ENCYCLOPEDIA OF

Microbiology

Volume 1 **A–C**

Editorial Advisory Board

Editor-in-Chief
Joshua Lederberg
Rockefeller University
New York, New York

ENCYCLOPEDIA OF Microbiology

Volume 1 A–C

ACADEMIC PRESS, INC.
A Division of Harcourt Brace & Company
San Diego New York Boston London Sydney Tokyo Toronto

This book is printed on acid-free paper. ∞

Academic Press, Inc.
1250 Sixth Avenue, San Diego, California 92101-4311

United Kingdom Edition published by
Academic Press Limited
24–28 Oval Road, London NW1 7DX

Library of Congress Cataloging-in-Publication Data

Encyclopedia of microbiology / edited by Joshua Lederberg
p. cm.
Includes bibliographical references and indexes.
ISBN 0-12-226891-1 (v. 1). -- ISBN 0-12-226892-X (v. 2). -- ISBN 0-12-226893-8 (v. 3). -- ISBN 0-12-226894-6 (v. 4)
1. Microbiology--Encyclopedias. I. Lederberg, Joshua.
QR9E53 1992
576'.03--dc20 92-4429
CIP

PRINTED IN THE UNITED STATES OF AMERICA

93 94 95 96 97 EB 9 8 7 6 5 4 3 2

Contents

C

Please refer to Volume 4 for the following information:

Preface

For the purposes of this encyclopedia, microbiology has been understood to embrace the study of "microorganisms," including the basic science and the roles of these organisms in practical arts (agriculture and technology) and in disease (public health and medicine). Microorganisms do not constitute a well-defined taxonomic group; they include the two kingdoms of Archaebacteria and Eubacteria, as well as protozoa and those fungi and algae that are predominantly unicellular in their habit. Viruses are also an important constituent, albeit they are not quite "organisms." Whether to include the mitochondria and chloroplasts of higher eukaryotes is a matter of choice, since these organelles are believed to be descended from free-living bacteria. Cell biology is practiced extensively with tissue cells in culture, where the cells are manipulated very much as though they were autonomous microbes; however, we shall exclude this branch of research. Microbiology also is enmeshed thoroughly with biotechnology, biochemistry, and genetics, since microbes are the canonical substrates for many investigations of genes, enzymes, and metabolic pathways, as well as the technical vehicles for discovery and manufacture of new biological products, for example, recombinant human insulin.

Within these arbitrarily designated limits, let us consider the overall volume of published literature in microbiology, where to find its core, and strategies for searching for current information on particular topics. Most of the data for this preface are derived from the 1988 Journal Citation Reports Current Contents (T) of the Institute for Scientific Information (ISI). Table I lists the 53 most consequential journals in microbiology, assessed by citation impact factor, the average number of literature citations per article published in a given journal. Table II presents that list sorted by the total number of articles printed in each journal in 1988. Table III shows the distribution of journals citing the *Journal of Bacteriology* and the distribution of journals cited in it.

Obviously, the publications of the American Society for Microbiology (indicated by AMS in the tables) play a commanding role. The society is now making its journals available in electronically searchable form (on optical disks), which will greatly facilitate locating and retrieving the most up-to-date information on any given subject. In addition, interdisciplinary journals such as *Nature (London), Science,* and the *Proceedings of the National Academy of Sciences, U.S.A.* are important sources of prompt news of scientific developments in microbiology. It is difficult to assess how much of their total publication addresses microbiology. As seen in Table III, the bibliographies in the *Journal of Bacteriology* cite half as many articles from the *Proceedings* (2348) as from the *Journal of Bacteriology* itself (5708). The 7038 articles indicated in Table II probably reach some 10,000 per year when these interdisciplinary and other dispersed sources are taken into account. An equal number might be added from overlapping aspects of molecular biology and genetics. To find and read all these titles would tax any scholar, although it could be done as a near full-time occupation with the help of the weekly Current Contents (T) of the ISI. To start afresh, with perhaps a decade's accumulation of timely background, would be beyond reasonable human competence. No one person would intelligently peruse more than a small fraction of the total texts.

The "Encyclopedia of Microbiology" is intended to survey the entire field coherently, complementing material that would be included in an advanced undergraduate and graduate major course of university study. Particular topics should be accessible to talented high school and college students, as well as graduates involved in teaching, research, and technical practice of microbiology.

Even these hefty volumes cannot embrace all current knowledge in the field. Each article does provide key references to the literature available at the time of writing. Acquisition of more detailed and up-to-date knowledge depends on (1) exploiting the review and monographic literature and (2) bibliographic retrieval of the preceding and current research literature. To make greatest use of review literature and monographs, the journals listed in Table II are invaluable. Titles such as *Annual Reviews* should not be misunderstood: these journals appear at annual intervals, but 5 or 10 years of accumulated research is necessary for the inclusion of a focused treatment of a given subject.

To access bibliographic materials in microbiology, the main retrieval resources are Medline, sponsored by the U.S. National Library of Medicine, and the Science Citation Index of the ISI. With governmental subsidy, Medline is widely available at modest cost: terminals are available at every medical school and at many other academic centers. Medline provides searches of the recent literature by author, title, and key word, and offers on-line displays of the relevant bibliographies and abstracts. Medical aspects of microbiology are covered exhaustively; general microbiology is covered in reasonable depth. The Science Citation Index must recover its costs from user fees, but is widely available at major research centers. It offers additional search capabilities, especially by citation linkage. Therefore, starting with the bibliography of a given encyclopedia article, one can quickly find (1) all articles more recently published that have cited those bibliographic reference starting points and (2) all other recent articles that share bibliographic information with the others. With luck, one of these articles may be identified as another comprehensive review that has digested more recent or broader primary material.

On a weekly basis, services such as Current Contents on Diskette (ISI) and Reference Update offer still more timely access to current literature as well as abstracts with a variety of useful features. Under the impetus of intense competition, these services are evolving rapidly, to the great benefit of a user community desperate for electronic assistance in coping with the rapidly growing and intertwined networks of discovery. The bibliographic services of Chemical Abstracts and Biological Abstracts would also be potentially invaluable; however, their coverage of microbiology is rather limited.

In addition, major monographs have appeared from time to time—"The Bacteria," "The Prokaryotes," and many others. Your local reference library should be consulted for these volumes.

Valuable collections of reviews also include *Critical Reviews for Microbiology, Symposia of the Society for General Microbiology, Monographs of the ASM,* and *Proceedings of the International Congresses of Microbiology.*

The articles in this encyclopedia are intended to be accessible to a broader audience, not to take the place of review articles with comprehensive bibliographies. Citations should be sufficient to give the reader access to the latter, as may be required. We do apologize to many individuals whose contributions to the growth of microbiology could not be adequately embraced by the secondary bibliographies included here.

The organization of encyclopedic knowledge is a daunting task in any discipline; it is all the more complex in such a diversified and rapidly moving domain as microbiology. The best way to anticipate the rapid further growth that we can expect in the near future is unclear. Perhaps more specialized series in subfields of microbiology would be more appropriate. The publishers and editors would welcome readers' comments on these points, as well as on any deficiencies that may be perceived in the current effort.

My personal thanks are extended to Kathryn Linenger at Academic Press for her diligent, patient, and professional work in overseeing this series; to my coeditors, Martin Alexander, David A. Hopwood, Barbara H. Iglewski, and Allen I. Laskin; and above all, to the many very busy scientists who took time to draft and review each of these articles.

Joshua Lederberg

Table I The Top Journals in Microbiology Listed by Impact Factor

Citation impact rank	Journal title	Number of articles published in 1988	Citation impact rank	Journal title	Number of articles published in 1988
1	*Microbiol. Rev.*	28	28	*FEMS Microbiol. Lett.*	365
2	*Adv. Microb. Ecol.*	10	29	*Am. J. Reprod. Immunol.*	50
3	*Annu. Rev. Microbiol.*	29	30	*Infection*	103
4	*FEMS Microbiol. Rev.*	13	31	*Can. J. Microbiol.*	236
5	*Yeast*	NA	32	*Curr. Microbiol.*	87
6	*J. Bacteriol.*	915	33	*J. Appl. Bacteriol.*	125
7	*Mol. Microbiol.*	94	34	*J. Microbiol. Meth.*	34
8	*Antimicrob. Agents Ch.*	408	35	*B. I. Pasteur*	20
9	*Rev. Infect. Dis.*	213	36	*ZBL Bakt. Mikr. Hyg. A*	164
10	*CRC Crit. Rev. Microbiol.*	12	37	*Ann. Inst. Pasteur Mic.*	58
11	*Syst. Appl. Microbiol.*	52	38	*Vet. Microbiol.*	104
12	*Int. J. Syst. Bacteriol.*	83	39	*Acta Path. Micro. Im. B*	NA
13	*J. Antimicrob. Chemoth.*	352	40	*Protistologica*	NA
14	*Appl. Environ. Microb.*	588	41	*Med. Microbiol. Immun.*	37
15	*J. Clin. Microbiol.*	619	42	*Diagn. Micr. Infec. Dis.*	60
16	*Adv. Appl. Microbiol.*	8	43	*Int. J. Food Microbiol.*	66
17	*Curr. Top. Microbiol.*	53	44	*J. Gen. Appl. Microbiol.*	27
18	*Arch. Microbiol.*	173	45	*Microbiol. Immunol.*	122
19	*J. Gen. Microbiol.*	367	46	*Lett. Appl. Microbiol.*	81
20	*Enzyme Microb. Tech.*	108	47	*Gen. Physiol. Biophys.*	57
21	*Eur. J. Clin. Microbiol.*	161	48	*A. Van Leeuw. J. Microb.*	51
22	*FEMS Microbiol. Ecol.*	42	49	*Symbiosis*	14
23	*J. Med. Microbiol.*	124	50	*Comp. Immunol. Microb.*	27
24	*J. Infection*	68	51	*Microbios.*	61
25	*Eur. J. Protistol.*	37	52	*ZBL Bakt. Mikr. Hyg. B*	76
26	*Microbiol. Sci.*	70	53	*J. Basic Microb.*	69
27	*Appl. Microbiol. Biot.*	270			

NA, Not available.

Table II Microbiology Journals Listed by Total Number of Articles Published per Year (1988)

Journal title	Number of articles published in 1988	Journal title	Number of articles published in 1988
J. Bacteriol.	915	*Int. J. Food Microbiol.*	66
J. Clin. Microbiol.	619	*Microbios.*	61
Appl. Environ. Microb.	588	*Diagn. Micr. Infec. Dis.*	60
Antimicrob. Agents Ch.	408	*Ann. Inst. Pasteur Mic.*	58
J. Gen. Microbiol.	367	*Gen. Physiol. Biophys.*	57
FEMS Microbiol. Lett.	365	*Curr. Top. Microbiol.*	53
J. Antimicrob. Chemoth.	352	*Syst. Appl. Microbiol.*	52
Appl. Microbiol. Biot.	270	*A. Van Leeuw. J. Microb.*	51
ZBL Bakt. Mikr. Hyg. A	240	*Am. J. Reprod. Immunol.*	50
Can. J. Microbiol.	236	*FEMS Microbiol. Ecol.*	42
Rev. Infect. Dis.	213	*Med. Microbiol. Immun.*	37
Arch. Microbiol.	173	*Eur. J. Protistol.*	37
Eur. J. Clin. Microbiol.	161	*J. Microbiol. Meth.*	34
J. Appl. Bacteriol.	125	*Eur. J. Protistology*	29
J. Med. Microbiol.	124	*Annu. Rev. Microbiol.*	29
Microbiol Immunol.	122	*Microbiol. Rev.*	28
Enzyme Microb. Tech.	108	*J. Gen. Appl. Microbiol.*	27
Vet. Microbiol.	104	*Comp. Immunol. Microb.*	27
Infection	103	*B. I. Pasteur*	20
Mol. Microbiol.	94	*Acta Path. Micro. Im.*	18
Curr. Microbiol.	87	*Symbiosis*	14
Int. J. Syst. Bacteriol.	83	*FEMS Microbiol. Rev.*	13
Lett. Appl. Microbiol.	81	*CRC Crit. R. Microbiol.*	12
Microbiol. Sci.	70	*Adv. Microb. Ecol.*	10
J. Basic Microb.	69	*Adv. Appl. Microbiol.*	8
		Total	7038

Table III.A Distribution of Journals Cited in *Journal of Bacteriology*, 1979–1988

Journal cited	Number of citations	Journal cited	Number of citations
J. Bacteriol.	5708	*Genetics*	183
P. Natl. Acad. Sci. U.S.A.	2348	*Can. J. Microbiol.*	139
J. Biol. Chem.	1698	*Arch. Biochem. Biophys.*	127
Mol. Gen. Genet.	1157	*Virology*	123
J. Mol. Biol.	1148	*Bacteriol. Rev.*	118
Gene	902	*Cold Spring Harb. Sym.*	110
Nature (London)	820	*Antimicrob. Agents Ch.*	109
Nucleic Acids Res.	804	*Escherichia Coli Sal.*	95
Cell	802	*Plant Physiol.*	80
J. Gen. Microbiol.	701	*J. Biochem.-Tokyo*	78
Infect. Immun.	478	*J. Virol.*	78
Methods Enzymol.	434	*Mol. Cell. Biol.*	68
Anal. Biochem.	411	*J. Infect. Dis.*	67
Biochim. Biophys. Acta	401	*Bio-Technol.*	61
Eur. J. Biochem.	376	*Exp. Gene Fusions*	60
Mol. Cloning Laboratory	363	*Trends Biochem. Sci.*	60
Microbiol. Rev.	361	*Mutat. Res.*	59
Arch. Microbiol.	347	*Syst. Appl. Microbiol.*	55
Embo J.	327	*Phytopathology*	51
Biochemistry-U.S.	310	*Adv. Bacterial Genet.*	50
Science	301	*Photochem. Photobiol.*	50
Appl. Environ. Microb.	294	*Biochimie*	49
FEMS Microbiol. Lett.	257	*J. Exp. Med.*	48
Exp. Mol. Genetics	234	*Agr. Biol. Chem. Tokyo*	47
Plasmid	234	*Int. J. Syst. Bacteriol.*	44
Biochem. Bioph. Res. Commun.	224	*FEMS Microbiol. Rev.*	43
FEBS Lett.	213	*J. Clin. Microbiol.*	42
Biochem. J.	207	*Curr. Microbiol.*	41
Annu. Rev. Microbiol.	194	*J. Cell Biol.*	41
Annu. Rev. Biochem.	188		
Annu. Rev. Genet.	187	All other (1301)	4311

(continues)

Table III.B (*continued*) Distribution of Journals Citing *Journal of Bacteriology*, 1979–1988

Journal citing	Number of citations	Journal citing	Number of citations
J. Bacteriol.	5708	*Curr. Genet.*	117
J. Biol. Chem.	1119	*FEMS Microbiol. Rev.*	115
J. Gen. Microbiol.	963	*J. Basic Microb.*	115
Mol. Gen. Genet.	896	*J. Antimicrob. Chemoth.*	112
Appl. Environ. Microb.	890	*Microb. Pathogenesis*	110
Microbiol. Rev.	759	*Science*	104
Infect. Immun.	663	*Ann. Inst. Pasteur Mic.*	101
FEMS Microbiol. Lett.	648	*Methods Enzymol.*	99
Gene	599	*ZBL Bakt. Mikr. Hyg. A*	98
P. Natl. Acad. Sci. U.S.A.	588	*A. Van Leeuw. J. Microb.*	95
Can. J. Microbiol.	579	*Annu. Rev. Biochem.*	94
Arch. Microbiol.	484	*Plant Physiol.*	88
Mol. Microbiol.	452	*J. Infect. Dis.*	86
J. Mol. Biol.	434	*J. Med. Microbiol.*	85
Nucleic Acids Res.	431	*Folia Microbiol.*	79
Biochim. Biophys. Acta	378	*Genetika*	79
Eur. J. Biochem.	350	*Gene Dev.*	78
Antimicrob. Agents Ch.	340	*Microbios.*	77
Annu. Rev. Microbiol.	316	*Arch. Biochem. Biophys.*	75
Cell	246	*Biotechnol. Bioeng.*	73
Biochimie	238	*Nature (London)*	69
Biochemistry-U.S.	236	*Syst. Appl. Microbiol.*	69
Plasmid	236	*Zh. Mikrob. Epid. Immun.*	67
Embo J.	234	*J. Antibiot.*	66
J. Clin. Microbiol.	214	*Annu. Rev. Genet.*	65
Genetics	201	*Microbiol. Immunol.*	65
Adv. Microb. Physiol.	199	*J. Biochem.-Tokyo*	64
Agr. Biol. Chem. Tokyo	198	*Microbial Ecol.*	60
Mol. Cell. Biol.	197	*Plant Soil*	58
CRC Crit. R. Microbiol.	194	*Anal. Biochem*	56
Curr. Microbiol	193	*Annu. Rev. Cell Biol.*	55
Appl. Microbiol. Biot.	183	*Biotechnol. Lett.*	54
J. Appl. Bacteriol.	169	*Adv. Microb. Ecol.*	53
Mutat. Res.	160	*Enzyme Microb. Tech.*	53
Biochem. Bioph. Res. Commun.	152	*Curr. Sci. India*	52
Rev. Infect. Dis.	141	*Eur. J. Clin. Microbiol.*	51
Biochem. J.	137	*J. Theor. Biol.*	51
Microbiol. Sci.	135	*Bot. Acta*	50
Int. J. Syst. Bacteriol.	128	*Photochem. Photobiol.*	50
FEBS Lett	125		

How to Use the Encyclopedia

This encyclopedia is organized in a manner that we believe will be the most useful to you, and we would like to acquaint you with some of its features.

The volumes are organized alphabetically as you would expect to find them in, for example, magazine articles. Thus, "Foodborne Illness" is listed as such and would not be found under "Illness, Foodborne." If the first words in a title are not the primary subject matter contained in an article, the main subject of the title is listed first (e.g., "Heavy Metals, Bacterial Resistances," "Marine Habitats, Bacteria," "Method, Philosophy," "Transcription, Viral"). This is also true if the primary word of a title is too general (e.g.,"Bacteriocins, Molecular Biology"). Here, the word "bacteriocins" is listed first because "molecular biology" is a very broad topic. Titles are alphabetized letter-by-letter so that "Cell Membrane: Structure and Function" is followed by "Cellulases" and then by "Cell Walls of Bacteria."

Each article contains a brief introductory Glossary wherein terms that may be unfamiliar to you are defined *in the context of their use in the article*. Thus, a term may appear in another article defined in a slightly different manner or with a subtle pedagogic nuance that is specific to that particular article. For clarity, we have allowed these differences in definition to remain so that the terms are defined relative to the context of each article.

Articles about closely related subjects are identified in the Index of Related Titles at the end of the last volume (Volume 4). The article titles that are cross-referenced within each article may be found in this index, along with other articles on related topics.

The Subject Index contains specific, detailed information about any subject discussed in the *Encyclopedia*. Entries appear with the source volume number in boldface followed by a colon and the page number in that volume where the information occurs (e.g., "DNA repair by bacterial cells, **2**:9"). Each article is also indexed by its title (or a shortened version thereof), and the page ranges of the article appear in boldface (e.g. "Lyme disease, **2:639–646**" means that the primary coverage of the topic of Lyme disease occurs on pages 639–646 of Volume 2).

If a topic is covered primarily under one heading but additional related information may be found elsewhere, a cross-reference is given to the related material. For example, "Biodegradation" would contain all the page numbers where relevant information occurs, followed by "*See also* Bioremediation; Pesticide biodegradation" for different but related information. Similarly, a "*See*" reference refers the reader from a less-used synonym (or acronym) to a more specific or descriptive subject heading. For example, "Immunogens, synthetic. *See* Vaccines, synthetic." A *See under* cross-reference guides the reader to a specific subheading under a term. For example, "Mixis. *See under* Genome rearrangement."

An additional feature of the Subject Index is the identification of Glossary terms. These appear in the index where the word "defined" (or the words "definition of") follows an entry. As we noted earlier, there may be more than one definition for a particular term, and as when using a dictionary, you will be able to choose among several different usages to find the particular meaning that is specifically of interest to you.

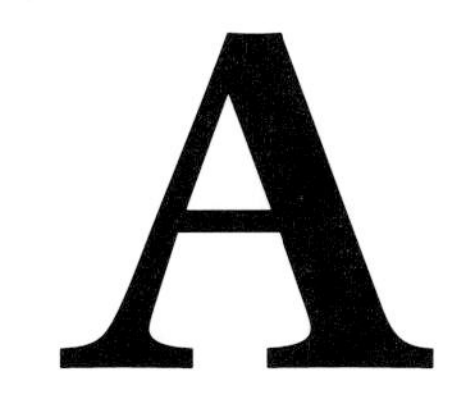

Acetogenesis and Acetogenic Bacteria

Harold L. Drake
Universität Bayreuth

Glossary

Autotroph Organism that uses CO_2 as its sole source of carbon

Chemolithoautotroph Organism that uses CO_2 as its sole source of carbon and an inorganic substrate for energy

CO_2 fixation Process in which CO_2 is covalently bonded to an organic compound

Heterotroph Organism that uses organic forms of carbon

Methanogens Obligate anaerobes that form methane as their end product

Obligate anaerobe Organism that cannot grow in the presence of oxygen

Sulfur-reducing bacteria Obligate anaerobes that use inorganic forms of sulfur as terminal electron acceptors

Syntrophy Relationship between two organisms in which each syntrophic partner is nutritionally dependent on the other

ACETOGENESIS is a term used to describe the metabolic processes by which certain bacteria form acetate from CO_2. There are three known processes by which acetogenesis occurs: the acetyl-CoA pathway, the glycine synthase–dependent pathway, and the reductive citric acid cycle. This article focuses on the acetyl-CoA pathway and those bacteria that are dependent on this pathway for the conservation of energy and the synthesis of biomass. These bacteria are defined in the present statement as acetogenic bacteria, or acetogens for short. Acetogens are obligate anaerobes and usually form acetate as their main end product. Many are chemolithoautotrophs. Ecologically, they play important roles in the turnover of carbon and energy in anaerobic habitats that range from soils and sediments to the gastrointestinal tract of many animals, including termites and humans. Three identifying features of acetogenic bacteria are (1) the use of chemolithoautotrophic substrates [carbon dioxide (CO_2) plus molecular hydrogen (H_2), or carbon monoxide (CO) alone] as sole sources of carbon and energy under strictly anaerobic conditions; (2) the capacity to convert certain sugars (e.g., glucose) stoichiometrically to acetate during heterotrophic growth; and (3) the use of aromatic compounds under acetogenic conditions. However, any single acetogen may not display all three of these metabolic features. Although other bacterial groups (e.g., methanogens and sulfur-reducing bacteria) also make use of the acetyl-CoA pathway for the synthesis of biomass or the oxidation of carbon, these bacterial groups use pathways that are slightly different from the acetyl-CoA pathway of acetogens, are dependent on other energy-yielding processes, and are not acetogenic.

It is nonetheless important to understand that the acetyl-CoA pathway is not restricted to a single bacterial group and is widespread in nature.

I. Acetogenic Bacteria

A. Definition of the Term "Acetogenic Bacterium"

When should the term "acetogenic bacterium" be used to describe an organism? There are many acetate-forming processes used by bacteria, and any bacterium that forms acetate could be termed an acetogen or an acetogenic organism. However, such broad usage of the term "acetogen" does not adequately distinquish between the different processes by which acetate is formed. For the purpose of this article, the following definition has been applied:

> An acetogenic bacterium is an obligately anaerobic bacterium that uses the acetyl-CoA pathway for (i) the synthesis of acetyl-CoA, (ii) the conservation of energy, and (iii) growth.

The following organisms are excluded by this definition: (i) purinolytic clostridia (e.g., *Clostridium acidiurici*) and other organisms that use the glycine synthase pathway for the formation of acetate; (ii) bacteria that employ the reductive citric acid cycle; and (iii) all other acetate-forming bacteria.

Nonetheless, this definition may not be as straightforward as it appears. For example, even though the capacity to form acetate as the sole or major product is viewed as synonymous with acetogenesis, the formation of acetate *per se* is not even required by the above definition. In other words, the fate of acetyl-CoA is less important than how it is formed. Even though the production of acetate by a true acetogen is an important observation, it is not the primary reason why these organisms should be defined as acetogens: the production of acetate is a manifestation of the reason, not the reason itself. To understand this point more clearly, several points can be examined:

1. Per the above definition, it follows that whether an acetogenic bacterium is chemolithoautotrophic (capable of growth at the expense of H_2/CO_2 or CO) or heterotrophic (dependent on organic carbon) is not a critical point of classification. Even true acetogens may not be capable of chemolithoautotrophic growth because they lack hydrogenase or autotrophic anabolic processes (e.g., *Clostridium magnum*). It should also be noted that many actogens isolated as strict heterotrophs have been later shown to be capable of chemolithoautotrophic growth. A classic example of this is *Clostridium thermoaceticum*. [*See* HETEROTROPHIC MICROORGANISMS.]

2. Some acetogens that are only shown to form acetate as a reduced end product during initial characterizations and thus termed "homoacetogens" may be later found to form alternative end products. An excellent example of this is *Acetobacterium woodii*. Originally shown to only form acetate, this acetogen has recently been found to reduce acetyl-CoA to ethanol under conditions of phosphate limitation. Another similar example is the capacity of *C. thermoaceticum* to form H_2 during CO-coupled growth. Furthermore, certain acetogens may never form acetate as their sole fermentation product, yet they are obligate anaerobes and use the acetyl-CoA pathway as defined above. For example, butyrate is formed by condensation of 2 acetyl-CoA's by the acetogens *Eubacterium limosum* and *Butyribacterium methylotrophicum* and under certain conditions can be the predominant end product. Can one safely apply the term "homoacetogen"?

3. Many anaerobic organisms use the acetyl-CoA pathway for purposes that are dissimilar to that of acetogenic bacteria; that is, the pathway is not used in the direction of acetyl-CoA synthesis or it is used primarily in an assimilatory fashion for biosynthesis [e.g., autotrophic methanogens and sulfur-reducing bacteria that use the pathway for assimilation of cell carbon, or organisms that may oxidize organic carbon via a reverse-type acetyl-CoA pathway (e.g., species of *Desulfobacterium* and *Methanosarcina*)]. Furthermore, portions of the acetyl-CoA pathway may (i) exist in organisms in a cryptic state and have no (known) function in the cell or (ii) perform non-acetogenic functions. Such organisms need not be considered acetogens (see Section II.E).

4. The above definition does not exclude the possibility that acetogens can also use other forms of energy conservation. Thus, although CO_2 is considered the primary terminal electron acceptor of acetogens, recent studies have shown that alternative acceptors might also be used in concert with CO_2 and give rise to other end products (e.g., the reduction of aromatic acrylate groups and fumarate appear to be coupled to the conservation of energy by *A. woodii* and *Clostridium formicoaceticum*, respectively, and yield products other than acetate).

To summarize these points, what is most critical in determining if an anaerobic bacterium is an acetogen per the above definition is demonstrating that it uses the acetyl-CoA pathway in the direction of acetyl-CoA synthesis and that this function is coupled to both the conservation of energy and growth. That an anaerobe grows chemolithoautotrophically and forms acetate as its sole product is extremely good evidence that the organism is indeed an acetogen. Equally compelling is the observation that acetate is the sole product obtained from the fermentation of certain carbohydrates. But, in general, and especially in cases where such results are not obtained, these lines of evidence should ideally be supplimented with additional tests to verify that the acetyl-CoA pathyway is used per the above definition. Perhaps the most conclusive evidence of this type would be demonstrating that acetyl-CoA synthase (often referred to as carbon monoxide dehydrogenase, as discussed in section II.B) is functional in the direction of acetyl-CoA synthesis.

Nonetheless, as many new and unusual organisms are isolated, one must also leave room for the possibility that the above definition is not absolutely appropriate for all acetogenic isolates. For example, a recent isolate appears to have the capacity to use a reversible acetyl-CoA pathway, depending on the conditions of growth. In such cases, the organism is only acetogenic when the pathway is used in the direction of acetate synthesis.

B. Origins of Acetogenic Bacteria

Approximately 40 different species have been isolated from extremely diverse anaerobic habitats (Table I). Given the diverse origins of isolates to date, it seems likely that acetogens occur in most anaerobic habitats. Although most isolates to date are mesophilic, psychrophilic and thermophilic species have also been isolated.

Most isolates are rod shaped, but coccus forms have also been studied. Staining properties vary, and both gram-positive and gram-negative species have been reported. Some acetogens have flagella and are therefore motile. Some form spores that remain viable for long periods; the thermophilic spore-forming species are fairly resistant to high temperatures. Thus, the ultrastructural features of acetogens are highly variable. The guanine-plus-cytosine (G+C) content of the genome of acetogens is also highly variable between species.

C. Isolation and Growth of Acetogenic Bacteria

Acetogens are strict anaerobes, and attempts to isolate or cultivate these organisms must take this fact into account. The basic Hungate technique is thus widely used for the cultivation of acetogenic bacteria. The Hungate technique is also used to cultivate other obligately anaerobic bacteria. Many versions of the basic technique are in use, but all have the same basic protocol: preparation of sterile anaerobic media and transfer of cultures under strictly anaerobic conditions. The basic approach involves the use of oxygen-free gases and air-tight culture vessels. Alternatively, anaerobic cultivation can also be achieved in anaerobic glove boxes, chambers that provide oxygen-free environments for the study and growth of obligate anaerobes.

Although all acetogens use the acetyl-CoA pathway, they vary in their capacity to use different substrates. In general, various carbohydrates (e.g., glucose, xylose, and cellobiose), alcohols (e.g., methanol and ethanol), methoxylated aromatic compounds (e.g., vanillate and trimethoxybenzene), and H_2/CO_2 or CO are growth-supportive substrates for acetogens and can be used to enrich for acetogens. Recent studies indicate that certain halogenated compounds can also be used as substrates for enrichment.

Not many acetogens have been studied relative to nitrogen metabolism, but *C. formicoaceticum* has been shown to catalyze the fixation of nitrogen (N_2). It is thus likely that other acetogens also have this capacity and are therefore able to assimilate N_2 during growth under certain conditions.

In most laboratory studies, growth of acetogens is under undefined conditions (i.e., in complex, undefined growth medium), and supplemental CO_2 is often supplied in the gas phase or in the form of dissolved carbonates (e.g., bicarbonate). Although many metabolic processes, including acetogenesis under certain conditions, give rise to CO_2, studies under defined conditions have shown that some acetogens need supplemental levels of CO_2 for growth. The reason for this requirement may be very complex, but it is likely due to the fact that CO_2 is the normal terminal electron acceptor for acetogenic bacteria and that under laboratory conditions the cell is unable to generate CO_2 in adequate amounts or rates. This fact must be taken into account when designing a growth medium for acetogens.

Table I Acetogenic Bacteria Isolated to Date[a]

Acetogen	Year	Source of isolate	Gram stain	Cell morph.	Growth temp.[b]
Acetitomaculum ruminis	1989	Rumen fluid, steer	+	Rod	Meso
Acetoanaerobium noterae	1985	Sediment	−	Rod	Meso
Acetobacterium carbinolicum	1984	Freshwater sediment	+	Rod	Meso
Acetobacterium malicum	1988	Freshwater sediment	+	Rod	Meso
Acetobacterium wieringae	1982	Sewage digester	+	Rod	Meso
Acetobacterium woodii	1977	Marine estuary	+	Rod	Meso
Acetobacterium sp. *B10*	1989	Waste water	+	Rod	Meso
Acetobacterium sp. *KoB58*	1988	Sewage sludge	+	Rod	Meso
Acetobacterium sp. *HP4*	1989	Lake sediment	+	Rod	Psychro
Acetobacterium sp. *MrTac1*	1987	Marine sediment	+	Rod	Meso
Acetobacterium sp. *OyTac1*	1987	Freshwater sediment	+	Rod	Meso
Acetobacterium sp. *AmMan1*	1991	Freshwater sediment	+	Rod	Meso
Acetogenium kivui	1981	Lake sediment	−	Rod	Thermo
Acetonema longum	1991	Wood-eating termite, gut	−	Rod	Meso
Butyribacterium methylotrophicum	1980	Sewage digester	+	Rod	Meso
Clostridium aceticum	1936	Soil	−	Rod	Meso
Clostridium sp. *CV-AA1*	1982	Sewage sludge	−	Rod	Meso
Clostridium fervidus	1987	Hot spring	−	Rod	Thermo
Clostridium formicoaceticum	1970	Sewage	−	Rod	Meso
Clostridium ljungdahlii	1990	Animal waste	+	Rod	Meso
Clostridium magnum	1984	Freshwater sediment	−	Rod	Meso
Clostridium mayombei	1991	Soil-feeding termite, gut	+	Rod	Meso
Clostridium pfennigii	1985	Rumen fluid, steer	+	Rod	Meso
Clostridium thermoaceticum	1942	Horse manure	+	Rod	Thermo
Clostridium thermoautotrophicum	1981	Hot spring	+/−	Rod	Thermo
Eubacterium limosum	1981	Rumen fluid, sheep	+	Rod	Meso
Peptostreptococcus productus	1984	Sewage digester	+	Coccus	Meso
Sporomusa acidovorans	1985	Distillation wastewater	−	Rod	Meso
Sporomusa malonica	1989	Freshwater sediment	−	Rod	Meso
Sporomusa ovata	1984	Silage	−	Rod	Meso
Sporomusa sphaeroides	1984	River mud	−	Rod	Meso
Sporomusa termitida	1988	Wood-eating termite, gut	−	Rod	Meso
Syntrophoccus sucromutans	1986	Rumen fluid, steer	−	Coccus	Meso
Unclassified					
AOR	1988	Thermophilic digester	+	Rod	Thermo
MC	1991	Sewage digester sludge	+	Coccus	Meso
Strain 22	1976	Sewage sludge	+	Rod	Meso
Strain X-8	1982	Vegetable wastewater	−	Rod	Meso
TH-001	1985	Sewage sludge	−	Rod	Meso

[a] Bacteria that appear to use the acetyl-CoA pathway for the synthesis of acetate and growth.

[b] General temperature preference indicated as psychrophilic (psychro), mesophilic (meso), and thermophilic (thermo). Optimum growth temperatures for these three temperature groups are 16–18°C, 31–34°C, and 58–62°C, respectively.

Table II Examples of Acetogens That Have Specific Vitamin Requirements for Growth

Acetogen	Essential vitamin(s)
Acetobacterium woodii	Pantothenate
Butyribacterium methylotrophicum	Pantothenate
Clostridium aceticum	Biotin, pantothenate, pyridoxamine
Clostridium formicoaceticum	Pyridoxine
Clostridium thermoaceticum	Nicotinic acid
Clostridium thermoautotrophicum	Nicotinic acid
Eubacterium limosum	Biotin, lipoic acid, pantothenate
Sporomusa sphaeroides	Vitamin B_{12}, folic acid, nicotinic acid

Many acetogens have specific vitamin requirements for growth (Table II). Various metals are also required for growth because many metal-containing proteins or cofactors are essential to acetogenesis. For example, acetyl-CoA synthase contains nickel, iron, and zinc, and formate dehydrogenase contains iron, molybdenum, selenium, and tungsten. Thus, it is common to grow acetogenic bacteria in a medium that is enriched in vitamins and trace metals. Furthermore, sodium is required for energy conservation by some acetogens and is therefore usually present in most growth media. Cultivation of acetogens in metal-deficient medium will usually result in poorer growth or alteration in product formation.

Some acetogens common to gastrointestinal environments have not been cultivated under defined conditions because they require unknown growth substrates or factors inherent to such habitats. Such acetogens are often cultivated in media that is supplemented with filter-sterilized rumen fluid (fluid obtained from the rumen of sheep or cattle). Only a few acetogens examined to date have the capacity to manufacture all biosynthetic compounds (vitamins, cofactors, amino acids, etc.) required for growth and can therefore grow solely at the expense of a single growth-supportive substrate in the absence of supplemental vitamins, amino acids, etc. Examples of such acetogens include *Acetogenium kivui, C. magnum,* and *Sporomusa ovata.*

D. Classification of Acetogenic Bacteria

As illustrated in Table I, many acetogens have been isolated; all isolates to date are eubacteria. It is beyond the scope and purpose of the present statement to delineate detailed protocols for the classification of acetogenic bacteria. At present, it is common to key out a new isolate according to the following characteristics: (1) morphological appearance, (2) presence, shape, and location of spores, (3) Gram stain, (4) motility and type of flagella present, (5) temperature and pH requirements, (6) G+C content of the genome, (7) a detailed examination of growth-supportive substrates and products formed, and (8) comparative analysis of 16S ribosomal RNA (not routinely possible in most laboratories). By close examination of these properties, a new isolate can be assigned to an existing species or classified as a new species if dissimilar to those previously described. The number of known species of acetogens is becoming quite large, and it is likely that more definitive phylogenetic analysis by oligonucleotide cataloging of 16S ribosomal RNA will become increasingly more valuable in the near future.

II. Acetyl-CoA (Wood) Pathway and Acetyl-CoA Synthase

A. Brief Historical Perspective on the Acetyl-CoA Pathway and Acetogenic Bacteria

The history of acetogenic bacteria is intimately tied to autotrophic processes and mechanisms of CO_2 fixation. In the late 1930s and early 1940s when the first two acetogens were isolated (*Clostridium aceticum* and *C. thermoaceticum,* isolated in 1936 and 1942, respectively), pathways for the autotrophic fixation of CO_2 were not yet resolved, and the concept that heterotrophic organisms were capable of fixing CO_2 was just emerging. Thus, the isolation of these two organisms that were later shown to have an intriguing autotrophic process did not catch the eye of many investigators at the time of their isolation. In fact, it was not until 1970 that the isolation of the third acetogen, *C. formicoaceticum,* was even reported.

What was the feature that most attracted early investigators to study *C. aceticum* and *C. thermoaceticum?* What was the clue that suggested the existence of a new autotrophic process? The answer can be understood easily by considering the consequences of the following fermentation stoichiometries observed with *C. thermoaceticum* and *C. aceticum:*

$$C_6H_{12}O_6 \text{ (glucose)} \rightarrow 3\ CH_3COOH \quad (I)$$

$$4\ CH_3COCOOH \text{ (pyruvate)} + 2\ H_2O \rightarrow 5\ CH_3COOH + 2\ CO_2 \quad (II)$$

$$2\ CO_2 + 4\ H_2 \rightarrow CH_3COOH + 2\ H_2O \quad (III)$$

When the first acetogens were isolated, the carbohydrate fermentations studied to date usually yielded ethanol, acetate, or an alternative two-carbon product and equimolar quantities of CO_2 or other one-carbon products. However, it is seen in Reaction I that acetate was the sole product obtained during the fermentation of glucose. Hence the term "homoacetate fermentation." Early investigators realized that this was an unusual fermentation pattern because typical fermentation mechanisms dependent on the classic "3-3 split" of glucose via glycolysis (e.g., as observed with lactate or ethanol fermentations) could not account for the formation of 3 moles of acetate per mole of glucose consumed. The stoichiometry of Reaction II, observed during pyruvate fermentations, was taken as further evidence of a mechanism by which two one-carbon units (at the level of CO_2) derived from pyruvate were combined to form acetate by an unknown mechanism. Such a mechanism was strongly implied by the capacity of *C. aceticum* to fix CO_2 to the level of acetate via the oxidation of H_2 according to Reaction III.

When the first acetogens were isolated, the use of isotopes for elucidating biochemical pathways was a technique just being developed, and acetogenic bacteria were among the first organisms studied using such techniques. These early "tracer" studies (in the 1940s with ^{14}C and in the 1950s with ^{13}C) made it clear that a new autotrophic mechanism for the fixation of carbon was present in acetogenic bacteria. Despite these early studies that presented an intriguing target to the investigators that followed, the complete mechanism by which two moles of CO_2 are fixed autotrophically to form 1 mole of acetate would require another 40 or so years to resolve.

From the 1950s to the mid 1980s, the biochemical steps involved in the acetyl-CoA pathway were resolved, and acetogenic bacteria were unequivocally shown to use a unique metabolic process for autotrophic growth. Acetogens are strict anaerobes, and many of the key catalysts involved in the acetyl-CoA pathway are extremely unstable to air (oxygen) and difficult to study. This fact played a large role in the length of time required to resolve the pathway, and it was not until strictly anaerobic techniques were developed that the final steps were elucidated in the early to mid 1980s. [*See* ANAEROBIC RESPIRATION.]

Ironically, the biochemical steps of this autotrophic process were resolved with *C. thermoaceticum,* an organism that was isolated and studied as an obligate heterotroph. Only after the autotrophic properties of the pathway were resolved at the biochemical level was it shown that the organism was actually capable of true chemolithoautotrophic growth at the expense of either H_2/Co_2 or CO.

Resolution of the biochemical details of the acetyl-CoA pathway can to a large part be credited to the studies of Harland G. Wood (recently deceased) and many of his collaborators (in particular, Lars G. Ljungdahl), and many individuals refer to the acetyl-CoA pathway as the Wood pathway in acknowledgment of Wood's countless contributions to the resolution of the pathway. Although many years have elapsed since *C. aceticum* and *C. thermoaceticum* were isolated, it should not be forgotten that a unique autotrophic process was clearly indicated by the first physiological characterizations of these organisms. The first investigators of these organisms (Wieringa, Kamen, Fontaine, Barker, and others) laid an indispensible foundation to the imagination of subsequent investigators who were later able to apply more sophisticated techniques and resolve the biochemical details of the acetyl-CoA pathway.

B. General Overview of the Pathway

The central unifying characteristic of acetogenic bacteria is their capacity to use the acetyl-CoA pathway for both the conservation of energy and the synthesis of biomass. Although the pathway is composed of many reactions, the overall process can be conceptualized as a process in which two one-carbon units are combined in the formation of an acetyl group. This reaction is dependent on CoA, and acetyl-CoA is the immediate intracellular product formed. Under autotrophic conditions, these two one-carbon units can be derived from CO_2 (Fig. 1).

However, an important aspect of the pathway is that it can be used under both autotrophic and heterotrophic conditions. That is, the carbon and energy used in the formation of acetate and biomass

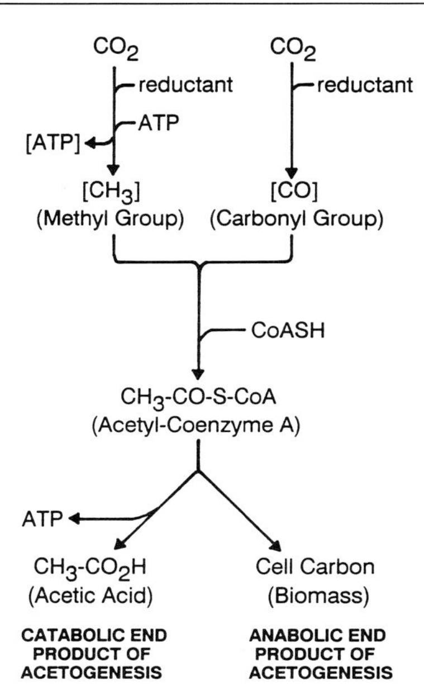

Figure 1 Simplified overview of chemolithoautotrophic acetyl-CoA pathway, illustrating the flow of carbon toward catabolic and anabolic end products. Under chemolithoautotrophic conditions, the reductant is derived from H_2 or CO, whereas under heterotrophic conditions, the reductant is derived from oxidation of organic compounds (see Fig. 2). Brackets around the methyl and carbonyl groups indicate that these groups are bound to specific cofactors or proteins. CoASH is coenzyme A. One ATP is consumed in formation of the methyl group, and formation of ATP indicates those portions of the pathway involved in conservation of energy. Energy conserved (ATP formed) during reductive synthesis of the methyl group likely occurs via membranous electron transport (indicated by [ATP]), whereas energy conserved in conversion of acetyl-CoA to acetate occurs via substrate-level phosphorylation by acetate kinase.

can be inorganic (e.g., H_2/CO_2) or organic (e.g., glucose) in origin. As illustrated in Fig. 2, the origins of the methyl and carbonyl groups used in the synthesis of acetyl-CoA can include many organic substrates.

The common interface to both the autotrophic and heterotrophic processes is the formation of acetyl-CoA from two one-carbon precursors, and this reaction is dependent on a special enzyme, acetyl-CoA synthase. This enzyme is also termed "carbon monoxide (CO) dehydrogenase" because it was first discovered on the basis of its ability to catalyze the following reaction:

$$CO + H_2O \rightarrow CO_2 + 2\,H^+ + 2\,e^- \qquad \text{(IV)}$$

The reducing equivalents (e^-) generated from CO can be detected with low redox potential, artificial electron acceptors (e.g., methyl viologen). The capacity of acetyl-CoA synthase/CO dehydrogenase to reduce redox dyes provides the basis for an easy spectrophotometric method for the detection and assay of this enzyme. Even though most investigators continue to use the term "CO dehydrogenase" instead of "acetyl-CoA synthase," it is important to recognize that this enzyme performs a function that is not equivalent to the CO dehydrogenases of aerobic organisms (e.g., *Pseudomonas carboxydovorans*) that also catalyze the oxidation of CO to CO_2 but do not use the acetyl-CoA pathway.

As illustrated in Figs. 1 and 2, the acetyl-CoA pathway performs two main functions for the cell. One function is catabolic and provides a mechanism for the conservation of energy. Both substrate-level phosphorylation and respiratory processes involving electron transfer and chemiosmotic mechanisms can be used (see Section III.B). The second function is anabolic and provides material for the formation of biomass via standard assimilatory metabolic processes. Obviously, this latter function becomes more critical to acetogens under increasingly minimal or autotrophic conditions because they are less able to obtain preformed organic compounds required for biosynthesis.

C. Acetyl-CoA Pathway

Figure 3 outlines the details of the pathway and illustrates how both heterotrophic and autotrophic processes are integrated during the degradation of glucose. In this figure, glycolysis and pyruvate-ferredoxin oxidoreductase (enzyme 12 in Fig. 3) are depicted as the heterotrophic origins of the reducing equivalents used in the autotrophic formation of the methyl and carbonyl groups. During glucose-dependent acetogenesis, glucose is converted to two molecules of pyruvate via glycolysis. Pyruvate is then oxidized to acetyl-CoA and CO_2, thus generating a total of eight reducing equivalents from glucose. Carbon dioxide is then used as a terminal electron acceptor, and these eight reducing equivalents are consumed in the formation of a third molecule of acetyl-CoA (acetate III in Fig. 3). All three molecules of acetyl-CoA are converted to acetate via phosphotransacetylase and acetate kinase. Thus, reducing equivalents generated during the oxidation of glucose are vented from the cell in the form of acetate.

All the enzymes of the pathway have been purified and characterized from *C. thermoaceticum* (Table III). Although it is assumed that the enzymes, cofactors, and steps of the pathway of other aceto-

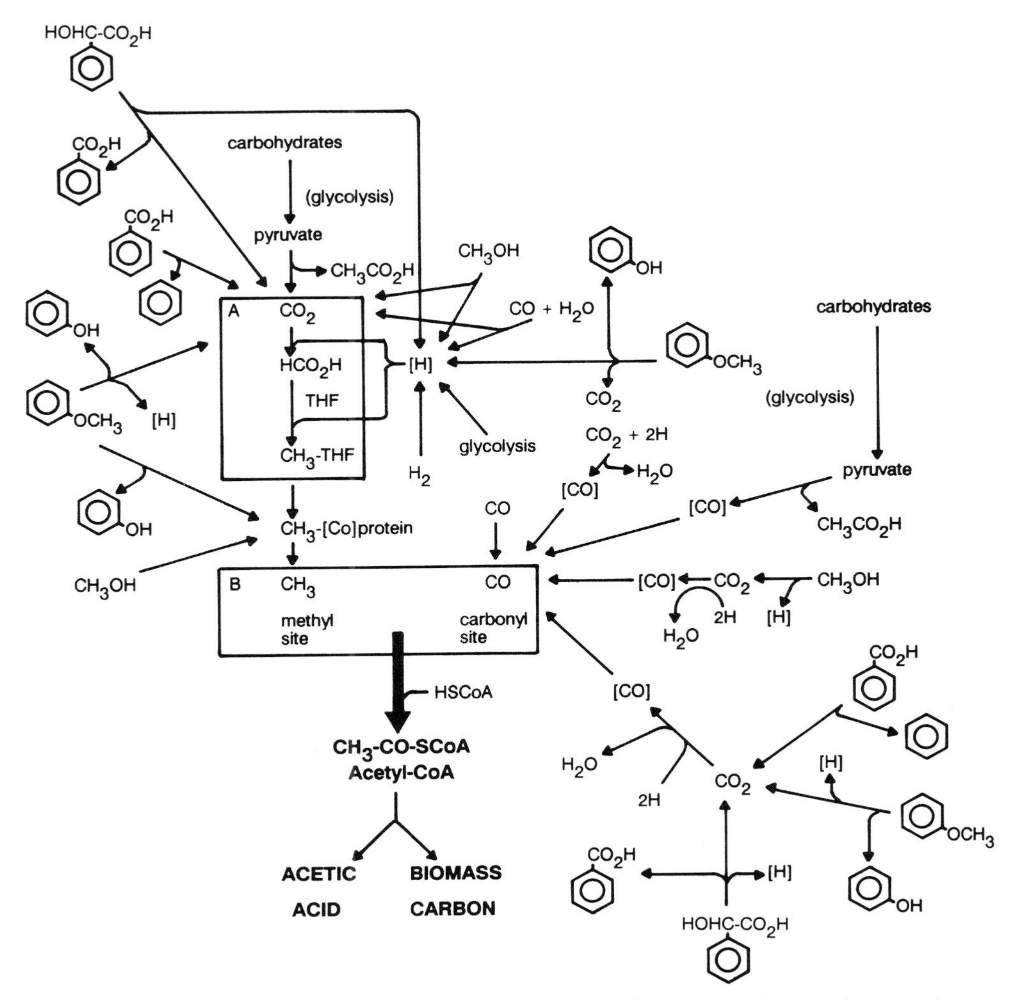

Figure 2 Overview of diverse origins for methyl and carbonyl carbons in the acetyl-CoA pathway used by acetogenic bacteria. Box A: Reductive tetrahydrofolate (THF) pathway. Box B: Acetyl-CoA synthase. Additional substituent groups of aromatic compounds are not shown. [Co]protein: corrinoid enzyme involved in methyl transfer. NOTE: Some of the reactions paths are hypothetical and remain to be proven.

gens are similar, there may be variations not yet apparent from the limited number of studies to date.

It is important to realize that reducing equivalents can be derived from many sources (Fig. 2) and that under strictly chemolithoautotrophic conditions, H_2 or CO are oxidized and drive the reductive synthesis of acetyl-CoA from CO_2 (yielding what is depicted in Fig. 3 as acetate III). Of course, when acetogenesis is under chemolithoautotrophic conditions, the reactions illustrated in Fig. 3 that lead to the formation of acetates I and II do not occur.

D. Acetyl-CoA Synthase and Autotrophic Formation of Acetyl-CoA

As illustrated in Fig. 3, acetyl-CoA synthase is involved in a series of complex reactions that culminate in the formation of the carbon-carbon covalent bond between the methyl and carbonyl carbon atoms. In this sense, acetyl-CoA synthase is the most important enzyme of the pathway in that it actually forms the autotrophically synthesized acetyl-CoA.

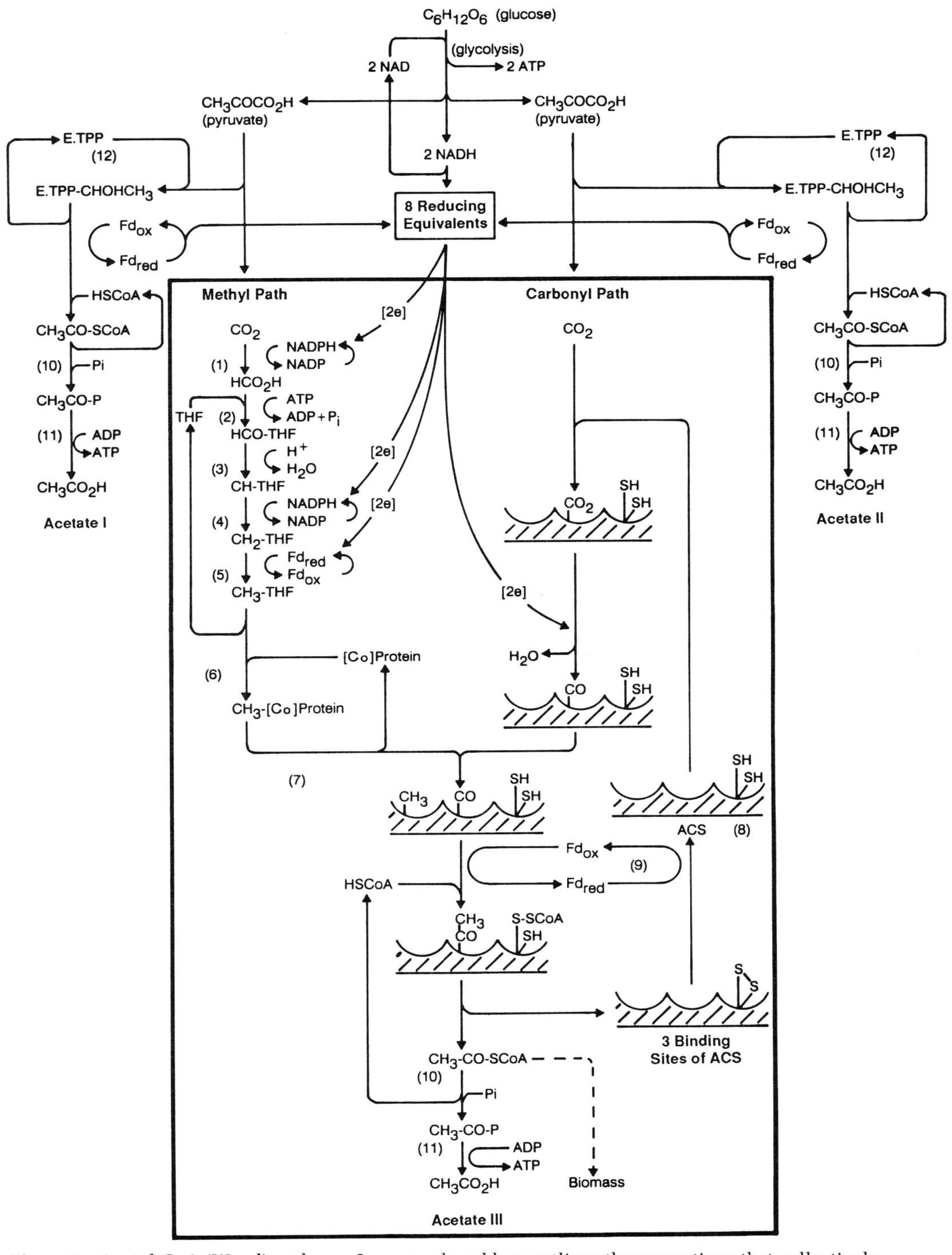

Figure 3 Acetyl-CoA (Wood) pathway. Large enclosed box outlines those reactions that collectively constitute the acetyl-CoA pathway. The Reactions outside the box are those reactions that result in (a) pyruvate oxidoreductase–dependent formation of acetates I and II and (b) glucose-dependent generation of reducing equivalents required for synthesis of acetate III via acetyl-CoA synthase. E.TPP, thiamine pyrophosphate of the pyruvate oxidoreductase complex; Fd, ferredoxin; ACS, acetyl-CoA synthase; Pi, inorganic phosphate; HSCoA, coenyzme A; THF, tetrahydrofolate; [Co]protein, corrinoid enzyme; [2e], 2 reducing equivalents. The broken line leading to biomass represents anabolic processes. See text and Table III for further information.

Table III Enzymes of the Autotrophic Portion of the Acetyl-CoA Pathway (box of Fig. 3)[a]

No. in Fig. 3	Enzyme	Molecular weight	Subunit composition	Function in pathway (illustrated in Fig. 3)
1	Formate dehydrogenase	340,000	$\alpha_2\beta_2$ (96,000 & 76,000)	Reduction of CO_2 to formate
2	Formyltetrahydrofolate (HCO-THF) synthetase	240,000	α_4	Conversion of formate to HCO-THF
3 & 4	Methenyltetrahydrofolate (CH-THF) cyclohydrolase and methylenetetrahydrofolate (CH_2-THF) dehydrogenase complex	55,000	α_2	HCO-THF converted to CH-THF, CH-THF reduced to CH_2-THF
5	Methylenetetrahydrofolate (CH_2-THF) reductase	125,900	α_2[b]	Reduction of CH_2-THF to CH_3-THF
6	Methyltransferase	58,900	α_2	Transfer of methyl unit from CH_3-THF to corrinoid enzyme
7	Corrinoid enzyme	89,000	$\alpha\beta$ (34,000 & 55,000)	Methylation of acetyl-CoA synthase
8	Acetyl-CoA synthase	440,000	$\alpha_3\beta_3$ (78,000 & 71,000) (81,730 & 72,928)[c]	Synthesis of acetyl-CoA
9	Acetyl-CoA synthase disulfide reductase	225,000	α_4	Reduction of disulfide of CoA binding site of acetyl-CoA synthase
10	Phosphotransacetylase	88,100	α_4	Conversion of acetyl-CoA to acetylphosphate
11	Acetate kinase	60,000	Not reported	Conversion of acetyl-CoA and ADP to acetate and ATP

[a] Enzymes indicated have been purified from *C. thermoaceticum*. Abbreviations: THF, tetrahydrofolate; CoA, coenzyme A.
[b] Recent results indicate methylenetetrahydrofolate reductase is an octomer (α_8) of identical subunits ($M_r = 36{,}143$ per subunit).
[c] As deducted from amino acid composition.

There are three binding sites of acetyl-CoA synthase (Fig. 3). One site accepts the methyl group from the methylated corrinoid enzyme. Another site is involved in the binding and formation of the carbonyl carbon. The third site appears to contain a disulfide bridge and is involved in the binding of CoA. The exact nature of the three sites is not yet clear, and precisely how these three binding sites interact in the formation of acetyl-CoA is still a matter of speculation. It appears that the binding sites of both the methyl and carbonyl carbons are very close together and contain the redox-active metals nickel and iron and that these metals are involved in catalysis. There is another binding site that contains cysteine and can reversibly bind the methyl group, but this site may not be directly involved in the formation of the carbon-carbon bond of the acetyl group.

The acetyl-CoA pathway can be divided into two reaction paths, both of which converge at acetyl-CoA synthase and the formation of acetyl-CoA. In the methyl reaction path, CO_2 acts as a terminal electron acceptor and is reduced. Six reducing equivalents are consumed in the overall reduction of CO_2 to the methyl level via formate and tetrahydrofolate-bound intermediates. A corrinoid-containing protein subsequently mediates the methylation of the methyl site of acetyl-CoA synthase.

The carbonyl reaction path is less complicated. CO_2 is bound by acetyl-CoA synthase, and two reducing equivalents are consumed in the formation of a carbonyl group (i.e., the CO-level precursor of the carboxyl group of acetate III). This reaction is reversible and is the reason why acetyl-CoA synthase was first discovered on the basis of its ability to oxidize CO to CO_2 (Reaction IV, above). When CO is available to the cell, CO_2 need not be reduced to the CO level; under these conditions, CO can be used directly by acetyl-CoA synthase as a preformed carbonyl unit. This is energetically favorable to the cell because the reduction of CO_2 to CO is thermodynamically unfavorable and requires energy. However, not all acetogens are able to growth with CO. The reasons for this are not clear but in some cases may be due to unknown toxic effects of CO rather than the inability to use CO.

Although Fig. 3 depicts the methylation of acetyl-CoA synthase as being subsequent to the formation of the carbonyl unit, the order in which the methyl or carbonyl sites are filled may be random, and methylation may occur before the filling of the carbonyl site by CO_2 or CO. Regardless of the order in which the methyl and carbonyl sites are filled, the methyl and carbonyl groups are combined to form an acetyl group that is subsequently transferred to CoA that is bound to the CoA-binding site of acetyl-CoA synthase. The CoA-binding site contains a disulfide bridge that undergoes a ferredoxin-dependent reduction before the binding of CoA. The acetyl-CoA that is formed is then subject to (i) conversion to acetate or (ii) consumption in anabolic processes required for the synthesis of cell material.

E. Variations of the Acetyl-CoA Pathway Used by Other Bacterial Groups

Acetogenic bacteria are not the only organisms that contain acetyl-CoA synthase; it and modified versions of the acetyl-CoA pathway are present in other bacteria, including certain archaebacteria (Table IV). The basic acetyl-CoA pathway and the enzyme acetyl-CoA synthase thus play diverse roles in nature. However, to date, only acetogens are known to use the pathway in the direction of acetate synthesis for both energy conservation (catabolism) and biosynthesis (anabolism). Those organisms that use the pathway in the direction of acetyl-CoA degradation or for only anabolic processes employ different enzyme systems and cofactors that accommodate alternative reactivity, and the details of these processes are outside the scope of the present statement. One organism that has not been classified (strain AOR) appears to have the capacity for the bi-directional use of acetyl-CoA and has been nicknamed "Reversibacterium." Detailed studies on such unusual isolates and their enzymes may shed light on how the pathway can be used for both the formation and degradation of acetyl-CoA.

The acetyl-CoA pathway has been described as the least complicated mechanism known for autotrophic growth. Because the basic acetyl-CoA pathway functions in two directions, depending on the organism, one wonders which directional usage of acetyl-CoA synthase evolved first. Did the capacity to oxidize acetyl-CoA (formed from acetate or other simple organic compounds) precede the capacity to

Table IV Potential Functions of Acetyl-CoA Synthase in Obligately Anaerobic Bacteria[a]

Type of function	Representative bacterial group	Representative organism	Typical reduced end product of metabolism	Interfaced to catabolic or anabolic processes	General function of acetyl-CoA relative to carbon flow
A. Acetyl-CoA forming	Acetogens	*Clostridium thermoaceticum*	Acetate	Both	CO_2 + [H] → [acetyl-CoA] → acetate + biomass
	Autotrophic methanogens	*Methanobacterium thermoautotrophicum*	Methane	Anabolic	CO_2 + [H] → [acetyl-CoA] → biomass
	Autotrophic S-reducing bacteria	*Desulfobacterium autotrophicum*	Sulfide	Anabolic	CO_2 + [H] → [acetyl-CoA] → biomass
B. Acetyl-CoA degrading	Acetate-oxidizing methanogens	*Methanosarcina barkeri*	Methane	Catabolic	acetate → [acetyl-CoA] → CO_2 + CH_4
	Acetate-oxidizing S-reducing bacteria	*Desulfotomaculum acetoxidans*	Sulfide	Catabolic	acetate → [acetyl-CoA] → CO_2 + [H]
C. Bidirectional	?(bidirectional acetogens?)	"Reversibacterium (strain AOR)[b]	Acetate or hydrogen	Both	acetate ⇄ [acetyl-CoA] ⇄ CO_2 + H_2

[a] Fermentative organisms (e.g., *Clostridium pasteurianum*) that have the capacity to oxidize CO via what appears to be acetyl-CoA synthase-dependent activity are not included, as such organisms do not have a known physiological function for this cryptic activity.

[b] An unclassified isolate capable of growth at the expense of either acetate or H_2. When growth is in the direction of H_2 synthesis, the partial pressure of H_2 must be kept very low; these conditions could occur in syntrophic association with H_2 consumers.

autotrophically synthesize acetyl-CoA? At the present time, one can only speculate. However, given the existence of the pathway in archaebacteria, it is tempting to speculate that acetyl-CoA synthase first evolved for the purpose of autotrophic CO_2 fixation and the synthesis of biomass under conditions that were nutritionally minimal and that the capacity to reverse this reaction or to couple this anabolic function to the conservation of energy was evolved subsequently. [*See* ARCHAEBACTERIA (Archaea).]

III. Physiology of Acetogenic Bacteria

A. Physiological Potentials and Regulation

Based on their potentials to use very diverse heterotrophic and chemolithoautotrophic substrates, acetogens must be viewed as a physiologically diverse group. This is in contrast to methanogens that appear to be more restricted by the number and type of substrates they can use.

In recent years, the number of known growth-supportive substrates for acetogens has increased significantly, and the stoichiometries for acetate formation of many of these substrates are outlined in Table V. Note that Fig. 2 illustrates at what point one-carbon units or reductant derived from many of these substrates might enter the acetyl-CoA pathway. Because the enzyme systems responsible for the integration of most substrates into the acetyl-CoA pathway have not been characterized, it is not yet clear how the flow of carbon and reductant is mediated from all growth-supportive substrates. Thus, many of the unique metabolic features of acetogens are yet to be unraveled.

Acetogens have been studied for their potential application in the production of chemicals. One feature that makes them attractive for application is their ability to form acetate as their sole product. With an inexpensive feedstock (carbon source),

Table V Representative Growth-Supportive Substrates and Overall Substrate-Product Stoichiometries of Acetogenic Bacteria That Form Acetate as Their Predominant Product[a]

Substrate	Overall stoichiometry for acetate production
Acetoin	$2\ CH_3COCHOHCH_3 + 2\ CO_2 + 4\ H_2O \rightarrow 5\ CH_3COOH + 4\ H^+$
Alcoxyethanols	$4\ RO\text{-}CH_2CH_2OH + 2\ CO_2 + 2\ H_2O \rightarrow 4\ ROH + 5\ CH_3COOH$
e.g., 2-methoxyethanol	$4\ CH_3OCH_2CH_2OH + 2\ CO_2 + 2\ H_2O \rightarrow 4\ CH_3OH + 5\ CH_3COOH$
e.g., 2-ethoxyethanol	$4\ C_2H_5OCH_2CH_2OH + 2\ CO_2 + 2\ H_2O \rightarrow 4\ C_2H_5OH + 5\ CH_3COOH$
2,3-Butanediol	$4\ CH_3CHOHCHOHCH_3 + 6\ CO_2 + 2\ H_2O \rightarrow 11\ CH_3COOH$
Carbon monoxide	$4\ CO + 2\ H_2O \rightarrow CH_3COOH + 2\ CO_2$
Citrate	$4\ C_6H_8O_7 + 2\ H_2O \rightarrow 9\ CH_3COOH + 6\ CO_2$
Formate	$4\ HCOOH \rightarrow CH_3COOH + 2\ CO_2 + 2\ H_2O$
Fructose	$C_6H_{12}O_6 \rightarrow 3\ CH_3COOH$
Glucose	$C_6H_{12}O_6 \rightarrow 3\ CH_3COOH$
Hydrogen + carbon dioxide	$4\ H_2 + 2\ CO_2 \rightarrow CH_3COOH + 2\ H_2O$
Hydrogen + carbon monoxide	$2\ H_2 + 2\ CO \rightarrow CH_3COOH$
Hydrogen + formate	$2\ H_2 + 2\ HCOOH \rightarrow CH_3COOH + 2\ H_2O$
Malate	$2\ HOOCCHOHCH_2COOH \rightarrow 3\ CH_3COOH + 2\ CO_2$
Methanol	$4\ CH_3OH + 2\ CO_2 \rightarrow 3\ CH_3COOH + 2\ H_2O$
Methoxylated aromatics	$4\ \text{aromatic-}[OCH_3] + 2\ CO_2 + 2\ H_2O \rightarrow 4\ \text{aromatic-}[OH] + 3\ CH_3COOH$
e.g., syringate	$2\ \text{syringate}[OCH_3]_2 + 2\ CO_2 + 2\ H_2O \rightarrow 2\ \text{gallate}[\text{-}OH]_2 + 3\ CH_3COOH$
e.g., vanillate	$4\ \text{vanillate}[\text{-}OCH_3] + 2\ CO_2 + 2\ H_2O \rightarrow 4\ \text{protocatechuate}[\text{-}OH] + 3\ CH_3COOH$
Methyl chloride	$4\ CH_3Cl + 2\ CO_2 + 2\ H_2O \rightarrow 3\ CH_3COOH + 4\ CHl$
Oxalate	$4\ HOOCCOOH \rightarrow CH_3COOH + 6\ CO_2 + 2\ H_2O$
Pyruvate	$4\ CH_3COCOOH + 2\ H_2O \rightarrow 5\ CH_3COOH + 2\ CO_2$
Xylose	$2\ C_5H_{10}O_5 \rightarrow 5\ CH_3COOH$

[a] No distinction is made (a) between acids and their dissociated salt forms or (b) between CO_2 and its carbonate or bicarbonate forms. See text for further information.

large amounts of acetate can be produced. Also intriguing is the ability of some acetogens to convert synthesis gas (CO and H_2) (produced from the gasification of coal, petroleum, or biomass) to acetate or butyrate. *E. limosum* and *B. methylotrophicum* can produce butyrate via the acetyl-CoA pathway and aldol condensation of acetyl-CoA:

$$10\,CO + 4\,H_2O \rightarrow [2\text{ acetyl} = \text{CoA}] \rightarrow CH_3CH_2COOH + 6\,CO_2 \qquad \text{(V)}$$

Current work in this area centers on developing methods (chemical or biological) for the conversion of acetogen-produced acetate and butyrate to other commercially useful products, including solvents, alcohols, methane, gasoline enhancers, biomedical polymers, biodegradable plastics, and environmentally safe road de-icer (calcium-magnesium acetate).

Very little is known of how acetogens regulate their metabolic potentials. Acetyl-CoA synthase, the corrinoid enzyme, and the methyltransferase of *C. thermoaceticum* co-exist on a gene cluster, but the potential regulation of this cluster is not known. Acetyl-CoA synthase was thought for many years to be a constituitive enzyme, but recent studies with *C. thermoaceticum* indicate the activity of this key enzyme is dependent on the growth substrate used to cultivate the organism. Furthermore, not all acetogens have been examined in much detail, and the regulation of acetyl-CoA synthase may vary between species.

Some enzymes responsible for the generation of reductant or use of aromatic substituent groups appear to be strongly regulated. For example, based on activity profiles, the hydrogenase and aromatic *O*-demethylase that generate utilizable reductant from hydrogen and *O*-demethylates aromatic methoxyl groups, respectively, are not produced or are at very low levels unless the acetogen is cultivated in the presence of these substrates. Furthermore, alternative sources of reductant can have a regulatory effect on the activity of these enzymes. For example, in *C. thermoaceticum,* CO induces hydrogenase and glucose represses the *O*-demethylase. Although it is clear that many catabolic processes are regulated at either the gene or enzyme levels, the biochemical mechanisms by which acetogens regulate their metabolic activities are yet to be resolved.

B. Bioenergetics of Acetogens

As shown in Fig. 3, 4 ATP is synthesized from glucose by substrate-level phosphorylation independent of the autotrophic acetyl-CoA pathway. The autotrophic pathway itself is unable to generate any net gain in ATP by substrate-level phosphorylation because the ATP gained at the level of acetate kinase is negated by the cell's need to activate formate during the ATP-dependent synthesis of formyltetrahydrofolate. In general, 1 mole of ATP can yield about 10 g cell biomass dry weight. However, acetogens (e.g., *C. thermoaceticum* and *A. woodii*) can have cell yields approximating 50 to 70 g cell biomass dry weight per mole of glucose or fructose. Because the maximum cell yield from glucose or fructose should only yield about 40 g of cell biomass dry weight and because under autotrophic conditions there is no net gain of ATP possible via substrate-level phosphorylation, the amount of ATP formed by substrate-level phosphorylation does not totally explain how acetogens grow at the expense of acetate synthesis. Because of these observations, acetogens were among the first anaerobes for which anaerobic oxidative phosphorylation was proposed.

Recent studies have shown that many acetogens are rich in cytochromes and other electron carriers and that many carriers are membrane-associated. Furthermore, ion porters and membranous ATPases have been studied, and it has been demonstrated that membrane-mediated electron transport can be coupled to the generation of a proton motive force in at least a few species. These observations strongly support the concept that acetogens catalyze electron transport-mediated phosphorylation.

How then is the oxidation of substrates and the reductive acetyl-CoA pathway coupled to oxidative phosphorylation? In theory, this question can be partially answered by evaluating the changes in the Gibbs free energies ($\Delta G^{\circ\prime}$, indicated in brackets in kJ per reaction) of the reactions involved in H_2/CO_2-dependent acetogenesis:

$$CO_2 + H_2 \rightleftarrows \text{formate} + H^+\ [\Delta G^{\circ\prime} = +3.4] \qquad \text{(VI)}$$
$$\text{formate} + \text{THF} + \text{ATP} \rightleftarrows \text{formyl-THF} + \text{ADP} + P_i\ [\Delta G^{\circ\prime} = -8.4] \qquad \text{(VII)}$$
$$\text{formyl-THF} + H^+ \rightleftarrows \text{methenyl-THF} + H_2O\ [\Delta G^{\circ\prime} = -4.0] \qquad \text{(VIII)}$$
$$\text{methenyl-THF} + H_2 \rightleftarrows \text{methylene-THF} + H^+\ [\Delta G^{\circ\prime} = -23.0] \qquad \text{(IX)}$$
$$\text{methylene-THF} + H_2 \rightleftarrows \text{methyl-THF}\ [\Delta G^{\circ\prime} = -57.3] \qquad \text{(X)}$$
$$CO_2 + H_2 \rightleftarrows CO + H_2O\ [\Delta G^{\circ\prime} = +20.1] \qquad \text{(XI)}$$
$$\text{methyl-THF} + CO + \text{CoA} \rightleftarrows \text{acetyl-CoA} + \text{THF}\ [\Delta G^{\circ\prime} = +21.8] \qquad \text{(XII)}$$
$$\text{acetyl-CoA} + P_i \rightleftarrows \text{acetyl phosphate} + \text{CoA}\ [\Delta G^{\circ\prime} = +9.0] \qquad \text{(XIII)}$$
$$\text{acetyl phosphate} + \text{ADP} \rightleftarrows \text{acetate} + \text{ATP}\ [\Delta G^{\circ\prime} = -13.0] \qquad \text{(XIV)}$$

overall for acetate formation:

$$2\,CO_2 + 4\,H_2 \rightarrow \text{acetate} + H^+\ [\Delta G^{\circ\prime} = -95.0] \qquad \text{(XV)}$$

As CO_2 is reduced, several oxidation–reduction reactions occur, and the one in which methylenetetrahydrofolate is reduced to methyltetrahydrofolate (Reaction X) is energetically posed at a level that is particularly favorable for the synthesis of ATP via oxidative phosphorylation. The enzyme that catalyzes this reaction (methylenetetrahydrofolate reductase) has been shown to be membrane associated in *Clostridium thermoautotrophicum*. Sodium-dependent, energy-conserving processes appear to be involved in some but not all acetogens, and some evidence suggests that Reaction X is sodium-dependent. However, the role of sodium remains unknown. Although a general model has been proposed for a membranous electron transport chain in *C. thermoautotrophicum* that terminates at methylenetetrahydrofolate reductase and extrudes protons used to drive the H^+-ATPase-dependent synthesis of ATP, a comprehensive statement that would apply to all acetogens is not possible at the present time. This is especially true in the case of acetogens that are capable of using alternative terminal electron acceptors (e.g., acrylates and fumarate) during the oxidation of certain substrates.

IV. Environmental Roles of Acetogenic Bacteria

Although acetogenic bacteria were initially considered a peculiar bacteriological group that used an unusual process for the fixation of CO_2, the study of acetogenic bacteria and the acetyl-CoA pathway has taken on new dimensions in recent years. Acetogenic bacteria have been shown to have many diverse metabolic potentials that link them closely to the turnover of carbon in anaerobic habitats. That they have such diverse capabilities should make them very competitive in their native habitats. In this regard, the advantageous use of aromatic compounds might provide them with growth-supportive substrates under conditions not conducive to the growth of many anaerobes. For example, *C. thermoaceticum* has recently been shown to use aromatic carboxyl groups as a CO_2-equivalent terminal electron acceptor for both the conservation of energy and the total synthesis of acetate under conditions that would not support the growth of many anaerobes. In addition, many acetogens are also able to use more than one growth-supportive substrate at a time and can grow mixotrophically. These general potentials would also make them very competitive relative to other organisms present in their environments.

To date, acetogens capable of using large polymers such as cellulose and lignin have not been isolated. However, they have been shown to be involved in the overall breakdown of such polymers in co-culture with other organisms. A recent isolate of *C. thermoautotrophicum* has been shown to degrade the plant polysaccharide inulin, $(C_6H_{10}O_5)_4$, and it is therefore likely that other larger polymers are subject to degradation or transformation by acetogens that have not yet been isolated. In addition, recent evidence indicates that acetogens can be linked to the turnover of halogenated compounds and may thus play a role in not only the cycle of carbon but also the bioremediation of contaminated soils and sediments.

Acetogens form nutritional partnerships with many diversely related organisms, and many specialized ecosystems are, in fact, strictly dependent on either acetogenic bacteria or biochemical variations of the acetyl-CoA pathway. For example, they occur in many gastrointestinal ecosystems. Certain termites are nutritionally dependent on acetate production by acetogens that reside in their intestinal tract, and it has been estimated that 3.3×10^6 tons of acetate are produced per day from the hindgut of these animals. This amount is several-fold higher than the amount of methane produced globally from CO_2 (estimated at 6×10^5 tons of CO_2 reduced to methane per day from all biogenic sources). It has also been estimated that acetogens common to the human gut produce nearly 5 million metric tons of acetate per year from H_2 and CO_2. In addition, some acetogens appear to participate in interspecies H_2 transfer and form syntrophic relationships with other bacteria; resolving such relationships will further our understanding of where acetogenic bacteria are found in nature and what specialized features allow them to grow syntrophically.

Because acetate is one of the primary substrates for methanogens, methanogenesis is, in many cases, coupled to the production of acetate, and wastewater recycling is likely coupled to degradative processes catalyzed by acetogens. Methanogens and sulfur-reducing bacteria play extremely critical roles in the carbon and sulfur cycles, and the use of the acetyl-CoA pathway by members of these bacteriological groups further accentuates the general importance of the pathway. [*See* METHANOGENESIS.]

Most studies to date have examined biochemical, physiological, or ecological properties of acetogens. Relatively little is known of the genetics or natural mechanisms of gene transfer used by acetogenic bacteria, and how these processes contribute to the metabolic interface between acetogens and their environments is not known. However, the presence of acetogens in various ecosystems is now being studied using group-specific DNA probes. Such techniques can be applied to study the natural populations, species diversity, and *in vivo* stability of acetogens. Although future studies must resolve more clearly the extent to which acetogens and other organisms that use variations of the acetyl-CoA pathway populate habitats, form communities, and impact on global cycles, it is clear that acetogens and the autotrophic CO_2-fixing pathway they use are far more important than originally conceived.

Bibliography

Breznak, J. A., and Blum, J. S. (1991). *Arch. Microbiol.* **156,** 105–110.

Breznak, J. A., and Kane, M. D. (1990). *FEMS Microbiol. Rev.* **87,** 309–314.

Daniel, S. L., Hsu, T., Dean, S. I., and Drake, H. (1990). *J. Bacteriol.* **172,** 4464–4471.

Daniel, S. L., Keith, E. S., Yang, H., Lin, Y.-S., and Drake, H. L. (1991). *Biochem. Biophys. Res. Commun.* **180,** 416–422.

Dörner, C., and Schink, B. (1991). *Arch. Microbiol.* **156,** 302–306.

Fuchs, G. (1986). *FEMS Microbiol. Rev.* **39,** 181–213.

Fuchs, G. (1989). *In* "Autotrophic Bacteria" (H. G. Schlegel and B. Bowien, eds.), pp. 365–382. Springer-Verlag, Berlin.

Geerligs, G., Schönheit, P., and Diekert, G. (1989). *FEMS Microbiol. Letters* **57,** 253–258.

Gottschalk, G. (1989). *In* "Autotrophic Bacteria" (H. G. Schlegel and B. Bowien, eds.), pp. 383–396. Springer-Verlag, Berlin.

Heise, R., Müller, V., and Gottschalk, G. (1989). *J. Bacteriol.* **171,** 5473–5478.

Hsu, T., Lux, M. F., and Drake, H. L. (1990). *J. Bacteriol.* **172,** 5901–5907.

Hugenholtz, J., Ivey, D. M., and Ljungdahl, L. G. (1987). *J. Bacteriol.* **169,** 5845–5847.

Hugenholtz, J., and Ljungdahl, L. G. (1989). *J. Bacteriol.* **171,** 2873–2875.

Lajoie, S. F., Bank, S., Miller, T. L., and Wolin, M. J. (1988). *Appl. Environ. Microbiol.* **54,** 2723–2727.

Lee, M. J., and Zinder, S. H. (1988). *Appl. Environ. Microbiol.* **54,** 124–129.

Lovell, C. R., and Hui, Y. (1991). *Appl. Environ. Microbiol.* **57,** 2602–2609.

Lu, W.-P., and Ragsdale, S. W. (1991). *J. Biol. Chem.* **266,** 3554–3564.

Ragsdale, S. W. (1991). *Crit. Rev. Biochem. Mol. Biol.* **26,** 261–300.

Roberts, D. L., James-Hagstrom, J. E., Garvin, D. K., Gorst, C. M., Runquist, J. A., Bauer, J. R., Haase, F. C., and Ragsdale, S. R. (1989). *Proc. Natl. Acad. Sci. USA* **86,** 32–36.

Terracciano, J. S., Schreurs, W. J. A., and Kashket, E. R. (1987). *Appl. Environ. Microbiol.* **53,** 782–786.

Thauer, R. K., Möller-Zinkhan, D., and Spormann, A. M. (1989). *Annu. Rev. Microbiol.* **43,** 43–67.

Traunecker, J., Preuβ, A., and Diekert, G. (1991). *Arch. Microbiol.* **156,** 416–421.

Wiegel, J., Carreira, L. H., Garrison, R. J., Robeck, N. E., and Ljungdahl, L. G. (1990). *In* "Calcium Magnesium Acetate" (D. L. Wise, Y. Levendes, and M. Metghalci, eds.) pp. 359–416. Elsevier, Amsterdam.

Wood, H. G. (1991). *FASEB J.* **5,** 156–163.

Wood, H. G., and Ljungdahl, L. G. (1991). *In* "Variations in Autotrophic Life" (J. M. Shively and L. L. Barton, eds.), pp. 201–250. Academic Press, San Diego.

Worden, R. M., Grethlein, A. J., Zeikus, J. G., and Datta, R. (1989). *Appl. Biochem. Biotechnol.* **20/21,** 687–698.

Yang, H., and Drake, H. L. (1990). *Appl. Environ. Microbiol.* **56,** 81–86.

Young, L. Y., and Frazer, A. C. (1987). *Geomicrobiol. J.* **5,** 261–293.

Acquired Immunodeficiency Syndrome (AIDS)

Jonathan Vogel and Gilbert Jay
Jerome H. Holland Laboratory

Glossary

Cell tropism When an organism such as a virus shows a special affinity for and may preferentially infect or replicate in a particular cell type or tissue

Cytopathicity Pertaining to or characterized by the pathological changes in cells

Dementia Organic mental disorder characterized by a general loss of intellectual abilities involving impairment of memory, judgment, and abstract thinking as well as changes in personality

Epitope Antigenic determinant

Homologous Corresponding or similar in structure, position, origin, and sequence

Pathogenesis Development of morbid conditions or of disease; more specifically, the cellular events and reactions and other pathologic mechanisms occurring in the development of disease

Replication Process of duplicating or reproducing, such as the replication of an exact copy of a polynucleotide strand of DNA or RNA, or the reproduction of new progeny viruses

Retroviruses Large group of RNA viruses including lentiviruses, so named because they carry the enzyme reverse transcriptase used to convert viral genomic RNA into double-stranded DNA

Splicing Removal or splicing out of intervening genomic sequences called introns and the joining or splicing together of exons during transcription of a gene

Transcription Synthesis of a single-stranded RNA with a base sequence complementary to one strand of a double-stranded DNA

THE ACQUIRED IMMUNODEFICIENCY SYNDROME, or AIDS, is a constellation of clinical manifestations that result from human immunodeficiency virus (HIV) infection. Although a long period of up to 10 years may be present between the time of infection with HIV and the actual onset of AIDS, it is generally believed that most people infected with HIV will eventually develop AIDS.

I. Introduction

The manifestations of AIDS are best understood by dividing them into three main groups: the immunodeficient state, AIDS-associated neurological problems termed AIDS–dementia complex, and AIDS-associated malignancies. As the moniker AIDS implies, the immunodeficiency associated with HIV infection can be profound and is the major cause of death. Because of the loss or dysfunction of the immune system, patients are no longer able to defend themselves against infectious agents such as bacteria, viruses, fungi, and protozoans and are overwhelmed with infectious complications. Many of these infectious agents are commensal organisms and in normal individuals would not be the cause of an infection. The neurological problems of AIDS–dementia complex are a common finding in HIV infection and consist of both cognitive (memory, calculations) and motor (coordination) abnormalities as well as behavioral problems. Finally, a high percentage of HIV-infected individuals will develop cancer at some point during the course of their disease, the most common being Kaposi's sarcoma and non-Hodgkin's lymphoma.

Although many questions about how HIV causes

AIDS remain unanswered, an astonishing amount of information about HIV has been accumulated in a relatively short time. This is due to the contributions of many different scientific disciplines, including epidemiology, molecular biology, virology, and clinical medicine, which have converged to give us a clearer picture of the characteristics of HIV. The story of HIV infection and the resulting AIDS epidemic is best told from these different perspectives. Correspondingly, we will first discuss HIV infection and AIDS from an epidemiological perspective. Next, we will discuss the molecular biology and structure of HIV and attempt to correlate the function and life cycle of HIV with the progression of the different disease manifestations of AIDS. The correlation of structure and function of HIV with disease progression is important because it allows us to attempt to define actual mechanisms. As an understanding of these disease mechanisms improves, precise treatment and therapeutic modalities can be defined and implemented. As we proceed, a number of questions should be kept in mind. How does infection with HIV cause the immunodeficiency of AIDS? After an individual is infected with HIV, why does it take up to 10 years to develop AIDS? If the virus is quiet or clinically latent for this long period of time, what activates it and causes active disease? How do we explain the role of HIV in important associated illnesses such as AIDS–dementia complex and AIDS-associated cancer? And what future trends do the epidemiologists predict for HIV infection and AIDS in the future? Some of the questions can be answered fairly accurately or at least with good educated guesses, whereas others still require a great deal of speculation. [*See* EPIDEMIOLOGIC CONCEPTS.]

II. Epidemiology of HIV Infection and AIDS

Epidemiological data are crucial in framing and defining an epidemic such as AIDS. By searching for common denominators in the characteristics of infected individuals, such as their sex, race, geographical distribution, age, behavior, and lifestyle, epidemiologists could define the risk factors most likely to be associated with HIV infection and development of AIDS. This contributed to the early realization that an infectious agent was involved and helped to define the three main routes of infection: sexual transmission from infected individuals, parenteral transmission via infected blood products, and perinatally from infected mother to child. These three main routes of infection still define the population groups at high risk for HIV infection. When analyzing this epidemiological data, it is important to remember that two different groups are being tracked: those individuals who already have the symptoms of AIDS and those individuals who are infected with HIV but do not yet have AIDS. In practice, it is much easier to obtain accurate data on individuals with AIDS than to determine how many people are infected with HIV but free of AIDS symptoms. There is no standardized or general screening of large populations to determine who is infected with HIV. Given the long and variable lag period between HIV infection and AIDS, and the imprecise knowledge of how many individuals are infected, only rough estimates can be made about the numbers of AIDS cases that will occur in the future. Another important feature to consider is that although cumulative epidemiological data are often presented (1981–1989), there are changing trends in the HIV epidemic in terms of who is being infected, what their age, sex, race, and ethnic group are, and where they live. Additionally, each risk group may have its own unique trend. Cumulative data can often mask or hide these recent and changing trends.

Between 1981 and 1989, more than 100,000 AIDS cases were reported to the Centers for Disease Control, with over 35,000 cases reported in 1989 alone. During the 1990s, the number of AIDS cases each year is expected to increase, with a cumulative AIDS case count expected to reach 390,000–480,000 in the United States by 1993. Death is the usual inevitable result once AIDS symptoms occur. The number of HIV-infected individuals nationwide is estimated at 1 million.

The global statistics are even more troubling. The number of people worldwide infected with HIV is estimated to be 6–8 million. Projection of the number of AIDS cases that will occur worldwide throughout the 1990s is at least 3 million, resulting from the up to 8 million people who are estimated to have already been infected with HIV as of mid-1990. The sub-Saharan countries of Africa have been particularly hard hit with cumulative estimated HIV infections of over 5 million and over 500,000 AIDS cases. The heterosexual spread of HIV infection in Africa and parts of Asia has continued to increase and is a major source of new HIV infections in the world today. The spread of HIV infection in Latin

America initially was confined to homosexual or bisexual men, intravenous drug users (IVDUs), or those receiving contaminated blood products. However, heterosexual spread has now become the dominant mode of transmission of virus in this region. Latin America has approximately 1 million HIV-infected individuals and 75,000 cumulative AIDS cases. These global trends portend the increasing role that heterosexual transmission will play in the spread of HIV infection in the United States.

A. Geographical Distribution of AIDS in the United States

The HIV epidemic is different today compared to 9 years ago, the geographic focus has widened, the profile of persons at greatest risk is changing, and the relative importance of different routes of transmission is shifting. Geographically, about two-thirds of all AIDS cases have come from five states: New York, New Jersey, Florida, Texas, and California. The rates are the highest in the most populous metropolitan areas with populations over 1 million. New York City continues to report the most cases (18,000) and, at the end of 1988, exceeded the combined total from San Francisco (6600), Los Angeles (6000), Houston (2600), and Newark (2500). Although these urban trends are expected to persist, the proportion of AIDS cases in urban areas under 500,000 population is beginning to increase, most notably in the Midwest and South Atlantic regions. Correspondingly, the proportion of recent AIDS cases from New York and New Jersey is beginning to decrease.

B. Trends in AIDS Risk Groups

As has been the case since the epidemic's earliest days, most AIDS patients are male (90%), although the proportion of adult women with AIDS has been steadily increasing from 7% in 1984 to 10% in 1989. The vast majority of adult AIDS patients are homosexual or bisexual men (61%), intravenous drug users (21%), or both (7%). Note that these two risk groups include sexual and parenteral routes of HIV infection, respectively. An additional 2% of AIDS patients are recipients of HIV-infected blood transfusions, and 1% are patients with blood disorders such as hemophilia who presumably received HIV-contaminated blood products. The last defined group of approximately 5% consists of individuals, male and female, who were infected through heterosexual contact with an HIV-infected partner. The remaining 3% of HIV-infected individuals have an unknown source or route of infection.

Homosexual and bisexual men were the first group in which AIDS was recognized in the United States. However, the trends of HIV infection are changing in this group. On a yearly basis, they now account for a smaller proportion of all AIDS cases, from 63% in 1985 to 57% in 1989. Because the risk of infection through sexual contact increases with the number of sex partners and the frequency of sexual practices such as receptive anal intercourse, changes in sexual practice may account for the reported drop of new HIV infections in some homosexual populations. Geographically, a declining proportion of AIDS cases in homosexuals is now being reported in New York City, Los Angeles, and San Francisco.

The parenteral spread of HIV by blood-contaminated needles for intravenous drug use is playing an increasing role in the spread of HIV, especially into the heterosexual population. The transmission of HIV has spread from IVDUs to their sex partners and their newborn children. The incidence of IVDU-associated AIDS cases is increasing and made up approximately 33% of all AIDS cases reported in 1989. This IVDU-associated group includes both heterosexual and homosexual IVDUs as well as the partners and offspring of IVDUs. Intravenous drug use is now the major path through which HIV is infecting heterosexual men and women in the United States and accounts for approximately 70% of the AIDS cases in which HIV infection was due to heterosexual contact. In most cases, this heterosexual contact is usually a partner who is an IVDU and infected with HIV. Correspondingly, HIV infection in prostitutes is strongly associated with a history of intravenous drug use by these prostitutes. Intravenous drug use by mothers also accounts for about 50% of perinatally infected newborns, so it should be apparent that the IVDU-associated route of transmission is affecting many different target populations. With regard to race and ethnic group analysis, the rate of IVDU-associated AIDS is higher for Blacks and Hispanics than for Whites. In fact, this higher incidence of IVDU-associated AIDS is the most significant contributing factor to the overall higher relative incidence of AIDS in Black and Hispanic groups, which have 40% of all reported AIDS cases. The infection rate among IVDUs, derived from studies conducted in drug treatment centers, is highest in New York City, Puerto Rico, and the metropolitan areas of the Northeast. In contrast to this rising rate, the rate of transmission of HIV infection through contaminated blood and blood products has been greatly reduced by routine screening of blood and blood products for HIV infection.

C. Natural History of HIV Infection

The studies of the natural history of HIV indicate that the interval between HIV infection and the manifestations of AIDS is variable but often long. In one study of HIV-infected homosexual men in San Francisco, the median period between infection and AIDS is estimated at 9.8 years, with only 15% of the men asymptomatic after 100 months. Other studies have shown that the incubation period varies by age and is shorter in infants and older adults than in young adults and adolescents. Once AIDS has been diagnosed, the average survival time has been approximately 12 months, but this varies and is influenced by which clinical manifestations or complications of AIDS the patient exhibits. Because AIDS disproportionately affects younger people (20–44-yr-old age group), deaths due to AIDS lead to substantial decreases in life expectancy, especially in areas of high AIDS incidence. It is hoped that early prophylactic treatment of infectious complications and new therapies will prolong the survival after diagnosis.

III. HIV Gene Structure and Regulation of Expression

A precise knowledge of the genetic structure of HIV is necessary to gain an understanding of how HIV infection eventually leads to the symptoms of AIDS. This knowledge will also provide the basis for the design of antiviral drugs and vaccine development.

A. Retroviral Life Cycle

HIV is a human retrovirus that contains a diploid single-stranded (ss) RNA genome (Fig. 1). The mature retrovirus particle contains an RNA-protein core surrounded by a lipid membrane embedded with viral proteins. All retroviruses have similar replicative life cycles. After the virus enters into the cell, the RNA genome is converted into a double-stranded (ds) DNA form by an enzyme produced by

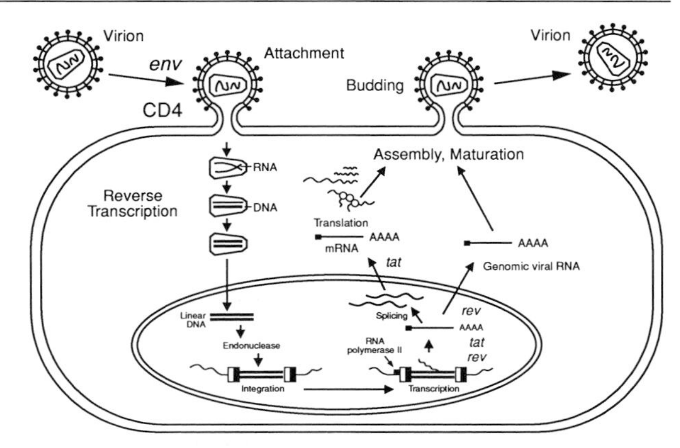

Figure 1 Life cycle of HIV.

the retrovirus, a process called reverse transcription. This dsDNA migrates to the nucleus, where it is integrated into the cellular chromosomal DNA by an enzyme named integrase, also made by the retrovirus. This integrated dsDNA is the proviral form of the retrovirus. To complete the replicative life cycle, the dsDNA must be transcribed into messenger RNA (mRNA) to produce necessary viral proteins as well as complete viral genomic ssRNA, which will be incorporated into progeny viral particles. [*See* RETROVIRUSES.]

On the basis of genetic and morphologic parameters, HIV is placed in the Lentivirinae subfamily of retroviruses. There is significant homology between HIV and the animal lentiviruses (e.g., visna virus, equine infectious anemia virus, caprine arthritis–encephalitis virus) both in genome organization and in the amino acid sequence of viral proteins. Parallels also exist between HIV and the animal lentiviruses in their biological properties such as their propensity to induce progressive wasting disease and disorders involving the immune and neurologic systems. Additionally, both HIV and the other lentiviruses demonstrate rapid genetic drift, which means infected individuals may contain different subtypes of HIV with different sequences.

B. The HIV Genome

The integrated proviral genome of HIV is 9700 bp long and has a typical structure for retroviruses consisting of long terminal repeat (LTR) regions flanking the genes coding for the major structural proteins, *gag*, *pol*, and *env* (Fig. 2). These structural proteins either form the viral particle or consist of enzymes that are critical to HIV replication, and all are incorporated into the viral particle. The LTR regions do not code for any genes but do contain regulatory sequences that are important for controlling viral gene expression. In addition to the above genes, HIV possibly contains eight additional genes that encode proteins that provide the virus with the ability to control its own expression at many different levels (Fig. 2). These regulatory genes are known as *tat, rev, nef, vif, vpu, vpr, vpt,* and *tev.* The ability of these different regulatory components to interact with each other will be discussed in the following section.

C. The Long Terminal Repeat Region

The identical 630-bp LTR regions are at both ends of the proviral genome and are divided into discrete functional units designated U3, R, and U5, extending 5′ to 3′. The U3 and R units of the LTR contain at least five elements or sequences which control the expression of all the structural and regulatory genes (Fig. 3). These sequences interact with both cellular transcription factors and HIV regulatory proteins and are known as cis-acting control elements. All HIV mRNA transcripts are initiated or started at the U3-R junction (Fig. 3). The five elements present in the U3 and R regions are the TATAA promoter region, Sp1-binding sites, the transcriptional enhancer, the negative regulatory elements (NREs), and the trans-acting responsive region (TAR region). The TATAA promoter region is present in most eukaryotic genes and is responsible for determining the proper initiation of transcription at the U3-R junction. The three Sp1 elements bind the cellular transcription factor Sp1. Factors that bind to the Sp1 region also may help to stabilize factors that bind to the adjacent TATAA and enhancer elements. The enhancer region is important because it contains an element that binds a cellular factor known as NFκB. This factor is produced by T lymphocytes when they are activated and, consequently, may play a role in turning on HIV gene

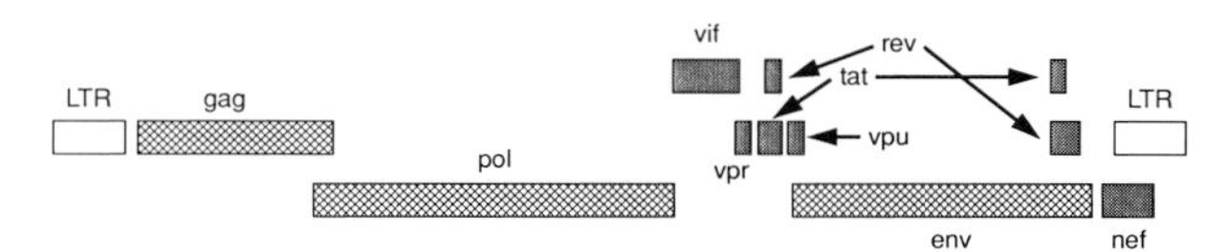

Figure 2 The genomic structure of human immunodeficiency virus (HIV), including the long terminal repeat (LTR) regulatory regions, the structural genes, and the regulatory genes.

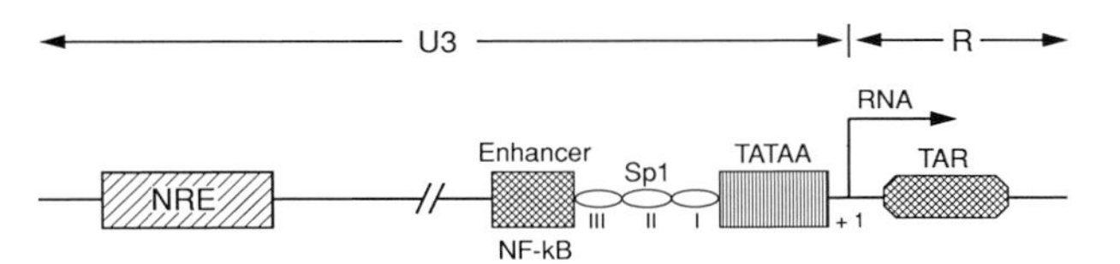

Figure 3 The organization of critical regulatory sequences in the U3 and R regions of the long terminal repeat. The initiation or start site for RNA transcription is shown at the junction of the U3 and R regions. NF-kB, NFκB cellular factor; NRE, negative regulatory element; TAR, trans-acting responsive region.

expression in these activated T cells. As we will discuss in further detail later, the activation of HIV in T lymphocytes may play an important role in the destruction of these cells and in the pathogenesis of HIV. The NRE appears to suppress the initiation of transcription because deletion of this sequence yields more replicating viruses in tissue culture and increased LTR gene expression. It may play a role in down-regulating HIV gene expression. [*See* T Lymphocytes; Transcriptional Regulation, Eukaryotes.]

Perhaps the most intriguing and important cis-acting regulatory element in the HIV LTR is the TAR region, which is present in the R unit immediately after the start site of mRNA transcription (Fig. 3). This region is absolutely critical for HIV replication and gene expression. The ability of the potent transactivating gene of HIV, the *tat* gene, to up-regulate or increase viral transcription is mediated through this target sequence. The exact mechanism of how the *tat* gene transactivates through the TAR region is an area of intense scientific interest. One of the confusing features of the TAR element is that, because it is located after the mRNA start site, this sequence is present in both the proviral DNA and the transcribed RNA. Therefore, it is not immediately clear if the Tat protein and other cellular factors interact with the DNA or mRNA sequences when they are up-regulating HIV gene expression. Much of the recent research suggests that Tat transactivates by acting on the TAR region on the RNA level. The TAR region can form a stable secondary structure known as a stem-loop, where the RNA folds back on itself, allowing complementary RNA base pairs to form hydrogen bonds. The formation of this secondary structure is critical for the TAR region, as mutations that disrupt this stem-loop structure abolish transactivation. Other features critical for function are its 5′ to 3′ orientation and its location relative to the RNA start site. The Tat protein and other cellular factors may bind to this TAR region stem-loop structure and enhance transcription of mRNA through this region, which is present in all transcribed HIV molecules. An alternative mechanism would be that Tat and cellular factors may bind to stabilize the RNA, thereby increasing its accumulation and accelerating the production of viral proteins. It is important to emphasize that the synthesis of all viral proteins, both structural and regulatory, depends on the Tat protein–TAR region interaction because the transcription of all viral mRNAs start in the LTR and contain the TAR region at their 5′ ends.

D. The Structural Genes

The *gag* gene encodes the precursor of the internal structural proteins of the virus, including the inner membrane protein (p17), major capsid protein (p24), and nucleocapsid proteins (p9, p7). The nucleocapsid proteins bind to the viral RNA genome when they are packaged into the viral particle.

The *pol* gene encodes a precursor protein, which contains four different enzymatic activities that are important for the HIV replicative life cycle: protease, reverse transcriptase DNA polymerase, ribonuclease H, and integrase. It should be noted that these enzymes are contained within the viral core and, consequently, can function as soon as the virus enters the cell. The role of the protease is the cleavage of the Gag and Pol precursor proteins into their respective subunits. Additionally, the final maturation step of the viral particle after it leaves the cell is the condensation of the protein-RNA core, and this step depends on viral protease activity. Both the reverse transcriptase and the ribonuclease H enzymes are required for the conversion of the viral genomic ssRNA into complementary dsDNA. The final enzymatic activity contained in the Pol precursor protein is the integrase, which has an endonuclease activity and is involved with the insertion of the HIV proviral dsDNA into cellular DNA.

The product of the *env* gene has been the focus of intense research interest for a number of reasons. The Env protein gets added to the viral particle as it buds out of the cell, and, because this is the exposed or coating protein of the viral particle (Fig. 1), the Env protein is a target for the immune system in individuals exposed to HIV. It therefore provides a target for antiviral drug and vaccine development, and extensive sequence comparisons and epitope mapping have been performed on the Env protein of different HIV strains. The Env protein also determines which cells the HIV particle will attach to once an individual is infected. The Env protein has a very high binding affinity for the CD4 antigen, which is present on immune cells, including a very important subset of T lymphocytes, known as helper T lymphocytes. These helper cells enhance the ability of the immune system to respond to foreign antigens and infectious organisms. Therefore, the Env protein may play a pivotal role in the toxicity of HIV toward T lymphocytes by targeting these cells. The Env protein is made as a single polypeptide chain. It is heavily glycosylated by cellular enzymes that add complex sugar side chains. After the protein is transported to the cell surface, it is cleaved into the N-terminal gp120 portion and the C-terminal gp41 protein. The gp120 protein is located entirely external to the membrane, whereas the gp41 protein is the transmembrane protein, which is partially in and partially out of the membrane.

Because of the extensive comparisons of the Env sequences of different strains of HIV, four conserved regions and five hypervariable regions have been identified within gp120. In general, the hypervariable regions are separated by conserved regions. These variable regions have characteristics of antigenic sites, and their sequence variation means that different strains of HIV will require different antibodies to neutralize or prevent infection of susceptible target cells. This is a problem for the development of a vaccine against the Env protein. The most important variable region of gp120 for eliciting type-specific neutralizing antibodies is in the third variable region known as V3. A number of other critical functional domains of gp120 and gp41 have been identified. A specific CD4-binding site that mediates the cell-tropism of HIV for $CD4^+$ cells is present in a conserved region of the C terminus of gp120. There is also a domain that specifies membrane-to-membrane fusion of an HIV-infected cell to other infected cells or to noninfected cells that carry the CD4 marker. The membrane-to-membrane fusion refers to the formation of giant multinucleated syncytia, which HIV-infected cells are known to form in tissue culture. This syncytial formation has been shown to be mediated by the Env protein, which is present on the surface of infected cells.

E. The Regulatory Genes

Although the exact function of each of the eight regulatory genes listed above are not clearly defined, much information has been acquired about the critical role that the *tat* and *rev* genes play in HIV expression and replication. To understand fully how the regulatory genes influence viral replication, one must understand how splicing of the RNA from HIV into different mRNA transcripts influences the relative amounts of different HIV proteins produced. The process of splicing occurs during the transcription of most cellular genes as well as the integrated HIV genome and can be thought of as the joining together of mRNA sequences from different portions of the genome to create new mRNA transcripts. The portion of the genomic DNA that becomes part of the mRNA transcript after splicing is known as the exon, and the part of the genome that is spliced out or does not become part of the mRNA transcript is termed the intron. The same genomic sequence could contain different combinations of exons and introns and, by different splicing patterns, could give rise to mRNA transcripts of different sizes and sequences and ultimately produce different protein products. This is the situation for HIV in that the integrated viral genome will give rise to mRNA transcripts of different sizes and sequences; some are unspliced and 9700 bp long, while others are shorter, depending on the amount of splicing that occurs. In general, the regulatory genes of HIV are present in multispliced mRNA transcripts (Fig. 2), whereas by comparison, the structural *gag* and *pol* genes are present in an unspliced full-length viral mRNA, and the *env* gene is present in a singly-spliced transcript. For example, through the splicing process, the *tat* and *rev* genes are made up of sequences or exons derived from different portions of the HIV genome as shown in Fig. 2. Regulating the relative amounts of these differentially spliced transcripts is important for HIV because this is how HIV can control the relative amounts of different viral proteins that are made. For HIV to have a successful replicative life cycle, different structural and regulatory proteins must be made in the correct amounts at the correct time. To summarize, the function of the known regulatory genes is to both maintain the correct rate of transcription and control the relative amounts of the different spliced RNA transcripts. [*See* RNA SPLICING.]

The *tat* transactivator gene is a positive feedback regulator that increases the rate of synthesis of all viral proteins and is necessary for HIV expression and replication. The TAR region, which is the cis-acting target sequence of the Tat protein, has already been described along with potential mechanisms for Tat action. The Tat protein is 86 amino acids long and is encoded by a multispliced transcript that contains the two coding exons indicated in Fig. 2. Three functional domains have been proposed based on mutational analysis and sequence comparisons: an acidic domain, a cysteine-containing domain, and a basic domain. Mutations or substitutions in the acidic domain and point mutations of the cysteines in the second domain are each capable of causing a loss of Tat function, whereas mutations in the basic region seem to interfere with its nuclear localization.

The *rev* regulatory gene is also a transactivating gene; however, in contrast to the *tat* gene, which up-regulates the expression of all HIV proteins, the Rev protein up-regulates the expression of structural proteins and down-regulates the expression of other regulatory genes. By controlling the relative amounts of the different spliced mRNA transcripts that encode the different HIV proteins, Rev can control which proteins are made. It will increase the transcription of the unspliced and singly spliced transcripts, which code for structural proteins, while inhibiting the transcription of the multispliced transcripts, which encode the regulatory proteins. Two cis-acting sequences are important in the Rev regulatory mechanism. The first are the multiple cis-acting repression sequences (CRS) within the structural *gag, pol,* and *env* genes, which have an inhibitory effect on protein expression. The CRS may prevent the unspliced structural gene transcripts from being used as a substrate for protein synthesis by retaining the viral RNA in a nuclear compartment where splicing and degradation can occur. Without the presence of Rev, unspliced RNA is not transported out of the nucleus. The inhibitory effect of the CRS can be overcome by a second cis-acting sequence, the Rev-responsive element (RRE), which is present in the gp41 portion of the *env* gene. The interaction of Rev with the RRE sequence overrides the inhibitory effect of the CRS on the synthesis of structural proteins and permits the structural gene mRNA that contains the CRS sequences to be transported out of the nucleus in an unspliced form and translated into protein. The RRE sequence is a highly conserved 234-bp sequence that can form a very stable stem-loop secondary structure, and this secondary RNA structure is probably bound directly by the Rev protein. The Rev protein appears to have at least two functional domains: a basic N-terminal domain and a C-terminal domain. Mutations in the N-terminal domain result in loss of function probably by interfering with nuclear localization, whereas specific mutations in the C-terminal end also cause a loss of Rev's ability to transactivate. To summarize, the interaction of the Rev protein with the RRE sequence is critical for the enhanced expression of structural proteins from unspliced HIV mRNAs. In Fig. 1, the effects of Tat and Rev are schematically depicted.

The exact functions of the other regulatory genes are not as well defined and, in contrast to the *tat* and *rev* genes, do not appear to be absolutely critical for HIV gene expression and replication. The *nef* (negative regulation factor) gene may play the role of a negative regulatory gene, which is capable of downmodulating viral gene expression and replication. There are reports that Nef mutants can replicate more rapidly in tissue culture and may have enhanced infectivity, suggesting that Nef can inhibit replication. There are also reports that the Nef protein inhibits expression from the HIV LTR through cis-acting elements in the NRE region of the LTR, although other data have failed to confirm this. The Nef protein is located in the cytoplasm but may also be embedded in the inner surface of the cell membrane. Several unique structural features of the Nef protein suggest that it has the potential to act as a signal-transducing factor by modifying cellular proteins that regulate transcription initiation. These features are the following. (1) Functional assays suggest that Nef both binds guanosine triphosphate (GTP) and has GTPase activities. Nef is also reported to be a protein kinase capable of autophosphorylation and is a substrate of protein kinase C. (2) Nef has amino acid sequence homologies with the Ras family of proteins, including regions that bind nucleotides such as GTP. (3) Nef can contain the fatty acid myristic acid at its N terminus, suggesting that this form of protein is localized at the inner membrane.

The *vif* (virion infectivity factor) gene is conserved among different HIV strains and encodes a 23-kDa protein. Proviruses that have a mutated or deleted *vif* gene are capable of both replication and cytopathicity in tissue culture assays but with reduced efficiency. The Vif protein may increase the efficiency of steps that occur after virus attachment but before the viral DNA is integrated into the host cell DNA. However, the exact mechanism of Vif is unknown.

The *vpu* gene is not required for HIV replication because mutants defective in the gene are capable of replicating. Functional analysis of the Vpu protein suggests that it enhances viral release from the cell by facilitating HIV assembly and budding of viral particles. Both Vpu and Vif appear to enhance the production of large amounts of infectious virus.

The *vpr* gene is highly conserved in all HIV strains as well as other lentiviruses. Mutational analysis indicates that Vpr is not required for HIV replication and cytopathicity in tissue culture. Nevertheless, Vpr seems to function as a transactivator and to activate the HIV LTR and accelerate viral replication. It is the only regulatory protein incorporated into the viral particle, suggesting that it may play an early role in the establishment of infection. Researchers believe that additional HIV genes such as the *vpt* and *tev* genes encode proteins, but the function of these genes is still unknown.

IV. Pathogenesis of HIV Infection

Following HIV infection, the progression to AIDS proceeds through relatively distinct clinical states. The initial clinical infection is often associated with symptoms that are typical for viral infections in general. Rapid viral replication presumably occurs during this initial infection, and cells with the CD4 cell-surface marker, such as helper T lymphocytes, are preferentially infected. Consequently, these cells may decline while still remaining within normal limits. With the development of an immune response to HIV, the initial infection ends and an asymptomatic period of clinical latency then follows. The term clinical latency only refers to a lack of clinical symptoms, not to a lack of expression of HIV genes or HIV replication, which presumably continues. The degree of HIV replication during this period is debated, but presumably the rate of replication is well controlled or suppressed and target cells of HIV infection, such as the $CD4^+$ cells, either remain stable or slowly decline during this period. The average length of this asymptomatic period can be quite long, up to 10 yr. The third and final period is associated with an increased rate of HIV replication and a breakdown in the immune system, associated with significant decreases in target $CD4^+$ T lymphocytes. Many of the symptoms associated with AIDS can develop during this period.

To answer some of the questions posed in Section I, the progression of clinical disease can be analyzed on several different levels. On a cellular level, the rate of spread of HIV through potential target cells and the effects on both those target cells and the individual need to be assessed. On a molecular level, we have described how the regulatory genes control the amount of HIV expression and replication, and any conditions or factors that cause these genes to switch to more active replication need to be assessed.

A. Effects of HIV on CD4$^+$ Target Cells

Because the CD4 surface molecule is the principle receptor for the Env protein, any cell bearing this receptor will be preferentially infected by HIV. The CD4$^+$ T lymphocytes are a principle target, and their depletion is closely correlated with clinical progression to AIDS. These cells can be destroyed by HIV through both direct and indirect mechanisms.

The direct mechanism of destruction has been observed in studies of HIV replication in T lymphocytes in tissue culture and presumably also applies to the *in vivo* destruction of CD4$^+$ T lymphocytes in infected individuals. During the period of rapid HIV replication, injury to the cell membrane can result as part of the budding process of the virus when HIV is coated with the Env protein embedded in the lipid membrane. Small holes in the cell membrane occur as a result, and the cell is no longer able to maintain ionic balance. Second, the production of large amounts of both viral RNA and protein can usurp the production of the usual cell mRNA and proteins, leading to a loss of normal cell function and eventual cell death. Finally, the interaction of the Env protein with the CD4 molecule has been proposed to play a role in direct cell destruction. Cells with active HIV replication have been shown to have decreased amounts of CD4 on their cell surface, and the Env protein may be responsible for this down-regulation by binding CD4 inside the cell or at the cell surface and preventing its proper deployment.

Indirect mechanisms also exist for the destruction of CD4$^+$ T lymphocytes. Many of these mechanisms evolve around the Env protein's ability, either in soluble form or as part of the cell surface of infected cells, to bind the CD4 molecule of uninfected CD4$^+$ T lymphocytes tightly. This allows infected T lymphocytes that have Env on their cell surface to fuse with multiple uninfected CD4$^+$ T lymphocytes and form giant multinucleated syncytium, which consists of many dying cells. Alternatively, immune mechanisms of destruction can result when soluble Env protein binds to the CD4 molecule of uninfected T lymphocytes and antibodies directed against the Env protein cause the destruction of these uninfected cells. Similarly, cytotoxic T lymphocytes, which are specific for killing cells with the Env protein on their surface, may target and kill uninfected CD4$^+$ T lymphocytes, which have bound and processed the Env protein. Other theoretical but unproven indirect mechanisms involve the selective depletion of cells that are critical for the proper functioning of CD4$^+$ T lymphocytes. For example, HIV infection and destruction of precursors of the CD4$^+$ T lymphocyte could lead to a depletion of this subset. Alternatively, infected T lymphocytes or nonlymphoid cells may release soluble substances that are toxic to uninfected CD4$^+$ T lymphocytes.

The loss of the critical subset of helper T lymphocytes will cause profound deficits in the immune system, eventually resulting in overwhelming infection. But HIV infection has also been demonstrated to cause functional abnormalities in the immune response even before any significant loss of the helper T cells. This functional impairment consists of poor T-cell proliferation in response to soluble antigens. The impairment may result from a noncytopathic HIV infection of T lymphocytes, which interferes with their normal cellular functions of presenting antigens and stimulating an immune response. Additionally, it has been shown that direct infection of the CD4$^+$ T cells is not required to cause the functional impairment. Exposure of these cells to the serum of HIV-infected patients also limits their ability to induce an immune response. Again, the culprit may be the soluble Env protein, which binds to the critical CD4 molecule that plays an important role when different cells of the immune system communicate with each other. Other cell types that play an important role in the immune system response also have abnormal responses to antigens or infectious agents. These include immunoglobulin-producing B cells, monocyte–macrophage cells, neutrophils, natural killer cells, and epidermal Langerhans cells. Although HIV has been reported to be present in all these different cell types, whether direct HIV infection or more indirect mechanisms are involved in the impaired function is still not clear.

As research progresses, investigators are becoming aware that an increasing number of different cell types and tissues may be infected with HIV. Many of these different cell types either express CD4 protein or contain CD4 mRNA, which is presumably the route of entry for HIV. However, increasing evidence indicates that other routes of virus entry exist that do not involve the CD4 receptor, because inefficient infection has been shown to occur in cell types lacking CD4. Cell contact or fusion may also cause the spread of HIV to $CD4^-$ cells. In general, entry of HIV via the CD4 cell-surface molecule is the most efficient mechanism, and the predominant cell types infected are the $CD4^+$ T lymphocytes and the $CD4^+$ monocytes–macrophages.

B. Genetic and Biological Diversity of HIV

The genetic and biological diversity of different HIV isolates is a feature that underlies all of our discussions about the pathogenesis of HIV infection and needs to be considered when assessing the different cell types and the different tissues that are infected with HIV. This genetic or sequence diversity of HIV strains is present in isolates from different individuals as well as in isolates taken from the same individual at different time points. Genetic variability is a feature of retroviruses since the replication of all retroviruses involves three separate conversions of nucleic acid, thus increasing the chance for mutational changes (Fig. 1). The genomic viral RNA is first transcribed into an RNA/DNA hybrid and then converted to dsDNA/DNA by the retroviral reverse transcriptase. Following integration, the proviral dsDNA/DNA is then transcribed into ssRNA for the new viral particle. The HIV reverse transcriptase has one of the highest error rates of all retroviruses and, like other viral reverse transcriptases, it does not have the ability to correct misreadings and incorrect incorporation of nucleotides that result from mismatched base-pairing during polymerization. Error rates can be as high as 1/1700 to 1/2000 per detectable nucleotide incorporated. Overall, sequence comparisons of different HIV strains demonstrate nucleic acid sequence divergence between 1.5 and 9.3% and amino acid variability from 2.2 to 14.25%. While these sequence differences can be present in any part of the genome, only mutations and changes in the nonessential regions of genes will be tolerated. The critical functional domains of HIV genes will be conserved. While the *env* gene contains highly conserved functional domains, the most striking sequence differences are seen in the previously mentioned variable regions of Env.

The propensity for genomic diversity in genes such as *env* may have profound implications for immune recognition of HIV and the development of vaccines against HIV. Vaccine development is complicated if the target of vaccines are variable regions that can easily mutate and no longer be recognized by antibodies stimulated by the vaccine. Most of the defined immunodominant epitopes detected by neutralizing antibodies in HIV are located in these variable Env regions. The normal immune response against HIV may continue to select for genetic variants and eventually result in variants that escape immune recognition. Additionally, the variants that escape detection may begin to predominate just as the immune system begins to be crippled by the selective loss of the $CD4^+$ T cells. Thus, the diabolical nature of HIV is evident in that it selectively depletes the T lymphocytes, which are responsible for directing the immune response against all the HIV variants.

Biological diversity also results from these different genetic variants, and different HIV strains may have distinct cell tropisms or predilections for infecting certain cell types. For example, some HIV strains may prefer to infect and replicate in T lymphocytes, whereas other HIV strains prefer the monocyte–macrophage cell lineage. Both strains may be present in the same individual at the same time. The biological variability of different strains also manifests itself by different replication rates and by a differing ability to cause cytopathicity and syncytial formation. These biological differences have been largely observed *in vitro* but do seem to correlate with the clinical state of the infected individual at the time the isolates were removed. The HIV strains isolated from patients with AIDS often replicate faster and have increased cytopathicity than do strains isolated from asymptomatic individuals. To summarize, the genomic diversity of HIV influences how the virus interacts with and avoids the immune system, and the biological variants present in HIV-infected individuals will determine the manifestations of disease.

C. Monocytes and Macrophages in HIV Infection

Monocytes and macrophages contain the CD4 cell-surface molecule and can also be infected by HIV. Several distinct characteristics of HIV replication in the monocyte–macrophage lineage may allow it to be an important reservoir of HIV in infected individuals. Unlike T lymphocytes infected in tissue culture, monocytes–macrophages are resistent to the lytic effects of HIV and can persistently produce HIV over time. Also, HIV replication in monocytes–macrophages results in the intracellular production of mature virions, which undergo budding from the membranes of intracytoplasmic vesicles rather than from the cell surface. The intracellular production of HIV with release from intracytoplasmic vesicles would allow infected monocytes–macrophages to escape immune surveillance because no virion proteins would be expressed on the cell surface. As a reservoir, infected monocytes–macrophages may play a role in spreading HIV to organ systems such as the lung and brain.

D. The Role of HIV in Neurological Disease

AIDS–dementia complex or AIDS encephalopathy is a common finding in HIV infection. However, the role of HIV in the mechanisms of neuropathogenesis remains obscure. HIV can be isolated from the brains and cerebrospinal fluids of infected individuals, but evidence indicating direct infection of neural cells *in vivo* is lacking. The predominant cell type in the HIV-infected brain is the macrophage. These infected macrophages may cause neuropathology by releasing cytokines or other factors that impair neuronal function. One potential candidate is tumor necrosis factor (TNFα).

E. Neoplasms Associated with HIV Infection

Epidemiological data show that HIV-infected individuals have an increased risk for developing cancer during their infection. The two neoplasms commonly associated with HIV infection are Kaposi's sarcoma and non-Hodgkin's lymphoma. However, other additional cancers have been associated with HIV including basal cell carcinoma; squamous cell carcinomas of the rectum, skin, head, and neck; melanoma; and hepatocellular carcinoma. Kaposi's sarcoma is a vascular tumor most commonly found in the skin but also present in other organs.

The most rapidly increasing neoplasm associated with HIV infection is non-Hodgkin's lymphoma (NHL). NHL may become an increasingly important complication of HIV infection throughout the 1990s. These NHL tumors are of the B-lymphocyte lineage and their high-grade, aggressive nature makes them difficult to treat.

The role of HIV in the development of cancer is not well understood, but several potential mechanisms may be involved either singly or in combination with each other. The functional impairment of the immune surveillance system by HIV may make it easier for cancers to develop. Although cancers such as Kaposi's sarcoma and non-Hodgkin's lymphoma can develop before any significant immunodeficiency or depletion of $CD4^+$ T cells, the subtle impairment of T-lymphocyte function that happens early in HIV infection may prevent adequate detection of these cancers. The role of viruses in AIDS-associated cancers also needs to be considered, especially since many HIV-infected individuals are infected with a large number of different viruses in addition to HIV. HIV or one of its genes could interact with and activate one of several DNA viruses such as the Epstein–Barr virus (EBV), cytomegalovirus, and hepatitis B virus (HBV) and together induce neoplasia. Alternatively, these viruses have been demonstrated to up-regulate HIV expression in tissue culture and may act as cofactors of HIV. The behavior and pathogenesis of these viruses may be altered by the HIV-induced immunodeficiency, and reactivation of latent virus could lead to neoplasia. The non-Hodgkin's lymphoma seen in HIV infection may actually represent three distinct disease entities, and EBV may play a role in at least one of the categories. Additionally, EBV may also play a role in the squamous cell carcinomas of the head and neck (nasopharyngeal carcinoma) seen in HIV infection. A final potential mechanism is that HIV genes may directly induce neoplasia in cells where they are expressed by interfering with important differentiated cellular functions and/or inducing cellular proteins, thus resulting in transformation. Both *in vitro* tissue culture work and *in vivo* transgenic mouse models suggest that the *tat* gene of HIV may play a role in inducing and enhancing the development of Kaposi's sarcoma lesions in HIV-infected individuals. It is possible that other HIV genes, such as the regulatory transactivators, will promote other types of cancer.

F. Activation of Latent HIV

The long clinical latency period that follows HIV infection suggests that HIV expression and replication are presumably restricted until some factor or change in biological conditions results in unrestricted viral expression. It is of obvious clinical and therapeutic importance to determine what these factors are because intervention could prevent the onset of AIDS. *In vitro* experiments suggest that factors or conditions that result in the activation of HIV-infected cells also may stimulate HIV replication. Part of the basic problem of HIV infection is that HIV is present in the T lymphocytes and monocytes–macrophages that are most likely to be activated in the body by foreign antigens and infectious agents. When this activation of cells occurs, HIV expression and replication is initiated, and additional T lymphocytes and monocytes–macrophages can become infected and subsequently activated, starting the entire cycle over again. The underlying genetic and biological diversity of HIV only aggravate the inability of the immune system to respond to this challenge. The end result is the destruction of T lymphocytes and the massive failure of the immune system.

The activation of HIV-infected cells results in the production of cellular factors capable of binding to the HIV promoter region and initiating HIV gene expression. The NFκB region of the LTR described above is an important target for these cellular factors. Thus, when T lymphocytes are activated by a variety of different mechanisms, the activation of HIV is mediated by cellular factors binding to the NFκB region. Cytokines that are produced by immunologically activated cells may play an important role in this up-regulation of HIV expression by activating HIV-infected cells. These include TNFα, granulocyte–macrophage colony-stimulating factor, and interleukin-6. Other co-infecting viruses such as EBV, cytomegalovirus, HBV, and herpes simplex virus type 1 may also be present in HIV-infected individuals. These viruses may activate the expression of HIV inside infected cells by two different mechanisms. The viral gene products may directly interact with the HIV LTR regulatory region or they may induce cellular proteins that activate HIV expression by binding the NFκB region of the LTR enhancer. Once again, cytokines and factors important for the normal functioning of the immune system induce HIV activation in infected immune system cells and cause their destruction. [*See* INTERLEUKINS; CYTOKINES IN BACTERIAL AND PARASITIC DISEASES.]

AIDS clearly is a complex constellation of clinical manifestations caused by an equally complex virus. A cure for AIDS is possible only with a clear understanding of the workings of this virus.

Bibliography

Gallo, R. C., and Jay, G. (eds.) (1991). "The Human Retroviruses." Academic Press, San Diego.

Sci. Am. (1988). **259,** 40–134.

Science (1988). **239,** 573–622.

Harawi, S. J., and O'Hara, C. J. (eds.) (1989). "Pathology and Pathophysiology of AIDS and HIV-Related Diseases." C. V. Mosby, St. Louis.

Adhesion, Bacterial

Ann G. Matthysse
University of North Carolina at Chapel Hill

Glossary

Biofilm Group of organisms growing as a layer on a surface

Fibronectin Serum protein that plays a role in the adhesion of mammalian cells to each other and to substrates

Fimbriae or pili Straight filamentous appendages composed of proteins that project from the surface of bacteria

Glycocalyx Loose layer composed largely of polysaccharide fibers surrounding a cell

Lectins Multivalent carbohydrate-binding proteins

MICROBIOLOGISTS ACCUSTOMED to the study of *Escherichia coli* grown in shaker flasks in the laboratory tend to regard bacteria that live in suspension as the norm. However, in the real world many important bacterial populations live attached to a substratum. In habitats ranging from the Alaska oil pipeline to the human intestine, many bacteria grow attached to the surface rather than suspended in the liquid that surrounds them. Adherent bacteria play an important role in ecology (e.g., in microbial mats and in biofilms). Adherent bacteria compose the normal flora found on the outer and inner (gastrointestinal) surfaces of humans and animals and the normal flora found on the surface of plants. In addition, many pathogenic bacteria depend on their ability to adhere to host surfaces to cause disease.

Despite the importance of adherent bacteria, the study of microbial adhesion has only recently received a significant amount of attention by microbiologists. This field, like many other aspects of microbiology, is rapidly uncovering new information. Therefore the information given here should be supplemented with reference to the literature. The mechanisms of microbial adhesion differ for fungi, viruses, and bacteria. This review will consider only the adherence of bacteria.

I. Description of Bacterial Adhesion

Adherent bacteria vary from isolated cells widely spaced on the substratum to thick bacterial mats. Both are visible in the light microscope, but usually only spacing between the bacteria and very little else can be distinguished if only individual bacteria are bound to the surface. In the case of microbial mats or multilayered complex bacterial aggregates, observation in the light microscope may be helpful in identifying the overall organization of the attached bacteria.

The determination of whether bacteria are truly bound to a substrate or are simply resting on it can be difficult in the light microscope. If the preparation is of a type to permit moving the liquid or substratum while observing the bacteria, this can often aid in distinguishing attached bacteria from those simply resting on the surface. Observation of adherent bacteria in the light microscope has the advantage that one can look at living material and thus avoid many of the artifacts that are present in fixed material. However, the major disadvantage of light microscope observation is the difficulty in obtaining reliable quantitative results. In general, observation of living preparations of attached bacteria in the light microscope is a useful first step in characterizing adherent bacteria.

The details of bacterial attachment are more readily visible in the scanning and transmission electron microscopes (SEM and TEM). These preparations must be fixed and metal coated or stained before observation. Very small, delicate structures such as the details of the glycocalyx or fimbriae are unlikely to be visible after metal coating for the SEM. The SEM allows the observation of a relatively large surface area and of samples of moderate depth and so is suitable for obtaining information about statistical distributions with regard to such parameters as bacterial orientation, occurrence of aggregates, relationship of bacteria of differing morphologies to each other, and relationship of the adherent bacteria to various surface structures of the substratum. A comparison of light microscope and SEM views of bacteria attached to carrot cells is seen in Fig. 1.

Sections of adherent bacteria viewed in the TEM can show great detail including the glycocalyx and fimbriae. Examining sectioned material in the TEM is labor-intensive, and it is seldom possible to do wide spatial or statistical sampling. A less labor-intensive technique involves negative staining of the bacteria. The presence of such structures as fimbriae can be determined easily using this technique. [*See* Electron Microscopy, Microbial; Glycocalyx, Bacterial.]

II. Measurement of Bacterial Attachment

The number of adherent bacteria can be estimated in the light or SEM. However, it is seldom possible to obtain accurate numbers using microscopic observations. For this purpose the bacteria can be counted as numbers of viable cells using plate counts of free bacteria and those removed from the sample after various washing procedures or those that grow after the sample has been mechanically or enzymatically disrupted. These methods estimate only the number of bacteria present that will grow on the medium used and that will survive the procedure used to release them. Dead bacteria or bacteria that are not viable under the conditions used will not be scored. Alternatively the bacteria can be labeled before they are added to the substratum to which they will adhere. This labeling can be with radioisotope such as ^{3}H-thymidine, with fluorescent dyes that may be linked to antibodies, or by the inclusion in the bacterium's genes of a *lux* gene cassette with a promoter that will fluoresce under the desired conditions. The numbers of free and adherent bacteria are then determined using a scintillation counter, spectrophotometer, or fluorimeter. Care must be taken that the signal coming from the adherent bacteria is not masked or quenched by the presence of large bacterial aggregates or by the substratum.

The change in adherent bacteria over time can be examined using several different techniques. If the sample is suitable, time-lapse microcinematography may be helpful. Numbers of free and attached bacteria can be determined after various time intervals. In examining such data, it should be remembered that both free and attached bacteria may grow and that the doubling time for adherent bacteria is not necessarily the same as that for free bacteria. Thus mathematical modeling of the time course of bacterial adhesion can be complex. Adherent bacteria may modify the surface to which they adhere so that rates of attachment may change with time. In addition, attached bacteria may themselves be altered, and this may affect subsequent attachment. An example is the case of *Agrobacterium tumefaciens*, which binds to plant cell surfaces. After binding, substances released from the plant cells cause the bacteria to elaborate cellulose fibrils. These fibrils are sticky, and fresh bacteria entering on the scene may bind to either the original plant cell surface or become entrapped in the fibrils elaborated by the previously bound bacteria. The later process results in the formation of large aggregates of bacteria bound to the plant cell surface. Mathematical modeling of a process such as this in which all the bacteria are capable of growth at different rates depending on their location (free, attached directly, or attached via cellulose fibrils) and in which the free bacteria can become bound either directly or indirectly is quite complex.

III. Mechanisms of Bacterial Adhesion

A. Charge Interactions

Bacteria can bind to surfaces nonspecifically via ionic interactions. The surface of most gram-negative and many gram-positive bacteria is negatively charged. Thus these bacteria will adhere to surfaces that have a positive charge. This property of bacteria is frequently made use of in preparing bacterial samples for SEM. The bacteria are allowed to adhere to a glass or other surface that has been coated with poly-L-lysine. The bacteria bind to the poly-L-lysine via an ionic interaction between the

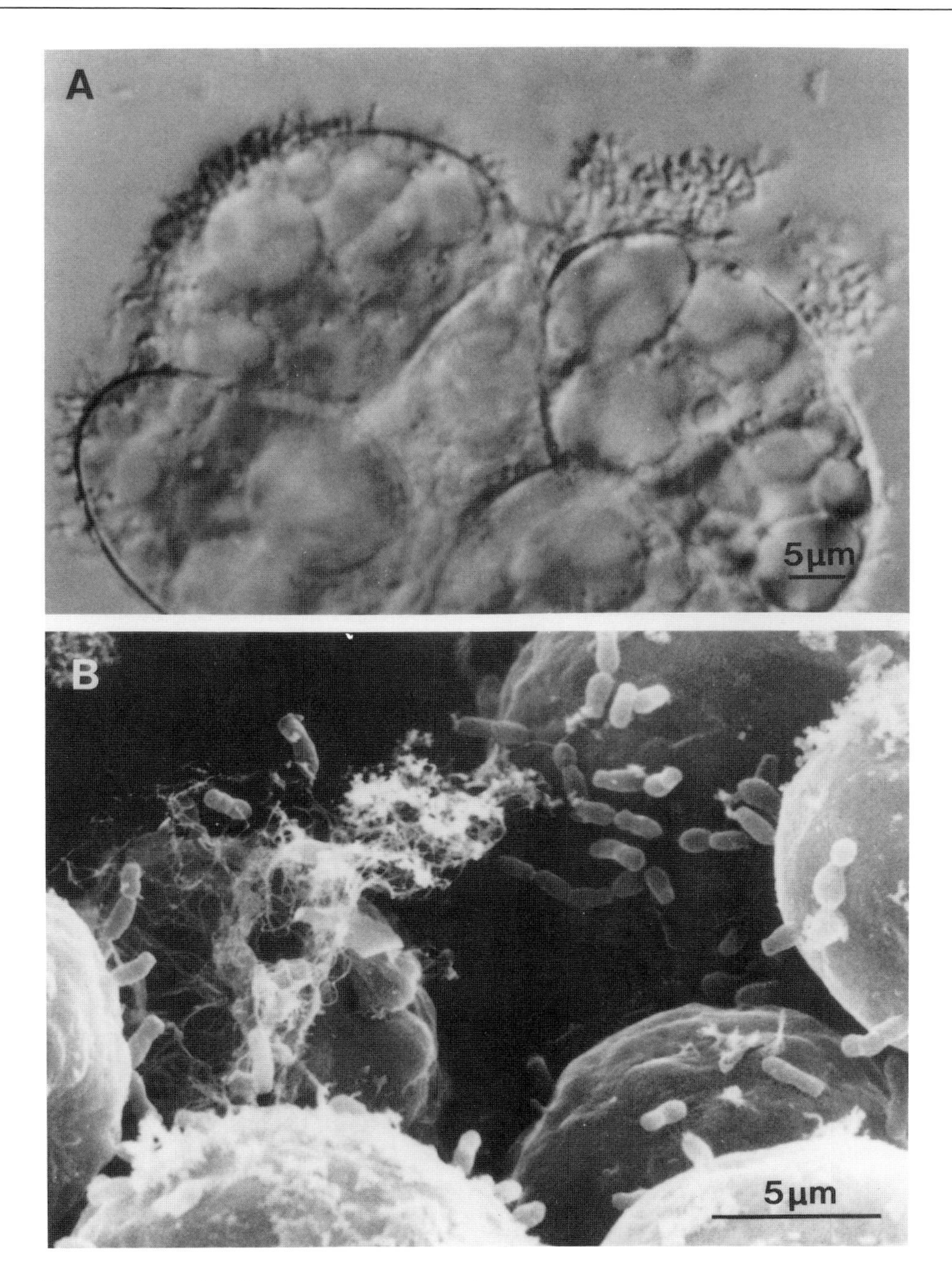

Figure 1 Bacterial adhesion as seen in the light and scanning electron microscopes. *Agrobacterium tumefaciens* attached to the surface of carrot suspension culture cells viewed (A) in the light microscope using living material and Nomarski optics and (B) in the scanning electron microscope using fixed and metal-coated material. Note that although the two views are similar, structures such as the fibrillar material holding the bacterial aggregates together are only visible in the scanning electron microscope.

positively charged polypeptide and the negatively charged bacterial surface. In general, this type of ionic interaction becomes weaker as the ionic strength of the medium is increased. Thus the effect on bacterial attachment of alterations in the ionic strength of the surrounding medium is one way of determining that an attachment may have an ionic component. However, one must remember that the bacteria (and possibly the substratum to which they are binding) are alive and will very likely exhibit a physiological response to changes in the medium.

B. Hydrophobic Interactions

Many surfaces encountered in nature by bacteria are hydrophobic. These may consist of droplets of oils or lipids, exposed surfaces of membranes including those of mammalian cells, and waxy surfaces including the cuticle of plants. Bacteria that themselves have a hydrophobic surface will tend to adhere to such surfaces. However, the surface of many bacteria is covered with exopolysaccharides and lipopolysaccharides, which are generally hydrophilic. Rough mutants of many bacteria show increased hydrophobicity. Projections from the bacterial surface such as fimbriae or pili are often hydrophobic and may aid in the binding of the bacteria to hydrophobic surfaces. Bacterial mutants that fail to make fimbriae often show decreased adherence to hydrophobic substances.

C. Lectins

Lectins are multivalent carbohydrate-binding proteins. They are made by bacteria, plants, and animals. Lectins have been shown to be involved in the binding of bacteria possessing type I fimbriae to surfaces having an exposed maltose residue. The lectin that binds the sugar is at the tip of the pilus. Other fimbriae have been shown to recognize other sugars. These fimbriae participate in the binding of pathogenic bacteria to host cells including the K88 pilus-mediated binding of enteropathogenic *E. coli* to the intestine, P pilus-mediated binding of uropathogenic *E. coli* to the bladder, the binding of *Actinomyces viscosis* via type II galactosyl-binding fimbriae to the surface of the oral mucosa, and the binding of *Streptococcus sanguis* to *A. viscosis* via a lactose-sensitive lectin on the type 2 fimbriae of *A. viscosis*.

Lectins produced by the plant and found on the surface of root hairs have been implicated in the initial binding of Rhizobia to their hosts. At least some of the specificity of the *Rhizobium*–legume interaction can be attributed to the plant lectin. *Rhizobium leguminosarum* binds to pea roots and forms nodules on peas. These bacteria do not bind to clover and do not form nodules on clover roots. When the pea lectin gene was introduced into clover roots growing in culture by gene transfer, the resulting clover roots bound *R. leguminosarum* and were able to form nodules with these bacteria.

D. Fibronectin Binding

Fibronectin and vitronectin are serum proteins that play a role in the adherence of mammalian cells to each other and to substrates. These molecules also serve as receptors for the attachment of some pathogenic bacteria. *Staphylococcus aureus,* various other staphylococci, streptococci, *Treponema pallidum,* and borrelia bind to fibronectin. The binding is specific and involves a bacterial surface protein. In at least some cases, the active binding site on this protein contains repeated domains. Synthetic peptides that mimic these domains inhibit bacterial binding.

Plants are not known to contain fibronectin; however, they do contain vitronectin. Some plant pathogenic bacteria may use this protein as a binding site.

E. Collagen Binding

Many of the same bacteria that bind to fibronectin also bind to collagen. In addition, several bacteria associated with oral disease bind to collagen, including *Streptococcus mutans* and *Bacteroides gingivalis*. Fibronectin also binds to collagen and is able to block the binding of *B. gingivalis*. It has been proposed that the extracellular proteases that are found in infected gingiva may digest the fibronectin and expose the collagen, thus allowing the binding of *B. gingivalis*.

F. Role of Fimbriae

The role of bacterial fimbriae (or pili) has already been mentioned. These projections from the bacterial surface appear to serve a role in the binding of bacteria to other bacteria in conjugation and in the binding of bacteria to surfaces during pathogenesis. Because fimbriae are long and the binding sites are often at the tip end, fimbriae may allow bacteria to attach to surfaces that hydrophobic or ionic interactions prevent them from approaching too closely. The actual binding interaction of fimbriae may involve lectins on the fimbriae and carbohydrates on the surface to which the bacteria bind. In addition, in some bacteria such as *A. viscosis,* fimbriae may mediate binding of the bacteria to proteins with no known involvement of sugars or lectins. Fimbriae are involved in the binding of enteropathogenic *E. coli* to the gut, of uropathogenic *E. coli* to the bladder, of *Neisseria gonorrhoeae* to epithelial cells, of *Streptococcus pyogenes* to epithelial cells, of *Pseudomonas syringae* pv. *phaseolicola* to the leaves of bean plants, and in a myriad of other bacterial adhesions. Fimbriae may also be hydrophobic and aid in the attachment of bacteria to hydrophobic surfaces.

G. Other Mechanisms

Binding of *Mycoplasma pneumoniae* to respiratory epithelial cells involves a structure at the elongated tip of the bacteria. Several surface proteins are found preferentially in this tip and are required for adherence. How these proteins function in attachment is not known.

The formation of swarms and fruiting bodies by *Myxococcus xanthus* involves the adhesion of the bacteria to each other. Pili and an extracellular matrix composed of thick fibrils both seem to be involved in this adhesion.

In *Vibrio parahemolyticus,* lateral flagellae appear to be involved in the adhesion to surfaces.

Lipoteichoic acid found on the surface of gram-positive bacteria appears to be involved in the adhesion of *S. pyogenes* and *Staphylococcus aureus* to epithelial cells. The gram-negative bacterium *Fusobacterium nucleatum* aggregates with streptococci. Two receptors seem to be involved. In one, a surface protein of *F. nucleatum* binds to lipoteichoic acid on the streptococci. The second may involve an interaction between surface proteins on the two bacteria.

Conjugation in *Streptococcus faecalis* involves the production of proteinaceous adhesins by the pheromone-induced donor cell. These adhesins may be involved in the formation of mating pairs.

In several bacteria, attachment to a substratum may be followed by the elaboration of surface structures, which results in tighter adhesion or in the subsequent binding of free bacteria from the surrounding medium. For example, *A. tumefaciens* binds to the surface of plant cells via outer membrane proteins. The plant cells release substance(s) that cause the bacteria to elaborate cellulose fibrils. These fibrils bind the bacteria tightly to the host cell surface. They are also sticky and entrap free bacteria, resulting in the formation of large aggregates of bacteria on the plant cell surface.

IV. Genetics of Microbial Adhesion

A. Genes Involved

In most cases the genes involved in microbial adhesion are found in operons that are concerned with the surface of the bacterium. In many cases, the operons specify the synthesis of a pilus. The F pilus of *E. coli* requires about 13 genes contained in one large operon for its assembly. The pilus may not exist as a stable rigid rod, but instead may be retracted and extended from the bacterial surface. If this model is correct, the attachment of the pilus to a surface and its subsequent retraction would result in closer contact of the bacterium with the surface. In bacterial conjugation, this closer contact is followed by the formation of unstable mating pairs and then stable mating pairs. This process is under the control of various *tra* genes. In many cases the genes for pili or fimbriae are carried on plasmids and thus can be easily transferred from one bacterium to another.

Genes for the synthesis of flagellae are also found in large operons. The synthesis of extracellular polysaccharides may involve only a few genes, as in the case of the acidic exopolysaccharide of *Rhizobium.* However, the synthesis of some extracellular polysaccharides such as cellulose, which is synthesized by *Acetobacter xylinum,* involves more than one operon and about 15 kb of chromosomal DNA.

The genes for outer membrane proteins are found in small operons in *E. coli.* Many of these proteins may have more than one role. For example, the *mal* gene, which is involved in the uptake of maltose and in the binding of bacteriophage lambda to the bacterial surface.

Many mutations may prevent bacterial adhesion without affecting specifically the gene products directly involved in bacterial attachment. For example, mutations that alter the entire bacterial surface will generally reduce the ability of the bacteria to adhere to a substrate. Such mutations include those in genes involved in the secretion of proteins through the plasma membrane, some deep rough mutations, and some mutations that affect the synthesis of the outer membrane or of the cell wall. In *M. pneumoniae,* mutations in the genes for several high-molecular-weight proteins (HMW 1, 2, 3, and 4) result in the failure of the major P1 adhesin to become localized at the tip of the bacterium so that the bacteria can no longer bind effectively to host cells. Mutations that result in the overproduction of exopolysaccharides or the production of larger more highly substituted lipopolysaccharides may also prevent the binding of bacteria by covering the bacterial binding site so that it is no longer accessible to the receptor to which it would normally bind.

B. Gene Regulation

Genes required for bacterial attachment may be regulated in response to changes in the environment. In addition they may be under a variable control such as phase variation so that only a fraction of the population expresses the attachment genes at any one time. Regulation of attachment genes is required so that a population of bacteria has the ability to move from one location to another. In some environments it may be advantageous to be attached, and in other environments or under other conditions it may be more advantageous to be free-living. For example, *E. coli* that adhere to the gut via K99 fimbriae do not make such fimbriae at temperatures below 20°C. At low temperatures the bacteria would presumably be outside the animal host, and the ability to adhere to intestinal epithelial cells would not be useful. At 37°C a portion of the bacterial population make fimbriae and thus can adhere to the intestinal epithelium.

In *V. parahemolyticus,* the production of polar flagellae is constitutive, but lateral flagellae are produced only when the bacteria are on a surface. Lateral flagellae are involved in adhesion to surfaces by this bacterium. The stimulus for their production appears to involve the obstruction of the rotation of the polar flagellae.

In some bacterial species, adhesion is regulated as a part of the life cycle of the bacteria. This is the case in the myxobacteria and in caulobacteria. In the myxobacteria *Stigmatella auranteaia* and *S. erecta,* there are two cohesion systems, one constitutive and one that is induced by Ca^{2+}. In *M. xanthus,* cell cohesion requires the S genes (about 10 loci), Mg^{2+}, Ca^{2+}, energy, and the products of the *asgA, asgB,* and *asgC* genes.

Phase variation in the production of fimbriae or adhesins provides the bacterium with a constant population of both adherent and nonadherent bacteria. This appears to be the advantage to the bacterium of the phase variation seen with type 1 fimbriae in *E. coli.* The switch occurs with a frequency of about 1 in 10^3 per generation, which is much higher than mutation rates. It involves the inversion of a small (314-bp) piece of DNA that contains a promoter. Borders of 9-bp inverted repeats are required, as well as at least five genes located elsewhere in the chromosome. In *N. gonorrhoeae,* phase variation in the serotypes of adherent pili is used to evade the host's immune system. The switching here is rather complicated, and more than 50 pilus serotypes are known as well as nonpiliated variants. [*See* Microbial Attachment, Molecular Mechanisms.]

IV. Role of Microbial Adhesion

A. Medical

The ability to adhere to host cells is a significant virulence factor for many human pathogens. In general, there is a correlation between the ability of the bacteria to adhere to a particular site and the extent to which that site is colonized by that bacterium. Bacterial mutants that lack the ability to bind to host tissues are usually reduced in virulence. For example, mutants of *M. pneumoniae* that lack the major adhesin P1 have decreased virulence in hamsters. Adherent *E. coli* have been shown to have both a growth advantage and increased toxicity when compared with isogeneic nonfimbriated strains. Enhanced toxicity has also been shown for adherent streptolysin-producing *S. pyogenes.*

Bacterial adhesion to human cells can result in colonization of the cell surface. However, in some cases, adhesion is followed by bacterial invasion of the host cell. The factors that determine whether bacterial adhesion results in surface colonization or invasion include the species of bacterium, the type of host cell, and the nature of the molecules involved in the binding reaction, including the number of sites and the tightness of the adherence. Bacteria such as *Yersinia enterocolitica* bind to integrins on the host cell surface via a bacterial invasin. This binding is a high-affinity reaction and promotes the invasion of the host cell by the bacteria.

Adherent bacteria may form biofilms on foreign materials present in the human body. The materials so colonized can range from contact lenses to catheters, suture threads, and surgical implants. Bacteria in these biofilms are typically more resistant to antibiotics than free-living bacteria and are very difficult to remove. [*See* BIOFILMS AND BIOFOULING.]

B. Dental

The complex associations of bacterial species seen on different tissues in the mouth appear to reflect differences in the adherence of the bacteria to different surfaces and also specificity in the adherence of bacterial species to each other. For example, there is a correlation between laboratory measurements of the ability of particular bacteria to adhere to the tongue, to buccal mucosa, or to teeth and the presence of those bacteria as part of the normal flora in that location. In addition, the associations of bacteria with each other in these locations are specific. A diagram showing some of these specific associations of bacteria on the surface of teeth is presented in Fig. 2. The particular association of bacterial species varies with the location in the mouth. *Veillonella* species from the tongue form coaggregates with *Streptococcus salivarius* and not with *Actinomyces* or *S. sanguis* from the surface of teeth. On the other hand, *Veillonella* species isolated form dental plaque form coaggregates with *Actinomyces* and *S. sanguis* and not with *S. salivarius* from the tongue. These bacterial associations on the tooth surface are critical in the formation of dental caries because of the production of acid by the bacteria in an environment in which the adherent bacteria cause its local accumulation. The formation of a layer of adherent bacteria is also important in the development of gingival disease in which the species present in the bacterial aggregate may interact with each other to

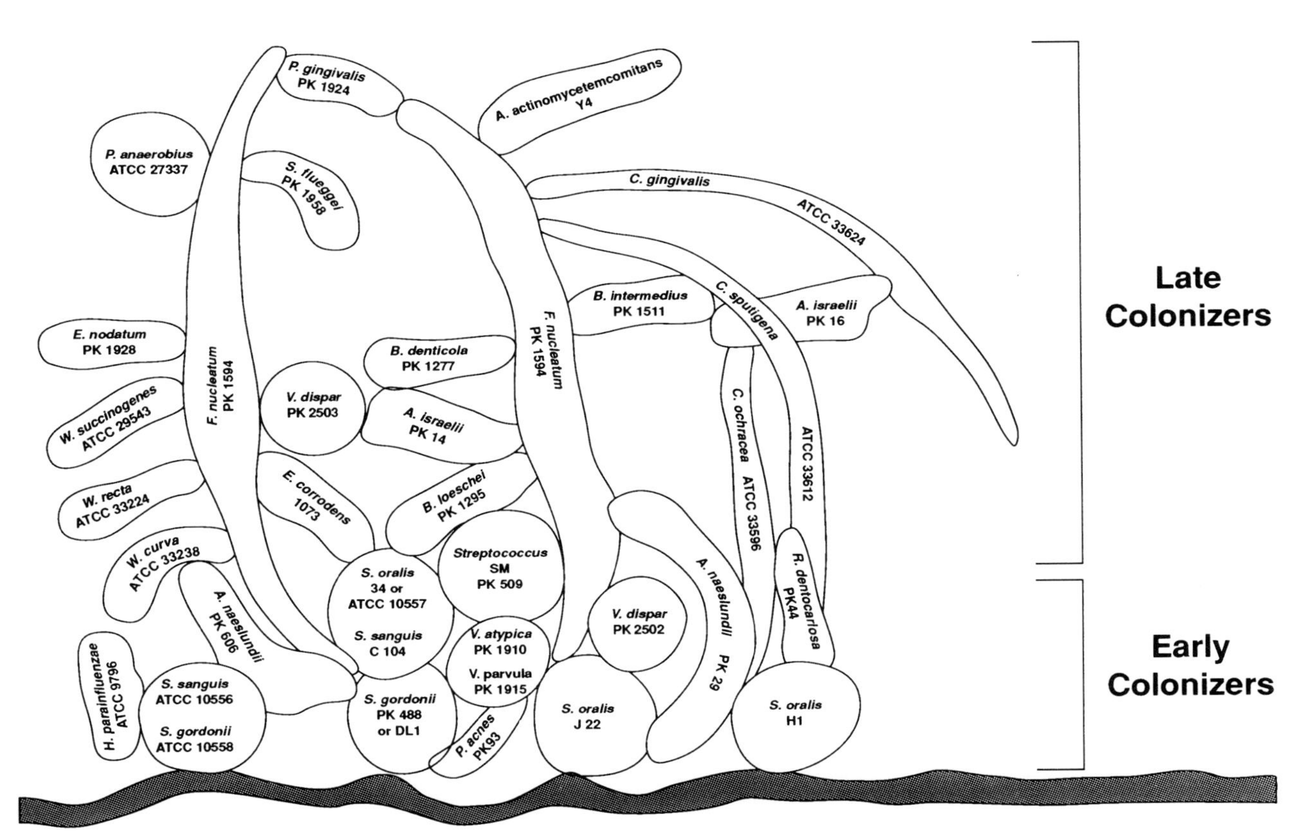

Figure 2 Diagram showing the complex interactions between various species of bacteria adhering to teeth. [From Kohlenbrander, P. (1991). Microbial adhesion to teeth. *In* "Microbial Cell–Cell Interactions " (M. Dworkin, ed.). American Society for Microbiology, Washington, D.C.]

expose underlying layers of host proteins to degradative action by the bacteria.

C. Plant Interactions

The best studied cases of the role of bacterial adherence in bacterial plant interactions are the interactions of Rhizobia with legumes and the interactions of *Agrobacterium* with host cells. The available evidence suggests that the ability of *Rhizobium* to bind to legume root hairs aids in the formation of nitrogen-fixing nodules. However, at present there is no noncontroversial evidence that the ability of the bacteria to bind to root hairs is required for nodule formation. However, the ability of both *A. tumefaciens* and *A. rhizogenes* to bind to host cells does appear to be required for the formation of crown gall tumors and hairy roots, respectively. All known mutants of *Agrobacterium* that fail to bind to host cells are avirulent.

Many potentially pathogenic bacteria normally reside on the surface of the plant. The mechanisms by which they adhere to the plant surface are largely unknown. However, in a few cases such as *P. syringae* pv. *phaseolicola,* the adherence has been shown to involve fimbriae. Bacteria that lack these fimbriae bind only poorly to the plant surface and thus would be expected to be reduced in the ability to cause disease in the field.

Bacteria that promote plant growth are also often observed to bind to the plant (usually root) surface. This binding may enhance their ability to promote the growth of the host plant.

D. Attachment to Inanimate Surfaces

Many bacteria are found attached to inanimate surfaces. Organic molecules and various salts may become concentrated at such surfaces, making them a nutritionally favorable site. Gram-negative marine bacteria are reported to respond to starvation by an increase in adhesiveness. In systems with flowing liquid such as streams or pipelines, adhesion is necessary for the bacteria to maintain themselves in a particular location. In many species of bacteria, adhesion may be an essential part of the life cycle. This is the case for myxobacteria and for caulobacteria. In some cases, bacteria attach to a surface to degrade it and use it as a food source. Examples are bacteria that degrade cellulose and adhere to the surface of cellulose fibers and bacteria that degrade components of oil droplets and adhere to the oil–water interface.

E. Biofilms and Microbial Mats

In most natural systems with adherent bacteria, the bacteria occur not as isolated colonies but as a biofilm covering the exposed surface. These biofilms are often made up of several species of bacteria that may be interdependent metabolically. Toxic compounds such as antibiotics do not usually penetrate a biofilm effectively, and so the bacteria are protected by this type of growth. [*See* BIOFILMS AND BIOFOULING.]

In evolutionary terms, microbial mats composed of one or more species of bacteria are very old. Some of the most ancient fossils are stromatolites, which are formed by microbial mats found in shallow bodies of water. Typically photosynthetic organisms are found in the top layers of such mats, with nonphotosynthetic decomposers making up the bottom layers. These mats can be found in both moderate and extreme environments such as under the ice in the antarctic. The biochemical and molecular genetic mechanisms involved in the formation of such mats have not been studied in much detail.

Bibliography

Costerton, J. W., Cheng, K. J., Geesey, G. G., Ladd, T. I., Nickel, J. C., Dasgupta, M., and Marrie, T. J. (1987). *Annu. Rev. Microbiol.* **41,** 435–464.

Doyle, R. J., and Rosenberg, M. (eds.) (1990). "Microbial Cell Surface Hydrophobicity." American Society for Microbiology, Washington, D.C.

Dworkin, M. (ed.) (1991). "Microbial Cell–Cell Interactions." American Society for Microbiology, Washington, D.C.

Isberg, R. R. (1991). *Science* **252,** 954–958.

Savage, D. W., and Fletcher, M. M. (eds.) (1985). "Bacterial Adhesion Mechanisms and Physiological Significance." Plenum Press, New York.

Agrobacterium and Plant Cell Transformation

Andrew N. Binns
University of Pennsylvania

Rolf D. Joerger
United States Department of Agriculture

John E. Ward, Jr.
Southern Methodist University

Glossary

Border sequences 25-bp direct, imperfect repeats that delineate the DNA transferred from Agrobacterium to the plant cell

Conjugation Transfer of DNA, via a single-stranded intermediate, between bacteria in a process that requires specific physical contact

Conserved Processes or molecules common to many members of a given group of organisms.

T-DNA DNA between the border repeats that is transferred from Agrobacterium to plant cells

AGROBACTERIUM TUMEFACIENS (originally called *Bacterium tumefaciens*) and *Agrobacterium rhizogenes* are gram-negative soil bacteria that can induce aberrant growths on a wide variety of dicotyledonous plants. The "crown gall" tumors induced by *A. tumefaciens* are capable of growing continuously in the absence of the inciting bacterium. This result indicates that the bacteria cause a stable change in the cellular heredity of the affected plant cells, resulting in their uncontrolled proliferation. During the 1970s several laboratories showed that the presence of a large (>200 kb) plasmid was required for virulence of Agrobacterium, and in 1977, bacterial DNA was demonstrated to be present in the genome of cells from bacteria-free crown gall tumors. During the past 15 years substantial progress has been made in understanding the cellular, biochemical, and molecular basis for this DNA transfer and the resultant tumorous growth. Here we will stress the observations that conserved processes, both in the bacterium and in the plant, are used in novel ways to elicit tumor formation. This discussion will focus on the following topics: (1) How do bacteria recognize that a competent transformable plant is available? (2) What is the mechanism used to deliver bacterial DNA into the plant cell? (3) How does the bacterial DNA, once inside the plant cell, ultimately cause tumorous growth? and (4) What are the current and future prospects for use of Agrobacterium-mediated delivery of foreign DNA?

I. Infection Process—General Features

A. Productive Interaction

Agrobacterium tumefaciens initiates unorganized growths, termed *crown gall tumors,* whereas *A. rhizogenes* initiates root-forming tissues, often referred to as *hairy roots,* at infected wound sites on susceptible plants. Pathogenicity is due to the presence of large plasmid residing in the bacteria. *Agrobacter-*

ium tumefaciens carries the Ti (tumor-inducing) plasmid, whereas *A. rhizogenes* harbors a plasmid referred to as Ri (root-inducing). In the natural soil environment, nonpathogenic forms of *Agrobacterium,* lacking Ti or Ri plasmids, vastly outnumber the plasmid-carrying forms. Only the latter are virulent, because the delivery of a portion of the Ti or Ri plasmids into the plant cell is required for the abnormal plant growth response. [*See* PLASMIDS.]

The basic infection cycle of these two pathogens is quite similar (Fig. 1). Bacteria infect a wound site on the plant and attach to plant cells, probably at specific sites on the plant cell wall. During this time molecules produced by the wounded plant tissues are recognized by the bacterium, which then synthesizes a DNA intermediate from the Ti (or Ri) plasmid. This intermediate is transferred into the plant cell, crossing numerous wall and membrane boundaries, and, once in the plant cell is integrated into the nuclear DNA. Transcription of the transferred DNA (T-DNA) produces RNA that is processed into typical polyadenylated mRNA and translated in the cytoplasm. Ultimately, these gene products of the T-DNA are responsible for the tumorous or rooty growth of the transformed plant cells.

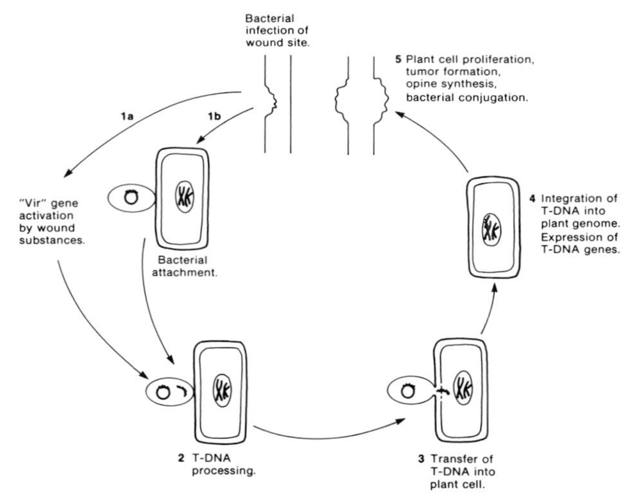

Figure 1 Summary of the Agrobacterium–plant cell interaction that leads to tumor formation. *Wavy line* represents T-DNA and its possible intermediates (see text for details). [From Binns, A. N., and Thomashow, M. F. (1988). *Annu. Rev. Microbiol.* **42,** 575–606.]

B. Genes Involved in Tumorigenesis

Genetic and molecular strategies have led to the identification of a variety of genes, both on the Ti (Ri) plasmid and on the bacterial chromosome, that are required for T-DNA transfer and tumor formation (Fig. 2). In general, the genes of *A. tumefaciens* and *A. rhizogenes* required for DNA transfer are structurally and functionally homologous, whereas the genes of the T-DNA can be quite divergent, particularly those involved in the growth response of the plant cell. For this reason, discussion of the DNA transfer events will refer to *A. tumefaciens* and the Ti plasmid almost exclusively, whereas discussion of the T-DNA and its function will include both of these pathogens.

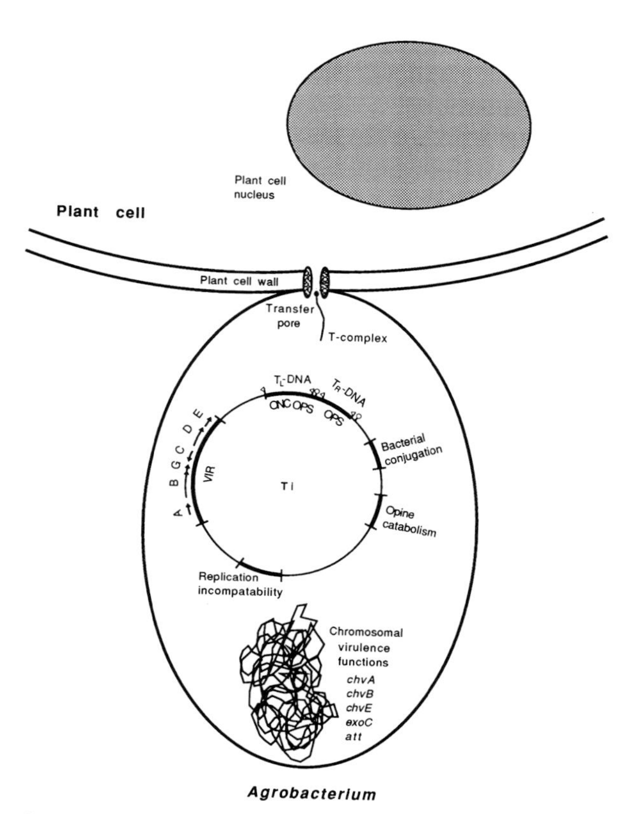

Figure 2 Schematic representation (not to scale) of Agrobacterium–plant cell interaction, showing Ti plasmid and chromosomal genes involved in DNA transfer and tumorigenesis and proposed T complex and DNA transfer pore. Octopine-type Ti plasmid is shown, indicating location and polarity of *vir* loci; location, general functions, and border repeats of T-DNA; and approximate locations of genes involved in bacterial conjugation, opine catabolism, and plasmid replication. Δ, border repeats; o, overdrive sequences.

Several loci on the bacterial chromosome have been identified that are required for virulence. Although most of these chromosomal virulence (*chv*) genes identified to date are involved in attachment of the bacterium to the plant cell, at least one gene (*chvE*) is involved in recognizing plant cells competent for transformation (see text below). Many of the genes involved specifically in DNA transfer from the Ti plasmid to the plant cell are located on the plasmid itself and are referred to as virulence (*vir*) genes. There are six critical *vir* operons, two of which (*virA* and *virG*) have only a single open reading frame, whereas the others have between 2 (*virE*) and 11 (*virB*) open reading frames (Table I). Because the protein products of the *vir* operons B, C, D, and E are involved in the production and transfer of T-DNA out of the bacterium (see text below), it is not surprising that their expression is highly regulated. The gene products of *virA* and *virG* are critical in this regulation. Certain Ti plasmids contain two additional *vir* operons (*virF* and *virH*) that may play a role in host range but have not yet been well characterized. In addition, *virA* and *virG* also control the expression of at least five virulence related proteins (VRPs), two of which are encoded by Ti plasmid genes and the others, presumably, by the bacterial chromosome. The function of the VRPs in the transformation process, if any, is unknown. [*See* CHROMOSOME, BACTERIAL.]

The genetic material transferred to plant cells by Agrobacteria encodes gene products involved in two distinct processes. First, as described above, the T-DNA encodes genes involved in the neoplastic growth of the transformed plant cells. These genes, often referred to as the oncogenes (*onc*) of the T-DNA, are involved in producing or responding to plant hormones that control cell proliferation. The second set of T-DNA genes is involved in the synthesis and transport of unusual amino acid or sugar conjugates, termed *opines*. The opines appear to be the biochemical basis for the evolution of the entire disease process: Plant cells are incapable of metabolizing the opines, whereas each inciting Agrobacterium carries opine catabolism loci that enable it to use the opine as a sole nitrogen and/or carbon source. Some of these opines will also induce conjugative transfer of the Ti or Ri plasmid to plasmid-free strains of Agrobacteria. Thus, Agrobacteria have evolved as natural genetic engineers that, via transformation, diverts the resources of the transformed plant cell to produce materials that are specifically used by the pathogen. Interestingly, recent studies have shown that certain other soil bacteria have also evolved the ability to use certain opines and can be found in natural crown gall tumors isolated in the field.

C. Host Range

Agrobacteria transform a wide range of dicotyledonous plants, but transform monocotyledonous plants, particularly the agronomically important Gramineae, either not at all or with vastly reduced frequency. The basis for host range variation probably reflects the complexity of the infection and transformation process itself. At many steps in this process, activities of both the plant and the bacterium are required for the ultimate transfer, integration, and expression of the T-DNA. Any inapproriate or missing activity would disrupt the transformation process. In some cases closely related plants, even different varieties of the same species, can vary more than 100-fold in their sensitivity to the same inciting Agrobacterium. However, the fact that diverse groups of plants *can* be transformed suggests that conserved plant processes are involved. As discussed below, current evidence indicates this is the case, although the entire set of processes (bacterial and plant) required for transformation has not been fully elucidated.

II. Signaling Involved in Initiating DNA Transfer

A. Plant Wound Response: Making the Plant Cell Competent

One of the earliest recognized and least understood process involved in Agrobacterium-mediated transformation is the plant wound response. A wound, and in particular a specific type of wound response, is required for successful transformation. Pioneering studies carried out in the 1950s showed that a "window of competence" occurs at wound sites: If Agrobacteria are present only before or after this critical period of time after wounding, tumors rarely develop, whereas if they are present during this time, extensive tumor formation occurs. More recent studies in which regenerating plant protoplasts are cocultivated with virulent bacteria confirm that a distinct time period after wounding (in this case, protoplast formation by enzymatic digestion) exists

Table I Summary of *vir* Gene Products

Locus	Size (kb)	ORFs[a]	Proteins: Size (kDa)	Proteins: Amount	Proteins: Location[b]	Function
virA	2.0	1	90	+	M	Plant signal sensor, protein kinase
virG	1.0	1	30	++	C	Transcriptional activator
virD	4.5	4	16, 47, 21, 75	+	C/M?	T-DNA border endonuclease (VirD1 and VirD2); pilot protein? nuclear localization? (VirD2)
virC	2.0	2	26, 23	+	C?	Processing of T-DNA (VirC1)
virE	2.0	2	7, 60.5	+++	C/M?	Single-strand DNA-binding protein (VirE2)
virB	9.5	11	26, 12, 11, 87, 23, 32, 5.5, 25, 32, 48, 38	+++	M	T-DNA transfer apparatus?

[a] Open reading frames predicted by DNA sequence.
[b] C, cytoplasm; M, membrane.

in which the damaged cells are particularly susceptible to transformation.

The biochemical processes that occur during a wound response have been studied in some detail. Immediately after wounding, most dicotyledonous plants respond with measures intended to deter or prevent microorganisms from becoming established at the wound site. Enzymes already present before wounding and others synthesized as a result of the wounding event produce antimicrobial compounds. These chemical defense mechanism are accompanied by the slower process of cordoning off the wound area. Entire cell layers can be converted into barriers of dead cells, whereas other cells dedifferentiate and begin to divide. Cell wall synthesis is initiated and existing walls are fortified. All these activities result in local acidification and the sustained release of characteristic chemicals.

Of main concern to virulent Agrobacteria are phenolic compounds, sugars, and sugar derivatives because these substances are signals to the bacteria (see text below) that indicate ongoing cell divisions and synthesis of cell wall polysaccharides and lignin. The "window of competence" described earlier is correlated to the period of maximal cell division at the wound site. Thus, it may be that the T-DNA is incorporated most efficiently into the plant cell genome (see text below) during this period of active cell cycling. Interestingly, most monocotyledonous plants have a quite different wound response. Although many of the phenolics and sugars are produced, there is a lack of cell division at the wound site. This may be one reason for the general inability of Agrobacterium to transform this class of higher plants.

B. Chemotaxis

Agrobacterium species are common inhabitants of soil, but they are particularly prevalent in the rhizosphere, where organic compounds released by the plant provide a rich source of nutrients. Organic acids such as succinate and *p*-hydroxybenzoate, certain amino acids (valine and arginine), and especially carbohydrates are powerful chemoattractants for *A. tumefaciens*. In laboratory experiments, *A. tumefaciens* cells can reach speeds of up to 60 μm/sec over straight runs of distances more than 500 μm when moving toward the source of such compounds. Chemotaxis is thought to guide Agrobacterium cells toward wound sites on plants and on toward cells within the wound area that are competent to receive and express T-DNA. Chemoattractants, including sugars and soluble phenolics, are released from wounded plant cells, and plant and microbial enzymes that degrade macromolecular plant constituents further replenish the pool of attracting molecules. Agrobacteria owe their motility to peritrichous flagella controlled by a chemotaxis system whose genetic determinants are chromosomally located since strains with or without a Ti plasmid exhibit chemotactic behavior. One of the genes known to be involved in chemotaxis is chvE, which encodes a periplasmic, sugar-binding protein that presumably interacts with chemotaxis receptors in the cell membrane. [*See* CHEMOTAXIS.]

C. *Vir* Gene Induction

The activation of the *vir* genes involved in the synthesis and transfer of the T-DNA into plant cells is carefully regulated and is dependent on the ability of the bacteria to sense phenolics and sugars synthesized during the plant wound response (Fig. 3). The VirA and VirG proteins form a two-component regulatory system that is responsible for this function. VirA senses signals from the environment that indicate the presence of cells that are potential targets for T-DNA transfer. Functionally and structurally, it is similar to other bacterial sensory proteins such as CheA, NtrB, and EnvZ, which are involved in chemotaxis, nitrogen responsiveness, and osmoticum responsiveness, respectively. VirA is an autophosphorylating protein and is thought to respond to the appropriate environmental signals by transferring a phosphate group from one of its histidine residues to an aspartate residue of VirG. The phosphorylated VirG than acts as a transcriptional activator. VirG is a protein located entirely in the cytoplasm, whereas VirA spans the cytoplasmic membrane. The cytoplasmic domain of VirA exhibits the autophosphorylation activity and interacts with VirG. The periplasmic domain, but possibly also other sections of the protein including the membrane-spanning and cytoplasmic domains, is the receptor site for signals from the environment.

Three types of signals are recognized by VirA. The most critical signal molecules are phenols, carring an *ortho*-methoxy group, that are related to lignin precursors. Substitutions at the *para* position of the phenol determines its level of *vir*-inducing activity. Comparisons between strong inducers such as acetosyringone and weaker inducers such as ferulic and syringic acid and acetovanillone suggest that

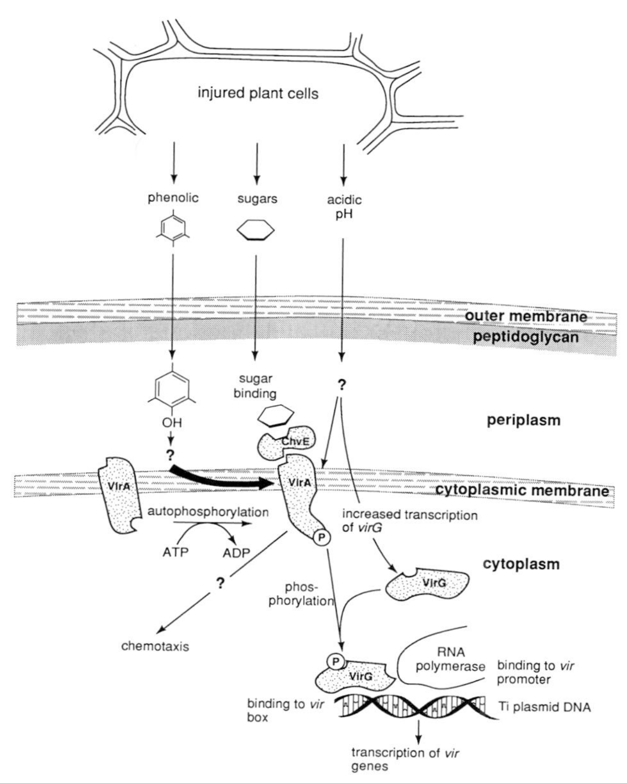

Figure 3 Summary of *vir* gene induction in response to wounded plant cells, indicating location and activities of various proteins and gene involved (see text for additional details).

molecules with *para* substituents that facilitate the dissociation of the proton from the hydroxl group are stronger inducers. Whether these phenolic compounds interact directly with VirA or through a phenolic-binding protein intermediate is not known. The nature of the structural changes in the VirA protein after the recognition of phenolic compounds is also not understood.

Certain pyranose sugars (e.g.,L-(+)-arabinose) and sugar derivatives (e.g., galacturonic acid) strongly enhance the inducing activity of phenolics. Such sugars are recognized by a sugar-binding protein, encoded by *chvE*. *ChvE*–sugar complexes are thought to interact with the periplasmic domain of VirA and to cause a conformational change that increases the sensitivity of VirA toward phenolic inducer molecules. Finally, the presence of inducing phenolics and sugars alone is not sufficient for *vir* induction. The periplasmic domain of VirA also is involved in the response to low pH (optimum, approximately 5.3). The means by which information on the pH of the environment is received by VirA is not known. The requirement for the detection of several signals from the environment before maximal VirA activity is achieved ensures that expression of the *vir* system is induced only under conditions promising success for tumorigenicity. The chemical and physical changes of VirA that result from these environmental conditions and that modulate the VirG-specific phosphonkinase activity of VirA still remain to be elucidated.

The amino acid sequence of VirG indicates the presence of several domains characteristic of DNA-binding proteins. One domain, located in the N-terminal half of the protein, is thought to interact with the transcription machinery of cell (sigma factor, RNA polymerase). As in other bacterial transcriptional activators [e.g., Che Y (chemotaxis), NtrC (nitrogen status), or OmpR (osmotic conditions)], an ATP-binding site and a conserved aspartate residue are present in VirG. This aspartate residue is phosphorylated by VirA. The C-terminal domain contains a helix-turn-helix DNA-binding motif and amino acid residues in this region of the protein contact specific bases on the DNA. VirG specifically binds to a 12-bp sequence, the *vir* box, located upstream or overlapping all the *vir* promoters. Although it is able to bind to this sequence regardless of its phosphorylation status, only phosphorylated VirG is a transcriptional activator. The assumption is that phosphorylation allows VirG to interact with, and activate, the transcriptional machinery. Finally, a *vir* box is located upstream of the *virA* promoter and three *vir* boxes are upstream from the transcription start site of *virG*. This suggests that transcription of the *vir* regulatory genes is autoregulated in a positive fashion on reception of the phenolic and sugar signals.

VirG is the rate-limiting protein for *vir* gene induction, and its transcription increases in response to the presence of inducing phenolics, an acidic environment, and phosphate starvation. Transcription occurs from two promoters on activation by phosphorylated VirG bound to *vir* boxes located upstream from the promoters. Transcription from promoter 1 increases in the presence of phenolic compounds and under acidic conditions. VirG promoter 1 is similar to *Escherichia coli* promoters inducible by phosphate starvation, and promoter 2 resembles *E. coli* heat shock promoters, which are inducible by a number of environmental stresses. Thus it appears that adverse environmental conditions (e.g., lack of certain nutrients, presence of inhibitors) increase the concentration of VirG and make the cell more responsive to signals that indicate the presence of plant cells that can be infected. As these signals activate the phosphokinase activity of VirA, phosphorylated VirG enhances transcription of its own promoter, and ultimately sufficient active VirG is present to activate transcription of entire *vir* regulon.

III. Bacterial Attachment to Plant Cells and DNA Delivery

A. Plant Components of Attachment

Physical interaction of the bacteria and plant cells is required for transformation and can take place even in the absence of *vir* gene induction (see Fig. 1). Binding appears to be a two-step process in which individual bacteria first loosely attach to the plant cell surface, often in a polar fashion. This initial, reversible loose attachment is followed by irreversible tight binding of the bacteria to the plant cell surface. During this second stage, the bacteria synthesize cellulose fibrils that may be involved in stabilizing the interaction with the plant surface. However, the cellulose fibrils are not essential for productive attachment because *cel* mutants lacking fibrils bind to plant cells and are fully virulent. Stability of the attachment has been demonstrated by monitoring the ability of attached, radiolabeled bacteria to leave the plant surface: Once the second phase of attachment has occurred (usually within 3 to 6 hr), bacteria remain associated with the plant cell and cannot be displaced by nonlabeled bacteria.

Attachment of Agrobacteria appears to occur at specific sites on the plant cell surface. This hypothesis is supported by the facts that (1) Ti plasmidless strains capable of attachment to plant cells can inhibit the binding of virulent strains and prevent tumor formation, and (2) quantitative binding assays show that Agrobacterium binding to plant cells is saturable, suggesting the presence of a limited number of attachment sites on the plant cell. It has been estimated that, at saturation, a few hundred bacteria are bound per cell. The exact nature and location of the plant cell receptor remains elusive. Soluble, pectin-enriched, protein-containing plant cell wall fractions (CWF) can, when incubated with Agrobacterium, prevent the bacteria from attaching to plant cells and can block transformation. This suggests that the CWF contains a moiety(ies) involved in the productive physical interaction between the plant and bacterial cells. The inhibitory activities of the CFW are eliminated by heat and acid treatment and partially reduced by protease treatment. Similarly, treatment of carrot suspension cells with protease, or extraction with the non-ionic detergent Triton X-100, causes them to lose the ability to bind *Agrobacterium*. Together, all these results suggest that the plant receptor molecule may be pectin-associated protein or glycoprotein.

B. Bacterial Components of Attachment

Efficient attachment of bacteria to plant cells requires the products of three *A. tumefaciens* chromosomal loci, *chvA, chvB, and exoC (pscA)*. Mutations in any of these constitutively expressed loci significantly reduce virulence and cause pleiotropic phenotypes suggestive of cell envelope changes (e.g., bacteriophage resistance, loss of motility, and calcofluor negative staining). Both *chvB* and *exoC* functions are involved in the synthesis of a low-molecular-weight, cyclic β-1,2-glucan molecule. The *exoC* gene product is a phosphoglucomutase that converts glucose 6-phosphate to glucose 1-phosphate, an intermediate in the synthesis of UDP-glucose. *chvB* encodes a 235-kDa inner membrane protein, which incorporates UDP-glucose into the β-1,2-glucan chain. Transport of β-1,2-glucan from the cytoplasm to its active site in the periplasm requires the *chvA* gene product, which shares amino acid similarity with both prokaryotic and eukaryotic export proteins. Genes homologous to the closely linked *chvA* and *chvB* loci are required for the plant cell attachment by *Rhizobium meliloti* as well, suggestomg these attachment function are conserved in those related bacteria. [*See* ADHESION, BACTERIAL.]

Synthesis and export of β-1,2-glucan to the periplasm is clearly a prerequisite for attachment of *A. tumefaciens* to plant cells. However, the exact role of this oligosaccharide in the attachment process is still unknown. Exogenously added purified β-1,2-glucan does not restore attachment or tumorigenesis to mutants lacking this glucan, suggesting that periplasmic glucan may be the form related to binding activity. Periplasmic β-1,2-glucan is known to play a role in osmotic protection, and it has been proposed that loss of the glucan might indirectly affect virulence by reducing the activity of cell surface virulence proteins under suboptimal osmotic conditions. However, in cocultivation experiments, medium osmolarity does not affect the virulence of *chv* mutants. Thus, β-1,2-glucan appears to affect attachment in a manner unrelated to its role in osmoadaptation. It should be noted that *chvA* and *chvB* mutant strains are not avirulent but rather are considerably reduced in virulence compared with wild type cells. Measurable attachment does occur with these mutants, and it is saturable. Such attachment of the *chv* mutants is also stable, as demonstrated by the fact that radioactively labeled bacteria do not dissociate from the plant cell over a 6-hr period. One interpretation of these results is that the *chv* mutants are deficient at "finding" or "uncovering" sites to which stable attachment occurs.

A second class of attachment deficient mutants (*att*) have been identified. The six *att* mutations identified to date all map to a single 12-kb chromosomal *Eco*RI fragment at a location different from that of *chv* and *exoC*. The *att* mutants display greatly reduced attachment and virulence on carrot cells but differ from wild-type cells only in the loss of one or two of three distinct membrane proteins. The role of these proteins in the attachment process is unknown.

C. DNA Delivery Process: A Conjugal Model

1. Production of the Transferred Intermediate

Once the *vir* genes have been activated and the bacteria are attached to the plant cells, the T-DNA transfer process begins by the excision of the T-DNA out of the Ti plasmid into a form transferable to plant cells. This process involves functions of the *virD, virC,* and *virE,* loci (see Table I; Fig. 4). Formation of the T-DNA intermediate also requires the border sequences, conserved *cis*-acting 25-bp imperfect direct repeats that delineate the ends of the T-DNA in all Ti and Ri plasmids. The borders function in an orientation-specific fashion, and deletion of the right border, or changing its orientation, causes a much greater decrease in tumorigenesis than does a similar alteration of the left border repeat. This indicates the right border is more important than the left and suggests a polarity of T-DNA transfer. T-DNA processing is initiated by an endonucleolytic cleavage reaction at the borders mediated by the *virD1* and *virD2* gene products. The VirD1 topisomerase first converts the Ti plasmid DNA near the border from a supercoiled to a relaxed

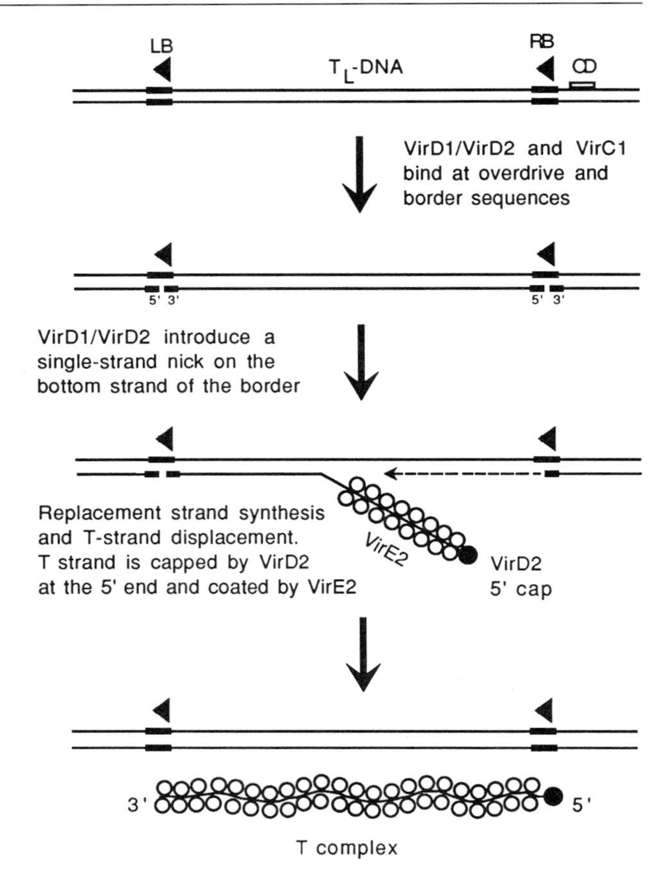

Figure 4 Strand displacement model and T-complex formation (see text for details).

form. A VirD2 endonuclease then introduces a site-specific nick on the bottom DNA strand of the border. In some Ti plasmids the VirC1 protein binds at a conserved 24-bp sequence, termed *overdrive,* located adjacent to the right border. The VirC1–overdrive complex stimulates VirD endonuclease binding and nicking at the border, perhaps contributing to the greater activity of the right border. The Agrobacterium host DNA replication machinery is then thought to initiate DNA synthesis at the 3′-OH end of the border cleavage site, using the top DNA strand as a template. Replacement synthesis continues until the left border cleavage sit is reached. The original bottom DNA strand between the borders is then released as a linear, single-strand T-DNA molecule, called the *T strand.* In response to cocultivation with plant cells, or *vir* gene induction with phenolics *in vitro,* the T strand is produced at about one copy per bacterial cell. The VirD2 protein remains covalently attached to the 5′ end of the T strand, which becomes coated along its length by

VirE2, a cooperative single-strand DNA-binding protein. The Vir proteins associated with the T strand (also called the *T complex*) may serve to protect the DNA molecule from bacterial and plant cell nucleases, endow the molecule with a conformation necessary for transport, or provide a "pilot" function for guiding the DNA out of the bacterial cell. In addition, recent results suggest that VirD2 might also function in the plant to target the T complex to the plant cell nucleus (see text below).

Considerable evidence now suggests that the T complex represents the T-DNA intermediate transferred to plants. A direct correlation exists between the level of T-complex production and the frequency of tumor formation, a result expected if the T complex is the biologically active intermediate. In addition, the fact that *A. tumefaciens* encodes a single-strand DNA-binding protein (VirE2) necessary for virulence on most plants also suggests that the transferred DNA is single-stranded. It must be pointed out, however, that *vir* gene induction leads to the formation of other potential T-DNA transfer intermediates. VirD-dependent double-strand T-DNA border nicks have been observed, and both double-stranded linear and circular T-DNA forms have been identified in *A. tumefaciens*. However, these types of molecules have generally been observed at a much lower frequency than the T complex. Further, the production of a linear T-DNA by double-strand breaks would be a nonconservative process, leading to loss of the T-DNA from the Ti plasmid. The T-complex formation model, however, postulates the retention of the original structure of the T-DNA on the Ti plasmid. The strand displacement model is also attractive because it helps explain the apparent polarity of border function in T-DNA transfer.

Analysis of the early events in T-DNA processing have lead to the intriguing hypothesis that T-DNA transfer to plants occurs by a mechanism analogous to bacterial conjugation. There are many similarities between the processes of T-DNA transfer and bacterial plasmid conjugal DNA exchange. These include (1) Direct surface–surface contact is required between the donor and recipient cells; (2) DNA transfer is initiated after a single-strand nick at the origin of transfer (e.g., Ti plasmid borders and the *oriT* sequences of conjugal plasmids); (3) A linear, single-stranded DNA molecule is transferred to the recipient in a 5′–3′ unidirectional manner; and (4) Single-stranded DNA-binding proteins are required. Additional evidence of a conjugative mechanism of T-DNA transfer comes from analysis of mobilizable, wide host range plasmids of *E. coli*. Certain mobilizable plasmids can use their own *oriT* and mobilization (*mob*) functions to generate a conjugal intermediate transferable to other bacterial by the transfer (*tra*) functions of conjugative plasmids residing within the same bacterium. In this case the *mob* gene products act in a DNA sequence-specific fashion to nick the plasmid at *oriT* and initiate mobilization. The *tra* gene products encode the transfer machinery and can use almost any DNA substrate in a nonsequence specific manner. Interestingly, when present in an *A. tumefaciens* strain harboring a disarmed Ti plasmid (containing the *vir* genes but lacking the T-DNA or its borders), the *oriT*/*mob* functions of the wide-host range, mobilizable IncQ plasmid RSF1010 can functionally substitute for T-DNA borders and generate a DNA molecule transferable to plants by the *vir* gene transfer machinery. This suggests that the normal T-DNA intermediate resembles the RSF1010 conjugal intermediate, and further, that the *vir* genes are functionally analogous to the *tra* genes of conjugative plasmids. These results strongly support a conjugal model of T-DNA transfer.

Recent work suggests that an evolutionary relationship exists between the processes of T-DNA transfer and conjugation. For example, a 12-bp sequence spanning the VirD cleavage site on the T-DNA border is also present in the *oriT* sequences of several wide-host range IncP conjugal plasmids. Furthermore, DNA sequence analysis revealed that the *mob* gene operon from the conjugal IncP plasmid RP4 contains four genes, the first two of which (*traJ* and *traI*) encode proteins responsible for nicking at the *oriT* sequence during plasmid mobilization. Interestingly, the organization of the genes in this RP4 *mob* operon very closely resembled that of the pTiA6 *virD* operon of *A. tumefaciens*. Although the *traJ*/*virD1* and *traI*/*virD2* gene pairs were not significantly homologous to each other, another of the RP4 *mob* gene products, TraG, did share considerable amino acid similarity with the predicted pTiA6 *virD4* gene product. These results are exciting in that they suggest an evolutionary relationship between the IncP conjugal transfer system and the Agrobacterium T-DNA transfer system. [*See* CONJUGATION, GENETICS.]

2. Movement of the Transferred Intermediate

The mechanism by which the T complex is transported to the plant cell across both the bacterial and plant cell membranes and walls remains poorly understood. In the best characterized bacterial conjugal transfer system, that of the *E. coli* F plasmid, DNA transfer requires an extracellular structure called the *F pilus*. The pilus first mediates donor and recipient cell binding, then undergoes retraction and disassembly to bring the membranes of the two cells in close contact. A membrane-spanning, multicomponent protein complex located at the pilus base is then thought to serve as a conduit for transfer of the conjugal DNA molecule into the recipient. As described above, the attachment of *A. tumefaciens* cells to plant cells is mediated by bacterial chromosomal gene products, and extracellular pili have not been implicated in virulence. By analogy with plasmid conjugal systems, however, it seems likely that a *vir*-specified membrane pore is formed to deliver the T complex into plant cells.

Products of the Ti plasmid *virB* operon are the best candidates to form a membrane-associated T-DNA transport apparatus. This follows from the fact that *virB* nucleotide sequence analysis predicts that most of the eleven *virB* gene products are membrane-associated (see Table I), and to date six VirB proteins have been localized to the *A. tumefaciens* inner membrane. In addition, *virB* functions are not involved in T-DNA processing, whereas genetic analysis has demonstrated that at least three *virB* genes (*virB9, virB10,* and *virB11*) are required for virulence. This suggests that VirB proteins mediate a late step or steps in T-DNA transfer process and is consistent with their proposed role in T-DNA transport.

The exact function of any VirB protein has yet to be determined. One of the VirB proteins essential for virulence, the 48-kDa *virB10* gene product, has been identified as a component of an inner membrane-spanning protein aggregate in *A. tumefaciens*. Thus, VirB10 might be a structural component of the T-DNA transport apparatus. In addition, both the predicted *virB4* and *virB11* gene products contain a consensus mononucleotide (NTP) binding site, suggesting that VirB protein phosphorylation may somehow play a role in the T-DNA transfer process. Of particular interest is the fact that VirB11 displays amino acid similarity, including an NTP binding site, with the *comG1* gene product of *Bacillis subtilis*. The ComG1 protein has been proposed to regulate the activities of membrane proteins involved in DNA uptake through the membrane of *B. subtilis* during the development of genetic competence. The similarity between these proteins suggests that the essential virulence function carried out by VirB11, a 38-kDa extrinsic membrane protein, has been conserved in two different processes involving DNA transport through membranes (i.e., T-DNA transfer and bacterial transformation). In addition, purified VirB11 protein has been shown to undergo autophosphorylation in vitro. Thus, VirB11 could regulate its own activity via phosphorylation or might act as a kinase to regulate the conformation or activity of another component(s) of the T-DNA transport apparatus. Interestingly, *virB11* mutants display a reduced frequency of Ti plasmid conjugal transfer between *A. tumefaciens* cells (see Section IV.C), suggesting that VirB11 might somehow influence plasmid conjugal exchange as well.

Finally, recent evidence indicates that *virB* gene products affect the substrate specificity of the T-DNA transfer machinery. It was found that the conjugal intermediate of an RSF1010-derivative plasmid (see Section III.C.1) greatly suppressed virulence in *A. tumefaciens* cells containing the octopine-type plasmid pTiA6. Presumably the RSF1010 conjugal intermediate (the "R complex") interfered with T-complex transfer to plants by competing for and/or saturating the VirB transport apparatus. Surprisingly, coordinate overproduction of the *virB9, virB10,* and *virB11* gene products partially relieved this oncogenic suppression and restored tumorigenicity but did not increase the transfer frequency of the RSF1010 plasmid to plant cells. This result indicates that overexpression of these three *virB* genes increases the efficiency of T-complex transfer to plants in comparison with that of the R complex. Further, this suggests that the T complex possesses physical characteristics or specific factors (e.g., a VirD2 5′ cap or VirE2 coat) that result in more efficient T-complex transport (or targeting to the transport machinery) in the presence of additional VirB9, VirB10, and VirB11 proteins.

IV. Inside the Plant Cell

A. T-DNA Movement and Integration

The means by which the T complex enters the plant cell, moves to the nucleus, and integrates into the nuclear DNA are unknown. Most models concerning T-complex movement center around the concept that the capping protein, VirD2, has amino acid homology to the nuclear localization signals of eukaryotic nuclear proteins. When a domain of this protein containing such a signal is fused to a reporter gene (e.g., β-glucuronidase), it can mediate transfer of the fusion protein to the nucleus of plant cells. Because VirD2 is covalently attached to the T complex, its movement to the nucleus could result in movement of the entire complex, although this has yet to be established experimentally. The fate of the VirE2 that coats the T strand is also unknown. Interestingly, co-inoculation experiments have shown that a nonvirulent strain carrying a wild-type *virE* gene (but no T-DNA) will cause tumor formation by a *virE* mutant strain containing wild-type T-DNA. This extracellular complementation requires attachment of the bacteria to the plant cell and *vir* gene activity (e.g., *virB*) and suggests a role for the VirE proteins inside the plant cell. As previously described, one important function could be to protect the single-stranded DNA of the T complex from nucleases.

Most analyses of the integration of T-DNA into host plant DNA have been carried out by studying the T-DNA–plant DNA junctions in the genome of transformed cells. This has revealed several consistent features of the integration process. First, the right border of the integrated T-DNA is generally (but not always) within a few base pairs of the right border nick site on the Ti plasmid. The left border of the integrated T-DNA is considerably less precise with the T-DNA–plant DNA junction often being 100 bp or more from the border repeat of the Ti plasmid. This suggests a polarity in the integration process that may reflect the fact that the 5′ (right border) end of the T complex is capped by the VirD2 protein whereas the 3′ end is not. Second, although the T-DNA is often found as a single insertion of a single copy of the T-DNA, it can also be found in much more complex configurations. For example, there can be multiple insertion events in a single plant cell and/or multiple tandem insertions at a single site. Third, the distribution of insertions of the T-DNA in the plant genome appears to be random, with the possible exception that insertion may be more frequent in sites of active plant gene expression. (Because of these attributes, Agrobacterium-mediated transformation has been used to mutagenize and "tag" plant genes with the T-DNA.) Finally, the plant sequences at both the left and right border of the inserted T-DNA have been compared with the sequence of this region before integration. These analyses showed that the target plant DNA can undergo several types of rearrangements, including small duplications and deletions. Outside of the possible functions of the VirD2 protein at the integration site, no known integration specific *vir* gene functions have been identified. This, along with the observation of deletions and duplications at the integration site, suggest that insertion occurs during periods of host plant directed DNA synthesis. Such a hypothesis is consistent with the fact that the highest frequency of transformation is correlated with cell division at the wound site.

Once integrated into the plant genome, the T-DNA appears to be used in a fashion identical to that of the host DNA: Transcription, formation of polyadenylated mRNA, RNA movement into the cytoplasm, and translation all are characteristic of eukaryotic cells. Although the genes of the T-DNA do not carry introns, genes that do carry introns can be introduced into the T-DNA (see Section V) and splicing occurs in the predicted fashion. One of the great mysteries of the entire process is how eukaryotic transcription and translation regulation evolved within Agrobacterium genes when selection for them is so indirect: This gene expression must occur *in the plant* before selective advantage for the bacterium can be derived (see Section IV.B). At this point there is no evidence that genes from some plant were "captured" in the manner so often observed in the case of animal retroviruses.

B. T-DNA Expression: Opine Production

One of the early fundamental findings concerning Agrobacterium-mediated transformation was that crown gall tumors contain novel metabolites whose production is specified by the inciting Agrobacterium strain rather than by the host plant. For example, strain B6 of *A. tumefaciens* induces tumors that produce octopine, a reductive condensation product of pyruvate and arginine, whereas strain T37 induces tumors that produce nopaline, derived from α-ketoglutarate and arginine. Numerous other opines have been identified, and in general, these compounds are made from an amino acid and a sugar or sugar acid. In the cases of both the Ti and Ri plasmids, the genes encoding the opine biosynthetic enzymes are found on the T-DNA and are expressed only in the transformed plant cells. Although the T-DNA genes involved in tumor formation are usually physically linked to the genes involved in opine production, some exceptions have been noted. For example, the octopine-type Ti plasmids carry two

T-DNAs, each flanked by the border repeats (see Fig. 2). The T-left (T_L) T-DNA carries genes responsible for both opine synthesis (*ops*) and tumorigenesis (*onc*), whereas the T-right (T_R) T-DNA carries genes encoding only opine production. Finally, molecular analysis of the genes encoding opine biosynthesis has shown that the mRNAs of these genes can be highly abundant and, in some cases, are regulated by the plant hormones involved in the control of cell division (see Section IV.C).

The opines are the compounds that form the theoretical basis for the selective advantage of the DNA transfer process. Plant cells cannot metabolize these molecucles even though they can be extremely abundant (nopaline has been found in tobacco tumors at 60 μmol/g dry weight). In contrast, the bacterial strains that induce the tumors carry catabolism genes on their Ti or Ri plasmids that are responsible for two critical processes: active transport of the opine specified by this strain into the bacterial cells, and cleavage of the opine into the parent compounds. These processes provide a unique source of carbon and nitrogen for the bacterium. Generally, the bacterial catabolism genes are not expressed unless their transcription is induced by the cognate opine. An intriguing observation is that certain of the opines can serve to induce conjugative transfer of the Ti plasmid to Ti plasmid-free strains of Agrobacterium. Thus, the opines not only promote the growth of bacteria carrying the appropriate Ti plasmid, but also stimulate the transfer of this plasmid to other bacteria, thereby stimulating its abundance in the general Agrobacteria population.

C. T-DNA Expression: Tumorigenesis and Hairy Root Formation

As previously described, the transfer of the T-DNA into plant cells results in the production of opines that can be advantageous to the inciting bacterium. However, the advantage of opine production alone for the bacteria would be limited to the frequency of transformation: The opines have no documented effect on plant cell survival and/or growth. Thus, each transformation event would result in the production of opines by only one plant cell or the limited number of cells derived from it during wound healing of the plant. This problem is eliminated by the fact that the T-DNA also carries what may best be described as plant oncogenes: Expression of these genes in the transformed plant cells results in their uncontrolled or abnormally regulated proliferation, and thus, tumor or hairy root formation.

The physiological and molecular basis of tumor formation as a result of *A. tumefaciens* is reasonably clear: The transformed cells produce compounds that stimulate cell division. Most dicotyledonous plant cells of the type susceptible to transformation can be stimulated to proliferate in vitro on a defined tissue culture medium that includes, in addition to a carbon source and various inorganic salts, two plant hormones, auxin and cytokinin. These hormones are highly conserved in the plant kingdom and are both derived from abundant and important metabolites. Auxin [e.g., indole-3-acetic acid (IAA)] is derived from tryptophan, and cytokinin (e.g., isopentenyl adenosine) is derived from adenosine. Each of these growth factors individually has marked effects on plant cell growth and differentiation, and synergistically they can stimulate continuous cell division.

A variety of plant pathogens (e.g., *Pseudomonas*) exert effects on plant growth or development by producing one or both of these hormones, which then affect the plant. Agrobacteria have evolved a different strategy. Several of the T-DNA genes encode proteins that synthesize auxin and cytokinin. Originally observed in genetic analysis as genes that affected the morphology and growth potential of the resultant tumor, molecular and biochemical analysis has precisely defined their activity. Two genes, originally referred to as *tms1* and *tms2* (becauses mutation in them results in tumor morphology that is shooty), encode a two-step biosynthetic pathway for the production of IAA. The *tms1* encodes a tryptophan 2-monooxygenase that converts tryptophan to indole-3-acetamide (IAM) and the *tms2* encodes an IAM hydrolase that converts IAM to IAA. Interestingly, these enzymes, and the genes encoding their synthesis, are highly homologous to those originally found in *Pseudomonas*. No similar genes or enzymes have been found in higher plants. Thus, a bacterially derived plant hormone biosynthetic pathway has evolved that can be expressed in higher plant cells. Similarly, a cytokinin, isopentenyl adenosine monophosphate, is synthesized from 5′ adenosine monophosphate and isopentenyl pyrophosphate by an ispentenyl transferase encoded by the *tmr* (tumor morphology rooty) gene of the T-DNA.

The ultimate phenotype of tumors initiated by *A. tumefaciens* is defined by a combination of T-DNA activities and host responses. Besides the hormone biosynthesis genes, other T-DNA genes have been shown to affect tumor growth or morphology. One example of this is the *tml* (tomor morphology large) gene, which, when mutated, results in strains that produce larger than normal tumors on some host plants. The molecular and biochemical basis of this phenotype is unknown. Finally, the response of different plants to the same T-DNA can vary enormously. For example, *Nicotiana tabacum* responds to mutant Agrobacteria missing the *tms* loci of the T-DNA by producing shoot-forming tumors, an expected result because of the absence of IAA production combined with the presence of cytokinin production (known to stimulate shoot formation). However, the related species *Nicotiana glutinosa* responds to the identical Agrobacterium strain by forming typical, unorganized crown gall tumors. These results demonstrate that host activities can supplement or replace T-DNA-specified hormone biosynthesis and result in a fully neoplastic cellular state. The physiological and molecular basis of this has not been fully defined.

In the transformed cells, the transcripts from the T-DNA hormone biosynthetic genes are extremely rare, particularly in comparison with the transcripts from the opine biosynthetic genes. This is likely to reflect the fact that the hormones are active at extremely low doses (10^{-6}–10^{-9} M) but can be toxic at higher doses. Thus, it is important for these genes to be expressed at a level that is compatible with survival of the transformed cell. One result of hormone production by the transformed cells is that they grow continuously, thus expanding the population of opine-producing cells. It should also be noted, however, that these hormones diffuse to surrounding cells and tissues and result in the proliferation of nontransformed cells. Thus, virtually all naturally occurring crown gall tumors are mixtures of transformed and nontransformed cells. The advantage of this to the bacteria may be that the larger the tumor, the greater a nutrient sink it becomes, attracting the reserves of the plant to itself rather than the usual sites requiring nutrition (e.g., meristems and young leaf primordia of the plant). Alternatively, it is possible that growth of the nontransformed cells essentially dilutes the concentration of the plant hormones that are known to be growth inhibiting at high concentrations.

Agrobacterium rhizogenes, as described earlier, does not induce tumor formation but rather causes the transformed tissues to form numerous roots, described as "hairy roots" in the early literature. As in the octopine type Ti plasmids, many Ri plasmids have two T-DNAs (also referred to as T_L and T_R), both flanked by border repeats. In this case both T-DNAs are involved in the aberrant growth response and opine production. The formation of roots in many cultured plant tissues is controlled by the *ratio* of exogenously supplied auxin and cytokinin, rather than their absolute concentrations. Thus, it was originally thought that the T-DNA of the Ri plasmid would simply involve auxin production, thereby increasing the auxin to cytokinin ratio, favoring root formation. Indeed, the T_R-DNA does carry genes encoding the same auxin biosynthetic pathway as that of the T-DNA from the Ti plasmid. However, genetic analysis has shown that several genes on the T_L-DNA, *rol* (root locus) *A*, *B*, and *C*, are also involved in the hairy root phenotype. These genes are particularly interesting because they seem to mediate root formation from specific tissues on a plant or from certain hosts: A particular *rol* gene may be necessary for hairy root formation in one host or tissue but not in another. Recent studies introducing the different *rol* genes into plant cells, alone or in combination, suggest that they drastically affect the ability of the transformed cells to form roots in response to auxin. For example, in certain hosts either *rolA* or *rolB* can act synergistically with *rolC* to evoke the hairy root phenotype. Interestingly, *rolC* can complement mutations in the cytokinin biosynthesis locus of the Ti plasmid T-DNA, suggesting it may be involved in some aspect of cytokinin activity. Additional studies have shown that the cells containing the *rol* genes are approximately 100 times more sensitive, on a molar basis, to the effects of auxin on the transmembrane potential of their cytoplasmic membrane. Although the precise biochemical activities of the *rol* genes have yet to be elucidated, it is clear that these genes have the capacity to affect hormonal response systems and should prove extremely useful in the analysis of hormonal signal transduction in plants.

V. Genetic Engineering with Agrobacterium

A. Vectors

The ability of Agrobacterium to transfer DNA into plant cells and the clear exploitation of plant nutrients (opine production) resulting from the expression of this DNA lead to the concept that this process could be modified to allow the insertion of any chosen piece of DNA into plant cells. The development of this capability was heralded by the findings that (1) mutation or deletion of the oncogenes within the T-DNA did not affect the ability of the T-DNA to be transferred, and (2) cells transformed by such "disarmed" T-DNA could be identified either by screening for opine production or by inserting a plant-selectable marker into the T-DNA. In many cases, entire fertile plants were regenerated from such transformed cells and the transferred genes were shown to be expressed and heritable in a Mendellian fashion.

These observations lead to the development of highly modified vector systems and transformation protocols. Two types of vectors are routinely used. The first are derivatives of the "disarmed" Ti plasmids that have most of the T-DNA between the border repeats deleted and replaced with, for example, a plant selectable marker and pBR322 plasmid DNA. The latter can serve as a site for recombination with an incoming pBR322 derivative containing the gene of interest. An alternate system takes advantage of the fact that the *vir* genes can work in *trans* to generate T complexes from any plasmid in Agrobacterium that carries the border repeats. Typically this strategy includes a disarmed Ti plasmid, lacking the T-DNA and borders but carrying the *vir* genes, and smaller plasmids that can replicate in both *E. coli* and *A. tumefaciens*. These "binary vectors" carry plant-selectable and/or scorable markers and various cloning sites within the border repeats.

B. Vector Utilization

Both types of vector strategies have been used effectively by numerous laboratories interested in constructing and using transgenic plants. Such studies have ranged from basic analysis of transcriptional control to novel methods for generating disease and pest resistance in agronomically important plant species. The reason for such enormous success has been that, in appropriate host plants, the frequency of transformation can be extremely high. For example, by co-cultivating individual regenerating plant protoplasts with *A. tumefaciens* carrying the appropriate vectors, it is possible to show that up to 30–50% of *nonselected* plant colonies derived from the protoplasts are transformed. More frequently used are transformation strategies in which tissue explants (e.g., leaf, stem, or root sections) are cocultivated with *Agrobacterium* and then immediately placed on selective medium that will also induce plant regeneration from the tissues. This allows for rapid production of transgenic plants from cells that have spent very little time being cultured *in vitro*. Because *in vitro* culture of plant cells often leads to increased spontaneous genetic variation, the rapid transformation–regeneration strategies can be very important.

Unfortunately for the genetic engineer, many plants, and in particular the agronomically important Gramineae (e.g., corn), either are not susceptible to transformation by Agrobacterium or have susceptible tissues that are not competent (at least using current protocols) to regenerate plants. As discussed earlier, the reasons for variation in the host range of DNA transfer from Agrobacterium can be as complex as the numerous steps involved in this plant–pathogen interaction. Interestingly, Agrobacterium can transfer DNA into a specific subset of corn cells, but this transfer does not result in stably transformed cells, at least none that have been identified. This suggests that some step(s) in the transformation process *after* DNA transfer into the plant cell is limiting.

VI. Conclusions

The successful colonization and transformation of plant cells by *A. tumefaciens* and *A. rhizogenes* is a remarkably complex process that is dependent on the integration of cellular activities of both the pathogen and the plant. Substantial progress has been made during the past 15 years describing the molecular basis of transformation, particularly as it relates to activities within the bacterium and the oncogenic functions of the T-DNA in the plant. However, numerous intriguing and important questions abound. In particular, the elucidation of the basic cell biology and biochemistry of the T-complex movement into the plant cells and its integration into the host genome are likely to reveal fundamental plant processes that have been exploited by the pathogen. Understanding these processes may also allow the manipulation of the system to transform plants that heretofore have been outside the previously defined host range of Agrobacterium.

Bibliography

Binns, A. N., and Thomashow, M. F. (1988). *Annu. Rev. Microbiol.* **42,** 575–606.

Braun, A. C. (1982). A history of the crown gall problem. *In* "Molecular Biology of Plant Tumors" (G. Kahl and J. Schell, eds.), pp. 155–210. Academic Press, New York.

Buchanan-Wollaston, V., Passiatore, J. E., and Cannon, F. (1987). *Nature* **328,** 172–175.

Heinemann, J. A. (1991). *Trends Genet.* **7,** 181–186.

Klee, H., Horsch, R., and Rogers, S. (1987). *Annu. Rev. Plant Physiol.* **38,** 467–486.

Melchers, L. S., and Hooykaas, P. J. J. (1987). *Ox. Surv. Plant Mol. Cell Biol.* **4,** 167–220.

Ream, W. (1989). *Annu. Rev. Phytopathol.* **27,** 583–618.

Zambryski, P. (1988). *Annu. Rev. Genet.* **22,** 1–30.

Zambryski, P., Tempe, J., and Schell, J. (1989). *Cell* **56,** 193–201.

Airborne Microorganisms

Linda D. Stetzenbach
Harry Reid Center for Environmental Studies–University of Nevada, Las Vegas

Glossary

Autochthonous Indigenous, naturally occurring in a particular ecosystem

Bioaerosol Suspension of microorganisms, microbial by-products, and/or pollen in the air

Building-related illness Specific adverse health reaction resulting from indoor exposure to pollutants

Opportunistic pathogen Organisms not usually considered pathogenic but may cause adverse health affects under certain conditions

Sick building syndrome Random symptoms of ill health reported by building occupants when no causative agent has been identified

AIRBORNE MICROORGANISMS include bacteria, algae, fungi, viruses, and protozoa. These organisms are not autochthonous to the aerosphere but are being passively transported from one place to another as a bioaerosol. Aerobiology is the study of the dispersal and survival of bioaerosols and the impact that airborne microorganisms have on the biosphere. Historically, as far back in time as Hippocrates, the science of aerobiology was limited to discussion of airborne organisms as "germ clouds" transmitting disease via the respiratory route. Recently, however, the scope of aerobiology has expanded to include not only pathogens but also opportunistic and nonpathogenic organisms. Aerobiology also concerns exposure to bioaerosols consisting of fragments of microbial cells and microbial metabolites resulting in adverse health effects and allergic reactions and the potential impact to the environment of aerosol-released, genetically engineered microorganisms.

I. Bioaerosols

A. Characteristics of Bioaerosols

An aerosol is a collection of particles suspended in a given body of air. The size range of particles in an aerosol is approximately 0.5–30 μm in diameter. This size range includes bacteria, fungal spores, viruses, and some parasites.

1. Microorganisms as Bioaerosols

Microorganisms are suspended as bioaerosols consisting of single-organism units (e.g., one cell, one spore) or as a cluster of several organisms. These bioaerosols can occur as liquid droplets or as dry particulates. Single fungal spores are the largest microorganisms in bioaerosols, measuring between 3 and 30 μm in diameter (Fig. 1), whereas viruses are the smallest (<0.1 μm). Suspended single bacterial cells are approximately 1 μm and amoebic cysts that remain airborne for any length of time following aerosolization are generally <15 μm in diameter. Microorganisms are often suspended as clumps of cells. The aggregation of cells into clumps results in larger droplets that are less likely to travel long distances, but survival may be enhanced as cells in the central portion are protected by cells along the periphery. Droplets containing liquid either within the microspace between the cells or surrounding them may also afford protection from environmental stresses. Similarly, the dry bioaerosol (consisting of particulates such as dust, skin scales, or plant material that microorganisms use as a raft) may also protect organisms from stresses.

2. Other Bioaerosols

Fragments of microbial cells (e.g., cell wall segments, flagella, genetic material) and by-products of

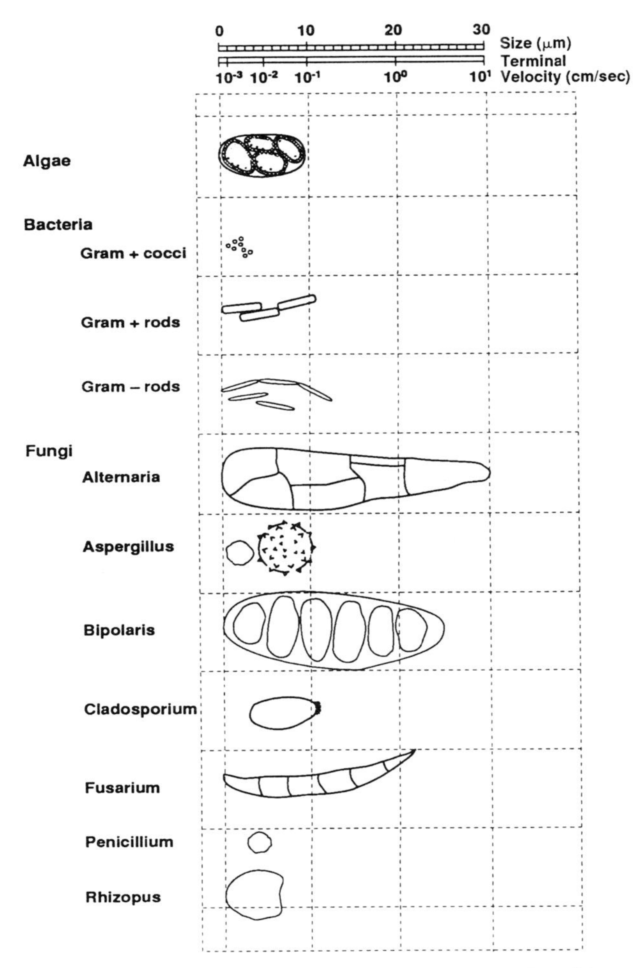

Figure 1 An illustration of some microorganisms found in bioaerosols and their approximate settling velocity as single units. (Courtesy of Shirley Burns, UNLV.)

metabolism (e.g., volatile organic compounds, toxins) may also be transported as bioaerosols.

B. Dispersal

The transport of microorganisms in the air is determined by the aerodynamic characteristics of the organism in the bioaerosol and the current environmental conditions.

1. Physical Properties

The settling of the bioaerosol from the air is influenced by the size, density, and shape of the microorganisms. In general, the terminal velocity of a suspended organism is inversely proportional to the square of its radius. Approximate settling times for some microorganisms commonly occurring in the air are listed in Fig. 1.

2. Environmental Conditions

The transport of microorganisms in the air is controlled by several environmental parameters. Figure 2 illustrates the effects of some of these variables on bacterial cells released via aerosol onto a plant leaf.

a. Temperature and Humidity

Bacterial cells may experience upward movement from plant leaves during the hot, dry periods of the day. High temperatures and low relative humidity (RH) result in unsettled air leading to dispersal of organisms into the atmosphere. Positive electrostatic charges on plant surfaces during sunny conditions also aid in dispersal because they repel negatively charged microorganisms, keeping them suspended in the air. Conversely, thermal inversion and stagnant air cause a settling of aerosols and result in decreased numbers of airborne cells. Raindrops will cause suspended organisms >2 μm in diameter to settle out, but plant pathogens have been dispersed from diseased crops onto healthy plants by rain impaction. Simulated raindrop studies show that the impact of a 2-mm drop of rain possesses enough energy to generate bioaerosols from bacterial infected potato stems and viable, infective

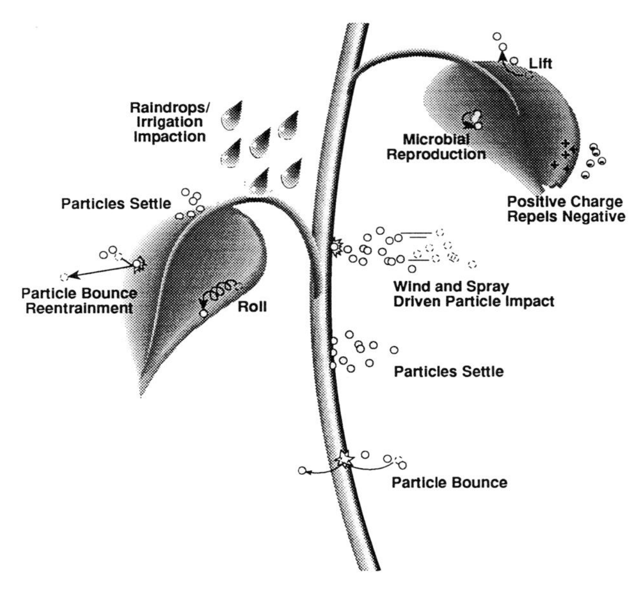

Figure 2 Environmental conditions affecting the dispersal of airborne bacteria on plant surfaces. (Courtesy of Shirley Burns, UNLV.)

bacterial cells may remain in the air for 60–90 min when transmission is enhanced by accompanying wind.

Humidity also is important for dispersal of fungal spores. Dry air has been correlated to high numbers of *Cladosporium* and *Alternaria* spores, whereas moist air, especially following a rain shower, contains larger numbers of mushroom spores and ascospores.

b. Wind

Once microorganisms are airborne, the primary meteorological condition influencing their dispersal outdoors is the wind. Both the wind speed and zones of turbulence influence dispersal. Transport of airborne cells is particularly important for aerial spray releases of genetically engineered microorganisms developed to enhance agricultural yield. The U.S. Environmental Protection Agency, therefore, suggests outdoor aerosol releases of microorganisms only when wind is <5 mph to minimize drift off the target plants. Wind is also critical to the dispersal of fungal spores outdoors. While some fungi have active spore release mechanisms, many genera found in the soil and on plant surfaces rely on wind to lift them into the aerosphere for dispersal.

Air currents may also transport airborne microorganisms through open windows and doors from the outdoors into home environments. Air flows from mechanically ventilated air-handling systems also may contribute to the dispersal of microorganisms in office, residential, and commercial indoor environments.

C. Survival of Airborne Microorganisms

Viability of airborne microorganisms and their ability to colonize at a new location or adapt to a different type of environment are affected by the conditions during transport, the constituents of the aerosol, and the type of microorganism. In general, bacteria and algae are more susceptible to stresses encountered during aerosolization and transport than fungal spores, enteric viruses, or amoebic cysts. Some bacteria develop nonvegetative forms that are quite resistant such as *Bacillus* sp. Fungal spores and amoebic cysts are nonvegetative forms that are resistant to environmental and physical stresses and can survive for long periods of time under adverse conditions.

1. Environmental Factors

a. Temperature and Relative Humidity

Survival of airborne bacteria and viruses is markedly affected by the air temperature and RH. Generally, die-off of airborne bacterial cells is geometric in nature with maximum death rates occurring in the first 0.7–1.0 sec after aerosolization. This almost instantaneous death is a result of rapid evaporation of the water associated with airborne droplets in high temperature/low humidity environments. Mid-range (60–80% RH) values have also been associated with poor survival rates for airborne bacteria. Similar conclusions have been reported for airborne viruses. Survival of some viruses (e.g., polio, influenza) is inversely proportional to air temperature and inactivation is rapid at mid-range RH values. Air-dried algal cells may remain viable for prolonged periods of time, and blue-green algae can survive high and low temperature extremes. Some fungi such as *Coccidioides immitis* have adapted to extreme temperatures. *C. immitis* survives only in hot, dry deserts of the Sonoran life zone of Mexico and the southwestern United States.

b. Radiation

Solar radiation is important to the survival of airborne microorganisms outdoors. Ultraviolet radiation at 250–260 nm is strongly germicidal, resulting in mutations in bacterial DNA, decreased algal survival, and inactivation of viruses. Yellow, orange, and red carotenoid pigments, however, help to prevent the formation of singlet oxygen caused by visible light, and organisms with these pigments are more resistant to sunlight exposure. Gamma and X-ray radiation may cause random breaks in DNA, inactivate microbial enzymes, and alter membrane permeability of bacteria. X-rays also will inactivate viruses. Levels of ionizing radiation in the environment, however, are low and not a hazard for airborne organisms.

c. Air Pollutants

Ozone as an air pollutant has shown antibacterial activity, particularly when other photochemical oxidants commonly associated with automobile exhaust, petroleum vapors, and olefins are present. The toxic effects of ozone are also increased when the RH is elevated. Exposure to a combination of sulfur dioxide and mid-range RH has been linked with inactivation of airborne virus. Atmospheric nitrogen dioxide has also been implicated in airborne viral inactivation.

2. Droplet Composition

Decreased inactivation has been shown when the viruses were aerosolized in nasal secretions, feces, or saliva compared to water or buffers. The protective effect may be associated with the protein material in body fluids, although in another study viruses aerosolized in media containing glucose and amino acids survived better than those sprayed from nasal secretions.

Addition of low concentrations of some chemicals to spray suspensions affords protection to aerosolized bacterial cells by decreasing the rate of evaporation and providing a protective layer around the cell wall. Similarly, microorganisms suspended from sewage are often protected due to components in the effluent. Betaine, inositol, and other osmoprotectants stored by bacterial cells prior to aerosolization or supplied in the collection medium during aerosolization have been reported to increase survival. Some chemical additives, however, may set up a hypertonic state around the suspended cell. Water is then forced out of the cell, resulting in death.

Droplet size is also a significant factor in the survival of airborne bacterial cells, with decreasing droplet size correlated with decreased survival. Viability studies have shown 13% of airborne organisms in droplets <7 μm in diameter were viable compared to 30% for droplets >7 μm, and approximately 3% of droplets <4.7 μm in diameter were viable compared to $>20\%$ for larger droplets. Dry aerosols with particulate matter serving as rafts may also aid in survival of airborne bacteria, algae, and viruses. Particulates may protect these organisms from desiccation and shield them from radiation.

3. Microorganism

Gram-positive bacterial genera generally are more likely to survive in the airborne state than gram-negative bacilli. Percent retrieval of airborne *Staphylococcus aureus* was shown to be unaffected during prolonged suspension as a bioaerosol, yet gram-negative bacilli (*Escherichia coli* and *Proteus mirabilis*) exhibited marked decreases in viability (20–60% less).

In general, airborne viruses are more resistant than coliforms or fecal streptococci. Survival of airborne viruses, however, is highly variable and depends on the type of viral agent. As an example, enveloped viruses survive better in low humidity environments than nonenveloped viruses due to protection of the interior of the virus by the envelope.

II. Sources of Airborne Microorganisms

A. Outdoor Sources

Airborne concentrations of microorganisms are affected by the activities occurring on the land masses and watersheds. Resuspension of organisms from plant surfaces, raindrop impaction, and wind-blown soil contributes to large numbers of organisms to the atmosphere. Agitation of lakes, rivers, and reservoirs also produces aerosols containing microorganisms from the water column and sediments. Similarly, wave action generates bioaerosols of organisms in marine waters and beach sediments. Additional input of microorganisms to outdoor air occurs during agricultural practices and wastewater treatment. The rapid development of biotechnology and the potential use of genetically engineered microorganisms and their by-products in the environment may also serve as a source of airborne microorganisms.

1. Agricultural

Harvesting, tilling, and other common agricultural practices in the field contribute to high concentrations of airborne dust and microorganisms. During harvesting of grass-seed, the mean concentrations of bacteria and fungi have been calculated at the combine at 10^8 colony-forming units per cubic meter of air sampled (CFU/m^3). At this source strength, the contribution of these airborne microorganisms to the microbial count in the airshed of the valley exceeded 10^2 bacterial and fungal CFU/m^3 at a 1684-m elevation.

Swine houses and cattle feed lots have also been implicated as sources of airborne organisms. Dairy and food-processing facilities where surfaces are routinely rinsed down with a water hose have shown increases of 8–100% in the number of viable airborne cells following the rinsing procedures.

2. Municipal Waste Treatment Facilities

Domestic sewage contains approximately 10^7–10^9 coliforms and 1–10 enteric viruses per 100 ml. Depending on the level of treatment and resulting death of bacteria and inactivation of viruses, effluent at treatment plants may contain $>10^3$ coliforms and 0–0.3 viruses per 100 ml. Calculations of the mean bacterial emission rate at aeration sewage plants has been reported as 440 CFU/m^3/sec and droplets dispensed from spray equipment at sewage plants may contain viable bacteria in the respirable size range (1–5 μm in diameter). While airborne microorganisms have been detected at 1.2 km from sewage treatment plants, the initial bacterial and virus concentration in the air is reduced almost 90% due to dispersion and die-off, and increased enteric disease resulting from close proximity to sewage treatment plants has not been correlated to airborne microorganisms.

3. Reservoir Hosts and Soil

Numerous fungi are harbored and disseminated into the environment by birds and other reservoir hosts, and several respiratory illnesses are linked to exposure to these airborne organisms. *Histoplasma capsulatum* is disseminated from contaminated chicken, blackbird, or bat excreta and is endemic in the Mississippi, Ohio, and Missouri river valleys of the United States. *Cryptococcus neoformans* is also excreted from birds, mostly pigeons, and may reach very high numbers in their roosts in old buildings. *Cryptococcus neoformans* and *Blastomyces dermatitidis,* found in soil and old structures, are often respired into the lungs during renovation of buildings, and *Coccidioides immitis* is ubiquitous in southwestern desert soils.

Bacteria and viruses are also released into the environment by reservoir hosts. Transmission of *Bacillus anthracis,* the cause of anthrax, has been linked to inhalation or ingestion of the bacilli spores carried by infected animals. Woolsorter's disease is a hemorrhagic infection of the lungs caused by inhalation of *B. anthracis* spores from goat hair disseminated during processing. Although viruses are very host-specific, bat-infested caves were the site of an airborne infection of rabies virus by two investigators in Texas.

4. Biotechnological Application

Biotechnological products such as genetically engineered microorganisms are being developed to enhance domestic waste treatment, mining recovery, oil spill clean-up, and toxic waste disposal. The application of these types of products may involve aerosol mists from which airborne organisms are released into the environment. Experimental releases of naturally occurring bacteria onto potato and strawberry fields in California in 1987 demonstrated the drift of sprayed cells to distances >50 m from the target crops even under ideal conditions. The environmental impact of these airborne cells beyond the target plot, however, can be minimized by the use of nonpathogenic microbes and the presence of a fallow buffer zone surrounding the target crops. [*See* GENETICALLY ENGINEERED MICROORGANISMS, ENVIRONMENTAL INTRODUCTION.]

B. Indoor Sources

The terms sick building syndrome and building-related illness have been used to describe a variety of upper respiratory and flulike complaints associated with persons sequestered indoors such as office workers and school children. Because complaints of these type have been increasing since the "tightening" of buildings in an effort to save on energy costs, poor indoor air quality has been implicated as a probable cause. Temperature, humidity, chemical pollutants, and biological agents have been suggested as contributors to poor indoor air quality. The presence and the impact of airborne microorganisms on indoor air quality have been studied in residences, hospitals, nursing homes, and office settings, and in approximately 45–60% of those surveys an association of adverse health reactions resulting from airborne microbes was observed. Bacteria, algae, fungi, and fungal by-products can affect the indoor environment, and anywhere there is moisture is a potential source of microorganisms indoors. This includes the air-handling system and building materials (e.g., carpet, wall covering, ceiling tiles). Building occupants and their activity are also considered a source of indoor airborne microorganisms. Health care facilities are especially conscious of the exposure of patients to airborne microbial agents, but even people without medical confinement spend up to 22 hr/day inside and are, therefore, potentially exposed to a wide variety of indoor contaminants.

1. Air-Handling Systems

a. Portable Systems

Portable humidifiers, ventilators, and cool-mist vaporizers have been directly linked to illness indoors.

Bacteria, fungi, algae, and free-living amoeba have been isolated from water reservoirs of these systems and are often found in high concentrations. Operation of these devices results in the aerosolization of water droplets and microorganisms in the reservoirs, and attempts to decontaminate these portable units by the homeowners are often unsuccessful. The potential for growth and dispersal of microorganisms via heating and cooling systems in homes and large buildings is of concern because equipment is less accessible and maintenance is expensive.

b. Evaporative Cooling Systems

Evaporative cooling systems consist of fan-driven air flowing over water-soaked fibers. These systems are commonly used in homes in the southwestern desert climates of the United States, where the relative humidity is low and the daytime temperatures are high. Evaporative cooled residences may have air exchange rates 10 times higher than traditional air-conditioned rooms as large volumes of outside air are imported through the system. The evaporative cooling system is a low-cost, energy-efficient alternative to traditional air-conditioning systems, but the potential for transporting airborne microorganisms from the outdoors and growth of these organisms in the water-soaked pads and subsequent aerosolization has raised questions of their impact on indoor air quality.

c. Air-Conditioning Systems

In contrast to evaporative systems, air conditioners recycle up to 90% of indoor air and import only the remainder from the outside. The air is filtered and normally discharged at 5–6 air exchanges per room per hour for residential users. Recycled air, however, could contain high concentrations of airborne microorganisms due to growth in the ducting or building furnishings and poor exchange with fresh outside air. Fungal and bacterial cells also can grow on the downstream side of filters resulting in increased airborne counts. Water condensation collection areas and water-spray commercial air-handling systems are potential source areas for the amplification and growth of algae, bacteria, and parasites that could become airborne and dispersed throughout the system. Air-conditioning units with central electrostatic filtration, however, have been shown to have significantly fewer airborne fungal isolates than systems without these filters.

2. Indoor Building Materials and Furnishings

While air-handling systems may be a significant source for indoor air microbial contaminants, other building materials and furnishings may also provide a niche for bacteria, fungi, algae, and free-living protozoa.

Carpeting is used in buildings to muffle sound, cushion traffic, moderate cold floors, and enhance decoration of the living or working space. Carpets, however, wick moisture and collect dust, pet dander and hair, and other debris. This provides an ideal environment for the growth of bacteria and fungi deposited onto the carpet. *Alternaria* are the most common fungi isolated from carpet dust, and monitoring has shown that carpeting enhances the survival of this organism in the winter. Carpet composition has also been linked with fungal spore survival and growth. A greater occurrence of *Alternaria* and *Monilia* have been found in woolen carpeting compared to synthetics, and the deeper pile depths are sites for growth of *Cladosporium*, *Alternaria*, and *Monilia*. Similarly, drapery and upholstered furniture may be amplification sites for microbial growth. Ceiling tiles that have been wetted due to leaking pipes in plenum areas are primary sites for microbial growth. Painted surfaces such as walls and ceilings have also been shown to provide sites for the growth of saprophytic fungi.

Kitchen and bathroom areas of residences are excellent sites for bacterial growth. *Escherichia coli*, *Citrobacter freundii*, and *Klebsiella pneumoniae* have been isolated in high numbers from wet areas in indoor environments, especially around sink areas, refrigerator drip pans, and bathroom surfaces, whereas *Pseudomonas* and *Bacillus* species are frequently isolated in drier areas (e.g., flooring). The majority of these organisms were isolated during surface sampling, yet the potential for dispersal into the indoor air is evident.

Aquariums have been implicated as sources of algae and free-living parasites in indoor environments. These organisms are aerosolized due to bursting bubbles and agitation of the surface.

3. Occupants

Approximately 85–90% of airborne bacteria in indoor surveys of residences have been identified as gram-positive cocci. The most predominant genera (*Micrococcus*, *Staphylococcus*, and *Streptococcus*) are commonly associated with human activity. Dissemination of microorganisms from nasal and oral surfaces, skin, clothing, and hair of building occupants may be the sources of these airborne bacterial cells. Coughing has often been cited as a major vehicle for the transport of upper respiratory tract flora, and these organisms are also prevalent on clothing and dust. High ratios of these organisms compared to their numbers in outside air have been used as an indication of high occupancy rate and poor ventilation of commercial buildings and schools. [*See* GRAM-POSITIVE COCCI.]

Viruses are also disseminated indoors as a result of human activity. Some viruses may be inhaled and deposited in the oropharynx. These viruses may then replicate and be re-aerosolized during coughing and talking. Enteric viruses enter the body through the mouth and replicate in the intestines. Infective viral particles are then excreted in the feces of an infected individual for several weeks. Dissemination of enteric virus may occur during toilet flushing, which aerosolizes particles into the bathroom air.

4. Health Care Situations

While approximately 50% of hospital infections may be due to exposure in the community prior to the hospital stay, the remainder can be described as nosocomial (acquired during hospital stay). Although hospitals adhere to strict infection control procedures and routinely monitor for airborne organisms, increasing numbers of elderly people are living in nursing home and home health care facilities where maintenance and decontamination practices may not be as rigidly followed. Exposure to microorganisms that may act as opportunistic pathogens is, therefore, increased.

III. Bioaerosols Implicated in Building-Related Illnesses

A. Bacterial Sources

1. *Legionella*

The discovery of a previously unreported bacterium as the cause of numerous deaths in Philadelphia at the 1976 American Legion convention focused attention on the relationship of environmental microorganisms to adverse human health effects. The illness, called Legionnaire's disease, was not transmitted person-to-person but, rather, via a contaminated air-handling system. The organism causing the disease, *Legionella pneumophila*, has since been found to be a naturally occurring waterborne aerobic gram-negative bacillus commonly isolated from warm water environs. Pathogenic and nonpathogenic strains of *Legionella* sp. often associate with sediment, algae, waterborne parasites, and other waterborne bacteria in a synergistic relationship to resist environmental stresses and water treatment practices. In recent years, reported cases of Legionnaire's disease have been linked to aerosolization of the organism from shower spray heads, humidifiers, and cooling units.

2. Thermophilic Actinomycetes

A group of organisms called thermophilic actinomycetes have been implicated in numerous cases of hypersensitivity pneumonitis and other allergic reactions. Exposure to these organisms whose optimal growth temperature is $\geq$40°C may occur when handling decomposing organic matter such as compost or municipal garbage that results in aerial release of these organisms. *Micropolyspora faeni* is the thermophile most commonly associated with hypersensitivity pneumonitis. [*See* THERMOPHILIC MICROORGANISMS.]

3. Endotoxin

Endotoxin is a lipopolysaccharide found in the cell envelope of many gram-negative bacilli. Inhalation of airborne endotoxin has been associated with the presence of organic dusts and often results in severe respiratory complications and allergic reactions. Exposure to endotoxin has also been linked to reports of fever, cough, headache, and diarrhea, and it may be a major cause of illness in agricultural settings such as silage facilities, barnyards, and poultry-processing areas. Chilled-water spray humidification systems used in textile manufacturing facilities have also been implicated as sources of aerosolized endotoxin causing cases of humidifier fever and hypersensitivity pneumonitis. Some of the bacteria associated with airborne endotoxin exposure are *Pseudomonas, Cytophaga, Salmonella,* and *E. coli*.

B. Algal Sources

Skin testing for allergic reactions to algae and air sampling for airborne algal cells have resulted in an association of both green algae (*Chlorella* and *Chlorococcum*) and blue-green algae (*Schizothrix* and *Anabena*) with adverse human health effects. While the extent of allergic reactions due to algal exposure has not been fully investigated, house dust and aeration of aquariums and eutrophic lakes have been proposed as possible sources.

C. Fungal Sources

While over 400 species of fungi are recognized as human pathogens, few are capable of infection without an injury to host tissue or immunosuppression of host defenses by another disease-causing agent or therapy (e.g., antibiotics, steroids, drugs). Some fungi, however, can cause adverse human health effects through allergic and hypersensitivity reactions, and others produce toxic products as a result of their metabolism.

1. Spores and Vegetative Cells

Cladosporium is the most common fungus isolated in indoor air environments. This fungus was once thought to be only a reflection of outdoor fungal populations and not a human health concern. Recently, however, incidence of allergic reactions and sensitivity to this fungus has raised questions as to its role in adverse health effects in indoor environments. Other commonly isolated fungi implicated in allergic and adverse health reactions are some species of *Aspergillus, Alternaria, Penicillium,* and *Monilia. Aspergillus fumigatus* is a fungus with known, pathogenic effects including farmer's lung, a debilitating respiratory disease often contracted in barnyard settings where contaminated hay is handled. This fungus and many others have been associated with animal fur and dander and may be harbored in carpeting.

2. Mycotoxins

The spores of toxigenic fungi produce metabolic products—mycotoxins—that cause severe adverse health effects in humans. Aflatoxins produced by *Aspergillus flavus* and *Aspergillus parasiticus* are potent liver carcinogens, but accumulation of these fungi are most often seen in agricultural products (e.g., peanuts) and are rarely found in offices or homes. Toxigenic *Penicillium* and another mycotoxin-producing fungus, *Stachybotrys atra,* are found in contaminated buildings and have been associated with reports of illness. *Stachybotrys atra* grows on wetted cellulose (e.g., ceiling tiles, paper products) and has been isolated from jute-backed carpeting. Unfortunately, this fungus does not compete well with other fungi on isolation medium commonly used for indoor air surveys and may be overlooked during monitoring. Studies have demonstrated that inhalation exposure doses can produce toxicosis and low numbers of toxigenic fungal spores can result in severe health problems. [*See* MYCOTOXICOSES.]

D. Viruses

Viruses are inert outside of a living cell and must invade a specific host and mobilize metabolic processes in order to replicate themselves. Numerous viruses are transmitted via droplets and spread by the respiratory route from one person to another in the indoor environment. Large numbers of viral particles are produced in the nose and throat of infected individuals and are expelled during coughing and sneezing. These viral particles may be directly transmitted to a new host or cause contamination of surfaces that are later handled by a new host. Mumps, measles, chicken pox, polio, and the common cold are all viral diseases that can be transmitted through the air.

E. Parasites

Soil and water are natural habitats for free-living amoebae. Natural and artificially heated waters such as power plant discharges, lakes, and hot springs are likely sources of *Naegleria fowleri* and *Acanthamoeba*. These free-living amoebae have been cited as causing severe illness often resulting in death. Cooling system waters have also been implicated as potential sources of aerosolized *N. fowleri* and *Acanthamoeba*. Nasal mucosa is the initial site for *N. fowleri* infections and aerosolized *Naegleria* antigens may also lead to allergic reactions. *Acanthamoeba* have been isolated from eye infections, usually following injury, and some cornea infections have been traced to contact lens solutions contaminated with dust.

IV. Monitoring for Airborne Microorganisms

A. Methods for Detection and Enumeration of Airborne Bacteria and Fungi

Collection of viable airborne microorganisms requires apparatus that will minimize sampling stress and capture the particle size of the bioaerosol droplet efficiently. Enumeration of organisms per unit volume of air and/or quantitation of respirable-sized particles may also be desirable. Culture methods rely on growth of the organisms following collection and assume that the organisms will grow and produce classical characteristics within a specified time period. Airborne microorganisms, however, are stressed during transport and may not respond to traditional culture techniques.

Total count procedures rely on the uptake of a dye by the microorganisms and subsequent enumeration using a microscope. These procedures are used with samplers that either collect the sample onto a filter or into liquid. Acridine orange is a commonly used dye for fluorescent enumeration of cells, and this technique counts both viable and nonviable cells. Unfortunately, this stain does not distinguish between living and deal cells, nor does it permit identification of the organisms.

1. Traditional Aerobiological Samplers

a. Passive/Depositional Sampling

Passive sampling, such as open Petri dishes with culture medium or glass slides covered with a sticky film, relies on the settling out of particles from the air (Table I). This type of sampling has been widely used for pollen and large particulates, but small particles such as bacterial cells may remain suspended and not be enumerated. Passive sampling methods also do not permit quantitation of particles per unit volume of air.

Table I Aerobiological Sampling Devices for Monitoring of Bioaerosols

Type of sampling	Sampler	Sampling surface	Type of organism	Considerations
Depositional sampling	Petri dish	Agar	Bacteria Fungal spores Algae Amoebic cysts	Settling out of particles; not quantitative
	Glass slide	Sticky film	Bacteria Fungal spores Algae Amoebic cysts	Settling out of particles; not quantitative
Forced air-flow sampling	SAS high-volume sampler	Agar-filled plates	Bacteria Fungal spores	Small sampling surface; overload plates with high concentrations of fungi
	Biotest RCS sampler	Strip of agar-filled wells	Bacteria Fungal spores	Very small sampling wells; overload with high concentrations of fungi
	Andersen 2- or 6-stage sampler	Agar-filled plates	Bacteria Fungal spores Algae Amoebic cysts Viruses	Overload in high concentrations of organisms; size distribution of droplet
	All-glass impinger	Buffer-filled sampler	Bacteria Algae Viruses Antigens	Poor recovery of hydrophobic spores; can dilute in high-concentration situations
	Filtration sampling	Filters	Bacteria Fungal spores Algae Amoebic cysts Viruses Antigen	Desiccation of bacterial and algal cells; useful with microscopic and biotechnological detection methods
	Burkard spore trap sampler	Glass slide or sticky tape	Fungal spores	Time discrimination

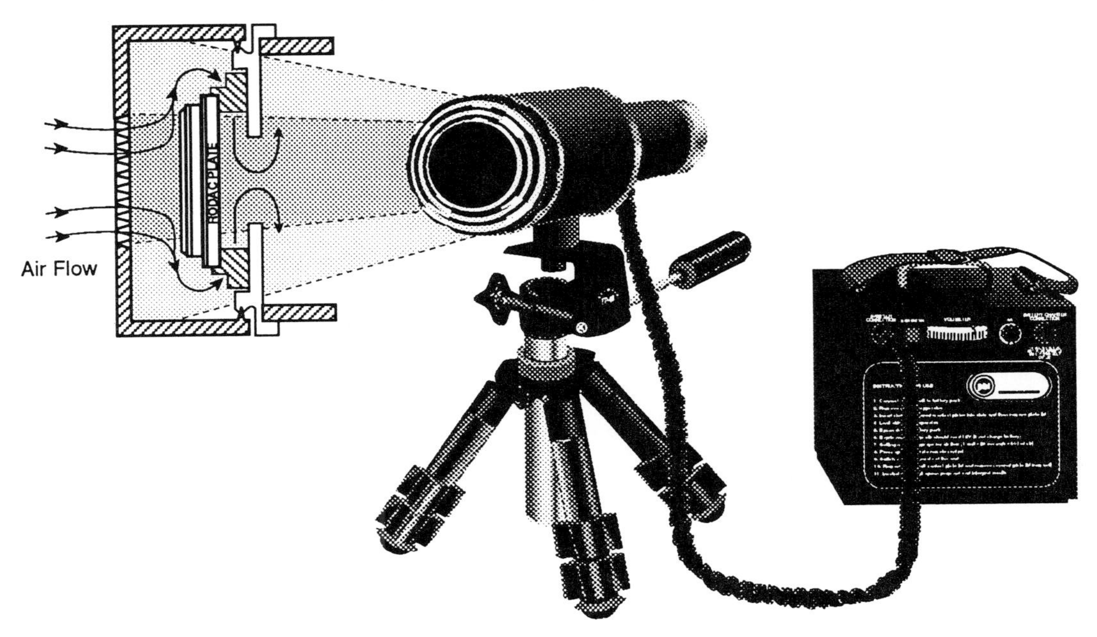

Figure 3 Surface Air System high-volume aerobiological sampler. (Courtesy of Shirley Burns, UNLV.)

b. Forced Air Flow Samplers

To enhance the detection and enumeration of bioaerosols, samplers with a mechanically induced air flow were designed. These samplers collect the microorganisms on agar surfaces, liquid buffer, filters, or tape (Table I).

Impactor samplers collect organisms onto an agar surface as air is drawn through an orifice. Portable impactor samplers such as the Surface Air System (SAS) and RCS samplers have a single orifice. The SAS (Fig. 3) samples at >140 liters/min and impacts cells onto either a 57- or 80-mm-diameter agar plate over a 20-, 40-, or 60-sec sampling time. The RCS sampler (Fig. 4) impacts cells onto agar-filled wells on a sampling strip. This sampler is designed to sample according to the volume of air through the device rather than a specified sampling time. The Andersen multistage sampler, with a vacuum pump attachment, directs air at 28 liters/min through a

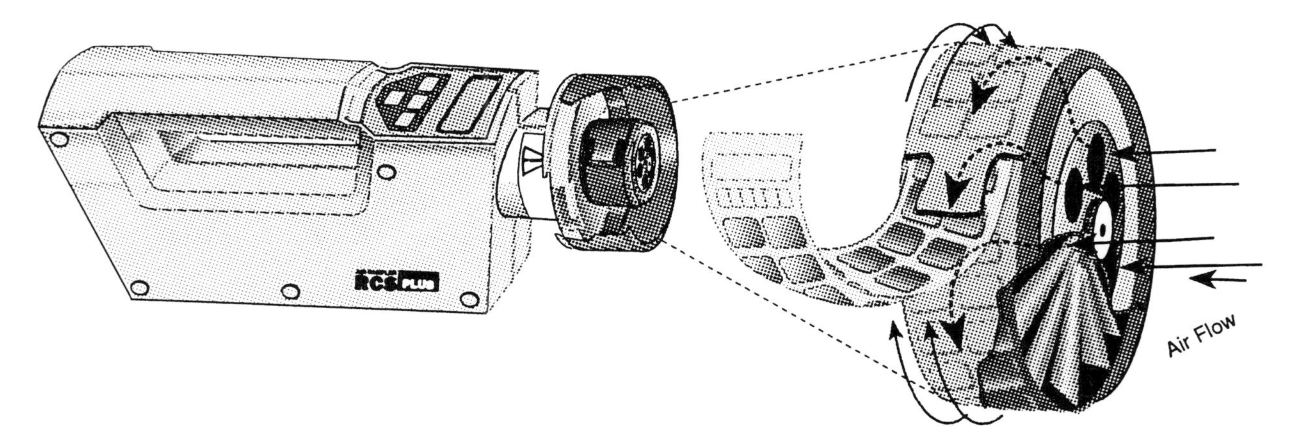

Figure 4 Biotest RCS Plus aerobiological sampler. (Courtesy of Shirley Burns, UNLV.)

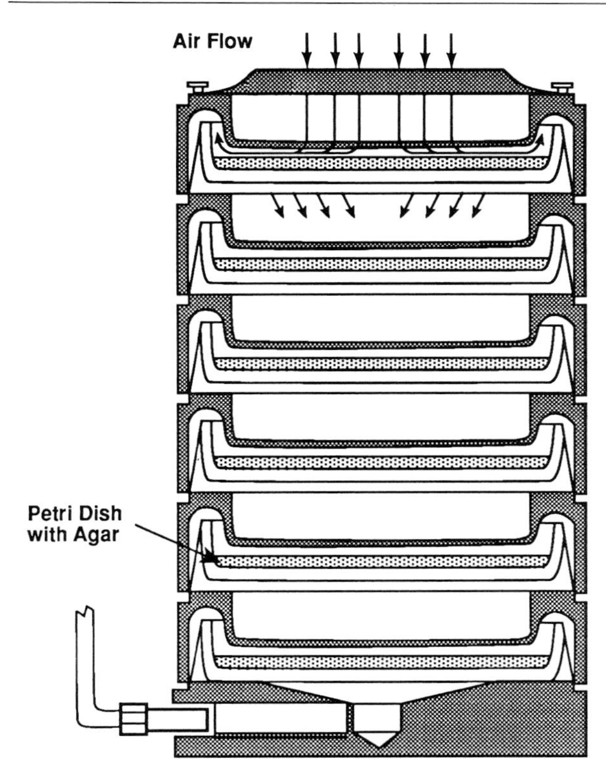

Figure 5 Andersen six-stage impactor sampler. (Courtesy of Shirley Burns, UNLV.)

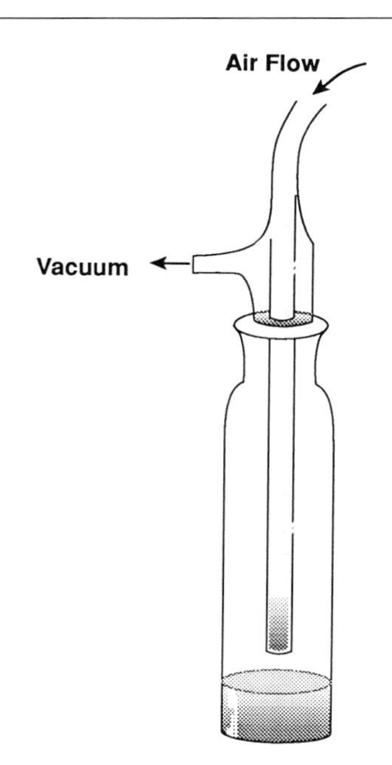

Figure 6 AGI-30 all-glass impinger sampler. (Courtesy of Shirley Burns, UNLV.)

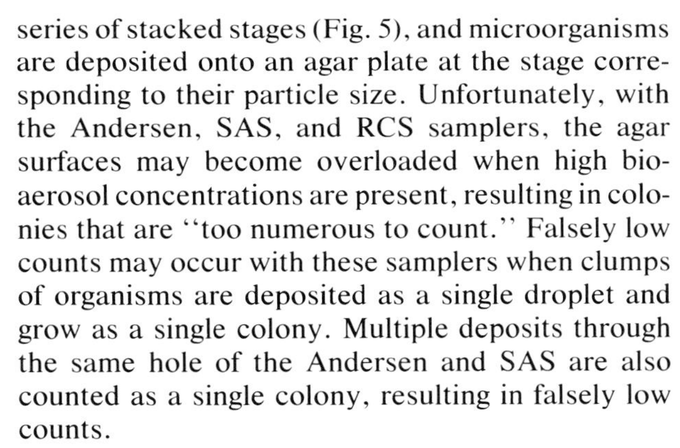

series of stacked stages (Fig. 5), and microorganisms are deposited onto an agar plate at the stage corresponding to their particle size. Unfortunately, with the Andersen, SAS, and RCS samplers, the agar surfaces may become overloaded when high bioaerosol concentrations are present, resulting in colonies that are "too numerous to count." Falsely low counts may occur with these samplers when clumps of organisms are deposited as a single droplet and grow as a single colony. Multiple deposits through the same hole of the Andersen and SAS are also counted as a single colony, resulting in falsely low counts.

Impinger samplers, such as the all-glass impinger (Fig. 6), were originally designed to simulate air flow through nasal passages and operate by drawing air through a curved tube and into a liquid buffer. The liquid collection fluid then can be used as an inoculum for culture media and examined by total count techniques. The buffer can be diluted prior to culturing to avoid overloading in high-density situations and subsamples can be plated onto several media to enhance the recovery of injured or fastidious organisms. Samplers with liquid buffer, however, require strict adherence to aseptic techniques to avoid contamination during the preparation, sampling, and subculturing steps.

The Burkard spore trap sampler (Fig. 7) is a slit sampler design that uses an adhesive-coated sliding glass slide or rotating drum apparatus to enable continuous sampling over 24 hr or 7 days. This sampler, however, is limited to total counts of fungal spores and pollen, and a highly trained technician is needed to identify the different spore and pollen types.

2. Biotechnology Based Assay

Collection of microorganisms onto filters is not widely used for airborne cells due to the loss of viability through desiccation. Biotechnology assay methods of samples collected onto filters or into liquid buffer, however, do not require growth of the organisms for detection and enumeration.

a. Gene Probes

Biotechnological methods to enhance the detection and quantitation of airborne bacterial contaminants

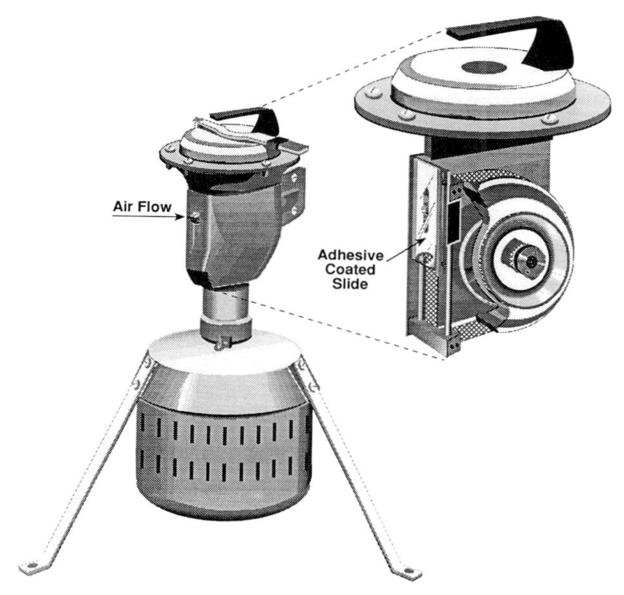

Figure 7 Burkard spore trap sampler. (Courtesy of Shirley Burns, UNLV.)

are currently being developed. DNA and RNA probes for specific genetic sequences combined with polymerase chain reaction technology should result in increased selectivity and sensitivity. [*See* POLYMERASE CHAIN REACTION (PCR).]

b. Immunoassay

Direct counting of organisms using antibodies is also rapidly developing. Immunoassay methods rely on development of antigen-specific antibodies combined with fluorescent, radioactive, or enzymatic detection methods. Immunoassay, therefore, could be used for monitoring microbial antigen and byproducts of microbial metabolism without the need to detect a living cell.

B. Methods for Detection of Other Bioaerosol Contaminants

1. Virus

Routing monitoring for airborne viruses is rarely done owing to difficulties in sampling and analysis. Sampling, however, has been accomplished using the SAS high-volume and Andersen samplers with freshly made trypticase soy agar plates. The plates are not incubated, as for bacterial and fungal cultures, but the agar is removed from the dishes and minced with viral culture medium in a vortex. The suspension is centrifuged and the supernatant fluid is inoculated into cell culture. High-volume liquid samplers and all-glass impingers have also been used for isolation of airborne virus with the liquid buffer sample inoculated onto cell culture or viewed with electron microscopy. Buffer-moistened filters have been used for the retrieval of viruses in monitoring of bathrooms. In all cases, the samples must remain moist because desiccation will quickly inactivate viruses.

2. Algae

Although airborne algae may be a significant allergin in indoor air, monitoring for these organisms is not currently routine. Samples for algae may be taken using liquid impingement as described for bacteria, but sampling of sources where the algae are concentrated and/or growing is often more productive. Samples of house dust diluted in Tris buffer can be inoculated onto agar with a soft agar overlay to quantify algae cells per gram of dust, and inoculation of water has been successfully used for standing water sites.

3. Endotoxin

Endotoxins are concentrated in dust, and the target of these toxins is the bronchoalveolar region of the lung. Since the biological activity of endotoxin does not depend on viability of the bacterial cell, sampling is not limited to culture-based methods. Filter cassettes are often used for the retrieval of respirable-sized dust particles followed by elution and extraction of the toxin from the dust. Suspected endotoxin in humidification systems has been monitored by sampling of a gelatinous film in the water reservoir.

4. Free-Living Amoebae

Water reservoirs and standing water near air-handling systems are the most likely sites for the detection of free-living amoebae impacting indoor air quality. Similarly, surface waters would be sites to sample in outdoor environments using grab samples or filtration techniques. Andersen samplers and liquid impinger samplers have been used for sampling airborne amoebae cysts. Liquid samples are concentrated by centrifugation and filtered through cellulose filters that are plated onto nonnutritive agar spread with a lawn of *E. coli*. The Andersen collection method utilizes a single agar plate spread with an *E. coli* lawn inserted into the sixth stage.

Bibliography

Cox, C. S. (1987). "Aerobiological Pathways of Microorganisms." John Wiley and Sons, New York.

Jacobs, R. R. (1989). *Appl. Ind. Hyg.* **4,** 50–56.

Henis, Y. (1987). "Survival and Dormancy of Microorganisms." John Wiley and Sons, New York.

Lighthart, B., and Kim, J. (1989). *Appl. Environ. Microbiol.* **55,** 2349–2355.

Morey, P., Feeley, J. C., and Otten, J. A. (1991). "Biological Contaminants in Indoor Environments." American Society for Testing and Materials, Philadelphia, Pennsylvania.

Sattar, S. A., and Ijaz, M. K. (1987). *CRC Crit. Rev. Environ. Control* **17,** 89–131.

Algal Blooms

Gerhard C. Cadée
The Netherlands Institute for Sea Research

Glossary

Amnesic shellfish poisoning (ASP) From domoic acid, a toxin derived from the diatoms *Nitzschia pungens* and *Nitzschia pseudodelicatissima*, common members of the phytoplankton community not known to produce toxins before 1988; causes disruption of normal neurochemical transmission in the brain, with abdominal cramps, memory loss, disorientation, and, in some instances, mortality

Coccolithophores Group of flagellate phytoplankton algae that produce internal calcium carbonate plates (coccoliths); widespread and often abundant in the oceans, at high concentrations (blooms) they cause a milky white appearance in water due to light scatter from cells and detached coccoliths

Cyanobacteria Cyanophyta, or blue-green algae, with prokaryotic, photosynthetic cells and a threadlike or nearly spherical shape; benthic and planktonic species, with many bloom-forming species in freshwater and brackish water and a few in the open ocean; many species can float due to gas vesicles or gas vacuoles and can bind gaseous nitrogen, making them strong competitors in the phytoplankton community and causing cyanobacteria blooms

Diarrhetic shellfish poisoning (DSP) Occurs after consumption of shellfish that has fed on toxin-producing algae (e.g., the dinoflagellate *Dinophysis acuminata, Dinophysis norvegica*); easily confused with gastroenteritis and general stomach upsets associated with eating shellfish or contaminated shellfish; has only been recognized as a poisoning for the past 10 years, and no mortalities have been recorded; toxins include okadaic acid and "dinophysistoxin"

Diatoms Unicellular algae protected by an external pillboxlike skeleton of amorphous silica; they occur in all wet environments, with high concentrations (blooms) forming in upwelling areas, and in temperate and (ant)arctic regions in seas and lakes.

Dinoflagellates Unicellular algae, with cellulose cell walls, in most species consisting of a number of polygonal plates but some are naked; two flagella in perpendicular grooves of this cell wall; some species are heterotrophic and without pigments; blooms of dinoflagellates are often red (red tides), and some species produce toxins

Dissolved organic carbon All organic carbon present in water that passes a filter of a 0.45-μm pore size (recently, more often 0.2 μm) and, thus, includes colloidal organic matter; what is retained on the filter is called particulate organic carbon

Eutrophication Both natural and artificial increase in nutrient load of a water body; involves increase in algal growth-promoting substances and leads to reduction in phytoplankton diversity and massive production of a few species, hence to an increase in algal blooms

Neurotoxic shellfish poisoning (NSP) Symptoms similar to, but milder than, diarrhetic shellfish poisoning due to the dinoflagellate *Ptychodiscus brevis;* no mortalities reported

Paralytic shellfish poisoning (PSP) Serious illness caused by eating shellfish that have consumed large quantities of some dinoflagellates such as *Gonyaulax catenella, Alexandrium tamarensis, Pyrodinium bahamense*, and *Gymnodinium* spp.; they produce potent nerve toxins, the poison acts very rapidly, no antidote has yet been discovered, and mortalities are common

Phytoplankton Community of free-floating, microscopic algae living in the water column;

largely confined to the euphotic layer, where light is sufficient for growth of the algae

ALGAL BLOOMS are loosely defined as discolorations of the water due to the occurrence of phytoplankton concentrations. Such concentrations are usually monospecific and are reported from all over the world in freshwater and in the sea. Growing evidence indicates that they occur now more often than before, certainly in freshwater and coastal areas, due to human-made changes in the environment, particularly the increase in nutrients nitrogen and phosphorus. Some of the bloom-forming algae produce toxic substances that may kill invertebrates and fish; consumption of mollusks that have ingested toxic algae may lead to illness in humans, sometimes followed by death. Most bloom-forming algae are nontoxic, and the high amounts of organic matter produced may be beneficial to the next steps in the food chain. However, during blooms, so much organic matter may be produced that remineralization during and after the bloom may consume all the oxygen present in the water, which may also lead to mass mortalities among invertebrates and fish.

I. History of Algal Bloom Research

Discolorations of waters probably due to algal blooms have been reported throughout history, and the fossil record suggests the presence of algal blooms in older periods; however, the invention of the microscope was necessary to determine the true nature of these discolorations. In his letters to the Royal Society, Anthonie van Leeuwenhoek was the first to describe organisms causing discolorations of water. On 7 September 1674, he described "green clouds" in a freshwater lake near his hometown of Delft, due to a filamentous "animalcule" now known as the green alga *Spirogyra*. In his letter of 9 February 1702, he described small, red, moving "animalcules" (now known as the flagellate *Haematococcus*) coloring the water red in his gutter. [*See* HISTORY OF MICROBIOLOGY.]

Not all discolorations of water are due to algal blooms. Other substances such as inorganic suspended particles or dissolved organic substances (*Gelbstoff*) may also discolor water. Particularly in freshwater, dissolved organic carbon may give the water a deep brown color, reflected by names such as Cola-creek (in Surinam). The Yellow Sea is named thus because of its high-suspended matter content and the Red Sea after its frequent blooms of the cyanobacterium *Oscillatoria* (*Trichodesmium* in older literature).

In 1845, Johannes Müller was the first to collect plankton with a net on Heligoland (called for a long time as Müller's nets; the name plankton was introduced later). In an attempt to estimate variations among places in the primary production of algal blooms, Victor Hensen introduced vertical net hauls from a 200-m depth to the surface to determine the amount of plankton in the sea. Used during the Plankton Expedition of 1899 at 126 stations distributed throughout the Atlantic Ocean, this method showed wide variations in the amounts of phytoplankton.

Hans Lohmann observed that the plankton nets used were not capable of collecting all the phytoplankton present, because the smallest cells, several microns in diameter, passed through the nets. Therefore, he introduced paper filters and centrifugation as methods to collect all phytoplankton. A more popular method to estimate the phytoplankton cell numbers has become the Utermöhl's settling method. Phytoplankton is concentrated (after fixation) by settling in special settling tubes with a thin glass bottom. The cells are counted using an "inverted" or plankton microscope, with the objective that one scans the bottom of the settling tubes from underneath. With this method, we can now identify and count the different species present. For the more recently discovered very small algae of $<1\ \mu m$, the light microscope is insufficient; either fluorescent or electron microscopes are necessary for these algae.

In addition to a number of pigments, all algae contain chlorophyll a. To estimate the total biomass of algae, chlorophyll a is measured by filtering a certain amount of water and then extracting the pigments from the algae collected with a suitable solvent (e.g., acetone). Green coloration of this extract is measured with a spectrophotometer or a fluorometer. Since its introduction some 50 years ago, the method has improved, but these improvements make comparison with the earlier data difficult. Because blooms are often (but not always) confined to the surface layer, remote sensing has become a powerful method of collecting synoptic data on the occurrence, size, and changes in time of algal blooms over wide areas of the ocean.

Although the study of phytoplankton has a relatively long history, as indicated by this historical survey, descriptions of phytoplankton blooms made before ca. 1900 are mostly qualitative or, at best, based on net phytoplankton samples, which do not collect all algae. Chlorophyll a measurements and the Utermöhl method came into use in the first half of the twentieth century, but only a few marine stations can boast long-term monitoring programs for phytoplankton. The best-known monitoring program, the Continuous Plankton Recorder Survey, was started by Alister Hardy in 1932. It consists of a plankton net in a torpedolike housing towed by ships along fixed lines from harbor to harbor. It is the longest monitoring program for a large sea area, starting first in the North Sea and now covering large areas of the North Atlantic. Its main drawback is its use of a plankton net with a relatively large mesh size (~280 μm), which retains only large phytoplankton cells and zooplankton. Nevertheless, its results nicely indicate parallel plankton changes in a wide area, thus stressing the importance of climatic changes on plankton. We must keep in mind that the absence of long phytoplankton monitoring data in most areas makes it difficult to investigate whether or not algal blooms are now more frequent and/or of longer duration than before.

II. Why Do Algal Blooms Occur?

Algal blooms are normal occurrences in the surface waters of lakes and seas. The spring phytoplankton bloom in temperate and arctic areas is a particularly well-known phenomenon. It is related to the increase in day length and the amount of solar radiation in spring coupled with the onset of water stratification and the high amounts of nutrients brought to the surface by vertical mixing in winter, leading to an exponential growth of phytoplankton. During the spring bloom, a succession of phytoplankton species usually may be observed because conditions (nutrients, light, temperature) change in time. Competition usually finally leads to a monospecific algal bloom at the spring peak, consisting of the colonial *Phaeocystis* in continental coastal North Sea waters.

Algal blooms are often predictable features following hydrobiological changes that increase the nutrient content of the upper water layers, where sufficient light promotes algal growth. This may occur after enhanced terrigenous runoff or after water column turnover or upwelling of deep water (both of which bring nutrient-rich water to the surface), coupled with light conditions that favor growth and proliferation of algae. If herbivore grazing by zooplankton cannot keep algal biomass low under these conditions (usually because zooplankton cannot respond quickly enough by an increase in numbers), algal cell numbers may increase to high numbers, thus causing algal blooms to develop and cause a discoloration of water. Algal cell numbers may attain millions per liter during blooms, with one billion produced in the recent "brown tides" of the small alga *Aureococcus anophagefferans* in Long Island Sound waters.

The discoloration may be green, brown, or red, depending on the algae responsible and, thus, on the algal pigments they contain. Some blooms, mainly caused by dinoflagellates, are red and are also known as red tides. Whitish algal blooms are due to the calcareous skeleton elements of blooming coccolithophores such as *Emiliania huxleyi* (Figs. 1 and 2). What is called an algal bloom depends on the area considered. In the nutrient-poor and clear, open ocean waters, a barely perceptible change in water color and a decrease in Secchi depth from 25 to 15 m (the depth to which a white disc lowered in seawater

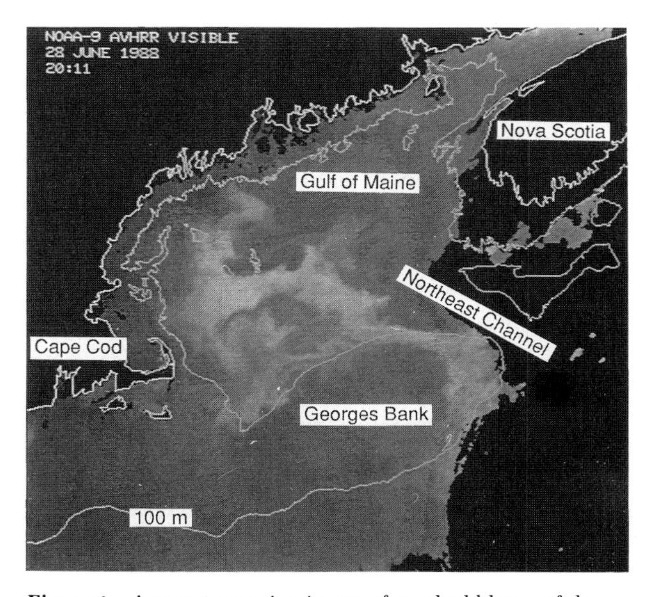

Figure 1 A remote sensing image of an algal bloom of the coccolithophore *Emiliania huxleyi* in the Gulf of Maine, 28 June 1988, obtained with an advanced very high resolution radiometer. The bloom formed over the deep waters of the Gulf and persisted for approximately 3 wk. [Original, courtesy of S. G. Ackleson (Lockheed, Houston) and P. M. Holligan (Plymouth Marine Laboratories).]

Figure 2 Scanning electron microscope picture of *Emiliania huxleyi*, Gulf of Suez. Magnification: 7000×. [Original by A. Kleijne, Free University, Amsterdam.]

can be seen) and chlorophyll a values of about 5 mg/m^3 may be called a bloom. In nutrient-rich coastal or lake waters, algal blooms are recognized when Secchi disc visibility drops to some decimeters or centimeters and chlorophyll reaches values above 10 mg/m^3. During exceptional blooms, even values above 100 mg/m^3 are encountered.

III. Effects of Algal Blooms and Their Toxicity

The poor underwater light-climate in lakes caused by excessive blooms (usually of cyanobacteria) causes drastic changes in the lake ecosystem. Due to this increased turbidity, larger plants (macrophytes) and the visual-hunting piscivorous fish have disappeared from most Dutch eutrophic lakes. The fish population has changed from piscivorous to water-filtering planktivorous species, leading to a virtual disappearance of the larger crustacean zooplankton.

Algal blooms may become a nuisance for humans when they lead to perceptible water-quality deterioration, including foul smell. This involves a loss of aesthetic and, hence, recreational values of the water. Tourists tend to avoid such areas. This has happened, for example, in the Baltic and the northern Adriatic Sea.

Some bloom-forming algae (particularly *Phaeocystis* and the coccolithophore *Emiliania huxleyi*) are significant producers of dimethylsulfide (DMS) in the sea. Liberated to the atmosphere, DMS is implicated in acid precipitation and the production of aerosols. The particles formed may act as nuclei in cloud formation and, thus, affect global climate.

Some bloom-forming algae produce toxins, which may lead to mass mortalities in fish and benthic organisms. One of the best-documented toxic blooms was the *Chrysochromulina polylepis* bloom in Scandinavian waters (May–June 1988), which caused a considerable setback to fish aquaculture industry in that area. Shellfish may accumulate the algal toxins when feeding on toxic algae. Consumption of these shellfish may lead to illness. Paralytic shellfish poisoning has caused the death of several people in the past; diarrhetic, amnesic, and neurotoxic shellfish poisonings are other illnesses due to consumption of shellfish that had accumulated algal toxins. Shellfish toxicity usually lasts over one month after the toxic algal bloom, but in some cases toxicity remains as long as 1 yr after the bloom. Economic losses in the mariculture of shellfish can be kept to a minimum through the combined efforts of an intensive monitoring program for the toxic algae or the toxins themselves, a ban on shellfish harvest as long as toxin levels are unacceptable, and the culture of species that rapidly release toxins (e.g., the blue mussel *Mytilus edulis*) or species known to avoid toxic dinoflagellates (e.g., most oysters, *Mercenaria,* scallops).

Mass mortalities during blooms are not always due to toxins. Respiration of the algae during the night and the breakdown of the large amounts of organic matter produced during blooms use large quantities of oxygen. This may lead to anoxic conditions and H_2S formation, which may kill organisms. Such mass mortalities occur regularly (e.g., in the upwelling area off southwest Namibia near Walvis Bay). Anoxic conditions related to algal blooms have also been reported in recent years from coastal areas where they previously occurred less often or not at all (e.g., the Baltic Sea, the northern Adriatic Sea, the German Bight).

IV. Are Algal Blooms Increasing?

For many areas, there is a lack of sufficient, long-term phytoplankton data to answer this question directly. Moreover, algal blooms now receive more attention than before, because more scientists are now engaged in research on (toxic) algal blooms. The increase in mariculture in coastal areas has made us more vulnerable to toxic algal blooms. Algal blooms make headlines in the news, particularly when they are combined with mass mortalities. There is a growing awareness that anthropogenic activity could be the cause. Human-made eutrophication has led to increased algal biomass and algal blooms in many lakes in populated and industrialized areas all over the world. Growing evidence indicates that the number and duration of algal blooms increased in many coastal areas directly influenced by human-made perturbations in nutrients (phosphorus and nitrogen from fertilizers, sewage dumping, sewage treatment plants effluents, detergents). An increase in algal blooms is documented for coastal areas in Hong Kong, South Africa, the Bay of Fundy in Canada, the Baltic Sea, Skagerrak, Kattegat, Dutch Wadden Sea, Black Sea, Seto Inland Sea, and the northern Adriatic Sea. In Tolo Harbor, Hong Kong, the number of red tides per year increased 8-fold between 1976 and 1986; the human population increased 6-fold in the same period, leading to a 2.5-fold increase in nutrient loading.

The human-induced increase in nutrients also changed nutrient ratios, leading to changes in phytoplankton composition. For the coastal North Sea, total algal numbers increased, but diatoms did not increase in the Helgoland and Dutch Wadden Sea (Marsdiep) data. In the Marsdiep series, we observed a considerable increase in the duration of *Phaeocystis* blooms (Fig. 3), a nontoxic colonial alga. *Phaeocystis* blooms are now recorded regularly all along the continental North Sea coast. The human-induced eutrophication increased the nitrogen and phosphorous, but not the silicate, content in the coastal waters. This explains why diatoms needing silicate for their skeleton did not increase. Thus, the human-induced changes in nutrient ratios are an important factor in changing the phytoplankton community. As phytoplankton forms the food for the next step in the food chain, such changes in nutrient ratios may indirectly affect higher trophic levels. The increase in *Phaeocystis* observed along the continental North Sea coast could not be found in the Continuous Plankton Recorder survey sampling the offshore North Sea. Here, even a decrease in *Phaeocystis* was observed. This indicates that eutrophication responsible for this increase in nondiatoms is restricted to the coastal zone, in accordance with the longshore current pattern in the North Sea. Also in other parts of the world, eutrophication effects in sea are confined to coastal zones.

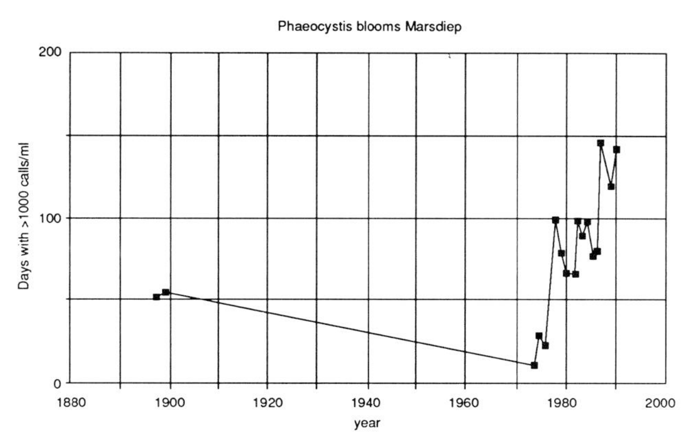

Figure 3 Increase in duration of *Phaeocystis* blooms in the westernmost inlet of the Dutch Wadden Sea (Marsdiep) measured as number of days with >1000 cells/cm³. Data for 1897 and 1899 estimated from P. T. Cleve. Increase in 1970s and 1980s due to eutrophication. [Adapted from Cadée, G. C. (1991). *Hydrobiol. Bull.* **24,** 111–118.]

V. Global Spreading of Bloom Species

The first report of the dinoflagellate *Gyrodinium aureolum* in European waters dates from 1966, when blooms along the Norwegian coast from Oslo to Bergen caused sea trout mortality. Since 1966, it has spread throughout north European waters.

The dinoflagellate *Gymnodinium catenatum* was known only from southern Californian waters. Recently it caused paralytic shellfish poisoning toxicity in shellfish reared in Portugal, Spain, Japan, and Tasmania.

These and similar examples suggest that certain toxic species are spreading globally and regionally. The spread of nontoxic species has also been documented (e.g., the introduction of the diatom *Biddulphia sinensis* from Asiatic waters to northwestern Europe early this century). Ballast water of large cargo ships has been proposed as an important agent in global dispersion either of cells or of resting stages (such as dinoflagellate cysts). Once established within a region, local current patterns will help their regional distribution. This has also been documented for many introduced invertebrates with a planktonic larval stage.

More enigmatic is the toxicity of some worldwide-occurring bloom-forming algae that before 1988 were not reported to be toxic. Examples are the

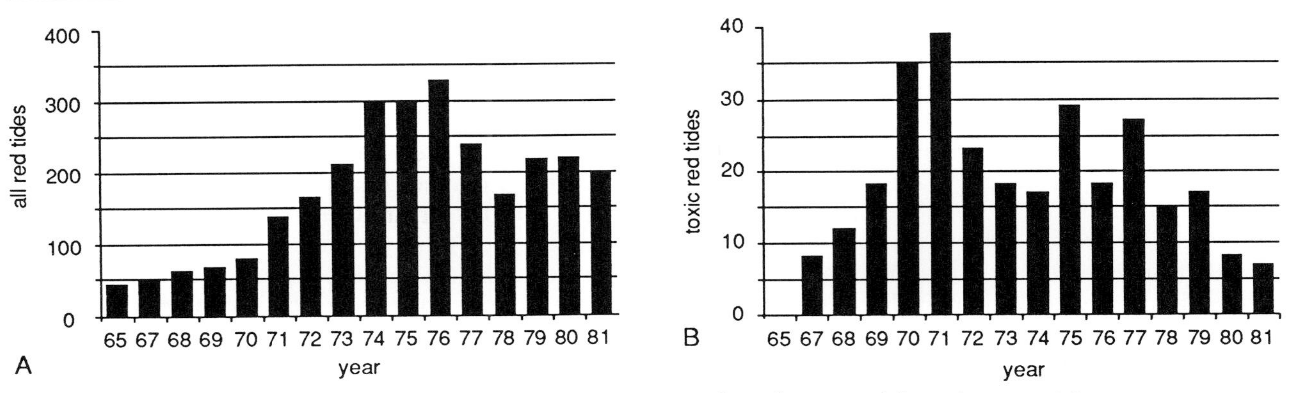

Figure 4 Numbers of red tides per year in Seto Inland Sea (Japan). Total number (A) and those damaging fisheries (B) separately, increasing first with increasing eutrophication and decreasing after control measures to decrease eutrophication. [Adapted from Prakash, A. (1987). Rapp. P.-V. Réun., *Cons. Int. Explor. Mer* **187,** 61–65.]

diatoms *Nitzschia pungens, f. maltiseries* and *Nitzschia pseudodelicatissima,* which produce domoic acid leading to amnesic shellfish poisoning. Toxicity of some algae has recently been related to infection by viruses and bacteria. Such an infection might explain toxicity of species that formerly were nontoxic as well as toxicity differences in different populations of one phytoplankton species.

VI. The Future

It is widely accepted that reduction of nutrient loading may help to improve water quality and decrease algal blooms. A promising example is provided by the Seto Inland Sea (Fig. 4), where algal blooms (red tides of different dinoflagellates) increased from 1965 to the early 1970s. These red tides caused fish kills, destruction of marine life, economic losses, dangers to public health, and quality of the environment. Public pressure resulted in control measures in the late 1970s and, since that time, in a marked decrease in red tides.

Restoration of water quality, however, is not always simple. In Dutch lakes, it is impeded by the large amounts of phosphorous stored in the lake sediments and the altered structure of the food web: The large stocks of planktivorous fish remove the zooplankton and, thus, decrease the grazing pressure on phytoplankton. As a result, algal blooms persist despite the decreased nutrient inputs. Large experiments in some lakes are now in progress whereby lake sediments are removed or phosphorus in the sediment immobilized, and the original fish-stock is restored to accelerate the process of water quality improvement and the decrease of algal blooms. The first results are promising, but such an expensive biomanipulation is, of course, impossible for all eutrophicated lakes. In these lakes, restoration of water quality and decrease of algal blooms after reduction of the nutrient load will take many years.

Bibliography

Anderson, D. M., White A. M., and Baden, D. G. (eds.) (1985). "Toxic Dinoflagellates." Elsevier, New York.

Cadée, G. C. (1990). *EC Water Pollution Res. Rep.* **12,** 105–112.

Cosper, E. M., Bricelji, V. M., and Carpenter E. J. (eds.) (1990). *Coast. Estuar. Stud.* **35,** 1–797.

Donk, E. van, and Gulati, R. D. (eds.) (1989). *Hydrobiol. Bull.* **23,** 1–99.

Fukuoka, Y., Takano, H., Chihara, M., and Matsuoka, K. (1990). "Red Tide Organisms in Japan, an Illustrated Taxonomic Guide." Uchida Rokakuho, Tokyo.

Granéli, E., Sundström, B., Edler, L., and Anderson, D. M. (eds.) (1990). "Toxic Marine Phytoplankton." Elsevier, New York.

Keller, M. D., Bellows, W. K., and Guillard, R. R. L. (1989). Dimethyl sulfide production in marine phytoplankton. *In* "Biogenic Sulfur in the Environment" (E. S. Satzman and W. J. Cooper, eds.). *Am. Chem. Soc. Symp. Ser.* **393,** 167–182.

Okaichi, T., Anderson, D. M., and Nemoto, T. (eds.) (1989). "Red Tides, Biology, Environmental Science and Toxicology." Elsevier, New York.

Parker, M., and Tett, P., eds. (1987). *Rapp. P.-V. Réun., Cons. Int. Explor. Mer* **187,** 1–114.

Pearl, H. W. (1990). *Limnol. Oceanogr.* **33,** 823–847.

Shumway, S. E. (1990). *J. World Aquacult. Soc.* **21,** 65–104.

Smayda, T. J. (1989). *Coast. Estuar. Stud.* **35,** 449–483.

Alkaline Environments

W. D. Grant
University of Leicester

Glossary

Alkaliphile Organism with a pH optimum for growth in excess of pH 8; obligate alkaliphiles are incapable of growth at neutrality; alkalitolerant or alkalitrophic organisms are capable of growth at high pH but are also capable of growth under neutral or acidic conditions

Haloalkaliphile Organism with a pH optimum for growth in excess of pH and a requirement for >10% w/v NaCl in the growth medium

Halobacteria Obligately aerobic red-pigmented obligately halophilic (i.e., a requirement for >10% w/v NaCl in the growth medium) archaeobacteria belonging to the order Halobacteriales

Soda lake Alkaline lake where the alkalinity is due to large amounts of Na_2CO_3 (or complexes of this salt)

STABLE ALKALINE ENVIRONMENTS (pH >10) are not common and are caused by an unusual combination of ecological, geographical, and climatic conditions that mitigate against the significant buffering capacity of atmospheric CO_2. Alkaliphiles are defined as organisms that grow at alkaline pH, with pH optima for growth in excess of pH 8, usually between pH 9 and 10. Some are capable of growth at pH 11.5, and obligate alkaliphiles are incapable of growth at neutrality. Alkalitrophic (or alkalitolerant) organisms are defined as being capable of growth at high pH, but with pH optima for growth in the acid or neutral region of the pH scale. A number of man-made unstable alkaline environments are produced by industrial processes including cement manufacture, the disposal of blast furnace slag, electroplating, food processing, and the manufacture of paper and board, all of which introduce alkali (usually NaOH) either as a part of the process or, as in the cases of the cement and blast furnace waste, as a consequence of the generation of $Ca(OH)_2$. Alkaliphiles are widely distributed in such human-made environments and can be found in almost any environment including soils, even in those that would not be considered to be particularly alkaline. It is assumed that in such environments transient alkaline environments are generated by biological activity allowing survival and limited growth of a small alkaliphile population. These environments do, however, yield a restricted range of alkaliphile types, notably endospore-forming bacilli. On the other hand, naturally occurring stable alkaline environments yield a much more diverse population of alkaliphiles. Soda lakes and deserts constitute the major natural alkaline habitat. These Na_2CO_3-dominated environments have pH values of up to 11.5, probably the highest stable pH values on earth.

I. The Genesis of Soda Lakes and Soda Deserts (Low Ca^{2+} Environments)

Soda lakes and soda deserts represent the major type of naturally occurring highly alkaline environments. Such sites are widely distributed (Table I), although most of the detailed biological analyses have been limnological rather than microbiological. Soda lakes and deserts, as the name implies, are characterized by the presence of large amounts of Na_2CO_3 (usually as Na_2CO_3 10 H_2O or $Na_2CO_3 \cdot NaHCO_3$ 2 H_2O) and are significantly depleted

Table I Worldwide Distribution of Soda Lakes and Soda Deserts

Region	Lakes
North America	
Canada	Manito
United States	Alkali Valley, Albert Lake, Lake Lenore, Soap Lake, Big Soda Lake, Owens Lake, Mono Lake, Searles Lake, Deep Springs, Rhodes, Marsh, Harney Lake, Summer Lake, Surprise Valley, Pyramid Lake, Walker Lake
Central America	
Mexico	Texcoco
South America	
Venezuela	Langunilla Valley
Chile	Antofagasta
Europe	
Hungary	Lake Feher
Yugoslavia	Pecena Slatina
Russia	Kulunda Steppe, Tanatar Lakes, Karakul, Araxes plain, Chita, Barnaul, Slavgerod
Asia	
Turkey	Van
India	Lake Looner, Lake Sambhar
China	Qinhgai Hu; Sui-Yuan, Heilungkiang, Kirin, Jehol, Chahar, Shansi, Shensi, Kansu
Africa	
Libya	Lake Fezzan
Egypt	Wadi Natrun
Ethiopia	Lake Aranguadi, Lake Kilotes, Lake Abiata, Lake Shala, Lake Chilu, Lake Hertale, Lake Metahara
Sudan	Dariba lakes
Kenya	Lake Bogoria, Lake Nakuru, Lake Elmentieta, Lake Magadi, Lake Simbi, Lake Sonachi
Tanzania	Lake Natron, Lake Embagi, Lake Magad, Lake Manyara, Lake Balangida, Basotu Crater, Lakes, Lake Kusare, Lake Tulusia, El Kekhooito, Momela Lakes, Lake Lekandiro, Lake Reshitani, Lake Lgarya, Lake Ndutu, Lake Rukwa North
Uganda	Lake Katwe, Lake Mahega, Lake Kikorongo, Lake Nyamunuka, Lake Munyanyange, Lake Murumuli, Lake Nunyampaka
Chad	Lake Bodu, Lake Rombou, Lake Dijikare, Lake Momboio, Lake Yoan
Australia	Lake Corangamite, Red Rock Lake, Lake Werowrap, Lake Chidnup

in Mg^{2+} and Ca^{2+}. The best-studied soda environments are the highly alkaline lakes of the East African Rift Valley, where both detailed limnological and, unusually, microbiological analyses have been carried out over many years.

A number of theories concern the generation of the alkalinity. The East African Rift Valley is an area of active vulcanism dominated by alkaline trachyte lavas (high Na^+, low Mg^{2+}, Ca^{2+}), and the simplest theory would suppose the contribution of Na_2CO_3 via vulcanism—At least one active volcano generates a soda-rich lava flow in the southern path of the Rift Valley.

An alternative theory links alkalinity to sulfate reduction in anaerobic basins. Anaerobic basins are characterized by sulfate reduction, which mineralizes sinking algal organic matter and generates alkalinity (i.e., the sum of the charges of all of weak acids, $HCO_3^- + CO_3^{2-} + OH^-$, etc.) in the following way:

$$53\ SO_4^{2-} + C_{106}H_{263}O_{110}N_{16}P_1 + 14\ H_2O \rightarrow 53\ H_2S + 106\ HCO_3^- + HPO_4^{2-} + 16\ NH_4^+ + 15\ OH^-$$

Anaerobic basins were common features in the earth's history, although today the Black Sea represents the only example of a large anoxic basin. Bacterial sulfate reduction in surrounding swamps has thus been proposed to contribute toward alkalinity in certain soda lakes such as those of the Wadi Natrun depression in Egypt and certain Hungarian sites.

However, these mechanisms probably do not explain the genesis of alkalinity in most soda lakes. Most likely, geological and climatic features, particularly the absence of Ca^{2+} and Mg^{2+} in the surrounding topography together with distinct isolated drainage basins and high rates of evaporative concentration, are responsible for the formation of soda lakes in many parts of the world.

The East African Rift Valley is a typical low Ca^{2+}

Mg^{2+} area characterized by a high Na^+ geology, a series of closed basins, and as a result a considerable number of soda lakes have developed along the floor of the valley. Evaporative concentration has played a significant role in the genesis of these lakes and the lakes are both alkaline and saline. Typical total salinities range from around 5% (w/v) in some of the northern lakes such as Lake Bogoria to saturation (33% w/v) on parts of the southern lakes (lakes Magadi and Natron). Table II illustrates typical ion comparisons of a range of these lakes from the Rift Valley and includes analyses carried out on non-Rift Valley soda lakes for comparison. All of the lakes are devoid of significant amounts of Ca^{2+} and Mg^{2+}, and pH values of 11.5 have been recorded. However, pH measurements under alkaline, very saline conditions, are error-prone. Such lakes have great buffering capacity and maintain their high pH in seasons with widely differing rainfall.

Researchers postulate that in surface and near-surface zones, weathering and biological activity produces CO_2-charged surface water and, thus, a HCO_3^{2-}/CO_3^{2-} solution of minerals is produced, reflecting the surrounding geology (in this case, Na^+-dominated). In closed basins, where intense evaporative concentration takes place, if the concentration of HCO_3^{2-}/CO_3^{2-} greatly exceeds that of any Mg^{2+} and Ca^{2+}, these are removed from solution as insoluble carbonates, and alkalinity develops as a consequence of the shift in the $CO_2/HCO_3^-/CO_3^{2-}/OH^-$ equilibrium as concentration progresses. In normal surface waters, Ca^{2+} and Mg^{2+} concentrations generally greatly exceed that of HCO_3^-/CO_3^{2-} and the precipitation of carbonates has a buffering action (as in the marine environment).

Figure 1 illustrates the schematic representation of the proposed mechanisms involved in the formation of Lake Magadi, an alkaline and very saline lake some 200 km south of Nairobi at the lowest point in the Rift Valley (200 m). However, the generation of alkalinity is undoubtedly a complex process and other factors may be involved (e.g., in the Lake Magadi area, subterranean reservoirs probably contribute to the final ionic composition of the lake).

In certain soda lakes, calcium-rich ground water, which may have traveled a considerable distance, seeps into the alkaline lake water, giving rise to localized calcite ($CaCO_3$) precipitation in the form of columns or structures. These "tufa columns" may be of considerable size. Mono Lake in California exhibits impressive formations of these structures and recently similar structures 40 m high and extending down to a depth of 100 m have been described in Lake Van in Eastern Turkey where cyanobacteria contribute to further permineralization via the biogenic precipitation of aragonite, another crystalline form of $CaCO_3$. Stromatolithic structures like these were widespread in Proterozoic shallow seas early in the earth's history. It has been argued that these stromatolite-dominated pre-Cambrian oceans may have been soda oceans since this period predates the widespread mobilization and ubiquitous deposition of Ca^{2+} by eukaryotic marine algae.

II. The Genesis of High Ca^{2+} Alkaline Environments

Alkaline Ca^{2+}-bearing ground waters have been identified in various geological locations in California, Oman, Yugoslavia, Cyprus, and Jordan. This kind of ground water is similar to the effluent produced by cement manufacturing and casting, where the alkalinity (pH >11) is generated by the aqueous

Table II Hydrochemistry of Selected Soda Lakes (m*M*)

	Lake Bogoria (Kenya)	Lake Magadi (Kenya)	Wady Natrun-Zugm (Egypt)	Sambhar Lake (India)
Na^+	652	2000	6177	1630
K^+	5	16	58	12
Cl^-	85	400	4360	572
CO_3^{2-}	220	582	1120	10
SO_4^{2-}	not done	not done	235	22
Mg^{2+}	<0.1	<0.1	<0.1	<0.1
Ca^{2+}	<0.1	<0.1	<0.1	<0.1
pH	10.5	10.9	11.0	9.0

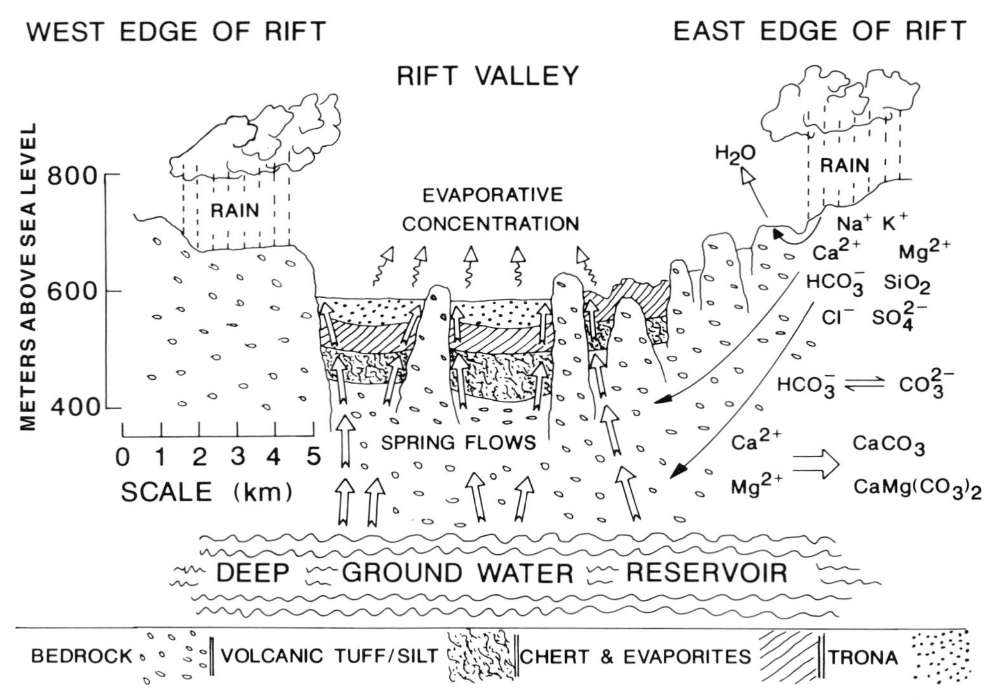

Figure 1 Schematic representation of the possible mechanisms involved in the formation of alkaline lakes (Lake Magadi). [From Grant, W. D., Mwatha, W. E., and Jones, B. E. (1990). *FEMS Microbiol. Rev.* **75,** 235–270.]

phase being in equilibrium with solid phase $Ca(OH)_2$. The hydrochemical conditions are inferred to resemble those in cement pore waters and are, thus, of interest as a model system in which to assess any long-term microbial or chemical activity that might eventually compromise the long-term integrity of concrete structures in contact with water.

Unlike the situation in soda lakes, the alkalinity is generated in the virtual absence of CO_3^{2-}, which is removed from solution in the presence of excess amounts of Ca^{2+} and Mg^{2+} in the surrounding topography. As in the initial stages of soda lake genesis, CO_2-charged surface waters decompose surrounding minerals. In this case, the chemistry is determined by the low temperature subsurface weathering of two primary minerals (olivine and pyroxene). In near-surface zones, these minerals are decomposed in the manner depicted in Fig. 2, resulting in Ca^{2+} and OH^- being released into solution. The process, known as serpentinization thus produces a $Ca(OH)_2$-dominated brine where solid phase $Ca(OH)_2$ is in equilibrium with the soluble phase. Unlike the situation for soda lakes, the chemistry is dilute since the solubility of $Ca(OH)_2$ is extremely low under most conditions (around 10 mM) and such waters have an extremely low buffering capacity when separated from solid phase $Ca(OH)_2$ and exposed to atmospheric CO_2. Some of the consequences of the pathways shown in Fig. 2 are that CO_3^{2-} rapidly becomes removed from solution as carbonates, Mg^{2+} is largely removed by the genesis of the mineral serpentine, and conditions are profoundly reducing due to the release of Fe^{2+} and the generation of hydrogen by the hydrolysis of transient metal hydroxides. Na^+ and Cl^- are introduced by the teaching of other minerals entrapped in the primary minerals. Table III indicates the ionic composition of $Ca(OH)_2$-dominated ground waters in Oman. The chemistry is dilute and comparable to that seen for oligotrophic waters, again unlike the situation for soda lakes, where eutrophic conditions dominate.

III. Alkaliphile Taxa from Neutral Environments

Neutral (i.e., neutral when subject to bulk pH measurements) soils almost invariably yield gram-positive endospore-forming rods that are classified in the genus *Bacillus*. It should be stressed that culture procedures are generally carried out in nutrient-

$$MgFeSiO_4 + CO_2 + H_2O \longrightarrow Mg^{2+} + HCO_3^- + H_4SiO_4 + Fe^{2+} + OH^-$$

OLIVINE

$$MgCaFeSiO_3 + CO_2 + H_2O \longrightarrow Mg^{2+} + Ca^{2+} + HCO_3^- + H_4SiO_4 + Fe^{2+} + OH^-$$

PYROXENE

$$Mg^{2+} + H_4SiO_4 + H_2O \longrightarrow Mg_3Si_2O_5(OH)_4 + H^+$$

SERPENTINE

$$MgFeSiO_4 + MgCaFeSiO_3 + H_2O \longrightarrow Mg_3Si_2O_5(OH)_4 + Fe^{2+} + Ca^{2+} + OH^-$$

$$Fe(OH)_2 \longrightarrow Fe_2O_3 + H_2 + H_2O$$

Figure 2 Geochemical evolution of Ca^{2+} springs. [Modified from Bath, A. H., Cristofi, N., Neal, C., Philip, J. C., McKinley, I. G., and Berner, U. (1987). Trace element and microbiological studies of alkaline ground waters in Oman, Arabian Gulf: A natural analogue for cement pore waters. Reports of the Fluid Processes Research Group of the British Geological Survey FLPU 87-2.]

rich media-made alkaline with Na_2CO_3, under aerobic conditions—other growth conditions are essentially unexplored. The spore-forming isolates are often pigmented, usually yellow or red, and grow over the pH range 8–11.5. The rigorous examination of pH optima for growth and the examination of biochemical characteristics requires the automated addition of alkali to the growth medium, to avoid the buffering effect of atmospheric CO_2 alluded to earlier. To date, a very large number of alkaliphilic bacilli have been isolated largely because of interest in the biotechnological applications of alkali-stable enzymes. Organisms of this type were first isolated 50 years ago and most bear names that have no taxonomic standing. It is clear that the alkaliphilic bacilli, although usually described as *Bacillus alkalophilus* or *Bacillus firmus*, comprise an extremely diverse group, probably representing several major taxa within the *Bacillus* taxon complex, and a considerable amount of work will be necessary to clarify the systematics of the group. These bacilli will also grow in media made alkaline by contact with solid phase $Ca(OH)_2$, so there is no *a priori* requirement for Na_2CO_3 as the alkalinizing agent, although the organisms do have a requirement for Na^+ ion. Organisms of this type, being simple to grow, have also been extensively used as model systems with which to test the basic tenets of the chemiosmotic theory of energy generation since they are obviously capable of energy transduction despite having a "reversed" pH gradient across the cell membrane. [*See* ALKALIPHILES.]

Table III Hydrochemistry of Selected Ca^{2+} Springs in Oman (mM)

	Nizwa Jill	Karkin	Bahla
Na^+	9.5	11.2	8.2
K^+	0.23	0.9	0.21
Cl^-	8.1	9.9	7.7
CO_3^{2-}	0	0	0
SO_4^{2-}	0.01	0.02	0.002
Mg^{2+}	<0.1	<0.1	<0.1
Ca^{2+}	1.2	1.6	1.38
OH^-	4.9	6.1	4.9
pH	11.2	11.4	11.4

[From Bath, A. H., Cristofi, N., Neal, C., Philip, J. C., McKinley, I. G., and Berner, U. (1987). Trace element and microbiological studies of alkaline ground waters in Oran, Arabian Gulf: A natural analogue for cement pore waters. Reports of the Fluid Processes Research Group of the British Geological Survey FLPU 87-2.]

Nonsporing alkaliphilic isolates can also be obtained from soils and particularly from waters, although at a much lower frequency as compared to sporing isolates. These include *Pseudomonas* spp., *Paracoccus* spp., *Micrococcus* spp., *Aeromonas* spp., *Corynebacterium* spp., and *Actinopolyspora* spp., including psychrophilic representatives and even alkaliphilic fungi, but characterizations are often incomplete. Human-made alkaline environments presumably select particular alkaliphiles from adjacent soils and waters. Isolates from such environments include the gram-positive *Exiguobacterium aurantiacum* from potato processing waste and a gram-negative rod described as an *Ancylobacterium* sp. from Kraft paper and board process effluents.

IV. Alkaliphile Taxa from $Ca(OH)_2$-Dominated Ground Waters?

Microbiological analyses are largely confined to a series of Ca^{2+} springs in Oman. Organisms isolated under largely aerobic conditions were similar to those in normal environments including *Bacillus* spp., *Vibrio* spp., *Flavobacterium* spp., *Pseudomonas* spp., and members of the enterobacteria. However, the low numbers encountered (10^4 ml^{-1}) could have been introduced as surface contaminants from adjacent animals, and no evidence indicates that the organisms were actually indigenous to the ground waters, particularly since few were obligately alkaliphilic. Logically, such environments might be expected to harbor oligotrophic anaerobes. Alkaline springs elsewhere with similar chemistries have also yielded isolates that were assigned to extant taxa, although again no evidence indicates that the organisms were actually growing *in situ*. Accordingly, the question of whether or not organisms actually proliferate in these environments remains to be established.

V. Alkaliphile Taxa from Soda Lakes and Soda Deserts

Soda lakes are probably the most productive naturally occurring aquatic environments in the world. Whereas the mean gross primary productivity for streams and lakes is calculated to be around 0.6 g cm^{-2} day^{-1}, soda lakes exhibit productivities >1 order of magnitude greater than this value, as a consequence of relatively high surface temperatures (30°–35°C), high intensity of illumination due to tropical or semitropical locations, and virtually unlimited reserves of HCO_3^- for photosynthesis. These lakes support large standing crops of a diverse range of microorganisms that are undoubtedly proliferating *in situ*.

Soda lakes often exhibit blooms of phototrophs that are sufficiently dense to color the water's various shades of green, orange, or red, depending on the organism(s) involved. These lakes are unique in that the productivity is composed almost entirely of prokaryotes, notably cyanobacteria. The more dilute lakes (total salinities between 5 and 15% w/v) often exhibit permanent or seasonal populations of cyanobacteria, generally consisting of only one or two species. Cell densities may reach up to 13,000 cyanobacterial filaments ml^{-1}, reflected in chlorophyll levels of up to 200 μg ml^{-1} of lake water. [*See* CYANOBACTERIA, MOLECULAR GENETICS.]

Cyanobacteria commonly participating in bloom formation include the filamentous *Cyanospira* (*Anabaenopsis*) spp. and unicellular examples such as *Chroococcus* spp.; *Pleurocapsa* spp. are implicated in the tufa stromatolites seen in Lake Van. The organisms are often poorly characterized, identification often being made on the basis of microscopic examination alone. There are considerable difficulties in obtaining axenic cultures of these cyanobacteria due to adhering bacterial cells.

In many cases, cyanobacterial blooms may consist almost entirely of organisms variously known as *Spirulina* or *Arthrospira* spp. These filamentous cyanobacteria grow in the form of tight right- or left-handed helices, and there is confusion over the correct name and the number of species. East African lakes are reported as containing *Spirulina platensis,* whereas other lakes may harbor *Spirulina maxima,* although the differences between the two types are not clear and the same organism may have a worldwide distribution. The organisms are gas vacuolate and accumulate as surface scums along the shorelines of these less saline lakes. Enormous numbers of lesser flamingos feed almost exclusively on these large (10–20 × 100–400 μm) cyanobacterial filaments, and up to 2×10^6 birds have been recorded on Lake Nakuru, one of the Rift Valley Lakes, such a population consuming up to 200 tons of algal biomass each day. *Spirulina* also has a long history as a human food source, originally consumed as food supplements by the Aztecs (tecuilaitl) and currently part of the diet of the native population in the Kanen region near Lake Chad. In recent times, *Spirulina* has obtained the status of a health food, having a very favorable amino acid spectrum and a low nucleic acid content, its use ranging from food supplements for Olympic athletes to use as a space diet food. There is now increasing interest in the culture of *Spirulina* under natural conditions as a producer of fine chemicals, such as essential fatty acids, water-soluble vitamins, and β-carotene, both for possible human use and for the rearing of fish and poultry—in particular, the high concentration of β-carotene results in the enhancement of pigmentation, which is often desirable. The photosynthetic yield of *Spirulina* surpasses that of anything in the terrestrial environment, and there is a considerable effort directed toward optimizing yield conditions under various culture conditions. In general, the organism has great potential: it grows under conditions that exclude most other organisms, particularly pathogens, and it can be readily recovered due to its buoyancy provided by gas vacuoles.

The most dilute soda lakes also harbor dense, stable populations of nonphototrophic bacteria (total counts using epifluoresence microscopy indicate 10^7–10^8 bacteria ml^{-1}). Viable counts of aerobic organotrophic bacteria on high-nutrient media similar to those used for isolating alkaliphiles from soil, range from 10^5 to 10^6 colony-forming units/ml, the majority being obligately alkaliphilic. Counts made on a range of dilute soda lakes in the Rift Valley indicate that the population is stable throughout the year, apparently independent of seasonal *Spirulina* blooms and significant changes in rainfall. Each lake had a unique population not shared by any other and, unlike the situation for soils, the populations are largely composed of gram-negative types, as determined by quinone composition and sensitivity to lysis by 3% (w/v) KOH. The taxonomic identity of these isolates remains to be established, but it is clear that there are several new gram-negative taxa probably at genus level.

Work so far has been largely confined to aerobic organotrophs, although a start has been made on anaerobes including the observation that anoxygenic phototrophic bacteria of the genus *Ectothiorhodospira* may make a significant contribution to primary productivity in some of the lakes at particular times. Significant blooms of these organisms (usually the red-pigmented *Ectothiorhodospira mobilis* and *Ectothiorhodospira vacuolata*) have been seen together with cyanobacterial blooms in some of these lakes. It is now clear that the soda lake is the natural habitat for these characteristic prototrophs, although early isolations were made from nonalkaline environments. *Ectothiorhodospira* spp. have a role to play in the sulphur cycle in these lakes because they utilize H_2S as an electron donor in photosynthesis. Because the lakes contain black anoxic sediments, it is to be assumed that as-yet unidentified alkaliphilic sulfate-reducing bacteria complete the reductive part of the cycle. In principle, one might expect to find alkaliphilic representatives of key organisms participating in element cycles in neutral environments.

Methanogenesis is also a feature of these anoxic sediments and enrichment cultures indicate that methylamine-utilizing methanogens probably of the genus *Methanohalophilus* are widely distributed in such lakes. [*See* METHANOGENESIS.]

Highly saline (>20% NaCl w/v) and alkaline environments such as Lake Magadi in the Rift Valley, Owens Lake in California, the Wadi Natrun lakes in Egypt, and a wide range of saline soda lakes and soils in Tibet, Mongolia, Pakistan, India, and Russia harbor a different population of prokaryotes. The lakes are often colored red by large numbers (10^7–10^8 ml^{-1}) of haloalkaliphilic archaeobacteria (archaea). These organisms are currently classified into two genera, the rod-shaped isolates in the genus *Natronobacterium* with three species *N. pharaonis*, *N. gregoryi*, and *N. magadii*, and the coccoid types in the genus *Natronococcus*, which is currently a monospecific genus with *Natronococcus occultus* as the sole species. It is quite clear that representatives of these genera are to be found in soda lakes and soda deserts throughout the world and that new types will be described. The organisms are members of the order Halobacteriales and are halobacteria *sensu strictu* but represent distinct lines within the halobacterial phenotype. No overlap exists between the ecological niche occupied by these organisms and those occupied by other halobacteria. These organisms clearly dominate highly saline alkaline environments in the final stages of concentration before the crystallization of salts occurs. The red pigmentation that these organisms impart to the environment is a consequence of the possession of C_{50} carotenoids, in common with other halobacteria, and the color is particularly pronounced where solar evaporation ponds are used to crystallize NaCl from soda brines. In this respect, these ponds resemble neutral salterns where neutrophilic halobacteria color the brines. Examples of such alkaline salterns are to be found at Lake Magadi where a complex cycle of precipitation and solution of both NaCl and Na_2CO_3 occurs in the salt-making process, quite different from the nonalkaline marine salterns more commonly seen throughout tropical and subtropical areas of the world.

Alkaliphilic and halotolerant eubacteria are also to be found in these highly saline environments but do not constitute a significant fraction of the population on or about saturation. However, large numbers of different types of these bacteria can be isolated from soda lakes where the salinity is around 20% (w/v). In environments like this, they are at least as numerous as haloalkaliphilic archaeobacteria and, thus, the situation is comparable to that seen in nonalkaline saline environments where eubacteria make up a significant fraction of the environment, coexisting with archaeobacteria at salinities between 10 and 20% (w/v). These alkaliphilic and possibly halophilic eubacteria certainly represent new taxa but remain to be unequivocably classified.

In common with the more dilute soda lakes, the saline lakes also harbor *Ectothiorhodospira* spp., in this case halotolerant species such as *Ectothiorhodospira halochloris* and *Ectothiorhodospira halophila,* which are probably responsible for primary productivity in these lakes as the salt concentration approaches saturation. Haloalkaliphilic archaeobacteria thus represent secondary producers in these lakes, particularly because none of the examples examined to date have been shown to be able to harness light energy via the bacteriorhodopsin system. Methanogenesis is also seen in highly saline soda lakes and a number of alkaliphilic and halophilic methanogenic archaeobacteria have been isolated. A new genus, *Methanohalophilus,* has been created to accommodate such isolates and other primarily halophilic isolates from nonalkaline saline lakes. Other haloalkaliphilic bacteria, including sulfate-reducing bacteria have yet to be described, but these undoubtedly exist. The eventual isolation and study of such should add further important aspects to our understanding of the biology of the remarkable inhabitants of the most alkaline naturally occurring environments on earth.

Bibliography

Bath, A. H., Cristofi, N., Neal, C., Philip, J. C., McKinley, I. G., and Berner, U. (1987). Trace element and microbiological studies of alkaline ground waters in Oman, Arabian Gulf: A natural analogue for cement pore waters. Reports of the Fluid Processes Research Group of the British Geological Survey FLPU 87-2.

Grant, W. D., and Larsen, H. (1989). Order *Halobacteriales. In* "Bergey's Manual of Determinative Bacteriology," Vol. 3 (J. T. Staley, M. P. Bryant, N. Pfennig, and J. G. Holt, eds.), pp. 2216–2234. Williams and Wilkins, Baltimore.

Grant, W. D., Mwatha, W. E., and Jones, B. E. (1990). *FEMS Microbiol. Rev.* **75,** 235–270.

Grant, W. D., and Tindall, B. J. (1986). The alkaline, saline environment. *In* "Microbes in Extreme Environments" (R. A. Herbert and G. A. Codd, eds.), pp. 25–54. Academic Press, London.

Hardie, L. A., and Eugster, H. P. (1970). *Mineral. Soc. Am. Spec. Pap.* **3,** 273–296.

Tindall, B. J. (1988). Prokaryotic life in the alkaline saline athalassic environment. *In* "Halophilic Bacteria," Vol. 1 (F. Rodriguez-Valera, ed.), pp. 31–70. CRC Press, Boca Raton, Florida.

Alkaliphiles

Tetsuo Hamamoto and Koki Horikoshi
The RIKEN Institute

Glossary

Alkaline enzyme Enzymes whose optimal working pH is alkaline (>10); alkaliphilic microorganisms produce various extracellular alkaline enzymes (e.g., cellulases, amylases, proteases); in contrast, the internal enzymes produced by alkaliphiles usually function at neutral pH, a reflection of an internal neutral pH

Alkaliphile Microorganisms capable of growth at alkaline pH (>8) but not at neutral pH (7)

Alkalitroph Microorganisms capable of growth at alkaline pH (10 or more) as well as at neutral pH (7)

Na^+/H^+ antiport system Carrier that exchanges Na^+ and H^+ through the membrane; transports Na^+ and H^+ in opposite directions through the membrane

ALKALIPHILIC MICROORGANISMS consist of two main physiological groups of microorganisms: nonhalophilic alkaliphiles and haloalkaliphiles. The nonhalophilic alkaliphiles require an alkaline pH of 8 or more for their growth and have an optimal growth pH of around 10. On the other hand, the haloalkaliphiles require alkaline pH (>10) and also high salinity (up to 33% NaCl + Na_2CO_3). The haloalkaliphiles have been found in extremely alkaline saline environments such as soda lakes and, thus, are considered to be extremophiles. In contrast, the nonhalophilic alkaliphiles have been isolated mainly from neutral environments. Indeed, most of the known alkaliphiles have been isolated from soil samples of neutral pH, sometimes even from acidic soil samples. However, they should also be considered as extremophiles, even though they can be isolated from *normal* neutral environments because of their alkaliphilicity. The only difference in procedure compared with the isolation of neutrophiles is that the alkaline (pH 10) medium should be used for the isolation. They were found in neutral environments because, presumably, they have alkalinized the environment right outside their cell surface by their metabolic activity. Strictly speaking, two categories of bacteria grow at alkaline pH: the alkaliphiles, which grow only at alkaline pH, and the alkalitrophs, which are also capable of growing at neutral pH (Fig. 1). The nonhalophilic alkaliphiles show a number of interesting physiological characteristics and, in addition, have given rise to significant biotechnological applications.

I. Alkaliphilic Microorganisms and Alkaline Environments

A. Extremely Saline Alkaline Environments and Alkaliphiles

Very few naturally occurring alkaline environments are located on earth. The most remarkable examples of such environments are soda deserts and soda lakes. Extremely alkaline lakes (e.g., Lake Magadi in Kenya, the Wady Natrun in Egypt) can be regarded as the most stable alkaline environments on earth, with a pH of 10.5–11.0. In these lakes, Mg^{2+} and Ca^{2+} precipitate as carbonate salts; thus, very low levels of these ions are present in a soluble form. In these environments, sodium carbonate is the major source of alkalinity; accordingly, this chemical is usually used in the isolation of microorganisms that are capable of growth under alkaline conditions. Those organisms isolated from alkaline and highly saline environments such as soda lakes require high salinity, which is achieved by adding NaCl to the isolation medium. These include representatives of

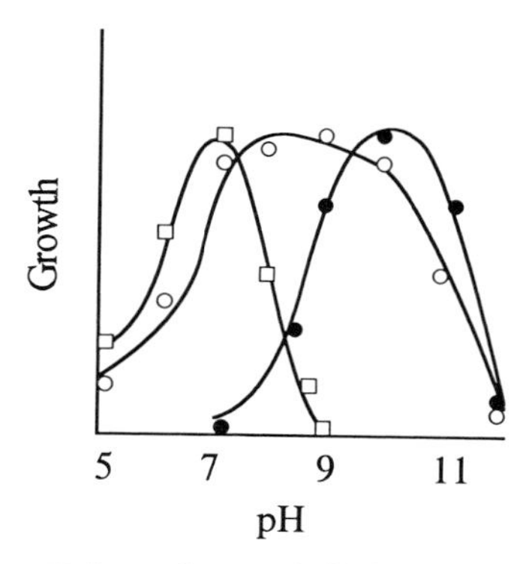

Figure 1 The pH dependency of alkaliphilic microorganisms. Typical growth–pH dependency of neutrophilic (squares), alkalitrophic (open circles), and alkaliphilic (closed circles) bacteria are shown.

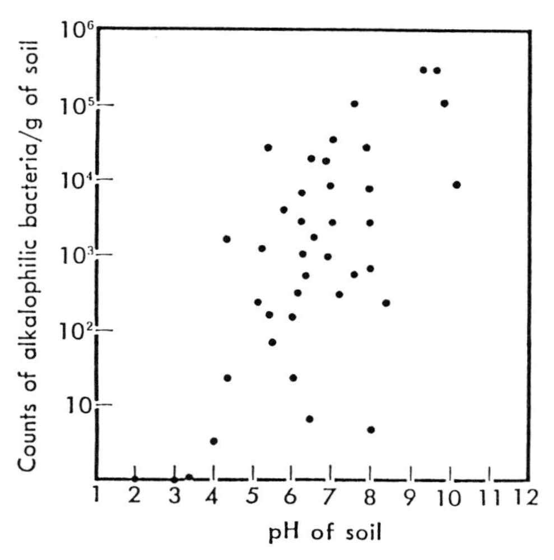

Figure 2 Schematic diagram of occurrence of alkaliphilic microorganisms in various pH environments. Circles indicate the numbers of the alkaliphiles isolated from the samples at the corresponding pH. [From Horikoshi, K., and Akiba, T. (1982). "Alkalophilic Microorganisms." pp. 11. Springer-Verlag, Heidelberg, Germany.]

the archaebacteria. These microorganisms occasionally reach numbers of 10^7–10^8/ml in soda lakes. Because these organisms usually require a considerable amount of organic matter as nutrients, soda lakes are considered to be rich in organic materials. The origin of the organic materials that allow these heterotrophic microorganisms to grow to high density in the lakes has not been clearly explained; however, one possible explanation is that salinities of parts of the lake decrease due to rainfall, which causes temporary growth of less halophilic autotrophic organisms, and then, as the salinity increases without further supply of rain and by intensive evaporation of lake water, those organisms die and the organic debris left in the water can be utilized as nutrients for haloalkaliphilic microorganisms. In support of this explanation, extraordinary occasional blooms of a photosynthetic bacterium such as *Ectothiorhodospira* spp. have been reported in Lake Magadi and the Wadi Natrun. [*See* ALKALINE ENVIRONMENTS.]

B. Nonsaline Alkaline Environments and Alkaliphiles

Alkaliphilic microorganisms have been readily isolated from the neutral environments surrounding us as well as from the alkaline environments. In other words, alkaliphilic microorganisms coexist ubiquitously with neutrophilic microorganisms as well as occupy specific extreme environments in nature. Figure 2 illustrates the relationship between the occurrence of alkaliphilic microorganisms and the pH of the samples of their origins. To isolate alkaliphiles, an alkaline medium must be used. Sodium carbonate is generally used to adjust the pH to around 10, because they usually require sodium ions. Table I shows these general-purpose alkaline media for the isolation of alkaliphiles. The frequency of alkaliphilic (including alkalitrophic) microorganisms in neutral, ordinary soil samples is considered to be 10^2–10^5/g of soil, which corresponds to $\frac{1}{100}$–$\frac{1}{1000}$th of the population of neutrophilic microorganisms. Many alkaliphilic microorganisms have been isolated, from bacteria belonging to the genera *Bacillus, Micrococcus, Pseudomonas,* and *Streptomyces* and to eukaryotes such as yeasts and fungi. In particular, a large number of alkaliphilic bacilli have been intensively isolated. Although some of

Table I Media Used for Nonhalophilic Alkaliphilic Bacteria Isolation

	Horikoshi-I (g/liter)	Horikoshi-II (g/liter)
Glucose	10	
Soluble starch		10
Polypeptone	5	5
Yeast extract	5	5
K_2HPO_4	1	1
$MgSO_4$ 7 H_2O	0.2	0.2
Na_2CO_3	10	10

these bacilli have turned out to be alkalitrophs rather than real alkaliphiles, the discovery of these organisms provided suitable tools for investigating membrane function and the cellular physiology of alkaliphilic microorganisms. Furthermore, these organisms have also found significant industrial application.

The reasons why alkaliphilic microorganisms are found in neutral, sometimes even acidic, environments, could be complex; however, there are two suggestive experimental results. First, alkaliphilic microorganisms are capable of altering the pH of their surrounding environments from neutral to alkaline. This has been proven by cultivation of alkaliphiles in various pH media. The final pH of these cultures changed to a pH that is optimal for the alkaliphiles (such as pH 9). Second, sequential transition of pH by several different microorganisms has been observed. One of the model experiments showed a pH transition from 5 to 9. A fungus, *Aspergillus oryzae* (optimal pH for growth of 4–5), a bacterium, *Bacillus circulans* (optimal pH for growth of 6–8), and an alkaliphilic *Bacillus* species were inoculated in a medium at pH 5. Initially, *A. oryzae,* which grows at acidic pH, began to grow at pH 5. After cessation of growth, the pH of the medium turned to neutral due to autolysis of the fungus. *Bacillus circulans,* which then began to grow at neutral pH, produced a fungal cell-wall lytic enzyme and increased the pH to 8, which allowed the growth of alkaliphiles. This experiment thus demonstrated the overall growth of alkaliphiles in a medium at an initial pH of 5.

II. Physiology of Alkaliphiles

A. Internal pH

The optimal growth pH for most alkaliphiles is around 10, which is the most significant difference from well-investigated neutrophilic microorganisms. Differences between internal pH and environmental pH affect the transportation of ions and substances through the cell membrane, one of the important parameters in maintaining vital biological activities. Because directly measuring the internal pH of microorganisms is difficult, several indirect methods have been applied. One of the methods is to measure the distribution in the cells of compounds that are not actively transported by cells and that have an ionized, pH-dependent form. A weakly alkaline compound, methylamine can be used for this purpose under alkaline pH. Calculating the distribution of methylamine showed that the alkaliphiles had a lower internal pH than that in the external environment (Table II). The internal pH was maintained at around 8, despite high external pH (8–11). In addition to this method, internal pH can also be estimated from the optimal pH of intracellular enzymes. For example, α-galactosidase from alkaliphile *Micrococcus* sp. No. 31-2 had its optimal catalytic pH at 7.5, suggesting that the internal pH is around neutral (Table III). Furthermore, cell-free protein synthesis machinery consisting of ribosomes and transfer RNA from alkaliphiles optimally incorporated amino acids into protein at pH 8.2–8.5, which was only 0.5 pH units higher than that of neutrophilic *B. subtilis*. Therefore, one of the key features in alkaliphilicity is associated with the cell surface, which discriminates between the intracellular neutral environment and the extracellular alkaline environment.

B. Na^+ Ion and Membrane Transport

As already mentioned, alkaliphilic microorganisms grow vigorously at pH 9–11 and require Na^+ for their growth. The presence of sodium ions in the surrounding environments has proven to be essential for effective solute transport through the membranes of alkaliphilic *Bacillus* spp. According to the chemiosmotic theory, the proton motive force in the cells is generated by excreted H^+ derived from adenosine triphosphate (ATP) metabolism by AT-

Table II Internal and External pHs of Alkaliphilic Microorganisms

	Optimal growth pH	Internal pH[a]	External pH
Bacillus alcalophilus	10.5	7.4	7.0
		8.0	8.0
		7.6	9.0
		8.6	10.0
		9.2	11.0
Bacillus sp. No. 8-2	9.5	7.8–8.2	7–10

[a] Internal pHs of the alkaliphiles were measured by the distribution of methylamine.

[From Horikoshi, K., and Akiba, T. (1982). "Alkalophilic Microorganisms." pp. 51. Springer-Verlag, Heidelberg, Germany.]

Table III Optimal pH of the Intracellular Enzymes Produced by Alkaliphiles

Enzyme	Optimal pH	Stable pH range	Producer
α-Galactosidase	7.5	7.5–8.0	*Micrococcus* sp. No. 31-2
β-Galactosidase	6.5	5.5–9.0	*Bacillus* sp. No. C-125

Pase, or respiration. Then, the H^+ is reincorporated into the cells with cotransport of various substrates. In the case of Na^+-dependent transport systems, the H^+ is exchanged with Na^+ by Na^+/H^+ antiport systems, thus generating a sodium motive force, which drives substrates accompanied by Na^+ into the cells. The incorporation of α-aminoisobutyrate (AIB; used as a nonmetabolizable amino acid analog) increased twice as the external pH shifted from 7 to 9, and the presence of sodium ion significantly enhanced the incorporation; 0.2 *M* NaCl produced an optimum that was 20 times the rate observed in the absence of NaCl. Other cations, including K^+, Li^+, $NH_4{}^+$, Cs^+, and Rb^+ showed no effect, nor did their counter anions. [*See* Ion Transport; ATPases and Ion Currents.]

C. *In Vitro* Membrane Transport Systems

It is generally accepted that Na^+ enhances the affinity between Na^+ carrier proteins and substrates in active transport systems. These carrier proteins are also required in the passive excretion of the substrates from the cells. In alkaliphiles, the incorporation of AIB and serine was shown to be enhanced by the presence of a Na^+ concentration gradient through the membrane as well as by the membrane electron potential in experiments with *in vitro* membrane particle systems. This suggests that sodium motive force and sodium ion-dependent membrane transport systems function both in the incorporation of nutrients and in internal pH adjustment, whereas neutrophiles utilize a proton motive force in the same way. Because alkaline environments contain high concentrations of Na^+ ions, in contrast to neutral environments, alkaliphiles may have Na^+/H^+ antiport systems that share the same subunits with nutrient transport systems.

D. Flagella of Alkaliphiles

Some alkaliphiles, such as *Bacillus alkalophilus, Bacillus firmus* RAB, and *Bacillus* sp. No. 8-1 are motile with flagella. This flagella-induced motility is considered to be driven by a sodium motive force instead of a proton motive force, as shown by neutrophiles. These alkaliphiles are most motile at pH 9.0–10.5, whereas no motility is observed at pH 7; in addition, they require Na^+ for motility. Even in the presence of Na^+, motility is inhibited after abolishing their membrane potential by adding ionophoric substances. [*See* Flagella; Motility.]

III. Cell Walls of Alkaliphiles

A. Cell-Wall Composition

Despite the external alkaline pH, the intracellular pH of alkaliphilic microorganisms is neutral (pH 7.5–8.0). Therefore, the pH difference must be due to cell-surface components, namely, the cytoplasmic membrane and the cell wall. Since the protoplasts of alkaliphilic *Bacillus* strains lose their stability in alkaline environments, the cell wall may play a role in protecting the cell from alkaline environments. Components of the cell walls of several alkaliphilic *Bacillus* spp. have been investigated in comparison with those of the neutrophilic *B. subtilis*. In addition to the peptidoglycan, which was essentially similar in composition to that of *B. subtilis,* alkaliphilic *Bacillus* spp. contain acidic compounds. Hydrolytic analysis of the acidic compounds yielded galacturonic acid, gluconic acid, glutamic acid, aspartic acid, and phosphoric acid. This was a clear relationship between their cell-wall composition and their growth–pH characteristics. They could be divided into three groups, as described in Table IV. [*See* Cell Walls of Bacteria.]

B. Composition of the Peptidoglycans

The peptidoglycans of alkaliphilic *Bacillus* apparently are similar to that of *B. subtilis;* however, their composition was characterized by the excess of hexosamines and amino acids in the cell walls. Glucosamine, muramic acid, D- and L-alanine, D-glutamic acid, *meso*-diaminopimeric acid, and acetic acid were found in hydrolysate. Although some variation in the amide content among the peptidoglycans forms alkaliphilic *Bacillus* strains, the variation in

Table IV Cell-Wall Composition of Alkaliphilic *Bacillus* Strains

Group	Components of cell walls	Growth pH	Ion requirement
1	High in glucuronic acid, teichuronic acid, and hexosamine	No growth at pH 7	Na^+ (essential)
2	High in glutamic acid, asparctic acid, galacturonic acid, and glucuronic acid	Capable of growth at pH 7	Na^+ (essential)
3	Presence of phosphoric acid, similar to *Bacillus subtilis*	Capable of growth at pH 7 and 10 in the presence of Na^+ and K^+	

pattern was similar to that in neutrophilic *Bacillus* species.

C. Acidic Polymers in the Cell Walls

As described in Table IV, the alkaliphilic *Bacillus* strains in group 1 contain teichuronic acid in the cell walls and cannot grow at neutral pH; on the other hand, the cell walls of group 2 strains contain large amounts of acidic amino acids and uronic acids. In contrast, the walls of group 3 organisms contain teichoic acid. Although most of the strains in group 2 can grow at neutral pH, the amount of acidic amino acids and uronic acids found in cell walls from bacteria grown at neutral pH was much smaller than the amount found in those from an alkaline pH. Therefore, the acidic components in the outermost layer of the group 2 bacteria probably help to support growth at alkaline pH. The negative charges on the acidic nonpeptidoglycan components may enable the cell surface to adsorb sodium and hydronium ions and repulse hydroxide ions and, as a consequence, enable the cells to grow in alkaline environments.

IV. Genetic Analysis of Alkaliphily

A. Characteristics of Alkali-Sensitive Mutants

Two alkali-sensitive mutants of alkaliphilic *Bacillus* No. C-125 have been obtained by chemical mutagenesis. Because strain No. C-125 grows well in Horikoshi minimal medium (supplemented with 0.2% glutamate and 0.5% glycerol as nutrients in place of glucose–starch polypeptone and yeast extract) over the pH range of 7.5–10.0, this strain is convenient for obtaining mutants defective in alkaline growth. Moreover, the strain closely resembles *B. sutilis,* and genes from the strain are known to be expressed well in *Escherichia coli.* Therefore, the strain is considered to be suitable for genetic analysis in *E. coli.* The two mutants showed different properties from each other. One of the mutants, No. 18224, cannot grow at pH >9. The internal pH of the mutants, calculated by methylamine incorporation across the membrane, was 8.7 in the presence of Na_2CO_3. This value is close to that of the parent strain No. C-125 (pH 8.6). However, this mutant cannot maintain a low internal pH in the presence of K_2CO_3. It was also found that the parent alkaliphilic strain can maintain a low internal pH in the presence of Na_2CO_3 but not in the presence of K_2CO_3, whereas another mutant, No. 38154, was unable to sustain low internal pH in the presence of either Na_2CO_3 or K_2CO_3. The internal pH was 10.4, which was the same as that for all strains in the presence of K_2CO_3. To summarize, it is suggested that mutant No. 38154 was defective in the regulation of internal pH, whereas mutant No. 18224 apparently showed normal regulation of internal pH values; at the same time, Na^+ ion plays some role in the pH homeostasis mechanism, either directly or indirectly. In the case of mutant No. 18224, its inability to grow at alkaline pH is due not to its inability to generate a pH gradient but, rather, to a defect in some other function.

B. Molecular Cloning of DNA Fragments Conferring Alkaliphily

Attempts to identify the genetic basis of alkaliphily revealed the existence of two different DNA factors that confer alkaliphily on respective mutants. Original alkaliphily-conferring recombinant plasmids, pALK11 and pALK2, contained 1.4 and 2.0 kb of DNA fragments from strain No. C-125, respectively. Plasmid pALK11 conferred alkaliphily on mutant No. 18224 but not on mutant No. 38154. On the other hand, mutant No. 38154, which cannot regulate its internal pH in alkaline media, recovered pH homeostasis by the introduction of pALK2. However, the introduction of pALK11 had no effect on recovery of pH in mutant No. 38154. These results indicate that at least two independent factors are involved in alkaliphily, even though the precise functions of these two DNA fragments are not yet clear. Analysis on the locations of these original two fragments on the chromosome of No. C-125 demonstrated the overlap of the fragments (Fig. 3). It has been proven that the two factors contained in these DNA fragments are independent of each other by trimming the DNA fragments. In addition, several alkaliphilic *Bacillus* strains contain DNA regions homologous to these two DNA fragments, which are not present in the chromosome of neutrophilic *B. subtilis*. Therefore, these genetic elements may be common to all alkaliphilic *Bacillus* spp.

V. Alkaline Enzymes

A. Occurrence of Alkaline Enzymes

Published in 1971, the first report concerning an alkaline enzyme described an alkaline protease produced by *Bacillus* sp. No. 221. Strain No. 221 showed significant growth even at pH 11 but not at pH 7. The alkaline protease had an optimal pH for activity at pH 11.5–12.0 and showed substantial remaining activity at pH 13. Since then, a number of enzymes such as proteases, amylases, and xylanases have been discovered. Table V lists the alkaline enzymes, producer strains, and optimal pH for the enzymes.

B. Alkaline Proteases

Alkaliphilic *Bacillus* No. 221 (ATCC21522) produces an alkaline protease that can be readily crystallized because of the low content of contaminating proteins, one of the advantages of alkaline enzymes. This enzyme possesses an optimal pH of 11.5–12.0,

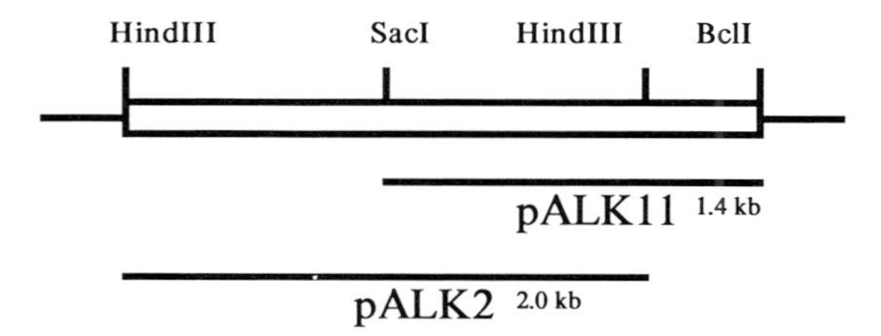

Figure 3 Map of DNA fragments in chromosomal DNA that confer alkaliphily on alkali-sensitive mutants of alkaliphilic *Bacillus* strain C-125. The box represents the DNA fragment of the particular region (ca. 2.5 kb) in the chromosomal DNA from alkaliphilic *Bacillus* strain C-125. Restriction enzyme sites are shown above the corresponding cleavage sites. Regions of DNA corresponding to the cloned DNA fragments in pALK11 and pALK2 are indicated with lines.

Table V Characteristics of Alkaline Enzymes

Producer	Active pH	Stable pH	Characteristics
Protease			
No. 221	11.5–12.0	4–11	Used in detergents
No. 8-2	10.5–11.0	6–9	
No. D-6	10.5–11.0	4–12	Contains two enzymes
Amylase			
No. A-40-2	10.5	7–9	Liquefying type
No. H 167	10–11	6–11	Produces maltohexaose
No. 17-1	4.5–10.0	6–10	Produces cyclodextrins
No. 38-2	4.5–9.0	6–9	Produces cyclodextrins
Pectinase			
No. P-4-N	10.0	5–9	Thermostable
Pullulanase			
No. 202-1	9.0	6–10	
Cellulase (carboxymethylcellulase)			
No. N-4	6–11	5–11	Used in detergents
No. 212	6–8	5–11	
No. 1139	9	5–11	
Xylanase			
No. C-59-2	5.5–9.0	5–9	
No. C-125	6–10	4–12	

and it is stable at 60°C for 10 min. In particular, the enzyme is stable at alkaline pH as high as pH 13. Moreover, the specific activity is almost eight times higher than that of subtilisin. Interestingly, the isoelectric point of the enzyme is also highly alkaline (pH 11). Attempts to improve the properties of biological detergents have been going on for nearly 20 years, and many alkaline proteases are now practically employed in a number of detergent formulations.

C. Alkaline Amylases

Many alkaline amylases have been isolated from alkaliphiles. One of them, an alkaline amylase from alkaliphilic *Bacillus* No. A-40-2 (ATCC21592), has an optimal pH of 10.0–10.5, and it also shows 50% of the activity remaining at pH 9.0 and 11.5. This enzyme hydrolyzes about 70% of starch in the manner of an α-amylase.

D. Cyclodextrin Glucanotransferase

Alkaliphilic *Bacillus* No. 38-2 (ATCC21783) produces a cyclodextrin glucanotransferase with excellent properties. Although this enzyme is not an alkaline enzyme in the strict sense, it is stable up to pH 10. The optimum pH for the activity is significantly broad, ranging from pH 4.5 to 10.0. In addition, the conversion ratio from starch to cyclodextrins reaches 75–80%, exceeding that of the conventional enzyme from *Bacillus macerans*, which achieves only 50%. These properties enabled direct crystallization of cyclodextrin from the primary enzyme reaction mixture with starch for the industrial production.

E. Alkaline Cellulase

Cellulases, which degrade cellulose derivatives, were not found until alkaline cellulases [carboxymethylcellulases (CMCases)] were found. A number of alkalophilic *Bacillus* spp. produce CMCase, which degrades CMC at pH 9–10. Alkaliphilic *Bacillus* No. N-4 produces several cellulases. Crude cellulase preparations exhibit a broad optimal pH (4.5–11.0). These complex enzymes were analyzed by cloning of the genes encoding for the constituent enzymes in *E. coli*. Another alkaliphile, *Bacillus* No. 1139, produces a real alkaline cellulase that has an optimum pH of 9 but is not active at pH 7. [*See* CELLULASES.]

Bibliography

Grant, W. D. (1987). *Microbiolog. Sci.* **4,** 251–255.

Horikoshi, K., and Akiba, T. (1982). "Alkalophilic Microorganisms." Springer-Verlag, Heidelberg, Germany.

Horikoshi, K., and Grant, W. D. (1991). "Superbugs." Springer-Verlag, Heidelberg, Germany.

Krulwich, T. A., and Guffanti, A. A. (1989). *Annu. Rev. Microbiol.* **43,** 435–463.

Kudo, T., Hino, M., Kitada, M., and Horikoshi, K. (1990). *J. Bacteriol.* **172,** 7282–7283.

Anaerobic Respiration

Valley Stewart
Cornell University

Glossary

Cytochrome Cytoplasmic membrane-bound hemoprotein, involved in electron transfer between respiratory chain components

Electron acceptor Small inorganic or organic compound that is reduced to complete an electron transport chain

Electron donor Small inorganic or organic compound that is oxidized to initiate an electron transport chain

Oxidative phosphorylation Adenosine triphosphate synthesis coupled to protonmotive force

Protonmotive force Chemical (ΔpH) and electrical ($\Delta\Psi$) gradient formed across the cytoplasmic membrane during respiration

Quinone Lipid-soluble electron carrier that mediates electron transfer between respiratory enzymes

Respiration Generation of protonmotive force by the coupled oxidation and reduction of substrates

Standard redox potential Calculated electromotive force of an electron donor/acceptor couple in an activity ratio of 1.0 at 25°C ($E^{\circ\prime}$ or $E_{m,7}$)

ANAEROBIC RESPIRATION provides the ability to conserve energy [adenosine triphosphate (ATP) synthesis] or perform work (solute transport) through production of a protonmotive force in the absence of molecular oxygen. Distinct respiratory chains, assembled in the cytoplasmic membrane, consist of a dehydrogenase, a reductase, and intermediate electron carriers such as cytochromes and quinones. Electrons derived from electron donor oxidation are transferred sequentially through the intermediate carriers to the reductase complex, where they are used to reduce the electron acceptor. Concomitant with this electron transfer is a net transfer of protons from the cytoplasm to the exterior face of the cytoplasmic membrane. The resultant protonmotive force is used to drive energy-dependent processes such as ATP synthesis and active transport. Respiration is substantially more efficient than fermentation, so a respiring culture of a given organism will form more cell mass and grow faster than a fermenting culture of the same organism. Anaerobic respiration is identical in principle to aerobic respiration, except that molecular oxygen is not used as the terminal electron acceptor.

I. Architecture of Anaerobic Respiratory Chains

A. Thermodynamic Considerations

Respiration results in the formation of a protonmotive force (proton electrochemical gradient), which consists of two components: a chemical gradient, the asymmetric distribution of protons across the cytoplasmic membrane (ΔpH; alkaline inside); and the electrical gradient, the asymmetric distribution of charge across the cytoplasmic membrane ($\Delta\Psi$; negative inside). The magnitude of the protonmotive force must be sufficient to drive ATP synthesis from adenosine diphosphate (ADP), which has a standard free energy ($\Delta G^{\circ\prime}$) of approximately 10–12 kcal/mol.

Respiration consists of two coupled half-reactions linked by an electron transport cascade. The electron donor is oxidized, and electrons derived from this oxidation are transferred sequentially through a series of electron carriers (such as iron-sulfur centers, cytochromes, and quinones) and used to reduce the electron acceptor. For example, in formate → nitrate respiration, the electron donor couple is HCO_2^-/CO_2 (formate/carbon dioxide), while the electron acceptor couple is NO_3^-/NO_2^- (nitrate/nitrite). The standard redox potentials of these couples are approximately −430 and +430 mV, respectively, giving a standard redox span of 860 mV. Since $\Delta G^{\circ\prime} = -nF\,\Delta E^{\circ\prime}$ (where n is the number of electrons transferred and F is the Faraday constant, 23.06 kcal/V·mol), these coupled half-reactions generate −39.6 kcal/mol under standard conditions. In theory, this is sufficient to drive the synthesis of at least 3 mol of ATP from ADP. Of course, living organisms operate their energy metabolism far from "standard conditions," and much of the calculated available energy is lost as heat, so these calculations provide only crude estimates of *in vivo* ATP synthesis. However, consideration of standard redox potentials is useful in estimating the relative efficiencies of different respiratory chains.

A wide variety of bacteria are capable of anaerobic respiration, including both strict anaerobes and facultative aerobes (such as *Escherichia coli*). Indeed, some organisms considered to be aerobes, such as *Pseudomonas aeruginosa,* grow well anaerobically by respiring nitrogen oxides as terminal electron acceptors. The diversity of bacteria that respire anaerobically is reflected in the diversity of compounds that serve as electron donors or acceptors for respiration; some examples are shown in Table I. The principal requirement for assembling a satisfactory respiratory chain is that the redox span between the donor couple and the acceptor couple be of sufficient magnitude to support ATP synthesis. In some cases, redox pairs may serve as electron donors or acceptors. For example, succinate (oxidized to fumarate) serves as an electron *donor* for respiration with nitrate as electron acceptor, while fumarate (reduced to succinate) is an electron *acceptor* for respiration with glycerol-3-phosphate as electron donor.

Table I Approximate Standard Redox Potentials of Electron Donor and Acceptor Couples and Respiratory Chain Components

Couple	$E^{\circ\prime}$ (mV)
N_2O/N_2	+1360
NO/N_2O	+1180
O_2/H_2O	+820
NO_3^-/NO_2^-	+430
NO_2^-/NH_4^+	+360
NO_2^-/NO	+350
Dimethylsulfoxide/dimethylsulfate	+160
Trimethylamine *N*-oxide/trimethylamine	+130
Ubiquinone/ubiquinol	+110
Fumarate/succinate	+33
Menaquinone/menaquinol	−74
Dihydroxyacetone phosphate/glycerol-3-phosphate	−190
Pyruvate/lactate	−190
S°/HS^-	−270
$NAD^+/NADH$	−320
H^+/H_2	−410
CO_2/HCO_2^-	−430
SO_4^{2-}/HSO_3^-	−520

[Values extracted from Thauer, R. K., Jungerman, K., and Decker, K. (1977). Energy conservation in chemotrophic anaerobic bacteria. *Microbiol. Rev.* **41**, 100–180.]

B. Proton-Translocating ATP Synthase

How is protonmotive force used to drive ATP synthesis? This amazing feat is performed by the proton-translocating ATP synthase, also termed ATPase. This enzyme complex consists of three integral membrane proteins, which form the F_O ("O" for oligomycin binding) portion of the complex, and five cytoplasmic proteins, which form the F_1. The F_1 and F_O are attached to each other by a short stalk; in electron micrographs, the ATP synthase complex resembles a mushroom, with the bulbous F_1 "cap" meeting the membrane "soil" at the F_O base.

The F_O portion of ATP synthase forms a proton channel, through which extracytoplasmic protons flow down the proton electrochemical gradient back into the cytoplasm. The release of energy during this electron flow is captured by the F_1 portion and used to synthesize ATP from ADP and inorganic phosphate. Estimates of the H^+/ATP ratio vary, but most place the value close to three. This capture of energy for ATP synthesis is termed oxidative phosphorylation.

During fermentation, when electron acceptors are not available for respiration to generate protonmotive force, ATP is generated by substrate-level phosphorylation. Under these conditions, the ATP synthase operates in reverse, hydrolyzing ATP to ADP to drive the extrusion of protons into the extracytoplasmic space. This also results in the generation of a protonmotive force, necessary to drive solute uptake and motility. This illustrates why respiration provides so much more energy for metabolism and growth than does fermentation.

C. Composition of Anaerobic Respiratory Chains

An essential requirement for respiration is the presence of an intact cytoplasmic membrane. Generation of protonmotive force involves establishment of proton and electrical gradients across the membrane. Thus, respiratory chain components are usually associated with the cytoplasmic membrane and are asymmetrically placed with respect to the membrane. This asymmetry is established such that electron transfer between components results in the net consumption of protons from the cytoplasmic compartment and the net release of protons into the extracytoplasmic compartment (periplasm in gram-negative bacteria). Most often, dehydrogenase components (generators of electrons) are tightly associated with the cytoplasmic membrane; reductase components (consumers of electrons) may be membrane-associated, or they may be soluble enzymes in the extracytoplasmic space. For example, the two large subunits of *E. coli* respiratory nitrate reductase are associated with the cytoplasmic face of the inner membrane, while the third subunit (cytochrome b_{556}) is an integral component that spans the inner membrane. By contrast, the *E. coli* respiratory nitrite reductase (cytochrome c_{552}) is a soluble protein located in the periplasmic space. In both cases, however, electron acceptor reduction is associated with the net translocation of protons into the periplasm.

The cytoplasmic membrane contains lipid-soluble electron carriers, quinones, that mediate electron transfer between dehydrogenase and reductase components of respiratory chains. Bacteria contain at least two types of quinones: benzoquinones, such as ubiquinone, and naphthoquinones, such as menaquinone and demethylmenaquinone. Quinone reduction to quinol is coupled to substrate oxidation by respiratory dehydrogenases, and quinol oxidation to quinone is coupled to substrate reduction by reductase components (Fig. 1). Thus, these small molecules act as electron shuttles between dehydrogenases and reductases. Many species of bacteria synthesize more than one type of quinone and menaquinone. The participation of a given quinone in a particular electron transport chain depends to a large extent on thermodynamics, as the different quinones have different standard redox potentials (Table I). For example, in *E. coli,* the glycerol-3-phosphate → fumarate respiratory chain uses menaquinone, while the glycerol-3-phosphate → nitrate respiratory chain probably uses both menaquinone and ubiquinone.

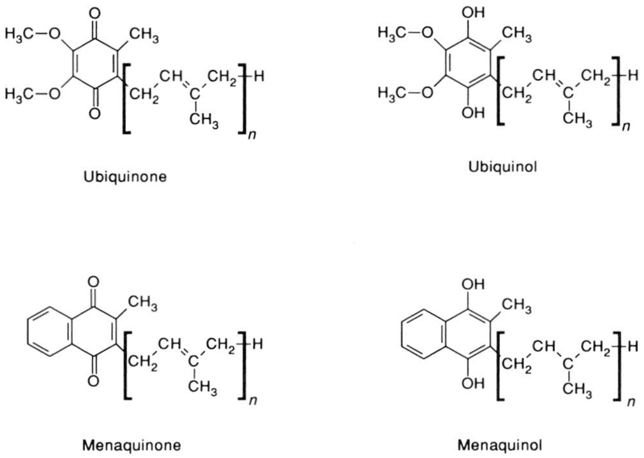

Ubiquinone

Ubiquinol

Menaquinone

Menaquinol

Figure 1 Structures of representative quinones. A typical benzoquinone is ubiquinone, and a typical naphthoquinone is menaquinone. Both molecules undergo reversible oxidation and reduction between the quinone and quinol forms during electron transport. The length of the isopentenyl side chain (n) varies from 4 to 10 in different species.

Bacterial respiratory chains are extremely flexible and generally consist of dehydrogenase and reductase modules that interchangeably interact with other components to form specific respiratory chains under specific environmental conditions. Because quinones serve as intermediate electron carriers, there is no requirement that dehydrogenases and reductases physically associate with each other. Indeed, oxidation of a variety of substrates can contribute electrons to a common quinone pool, which then transfers electrons to the terminal reductase. A given organism has the genetic capacity to synthesize a variety of different dehydrogenase and reductase components, which can be "mixed and matched" to form specific respiratory chains depending on the availability of substrates. Thus, if an organism is grown in medium containing lactate and nitrate, for example, lactate dehydrogenase and nitrate reductase will be synthesized and constitute an overall lactate → nitrate respiratory chain.

D. Mechanisms of Proton Translocation

Understanding the precise mechanism of proton translocation represents one of the principal challenges in studies of respiration. In general, two types of proton translocation mechanisms are envisioned.

The first may be termed a proton pump, in which protons from the cytoplasm are physically translocated into the extracytoplasmic space. The mechanism(s) by which this might be achieved are obscure, but it seems likely that the proton pump components form a specific pore or channel in the cytoplasmic membrane, perhaps analogous to that formed by the proton-translocating ATP synthase. An important element of this type of mechanism has to do with stoichiometry: In principle, the ratio of protons translocated to electrons consumed could exceed one. Such ratios could allow for more efficient energy conservation than scalar mechanisms (see later), where the ratio is precisely one, because more protons translocated per mol of electron acceptor reduced results in more ATP synthesized per mol of electron donor reduced.

The second mechanism for proton translocation is the so-called scalar mechanism. In this scheme, electron acceptor reduction consumes two protons from the cytoplasm, while quinol oxidation results in the release of two protons into the extracytoplasmic space. Thus, a given proton is not translocated into the extracytoplasmic space, but rather the net consumption and generation of protons results in the formation of protonmotive force. Key to this scheme is the physical separation of the sites for quinol oxidation and acceptor reduction, which must be on opposite sides of the cytoplasmic membrane. This mechanism has been established for a number of terminal reductase components, including *E. coli* respiratory nitrate reductase, which is diagrammed in Fig. 2.

II. Examples of Anaerobic Respiratory Chains

The variety of anaerobic respiratory chains, coupled with the stunning diversity of bacterial species and

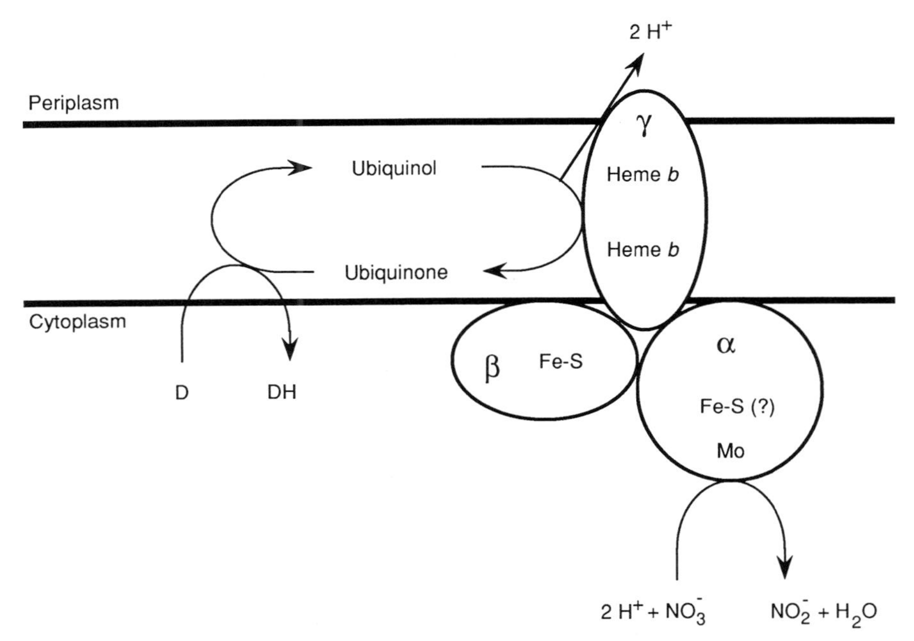

Figure 2 Organization and mechanism of *Escherichia coli* respiratory nitrate reductase. Ubiquinol is oxidized to ubiquinone by cytochrome b_{556}^{Nar} (γ-subunit) of nitrate reductase, releasing two protons into the periplasm. Electrons are passed from cytochrome b_{556}^{Nar} to iron-sulfur centers in the β-subunit, and thence to the molybdenum center in the α-subunit. Nitrate is reduced to nitrite, with the consumption of two protons from the cytoplasm. Thus, the reaction results in the net transfer of two protons from the cytoplasm to the periplasm for each pair of electrons passed to nitrate. Ubiquinone is rereduced to ubiquinol by any of a variety of electron donors such as formate or NADH. Subunit sizes are not to scale, and their interactions are hypothetical. D, oxidized electron donor; DH, reduced electron donor; Fe-S, iron-sulfur cluster. [Modified from Stewart, V. (1988). Nitrate respiration in relation to facultative metabolism in enterobacteria. *Microbiol. Rev.* **52**, 190–232.]

habitats, precludes an exhaustive listing of all the different modes of anaerobic respiration that have been characterized. The following section focuses on a small number of well-studied respiratory chains, to provide examples of some of the general considerations discussed earlier. [*See* SULFUR METABOLISM; NITROGEN CYCLE.]

A. Formate → Nitrate Respiration in *Escherichia coli*

The formate → nitrate respiratory chain in *E. coli* is one of the best-studied respiratory pathways in bacteria. This relatively simple chain consists of formate dehydrogenase-N, quinone (both ubiquinone and menaquinone), and respiratory nitrate reductase. Formate (HCO_2^-) is an anaerobic product of glucose metabolism, nitrate (NO_3^-) is widely distributed in soils and water, and the energy yield from formate → nitrate respiration is substantial (see earlier). This is a significant anaerobic respiratory chain, and it is widely distributed in diverse bacterial species.

Formate dehydrogenase-N consists of three subunits: α (113 kDa), which contains molybdenum cofactor, selenocysteine, and possibly an iron-sulfur center; β (32 kDa), which contains four iron-sulfur centers, and γ (21 kDa; cytochrome b_{556}^{Fdn}), which contains a single heme *b* ligand. Respiratory nitrate reductase has a similar three-subunit structure: α (139 kDa), which contains molybdenum cofactor and possibly an iron-sulfur center; β (58 kDa), which contains four iron-sulfur centers; and γ (26 kDa; cytochrome b_{556}^{Nar}), which contains two heme *b* ligands. Molybdenum cofactor consists of molybdenum complexed with molybdopterin guanine dinucleotide and is essential for the reactions catalyzed by a variety of anaerobic respiratory enzymes. Selenocysteine is found in a variety of redox enzymes and also plays role in catalysis. Iron-sulfur centers and cytochromes are involved in electron transfer.

The measured midpoint potential of cytochrome b_{556}^{Fdn} is approximately −110 mV, whereas those of the diheme cytochrome b_{556}^{Nar} are approximately +10 and +125 mV. Thus, very low-potential electrons derived from the oxidation of formate to carbon dioxide are transferred through increasingly higher-potential components (molybdenum-selenium center, iron-sulfur centers, cytochrome b_{556}^{Fdn}, quinone, cytochrome b_{556}^{Nar}, iron-sulfur centers, and a molybdenum center) in order to reduce nitrate to nitrite, a high-potential reaction (Table I). (It is difficult to compare the measured midpoint potentials of the cytochrome species with the "standard" redox potentials of the quinone/quinol couples as shown in Table I, but the overall trend to increasingly higher potential is clear.) It is not known how the formate → quinone portion of this respiratory chain is involved in proton translocation, but the scalar movement of protons associated with the quinol → nitrite portion is well documented (Fig. 2).

The structural genes for formate dehydrogenase-N and nitrite reductase are organized in two unlinked operons, *fdnGHI* and *narGHJI*, respectively, where genes *G, H,* and *I* encode the α-, β-, and γ-subunits of the two enzymes, respectively. (The role of the *narJ* gene product is not known, but this polypeptide does not seem to be associated with nitrate reductase.) Regulation of *fdn* and *nar* operon expression is described later.

B. Denitrification in *Pseudomonas stutzeri*

Denitrification consists of a series of anaerobic respiratory reactions involving nitrogen oxides. The overall pathway of denitrification is

$$\underset{\text{nitrate}}{NO_3^-} \rightarrow \underset{\text{nitrite}}{NO_2^-} \rightarrow \underset{\text{nitric oxide}}{NO} \rightarrow \underset{\text{nitrous oxide}}{N_2O} \rightarrow \underset{\text{dinitrogen}}{N_2} \quad (1)$$

Each reduction step is associated with proton translocation, and each is thermodynamically favorable for energy conservation (Table I). These steps are each catalyzed by a distinct enzyme complex (see later), and a given bacterial species may or may not have the capacity to perform all the steps of denitrification. By definition, denitrification results in the formation of gaseous products (nitric oxide, nitrous oxide, or dinitrogen). Thus, enterobacteria such as *E. coli* are not denitrifiers, because the nitrite generated by nitrate respiration is further reduced to ammonium.

The role of nitric oxide in denitrification has been debated for many years, with some workers holding that nitrite is directly converted to nitrous oxide by nitrite reductase, without nitric oxide as a free intermediate:

$$\underset{\text{nitrate}}{NO_3^-} \rightarrow \underset{\text{nitrite}}{NO_2^-} \rightarrow \underset{\text{nitrous oxide}}{N_2O} \rightarrow \underset{\text{dinitrogen}}{N_2} \quad (2)$$

The purification of nitric oxide reductase from *P. stutzeri* (see later) and *Paracoccus denitrificans* helped to establish the veracity of Scheme (1) in these organisms. Recent isotope tracer studies indicate that Scheme (1) operates in a majority of denitrifiers. However, these same studies suggest that Scheme (2) is the principal denitrification pathway in *Achromobacter cycloclastes* and *Pseudomonas aureofaciens*.

Pseudomonas stutzeri has emerged as a favorite organism for denitrification research, because of its vigorous denitrification pathway and its relative amenability to genetic studies. Thus, results obtained with *P. stutzeri* serve to illustrate general principles concerning the enzymology and genetics of denitrification. [*See* NITROGEN CYCLE.]

Respiratory nitrate reductase from *P. stutzeri* contains at least two subunits of 112 and 60 kDa and is tightly associated with the cytoplasmic membrane. The enzyme, as purified by heat release from membranes, is not associated with a cytochrome *b* subunit, but by analogy with other respiratory nitrate reductases it seems likely that such a cytochrome is involved in the *in vivo* function of the enzyme. The enzyme contains molybdenum cofactor and iron-sulfur clusters, and except for the relatively small size of the α-subunit, it is probably similar to *E. coli* respiratory nitrate reductase.

Respiratory nitrite reductase from *P. stutzeri* is cytochrome cd_1, a dimer of identical 62-kDa subunits; the enzyme is located in the periplasm. Biochemical studies have suggested that electron transport from quinone is mediated by periplasmic cytochrome c_{551}. However, the structural gene for nitrite reductase (*nirS*) is in an operon with at least two other genes, *nirT* and *nirB*, which encode a presumed cytochrome *c* and cytochrome c_{552}, respectively. The structural gene for cytochrome c_{551}, *nirM*, is immediately downstream of the *nirSTB* operon. All of these proteins are located in the periplasm. Thus, electron transfer to nitrite probably involves multiple cytocrome *c* species. Nitrite reductase in some other organisms is a periplasmic copper protein, rather than cytochrome cd_1.

Nitric oxide reductase from *P. stutzeri* contains at least two subunits of 38 and 17 kDa and is tightly associated with the cytoplasmic membrane. The two subunits are cytochrome *b* and cytochrome *c*, respectively.

Nitrous oxide reductase from *P. stutzeri* is a dimer of identical 71-kDa subunits; the enzyme is located in the periplasm. This enzyme requires copper for activity. The structural gene, *nosZ*, is located upstream of genes whose products are involved in copper acquisition and processing, and downstream of the cluster of genes encoding nitrite reductase and associated cytochromes. Thus, the genes for these two enzymes and associated components form a "supraoperonic cluster" typical of some *Pseudomonas* catabolic genes. To date, the location and organization of the genes for nitrate reductase and nitric oxide reductase have not been reported.

C. Glycerol-3-Phosphate → Dimethylsulfoxide Respiration in *Escherichia coli*

Glycerol and glycerol-3-phosphate are degradation products of phospholipid metabolism and are, thus, found in a variety of environments. In bacteria such as *E. coli*, glycerol cannot be fermented but, rather, must be converted to glycerol-3-phosphate and then oxidized. This oxidation requires an electron acceptor and a functioning respiratory chain. Dimethylsulfoxide, trimethylamine *N*-oxide, and other similar compounds are found in a variety of anaerobic environments and serve as relatively efficient electron acceptors for a variety of bacteria (Table I).

Escherichia coli synthesizes two glycerol-3-phosphate dehydrogenases, one of which is used for aerobic respiration, and the other for anaerobic respiration. The anaerobic enzyme contains three subunits of 62 kDa (subunit A), 43 kDa (subunit B), and 44 kDa (subunit C). Subunit A binds flavin adenine dinucleotide (FAD), whereas subunit B may bind flavin mononucleotide (FMN); these two subunits constitute the catalytic dimer. Subunit C is tightly associated with the cytoplasmic membrane and contains iron-sulfur clusters involved in mediating electron flow from the flavin subunits to quinone. The three subunits are encoded by the *glpABC* operon.

At least three *E. coli* enzymes have dimethylsulfoxide reductase and/or trimethylamine *N*-oxide reductase activity. One enzyme has been well characterized and shows good activity on a broad variety of substrates including diverse sulf-oxides and *N*-oxides. This enzyme contains three subunits of 87 kDa (subunit A), 23 kDa (subunit B), and 31 kDa (subunit C). Subunit A binds molybdenum cofactor. Subunit B contains four iron-sulfur centers, and subunit C is a membrane anchor subunit. The three subunits are encoded by the *dmsABC* operon.

The glycerol-3-phosphate → dimethylsulfoxide (trimethylamine *N*-oxide) respiratory chain is thus striking in its apparent lack of associated cytochromes. Presumably, electrons are transferred directly between the iron-sulfur centers and quinone.

The mechanism of proton translocation by this respiratory chain is unknown.

This pathway illustrates the modular nature and great flexibility of anaerobic respiratory chains. Synthesis of glycerol-3-phosphate dehydrogenase and dimethylsulfoxide reductase requires anaerobiosis; synthesis of the former enzyme is induced by glycerol, whereas that of the latter is repressed by nitrate (see later). Both of these enzyme complexes will interact with other respiratory enzymes to form other respiratory chains, depending on the availability of substrates. Thus, *E. coli* can also elaborate a glycerol-3-phosphate → fumarate respiratory chain, or an reduced nicotinamide adenine dinucleotide (NADH) → dimethylsulfoxide respiratory chain, using these same enzymes.

D. Hydrogen → Fumarate Respiration in *Wolinella succinogenes*

Hydrogen gas is a metabolic end-product found in anaerobic environments. The low standard redox potential of the H^+/H_2 couple makes hydrogen an effective electron donor for respiration, and thus hydrogen is an important electron donor for numerous species.

In the strict anaerobe *W. succinogenes*, the hydrogen → fumarate respiratory chain consists of hydrogenase, menaquinone, and fumarate reductase. Hydrogenase is tightly associated with the cytoplasmic membrane, contains two subunits of 60 kDa and 30 kDa, and requires nickel for activity. The large subunit probably contains iron-sulfur clusters and nickel. Fumarate reductase is also a membrane-associated complex and contains three subunits of 73 kDa (subunit A), 27 kDa (subunit B), and 30 kDa (subunit C). Subunit A contains covalently bound F A Dinucleotide, whereas subunit B contains iron-sulfur centers. Subunit C, an integral membrane protein, is a diheme cytochrome *b*, with heme midpoint potentials of −20 and −200 mV. The three subunits are encoded by the *frdCAB* operon.

III. Regulation of Respiratory Enzyme Synthesis in *Escherichia coli*

A. Regulation by Oxygen

Anaerobic respiratory enzymes are only synthesized during anaerobic growth, as they serve no function when oxygen is available as terminal electron acceptor. This oxygen regulation is mediated by a transcriptional activator protein, FNR (fumarate and nitrate reductase). Synthesis of all known anaerobic respiratory enzymes, including formate dehydrogenase-N, anaerobic glycerol-3-phosphate dehydrogenase, nitrate reductase, and dimethylsulfoxide reductase, is controlled by FNR. However, *fnr* mutants still grow anaerobically by fermentation, so distinct regulatory circuits are involved in oxygen regulation of fermentation enzyme synthesis.

FNR is 30 kDa and shares sequence and functional similarity with CRP (cyclic adenosine monophosphate receptor protein), required for expression of carbon catabolic operons. Four cysteyl residues of FNR are essential for activity, three near the amino-terminus and one near the middle of the polypeptide chain. Current evidence suggests that these cysteyl residues are involved in coordinating an atom of iron, whose redox status signals the presence or absence of oxygen. In this model, Fe^{2+} (ferrous iron) would be bound to FNR and promote or stabilize the active conformation. Upon shift to aerobiosis, the Fe^{2+} would be oxidized to Fe^{3+} (ferric iron), which might dissociate from FNR or perhaps destabilize the active conformation. [*See* Catabolite Repression.]

The DNA-binding site for FNR has been determined by mutational analysis of several operons, and the consensus sequence is TTGAT-N_4-ATCAA (where N is any nucleotide). This is quite similar to the consensus binding sequence for CRP (TGTGA-N_6-TCACA), with a common core of TGA-N_6-TCA. Indeed, a variety of experiments demonstrate that CRP and FNR probably activate transcription initiation by similar mechanisms. The center of the FNR-binding site is located 41–43 bp upstream of the transcription initiation site in all cases examined.

Recent work with *P. aeruginosa* has uncovered a homolog of FNR, termed ANR, which is required for expression of genes for denitrification and other anaerobic processes. Thus, structural and functional equivalents of FNR are probably present in a wide diversity of facultative anaerobes.

B. Regulation by Nitrate

A second level of anaerobic respiratory gene regulation is exerted by nitrate, the most energy-yielding of the anaerobic electron acceptors in *E. coli* (Table I). During anaerobic growth, nitrate induces the synthesis of enzymes required for nitrate respiration (formate dehydrogenase-N and nitrate reductase) and represses the synthesis of other respiratory enzymes (such as dimethylsulfoxide reductase). This dual regulation by oxygen and nitrate ensures that cells will only respire with the energetically most favorable compounds, even when faced with a mixture of terminal electron acceptors.

Nitrate regulation is mediated by a transcriptional regulatory protein, NarL, which serves to activate expression of the *fdnGHI* and *narGHJI* (formate dehydrogenase-N and nitrate reductase) operons and repress expression of the *dmsABC* (dimethylsulfoxide reductase) and other anaerobic respiratory operons. The binding site(s) for NarL and its mode of action remain to be established. To date, homologs of NarL have not been described in bacteria other than the enterobacteria, so the mechanism(s) of nitrate regulation in other groups of bacteria remains unknown.

Most known targets of NarL action are also controlled by FNR, suggesting perhaps a functional coupling between the two regulators. However, while most FNR-controlled genes are activated (a few cases of FNR repression are known), NarL-controlled genes are roughly equally divided between those which are activated and those which are repressed by nitrate. Again, some FNR-independent fermentation enzymes are only synthesized in the absence of nitrate, and this nitrate regulation is mediated by an unknown distinct mechanism.

NarL is a response regulator protein of so-called two-component regulatory systems involved in bacterial signal transduction. The second component, NarX, is a cytoplasmic membrane-bound protein that probably senses the presence of nitrate in the periplasm, whereupon it transfers a phosphate group to NarL, activating it for its roles in regulating gene expression. A hypothetical second sensor, termed NarQ, may also interact with NarL functionally *in vivo*. NarX and NarQ may be functionally redundant.

Bibliography

Collins, M. D., and Jones, D. (1981). Distribution of isoprenoid quinone structural types in bacteria and their taxonomic implications. *Microbiol. Rev.* **45,** 316–354.

Egan, S. M., and Stewart, V. (1991). *J. Bacteriol.* **173,** 4424–4432.

Harold, F. M. (1986). "The Vital Force: A Study of Bioenergetics. W. H. Freeman, New York.

Jüngst, A., Braun, C., and Zumft, W. G. (1991). *Mol. Gen. Genet.* **225,** 241–248.

Körtner, C., Lauterbach, F., Tripler, D., Unden, G., and Kröger, A. (1990). *Mol. Microbiol.* **4,** 855–860.

Lin, E. C. C., and Kuritzkes, D. R. (1987). Pathways for anaerobic electron transport. *In* "*Escherichia coli* and *Salmonella typhimurium*. Cell and Molecular Biology" (F. C. Neidhardt, J. L. Ingraham, K. B. Low, B. Magasanik, M. Schaechter, and H. E. Umbarger, eds.), pp. 210–221. American Society for Microbiology, Washington, D.C.

Maloney, P. C. (1987). Coupling to an energized membrane: Role of ion-motive gradients in the transduction of metabolic energy. *In* "*Escherichia coli* and *Salmonella typhimurium*. Cell and Molecular Biology" (F. C. Neidhardt, J. L. Ingraham, K. B. Low, B. Magasanik, M. Schaechter, and H. E. Umbarger, eds.), pp. 222–243. American Society for Microbiology, Washington, D.C.

Spiro, S., and Guest, J. R. (1990). *FEMS Microbiol. Rev.* **75,** 399–428.

Stewart, V. (1988). *Microbiol. Rev.* **52,** 190–232.

Thauer, R. K., Jungerman, K., and Decker, K. (1977). *Microbiol. Rev.* **41,** 100–180.

Ye, R. W., Toro-Suarez, I., Tiedje, J. M., and Averill, B. A. (1991). *J. Biol. Chem.* **266,** 12848–12851.

Antibiotic Resistance

Roy H. Mosher and Leo C. Vining
Dalhousie University

Glossary

Acyl Chemical substituent derived from an organic acid

Catabolite Repression Reduced expression of a gene caused by the negative effect on transcription of a readily assimilated nutrient

Enzyme Protein catalyst that acts specifically to enhance the rate of a biochemical reaction

Hydrophilic Having an affinity for water

Induction Increased expression of a gene caused by stimulation by a substance in the cellular environment

Peptidoglycan Biopolymer found in the cell walls of bacteria; it consists of relatively long linear chains of sugars (the glycan part) that are connected to one another by shorter chains of amino acids (the peptido part)

Ribosome Subcellular particle consisting of ribonucleic acids and proteins; ribosomes are the site of protein synthesis and in bacteria consist of 30S and 50S subunits

THAT THE DEVELOPMENT OF ANTIBIOTIC RESISTANCE would be a problem in medicine became apparent soon after the introduction of penicillin into clinical practice. Because of their inherent ability at low concentrations to inhibit or kill other microorganisms, antibiotics were recognized by Alexander Fleming and other pioneers to have great potential value for suppressing the pathogenic microorganisms responsible for infectious diseases in humans. The purification of penicillin from culture broths of the fungus *Penicillium notatum* and discovery that in its refined form the substance was nontoxic to animals allowed the potential of antibiotic therapy to be realized. With the success of penicillin in controlling infections caused by many gram-positive bacteria, the screening of soil and other sources of antibiotic-producing microorganisms intensified and yielded a succession of new agents (e.g., chloramphenicol, streptomycin, tetracyclines, and erythromycin), widening the range of antibiotic therapy. However, with increasing frequency, bacterial infections that should have been suppressed by one of these antibiotics were found to resist treatment. In general, an increased incidence of resistant organisms in a population exposed to an agent that affects growth and reproduction can be attributed to the survival and proliferation of a small fraction of the original population. The resistant survivors are genetic variants, different in structure, physiology, or behavior from the majority. The introduction of an antibiotic into medical practice will inevitably have such an effect. By creating a hostile environment for the susceptible majority of the pathogen population, it fosters the selective survival and multiplication of forms against which the antibiotic has little or no effect. The discovery that susceptible pathogenic bacteria had developed resistance to penicillin, although predictable, was nevertheless a cause for anxiety. It seemed possible that each promising new antibiotic would quickly lose its usefulness as resistance to it spread among disease-causing microorganisms. The danger this posed to the successful control of infections has provided an impetus for research into the mode of action of antibiotics and the mechanisms by which resistant organisms can circumvent their effects.

I. Antibiotic Activity and Resistance Mechanisms

To find a useful place in medicine, an antibiotic must not only be highly toxic to a disease-causing microorganism but must also be nontoxic, or relatively so, in the human patient. In the normal course of discovering therapeutically useful new antibiotics, large numbers of potential candidates are screened for a favorable combination of antibiotic potency and low animal toxicity. The few promising substances are then subjected to more intensive testing, which again yields only a small proportion of safe and effective drugs; however, the critical selective toxicity that establishes whether an antibiotic will be of value in treating a diseased host depends a great deal on the biochemistry of its toxic action. Useful antibiotics interfere in essential biochemical processes in the microbial pathogen that either do not exist or are sufficiently different to escape interference in the host. Research into the ways in which antibiotics achieve this selective toxicity has provided us with insight not only into the differences between microorganisms and more complex differentiated organisms, but also into the mechanisms by which susceptible microorganisms can acquire resistance to antibiotics.

Most antibiotics exert their effects by interfering with a cellular function, and the mechanisms by which microorganisms acquire resistance reflect this general mode of action. Antibiotics act on precise biochemical targets, and as these have been identified through mode of action studies, it has become clear that there are four general ways in which a susceptible organism becomes shielded from lethal attack. These are (1) a change in molecular structure at the specific location (target site) where the antibiotic acts; (2) acquisition of a biochemical activity that chemically modifies and thereby inactivates the antibiotic; (3) a change in the permeability of the cell envelope, preventing the antibiotic from reaching its target site; and (4) aquisition of an additional enzyme that can be used to bypass the inhibited reaction. Given enough time, any large population of microorganisms will evolve to deploy one or more of these resistance mechanisms, thereby neutralizing the effects of an antibiotic. This article will describe some of the more clinically important examples of resistance acquired through each of these mechanisms. It will focus on those antibiotics for which resistance mechanisms are best understood. By and large, these substances belong to the more widely used and therefore best known groups, represented by the penicillins, tetracyclines, aminoglycosides, macrolides, and chloramphenicol. We will relate the biochemical action responsible for the selective toxicity of each antibiotic to the mechanisms through which resistance is acquired. We will then discuss the spread of resistance and the possible origins and evolution of antibiotic resistance in microorganisms.

A. Target Site Modification

Some types of antibiotic that are quite dissimilar in chemical structure and biosynthetic origin nevertheless have much in common when considered from the viewpoint of their mode of action. One such group comprises the antibiotic families known as macrolides, lincosamides, and streptogramins, which are referred to collectively as the *MLS antibiotics*. The macrolides, of which erythromycin is an example, feature a large lactone ring, which typically contains 12–16 carbon atoms and carries a variety of substituents that include sugars and amino sugars. Besides erythromycin, the macrolide antibiotics that will be mentioned below are oleandomycin, tylosin, and spiramycin. The lincosamide antibiotics are structurally quite different from the macrolides and consist of a substituted five-membered nitrogen-containing ring connected through an amide bond to an unusual sulfur-containing sugar. In the lincosamide family are the potent antibacterial agents lincomycin and clindamycin. The streptogramin B antibiotics form a third family structurally unrelated to either the macrolides or the lincosamides. They are small peptides with five to seven amino acids cyclized by an ester bond. All MLS antibiotics inhibit protein synthesis in sensitive bacteria by binding to the large (50S) subunit of ribosomes.

In certain bacteria individual members of the MLS antibiotic group can elicit an inducible resistance that enables the organism to withstand the effects of all types of MLS antibiotics. This inducible phenotype, generally referred to as *MLS resistance,* was first observed in *Staphylococcus aureus,* which is moderately resistant to erythromycin but normally quite sensitive to spiramycin. After it had been exposed to a low concentration of erythromycin, *S. aureus* was very resistant to spiramycin, as well as to lincosamide and streptogramin B antibiotics. MLS resistance has proven to be quite common in bacteria and has been found in some antibiotic-producing *Streptomyces*. It is now known to be inducible by only a select group of macrolides, including erythromycin and oleandomycin; other macrolides, lincosamides, and streptogramin B are ineffective.

Clues to the biochemistry behind MLS resistance

came from evidence that the ribosomes of erythromycin-induced *S. aureus* failed to bind MLS antibiotics. In contrast, ribosomes from susceptible uninduced bacteria bound MLS antibiotics readily. The difference was related to altered properties of the 50S ribosomal subunit, which on further analysis was traced to a difference in its 23S ribosomal RNA (rRNA) component. In resistant strains, but not in susceptible ones, a specific adenine residue, A2058, within the nucleotide sequence GAAAG of the 23S rRNA was dimethylated. That this modification was responsible for MLS resistance was supported by ribosomal reconstitution experiments. Components of the 50S ribosomal subunit from sensitive bacteria were separated from the 23S rRNA fraction, purified, and used to reconstitute the subunit by combining them with 23S rRNA extracted and purified from either MLS-sensitive bacterial cells or from cells induced by erythromycin and now resistant to MLS antibiotics. When the 23S rRNA was from MLS-induced cells, the reconstituted 50S ribosomal subunits were unable to bind MLS antibiotics; in contrast MLS antibiotics readily bound to subunits reconstituted from the 23S rRNA of sensitive cells. [*See* RIBOSOMES.]

An enzyme that catalyzes the methylation of 23S rRNA is present in the cells of bacteria that have acquired MLS resistance; comparable methylating activity is absent from MLS-sensitive species. In MLS-inducible bacteria, the methylase is present only when cells are grown in media with low levels of erythromycin. A similar rRNA methylase has been extracted as well as from cells of the erythromycin-producing species, *Streptomyces erythraeus* (now *Saccharopolyspora erythraea*). The enzymes from MLS-resistant cells and from *S. erythraea* each dimethylate nucleotide A2058 of purified 23S rRNA. However, only the methylase from MLS-inducible bacteria is able to methylate the rRNA when it is present in intact 50S ribosomal subunits. It has been suggested that the properties of the methylases from the two sources reflect different physiological strategies. The MLS-inducible bacteria respond rapidly when exposed to exogenous sources of antibiotic by modifying their existing sensitive ribosomes. In contrast, *S. erythraea* probably produces some erythromycin throughout its life cycle and must constantly protect its ribosomes. This is achieved through constitutive expression of resistance to erythromycin. The methylase is always present and can modify the 23S rRNA before it has been incorporated into the 50S ribosomal subunit.

The mechanism by which the rRNA methylase is induced in MLS-resistant bacteria appears to involve translational attenuation similar in some respects to the well-known transcriptional attenuation that regulates the biosynthesis of amino acids such as tryptophan. In the absence of erythromycin, the messenger RNA (mRNA) containing the coding sequence for the methylase forms a base-paired structure just in front (upstream) of the coding region. This structure prevents translation of the coding region into protein by ribosomes. Thus, although the mRNA for the methylase in MLS-inducible bacteria is always formed, it adopts an inactive state in uninduced cells. If low levels of erythromycin are present, the antibiotic will occasionally bind to ribosomes attached to the mRNA upstream of the region encoding the methylase. The ribosomes then stall, and in doing so destabilize the base-paired structure that had prevented translation. Normal (erythromycin-free) ribosomes can then attach at the coding region and translate the message. The net result is that low levels of antibiotic allow some of the mRNA to be translated; the methylase produced then causes the ribosomes to be methylated and so prevents further inhibition by MLS antibiotics. [*See* TRANSCRIPTION ATTENUATION.]

Penicillin, discovered as a metabolite of the fungus *Penicillium notatum* by Fleming in 1927, was the first antibiotic to be used clinically on a large scale and was extremely effective against the gram-positive disease-causing staphylococci and streptococci, as well as against the gram-negative coccobacilli responsible for gonorrhea and meningitis. However, it was largely ineffective against other more common gram-negative bacterial pathogens. Penicillin is a member of the β-lactam family, named for its common chemical feature, a characteristic four-membered cyclic amide ring. In penicillin, this is fused with a sulfur-containing thiazolidine ring and has an acyl side chain attached via an amide substituent (Fig. 1A). The acyl side chain is variable in penicillin and can be altered by changing the availability of precursor substances in the nutrient medium. The variously acylated penicillins thus formed differ somewhat in the spectrum of their antimicrobial action and in their sensitivity to acidic conditions. Penicillins G and V, produced by feeding cultures phenylacetate and phenoxyacetate, respectively, have better antibiotic spectra than other natural penicillins; penicillin V is less affected by stomach acids and so can be given orally.

This method of directing β-lactam synthesis with

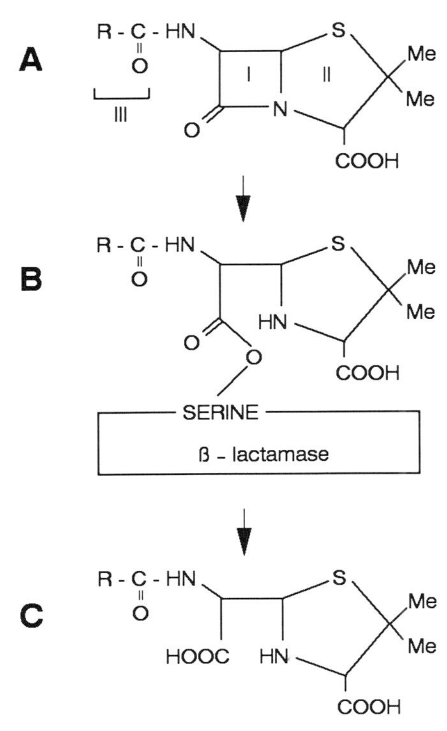

Figure 1 (A) Structure of penicillin showing β-lactam ring (I), five-membered thiazolidine ring (II), and variable acyl side chain (III). (B) Acyl-enzyme intermediate formed by reaction between serine at the active site of β-lactamase and the carbonyl group of β-lactam ring in penicillin. (C) Antibiotically inactive product (penicilloic acid) formed by action of β-lactamase on penicillin.

precursors is limited, however, by the flexibility of the fungal enzymes involved in β-lactam synthesis. Fermentative modification is now augmented by chemical and enzymatic procedures for removing the phenylacetate side chain of penicillin G and replacing it with a wide range of acyl groups. This "semi-synthesis" of valuable new penicillins such as methicillin has been extended by modifying the thiazolidine ring and has yielded new generations of β-lactams with not only broader spectra of antibacterial activity but activity against resistant organisms. The semisynthetic antibiotics and newly discovered natural β-lactams act against hitherto unaffected gram-negative bacteria and also against gram-positive forms that have become resistant to penicillin G. As a result of these steady improvements, β-lactams have remained one of the most clinically valuable and therefore most widely used types of antibiotics.

The antibacterial effect of β-lactams is largely due to their ability to bind covalently to cell wall biosynthetic enzymes on the external face of the cytoplasmic membrane. These enzymes have become known as penicillin binding proteins (PBPs). Eubacteria possess from three to seven PBPs (*Escherichia coli* possesses five), which vary in molecular weight and enzymatic activity. Most of the large PBPs are transpeptidases and some are also transglycosylases; the smaller PBPs are usually DD-carboxypeptidases. These enzymes catalyze polymerization of the peptidoglycan network that comprises an important part of the bacterial cell wall. The transglycosylase adds each newly formed wall-building unit, which consists of a disaccharide-pentapeptide, to a growing peptidoglycan chain by linking a sugar of the disaccharide-pentapeptide to the free sugar at the end of a peptidoglycan chain. The transpeptidase cross-links the peptidoglycan chains by connecting their pentapeptide side chains. The cross-linking reaction connects the thousands of peptidoglycan chains in the bacterial cell wall into an extremely strong, chemically resistant mesh.

The β-lactams specifically inhibit the ability of PBPs to bring about transpeptidation reactions. It is believed that the molecular configuration of the β-lactam mimics the shape of the terminal D-alanyl-D-alanine portion of an uncross-linked pentapeptide side chain, which is the natural substrate of the transpeptidases. The transpeptidase hydrolyzes the peptide bond between the two D-alanine residues and covalently bonds to the penultimate residue, releasing the other. The penultimate D-alanine is then transferred from the enzyme to the basic amino acid of a neighboring side chain, generating a cross-link and restoring the active site of the enzyme. When instead of the natural substrate, a β-lactam interacts with the transpeptidase, the β-lactam is hydrolyzed and the ring-opened structure covalently bonds with the transpeptidase. Unlike the penultimate D-alanine, the covalently bound β-lactam cannot be transferred to the neighboring side chain; it remains bound to the active site of the enzyme, which is thereby inactivated. By blocking transpeptidase activity, β-lactams inhibit cross-linking of the bacterial cell wall and weaken it severely. The consequences depend on the function of the PBP inhibited and its importance in maintaining cell wall integrity. For example, β-lactams that bind specifically to PBPs 1A and 1B of *E. coli* invariably cause cell lysis, whereas those that bind to either PBPs 2 or 3 usually stop cell division and cause abnormally shaped cells.

Some clinically important bacteria isolated recently exhibit relatively high levels of resistance to a variety of β-lactams and possess modified PBPs. Most of the modifications reduce the affinity of the PBPs for β-lactams without significantly altering their affinity for the natural substrate. The modifications are usually due to simple base mutations in genes coding for PBPs; as a result, specific amino acid residues within the active site of the PBP are altered. Because studies with isolated PBPs have shown that even small alterations in the active site can cause the protein to become thermolabile, Brian Spratt has proposed that two types of change occur. A mutation changing an amino acid in the active site is accompanied by another mutation stabilizing the protein so that the viability of the PBP is maintained. In these circumstances, formation of a modified PBP in a resistant bacterium is the result of sequential mutations and natural selection. The latter would be enhanced by heavy use of β-lactams, as is required, for example, in patients suffering from the frequent, severe respiratory infections associated with cystic fibrosis.

The clinically important tetracycline family of broad spectrum antibiotics includes tetracycline, chlortetracycline, and oxytetracycline, all of which contain a characteristic linearly fused four-ring system. Tetracyclines are potent inhibitors of bacterial protein synthesis. Although the mechanism by which they act is still not fully known, it is clear that they bind strongly but reversibly to a site on the 30S subunit of the bacterial ribosome and thereby interfere with the binding of incoming transfer RNA (tRNA) carrying a new amino acid to the site of polypeptide assembly.

Resistance to tetracyclines has been observed in a variety of bacterial species, but no naturally occurring strain that owes its resistance to modification of its ribosomes has been reported. Instead, recent work has demonstrated the presence in various bacteria of a resistance mechanism that protects ribosomes from the inhibitory action of tetracyclines. Although the protected ribosomes purified from such bacteria may not bind lesser amounts of tetracycline than those from sensitive strains, they are significantly less inhibited by tetracycline in protein synthesis assays carried out with cell extracts. Resistant ribosomes become sensitive to tetracyclines if they are washed in a high-salt solution, and sensitive ribosomes become resistant if co-incubated with a ribosome-free cell extract from a resistant bacterium. The genes for tetracycline resistance via ribosomal protection have been cloned from a number of bacteria, both gram-positive and gram-negative, and their nucleotide sequences have been determined. In general, they encode a hydrophilic protein with regions that are highly homologous to portions of the protein synthesis elongation factors Tu and G. This homology appears to be restricted to and centered around the GTP-binding portions of the proteins. Because the genes from diverse species of bacteria show significant and extensive nucleotide sequence similarity, it has been suggested that they are distributed by horizontal transfer and may have originated in the tetracycline-producing *Streptomyces*. Consistent with this idea is evidence that the oxytetracycline producer, *Streptomyces rimosus*, uses a ribosomal protection mechanism to avoid inhibition by its own metabolite.

B. Antibiotic Inactivation

Resistance mechanisms involving modification and inactivation of antibiotics are relatively common, one of the best known and probably the most frequent cause of resistance in clinically important bacteria being the inactivation of β-lactam antibiotics by enzymatic hydrolysis of the β-lactam ring. Such an event was observed soon after the introduction of β-lactams in 1940, when Abraham and Chain found that an extract of *E. coli* could inactivate penicillin. β-Lactamases have subsequently been discovered in almost every species of bacteria examined and are perhaps the most ubiquitous enzymes in the prokaryotic world.

The bioactive properties of even the most chemically complex β-lactam antibiotics are linked to the integrity of their four-membered β-lactam ring. Although their specificity and range of catalytic activity varies from enzyme to enzyme, all β-lactamases bring about essentially the same reaction—hydrolytic cleavage of the β-lactam ring at its amide bond. Hundreds of different β-lactamases have been discovered, but their active site structure and mechanism of bond scission divide them into only two categories. The first, and most common, is serine hydrolases; they have the amino acid serine as an essential component in their active site. The second and much less common category of β-lactamases is metalloenzymes, which generally possess a zinc atom coordinately bonded to three histidine residues within their active sites. The zinc atom is essential for enzyme activity.

The serine-based enzymes have been studied extensively, and many aspects of the mechanism by which they catalyze hydrolysis of their substrate have been elucidated. The first step is formation of an acyl-enzyme intermediate (Fig. 1B). The hydroxyl group of the active-site serine attacks the amide bond of the β-lactam and opens the ring by forming an ester linkage with the amide carbonyl group. The acyl-enzyme intermediate is then hydrolyzed to regenerate the active-site serine and a biologically inactive ring-opened product (Fig. 1C). The zinc-based enzymes have received less attention than their serine-based cousins, and the mechanism by which they catalyze β-lactam hydrolysis is not so clearly established. Unlike the serine enzymes, they do not appear to form acyl-enzyme intermediates.

The catalytic action of the serine hydrolase β-lactamases suggests an evolutionary connection with the PBPs responsible for cell wall synthesis. When they encounter a β-lactam, both form acyl-enzyme intermediates in which an ester bond links the serine residue in their active site to the ring-opened antibiotic. Unlike the PBPs, β-lactamases readily hydrolyze the serine ester intermediate and regenerate the active site for catalytic inactivation of additional β-lactam molecules. They may have arisen by duplication of PBP genes followed by mutations that allowed more efficient hydrolysis of the serine ester intermediate. An extracellular DD-carboxypeptidase of *Streptomyces* R61 may bridge the two systems in exhibiting transpeptidase activity as well as significant active-site relatedness to the chromosomally encoded β-lactamases of *E. coli*.

Chloramphenicol is a broad spectrum bacteriostatic antibiotic produced by *Streptomyces venezuelae*. It inhibits protein synthesis by binding to the 50S subunit of prokaryotic ribosomes and interfering with the formation of a peptide bond between an incoming amino acid and the existing polypeptide chain. Although the chloramphenicol was one of the first antibiotics to be brought into general use after penicillin, prolonged treatment can cause a fatal and irreversible aplastic anemia in 1 of approximately 25,000 people; therefore its clinical use is now limited to severe cases of rickettsial infections, typhus, and bacterial meningitis. The most common cause of chloramphenicol resistance in clinically important bacteria is the presence of chloramphenicol acetyltransferase (CAT). This enzyme has been discovered in both gram-positive and gram-negative bacteria and catalyzes the transfer of an acetate group from acetyl coenzyme A (acetyl CoA) to the hydroxyl group at carbon-3 of chloramphenicol (Fig. 2). The 3-acetoxychloramphenicol formed undergoes a spontaneous rearrangement in which the acetate group is transferred from carbon-3 to carbon-1. This allows CAT to reacylate carbon-3 and form 1,3-diacetoxychloramphenicol. Because monoacetoxy chloramphenicol is unable to bind to prokaryotic ribosomes and therefore lacks antibiotic activity, the second acylation has no bearing on resistance.

In contrast to the action of β-lactamase on β-lactams, CAT's action on chloramphenicol is energy expensive and the acetyl CoA it consumes is a valuable metabolic intermediate. Probably for this reason, the expression of CAT is closely regulated in most gram-positive bacteria. The mechanism of regulation is still being investigated but in some cases appears to involve translational attenuation, as in MLS resistance. In *E. coli,* CAT appears to be regulated by catabolite repression, although in other gram-negative bacteria, CAT genes are nearly always expressed constitutively. [*See* CATABOLITE REPRESSION.]

The aminoglycosides are a large group of antibiotics that typically consist of an aminocyclitol such as streptamine (in streptomycin) or deoxystreptamine (in neomycin and kanamycin) linked to one or more amino sugars. They have a broad spectrum of activity, inhibiting protein synthesis in both gram-positive and gram-negative bacteria by binding to the 30S subunit of ribosomes. Clinical applications have been limited by their ototoxicity (tendency to damage the 8th intercranial nerve, often resulting in deafness) and effect on kidney function (nephrotoxicity); however, if large doses and prolonged therapy are avoided, they are valuable agents for treating bacterial infections that resist other antibiotics. Streptomycin, an aminoglycoside introduced during the late 1940s, was the first effective drug for treating tuberculosis. The most common mechanism of resistance to an aminoglycoside is via covalent modification of the antibiotic. Clinical isolates of resistant bacteria produce a wide variety of modifying enzymes, but these generally fall into three categories: (1) phosphotransferases, which catalyze the addition of a phosphate group from ATP; (2) nucleotidyltransferases, which catalyze addition of the adenylate moiety from ATP; and (3) acetyltransferases, which catalyze addition of the acetyl group from acetyl CoA. One strategy used to develop aminoglycoside antibiotics that escape the activity of such enzymes is to remove or mask the functional group on the molecule to which the donor phosphate, nucleotide, or acetyl group is transferred. This needs

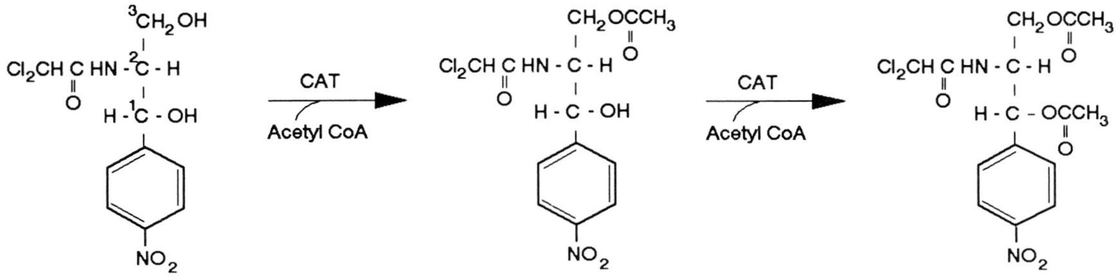

Figure 2 Inactivation of chloramphenicol through action of chloramphenicol acetyltransferase (CAT).

to be done with discretion to avoid loss of antibiotic activity.

Modification of aminoglycosides by inactivating enzymes nearly always results in the antibiotic losing affinity for its ribosomal target. However, in cultures of resistant bacteria, little extracellular antibiotic is actually modified. Most studies have shown that aminoglycoside-modifying enzymes are closely associated with the cytoplasmic membrane. The antibiotic is evidently modified only as it is taken up by the bacterial cell, and the modifying enzyme competes for the aminoglycoside with the cellular uptake mechanism. The efficiency of modifying enzymes in conferring resistance is closely related to their affinity for the antibiotic; those with high affinities successfully inactivate the antibiotic before it reaches its target, whereas those with low affinities compete less well with uptake and confer only a low level of resistance. Another aspect of this form of resistance is the mechanism by which aminoglycosides affect protein synthesis. There is evidence that the ability of a cell to synthesize protein when it is first exposed to streptomycin is only partially affected, but the synthesis and incorporation into the cellular membrane of aberrant proteins resulting from the initial effect enhances the permeability of the membrane for the aminoglycoside and accelerates uptake of the antibiotic. It would be predicted, therefore, that a modified aminoglycoside in failing to bind to ribosomes also fails to accelerate the uptake of unmodified antibiotic.

C. Cellular Exclusion

Cellular exclusion is a potentially powerful mechanism for generating antibiotic resistance. The gram-negative cell is surrounded by two membranes, the inner of which is the cytoplasmic membrane. The outer membrane is composed chiefly of lipopolysaccharide and lipoprotein; smaller quantities of proteins commonly known as porins are also present. Porins belong to a class of proteins located in the outer membrane, and as their name suggests, assemble there to form pores. These pores allow the passive transport of hydrophilic substances such as sugars and amino acids and provide a route for most β-lactams to gain entrance to the space between the inner and outer gram-negative membranes. Because the hydrophilic channels formed by porin proteins can vary in size and number, the outer membrane is an effective sieving agent. In some β-lactam-resistant bacteria the number of porin molecules seems to have been reduced; in others, the number of porin molecules is unchanged but the diameter of the hydrophilic channels has been reduced. In both of these circumstances the amount of antibiotic penetrating the outer membrane is drastically decreased.

Tetracycline passes passively through the outer membrane of gram-negative bacteria by means of porin channels, and in all tetracycline-sensitive bacteria, the antibiotic is taken through the cytoplasmic membrane by both energy-dependent and energy-independent mechanisms. One of the major mechanisms of tetracycline resistance in both gram-negative and gram-positive bacteria is thought to be mediated by an efflux system. Most bacteria possess low-affinity efflux systems, which pump out tetracycline. In many tetracycline-resistant bacteria this efflux system is bolstered by the addition of an inducible energy-dependent high-affinity system that protects the usually vulnerable ribosomes by efficiently pumping tetracycline out of the cell as fast as it enters. The membrane proteins that make up this system are usually encoded by genes located on a plasmid or transposon.

It is now apparent that antibiotic resistance in bacteria is a complex phenomenon and is often due to more than one mechanism. In fact a number of bacteria possess multiple lines of defense against individual types of antibiotic. In gram-negative bacteria, resistance to β-lactams is multifactorial, beginning with a relatively impermeable outer membrane. If the antibiotic is able to penetrate the outer membrane, it might then encounter a periplasmic β-lactamase. In the event that both of these protective mechanisms fail, the bacterium may also possess PBPs with low affinities for the β-lactam. Another example of this type of resistance network can be seen in tetracycline-resistant gram-positive bacteria. Many of these organisms possess not only an efflux system but also a ribosomal protection mechanism that prevents an antibiotic penetrating to the cytoplasm from binding to its target. As a general rule, therefore, antibiotic resistance systems in bacteria are two- or even three-tiered in their degree of complexity.

D. Target Site Bypass

Target site bypass occurs most commonly in mammalian cells during treatment with drugs that inhibit enzymes required for cell replication. Resistant cells elaborate an enzyme that catalyzes the same reaction as the inhibited enzyme but has a much lower affinity for the drug.

In prokaryotes, target site bypass as a mechanism for evading the effects of antibacterial drugs is relatively uncommon. However, it is found in strains of *Staphylococcus aureus* that have become resistant to methicillin. Exposure to methicillin inactivates the four binding proteins (PBPs) normally present, but causes the formation of a new PBP with sharply reduced affinity for β-lactams. In the absence of methicillin, the new PBP is undetectable. The gene for it has been identified on an unstable segment of DNA in the *S. aureus* chromosome, and it is believed to have been obtained through genetic exchange with a related bacterium. High-level expression of the gene during exposure to methicillin facilitates methicillin-resistant peptidoglycan synthesis, thus compensating for inactivation of the four normal methicillin-blocked PBPs.

Target site bypass can be responsible for sulphonamide resistance in bacteria. Sulphonamides competitively bind to the active site of the enzyme dihydropteroate synthetase, inhibiting folic acid synthesis and thus bacterial growth. Sulphonamide-resistant bacteria often possess two dihydropteroate synthetases, one of which has a very much lower affinity for sulphonamides. The gene for the resistant enzyme is usually located on a mobile piece of DNA, either a transposon or an R plasmid. The mechanism of bacterial resistance to trimethoprim, a drug that inhibits folic acid biosynthesis by binding competitively to dihydrofolate reductase can also be classed as target site bypass. The resistant organism acquires a second dihydrofolate reductase with low affinity for trimethoprim, and again the additional gene is located on a mobile DNA element.

II. Spread of Antibiotic Resistance

When resistant strains of bacteria first appeared, the most commonly heard explanation was that in large bacterial populations the spontaneous rate of mutation (one in every 10^5–10^7 individuals) is high enough to generate a sufficient variety of altered biochemistries to confer resistance on a few individuals. These flourish in the toxic environment and go on to create a population of antibiotic-resistant progeny. However, in Japan during the 1950s and 1960s, many outbreaks of dysentry caused by *Shigella flexneri* were found to be caused by strains possessing multiple antibiotic resistance. These bacteria exhibited high resistance to at least four mechanistically distinct antibacterial agents: tetracycline, chloramphenicol, streptomycin, and sulfonamide. The resistant populations contained an unusually low level of sensitive individuals, and strains of *E. coli* isolated from patients infected with *S. flexneri* possessed an identical antibiotic resistance phenotype.

The sudden appearance of multiple antibiotic resistance in a bacterial population was inconsistent with the spontaneous mutation hypothesis, and coupled with the low level of sensitive individuals and the isolation of both *S. flexneri* and *E. coli* strains with identical resistance phenotypes, suggested that resistance genes could spread within and between bacterial populations. Molecular analysis of resistant strains established that the genes were on a plasmid—a small circular DNA molecule separate from the chromosome and sometimes capable of transferring independently from cell to cell. Plasmids conferring resistance, called R plasmids or R factors, have since been discovered in numerous gram-negative and gram-positive bacteria. They can be grouped into families according to their size, antibiotic resistance traits, and incompatibility characteristics (i.e., their ability to coexist in the same host). [*See* PLASMIDS.]

Although R plasmids have a range of structures, in general they consist of two components, a resistance transfer factor (RTF) and the resistance determinant (r-det). The RTF of R plasmids carries genes necessary for conjugation, including those directing the formation of small appendages (pili) on the outer surface of fertile bacteria that bring two cells into contact so that plasmid transfer can take place. In many respects RTFs are similar in structure and function to the well-characterized sex (F) plasmid of *E. coli*. The r-det portion of R plasmids is a highly variable component encoding anywhere from one to eight different resistance genes. Both the RTF and the r-det components can replicate independently of each other; however, the extent to which they do so depends on the host in which they reside. Antibiotic-resistant bacteria possessing only the r-det portion of an R plasmid are, as might be expected, incapable of conjugatively transferring their resistance genes.

The variability of the r-det portion of R plasmids is a source of concern because it implies that bacteria can acquire new antibiotic resistances relatively easily. Even more disturbing is the evidence that variability is due to the location of nearly all r-det antibiotic resistance genes on mobile regions of DNA called *transposons*. Transposons are pieces of DNA capable of moving from one DNA molecule to another and for this reason are often referred to as *jumping genes*. Most of the transposable elements associated with R plasmids are "composite" transposons, which consist of a region of DNA containing one or more antibiotic resistance genes bordered by a pair of insertion sequences (ISs). Insertion sequences, also called *simple transposons*, can move from one DNA molecule to another, inserting into a new "target" site, but leaving a copy of themselves at their original position. They have been found in nearly all bacterial species examined. When two ISs come in close proximity to each other they are sometimes capable of transposing not only their own DNA but the DNA that lies between them. Composite transposons are, therefore, pieces of DNA that have become mobile through association with two ISs. [*See* TRANSPOSABLE ELEMENTS.]

The discovery of R plasmids carrying multiple resistance genes that can be transferred from one microorganism to another has influenced physicians' use of antibiotics as therapeutic agents. It has discouraged prolonged or unnecessary treatment because antibiotic therapy inevitably favors the selection of antibiotic-resistant variants and the generation of antibiotic-resistant populations. However, when grown in the absence of antibiotic under laboratory conditions, bacteria that possess R plasmids often compete poorly with plasmid-minus bacteria. Furthermore, R plasmids are generally unstable and tend to be lost from the bacterial cell in the absence of antibiotic selection. Because the infectious nature of R plasmids is partially compensated for by their inherent instability and the apparent "metabolic load" they place on their host, judicious use of antibiotics should greatly reduce the risk of a rapid spread of resistant pathogens. In terms of the total volume of antibiotics delivered into the environment, uses in agriculture are as important as those in human medicine and therefore also feature prominently in concerns about the spread of resistant strains. The incorporation of low levels of antibiotics into feedstuffs for poultry and animals not only helps to maintain good health and productivity but also increases the efficiency of food utilization. However, the practice has been viewed by many as a potential danger to the future of antibiotic therapy in human medicine. There is fear that resistant strains of bacteria pathogenic in animals might emerge as a human health hazard, but even greater concern has been expressed over the possible transfer to human pathogens of mobile resistance genes acquired by bacteria exposed to antibiotics in the animal feedstuffs. The magnitude of the danger is controversial, but it can be minimized by developing for veterinary use and animal feed supplementation, antibiotics different from and lacking cross-resistance to those used in human medicine.

A better understanding of R plasmids and their role in resistance has led to more appropriate regimens of antibiotic therapy. Elucidation of the resistance mechanisms associated with R plasmids has also encouraged the search for ways in which these mechanisms can be bypassed. Pharmaceutical companies have synthesized new penicillins resistant to many of the plasmid encoded β-lactamases and have discovered new compounds that directly interact with enzymes involved in antibiotic resistance. An example of such a compound is clavulanic acid, which binds to the active site of β-lactamases and inhibits their function. The molecular cloning and characterization of genes for antibiotic biosynthesis in streptomycetes and fungi promise a future in which genetically engineered hybrid antibiotics with chemical structures not previously known in nature may circumvent the reservoir of resistance mechanisms that have accumulated over time in bacteria.

III. Evolution of Antibiotic Resistance

Antibiotics belong to a large group of natural substances called secondary metabolites, a group that consists of plant, animal, and microbial metabolites, with no direct role in the essential biochemical processes that support growth and reproduction. Although not directly involved in the primary metabolic network, secondary metabolites are nevertheless of vital importance to organisms in their interactions with the environment and with other organisms. They include a diverse array of attractants, repellents, stimulants, and toxins that determine the relationships between organisms within ecosystems. Antibiotics represent a subset of secondary metabolites, limited by the original definition of the term to those that are produced by microorganisms and are capable of killing or inhibiting the growth of other microorganisms. Because the biochemical evolution of antibiotic production in the natural world has taken place in a situation where resistance to the effects of an antibiotic would confer a selective advantage on competitors, it can reasonably be expected that such resistance has a long evolutionary history.

Streptomycetes produce more than 70% of the clinically important antibiotics. In these bacteria the genes for antibiotic synthesis are usually clustered with resistance genes that offer protection from the autotoxic product. These resistance genes in the antibiotic producers have long been suspected of providing the genetic reservoir for antibiotic resistance genes in other bacteria. The discovery of transposable antibiotic resistance genes and subsequent realization that many of the mechanisms encoded by these genes are similar to the resistance mechanisms used by the antibiotic-producing streptomycetes strengthened this view. The streptomycetes also possess a number of mobile genetic elements including plasmids, bacteriophages, and insertion sequences. If the possibility of interspecies transfer is accepted, streptomycetes might indeed have been the source of antibiotic resistance genes in clinically important bacteria. Mobilization of the resistance genes by IS elements and their insertion into plasmids capable of infectious transfer may have introduced the genes into a succession of bacteria in which they became part of the host genome. Nascent R plasmids transferred from bacterium to bacterium under conditions where selection was exerted by the presence of an antibiotic in the environment would pick up the mobile resistance gene by transposition; in this way R plasmids carrying multiple resistance factors would be formed.

Although this scenario is largely speculative, there is some evidence in its favor. Biochemical and genetic data have shown that the enzymes and proteins encoded by R-plasmid resistance determinants are, in general, significantly similar to those found in streptomycetes. In contrast, antibiotic-resistant bacteria generated from originally sensitive strains by mutagenesis under laboratory conditions rarely possess any of the mechanisms encoded by R plasmids. That genetic transfer between streptomycetes and gram-negative enterics might be possible is suggested by a recent observation under laboratory conditions of plasmid-mediated conjugation between these two groups of bacteria. Finally, some cultures of bacteria preserved and stored before antibiotics were discovered and introduced into medicine possess R plasmids: This suggests that bacteria obtained antibiotic resistance genes by mechanisms that existed before the strong natural selection that was imposed by therapeutic use of antibiotics.

Bibliography

Bryan, L. E. (ed.) (1989). "Microbial Resistance to Drugs. Handbook of Experimental Pharmacology," Vol. 91. Springer-Verlag, New York.

Cundliffe, E. (1989). *Annu. Rev. Microbiol.* **43,** 207–233.

Davies, J. (1986). *FEMS Microbiol. Rev.* **39,** 363–371.

Franklin, T. J., Snow, G. A., Barrett-Bee, K. J., and Nolan, R. D. (1989). "Biochemistry of Antimicrobial Action," 4th ed. Chapman and Hall, New York.

Gale, E. F., Cundliffe, E., Reynolds, P. E., Richmond, M. H., and Waring, M. J. (1981). "The Molecular Basis of Antibiotic Action," 2nd ed. Wiley, Toronto.

Kirby, R. (1990). *J. Mol. Evol.* **30,** 489–492.

Lupski, J. R. (1987). *Rev. Infect. Dis.* **9,** 357–368.

Salyers, A. A., Speer, B. S., and Shoemaker, N. B. (1990). *Mol. Microbiol.* **4,** 151–156.

Waley, S. G. (1988). *Sci. Prog.* **72,** 579–597.

Antifungal Agents

Paul D. Hoeprich
University of California, Davis

Glossary

Antimicrobic Compound of natural, semisynthetic, or synthetic origin that inhibits or kills free-living, commensal, or pathogenic microorganisms while causing little or no injury to the host

Dematiaceous Fungi that are darkly pigmented, to a variable degree, in their spores, conidia, and hyphae, i.e., in any structure

Dermatophytes Fungi that cause superficial infections of keratinized structures, namely *Trichophyton* spp., *Microsporum* spp., and *Epidermophyton floccosum*

Fungicidal Lethal in antifungal activity; by testing *in vitro*, the number of colony-forming units is reduced to ≤1% of the starting inoculum under specified conditions

Fungistatic Inhibits fungal replication; by testing *in vitro*, there is insignificant increase in the number of colony-forming units under specified conditions

Fungus Life form that is eukaryotic, achlorophyllous, heterotrophic, obligate, or facultatively aerobic, capable of sexual reproduction, and composed of either yeasts (unicellular ovoids or spheres 3–8 μm in diameter) or hyphae (branching filaments 2–10 μm in diameter)

Minimal inhibitory concentration Lowest concentration of an antimicrobic that stops replication

Minimal lethal concentration Lowest concentration of an antimicrobic that is lethal

Mycosis Infectious disease caused by a fungus; classified as superficial (present on or in skin or mucous membranes, without penetrating through underlying subcutaneous or submucosal tissues), deep (involving organs deep to the portal of entry), or blood-borne (a fungemia that may portend dissemination)

Nonsystemic agent Compound that exerts its effect locally, only in the region to which it is applied, having no effect on organs, or tissues elsewhere

Systemic agent Compound that is absorbed from the gut or a parenteral locus of administration, with distribution to many organs and tissues

ANTIFUNGAL AGENTS may be defined as compounds that cause fungi to fail to thrive. Such a broad definition sweeps from compounds that kill all forms of life to agents that appear to affect only fungi adversely. While the former have public health and industrial applications, the latter are of particular value in the systemic treatment of mycoses in humans.

I. Introduction

Superficial fungal infections of the skin, mucous membranes, and dermal appendages are so common that virtually anyone who succeeds to adulthood has experienced this kind of mycosis. They are caused most often by the dermatophytes, *Malassezia furfur*, or *Candida* spp. Of these, only the *Candida* spp. have the capacity to invade and become life-threatening; however, as all superficial mycoses are annoying, a plethora of nostrums vie with a few effective remedies for market share.

Deep mycoses were once thought to be rare or peculiar to endemic loci. While sharpened diagnostic capabilities have belied rarity, the burgeoning of deep mycoses is linked directly to the now common occurrence of various states of diminished host defenses, including immunodeficiencies. Thus, the prevalence of fungal infections is largely consequent on therapeutic successes in diverse areas of medicine; many patients are preserved in a state of hypervulnerability, as those with diabetes mellitus, cancer chemotherapy, extensive surgery, and transplantation of organs. Nonfungal infectious diseases are also implicated. There is nearly universal occurrence of mycoses in patients with the acquired immunodeficiency syndrome, AIDS. Moreover, the very effectiveness of an array of antibacterial antimicrobics has cast into relief the shortcomings of available antifungal agents (e.g., the death of a neutropenic patient from general aspergillosis after cure of staphylococcal sepsis). [*See* ACQUIRED IMMUNODEFICIENCY SYNDROME (AIDS).]

The treatment of deep mycoses began with the discovery of amphotericin B in 1953. Thereafter, the pace of discovery of additional antifungal agents has, in comparison with other anti-infectives, been slow, indeed: flucytosine in 1957; imidazole derivatives in the 1980s (now obsolete as triazole derivatives are licensed or under clinical evaluation); polypeptides (first reported in 1977 and 1980, with a semisynthetic lipopeptide recently in clinical trial); and allylamines in 1978, with a perorally administrable derivative under clinical study. Whereas early discoveries were serendipitous (Table I), design of compounds to attack specific sites of fungal metabolism has marked recent efforts (Table II).

Table I Serendipitous Antifungal Antimicrobics[a]

Year	Compound	Source
1939 (1958[b])	Griseofulvin	Class exercise in microbial chemistry in England
1950	Nystatin[c]	Soil at dairy farm in Virginia
1953	Amphotericin B[c]	Rotting vegetation on banks of Orinoco River, Venezuela
1957	Flucytosine	Halogenated pyrimidines synthesized as putative anticancer agents
1958	Ambruticin[c]	Soil at unspecified origin
1961	Saramycetin[c]	Soil at Roche Park, Nutley, New Jersey
1984	Naftifine[c] Terbinafine[c]	Evolved from synthesis aimed at producing compounds active in the central nervous system

[a] Compounds either discovered in nature or made by synthesis not specifically aimed at producing anti-infective drugs.
[b] Year introduced into clinical use.
[c] Potentially fungicidal.

II. Nonsystemic Agents

Nonsystemic agents must be applied directly to the site of infection to make contact with the infecting fungus. Hence, they are employed to treat superficial mycoses.

A. Indirect-Acting Agents

Indirect-acting agents are antifungal by altering the immediate microenvironment of the host to compromise fungal growth.

Keratolytic drugs destroy or remove the substrate for certain fungal parasites (e.g., salicylic acid to treat dermatophytoses). Anhidrotic agents (aluminum and zinc salts) and astringent preparations (borate and bicarbonate) retard the growth of dermatophytes through drying and diminishing inflammation.

B. Direct-Acting Agents

Direct-acting agents may be either fungistatic or fungicidal.

1. Nonspecific, High-Level Disinfectants

Nonspecific agents include the high-level disinfectants such as elemental iodine, ethanol, and hypochlorite; all may be irritating to tissues.

2. Specific Agents

Specific agents are used primarily for topical antifungal therapy.

a. Organic Acids

Organic acids (e.g., benzoic, propionic, caprylic, undecylenic) are primarily fungistatic and are the least effective agents of this group.

b. Thiocarbamate—Tolnaftate

The thiocarbamate compound tolnaftate may be fungicidal to dermatophytes but has no effect on *Candida* spp.

Table II Designer Antifungal Antimicrobics[a]

Year	Compound	Mechanism of action
1967	Clotrimazole	Inhibition of microsomal cytochrome P-450-dependent lanosterol 14-α-demethylase system, decreasing synthesis of ergosterol with accumulation of 14-methylated intermediates, affecting surface enzyme activities, fluidity, permeability
1969	Miconazole	
1977	Ketoconazole	
1982	Fluconazole	
1984	Itraconazole	
1988	Genaconazole	
1989	Saperconazole	
1981	Amorolfine[b]	Inhibition of synthesis of ergosterol at two sites leading to accumulation of unnatural sterols; abnormal cell wall synthesis
1984	Cilofungin[b]	Inhibition of formation of β-1,3 glucan in fungal cell wall

[a] Compounds not found in nature that were synthesized specifically to attack fungi.
[b] Potentially fungicidal.

c. Polyene—Nystatin

Polyenes used topically include nystatin, amphotericin B (AmB), and natamycin. All are fungicidal against most of the causes of superficial mycoses, but they do not penetrate well into sites of infection, and they may irritate tissues. While a more complete discussion of polyenic antifungals follows in Section II, nystatin is considered here because it is the principal polyene used for topical application.

Nystatin is a degraded heptaene elaborated by a strain of *Streptomyces noursei*. It is insoluble in water at physiologic pH, is too toxic for systemic use, and is not absorbed through intact skin or mucous membranes.

Preparations of nystatin include a suspension (100,000 units/ml), tablets (500,000 units), and cream and ointment formulations (100,000 units/g). A wide variety of fungi is susceptible to nystatin. While the drug is fungicidal in the immediate vicinity of application, its efficacy is limited by poor diffusion. Nystatin is presumed to act against fungi by combining with cell membrane ergosterol.

Nystatin is applied topically to the mucous membranes of the mouth, oropharynx, gut, and female genital tract. Systemic distribution after topical application has not been detected. Perorally (PO) administered nystatin is passed in the feces. Nausea, vomiting, and diarrhea have been associated with PO administration of large doses.

d. Azole Derivatives

Azole derivatives (clotrimazole, miconazole, econazole, tioconazole) are broadly antifungal. As applied in concentrations of 1–2%, they may be fungicidal and yet are well tolerated, even by mucous membranes (see Section II).

e. Amorolfine

Amorolfine (AMF) is a morpholine derivative with activity against fungi pathogenic for humans. Its origins go back to the 1960s when dimethylmorpholine compounds were synthesized for use against fungal pathogens of crop plants. Successful application of derivatives of 2,6-dimethylmorpholine as agrifungicides led to synthesis of compounds active against fungal pathogens of humans. AMF is the first such product to come to investigational evaluation in humans.

AMF, as the hydrochloride salt, is a colorless solid that is water-soluble. Its antifungal activity is configuration-dependent; the cis-form of the 2,6-dimethylmorpholine nucleus must be coupled with the (+)-*S*-enantiomer side chain characteristic of AMF (Fig. 1).

For topical applications, creams (0.125, 0.25, and 0.50%), lacquers (2 and 5%), and tablets (50 and 100 mg) have been used.

By testing *in vitro,* AMF is active against the dermatophytes, *Candida* spp., dematiaceous fungi, *Histoplasma capsulatum, Coccidioides immitis, Blastomyces dermatitidis, Sporothrix schenckii, Pseudallescheria boydii,* and *Alternaria* spp; *Aspergillus* spp. and the zygomycetes are resistant.

AMF was effective in the treatment of experimental dermatophytosis and vaginal candidosis. In humans, limited trials showed good effect from topical treatment of dermatophytosis, vaginal candidosis, and, possibly, onychomycosis.

The primary mechanism of the antifungal action of the morpholine derivatives appears to be inhibition of $\Delta^8 \rightarrow \Delta^7$-sterol isomerase and Δ^{14} reductase (Fig. 2). While the degree of inhibition of these enzymes may vary with different species of fungi, overall the production of ergosterol is decreased, unnatural demethylsterols accumulate, and there is hyperfluidity of the cell membrane. As a secondary phenomenon, irregular thickening of fungal cell walls has been observed, perhaps reflecting deposition of chitin, a finding associated with fungicidal action. Resistance has not been observed to develop in susceptible fungi exposed to AMF—possibly because two se-

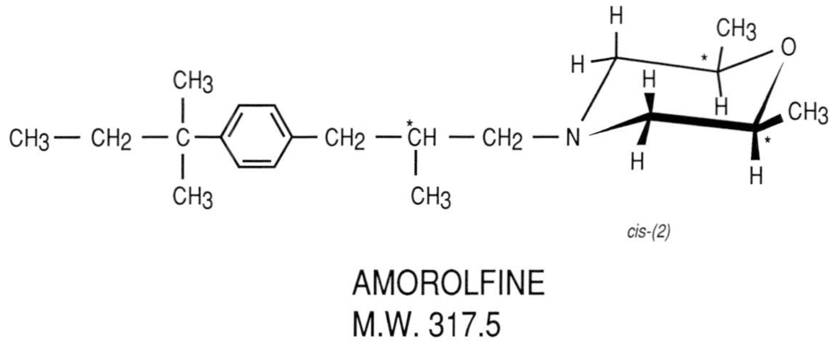

CH3—CH2—C(CH3)(CH3)—C6H4—CH2—*CH(CH3)—CH2—N

cis-(2)

AMOROLFINE
M.W. 317.5

Figure 1 Amorolfine has useful antifungal activity when the cis-form of the 2,6-dimethylmorpholine nucleus is coupled with the (+)-S-enantiomorph of the phenylpropyl side chain. *, site of optical activity.

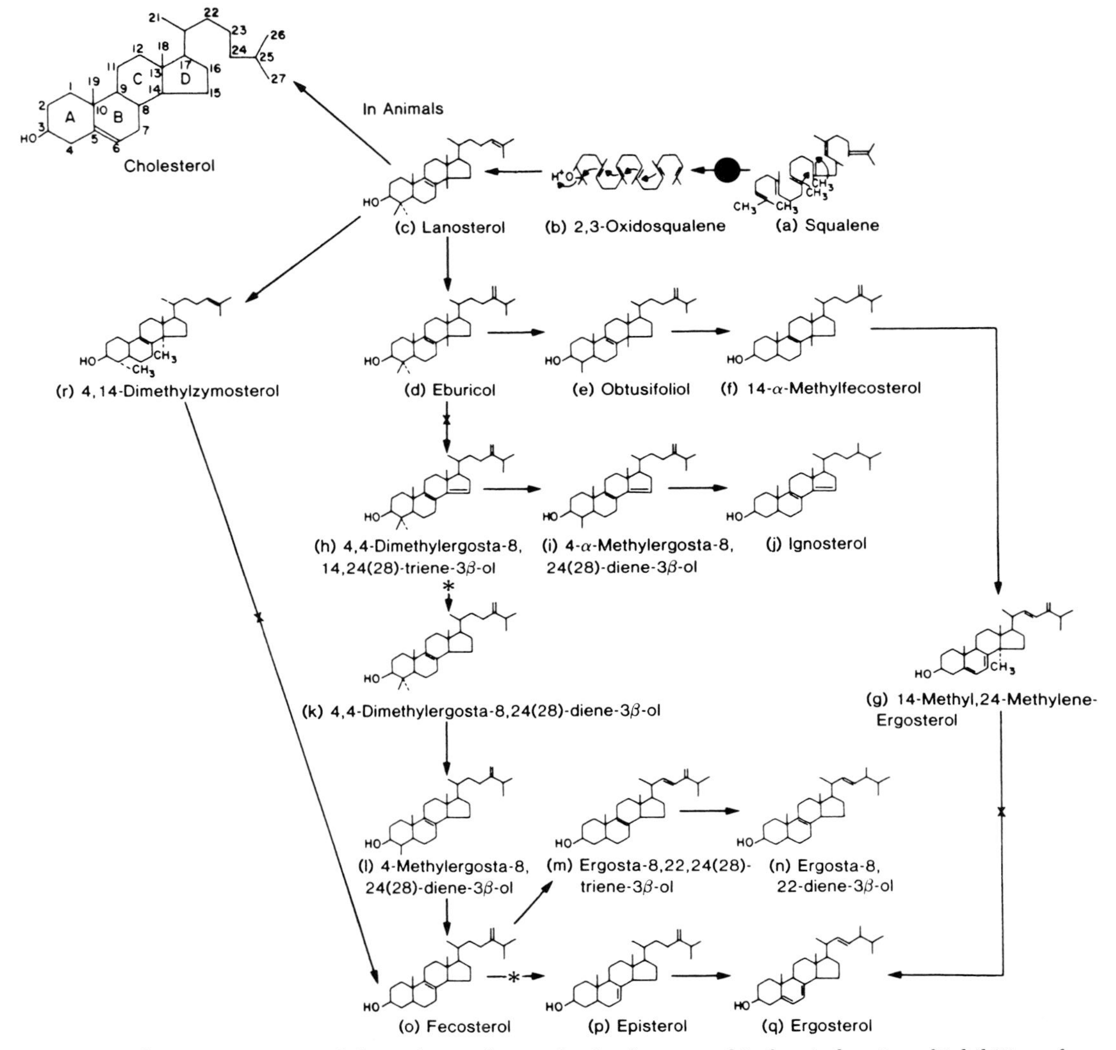

Figure 2 In this representation of the pathways for synthesis of ergosterol in fungi, the sites of inhibition of antifungal antimicrobics are marked as follows: ●, allylamines; *, amorolfine; × and *, azoles. [Modified from Rippon, J. W. (1986). A new era in antimycotic agents. *Arch. Dermatol.* **122,** 399–402. Copyright 1986, American Medical Association.]

quential sites in the same synthetic pathway are affected.

There was virtually no absorption from sites of topical application. No data exist regarding either systemic distribution or means of elimination. Topical application to the skin, mucous membranes, and nails was well tolerated.

f. Naftifine

Naftifine (Fig. 3) is the first allylamine antifungal agent to come to clinical trial. It was inadvertently synthesized in a program aimed at producing drugs active in the central nervous system; its antifungal activity was discovered in the course of general screening of novel compounds for bioactivity.

The allylamine antifungal agents are highly lipophilic, synthetic compounds derived from heterocyclic spironaphthalenones. From evaluation of many derivatives, it was shown that the allylamine double bond must have the trans orientation for good activity.

The primary target of the allylamines is squalene epoxidase (Fig. 2). In susceptible fungi, the inhibition is specific, noncompetitive, and reversible and results in failure of synthesis of ergosterol leading to fungal death through accumulation of squalene. Rat and guinea pig squalene epoxidases are 2–3 orders of magnitude less sensitive than the candidal enzymes and are inhibited competitively and reversibly. Cytochrome P-450 enzymes are unaffected by the allylamines.

Naftifine is an investigational drug in the United States; it is available only as a 1% cream for topical application. The dermatophytes are susceptible (MICs) [minimal inhibitory concentrations = 0.05–0.50 μg/ml], whereas other pathogenic fungi are relatively resistant. Clinical efficacy in the dermatophytoses may relate in part to the ease of penetration, with accumulation, of naftifine in the outer layers of the skin and mucous membranes.

g. Miscellaneous

Ciclopiroxolamine, haloprogin, and acrisorcin are among the many other agents that may be effective on topical application.

III. Systemic Agents

A. Griseofulvin

Elaborated by *Penicillium* spp., griseofulvin was introduced into medical therapy in 1958, after almost a decade of use to prevent wilting of vegetable produce. It is a water-insoluble phenol-ether compound with the structure shown in Fig. 4.

Ultramicrosized crystals of griseofulvin partially dissolved in polyethylene glycol are formulated in tablets containing 125, 165, 250, or 330 mg. Preparations of microsized crystals (250 or 500 mg tablets) are also available but are less well absorbed. Griseofulvin is fungistatic and is effective only against the dermatophytes.

As keratinized and cornified dermal structures become laden with griseofulvin, they are rendered resistant to parasitization by dermatophytes through

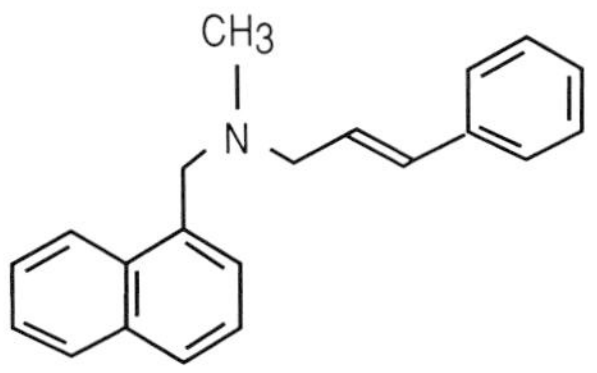

CH3 N NAFTIFINE M.W. 287.4

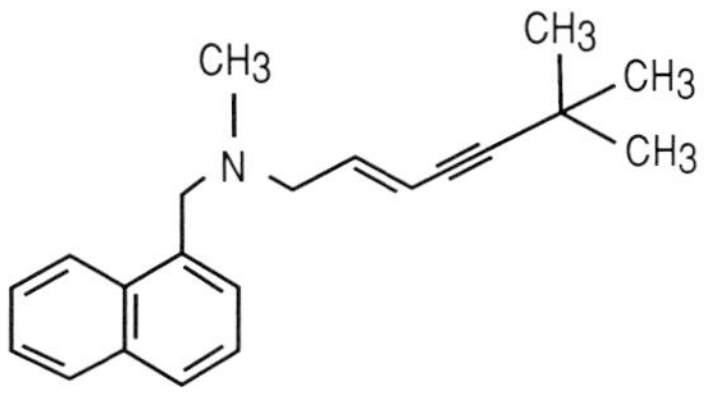

CH3 CH3 CH3 CH3 N TERBINAFINE M.W. 291.4

Figure 3 Naftifine and terbinafine are allylamine antifungal agents currently in clinical trial.

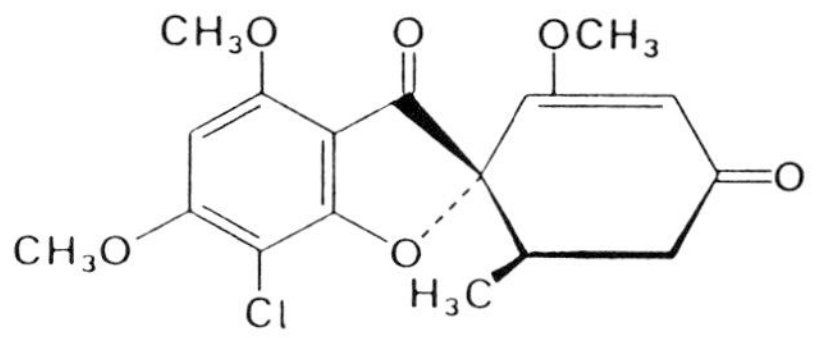

CH_3O O OCH_3 O CH_3O O Cl H_3C GRISEOFULVIN M.W. 352.5

Figure 4 Griseofulvin is a phenol-ether compound with antifungal activity confined to the dermatophytes.

unknown mechanisms. Griseofulvin has an antimitotic effect on susceptible fungi causing production of multinucleate giant cells.

Griseofulvin was the first PO administrable, systemic, antifungal agent. Absorption from the gut may be favored by taking the drug after eating, particularly after a fatty meal. In some patients, treatment with griseofulvin augments the catabolism of coumadin. Keratin-producing cells take up the drug and deposit it in keratin as it is formed. In addition, griseofulvin is delivered to the stratum corneum in eccrine sweat. The drug is eliminated as keratinized structures are sloughed. Adverse reactions are rare; most are trivial and disappear as treatment is continued.

B. Polyene

1. Amphotericin B

a. Deoxycholate Complex

While many polyenes are active against fungi, AmB is the only polyenic agent in use for the systemic therapy of mycoses. Elaborated by strains of *Streptomyces nodosus* selected for production of AmB, the compound is water-insoluble if the pH is >2 and <11. Unstable at the extremes of pH and degraded by light, heat, and oxygen, AmB is a heptaenic macrolide with the structure shown in Fig. 5. Its deep yellow color is contributed by the rigid planar, hydrophobic portion of the molecule that is linked through an oxygen bridge to a flexible, polyhydroxylated, lipophobic region to which is attached a microlactone ring subtending at C-18 the one carboxyl of the compound. Mycosamine, an amino sugar also found in other antifungal polyenes, is glycosidically linked to the hydroxyl at C-19 on the macrolactone ring. Amphotericity is contributed by the carboxyl and the primary amine; an intact primary amine is essential to antifungal activity, whereas the carboxyl is not.

b. Preparations

i. Amphotericin B–Deoxycholate Complex AmB–deoxycholate complex (AmB-DC) is the form

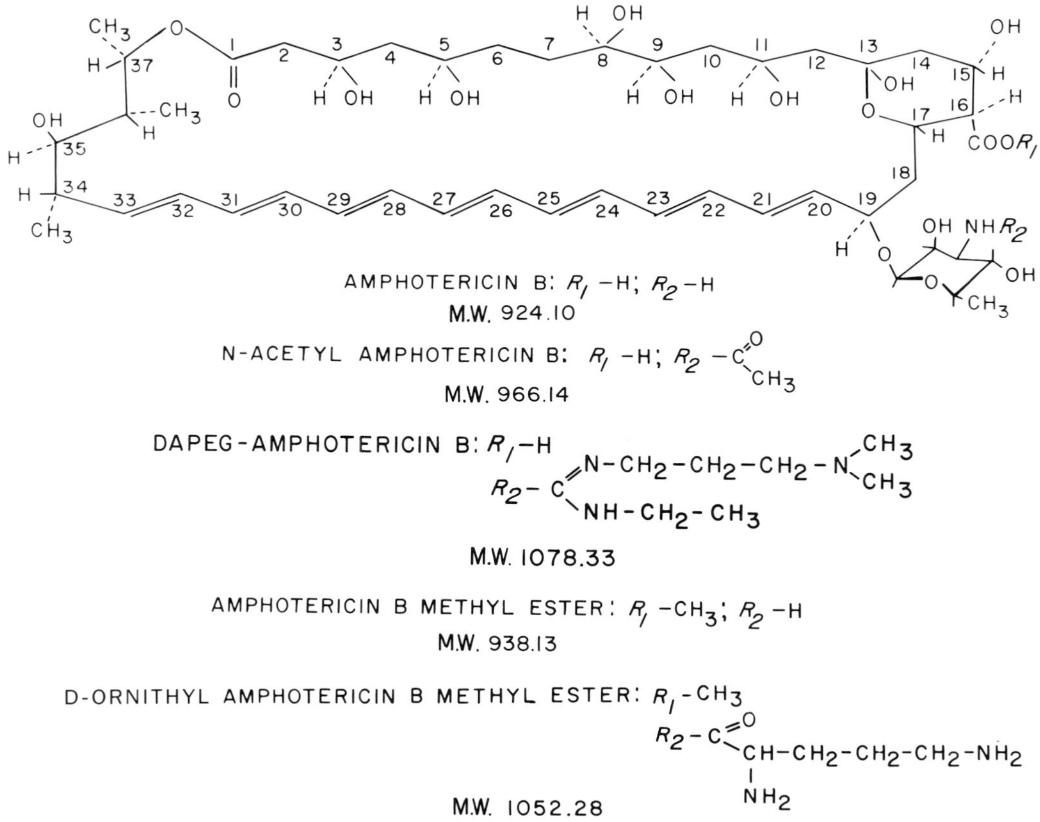

AMPHOTERICIN B: R_1 –H; R_2 –H
M.W. 924.10

N-ACETYL AMPHOTERICIN B: R_1 –H; R_2 –C(=O)–CH_3
M.W. 966.14

DAPEG-AMPHOTERICIN B: R_1–H
R_2– C(=N–CH_2–CH_2–CH_2–N(CH_3)$_2$)–NH–CH_2–CH_3
M.W. 1078.33

AMPHOTERICIN B METHYL ESTER: R_1 –CH_3; R_2 –H
M.W. 938.13

D-ORNITHYL AMPHOTERICIN B METHYL ESTER: R_1 -CH_3
R_2 - C(=O)–CH(NH_2)–CH_2–CH_2–CH_2–NH_2
M.W. 1052.28

Figure 5 Amphotericin B has been subject to chemical modification since the 1950s. The goals of decreased toxicity without loss of antifungal activity have yet to be achieved.

used for parenteral therapy in the United States. The commercial preparation is a dry powder consisting of 50 mg AmB, 11 mg sodium deoxycholate, and 25.2 mg sodium phosphates; it is supplied in glass, rubber-stoppered vials sealed under nitrogen. The addition of 10 ml sterile water for injection yields, with shaking, a clear, yellow, stable colloidal dispersion of AmB-DC. In other countries, 2-aminoglucose has been used to stabilize AmB in aqueous colloidal form.

ii. Amphotericin B–Liposomal AmB–liposomal (L-AmB) has been prepared in several formulations, all aimed at abating the toxicity of AmB-DC. In principle, AmB (without DC) is intercalated into the walls of liposomes during the preparation of these biocompatible vesicles from phospholipids (usually, phosphotidylcholines and/or phosphotidylglycerols), with or without sterols (either cholesterol or ergosterol). The preparation of liposomes is simple, but many variables affect size, extent of layering of walls, and efficiency of incorporation of AmB. Toxicity appears to increase with increase in liposomal size, and the influence of incorporation of sterols on toxicity is controversial. Pharmacokinetics and efficacy also vary with liposomal size and constituents. There may be some difficulty in batch-to-batch replication of preparations of L-AmB.

iii. Semisynthetic Derivatives

(a.) N-Acetyl N-acetyl AmB (Fig. 5) is of historical significance, because it was prepared in 1960, shortly after AmB was discovered. While the intent of achieving a water-soluble derivative was fulfilled, the compound had virtually no antifungal activity, thereby demonstrating the essentiality of the primary amine to antibiotic utility.

(b.) DAPEG N-(N^1-(3-dimethyl-aminopropyl) N^{11}-ethyl guanyl amphotericin B. (DAPEG-AmB), a guanido derivative of AmB in which the carboxyl is also preserved (Fig. 5), and water-soluble under physiologic conditions. Other differences were found in experimental animals: less toxicity, entry into the central nervous system (CNS), and prompt excretion in the urine. However, it was so much less active than the parent AmB [testing *in vitro* and *in vivo* (experimental murine mycoses)] that work with it ceased.

(c.) Methyl Ester AmB methyl ester (AME; Fig. 5) is the only chemical derivative of AmB to come to clinical trial. Water-soluble at physiologic pH, greatly reduced in nephrotoxicity, and better tolerated than AmB-DC, AME was thought to have been more neurotoxic, a matter fundamentally confused by the fact that all of the preparations used in therapy were mixtures containing 35–67% AME, with 2–8% AmB, and various amounts of six to seven multimethylated derivatives. Evaluation of pure AME in neural cell cultures and in experimental animals has shown that AME is actually about one-tenth as neurotoxic as AmB.

(d.) d-Ornithyl Methyl Ester AmB d-ornithyl methyl ester offered increased cationicity through peptide bonding of the d-isomer of the dibasic amino acid, ornithine, to the primary amine of AME (Fig. 5). The solubility and antifungal activity of AME were retained; however, in experimental animals the compound had greater nephrotoxicity (rats) and was neurotoxic (beagle dogs).

c. Antifungal Spectrum

When tested *in vitro* using AmB-DC, many of the fungi that cause systemic mycoses are inhibited by concentrations equal to or less than those attained in the blood during therapy; i.e., the MICs for clinical isolates of some *Aspergillus* spp., *Blastomyces* spp., *Candida* spp., *C. immitis, Cryptococcus neoformans, Histoplasma* spp., *S. schenckii,* and the zygomycetes are usually $\leq$1.5 μg/ml; however, susceptibility varies. Many strains of *Aspergillus* spp. are not susceptible; *Candida lusitaniae* are generally resistant, whereas the other pathogenic *Candida* spp. are usually susceptible; and *P. boydii, Fusarium* spp., *Geotrichum* spp., and many of the dematiaceous fungi are often resistant. Although the minimal lethal concentrations (MLCs) of AmB-DC are usually only modestly higher than the MICs, lethal concentrations generally are not safely attainable in patients.

The spectrum of antifungal activity of AmB derivatives is essentially unchanged from that of the parent compound, although the potency of certain derivatives may be less.

d. Mechanisms of Action

AmB and the other antifungal polyenes interact with sterols in the cell membranes of eukaryotic cells to cause destabilization. With minimal injury, as manifested by leakage of intracellular cations, the damage is reversible; with more severe injury, loss of larger molecules (e.g., nucleoproteins) may culminate in cell death. Cholesterol, the membrane-stabilizing sterol characteristic of human cells [and the cells of other mammals, helminths, protozoa, some mycoplasmas, and certain viruses (those enveloped with constituents of the outer membrane of a mammalian host cell)] may serve as a ligand for AmB. However, binding by ergosterol, the cell membrane sterol characteristic of fungi, appears to be stronger—to an extent permitting therapy. Rarely, resistance may develop during therapy, an occurrence marked by the disappearance of ergosterol from the fungal cell membrane, a reduction in the binding of AmB by the fungal cell membrane, and a decrease in the pathogenicity of the resistant fungi as tested in hypercorticoid mice.

It has been suggested that AmB may also cause oxidative damage to the fungal cell membrane and, in some way, also stimulate the immune response of the host.

e. Pharmacologic Properties

i. Routes of Administration None of the preparations of AmB, or derivatives of AmB, are absorbed well enough from the gut to permit systemic therapy. Intravenous (IV) injection is the usual route of parenteral administration [subcutaneous (SC), intramuscular, or intraperitoneal injections are too irritating for use].

Intrathecal (IT) injection may be necessary to attain therapeutic concentrations in the cerebrospinal fluid (CSF), because there is meager penetration from the blood into the CNS.

Subconjunctival and intravitreal injection into the eye are required to treat intraocular mycoses, as entry from the blood is too poor to be therapeutically useful.

Instillation of low concentrations into serous cavities, joint spaces, and the urinary bladder may also be carried out.

ii. Distribution Using ^{3}H-AmB complexed with ^{14}C-DC in rhesus monkeys, 24 hr after the IV injection of a dose of 1 mg/kg body weight, there was dissociation of the complex with the DC collecting in the liver and bile; in descending order, the AmB went to the kidneys, liver, spleen, adrenals, lungs, thyroid, heart, somatic muscle, pancreas, brain, and bone. AmB was barely detectable in the CSF, CNS, and aqueous and vitreous humors, and only meager concentrations were detectable in the urine.

When normal sheep were given AmB-DC in a dose of 1 mg/kg body weight by IV injection, the drug appeared promptly in pulmonary lymph and disappeared at an approximately exponential rate from both pulmonary lymph and venous blood. Apparently, neither the colloidal state of the injected AmB-DC nor binding by plasma proteins interfered with the passage of AmB from the blood into the interstitium of the lungs and, thence, into the pulmonary lymph.

After IV injection of L-AmB into nonhuman mammals, AmB was found primarily in the major organs of the reticuloendothelial system—liver, spleen, kidneys, and lungs—much as occurs with AmB-DC. However, there was greater penetration of the CNS, possibly as a result of uptake and carriage in phagocytic cells.

The distribution of AME in nonhuman primates using doubly radiolabeled drug (^{3}H in the macro- and microlactones and mycosamine, ^{14}C in the methyl ester) was much the same as for AmB except that relatively higher concentrations were found in the lungs and kidneys.

f. Elimination

About 5% of AmB-DC appears in the urine as antifungally active drug during the 24 hr postdose. In rhesus monkeys, urinary excretion was about the same as in humans; the major route of excretion was in the bile.

Injection of L-AmB into experimental animals was followed by markedly slower elimination than is characteristic with AmB-DC (particularly with preparations rich in small, positively charged vesicles high in lipid content).

About half of a dose of AME is excreted in the urine, with the remainder eliminated in the bile.

g. Adverse Reactions

Adverse reactions to any preparation of AmB are presumed to result mainly through interaction with cholesterol in the membranes of cells, causing perturbations of regulatory functions. However, in the case of AmB-DC, the DC itself may derange the integrity of cell membranes through a detergent effect that is probably of greatest import following subarachnoid injection.

i. Intravenously Administered Amphotericin B-Deoxycholate AmB-DC Reversible adverse reactions consist of chills, fever, headache, nausea, and vomiting. Hypotension is uncommon, usually asymptomatic, and not often engendered after the first or second dose. Anorexia and malaise accompany the other adverse reactions, becoming increasingly prominent with prolonged therapy.

Anemia develops in at least 75% of patients, sometimes with thrombocytopenia. It results primarily from direct suppression of erythropoiesis (and platlet formation) but is due in part to renal failure. Hemolysis from direct interaction between erythrocytes and AmB is unlikely to be contributory, because much higher concentrations than are attained in therapy are necessary by testing *in vitro*.

Nephropathy is virtually always produced but varies in severity from patient to patient. Generally, if renal function was normal at the outset of treatment and the total dose was ≤2 g (about 30 mg/kg body weight in an adult), <15% of patients will have detectable, not crippling but permanent, renal damage; if the total dose was ≥5 g (about 75 mg/kg body weight in an adult), approximately 80% of patients will have permanent, severe damage. Destruction is maximal in the tubules (hypokalemia, hyposthenuria, and diminished capacity to excrete acid), but the glomeruli are also involved (decreased creatinine clearance, with or without azotemia). Necrosis and calcification of the renal tubules is the histologic hallmark of kidneys rendered nonfunctional by treatment with AmB-DC. Renal hypoperfusion is contributory to nephrotoxicity. Within minutes after beginning IV injection of AmB-DC, the renal blood flow is reduced and the production of urine falls—effects that occur despite maintenance of the systemic blood pressure and recur with every dose. Such hypoperfusion appears to have particular impact in the relatively poorly vascularized renal medulla. Here, oxygen-requiring cation transport enzymes are pushed to maximal activity to conserve cations, especially Na^+, because the direct toxic effect of AmB on renal tubular cell membranes is expressed principally by loss of sodium. Anoxic necrosis of renal tubular epithelium may result.

The neurotoxic potential of AmB-DC is well documented. IV injection has been associated with hyperthermia, hypotension, confusion, incoherence, delerium, depression, obtundation, psychotic behavior, tremors, convulsions, blurring of vision, loss of hearing, flaccid quadriparesis with degeneration of the myelin in the brachial plexus, akinetic mutism, and diffuse cerebral leukoencephalopathy. Intrathecal injection may cause radiculitis, arachnoiditis, pareses (mono- or para-), hypesthesias, paresthesias, urinary retention, fever, impairment of vision, loss of hearing, and delerium. Injection of AmB-DC into serous cavities, joint spaces, or the urinary bladder is usually without adverse reaction.

ii. Amphotericin B–Liposomal L-AmB injected IV causes less severe immediate and short-term adverse reactions than AmB-DC; however, long-term evaluation of toxicity has yet to be carried out.

iii. Amphotericin B-Methyl Ester IV and IT administration of AME results in less severe immediate adverse reactions, and in far less severe nephrotoxiciticy, as compared with AmB-DC. The neurotoxic potential of pure AME has yet to be studied in humans. In neural cell cultures and in rats, pure AME is about one-tenth as neurotoxic as AmB-DC; contaminating pure AME with AmB-DC results in neurotoxicity.

2. Hamycin

Elaborated by a strain of *Streptomyces pimprina,* hamycin is a heptaenic polyene that is insoluble in water at physiologic pH. There has been limited investigational study of this compound since its discovery in 1961. Hamycin has been categorized as an "orphan" drug.

Individual investigators have made their own preparations for investigational use. A proprietary product, labeled JAI-hamycin, has been reported to be water-soluble under physiological conditions. The antifungal spectrum of hamycin is virtually the same as that of AmB. Hamycin is presumed to act on susceptible fungi through combination with ergosterol in the cell membrane.

Hamycin, in various preparations, has been applied topically, given perorally, and injected. Absorption from the gut is limited. No data exist regarding extraenteric distribution. After PO administration, it is presumed that hamycin is eliminated in the feces. Nausea, vomiting, and diarrhea may occur following PO administration.

C. Pyrimidine—Flucytosine

Flucytosine [5-fluorocytosine (5FC)] was one of several mock pyrimidines synthesized as candidate anticancer agents. It was found to have antifungal activity when a favorable effect was noted in experimental murine candidosis and cryptococcosis. A stable white powder that is moderately water-soluble (about 15 g/liter), 5FC exists in tautomeric form at physiologic pH (Fig. 6).

Tablets containing either 0.25 or 0.50 g are commercially available. An investigational solution for IV administration contains 1.0% 5FC in 0.09% NaCl solution.

Initially, the results of suceptibility testing *in vitro*

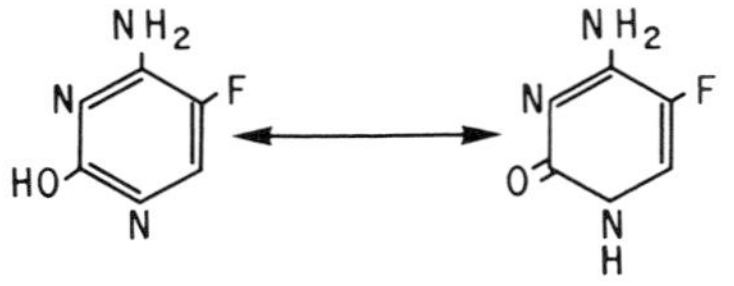

5-FLUOROCYTOSINE
M.W. 129.1

Figure 6 Flucytosine (5-fluorocytosine) exists in tautomeric form under physiologic conditions.

indicated lack of activity, in contrast to effectiveness *in vivo* in candidosis and cryptococcosis in experimental animals and in humans. The paradox was resolved when it was shown that conventional, complex, and undefined culture media contained pyrimidines antagonistic to 5FC. Native resistance to 5FC (growth in $\geq$25 μg/ml) may vary from one geographic region to another and may be increasing in frequency—about 10–15% of isolates of *Candida albicans* and about 3–5% of isolates of *Cryptococcus neoformans*. With *C. albicans*, there is a distinct relationship to serogroups in that about 15% of Group A and about 95% of Group B strains are natively resistant, whereas with *C. neoformans*, no serogroup relationship to resistance has been found. Variation in susceptibility may be even greater among non-*albicans Candida* spp. Fungistasis is the usual effect; however, most isolates of *Torulopsis glabrata* are killed at clinically relevant concentrations, with native resistance in about 5% of isolates. Other pathogenic fungi are generally resistant to 5FC.

No evidence indicates that 5FC is itself cytotoxic. However, some bacteria (*Salmonella typhimurium* and presumably other genera of the class Enterobacteriaceae) and susceptible fungi internalize 5FC by active transport and then through the action of cytosine deaminase, convert it into 5-fluorouracil (5FU). 5FU is potentially lethal as fluorouridine monophosphate, an inhibitor of thymidylate synthase; in addition, the formation of either fluorouridine triphosphate or fluorocytidine triphosphate may lead to the production of abnormal RNA that either blocks the synthesis of proteins or results in the synthesis of faulty proteins. The safety of treatment with 5FC reflects the lack of cytosine deaminase in humans. However, the microbial population of the colons of humans almost surely abounds in microorganisms that are capable of deaminating 5FC; no formal studies of this question have been reported (fortunately, 5FU is virtually not absorbed from the normal colon).

Native resistance to 5FC has not been studied systematically either in fungi or bacteria. However, it is thought that microorganisms may attain resistance by (1) not taking up 5FC, (2) lacking cytosine deaminase, (3) having an altered capacity to phosphorylate uridine monophosphate, or (4) overproducing pyrimidines. Development of resistance during monotherapy with 5FC in a patient with infective endocarditis caused by *Candida parapsilosis* was shown to have resulted from mutational deletion of cytosine deaminase.

5FC is usually administered PO. IV injection has been used in a few patients, apparently with success.

As a small molecule, 5FC is present in all body water, encountering no barriers. The differences in concentrations, as referred to contemporaneous blood, probably reflect a temporal lag in equilibration: peritoneal fluid, 100%; CSF, 75%; synovial fluid, 60%; aqueous humor, 20%; and bronchial secretions, 76%.

5FC and its metabolites are virtually entirely eliminated in the urine. About 90% of a dose given PO appears as an antifungally active drug in the urine, attaining concentrations 10–100 times those in the serum. Of probable relevance to toxicity in humans was the finding that about 1% of a PO dose of ^{14}C-labeled 5FC was present in the urine in the form of α-fluoro-β-ureido-propionic acid, a metabolite of 5FU most likely derived from deamination of 5FC by enteric microorganisms, followed by absorption and catabolism.

Adverse reactions to 5FC are not only uncommon but also are usually of no great potential for morbidity—anorexia, nausea, vomiting, diarrhea, abdominal pain in about 6%, elevations of transaminases, alkaline phosphatase (and quite rarely, hyperbilirubinemia, with or without hepatomegaly) in about 5%. Potentially more serious is depression of the bone marrow with leukopenia, thrombocytopenia (and rarely, agranulocytosis, anemia, or pancytopenia) in about 5%. These values are approximations because the patients often have either underlying diseases that have already compromised the function of the gut, liver, and bone marrow or have been treated with drugs that injure these organs; moreover, such patients are often under treatment with nephrotoxic antimicrobics, frequently including AmB, that impair the excretion of 5FC and metabolites of it. Disturbed enteric motility, especially slowing, permits a shift in the colonic microorganisms mouthward, greatly increasing the probability of occurrence of microbial intraenteric deamination of 5FC to yield 5FU. Concomitant erosion or other injury to the enteric mucosa may facilitate absorption of 5FU; the net effect is increased toxicity.

D. Azole Derivatives

All of the antifungal azole derivatives are the products of chemical syntheses. Less toxic than AmB, the antifungal azoles are broadly active. Of the many compounds that have been synthesized, six have been selected for discussion (Fig. 7). Both of the imidazoles, miconazole and ketoconazole, are licensed in the United States; of the triazoles, fluconazole is licensed and itraconazole and saperconazole are investigational, whereas genaconazole (SCH 42427) is no longer under study.

The antifungal azoles act through inhibition of the microsomal cytochrome P-450-dependent lanosterol 14-alphademethylase system (Fig. 2). To an extent that is dose-dependent, the biosynthesis of ergosterol is decreased and 14-methylated intermediary sterols accumulate. Both the liquidity and the permeability of the fungal cell wall–membrane complex

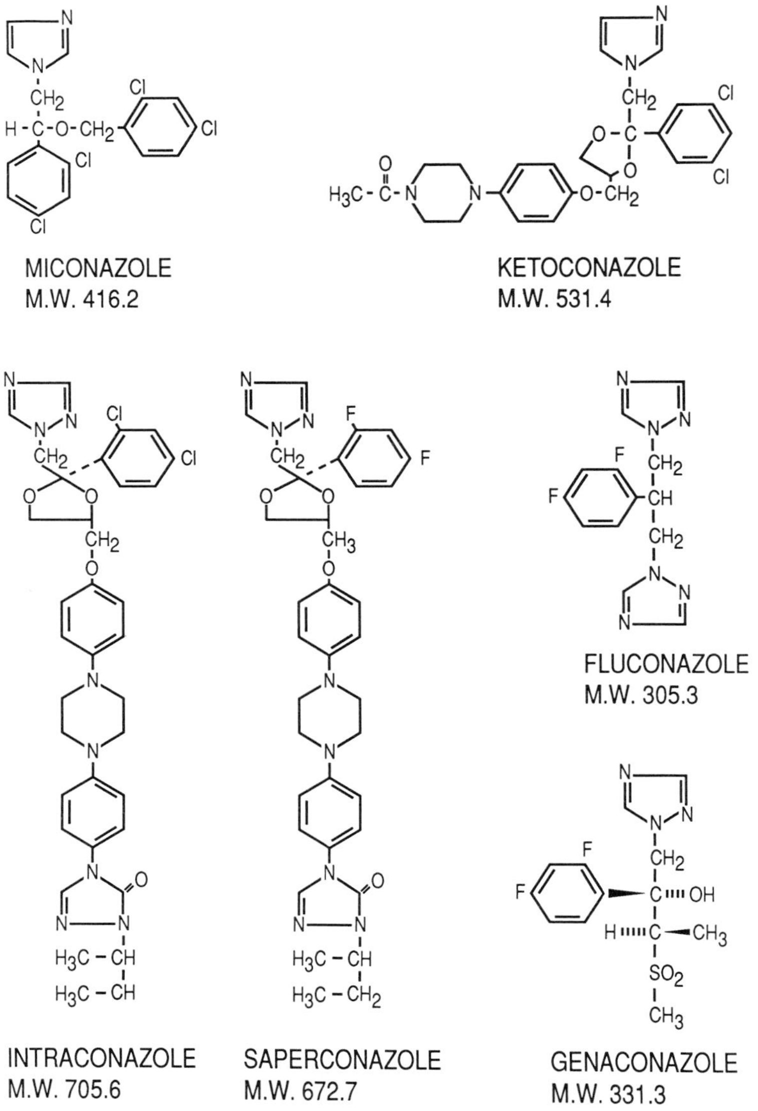

MICONAZOLE
M.W. 416.2

KETOCONAZOLE
M.W. 531.4

FLUCONAZOLE
M.W. 305.3

INTRACONAZOLE
M.W. 705.6

SAPERCONAZOLE
M.W. 672.7

GENACONAZOLE
M.W. 331.3

Figure 7 Of the many antifungal azole derivatives, the two licensed imidazoles, miconazole and ketoconazole, may be superceded by triazole derivatives. Fluconazole and itraconazole have been most widely used of the triazoles; although more potent and possessed of more favorable pharmacokinetics, genaconazole and saperconazole are burdened by carcinogenic potential expressed in rodents given large doses for long periods.

suffer, the activities of surface enzymes (chitin synthetase, lipid metabolism, oxidative enzymes) are disarrayed, and there is retention of metabolites (e.g., glucose). Disordered chitin synthetase activity is of particular interest because foci of growth are affected most dramatically—sites of abcission of buds in yeast forms, and hyphal tips and septa in mycelial forms. The fungi survive, but the damage they sustain may heighten their susceptibility to destruction by phagocytes. Because depletion of ergosterol in fungi exposed to the antifungal azoles should lead to a decrease in cell membrane sites for the interaction of polyenic antifungal agents, it is reasonable to expect that the azoles might antagonize AmB. This is indeed the case, *in vivo* as well as *in vitro*.

Unfortunately, none of the antifungal azoles is lethal in action against any of the pathogenic fungi at concentrations relevant to therapy. This is a distinct limitation because most systemic fungal infections are consequent on diminution of host defenses; hence, optimal therapy often requires fungicidal drug action.

Evaluation of the antifungal activity of azole compounds by testing *in vitro* has been difficult. Not only are there the usual variables of inocula, culture media, temperature, and duration of incubation, but also the drugs themselves are influenced differently by these variables. The preference for testing *in vivo* in experimental animals provides other problems such as species, inoculum, dosage, route of administration, length of therapy, period of post-therapy observation, and the impact of investigator-induced immunosuppression.

In the course of nearly two decades, there has been evolution from the first imidazoles with clinically useful antifungal activity to the triazoles currently under investigation. At least three goals were pursued: (1) the design of drugs with high potency, i.e., greater affinity for fungal cytochrome P-450 enzymes than for human cytochrome P-450 enzymes; (2) PO administrable agents, also adaptable for IV injection, that are slowly eliminated, permitting once-a-day treatment; and (3) freedom from significant adverse effects. Remarkable success has been achieved as is testified by comparing the first imidazole to come to clinical trial, clotrimazole (a drug with formidable gastrointestinal toxicity and induced hepatic catabolism) with fluconazole, the first licensed triazole (a relatively benign drug). However, it must be kept in mind that the cytochrome P-450 enzymes are a family of enzymes; they are as vital to mammals, including humans, as to fungi. Chemical manipulations of azoles to yield compounds ever more potent as inhibitors of the fungal P-450 enzymes may at the same time provide drugs that exert subtle but crucial effects on mammalian P-450-mediated functions. Immediate or short-term adverse effects may not be detectable or may be dismissed as inconsequential. However, the conjunction of efficient absorption from the gut, slow elimination from the body, and long-term administration—often necessary in antifungal therapy—may set the stage for serious, even mortal toxicity, as illustrated by the carcinogenicity of genaconazole and the tumorogenicity of saperconazole on chronic, high-dose administration to rats.

1. Miconazole

Miconazole (MON) is one of a series of β-substituted 1-phenethylimidazoles synthesized in the late 1960s (Fig. 7; Table III). It is a white powder that is insoluble in water but is soluble in a variety of organic solvents.

MON is active against most isolates of *Candida* spp., *T. glabrata, C. neoformans, C. immitis,* and about one-third of *Aspergillus* spp. In addition, the dermatophytes, *S. schenckii, B. dermatitidis, H. capsulatum,* and *P. boydii* are susceptible. MON is fungistatic.

Each milliliter of the commercial preparation for IV injection consists of 10 mg of a colloidal suspension of MON, stabilized by 0.115 ml of Cremaphor EL—polyethoxylated castor oil (a mixture of ricinoleic acid polyglycol ester, glycerol polyglycol ethers, and polyglycols), 1.82 mg of methylparaben, and 0.18 mg of propylparaben).

For the treatment of systemic mycoses, MON must be injected IV. It is no longer injected IT to treat fungal meningitis.

There is insignificant entry of MON into the urine, CSF, or sputum, although it penetrates into pus and synovial fluid.

Only about 12% of the dose of MON can be detected in antifungally active form in the feces. Most of the drug is degraded in the liver (without inducing augmented catabolic capacity as therapy is continued) to yield metabolites that appear in the urine.

The Cremaphor EL component of the formulation may cause the frequent thrombophlebitis, rouleauxing of erythrocytes, hyperlipidemia, and the uncommon anaphylaxis. Pruritus, anemia, thrombocytosis, hyponatremia, and nausea are toxicities probably attributable to the miconazole itself.

Table III Some Properties of Selected Antifungal Azole Derivatives

	Miconazole	Ketoconazole	Fluconazole	Genaconazole	Itraconazole	Saperconazole
Molecular weight	531.4	533.0	305.3	331.3	705.5	673.0
$T_{1/2}$ in blood (hr)	≤24	≥8	≥24	≥60	≥15	≥20
Antisteroidogenic	?	4+	0	0	0	?
H^+ for absorption	n.a.	3+	0	0	3+	?
Enter cerebrospinal fluid/central nervous system/eye	0	±	4+	4+	±	?
Enter urine	0	0	4+	4+	0	?
Hepatic catabolism	0	0	0	?	0	?
% Protein binding	≥80	≥90	≥20	?	≥90	?

n.a., not applicable; ?, data not available.

2. Ketoconazole

Ketoconazole (KET) is a white powder that is insoluble in water at physiological pH but does dissolve in 0.1 *N* HCl. The structural formula is given in Fig. 7, and several properties are listed in Table III. The antifungal spectrum of KET is virtually identical with that of MON.

KET is available only as 200-mg tablets for PO administration. The drug cannot be absorbed unless it is dissolved in an acid medium; hence, patients with achlorhydria and those rendered functionally hypochlorhydric or achlorhydric by treatment with histamine type 2 (H-2) blockers or antacids, may not respond to PO therapy. In such patients, each dose may be dissolved in 0.1 *N* HCl and taken through a glass tube or plastic straw; H-2 blockers should be withheld for 8–12 hr prior to the administration of a dose, and antacid therapy should also be withheld.

Peak concentrations in the blood are attained 2–4 hr after ingestion of a dose and range from 2 to 20 μg/ml after doses of 200–1200 mg. KET does not attain effective concentrations in the urine, CSF (except with meningitis when the concentration may approach 15% of the contemporaneous blood), CNS, or eye.

Following noninducible hepatic catabolism, the inactive metabolites of KET are mostly excreted in the bile, and the remainder is passed in the urine.

The severity of nausea and vomiting increases in direct proportion to increases in dose. Also directly related to dosage is the occurrence of endocrinopathies manifested in males as gynecomastia, loss of libido, and oligospermia; in females, irregular menses and amenorrhea may occur. Hepatitis is uncommon and is usually reversible, although fatalities have been reported.

3. Fluconazole

Fluconazole (FLU) is a water-soluble triazole (Fig. 7; Table III). Tablets containing 50, 100, or 200 mg of FLU are commercially available for PO administration. For IV administration, isosmotic solutions containing 2 mg/ml are available. Both PO and IV administration are practical.

Absorption of FLU from the gut is independent of gastric acidity and is virtually quantititative. Unlike other antifungal azoles, FLU does not undergo hepatic catabolism, and antifungally active drug is present in both the urine and the feces. There is excellent entry into the CSF, even in the absence of meningeal inflammation, with concentrations >60% of contemporaneous concentrations in the blood.

In excess of 60% of FLU is excreted in the urine as unchanged, active drug; most of the remainder appears in the feces.

FLU may cause nausea in 4%, vomiting in 2%, abdominal pain in 2%, diarrhea in 1%, headache in 2%, and skin rash in 2% of patients. Asymptomatic rises in hepatocellular enzymes may be detected in 1% of patients. Thus far, severe hepatotoxicity and exfoliative dermatitis have been rare. No interference with steroidogenesis has been detected.

Drug–drug interactions result in increased prothrombin time with coumadin, increased concentrations of phenytoin, increased concentrations of cyclosporine, hypoglycemia with sulfonylurea agents, and increased concentrations of FLU with rifampin.

4. Genaconazole

Genaconazole (GEN; SCH 42427) is the SS,+ enantiomorph of the racemic SCH 39304; it was selected for study because the RR,− enantiomer (SCH 42426) had virtually no antifungal activity. Somewhat analogous in structure and properties to FLU (Fig. 7; Table III), the initial work was carried out with the racemic mixture. The preparations for investigational use included formulations for topical, PO, and IV use.

GEN displayed broad antifungal activity against *Aspergillus* spp., *B. dermatitidis, Candida* spp., *C. immitis, Cryptococcus* spp., *Fonseca* spp., *Fusarium* spp., *H. capsulatum, Paracoccidioides brasiliensis,* and *Trichophyton* spp. and the other dermatophytes. In comparison with other triazoles, GEN appeared to combine superior antifungal activity with excellent pharmacologic properties. However, investigation with GEN was halted when chronic, high-dose administration to rodents was associated with the development of hepatocellular carcinomas.

5. Itraconazole

Itraconazole (ITR) is a lipophilic triazole that is insoluble in water (Fig 7; Table III). It is an investigational drug.

Although the reproducibility of *in vitro* tests of susceptibility to ITR is not satisfactory, some *Aspergillus* spp. and many of the dematiaceous fungi, as well as the range of pathogenic fungi listed for MON, are apparently susceptibile.

Capsules containing 50 and 100 mg of ITR have been used. ITR has been given to patients solely PO. The absorption of ITR is favored by ingestion with meals; an acid environment is necessary for absorption.

Entry of ITR into the CSF, tracheobronchial secretions, and sputum is too meager to have therapeutic effect. Similarly, $<1\%$ of ITR is excreted in antifungally active form in the urine. In rats, ITR undergoes noninducible catabolism by the liver, yielding inactive metabolites that appear primarily in the feces and, to a lesser extent, in the urine.

From preliminary experience, ITR is less likely to cause nausea and vomiting than KET. Suppression of steroidogenesis in the testes or adrenal cortex may occur with high dosage.

6. Saperconazole

Saperconazole (SAP) is a water-insoluble, lipophilic investigational triazole that is identical in structure to ITR except for replacement of chlorine with fluorine (Fig. 7), a change that should enhance the electronegativity of the compound.

Capsules containing 100 mg have been employed for investigational use in volunteers. Preparations for topical application and for IV injection have been evaluated in experimental mycoses in nonhumans.

The utility of testing the susceptibility of fungi to SAP *in vitro* has yet to be determined. However, from preliminary results, SAP appears to surpass the activity of earlier azole derivatives, including ITR, against *Aspergillus* spp. particularly, as well as against *Phaehyphomyces* spp. and many of the dematiaceous fungi.

Absorption of SAP after PO administration is enhanced if the drug is taken after eating. In volunteers, peak concentrations of 1 μg/ml were attained 1–2 hr after a 200-mg dose. Nausea, vomiting, dizziness, vertigo, and vulvar itching have been reported in patients treated with SAP.

While few data are available, SAP apparently shares many properties with ITR (Table III). Moreover, the affinity of SAP for fungal P-450 cytochrome enzymes appears to be several orders of magnitude greater than for analogous mammalian enzymes. However, chronic administration of high doses to rats has been associated with the appearance of ovarian granulosa-theca cell tumors.

E. Peptide

1. Cilofungin

Cilofungin (CIL) is a semisynthetic, amphiphilic lipopeptide (Fig. 8) derived from echinocandin B. It is a white powder that is stable if stored in the absence of water.

For IV administration, CIL, at a final concentration of 25 mg/ml, is dissolved in a vehicle consisting of 26% polyethylene glycol (v/v) in distilled water containing per 100 ml the following: 1.53 g NaH_2PO_4, 1.58 g Na_2HPO_4, and 1.8 g NaCl; the pH is set at 6.80.

Candida spp. are susceptible to CIL. In side-by-side testing, CIL was as active as AmB against *C. albicans* and *Candida tropicalis;* however, *C. parapsilosis* and *Cryptococcus* spp. were resistant. Some isolates of *Aspergillus* spp. may be susceptible, and CIL was effective in the treatment of experimental pneumocystosis in rats.

Actively growing *Candida* spp. are lysed, apparently as a consequence of inhibition of synthesis of the $\beta_{1,3}$ glucan component of the cell wall by CIL. The action is fungicidal.

CIL is not absorbed from the gut. It has been injected IV in tests in experimental mycoses and in humans. After IV injection, CIL does not enter the CSF, aqueous humor, or urine in antifungally active

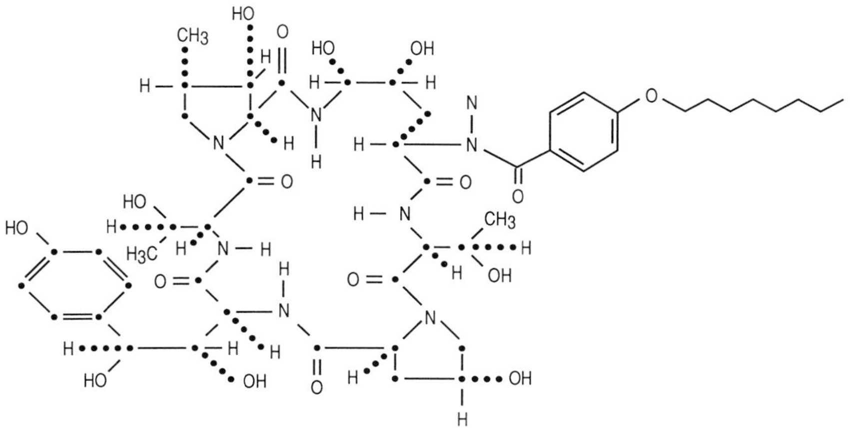

CILOFUNGIN
M.W. 1030.2

Figure 8 Cilofungin is a semisynthetic polypeptide that has limited antifungal spectrum but has fungicidal potential.

form. No data have been published on the elimination of CIL.

At 100 mg/kg body weight per dose, the highest dose tested in dogs, CIL caused immediate adverse reactions suggestive of histamine release. In addition, there were elevations in the serum concentrations of hepatocyte-origin enzymes, hepatomegaly, and centrilobular fatty vacuolation. Thrombophlebitis was present at the sites of IV injection. In humans, severe metabolic acidosis was provoked in some patients by the IV administration of CIL, an adverse effect attributed to the polyethylene glycol necessary to solubilize the drug. Clinical evaluation has been halted.

2. Saramycetin

Saramycetin is a cyclic thiazolyl peptide with a molecular weight of 1452. It is produced by *Streptomyces saraceticus*. At present, saramycetin is categorized as an "orphan" drug.

Saramycetin was prepared as a solution in 0.9% NaCl for injection. Antifungal activity was not apparent by conventional testing *in vitro*.

In experimental mycoses, saramycetin was effective against *H. capsulatum, B. dermatitidis, C. immitis,* and *S. schenckii*. It was not effective in murine candidosis.

When used to treat a small number of patients, saramycetin was effective in blastomycosis, histoplasmosis, sporotrichosis, aspergillosis, and zygomycosis.

Susceptible fungi undergo changes of their cell walls that culminate in lysis on exposure to saramycetin.

Saramycetin has been administered by SC injection. No data regarding distribution are available. The route of elimination of saramycetin is not known. The primary adverse reaction was hepatic dysfunction. In addition, inflammation at the sites of SC injection, eosinophilia, and fever were noted.

F. Allylamine

Terbinafine (TER) is unique in the presence of an acetylene function conjugated to the allylamine portion of the molecule (Fig. 3; general properties are discussed in Section II.B). TER is an investigational drug in the United States.

For PO administration, TER was prepared as 250-mg tablets. By testing *in vitro,* the dermatophytes are exquisitely susceptible to TER. Among other fungi, *Aspergillus fumigatus, Aspergillus flavus, Aspergillus niger, Candida* spp., and *S. schenckii* appear to be susceptible. Only PO administration of TER has been reported. In volunteers, PO doses of 500 mg yielded peak concentrations in the serum of 2 μg/ml; the half-life was 11.3 hr. No data regarding distribution in the body or means of elimination have been published. Adverse reactions have not been reported.

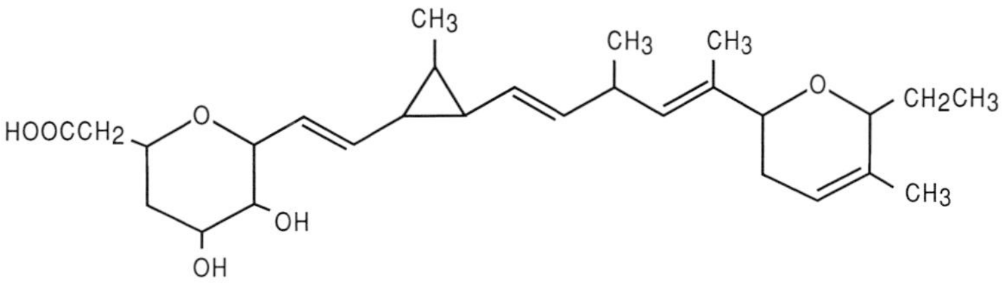

CH3 CH3 CH3
HOOCCH2 O O CH2CH3
OH CH3
OH

AMBRUTICIN
M.W. 474.3

Figure 9 Ambruticin, a cyclopropyl-polyene-pyran acid, is currently classified as an "orphan" drug.

G. Miscellaneous

Ambruticin (BRT), a cyclopropyl-polyene-pyran acid (Fig. 9) is the major component among several antimicrobics produced by *Polyangium cellulosum* subsp. *fulvum*. The data regarding BRT were derived from observations *in vitro* or *in vivo* using nonhuman animals. At present, BRT is categorized as an "orphan" drug.

BRT has been prepared for topical and PO administration. The MICs of BRT against *C. immitis, B. dermatitidis, S. schenckii,* dematiaceous fungi, the dermatophytes, and zygomycetes were lower or the same as those of AmB. With *H. capsulatum, Aspergillus* spp., and *P. boydii,* the MICs were higher than those of AmB. Generally, the yeastlike fungi were resistant to BRT.

BRT was effective in the treatment of experimental mycoses caused by fungi susceptible to it by testing *in vitro*. Cure of mice infected with *C. immitis* by treatment with BRT given PO was most notable.

Mechanisms of action that have been suggested include interference with the synthesis of RNA, inhibition of the uptake of amino acids, and blockage of carbohydrate metabolism. No evidence was found for derangement of the permeability of fungal cell membranes. Susceptible fungi are killed by BRT; development of resistance was not detected either *in vitro* or *in vivo*.

The pharmacologic properties of BRT were studied only in nonhuman animals using preparations for topical and PO administration. Absorption from the gut was rapid, with general distribution except to the CNS and CSF. The principal route of elimination was in the bile.

BRT was well tolerated by mice in doses as high as 150 mg/kg/day, PO, for 30 days. The acute LD_{50} in mice was >1000 mg/kg.

Bibliography

Bodey, G. P. (guest ed.) (1989). *Eur. J. Microbiol. Infect. Dis.* **8,** 323–375, 402–490.

Drutz, D. J. (guest ed.) (1988). *Infect. Dis. Clin. N. Am.* **2,** 779–954; **3,** 1–133.

Fromtling, R. A. (ed.) (1987). "International Telesymposium on Recent Trends in the Discovery, Development and Evaluation of Antifungal Agents. J. R. Prous Science Publishers, Barcelona, Spain.

Hoeprich, P. D. (1989). Chemotherapy for systemic mycoses. *In* "Progress in Drug Research" (E. Jucker, ed.), pp. 317–351. Birkhauser Verlag, Basel.

Kucers, A., Bennett, N. McK., and Kemp, R. J. (eds.) (1987). "The Use of Antibiotics." William Heinemann Medical Books, London.

St. Georgiev, V. (ed.) (1988). *Ann. N.Y. Acad. Sci.* **544,** 1–613.

Antigenic Variation

Carolyn M. Black and Alison C. Mawle
Centers for Disease Control

Glossary

Antigenic drift Antigenic variability caused by accumulation of point mutations in nucleic acid that result in changes in the encoded protein

Antigenic polymorphism Antigenic variability determined by different alleles of the same gene in different strains

Antigenic shift Antigenic variability caused by genetic recombination between different segments of nucleic acid

Immunodominant epitope Predominant unit seen by the immune system on a given protein

Neutralizing antibody Antibody that, when bound to a microorganism, renders it incapable of causing an infection *in vitro*

Phase variation Turning expression of a protein off and on; can only occur for proteins nonessential for survival

ANTIGENIC VARIATION is the rapid change in surface-expressed structures that occurs in many diverse microorganisms as a means of adaptation to environmental conditions and for the purpose of selective advantage. These adaptations may be a response to changes in environmental conditions that affect nutrition, to colonization of surfaces, or to the assault of the immune system of a vertebrate host. Adjustments or replacements in the repertoire of surface-expressed antigens can occur as a result of strategic regulation mechanisms or at random. Random changes are classically termed antigenic drift, especially in the influenza field, and also have been referred to in the literature as antigenic variability. This process consists of gradual and cumulative mutations in the genes that result in altered amino acid composition and structure of surface proteins; this process can take place more quickly in some organisms than in others. An example of this mechanism exists in viruses that harbor error-prone or "sloppy" RNA polymerases that make nucleotide base substitutions at high frequency.

Nonrandom control over changes in surface components can occur in several different forms. Shedding of antigens to evade the immune response (*Leishmania*) is an elaborate example of one form of change that has been referred to as antigenic modulation. Acquisition of host molecules on surfaces (*Schistosoma*) and mimicry of host antigens (*Streptococcus*) are further instances of highly evolved tactical systems. Antigenic variation in the strict sense, however, is defined by the ability of a microbe to alter its surface structures using a specific mechanism that involves multiple nonallelic genes and, usually, some type of recombination or gene rearrangement. Historically, the term "antigenic shift" has been analogous to that of "antigenic variation" to indicate strategic control over change through a process of DNA recombination between genes; more recently, the intended definition of the two terms has diverged. Antigenic shift is defined currently as a change occurring through a process of recombination between DNA of separate strains of bacteria during a mixed infection, whereas antigenic variation is the result of recombination of DNA within a single strain. Since this distinction has not yet been made for many disease agents, this discussion will include both processes.

Since we have emphasized only those organisms that have been well studied, this article is by no means an exhaustive account of the subject. We refer the reader to the reviews listed in the bibliography for further detail on individual topics and more complete reference lists.

I. Rules of Host–Parasite Interactions

The most efficient parasites exist in a delicate state of balance with the host that must prevent excessive injury to the host and still insure growth, or at least persistence, of the parasite. This compromise is maintained by the parasite through some means of evasion or counteraction of the host defense mechanisms that constitute the immune system. Many diverse and elegant systems have evolved in microbial pathogens to accomplish this task, including establishment of an intracellular state, masking by mimicry or acquisition of host molecules, and altering target molecules to simply outpace the host response. This discussion will focus on the latter system of antigenic variation.

The purpose of antigenic variation in this sense is to provide a means of avoiding recognition and, subsequently, destruction by antibodies made by the host in response to initial stimuli. Antibody responses are mounted days to weeks after the initial stimulus. Early antibodies are of lower affinity for antigen than late antibodies; thus, the multiplying parasite with a far quicker generation time than that of high-affinity antibodies has ample opportunity to activate mechanisms for switching antigen targets and escaping recognition. However, some microbes exhibit heterogeneity of surface targets at the time of infection and before specific antibodies are formed. In many cases, whether antibody responses affect the frequency of variation of microbial antigens or merely select for existing or spontaneously occurring variants is not known.

Evasion of host antibodies is thus not the only operational function of antigenic variation. Diversity of surface molecules also serves to maximize opportunities for attaching to host cells as a first step in colonization or to induce endocytosis in the case of intracellular parasites, or may simply expand the repertoire of nutritional substrata that can be used by the parasite. The remainder of this discussion will examine some of the most well-studied and nonrandom mechanisms of antigenic variation employed by parasites, bacteria, viruses, and fungi.

II. Antigenic Variation in Parasites

Parasitic infections are a huge problem in both humans and animals, primarily in the developing world. An estimated 500–1000 million people suffer from a parasitic disease; the cost is enormous in both human and economic terms. The life cycle of a parasite occurs in at least two different hosts; the cycle of infection and transmission between them involves several distinct parasitic forms.

Parasites survive by existing in uneasy equilibrium with their host. In order to survive, they must avoid attack by the host's immune system and they must avoid killing their host at least long enough to move into their carrier species. A successful parasite does not kill its host, but continues to replicate within it and be transmitted to the next vector. To achieve this success, a balance is struck between the parasite and the immune system of the host; the immune response to the parasite must be sufficiently effective to prevent the host's being overwhelmed, but sufficiently ineffective to allow the parasite to survive. Many different strategies have been adopted by different organisms to achieve this end, including evading the immune system by becoming intracellular and, thus, avoiding any neutralizing antibodies, shedding of antigen from the cell surface to block antibody, and variation of antigenic determinants on the surface of the organism, thus rendering it insensitive to any immune defense already mounted. The success of this strategy depends on the fact that the time taken to mount an effective immune response is many times longer than the replication time of the parasite. In addition, this strategy makes vaccine development very difficult; to date, no effective vaccine exists against any major parasitic disease of humans.

A. *Trypanosoma* Species

1. Clinical Significance

Trypanosomes are the causative agents of sleeping sickness in both Africa and South America. Although the organisms on each continent both belong to the genus *Trypanosoma,* the species use different strategies to evade the immune response. The South American trypanosome, *T. cruzi,* survives as a result of its ability to invade cells and evade the immune response by replicating in an immunologically privileged site; this organism will not be discussed further. *Trypanosoma brucei,* however, demonstrates true antigenic variation.

The hosts of the African trypanosome include humans and game and domestic animals. *Trypanosoma congolense* and *T. vivax* are major animal pathogens, whereas members of the *T. brucei* group cause sleeping sickness in humans. *Trypanosoma b. rhodesiense* causes acute sleeping sickness, generally in East Africa, and *T. b. gambiense* is responsible for a chronic form of the disease, primarily in West Africa. Human infections result in a relapsing parasitemia. Annually, 25,000 new infections are reported; the drug treatment available is only partially effective and has deleterious side effects. *Trypanosoma b. brucei* does not infect humans, although it has a wide host range and can infect virtually all domestic animals. The mechanism of antigenic variation has been studied most extensively in the *T. brucei* group.

2. Life Cycle

Trypanosoma brucei is a protozoan and has a relatively simple life cycle in tsetse flies and mammals (Fig. 1). The parasite is ingested by the fly in a nondividing stumpy form that circulates in the blood of the host animal. In the fly gut, this form multiplies and develops into the trypomastigote stage, which then migrates to the salivary glands. Here the organism develops into the epimastigote, which is attached to the gland as a monolayer; on further division, the free-swimming metacyclic form arises. This stage is infectious for mammals and is injected into the bloodstream when the fly bites. This form then divides to become the long slender form that can invade the intercellular tissue fluids and, in turn, transform to the stumpy form that reinfects the fly. The mammalian form has a surface glycoprotein coat that undergoes antigenic variation and is the primary defense of the organism against the host immune response.

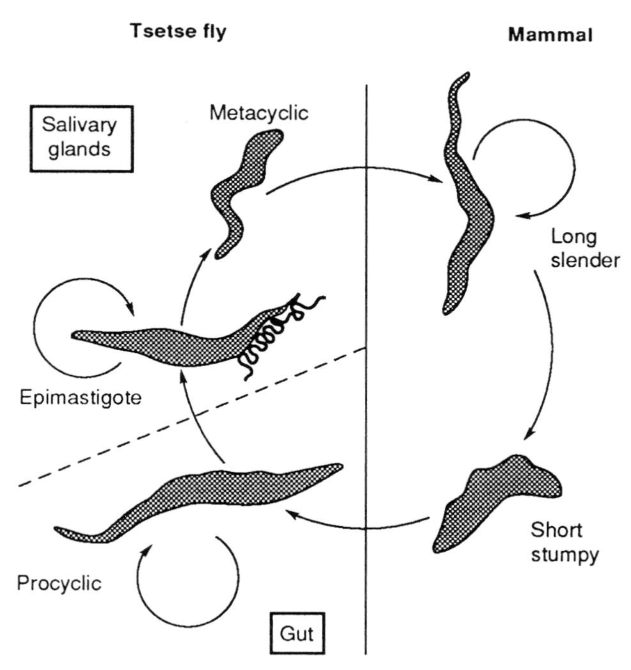

Figure 1 The life cycle of *T. brucei*. The main role of each stage is indicated. The long slender procyclic and epimastigote stages multiply by division and develop to the next stages, whereas the transient short stumpy and metacyclic stages are preadapted for transmission between hosts and subsequent development. [Reproduced from Barry, J. D. (1989). *In* "New Strategies in Parasitology" (K. P. W. J. McAdam, ed.), pp. 101–120. Churchill Livingstone, Edinburgh.]

3. Molecular Basis of Antigenic Variation

The entire surface coat of the bloodstream form of *T. brucei* is made of a single type of molecule, the variant surface glycoprotein (VSG). There are $>10^7$ VSG molecules in the coat; the importance of the VSG lies in the fact that the organism is able to vary the protein expressed via a switching mechanism. Only one VSG is expressed at a time by each trypanosome, except during switching, when double expression of the old and new forms can be detected. At any given time, the vast majority of the parasites express the same VSG. This homogeneity is achieved in two ways. First, the rate of switching is low and, second, the switch occurs in a certain order that avoids population heterogeneity.

Approximately 10^3 genes in the genome of *T. brucei* code for VSGs, most of which can be expressed. The genes are located predominantly in tandem arrays within chromosomes, and are known as basic copy genes. In this form the genes are never expressed; for expression to occur, these genes must be copied into specific expression sites that are located at telomeres. This process occurs by gene conversion, and destroys the VSG gene already located there (Fig. 2).

A second set of basic copy genes is already located at telomeric sites. The existence of approximately 100 minichromosomes in the trypanosome genome insures a large number of telomeres. Trypanosome chromosomes can be divided into four size classes; only the two largest contain structural genes. The two smallest classes contain only VSGs and standard telomere repeats. Since not all telomeric sites are active expression sites, telomeric VSGs can be activated by two different mechanisms. Like nontelomeric basic copy genes, they can be duplicated into an active site, destroying the existing VSG; however, this appears to be uncommon. Alternatively, they can be expressed by the simultaneous activation of a silent telomeric site and the inactivation of the currently expressing site.

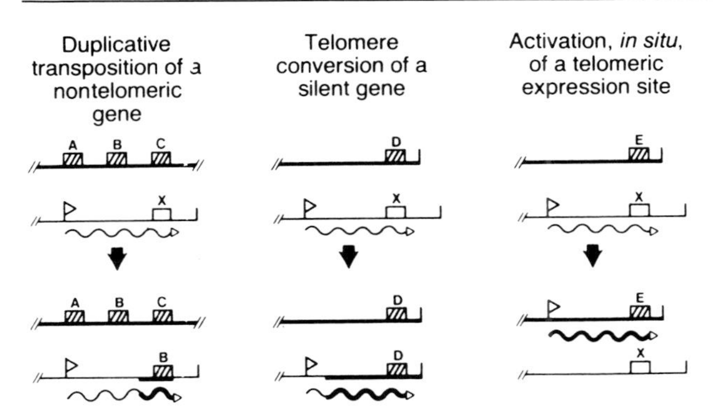

Figure 2 The various pathways for activating a variant-specific surface glycoprotein gene in *T. brucei*. Boxes A, B, and C are nontelomeric genes; X is a telomeric gene. Duplicative transposition of B displaces X and brings B under control of the expression-site promoter (flag). The vertical bar marks the end of a chromosome. [Reproduced from Borst, P. (1991). Molecular genetics of antigenic variation. *In* "Immunoparasitology Today" (C. Ash and R. B. Gallagher, eds.), pp. A29–A33. Elsevier Trends Journals, Cambridge.]

These two events appear to be mechanistically independent of one another and their coordination in the organism is not understood.

The rate of gene switching is high, 10^{-2}–10^{-3} per trypanosome per generation, implying that this system is driven and not simply the result of a background level of gene rearrangement. The molecular basis of this mechanism is not known.

The order of VSG gene expression is clearly a complex, highly regulated process, although the manner in which this order is achieved is still speculative. The order in which variants appear seems to be determined by the relative frequency at which that gene is turned on. Thus telomeric genes in active expression sites should be expressed at the highest frequency. Products of genes having high homology with these regions should also be common, since homologous genes can easily undergo gene conversion into the active site. Genes having poor homology with the expression site would be expressed later because of the relative infrequency of a conversion event. It should be emphasized that these ideas are speculative, and the order of gene expression in the organism is not a precise event.

VSGs are not expressed in the tsetse fly; expression resumes as the infective metacyclic stage develops. Expression at this stage differs from that in the bloodstream because only a subset of the VSG repertoire is used. This subset is known as the metacyclic variable antigenic type (M-VAT) repertoire, and is very predictable, suggesting a different selection mechanism for this form than for the bloodstream form.

4. Immune Response to *Trypanosoma brucei*

Infection with trypanosomes induces a very effective humoral response that results in the elimination of the parasites carrying the VSG to which the response is made. The few parasites that remain have already undergone spontaneous switching and therefore expand until they in turn are eliminated by an antibody response. Thus, if a patient with sleeping sickness is monitored, successive waves of parasitemia are seen, each of which is antigenically distinct from the last. Each wave of parasitemia is estimated to end by the destruction of 99.9% of parasites present. The success of this survival strategy depends on the antigenic distinction of each successive VSG from the previous one, so there are no cross-reacting antibodies. The population is homogeneous with respect to the glycoprotein coat and the coat is very densely packed on the surface of the organism, preventing accessibility of any conserved structural protein to the immune system. Antigenic switching is not driven by the production of specific antibody, but by a relatively high level of spontaneous variation arising in the population. This variety insures a small fraction of survivors once a specific response is mounted. There is no evidence that the immune response to VSGs is impaired in any way: the success of the strategy depends on the tremendous antigenic variation available in the glycoprotein coat.

The enormous variability of the VSG means that approaches to vaccine design are very difficult. It has been suggested that the relatively conserved M-VATs of the metacyclic infective form might provide the basis for a vaccine. Another approach has been to look for conserved proteins that might be available to the immune system. The trypanosome endocytoses molecules that are required for its growth through a specialized area at the base of the flagellum known as the flagellar pocket. Mice immunized with a purified fraction of this area have been partially protected from subsequent infection. At present, this approach holds what little hope there is for a vaccine against this devastating disease.

B. *Plasmodium* Species

1. Clinical Significance

Malaria has a reputation as the greatest scourge of humankind, and has been known since ancient times. The parasite replicates in the red blood cells. The widespread occurrence of the disease is responsible for maintaining genetic polymorphisms such as sickle-cell anemia in susceptible populations, since heterozygotes for these genes are less susceptible to infection. Malaria in humans is caused by four different species: *Plasmodium vivax, P. malariae, P. ovale,* and *P. falciparum.* These species vary in their geographical distribution and in the severity of disease. *Plasmodium vivax* is found predominantly in Asia and accounts for about 43% of malaria cases in the world. The disease is one of relapsing fever and does not usually cause death. *Plasmodium falciparum* is found predominantly in tropical areas and accounts for 50% of all malaria. The disease is much more severe and is responsible for the greatest number of malarial deaths, due to complications that develop as a result of clogging of the capillaries. Because of the severity of the disease it causes, most research has focused on *P. falciparum;* thus, the following discussion will concentrate on this organism.

2. Life Cycle

The life cycles of all four species of malaria are similar. The insect vectors are mosquitoes of the genus *Anopheles* and only the female, which requires a blood meal for development of her eggs, transmits the parasite. Once infected, a mosquito can transmit the parasite for life.

The stages of the life cycle that occur in humans are asexual, whereas the insect stage undergoes sexual reproduction (Fig. 3). The parasite is injected into the bloodstream as a sporozoite, which immediately migrates to the liver and invades the parenchyma cells. There it replicates prolifically and undergoes maturation to the merozoite form (~30,000 merozoites per sporozoite). The merozoites are released into the bloodstream, where they invade the red blood cells (RBC) and undergo successive rounds of replication. Eventually the RBCs rupture and release merozoites and metabolic waste into the bloodstream. The rupture of parasitized RBCs is synchronized to a characteristic time interval for each species. This synchronization is the cause of bouts of fever at regular intervals. A few merozoites develop into gametocytes in the erythrocytes. Gametocytes are the sexual stage of the parasite; no further development occurs until they are ingested by a mosquito. Male and female gametes combine in the gut of the mosquito and the sporozoite develops from the zygote. Newly formed sporozoites migrate to the salivary glands and the cycle begins again.

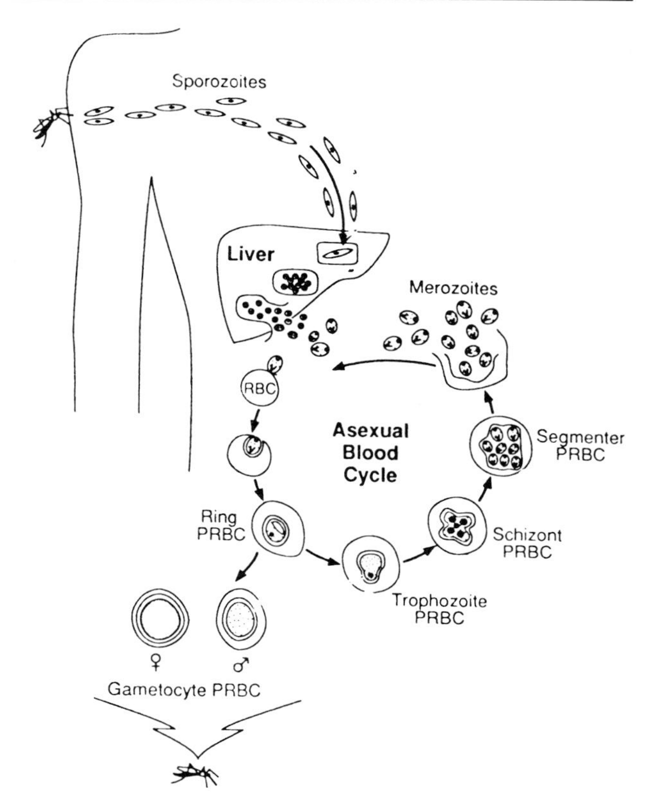

Figure 3 The life cycle of *P. falciparum* malaria in humans. RBC, red blood cell; PRBC, parasitized red blood cell. [Reproduced from Howard, R. J. (1989). *Blood* **74**, 533–536.]

3. Antigenic Variation in Malaria

The parasitic infection in malaria is characterized as predominantly intracellular, with a brief exposure of the sporozoite and merozoite forms in the blood; thus, very little exposure to the immune system takes place. This is in contrast to *T. brucei,* which exists only in the blood and extracellular fluid. Thus, the mechanisms of defense of the plasmodia are somewhat different.

Several protein antigens have been cloned and sequenced from the different asexual blood stages. Although the proteins have been shown to be antigenically distinct for different strains of *P. falciparum,* the organism carries only one gene copy per cell. Thus, the genes are polymorphic among strains, and each strain exhibits only one antigenic type. Such polymorphism is not true antigenic variation in the sense of gene switching within one organism, but because the antigenic variation among strains contributes substantially to the ability of the organism to evade the immune response, these proteins will be discussed in the following sections.

a. S-Antigens

S-Antigens are heat-stable antigens that are released into the serum of infected individuals. They are secreted into the vacuole that surrounds the developing merozoites (schizont) and are released into the bloodstream when the RBC ruptures. The diversity of these antigens is such that they can be used as stable markers for each individual parasite strain. At least five different S-antigen genes have been cloned, each of which contains a single exon with a large central block of repeats. The length and number of the repeats can vary widely. The first S-antigen cloned, FC27, has an 11-amino-acid sequence repeated approximately 100 times, whereas a subsequent isolate described, NF7, has an 8-amino-acid sequence seen in about 40 repeats. In addition, this gene has two 8-amino-acid repeats created by a substitution of leucine for arginine at one of the positions in the repeat. However, the repeats from different strains are related at the nucleotide level, and appear to have evolved from a common precursor. Antibodies directed against the S-antigen can be found in infected individuals; commonly, antibodies against more than one strain can be detected. Studies using polyclonal antisera with multiple specificities show a very large number of S-antigens that are widely dispersed. These results have been confirmed for individual S-antigen serotypes such as FC27 using specific monoclonal antibodies. However, the fact that S-antigens are expressed neither on the surface of the parasite nor on the infected cell means that the immune response to this protein does not confer any protection against the disease to the individual. Instead, the protein provides an immunological smoke-screen in which an immune response is made to the most abundant protein.

b. Merozoite Surface Antigens

Two different although structurally related antigens, merozoite surface antigens (MSA) 1 and 2, have been isolated from the surface of the merozoite. Both are membrane proteins, and exhibit considerable structural diversity. As for S-antigens, there is only one gene copy per organism, so MSA 1 and 2 also provide an example of antigenic polymorphism. Several alleles of each gene have been cloned and the basis of antigenic diversity is understood (Fig. 4). The polypeptide for MSA 1 can be divided into 17 blocks that are classified as variable, semiconserved, or conserved. The structure of these blocks

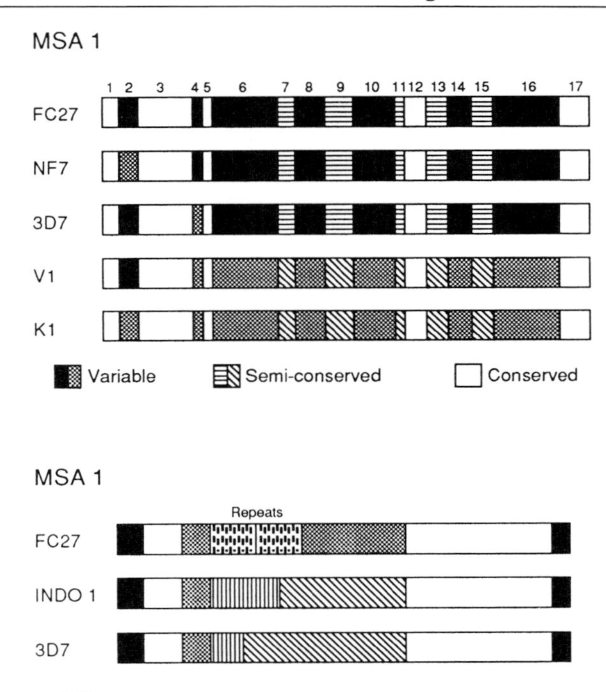

Figure 4 Structural diversity in two merozoite surface antigens of *P. falciparum*. [Reproduced from Anders, R. F., and Smythe, J. A. (1989). *Blood* **74**, 1865–1875.]

suggests that the different alleles have been generated by limited intragenic recombination at the amino terminus of the molecule. In addition, at least one form of variable block 2 contains tripeptide repeats.

MSA 2 is a much smaller molecule than MSA 1, although it exhibits a similar structural organization (Fig. 4). MSA 2 molecules contain repeat structures that vary among isolates, as do the repeats of the S-antigens. Thus, isolate FC27 contains two copies of a 32-amino-acid repeat, whereas two other isolates contain differing numbers of the same 4-amino-acid sequence. However, both the amino and carboxy terminus of this molecule are highly conserved.

c. Antigens Associated with the Erythrocyte Membrane

Several different proteins have been identified that are associated with the erythrocyte membrane during the asexual stages. Genes encoding several of these proteins have been cloned and sequenced, and contain repeats that are polymorphic in size among different isolates. The known proteins include the mature parasite-infected surface antigen (Pf EMP 2 or MESA) and two histidine-rich proteins (HRP1 and HRP2). The contribution of these molecules to antigenic diversity of the parasite is not yet clear.

Another protein, Pf EMP 1, is polymorphic in size and exhibits considerable antigenic diversity. Since the gene encoding this protein has not yet been cloned, the molecular basis for this diversity is not known. Experimental evidence suggests that plasmodia can undergo true antigenic variation. This molecule has been suggested as a possible candidate for such a process, which could explain the size variation seen in the molecule.

d. Circumsporozoite Protein

The circumsporozoite protein (CSP) constitutes the outer coat of the sporozoite stage of the organism. It also contains a region of oligopeptide repeats but, unlike the other proteins described, these repeats are highly conserved and compose the major B-cell epitope on this molecule. However, vaccines using the repeat are poorly protective; further studies have shown that T-cell epitopes on this molecule are polymorphic and constitute the important determinant of the immune response to this molecule.

4. Immune Response and Vaccine Design

Infection of naive subjects with *P. falciparum* invariably leads to clinical disease with a high rate of fatality if untreated. However, in endemic areas a partial immunity does develop with age, since the blood parasite count drops and clinical disease is mild. This phenomenon is dependent on reinfection, since individuals who move from an endemic to a nonendemic area may suffer severe illness on their return. The long period of time required to acquire this natural immunity has two possible explanations that are not mutually exclusive. Either the parasite may be poorly immunogenic, as far as protective immunity is concerned, or the immune response may be strain-specific, and a certain amount of time is required to develop an effective immune response to many different strains. The data available on proteins from the parasite suggest that both mechanisms operate.

Vaccine development has been hampered because of these two mechanisms. The existence of naturally acquired immunity suggests that a vaccine should be possible, but the antigenic diversity exhibited by the organism makes design difficult. Early studies showed that irradiated sporozoites could protect against infection in animal models. A great deal of effort has been expended on developing a vaccine based on the conserved repeat of CSP. However, as noted earlier, this approach has proved less effective than was hoped. The conserved regions of MSA currently hold some promise for vaccine development.

C. Other Parasites: *Giardia* Species

Evidence of antigenic variation has been found in the protozoan *Giardia lamblia*, both *in vitro* and *in vivo*. *Giardia* is a common intestinal parasite of humans and animals, commonly transmitted by exposure to water that has been contaminated with infected feces. Variation of *Giardia* antigens has been observed in a 64-kDa cysteine-rich major surface protein. Variants have been found in cloned populations prior to selection processes; thus, variation does not appear to be induced. The genetic mechanism responsible is not yet understood, but involves some form of gene rearrangement and occurs at a relatively rapid rate (10^{-3} per cell division). Since the human antibody response is directed primarily against the initially expressed surface protein, variation of the 64-kDa protein is believed to contribute to the chronicity of infections and the lack of protection from reinfection.

III. Antigenic Variation in Bacteria

All bacteria undergo gradual and random mutations in DNA sequence that eventually serve to alter structural components. This process is known as antigenic drift. Some clinically significant examples of bacterial structural components that have been subject to this process are the M protein of *Streptococcus* species and the lipopolysaccharide of *Salmonella* species. In addition, some bacteria contain all the information necessary for successive expression of multiple different antigenic determinants of the same protein. In these cases, the genetic information necessary for expression of serotype is not lost with the change from one serotype to another. The following examples represent some of the most extensively studied systems.

A. *Borrelia* Species

The spirochete *Borrelia* is similar in many respects to the trypanosomes. *Borrelia recurrentis* causes relapsing fever, a tick-borne disease that is endemic in many parts of the world. Relapsing fever is characterized by a recrudescent septicemia in which isolates from successive bacteremic stages express unique surface antigens not seen in previous stages. Recovery from fever is mediated by the appearance of serum antibodies directed against determinants expressed by the currently predominant spirochete. Since *Borrelia* undergoes antigenic variation, recovery does not constitute immunity or cure; the fever returns with septicemia caused by spirochetes expressing a different antigen. In studies of mice infected with *B. hermsii,* at least 26 different serotypes have been produced from a single clone of this microorganism. The immunodominant antigens that dictate serotype are contained in a surface-exposed membrane protein that varies around 40-kDa in molecular mass and is known as the variable major protein (VMP). The genes that encode different VMPs may or may not be expressed and are contained on multicopy linear 30-kb "storage" plasmids. The silent gene copies on plasmids lack the sequence information required for expression, presumably the promoter region. The current models for VMP variation are shown in Fig. 5. To switch from one VMP type to another, site-specific recombination occurs between silent copies of genes on storage plasmids and expression sites on other plasmids by a process called duplicative transposition. Whether this takes place as a single-site recombination or as a double recombination that retains the original 3′ end of the expression plasmid is not yet clear. VMP switching occurs at a rate of 10^{-4} to 10^{-3} per generation.

Borrelia burgdorferi, the etiologic agent of Lyme disease, also contains small linear plasmids, but apparently does not use this kind of surface protein switching mechanism. Variability in its outer membrane protein, OspB, has been attributed to antigenic drift.

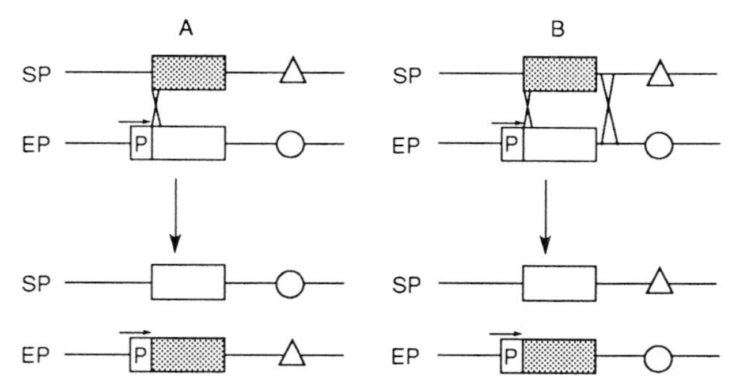

Figure 5 Models for *Borrelia*VMP variation. (A) Single site-specific recombination reaction that replaces all the sequences 3′ to the expressed VMP gene in the expression plasmid (EP) with those that are 3′ to the new VMP gene in the storage plasmid (SP). (B) Double site-specific recombination reaction that replaces the sequences immediately 3′ to the expressed VMP gene with the sequences immediately 3′ to the new VMP gene from the SP; however, another undefined recombination site occurs before the end of the SP and EP. The triangle and circle represent undefined genetic elements in the SP and EP 3′ to the expression site. P represents the promoter of the EP; the arrow indicates the direction of transcription. [Reproduced from Seifert, H. S., So, M. (1988). *Microbiol. Rev.* **52**, 327–336.]

B. *Neisseria* Species

Neisseria species are gram-negative cocci that cause gonorrhea, a sexually transmitted disease of epidemic proportions throughout the developed world, and are also an important cause of meningitis. Antigenic variation in *Neisseria* has been the subject of intense interest because it plays a major role in pathogenesis of the organism. At least two surface proteins that are immunodominant antigens are involved in both phase (turning expression on and off) and antigenic variation. These proteins are the pilins, subunits of pili that are essential to virulence because they mediate attachment to host cells, and the opacity-associated protein, which is also involved in adhesion to host surfaces and implicated in aiding colonization through self-adhesion and resistance to serum bactericidal factors and immune cells. [*See* Adhesion, Bacterial.]

1. Pilin Subunit Proteins

As in *Borrelia* spp. and the trypanosomes, multiple gene loci exist in the genome for the pilin subunit proteins. Antigenic variation takes place through recombination between copies of silent and expressed genes. Aspects that are unique to *Neisseria* include loss of pilins through phase variation as a result of the same mechanism. This contrasts with the trypanosome coat protein and *Borrelia* VMP, each of which are essential for survival, making phase variation impossible.

The pilin genes include one or two expression genes (*pil* E) and an array of silent genes (*pil* S). The silent genes lack not only promoter sequence information but also amino-terminal invariant regions of pilin. The expression sites supply both sets of information to make the complete and expressed sequence. Phase variation resulting in no expression of pili (P^- variants) comes from recombination with nonfunctional *pil* S genes containing missense or nonsense coding sequences that express pilin components that are incapable of forming stable pili at the membrane surface. P^- variants can revert to P^+ after a second gene conversion through recombination with a different *pil* S gene. The model for the genetic mechanism that functions for both antigenic and phase variation is shown in Fig. 6. Pilin genes contain six variable sequences called "minicassettes" that are separated by conserved sequences. The conserved regions are thought to mediate nonreciprocal exchange of sequence information between silent and expressed pilin genes, and probably encode essential components of the pilin protein. Since the expressed gene acquires, through intragenic recombination, some but not all of its minicassettes from any of the *pil* S loci, this system is capable of vast combination possibilities.

Another aspect unique to *Neisseria* is that pilin antigen switching can take place in more than one way. *Neisseria* tends to undergo autolysis spontaneously and P^+ cells are naturally competent for transformation. Therefore, pilin gene conversion can occur as a result of transformation through recombination with species-specific exogenous DNA from lysed cells. In fact, this is thought to be the most common method of pilin antigen switching, since it occurs at a frequency of 10^{-3} per cell division.

The clinical and pathogenic significance of such extreme potential for variation in pilin genes is evidenced by observations that gonococcal isolates from sexual partners and from different anatomical sites on the same patient commonly differ in pilus type. Recent evidence suggests that antigenic variation of the pilin proteins may serve the organism not only by evading immune responses but by altering the capacity of *Neisseria* to bind to different host cell types, possibly mediated by cell-specific receptors.

2. Opacity-Associated Protein

The opacity-associated protein (OPA) also has been referred to as P.II in the gonococci and as class 5 in the meningococci. In contrast to what has been described for other bacterial systems, phase variation of OPA is accomplished through an unusual control mechanism. From 2 to 12 variant copies of the *opa* gene exist in the genome, depending on the species; unlike the pilin genes, all copies are constitutively transcribed. However, all copies are not translated because of an expression control mechanism mediated by a repeating sequence or coding repeat known as CR. The coding repeat is part of the membrane transport leader sequence of the *opa* gene and consists of repeating units of the sequence CTCTT, as shown in Fig. 7. The repeated sequence encodes the hydrophobic core of the leader peptide of the OPA protein. The number of coding repeat units is

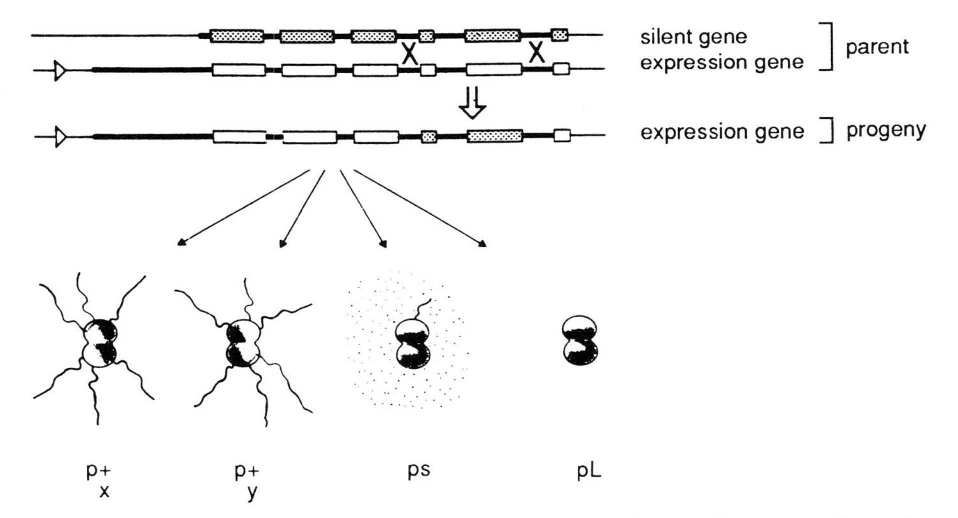

Figure 6 Generation of variant pilin forms. Intragenic recombination between a silent *pil* copy and the expressed copy leads to a new combination of minicassettes in the expressed copy of the progeny variant. In some instances (not shown) additional minicassettes from tandemly arranged silent copies may be incorporated into the expressed gene, generating an extended expression locus (L-type). Depending on the sequence arrangement in the expressed gene, different types of pilin molecules are produced, variant P^+ pilin, S-pilin, or L-pilin. [Reproduced from Meyer, T. F., Gibbs, C. P., and Haas, R. (1990). *Ann. Rev. Microbiol.* **44,** 451–477.]

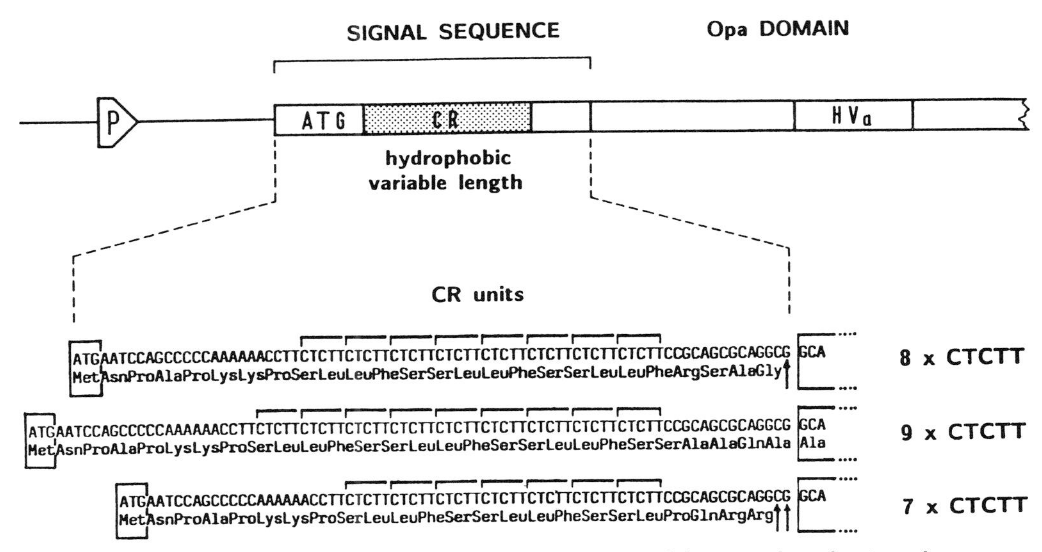

Figure 7 Location of the coding repeat (CR) in the *opa* genes and the control mechanism of *opa* gene expression. The number of CR units determines the reading frame of *opa* genes; if the reading frame is not correct (e.g., eight or nine CR units), no OPA protein will be produced. [Reproduced from Meyer, T. F., and van Putten, J. P. M. (1989). *Clin. Microbiol. Rev.* **2** (Suppl.), S139–S145.]

variable in different copies of *opa* genes (ranging from 7 to 28 repeats); these differences in length of the coding repeats determine the reading frame of each mRNA fragment. Addition or subtraction of one or two coding repeats causes a frameshift that results in a nonfunctional truncated protein. Switching *opa* genes in and out of frame occurs at a rate of 10^{-2} per cell division. Antigenic variation of OPA is accomplished by successful expression of variant copies of *opa* genes. Thus, the amount of variation is mostly restricted to genetic information that is already available in copies in the genome and is rarely altered by inter- or intragenic recombination events.

C. Phase Variation in *Haemophilus* and *Escherichia coli*

Type b strains of *Haemophilus influenzae* are notorious agents of meningitis and other serious infections. Extracellular production of a polysaccharide capsule is strongly associated with virulence of these strains and, accordingly, capsule polymers have been employed effectively as vaccine components. However, loss of capsule production yields a selective advantage for this agent at certain times, for example, during adherence to epithelial cells. The gene family required for capsule production is in a locus called *cap*. This locus contains the gene *bex*A (b capsule expression), which encodes a protein required for extracellular transport of the capsule. Spontaneous deletions of the region containing *bex*A result in strains that are phenotypically nonencapsulated although the capsule antigen is made and accumulates intracellularly. In contrast to other organisms that use phase variation, nonencapsulated variant strains of *H. influenzae* are apparently unable to return to their original state of extracellular capsule production.

Most *Escherichia coli* strains are capable of undergoing phase variation by a switching mechanism that turns production of type 1 pili on and off. Type 1 pili function by adhering to mannose molecules on host cells but also make bacterial cells more susceptible to phagocytic uptake and clearance mechanisms of the host. The switch in phase involves a 314-bp DNA segment that exists immediately upstream of the pilin gene (*fim*A). The segment encodes a promoter sequence that undergoes site-specific recombination, resulting in inverted (and nonfunctional) orientation of the promoter in the locus. The recombination is mediated by products of two additional genes in the locus (*fim*B and *fim*E). FimB regulates switching from off to on (pil^+) and FimE regulates switching from on to off (pil^-). This process of pilin phase variation occurs at a frequency of 10^{-3} per cell division.

D. Streptococcal M Protein Variation

Group A streptococci are the causative agents of a spectrum of disease ranging from acute respiratory to severe multiorgan illnesses. Group A streptococci have an impressive array of virulence factors available for gaining advantage over the host immune response, including exotoxins that also act as immunomodulating agents, molecules that mimic host antigens, and the M protein. The M protein is a surface structure with strong antiphagocytic properties that make it primarily responsible for the ability of Group A streptococci to persist in host tissues. In contrast to many other bacterial systems of antigenic variation, the M protein undergoes variation through intragenic recombination of inexact repeat sequences located in tandem on the single copy gene. Sequences encoding antigenic determinants in the M protein may be duplicated or deleted by such recombination events. This recombination produces M protein variants of different sizes, although all are functionally antiphagocytic. At present, over 80 different serotypes of M protein have been identified. Since resistance to infection is associated with the presence of type-specific antibodies, an individual can be infected with many different types of Group A streptococcus over the course of a lifetime.

IV. Antigenic Variation in Viruses

As described for bacteria, all viruses undergo continuous random mutations in their nucleic acid sequence that eventually alter their structural components (antigenic drift). However, this process occurs much more frequently in some viruses than others, suggesting an active mechanism. By definition, this process occurs more rapidly in viruses with RNA genomes than in DNA viruses, since RNA polymerases are less accurate than DNA polymerases. Antigenic variation does not seem to be a major mechanism of immune evasion in the DNA viruses studied so far. However, in RNA viruses there are marked differences in the rate of accumulation of mutations that affect antigenic structure. Poliovirus, a positive-strand RNA virus, can be divided into three different serotypes that, although they differ antigenically from each other, have remained individually stable over time. In contrast, lentiviruses and influenza both accumulate mutations that affect protein structure at a rate that makes vaccine design very difficult; these examples will be discussed further. In addition, influenza A virus undergoes a process known as antigenic shift, in which portions of its genome are exchanged with portions of genome from influenza viruses that infect other species, creating a new antigenic subtype.

A. Influenza

1. Clinical Significance

Influenza causes an acute respiratory illness that is responsible for significant morbidity and mortality in susceptible populations. Illness is particularly severe in the elderly and the young. Epidemics occur relatively frequently; periodically, world-wide pandemics occur. The last of these, in 1918–1919 killed 20–40 million people. [*See* INFLUENZA.]

2. Structure and Host Range

Influenza is a negative-strand RNA virus. Its genome is divided into eight separate RNA segments that encode 12 known viral proteins. The two surface proteins of the virus are hemagglutinin (HA) and neuraminidase (NA); these proteins undergo antigenic variation. Influenza viruses are divided into types A, B, and C based on antigenic differences in the nucleoprotein and the matrix protein. All three types undergo antigenic drift, but only influenza type A exhibits antigenic shift. Types B and C infect only humans, although recently an influenza C isolate has been recovered from swine. In contrast, influenza A has a wide host range and infects swine, horses, and seals, as well as a large variety of birds.

3. Antigenic Shift

Antigenic shift occurs when a new subtype of the hemagglutinin or nucleoprotein gene is introduced into the human population. This can happen when two different influenza viruses infect the same cell and reassortment takes place between the RNA segments of their genomes. In this century, three different hemagglutinin subtypes—H1, H2, and H3—and two different nucleoproteins—N1 and N2—have been described. In 1957, the H1N1 subtype was replaced by H2N2; in 1968, H2N2 was replaced by H3N2. Sequence analysis showed that H3 hemagglutinin was 98% homologous with the same protein from an avian influenza strain. Genetic reassortment can be readily demonstrated experimentally; it is reasonable to suppose that this mechanism is responsible for the generation of new subtypes of influenza A. The lack of antigenic shift in subtypes B and C may reflect the lack of an animal reservoir.

4. Antigenic Drift

Antigenic drift occurs by the accumulation of a series of point mutations. These mutations lead to sufficient changes in amino acid sequence that the virus is no longer recognized by the initial host immune response. Antigenic drift occurs in all three types of influenza and is most marked in HA and NA, the two viral surface molecules. Antigenic drift can be mimicked *in vitro* using monoclonal antibodies directed at HA; variants emerge at a frequency of 10^{-4}–10^{-5}. The majority of these variants has a single amino acid substitution in the HA1 polypeptide chain. Sequence analysis of naturally occurring viruses confirms that the majority of changes maps to this chain. However, naturally occurring variants generally have more than one amino acid substitution. The frequency of isolation of variants in NA is similar to that of HA.

5. Immune Response

Immunity to a given influenza subtype is long-lived and directed primarily at the HA and NA proteins. Thus, individuals infected with a given subtype are immune if it reemerges years later. Antibody alone is protective, although an efficient cell-mediated response is made. However, protection may be incomplete as antigenic drift modifies the virus.

B. Human Immunodeficiency Virus

1. Life Cycle and Clinical Significance

Human immunodeficiency virus (HIV-1) is a member of the lentivirus family of retroviruses, and the causative organism of acquired immune deficiency syndrome (AIDS). HIV-1 is an RNA virus that, after entry into the cell is transcribed to DNA by the enzyme reverse transcriptase. The DNA becomes incorporated into the host genome, where it is replicated whenever the host cell divides. This is considered a latent phase. The viral genome is 10 kb and is one of the most complex retroviral genomes known. In addition to the classical envelope (*env*), nucleoprotein (*gag*), and polymerase (*pol*) genes, the virus also has at least five regulatory proteins that control its replicative state in the cell. Transcription of the integrated DNA results in the formation of new virus particles.

The receptor for the virus is the CD4 molecule, which is found predominantly on the helper T lymphocyte population of the immune system. These cells are primarily infected by HIV-1 and their loss leads to the devastating immune deficiency seen in AIDS. The envelope protein (gp120) binds directly to the CD4 molecule; the regions responsible for this interaction have been mapped on both. [*See* ACQUIRED IMMUNODEFICIENCY SYNDROME (AIDS).]

2. Antigenic Variation

The variation seen among different strains of HIV-1 is enormous; no two isolates are alike and, even within a given individual, several different strains are observed. The bulk of this variation is seen in the envelope protein; the other proteins remain relatively conserved. Five variable regions have been identified in the envelope gene where the rate of nucleotide substitutions is substantially greater than that in other regions of the molecule. One of these, V3, contains a neutralizing epitope that is strain-specific.

The variation in the envelope protein appears to be driven in response to selection pressure by antibody. Neutralizing antibody is induced early and appears to limit viral spread initially. However, this response selects for variants that are poorly neutralized; since the major neutralizing epitope is strain-specific and highly variable, selection occurs rapidly. This selection pressure also seems to select for more rapidly growing strains of virus, although this mechanism is poorly understood.

Other lentiviruses that chronically infect animals have similar characteristics. In equine infectious anemia (EIAV), sera from chronically infected horses can neutralize isolates from previous disease episodes but not from subsequent ones. Similar observations have been made in sheep infected with Visna virus; antigenic variants have been described in goats infected with caprine arthritis–encephalitis virus (CAEV). Thus, antigenic variation appears to be a major mechanism of escape from the immune response in this family of viruses.

3. Immune Response and Vaccine Design

As described earlier, the antibody response to HIV-1 is vigorous but ineffective in controlling the disease. In addition, the major neutralizing epitope is strain-specific, making design of a vaccine based on the envelope protein difficult. Currently, much effort is being devoted to developing a more broadly reactive antibody response to the envelope protein. However, the other structural proteins of the virus, gag and pol, are much more conserved and elicit a good, broadly reactive, cell-mediated response. Recent evidence suggests that cell-mediated immunity in the form of cytotoxic T cells is important in controlling disease progression. Thus, a vaccine that stimulates virus-specific cytotoxic T cells may be more promising. At the present time, an effective vaccine for HIV-1 is a long way off. [*See* T LYMPHOCYTES.]

V. Antigenic Variation in Fungi: *Candida*

Candida albicans is the single most common fungal pathogen for humans. This dimorphic fungus is responsible for infections that range from the extremely common mucous membrane infections (oral in newborns, vaginal in adults) to the potentially fatal systemic candidiasis that has become increasingly associated with an immunocompromised state. The dimorphic life cycle of this microbe presents unique opportunities for antigenic variation. As the cells transform from mycelial phase to yeast phase, there are profound qualitative and quantitative changes in the expression of cell-surface antigens. Only a small number of proteins from cell walls of one phase is recognizable by serum antibodies prepared against the other phase. Relatively little is known about what stimulates phase transition *in vivo,* but there is evidence that the mycelial phase cells are more invasive of host cells and more effective in eliciting host immune responses than the yeast phase cells. It is not unusual to find both phases existing simultaneously in infected host tissues. Analysis of *Candida* at the genetic level has been impaired for reasons that include lack of precision in identification methodology, its polyploid nature, and the absence of chromosome condensation during metaphase. Epidemiologic studies of this pathogen have historically used identification parameters such as patterns of phenotype characteristics or "biotyping" of strains. This system has fallen out of favor since the discovery that a strain of *C. albicans* can switch reversibly and at a high frequency among at least seven different phenotypes defined by colony morphology. This switching of phenotype is also associated with changes in virulence and is thought to be mediated through some kind of chromosome reorganization process.

VI. Conclusions

In this article we have emphasized the strategies of the parasite in maintaining and gaining advantage in the host–parasite relationship. Variation of exposed antigens can serve the microbe by permitting it to escape immune recognition or by allowing it to take advantage of newly encountered environments. Antigenic variation can be a rapid and nonrandom process that allows the parasite to adapt immediately to new challenges, or a more gradual and random process that eventually defines a new environmental niche (or host) for the parasite. The active role in this relationship can be taken alternately by the parasite (activating regulatory mechanisms) or by the host (selecting the most successful among variants already present). The host–parasite relationship is thus always dynamic, with a continually shifting balance of opposing strategies. It should be noted that the system of immunoglobulin diversification in mammals has evolved to be remarkably similar to antigenic variation. In each case, genetic recombination is the principal mechanism employed to achieve diversified or altered products. Each system is likely to have provided the evolutionary impetus for the other.

Bibliography

Anders, R. F., and Smythe, J. A. (1989). *Blood* **74,** 1865–1875.

Borst, P. (1991). Molecular genetics of antigenic variation. *In* "Immunoparasitology Today" (C. Ash and R. B. Gallagher, eds.), pp. A29–A33. Elsevier Trends Journals, Cambridge.

DiRita, V. J., and Mekalanos, J. J. (1989). *Ann. Rev. Genetics* **23,** 455–482.

Goudsmit, J., Back, N. K. T., and Nara, P. L. (1991). *FASEB J.* **5,** 2427–2436.

Seifert, H. S., and So, M. (1988). *Microbiol. Rev.* **52,** 327–336.

Wilson, I. A., and Cox, N. J. (1990). *Ann. Rev. Immunol.* **8,** 737–771.

Antiviral Agents

George J. Galasso
National Institutes of Health

Glossary

Debilitating disease Ailment that causes severe loss of strength and ability

Empirical antiviral drug development Design of antiviral agents through screening programs followed by chemical modification of agent to develop more effective compounds

Open studies Evaluate a drug with no controls; the investigator and clinician are aware of the drug and dosage used

Papillomatosis Development of multiple benign epithelial growths such as warts

Prophylactic Treatment regimen used to prevent disease

Randomized, double-blind, placebo-controlled clinical trial Evaluates the efficacy of a drug where patients are randomly placed on a placebo or the test drug and neither the physician nor the patient is aware which is being used; all patients receive the standard treatment; if a drug does exist for the treatment of the disease, it is used instead of the placebo

Rational antiviral drug development Design of antiviral agents to inhibit virus at any step in its replicative function, including gene targeting by antisense based on molecular structure of the virus particle or receptor site on the host cell

Therapeutic Treatment regimen used to treat disease

VIRAL DISEASES are among the greatest cause of morbidity and mortality in the world. Vaccines have been most helpful in the prevention and even eradication (smallpox) of some of these diseases; however, treatment of viral infections has not been very successful. Viruses are metabolically inert substances in the extracellular fluids of the host system. They replicate by entering a host cell and utilizing the cellular components for replication. Therefore, unlike antibiotics, which can attack bacteria as they replicate extracellularly, antiviral agents must be capable of entering the host cell and inhibiting viral replication without affecting the host cell. Despite this great obstacle, considerable advances have been made and exciting developments are expected. [*See* SMALLPOX.]

I. Development of Antiviral Agents

Early attempts at finding useful antivirals were disappointing and consistent with this pessimism. In 1957, interferon (IFN) was first described as a natural product that inhibits viral replication *in vitro* and *in vivo*. Because patients recover from some viral infections prior to the development of antibody, it was assumed that the IFN produced early in infection was the body's first line of defense. There was early expectation that the "penicillin of viruses" had been found. Initially, production of this material in quantities sufficient for clinical trials proved elusive. Chemical inducers to promote host production were identified, but they proved too toxic for clinical application. When it was discovered that IFN could be produced *in vitro* in human leukocytes, sufficient material was made for a few limited clinical studies. The results of the initial studies were disappointing, particularly when coupled with the toxicity observed at the desired concentrations. Another promising early antiviral agent was methisazone. It showed some activity against poxviruses and was field-tested against smallpox in 1963. The results were inconclusive and revealed considerable toxicity. These early failures added to the general pessi-

Encyclopedia of Microbiology, Volume 1

mism toward the potential of antiviral agents. [*See* INTERFERON.]

The skeptics further indicated that treatment of viral diseases must overcome another burden—the need for rapid and reliable diagnostic assays. The pathogen associated with bacterial infections could usually be identified within 24 hr, whereas viral isolation and identification usually took 1 wk or more. How a treatment regimen could be properly administered in time to prevent viral replication and the ensuing damage was difficult to conceive. Nevertheless, some persistent stalwarts made small advances.

The first antiviral of real importance, idoxuridine, was synthesized by Prusoff in 1959. In 1962, Kaufman demonstrated that this compound could be successfully used topically for the treatment of herpetic keratitis. Although this finding was impressive, the skeptics were not convinced. The drug was toxic, and its success was due in part to the localized nature of the infection. The eye is an isolated organ that is well suited to topical treatment; therefore, the problem of systemic toxicity is avoided. Idoxuridine was the first antiviral agent to be approved by the Food and Drug Administration (FDA) for clinical use. Amantadine was subsequently found to be effective prophylactically against influenza A and was approved by the FDA in 1966. However, because of unwarranted concerns about toxicity and lack of interest in prophylactic drugs, amantadine did not generate the usage it deserved. The turning point came in 1977 with the advent of vidarabine. It was the first effective systemically administered antiviral drug to treat a serious ongoing viral disease, herpes encephalitis. Skeptics became cautious believers, and research in the field began to expand.

All of the antiviral agents currently approved by the FDA (Table I) were found serendipitously, due mostly to an empirical approach. Most of the compounds emanated from the various anticancer screening programs. If a compound was found to have some anticancer activity, it would be tested for antiviral activity. If positive, it would be given to the chemist to attempt various structural modifications to reduce toxicity and increase activity.

The growth in the field of molecular biology contributed greatly to our understanding of viral composition and replication. This knowledge can now be applied to a rational approach for the inhibition of virus and the development of antiviral agents. Specific steps in viral replication can be identified for which inhibitors can be developed. Specific viral nucleic acids, viral receptors, viral-encoded enzymes, proteins, etc., essential for viral attachment, penetration, uncoating, replication, assembly, and release, can also be identified. By selecting specific targets in the process of viral replication, scientists should be able to develop antiviral agents that are less toxic to the host, and, by better understanding the mechanism of the interaction, investigators

Table I Antiviral Agents Approved by the FDA (United States)

Drug	Indication	Date approved
Topical idoxuridine	Herpes keratitis	6/63
Topical trifluorthymidine (trifluridine)	Herpes keratitis	4/80
Topical vidarabine (Vira-A)	Herpes keratitis	11/76
Vidarabine (Vira-A)	Herpes encephalitis	10/78
	Neonatal herpes	
	VZV in immuno-compromised	
Amantadine (symmetrel)	Influenza A	10/66
Ribavirin (virazol)	Respiratory syncytial virus infection in the neonate	12/85
Acyclovir (zovirax)	Primary genital herpes and suppression	3/82
	VZV in immuno-compromised	3/90
	Herpes encephalitis	3/90
	VZV (shingles)	3/90
Zidovudine (zidovidine, retrovir)	AIDS	3/87
	CD4+ T cells <500/mm^3	3/90
Interferon-α 2B (intron A)	Genital papilloma	11/88
	Chronic hepatitis C	2/91
Ganciclovir (cytovene)	Cytomegalovirus retinitis in immuno-compromised	6/89
Foscarnet	Cytomegalovirus retinitis in immuno-compromised	10/91
Dideoxyinosine* (ddI)	AIDS	10/91

[a] Conditional approval.
VZV, varicella zoster virus.

should be able to develop more effective synergistic combinations. [*See* DNA REPLICATION.]

Tremendous strides have been made in the rational design of antiviral agents due to the international effort to combat acquired immunodeficiency syndrome, or AIDS. This fatal disease provided the added incentive for collaborations of chemists, physicists, molecular biologists, computer modelers, and virologists working for a common goal. Although the desirability of the rational approach for the development of antivirals has long been recognized, the impetus and the funds necessary to implement the approach were lacking until this unfortunate pandemic. The rational approach is now beyond theory. [*See* ACQUIRED IMMUNODEFICIENCY SYNDROME (AIDS).]

Crystallographic techniques and computer modeling have made it possible to determine the three-dimensional structure of some viruses at the atomic level. The attachment site of the viral particle to the host cell can be visualized and compounds can be synthesized that could react with these attachment sites and interfere with infection. The potential of this approach has been demonstrated. A known viral inhibitor was shown to fit into a cavity of the rhinovirus attachment site. Studies were then initiated to determine if compounds could be synthesized to form a better fit and be more active. Such studies have also been effective in understanding the mechanism for the development of resistance to some antiviral agents. In a recent development, protease, an essential component of human immunodeficiency virus (HIV) replication, was identified. After its three-dimensional structure was determined, chemical compounds similar in structure were designed to mimic the protease and halt HIV production. These, in turn, were tested *in vitro* for viral inhibition. Early results have been encouraging.

A novel approach to viral inhibition is to develop antisense genes or oligonucleotides that would bind with specific messenger RNA encoded to produce viral protein, thus preventing viral replication. This complex may then be destroyed by cellular enzymes. Similar antisense oligonucleotides may also bind to DNA for similar results. This approach, although intellectually satisfying, is not without its own set of problems. Cells are not hospitable to foreign oligonucleotides and rapidly degrade them. Stoichiometric problems also need to be overcome. In the normal process of viral replication, the viral genome in infected cells is extensively amplified. Therefore, delivery of sufficient antisense materials to the target cell to inhibit the virus adequately may prove difficult. This approach must overcome substantial problems of drug delivery and stability before it becomes feasible.

We are now at the stage where large numbers of agents are ready for clinical evaluation, particularly against HIV. The two-pronged approach to identifying new agents—the traditional empirical approach of screening large numbers of potential compounds and the rational approach of targeting specific steps in the viral replicative cycle—will continue to work in parallel. The advances against HIV are relevant to efforts to combat other viruses.

One final introductory note is to acknowledge the importance of animal models in the development of antiviral agents. The role they have and continue to play could justify an article by itself. *In vitro* studies can identify viral inhibition and cell toxicity, but the role of pharmacokinetics, general toxicity, chemical inactivation by body fluids/enzymes, ability to reach target organs, etc., can only be determined *in vivo*.

II. Viral Diseases

A. Rhinoviruses

The common cold is the most frequent cause of absenteeism from work, thus resulting in significant economic loss. The treatment of the common cold has been a prime target for pharmaceutical companies interested in antivirals, yet effective therapy remains elusive. The reason is twofold. The common cold may be caused by several viruses or several strains of the same virus (such as the rhinovirus, with >100 strains). It is also a mild and short-lived infection. Therefore, to be successful, the antiviral would have to have a broad spectrum, be free of toxicity, and be easily administered at the earliest onset of symptoms. Ideally, it should also be an over-the-counter product. These are difficult obstacles to overcome; however, because of the potential market, attempts at solving the problem continue. [*See* RHINOVIRUSES.]

The prime target for the common cold has been rhinoviruses, the most common etiological agent. It has been demonstrated that the major cell receptor site is shared by 90% of the rhinovirus strains and is identical to the intracellular adhesion molecule (ICAM-1). Therefore, soluble ICAM-1 may be a successful broad-spectrum inhibitor of rhinovirus attachment. The gene for this receptor has been mapped and large quantities of the synthetic receptor have been made. If shown to be clinically effective, this material could be used prophylactically in high-risk patients and family contacts of an index case.

Isoxazole derivatives have been shown to inhibit rhinoviruses. The previously discussed crystallographic techniques demonstrated that the inhibition was the result of the compound filling a cavity of the virus receptor site. Based on these structural studies, further chemical synthesis of similar compounds was initiated to develop more potent and broader-spectrum compounds. One such compound, 5-[5-[2,6-dichloro-4-(4,5-dihydro-2-oxazolyl)phenoxy] pentyl]-3-methylisoxazole (WIN 54954), showed a 20-fold improvement in spectrum and potency over the previous compound. It is also effective against other picornaviruses such as coxsackieviruses and echoviruses both *in vitro* and *in vivo* and represents a good candidate for clinical studies. In addition, it demonstrates the viability of the rational approach to antiviral development.

A series of other compounds from a number of different pharmaceutical firms, including chalcones, enviroxime, flavans, and ribavirin, have been examined singly and in combination (for synergy) with encouraging results. These *in vitro* studies led to a follow-up clinical study with one of the most promising compounds, 3-methoxy-6-[4-(3-methylphenyl)-1-piperazinyl] pyridazine, which inhibited >70% of the rhinovirus serotypes *in vitro*. Some efficacy was observed in the prevention of colds, but the results were not impressive.

The first clinical study of IFN was against the common cold. The world supply of human IFN was acquired by the Common Cold Unit of the Medical Research Council in England for the study. Although these early studies were encouraging, the use of IFN for this purpose was not deemed practical. With the advent of recombinant nucleic acid technology, synthesis of IFN was achieved and attempts at treatment/prevention of the common cold were reinitiated. IFN-α has been demonstrated to be effective in postexposure prophylaxis studies in family settings, yet practical application continues to be a problem. The major toxicity is pyrogenicity, which is an unacceptable side effect for a viral disease with mild symptoms. IFN-β, although less toxic, was also studied and found ineffective. This emphasizes the need for *in vivo* studies and the importance of animal models wherever available, since early *in vitro* studies had shown IFN to be efficacious. *In vivo* efficacy often does not directly translate into clinical efficacy.

One of the major problems in developing a viral inhibitor for the common cold is reaching the target cells in the nasal epithelium. Ideally, the antiviral would be administered as a nasal spray; therefore, it must penetrate the nasal secretions and not be washed away by mucocilliary clearance mechanisms. This presents an additional substantial complication to the development of an effective agent. Although the common cold continues to prove elusive, progress continues to be made.

B. Influenza

Influenza is a more serious respiratory infection and one for which effective antivirals have been available but underutilized. Amantadine was shown to be an effective prophylactic agent against influenza A, the most common form of influenza, in the mid-1960s. However, some minor side effects (such as nervousness, insomnia, and dizziness, as well as nausea in 10% of the patients) coupled with a lack of interest in prophylactic drugs hindered its proper use in the clinic. With the threat of swine flu in the mid-1970s, further studies were done with both amantadine and its analog rimantadine. It was shown conclusively that both drugs were effective both prophylactically and therapeutically against influenza A. Although minor side effects in a small percent of patients treated with amantadine were observed, rimantadine was essentially free of side effects and comparable to the placebo in this regard. Amantadine was slightly more effective, but both drugs were about 80% effective in prevention of disease and nearly 100% effective in reduction of illness and virus shedding in infected individuals when treated early in the disease. [*See* INFLUENZA.]

Recent studies have demonstrated that resistant viruses can be isolated from some treated patients. If the index case in a family setting is treated and the other members of the family are put on prophylaxis, the drug may not prove effective in preventing disease. This is due to emergence of resistant virus from the treated patient. If, however, the patient is not treated, prophylaxis of the family members is effective. Doses as low as 100 mg rimantadine daily are highly effective in prevention of influenza A. Early studies with 200 mg were devoid of toxic side effects. Once influenza A is identified in a community, prescribing rimantadine at a daily dosage of 100 mg for 5–6 wk would be useful for prevention of disease in a defined population. This regimen is currently under consideration by the FDA. Combined treatment and prophylaxis in a closed setting should probably be avoided in view of the previously cited family study. The need remains for a safe, orally available drug effective against both influenza A and B.

C. Respiratory Syncytial Virus

A serious problem in neonates requiring hospitalization is pneumonia and bronchiolitis caused by respiratory syncytial virus. The use of aerosolized ribavirin for 3–6 days was effective in reducing the course of the illness. The drug brings about more rapid improvement of rales, cough, and lethargy in treated infants, as well as reduced viral shedding and improved arterial oxygen saturation. This treatment continues to be studied for long-term effects, although no toxicity has been observed to date. The mode of administration also needs further improvement. Because the drug tends to precipitate and clog the aerosolizing machine, care must be taken to assure that a proper dosage is being administered.

D. Arenaviruses

Ribavirin has its greatest potential against diseases caused by bunyaviruses and arenaviruses, including Lassa fever and a variety of hemorrhagic fevers. Clinical studies have been completed against Lassa fever and hemorraghic fever with renal syndrome (HFRS). Lassa fever was studied in Sierra Leone, where investigators reported striking results when patients were treated with ribavirin within 6 days from onset of fever. In a disease with a mortality rate of $>75\%$ in untreated patients, the mortality rate dropped to $<10\%$ using ribavirin.

Clinical studies against HFRS, caused by hantaan virus, have been done in the People's Republic of China. In a prospective, randomized, double-blind placebo-controlled study, ribavirin treatment significantly reduced mortality from 10 of 118 in the placebo group to 3 of 126 in the treatment group. Ribavirin was also successful in reducing the number of patients progressing to hemorrhagic disease, shortening the duration of fever and other clinical signs.

E. Hepatitis

Viral hepatitis is caused by at least five viruses, hepatitis A, B, and D and non-A and non-B viruses, tentatively called hepatitis C and E. Hepatitis B and D and the classical non-A and non-B can lead to chronic hepatitis. Strong evidence indicates that chronic hepatitis can lead to hepatocellular carcinoma. Chronic hepatitis has presented a considerable challenge of worldwide importance. It is a common and often progressive liver disease, particularly in Asia, for which no adequate treatment exists. Chronic hepatitis B was one of the first infections to be tested with IFN. Early studies with leukocyte-derived IFN and vidarabine showed some promise, but disease returned with cessation of treatment. Some of these early studies demonstrated that corticosteroids enhanced the effectiveness of the antiviral therapy. It was postulated that the resulting immunosuppression caused by corticosteroids could stimulate viral replication in patients with chronic hepatitis. If latent virus could be induced, the antiviral treatment could be more effective. Use of corticosteroids such as prednisone to enhance viral replication is controversial because it induces an exacerbation of disease. However, preliminary evidence indicates that a short course of prednisone followed by IFN leads to a high rate of remission in the disease. A recent study in chronic hepatitis B patients compared the effects of IFN treatment alone, IFN following prednisone, and untreated controls. Patients received 5 million units of recombinant IFN-α per day for 16 wk after 6 wk of placebo or 5 million units of recombinant IFN-α per day for 16 wk after 6 wk of treatment with prednisone. Both treatments reduced viral markers in the patients' serum. There was no significant difference between the treated groups, but there were significant differences between the treated and the untreated groups. These studies demonstrated that the strongest independent predictor of a response to treatment was the level of viral DNA in the serum at initiation of therapy. Therefore, it is an indication that the use of prednisone, which would enhance serum viral DNA levels, needs further study. Studies with significant follow-up periods indicate that IFN has potential in the treatment of these patients and may triple the spontaneous seroconversion rate normally observed in these patients. [*See* HEPATITIS.]

Similar beneficial results with IFN treatment have been observed in hepatitis C. Results were sufficiently efficacious to warrant approval by the FDA for this indication. However, further studies are warranted to address relapses that occur after cessation of treatment. This remains a problem in at least half the patients showing benefit.

F. Herpesviruses

The most extensive experience with antivirals has been against herpesviruses. As noted earlier, the first antiviral drug successfully used against a viral infection was topical idoxuridine against herpes keratitis. The success was largely due to the nature of the disease, a surface infection in an isolated organ, the eye, that could be treated topically. Herpes keratitis was later successfully treated topically with other drugs such as vidarabine, trifluridine, acyclovir, and IFN. The drug of choice appears to be trifluridine, but it is also one of the most expensive. [*See* HERPESVIRUSES.]

The first successful treatment of a serious ongoing systemic viral disease was for herpes encephalitis with vidarabine. It was also shown to be effective for other herpes infections. This drug was subsequently replaced by acyclovir for most herpes infections; it proved more effective and can be administered orally. Vidarabine has a disadvantage in that it must be administered intravenously by slow-drip infusion and has not yet been formulated for oral administration.

Similarly, vidarabine proved efficacious in the treatment of varicella-zoster in immunocompromised patients but was soon replaced by acyclovir. Acyclovir was also shown to have some beneficial effect in early treatment of otherwise healthy children with varicella. In oral dosages ranging from 10 to 20 mg/kg (depending on age) administered four times daily, patients receiving the drug experienced fewer lesions, quicker cutaneous healing, and less fever.

A direct comparison of acyclovir and vidarabine was made against neonatal herpes. The results showed both drugs to be effective, with no clear clinical advantage for either drug. Acyclovir still has the advantage of ease of administration. Thus, acyclovir has a clear advantage over vidarabine against encephalitis; however, they are equally effective against neonatal herpes. The success of antivirals against herpesviruses (see Table I) is due to an enzyme highly specific to herpes virus, thymidine kinase. Pyrimidine and purine nucleoside analogs, such as acyclovir, can be selectively phosphorylated to the active compound by the viral thymidine kinase in infected cells. The phosphorylated antivirals can then enter into the metabolic pathway of viral precursors, thereby interfering with viral DNA synthesis. The acyclovir triphosphate is also a selective inhibitor of the viral polymerase.

Acyclovir has proven efficacious in the treatment of first and recurrent episodes of genital herpes. First clinical episodes of the infection were shown to be effectively treated with 200 mg five times daily for 5–10 days. The drug significantly reduces viral shedding and shortens time to healing. However, treatment of first episodes does not prevent subsequent recurrences.

Treatment of severe recurrences with acyclovir (200 mg five times daily for 5 days) does provide relief by reducing healing time and associated parameters of disease. Additional studies have demonstrated that individuals who suffer severe and frequent recurrences could be successfully put on suppressive treatment, 400–1000 mg of oral acyclovir for several months. This could reduce the number of recurrent episodes but not eliminate the infection in these patients. A recent report described the drug as well tolerated and safe even when used continuously for 4 yr. Dosages could effectively be reduced to one daily dose of 800 mg. Because acyclovir is a nucleoside analog, there is always concern about its potential carcinogenic, mutagenic, and teratogenic effects; none have been observed to date in animals or humans. There have been 77 episodes in which pregnant women have taken acyclovir in the first trimester with no subsequent abnormalities.

Because acyclovir was associated with some evidence of testicular atrophy in rats and decreased spermatogenesis in dogs, a study of sperm production was made in patients with prolonged treatment. No effects were seen in sperm count, motility, or morphology. There was some concentration of acyclovir in the sperm. The functional reproductive capacity of the sperm could not be determined from this study nor could the significance of the accumulated acyclovir on potential embryos be determined.

An alternate treatment also found to benefit patients with genital herpes in reducing time of healing, crusting, viral shedding, and symptom duration was topical treatment with IFN-α (10^6 units/g with 1% nonoxynol-9 in 3% methylcellulose). Early application was critical.

Although not practical for the average patient with herpes labialis (cold sores), acyclovir may be of some value to those who suffer from severe lesions. In these cases, the 37% decrease in duration of pain and 27% decrease in healing time may be of value.

Although cytomegalovirus (CMV) usually causes a common, benign infection, it is a serious pathogen in fetuses and in immunocompromised cancer, organ and bone marrow transplant, and AIDS patients. The most common forms of disease in these patients is pneumonia, gastrointestinal disease, and retinitis; the latter two are most common in AIDS patients. CMV is a herpesvirus but is not as sensitive to acyclovir as herpesvirus 1 and 2 (HSV-1 and HSV-2; the former usually associated with herpes labialis and the latter with genital herpes). The lack of sensitivity is attributed to the lack of thymidine kinase in CMV. However, acyclovir did reduce the incidence of infection to 36% compared to 61% in the placebo group in a randomized, prophylactic, double-blind, placebo-controlled study in recipients of renal transplants. Similar results were obtained in allogeneic bone marrow transplantation, where CMV disease occurred in 22% of the treated patients compared to 38% in the controls.

A more promising compound is ganciclovir didehydroxy propoxymethyl guanine (DHPG), which appears to be as equally effective and tolerated in pediatric populations as it is in adults for the treatment of CMV infections. It differs from acyclovir by the addition of a single carboxyl group and is 100 times more active against CMV in tissue culture than acyclovir. It has been widely tested in AIDS patients, where CMV retinitis poses a severe problem. Ganciclovir has proven clinically effective in stabilizing the disease in >80% of the patients; however, relapses are common and patients require maintenance treatment.

Foscarnet has recently been approved for the same indication. Studies with foscarnet yielded the unexpected result that AIDS patients who were taking foscarnet lived an average of four months longer than patients taking ganciclovir.

The prognosis for CMV pneumonia in renal transplant patients treated early with ganciclovir is excellent; however, similar results were not obtained in treatment of bone marrow transplant patients with CMV pneumonia. The best results utilizing ganciclovir were for the treatment of CMV gastrointestinal disease regardless of the underlying cause of immunosuppression. In separate studies, 75, 83, and 92% of patients improved with treatment; however, further work is necessary before definite conclusions can be reached on the role of this drug. The key to efficacy appears to be time of treatment; for the drug to work, it must be administered early in the infection. In a recent study to treat gastroenteritis in bone marrow transplant patients, 2 wk of treatment was not sufficient to achieve clinical improvement, although viral replication was suppressed.

Foscarnet appears to be a second promising drug in CMV disease, but it must be administered by continuous or intermittent infusion. Further studies with both drugs are necessary to arrive at optimal treatment regimens.

The phosphorate nucleoside analog (s)-1-(3-hydroxy-2-phosphoryl methoxy propyl) cytosine has broad-spectrum activity against the herpesvirus group. Studies in animals have shown it to be more effective than ganciclovir against CMV infections and more effective than acyclovir against HSV 1 and HSV 2 infections. In addition to being more potent, it also has a longer half-life and a longer duration of activity. Preliminary studies are encouraging for its potential clinical use.

G. Papillomavirus

It is often surprising that genital warts (*Condyloma acuminata*) caused by papillomavirus is a bigger problem than genital herpes. They are both common chronic infections implicated in genital dysplasia and carcinoma (the evidence for herpes is weaker than for papillomavirus). However, genital warts are constantly present with rare periods of regression, as compared to genital herpes, which is a recurrent infection with periods of quiescence. Genital warts are caused by many distinct types of papillomaviruses. Other types cause palmar, plantar, or common cutaneous warts. A wide range of topical and ablative therapies for genital warts have been attempted, including podophyllin, trichloroacetic acid, electrodesiccation, neutral red plus ultraviolet light, electrocautery, and surgery, all with little benefit. Crude IFN was used as early as 1975. These early studies utilized lyophilized preparations applied topically; subsequent studies used purified IFN or recombinant IFN-α in an ointment. Dosages varied from 4×10^3 to 2×10^6 units/g. Topical treatment offers ease of application, but penetration is a problem, and studies proved negative for clinical efficacy. Other studies applied IFN by intramuscular or intralesional administration. Results were generally positive, averaging approximately 30% clearance, but with some recurrence and adverse reactions. Reichman and his associates performed a double-blind, placebo-controlled study evaluating three different IFN-α preparations administered subcutaneously. All three showed some positive effect in treating warts, although the lymphoblastoid IFN appeared to have a more positive effect than the two recombinant preparations. However, the authors recommend further studies for longer periods of treatment to obtain statistically significant results.

Topical 5-fluorouracil has also been used for topical treatment of genital warts. Open studies (uncontrolled trials) have claimed beneficial results with clearance in 10–95% of patients. However, considerable toxicities were observed, including local irritation, vulvitis, ulcerative balanitis, and meatitis.

Papillomavirus is also responsible for recurrent respiratory papillomatosis, a severe debilitating disease. In this disease, a growth occludes the larynx. Early studies showed dramatic reduction in the size and frequency of these growths. A recent randomized multicenter study in 123 patients evaluated the role of IFN-α in preventing the recurrence of the lesion following surgical removal of the growth. The 2-yr study compared these patients to patients who received surgery alone. During the first 6 mo, there was a significant beneficial effect with IFN. These patients experienced a lower rate of papilloma growth. However, with time, this effect decreased, leading the investigators to conclude that there was no real value in the treatment with IFN. Whether a different regimen of treatment might accentuate or prolong the early beneficial effects remains to be shown.

H. Human Immunodeficiency Virus

A remarkable amount of progress has been made in understanding the molecular biology of the causative agent of AIDS. It has also been the single greatest influence in renewing interest in antiviral research. Zidovidine (AZT), an agent originally synthesized in a cancer chemotherapy program and known to have antireverse transcriptase activity, was tested against HIV-1 and found to inhibit the virus. AZT was quickly put into clinical trial and licensed for the treatment of AIDS in 1986. An impressive effort by laboratory investigators, clinicians, and the FDA brought the drug to the clinic in record time. This record achievement could not have been accomplished without the rapid involvement and collaboration of government, academia, and scientists.

The initial studies showing the efficacy of AZT in prolonging life in AIDS patients was done with dosages of 1500 mg/day. This level resulted in significant adverse effects. Subsequent studies have been done to determine whether or not reduced, less toxic doses would be effective and whether or not asymptomatic HIV-1-positive individuals would benefit.

To evaluate the benefits of a lower dosage in patients, a randomized controlled trial of a reduced daily dose of AZT was conducted in AIDS patients who had had a first episode of *Pneumocystis carinii* pneumonia. Patients were given either a dose of 250 mg orally every 4 hr, as the standard medication, or a dose of 200 mg orally every 4 hr for 4 wk, followed by half that dose thereafter. No therapeutic differences were seen between the two groups, with less toxicity observed with the lower dosage. As a result of this multicenter study, the investigators recommend a daily dose of 600 mg, although they conclude that the optimum dosage has not yet been determined.

AZT has also proven effective in prolonging the life and delaying the progression of the disease in asymptomatic patients; however, the long-term cost of such therapy is considerable. In an attempt to reduce both toxicity and cost, a study was performed to determine the efficacy of lower dosages in this population. Three daily doses were tested: 1500, 600, and 300 mg. As in an earlier study, which saw no difference between 1500 mg/day and 500 mg/day, these investigators could determine no difference in any dosage used; therefore, the lowest dose is recommended. However, the authors indicate that further studies should be done to determine whether or not an even lower dose may be advisable. This study also was designed to determine whether or not acyclovir would have a synergistic effect when given in combination with AZT. No benefits could be determined from the combination.

Ongoing studies in a pediatric population (5 mo to 13 yr) indicate that AZT may be as effective in this population as it is in adults.

The effectiveness of AZT has led to the screening of other nucleoside analogs. Two of these compounds show particular promise—2′,3′-dideoxyinosine (ddI) and 2′,3′-dideoxycytidine (ddC). The first, ddI, appears to be effective in patients with AIDS or advanced AIDS-related complex as measured by T4 cell counts and p24 antigen levels as well as survival and toxicity. Although toxic effects such as peripheral neuropathy, pancreatic inflammation, hepatitis, and seizures have been observed in patients treated with ddI, these side effects are observed less frequently when the dosage is in the 3.2–9.6-mg/kg/day range. A survival rate of 88% was observed after 21 mo of ddI therapy. This study was performed in patients intolerant to AZT or who are clinically deteriorating while on AZT, indicating that ddI could be a useful alternative. The second compound, ddC, was also promising in early clinical studies, but it is also associated with considerable toxicity such as severe peripheral neuropathy. However, it may offer additional hope for alternate or combination therapy if ongoing clinical trials can identify an acceptable therapeutic modality. In early clinical trials, didehydrodeoxyuridine showed neurological toxicities similar to those seen with ddC and ddI at a dosage of 1–12 mg/kg daily. However, increases in CD4 cell counts and significant decreases in HIV p24 antigen levels were observed even at the lowest dosage levels. Studies continue to determine lowest effective dosages and to determine extent of adverse effects.

In October 1991, the FDA granted conditional approval to ddI for the treatment of AIDS. Until further information is available, conditional approval based on surrogate endpoints has been granted for use in adult and pediatric AIDS patients who are intolerant or whose health has significantly deteriorated while on AZT.

Other drugs showing some early promise against HIV-1 *in vitro* include azidodeoxyuridine. However, no signs of efficacy were observed in patients with CD4 cell counts of 200–400, even with dosages of 48 mg/kg. Acemannan has also shown beneficial effects in preliminary studies, with improvement in several clinical factors including p24 antigen levels, CD4 cell counts, weight gain, and reduced opportunistic infections. Although there was some early promise for ribavirin, a recent study was unable to show a clinical benefit on virologic or immunologic indicators with daily dosages up to 1600 mg. High dosages were associated with adverse effects. Although IFN-α may play a role in slowing down progression of HIV-1 disease, it is not appropriate for long-term treatment. Its optimal benefits may be seen in combination therapies.

Inhibitors of HIV-1 α-glucosidase-1 such as castenospermine, tunicamycin, and *N*-butyl deoxynojirimycin, which can interfere with viral glycosylation, appear to prevent the glycoprotein from migrating to the cell surface for incorporation into the budding virions. An inherent problem, in addition to not being as active as AZT, is that such compounds also inhibit normal cell function.

Based on the knowledge of the structure and action of the HIV-1 protease, a number of inhibitors have been developed. Although HIV protease has many features in common with other human aspartic proteinases, it appears to have sufficient viral specificity to make it a potential antiviral agent target. The compounds developed appear to be effective and highly selective. This selectivity has the potential for low toxicity and minimal effect on the related human proteases.

Numerous other promising antivirals are directed against HIV-1. Among the more promising are 2-hydroxyethoxymethyl-6-phenylthiothymine (HEPT) and tetrahydro-imidazo-benzodiazepinthione derivatives (TIBO). They are very specific for HIV-1 and do not inhibit HIV-2. Both compounds inhibit reverse transcriptase, but the specific mode of action remains to be determined. Their structures are unrelated to other available antiviral agents. *In vitro* studies show they are considerably more effective than AZT, ddI, or ddC. The amount of drug required to get viral inhibition is 6000 times less than the amount of drug that will cause toxicity, a very favorable therapeutic index. A review by Myers describes clinical studies of antiviral agents currently under way in AIDS patients.

III. Resistance

An ever-increasing concern with the use of antiviral agents is the development of resistant viruses with the potential for more potent pathogenicity. Although resistant viruses have been demonstrated, particularly in treated immunocompromised patients, they were considered to be rare and less pathogenic than the wild type. In some instances, the resistant strains are overtaken by the susceptible virus when grown *in vitro*. Nevertheless, the concerns with resistance have increased with the extensive use of antivirals in AIDS patients. Progressive disease due to acyclovir-resistant herpese simplex virus and ganciclovir-resistant CMV in AIDS patients has been observed. Studies have shown that 7.6% of AIDS patients, treated with ganciclovir for >3 mo, excreted resistant virus; However, when these patients experience a recurrence, it is usually with a new and sensitive virus. This may present a problem, but it is not one that cannot be overcome or controlled with judicious use of multiple antivirals (e.g., antivirals that work by a different mechanism). Resistance to all the currently available antiviral agents has been shown to result from single nonlethal mutations. The widest experience to date has been with acyclovir. Most strains isolated have reduced pathogenicity. After more than 5 yr of widespread use, resistant strains have not proven significant even with long-term prophylactic use in immunocompetent patients. However, the wide use of antivirals in immunocompromised patients is still in its infancy, and long-term surveillance is necessary. Limited information is available on frequency of occurrence, mechanism of action, increased virulence, and clinical effect due to resistance. Studies of AZT-resistant strains showed no change in clinical symptoms in a limited number of patients and were successfully treated with alternate antivirals. [*See* ANTIBIOTIC RESISTANCE.]

IV. Drug Combination

As discussed previously, an inherent problem with antiviral agents is their potential toxicity. Drugs that are highly viral-specific are potentially less toxic but have a greater likelihood for the development of resistance. A method to circumvent these problems is to use drugs either in combination or alternately. The former provides the opportunity to use two potentially toxic drugs in reduced dosages, avoiding toxicity and, in some cases, producing synergy. The latter provides a mechanism to avoid side effects or make the drug endurable. Combination therapy also renders development of resistance less likely, particularly if the two drugs act at different sites of viral replication. Strategies using antivirals in combination with other agents (granulocyte–monocyte colony-stimulating factor) to control toxicity are also under study. An extensive review of the subject is presented by Shinazi. Shinazi and his associates were among the first to show the practicality of this approach, *in vitro,* utilizing acyclovir in combination with vidarabine and IFN against herpesviruses.

Dideoxycytosine was shown to be highly efficacious in inhibiting HIV *in vitro;* however, in clinical studies, it was shown to cause neuropathy. Instead of discarding this drug, a small pilot study was done alternating ddC with AZT (two drugs with different manifestations of toxicity: AZT, showing hematologic toxicity, and ddC, showing neurologic toxicity). Preliminary reports cite a decrease in p24 and an increase in CD4 cells. An expanded study using about 200 patients will soon be underway. Other studies in clinical trial include AZT with IFN and AZT with acyclovir. *In vitro* studies combining CD4 with AZT, ddI, or ddC and AZT with foscarnet also show promise against HIV. Ganciclovir and foscarnet against CMV also show promise *in vitro.*

Attempts have also been made to evaluate the beneficial and/or synergistic effect of three drugs in combination. A study utilizing AZT, soluble CD4, and recombinant IFN-α A in two- and three-drug combinations reported a greater positive effect with the three-drug combination than with two. The investigators reported nearly complete virus suppression *in vitro* over a 28-day period without cellular toxicity.

Caution must be used when administering drug combinations, because not all drug combinations are effective. Sometimes a combination regimen could increase the toxicity of one or both of the drugs or the response of one drug may nullify the other. In the case of the AZT–ribavirin combination for AIDS, it was assumed that because AZT is a nucleoside analog and ribavirin is not, and because the site of toxicity is different in each, this pair of drugs could prove to be a winning combination by attacking different viral targets. The results proved to the contrary when increased toxicity was observed. A similar situation also occurred when combining AZT with DHPG to inhibit HIV and its companion opportunistic infection of CMV; 82% of the treated patients developed severe life-threatening hematological toxicity.

Therefore, although combination has many beneficial attributes, each attempt requires a clear understanding of the pharmacokinetics of the combinations. Drugs that have been widely tested independently require further efficacy and safety clinical studies. This is not an easy task, particularly if the drugs are not manufactured by the same company. Such a task is more easily performed through independent or government-sponsored trials.

V. Drug Delivery

The increased activity in antiviral agent research is beginning to yield a number of promising compounds, but the early concern continues—a number of agents administered at levels necessary to cause a beneficial effect are toxic. Similarly, getting the agents to the target cells remains a problem. Many are attacked by the body's defense mechanisms and enzymes before they can take effect. Therefore, an ancillary field of antiviral research is drug delivery—devising new methods to get small amounts of drugs efficiently to the target cells. In addition to reducing toxicity, slow and prolonged release of the drug is desirable to meet the increased need for long-term maintenance therapy in patients with AIDS, genital herpes, and chronic hepatitis. A variety of techniques are being tested such as embedding the drug in fat globules (liposomes) for slow release, coating with other biodegradable polymers, slow-release skin patches and implants, implantable pumps, and conjugating with cell/organ-specific monoclonal antibody. Some of these approaches are already in use, such as rate-controlled transdermal patches. The patches have the advantage of maintaining constant, level drug delivery. The limiting factor is the molecular size of the drug; however, new biodegradable polymers capable of containing and releasing more complex chemicals are becoming available.

Early studies with liposomal encapsulated drugs have been discouraging because of the fragility of the liposomes. However, progress has been made with the multivescular liposomes using different formulations. The half-life of drugs has been extended from 1 to 23 hr by liposomes.

Although conjugating toxins to monoclonal antibodies is theoretically attractive, its practicality remains to be proven. Toxins such as *Pseudomonas* exotoxin or plant toxin ricin as well as antivirals could be conjugated to monoclonal antibody for delivery to infected cells. However, problems such as potential immune sensitivity, cross-reactivity, and directing the complex to the target organ remain a concern. In addition, the technology for conjugation has not been perfected to a level that would prevent toxins from detaching in transit and causing side effects.

VI. Conclusion

The 1980s produced significant progress toward the development of antiviral agents. The earlier skepticism began to give way with the advent of vidarabine; soonafter, additional antivirals became licensed, notably acyclovir. The advent of AIDS was the impetus for increased efforts in the field, with an explosion of activity and results. We are no longer faced with the question "Where are the drugs to test in clinical trials?" The question now is "Which drug should we choose to go into clinical trial?" Most of these are directed against AIDS, but the knowledge gained will contribute significantly to other viral diseases.

Acknowledgment

Grateful thanks to Sue Ohata for her considerable contributions to this text.

Bibliography

Crumpacker, C. S., II (1989). *N. Engl. J. Med.* **321,** 163–172.

Fischl, M. A., Parker, C. B., and Petinelli, C., *et al.* (1990). *N. Engl. J. Med.* **323,** 1009–1014.

Galasso, G. J., Merigan, T. C., and Whitley, R. J. (eds.) (1990). "Antiviral Agents and Viral Diseases of Man," 3rd ed. Raven Press, New York.

Hazeltine, W. A. (1990). *J. AIDS* **2,** 311–334.

Hirsch, M. S. (1990). *J. Infect. Dis.* **161,** 845–857.

Myers, M. W. (1990). *Rev. Infect. Dis.* **12,** 944–950.

Reichman, R. C., Oakes, D., and Bonnez, W., *et al.* (1990). *J. Infect. Dis.* **162,** 1270–1276.

Sacks, S. L., *et al.* (1990). *J. Infect. Dis.* **161,** 692–698.

Shinazi, R. F. (1991). Combined chemotherapeutic modalities for viral infections: Rationale and clinical potential. *In* "Synergism and Antagonism in Chemotherapy (T.-C. Chan and D. C. Rideout, eds.), pp. 109–181. Academic Press, New York.

Sperber, S. J., and Hayden, F. G. (1988). *Antimicrob. Agents Chemother.* **32,** 409–419.

Archaebacteria (Archaea)

Kenneth M. Noll
University of Connecticut

Glossary

Archaea New domain composed of organisms previously called archaebacteria; plural archaea; singular, archaeon; possessive, archaeal

Bacteria New domain composed of organisms previously called eubacteria; plural bacteria; singular, bacterium; possessive, bacterial

Crenarchaeota One of two kingdoms of the domain Archaea; includes thermoacidophiles and strictly anaerobic thermophiles; plural crenarchaeotes; singular, crenarchaeote; possessive, crenarchaeotal

Domain New taxonomic division above kingdom level. Recognizes three major forms of organisms distinguished by analyses of macromolecular sequences

Eucarya New domain of organisms composed of eukaryotes; plural eucarya; singular, eucaryon; possessive, eucaryal

Eukaryotes Organisms that maintain their genome within a defined nucleus; not used in any formal taxonomic system but now may be used to refer to members of the domain Eucarya

Euryarchaeota One of two kingdoms of organisms of the domain Archaea; includes halophiles, methanogens, and some strictly anaerobic thermophiles, plural euryarchaeotes; singular, euryarchaeote; possessive, euryarchaeotal

Phylogeny Evolutionary history of a group of organisms

Prokaryote Organisms that maintain their genome dispersed throughout the cytoplasm; not used in any formal taxonomic system but was previously used interchangeably with the term bacteria

Taxonomy Study of the classification of organisms according to standard rules. Most modern taxonomies are phylogenetic (or "natural"); they attempt to group organisms according to evolutionary descent

ARCHAEBACTERIA are a group of prokaryotic microorganisms that are only distantly related to eukaryotes and the other prokaryotes. A recent proposal suggests that this group be elevated to a new taxonomic division above the kingdom level, a domain, to stress their distinctive evolutionary lineage. This domain is called the Archaea. The other prokaryotes are placed in the domain Bacteria, and eukaryotes are placed in the domain Eucarya. Although they resemble bacteria morphologically, archaea and bacteria differ in a variety of molecular features. Three major physiological groups comprise the Archaea. One group contains organisms that live at very high temperatures and reduce elemental sulfur to hydrogen sulfide. Most are strict anaerobes, but those that will grow aerobically also live in extremely acidic environments where they oxidize elemental sulfur to sulfuric acid. The second group is the methanogens. These archaea are strict anaerobes that convert carbon dioxide and simple organic molecules to methane. The final group is the extreme halophiles: those organisms that require high salt for growth. Some grow in highly alkaline, hypersaline environments. Because of their distant relationship to the Bacteria and Eucarya, the Archaea offer a third perspective on the evolution of biological processes and the origin of life.

I. Bacterial Phylogeny and the Archaea

A. Traditional Taxonomic Systems

Prior to the recognition of the Archaea as a group, these organisms were often considered unusual variants within traditional bacterial groups. With the recognition that these "funny bugs" are closely related to one another to the exclusion of all other organisms, the underlying explanation for these novel physiological features became apparent. Rather than being adaptations to unusual environments, these features are the result of inheritance from a common ancestor that harbored these traits. As these organisms have been studied in greater detail and new archaea have been discovered, the distinctive molecular and physiological characteristics of the Archaea have been revealed.

To fully appreciate the unique characteristics that describe the Archaea, it is important to understand how the group was discovered. Phylogenetic groups are delineated on the basis of shared characteristics. In taxonomic systems of plants and animals, which attempt to group organisms based on proposed evolutionary relationships, these shared characteristics are morphological features. For prokaryotes, taxonomic and phylogenetic schemes based on cell morphology have not been informative. Physiological characteristics, such as metabolic end products and distinctive enzymatic activities, offer another means to define prokaryotic groups and are often used to describe and identify newly-isolated organisms. Physiological criteria alone, however, have failed to produce phylogenetically coherent groupings. [*See* TAXONOMIC METHODS.]

B. Molecular Phylogeny

To define natural taxonomic groups, one can examine the sequences of the monomers of macromolecules isolated from a wide variety of organisms. In 1965, Emile Zuckerkandl and Linus Pauling described these macromolecules as "molecular chronometers." The sequence of the monomers (amino acids for proteins and nucleotides for nucleic acids) are determined through chemical means. The sequences of two identical macromolecules isolated from different organisms are compared and their overall similarity assessed. The similarity of the sequences is a measure of the evolutionary relationship of the organisms. Convergent evolution (similarity arising from accumulated changes in distantly-related molecules) is highly unlikely for such large molecules. From these similarity comparisons, phylogenetic diagrams ("trees") can be constructed to show the evolutionary relationship among the organisms analyzed. The branch points or nodes of these diagrams indicate times at which two groups diverged from a common ancestor.

Carl Woese and his colleagues have analyzed the sequences of the small subunit ribosomal RNAs (the 16S RNA) isolated from hundreds of organisms to assess their relationships. From these analyses, Woese discovered that methanogens are no more related to other prokaryotes than they are to eukaryotes. The term "archaebacteria" was coined to describe this new group of organisms. The name was chosen as a trivial name (not to be used in a strict taxonomic sense) to reflect the presumed ancient divergence of this group from a lineage that gave rise to all other organisms. The bacteria were called "eubacteria" (or "true bacteria") to distinguish them from archaebacteria. Subsequently, they found the halophiles, *Sulfolobus,* and *Thermoplasma* were specifically related to the methanogens (Fig. 1). Since recognition of the Archaea, new members have been found, including many extremely thermophilic anaerobes (Table I). Many new bacteria and eucarya have also been discovered that, in the light of the rRNA-derived phylogeny, enrich our understanding of the nature of the earliest ancestors and the progression of early evolutionary events. [*See* THERMOPHILIC MICROORGANISMS.]

Evidence is beginning to accumulate that suggests the Archaea are more closely related to the Eucarya than they are to the Bacteria. This would mean that a common ancestor of the Archaea and the Eucarya and an ancestral bacterium diverged prior to the divergence of the Archaea and the Eucarya. There are no known eucarya that are as closely related to this common ancestor as are some of the Archaea. Until such organisms are found (if they still exist), the nature of this common ancestor will remain mysterious.

C. A Taxonomic Revision

This newly recognized group of organisms has posed a problem for traditional taxonomy. The distinctiveness of bacteria has long been recognized in the terms eukaryote and prokaryote. The traditional five-kingdom taxonomy proposed by R. H. Whittaker in 1959 places the single prokaryotic kingdom

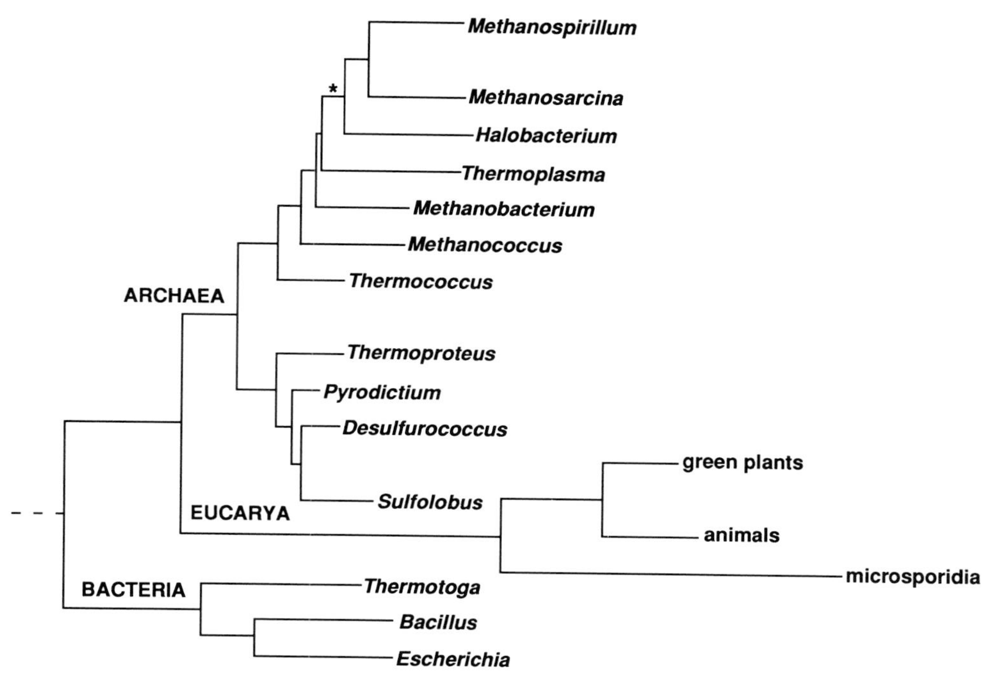

Figure 1 Schematic phylogenetic diagram depicting the relationships among the major genera of Archaea and the relationships among the three domains. Branch lengths (indicated by horizontal distances from nodes) are proportional to the evolutionary distances determined by comparative analyses of 16S rRNA sequences. Vertical distances are arbitrary. The dotted line indicates the "root" of the tree (i.e., the position of the common ancestor of all life). The asterisk is placed at the position from which *Archaeoglobus* was found to diverge based on sequence analyses. [From Woese, C. R. (1987). *Microbiol. Rev.* **51,** 221–271; and Woese, C. R., Kandler, O. and Wheelis, M. L. (1990). *Proc. Natl. Acad. Sci. USA* **87,** 4576–4579.]

(Monera) on an equivalent taxonomic level with the four eukaryotic kingdoms: animals (Animalia), green plants (Plantae), fungi (Fungi), and single-celled protozoans (Protista). The molecular phylogeny revealed by Woese, however, not only made the prokaryote distinction more apparent, but also demonstrated that the kingdom Monera is not composed of a single group of organisms.

In 1990, Woese, Kandler, and Wheelis proposed that a new taxonomic division above the kingdom level be introduced to recognize the newly emerging phylogeny, thus restoring the traditional taxonomy to a natural system as envisioned by Linnaeus. This new division is called a domain and the new domains are the Archaea (archaebacteria), Bacteria (eubacteria), and Eucarya (eukaryotes). Not only does this new system allow for future assignments to be made on a phylogenetic basis, but it recognizes that archaea are *not* bacteria (as the term archaebacteria implies). The Archaea are further divided into two kingdoms following the two branches revealed by the 16S rRNA phylogeny. The kingdom Euryarchaeota (informally the euryarchaeotes) is composed of the halophiles, methanogens, *Thermoplasma, Archaeoglobus,* and *Thermococcus.* The kingdom Crenarchaeota (informally the crenarchaeotes) includes aerobic and anaerobic extreme thermophiles, all of which metabolize elemental sulfur (Table I). This new taxonomy will be used throughout this article to reinforce the idea that these organisms are not bacteria in the traditional sense of the word.

II. Major Physiological Groups of the Euryarchaeota

A. Extremely Halophilic Euryarchaeotes

1. Hypersaline Environments

The extremely halophilic ("salt-loving") archaea are found in environments characterized by very high salt concentrations, often high light intensity,

Table I Characteristics of Selected Archaea

	Optimal growth conditions					Sulfur reduction[a]	Lipids $C_{20}:C_{40}$[b]	Cell wall type[c]	Year isolated
	Temp. (°C)	pH	NaCl (M)	Aerobic	Anaerobic				
EURYARCHAEOTES									
Halobacterium halobium	40	7.3	4	+	−	−	1:0	SL	1931
Halococcus morrhuae	34	7.2	4	+	−	−	1:0	HPS	1880
Natronobacterium gregoryi	37	9.5	3	+	−	−	1:0	SL?	1984
Natronococcus occultus	38	9.5	3.5	+	−	−	1:0	ND	1984
Methanosarcina barkerii	35	7.0	0.01	−	+	+	1:0	HPS	1978
Methanospirillum hungatei	34	7.0	0.01	−	+	ND	4:6	Sh	1974
Methanomicrobium mobile	40	6.5	0.01	−	+	ND	+	SL	1968
Methanobacterium formicicum	37	7.0	0.01	−	+	ND	+	PM	1947
Methanothermus fervidus	83	6.5	0.01	−	+	+	1.4:8.7	PM/SL	1981
Methanococcus vannielii	38	8.0	<0.5	−	+	ND	1:0	SL	1951
Methanopyrus strain AV 19	98	6.5	0.25	−	+	ND	+	ND	1989
Thermoplasma acidophilum	60	1.5	tr	+	+	+	1:9	none	1970
Archaeoglobus fulgidus	83	7.0	0.3	−	+	−	+	SL	1987
Pyrococcus furiosus	100	7.0	0.3	−	+	+	+	SL	1986
CRENARCHAEOTES									
Sulfolobus acidocaldarius	73	1.5	tr	+	+	+	0:1	SL	1972
Pyrodictium occultum	105	6.5	0.03	−	+	+	+/+	SL	1983
Thermoproteus tenax	88	5.5	tr	−	+	+	+/+	SL	1981

[a] Reduction of elemental sulfur to hydrogen sulfide by growing cells.

[b] The ratio of diether to tetraether lipids in the cell membrane. +, Ether-linked lipids are present, but no quantitative information is published; +/+, Both lipid types are present, but quantitative information is not published.

[c] SL, Surface layer composed of proteinaceous subunits; HPS, Heteropolysaccharide sacculus; Sh, Individual cells surrounded by a proteinaceous layer and cells are held together by protein fibrils (sheath); PM, Sacculus composed of pseudomurein.

tr, trace; ND, not determined.

and sometimes high alkalinity. Inland seas, such as the Great Salt Lake and the Dead Sea, provide salt-rich habitats for these organisms. Halophilic archaea also cause spoilage of fish, meat, and animal hides that are stored in salt. In areas in which seawater slowly evaporates (such as in the production of sea salt), light intensity and salinity are very high. These waters often appear red because of the high density of halophilic archaea, which produce red carotenoid pigments to protect them from the light. Caustic salts in some arid, low-lying areas cause halobacterial habitats to become highly alkaline (pH to 11). [*See* ALKALINE ENVIRONMENTS.]

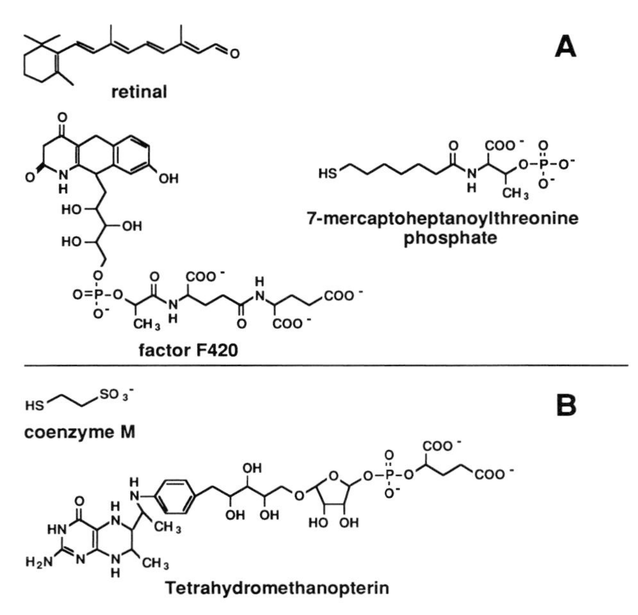

Figure 2 Cofactors of archaea. (A) Cofactors that mediate transport of electrons, protons, or both. Retinal participates in proton translocation in halophiles, 7-mercaptoheptanoylthreonine phosphate donates reducing equivalents for methane formation, and factor F420 is an electron carrier in methanogens and *Archaeoglobus*. It is also present in halophiles and thermoacidophiles where its function is unknown. (B) Cofactors that carry a carbon atom. All three carry a carbon atom for methane formation. Tetrahydromethanopterin and methanofuran function in carbon dioxide fixation in *Archaeoglobus*.

2. Intermediary Metabolism

All the extreme halophiles are aerobes, and only a few are capable of anaerobic growth by nitrate reduction. Peptides and amino acids are most commonly used as carbon sources, although sugars are catabolized by a few isolates. They generate energy by oxidative phosphorylation using a respiratory chain composed of cytochromes.

Those that catabolize sugars do so by an unusual pathway (Fig. 2). This pathway is a modification of the Entner-Doudoroff glycolytic pathway. In this modified pathway, glucose is not phosphorylated before it is oxidized to gluconate. Instead, 2-keto-3-deoxygluconate is phosphorylated before it is hydrolyzed. This does not change the net yield of adenosine triphosphate (ATP) compared to the Entner-Doudoroff pathway. This modified pathway is also used by some bacteria for growth on gluconate.

3. Photometabolism

Many, though not all, of the species of *Halobacterium* use light energy to drive cellular metabolism when oxygen levels drop too low to support respiration. Unlike the photosystems found in photosynthetic bacteria and plants, the *Halobacterium* photosystem does not use chlorophylls to capture light. Instead, retinal, the same photopigment found in human eyes, captures the energy of light and converts it to electrical and chemical energy that the cell uses for growth (Fig. 3).

Bacteriorhodopsin, the retinal-containing protein isolated from *Halobacterium halobium,* is among the most thoroughly studied proteins. Bacteriorhodopsin converts light energy to electrochemical energy by "pumping" protons from inside the cell to the outside. This creates an electrical potential across the membrane that the cell can use for other energy-requiring processes. Unlike complex, multicomponent proton pumping systems found in other organisms, bacteriorhodopsin is a relatively simple protein that can be purified in large quantities for study. Thus, it has become an important subject for investigations of bioenergetic mechanisms.

B. Methanogenic Euryarchaeotes

1. Habitats

While most archaea can be said to inhabit "extreme" environments occupied almost exclusively by other archaea, most methanogens live in more benign habitats. Methanogens are very sensitive to oxygen and occupy highly anaerobic niches, such as freshwater and marine sediments, sewage, digestive tracts, marshes, and landfills. They inhabit these

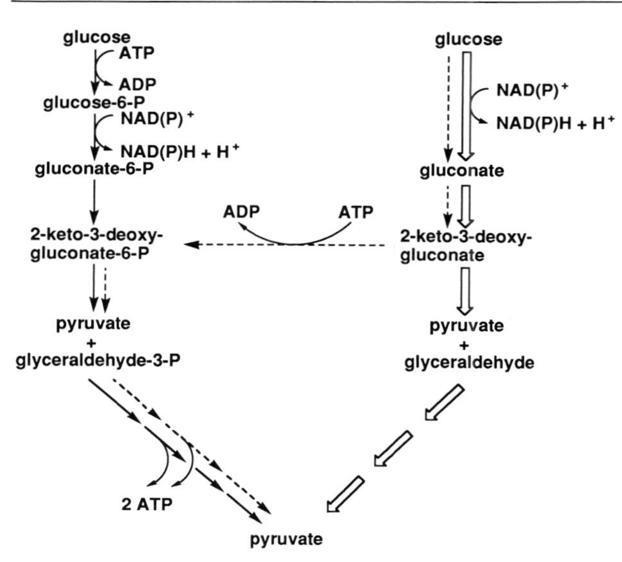

Figure 3 Glucose catabolism in archaea. Entner-Doudoroff pathway (→). Modified pathway found in *Halobacterium* (--→). Modified pathway found in *Sulfolobus* and *Thermoplasma* (⇒). The conversion of glyceraldehyde to pyruvate has only been shown in *Thermoplasma*.

areas along with a wide variety of bacteria, some of which depend directly on the methanogens to facilitate their metabolic processes. Some methanogens live among members of the other groups of Archaea and so do live in "extreme" environments. Both halophilic and extremely thermophilic methanogens have been discovered. [*See* METHANOGENESIS.]

The methanogenic archaea are solely responsible for biogenically produced methane. They use a rather limited range of carbon substrates for methanogenesis. Acetate, formate, carbon dioxide, methanol, and methylamines can each be reduced to methane. In nature, acetate and carbon dioxide (plus hydrogen) are the principle substrates. Methanogens serve as the last link in the anaerobic food chain because they convert these waste products from other organisms into cell material and an insoluble gas, much of which escapes into the atmosphere.

2. Methanogenesis

The study of the biochemistry of methane formation has yielded a number of novel biochemical mechanisms. The most detailed studies have been carried out with *Methanobacterium thermoautotrophicum*. This thermophile grows optimally at about 65°C and uses molecular hydrogen to reduce carbon dioxide to methane in a multistep reaction. The carbon atom of carbon dioxide is carried by unusual cofactors (methanofuran, methanopterin, and coenzyme M) as it is reduced to methane by electrons derived from hydrogen (Fig. 2). These electrons are carried by other novel cofactors (factor F420 and 7-mercaptoheptanoylthreonine phosphate). Another unique cofactor (factor F430) is a nickel-containing tetrapyrrole that plays an unknown role in the final step of methane formation. [*See* METHANOGENESIS, BIOCHEMISTRY.]

3. Intermediary Metabolism

Carbon dioxide serves as the major carbon source for a number of methanogens. Carbon dioxide is directed toward the synthesis of cellular components through a central autotrophic pathway, an important component of which involves the synthesis of acetate from two molecules of carbon dioxide. The methyl portion of acetate is synthesized via the methanogenic pathway. The carboxyl portion of the acetate is derived from a second molecule of carbon dioxide that is reduced to a carbonyl group by the carbon monoxide dehydrogenase enzyme complex. The methyl group and the carbonyl group are then joined by this same enzyme complex. This mechanism is similar to that used by certain bacterial anaerobes that synthesize and excrete acetic acid derived from hydrogen and carbon dioxide. This bacterial acetogenic pathway, however, uses cofactors found in all bacteria and eucarya. Thus, a similar biochemical mechanism for the reduction of carbon dioxide with hydrogen may have arisen independently among the Archaea and Bacteria.

This acetate is then converted to pyruvate and then oxaloacetate by the successive incorporation of two more molecules of carbon dioxide. The oxaloacetate then feeds into an incomplete tricarboxylic acid cycle. Sugars are synthesized from pyruvate by reversing the Embden-Meyerhoff glycolytic pathway. Methanogens cannot use sugars as a source of carbon and so do not contain the necessary enzymes for its catabolism.

C. Extremely Thermophilic Euryarchaeotes

1. *Archaeoglobus*

Archaeoglobus fulgidus was isolated in 1987 in the laboratory of Karl Stetter from a shallow marine volcanic area off the coast of Italy. It is a strict anaerobe that grows optimally at 83°C. It is the first archaeon found to derive energy from the reduction of sulfate to sulfide (a process called dissimilatory sulfate reduction). Dissimilatory sulfate reduction is carried out by at least three different groups of bacteria, some of which are moderate thermophiles. Glucose, lactate, pyruvate, 2,3-butanediol, fumarate, formate, and molecular hydrogen can serve as sources of electrons for the reduction of sulfate and thiosulfate. Elemental sulfur is not reduced.

Comparative analyses of the 16S rRNA from *A. fulgidus* and other archaea indicate that it is an euryarchaeote. Its phylogenetic position relative to the methanogens is uncertain, however. Comparison of 16S rRNA sequences suggests its ancestor diverged prior to the appearance of the methanogens. Its sequence contains nucleotides in particular positions (so called "signature sequences") that are characteristic of the halophiles and the "Methanomicrobiaceae" (methanogens most closely related to the halophiles). Comparative analyses of 23S rRNA sequences also support its closer relationship to these methanogens. It is not yet clear whether *Archaeoglobus* retains characteristics derived from a methanogenic ancestor or whether it retains characters that were present in an organism ancestral to the methanogens.

Some interesting observations have been made concerning the physiology of *Archaeoglobus* that may shed light on its phylogenetic position. When cells of *A. fulgidus* are examined under ultraviolet light, they fluoresce with the same blue-green color characteristic of the methanogens. This color is caused by factor F420, the same cofactor that causes methanogens to fluoresce. *Archaeoglobus fulgidus* also contains methanofuran and tetrahydromethanopterin, which are involved in one-carbon metabolism as they are in methanogens. It does not contain coenzyme M or factor F430, which are specifically involved in methanogenesis. It also contains the enzymes involved in the novel carbon dioxide fixation pathway found in methanogens. The presence of these shared metabolic capabilities support the close relationship between *Archaeoglobus* and the methanogens, but they do not support either hypothesis regarding their relationship.

2. Thermoplasma

Thermoplasma is a facultatively anaerobic thermoacidophile growing optimally at 60°C at pH 1.5. The first isolate of *Thermoplasma* was discovered in a self-heating pile of debris left from coal mining. Aerobically it catabolizes sugars, and anaerobically it grows by reducing elemental sulfur to hydrogen sulfide while catabolizing complex carbon sources. It grows better anaerobically, and cultures of organisms are more stable under anaerobic conditions than under aerobic conditions.

Thermoplasma does not have a cell wall external to its cell membrane. This property resulted in its original classification among the bacterial group of mycoplasmas. Those organisms also lack cell walls, but only grow in close association with plants or animals from which they derive essential growth factors. *Thermoplasma* was unusual among these organisms for both its extreme growth conditions and its ability to live without a host. Analyses of its 16S rRNA sequence and other molecular features clearly demonstrate that *Thermoplasma* is an archaeon.

Thermoplasma also uses a modified Entner-Doudoroff pathway for glucose catabolism (Fig. 3). Unlike that found in the halophiles, this pathway does not yield net ATP synthesis. *Thermoplasma* apparently uses pyruvate to form acetyl-CoA, which is then used to form acetate and ATP through the action of an acetyl-CoA synthetase. It is not clear whether or not *Thermoplasma* is capable of coupling proton extrusion via membrane-associated electron transport with ATP synthesis nor is anything known about its metabolism under anaerobic conditions.

3. Thermococcus/Pyrococcus

Members of the genera *Thermococcus* and *Pyrococcus* are very closely related and differ mainly in their optimal growth temperatures. *Thermococcus* grows optimally at 88°C while *Pyrococcus* grows optimally at 100°C. They are both obligate anaerobes that grow at the expense of sugars and complex carbon sources by reducing elemental sulfur to hydrogen sulfide. They will grow to low cell densities without elemental sulfur, but the accumulation of hydrogen as an end product of their catabolism inhibits further growth. If this hydrogen is removed by continuous flushing of cultures during growth, growth can continue. It is thought that elemental sulfur serves to "detoxify" the hydrogen by preventing its accumulation through the generation of hydrogen sulfide.

These organisms belong to the shortest and deepest branch of the archaeal lineage. This suggests that they have diverged little from the common ancestor of the Archaea.

The physiology of these organisms has not been studied in detail, but a few enzymes have been isolated from *Pyrococcus furiosus*. *Pyrococcus furio-*

sus contains proteases and amylases that are active at temperatures up to 100°C and are very stable at lower temperatures. These make them attractive for industrial purposes. There is no evidence about which pathway is used for glucose catabolism.

III. Major Physiological Groups of the Crenarchaeota

A. Thermoacidophilic Crenarchaeotes

1. Habitats

The thermoacidophilic crenarchaeotes belong to the genera *Sulfolobus, Acidianus,* and *Acidothermus*. These organisms are obligate or facultative aerobes that grow above 60°C at a pH of <4. These conditions are found in nature in continental hot springs where sulfide-rich, geothermally-heated water comes into contact with air. The sulfide oxidizes to elemental sulfur in air, and microorganisms oxidize the elemental sulfur further to sulfuric acid. These acidophiles are also used in the leaching of metals from low-grade ores rich in metal sulfides. [*See* SULFIDE-CONTAINING ENVIRONMENTS.]

2. Physiology

Members of the genus *Sulfolobus* are heterotrophs that ferment complex carbon sources or respire aerobically. Some are also autotrophs that derive electrons from the oxidation of elemental sulfur to sulfuric acid under aerobic conditions or hydrogen under anaerobic conditions. *Sulfolobus* uses a reductive citric acid cycle and not a Calvin cycle to fix carbon dioxide. Some isolates have been found to grow by reducing elemental sulfur to hydrogen sulfide under anaerobic conditions. Thus, members of all genera of crenarchaeotes are capable of anaerobic growth by elemental sulfur reduction, but all the thermoacidophiles are also capable of aerobic growth.

Sulfolobus uses the same modified Entner-Doudoroff pathway for glucose catabolism that is used by *Thermoplasma*. Unlike *Thermoplasma*, however, ATP can be generated through a membrane-bound adenosinetriphosphatase (ATPase) under aerobic conditions. The mechanism of synthesis of ATP during anaerobic growth has not been elucidated. [*See* ATPASES AND ION CURRENTS.]

B. Strictly Anaerobic Crenarchaeotes

1. Habitats

The majority of the crenarchaeotes are strict anaerobes that reduce elemental sulfur to hydrogen sulfide and grow under mildly acidic conditions (pH 4.5–6.5) at temperatures >85°C. The many genera that compose this group are differentiated on the basis of cell morphology, optimal growth temperature, and growth substrates. They have been isolated from terrestrial, marine, and fresh water geothermal environments around the world where they often grow in close association with one another.

Pyrodictium occultum was the first organism isolated that has an optimal growth temperature >100°C (optimal growth at 105°C). Isolates were obtained from submarine volcanic areas off the coast of Italy. Subsequently, other species have been found in other deep marine geothermal areas. These environments are unusual in that the overlying water column creates high hydrostatic pressure allowing the heated seawater to reach temperatures above 100°C without boiling. Other archaea that grow optimally at or above 100°C have been found in these environments. Thus far, no bacteria have been discovered that grow at these temperatures.

2. Physiology

The intermediary metabolism of most of the strictly anaerobic crenarchaeotes has not been studied in detail. Some aspects of carbon metabolism in *Thermoproteus* have been examined. *Thermoproteus* is an obligate anaerobe that grows either autotrophically using hydrogen to reduce carbon dioxide or heterotrophically on sugars, simple alcohols, and organic acids with the reduction of elemental sulfur. One species, *Thermoproteus neutrophilus,* only grows autotrophically. It uses a reductive citric acid cycle to fix carbon dioxide. Interestingly, the addition of acetate to a culture represses carbon dioxide fixation, and the pathway becomes noncyclic through the loss of fumarate reductase activity. The nature of this loss is unknown.

The high optimal growth temperature of *Pyrodictium* has aroused interest in the thermostability of its enzymes. A hydrogenase from *Pyrodictium brockii* has been isolated that is maximally active in temperatures >90°C. Thermostable enzymes from *Pyrodictium* are interesting for studies of protein structure and stability as well as for potential biotechnological applications.

IV. Molecular Biology of the Archaea

A. Genetic Organization

1. Chromosome Size

The genomes of Archaea range from 0.84 to 2.3×10^9 daltons. These are smaller than that of *Escherichia coli*, but larger than that of many mycoplasmas. Archaeal chromosomes examined to date are circular as in most bacteria.

2. Operons

Bacterial genes are frequently clustered in operons, and at least some genes of archaea are also organized in operons. Although few operons have been characterized in detail, polycistronic mRNA transcripts have been identified in genes from methanogens and *Sulfolobus*. In addition, adjacent, polypeptide-encoding genes in operon-like structures have been observed in the halophiles.

Stable RNAs (16S, 23S, 5S, and tRNAs) are also found in operons. The organization of these differ from those found in bacteria, however. The typical bacterial 16S-23S-5S order is found among all methanogens and halophiles, and most have a $tRNA^{Ala}$ gene between the 16S and 23S genes as do bacteria. Unlike bacteria, a $tRNA^{Ile}$ gene is not linked to this $tRNA^{Ala}$ gene. Among the thermophiles (both euryarchaeotes and crenarchaeotes), the 5S gene is frequently found outside the 16S-23S operon. In *Thermoplasma*, the 16S and 23S rRNA genes are also unlinked.

3. Introns

Introns are usually considered features of eucaryal gene structure (although the *E. coli* T-even phages contain introns). Introns are also found in some tRNA genes from archaea. Introns are present in an rRNA gene from *Desulfurococcus* and in tRNA genes from *Halobacterium volcanii, Sulfolobus solfataricus*, and *Thermoproteus tenax*. The tRNA introns of archaea are small as are those found in the nuclear-encoded tRNAs of eucarya. The tRNA introns from eucaryal nuclei, *H. volcanii*, and *S. solfataricus* are found near the anticodon of the tRNA transcript. Among *Thermoproteus* tRNA genes, introns are found near the anticodon, in the anticodon, and on the anticodon stem.

B. DNA Replication

1. DNA Polymerases

DNA polymerases from archaea have been found that bear similarities to both eucaryal and bacterial DNA polymerases. Aphidocolin, an inhibitor of eucaryal α-type DNA polymerases, inhibits the DNA polymerases of some methanogens and halophiles. DNA polymerases from *M. thermoautotrophicum* and *S. acidocaldarius* more closely resemble those found in bacteria.

2. Topoisomerases

A novel DNA topoisomerase, reverse gyrase, is present in *S. acidocaldarius*. It is an ATP-dependent, type I topoisomerase that introduces positive supercoils into the DNA. A positively-supercoiled plasmid has been isolated from this organism. Other archaea also possess reverse gyrase activity and thermophilic bacteria of the closely related genera *Thermotoga, Thermosipho*, and *Fervidobacterium* also possess reverse gyrase. Topoisomerase II is also present in *S. acidocaldarius, H. halobium*, and methanogens. This activity is inhibited by agents effective against both bacterial and eucaryal topoisomerases.

3. DNA-Binding Proteins

DNA-binding proteins and histone-like proteins have been observed in all groups of archaea. A histone-like protein from *Thermoplasma acidophilum* has been studied in detail. It is a basic protein that condenses double-stranded DNA into globular structures that substantially increase the thermal stability of the bound DNA. It shares amino acid sequence identity with both the eucaryal histones and the histone-like proteins of *E. coli*. A DNA-binding protein isolated from *Methanothermus fervidus* shows significant amino acid sequence identity with eucaryal histones. It also forms nucleosome-like structures and increases the thermal stability of the DNA to which it binds.

C. Transcription

1. RNA Polymerase

The DNA-dependent RNA polymerases (RNAP) of archaea contain from 7 to 12 component polypeptides as compared with 4 in *E. coli* and 9 or more in eucarya. They have a single RNAP (eucarya have three, bacteria one) that is insensitive to antibiotics that inhibit bacterial RNAP. Based on subunit composition, these RNAPs fall into two major classes: *Thermococcus, Thermoplasma,* and the crenarchaeotes contain three large subunits designated BAC (subunit B the largest, C the smallest). The methanogens and halophiles contain four large subunits (AB′B″C). Immunological and genetic evidence demonstrate that these subunits are homologous and the split B subunit of the the methanogens and halophiles arose from the single B subunit of the thermophiles. The genes are arranged as B(or B″B′)-A-C. This arrangement is the same as the homologous genes in *E. coli:* β(homologous with B)-β' (homologous with A and C). It is also the same arrangement as eucaryal RNAP I genes: B(homologous with B)-A(homologous with A and C). *Archaeoglobus* also has three large subunits, but the A and C subunits are fused and the B subunit is split (A + CB′B″).

2. Promotor Sequences

In bacteria, the synthesis of mRNA begins with RNAP binding at promotors that are composed of two regions of conserved sequences located 35 base pairs and 10 base pairs upstream of genes. In eucarya, there is no similar RNA polymerase-binding site, and the RNAP requires associated protein transcription factors for activity. Most protein-encoding eucaryal genes are preceded by an AT-rich region 25 base pairs upstream of the gene. This region may align the RNAP II correctly. In archaea, an AT-rich region is also present 25 base pairs upstream of many genes, and, in methanogens, it has been demonstrated that RNAP can bind to this region unassisted. These methanogen RNAPs did require additional factor(s) to initiate transcription, however. Thus, the initiation of transcription in archaea shares features found in both bacteria and eucarya.

3. Termination Sequences

The termination of transcription of protein-encoding archaeal genes often occurs through the the formation of "hairpin-loop" structures in the mRNA that cause the RNAP to disassociate from the template. This is similar to a process in *E. coli* called "rho-independent" termination. In archaeal stable RNA genes, termination occurs at sites indicated by specific nucleotide sequences that in some cases are similar to those used by eucaryal RNAP III.

D. Translation

1. Ribosome Structure

The archaeal ribosome is composed of subunits similar to those found in bacteria. It is a 70S structure composed of two subunits (52S and 31S) each containing rRNA (23S + 5S and 16S, respectively) and from 53 to 64 polypeptides. The secondary structure of the rRNAs are more similar to those of bacteria than eucarya, but distinctive archaeal features are present. The ribosomal proteins of the halophiles and some methanogens are acidic unlike the basic proteins found in bacteria. The crenarchaeotes and the thermophilic euryarchaeotes do not have acidic ribosomal proteins, and the proportion of acidic ribosomal proteins increases along the phylogenetic tree up to the halophiles. This is paralleled by an increase in the internal salt concentrations seen in some methanogens and the halophiles. These ions presumably serve to stabilize the ribosomes because ribosomes from the halophiles dissociate in low salt *in vitro.* Archaeal ribosomes are also insensitive to chloramphenicol and streptomycin, which inhibit bacterial ribosomes. They do bind anisomycin but not cycloheximide, which both bind to and inhibit eucaryal ribosomes. Archaea also contain an elongation factor, which is adenosine diphosphate (ADP)-ribosylated by diphtheria toxin, a property they share with eucaryal elongation factors. [*See* RIBOSOMES.]

2. tRNA

Transfer RNAs (tRNA) from archaea are structurally similar to those found in bacteria and eucarya. They differ from both domains in specific details of their structure and nucleotide composition, however. Archaeal tRNAs contain a number of uniquely modified bases, some of these containing sulfur. The normal triplet code is used by archaea, however, protein synthesis is initiated by methionine rather than N-formylmethionine as in bacteria.

E. Cell Envelope

1. Cell Walls

The cell walls of archaea are very different from those found in bacteria and eucarya. Many archaea (including all crenarchaeotes, thermophilic euryarchaeotes, halophiles, and *Methanococcus*) have an outer envelope (or S-layer) composed of hexagonally or tetragonally arranged protein or glycoprotein subunits that are easily disintegrated by mechanical shearing or detergents. In some cases, such as with *Thermoproteus,* the isolated S-layer retains the shape of the cell indicating strong bonding between the subunits. Members of the "methanobacteriales" and the halophiles have rigid cell walls composed of heteropolysaccharides. In *Methanobacterium,* the structure of this polymer is similar to that of bacterial murein and is called "pseudomurein." Pseudomurein differs from murein in that it (1) contains N-acetyl-L-talosaminuronic acid instead of N-acetyl-D-muramic acid; (2) the linkage between the glycan moieties is $\beta(1 \rightarrow 3)$ instead of $\beta(1 \rightarrow 4)$; and (3) the amino acids in the peptide moiety are different, and no D-amino acids are present. This latter property makes these archaea resistant to penicillin and D-cycloserine. *Thermoplasma* does not have a cell wall. Instead, cell integrity depends on a rigid cytoplasmic membrane composed of lipopolysaccharides and dibiphytanylglycerol tetraethers (see following discussion). [*See* CELL WALLS OF BACTERIA.]

2. Cell Membranes

The lipids that compose the cell membranes of archaea are perhaps the most unique feature of this domain. The dominant lipid structure found in both the polar and nonpolar lipids of the Archaea is based on the five-carbon, branched isoprene unit. In polar lipids, these units are linked "head-to-tail" to form chains of up to 20 carbons, and two of these chains can be linked "head-to-head" to form a 40-carbon chain. These hydrocarbon chains are linked to glycerol by ether rather than ester bonds as are found in bacteria and eucarya (Fig. 4). Thus, a 20-carbon diether bilayer membrane can be formed or a 40-carbon tetraether monolayer membrane can be formed. A combination of these two types is also possible and a high tetraether : diether ratio may be used to stabilize the membrane at high temperatures in the thermophiles. Halophiles contain only the diether, methanogens contain a mixture of diethers and tetraethers (depending on their phylogenetic position and growth temperature), and the thermophilic euryarchaeotes and crenarchaeotes contain predominantly tetraethers (see Table I). The nonpolar lipids are also composed of isoprenoid units of 15 to 30 carbons. The details of the biosynthesis of these lipids are under investigation (particularly the formation of the ether bond), but mevalonate has been identified as an intermediate in the biosynthesis as it is in bacteria and eucarya. This property led to the use of mevinolin (an inhibitor of the mevalonic acid pathway in eucarya) as an inhibitor of the growth of halophiles to select for mevinolin-resistant mutants that are useful for genetic studies. [*See* CELL MEMBRANE: STRUCTURE AND FUNCTION.]

V. Implications for Future Study

The recognition of the Archaea as a separate evolutionary lineage has had a tremendous impact on our view of how life has evolved and how cells function. Archaea have solved some of the same biological

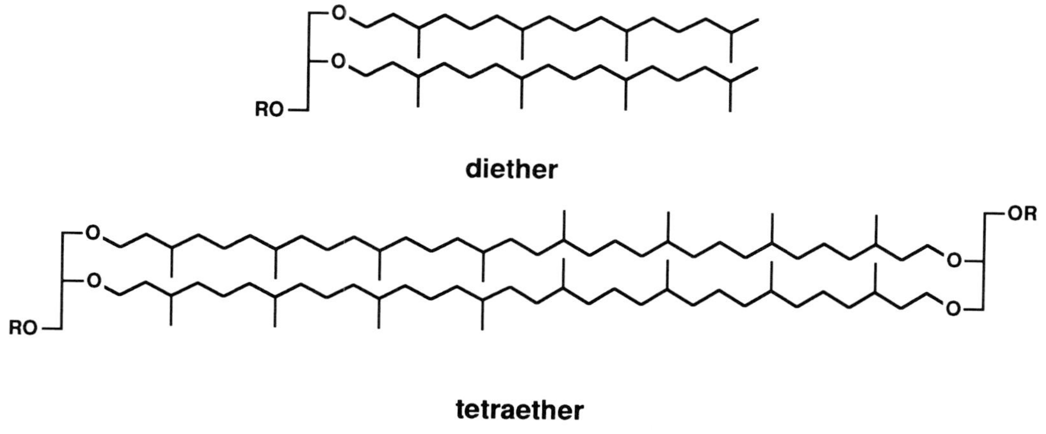

RO
O
O
diether

RO
O
O
O
O
OR
tetraether

Figure 4 Common glycerolipid structures of archaea. The diether is 2,3-di-*O*-phytanyl-*sn*-glycerol and the tetraether is 2,3,2′,3′-tetra-*O*-dibiphytanyl-di-*sn*-glycerol. The R groups include glycerol phosphate, phosphatidylglycerosulfate, and a variety of polysaccharides (some sulfated).

problems as bacteria and eucarya, but often in unique ways. For example, archaea employ novel methods of photometabolism, autotrophy, and cell envelope biosynthesis. Other features have not yet been found among the Archaea, such as multicellularity or resting stages (spores or cysts). As we learn more about the Archaea, we may begin to understand why organisms have evolved (or failed to evolve) particular metabolic capabilities. This will lead to a deeper understanding of the process of evolution at the molecular level.

It is interesting to note that the most slowly-evolving organisms of the bacterial lineage are thermophiles. The slowest of these belong the the genus *Thermotoga,* which physiologically resembles the most slowly-evolving archaeal lineages. *Thermotoga* are strict anaerobes, growing in temperatures of up to 90°C, that reduce elemental sulfur to hydrogen sulfide. They have unusual ether-linked lipids (though not isoprenoids) and have reverse gyrase activity. These factors have led Woese and others to speculate that the common ancestor of all life was a thermophile with similar features. From this, one might further speculate that the ancestral eucaryon was also a thermophilic anaerobe. If a slowly-evolving, anaerobic, thermophilic eucaryon were still alive, it would be very valuable to further define the Eucarya and to shed light on the mysteries of eucaryal evolution. As one begins to consider the organisms most closely related to the common ancestor of each of the three domains, one realizes that the features used to define each domain become few. Thus a true eucaryal ''living fossil'' may be difficult to distinguish from an archaeon.

Bibliography

Brown, J. W., Daniels, C. J., and Reeve, J. N. (1989). *CRC Crit. Rev. Microbiol.* **16,** 287–338.

Danson, M. J. (1988). *Adv. Microbiol. Physiol.* **29,** 165–231.

Jones, W. E., Nagle, D. P., Jr., and Whitman, W. B. (1987). *Microbiol. Rev.* **51,** 135–177.

Woese, C. R. (1987). *Microbiol. Rev.* **51,** 221–271.

Woese, C. R. and Wolfe, R. S. (1985). The bacteria. ''Archaebacteria'' Vol VIII, (C. R. Woese and R. S. Wolfe, eds.). Academic Press, Inc., Orlando, Florida.

Woese, C. R., Kandler, O., and Wheelis, M. L. (1990). *Proc. Natl. Acad. Sci. USA* **87,** 4576–4579.

ATPases and Ion Currents

Robert E. Marquis
University of Rochester

Glossary

Antiport Refers to counter-transport across a membrane catalyzed by an antiporter protein such that transport of one solute in one direction is coupled to transport of another solute in the opposite direction

Protonmotive force (Δp) Equal to $\Delta\Psi - z\Delta pH$, where $\Delta\Psi$ is the electrical potential across the cell membrane, ΔpH is the pH difference across the cell membrane, and z is equal to 2.3 RT/F (R is the gas constant, T the Kelvin temperature, and *F* the Faraday)

Symport Refers to co-transport across the cell membrane catalyzed by a symporter protein such that movement of one solute in one direction is coupled to movement of another solute in the same direction

ALL CELLS regulate the ionic composition of their cytoplasm by means of selective pooling of mineral ions such as K^+ and exclusion or excretion of other ions such as Na^+. Organic ions and other inorganic ions also are selectively taken up or extruded. The cell membrane is highly permeable to water but generally has much lower permeability to electrically charged species, primarily because of the lipid bilayer and hydrophobic proteins. However, the cell membrane also contains the major catalysts for selective movements of ions involved in transmembrane electrical currents. These currents are considered to be of fundamental importance in bioenergetics. In the late 1950s and early 1960s, Peter Mitchell formulated his chemiosmotic-coupling hypothesis based on earlier work of scientists such as E. J. Conway and H. Lundegardh. Mitchell proposed that respiratory electron transport resulted in extrusion of protons and build-up of electrical charge separation across the cell or mitochondrial membrane. The electrical energy involved in this charge separation could then be converted to chemical energy through the agency of the F_1F_0 adenosine triphosphatases (ATPases) or ATP synthases associated with the membrane. Specifically, the flow of protons back across the membrane was coupled to synthesis of ATP from adenosine diphosphate (ADP) and inorganic phosphate. Mitchell also developed the notion of the protonmotive force Δp equal to $\Delta\Psi$, the electrical potential across the membrane, minus $z\Delta pH$, where z is equal to 2.3 RT/*F* and ΔpH is the pH difference across the cell membrane. Here R refers to the gas constant (8.31 volt-coulombs $mol^{-1}\,K^{-1}$), T to the Kelvin temperature, and *F* to the Faraday (96,496 coulombs mol^{-1}). At a temperature of 25°C (298 K), z is equal approximately to 0.059 volts.

F_1F_0 or now simply F-ATPases can work also in the reverse direction for proton export coupled to hydrolysis of ATP. This export may be critical for maintenance of acid-base balances in acid-producing, fermentative organisms. Other ATPases may also function in proton transport; for example, vacuole V-ATPases serve to acidify the interiors of vacuoles or vesicles in a wide variety of cell organelles. E_1E_2 or P-ATPases, are commonly involved in transport of cations, and in contrast to F-ATPases or V-ATPases, they become phosphorylated during their catalytic cycles. Other ion transport systems may involve symporters or antiporters, which often are energetically coupled to ATPases. The energy for the work of solute transport may derive either from hydrolysis of ATP or from discharge of Δp across the cell membrane. The net result is a highly complex system of ion currents serving a variety of functions including respiratory or photosynthetic synthesis of ATP, osmoregulation, acid-base balance, motility, nutrient uptake, and product extrusion. [*See* ION TRANSPORT.]

I. F-ATPases (F_1F_0)

F-ATPases are considered to be evolutionarily derived from the same ancestor as vacuolar V-ATPases of primitive organisms. The current view is that primitive organisms were fermentative, and the primary function of their membrane-associated ATPases was in proton extrusion to keep the cell interior from becoming too acidified or to produce a potential for uptake of solutes. The ATP for proton extrusion would then be derived from fermentative reactions. With development of photosynthesis and respiration, F-ATPases functioned in the reverse direction as ATP synthases responsible for the bulk of ATP synthesized by the current biomass. For a human, this synthesis may be as much as a ton (1000 kg) per day, primarily by mitochondria with only small contributions from substrate-level phosphorylations. The other ATPases work mainly in the hydrolytic direction, and so, especially in eukaryotic cells, the ATP cycle may involve synthesis by F-ATPases and then hydrolysis by a variety of ATPases. For modern-day fermentative organisms, the major functions of the F-ATPases remain hydrolytic, especially in the transport of protons out of the cell to protect acid-sensitive cytoplasmic enzymes from damage and in development of ΔpH and Δp.

A. Molecular Structure

The F_1 or hydrolytic component of the enzyme protrudes from the inner face of the membrane and is made up of five subunits with a stoichiometry of $\alpha_3\beta_3\gamma\delta\epsilon$ (Fig. 1). The aggregate F_1 molecular mass is some 380 kDa, and the individual subunits (e.g., from *Escherichia coli*) have molecular masses of 55, 50, 31, 20, and 15 kDa, respectively. The β-subunit contains the catalytic site for hydrolysis. Current thinking is that the total F_1 complex has two active catalytic sites, and three sites on α-subunits have tightly bound nucleotide, which are not hydrolyzed during catalysis. The F_1 component of the enzyme can be removed from isolated membranes or derived F_1F_0 complexes by washing with low-ionic-strength buffers. The isolated F_1 then can catalyze ATP hydrolysis uncoupled from proton movements.

The F_0 component of the holoenzyme to which the F_1 is anchored is embedded in the membrane. In *E. coli,* it is made up of three polypeptides, a, b, and c, with respective molecular masses of 30, 17, and 8 kDa, and a stoichiometry of ab_2c_{6-12}. Thus, the aggregate mass of the F_0 component is 112–160 kDa, and the holoenzyme has an aggregate weight of some 492–540 kDa. The composition of the F_0 component varies much more among organisms than does that of the F_1 component. For example, the F_0 component from bovine heart mitochondria has eight or more subunits. The c-subunit of the *E. coli* enzyme is considered to form the proton channel through the membrane. However, genetic and biochemical data indicate that the a- and b-subunits must be present for coupled proton translocation to occur.

B. Isolation and Reconstitution

The F_1 component of the enzyme can be isolated from membranes by washing with low-ionic-strength buffers. Generally, a so-called osmolyte, such as glycerol, is included in the washing mix for conformational stabilization of the enzyme. Also, chaotropic agents such as thiocyanate can be used to enhance release of F_1.

The F_0 component is more difficult to isolate, and, generally, membranes must first be disrupted by detergents such as Triton X-100, octylglucoside, or deoxycholate. It is often most convenient to isolate the F_1F_0 holoenzyme, which can hydrolyze ATP, and then to separate the F_1 and F_0 components. The F_0 can then be incorporated into proteoliposomes for physiologic studies or can be separated into individual proteins by use of detergents such as sodium dodecyl sulfate.

F_1 components can be reattached to membranes previously stripped of F_1 or to isolated F_0 components to reconstitute the holoenzyme capable of reversible ATP hydrolysis coupled to proton translocation. The reattachment involves Mg^{2+} and the b-subunit of F_0, although a fully assembled F_0 appears to be required for reattachment. The δ- and ϵ-subunits of the F_1 are thought to be involved in the anchoring to F_0.

C. Genetics

Genetic aspects of the F-ATPase of *E. coli* have been explored in detail. The genes for the various subunits of the holoenzyme are arranged in an *atp* (*unc*) operon of approximately 7 kbp with a naturally

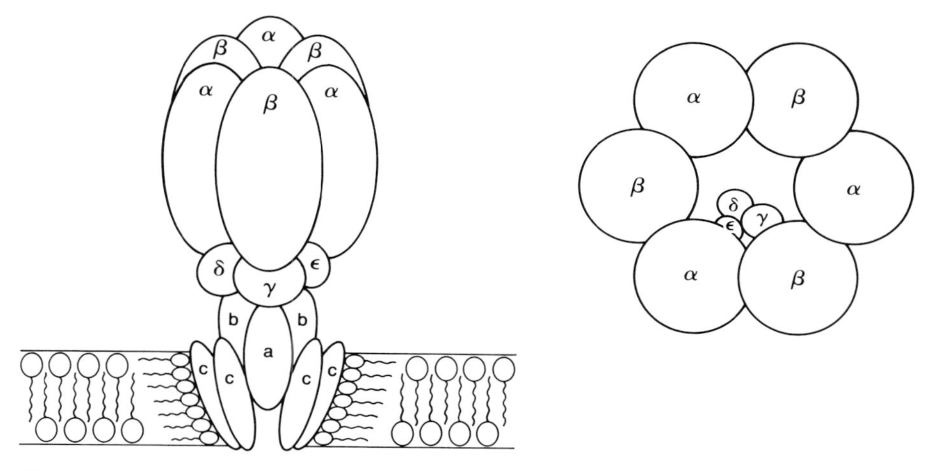

Figure 1 Diagram of the F-ATPase of *Escherichia coli*. Greek letters indicate F_1 proteins; latin letters indicate F_0 proteins. Phospholipids in the membrane (⌒∿∿) are associated with the enzyme.

occurring promoter and terminator. The gene order in terms of the protein subunits is a, c, b, δ, α, γ, β, and ϵ. The gene names are then *atp*A, *atp*C, etc. Between the promoter and the gene for the a-subunit, there is a gene designated *atp*I encoding a protein for which no well-defined function has been found. The *atp* operon is transcribed to yield a polycistronic message. This messenger RNA (mRNA) must then be translated in relation to the subunit stoichiometry of the enzyme to avoid synthesis of excess subunits. The control of translation is primarily at the points of translation initiation, although the molecular basis for control is only partly understood. There appears also to be differential mRNA stability in that the earlier parts of the polycistronic message are less stable, at least functionally. The system allows then for coordinate transcription of the genes but disparate translation. Assembly appears to be tightly coupled to synthesis so that free subunits or F_1 components generally are not found in the cytoplasm.

D. Catalytic Mechanism and Functions

The minimal structure for hydrolytic activity of F-ATPases appears in general to be a complex of $\alpha_3\beta_3\gamma$, which can be reconstituted from purified subunits. The γ-subunit appears to be stabilizing for the complex but not involved in catalysis. The enzyme exhibits extreme cooperativity, which has been interpreted in terms of alternating site cooperativity with the binding of ATP at a second unoccupied catalytic site leading to promotion of catalysis at the first or previously occupied catalytic site.

Adenine nucleotides can bind to at least two types of sites on the F_1 of *E. coli*. Noncatalytic sites, probably on the α-subunits, can bind ATP tightly without need for Mg^{2+}, while catalytic sites on the β-subunits can bind guanine or inosine nucleotides, ATP, or ADP in exchangeable forms in the presence of Mg^{2+}. The enzyme can catalyze hydrolysis of a variety of purine nucleoside triphosphates, although maximal hydrolytic rates are generally obtained with ATP. Mg^{2+} is the preferred cation; Ca^{2+} is less effective, and other divalent cations are largely ineffective. Substrate specificities and binding sites vary somewhat among F-ATPases from various sources. For the *E. coli* enzyme, overall K_m values for ATP are some 0.2–0.6 m*M*. Maximum rates of multisite catalysis are approximately 600 sec^{-1} for the F_1 enzyme and about half this value for the F_1F_0 enzyme. A proposed mechanism for ATP synthesis involves ADP and P_i binding tightly at one catalytic site on the β-subunit to form ATP, which remains tightly bound to the site. The energy for release then comes from the proton gradient, and binding of ADP and P_i to a second catalytic site also is required for release. This multisite mechanism is then effective in allowing ATP synthesis without product inhibition by cytoplasmic ATP, which in a metabolizing cell may be at a concentration of some 3–5 m*M*. As already indicated, ATP hydrolysis also is highly cooperative, and binding of ATP to a second catalytic site accelerates hydrolytic rates some 10^6-fold over the value of about 10^{-3} sec^{-1} for unisite catalysis.

Proton movements through the F_0 are thought to be mediated by proteolipid subunit c, which occurs in multiple copies in each F_0, and by subunit a. However, subunit b is required for a functional F_0. The c-subunit spans the membrane lipid bilayer in a hairpin form with the midchain bend protruding from the cytoplasmic side of the membrane. The a-subunits are considered to have four to six transmembrane helices. However, the exact structure of the H^+ channel in the F_0 and the mode of coupling to ATP synthesis or hydrolysis remain to be determined.

II. P-ATPases (E_1E_2)

P-ATPases differ from F-ATPases in that they form covalently phosphorylated intermediates during their reaction cycle. Thus, they pass through at least two distinct conformational states, E_1 and E_2, during catalysis, and the older E_1E_2 designation reflects the two-state cycle. The enzymes include a variety of cation-specific transporters in which the energy of ATP hydrolysis is coupled to transmembrane cation movements. Their actions are not reversible, and so ATP synthesis is not one of their functions. In general, the sites for covalent phosphorylation are distant, at least in terms of known amino acid sequences, from ion-binding sites, which may even be on other molecules of a P-ATPase complex. P-ATPases perform a variety of functions in eukaryotic cells as shown, for example, by the Na^+/K^+-, Ca^{2+}-, or H^+-transporting ATPases of plasma membranes and the Ca^{2+}-transporting ATPases of endoplasmic or sarcoplasmic reticula. However, P-ATPases also occur commonly in prokaryotic cells, and the inducible, K^+-transporting, Kdp ATPase of *E. coli* has been studied in detail genetically. Other P-ATPases of prokaryotes include an H^+-ATPase and possibly also a K^+-ATPase of *Enterococcus hirae;* Ca^{2+}-ATPases of *E. hirae, Streptococcus sanguis,* and *Lactococcus lactis;* the plasmid-encoded Cd^{2+}-ATPase; the Mg^{2+}-ATPase of *Salmonella typhimurium;* the FixI ATPase of *Rhizobium;* and Kdp-like ATPases of a variety of bacteria.

A. Molecular Structures

The Kdp, K^+-transporting ATPase of *E. coli* has three subunits with molecular masses of 59 kDa (KdpA), 72 kDa (KdpB), and 20.5 kDa (KdpC). However, P-ATPases may have only a single subunit, or sometimes two. The holoenzyme of Kdp appears to be composed of one molecule of each subunit but may involve higher-order complexes of the three-subunit modules. The genes for the enzyme are arranged in the *kdp*ABC operon. Expression of the operon depends on products of the *kdp*D and *kdp*E genes in an adjacent operon. The latter genes code for a cytoplasmic and a membrane protein of a sensor-effector system in which KdpE protein is thought to be phosphorylated by KdpD protein in response to decreases in cell turgor but not specifically changes in intracellular or extracellular K^+ levels. The function of the Kdp system is to scavenge K^+ in K-deficient environments. Therefore, the enzyme has a very high affinity for K^+ with a K_m of about 2 μM. The K^+-binding site is thought to be on the KdpA subunit, which is highly hydrophobic, is embedded in the membrane, and may form the K^+-conducting channel. The KdpB protein contains the phosphorylation site, an aspartyl residue in the sequence (Ser or Leu)-Asp-Lys-Thr-Gly-Thr-(Ile or Leu)-Thr. The protein has at least two or three membrane-spanning regions, which appear to serve mainly as membrane anchors. The functions of the KdpC protein are currently not well defined.

The Ca^{2+}-ATPase of sarcoplasmic reticulum with only a single subunit has some six to eight membrane-spanning regions. Thus, a single subunit serves for ATP binding (site closer to the N terminus), phosphorylation, energy transduction, and ion binding (site closer to the C terminus).

B. Isolation and Characterization

Because the P-ATPases are tightly integrated into the membrane, detergents or other membrane-disrupting agents are required for solubilization. The Kdp enzyme can be solubilized with nonionic detergents to yield a three-subunit complex, which can then be highly purified by means of ion-exchange and gel-filtration chromatography to allow for subsequent isolation and characterization of individual subunits.

C. Catalytic Mechanisms and Functions

Details of catalytic cycles have been worked out mainly with eukaryotic P-ATPases (e.g., for the plasma-membrane, Na^+/K^+ ATPase). Initially ATP is bound to the enzyme binding site, and then phosphate is transferred to the aspartyl residue of the phosphorylation site in association with binding of

3 Na^+ ions from the cytoplasm. The enzyme then undergoes a conformational change ($E_1 \rightarrow E_2$) associated with release of the 3 Na^+ ions into the environment. The enzyme becomes dephosphorylated and binds 2 K^+ ions from the environment. The 2 K^+ ions are next released into the cytoplasm in association with a conformational change ($E_2 \rightarrow E_1$) and binding of ATP so that the cycle can begin again.

The primary functions of the P-ATPases are in transmembrane ion transport, notably inorganic cations, and the transport may be for osmoregulation, maintenance of ionic homeostasis, development of ion gradients for energization of secondary processes, detoxification of cations such as Cd^{2+}, or replenishment of intracellular stores of ions such as Ca^{2+}. As indicated, the P-ATPases do not work reversibly; therefore, they are not involved in ATP synthesis. In fact, they may have a dependency on F-ATPases for their supply of ATP. Some of the P-ATPases can transport H^+, and this transport is of major importance in processes such as gastric acid secretion. The functions of H^+-transporting P-ATPases of bacteria are not well defined, especially in organisms such as *E. hirae,* which has high levels of F-ATPase for maintaining acid-base balances in acidic environments.

III. Other ATPases

A. Vacuolar V-ATPases

The vacuole or V-ATPases of eukaryotic cells, like F-ATPases, are not phosphorylated during their catalytic cycles. Moreover, they are multimeric enzymes with aggregate molecular masses of some 500 kDa. The holoenzyme from lysosomes or chromaffin granules includes a more hydrophilic complex made up of five different subunits, A–E, and a more hydrophobic complex, which is an integral part of the membrane. The hydrophobic complex is made up of an a-subunit of molecular masses 20 kDa, approximately 6 c-subunits of molecular masses 16 kDa each, and other subsidiary subunits. The body of the proton pore through the membrane appears to be made up of proteolipid c-subunits. The stoichiometry of the more hydrophilic, catalytic complex is A_3B_3CDE, and molecular masses of the individual subunits are, respectively, 70, 60, 41, 34, and 33 kDa. The A-subunits appear to have the catalytic sites, while the B-subunits appear to play regulatory roles. Yeast V-ATPases (e.g., that of *Saccharomyces cerevisiae*) are of a somewhat simpler construction with only A- and B-subunits in the hydrolytic component and only c-subunits in the proton-translocating, membrane component (Table I).

Procedures for dissociation of catalytic complexes from isolated membranes, for separation of a–c-subunit complexes from detergent solubilized membranes, and for separation of complexes into individual subunits are similar to those applied to F-ATPases.

Genes for the catalytic A-subunit and for the proteolipid c-subunit have been cloned from a number of organisms and sequenced. The A-subunit is highly conserved with about 62% sequence identity in plants and fungi. There is only some 25% overall identity between A-subunit sequences and those of the β-subunits of F-ATPases; however, there are highly conserved regions, such as the ATP-binding site (GXXGXGKT). The A-subunits differ from β-subunits in having an 88-residue, nonhomologous sequence near the N terminus and in having multiple cysteine residues with two or three of them in the proposed active site. Still, it is considered that F- and V-ATPases are genetically related but divergent families of proton transporters.

The 16-kDa c-subunit of the V-ATPases appears to be closely related to the 8-kDa c-subunit of F-ATPases with four transmembrane stretches, one of which contains the glutamate residue for binding of the inhibitor dicyclohexylcarbodiimide (DCCD). The generally accepted view is that c-subunits arose from a common ancestor but that there was a gene duplication for the V-ATPases.

V-ATPases act to produce the protonmotive force needed to energize vacular organelles including lysosomes, synaptic vesicles, clathrin-coated vesicles, the Golgi apparatus, and various plant and fungal vacuoles. The primary component of this protonmotive force is then the ΔpH component. Acidification of the interior of the vacuole is important for degradative processes and other metabolic activities of vacuolar organelles, including damage to phagocytized microbes in phagolysosomes. V-ATPases appear to be critical for survival of eukaryotic cells. Site-specific mutations in either the B-subunit of the catalytic complex or the c-subunit of the transmembrane complex result in strains of the yeast *S. cerevisiae* restricted in growth to a narrow pH range around 5.5. Mutagenic alterations in V-ATPases of higher prokaryotes can be lethal and not just growth-restrictive.

Table I Characteristics of Microbial ATPases Involved in Transmembrane Ion Currents

	Representative enzyme				
Characteristic	F-ATPase *Escherichia coli*	V-ATPase *Saccharomyces cerevisiae*	Kdp ATPase *Escherichia coli*	A-ATPase *Sulfolobus acidocaldarius*	Ars ATPase plasmid
Subunits					
Hydrophilic complex	$\alpha_3\beta_3\gamma\partial\epsilon$	A_3B_3	ABC	A_3B_3C	A_2B
Hydrophobic complex	ab_2c_{6-12}	ac_6	A	c	B
Approximate aggregate molecular mass (kDa)	500	576	152	400	170
Hydrolytic subunit	β	A	B	A	A
Proton or ion channel	a+c	c	A	C	B
Inhibitors	Azide DCCD	NO_3^- NEM DCCD	VO_4^{2-} DCCD	NO_3^-	
Reversible catalysis?	Yes	No	No	Yes	No
Functions	Synthase proton export	Proton pumping	K^+ uptake	Synthase proton export	Toxic metal export
Phosphorylated enzyme intermediate?	No	No	Yes	No	No

DCCD, dicyclohexylcarbodiimide; NEM, N-ethylmaleimide

B. Archaebacterial A-ATPases

The proton-translocating ATPases of the plasma membranes of archaebacteria have been found to resemble more closely V-ATPases than the F-ATPases of eubacterial membranes, and this finding has been taken as an indication that archaebacteria may be more closely related to eukaryotes than to eubacteria. The hydrolytic complex of the archaebacterial enzyme of *Sulfolobus acidocaldarius* is composed of three subunits with molecular masses of 64 (A), 54 (B), and 28 (C) kDa. The stoichiometry is A_3B_3C, and the catalytic sites are in the A-subunits. The proton-conducting complex is made up of the proteolipid c-subunit with molecular mass of about 10 kDa. Sequence analyses have shown that the archaebacterial A- and B-subunits have some 50% identity with A- and B-subunits of V-ATPases but only some 25% identity with β- and α-subunits of F-ATPases. [*See* ARCHAEBACTERIA (ARCHAEA).]

The hydrolytic complexes of archaebacterial or A-ATPases from *Halobacterium salinarium* or *Methanosarcina barkeri* have only two subunits, A and B. The genes for the subunits have been cloned and sequenced. They are contiguous, and it appears that they may form an operon with *atp*A read before *atp*B.

The A-ATPases can work reversibly, and their primary function is as an ATP synthase. However, since archaebacteria live in extreme environments, the hydrolytic, proton-extruding function may also be of major importance for survival.

C. Resistance ATPases

The abilities of certain bacteria to tolerate toxic heavy metals often depend on ATPase systems that can transport heavy metal ions out of the cell in association with ATP hydrolysis. Genes for these enzymes are often carried by resistance plasmids so that the resistance may be gained or lost with the plasmid or may be transferred to other bacteria. The genes may be located on transposons to increase further their mobilities. Many resistance ATPases (e.g., the Cd^+-ATPase) appear on the basis of their amino acid sequences and protein structures to be members of the P-ATPase family. The CadA AT-Pase of resistance plasmids appears from the nucleotide sequence of the cadA gene to be a polypeptide of 727 amino acids with six transmembrane spans. Details of the mechanism for Cd^+ transport out of the cell are currently being investigated.

In contrast, the oxyanion-transporting ArsA ATPase is related by sequence analysis to other bacterial ATPases only in its ATP-binding region. Thus, it appears to be distinctive and to represent yet another class of ATPases.

The Ars system involves three proteins: ArsA, ArsB, and ArsC. Ars A is the ATPase, ArsB is an integral membrane protein 429 amino acids in length, and ArsC is a small, soluble protein involved in substrate specificity. The system in *E. coli* confers resistance to arsenite, arsenate, and antimonite. The genes are carried by a conjugative, R-factor plasmid, which also has a regulatory *ars*R gene and *ars*D so that the system is inducible by its substrate. Regulation appears to be at the transcriptional level, and the genes *ars*ABC form an operon.

D. Na^+-Pumping F-ATPase

An interesting variant of F-ATPases, which normally translocate protons, has been discovered in the anaerobic bacterium *Propionigenium modestum*. This enzyme normally translocates Na^+ rather than H^+. Its overall structure is the same as that of the F-ATPase of *E. coli,* and there is little question that it is a member of the F-ATPase family. The organism also has a membrane-associated methylmalonyl-CoA decarboxylase. Decarboxylation of the substrate is coupled to extrusion of 2 Na^+ per molecule, and the Na^+ gradient developed can be used for synthesis of ATP with the Na-ATPase functioning in the synthase mode. The *P. modestum* enzyme can translocate H^+ but only at low Na^+ levels <1 m*M*. The finding of a Na^+-translocating F-ATPase is considered contrary to proton-wire models for proton translocation through the membrane and interpretable more in terms of binding sites allowing weak interactions of the cations with appropriately positioned oxygen or nitrogen atoms.

Mycoplasma organisms have membrane ATPases that appear to be variants of F-ATPases, but they are insensitive to inhibitors such as azide. The enzymes appear to function mainly in expulsion of Na^+, but the bacteria may have more than one type of ATPase involved in current flow across the membrane.

E. ATPases Involved in Other Membrane Functions

ATPases may be involved in energy coupling in various membrane functions. For example, the SecA protein of *E. coli* has been termed a translocation ATPase, and this activity is required for translocation of secreted proteins such as OmpA (outer membrane protein A). The proOmpA with leader sequence binds to SecA, and this binding stimulates ATP hydrolysis essential for protein translocation. Enzymes such as protein kinase generally have ATPase activities, but the activity may be only indirectly associated with transmembrane ion currents.

The energy source for flagellar rotation in bacteria is the transmembrane proton potential (Δp), and so motility is indirectly coupled to the action of F-ATPases. There has been recent indication that the flagellar operon of *Bacillus subtilis* codes for a protein with ATPase activity, and so ATPases may be more directly involved in motility in either assembly or functioning of flagella. ATP is known to be required for tactic responses of bacteria (e.g., chemotaxis in *E. coli*). [*See* MOTILITY; FLAGELLA.]

IV. ATPase Inhibitors

Inhibitors have been useful for studies of catalytic mechanisms and for distinguishing among the major classes of ATPases. DCCD is a general inhibitor of F-, V-, and P-ATPases. It reacts specifically with the carboxyl group of aspartyl-61 in transmembrane segment 2 of the c-subunit of F-ATPases. It blocks proton translocation and, thus, is inhibitory for the holoenzyme but not for the F_1 enzyme dissociated from the membrane complex. However, at high concentrations, DCCD can react with carboxyl groups in the F_1 enzyme, and selectivity of the inhibition depends on assay conditions.

There also are inhibitors specific for the various classes of ion-transporting ATPases. For example, azide inhibits F-ATPases, vanadate inhibits P-ATPases, and nitrate inhibits V- and A-ATPases. V-ATPases have active-site cysteine residues and so are sensitive to *N*-ethylmaleimide, whereas F-ATPases are insensitive to the agent. The antibiotic oligomycin inhibits F-ATPases of eukaryotes; in fact, early on they were called oligomycin-sensitive ATPases. Oligomycin is without effect on bacterial F-ATPases because they lack the OSCP (oligomycin sensitivity conferral protein) subunit in the F_0 membrane complex.

Fluoride is a general inhibitor for ATPases, but the actual inhibitor is now considered to be a metal–fluoride complex, most commonly AlF_4^-. This complex mimics phosphate, and inhibition appears to involve formation of ADP-AlF_4^- in the catalytic site. Both F- and P-ATPases are inhibited by AlF_4^-.

V. Regulation

ATPases that serve functions vital only in certain environments, such as the Ars ATPase, may be regulated at the transcriptional level so that synthesis is repressed when the enzyme is not needed but induced in response to substrate. ATPases such as the F-ATPases serve more generally vital functions and are subject to more restrictive regulation. Levels of enzyme activity may be modulated only within a relatively narrow range in response to environmental conditions. Mention has already been made of the regulation of synthesis of F-ATPase at the transcriptional level aimed at avoiding excess synthesis of subunits required in only a single copy in the assembled enzyme. Regulation may occur also at the transcriptional level. An example is the response of *E. hirae* to acid stress with resultant increased levels of F-ATPase to give the bacterium increased capacity to extrude protons. However, the increase constitutes only about a doubling or tripling of the activity in cells not exposed to acid stress. The capacity of an infectious organism to respond to acid stress may be important in allowing invading cells to resist acid killing in phagolysosomes. Adaptation to acid stress is important also in a variety of other situations. For example, it appears that the bacteria in dental plaque responsible for tooth decay can adapt to plaque acidification by increasing F-ATPase activity and, thus, increase their capacities to bring about acid erosion of enamel.

VI. Evolution

Attempts to gain a view of the evolution of ATPases have focused on the molecular properties of the enzymes and on sequence data. The progenitor enzyme is usually considered to be the closest to the V-ATPase. This ancestor to both the V- and F-ATPase families functioned to transport protons in association with ATP hydrolysis. Primitive organisms were anaerobic, and the major source of ATP would be from fermentative reactions. Modern-day V-ATPases then arose from the primitive forms. As O_2-producing photosynthesis developed, F-ATPases capable of ATP synthase activity also developed, in chloroplasts, eubacteria, and mitochondria. Also, A-ATPases with synthase function evolved to serve the energetic needs of archaebacteria. In addition, the P-ATPases appear to be related to F/V/A types, but it is thought that they diverged in their evolution from V-ATPases very early on, prior to the F/V divergence, and so it is difficult to define clearly evolutionary relationships.

VII. Symporters and Antiporters

Transmembrane currents commonly involve transport of charged molecules in association with hydrolysis of ATP. However, transmembrane currents may arise secondarily with energetic coupling to the protonmotive force. The systems may be electrogenic in that their activity results in development of charge across the membrane, or they may be electroneutral in that current flow in is balanced by current flow out.

For example, the uptake of galactose by many organisms involves phosphotransferase systems energized by phosphoenolpyruvate but may also involve H^+/galactose symport. The symport system is electrogenic in that there is a net flow of protons into the cell associated with sugar uptake. Moreover, if the cytoplasmic pH value is higher than the environmental value, then proton flow is favored energetically, and the organism can couple galactose uptake to a proton gradient established by the activities of F-ATPases.

Antiport systems often are electroneutral. One of the best examples is the Na^+/H^+ antiport system of alkaliphilic bacteria. This antiport is crucial for growth of the organisms in alkaline environments and serves to bring in protons extruded during processes such as respiration in exchange for cytoplasmic Na^+. The protons then serve to keep the internal pH value from becoming too alkaline. Other cells may have K^+/H^+ antiporters similar to those of eukaryotic cells.

An interesting antiport system discovered recently catalyzes exchange of arginine for ornithine. Arginine is the substrate of the arginine deiminase system. It is degraded to ornithine, 2 NH_3, and CO_2, and an ATP can be obtained from the reaction catalyzed by carbamate kinase. The cell can conserve this ATP because of the antiporter, which catalyzes uptake of the substrate arginine in exchange for the product ornithine. Thus, the substrate is brought into the cell without energy expenditure, and there is a net production of one ATP per molecule of arginine catabolized. This ATP can then serve to energize growth or other vital functions such as protection against acid damage.

Bibliography

Dibrov, P. A. (1991). *Biochim. Biophys. Acta* **1056,** 209–224.

Krulwich, T. A. (1990). "Bacterial Energetics." Academic Press, San Diego.

Nelson, N. (1989). *J. Bioenerg. Biomemb.* **21,** 553–571.

Nelson, N., and Taiz, L. (1989). *Trends Biochem. Sci.* **14,** 113–116.

Penefsky, H. S., and Cross, R. L. (1991). *Adv. Enzymol.* **64,** 173–214.

Rosen, B. P., and Silver, S. (1987). "Ion Transport in Prokaryotes." Academic Press, San Diego.

Senior, A. E. (1990). *Annu. Rev. Biophys. Chem.* **19,** 7–41.

Silver, S., Nucifora, G., Chu, L., and Misra, T. K. (1989). *Trends Biochem. Sci.* **14,** 76–80.

Tisa, L. S., and Rosen, B. P. (1990). *J. Bioeng. Biomemb.* **22,** 493–507.

Automation in Clinical Microbiology

Carl L. Pierson and Kenneth D. McClatchey
The University of Michigan

Glossary

Antigen capture Process of removing a specific antigen from a mixture of heterogeneous antigens using an antibody or lectin specific for the antigen or ligand

Antimicrobic susceptibility Measurement of an organism's level of resistance to one or more antimicrobics

DNA primers Pair of oligonucleotides that attach to the ends of a target DNA sequence to be repeatedly copied using a polymerase chain reaction

Growth-based techniques Method used for organism identification based on the ability of the organism to use specific nutrients for growth

Infrared spectrometry Method of measuring of the amount of a compound present that absorbs light emitted in the infrared range of wavelengths (e.g., CO_2)

Inoculum Suspension of organisms usually adjusted to a defined concentration

Light-emitting diode Semiconductive vacuum tube that allows unidirectional flow of current; during neutralization of charge, light is emitted within a defined wavelength

Luciferin Thiazoline carboxylic acid that serves as substrate for the enzyme luciferase; in the presence of ATP, luciferase oxidizes luciferin with the resultant emission of light

Pixel Smallest area of a field that can be detected as a discrete entity by an imaging device such as a video camera

Rapid identification Definitive identification of an organism, usually within a 8-hr work period

Saponification Process of releasing fatty acids from glycerides using alkai

AUTOMATED SYSTEMS for the clinical microbiology laboratory ideally would allow the introduction of a clinical specimen into an instrument along with the necessary patient identifiers and the tests to be run and would have the requested tests completed within 1 hr, including appropriate antimicrobic susceptibility tests. Such an instrument does not exist; however, definite progress has been made toward this end. The systems that are available have limited capabilities and, compared with automated instruments designed for use in clinical biochemistry, hematology, and immunology laboratories, should be classified as semiautomated at best because most require the operator to become involved with various stages of the operation.

There are several reasons for the delay in the development of successful diagnostic instruments for microbiology. The diverse nature of the specimens received frequently requires some initial preparation of the sample. Fluid samples such as blood, spinal fluid, and urine may be handled directly, but catheter tips, swabs, and various solid tissues usually require some initial handling to make them suitable for subsequent processing. The infecting organism may also be present in very low numbers, requiring the system to be extremely sensitive. In addition, there are many different species capable of causing infectious disease that the system would have to be able to detect and identify, frequently in a milieu of commensal organisms. With a few notable exceptions, microbiologists have relied on growth-based techniques that require an initial period of growth and isolation followed by inoculation into various test media and reagents to determine the organism's identification and susceptibility to antimicrobics. This process requires days to weeks to accomplish, depending on the microorganism. Clinicians have reluctantly accepted this as a problem inherent to the discipline and, therefore, may treat their patients empirically, using the laboratory test results as they become available to verify their original diagnosis or to modify therapy in accord with the laboratory finding when necessary. The goal of the clinical microbiologist is to make test results available to the clinician in a timely manner for use in initial decision making. We are hopeful that automation will soon provide the necessary accuracy and speed for this to occur.

The semiautomated instruments currently available for use by clinical microbiologists can be divided into two groups: (1) those that analyze the specimen directly after minimal preparation, and (2) those that require the initial isolation of the organism in sufficient quantity to be used for organism identification or antimicrobial susceptibility testing.

I. Semiautomated Instruments for Direct Specimen Analysis

A. Blood

Currently no instruments being used in clinical microbiology are capable of determining whether a blood specimen contains one or more organisms without the need to culture. Some preliminary work with flow cytometers has been performed, showing that this technique is capable of detecting microorganisms in blood provided that they are present in relatively high numbers (e.g., 10^3 CFU/ml), which rarely occurs in clinical blood specimens.

To date only two systems are being marketed specifically for culturing blood specimens: the Bactec and the BacT/Alert.

1. Bactec (Becton-Dickinson Diagnostic Instrument Systems, Sparks, Maryland)

There are several models of Bactec instruments available. Early models of the Bactec (e.g., the 460) are designed to detect ^{14}C-labeled CO_2 liberated into the headspace of the inoculated specimen bottle by the action of microbial enzymes on labeled substrates. At periodic intervals, the bottles must be manually removed from an off-line incubator and placed into a carrier on a chain-driven track that moves each bottle under a set of two sterile needles. The machine pushes the needles through the diaphram of each bottle and displaces the headspace gas into a chamber where the amount of liberated $^{14}CO_2$ is measured. Any measurement above a predetermined threshold is flagged by the instrument as a "positive" culture. Such bottles need to be sampled and stained for evidence of growth, usually by Gram stain or acridine orange stain, and subsequently subcultured if found to be stain-positive. Studies indicate that blind and terminal subcultures of "negative" bottles are of little use, and therefore, most laboratories have eliminated this procedure when using the Bactec.

Recently, Bactec models NR-660, NR-730, and NR-860 have been introduced. These systems use infrared spectroscopy to detect accumulated CO_2 in the headspace of inoculated bottles, thus eliminating the need for ^{14}C-labeled substrates. Clinical studies have shown these systems to be as accurate as the Bectec 460. These models are able to process each bottle faster (35 sec) than the 460 model. Model NR-860 has the capability of autoloading and reincubating the trays containing the bottles without manual assistance. These models are computerized and have data management programs available.

A variety of media is available with and without antimicrobic-absorbing resins. The blood volume capacity of the bottles has also been increased from the previous maximum of 5 ml to 10 ml for better recovery of organisms in low concentrations.

2. BacT/Alert (Organon Teknika, Durham, North Carolina)

This automated system also monitors inoculated blood culture vials for the accumulation of CO_2 but has obviated the need for bottle entry by incorporating a CO_2 absorbing sensor pad at the bottom of each vial, resulting in a color change produced by a pH indicator when microbial growth occurs. Each pad is monitored by a matching light-emitting diode and reflectometer. Colorimetric readings of each vial are taken every 10 min. A vial is flagged as "positive" when the instrument's computer indicates there has been a significant color change from green to yellow in the sensor pad. Each unit can accomodate up to 240 aerobic and anaerobic vials. The current vials accommodate up to 10 ml of blood. Each bottle has a unique bar code, allowing for random placement into the agitator/incubator unit. As many as four units can be interfaced to one main computer module. Although the system is being marketed, only one clinical trial using this system has been published to date. Several other clinical trials are ongoing as of this writing. Preliminary data indicate that this system is comparable with the Bactec systems in detecting microorganism causing bacteremia.

3. Investigational Blood Culture Systems

Two other blood culturing systems similar to the BacT/Alert have been introduced, but to date there has been little clinical information on their efficiency.

1. Bactec 9240, produced by Becton-Dickinson Diagnostic Instrument Systems, Sparks, Maryland, uses fluorescent sensors to monitor CO_2 accumulation without bottle entry.
2. Bio Argos produced by Diagnostics Pasteur, Marnes-La Coquette, France, uses infrared spectrophotometry to monitor CO_2 levels. Preliminary studies have shown it to be equivalent to the Bactec 460 in sensitivity.

It must be emphasized that these instruments are designed to detect microbial growth only; they do not assist the laboratory in determining the identification of the organism or its antimicrobic susceptibility, with the exception of the Bactec 460 for *Mycobacterium tuberculosis* described below (see Section I.C.1). [*See* CENTRIFUGATION AND FILTRATION: DETECTION OF BACTERIA IN BLOOD.]

B. Urine

Several instruments designed to detect "significant" levels of bacteriuria have come and gone because of either an unfavorable negative predictive value or impractical methods for routine use. Those that remain find their primary use in high-volume laboratories where most urine specimens received are negative. These instruments are designed to eliminate negative urine specimens so that the technologists can put their efforts toward processing those that are most likely positive. Currently, only one automated system is devoted to the screening of urine, the UTI-Screen Model 633. Three other systems have, as part of their diagnostic capabilities, the ability to screen urine.

1. UTI-Screen Model 633 (Los Alamos Diagnostics, Los Alamos, New Mexico)

This system quantitates the amount of microbial ATP present in urine by measuring the amount of light emitted in the presence of luciferin and luciferase.

$$\text{luciferin} + \text{ATP} \xrightarrow[\text{luciferase}]{Mg^{2+}} \text{luciferin-AMP} + \text{PPi} \quad (1)$$

$$\text{luciferin-AMP} + O^2 \longrightarrow \text{oxyluciferin} + CO^2 + \text{AMP} + \text{light} \quad (2)$$

If the amount of light released is above a predetermined threshold, the urine is declared positive. Thresholds can be adjusted by the operator to obtain the level of sensitivity required. The model 633 consists of a luminometer and a computer module. Urine is added to a vial containing a somatic cell lysing agent that releases and destroys the ATP derived from host cells during an initial 10-min off-line incubation period. Thereafter, the system automatically adds all subsequent reagents and prints out the relative light units and determines whether the number of units is above threshold. The system can accomodate up to 23 urine specimens per run and the batch run time is 7 min. "Positive" urine specimens can then be cultured for identification and susceptibility testing. Little published information is available on this system, but it appears that it can accurately detect microorganisms at a level of $\geq 10^4$ CFU/ml of urine. The company also produces a less automated instrument, the model 535, that works on the same principle.

The following systems can be used for direct detection of organisms in urine, but this function is only one component of their overall capability. The reader is referred to later discussions about these systems for more complete descriptions.

2. Vitek (Vitek Systems, Inc., Hazelwood, Missouri)

The Vitek consists of a filler/sealer module, reader/incubator, computer, data terminal, and printer. It is capable of identifying most clinically significant gram-positive and gram-negative aerobic and anaerobic bacteria and yeasts and performs antimicrobic susceptibility tests. The urine is used to inoculate a small plastic card containing numerous reagent wells in a vacuum filling module. The inoculated card is removed and inserted into the reader/incubator module, where each well of each card inserted is monitored hourly for relative changes in transmitted light through each well. The system determines the end points of each reaction and, based on the accumulated results, provides an interpretation of the results. The system is capable of identifying nine of the most common microorganisms causing urinary tract infections within 8–9 hr and indicates if the urine contained at least 50,000 viable organisms/ml within 6 hr. Test-negative urine specimens can be discarded without further workup. Test-positive urine specimens need to be cultured to obtain a pure culture of the significant organism(s) for subsequent antimicrobic susceptibility tests, if requested, and to identify organisms not identified by the urine card. These tests can be performed on this system by using appropriate cards for the purpose.

3. MS-2/Avantage (Abbott Laboratories, Irving, Texas)

This system consists of one or more Analysis Modules that are connected to a Control Module. Each Analysis Module serves as both an incubator and a shaker. It contains eight chambers into which one inserts an ampule carrier or a plastic cartridge. The system has programs to perform urine screening, organism identification from culture, and antimicrobic susceptibility testing. Each cartridge has 10 compartments through which light-emitting diodes direct light to a corresponding detector. Evidence of growth is detected by a relative change in transmitted light. Readings are taken every 5 min and stored in the computer storage unit.

Urine is added to a glass ampule containing a growth medium, and the inoculated ampules are placed into a carrier. The carrier is inserted into one of the chambers and tubidity measurements commence. The computer determines when the tubidity readings have changed sufficiently to declare a position positive and prints out a positive result. Positions without significant change after 5 hr of incubation are declared negative. Negative urine specimens can be discarded and positive urine specimens cultured for subsequent organism identification and susceptibility testing using this system, if desired.

4. Autobac Series II (Organon Teknika, Charleston, South Carolina)

This system consists of a photometer and one or more incubator/shaker modules that are linked to a computer for system and data management. An aliquot of urine is added to a eugonic broth in a disposable multichambered cuvette. The cartridge is placed in the incubator/shaker for 4 hr, removed and placed in the photometer for measurement of amount of forward light scatter. Readings above a given threshold are called positive, and these urine specimens can be selected and cultured.

C. Other Body Fluids and Exudates

1. Bactec TB System (Becton Dickinson, Sparks, Maryland)

The Bactec 460 model can be retrofitted with an exhaust hood, allowing it to be used for the detection of mycobacteria in many different clinical specimens by inoculating bottles containing a formulation of Middlebrook broth. Studies have shown that the average length of incubation time to detection of specimens containing *Mycobacterium tuberculosis* was reduced from a range of 20 to 25 days using conventional techniques to less than 15 days. Paranitroacetylamino-β-hydroxypropiophene (NAP) can be added to the bottles to facilitate the identification of *Mycobacterium*, which is selectively inhibited by this compound. The bottles may also be used to perform antimycobacterial susceptibility testing by adding test agents to the bottles followed by inoculation and monitoring the growth index for evidence of growth indicating drug resistance. To date, *Mycobacterium* identification and susceptibility testing cannot be performed on the infrared systems. [*See* MYCOBACTERIA.]

2. Vitek ImmunoDiagnostic Assay System (Vitek Systems, Inc., Hazelwood, Missouri)

The "VIDAS" automates the enzyme-linked immunosorbant assay methodology and is designed to be used for the direct detection of various antigens and antibodies. Each module contains 30 slots for reagent strips and companion "solid-phase receptacles." The slots are grouped into five sections containing six slots per section. Each section responds to an established protocol governed by a computer. Each reagent strip contains all the necessary reagents to perform an assay. The solid-phase receptacle resembles a plastic pipette tip. Its inner surface contains either specific antibody for capturing corresponding antigens in body fluids or specific antigens to capture corresponding antibodies from serum. [*See* BIOASSAYS IN MICROBIOLOGY.]

Either serum or a body fluid is added to the first position of the reagent strip, and the strip, along with its matching receptacle, is inserted into the appropriate position in the module. Accession information is entered through the computer keyboard for the selected test and the run started when ready. From this point on, all further processing of the test is performed automatically by the module. At the final step, a fluorogenic substrate is acted on by alkaline phosphatase to yield a fluorescent signal above threshold if the specimen is positive. Because the module is divided into five independent sections, at least five different assays can be run at the same time. Different assays using the same protocol can be run in the same section. Test results are available in 1–2 hr, depending on the assay. The specimen makes contact only with the disposable reagent strip and receptacle; therefore, there is no need to decontaminate anything within the module.

Currently, the system can do direct qualitative assays for respiratory syncytial virus, herpes simplex virus, *Mycoplasma pneumoniae, Chlamydia trachomatis, and Clostridium difficile* toxin A. Serum IgG antibody assays for rubella, *Toxoplasma,* HIV-1, and *Borrelia burgdorferi* are also available. Plans are underway to also include DNA probe tests as they become available.

II. Semiautomated Systems for the Identification and/or Antimicrobic Susceptibility Testing of Isolated Microorganisms

Several systems have been developed to expedite the identification and susceptibility testing of bacteria and yeasts once they have been isolated. The major components for these systems are (1) a main module into which inoculated plates or cards are placed and where the reactions are determined, (2) a computer that controls the main module functions and stores and interprets the data, and (3) a video terminal and/or printer. Most systems can be interfaced to the mainframe of a laboratory information system.

These systems require that the organism be in pure culture before use for inoculation. The inoculum density must be adjusted to a predetermined density that will vary depending on the type of reagents being used. Each system has disposable plasticware that is designed to be used only in that system. The plasticware contains reagents, either in hydrated or dehydrated form, that the system uses to determine organism identification and/or susceptibility to antimicrobial agents. The inoculated panels or cards are then manually placed into a module that serves as both an incubator and a reaction reader. Each panel has unique identifiers that are stored in the computer. At various times the instrument reads the reactions and reports the results to the computer. The computer monitors the reaction rates in each reaction chamber and determines when the test is complete. The completed biochemical and/or susceptibility profiles are compared with a stored data base to determine the probability of correct identification. If no good match exists, the system will report no identification, and supplementary tests may have to be performed to achieve an identification. [*See* IDENTIFICATION OF BACTERIA, COMPUTERIZED.]

To date there are three systems that are commercially available that approach full automation: the Vitek system, the autoSCAN-W/A, and the API Aladin. In addition, the MS-2/Avantage automatically performs antimicrobic susceptibility testing and can perform some microbial identifications. The Microbial Identification System uses fatty acid profiling to determine identification.

A. Vitek (Vitek Systems, Hazelwood, Missouri)

The Vitek system (previously called the AutoMicrobic System) consists of a filler/sealer module, a reader/incubator module, a computer with video display, and a printer. The system uses credit card–sized molded polystyrene "CARDS" that contain a series of microcuvettes connected by a distribution channel that extends to the edge of the CARD. Each microcuvette contains either a lyophilized substrate, biochemical, or antimicrobic that is used for organism identification or susceptibility testing. Numerous CARDs are available that enable the user to identify most commonly isolated gram-positive and gram-negative bacteria, anaerobes, and yeasts and to perform susceptibility testing. CARDs are also available for direct urine quantitation and common isolate identification as noted previously (see Section I.B.2).

Organisms are suspended in saline to a match a no. 1 McFarland turbidity standard. The short arm of an L-shaped transfer tube is inserted into the side opening of the CARD's distribution channel, and the long arm is immersed into the tube containing the organism suspension. This unit is then placed into a filler/sealer module, which cycles to evacuate the CARD and, with repressurization, forces the inoculum to enter the CARD through the distribution channel and fill each microcuvette. This is followed by a sealing step where the transfer tube is removed and the opening to the distribution channel sealed. After appropriate labeling and recording of the cytochrome oxidase test result on the gram-negative identification CARD, it is ready to be placed into a carousel in the reader/incubator module, which can accommodate up to either 60 or 120 cards, depending on the size of the unit. The module is equipped with a optical reader head that scans the microcuvettes once every hour until the test is determined by the computer to be complete based on spectrophotometric changes that occur in the microcuvettes. Test results can be completed as soon as 4 hr or take up to 18 hr for nonfermentative gram-negative bacilli. The spectrophotometric results are transmitted to a computer for storage and used to compare with a stored data base to determine the most probable organism identification. Antimicrobic susceptibility results, including antimicrobic minimum inhibitory concentrations and interpretations, are provided within 8 hr. Status reports on any CARD can be accessed anytime after the first reading, and final reports are generated on the printer when the tests are completed for each card. A data management system is also available to provide several types of reports, including epidemiology reports.

B. autoSCAN-W/A (Baxer Healthcare Corporation, West Sacramento, California)

The central component of the autoSCAN-W/A is the automated incubation–interpretation cabinet with functions controlled by an IBM PS2 computer. The system uses 96-well plastic microtiter trays containing substrates and/or antimicrobics. The company now offers 53 different standard panels with the opportunity to obtain customized panels. Although a limited number of frozen, hydrated panels are available, more effort is being placed on preparing lyophilized panels for room temperature storage and extended expiration times. Two panel types are available containing either chromogenic or fluorometric substrates. The dehydrated panels are hydrated and inoculated simultaneously by hand, using a prepared inoculum in growth medium. User-generated bar codes are affixed to each inoculated tray, and the tray is inserted into the cabinet onto an available shelf. Depending on the panel type, the instrument adds the necessary reagents at appropriate times to selected microtiter wells and takes both baseline readings and readings at preselected times of each reaction well using fiberopic technology. Chromogenic panels are read by photometery using up to six different wavelengths; fluorogenic panels are read by fluorometry. Depending on panel type, incubation times range from 2 hr for the fluorogenic identification panels to 48 hr for yeast identification panels. Susceptibility result times range from 3.5 hr for the fluorogenic panels to 24 hr for the chromogenic panels. The final readings are transmitted to the computer, and the panel results are compared with a data bank of stored profiles. The system can identify more that 150 different species of gram-positive and gram-negative bacteria and yeasts. The program prints a biocode for the isolate and determines the probability of correct identification. A probability of ≥85% is considered correct identification. The data management software allows selective reporting of antimicrobic susceptibility information.

C. API Aladin (Analytab Products, Plainview, New York)

This instrument contains an incubator/reader compartment equipped with a video camera, an associated IBM AT computer, and a printer. Analytab Products produces a line of panels that identify gram-positive and gram-negative aerobes, facultative anaerobes, nonfermenters, anaerobes, and yeasts as well as antimicrobic susceptibility panels. Many of these products have been used in manual procedures for several years. These products in modified trays can be used in the Aladin system and processed automatically.

The organism inoculum is prepared as a tubid suspension and loaded into each microcupule or well of the selected panel type, and the panel is hand-labeled before being placed into a disposable carrier. Each carrier can hold an identification panel or a combination of identification and susceptibility panels. The incubator/reader compartment has a capacity of 60 carriers. Once loaded into the compartment, the system software directs the elevator system and robotics that move the carrier to either the reagent dispensing station where reagents are added to the panels or to the reading station for monitoring test results. Each panel is monitor by a video camera that produces images in the form of a digitized matrix or pixels. Each well is represented by 200–300 pixels that are activated, depending on the test being performed and whether it is a colorimetric or turbidimetric measurement. Because the entire area of the microcupule or well is being monitored, the computer can selectively monitor specific areas that are of most interest for that test (i.e., either the entire well can be monitored or just one region of the well). The system software controls all environmental conditions, determines when the reaction time has expired, reports the test results, and automatically discards the carrier into a biohazard bag. That position then becomes available for a new loaded carrier. Results become available as early as 4 hr; nonfermenters take up to 24 hr.

D. MS-2/Avantage (Abbott Diagnostics, Irving, Texas)

The avantage consists of an analysis module, a control module, a video display terminal, and a printer. Each analysis module has eight chambers to accommodate a disposable polystyrene cartridge divided into a row of cuvettes. Each chamber has a series of matching light-emitting diodes and detector pairs that monitor the optical density of each cuvette position every 5 min until the test is completed. The control module is a microcomputer (Hewlett-Packard) that controls all functions of the system and through which commands are entered. The computer can handle up to eight analysis modules.

This system is capable of screening urine for significant numbers of organisms, identifying clinically significant gram-negative bacteria and yeasts, and performing antimicrobic susceptibility tests. The use of the system for urine screening has been described previously (see Section I.B.3). Organism identifications are performed by making a suspension of the test isolate and using a multiheaded pipettor to transfer the inoculum into a 20-cuvette identification cartridge that contains lyophilized biochemical substrates. The loaded cartridges are sequentially loaded into a specific position in an analysis module for accessioning and taking initial readings that are stored in the computer memory. The cartridges are incubated off-line for a 4–5-hr period, indole and oxidase reactions determined, and the cartridge reinserted for final reading and organism identification.

Susceptibility testing is a more automated procedure. A two-tiered cartridge is used that has a nonsegmented upper chamber and a lower chamber divided into 11 cuvettes with openings at the base into which antimicrobic-containing disks are inserted and mechanically sealed. Fifteen milliliters of eugonic broth is added to the upper chamber, and the broth is inoculated with the test organism. The cartridge is then placed into one of the analysis module chambers and accessioned. The module continuously agitates and incubates the cartridges. The optical density of the upper chamber is monitored, and when the OD reaches a preset value, the contents of the upper chamber are forced into the individual cuvettes in the lower chamber and thus exposed to the individual antimicrobics. Optical density readings are taken every 5 min thereafter until the computer determines that the change in optical density is sufficient to declare the organism sensitive, intermediately sensitive, or resistant to the drug. The computer also produces a calculated MIC for intermediate results. Growth curves can also be pulled up on the video monitor and printed. This system has also been used to perform endotoxin tests using limulus lysate gelation.

E. Microbial Identification System (Microbial Identification, Inc, Newark, Delaware)

The Microbial Identification System consists of a gas–liquid chromatograph (Hewlett-Packard 5890A series II) equipped with a flame ionization detector, an automatic sample injector, and a computer system that controls the system and provides data output at the completion of the analysis time. System libraries are available for the identification of various bacteria and yeasts.

Organisms are grown on routine culture media using standardized culture conditions. Pure growth is removed from the media and placed in an extraction tube where the organisms are saponified to release the cellular fatty acids. These acids are methylated to form methyl esters, which are highly volatile, and thus become suitable for gas–liquid chromatographic analysis. Total sample preparation time is about 90 min. The extracted fatty acid methyl esters are added to a sealed vial and placed into the autoinjector, and the accession number information is entered into the computer. The system automatically injects the extract onto the column, alters the temperature according to a preset program, and stores the analyte data. The computer identifies each species of fatty acid present according to retention time on the column, the amount present, and the relative ratios of certain fatty acids. This information is compared with patterns stored in the microbial library, and a best match is determined. If a good match is made, the organism is identified to either the genus, species, or subspecies level. Options are also available to prepare a personal library using specific isolates of interest. The system can analyze up to 50 isolates in a 24-hr period.

III. Future Systems

The science of clinical microbiology will become most useful to clinicians and patients when the infectious agent(s) can be identified directly from the infected body site within minutes or, at the most, a few hours accompanied by appropriate antimicrobial susceptibilities. This precludes reliance on organism isolation and metabolic characterization traditionally used to identify most bacteria and yeasts. Recent technologic advances have put this goal within reach.

A. Amplification of Nucleic Acids

Rapid progress is being made on identifying regions of microorganism DNA and RNA that are unique to a species or subspecies. These unique segments can be sequenced and DNA primers prepared for use in amplification using the polymerase chain reaction. By using this technique, many investigators have been able to detect specific microorganisms in tissues and body fluids at very low quantities within hours of specimen collection. The detection of any specific microorganism in body tissue or fluid is theoretically possible, even in the presence of a mixture of other microorganisms. By knowing the nucleic acid sequence of regions coding for enzymes that can inhibit the activity, directly or indirectly, of various antimicrobics, the organism's susceptibility to these drugs could be predicted without the need to perform the traditional susceptibility test. Automation has developed to the point where it is already being used, in part, for these amplification techniques. In the near future this powerful technology could be available to most clinical microbiology laboratories.

B. Flow Cytometry

Flow cytometric analysis of various body fluids for the presence of microorganisms has been reported with varying degrees of success. As few as 100 organisms/milliliter of blood have been detected using this technique. Targeted organisms in a liquid specimen are labeled with fluorescent dye-tagged monoclonal antibodies specific for their surface antigens. The specimen is slowly added to the center of a liquid vortex that causes the microorganisms to pass single file through lazer light beams. This intense light causes the fluorescent label to release a fluorescent signal that is picked up by a detector and converted to an electronic signal that is stored in the system's computer. The system can detect not only the presence and quantity of specific organisms but can also determine DNA, RNA, and protein content, cell size and shape, and G/C ratios, which have been used to separate different populations of organisms. Methods have also been developed that allow analysis of solid tissues for the presence of specific infectious agents. With the expansion of available hybridoma facilties and resulting monoclonal antibodies, it is possible to detect and quantitate organisms directly in clinical specimens. It may also be possible to use flow cytometry to monitor the effects of various antimicrobics on microorganisms by monitoring changes in cell membrane integrity, an effect that may occur within minutes after exposure to the drug. [*See* FLOW CYTOMETRY.]

The current limitation on the use of these and other potentially useful technologies for the clinical diagnosis of infectious diseases is primarily the limited supply of available specific reagents (e.g., primers, monoclonal antibodies) and the current expense of the systems and procedures. As these methods become fully automated, user-friendly, and less expensive we can expect to see them being implemented in clinical laboratories, with the resulting clinical benefit to patients.

Bibliography

Baker, J. S., Ederer, G. M., and Mundschenk, D. (1983). *Am. J. Med. Technol.* **49,** 727.

D'Amato, R. F., Bottone, E. J., and Amsterdam, D. (1991). Substrate profile systems for the identification of bacteria and yeasts by rapid and automated approaches. *In* "Manual of Clinical Microbiology," 5th ed. (A. Balows, W. J. Hausler, Jr., K. L. Herrmann, H. D. Isenberg, and H. J. Shadomy,

eds.), pp. 128–136. American Society for Microbiology, Washington, D.C.
Jorgensen, J. H. (1987). "Automation in Clinical Microbiology." CRC Press, Boca Raton, Florida.
Jorgensen, J. H., Johnson, J. E., Alexander, G. A., Paxson, R., and Alderson, G. L. (1983). *Am. J. Clin. Pathol.* **79,** 683.
Kelly, M. T. (1985). *Clin. Lab. Med.* **5,**. 91.
Pezzlo, M. T. (1984). *Lab Med.* **15,** 539.
Pfaller, M. A. (1985). *Diagn. Microbiol. Infect. Dis.* **3,** 15S–23S.
Pfaller, M. A., Sahm, D., O'Hara, C., Ciaglia, C., Yu, M., Yamane, N., Scharnweber, G., and Rhoden, D. (1991). *J. Clin. Microbiol.* **29,** 1422–1428.
Sasser, M., and Wichman, M. D. (1991). Identification of microorganisms through use of gas chromatography and high-performance liquid chromatography. *In* "Manual for Clinical Microbiology," 5th ed. (A. Balows, W. J. Hausler, Jr., K. L. Herrmann, H. D. Isenberg, and H. J. Shadomy, eds.) pp. 111–118. American Society for Microbiology, Washington, D.C.
Tilton, R. C. (1981). "Rapid Methods and Automation in Microbiology. "American Society for Microbiology, Washington, D.C.
Van Dilla, M. A., Langlois, R. G., Pinkel, D., Yaiko, D., and Hadley, W. K. (1983). *Science* **220,** 620.

B

Bacteriocins: Activities and Applications

Dallas G. Hoover
University of Delaware

Glossary

Bacteriocin Bacteriocidal peptide or protein produced by a bacterium
Colicin Bacteriocin produced by *Escherichia coli* or closely related enteric species
Conjugation Transfer of genetic material by cell–cell contact
Constitutive Noninducible enzymatic activity
Indicator strain Target organism sensitive to bacteriocin
Lactic acid bacteria Members of the genera *Lactococcus, Lactobacillus, Leuconostoc,* and *Pediococcus,* whose primary by-product is lactic acid
Lysogeny Incorporation of bacteriophage DNA into host genome; phage replication and cell lysis do not occur until phage DNA is activated or induced
Plasmid Self-replicating extrachromosomal DNA
Ribonuclease RNA-degrading enzyme

BACTERIOCINS are proteinaceous compounds produced by bacteria that are lethal against other bacteria. Normally, the bacterial cells producing the bacteriocin are immune to its antagonistic action. It is generally believed that bacteriocins give the producing culture a competitive advantage over other bacteria existing in the same ecological niche. Most of the information on bacteriocins has come from the study of colicins, bacteriocins produced by strains of *Escherichia coli* and closely related members of the Enterobacteriaceae. However, bacteriocins have been shown to be widespread throughout the prokaryotic world and quite diverse in their chemical and physical properties. Nisin, a bacteriocin produced by *Lactococcus lactis* subsp. *lactis,* has been approved for use as a preservative in pasteurized processed cheese spreads in the United States.

I. Historical Aspects

The definition for the term bacteriocin has gradually changed since the first bacteriocin was discovered by Gratia in 1925. Called "principle V," it was produced by one strain of *Escherichia coli* and found effective against another strain of *E. coli.* The term colicine was coined by Gratia and Fredericq in 1946, whereas bacteriocine was first used by Jacob and his co-workers in 1953. Because the colicins were the prototype bacteriocins comprising the bulk of the early work in this field, the key characteristics of the colicins were used to define bacteriocins in general.

As a consequence, the original concept of a bacteriocin was colicin-based (Table I). However, the accumulation of knowledge from "noncolicin" bacteriocins has resulted in a broadened description, whereby a bacteriocidal protein not active against the producing bacterium warrants the name bacteriocin. Normally, bacteriocins are given names according to the genus or species of the strain that produces them—for example, pediocins from *Pediococcus* or boticins from *Clostridium botulinum*.

In characterizing a bacteriocin, an important point of concern is the natural occurrence of other antibacterial products produced by bacteria. Antagonism between different types of bacteria have been studied and documented for many years. Other potential inhibitory products that can resemble bacteriocin activity include excreted metabolites such as lactic and acetic acids, lytic agents and degradative enzymes, "classical" or therapeutic low-molecular weight antibiotics, compounds that chelate or remove key nutrients, and defective bacteriophage. The bacteriocinlike activity of these bacterial by-products have sometimes resulted in instances of misidentification.

II. Colicins

A. Classification

The colicins are the most well-studied bacteriocins. In microbiology, colicins have acted as model systems in the study of protein secretion by bacteria as well as immunity and regulation of toxic compounds by bacteria. It has been estimated that 40% of all *E. coli* strains are colicin-positive (Col^+). The genes encoding for colicin synthesis and immunity are almost exclusively on *Col* plasmids that are easily transmissible to other strains of Enterobacteriaceae by conjugation or cell-to-cell contact. The activity of colicins is only effective against a narrow range of other cultures of enterobacteria. Colicins are classified into 20 groups based on their effects to specific colicin-insensitive indicator strains of enterobacteria; however, this classification scheme has had problems when colicin-insensitive indicator strains display resistance to more than one colicin and when the producing strain synthesizes more than one colicin. Such occurrences in nature are not uncommon. Colicin nomenclature reflects the colicin type and the bacterial strain that first demonstrated the colicin in question. For example, K-K12 indicates colicin type K and *E. coli* strain K12. The colicin types are subdivided according to immunity. Both colicins E2 and E3, perhaps the most thoroughly studied of the colicins, are members of the E group because they are ineffective versus the E-resistant colicin indicator strain, but exceptions exist for this classification protocol.

Table I Colicin-Based Definition of a Bacteriocin

- Proteinaceous compound
- Bacteriocidal action
- Narrow activity spectrum primarily against the homologous species
- Attachment to specific cell receptors
- Genes for biosynthesis and host cell immunity plasmid-associated
- Production by lethal biosynthesis

B. General Structure and Mode of Action

Some colicins consist of two proteins. The bacteriocidal activity of colicins E2 and E3 are associated with proteins of approximately 50,000 daltons; however, these particles are commonly attached to single peptides possessing molecular weights of 10,000. The association of the larger subunit with the smaller subunit enhances the antagonistic action of the bacteriocin, but more importantly the small peptide imparts immunity on those cells possessing that particular *Col* plasmid. The peptide does this by inhibiting the nuclease activity of the larger subunit. Both E2 and E3 are enzymes. E2 has endonuclease activity on DNA and E3 has ribonuclease (RNase) activity.

Cloacin DF13 is very similar to colicin E3; it also has RNase activity. DE13 is composed of proteins with molecular weights of 58,000 and 9000. Like E2 and E3, the larger subunit of DF13 is active against whole cells and contains the structural functionality for the complex's host specificity, receptor selection, and penetration through the outer and inner bacterial membranes.

The mode of bacteriocidal action of colicin E2 (and also E7 and E8) on sensitive *E. coli* cells is specific inhibition of DNA synthesis with subsequent degradation of DNA. *In vitro*, purified E2 shows deoxyribonuclease activity, but to demonstrate actual virulence against sensitive cells the colicin must have the capability to penetrate the cell envelope before it can interact with the bacterial chromosome. E2, E3, and DF13 are toxins because their amino acid sequences give them a structural form that allows them to cross the bacterial envelope.

Protein synthesis is blocked when sensitive *E. coli* cells are treated with E3 or DF13. After penetration through the bacterial surface, both bacteriocins act upon the 70S ribosome, whereby a single break near the 3′ terminus of the 16S RNA of the 30S ribosomal subunit results in the loss of a 49-nucleotide fragment with subsequent termination of protein synthesis. In addition, DF13-treated *Enterobacter cloacae* show leakage of intracellular potassium. Both E3 and DF13 appear to have endogenous RNase activity.

The mode of action for colicins A, E1, K, Ia, and Ib is the formation of ion-permeable channels in the membrane. A significant portion of the secondary phenomena observable in colicin-treated cells is apparently due to collapse of the membrane electrical potential, implicating the membrane as the primary site of activity for colicins A, E1, K, Ia, and Ib. It has been proposed that membrane depolarization is caused by insertion of the elongated protein molecules into the cytoplasmic membrane forming aqueous channels. Specifically inhibited by these bacteriocins are nucleic acid and protein syntheses and electon transport. Leakage of potassium is also evident, and in some cases magnesium and cobalt is lost as well. Leakage of these cations has been suggested as the primary cause of cell death. Factors such as decreased levels of adenosine triphosphate, a lack of accumulated substrates from active transport, and an inability to retain sufficient concentrations of enzyme cofactors probably cause inhibition of macromolecular synthesis. The formation of a single channel in the bacterial cytoplasmic membrane would account for these factors, including the single-hit killing of colicin-treated cells. A single channel with sufficient ion flow could depolarize a bacterial membrane within minutes.

C. Immunity

The immunity systems that exist to protect the bacteriocin-producing strains from their own toxic metabolites are quite specific. They involve inhibitor subunits that interact with the functional portions of the bacteriocin, thus inactivating the bacteriocin and protecting the producing strain that maintains its synthetic capacity while releasing the bacteriocin into the environment. The removal of the immunity protein apparently occurs during penetration through the cell surface and only active molecules enter the cell cytoplasm. Expression of these compounds occurs even when the bacteriocins are not induced; i.e., even when only a small percentage of the bacteriocin-capable cells are producing bacteriocin, all the cells will show evidence of producing free immunity protein. The regulation of these inhibitory proteins appears constitutive and under the direction of the appropriate immunity-determining plasmid.

D. Synthesis

The most common means of detecting bacteriocin activity is the use of solid growth media, usually either surface growth or a soft agar overlay. Therefore, the bacteriocin-producing bacteria growing in or on the agar medium must release the bacteriocins from the cell surface and the bacteriocin must be capable of freely diffusing into the surrounding medium to demonstrate antagonistic effect. This effect is a measurable zone of inhibition or halo of no growth surrounding the producing culture in a lawn of indicator strain growth. In the biosynthesis of colicins, the control mechanism is similar to that of temperate phages in lysogenic bacteria. In lysogeny, most bacterial cells are not producing phage particles, the production of the phage is suppressed, lysogenic bacteria are immune to the effects of similar phage, and the production of the phage is inducible. It is estimated that approximately 0.01% of a *Col*-positive population produces colicin when uninduced, although in an older population (stationary phase) the figure can be as much as 10%.

Presently no colicin-repressor protein analogous to the bacteriophage-repressor protein has been characterized. When a *Col*-positive population is induced (e.g., addition of mitomycin C, ultraviolet irradiation), at least 95% of the cells will be secreting colicin. With colicins such as E2, the colicin kills the producing cells, but for other colicins such as Ib, the producing cells are not killed. The mechanism for induction and reason for lethal biosynthesis are not understood. It appears that almost all bacteriocins from gram-negative bacteria are inducible, whereas most bacteriocins from gram-positive bacteria are not. Lactacin B produced by *Lactobacillus acidophilus* N2 is an example of an inducible bacteriocin from a gram-positive bacterium.

III. Methods for the Determination of Antagonistic Activity

A. Other Antagonistic Agents

Demonstration of inhibitive effects between cultures of bacteria is very common. In 1676, Antonie van Leeuwenhoek first documented antibiosis, whereby the product from one microorganism inhibited the growth of another. In 1877, Louis Pasteur and J. F. Joubert described the antagonistic effect of common bacteria from urine on *Bacillus anthracis*. From these early times in microbiology to the present, the actual mechanism of the biological effect has often been undetermined. Bacteriocins are but one class of compounds produced by bacteria that are antagonistic toward other bacteria. Besides the bacteriocins, possible inhibitors include the clinical or therapeutic low-molecular weight antibiotics, lytic agents, enzymes, defective bacteriophage, and other metabolic by-products such as ammonia, organic acids, free fatty acids, and hydrogen peroxide. Additional factors influencing microbial interactions are the competitive nature of nutrient acquisition between microbial populations and the unequal sensitivity of microbial groups toward environmental factors such as oxygen or low pH.

One of the earliest technical problems during the study of colicins was to distinguish them from bacteriophage. The action of bacteriophage mimics bacteriocin activity. In fact, many of the initial observations of bacteriocinlike activity originated with the study of lysogeny. Bacteriophage and colicin activities are similar in that both necessitate attachment to specific cell-surface receptors. In some cases, both share the same receptor. Also, inducers for phage production and subsequent cell lysis induce colicin production. Both prophage development and colicin production are lethal biosyntheses, and *Col*+ cells are immune to their homologous colicin just as prophage carriers are immune to their corresponding phage. A key differentiating characteristic is that bacteriocins do not carry genetic determinants necessary for self-replication within sensitive organisms. Only bacteriophage can be propagated on host cultures. A method of distinguishing the two phenomena is dilution to extinction. Increasing dilutions of preparations of bacteriocins gives diminishing zones of inhibition on indicator agar lawns, whereas dilution to extinction of bacteriophage preparations result in an appearance of individual plaques. Of course, there is the real chance that a culture has the dual capability of producing bacteriophage and bacteriocin. Additional means to elucidate the resemblance of effect of these two agents include pretreatments to exploit the greater resistance of bacteriophage to trypsin and the much greater resistances of bacteriocins to heat and ultraviolet irradiation. Use of a reverse-side–soft agar overlay technique will prevent contact of the indicator and producing strains that will limit antagonism by bacteriophage. Given the much broader activity spectra of bacteriocins produced from gram-positive bacteria and the narrow, often intraspecific host range of bacteriophage, the confusion between bacteriocin or bacteriophage effects is reduced. [*See* BACTERIOPHAGES.]

When dealing with bacteriocins from acid-producing strains, especially the lactic acid bacteria, one must be aware of metabolic by-products such as organic acids and hydrogen peroxide. Lactic acid, and sometimes acetic acid, can be produced in very high amounts. Inhibition by organic acids is due to the low pH and the undissociated forms of the acids that easily penetrate the bacterial cell causing death. Dialysis of the growth extract from producing strains prior to testing effectively removes these low-molecular weight compounds. Hydrogen peroxide is a common by-product of metabolism, especially by lactobacilli. Growth under anaerobic conditions or amendment of growth media with enzymes such as catalase and peroxidase will prevent the occurrence of hydrogen peroxide.

Other compounds producing bacteriocinlike properties are low-molecular weight inhibitors and bacteriolytic enzymes. Reuterin is a low-molecular weight (700), nonproteinaceous substance produced by *Lactobacillus reuterii* and found to be effective against gram-negative (*Pseudomonas, Salmonella,* and *Shigella*) and gram-positive (*Clostridium, Staphylococcus,* and *Listeria*) bacteria as well as yeast (*Candida*) and protozoa (*Trypanosoma*). Reuterin is apparently a fatty acid. Many bacteria can synthesize bacteriolytic enzymes. For example, members of the genus *Staphylococcus* have been shown to produce lytic enzymes such as virolysin, lysostaphin, lysozyme, endo-β-*N*-acetylglucosaminidase, and phage-associated lysin. Further ambiguity exists because some characterized bacteriocins, such as colicin M and pesticin A1122, generally kill by cell lysis. Given the disagreement among investigators as to what actually constitutes bacteriocin activity, the reoccurring confusion in classification can be expected.

B. The Assays

Any pure-culture laboratory method will only estimate the level of antagonism a bacteriocin-producing culture or purified bacteriocin actually delivers in its natural environment or in an application such as food preservation. Despite inherent limitations, the screening for bacteriocin activity using solid growth media is well accepted, relatively easy, cost-effective, and, more importantly, a usually accurate reflection of inhibitory potential. Most testing for bacteriocin production involves general growth media, whereby the producing strain is inoculated as a spot or streak on agar seeded with indicator bacteria. A positive response is usually determined by comparing the zone sizes of test and control strains. The zone of inhibition is a result of the diffusion of bacteriocidal protein through the agar preventing replication of the target cells. One should keep in mind that not only is the extent of the zone directly related to the rate by which the bacteriocin diffuses through the agar, but also by the rate of growth of the indicator organism. These two factors are therefore greatly influenced by incubation time and the level of the indicator population. For example, unusually large zones of inhibition may be due in large part to slow-growing indicator bacteria affected by low incubation temperature, minimal media, etc., whereas small zones may result when the bacteria grow rapidly.

Deferred methods often are more sensitive than direct or simultaneous tests and allow separation of the variables of incubation time and conditions of incubation for the producing and indicator strains. An example of a deferred agar method is the Kekessy–Piguet assay, where the producing and indicator strains can each be grown on different optimal media. For this method, the producing or test culture is spot-inoculated into a spread plate; after growth, the agar mass is aseptically dislodged with a spatula from the petri dish bottom and transferred to the dish lid by striking the inverted, closed disk onto the bench-top until the agar flips down onto the lid. A soft agar (0.7%) overlay seeded with an appropriate concentration of indicator bacteria is then poured over the inverted agar and, following solidification and reincubation, bacteriocin-positive strains will display a halo of clearing in the lawn around the original button of growth.

Some bacteriocins have been found to be produced only on solid media. Most cultures of *Salmonella typhi* will produce their noninducible bacteriocins on agar surfaces but not in a fluid substrate. Slopes of nutrient agar are prepared from which all condensation must be carefully removed; otherwise, the *S. typhi* cells exposed to water will display a greatly reduced level of bacteriocin production.

Cell-free growth supernatants can be examined for antagonistic activity in petri plates. Using this approach, wells can be cut into the seeded agar and the extract placed into the wells. Another approach is to absorb extract into sterile paper discs that can be placed onto seeded agar, incubated, and examined for zones of inhibition.

One can measure growth of indicator bacteria in liquid medium containing different concentrations of bacteriocin. For example, dialyzed, cell-free growth extracts of the producing culture can be diluted with sterile growth medium and inoculated with indicator bacteria. The lag time, doubling time, and final turbidity is measured spectrophotometrically. Liquid-based assays are superior to agar-based tests when examining nondiffusible bacteriocins.

Alternative methods have been developed for determining bacterial growth and sensitivity to antimicrobial agents. These newer procedures are based on a range of metabolic activities such as pH changes, changes in redox potential, radiometry, bioluminescence, microcalorimetry, and impedance or electrical conductivity.

The titer of a bacteriocin preparation is usually determined using the reciprocal of the highest dilution causing a measurable degree of inhibition to an indicator strain under standardized conditions. Defining the units of bacteriocin activity is normally based on the amount of antagonism in 1 ml of the preparation. One must always be aware of nonspecific inhibitory effects that can confuse results.

A common step included in some assays measuring bacteriocin activity has been killing the producing strain with chloroform. However, chloroform sterilization has also been shown to inactivate some bacteriocins that are chloroform-sensitive. Use of chloroform for this purpose has diminished in recent years.

Bacteriocins from gram-positive bacteria can occur in a polymeric form and, because of this, do not readily diffuse through agar media. Therefore, it is often valuable to incorporate a surfactant, such as Tween 20 or 80 into the agar. This eliminates the need for prediffusion, permitting a simple diffusion assay. This is especially useful for the estimation of nisin in foods, because such samples are susceptible to cloudiness from food particles. Many foods contain interfering substances that can bind the polymeric forms of nisin, altering the zones of inhibition so that a direct reading from a standard curve can cause serious error.

Whatever the method for determining bacteriocin activity, none will be perfect. The fact that bacteriocin synthesis is often plasmid-associated assures a certain degree of phenotypic instability based on the transient nature of many plasmids. Marked strain-to-strain variation in maximum yields of bacteriocin has been established in some instances, as with colicin D, and confusion can arise with the frequent occurrence of multiple bacteriocinogeny as well as the presence of inhibitors or inactivators in the test medium. It is also true that the factors controlling the synthesis of bacteriocins is poorly understood, especially in gram-positive bacteria.

IV. Application of Bacteriocins

A. Nisin

1. Properties

Nisin is an antibacterial polypeptide of approximately 3500 daltons produced by some strains of *Lactococcus lactis* subsp. *lactis*. Nisin is undoubtedly the best studied of the bacteriocins produced by gram-positive bacteria. Of all the bacteriocins, including the colicins, it has demonstrated the greatest commercial application, having first been used as a preservative agent in 1951. Nisin has gained worldwide acceptance as a food preservative, including in the United States, where nisin received Food and Drug Administration approval for use in pasteurized processed cheese spreads in 1988.

Until recently, nisin has often been described as an antibiotic, but this term is now more carefully avoided in referring to nisin and other bacteriocins from lactic acid bacteria. There is considerable and justifiable resistance to the use of therapeutic or clinical antibiotics as food additives (e.g., it is estimated that 8% of the U.S. population is strongly allergic to penicillin). Although nisin can be considered an antibiotic because of its substantial antibacterial effect, it has been deemed more politically appropriate to call nisin a bacteriocin, in light of its use in foods and its production by food-safe, cheese-making lactococci. Nisin occurs naturally in milk and has been consumed by humans for thousands of years in cultured dairy products with no documented toxicological problems. Also, nisin is not used therapeutically in human or veterinary medicine, and it is not used as an animal feedstuff additive or for growth promotion. Table II summarizes the requirements for bacteriocins intended for food use.

At one time, there was concern that nisin use might hide unsanitary manufacturing practices in the food industry, but due to the narrow target spectrum of nisin (it is ineffective against gram-negative, rapidly growing, spoilage bacteria as well as molds) this is not a valid issue. In addition, the effectiveness of nisin against sensitive gram-positive bacteria depends on the bacterial load. The antibacterial effectiveness of nisin drops as the population of the microflora increases. It has been shown that although nisin is effective in extending the shelf-life of canned soups, it has no efficacy when poor-quality starting materials were used.

Originally, the molecular weight of nisin was thought to be 7000; however, in 1967 researchers showed that the molecular weight was about 3500. Nisin normally occurs as a dimer. The formation of aggregates by bacteriocins from gram-positive bacteria is a rather common occurrence, whereas bacteriocins from gram-negative bacteria rarely form multimers. The configuration of the atoms in the nisin molecule is unusual and complex (Fig. 1). The amino acid and the synthetic production has been accomplished, although commercial production is by concentration and purification of culture supernatant. Nisin contains L-amino acids and the unusual S-amino acids, lanthionine, and β-methyllanthionine. Because of these amino acids, nisin is

Table II Requirements for Bacteriocins Intended for Use as Food Preservatives

- Proven safe for human consumption
- Economically acceptable cost
- Proven effective at relatively low concentrations
- No detrimental effect on organoleptic characteristics of the food
- Stable during storage and effective for the shelf-life of the food
- No medical uses

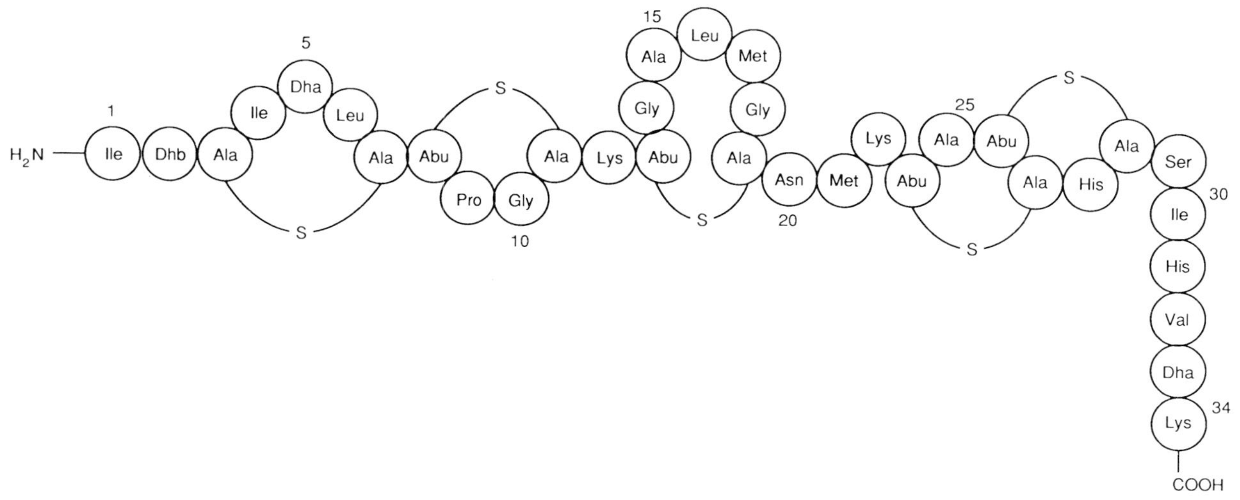

Figure 1 Structure of nisin.

sometimes referred to as a lantibiotic. These amino acids are common to nisin and subtilin (produced by *Bacillus subtilis*); nisin has a common ancestry with subtilin. There are five minor compositional variants of nisin: A, B, C, D, and E. The commercially available nisin is nisin A, which is the most biologically active of the nisins.

Nisin has no absorbance at 280 nm because it contains no aromatic amino acids. The solubility and stability of nisin depend on the pH of the aqueous solution. In solutions of dilute HCl at pH 2.5, its solubility is 12%, and it may be autoclaved without loss of activity. The solubility decreases to 4% at pH 5.0. At neutral and alkaline pH, nisin is practically insoluble and irreversible inactivation occurs even at room temperature. In low-acid foods at pH 6.1–6.9 and high-acid foods at pH 3.3–4.5, heating for 3 min at 121°C reduces the activity of nisin by 25–50%. Milk protects nisin from the effects of heat, but meat particles seem to increase sensitivity to heat. While refrigerated, purified nisin in powdered form is stable indefinitely, but nisin activity is gradually lost in foods. A reason for its inactivation in foods is the susceptibility of nisin to α-chymotrypsin and other endogenous proteolytic enzymes. Another factor is nisinase. Many lactic acid bacteria inactivate nisin by production of nisinase. Nisinase is apparently a dehydropeptide reductase. The presence of nisinase is a serious problem in cheesemaking and is a primary reason why nisin-producing cheese starter cultures cannot be used. Penicillinase at high levels has been reported to also inactivate nisin.

2. Antibacterial Activity

Nisin has no substantial effect against eukaryotic microorganisms and gram-negative bacteria, and the sensitivity of gram-positive bacteria varies. The organisms most sensitive to nisin are populations of *Lactococcus lactis* subsp. *cremoris,* whereas the closely related fecal streptococci, such as *Enterococcus faecalis* are most often resistant. Other asporogenous gram-positive bacteria that are inhibited by the presence of nisin include *Micrococcus, Staphylococcus, Mycobacterium, Corynebacterium, Listeria,* and *Lactobacillus*. What makes nisin so valuable commercially in food is its activity against the clostridia and bacilli. These endospore-forming bacteria are major spoilage organisms to foods, and *C. botulinum,* producer of the world's most deadly natural toxin, is of premier concern to food safety. *Bacillus coagulans* and *Bacillus stearothermophilus* (which produces the most heat-resistant spores known) are extremely nisin-sensitive. Small amounts of nisin increase the heat sensitivity of spores. This is why nisin treatment works so well in thermally processed canned foods. The heat treatment can be decreased because nisin makes the endospores more susceptible to the lethal effects of heating (e.g., thermal processing D values are reduced 50–60% with nisin).

The spores of the small-spored species of *Bacillus*

are more sensitive to nisin than those of the large-spored species. For instance, *B. subtilis* spores (small) are inhibited by about 5 IU/ml of nisin, whereas *Bacillus cereus* spores (large) require >100 IU/ml nisin for inhibition. Nisin appears to block pre-emergence swelling of spores.

In the *Clostridium* species, *C. sporogenes, C. butyricum* and *C. bifermentans,* the spores are significantly more sensitive to nisin than the vegetative cells. For *C. butyricum,* the sporicidal effect of nisin occurs after the spores germinate. Against *C. botulinum,* nisin is more effective against type E than type A. The amounts of nisin used to inhibit outgrowth of *C. botulinum* spores range from 100 to 5000 IU/ml, depending on the variety of *C. botulinum* tested, the spore load, and the medium examined.

In vegetative cells, nisin acts as a cationic surfactant that adsorbs strongly to the cytoplasmic membrane, disrupting the membrane by inactivation of sulfhydryl groups and inducing cell lysis. Resistant strains of *C. butyricum* have been shown not to adsorb nisin. This adsorption process is highly pH-dependent. Nisin blocks peptidoglycan synthesis at the cell wall.

3. Uses

Nisin is applicable as a food preservative in acidic foods and beverages, where nisin-sensitive, gram-positive bacteria require control. Nisin is very effective in cheeses, especially in pasteurized processed cheeses and cheese spreads. ''Blowing'' faults caused by *C. butyricum* and *Clostridium tyrobutyricum* are effectively controlled in processed cheese products at ranges anywhere from 100 to >500 IU/ml, depending on water activity, pH, the levels of sodium chloride and phosphate salts, and the legal limits set by the country of product origin. The use of nisin allows processed cheese spreads to have higher moisture contents and lower levels of salts while maintaining product stability. For dairy products, nisin is often sold mixed with skim milk powder. In many less industrialized countries, where the availability of refrigeration is limited, nisin is added to milk to increase its shelf-life and safety. In Egypt, nisin-amended whole milks can maintain satisfactory quality for 21 days at 37°C with the heat treatment or pasteurization process reduced by 80%. The use of nisin with reduced heat treatment makes powdered milk available for reconstitution in those parts of the world where water is unsafe and resources are scarce.

Eastern Europe adds nisin to canned vegetables. Companies in the Soviet Union and Poland produce nisin for large-scale commercial use by canneries. To prevent botulism, nonacid foods must receive a minimal heat treatment to destroy spores of *C. botulinum* type A, which are the most nisin-resistant of the bacterial endospores, but because the spores of spoilage clostridia and bacilli are more heat-resistant than those of *C. botulinum* type A, a greater heat treatment must accompany nisin use in these products. Nonetheless, the use of nisin significantly lowers energy consumption of the canning process and improves the texture, appearance, and nutritional quality of these products. [*See* TETANUS AND BOTULISM.]

Due to its narrow antibacterial activity spectrum, nisin plays a valuable role in the manufacture of alcoholic beverages, where lactic acid bacteria are important spoilage agents. Because nisin has no effect on fungi, nisin can be added to the alcoholic fermentation with no detrimental effects to commercial strains of wine or brewing yeast. Nisin inactivates the contaminating lactic acid bacteria of industrial yeast starter cultures so well that nisin treatment is replacing the traditional acid-washing techniques used to cleanse yeast strains because the use of nisin leaves the cultures of *Saccharomyces cerevisiae* with improved viability and unchanged fermentative performance.

The development of nisin-resistant strains of *Leuconostoc oenos* has led to successful pure culture malolactic fermentation in nisin-amended wines in the presence of undesirable lactic acid bacteria. Naturally occurring lactic acid bacteria can cause spoilage of the wine as well as inconsistent malolactate fermentations, resulting in a less desirable product. The application of nisin in this manner allows less sulfur dioxide to be used in the wine to prevent bacterial spoilage. Sulfites are another group of food additives where strong allergic reactions by consumers have been documented. Nisin has also been evaluated for use in fruit brandies and baked goods. [*See* WINE.]

Nisin has been examined for use in meats as a possible alternative or adjunct to nitrite in cured products such as cooked ham. Nisin and nitrite apparently have an additive effect against clostridia in some model meat systems (e.g., cooked ham slurry) where the nitrite level can be decreased to 40 ppm from the normally used 150 ppm without loss of preservative effect or color formation. However, reports in similar test systems have suggested the instability of nisin in stored refrigerated meats. More recent work presents evidence indicating that nisin does not work very well as a meat preservative. [*See* MEAT AND MEAT PRODUCTS.]

B. Other Bacteriocins for Use in Foods

In the marketplace, nisin is the only bacteriocin approved for use. All the other many bacteriocins can only be discussed with regard to future application. Given the historical association of nonpathogenic lactic acid bacteria with foods and beverages, it is understandable that their antibiotic proteins are being surveyed for use as preservative agents.

It is well known that members of the genera *Lactobacillus, Pediococcus,* and *Leuconostoc* synthesize bacteriocins and that *Lactococcus* produces other bacteriocins besides nisin. Diplococcin, a nonlanthionine-containing, 5300-dalton protein, is produced by some strains of *Lactococcus lactis* subsp. *diacetylactis* but has no commercial value due to its unstable nature and narrow inhibitory spectrum. It only inhibits other lactic acid bacteria. Other bacteriocins produced by lactococci include a group known as the lactostrepcins, which are from *L. lactic* subsp. *lactis* and are identified by numbers. The most well-studied is lactostrepcin 5, which has characteristics similar to those of diplococcin. Another bacteriocin, lacticin 481, has antagonistic activity against some clostridia.

Many species of *Lactobacillus* produce bacteriocins. Some of these characterized, named proteins are lactacins B and F, lactocins 27 and S, brevicin 37, caseicin 80, helveticin J, plantacin B, and sakacin A. Many of these derive their names from the producing species. Leucocin A-UAL 187 is produced by a variety of *Leuconostoc*. Pediocins A, PA-1, and AcH are produced by *Pediococcus*. Some of these, such as sakacin A, leucocin A-UAL 187, pediocins PA-1 and AcH, nisin, and a bacteriocin from *Carnobacterium pisciola* (*Lactobacillus carnis*), are effective against the foodborne pathogen *Listeria monocytogenes*. Although listeriosis is not as significant a food safety problem as salmonellosis, it is a relative newcomer whose infective dose remains undetermined. Foodborne listeriosis can be lethal, and the causative agent, *Listeria monocytogenes,* is commonly found in raw, animal-derived foods. Along with the inherent resistance of *Listeria* to extremes in temperature, acid, and salt, this organism has become a major concern in the food industry. Bacteriocins produced by lactic acid bacteria effective against *L. monocytogenes* offer a potential means of protecting our food supply against this infective organism. Thus, active research is ongoing to select and optimize conditions for this type of selective antibiosis. [*See* FOODBORNE ILLNESS.]

An advantage for commercial development that bacteriocins from gram-positive bacteria have over those from gram-negative bacteria is the ability to inhibit both gram-positive and gram-negative bacteria. The broader activity spectrum represents a greater range of applications. A good example is propionicin PLG-1 from *Propionibacterium thoenii,* a relative of the lactic acid bacteria and similar to the proprionibacteria used to put the "eyes" in Swiss cheese. Propionicin PLG-1 is not only inhibitive against lactic acid bacteria but also, more importantly, against deleterious gram-negative bacteria such as *Pseudomonas fluorescens, Pseudomonas aeruginosa, Vibrio parahaemolyticus,* and *Campylobacter jejuni.* Also, propionicin PLG-1 inhibits yeasts and molds such as *Aspergillus wentii, Candida utilis, Phialophora gregata, Saccharomycopsis fibuligera,* and *Trichoderma reesei.* Such versatility is highly desirable because the microflora present in foods and other products is so diverse.

Organisms such as *Staphylococcus aureus, Yersinia pestis, L. monocytogenes, C. botulinum, Streptococcus pyogenes,* and *Corynebacterium diphtheriae* all produce bacteriocins. However, given their established virulence as agents of severe human illnesses, the concept of putting a metabolic byproduct of their growth intentionally into a food or personal product simply is not acceptable—hence the screening of food-grade microorganisms for antibacterial or antimicrobial compounds.

C. Other Applications for Bacteriocins

Ambicin™ is a nontoxic bacteriocin formula that has been shown to kill bacteria that cause dental plaque on teeth and gums. It has been evaluated for intended use in mouthwash, toothpaste, soap, and other skin-care applications. It will be marketed as a safe and natural alternative to antibiotics and chemical germicidals. It is also effective in the prevention of bovine mastitis, a dairy cow disease of the udder that annually causes losses of $35 billion worldwide to the dairy industry. The bacteriocin formulation is rapid-acting and broad spectrum in activity and leaves no toxic residue to contaminate milk supplies; in addition, it shows no skin irritation common to chemical germicidals, which can dry and crack the teat skin leading to persistent intramammary infections. In less than 5 years, the market for mastitis teat dips and sprays worldwide is expected to reach $130 million.

Bacteriocins have application in biotechnology as genetic markers and in maintaining plasmid-bearing cells in continuous culture. As genetic markers, genes for bacteriocin synthesis and immunity can be used to tag or identify other plasmid-associated genes in gene transfer studies. That way, assuming the desired nearby gene or genes are cotransferred with the bacteriocin DNA, evidence of antibiosis by clones of the recipient strain can be used to screen for successful gene transfer. This can be easily visualized by indicator-seeded agar overlay of the recipient colonies. The zones of inhibition encircle those clones that suggest successful gene transfer; these can be isolated, grown, and examined for possession and expresssion of the desired properties. [*See* BACTERIOCINS, MOLECULAR BIOLOGY.]

By definition, plasmids are self-replicating extrachromosomal elements. Problems can arise from spontaneous plasmid loss, whereby segregational instability results in a lack of coordination with chromosome replication and cell division (i.e., one of the daughter cells lacks the plasmid and can outgrow those that do). Also, structural instability can arise from changes in plasmid DNA—such as insertions, deletions, or rearrangements—that prevent the desired gene from being expressed. Such loss is a critical issue in large-scale industrial fermentations where desirable traits are encoded on plasmids in genetically modified commercial strains. Stability refers to the persistence of cells that can express the plasmid-inserted gene over a long enough number of generations to be industrially productive. As already discussed, many naturally occurring plasmids give their host the ability to produce bacteriocins and to be immune to their effects. These plasmids can be used as vectors for genes encoding desirable traits. Therefore, in commercial bioreactors, plasmid-free cells arising from plasmid loss during cell division will possibly be killed by those cells retaining the plasmid, thus ensuring continued expression of the desired trait. It has been suggested that this kind of phenomenon is responsible for survival of such plasmids in natural environments. [*See* PLASMIDS; CATABOLIC PLASMIDS IN THE ENVIRONMENT.]

Bibliography

Banks, J. G., Board, R. G., and Sparks, N. H. C. (1986). *Biotechnol. Appl. Biochem.* **8,** 103–147.

Daeschel, M. A. (1989). *Food Technol.* **43**(1), 164–167.

Delves-Broughton, J. (1990). *Food Technol.* **44**(11), 100–117.

Klaenhammer, T. R. (1988). *Biochimie* **70,** 337–349.

Klaenhammer, T. R. (1990). "Proceedings of the 6th International Symposim of the Genetics of Industrial Microorganisms," Vol. I, pp. 433–445. Genetics of industrial Microorganisms 90, Strasbourg.

Bacteriocins, Molecular Biology

S. F. Barefoot, K. M. Harmon, D. A. Grinstead, and C. G. Nettles
Clemson University

Glossary

α-Helix Rodlike protein (or DNA) structure; tightly coiled protein backbone forms inner rod, and side chains extend outward in a helical array

β-Sheet Denotes a sheetlike protein conformation consisting of fully extended polypeptide chains laid side-to-side

Constitutive Refers to continuously expressed or unregulated genes

Genotype Genetic characteristics of an organism; denoted by italicized, lowercase letters (e.g., *bac* represents a gene for bacteriocin production)

Inducible Describes genes that are turned on when the appropriate signal is present; this signal inactivates the repressor and prevents its binding to the operator

Lex A protein Responsible for preventing or repressing synthesis of enzymes that repair damaged DNA and synthesis of colicins

Open reading frame Long stretch of triplet codons that is not interrupted by a stop codon; a protein-coding region

Operator DNA site at which binding of a repressor prevents synthesis of RNA

Operon Collection of functionally related genes subject to common regulation and transcribed into a single messenger RNA molecule

Phenotype Physical property of an organism; denoted by a capitalized, three-letter word (e.g., Bac represents bacteriocin expression)

Polycistronic DNA sequence that codes for the synthesis of several, usually related, proteins

Promoter DNA-binding site for RNA polymerase; start signal for RNA synthesis

SOS box Binding site for the Lex A repressor protein

Surfactant Surface active agent; decreases surface tension and increases fluid wettability

Vector Vehicle, often a plasmid, for cloning DNA; typically contains a selective marker (e.g., antibiotic resistance), an origin of replication compatible with the host, and a gene inactivated by insertion of passenger DNA

BACTERIOCINS are classically defined as bactericidal proteins with activity restricted to species closely related to the producer culture. More recently, that definition has been broadened to include all bactericidal proteins. Excluded are bacteriophages (bacterial viruses) and bacteriophage components that sometimes have been classified as particulate bacteriocins. Only bacteriocins that are soluble proteins will be discussed in this article. Bacteriocins have been studied since bacterial inhibition of closely related species was first observed in the late nineteenth century. The molecular biology of bacteriocins, or their physical, biochemical, and genetic characterization, has received considerable attention during recent years.

Bacteriocins are produced by diverse genera including *Bacillus, Bacteroides, Brucella, Carnobacterium, Caulobacter, Citrobacter, Clostridium, Corynebacterium, Enterobacter, Escherichia, Halobacteria, Klebsiella, Lactobacillus, Lactococcus, Leuconstoc, Listeria, Micrococcus, Mycobacterium, Neisseria, Pasteurella, Pediococcus, Propionibacterium, Proteus, Pseudomonas, Salmonella, Sarcina, Serratia, Staphylococcus, Streptococcus,* and *Vibrio*. Bacteriocins from most genera have been purified and biochemically characterized. Within the last 10 years, much information has become available about their actions against sensitive cells, their cellular targets, and the basis for bacteriocin insensitivity. Knowledge of the determinants and genetic regulation of bacteriocin induction, synthesis, and excretion has increased. Most recently identified genes encoding bacteriocins remain plasmidal; however, chromosomally encoded bacteriocins have been identified. Moreover, both plasmid and chromosomal genes in a few organisms may function in release, regulation, and immunity.

Molecular biology techniques have provided information about bacteriocins from both gram-positive and gram-negative species. However, colicins, the prototype bacteriocins produced by *Escherichia coli* and other *Enterobacteriaceae,* remain the most extensively characterized. This article will concentrate on the molecular biology of three groups of bacteriocins: (1) the colicins, (2) the low-molecular weight microcins produced by *E. coli,* and (3) the heterogeneous bacteriocins produced by lactic acid bacteria. For a listing of bacteriocins, producer organisms, and characteristics relevant to this discussion, see Table I.

I. Production and Purification of Bacteriocins

A. Conditions for Bacteriocin Production

Production conditions for bacteriocins apparently are unique to each bacterial species and/or strain. Many bacteriocins (e.g., plantaricin, a *Streptococcus salivarius* bacteriocin) are produced only on solid media and, as a consequence, must be solubilized by freeze–thaw elution or other methods prior to purification. Other bacteriocins are released into liquid cultures simplifying preparations for purification. Some bacteriocins are cell-bound or intracellular. How they are released is unclear. Examples include a *Bacteroides fragilis* bacteriocin, marcescin JF246, staphylococcin 414, enterocin EIA, and pesticin. Sonication or osmotic shock facilitates release of these bacteriocins from producer cells, allowing for further studies.

Changes in pH and elimination or addition of nutrients to growth media may influence both the amount of bacteriocin produced and the ease of purification. For example, maximum production of lactacin B occurs when producer cells are propagated in complex broth media at a constant pH of 6.0.

Synthesis of colicins and many other bacteriocins is inducible. Production of colicins is induced or increased by exposure of producer cultures to mitomycin C or ultraviolet light. Bacteriocins produced by other gram-negative species including klebocin, pyocin, vibriocin, and a *Pseudomonas solanacearum* bacteriocin are released into the culture medium after producer cells are exposed to mitomycin C and/or ultraviolet light. Similar conditions induce formation of bacteriocins produced by gram-positive species including the megacins of *Bacillus megaterium,* several clostridial bacteriocins, and the recently reported caseicin 80. The advantages conferred by inducibility for subsequent purification procedures include an increase in concentration of bacteriocin in the starting material, readily reproducible production conditions, and the potential for minimizing contaminating proteins.

In contrast, the low-molecular weight microcins and many bacteriocins produced by gram-positive species are not inducible. Examples include staphylococcin, pep 5, propionicin PLG1, boticin $S5_1$, nisin, lactocin LP27, the pediocins, and most bacteriocins produced by lactic acid bacteria. Treatment of producer cells with mitomycin C or ultraviolet light typically does not increase production of these bacteriocins; other approaches must be utilized to optimize their production.

Bacteriocin inducibility, or lack thereof, has not been assessed for other gram-negative and gram-positive cultures. These organisms (e.g., *Bacteroides fragilis, Lactobacillus acidophilus, Lactobacillus helveticus, Lactococcus lactis* subsp. *diacetylactis, Leuconostoc gelidum, Propionibacterium jensenii*) produce bacteriocins during growth in broth and/or agar media in the absence of inducing agents; therefore, production presumably is constitutive. Whether or not induction would increase yields for these cultures, as it does for colicins, remains undetermined.

Most efforts to increase bacteriocin concentration in crude preparations have addressed inducibility or manipulation of culture conditions. However, several authors have speculated that bacteriocin production provides a competitive advantage against closely related organisms seeking the same ecological niche. If so, the presence of bacteriocin-sensitive cells might be expected to elicit or enhance production by the producer cultures. This premise was examined in a recent preliminary, but interesting, study. The bacteriocin producer, *L. acidophilus* N2, was cocultured with a streptomycin-resistant variant of a closely related, bacteriocin-sensitive indicator, *Lactobacillus delbrueckii* subsp. *lactis* (also known as *Lactobacillus leichmannii*) ATCC 4797.

Table I Properties of Selected Colicins, Microcins, and Bacteriocins from Lactic Acid Bacteria

Bacteriocin						
Type	Designation	Producer	Size (1000 dalton)	Gene locus[a]	Mode of action[b]	Inducibility
Colicin	A	*Escherichia coli*	63	1a	I	+
	B	*E. coli*	76/90	1b	I	+
	D	*E. coli*	90	1a	IV	+
	E1	*E. coli*	56	1a	I	+
	E2	*E. coli*	66	1a	II	+
	E3	*E. coli*	66	1a	III	+
	E4	*E. coli*	64	1a	n.d.[c]	+
	E5	*E. coli*	66	1a	III	+
	E6	*E. coli*	66	1a	III	+
	E7	*E. coli*	65	1a	II	+
	Ia	*E. coli*	80	1b	I	+
	Ib	*E. coli*	80	1b	I	+
	K	*E. coli*	70	1a	I	+
	L	*E. coli*	64	n.d.	I	+
	M	*E. coli*	23	1b	V	+
	N	*E. coli*	39	1a	VI	+
	V	*E. coli*	4	1b	I	–
Cloacin	DF13	*Enterobacter cloacae*	56	1a	III	+
Microcin	A15	*Escherichia coli*	<1	1	IV	–
	B17	*E. coli*	4–5	1	VII	–
	C7	*E. coli*	<1	1	IV	–
	D15	*E. coli*	<1	1	X	–
	D140	*E. coli*	<1	1	X	–
	E492	*E. coli*	<1	1	X	–
	H47	*E. coli*	n.d.	2	n.d.	–
	15m	*E. coli*	<1	1	IV,XI	–
Lactacin	B	*Lactobacillus acidophilus*	6.5	2	n.d.	n.d.
	F	*L. acidophilus*	2.5	1	n.d.	n.d.
Caseicin	80	*L. casei*	40–42	n.d.	n.d.	+
Bacteriocin		*L. fermentum*	n.d.	n.d.	n.d.	–
Helveticin	J	*L. helveticus*	37	2	n.d.	n.d.
Lactocin	LP 27	*L. helveticus*	12.4	n.d.	IV, X	–
Plantaricin	A	*L. planatarum*	8	2	n.d.	n.d.
Sakacin	A	*L. sake*	n.d.	1	n.d.	n.d.
Nisin		*Lactococcus lactis* ssp. *lactis*	3.5	1 and 2	VIII	–
Diplococcin		*L. l.* ssp. *cremoris*	3.5	1	IX	n.d.
Lactostrepcins		*L. I.* ssp. *lactis* and *diacetylactis*	n.d.	2	IX, X	n.d.
Pediocin	PA1	*Pediococcus acidilactici*	16.5	1	n.d.	–
	AcH	*P. acidilactici*	2.7	1	X	–
	A	*P. pentosaceus*	n.d.	1	n.d.	–

[a] 1, plasmid; 1a, small high copy number plasmid; 1b, large low copy number plasmid; 2, chromosomal.
[b] I, pore formation; II, functions as a deoxyribonuclease; III, functions as an ribonuclease; IV, protein synthesis inhibitor; V, peptidoglycan synthesis inhibitor; VI, causes lysis; VII, blocks DNA replication; VIII, cationic surfactant; IX, terminates DNA, RNA, and protein synthesis; X, causes adenosine triphosphate and ion efflux and disrupts cellular membrane; XI, inhibits RNA synthesis.
[c] n.d., not determined.

The producer was propagated alone and with the sensitive indicator in a complex broth (initial pH 6.5) without pH control; bacteriocin activity was detected only in associative cultures. Effects of pH were eliminated by cultivating the producer alone or with the sensitive indicator in the broth at 37°C maintained at pH 6. Bacteriocin appeared 4–6 hr earlier in associative cultures but was detected at equal concentrations in both. Other experiments indicated that bacteriocin-enhancing activity was associated with indicator cells but not with spent indicator culture medium. Susceptibility of the enhancing activity to proteolytic, but not to cell wall-degradative, enzymes suggested participation of a protein. These results have not been confirmed, and unequivocal proof requires identification of the enhancing agent. Such studies may provide both support for the hypothesized role of bacteriocins in nature and a means for facilitating bacteriocin production. From the preceding discussion, it is clear that multiple factors influence production of each bacteriocin; selecting optimal conditions requires careful evaluation of several parameters.

Prior to purification, enzymatic treatments may assist in defining the chemical characteristics of bacteriocins in crude preparations. Inactivation by proteolytic enzymes is typical of these antimicrobial proteins. Bacteriocins also may be assessed for their sensitivity to other enzymes including lysozyme, phospholipase, lipase, amylase, deoxyribonuclease, ribonuclease, catalase, or peroxidase. Unequivocal results require that sensitivity to these enzymes be differentiated from inactivation by nonenzymatic means such as adsorption.

B. Purification of Bacteriocins

Development of bacteriocin purification protocols requires sensitive and reproducible methods for assessing both activity and protein concentration. Activity most commonly is determined by a critical dilution end-point technique in which the bacteriocin preparation is titered against a known population of cells of a sensitive indicator species. The titer commonly is defined as the reciprocal of the highest dilution inhibiting the indicator cell lawn and reported in activity units per milliliter. Protein concentration may be assessed by numerous techniques, including dye-binding methods (e.g., the Bradford), methods specific for peptide bonds and phenolic amino acids (Lowry), and monitoring absorbance at 260 and 280 nm.

The effects of purification on bacteriocin stability must be assessed at each step. Stability may be affected by numerous factors including heat, pH, solvents, and storage conditions. The tolerance of bacteriocins to heat may depend on the stage of purification, pH, presence of culture medium and other protective components, or other factors. For example, both crude and pure preparations of the lactostrepcins, diplococcin, pediocins, and the lactacins B and F are stable at high temperatures (100°–121°C) at acid pH and are sensitive to proteolytic contaminants at neutral pH. Activity of some gram-negative bacteriocins, including colicins M, Ia, Ib, and marcescin, is retained during storage in ethidium bromide, acriflavin, acridine orange, or bovine serum albumin. Colicins and some other bacteriocins are stable in nonionic detergents. Many low-molecular weight (<12,000) bacteriocins including lactacin B, lactacin F, and pediocin AcH are stable in chaotropic agents such as sodium dodecyl sulfate (SDS), urea, and guanidine HCl. A *B. fragilis* bacteriocin, lactocin LP27, lactacin B, and lactacin F initially were isolated as large-molecular weight complexes; the use of SDS or urea to dissociate them from large-molecular weight contaminants facilitated their purification. Microcins are stable and soluble in organic solvents; these characteristics may be used to separate them from larger proteins.

Purification of bacteriocins uses many techniques associated with purifying other proteins. Microcins typically are isolated by extraction with methanol and activated charcoal. Crude preparations of colicins have been obtained by centrifugation and ammonium sulfate precipitation. Further purification may be accomplished by gel filtration and ion-exchange chromatography, followed by high-pressure liquid chromatography (HPLC). Crude preparations of gram-positive bacteriocins have been obtained from the liquid released by freezing and thawing of agar cultures or from cell-free supernatants by precipitation with acetone, ethanol, or ammonium sulfate, with subsequent dialysis or ultrafiltration. Chromatography on ion-exchange resins and preparative isoelectric focusing have been employed for further purification. For very low titer bacteriocins, lyophilization, retention on ultrafiltration membranes, or dialysis against polyethylene glycol may provide an initial concentration step. Purification of bacteriocins to homogeneity nearly always requires a combination of protein separation techniques. For example, purification of lactacin F required three steps: treatment of culture supernatants with ammonium sulfate, then gel filtration of active fractions, and finally reversed phase HPLC. [*See* PROTEIN PURIFICATION.]

Molecular weight determination of bacteriocins may be performed throughout protein purification procedures. Gel filtration has been applied to the separation and sizing of marcescin, streptocins, enterocin, mutacin, colicins, pertucin, and megacin. SDS–polyacrylamide gel electrophoresis (SDS-PAGE) commonly is applied to size bacteriocins and provides a good estimation of molecular weight. Sedimentation equilibrium analysis, thin-channel ultrafiltration, and dialysis in semipermeable membranes also provide size estimations and have been used in conjunction with gel filtration or SDS-PAGE.

II. Molecular Properties and Modes of Action

A. Colicins

Extensive structural information to date is available only for the colicins, nisin, and the antibiotic subtilin. The colicins are elongated, single polypeptide chains; each chain consists of three regions. The structure of colicins was deduced by limited digestion with proteolytic enzymes and by analysis of the proteins following mutations in structural genes. Analyses of fragments from colicins A, B, E1, E2, E3, DF13, Ia, Ib, K, and N demonstrate similar functional and structural domains; differences reside in their amino acid sequences. The hydrophobic amino termini apparently are responsible for translocation of colicins across the cell membrane (Fig. 1). The lysine-rich carboxy termini mediate lethal activity. Between these two domains lies an area associated with colicin binding to the outer membrane receptors of sensitive cells. At low pH (5.0–5.3), the carboxy terminus of some pore-forming colicins consists of several α-helices in hairpin turns. β-pleated sheet secondary structure occurs at higher pH (5.5) but is unstable and reverts easily to the α-helical form.

The action of colicins against sensitive cells is well established and occurs in two steps. The proposed model has been deduced from the effects of colicins on sensitive cells and generally is accepted as true for other bacteriocins. First, the bacteriocin molecule attaches to a sensitive cell at a specific receptor site. This step is reversible because treatment with

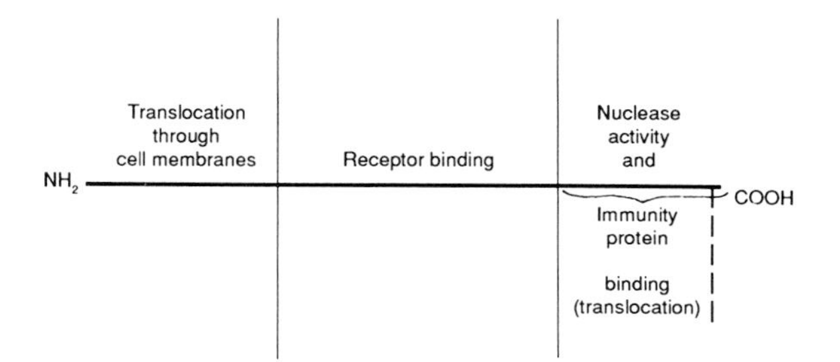

Figure 1 Functional domains of colicin E2, E3, and DF13 molecules. [Redrawn from Jakes, K. S. (1982). The mechanism of action of colicin E2, colicin E3, and cloacin DF13. *In* "Molecular Action of Toxins and Viruses" (P. Cohen and S. van Heyningen, eds.), pp. 131–167. Elsevier, New York.]

proteolytic enzymes degrades colicins and other bacteriocins and leaves the sensitive cell unscathed. Shortly thereafter, in a second irreversible step, the bacteriocin penetrates the cell membrane.

Alternatively, colicin entry and action may be visualized as occurring in three stages (Fig. 2). First, colicins adsorb to receptors; binding of the molecule to the receptor is favored by the shape of the colicin. Second, colicin adsorption results in a conformational change in the polypeptide chain, and α-helices associated with the carboxy terminus form a pore or channel in the membrane. The amino terminus then translocates across the cell membrane. Third, after entering the periplasmic space, the polypeptide realigns so that the carboxy terminus may penetrate the inner membrane and exert additional lethal activity, if any.

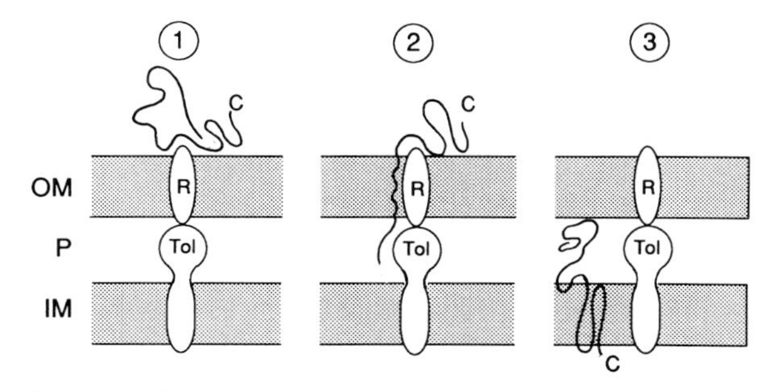

Figure 2 Three putative stages involved in colicin A entry and insertion into the cytoplasmic membrane. c, colicin carboxy terminus; IM, inner or cytoplasmic membrane; OM, outer membrane; P, periplasm; R, receptor protein; TOL, colicin transport protein. [Redrawn from Baty, D., Pattus, F., Parker, M., Benedetti, H., Feenette. M., Bourdinand, J., Cavard, D., Knibiehler, M., and Lazdunski, C. (1990). Uptake across the cell envelope and insertion into the inner membrane of ion channel-forming colicins in *Escherichia coli. Biochimie* **72,** 123-130.]

Penetration of some colicins affects the integrity of the cell membrane; examples include colicins A, DF13, E1, K, Ia, and Ib. These colicins are termed pore-formers because they form voltage-dependent channels in membranes. Their addition to sensitive cells results in an increase in conductance due to leakage of ions; the channels thus formed destroy the energy potential of the sensitive cell.

Once inside the cell membrane, other colicins, including B, D, E2, E3, M, and V, elicit specific bactericidal effects. Colicin B interferes with DNA, RNA, and protein synthesis; inhibits oxidative phosphorylation; and prevents galactose uptake. Colicin D specifically inhibits protein synthesis. Colicin E2 acts as an endonuclease and attacks cellular DNA. Colicin E3 is an endoribonuclease. Colicin M damages several sites in the outer membrane of sensitive cells causing lysis. Colicin V inhibits DNA, RNA, and protein synthesis in sensitive cells.

B. Microcins

To date, structural information for microcins is limited to their amino acid sequences. Little is known about microcin secondary protein structure. Their size (typically 1000 daltons) and stability to heat, chaotropic agents, and organic solvents suggest that little, if any, secondary or tertiary microcin structure exists.

The modes of action for microcins have been established. Microcin (Mcc) A15 blocks protein synthesis by inhibiting synthesis of methionine. MccB17 inhibits DNA replication and induces the SOS system, whereas MccD140 and MccD15 disrupt transmembrane electrical potential and deplete available adenosine triphosphate (ATP). MccE492 affects the electrical potential of sensitive cells, whereas MccC7 and Mcc15m interfere with protein synthesis; Mcc15m also inhibits RNA synthesis.

C. Bacteriocins of Lactic Acid Bacteria

There is considerable information about the structure of the broad-spectrum lactococcal bacteriocin nisin (Fig. 3). With subtilin, pep 5, and epidermidin, nisin is termed a lantibiotic because it contains unusual amino acids and five cyclic structures. The N-terminal amino acid in nisin is isoleucine; the dipeptide dehydroalanyllysine is the carboxy terminus. Nisin consists of 34 amino acids, including dehydrobutyrine (β-methyldehydroalanine), dehydroalanine, and α-aminobutryric acid (ABA).

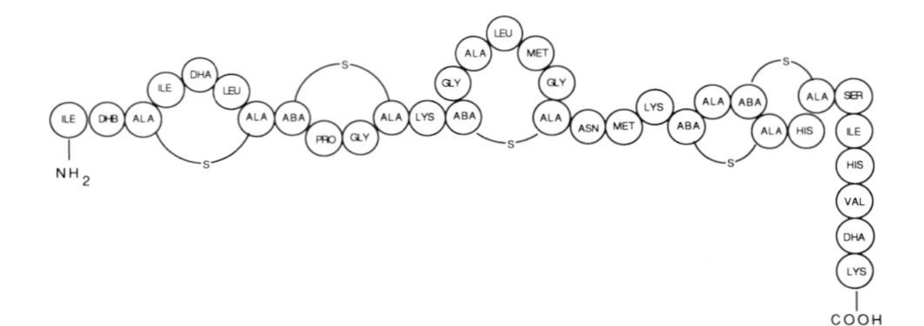

Figure 3 Structure of nisin. ABA, α-aminobutyric acid; DHA, dehydroalanine; DHB, dehydrobutyrine; standard three-letter abbreviations are used for amino acids. [Redrawn from Kaletta, C., and Entian, K.-D. (1989). Nisin, a peptide antibiotic: Cloning and sequencing of the *nisA* gene and posttranslational processing of its peptide source. *J. Bacteriol.* **171,** 1597–1601.]

The cyclic structures associated with nisin arise from the sulfide bridging of the amino acids. The amino acid lanthionine is formed when two alanine residues are linked by a sulfide bond (Ala-S-Ala). β-methyllanthionine results from the sulfide-bridging of ABA with alanine (ABA-S-Ala). Therefore, the cyclic structures of nisin include a single lanthionine residue, followed by four β-methyllanthione residues. Unlike the colicins, the secondary structures of lantibiotics and their effect on the action of these bacteriocins have not been determined.

With the exception of lantibiotics, little information is available about the lethal action of bacteriocins of lactic acid bacteria and other gram-positive species. Nisin and the other lantibiotics act as cationic surfactants or detergents. They inhibit cell wall synthesis, destroy membrane integrity, and cause efflux of cations and amino acids, ultimately resulting in cell lysis. An additional role for nisin is its sporostatic action; it prevents shedding of the spore wall and the spore swelling that prefaces outgrowth.

Other bacteriocins with established modes of action include lactostrepcin, diplococcin, lactocin LP27, and pediocin AcH. Lactostrepcin causes leakage of ATP and potassium ions, resulting in loss of cellular integrity, drop in intracellular pH, inhibition of uridine transport, and prevention of DNA, RNA, and protein synthesis. Diplococcin specifically inhibits DNA, RNA, and protein synthesis; lactocin LP27 terminates protein synthesis in sensitive lactobacilli. The treatment of sensitive cells with pediocin AcH affects membrane integrity, resulting in leakage of potassium ions and cytoplasmic materials. How cellular effects are related to the structures of these bacteriocins has not been determined.

III. Bacteriocin Insensitivity

Insensitivity to bacteriocins is achieved by three general mechanisms: immunity, resistance, and tolerance. Immunity results when cells produce a protein that protects against a corresponding bacteriocin. The gene encoding production of the immune protein often is linked closely with the gene encoding the bacteriocin. Bacteriocin-producing cells typically are immune to their own products.

Resistance is a second mechanism of bacteriocin insensitivity. Adsorption to specific receptors on the cell surface is required for bacteriocins to kill sensitive cells. Resistance results from loss of ability to adsorb bacteriocins and often is caused by a mutation in the sensitive cell that removes or alters the bacteriocin receptor. Bacteriocin receptors often serve dual functions as adsorption sites for molecules important to cell survival. As a consequence, mutations resulting in bacteriocin resistance also may be deleterious to resistant cells.

Tolerance is a third type of bacteriocin insensitivity. Tolerant cells adsorb bacteriocins but are not killed by them. Mechanisms of tolerance often involve mutations in sensitive cells. Tolerance (TOL) mutations may involve changes in functions important to cell survival such as transport of macromolecules across the cell membrane. Like resistance mutations, TOL mutations are often deleterious to the health of the cell.

A. Colicin Immunity

Immunity proteins may be divided into two groups for convenience of discussion: (1) immunity proteins of pore-forming colicins and (2) those of colicins that cleave DNA or RNA or inhibit protein synthesis. Immunity proteins of the pore-forming colicins A, E1, Ia, Ib, B, and N have been well studied.

Colicin A immunity protein (Cai) has been particularly well characterized and provides an adequate model for other immunity proteins of the pore-forming bacteriocins. The Cai protein is a minor component of the cell membrane. Cai is arranged in a zigzag fashion across the cell membrane (Fig. 4). Cai contains four separate hydrophobic regions consistent with its crossing the cell membrane four times. Its N and C termini and a loop of hydrophilic

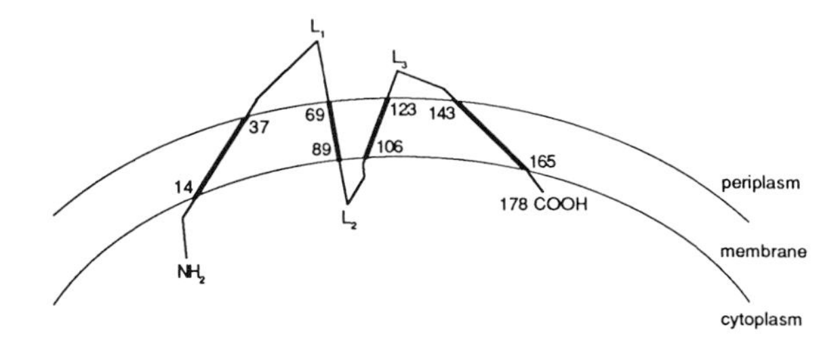

Figure 4 Hypothetical topology of the Cai polypeptide chain in the cytoplasmic membrane. [Redrawn from Geli, V., Baty, D., Crozel, V., Morlon, J., Lloubes, R., Pattus, F., and Lazdunski, C. (1986). *Mol. Gen. Genet.* **202,** 455–460.]

amino acids designated L2 are located in the cellular cytoplasm. Two hydrophilic loops termed L1 and L3 are located in the periplasm. Direct interaction at the cell membrane between Cai and the colicin A protein (Caa) has yet to be proven; consequently, precisely how cellular protection is conferred by immunity proteins is not known. However, these immunity proteins are necessary only to protect the cells from exogenous colicin. Pore-forming colicins produced by the cells are not toxic to the producer cells until they have been secreted.

The second group of colicins, typified by colicin E3, include nucleases and inhibitors of protein synthesis. The mechanism by which colicin E3 immunity protein (CE3i) protects the cell from colicin E3 involves a direct interaction between the immunity protein and colicin. This interaction takes place in a 1 : 1 molar ratio; i.e., one molecule of CE3i inactivates one molecule of colicin. Inactivation of colicin E3 is reversible, suggesting that there is no actual breakdown or digestion of the colicin by CE3i.

B. Colicin Resistance

The inability of some cells, generally mutants, to adsorb colicins is termed resistance. Colicins recognize and adsorb to specific receptors on the cell surface. The receptor sites for colicins are often vital for cell survival. For example, colicins E1, E2, and E3 adsorb to the vitamin B_{12} -binding site, and the receptor for colicin K may be involved in the uptake of nucleosides.

An examination of resistant mutants reveals that colicins may be classified by resistance into two groups: A and B. Group A includes colicins A, E1, E2, E3, K, L, N, S4, and X; group B includes colicins B, D, G, H, Ia, Ib, M, Q, S1, and V. Resistance to a colicin in group A may be accompanied by resistance to other group A colicins but not to a group B colicin. Mutants resistant to group A colicins commonly exhibit increased sensitivity to antibiotics, detergents, and surfactants. Mutants resistant to group B colicins, however, neither show increased sensitivity to antibiotics, detergents, and surfactants nor exhibit appreciable cross-resistance.

The mechanisms by which a sensitive cell becomes resistant vary. Changes in the binding site, elimination of the binding site, or production of a substance that competes with the colicin for the site are possible colicin resistance mechanisms. Whatever the mechanisms for resistance, colicins obviously have a powerful method to prevent the development of resistant mutants in nature. Their adsorption to receptors that are vital to cell survival minimizes the possibility of a resistant mutant being viable in nature; thus, the continued effectiveness of colicins is ensured.

C. Colicin Tolerance

A third mechanism by which cells can become insensitive to bacteriocins is tolerance. Tolerance refers to the ability of some cells, usually mutants, to bind a colicin without being killed by it. Some or all of the colicin molecule must enter the target cell to kill it. Several transport proteins are involved in this entry process. Typically, colicin transport proteins are present in the cell at low levels, and their exact functions within the cell are not understood fully. It is believed that they participate in transporting other large molecules across the cell membrane. TOL mutants, like resistant mutants, often grow poorly. This observation suggests that colicin entry mechanisms are essential for cell survival. If so, then the effectiveness of colicins is ensured by the use of transport mechanisms that make the survival of TOL mutants unlikely.

D. Insensitivity to Microcins

As they do for colicins, microcin immunity, resistance, and tolerance refer to similar mechanisms of insensitivity. However, the three phenomena are considerably less well characterized for microcins. The few microcin immunity mechanisms (primarily those for MccB17 and MccC7) that have been characterized differ considerably from those for colicins.

In particular, immunity to MccB17 has been well characterized and has several unique aspects. Unlike colicin-immune cells, which are unaffected by colicins, MccB17-immune cells exhibit similar damage to MccB17-sensitive cells. The primary difference between MccB17-sensitive and -insensitive cells is that damage to sensitive cells is irreversible and leads to cell death; in immune cells, damage is reversible and does not cause death.

This "incomplete" microcin immunity may be explained as follows. The three genes, *mcbE, mcbF,* and *mcbG,* responsible for immunity to MccB17, are located in an operon with the genes for MccB17 production. Two products, McbE and McbF, act in concert to confer partial immunity to MccB17. Either McbE or McbF singly provides no immunity to MccB17; however, cells harboring both *mcbE* and *mcbF* display 50 times the immunity of the wild type to MccB17. Cells harboring *mcbG* alone exhibit immunity roughly comparable to cells carrying *mcbE* and *mcbF.* Cells carrying all three genes possess 2000 times greater immunity than cells harboring none. It is interesting that McbE and McbF also are involved in export of MccB17, leading to the theory that these proteins form a pump to transport MccB17 out of the cell. This possibility explains "incomplete" immunity to MccB17. Cells possessing the MccB17 pump would be sensitive to excess exogenous microcin; upon its removal, the pump would eliminate endogenous microcin and restore cellular viability.

Immunity to MccC7 still is not well defined. Two separate regions of what probably constitutes an operon for production of MccC7 are required for immunity. Immunity to MccC7 is not sufficiently understood to allow for conclusions about specific mechanisms. However, the involvement of two separate DNA fragments in immunity suggests a system similar to that for MccB17.

Resistance and tolerance to microcins are not well characterized. Some evidence indicates involvement of the outer membrane porin protein F (OmpF) in uptake of microcin MccB17; mutations in the *ompF* or *ompR* genes result in decreased OmpF and decreased sensitivity to MccB17. A mutation in a third gene, *smbA,* results in cells that are insensitive to MccB17. The role of *smbA* is not understood; cells lacking *smbA* appear to be as healthy as wild-type cells. The most likely possibility is that the *smbA* gene product is involved with microcin uptake or binding. Until the precise roles of OmpF and SmbA are determined, it will not be possible to determine whether mutations in the corresponding genes result in tolerance or resistance.

E. Insensitivity to Bacteriocins Produced by Lactic Acid Bacteria

Insensitivity to bacteriocins produced by lactic acid bacteria is less understood than insensitivity to microcins. Genes for immunity to bacteriocins of this group usually are located near genes for production. This finding suggests that the determinants for immunity to bacteriocins in this group are similar, at least in location, to other immunity proteins. The precise structures and modes of action of these immunity proteins are yet to be detemined.

Resistance and tolerance to this group of bacteriocins have been poorly studied. Although insensitive mutants have been identified, the nature of the responsible mutations has not been determined. It seems reasonable to assume that mechanisms for resistance or tolerance to bacteriocins of lactic acid bacteria are similar to mechanisms for insensitivity to colicins.

IV. Genetics and Regulation of Bacteriocin Production

A. Colicins

Plasmid DNA carries the determinants encoding production of and immunity to most colicins. Genetic organization for colicins also may be divided into two groups. The first group is typified by the pore-forming colicins. The colicin structural gene and the lysis gene are in an operon and are oriented from left to right. The immunity gene is located between these two genes and is oriented in the opposite direction. Expression of the operon in growing cells is controlled by the SOS system that regulates error-prone DNA repair in *E. coli*. SOS regulation involves the Lex A and Rec A proteins. The operon promoter, generally very strong for pore-forming colicins, contains a *lex* operator sequence consisting of two overlapped SOS boxes to which Lex A protein binds, preventing synthesis of enzymes that repair damaged DNA. Lex A protein also prevents synthesis of colicin and lysis proteins. Exposure to DNA-damaging agents or inhibitors of DNA replication results in generation of an inducing signal responsible for reversible activation of the Rec A protein to its protease form. The activated Rec A protease cleaves the Lex A repressor, resulting in derepression of the SOS regulated genes, and subsequent production of the colicin and lysis proteins. Because the immunity gene is oriented in opposition to the structural and lysis genes, it is not under SOS control. Instead, it has its own weak promoter and is expressed constitutively. Transcription of the operon strongly interferes with transcription of the immunity gene, possibly from collision of the two RNA polymerase molecules moving in opposite directions. A reason for this interference may be that, because the immunity proteins for these types of colicins are aimed at external colicins, the small amount produced as a result of consitutive expression of the gene is sufficient for cell survival. [*See* RECA.]

For some of the colicins in this group, two termination sites have been found. Some read-through of the first terminator is essential to produce messenger RNA coding for the lysis protein. However, the first terminator appears important in controlling the rate of synthesis of the lysis protein. Overproduction of lysis protein may result in early cell death. In colicin A, the first terminator also acts as a terminator for the immunity gene, except it is located on the opposite DNA strand.

The genetic organization for the second group (colicins E2, E3, and DF13) is such that the structural, immunity, and lysis genes are contained in a single operon. The three gene products are released from the cell in a stoichiometric complex. Like the operon of pore-forming colicins, expression of this operon is SOS-inducible. In contrast, however, because the immunity gene is part of the operon, the immunity protein also is produced by the SOS response. For at least one of the colicins belonging to this group, a separate promoter has been found for the immunity gene. This separate promoter allows a small amount of the immunity protein to be constitutively produced to protect the cell against exogenous colicin. Higher-level production as a result of the SOS response is a protective mechanism against newly synthesized endogenous colicin. For these colicins, only one terminator has been found downstream from the lysis gene.

In addition to the SOS response, at least two other factors appear to regulate expression of colicin operons. First, catabolite repression has an effect on colicin production (i.e., in glucose medium, production of most colicins is reduced). Typically, in such regulation, intracellular concentration of cyclic adenosine monophosphate (cAMP) is elevated when a carbon source is unavailable. Catabolite activator protein (CAP) and cAMP form a complex that acts as a positive regulator. This regulator must be bound to the promoter region for transcription to occur. Thus, nutrient availability helps regulate metabolic activities. In some colicin operons, the operator contains a sequence homologous to that of the lac and gal operators, which is responsible for binding the cAMP–CAP complex. This region presumably has the same purpose in colicin operators. [*See* CATABOLITE REPRESSION.]

The other control factor relates to the degree of DNA supercoiling. An increase in the degree of supercoiling (as occurs in wild-type *E. coli* when grown in partial or complete anaerobiosis) increases expression of the operon and, hence, the amount of colicin and lysis proteins produced. If the immunity gene is part of the operon, its transcription is also increased. When the immunity gene is located between the structural and lysis genes in the opposite direction, it appears to be unaffected by changes in the degree of DNA supercoiling.

B. Microcins

Genetic determinants for production of most microcins are located on plasmid DNA. Frequently, microcin production also requires products encoded by the chromosome. The best-studied microcins are MccB17 and MccC7. Unlike colicins, production of these and other microcins is not induced by the SOS system.

MccB17 production requires four plasmid genes (*mcbA, mcbB, mcbC,* and *mcbD*). McbA is the precursor of MccB17. McbB, McbC, and McbD are necessary to process inactive precursor into an active form inside the cell. Immunity is mediated by the plasmid genes *mcbE, mcbF,* and *mcbG.* The major promoter of the MccB17 operon is growth-phase regulated (active in stationary phase only) and depends on the product of the chromosomal *omp-R* gene. (Omp-R is involved in the production of OmpC and OmpF, outer membrane porin proteins). The promoter is located upstream of *mcbA* and is responsible for production of McbA, McbB, McbC, and McbD. Another promoter also is responsible for low-level basal expression of *mcbD.* Transcription from both promoters presumably extends through the immunity genes. Immunity to MccB17 is expressed constitutively throughout both exponential and stationary phases.

Another chromosomal gene product that affects transcription of the MccB17 operon is MprA (microcin production regulation). MprA causes a reduction in production of MccB17 by preventing transcription from the major promoter. Other chromosomal gene products may also affect MccB17 production.

MccC7 production has several features in common with MccB17. Production and immunity require the activity of several adjacent genes in a polycistronic operon, with the promoter on the left side. Four regions have been identified, and at least six plasmid genes are responsible for production and immunity. As for MccB17, transcription of MccC7 genes is regulated by chromosomal gene products. The gene product of *appR* (involved in the production of acid phosphatase) activates transcription of two adjacent regions of the MccC7 operon. Also, as in MccB17, transcription of the MccC7 genes is inhibited by MprA.

In contrast to other microcins, the genes for MccH47 production and immunity are chromosomal. A continuous region of about 10 kb is required for expression of microcin production.

C. Lactic Acid Bacteria

Genetic aspects of bacteriocins produced by the lactic acid bacteria have not been as throughly investigated as those of colicins and microcins. For some bacteriocins, production and/or immunity have been linked to plasmid DNA. For others such as lactostrepcins (acid bacteriocins of lactic acid bacteria), investigations have failed to implicate plasmid DNA involvement. For still others (e.g., production of bacteriocins by *Leuconostoc* ssp.), genetic studies have not yet been conducted.

Production of and immunity to the *Lactobacillus* bacteriocins, sakacin A, lactocin S, and lactacin F, have been linked to plasmid DNA. In contrast, production of helveticin J is encoded chromosomally. Cloning and sequencing confirmed a chromosomal locus for the helveticin J (*hlv*) gene(s); the *hlv* structural gene appears to be located within an operon similar to the colicins.

Investigation of bacteriocin production and immunity in the pediococci often has revealed an association with plasmid DNA. Production of the following bacteriocins has been linked to plasmid DNA: pediocin PA-1 and bacteriocins of *Pediococcus cerevisiae* FBB-63, *P. pentosaceus* FBB-61, and *P. acidilactici* strains E, F, and H. The genes for immunity to pediocin PA-1 and the bacteriocins produced by *P. acidilactici* E and F are not located on the plasmids associated with production.

Plasmid DNA also has been implicated in bacteriocin production by *L. lactis* subsp. *diacetylactis* WM_4 and diplococcin production by *Lactococcus lactis* subsp. *cremoris*. Bacteriocin production by *L. lactis* subsp. *cremoris* 9B4 is plasmid-mediated; the determinants for both its production and immunity reside on a 60-kb plasmid. This plasmid has at least two regions coding for activity of and immunity to two different bacteriocins. One region has three open reading frames: the first two are involved with bacteriocin activity, the last with immunity. The other region has two open reading frames: the first is responsible for bacteriocin activity, the other for immunity. Both regions appear to be transcribed as operons.

The most thoroughly studied of the bacteriocins produced by *L. lactis* is nisin. Nisin production has been cloned from both plasmid and chromosomal DNA. The nucleotide sequences of the two cloned genes are identical. It is believed that the genes for both production and immunity are in an operon of at least 8.5 kb. Nisin is synthesized as a prepropeptide, which undergoes posttranslational processing to generate the mature, active antibiotic. Processing consists of dehydration of serines and threonines to their dehydro forms. Some of these react with cysteine residues to form thio-ester linkages, accounting for the unusual amino acids found in nisin. The leader peptide then is cleaved, and the mature antibiotic is exported from the cell. The structure of nisin, therefore, is dictated by the sequence of the structural gene, in contrast to other peptide antibiotics that are synthesized by nonribosomal pathways and involve multistep enzymatic pathways.

V. Future Directions in the Molecular Biology of Bacteriocins

Bacteriocin production and immunity may be exploited in various ways. Although many of the bacteriocins produced by the lactic acid bacteria inhibit only closely related microorganisms, nisin and some pediocins have a broader inhibitory spectrum that extends to unrelated gram-positive bacteria. Organisms inhibited by these substances include foodborne pathogens and spoilage microorganisms such as species of *Clostridium, Bacillus,* and *Staphylococcus*. A number of bacteriocins are active against *Listeria monocytogenes*. The application of nisin for control of clostridia and other foodborne pathogens has been investigated extensively. Its use as a food preservative has been permitted in Europe for several years. The wealth of biochemical, genetic, and usage information for nisin recently led the U.S. Food and Drug Administration to approve its use in processed cheese spreads. The application of both other bacteriocins and their producer cultures in foods to increase quality and safety continues to be assessed. In addition to applications information, complete biochemical and genetic data are prerequisites for the use of bacteriocins in foods. [*See* BACTERIOCINS: ACTIVITIES AND APPLICATIONS.]

In addition, the application of genetic techniques should be useful in developing strains that produce high levels of these inhibitory substances. Incorporation of these producer organisms in food products would be expected to result in greater protection against undesirable microorganisms in food. This goal could be achieved by using various types of genetic manipulation. For example, cloning the genes for production and immunity onto a high copy number plasmid should result in an increase in gene product. Factors affecting transcription, such as promoter strength and positive and negative regulators, could be altered to optimize production. It may also be possible to genetically alter the structure of certain bacteriocins to broaden their activity spectra.

Another use of bacteriocin genes is in the creation of cloning vectors. First, bacteriocin production and/or immunity genes from lactic acid bacteria could be used as markers for food-grade cloning vectors. Additionally, incorporation of colicin promoters into cloning vectors allows for controlled expression through the use of inducing agents. Another application is the use of weakly activated lysis genes (from colicins) to develop excretion vectors. Weak activation permits controlled release of material into the culture medium without causing cell death and facilitates purification of the substances being produced. One other benefit of having bacteriocin production and immunity encoded on a cloning vector is the resulting increase in stability of the plasmid. The presence of low levels of bacteriocin in the medium provides selective pressure for cells to maintain the plasmid vector and, hence, the cloned genes.

Acknowledgment

This work is technical contribution No. 3200 of the South Carolina Agricultural Experiment Station, Clemson University.

Bibliography

Baquero, F., and Moreno, F. (1984). *FEMS Microbiol. Lett.* **23,** 117–124.

Cramer, W. A., Cohen, F. S., Merrill, A. R., and Song, H. Y. (1990). *Molec. Microbiol.* **4,** 519–526.

Daeschel, M. A. (1990). Applications of bacteriocins in food systems. *In* "Biotechnology and Food Safety" (D. D. Bills and S. D. Kung, eds.), pp. 91–103. Butterworth-Heinemann, Boston.

Klaenhammer, T. R. (1988). *Biochimie* **70,** 337–349.

Konisky, J. (1982). *Annu. Rev. Microbiol.* **36,** 125–144.

Lazdunski, C. J. (1988). *Biochimie* **70,** 1291–1296.

Lazdunski, C. J., Baty, D., Geli, V., Cavard, D., Morlon, J., Lloubes, R., Howard, S. P., Knibiehler, M., Chartier, M., Varenne, S., Frenette, M., Dasseux, J. L., and Pattus, F. (1988). *Biochim. Biophys. Acta* **947,** 445–464.

Pugsley, A. P. (1984). *Microbiol. Sci.* **1,** 168–175, 203–205.

Schillinger, U. (1990). Bacteriocins of lactic acid bacteria. *In* "Biotechnology and Food Safety" (D. D. Bills and S. D. Kung, eds.), pp. 55–74, Butterworth-Heinemann, Boston.

Walker, G. C. (1984). *Microbiol. Rev.* **48,** 60–93.

Bacteriophages

Hans-Wolfgang Ackermann
Laval University

Glossary

Bacteriophage Virus that replicates in a bacterium; literally "eater of bacteria"

Capsid Protein coat surrounding the nucleic acid of a virus

Envelope Lipoprotein membrane surrounding a virus capsid

Genome Complete set of genes in a virus or a cell; in viruses, it consists of either DNA or RNA

Host range Number and nature of organisms in which a virus or group of viruses replicate

Prokaryote Type of cell whose DNA is not enclosed in a membrane

Restriction endonuclease Enzyme that recognizes a specific base sequence in double-stranded DNA and cuts the DNA strand at this or other sites

Superinfection Infection of a virus-infected host by a second virus

Virion Complete infectious virus particle

BACTERIOPHAGES, or "phages," are viruses of prokaryotes including eubacteria and archaebacteria. They were discovered and described twice, first in 1915 by the British pathologist Frederick William Twort and then in 1917 by the Canadian bacteriologist Félix Hubert d'Hérelle working at the Pasteur Institute of Paris. With about 3500 isolates of known morphology, phages constitute the largest of all virus groups. Phages are tailed, cubic, filamentous, or pleomorphic and contain single-stranded or double-stranded DNA (dsDNA), or RNA. They are classified into 12 families. Tailed phages are far more numerous than other types, are enormously diversified, and must be very old in geological terms.

Bacteriophages occur in over 100 bacterial genera and many different habitats. Infection results in phage multiplication or the establishment of lysogenic or carrier states. Bacterial genes may be transmitted in the process. Some phages (e.g., T4, λ, MS2, fd, or ϕX174) are famous experimental models. Phage research has led to major advances in virology, genetics, and molecular biology (concepts of lysogeny, provirus, induction, transduction, eclipse; DNA and RNA as carriers of genetic information; discovery of restriction endonucleases). Phages are used in phage typing and genetic engineering, but the high hopes set on phage therapy have been disappointed. In destroying valuable bacterial cultures, some phages are a nuisance in the fermentation industry.

I. Isolation and Identification of Phages

A. Propagation and Maintenance

1. Propagation

On solid media, phages produce clear, lysed areas in bacterial lawns or, if sufficiently diluted, small holes, or plaques, each of them corresponding to a single viable phage. In liquid media, phages sometimes produce complete clearing of bacterial cultures. Phages are grown on young bacteria in their logarithmic phase of growth, usually in conditions that are optimal for their hosts. Some phages require divalent cations (Ca^{++}, Mg^{++}) or other cofactors. Phages are propagated by three types of techniques: (1) in liquid media with host bacteria, (2) on agar surfaces inoculated with a monolayer of bacteria, and (3) in agar double layers consisting of normal bottom agar covered with a mixture of soft agar (0.3–0.9%), phages, and bacteria. Phages are harvested after a suitable incubation time, generally 3 hr for liquid cultures and 18 hr for solid media.

Phages from agar cultures are extracted with buffer or nutrient broth. Phage suspensions, or lysates, are sterilized, best by filtration through membrane filters (0.45-μm pore size), and titrated. Sterilization by chloroform or other agents is of questionable value.

2. Storage

No single technique is suitable for all phages. Many phages can be kept as lysates at +4°C or in lyophile, but others are quickly inactivated under these conditions. Lysates should be kept without additives such as thymol or chloroform. The best procedures seems to be preservation at −70°C. Phages may also be preserved in liquid nitrogen, by drying on filter paper, and, in the case of endospore-forming bacteria, by trapping phage genomes in spores. Ideally, any phage should be preserved by several techniques.

B. Isolation of Phages

1. Isolation from Nature

All samples must be liquid. Soil and other solid material are homogenized and suspended in an appropriate medium, and solids and bacteria are removed, usually by filtration preceded or not by centrifugation. Very rich samples can be assayed directly on indicator bacteria. In most cases, phages must be enriched by incubating the sample in a liquid medium inoculated with indicator bacteria. The culture is then filtered and titrated and phages are purified by repeated cloning of single plaques. Large samples must be concentrated before enrichment. This is done by centrifugation, filter adsorption and elution, flocculation, or precipitation by polyethylene glycol 6000. Adsorption–elution techniques may involve strongly acidic or alkaline conditions that inactivate phages. [*See* CENTRIFUGATION AND FILTRATION: DETECTION OF BACTERIA IN BLOOD.]

2. Isolation from Lysogenic Bacteria

Many bacteria produce phages spontaneously, and these phages may be detected by testing culture filtrates on indicator bacteria. It is generally preferable to induce phage production by mitomycin C, ultraviolet (UV) light, or other agents. A suspension of growing bacteria is exposed to the agent (e.g., 1 μg/ml of mitomycin C for 10 min or UV light for 1 min), incubated again, and then filtered. After mitomycin C induction, the bacteria should be separated from the agent by centrifugation and transferred into a fresh medium. Bacteriocins (see Section II.C), which are a source of error, are easily identified because they cannot be propagated and do not produce plaques when diluted.

C. Concentration and Purification

Small samples of <100 ml are usually concentrated by ultracentrifugation (about 60,000 *g* in swinging-bucket rotors), followed by several washes in buffer. Fixed-angle rotors allow considerable reduction of the *g* force, large phages sedimenting at as little as 10,000 *g* for 1 hr. Further purification may be achieved by centrifugation in a CsCl or sucrose density gradient. Large samples pose problems of contamination, aeration, and foaming. Preparation schedules are often complex: (1) pretreatment by low-speed centrifugation and/or filtration, (2) concentration, mostly by precipitation with polyethylene glycol, and (3) final purification in a density gradient or by ultracentrifugation.

D. Identification

Phage identification relies greatly on the observation that most phages are specific for their host genus; however, enterobacteria, in which polyvalent phages are common, are considered in this context as a single "genus." Phages are first examined in the electron microscope. This usually provides the family diagnosis and often indicates relationship on the species level. If no phages are known for a given host genus or only phages of different morphology, the new isolate may be considered as a new phage. If the same host genus has phages of identical morphology, they must be compared to the isolate by DNA–DNA hybridization and/or serology. Further identification may be achieved by determining restriction endonuclease patterns or constitutive proteins.

II. Phage Taxonomy

A. General

For d'Hérelle, only one phage had many races, the *Bacteriophagum intestinale*. Early attempts at classification by serology, host range, and inactivation tests showed that phages were highly diversified, but these attempts proved premature. Modern taxonomy started in 1962 when a system of viruses based on the properties of the virion and its nucleic acid was introduced by Lwoff, Horne, and Tournier. In 1967, phages were grouped into six basic types on the basis of morphology and nature of nucleic acid.

Other types were later established, and this process is likely to continue if more archaebacteria and other "unusual" microbes are investigated for the presence of phages. The International Committee on Taxonomy of Viruses presently recognizes 12 phage families and 12 phage genera. Their morphology is illustrated in Fig. 1 and their basic characteristics and host ranges are listed in Tables I–III. The most important family criteria are type of nucleic acid, particle shape, and presence or absence of an envelope. As in other viruses, family names end in *-viridae* and genus names in *-virus*. Species are designated by the vernacular names of their best-known (or only) members (e.g., T4 or λ.)

B. Phage Families and Genera

1. Tailed Phages

With about 3350 observations, tailed phages comprise >95% of phages and are the largest virus group known. They contain a single molecule of dsDNA and are characterized by a helical tail, a specialized structure for the transfer of phage DNA into host bacteria. Tailed phages fall into three families:

1. Myoviridae: phages with long complex tails consisting of a core and a contractile sheath (26% of tailed phages);
2. Siphoviridae: phages with long noncontractile, more or less flexible tails (59%); and
3. Podoviridae: phages with short tails (15%).

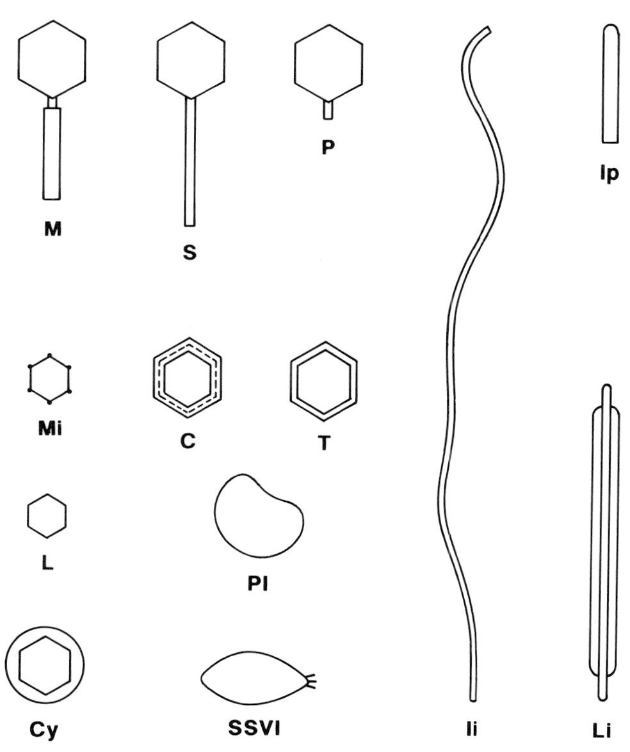

Figure 1 Morphology of phage families. C, Corticoviridae; Cy, Cystoviridae; Ii, Inoviridae, *Inovirus* genus; Ip, Inoviridae, *Plectrovirus* genus; L, Leviviridae; Li, Lipothrixviridae; M, Myoviridae; Mi, Microviridae; P, Podoviridae; Pl, Plasmaviridae; S, Siphoviridae; SSVI, SSV1 group; T, Tectiviridae. [Modified from Ackermann, H.-W. (1987). *Microbiol. Sci.* **4,** 214–218. With permission of Blackwell Scientific Publications Ltd., Edinburgh, Scotland.]

Phage capsids, usually named heads, are icosahedra or derivatives thereof. Capsomers are rarely visible. Elongated heads are relatively rare but occur in all three families. Heads and tails vary enormously in size and may have facultative structures such as head or tail fibers, collars, base plates, or terminal spikes (Fig. 2). The DNA is coiled inside the head. Its composition generally reflects that of the host bacterium, but it may contain unusual bases such as 5-hydroxymethylcytosine. Genetic maps are complex and include about 150 genes in phage T4 (possibly more in larger phages). Genes for related functions cluster together. Up to 40 proteins have been found in phage coats (T4). Lipids are generally absent but have been reported in a few exceptional cases. Response to inactivating agents is variable and no generalization is possible here. Despite the absence of lipids, about one-third of tailed phages are chloroform-sensitive, making chloroform use in phage isolation a dangerous procedure. Most properties of tailed phages appear as individual or species characteristics. Accordingly, genera have not yet been established, but about 220 species are recognizable, mostly on the basis of morphology, DNA–DNA hybridization, and serology.

2. Cubic, Filamentous, and Pleomorphic Phages

The cubic, filamentous, and pleomorphic phage group (Table IV) includes nine small phage families that correspond to <5% of phages, differ greatly in nucleic acid nature and particle structure, and sometimes have a single member. Host ranges are consequently narrow. Capsids with cubic symmetry are, with one exception, icosahedra or related bodies. Filamentous phages have, according to present knowledge, helical symmetry. Particles may or may not be enveloped. As in other viruses, the presence of lipids is accompanied by low buoyant density and high sensitivity to chloroform and ether.

Table I Main Properties and Frequency of Phage Families

Shape	Nucleic acid	Family	Genus	Particulars	Example	Members[a]
Tailed	DNA, ds, L	Myoviridae	—	Tail contractile	T4	874
		Siphoviridae	—	Tail long, noncontractile	λ	1,975
		Podoviridae	—	Tail short	T7	498
Cubic	DNA, ss, C	Microviridae	*Microvirus*	Conspicuous capsomers	ϕX174	38?
			Spiromicrovirus		SV4	
	DNA, ds, C, S	Corticoviridae	*Corticovirus*	Complex capsid, lipids	PM2	2?
	DNA, ds, L	Tectiviridae	*Tectivirus*	Double capsid, pseudo-tail, lipids	PRD1	14
	RNA, ss, L	Leviviridae	*Levivirus*		MS2	35
			Allolevirirus		Qβ	
	RNA, ds, L, M	Cystoviridae	*Cystovirus*	Envelope, lipids	ϕ6	1
Filamentous	DNA, ss, C	Inoviridae	*Inovirus*	Long filaments	fd	47
			Plectrovirus	Short rods	L51	
	DNA, ds, L	Lipothrixviridae	*Lipothrixvirus*	Envelope, lipids	TTV1	4?
Pleomorphic	DNA, ds, C, S	Plasmaviridae	*Plasmavirus*	Envelope, lipids, no capsid	MVL2	4?
	DNA, ds, C, S	SSV1 group[b]	SSV1 group[b]	Lemon-shaped	SSV1	1

[a] Excluding phagelike bacteriocins and known defective phages; computed 1 December 1990.
[b] Provisional name.
—, none established; C, circular; ds, double-stranded; L, linear; M, multipartite; S, supercoiled; ss, single-stranded.
[Modified from Ackermann, H.-W. (1987), *Microbiol. Sci.* 4, 214–218. With permission of Blackwell Scientific Publications Ltd., Edinburgh, Scotland.]

Table II Dimensions and Physiochemical Properties

Phage group or family	Virion					Nucleic acid		
	Particle size (nm)	Tail length (nm)	Weight (MDa)	Buoyant density	Lipids (%)	Content (%)	Molecular weight (kb)	G+C (%)
Tailed phages								
Average	63[a]	153	100	1.49	–	45	81	48
Range	38–160[a]	3–570	29–470	1.4–1.54	–	25–62	17–760	27–72
Microviridae	27	–	7	1.41	–	26	2.6	44
Corticoviridae	60	–	49	1.28	13	13	9.0	43
Tectiviridae	63	–	70	1.28	16	14	15.1	51
Leviviridae	24	–	4	1.43	–	31	1.9	51
Cystoviridae	75	–	90	1.27	23	12	16.1	58
Inoviridae								
Inovirus	760–1, 915 × 6	–	12–23	1.30	–	6?–21	2.9–4.7	40–60
Plectrovirus	85–250 × 14	–		1.37	–		4.4–8.5	
Lipothrixviridae	400 × 40	–			+		16.0	
Plasmaviridae	80	–			11		11.6	32
SSV1 group[b]	100 × 60	–			–		15.5	

[a] Isometric heads only.
[b] Provisional name.
+, present; –, absent; buoyant density is g/ml in CsCl; G+C, guanine–cytosine content.
[Modified from Ackermann, H.-W. (1987). *Microbiol. Sci.* **4**, 214–218. With permission of Blackwell Scientific Publications Ltd., Edinburgh, Scotland.]

Table III Comparative Biological Properties

Phage group or family	Adsorption site	Fixation structure	Infection by:	Assembly by:	Start of assembly	Inclusion bodies	Release	Nature of infection
Tailed phages	Cell wall, capsule, pili, flagella	Tail	Nucleic acid	Nucleoplasm, cell periphery	Capsid	+ or −	Lysis	Virulent or temperate
Microviridae[a]	Cell wall	Spikes	Nucleic acid	Nucleoplasm	Capsid	−	Lysis	Virulent
Corticoviridae		Spikes	Nucleic acid	Plasma membrane	Capsid	−	Lysis	Virulent
Tectiviridae	Pili, cell wall	Spikes, "tail"	Nucleic acid	Nucleoplasm	Capsid	−	Lysis	Virulent
Leviviridae	Pili	A protein	Nucleic acid	Cytoplasm	Nucleic acid	+ or −	Lysis	Virulent
Cystoviridae	Pili	Envelope	Capsid	Nucleoplasm	Capsid	−	Lysis	Virulent
Inoviridae								
Inovirus	Pili	Virus tip	Virion	Plasma membrane	Nucleic acid	−	Extrusion	Steady-state[b]
Plectrovirus	Plasma membrane	Virus tip	Virion?	Plasma membrane	Nucleic acid?	−	Extrusion	Steady-state
Lipothrixviridae	Pili	Virus tip					Lysis	Virulent or temperate
Plasmaviridae	Plasma membrane	Envelope	Nucleic acid?	Plasma membrane	Nucleic acid	−	Budding	Temperate
SSV1 group[b]		Spikes					Extrusion	Temperate

[a] Data are for *Microvirus* genus only.
[b] Provisional name.
[c] Two temperate phages have been reported.
+, present; −, absent.
[Modified with permission, from Ackermann, H.-W., and DuBow, M. S. (1987). "Viruses of Prokaryotes," Vol. I, p. 75. Copyright CRC Press, Boca Raton, Florida.]

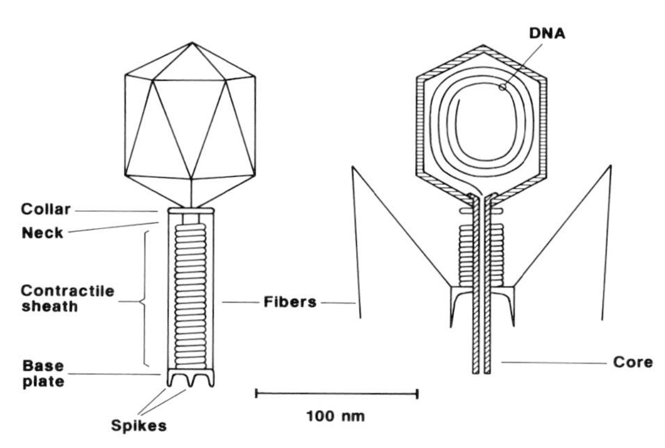

Figure 2 Schematic representation of phage T4 with extended tail and folded tail fibers (left) and sectioned with contracted tail (right). [Modified from Ackermann, H.-W. (1985). Les virus des bactéries. *In* "Virologie médicale" (J. Maurin, ed.), p. 200. With permission of Flammarion Médecine-Sciences, Paris.]

a. Cubic DNA Phages

i. Microviridae The genus *Microvirus* includes the phage ϕX174 and its relatives and all phages of enterobacteria and is characterized by large capsomers. Recently, the genus *Spiromicrovirus* was added. Similar, not-yet classified phages occur in *Chlamydia* and *Bdellovibrio*.

ii. Corticoviridae The only certain member of the family Corticoviridae is a maritime phage, PM2. Its capsid consists of two protein shells and a lipid bilayer sandwiched in between. A similar phage was isolated from seawater, but it is little known.

iii. Tectiviridae Phages of the family Tectiviridae are characterized by a double capsid and a unique mode of infection. The outer capsid, rigid and apparently proteinic, surrounds a thick, flexible membrane that seems to consist of lipoprotein. Upon adsorption to bacteria or chloroform treatment, this inner coat becomes a taillike tube of about 60 nm in length, obviously a nucleic acid ejection device. Tectiviruses may have apical spikes.

b. Cubic RNA Phages

i. Leviviridae Leviviruses resemble enteroviruses and have no morphological particulars. Most of them are plasmid-specific coliphages that adsorb to F or sex pili and have been divided, by serology and other criteria, into two genera. Several not-yet classified leviviruses are specific for other plasmid types (C, H, M, etc.) or occur outside of the enterobacteria family.

ii. Cystoviridae The single member of the family Cystoviridae is unique in several ways. It is the only phage to contain dsRNA and RNA polymerase and the only virus to have a dodecahedral capsid. The RNA is multipartite and consists of three molecules.

c. Filamentous Phages

i. Inoviridae The Inoviridae family includes two genera with widely differing host ranges and similarities in replication and morphogenesis that seem to derive from the single-stranded nature of phage DNA rather than a common origin of these phages. Despite the absence of lipids, viruses are chloroform–sensitive. The *Inovirus* genus includes 31 phages that are long, rigid, or flexible filaments of variable length. They are limited to a few Gram-negative bacteria only, sensitive to sonicaiton, and resistant to heat. Many of them are plasmid-specific. The *Plectrovirus* genus includes 16 isolates. Phages are short, straight rods and occur in mycoplasmas only.

ii. Lipothrixviridae In the family Lipothrixviridae, only one certain member is known, phage TTV1 of the archaebacterium *Thermoproteus tenax*. It is characterized by the combination of a lipoprotein envelope and rodlike shape. Three similar types of particles of different size have been observed, but they have not been propagated.

d. Pleomorphic Phages

i. Plasmaviridae In the family Plasmaviridae, only one certain member is known, *Acholeplasma* virus MVL2 or L2. It contains dsDNA, has no capsid, and may be called a nucleoprotein condensation with a lipoprotein envelope. Three similar isolates are known, but one of them contains single-stranded DNA and their taxonomic status is uncertain.

ii. SSV1 Group The SSV1 group does not yet have a scientific name and consists of a single member, SSV1, which is produced upon induction by the archaebacterium *Sulfolobus shibatae*. Particles are lemon-shaped with short spikes at one end. The coat consists of two hydrophobic proteins and is disrupted by chloroform. No lipids have been detected. SSV1 has not been propagated for absence of a

Table IV Occurrence and Frequency of Tailed and Cubic, Filamentous, and Pleomorpic (CFP) Phages[a]

Section	Bacterial group according to Bergey's Manual[b]	Phages		
		Tailed	CFP	Total
I, 1	Spirochetes	5		5
2	Aerobic/microaerophilic, motile, helical/vibrioid, gram-negative bacteria	19	9	28
4	Aerobic gram-negative rods and cocci	542	13?	555?
5	Facultatively anaerobic gram-negative rods	756	81	837
6	Anaerobic gram-negative straight, curved, and helical rods	13		13
7	Dissimilatory sulfate- or sulfur-reducing bacteria	2		2
8	Anaerobic gram-negative cocci	2		2
9	Rickettsias and chlamydias	1	1	2
10	Mycoplasmas	12	20	32
II, 1	Gram-positive cocci	686		686
2	Endospore-forming gram-positive bacteria	493	8	501
3	Nonspore-forming gram-positive rods of regular shape	176		176
4	Nonspore-forming gram-positive rods of irregular shape	168		168
5	Mycobacteria	74		74
6	Nocardioforms	22		22
III, 1	Anoxygenic phototrophic bacteria	10		10
2	Oxygenic phototropic bacteria	19		19
3	Gliding, fruiting bacteria	36	2	38
4	Gliding, nonfruiting bacteria	12		12
5	Sheathed bacteria	1		1
6	Budding and/or appendaged bacteria	112	8	120
7	Chemolithotrophic bacteria	2		2
8	Archaebacteria	9	5?	14?
IV, 2	Filamentous bacteria that form true sporangia	2		2
3	*Streptomyces* and similar genera	157		157
4	Filamentous bacteria of uncertain taxonomic position	16		16
Total		3,347	147?	3,494?

[a] Excluding phagelike bacteriocins and known defective phages; computed 1 December 1990.
[b] "Bergey's Manual of Systematic Bacteriology," Vols. 1–4 (J. G. Holt, editor-in-chief), Baltimore, Maryland (1984, 1986, 1989, 1990).

suitable host. It persists in bacterial cells as a plasmid and as an integrated prophage (see Section IV.B.1).

C. Plasmids, Episomes, and Bacteriocins

Plasmids are extrachromosomal genetic elements that consist of circular or linear dsDNA and replicate independently of the host chromosome. Certain prophages behave as plasmids, but phages and plasmids are sharply differentiated; contrary to plasmids, phages have a coat and genomes of uniform size, occur free in nature, and generally lyse their hosts. The term episome, which is no longer useful, designates plasmids and prophages that can integrate reversibly into host DNA. *Bacteriocins* are antibacterial agents that are produced by bacteria, require specific receptors, and kill other bacteria. "Particulate," or high-molecular weight, bacteriocins are defective phages (e.g., contractile or noncontractile tails without heads). "True," or low-molecular weight, bacteriocins are a mixed group of entities, including enzymes and phage tail spikes.

[*See* Plasmids; Bacteriocins: Activities and Applications; Bacteriocins, Molecular Biology.]

D. Origin and Evolution of Bacteriophages

Phages are probably polyphyletic in origin and came into being at different times. This is indicated by seemingly unbridgeable fundamental differences among most phage families and by their host ranges. Phages may have derived from cell constituents that acquired a coat and became independent [e.g., leviviruses from messenger RNA (MRNA); filamentous inoviruses from plasmids]. Tailed phages are obviously phylogenetically related and apparently are the oldest phage group of all. Their occurrence in eubacteria, notably cyanobacteria, and in archaebacteria suggests that they appeared before their hosts diverged, thus at least 3 million years ago. Phage groups linked to aerobic bacteria may have emerged when or after the atmosphere became oxygenated. In some cases, nature repeated itself. Convergent evolution is evident in the pseudo-tails of tectiviruses and perhaps in the general resemblance of *Inovirus* and *Plectrovirus* phages.

Microviruses, tectiviruses, and leviviruses show little or no morphological differentiation, possibly because of constraints imposed by capsid size or of a relatively young geological age of these phages. Inoviruses differentiated by elongation. By contrast, tailed phages are extremely diversified and must have an eventful evolutionary history. In terms of structural simplicity and present-day frequency, the archetypal tailed phage seems to be a *Siphovirus* with an isometric head, from which the other types evolved. The diversification of tailed phages is attributed, besides to point mutation and uniparental reproduction, which are found in all viruses, to two main factors: the exchange of gene blocks, or modular evolution, and the frequency of lysogeny (see Section IV.B.1), which perpetuates prophages and makes them available for recombination with superinfecting phages. Other avenues are gene rearrangement (deletions, duplications, inversions, transpositions) and recombination with plasmids or the host genome. On the other hand, morphological properties may be highly conserved and some phages appear as living fossils, indicating phylogenetic relationships of their hosts.

III. Phage Occurrence and Ecology

A. Distribution of Phages in Bacteria

Phages have been found in over 100 bacterial genera distributed all over the bacterial world: in aerobes and anaerobes, actinomycetes, archaebacteria, cyanobacteria and other phototrophs, endospore formers, appendaged, budding, gliding, and sheathed bacteria, spirochetes, mycoplasmas and chlamydias (Table V). Phagelike particles of the podovirus type have even been found in endosymbionts of paramecia. However, tailed phages reported in cultures of green algae and filamentous fungi are probably contaminants.

Most phages have been found in a few bacterial groups: enterobacteria (over 650 phages), bacilli, clostridia, lactococci, pseudomonads, staphylococci, and streptococci. This largely reflects the availability and ease of cultivation of these bacteria and the amount of work invested. About half of phages have been found in cultures of lysogenic bacteria. Tailed phages predominate everywhere except in mycoplasmas. In archaebacteria, they have been found in the genus *Halobacterium* only and not yet in methanotrophs and extreme thermophiles. Siphoviridae are particularly frequent in actinomycetes, coryneforms, lactocci, and streptococci. Myoviruses and podoviruses are relatively frequent in enterobacteria, pseudomonads, bacilli, and clostridia. This particular distribution must have phylogenetic reasons.

B. Phage Ecology

1. Habitats

Phages have essentially the same habitats as their hosts; indeed, their most important habitat is the lysogenic bacterium because it protects prophages from the environment and frees them from the need to find ever new bacteria for propagation. In nature, phages occur in an extraordinary variety of habitats ranging from Icelandic solfataras to fish sauce, fetal calf serum, and cooling towers of thermal power stations. They are found on the surfaces and in normal and pathological products of humans and animals, on plants, and in food, soil, air, and water, especially sewage. Body cavities with large bacterial populations, such as intestines and rumen, are ex-

Table V Host Range of Phage Families

Phage group or family	Host range
Tailed phages	Eubacteria and archaebacteria
Microviridae	Enterobacteria, *Chlamydia*, *Spiroplasma; Bdellovibrio?*
Corticoviridae	*Alteromonas*
Tectiviridae	i. Enterobacteria, *Acinetobacter*, *Pseudomonas*, *Vibrio* ii. *Bacillus*
Leviviridae	Enterobacteria, *Caulobacter*, *Pseudomonas*
Cystoviridae	*Pseudomonas*
Inoviridae	
Inovirus	Enterobacteria, *Pseudomonas*, *Vibrio*, *Xanthomonas*
Plectrovirus	*Acholeplasma*, *Spiroplasma*
Lipothrixviridae	*Thermoproteus*[a]
Plasmaviridae	*Acholeplasma*
SSV1 group[b]	*Sulfolobus*[a]

[a] Archaebacterium.
[b] Provisional name.
[Modified, with permission, from Ackermann, H.-W., and DuBow, M. S. (1987). "Viruses of Prokaryotes." Vol. I, p. 179. Copyright CRC Press, Boca Raton, Florida.]

tremely rich in phages. According to their habitat, phages may be acido-, alkali-, halo-, thermo-, or psychrophilic. These properties are not linked to particular phage groups but, rather, appear as individual adaptations. Psychrophilic phages are often temperature-sensitive and occur frequently in spoiled, refrigerated meat or fish.

2. Geographical Distribution

Except for phages from extreme environments, phage species generally seem to be distributed over the whole earth. This is suggested by (1) electron microscopical observations of rare and characteristical phage morphotypes in different countries and (2) global occurrence of certain lactococcal phage species in dairy plants and of RNA coliphages in sewage. Unfortunately, most data are from developed countries.

3. Frequency of Phages in Nature

Sizes of phage populations are difficult to estimate because plaque assays and enrichment and (most) concentration techniques depend on bacterial hosts; they therefore only detect phages for specific bacteria and environmental conditions. Consequently, phage titers vary considerably—for example, for coliphages between 0 and 10^9/g in human feces and between 0 and 10^7/ml in domestic sewage. Titers of actinophages in soil vary between 0 and 10^5/g. A few purely electron microscopic phage counts, which do not allow phage identification, indicate that total phage titers may be as high as 10^{10}/ml in sewage and 10^9/ml in rumen.

4. Persistence of Phages in the Environment

Phage survival in nature is studied largely with the aim of using phages as indicators of contamination. The principal experimental models are cubic RNA phages (MS2, f2) because of their resemblance to enteroviruses, other coliphages (ϕX174, T4), and cyanophages. The indicator value of phages has not been conclusively proven and still lacks a solid statistical basis, but considerable data on phage ecology have been obtained. Phages appear as parts of complex ecosystems including various competing bacteria. Their numbers are affected by factors governing bacterial growth, notably nutrient supply and, in cyanobacteria, sunlight. The lowest bacterial concentration compatible with phage multiplication seems to be 10^4 cells/ml. In addition, phage counts are affected by association of phages with solids and colloids (e.g., clay), presence of organic matter, concentration and type of ions, pH, temperature, UV and visible light, type of water (e.g., seawater), and nature and phage sensitivity of bacteria. Finally, phage titers depend on intrinsic phage properties such as burst size (see Section IV.A.4) and host range. No generalization is possible and each phage seems to have its own ecology.

IV. Phage Physiology

A. The Lytic Cycle

The lytic cycle, also called vegetative or productive, results in the production of new phages. Phages undergoing lytic cycles only are virulent. Lytic cycles consist of several steps and show considerable variation according to the type of phage (Table IV).

1. Adsorption

Phages encounter bacteria by chance and adsorb to specific receptors, generally located on the cell wall, but also on flagella, pili, capsules, or the plasma membrane. Adsorption sometimes consists of a reversible and an irreversible stage and may require cofactors (see Section I.A.1).

2. Infection

In most phage groups, only the viral nucleic acid enters the host and the shell remains outside. The mechanism of this step is generally poorly understood. In the *Inovirus* genus and in cystovirus $\phi 6$, the capsid penetrates the cell wall but not the plasma membrane. In phages with contractile tails, the cell wall is degraded by phage enzymes located on the tail tip. The sheath then contracts (Fig. 2) and the tail core is brought in contact with the plasma membrane.

3. Multiplication

The interval from infection to the release of new phages is called the latent period. It depends largely on the physiological state of the host and varies between 20 min and 30–40 hr. After infection, normal bacterial syntheses are shut off or modified. Phage nucleic acid is transcribed to mRNA using host and/or phage RNA polymerases. The RNA of leviviruses acts as mRNA and needs no transcription. In tailed phages, gene expression is largely sequential. Host syntheses are shut off first and structural genes are expressed last. According to present knowledge, replication of phage DNA and RNA is semiconservative, each strand of a double helix acting as a template for the synthesis of a complementary strand. In phages with single-stranded nucleic acid, double-stranded replicative forms are produced. In tailed phages, replication starts at fixed sites of the DNA molecule, is bidirectional, and generates giant DNA molecules, or concatemers, which are then cut to fit into phage heads. Translation is generally poorly known in phages. Microviridae, Leviviridae, the *Inovirus* genus, and the SSV1 type have overlapping genes that are translated in different reading frames, allowing the synthesis of different proteins from the same DNA or RNA segment. Lipids, if present in phages, are of variable origin. Phospholipids are specified or regulated by phages and fatty acids seem to derive from the host.

The assembly of new phages is called maturation. Phage constituents assemble spontaneously or with the help of specific enzymes. In most phage families, the nucleic acid enters a preformed capsid; in others, the capsid is constructed around the nucleic acid. In tailed phages, assembly is a highly regulated process with sequentially acting proteins and separate pathways for heads and tails, which are finally joined together. The envelope of plasmaviruses is acquired by budding, but that of cystovirus $\phi 6$ is of cellular origin. The assembly of tailed phages often results in aberrant particles including giant or multitailed phages and structures consisting of polymerized head or tail protein, called polyheads, polytails, or polysheaths. Inoviruses produce particles of abnormal length. Leviviruses and some tailed phages produce intracellular crystalline inclusion bodies.

4. Release

Phages are liberated by lysis, extrusion, or budding. Lysis occurs in tailed and cubic phages and in the Lipothrixviridae. Bacterial cells are weakened from the inside and burst, liberating some 20–1000 phages (often 50–100). Exceptional burst sizes (up to 20,000) have been recorded in leviviruses. Extrusion is observed in inoviruses and the SSV1 type. Phages are secreted through the membranes of their surviving host. Budding is found in plasmaviruses. Cells are not lysed and produce phages for hours. Progeny sizes for budding and extruded phages have been estimated at 130–1000 per cell.

B. The Temperate Cycle

1. Lysogeny

In phages called temperate, infection results in a special equilibrium between phage and host. The phage genome persists in a latent state in the host cell, replicates more or less in synchrony with it, and may be perpetuated indefinitely in this way. It behaves as a part of the bacterium. If this equilibrium breaks down, either spontaneously or after induction, phages are produced as in a lytic cycle. A bacterium harboring a latent phage genome or prophage is called lysogenic because it has acquired the ability to produce phages. Polylysogenic bacteria may carry up to five different prophages. Defective lysogeny is the perpetuation of temperate phages that are unable to replicate and often consist of single heads or tails.

Most temperate phages are tailed, but some members of the *Inovirus* genus, Lipothrixviridae, Plasmaviridae, and the SSV1 type can also lysogenize (Table III). Lysogeny is near-ubiquitous and occurs in eubacteria, including cyanobacteria, and in archaebacteria. Its frequency in a given bacterial species varies between 0 and 100% (often around 40%) according to the species, induction techniques, and number of indicator strains. Mitomycin C and UV light are the principal inducing agents (see Section I.B.2). Many others are known, notably antitumor agents, carcinogens, and mutagens. They often act by damaging host DNA or inhibiting its synthesis.

The behavior of coliphage λ is particularly well understood. After infection, the λ genome forms a circle and some λ proteins are immediately synthesized. They direct the cell to make a choice between the lytic and the temperate cycle. If a certain λ protein prevails, the λ genome integrates via a crossover at a specific site of the host DNA. It is then replicated at every bacterial division and makes the bacterium immune against superinfection by related phages. Spontaneous or induced excision of the λ prophage leads to normal phage replication. Other phages have several integration sites, or their DNA does not form circles and integrates at random. Still others persist as plasmids, perhaps in association with the plasma membrane. Their replication is less regular than that of integrated prophages. Phage λ has a prophage and a plasmid stage.

2. Pseudolysogeny and Steady-State Infections

In pseudolysogenic bacteria, only part of a culture is infected with phages and an equilibrium exists between free phages and noninfected, phage-sensitive bacteria. Phage-free strains can be obtained by simple cloning or by cultivating the bacteria in antiphage serum. In steady-state infections, the whole culture is infected, but cells are not lysed and produce phages continously (see Section IV.A.4).

3. Transduction and Conversion

Transduction is transfer of host DNA by viruses and is normally a rare event. In generalized transduction, fragments of bacterial DNA are packaged by accident into phage heads and transferred to a new bacterium. Any host gene may be so transferred and the implicated phages may be virulent or temperate. Specialized transduction is done by temperate phages that can integrate into host DNA (e.g., λ). If the phage DNA is not properly excised, bacterial genes adjacent to the prophage site may be packaged into phage heads along with normal genes. The resulting particle has a defective genome and may be nonviable. In conversion, bacteria acquire new properties through lysogenization by normal temperate phages. Conversion is a frequent event, affecting the whole bacterial population that has been lysogenized. The new properties are specified by phage genes and include new antigens, antibiotic resistance, colony characteristics, and toxin production (e.g., of diphtheria or botulinus toxin). They will disappear if the bacterium loses its prophage. Both transduction and conversion have been found in tailed phages only.

V. Phages in Applied Microbiology

A. Therapeutic Agents, Reagents, and Tools

1. Therapy and Prophylaxis of Infectious Diseases

Phage therapy started with high hopes and was strongly advocated by d'Hérelle himself. Phages were, enthusiastically and uncritically, applied in many human and animal diseases and spectacular results were reported as well as failures. When antibiotics became available, phage therapy was practically abandoned. The main reasons were the host specificity of phages and the rapid appearance of resistant bacteria. However, surprisingly good results were reported in this decade from Poland. They suggest that phage therapy is a viable therapeutic alternative in antibiotic-resistant infections, especially by pyogens (wounds, osteomyelitis, furunculosis, abscesses, septicemia). More basic research is needed, for example, on inactivation of phages by body fluids. Phage prophylaxis of infectious diseases was also attempted. Despite encouraging results in cholera prevention, it is of historic interest only. Phage control of plant diseases has not been recommended. [*See* ANTIBIOTIC RESISTANCE.]

2. Identification and Classification of Bacteria

Early attempts to use phages for bacterial identification were abandoned because no phage lyses all strains of a bacterial species and no others. A few diagnostic phages are still used as screening agents in specialized laboratories, for example, for rapid identification of *Bacillus anthracis*, members of the genus *Salmonella*, or the biotype El Tor of *Vibrio cholerae*. By contrast, phage typing is an important technique in epidemiology. In analogy with the antibiogram, bacteria are tested against a set of phages and subdivided into resistance patterns or phage types. Briefly, a continuous layer of bacteria is created on an agar surface, phage suspensions are deposited on it, and results are read the next day. Phage typing is invaluable for subdividing biochemically and serologically homogeneous bacterial species. Besides international typing schemes for *Salmonella typhi* and *Salmonella paratyphi* B, there are typing sets for most human, animal, and plant pathogenic bacteria. Because of their host specificity, phages are also valuable tools in bacterial taxonomy. Phage host ranges were a major argument in reclassifying *Pasteurella pestis* as an enterobacterium of the genus *Yersinia*. [*See* TAXONOMIC METHODS.]

3. Genetic Engineering

Phages have made many contributions to recombinant DNA technology. Restriction endonucleases were first identified in a phage host system and the DNA ligase of phage T4 is used to insert foreign DNA into viral or plasmid vectors. In addition, phages have several major applications:

1. Phage λ, derivatives of it, cosmids (hybrids between λ DNA and plasmids), and phage P1 are used as cloning vectors. Recombinant DNA (vector plus foreign DNA) is introduced into phage proheads. After completion of phage assembly, it can be injected into bacteria.
2. Filamentous coliphages of the *Inovirus* genus are used for DNA sequencing. Foreign DNA is introduced into the double-stranded replicative form of these phages.
3. Phage Mu DNA, able to integrate at random into any gene, is used to create mutations and to displace genes to other locations. [*See* RECOMBINANT DNA, BASIC PROCEDURES.]

4. Other Applications

1. Destruction of unwanted bacteria in bacterial and cell cultures, milk, meat, and freshwater (e.g., of cyanobacteria in "algal blooms").
2. Assay of antivirals, disinfectants, air filters, and aerosol samplers.
3. Detection of fecal pollution in water and of carcinogens, mutagens, and antitumor agents.
4. Tracers of water movements (surface water and aquifers).

B. Phages as Pests

In industrial microbiology, phages may destroy valuable starter cultures or disrupt fermentation processes. Phage interference has been reported in various branches of the fermentation industry, notably in the production of antibiotics, organic solvents, and cheese. In the dairy industry, phage infection is considered as the largest single cause of abnormal fermentations and a great source of economic losses. Phages derive from raw material, plant environment, or phage-carrying starter cultures. They are disseminated mechanically by air and may persist for months in a plant. Phage control is attempted by (1) preventing contamination by cleanliness, sterilization of raw material, sterile maintenance of starter cultures, and use of phage-free starters, (2) disinfection by heat, hypochlorites, UV light, and other agents, or (3) impeding phage development by starter rotation, use of genetically heterogeneous starters, and phage-inhibiting media. A recent approach is to construct phage-resistant starters by genetic engineering.

Bibliography

Ackermann, H.-W., and DuBow, M. S. (1987). "Viruses of Prokaryotes," Vol. I. "General Properties of Bacteriophages." CRC Press, Boca Raton, Florida.

Ackermann, H.-W., and DuBow, M. S. (1987). "Viruses of Prokaryotes," Vol. II. "Natural Groups of Bacteriophages." CRC Press, Boca Raton, Florida.

Casjens, S., and Hendrix, R. (1988). Control mechanisms in dsDNA bacteriophage assembly. *In* "The Bacteriophages," Vol. 1 (R. Calendar, ed.), pp. 15–91. Plenum Press, New York.

Francki, R. I. B., Fauquet, C. M., Knudson, D. L., and Brown, F. (eds.) (1991). "Classification and Nomenclature of Viruses. Fifth Report of the International Committee on Taxonomy of Viruses." *Arch. Virol.*, suppl. 2, 1–450.

Goyal, S. M., Gerba, C. P., and Bitton, G. (eds.) (1987). "Phage Ecology." John Wiley & Sons, New York.

Maniloff, J. (1988). *CRC Crit. Rev. Microbiol.* **15,** 339–389.

Van Duin, J. (1988). The single-stranded RNA bacteriophages. *In* "The Bacteriophages," Vol. 1 (R. Calendar, ed.), pp. 117–167. Plenum Press, New York.

Wünsche, L. (1989). *Acta Biotechnol.* **5,** 395–419.

Zillig, W., Reiter, W.-D., Palm. P., Gropp, F., Neumann, H., and Rettenberger, M. (1988). Viruses of archaebacteria. *In* "The Bacteriophages," Vol. 1 (R. Calendar, ed.), pp. 517–558. Plenum Press, New York.

B Cells

Ian M. Zitron
Wayne State University School of Medicine

Glossary

Antigen Molecule that can be bound specifically by antibodies (see also Determinant or Epitope)

Class II MHC antigens Cell membrane glycoproteins that are crucial to immune regulation. They are expressed on macrophages, dendritic cells, and B cells, where they form an integral part of the complex by which peptide antigens are presented to T helper cells.

Cluster of differentiation (CD) antigen As cells of the lymphoid system develop, they express a variety of molecules that are identifiable by specific antibodies. Some of these antigenic molecules are specific to a particular cell type, whereas others are shared by a number of cell types. In an attempt to standardize nomenclature, many of these antigenic molecules have been given CD numerical designations. For some of these CD antigens a function is known, whereas the function of others is still under investigation. Over 70 CD antigens have been defined.

Combinatorial joining Process by which a relatively small number of gene segments can be recombined to give a large number of full-length functional genes. This is the mechanism by which diversity is generated in the variable region genes for both immunoglobulins and T cell receptors. The term is also applied to the independent assortment of full-length heavy and light chains, so the variable regions of any heavy chain and any light chain can associate to form a combining site. In all cases, the principle is the same. Assume that a complete molecule is composed of two components, a and b; let the available number of each be n_a and n_b; then the total number of possible complete molecules is given by the product ($n_a \times n_b$).

Complement System of plasma proteins. These proteins function as a cascade. When antibodies of certain classes are bound to the surface of a cell the antibodies can initiate the cascade, which results in the lysis of that target cell. The process is called complement fixation. The antibody-dependent form of this process is called the classical pathway; another antibody-independent pathway, called the alternate pathway, also exists.

Determinant or epitope Antigen or immunogen are designations usually applied to large molecules. From the perspective of the lymphocytes, however, neither of these words is sufficiently precise, since the antigen-specific receptors of a single cell are capable of binding only a small region on the larger antigen. An individual small region of a molecule to which an antibody can bind is called a determinant or epitope.

Fluorescein One example of a group of molecules called fluorochromes. Such molecules absorb and emit light at characteristic wavelengths. The emitted light is in the visible spectrum, thus appearing colored. Fluorescein, for example, emits green light. Therefore, if a population of cells is stained with a fluorescein-conjugated antibody, those cells to which the antibody binds will appear green when appropriately illuminated and examined. Other fluorochromes are also in regular use and include red-, orange-, and blue-emitting molecules.

G proteins GTP-binding regulatory proteins that function as an important part of the signalling pathway in a number of membrane receptor systems

Hapten In order to form an antigen–antibody

complex that will precipitate, both the antibody and the antigen used must be at least bivalent. A hapten is a small monovalent molecule that may be bound by specific antibody, but will not form a precipitating complex. Hapten molecules are also capable of inhibiting precipitate formation when added to a mixture of antibody and bivalent antigen. When covalently coupled to a larger carrier, such as a protein, multivalent presentation can be achieved. In such a situation, the hapten groups become immunogenic and the hapten–carrier molecule can be precipitated by antihapten antibodies.

Immunogen Molecule that is capable of eliciting an immune response when introduced into an immunologically naive animal

Phagocytic cell or phagocyte From the Greek, meaning "eating cell." A designation given to cells such as macrophages, which can engulf and destroy particles such as bacteria. The engulfment process is called phagocytosis.

Phenotype Collective expression of the genome. In the context of B-cell development, this term is usually used to indicate the set of expressed molecules that defines a particular stage of development.

Repertoire Total number of distinct binding sites that an individual organism can generate, either in antibody or T-cell receptor combining sites

Retrovirus Virus that has an RNA genome and an enzyme, RNA-dependent DNA polymerase (reverse transcriptase), that is capable of transcribing a DNA copy of the viral RNA. Replication of retroviruses requires the production of such a DNA copy and the integration of it into the host cell's genome. An example of such a virus is human immunodeficiency virus type I, the etiological agent of AIDS.

Titer Measure of the concentration of specific antibodies in a serum sample. Titer is frequently expressed as the reciprocal of the highest dilution of the serum at which specific antibodies can be detected.

B CELLS are lymphocytes and can be activated from their resting state to become the effector cells of humoral immunity. Within a population of lymphocytes one can identify individual lineages and subpopulations by the presence of characteristic membrane molecules. B cells are most readily identified by the presence of membrane immunoglobulin (mIg) molecules. Resting B cells are morphologically identifiable as small cells (in the so-called G_0 phase of the cell cycle) of about 7 μm diameter, containing only a thin rim of cytoplasm. When an animal is immunized, the resting B cells are triggered to undergo both proliferation and differentiation. These events ultimately give rise to the terminally differentiated plasma cells that secrete antibodies. Antibodies are the effector molecules of humoral immunity that, because they are secreted, may act systemically, quite distant from their site(s) of production. Binding of foreign molecules by antibodies is very specific. If a preparation of purified antibodies is tested for its binding specificity, it can be shown to bind only those molecules used for immunization and very closely related molecules.

Antibodies are immunoglobulins. The words "antibody" and "immunoglobulin" are not synonymous, although they frequently appear to be used interchangeably. Although all antibodies are immunoglobulins, and we believe all immunoglobulins to be antibodies with binding specificity for something, the problem is that the universe of "somethings" is so enormous that we cannot possibly test the latter statement. Consequently, the word "antibody" should be used only when the binding specificity is known, whereas the word "immunoglobulin" should refer to the entire class of proteins, regardless of whether or not the binding specificity is known.

Immunoglobulins are glycoproteins with the common structural organization of a four-chain assembly of two large (heavy or H) and two small (light or L) chains; this assembly is often referred to as the basic H_2L_2 unit (Fig. 1). Within each H_2L_2 unit, the two H chains are identical to each other, as are the two L chains. Sequencing of purified H and L chains from many different antibodies has revealed that, in each chain, an N-terminal stretch of approximately 107 amino acids shows enormous sequence diversity; this is the *variable* (V) region. In contrast, the remainder of the molecule shows very limited diversity when the immunoglobulins are obtained from a single species; this is the *constant* (C) region of the molecule. With respect to function, antibody molecules demonstrate exquisite specificity of binding to their particular antigen. The antigen-binding function is the property of the combining sites, each of which is composed of one H chain variable region (VH) and one L chain variable region (VL). The two V regions fold together, forming a surface complementary to the antigen. The tremendous sequence

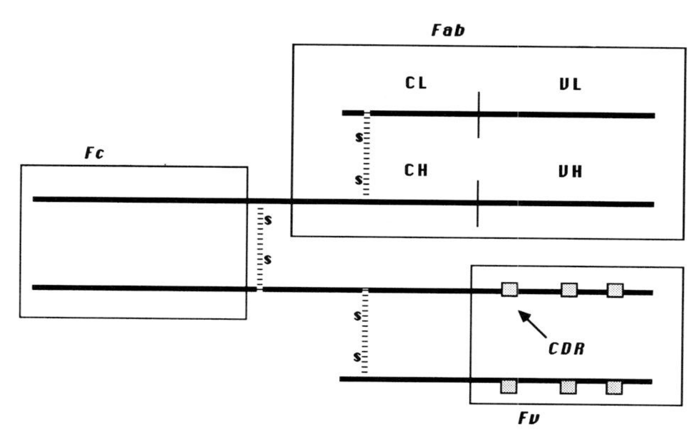

Figure 1 Basic four-chain immunoglobulin structure. The two halves of each molecule are identical, each consisting of one heavy and one light chain, shown by thick solid lines. The chains are linked by disulfide bonds, shown as hatched vertical lines. The heavy and light chains each consist of two regions, a variable region (labeled VH and VL) and a constant region (labeled CH and CL). Each variable region contains three stretches of amino acid sequences called the complementarity determining regions (CDRs). In the intact molecule, or an antigen-binding fragment derived therefrom, the VH and VL fold together so that the CDRs come together to form a surface that makes contact with the antigen molecule, effecting binding. The three boxed regions shown (labeled Fab, Fv, and Fc) are fragments that can be obtained by enzymatic digestion. Whereas an Fab or Fv fragment is monovalent, that is, has only one binding site, the four-chain Ig molecule is bivalent. Abbreviations: Fab, antigen-binding fragment; Fv, variable fragment; Fc, crystallizible fragment.

diversity observed in V regions explains much of the ability of the immune system to generate the large number of distinct combining sites (on the order of 10^8–10^9) that are known to exist. (See Table II and Section II.C.2 for a discussion of the genetic basis of variable region diversity.) Mutations arising during the proliferative phase of the response also contribute to the size of the repertoire. Each H_2L_2 unit contains two identical binding sites, that is, the antibody molecule is bivalent. Antibodies also exhibit a number of distinct biological properties, such as the fixation of complement or the ability to cross the placenta. The biological properties are much more limited and are properties of the C regions of the H chains. Both sequencing and serological studies have shown that there are a limited number of classes of immunoglobulins: five H chain classes in mammals, with different numbers of subclasses depending on the species. The individual classes of H chains are designated by (lower case) letters of the Greek alphabet; the class of the complete immunoglobulin molecule is indicated by the (upper case) English alphabet equivalent. For example, immunoglobulins of the IgG class have γ heavy chains. The class and biological properties of an immunoglobulin molecule are dictated solely by the H chain C region. However, two classes of L chain C region, κ and λ, are also identifiable. Some of the characteristics of immunoglobulins are shown in Table I.

Table I Characteristics of the Immunoglobulin Heavy Chain Classes[a]

Immunoglobulin	Heavy chain class	Molecular form	Molecular mass (kD)	Predominance
IgM	μ	$(\mu_2L_2)_5$[b]	900	Primary responses
mIgM	μ	μ_2L_2	180	Plasma membrane of B lineage cells from immature to mature virgin stage
IgG	γ[c]	γ_2L_2	160	Anamnestic responses; highest concentration of Ig classes in serum; membrane receptor on IgG memory cells
IgA	α	$(\alpha_2L_2)_2$/SC[d]	320	Secretions (some serum)
IgE	ε	ε_2L_2	200	Found on mast cells; lowest serum concentration of any of the Ig classes
IgD	δ	δ_2L_2 and δL	185	Mature virgin B cells; very low serum concentration

[a] In addition to the heavy chain classes, there are two classes of light chains: κ and λ. Each of the heavy chain classes may associate with either one of the light chain classes. For an individual heavy chain class, molecules of the form $H_2\kappa_2$ and $H_2\lambda_2$ exist.
[b] The pentameric secreted IgM molecules is held together by a small protein molecule called J chain. This is synthesized by the antibody-producing cell, in addition to the chains of the IgM. J chain is also found in other polymeric molecules, such as dimeric IgA.
[c] Subclasses of heavy chains exist. For example, in the mouse there are four γ subclasses designated γ1, γ2a, γ2b, and γ3. The Ig molecules in which these appear are called, respectively, IgG1, IgG2a, IgG2b, and IgG3.
[d] To enter secretions, IgA must be transported across an epithelial surface. The transport mechanism involves an additional molecule called secretory component (SC), which is produced by the epithelial cells.

I. Introduction

In response to the introduction of foreign molecules, such as the products of viral or bacterial infection, the acquired immune response is capable of mounting two separate and distinct effector responses. These responses are designated humoral and cell-mediated immunity (CMI). These names have their historical origins in a dispute that raged a century ago between two schools, the "Humoralists" and "Cellularists." The former held the view that soluble "humors" (which we now know to be antibodies) were the principal protective factors. In contrast, the latter argued that protection was largely the domain of cells, particularly phagocytic cells. We now know that both schools were correct, at least in part. Protective immunity is a consequence of the activity of lymphocytes, in collaboration with other cells. Both humoral immunity and CMI contribute to immune protection; one or the other of these is more efficient, depending on the insult or pathogen introduced into the body.

Any discussion of the immune response must include the Clonal Selection Hypothesis originally proposed by Macfarlane Burnet. Briefly, Burnet proposed that during the development of the immune system many independent clones (families) of lymphocytes arise. The members of each clone are characterized by the expression on their plasma membranes of receptors of one, and only one, antigen-binding specificity. Most importantly, the developmental steps leading to receptor expression are independent of encounter with the molecule(s) to which the receptors show specific binding, that is, they are *antigen independent*. Consequently, when a foreign substance is introduced, lymphocytes bearing receptors with the appropriate binding specificity are already present in the pool of cells. The lymphocytes with receptors having the "best fit" for the foreign molecule are selected for response and become activated (see Section III.A) as a direct consequence of the receptor binding events. Those lymphocytes that are not selected by antigen remain quiescent. Figure 2 depicts clonal selection. In the illustration, the receptors on clones 1 and 2 are incapable of binding the antigen, whereas those on clones 3 and 4 bind the antigen, permitting selection to occur. Clones 3 and 4 differ, however, by binding distinct portions of the antigen (see subsequent discussion of epitopes).

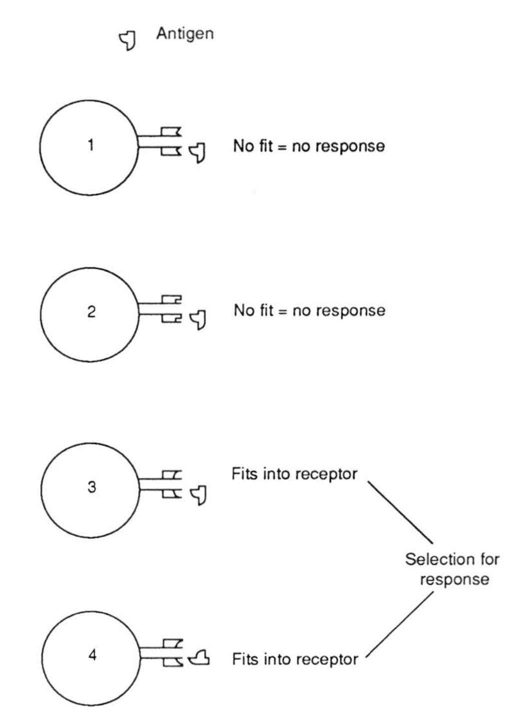

Figure 2 The Clonal Selection Hypothesis, originally proposed by Macfarlane Burnet. The antigen molecule is shown having an irregular shape. The receptors on clones 1 and 2 are not complementary to the antigen. Thus no binding occurs and the cells are unaffected by the presence of antigen. The receptors on clones 3 and 4 are complementary to the antigen, although in different orientations. Binding selects the cells in these clones for response.

Two crucial characteristics of the immune system are *specificity* and *memory*. The former was just described and is a direct function of the large diversity in the pool of combining sites that is clonally distributed in the lymphocyte population, that is, a single clone of lymphocytes has receptors with only a single binding specificity. The latter is much like the memory observed in the functioning of the nervous system: an ability to recognize, and respond to, the second (or third, etc.) exposure to a foreign substance in a way that is both quantitatively and qualitatively different from the response to the first, or primary, exposure. Both specificity and memory are evident in the function of B cells; together they form the underlying basis for the efficacy of vaccination. Figure 3 shows both specificity and memory in a single experiment. In this experiment, primary immunization with X elicits a relatively small response of short duration: the anti-X primary response. If, after the primary serum antibody has declined to an undetectable level, the animal is reimmunized (boosted) with the same antigen, a secondary response to X is elicited. The secondary response dif-

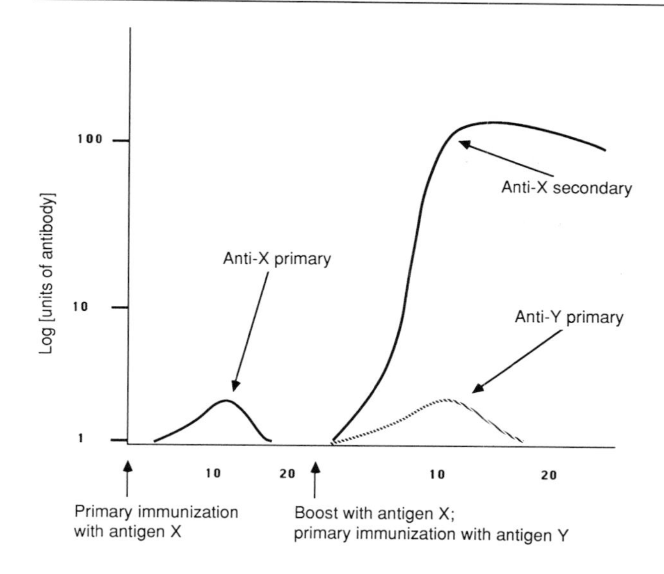

Figure 3 Primary versus secondary humoral responses. The responses differ in both magnitude and kinetics, as well as in class of antibody predominating (see text). These differences reflect immunological memory. Immunization with a non-cross-reacting antigen (Y) at the time of secondary immunization with X elicits a primary response to Y, indicating that the memory is antigen-specific.

fers from the primary in that its kinetics are faster, the serum antibody concentration attained is higher (note the logarithmic scale of the ordinate), and the decay phase is prolonged. Not shown in the figure are the classes of antibodies produced in the two responses. The primary response is almost exclusively IgM, whereas the secondary is IgG. These differences between the primary and secondary responses reflect immunological memory. Further, if the animal is immunized with a non-cross-reacting antigen, Y, at the time it is boosted with X, the anti-Y response will show the pattern of a primary response. Thus the memory that was generated by primary immunization with X is specific for X and does not reflect a generalized perturbation of the immune system affecting responses to other antigens. (See Section III.A.1 for an extended discussion of primary and secondary humoral responses.)

To elicit an immune response, a molecule must be several thousand daltons in molecular mass. In terms of B-cell recognition, even a molecule of this relatively small size may have several distinct sites that B cells can recognize. Each of these sites is called a *determinant* or *epitope*. B-cell responses to most immunogenic molecules are polyclonal, that is, a single molecule can elicit B-cell responses to each of its individual epitopes. All the antibodies produced will bind the whole molecule, but the population will, in fact, be composed of a collection of individual epitope-specific antibodies. Many immunogens are large molecules with many distinct epitopes. For example, many viral glycoproteins are globular molecules that bear both protein and carbohydrate epitopes. Moreover, an individual epitope can elicit responses from more than one clone of B cells. Whereas all the clones responding to a particular epitope may have specificity for it, they need not have identical combining sites and their combining sites may not bind the epitope in the same orientation or with the same affinity. Epitopes fall into two distinct categories: sequential and conformational. The former are epitopes formed by a contiguous stretch of amino acids or monosaccharides in a larger molecule. The latter depend on the higher order structure of the immunogen, since they depend on the folding of the molecule to form the epitope. B-cell receptors and antibodies can recognize both types of epitope. Polyclonal responses often show a predominance of antibodies that recognize conformational epitopes. Figure 4 depicts a protein molecule in which the solid dark line represents the polypeptide backbone. The sites on the molecule indicated as p, q, r, y, and z represent individual epitopes. Epitopes y and z are shown to involve limited contiguous stretches of amino acids; they are, therefore, sequential epitopes. In contrast, p, q, and r each involve contributions from more than one portion of the molecule; their existence depends on the folding of the polypeptide chain. These are conformational epitopes.

Because antibodies are large globular molecules, they cannot penetrate living cells. Thus their protective function is restricted to the extracellular milieu. This milieu includes the plasma and the interstitial fluids, which contain antibodies secreted by the *systemic* immune system. In addition, a *mucosal* immune system functions in parallel. Its role is to protect the mucosal surfaces of the body. These include the gastrointestinal, respiratory, and urogenital tracts; the eye; and glandular tissues such as the mammary glands. In many instances, effective protection results from an interplay of the two systems. The role of antibodies in protecting the human fetus and newborn is an example of this interplay. During intrauterine development, IgG produced by the maternal systemic immune system crosses the placenta, entering the fetal circulation. At birth, the neonatal immune system is insufficiently mature to

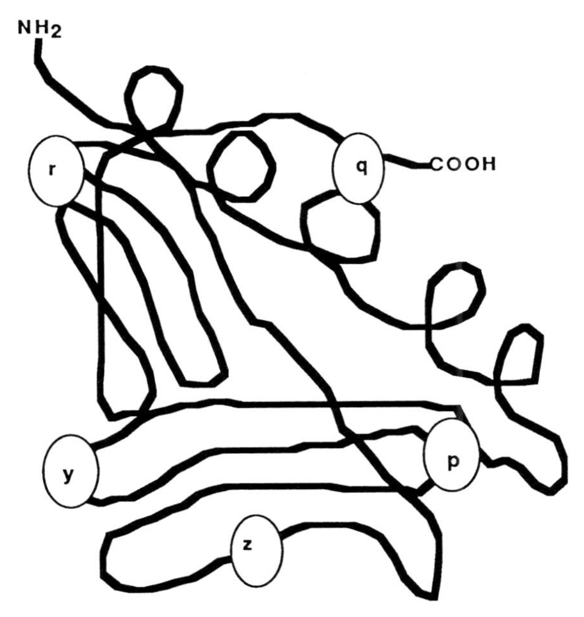

Figure 4 An imaginary protein molecule, showing both sequential and conformational epitopes. The amino acid backbone is shown as the thick solid line. The sites labeled y and z involve only short stretches of contiguous amino acids; these are sequential epitopes. The sites labeled p, q, and r indicate epitopes that depend on the folding of the native protein for their existence, since they involve amino acids from different regions of the backbone that are brought together only when the molecule is properly folded. These are conformational epitopes.

produce its own IgG; the maternally derived antibodies provide systemic protection for the first 6–9 months of life. In addition, the maternal mucosal immune system produces IgA antibodies that appear in the milk. A breast-fed infant thus obtains maternal antibodies that are protective of its gastrointestinal tract for as long as it continues to breast-feed. Both of these processes are examples of *passive immunization,* in which the immune system of the recipient is not responding per se, but effector molecules are being transferred and are conferring the protection.

II. Evolution and Ontogeny

A. Evolutionary Appearance of Humoral Immunity

The ability to distinguish between self and non-self and to react against the latter may be identified in organisms as primitive as sponges. Thus a system that may have been the precursor of the acquired immune response appeared very early in evolution. Extensive studies of invertebrates have shown that the immunity they manifest appears to resemble CMI rather than humoral immunity. Thus, cellular responses evolutionarily antedate antibody responses. Only when one studies the vertebrates is humoral immunity per se identifiable. Examining vertebrates, fish have been shown to express only one class of Ig, which resembles IgM. Diversification of other heavy chain classes occurred later in evolution: chickens, frogs, and toads have three classes. The immune systems of mammals have evolved to produce five heavy chain classes with, depending on the species, varying numbers of subclasses.

B. B-Cell Ontogeny: Primary Lymphoid Organs

All the cellular elements of the blood and lymphoid system originate from pluripotential hematopoietic stem cells. These cells are self-renewing, that is, they can divide and give rise to daughter cells identical to the parent or, alternatively, give rise to cells that become progenitors committed to any one of the four developmental lineages found in blood and lymphoid tissue: erythroid, lymphocytic, granulocyte/macrophage, and megakaryocyte/platelet. The direction of commitment appears to be under the control of growth factors, which direct cellular development along particular pathways. The lymphoid progenitors give rise to both B cells and another distinct lymphocyte population, the T cells. The organ sites at which lymphocyte development occurs are referred to as the *primary lymphoid organs*.

Studies of B-cell development were helped immensely by work performed in birds, which possess a discrete organ located just above the cloaca, the bursa of Fabricius, in which B-cell development occurs. Removal of the bursa immediately after hatching renders birds incompetent with respect to humoral immunity without compromising their ability to mount CMI. One of the origins of the name "B cells" is from the abbreviation B for bursa. Ontogenetic studies in birds showed that cells developing in the bursal follicles stained positively with fluorescein-conjugated antibodies specific for the Ig heavy chain classes. The earliest mIg-positive cells bore mIgM only (referred to hereafter as mIgM-positive cells). As development proceeded, mIgG-positive cells could be identified and, finally, mIgA-positive cells. Thus the ontogenetic sequence of mIg expression was defined. The sequential relationship between the various mIg-bearing populations and their ability to secrete antibodies of those classes was defined by showing that injection of anti-μ antibodies (specific for IgM and, thus, mIgM-positive cells) eliminated all humoral responses, presumably by their effects on the mIgM-bearing cells. In contrast, injection of anti-γ antibodies left intact the ability to make IgM, but abolished the ability to make IgG and IgA. As might be expected, administration of anti-α abolished IgA production, but spared IgM and IgG. These experiments thus demonstrated that the mIgM-positive cells gave rise to the mIgG-bearers and that these gave rise to the mIgA-positive B cells, thus explaining the sequential appearance of the heavy chain classes.

The work in birds led to similar studies in mammals. However, in the absence of a discrete organ comparable to the bursa, it was necessary to ask whether a mammalian bursa-equivalent existed or whether B-cell development occurred at multiple sites. These studies have shown that the site of B-cell ontogeny varies according to the stage of mammalian development. Lymphocyte production in the early embryo occurs first in the yolk sac. As development proceeds, the fetal liver predominates, followed by the spleen during the neonatal period. Finally, the bone marrow is the primary lymphoid organ for B-cell development in the postneonatal mammal.

C. Discrete Developmental Steps May Be Identified at Both Protein and DNA Levels

Most studies of B-cell development have been carried out with mouse cells. The following description summarizes results obtained in the mouse. When comparable studies have been performed with cells from other species, the pattern of development has been found to be very similar, although the nomenclature of some of the molecules and names of the reagents used may differ. Work performed in a number of laboratories has defined the sequence of events that occurs as cells mature along the B-cell pathway and has allowed investigators to determine the factors that regulate the process. Instrumental to these studies have been the development and refinement of *in vivo* and *in vitro* systems in which cells from the primary lymphoid organs can be grown and examined as they develop, coupled with the availability of monoclonal antibodies that recognize specific cell-surface molecules (see Section V). The various stages of development are recognized by the sets of molecules that the cells express and, from the pre-B cell stage onward, the DNA rearrangements in the Ig gene loci. Using these approaches, a generalized scheme of B-cell ontogeny has been obtained, beginning with an early B-cell precursor, the pro-B cell, and ending with the mature virgin B cell. The word "virgin" is used to denote a cell that is immunocompetent but has not yet encountered antigen. Figure 5 presents a generalized scheme of B-cell development, beginning with the pro-B cell, the earliest identifiable stage in the pathway, and culminating with the mature virgin B cell. For each stage, there is an indication of the molecular and genetic events characteristic of the stage. Additionally, the diameter of the arc depicting each stage is an indication of its relative size. The increase in arc size from pro-B to immature B cell indicates expansion in size from one stage to the next. Expansion ceases at the immature B cell stage; hence, the arcs depicting immature and mature B cells are the same size.

1. Pro-B Cells Coexpress Two Membrane Proteins Indicative of the B-Cell Lineage

The first cell in the scheme, the pro-B cell, may be recognized by its membrane expression of two molecules, designated CD45R and J11d. The Ig loci in the DNA are unrearranged in their germ-line configuration.

CD45R and J11d also illustrate the point that, although in some instances we can ascribe a function to the molecules that we use to identify a particular cell type, in other instances the function remains undefined. CD45R is a phosphotyrosine phosphatase (PTPase), which is thought to have an important role in regulating signal delivery to the B cell (see Section III.A.3). In contrast, no function has been ascribed to the molecule recognized by the J11d antibody, although it has been shown to be a heavily glycosylated integral membrane protein and a member of the family of molecules called *heat stable antigens*. Lack of information about the function of a particular molecule does not, however, compromise its utility as a marker of development and actually serves as a springboard for further investigation.

Expression of these two molecules is maintained

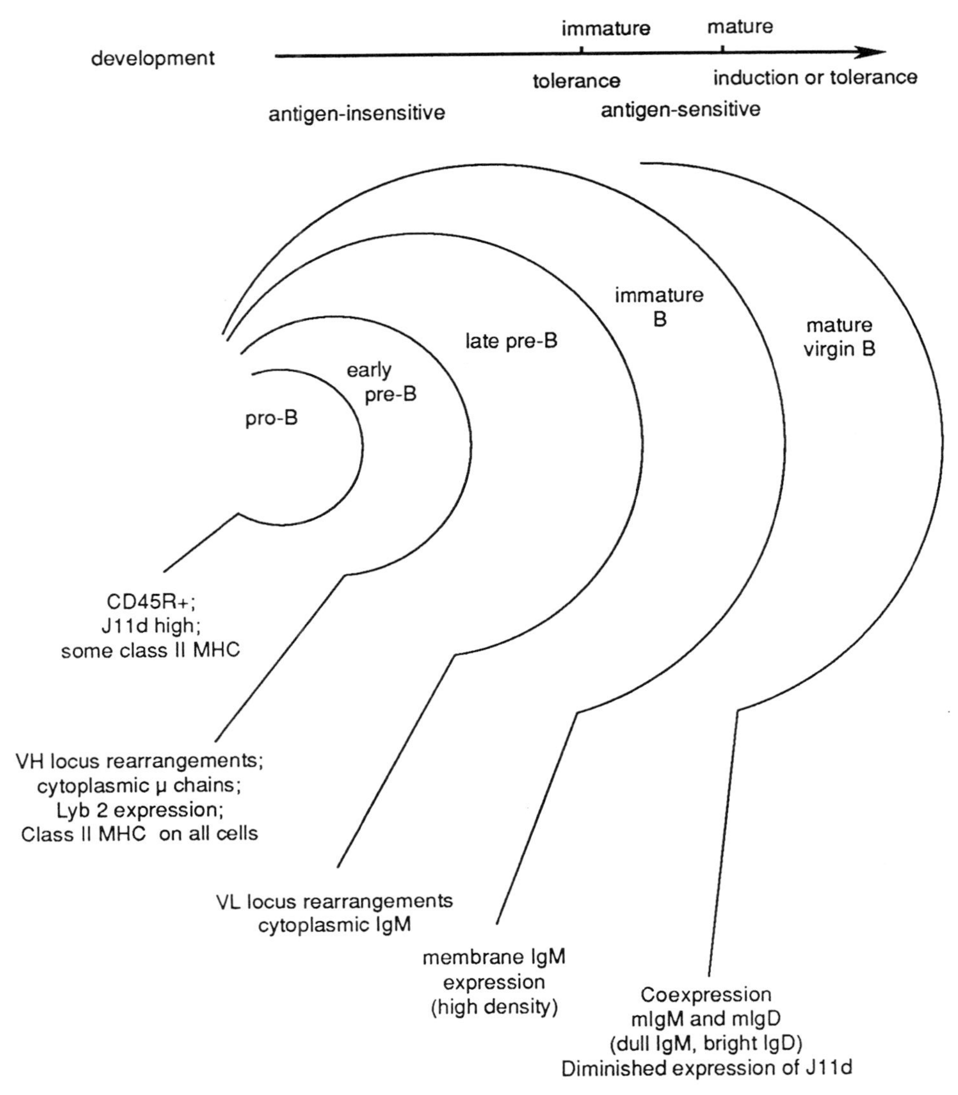

Figure 5 Generalized scheme of antigen-independent B-cell development. Development proceeds from left to right. Moving from one stage to the next, an increase in the size of the arc indicates that proliferation occurs. Once cells reach the immature B-cell stage, antigen-independent proliferation ceases. Cells are antigen-insensitive until they first express mIgM at the immature stage. At this point, antigen contact results in tolerance. Mature B cells respond to antigen either by induction or by becoming tolerant.

throughout the subsequent developmental stages. There is, however, some heterogeneity observed in the expression of the J11d antigen. B cells that have just entered the periphery but have not yet encountered a stimulus show high levels of expression of J11d. With time (a matter of days), in the absence of antigen contact, J11d expression diminishes. The functional correlation of J11d expression and B-cell response is considered later (see Section III.A.3).

2. The Organization and Rearrangement of Immunoglobulin Genes

Immunoglobulin genes are expressed only in cells of the B lineage. As indicated earlier, the B-cell repertoire contains on the order of 10^8–10^9 individual combining sites. Much of this repertoire is generated by rearrangements in the Ig gene loci, a process that occurs in B cells as they develop from the precursor to the immature B cell. Three clusters of genes encode Ig chains, one cluster each for H, κ, and λ. The three clusters are unlinked, existing on separate chromosomes. In each cluster one can identify discrete sets of genes. In the H chain cluster are four sets: V segment, D segment, J segment, and constant region genes. The κ and λ clusters contain three sets: V segment, J segment, and constant region genes. Rearrangements occur during B-cell development which give rise to genes that encode full-length Ig chains. The variable region domain of the heavy chain protein is encoded by three gene segments, one V, one D, and one J, that are brought together by recombination into a position 5′ to the gene encoding the μ chain constant region. The diversity in VH regions is an outcome of combinatorial joining, because of the multiple individual genes encoding the segments. On a single chromosome, then, the potential number of full-length V domain genes is the product of the numbers of genes for each of the individual segments. A similar process occurs during rearrangement at the light chain loci, again generating considerable diversity. Table II shows the number of genes in the H and κ loci of the mouse. The data indicate how a relatively small number of genes can be utilized to generate a large number of individual combining sites by a process of combinatorial joining. The total number of sites is actually greater than the product of the products because of additional mechanisms such as imprecise joining during recombination, the ability of the D segment genes to be read in all three frames, and somatic mutation, which occurs in proliferating B cells subsequent to antigen-driven stimulation.

DNA rearrangements can be identified by isolating DNA from B-cell precursors or, more easily, from a tissue source, such as liver, that reflects non-B cells. If this DNA is digested with restriction endonucleases, and the sizes of DNA fragments are determined by Southern blotting and the use of hybridization probes for individual gene segments, one can see patterns characteristic of the unrearranged germ-line organization. DNA from B cells, isolated and manipulated in parallel, gives hybridization signals on DNA fragments of different sizes, thus indicating that the genome has undergone rearrangement. The use of hybridization probes for non-Ig genes indicates that the rearrangement events are restricted to the Ig gene loci. The rearrangements in the Ig loci are, of course, genetic recombinations. The question arises, therefore, of how the recombinations are catalyzed. Two genes recently have been identified, the products of which appear to be responsible for recombinations in lymphoid cells. The genes have been designated *RAG-1* and *RAG-2; RAG* is the acronym for Recombination Activating Gene.

B cells are, like all other somatic cells, diploid and contain two copies of each of the immunoglobulin gene loci. However, an individual B cell expresses only one combining site in its Ig molecules. This phenomenon is called *allelic exclusion*. Successful rearrangement on one chromosome suppresses rearrangement of the allelic genes on its homolog. For Ig H chains, successful rearrangement to form a VH domain gene excludes only the other H chain. For L chains, however, successful rearrangement of one locus excludes not only its homolog, but the other two L chain loci. Thus, if a full-length Vκ gene is

Table II Combinatorial Joining Permits the Generation of Large Numbers of Immunoglobulin Variable Regions from a Small Number of Individual Genes

Gene	V segments	D segments	J segments	Sum	Product
Heavy chain locus	500	15	4	519	30,000
κ Light chain locus	200	NA[a]	4[b]	204	800
Sum of sums				723	
Product of products					2,400,000[c]

[a] NA, not applicable. D segments do not occur in immunoglobulin light chains.

[b] There are, in fact, five J segment genes in the κ locus. One of these is a pseudogene, so only four are expressed at the protein level.

[c] This number is a minimal estimate of diversity. Other mechanisms exist to increase the number of combining sites. These mechanisms include imprecise joining during the recombinational events, the ability of D segment genes to be read in three frames, and somatic mutation.

Abbreviations: V, variable; D, diversification; J, joining (not to be confused with the J chain found in polymeric immunoglobulins).

assembled, rearrangements in the other Vκ and in both Vλ loci are prevented.

3. Development from the Pre-B to the Mature B Cell

The next stage of development, the pre-B cell, is the first in which the Ig gene loci show DNA rearrangements. These rearrangements occur sequentially; the H chain locus shows changes before the L chain. At the level of protein expression from these loci, early pre-B cells can be shown to contain free cytoplasmic μ chain molecules, but no light chains. The late pre-B cell contains both cytoplasmic μ and light chains. Temporally, though not necessarily causally, associated with the Ig rearrangements, pre-B cells also express yet another surface protein, Lyb2.

Immature B cells, the cells formed in the next developmental step, are the first cells in which mIg expression can be identified. This is the first point in the B lineage at which the cells are antigen sensitive. However, in contrast to mature B cells, binding of specific antigen by immature B cells may result in tolerance, that is, the cells are rendered unresponsive. (See Section III.C for a description of B cell tolerance.) Prior to the immature B-cell stage, only a fraction of the cells expresses class II MHC gene products. At this stage, however, essentially 100% of the cells exhibit high levels of membrane expression of these molecules, which are essential for recognition by helper T cells (see subsequent text).

The final step in the pathway is the appearance of mature virgin B cells. The principal phenotypic change that occurs here is the coexpression of a second mIg class, mIgD. Mature B cells, then, have the phenotype $mIgM^{+}mIgD^{+}$. On an individual cell, the mIgM and mIgD have the same binding specificity. They are identical in their light chains and in the variable regions of their heavy chains and differ only in their heavy chain constant regions. The molecular mechanism by which immunoglobulins of these two distinct classes can be expressed on a single cell involves alternative pathways for processing a long primary messenger RNA transcript.

Mature virgin B cells migrate from the primary lymphoid organ to the periphery, where they are ready for encounter with antigen and, under the appropriate circumstances, the elaboration of a humoral response. The lifespan of these cells, once they have entered the periphery, is relatively short, only a few days, if they are not selected by antigen encounter. The mechanisms responsible for the limited lifespan of unstimulated mature B cells are unknown. New B cells are continuously being generated in the primary lymphoid organs at the rate of several million cells per day. Thus, over a short period of time, B cells encompassing the entire repertoire enter the peripheral compartment and are available to mount a protective immune response.

As noted earlier, the events in lymphocyte ontogeny leading to receptor expression are antigen independent. This is also true of the final step, from immature to mature B cell. However, this should not suggest that external influences are irrelevant to the process of B-cell development. In fact, a variety of growth and maturational factors have been identified that are involved in driving B-cell ontogeny. Among these are at least two T cell products, called interleukin-3 (IL-3) and interleukin-7 (IL-7), as well as other incompletely defined factors produced by bone marrow stromal cells. Developing B cells, therefore, express functional receptors for growth factors from very early in the lineage. These receptors are crucial for successful progress and more detailed definition of the factors, their receptors, and the modes of signaling through these receptors are at the forefront of research on B cells. The lymphocyte-derived growth and maturational factors (lymphokines) play important roles throughout the immune system (see Section III.A.2 and Table VI). [*See* INTERLEUKINS.]

4. CD5-Positive B Cells: A Discrete Subpopulation with an Autoreactive Bias

The foregoing sections have dealt with the development of those B cells that are found in the spleen and lymph nodes. There is, however, an additional and discrete subpopulation of B cells with a unique anatomical localization and set of characteristics. These cells are B cells, as shown by their immunoglobulin gene rearrangements and the presence of membrane Ig, but differ from the more conventional splenic and lymph node B cells by their expression of the membrane molecule designated CD5. CD5 (also known as Ly1) had been described originally as a unique marker of T lymphocytes, so the observation that it also was expressed on a population of mIg-positive lymphocytes was somewhat surprising. There is, however, no doubt that the population identified is, indeed, a true B cell population, albeit with some characteristics that distinguish its members from the B cells found in the secondary lymphoid organs. Some of the characteristics by which one can discriminate between CD5-positive and CD5-negative (conventional) B cells are shown in Table III.

Mature virgin B cells have been defined as bearing both mIgM and mIgD on their membranes. CD5-positive B cells express very little IgD. In the mouse, use of the two light chain classes κ and λ is in the ratio 95:5. This is true of B cells both in the secondary lymphoid organs and in serum immunoglobulin. In contrast, the CD5-positive population shows a fourfold greater frequency of λ-positive cells. CD5 is coexpressed with another antigen, Mac-1, that is not normally found on B cells. The significance of Mac-1 expression is unknown. One of the most striking aspects of CD5-positive B cells is their anatomical localization. They are at low or undetectable levels in the spleen and lymph nodes, but constitute nearly 20% of the B cells in the peritoneal cavity. Moreover, in contrast to the CD5-negative population, which is derived from bone marrow precursors in the adult, the precursors of CD5-positive B cells are located in the peritoneal cavity, suggesting that this may be a self-renewing population that diverges at about the time of birth, since in the fetus and newborn precursors for both CD5-positive and CD5-negative B cells may be detected in the liver.

Table III Comparison of Murine CD5− and CD5+ B Cells

Characteristic	CD5− cells	CD5+ cells
Membrane phenotype		
mIg expression	MIgM+ mIgD+	mIgM+ mIgD− or low
Light chain usage	~5% λ+ and 95% κ+ in peritoneal cavity[a]	~20% of peritoneal cavity μ+ cells are λ+
CD5/Mac-1 expression	Negative	Positive
B220 expression	Positive	Positive
Anatomical location		
Precursors		
Fetus and newborn	Liver/spleen	Liver
Adult	Bone marrow	Peritoneal cavity
Functional cells	Spleen/lymph nodes	Peritoneal cavity
Lifespan	Few days unless stimulated by antigen	Long and/or self-renewing

[a] The ratio of κ to λ light chain usage in peritoneal cavity CD5− B cells is identical to that found in splenic B cells.

A potentially important functional significance for the CD5-positive B cells has been suggested by two lines of evidence. The first line involved examination of the antibodies produced by these cells, which showed that a significant proportion of the antibodies were specific for self determinants, that is, had the potential to be autoantibodies. The second line involved mice with a specific mutation, called Motheaten (*me*), that profoundly affects immune function. Although the mutation affects both B and T cells, mice with this mutation have B cells only of the CD5-positive phenotype. Examination of the antibodies synthesized by *me* B cells showed that these were also strongly skewed toward the representation of autoreactive binding sites. Thus the CD5-positive population may represent a reservoir for many of the autoantibodies that exist. Whether such autoantibodies are simply a potential source of pathology (see Section IV.B.2), or whether they have some subtle autoregulatory function (Section III.B), must still be determined.

III. Antigen-Dependent Events

The recognition of antigen by a mature virgin B cell has two possible outcomes: the cell can be induced, initiating the events that lead to antibody production, or it can be tolerized. The principal factors that determine the outcome are the number and nature of the signals delivered to the cell. Briefly, if only one signal is delivered by the binding of antigen by mIg, the outcome is tolerance. If, on the other hand, a mIg-mediated signal is accompanied by, or closely associated in time with, signal(s) from other cell types, the outcome is induction.

A. Induction

In this section, we will consider inductive events at three levels: first, the level of the whole animal; second, with respect to the cellular cooperation that is obligatory for humoral responses to protein antigens; and third, the cellular and biochemical events that occur in the B cells proper and are initiated on binding of antigen by the mIg receptors.

1. Humoral Responses at the Whole Animal Level

In the periphery, B cells are largely sessile, residing mainly in specific areas of the secondary lymphoid organs (spleen, lymph nodes, Peyers patches), in the so-called the B-cell areas of these tissues. At these sites, the majority of the cells show the $mIgM^+mIgD^+$ phenotype and small size characteristic of mature virgin cells. Humoral immune responses take place in the secondary lymphoid organs. The teleological reason for this appears to be that the primary organs, being the sites of lymphocyte development, contain immature cells that are very susceptible to tolerance induction should they bind antigen via their receptors. The secondary organ in which the response actually takes place is dictated by the route of entry of the immunogen. Immunogens entering the bloodstream directly are delivered to the spleen, which is the site of response. Immunogens entering the tissues are delivered, via the afferent lymphatics, to the lymph nodes draining the immunization site; for example, an infection of a finger will lead to enlarged tender lymph nodes in the elbow of that arm. These two routes of immunization will activate components of the systemic immune system. In contrast, immunogens delivered to the gastrointestinal (GI) tract, for example, those in food, will be delivered to the Peyers patches and activate components of the mucosal immune system, leading to IgA in secretions. It should be emphasized that both B and T cells in the secondary organs are affected by the delivery of immunogen and that the antigen-driven events that one observes at the whole animal level result from the interplay of multiple cell types.

The response of an animal to its first exposure to a particular immunogen is called the primary response. Responses to subsequent encounters with the same antigen (secondary, tertiary, etc.) are grouped together under the heading *anamnestic* (not forgetting) responses. Primary and anamnestic responses show similarities in their overall patterns, but also differ from each other both quantitatively and qualitatively (see Fig. 3). In terms of kinetics, both show a lag period between immunization and detectable levels of circulating antibody; in the primary response, this lag is about 7 days, whereas in anamnestic responses the lag period is considerably shorter. The lag period is followed by a phase of rapid increase in antibody concentration, similar to the logarithmic phase of growth observed in a bacterial culture, which leads to a plateau. The slope of the rising portion of the curve is generally steeper in anamnestic responses than in a primary response and the height of the plateau is vastly greater in an anamnestic response. Finally, a decay phase is apparent in both, indicating the cessation of high-rate antibody production. The decay phase in a primary response is quite rapid, providing an indication of the relatively short half-life of the antibody molecules produced. In contrast, the decay phase in an anamnestic response is very slow, and antibodies are detectable for many years. In addition to the kinetic and quantitative differences, primary and anamnestic responses are dominated by different classes of antibodies. The predominant class in the primary response is IgM. In contrast, anamnestic responses, at least of the systemic immune system, are dominated by IgG antibodies; some serum IgA is detectable. The sequential appearance of IgM, IgG, and IgA antibodies of the same binding specificity is called *class-switching* and reflects genetic alterations in the clones of B cells responding to the immunogen (see Section III.A.2). It is clear, then, that primary immunization not only elicits a largely IgM response, but affects the responding population of B cells in a way that increases their number and also permits them to respond more rapidly and with antibodies of different heavy chain classes on secondary immunization. Table IV provides a comparison of some of the differences between the mature virgin (primary) B cells, which respond to the first exposure to antigen, and the memory (secondary) B cells, which are produced as a result of this exposure.

Antibody affinity is a measure of the strength of antigen binding. In general, IgM responses are of lower affinity than IgG responses. The average affinities in an IgG response rise with time, that is, the polyclonal antibody population gradually shifts from a predominance of low affinity to a predominance of higher affinity antibodies. This process is called *affinity maturation*. In physical terms, a low affinity antibody requires the presence of a higher concentration of antigen than one of high affinity for detectable binding to occur. In biological terms, if a humoral response is of low average affinity then there will come a point at which the *in vivo* concentration of antigen will fall to a level at which essentially no binding occurs. This will happen although circulating antigen may still be present. Thus such antibodies will no longer have a protective effect. Teleologically, then, affinity maturation is advantageous because, as the average affinity of the IgG population rises, the antibodies are able to bind and eliminate antigen at lower and lower residual antigen concentrations.

At a cellular level, affinity maturation is the result of selection. As the mature B cells proliferate, in the first phase of their response to antigen, mutations occur in their immunoglobulin variable region genes. A fraction of these mutations will result in

Table IV Comparison of Mature Virgin (Primary) B Cells and Memory (Secondary) B Cells

	Primary B cells	Memory B cells
Membrane Ig phenotype	mIgM and mIgD[a]	mIgG, mIgE, or mIgA[b]; essentially no mIgM or mIgD
J11d expression	Positive[c]	Negative
Peanut agglutinin (PNA)[d]		Positive early Negative late
High endothelial venule homing marker[e]	Negative	Positive
Recirculation	Little or none	Yes
Ig variable regions	Unmutated	Accumulated mutations
Antibody affinity	Low–intermediate	Intermediate–high
Precursors of primary AFC[f]	$mIgD^{low-high}$/$J11d^{high}$ and $mIgD^{int-high}$/$J11d^{low}$	
Precursors of memory cells[f]	$mIgD^{high}$/$J11d^{low}$ only	

[a] Virgin B cells in the secondary lymphoid organs show a spectrum of mIgM and mIgD expression. The initial cells express high IgM/low IgD; with time, the amount of mIgM decreases, while that of mIgD increases.

[b] Most memory B cells express on their surfaces Ig of the isotype that they will secrete upon stimulation; for example, an IgG memory cell is mIgG+mIgM−D−E−A−.

[c] Primary B cells that have newly entered the secondary lymphoid organs express high levels of J11d; this level diminishes with time (see Section II.C.1).

[d] PNA is one example of a group of proteins called lectins. Many lectins are of plant origin. They are characterized by highly specific carbohydrate-binding activity. In the case of PNA, the binding is specific for terminal galactose residues, present on glycoproteins. Germinal center B cells are strongly PNA positive. Early after immunization this population also contains most of the memory B cells. At later times, however, the antigen-specific memory B cells become PNA negative.

[e] The ability to bind to, and migrate through, high endothelial venules is necessary for lymphocytes to exit the circulation and enter secondary lymphoid organs. As such, they are functional markers of recirculating cells. These properties are due to the presence of specific molecules which may be detected on those cells which are capable of efficient recirculation. Such molecules have been detected using antibodies and have been characterized biochemically.

[f] See text.

cells that express mIg with higher affinity than that of the parental cell. As the free antigen concentration falls, due to binding by secreted antibodies and the clearance of antigen–antibody complexes, it will reach a point at which only the B cells with higher affinity mIg receptors can be triggered. Some of these cells will have been present in the initially immunized population, whereas others will have arisen as a result of mutation. These higher affinity clones of cells will be the only ones that can produce secreted antibodies. Thus, with time, the average affinity of the response will rise.

As already described, primary immunization affects the B-cell population in two ways. One is to elicit a primary antibody response, which is largely IgM. The other is to generate a population of memory (secondary) B cells that respond with Ig classes other than IgM on restimulation. An important question arises from the observation that primary immunization yields two cells with very different properties. Do both primary antibody-forming cells (AFCs) and memory cells arise from the same virgin B cells or can one identify in the virgin B-cell pool some cells with a propensity to give rise to primary AFCs and others that gives rise to memory cells? Two alternative models proposed to explain this process were termed "symmetric division" and "asymmetric division." These are shown in Fig. 6 (A, B). Experiments performed in the 1970s seemed to support the asymmetric division model. Antigenic stimulation of virgin B cells was proposed to initiate proliferation, so some of the daughter cells yielded primary AFCs while others became B memory cells. The central idea of the model, in its most extreme form, was that the virgin B cells represented a single compartment and that whether a postmitotic daughter cell became a primary AFC or a memory cell was a result of the quality or quantity of other signals, probably from T cells. With the availability of new techniques and antibodies to membrane molecules, however, it has proven possible to examine the virgin B-cell compartment in ways unavailable in the 1970s. The result of these experiments has been to identify cells with different response characteristics. Two of the most valuable membrane molecules in this regard are mIgD and J11d. Neither of these molecules are expressed on memory B cells and coexpression identifies virgin B cells. Analysis of the virgin B-cell compartment shows a gradient of expression of each, but these gradients are reciprocal and isolation of cells that can be distinguished by their relative levels of these two markers yields populations that show distinct functional characteristics on immunization. The illustration in Fig. 6 (C) summarizes the findings. The earliest immigrants into the secondary lymphoid organs express low levels of mIgD but high levels of J11d. With time of residence in the secondary organs, mIgD expression rises while that of J11d falls. The J11d-positive B-cell population was separated to obtain one pool of cells that contained the most intensely positive cells ($J11d^{high}$) and another pool that contained the most

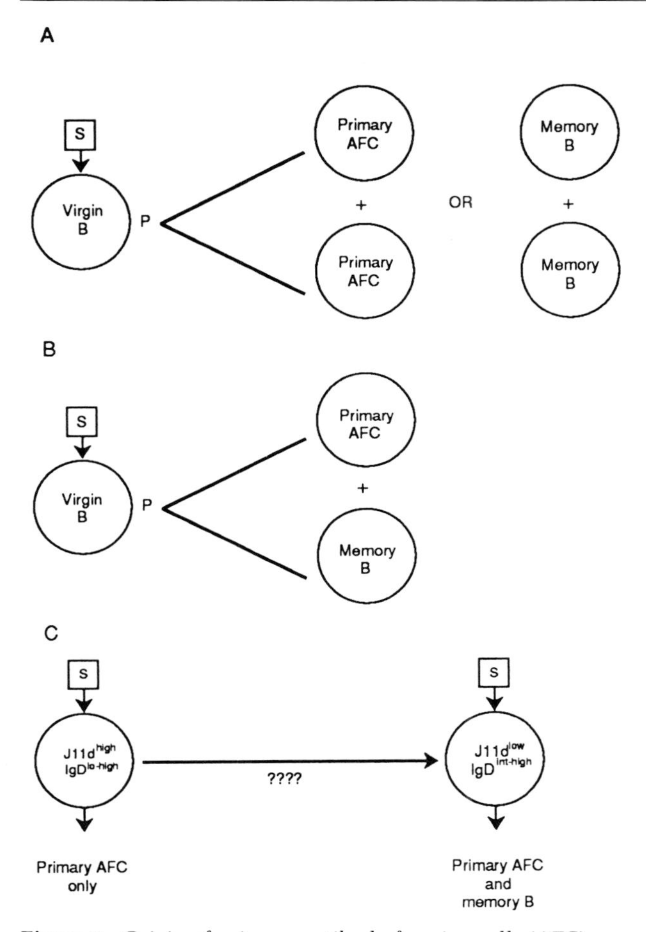

Figure 6 Origin of primary antibody-forming cells (AFC) and memory cells generated by initial exposure to antigen.(A) Symmetric division model. Stimulation (antigen plus additional signals) gives rise to proliferation. After each division, both postmitotic daughter cells are either primary AFC or memory cells. (B) Asymmetric division model. Stimulation causes proliferation, but after division one daughter cell gives rise to a primary AFC, whereas the other becomes a memory cell. (C) Summarized data from recent experimental work that shows that the virgin B-cell pool is separable, on the basis of J11d and mIgD surface densities, into two functionally distinct pools. One of these, J11d^{high}mIgD$^{low-high}$, yields only primary AFC on stimulation. The other, J11d^{low}mIgD$^{int-high}$, yields both primary AFC and memory cells. The question marks indicate that the responses of the cells between these two extremes have not been investigated. Abbreviations: high, high density; low–high, low to high density range; int–high, intermediate to high density range.

weakly positive cells (J11d^{low}). For the purpose of this study, the intermediate population was discarded. Further analysis showed that, in terms of mIgD expression, the J11d^{high} population contained cells that showed a spectrum from low to high intensity. In contrast, the J11d^{low} population contained cells with intermediate to high intensity of mIgD but lacked an mIgD^{low} component. Functional testing of the populations showed that each contained cells capable of giving rise to primary AFCs when appropriately stimulated. However, the J11d^{low}/mIgD$^{int-high}$ cell pool was the source of memory B cells. One conclusion that can be drawn from these studies is that, at least for the J11d^{high} population, the asymmetric division model does not hold, since these cells appear to be the source only of primary AFCs, regardless of the level of mIgD expression. The J11d^{low} cells may give rise to daughter cells that diverge postmitotically, or the population may contain separate subpopulations that give rise to primary AFCs or memory cells. Further analysis and separation of the J11d^{low} cells is necessary to resolve this issue. There is also the issue of those cells that express intermediate levels of J11d. Finally, even if primary AFCs and memory B cells can be shown to be derived from virgin cells that express different amounts of J11d (and other membrane molecules), do these represent distinct populations in the virgin B-cell pool, or are they different points on a continuum of development in the secondary lymphoid organs? All these questions come closer to resolution as our knowledge of B-cell biology increases. However, they still remain challenges to us in terms of our understanding and, thus, our ability to manipulate the humoral immune system.

2. Cellular Cooperation and the Requirement for Helper T Cells

The absence of an obvious bursa-like organ in mammals (see previous text) was the impetus for experiments to determine the origin of the cells responsible for antibody production. In one such series of experiments, Claman and his colleagues transplanted lymphoid cells between genetically identical mice. The recipient animals were irradiated prior to the transplant. The dose of irradiation administered destroyed the recipients' own immune and hematopoietic systems. Under these circumstances, animals that received injections of cells that could reconstitute the hematopoietic system could survive long-term and could function essentially as living culture systems in which to grow cells of interest. Such an experimental system is called an *adoptive transfer*. The use of donor animals genetically identical to the recipients obviated any problems of mismatching the tissue transplants. Recipient animals were immunized with sheep erythrocytes (SE), a strong immunogen, and bled periodically for the measurement of anti-SE antibodies in serum. Figure

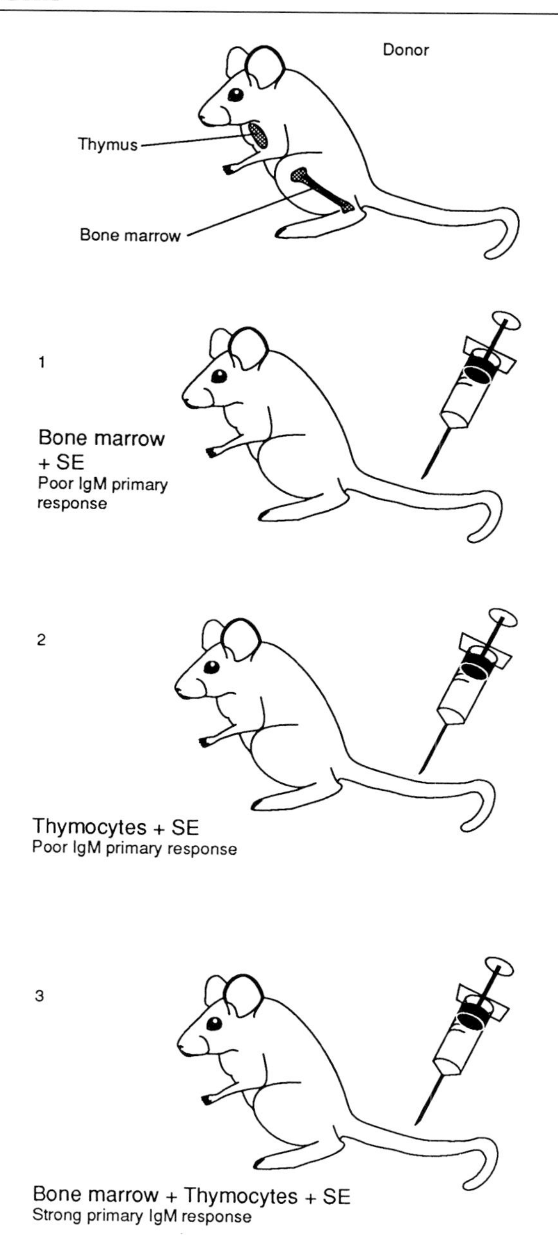

Figure 7 Adoptive transfer experiment showing cellular cooperation in the primary response to sheep erythrocytes. The donor animal is shown at the top of the illustration; tissue sources of the transferred cells (thymus and femoral bone marrow) are indicated. The three panels show individual groups of irradiated mice, the cells injected, and the experimental outcomes. Neither bone marrow cells (group 1) nor thymocytes (group 2) alone were adequate to generate a primary response to sheep erythrocytes. Only animals that received both marrow cells and thymocytes (group 3) showed a vigorous anti-sheep erythrocyte primary response. An additional control group of animals (not shown) received a mixture of bone marrow plus thymus transferred without antigen challenge; these animals gave no response. Abbreviation: SE, sheep erythrocyte.

7 presents a simple schematic of the experiment and the outcomes for the various groups of animals. The donor is shown at the top of the illustration. Three separate groups of recipients are shown individually.

Injection of bone marrow cells alone was shown to be capable of reconstituting the mice so they survived. However, they made few or no anti-SE antibodies (Group 1). When thymocytes were transplanted alone, they did not give rise to antibody production (Group 2). In contrast, animals that received bone marrow cells plus cells from the thymus (thymocytes) were not only reconstituted but also made significant amounts of anti-SE antibodies (Group 3). Experiments of this sort provided the first indication that more than one cell type, as defined by tissue of origin, was necessary for the humoral response to SE, that is, some form of "cooperation" between different cell types was involved, an astounding proposition at the time that the experiment was reported, but one which has been amply confirmed.

Subsequent experimental work carried out by many groups of investigators employed soluble proteins as antigens (SE is particulate) and, in most instances, concentrated on secondary IgG responses. A valuable twist was added to these experiments by using the soluble proteins covalently coupled to small molecules that as a general group, are referred to as *haptens*. The antibodies assayed in these experiments were those showing specificity for the hapten molecules; the larger proteins acted as *carriers*. The design and outcomes of an experiment of this type is shown in Fig. 8. The details of the experiment are given in the legend. The results of experiments such as this one confirmed that two separate specifically immunized populations of lymphocytes were required for a secondary antihapten antibody response. It was necessary for one population to be specifically immune to the hapten and the other to be immune to the protein carrier. The hapten-immune cells provided the source of antihapten antibody secretors; these were B cells. In contrast, the carrier-immune cells, although necessary for the antihapten antibody response, did not produce antihapten antibodies and were shown to be a distinct subpopulation of lymphocytes. Through the use of cell-surface markers, the carrier-specific cells were shown to be lymphocytes that had matured in the thymus and were designated T cells. The process by which immune T cells function in the generation of a humoral response has been called, by various investigators, "cooperation," "collaboration," or

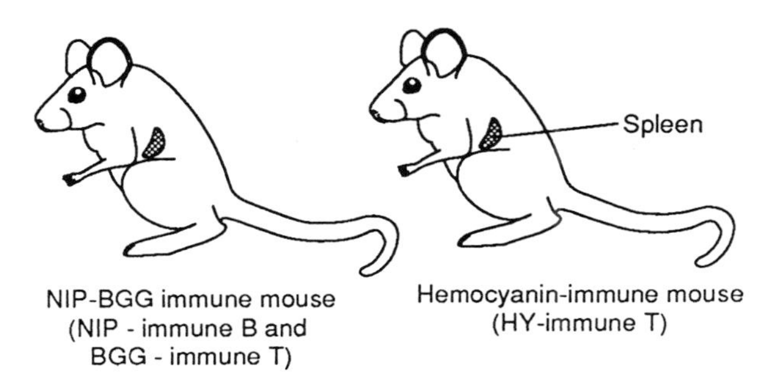

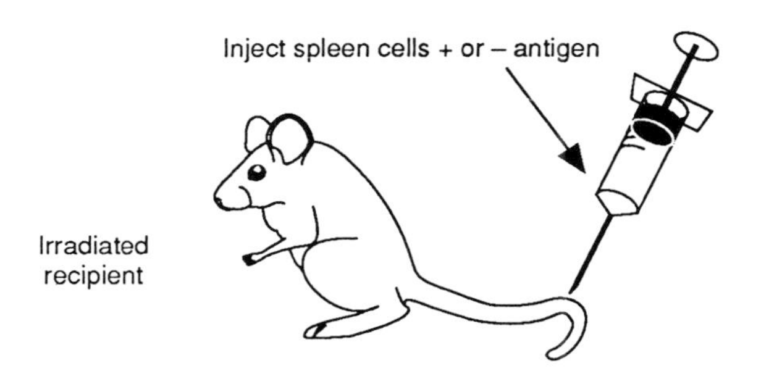

Spleen cells injected	antigen	anti-NIP IgG response
1 NIP - BGG	a. none	-
	b. NIP-BGG	+++
	c. NIP-HY	-
2 Hemocyanin	a. none	-
	b. NIP-BGG	-
	c. NIP-HY	-
3 NIP-BGG + Hemocyanin	a. none	-
	b. NIP-BGG	+++
	c. NIP-HY	+++
	d. NIP-OVA + HY	-

Figure 8 Adoptive transfer of a secondary anti-hapten (NIP) IgG response. The immunized donors are shown at the top of the illustration. One donor was immunized with the hapten conjugated to BGG; the other was immunized with unconjugated HY. Those cells that are functionally important for this experiment are indicated. Spleen cells were transferred into irradiated genetically identical recipients, with or without antigen challenge. The table shows the groups and subgroups within the experiment. Transfer of cells in the absence of antigen challenge did not give rise to a response. Response was, therefore, antigen dependent. Challenge with NIP–BGG elicited anti-NIP IgG only in animals with NIP–BGG-immune spleen cells in the transferred population. NIP–HY challenge gave rise to anti-NIP IGg only in animals that had received spleen cells from both sources. The HY-immune spleen cells thus lacked a population that could make anti-NIP IgG regardless of challenge and the NIP–BGG-immune cells, although containing a potential source of anti-NIP IgG, could only express this when challenged with the hapten–carrier with which they were originally immunized. Only when the two immune spleen cell populations were mixed did NIP–HY challenge give rise to anti-NIP IgG. The NIP–BGG-immune population contained the anti-NIP B memory cells and the HY-immune population contained the HY-immune Th cells. The failure to detect a secondary anti-NIP response after challenge with a mixture of NIP coupled to a third non-cross-reactive protein, OVA, plus HY indicated that the hapten and carrier had to be covalently linked for the two antigen-specific cell populations to interact. Abbreviations: NIP, nitroiodophenacetyl; BGG, bovine gamma globulin; HY, hemocyanin; OVA, ovalbumin.

"help." The last of these terms is now the most common. The cells themselves are referred to as T helper (Th) cells. [*See* T Lymphocytes.]

Humoral responses to many immunogens have an absolute requirement for Th cell activity. These are *thymus-dependent* (TD) immunogens. Almost all proteins and glycoproteins are TD immunogens. In contrast, immunogens have been identified that elicit significant humoral responses in the absence of Th cell activity; these are designated *thymus-independent* (TI) immunogens. Bacterial polysaccharides are good examples of TI immunogens. These often are large molecules that have repeating structural elements; this pattern of repeating structures is found in many TI immunogens. Another characteristic that is often associated with thymus independence is mitogenicity, the ability of a substance to stimulate lymphocytes to respond without regard to the cells' specificity for antigen, that is, mitogens stimulate lymphocytes through mechanisms that bypass their antigen-specific receptors. TI immunogens are able to provide sufficient signals for the B cells to undergo all the activation events that ultimately give rise to antibody secretion.

The mechanism of Th cell function in TD responses is an elegant example of biological regulation. Both B and Th cells possess receptors specific for determinants on the immunogen. The individual specificity of each of the lymphocytes is crucial for holding the two cell types in sufficient proximity, for an adequate period of time, to enable "help" to occur. However, there must be mechanisms by which Th and B cells can come into contact. These contacts are governed by *adhesion molecules*. They consist of a series of receptor molecules on the Th cell that recognize and bind to specific ligands on the B cell. These molecules are quite distinct from the antigen-specific receptors on either of the lymphocytes. The B cell–Th cell pairs formed have been called *conjugates* and the process has been called *conjugate formation*. In the absence of antigen, or prior to its recognition by the Th cell, conjugate formation permits the Th cell to "scan" the B cell for the peptide/class II MHC complex that its receptors will bind. These interactions provide sufficient contact for the T cell receptors (TCRs) on the Th cell to bind any peptide/class II MHC on the B cell for which they are specific. The interactions are reversible and, should the B cell not be presenting an appropriate complex, the Th cell releases and is able to "scan" another B cell. Table V lists the receptor–ligand combinations that enable antigen-independent "scanning" to occur. In the event of specific recognition, the other interactions in Table V come into play and the conjugate is stabilized. The most important of these interactions are the TCR–peptide/class II MHC and CD4–class II MHC interactions. Although both involve recognition of B-cell class II MHC, there is a distinct site on the class II molecule for each recognition event. Also indicated in Table V are interactions that play a role in the antigen-independent phase, but show differences in strength and duration as a consequence of recognition by the Th cell. One can, therefore, identify molecular mechanisms by which Th and B cells can be brought into apposition for the purposes of recognition and signaling that are needed for "help" to be delivered to the B cells. Figure 9 is a schematic of the steps in conjugate formation, beginning with the specific binding of the immunogen by the B cell's mIg. (See Section III.B for a discussion of B cells and antigen presentation to T cells.) The immunogen–mIg complex is endocytosed by the B cell and degraded into peptides in lysosomes (A). This is referred to as antigen processing. Peptide fragments are then re-expressed on the B-cell plasma membrane in association with newly synthesized class II MHC molecules. Figure 9 (B) shows the "scanning" process that precedes and is independent of recognition by the TCR, but allows the initial conjugate to form. In the event of TCR recognition, stabilization occurs (C). TCR binding of the peptide/MHC complex for which it is specific initiates Th cell activation. Among the consequences of this event is the synthesis and secretion of lymphokines that mediate the molecular events in "help."

Table V Interacting Molecules that Govern Conjugate Formation between Th Cells and B Cells

Antigen	Th-cell membrane molecule	B-cell membrane molecule
Absent[a]		
	LFA-1	ICAM-1
	CD4	Class II MHC
	CD2	LFA-3
	CD28	B7
Present[b]		
	TCR[c]	Peptide/class II MHC
	CD4	Class II MHC
	LFA-1[d]	ICAM-1
	CD2[e]	LFA-3
	CD28[e]	B7

[a] In the absence of antigen, these interactions are transient. They do not depend on antigen recognition, but permit it to occur.

[b] Interactions in the presence of antigen, so-called cognate recognition, are the most efficient at holding Th–B cell conjugates together.

[c] The TCR recognizes the combination of peptide and class II MHC antigen. The CD4 on the TH cell binds to a distinct site on the B-cell class II MHC.

[d] Although this interaction is also part of the antigen-independent phase, its strength increases transiently after Th-cell recognition of peptide/class II MHC. However, the increased binding persists for only a matter of minutes.

[e] The increase in binding due to this interaction occurs later and persists longer than that due to LFA-1/ICAM-1.

Abbreviations: LFA, lymphocyte function-associated (antigen); ICAM intercellular adhesion molecule; CD, cluster of differentiation; MHC, major histocompatibility complex; TCR, T cell receptor.

One can superimpose the specific experiment shown in Fig. 8 onto the schematic in Fig. 9 by initially focusing on the animals in Subgroup 1b (recipients of NIP–BGG-immune spleen cells, challenged with NIP–BGG). If the B cell in Fig. 9(A) is NIP-specific, then its receptors will bind the NIP–BGG and the complex will be endocytosed and degraded. This will result in BGG-derived peptides being presented on the B cell's surface, in a complex with class II MHC. If the Th cell in Fig. 9(B) is specific for a BGG–peptide/class II MHC complex, then the "scanning" process will result in conjugate formation, which will be stabilized by TCR recognition (C). The entire experiment shown in Fig. 8 can be understood in terms of the presence or absence of the appropriate antigen-specific lymphocytes combined with the processing and presentation of the antigens used for challenge.

The molecular mechanism of "help" is via soluble products of Th cells called *lymphokines*, sometimes also referred to as Interleukins. These are secreted glycoproteins, usually of relatively low molecular mass, in the range 15,000–20,000 daltons, that function as local hormones in regulating immune re-

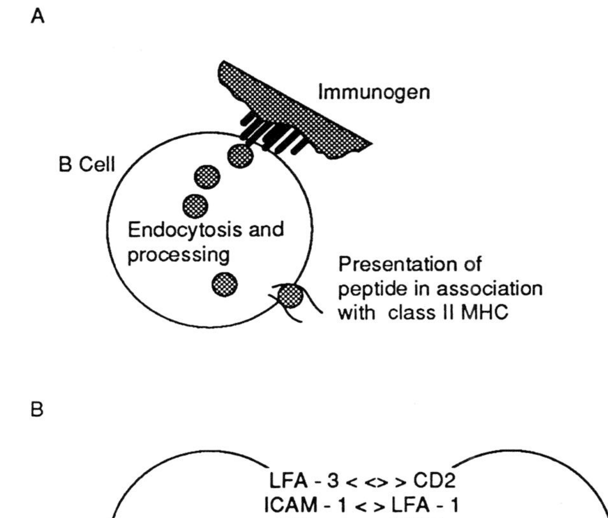

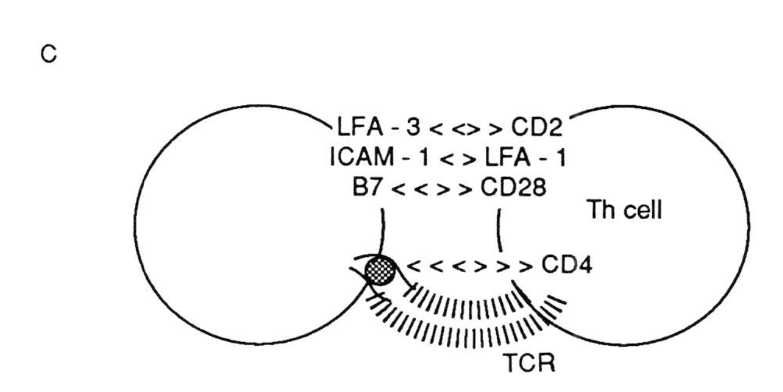

Figure 9 Three stages in conjugate formation enabling a specific Th cell to act on a B cell. (A) The B cell binds immunogen via its mIg receptors. The complex is internalized, digested in lysosomes, and peptide fragments presented on the B cell surface in association with class II MHC. (B) The Th cell is able to "scan" the B cell due to antigen-independent interactions. Conjugate formation at this stage is easily reversible, although it persists for a time sufficient for the TCR recognition to occur. (C) In the event of TCR recognition, the conjugate is stabilized. A further result of this is the activation of the Th cell, initiating the process of "help."

sponses. They have very short half-lives *in vivo,* on the order of seconds to minutes, and are enormously potent biological mediators. Lymphokines have no specificity for antigen. Antigen specificity is a characteristic of lymphocyte receptors; once the Th cells have been triggered by antigen recognition, the provision of "help" is mediated by these non-antigen-specific factors. On secretion by the Th cells, the lymphokines diffuse away in all directions. They act on B cells as well as other Th cells in close proximity to the activated Th cell. The stable interactions resulting in conjugate formation have the effect of retaining the specific B cells closest to the activated Th cells, thus causing them to be exposed to the highest concentrations of lymphokines.

Lymphokines are one class of small soluble glycoproteins that function as local hormones in the immune system. The name lymphokine indicates that they are products of lymphocytes. The other class of such molecules is the *monokines,* indicating that they are produced by cells of the monocyte/macrophage lineage. The generic name for such molecules is *cytokines.* Both lymphokines and monokines have a variety of effects on B cells. These include the stimulation of proliferation (cell division), immunoglobulin heavy chain class-switching, and the drive to terminal differentiation to plasma cells. Some lymphokines are also involved in the antigen-independent events in B-cell development. The cytokines involved in the development and triggering of B cells are listed in Table VI.

3. Cellular and Biochemical Events

Mature virgin B cells that have not yet been selected by antigen are small resting cells that express both mIgM and mIgD. Membrane Ig cross-linking by antigen, coupled with the lymphokines produced by activated Th cells, drives the B cells first through several rounds of proliferation and then to differentiation into plasma cells. At a morphological level, the resting B cell is small and quiescent, having only a thin rim of cytoplasm and few organelles; the actively dividing B cell, the lymphoblast, is larger (12 μm diameter) and retains the appearance of a lymphocyte, although it has lost mIgD expression. The plasma cell is larger still, about 16 μm diameter, with an eccentrically placed nucleus and a cytoplasm full of rough endoplasmic reticulum and Golgi apparatus structures, consistent with its function as a high-rate protein synthesizer.

Response may be initiated by the binding of antigen to either mIgM or mIgD alone, or by simultaneous binding to both. Binding cross-links the mIg; this event initiates a cascade of biochemical and genetic events. Figure 10 is a depiction of the most important biochemical changes initiated by the cross-linking of mIg by antigen. As noted earlier, the mIgM and mIgD on an individual cell have identical binding specificity and differ only in their heavy chain constant regions. Both of these immunoglobulins are integral membrane proteins, each of them having a hydrophobic transmembrane domain at the

Table VI Cytokines and Their Effects on Cells of the B Lineage

Cytokine	Source	Target cell	Effect
IL-1	Macrophages	Activated B cell	Early G_1
IL-2	Activated Th1[a]	Activated B cell	Proliferation
IL-3	Activated T cell	Pro-B cell	Proliferation and maturation
IL-4	Activated Th2[b]	Resting B cell	Increased class II MHC
		Activated B cell	Proliferation (overcomes block in early G_1); class-switching to IgG1 and IgE
IL-5	Activated Th2[b]	Activated B cell	Proliferation; class-switching to IgA
IL-6 (IFN-β 2)	Monocytes	Cycling B cell	Exit cycle and terminal differentiation
IL-7	Activated T cell	Pro-B cell	Proliferation and maturation
IFN-γ (immune interferon)	Activated Th1[a]	Activated B cell	Antagonize effects of IL-4; class-switching to IgG2

[a] Th cells have been divided into subpopulations, based on their lymphokine secretion profiles. These subpopulations have been designated Th1 and Th2. Some lymphokines are secreted by cells of both subpopulations, whereas others are subpopulation specific. IL-2 and IFN-γ are markers of the Th1 cells.

[b] IL-4 and IL-5 are the hallmark lymphokines secreted by Th2 cells. Abbreviations: IL, interleukin; IFN, interferon.

C-terminal end of the heavy chain. This domain is the region of the molecule that spans the lipid bilayer of the plasma membrane. The final three C-terminal amino acids of both the μ and δ chains constitute the cytoplasmic domain, that part of the molecule that protrudes into the cell beyond the inner limit of the plasma membrane. The cytoplasmic domains of the two are identical: lysine-valine-lysine-COOH (shown as KVK in the single letter code in the figure). The extremely short cytoplasmic domain is somewhat unlikely to be the sole site on the immunoglobulin molecule responsible for the delivery of signals into the cell. It suggests that mIg-mediated signaling might occur via an association of the transmembrane domains of the heavy chains with other molecules in the membrane structure. Biochemical analysis has confirmed that both mIgM and mIgD are noncovalently associated with other membrane proteins and these are thought to be responsible for the delivery into the cell of the information that the mIg has been cross-linked by antigen binding. The proteins identified have been shown to be heterodimers (proteins consisting of two nonidentical chains); the individual chains have been designated α and β. The activating signals involve the function of several membrane-associated enzymes. One very early event is depolarization of the B-cell plasma membrane. The change in membrane potential is detectable within a matter of seconds of mIg cross-linking. Another early step in the pathway is dependent on the activity of a G protein. This protein stimulates phospholipase C (PLC), which acts upon the membrane phospholipid *phosphatidyl inositol-4,5-biphosphate* (PIP_2) to cleave this into *inositol 1,4,5-triphosphate* (IP_3) and *diacyl glycerol* (DAG). The two PLC reaction products diffuse further into the cell where they act as second messengers. IP_3 mobilizes calcium from intracellular stores, thus raising the free cytoplasmic calcium concentration, whereas DAG activates the enzyme *protein kinase C* (PKC), which catalyzes the phosphorylation of a number of cellular proteins on serine and threonine residues. An additional, and quite independent, pathway of protein phosphorylation, this time on tyrosine residues, is simultaneously stimulated by mIg cross-linking. Tyrosine phosphorylation is particularly interesting, since it represents only about 1% of the total phosphorylated amino acids in the cell; phosphoserine and phosphothreonine represent the other 99%. Enzymes with the ability to phosphorylate tyrosine specifically, *protein tyrosine kinases* (PTKs), were first discovered in transforming retroviruses, which cause malignant transformation of cells. There is now a vast amount of evidence that indicates that the activity of tyrosine kinases is important both for the transforming properties of such viruses and for the regulation of normal untransformed cells. The genes for the tyrosine kinases found in normal cells essentially have been co-opted and taken out of their normal regulatory environment by the transforming retroviruses. These genes represent one class of so-called *oncogenes;* the normal cellular genes are designated cellular proto-oncogenes (c-*onc*), whereas their viral counterparts have been termed viral oncogenes (v-*onc*). Although the PTK responsible for the tyrosine phosphorylation induced by mIg cross-linking has not been characterized, the products of three cellular oncogenes are candidates, since they are expressed in B cells. These oncogenes are called *lyn, hck,* and *blk.*

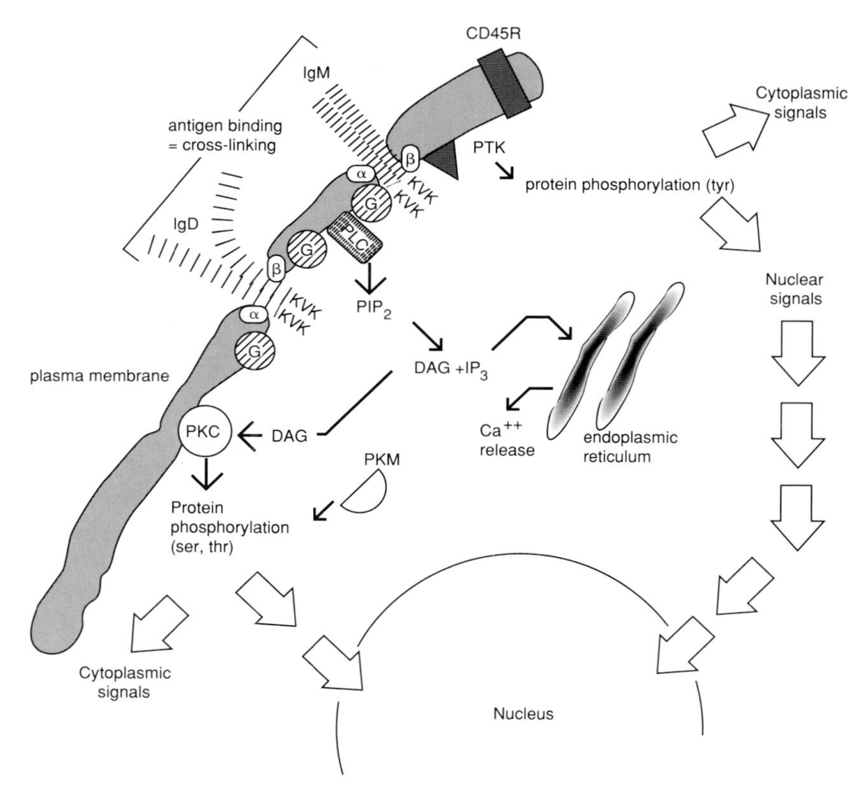

Figure 10 Biochemical events initiated in B cells by mIg cross-linking. The illustration depicts the cross-linking of mIgM and mIgD as the initiating stimulus. Cross-linking of mIgM alone or mIgD alone has the same effects. The mIg molecules are associated in the membrane with a heteodimeric protein; the two chains of this molecule are identified as α and β. There is no information available on the orientation of these proteins with respect to the mIg molecule; their location on each side of each mIg molecule is purely for ease of depiction. One signaling pathway is initiated by the action of a G protein. This activates PLC, which cleaves PIP_2 into DAG and IP_3. The former activates PKC, which exists in both a membrane-associated form and a soluble catalytic form, PKM, present in the cytoplasm. Both forms phosphorylate proteins on serine and threonine residues. IP_3 acts to release calcium ions from the endoplasmic reticulum, thus raising the intracellular Ca^{2+} concentration. A second, G-protein-independent pathway involves the activation of PTK. PTK phosphorylates proteins on exposed tyrosine residues. CD45R is thought to act as a regulatory element by removing phosphate groups from protein tyrosines, thus reversing the effects of PTK activity. The G-protein/PLC/PKC pathway alone is inadequate to cause the B cell to enter G_1. The combination of the two pathways may be adequate for the cell to begin to cycle and become receptive to Th-derived signals. Abbreviations: G, G protein; PLC, phospholipase C; PKC, protein kinase C; PKM, soluble catalytic fragment of PKC; PTK, protein tyrosine kinase; PIP_2, phosphatidylinositol-4,5-biphosphate; DAG, diacyl glycerol; IP_3, inositol 1,4,5-triphosphate; ser, serine; thr, threonine; tyr, tyrosine.

The phosphorylation of cellular proteins by either PKC or PTK is obviously important in the activation process. However, there is no unique substrate for either of the enzymes; each phosphorylates a number of distinct protein molecules. It is also important to note that the transmission of signals from the membrane to the nucleus requires input from both of these pathways, since activation of the phosphatidylinositol pathway alone is insufficient to cause full activation. It is obvious that the signal(s) must reach the nucleus, since the cell expresses new messenger RNA transcripts as part of its proliferation and differentiation program. Among the new proliferation-associated transcripts detected in B cells are those of the cellular oncogenes, c-*myc* and c-*fos*. The protein products of both c-*myc* and c-*fos* are not PTKs but DNA-binding proteins. Activation of c-*myc* is clearly important for proliferation although, once again, the precise mechanism is unclear. This time the question is which other gene(s) are regulated by the binding of the protein to DNA, leading to proliferation.

One of the early markers of B cells, CD45R, has been identified as a PTPase (see Section II.C.1). It has been suggested that the enzymatic activity of this molecule acts in counterpoint to the action of PTK, removing the phosphate from tyrosine residues and reversing the PTK-mediated signal.

The proliferative phase of the B-cell response is most easily understood in the context of an activation–progression model of the cell cycle. This model was first developed from studies of fibroblast growth, but is applicable to the behavior of lymphocytes. Briefly, a resting cell requires at least two separate signals to divide. The first of these, the activation signal, causes the quiescent cell to enter the cycle from the resting state. In order to transit the cycle, however, the activation signal alone is inadequate; the cell requires a progression signal for this. The two signals alone may each have detectable effects on the cell, but only the combination is adequate to cause it to divide. Figure 11 depicts this model as applied to B cells. In the simplest form of the model, mIg cross-linking provides the activation stimulus. Resting B cells are in G_0. Membrane immunoglobulin cross-linking activates them into the G_1 phase of the cycle. However, mIg-mediated signals alone are insufficient to drive the cell into the DNA synthesis (S) phase and beyond. Progression

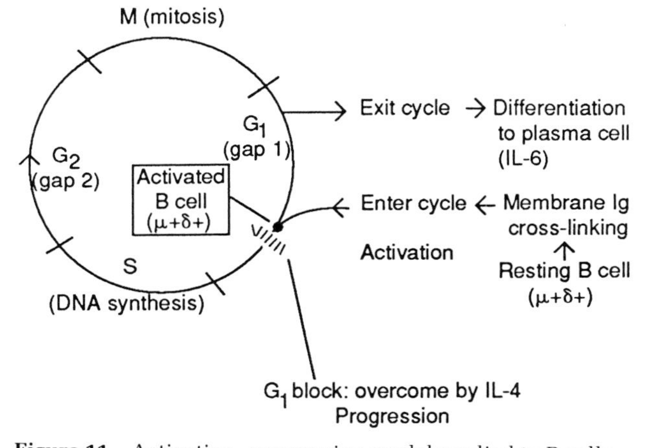

Figure 11 Activation–progression model applied to B cells. See text for explanation.

through the cell cycle is driven by IL-4. B cells activated by antigen-binding encounter a block in early G_1. IL-4 drives them past this block, so they complete G_1, transit S phase, and complete the cycle by going through G_2 and mitosis (M). Once back in G_1, the daughter cells of mitosis may then be driven to continue proliferation by IL-4. An alternative to this model is that cells that have undergone mitosis and returned to G_1 are caused to exit the cycle and differentiate by IL-6. Other cytokines, such as IL-1 and IL-5, have been shown to be functional in such a system. Thus, the model in Fig. 11 should be regarded as a minimal model.

IL-4 also has other important effects on B cells. One of these is to induce class-switching (see Table VI). The phenomenon of class-switching in the humoral response has been described at the whole organism level (Section III.A.1). The availability of purified and recombinant lymphokines has permitted experimental work that has shown that various lymphokines act on B cells to direct class-switching in a very directed manner. The word "directed" has a deliberate meaning here. For example, murine B cells exposed to IL-4 give rise to plasma cells that secrete IgG1 and IgE, but not the other IgG subclasses. In contrast, exposure to immune (or γ) interferon (IFN-γ) antagonizes IgG1 and IgE production, but promotes switching to IgG2. Interleukin-5 (IL-5) promotes switching to IgA and so may be of great importance in secretory immune responses.

Heavy chain class-switching occurs as a result of irreversible changes in the DNA of the B cells. The process has been described by a model called loop-and-excision, depicted in Fig. 12. The heavy chain constant region genes are arrayed tandemly, downstream (to the 3′ side) of the assembled gene that encodes the heavy chain variable domain, shown as VDJ in the figure. Upstream (to the 5′ side) of each of the heavy chain constant region genes except δ is a sequence called a switch sequence; only the switch sequences immediately upstream of the μ and γ1 constant region genes are shown (sμ and sγ1). Class-switching involves recombination between two switch sequences, one upstream of the heavy chain gene which the B cell is expressing *prior to* the switch (in this instance μ) and the other upstream of the gene which the cell will express *after* the switch (in this instance γ1). The two switch sequences form a stem and the DNA between the two forms a loop. The recombination event involves cutting the DNA in the stem, followed by religating. The DNA that formed the loop is lost. The result of the process is to bring the assembled heavy chain variable region

A

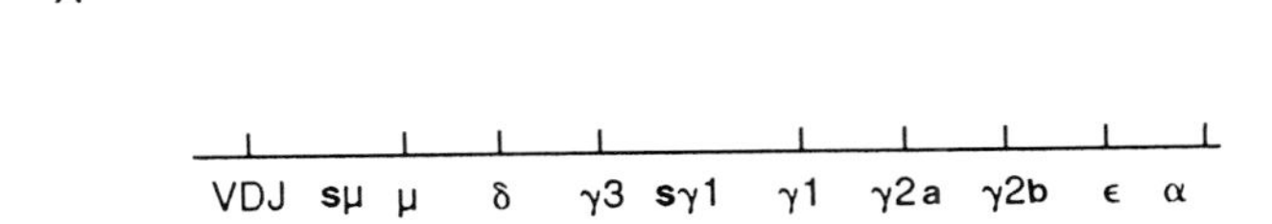

B

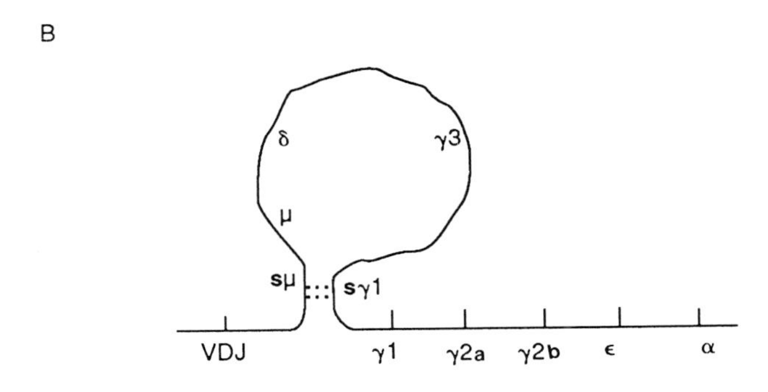

C

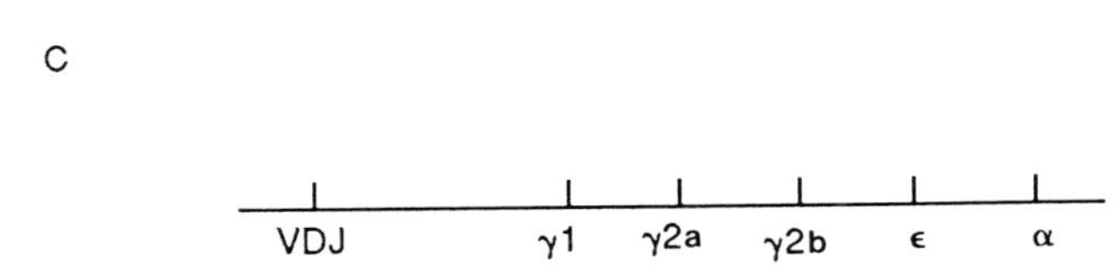

Figure 12 Loop-and-excision model for class-switching in B cells. (A) Assembled gene for the VH domain (VDJ) and germ line organization of the constant region genes. (B) Stem formed between the switch sequences immediately upstream of μ and $\gamma1$. The constant region genes between these are on the loop. (C) Organization of the locus after the stem has been cut and the DNA religated. An mRNA transcript initiating 5′ to VDJ would permit translation of a $\gamma1$ heavy chain, thus allowing the B cell to secrete IgG1.

gene immediately adjacent to the heavy chain constant region gene with which it will be expressed. In the illustration, the constant region genes for μ, δ, and $\gamma3$ have been lost and VDJ is immediately upstream of the $\gamma1$ constant region.

In order to undergo terminal differentiation to plasma cells, the B cells must exit cycle and express a different set of genes. This process is driven by interleukin-6 (IL-6). One of the earliest genes to be expressed in this phase is that for *J chain,* which is expressed in B cells en route to secretion of polymeric immunoglobulin, such as IgM. Serum IgM, the principal immunoglobulin class in the primary response, is secreted as a pentameric molecule, $(\mu_2L_2)_5$, of about 900,000 daltons. J chain (J = joining) is a small protein produced by B cells which functions to hold the IgM pentamer together. Another characteristic of the primary B-cell response is a change in the processing of the mRNA for the μ heavy chain. The resting cell expressed IgM as an integral membrane protein, with a hydrophobic transmembrane domain at its C terminus. The IgM secreted by plasma cells lacks this transmembrane domain, which is replaced by a different C terminus. This change in domains occurs at the level of RNA, a process of alternative splicing giving rise to two distinct mRNA species.

Thus, at the cellular level, the biochemical and genetic events associated with the humoral response are driven by a series of inputs to the B cell, starting with the cross-linking of mIg by antigen binding, but also requiring the active participation of Th cells by way of their secreted products. Although this is undoubtedly a complex process, the involvement of more than one cell type and the use of potent but short-range mediators in cellular communication provide a situation in which B-cell responses can be exquisitely regulated.

At several points in the foregoing discussion, it has been indicated that IgM and IgD are coexpressed as membrane immunoglobulins on the mature virgin B cell. Most of the mIgD is lost as the B cell is activated into cycle and only under very unusual circumstances does IgD appear as a secreted immunoglobulin. This has given rise to the hypothesis that mIgD exists solely as a receptor molecule on virgin B cells. There is a large body of experimental evidence in support of this idea, but the further question is begged: does mIgD have a unique function as a receptor, distinct from that of mIgM? After all, if a cell has mIgM, why does it need mIgD too? Membrane IgD is one of the hallmarks of mature B cells and a number of hypotheses have been proposed to explain the ostensible need for coexpression by mature cells. Although it has been demonstrated that mIgM and mIgD are transducing receptors, both capable of delivering signals into the cell, only very limited evidence indicates that in mature B cells the receptors show any differences in signaling capacity. Some information has been provided, however, by studies of B-cell lymphomas that have a phenotype suggestive of a transitional stage from immature to mature B cell, so they express low levels of mIgD and high levels of mIgM. In these cells, there does appear to be a difference between the two isotypes in terms of signaling. Thus one unique role for mIgD may exist at the earliest time point of its expression.

B. B Cells as Antigen-Presenting Cells

The interaction between helper T cells and antigen-presenting cells (APCs) is crucial to most immune responses. Helper T cells do not respond to free antigen but their receptors bind only peptides processed from protein antigens, presented to the Th cells in association with class II MHC molecules on the surface of the APC. A large body of evidence has been accumulated showing that macrophages and dendritic cells are potent APCs. Both of these cell types express high levels of class II MHC molecules on their membranes. B cells also express class II MHC molecules and recent work has demonstrated that this cell type can function as an extremely efficient APC for helper T-cell activation.

The efficiency of antigen-presenting activity of B cells is a function of both the state of activation of the B cell and the antigen-binding specificity of its mIg receptors. If the APC function of a population of B cells is assessed, resting B cells are relatively inefficient. In contrast, B cells that have been activated into cycle act as efficient APCs. This is due, at least in part, to the higher level of class II MHC expressed by activated cells compared with resting B cells. An increase in the expression of class II MHC molecules is one of the effects of activation of B cells by antigen. Further, if one asks whether the specificity of the B cell's mIg is of importance, it is clear that the most efficient of all are activated B cells that are presenting antigen for which their receptors are specific. Thus, not only does binding of antigen by mIg activate the B cell into cycle, but the antigen–mIg complex is endocytosed, leading to degradation of the antigen and representation of processed peptides in association with class II MHC on the B-cell surface. This provides an extremely efficient way of bringing antigen-specific B cells and helper T cells of the appropriate specificity into close apposition.

Where in the immune response does the antigen-presenting function of B cells play an important role, especially since, in an unimmunized B-cell population, the frequency of cells specific for a particular antigen is 10^{-3}–10^{-4}? Are there a sufficient number of specific B cells to play a major role in T helper cell activation in a primary immune response? There is no unambiguous answer to these questions, although evidence has been published that indicates that B cells are essential for the primary immunization of T cells in lymph nodes. However, the increase in frequency of specific B cells, as a consequence of immunization, suggests that there may at least be a sufficient number of cells present to play an important role in secondary and subsequent responses of T helper cells.

C. Regulation of the Humoral Response

For the purpose of biological economy, it is almost as important to turn B cells off as it is to turn them on. At least four mechanisms exist for this purpose. The first of these is decreasing antigen concentration. Antigen drives the system, eventually giving rise to the secretion of antibodies. These bind antigen and the resulting immune complexes are cleared by the cells of the *reticuloendothelial system* (RES). The phagocytic cells of the RES are capable of binding immune complexes, since these contain aggregated IgG molecules. Binding occurs via receptors specific for the Fc regions of aggregated IgG; these are called Fcγ receptors (FcRs). The phagocytes bind, engulf, and digest the immune complexes. Thus the effector arm of the B-cell response contributes directly to regulation by initiating the removal of the activating stimulus.

Soluble immune complexes, prior to clearance, also have a regulatory role. An immune complex with IgG antibodies is unlikely to have all the determinants on the antigen occupied by antibody molecules. These "spare" antigenic determinants may be recognized and bound by mIg molecules on antigen-specific B cells. This process might be expected to activate the B cells, as described earlier. However, the presence of the IgG in the immune complex permits recognition by FcRs on B cells. Occupancy of B-cell FcR prevents activation of phosphatidylinositol metabolism and so blocks the expression of the c-*myc* oncogene. This has the effect of rapidly stopping proliferation.

A third mechanism of regulation has been proposed by Jerne and has been called "network regulation." This depends on the immunogenicity of antibody molecules themselves as the regulatory drive. The variable regions of antibody molecules contain determinants that can be immunogenic in the animal in which they are made. These are the *idiotypic* determinants and may be thought of as the "signatures" of individual clones, or sets of clones, of B cells. One of the consequences of a humoral response is, therefore, the elicitation of a population of anti-idiotypic antibodies that recognizes the variable regions of those antibodies that were produced in response to the original immunogen. Anti-idiotypic antibodies can have profound regulatory effects on those B cells that have the appropriate idiotypic determinants on their membrane immunoglobulin receptors, either by sterically hindering further antigen binding or by signal delivery. Thus, network regulation may, depending on the characteristics of the individual anti-idiotypic antibody species, either down-regulate or up-regulate humoral responses. Indeed, one of the most intriguing prospects suggested by the network regulation model is the use of selected anti-idiotypic antibodies, those that are exactly complementary to the binding site of antibodies, as substitutes for antigen in vaccines. This could be extremely valuable in instances in which sufficient antigen is unavailable or in which antigen is toxic.

Finally, a regulatory mechanism involving a subpopulation of T cells has also been shown to affect B-cell responses. These are the so-called T suppressor (Ts) cells. Far less is known about Ts than Th cells and some controversy exists about their precise identity and mechanism of action. However, it has been shown that T cells can transfer the capacity to down-regulate immune responses in an antigen-specific manner from one animal to another. Although much remains to be learned about the biology of Ts cells, at a functional level they can be shown to have profound effects, both *in vivo* and *in vitro*.

D. B-Cell Tolerance

Tolerance is defined as antigen-specific unresponsiveness. It has been shown to occur in both B and T cells. The need for a process of tolerization (tolerance induction) comes from the mechanism by which diversity is generated in lymphocytes: the combinatorial joining of individual gene segments to yield the full-length variable regions of individual Ig molecules and T cell receptors. In accordance with the clonal selection hypothesis, the recombinational events and receptor expression occur without any extracellular signals dictating the repertoire that is generated. Thus, binding sites with specificity for self components may arise during the process. In addition, mutations accumulate in the Ig variable region genes after antigen-driven triggering of B cells. These provide an additional source of self-reactive cells. Should self-reactive lymphocytes generated by either of these processes be allowed to respond in an uncontrolled manner, severe autoimmune disease might result (see Section IV.B.2). Tolerance mechanisms have evolved as one important way of limiting immune attack on self components.

B cells can become tolerized at either the immature or mature stage of development. Whole animal studies showed that the injection of deaggregated protein antigens could induce tolerance in B cells (and T cells), reflected in an inability to respond to a subsequent immunogenic dose of the same antigen. The tolerant state persisted in B cells for up to two months, reflecting effects on both the mature peripheral cells and those maturing in the bone marrow. Subsequent *in vitro* studies showed, again, that both immature and mature cells could be tolerized, but the former were much more susceptible than the latter, as assessed by the concentration of tolerogen required in the cultures. At the cellular level, two mechanisms have been proposed by which lymphocytes become tolerized; these have been called *clonal abortion* (or deletion) and *clonal anergy*. In the former model, the cells are killed as a consequence of their interaction with the tolerizing molecule, whereas in the latter they persist in an unresponsive state.

In the past three years, genetic engineering techniques have been employed to construct *transgenic* mice, animals that have had introduced into their germ lines genes that encode and express proteins that are normally foreign to the host. This has allowed investigators to study the effects on the immune system of protein molecules that behave, in terms of expression, as if they were self. Such an approach more accurately reflects the processes that occur naturally to prevent self-reactivity. This approach has confirmed that both immature and mature B cells can be tolerized and has shown that both mechanisms of tolerization, abortion and anergy, occur. One of the factors dictating which of the two tolerization outcomes results appears to be the form in which the tolerizing epitope is presented. Membrane-associated molecules, which would presumably have an oligo- or polyvalent presentation, result in clonal abortion. In contrast, a soluble tolerogen, which would presumably be monovalent, gives rise to anergy. Regardless of the mechanism by which tolerization occurs, the processes can be put into the context of the number of signals delivered to the cell. Tolerance results from the delivery of only one signal, through the mIg receptor, in the absence of other signals necessary for the B cell to continue through the activation process. Clones of B cells that have been aborted are, by definition, dead, so there is no way in which they can be examined, nor is there any question about whether they can, at some later time, respond. Anergic B cells, however, persist, albeit unresponsive to antigen. The use of transgenic mice has permitted the identification of these cells and they have proven to have an unusual cell membrane phenotype with respect to mIg expression: they have markedly reduced mIgM but essentially normal expression of mIgD. Evidence has recently been presented to indicate that, under certain conditions of stimulation, the anergic state can be reversed in B cells. The full development of this work may provide important insights into the development of antibody-mediated autoimmune diseases.

IV. Pathology

A. Deficits: Immunodeficiencies

A number of immunodeficiencies have been described in which B cells are affected. One of the most profound is severe combined immunodeficiency (SCID), in which the lymphocytic progenitor cells are affected, leading to the absence of both B and T cells. Other deficiencies, however, are specific to the B-cell compartment. The classical example of these is Bruton's X-linked agammaglobulinemia. As its name suggests, this is a genetic defect with a locus on the X chromosome. Hence females (sex chromosome genotype XX) are carriers, whereas the affected individuals are males (sex chromosome genotype XY). The symptoms of the immunodeficiency often become profound at 12–18 months of age, with a history of frequent and severe bacterial infections. This condition, incidentally, provides an experiment of nature that demonstrates the biological role of humoral immunity in providing protection against extracellular bacteria and their products. The immunodeficiency results from a failure in the maturation of B cells, so very few mature B cells are present, although pre-B cells may be identified. The result is the almost total absence of immunoglobulin from the serum.

The most common immunodeficiency detected is selective IgA deficiency, in which only IgA production is absent. It occurs at a frequency of $\frac{1}{400}$–$\frac{1}{800}$. IgA is absent from both serum and secretions and, given the function of this class of immunoglobulin in the secretory immune system, it is rather surprising that the deficiency does not appear to have more profound effects. It is, however, most commonly identified quite by chance and does not appear to predispose to an increased frequency or severity of infections of mucosal surfaces. In contrast to Bruton's agammaglobulinemia, selective IgA deficiency may not reflect a defect in the B-cell compartment per se, but may be due to malfunction of the normal regulatory processes governing IgA production.

B. Excesses

There are a number of situations in which the excessive, or inappropriate, production of antibodies can have pathological consequences. Among these are *immediate hypersensitivity reactions* and *antibody-mediated autoimmune diseases*.

1. Immediate Hypersensitivity Reactions

Immediate hypersensitivity reactions are more commonly known as allergic reactions. They range in severity from allergic rhinitis, or hay fever, with its symptoms of a stuffy nose and red watering eyes, to hypersensitivity to insect venoms, which can lead to death within a few minutes of a sting. In both cases, specific antibodies of the IgE class are responsible for the symptoms. Mast cells have on their membranes a receptor that is specific for the Fc region of IgE, the FcεRI. IgE antibodies are very avidly bound by these receptors. Introduction of antigen cross-links the bound IgE molecules, initiating the process of mast-cell degranulation. A variety of extremely potent biological mediators, both preformed and stored in granules and newly synthesized molecules, are released from the activated mast cells. One of the most important of these mediators is histamine, which causes vasodilation, increased capillary permeability, and bronchoconstriction. When the antigen is delivered locally to the mucosal surfaces of the eye and upper respiratory tract, as occurs with pollen grains, then the symptoms of hay fever result. Antigen delivered to the mast cells in the lower respiratory tract can lead to an asthmatic attack, with the characteristic wheezing and profound difficulty in breathing. If, however, the antigen is delivered by injection, which is what essentially occurs in the case of an insect sting, then a massive systemic release of histamine can occur, causing an *anaphylactic reaction*. Such a reaction can cause death within a few minutes.

All individuals make antigen-specific IgE. However, allergic individuals appear to make significantly higher quantities of this class of antibody. The control of class-switching in B cells, in particular those events leading to IgE production, are being defined. IL-4 has been shown to be the principal drive to switching to IgE. The other regulatory events controlling the amount of IgE produced are under study and it is not yet clear where in the sequence the disregulated events occur in allergic individuals, giving rise increased IgE production.

2. Antibody-Mediated Autoimmune Diseases

Although the protective function of the humoral immune response is against foreign molecules, conditions arise in which antibodies reactive to self components are produced. These are termed autoantibodies. The presence of autoantibodies in an individual's serum is not abnormal. Most autoantibodies cause no detectable symptoms. However, there are instances in which they cause pathological effects; these are the antibody-mediated autoimmune diseases. There are many such diseases and they differ in the range and specificity of the autoantibodies produced and thus the number and types of tissues affected. At one end of the spectrum is systemic lupus erythematosus (SLE), a disease in which there is no obvious single organ-specific population of autoantibodies. Antibodies that react with self erythrocytes, DNA, and RNA are all found in SLE patients. Myasthenia gravis (MG) is an organ-specific disease representative of the other end of the spectrum. In MG, the target molecule of the autoantibodies is the α chain of the nicotinic acetylcholine receptor and the antibodies disrupt neuromuscular transmission, leading to muscle weakness and easy fatigability.

C. Malignancies

Tumors have been identified that appear to be the malignant counterparts of cells throughout the B-cell developmental scheme, from pre-B cells to plasma cells. Such tumors arise in humans, but there are also animal models, particularly in mice and birds, that have been extensively studied. This work has been particularly instructive in revealing the role of oncogenes and how disregulation of these can lead to malignancy.

Pre-B and B-cell lymphomas and leukemias show the appropriate phenotypes permitting their classification. Myelomas (plasmacytomas) are tumors that have the morphology of plasma cells and frequently secrete immunoglobulin molecules. The immunoglobulin secreted by myeloma cells may be detected as a pronounced monoclonal spike, if one uses electrophoresis to analyze serum proteins from humans or animals with myelomas. It should be pointed out that, regardless of the developmental stage of the tumor cells, we do not know the identity of the cell that actually underwent transformation. Viral transformation systems, such as the Abelson murine leukemia virus, are known in which the target of the virus is a pre-B cell but the malignancy appears as a disease of B cells.

Studies of B-cell malignancies have revealed at least three mechanisms by which transformation occurs. These are the introduction of a transforming v-*onc* gene into the genome of the target cell; insertional activation, in which a viral regulatory element becomes inserted into the host cell's genome so it subverts the normal regulation of a c-*onc* gene; and chromosomal translocation, in which there is a physical rearrangement that results in the activation of a c-*onc* gene, without any evidence for the direct participation of a virus.

The retrovirus Abelson murine leukemia virus provides an example of the first mechanism. This virus contains an oncogene, v-*abl*, the protein product of which is a protein tyrosine kinase. Expression of v-*abl* results in transformation. However, as noted earlier, although the target cell for this virus is at the pre-B cell stage, the transformed cell can frequently progress along its developmental pathway so the tumor appears to be a B cell. A number of other examples of such a mechanism of transformation have been described.

One of the first examples of insertional activation described was the development of bursal lymphomas in birds by another retrovirus, avian leukosis virus. Rather than introducing a v-*onc* gene into the target cell genome, the viral element inserted was a promoter sequence. When the DNA from tumors was analyzed, it was found that the viral promoter had been introduced close to the endogenous c-*myc* gene. This had the result of disregulating expression of the endogenous gene, resulting in uncontrolled proliferation.

The first two mechanisms described very clearly implicate retroviruses in the etiology of B-cell malignancies. It should be noted that the examples given have been of tumors in, respectively, mice and birds. There is no convincing evidence to date of a retroviral etiology for B-cell tumors in humans, although one cannot exclude this as a formal possibility.

The third mechanism of transformation is found in human tumors, as well as in nonhuman systems. This is chromosomal translocation. In this process, an endogenous c-*onc* gene is activated by being brought into physical proximity with one of the immunoglobulin gene loci. One example of this process is observed in *Burkitt's lymphoma,* a B-cell lymphoma that is prevalent in parts of Africa. Analysis of chromosome spreads from the malignant cells of Burkitt's lymphoma patients showed that in about

90% of the cases a translocation was identified involving chromosomes 14 and 8. In 5%, the translocation involved chromosomes 2 and 8, and in the final 5%, chromosomes 22 and 8 were involved. Mapping studies have shown that the immunoglobulin heavy chain locus in humans is on chromosome 14. There are two separate light chain loci; these are on chromosomes 2 and 22. The c-*myc* gene maps to chromosome 8. In this transformation mechanism, then, the regulation of a gene that is intimately involved in proliferation, c-*myc*, is being disrupted and replaced by regulatory elements that are active in cells of the B lineage. Interestingly, a human herpes virus, Epstein-Barr virus, has been associated with Burkitt's lymphoma, but there is no direct evidence to indicate that this virus is the etiological agent. In murine myelomas, the translocations observed also involve the immunoglobulin and c-*myc* loci.

V. Monoclonal Antibodies

A. Introduction

As indicated earlier, humoral responses are polyclonal. This provides obvious advantages for the survival of the intact organism. Further, even in a serum sample that exhibits a high titer of specific antibodies, 90% of the immunoglobulin molecules demonstrate no binding specificity for the antigen of interest. Both of these characteristics are disadvantageous if one requires large amounts of highly purified monospecific antibodies for experimental or clinical use. In the 1970s, Milstein and Kohler devised a procedure whereby the application of somatic cell hybridization techniques allowed investigators to generate hybrid cell lines that could grow indefinitely in culture and produce only one species of antibody. These are called monoclonal antibodies; they are sometimes referred to as hybridomas. Figure 13 illustrates the general scheme for the production of a monoclonal antibody. B cells are taken from an immunized mouse and hybrids are formed between these cells and myeloma tumor cells that have been adapted to growth *in vitro*. In this procedure, the immune B cells provide both the genetic information for the synthesis of the desired antibody and an enzyme required for surviving the selection procedure, while the myeloma cells provide immortality. The hybridization step is performed *in vitro;* any cells from the immune animal that fail to hybridize die in a few days. The survivors, should there be no selective pressure applied, are either hybrids or myeloma cells that have not fused. A critical step in separating these two cell types is to kill the unhybridized myeloma cells by taking advantage of the HAT (hypoxanthine–aminopterin–thymidine) selection system. In the presence of HAT, cells with a deficiency in the enzyme hypoxanthine–guanine phosphoribosyl transferase (HGPRT) die whereas cells containing the enzyme are able to grow. An HGPRT-deficient variant of the myeloma cells is used for the hybridization. The B cells from the immune mouse contain the enzyme; thus all hybrids between B cells and myeloma cells will be resistant to HAT and able to grow in culture. After hybridization and the application of selection, the cells are seeded into tissue culture wells at limiting dilution (a procedure called cloning). Each individual culture is then screened for the production of specific antibody. Those that are identified as positive may be expanded and used as a source of monoclonal antibodies. Large-scale production can be achieved either in tissue culture or by growing the cells as tumors intra-abdominally in mice.

B. Uses in Diagnosis, Research, and Therapy

It was immediately apparent that monoclonal antibodies could have a tremendous impact in medicine, particularly if antibodies could be generated that would unambiguously discriminate between different human cell types and subpopulations. Among the first to be obtained were antibodies that reacted specifically with the human CD antigens CD4 and CD8. These antigens are expressed by T cells and, in the periphery (secondary lymphoid organs and blood), show essentially mutually exclusive expression. The $CD4^{+}CD8^{-}$ cells are the T helpers; the $CD4^{-}CD8^{+}$ cells are the cytotoxic and suppressor T cells. When acquired immunodeficiency syndrome (AIDS) was first recognized as a novel disease, investigators quickly identified the underlying cause of the immune deficiency as profound loss of the CD4+ subpopulation. In uninfected individuals, the ratio of CD4+ : CD8+ cells in the peripheral blood mononuclear cell population is about 1.4 : 1. Individuals suffering from AIDS have such a profound loss of the CD4+ cells that the ratio can fall to a value close to zero. Thus, the investigative and diagnostic utility of these monoclonal antibodies is clear. Further, when *human immunodeficiency virus type I* (HIV-1), the virus that causes the disease, was isolated, anti-CD4 monoclonal antibodies were used to

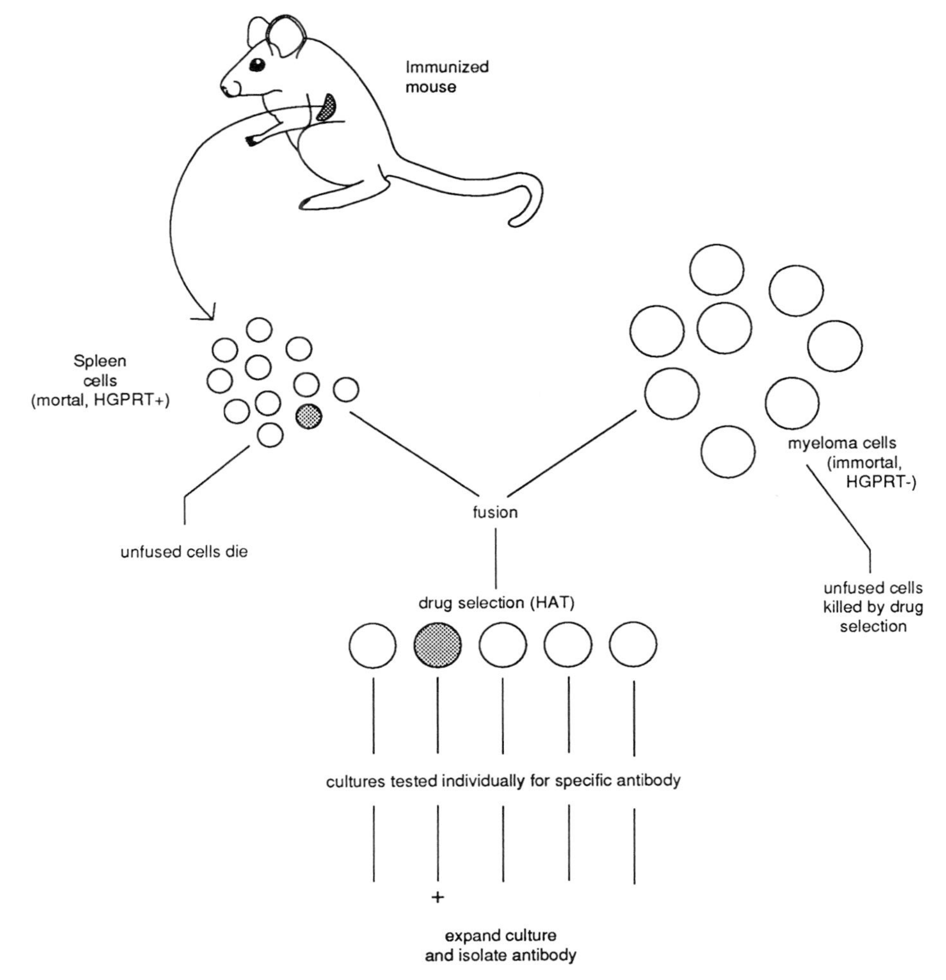

Figure 13 Scheme for the production of monoclonal antibodies. The spleen cell suspension from the immunized mouse contains B cells of the desired specificity only at low frequency (shown as the hatched cell). After fusion, unfused spleen cells die and unfused myeloma cells are killed by the selection procedure. The hybrids survive and a small fraction are of the specificity of interest. The hybrids are grown *in vitro* at limiting dilution and the cultures are tested individually for the presence of the desired antibody (+). Positive cultures are expanded for large scale production and isolation of antibodies.

demonstrate that the tropism of the virus for this particular subpopulation was due to the specific binding of the viral envelope glycoprotein, gp120, to the CD4 molecule itself. This was one of the first cellular receptors for a virus to be identified. [*See* Acquired Immunodeficiency Syndrome (AIDS).]

Monoclonal antibodies are now being administered for therapeutic purposes, at least at the level of clinical trials. One of these is anti-CD4. The aim here is not to reproduce the profound immunodeficiency seen in AIDS, but to identify a dose of antibody so the T cells that are active in certain autoimmune diseases can be controlled without rendering the patient incapable of responding to the microorganisms to which we are all exposed in normal life.

C. Human and Chimeric Monoclonal Antibodies

Most monoclonal antibodies have been produced using spleen cells from immunized mice as the source of the B cells. The antibodies produced will, therefore, be of mouse origin. For diagnostic and research purposes, this presents no problem. However, when such antibodies are administered to a member of a different species, humans for example, they elicit a strong immune response, ironically enough a humoral response. These antimouse immunoglobulin antibodies are not only able to neutralize any subsequently administered mouse monoclonal antibodies, thus rendering them ineffective, but the immune complexes formed can themselves be potentially dangerous. Therefore strategies have been devised to produce monoclonal antibodies that are less "foreign" to the intended recipient.

Since a major clinical use for monoclonal antibodies is likely to be in human disease, one approach is to make human monoclonals, using immune human B cells as the source of antibodies. This has necessitated the development of appropriate cell lines for hybridization that permit the retention of human chromosomes in the hybrids. An additional consideration has been the immunization protocol to obtain the human B cells to hybridize. *In vitro* immunization techniques may be of great use here.

An alternative approach has been to "humanize" mouse monoclonal antibodies. Although epitopes in both the V and C regions of immunoglobulins are strongly immunogenic across species barriers, the binding specificity of the antibody resides in only a relatively small percentage of the total molecular mass; for IgG, only about 6% of the amino acid residues is involved in the binding site. Using recombinant DNA techniques, investigators have isolated cDNA clones for the entire H and L chains of monoclonal antibodies of mouse origin and replaced the constant regions with human sequences. Such chimeric antibodies can then be re-expressed. The resulting monoclonal antibody retains the binding specificity of the original mouse antibody, but much of the molecule has now been rendered human and thus less strongly immunogenic to humans.

VI. B Cells and Their Products: Applications in Microbiology and Medicine

A. Vaccines and Immunization

The active stimulation of the immune system, with the development of circulating antibodies and immunological memory, is the basis of vaccination. Vaccines have been strikingly effective in combating many infectious diseases. Among the best examples are smallpox, which has been eradicated worldwide, and poliomyelitis, the incidence of which has been markedly reduced in those countries with the economic ability to introduce large-scale immunization programs. Such programs, coupled with such public health measures as improved sanitation and drinking water supplies, are very cost-effective ways of maintaining and improving the health status of the population. Worldwide childhood immunization programs, which would cost only pennies per individual, might seem expensive in aggregate cost, given the number of individuals involved, but are well worth the expenditure in overall economic terms. [*See* SMALLPOX.]

B. Serological Diagnosis and Epidemiology

In the early stages of the infection of an unimmunized individual, the serum antibody titer to the pathogen is low. As antibodies are produced, these are able to neutralize the pathogen but their titer continues to rise. A classical and time-honored method for diagnosis is to compare the titers of serum samples taken from a patient during the course of the infection (an "acute" serum) and several weeks later, after recovery (a "convalescent" serum). An increase of at least fourfold in specific titer, from acute to convalescent sample, is considered strong evidence of infection by the pathogen in question. In the case of bacterial infection, this might seem a somewhat academic point, if the organism can be cultured and identified. However, there are instances in which the organism cannot be identified at the time of infection and such serological evidence can only be obtained a significant time after the event. An example of such *ex post facto* diagnosis occurred after the identification of *Legionella pneumophila,* the organism causing Legionnaire's disease. Some years previously, an outbreak of pneumonia had occurred in Pontiac, Michigan. The etiologic agent was not identified at the time. However, serum samples were stored and, when the *Legionella* organisms were subsequently isolated, it became possible to demonstrate that organisms of this species had been responsible for the Pontiac episode.

C. Therapeutic Uses of Antibodies

1. Passive Immunization with Immune Globulin

Naturally occurring passive immunization in the protection of the fetus and infant by maternal antibodies was described earlier. There are clinical situations in which immune globulin is administered to patients to provide the protection that their immune systems are unable to produce. One such example is Bruton's X-linked agammaglobulinemia (see previous text). Patients are maintained by the administration of pooled human immune globulin. This immune globulin is deliberately not selected from hyperimmune sera raised against particular antigens, since the intent is to provide protection against as wide a range of pathogens as possible.

2. Use of Monoclonal Antibodies to Target Specific Cells: "Magic Bullets?"

In contrast to the use of pooled immunoglobulin is the use of monoclonal antibodies as delivery systems to target specific cells, such as tumor cells. The antibodies can be used to deliver compounds such as radioisotopes, drugs, or toxins specifically to the tumor cells, for diagnosis or therapy. The underlying basis for these approaches is to employ the exquisite specificity of the antibodies to deliver these compounds directly to their targets, thus achieving very high local concentrations. Systemic administration of these compounds at doses sufficient to achieve the same concentration at the site of a tumor would be either ineffective at identification, in the case of tumor imaging with radioisotopes, or lethal to the patient, in the case of toxins.

For diagnostic use, the antibodies can be labeled with radioisotopes, the emission from which can be detected by whole-body scanning (imaging) techniques. The antibodies are injected, circulate throughout the body, and bind to cells in the primary and metastatic tumors. The therapy of tumors is taking a similar approach, although additional procedures are being developed by which the antibodies are used to deliver toxins or drugs. One toxin–antibody conjugate that shows promise uses ricin, an extremely potent toxin obtained from castor beans, linked to antibodies specific for malignant B cells. It is obvious that the antibodies used in these approaches must be very carefully chosen to react only with tumor cells, since cross-reactivity with nontumor cells would lead to false positives in diagnosis and destruction of normal cells in therapy.

Acknowledgments

I thank Amelia Afsari, Steve King, Myron Leon, and Victoria Zitron for their careful reading of, and astute comments on, this article during the course of its preparation.

Bibliography

Ales-Martinez, J. E., Cuende, E., Martinez, A. C., Parkhouse, R. M. E., Pezzi, L., and Scott, D. W. (1991). *Immunol. Today* **12,** 201–205.

Goodnow, C. C., Adelstein, S., and Basten, A. (1990). *Science* **248,** 1373–1379.

Hardy, R. R. (1990). *Sem. Immunol.* **2,** 197–206.

Noelle, R. J., and Snow, E. C. (1991). *FASEB J.* **5,** 2770–2776.

Paul, W. E., and Ohara, J. (1987). *Ann. Rev. Immunol.* **5,** 429–459.

Reth, M., Hombach, J., Wienands, J., Campbell, K. S., Chien, N., Justement, L. B., and Cambier, J. C. (1991). *Immunol. Today* **12,** 196–201.

Rolink, A., and Melchers, F. (1991). *Cell* **66,** 1081–1094.

Springer, T. A. (1990). *Nature* (*London*) **346,** 425–434.

Vitetta, E. S., and Thorpe, P. E. (1992). Immunotoxins. *In* "Biologic Therapy of Cancer: Principles and Practice." (V. DeVita, F. Hellman, and S. Rosenberg, eds.) Lippincott. (In press.)

Beer

Mark A. Harrison
University of Georgia

Glossary

Bottom-fermenting yeasts Yeasts used to produce the type of beer known as lager

Chill-proofing Use of proteases to prevent haze development when beer is chilled to refrigerated temperatures

Hops Dried flowers of the female *Humulus lupulus* plant that contribute flavor and antibacterial compounds to beer

Malt Major raw material used in brewing that provides the appropriate substrate and enzymes needed to yield wort

Top-fermenting yeasts Yeasts used to produce the type of beer known as ale

Wild yeasts Yeasts that are present in the brewing process that were not introduced purposely nor tolerated for a specific purpose during brewing

Wort Liquid that remains after mash is strained containing soluble fermentable compounds

BEER is defined in the Bavarian Purity Law of Germany as a fermented alcoholic beverage made of malted cereals, water, hops, and yeast. This is the classical definition and has been enforced in Germany since the sixteenth century. Many countries, however, now allow additional substances to be used in this product. For instance, various enzymes and antifoaming agents are used by some brewers during the fermentation process. Others supplement expensive malt with unmalted cereals such as corn, rice, or wheat, which contribute to beer flavor while reducing processing costs.

Beer can be classified into different types by its flavor and color. Two basic types are lagers and ales. Examples of lagers are bock, pilsners, munich, and weissbier, whereas porter and stout are examples of ales. Lagers usually contain 3–5% alcohol by weight and have a lower hop content compared to ales. Bock beer also has a rather high alcohol content but is heavy and dark in color and traditionally only made at certain times of year. Pilsners are pale in color with a medium hop flavor and aroma, no sweetness, and an alcohol content of 3.0–3.8%. Munich beer is a full-bodied, dark brown brew with a sweet, maltlike flavor and 3–5% alcohol. Weissbier is made from wheat and barley. It has a tart taste with noticeable malt flavor and rich foam characteristics. Ales are pale in color and are tart-tasting with a noticeable hop flavor. They normally contain 4–6% alcohol. Porters are full-bodied beverages that are darker brown in color and have a sweet, malty flavor. Porters contain less hops than other ales and have an alcohol content of approximately 5%. Stout is a darker ale with an alcohol content of 5–6% and a rich, sweet, malt flavor. This ale has a strong, bitter hop taste. Light (or lite) beers are a twentieth century phenomenon. Since their introduction to the market, these bland-tasting beers have become very popular due to their reduced caloric content.

I. History of Brewing

Evidence shows that brewing beer was a popular practice in Mesopotamia before 6000 B.C. Beers offered people a flavorful alternative to drinking water and were often thought to possess therapeutic properties. Brewing beer became popular in other areas of the Middle East including Ancient Egypt and Israel. Brewing practices had spread to Rome during Caesar's reign and into other parts of Europe. The Germanic people probably learned the brewing process from the Celts. Over the centuries, it was discovered that adding hops and other spices would improve the flavor of beer. Since that time, hops have become essential in beer due largely to the contribution of flavor.

Beer was made in North America by the colonists. During the sixteenth and seventeenth centuries, breweries were established in Virginia and New England. Barley did not prosper in the New England climate so a variety of unusual ingredients were fermented including pumpkins, maple sugar, persimmons, and apples. Two factors later contributed to the further development of the North American brewing industry. Pennsylvania was found to be a good barley and hop production area, and the immigration of Germanic and Dutch brewmasters bolstered the American industry.

During this entire period, brewing was basically a hit-or-miss process. Individuals experienced in brewing recognized that using old brewing vessels yielded better products than new ones. We now realize that the cracks, crevices, and pores present in the older vessels, but lacking in the newer, harbored the yeasts and bacteria responsible for the fermentation. Several theories were developed in the nineteenth century in an attempt to explain the changes that occur during fermentation. Much of the debate centered around the issue of whether fermentation was a purely chemical process or a biological process. The issue was largely settled by Pasteur in the 1860s and 1870s when he published reports concluding that fermentation was due to the actions of yeast.

In 1883, Emil Christian Hansen established the method of using pure yeast cultures to produce beer at the Carlsberg Brewery in Copenhagen, Denmark. He had demonstrated previously that the culture used in brewing was often a mixed culture and that the metabolism of wild yeasts caused many of the defects in improperly processed beer. Over the next several years, the practice of using pure yeast cultures to produce beer became more widely accepted. The purpose and function of yeast enzymes in the fermentation process was shown by Buchner in 1897. It was found that cell-free extracts contained enzymes that could ferment sugars.

II. Brewing

The conversion of cereals into beer is not a direct process. The cereals used in beer production do not contain sufficient quantities of fermentable sugars. These cereals must first undergo modification during the malting and mashing steps to yield carbohydrates that yeast can convert during the fermentation step into ethyl alcohol and carbon dioxide. Freshly produced beer can then be aged for flavor development before it undergoes finishing steps, which can include filtering, pasteurization, and packaging. Each of these steps are examined more closely in the following section and in Fig. 1.

A. Ingredients

The basic ingredients in beer are water, malted cereals, hops, and yeast. Water comprises 90–95% of the content of finished beer, and its quality can influence the flavor of beer. Barley is the most common cereal used in the Americas and Europe to produce malt, although small volumes of beer are made from other cereal grains. Better beers are produced with clean barley that was properly dried after harvest. Overheating barley during drying can lead to its becoming unacceptable for malting because its germination potential is adversely affected. Barley contains a high starch content suitable for conversion to fermentable carbohydrates and a sufficient protein content to support yeast growth and contribute to forming beer foam. In addition, it contributes unique flavor components. Hops are the dried flowers from the female hop (*Humulus lupulus*) plant and contribute flavor and antibacterial compounds to beer. Yeasts are the predominant fermentation organism used to make beer worldwide. In some instances, bacteria may contribute certain characteristics to some regional beers. Malt adjuncts such as corn, rice, wheat, sorghum grain, soybeans, cassava, potatoes, sugars, and syrups may be used in some formulations. The adjuncts are all starch- or sugar-containing substrates that contribute fermentable carbohydrates. They also contribute flavor characteristics to produce distinctive varieties of beers. While enzymes other than those in the malt and those contributed by the yeast are not needed to produce beer, some brewers use additional enzymes to impart desired characteristics to their product.

B. Malting

The main objective of malting is to produce an ample supply of enzymes in grain that will degrade starch, proteins, fats, and other components of grain. The subsequent enzymatic changes provide fermentable sugars from starch and substances needed to support yeast growth (e.g., amino acids, fatty acids) from the other substrates. To produce malted bar-

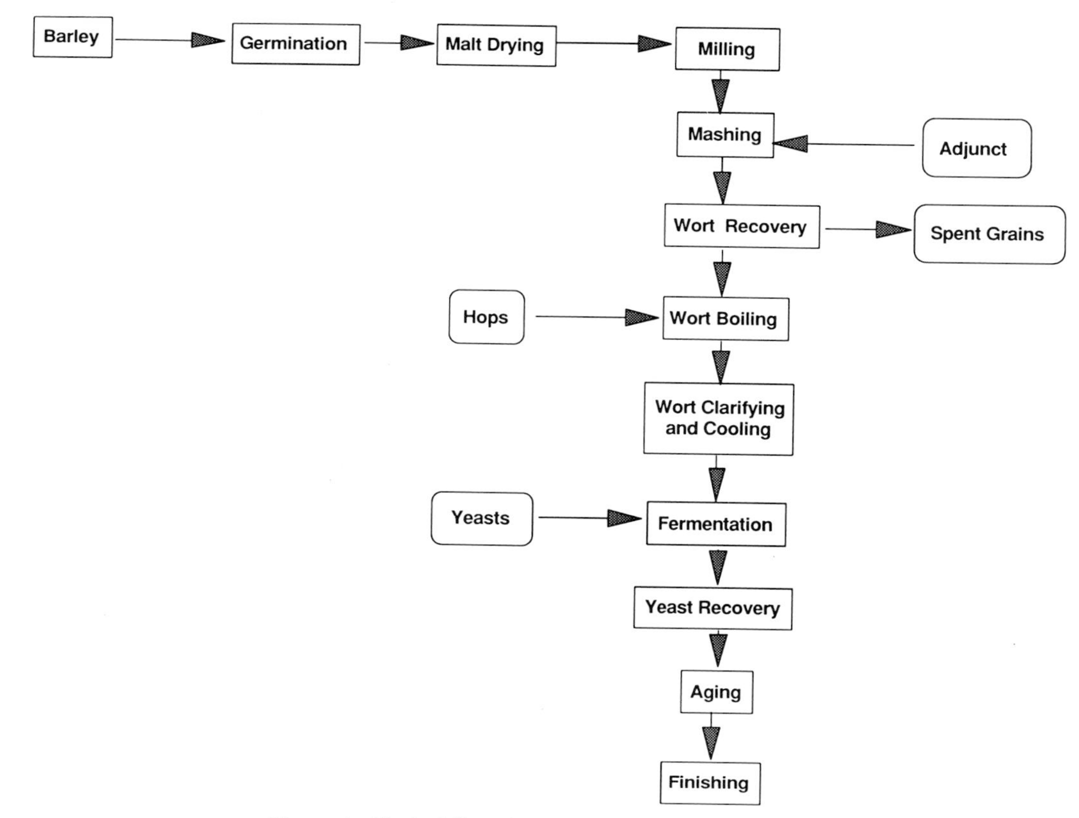

Figure 1 Typical flow diagram for the brewing process.

ley, barley grains are first steeped in 10–15°C water and then germinated at 15–20°C over 3–7 days. After the barley germinates, the sprouts are removed, leaving a medium rich with α-amylase, β-amylase, proteases, and their respective substrates. The malt is dried to about 5% moisture and ground. Grinding exposes the starchy endosperm of the grain, which makes the carbohydrates more available.

C. Mashing

During the mashing step, most of the nonsoluble, unfermentable carbohydrates and proteins are hydrolyzed into soluble fermentable materials by the enzymes present in the malt. To accomplish this, the ground malt is mixed with water and placed into a mash tun. To enhance protease action during the initial mashing period, the temperature of the mixture is maintained between 38 and 50°C. After a period of time, the temperature of the mix is raised to 65–70°C to enhance amylase activity. Within a few hours, the process is complete, and the temperature is increased to at least 75°C to inactivate the enzymes.

While the amylases are active, they degrade the starch contributed by the grain. The β-amylase splits off the disaccharide maltose from the amylose portion of the starch. Larger portions of starch are split off by yielding sections that are then acted upon by β-amylase. The products that result from the action of the two amylases known as dextrins also undergo additional enzymatic changes. Branch linkages of the amylopectin portion of starch are broken by debranching enzymes while amyloglucoside removes single glucose residues from the dextrins. Alterations in the color are also noted (changing from light to dark amber) during mashing.

The normal pH of malt is approximately 5.8 and is not acidic enough for optimum enzyme activity. To achieve optimum activity, the pH can be reduced to approximately 5.2 for lager production. The pH is usually more acidic for ale production. Adjustment of the acidity, if desired, can be accomplished by addition of acid, usually lactic acid, or by bacterial fermentation. While the lactic acid bacteria usually are undesirable contaminants, *Lactobacillus delbrueckii* has been used to accomplish this pH reduction in the past. This thermophilic bacterium converts sugars to lactic acid efficiently at temperatures of 42–51°C. Because the variety of microorganisms that can grow at these temperatures is small, it is easier to maintain a pure culture bacterial fermentation as well as to reduce the possibility of contamination by other microbes. Processors may find, however, that the bacterial modification requires much greater supervision and, if not controlled, may contribute to beer spoilage. [*See* THERMOPHILIC MICROORGANISMS; ALKALINE ENVIRONMENTS.]

After the naturally occurring and any added enzymes are inactivated, the solids settle out, leaving the wort. Wort contains the soluble compounds and is separated from the solids before it is transferred into the brew kettle. The spent grain can be used in animal feed.

The blend of grains used and the degree of enzymatic activity will influence the composition of the wort. These differences are some of the factors contributing to the characteristics noted in beers from different breweries.

D. Wort Processing

Wort is boiled in the brew kettle along with added hops for up to 2.5 hr. Boiling serves several purposes. It stops any further enzymatic action in the wort, causes any unhydrolyzed proteins to precipitate, extracts flavor compounds from the hops, and concentrates and sterilizes the wort. Among the compounds extracted from the hops are the essential oils humulone (α-bitter acid), lupulone (β-bitter acid), and tannin. Important flavor characteristics are contributed by the oils humulone and lupulone. In addition humulone and lupulone have some antimicrobial properties. The wort is then separated from the spent hops, cooled rapidly, and placed into a fermentation vessel. The spent hops may be used in fertilizer.

Light beers are made by reducing the unfermentable dextrin content in wort before fermentation occurs. The enzyme glucoamylase is added to the wort to hydrolyze most of the dextrins to glucose. During fermentation, the yeasts can ferment the glucose to alcohol. Thus, the amount of carbohydrates present that could contribute to the caloric content of the product is decreased.

E. Fermentation

At this point in the process, the wort is inoculated with brewers' yeast and fermented. During fermentation, the yeast produces alcohol, carbon dioxide, and some additional flavor constituents. The inoculation step also goes by other names such as pitching and seeding.

The fermentation room must be maintained in a clean manner to reduce possible contamination problems and should be kept at a constant temperature and humidity to maintain the desired growth rate for the yeast. Fermentation vats can be glass-lined or constructed of wood, stainless steel, or aluminum. Wooden vats pose problems in cleaning and disinfecting that are not experienced with vessels constructed of any of the other materials.

The strain of yeast used depends on what type of beer is desired. Lagers are produced using bottom-fermentating yeasts while ales are produced using top-fermenting yeasts. These yeasts have traditionally been referred to as *Saccharomyces carlsbergensis* (*uvarum*) and *Saccharomyces cerevisiae,* respectively. Many brewers consider these as separate species, although fungal taxonomists do not recognize them as distinct species. Nevertheless, they do behave slightly different during the fermentation process.

The temperature of fermentation for lagers produced by the bottom-fermenting yeasts is usually in the range of 6–15°C and takes 7–12 days. During fermentation, the yeasts tend to flocculate and settle to the bottom of the fermentation vat. These yeasts can be collected from the bottom of the vat for reuse in subsequent fermentations. By varying the fermentation temperature, slightly different versions of lagers can be produced.

Ales are traditionally produced using top-fermenting yeasts and incubation temperatures of 18–22°C over 5–7 days. These yeasts tend to form loose clumps of cells that are carried to the top of the fermenting liquid adsorbed to bubbles of carbon dioxide. These yeast cells can be collected from the surface for reuse with the next fermentation batch.

Since the mid-1800s, the use of bottom-fermenting yeasts has increased worldwide, whereas the use of top-fermenting yeasts has decreased over much of this time. In some areas of Europe and North America, the use of top-fermenters has regained some of its popularity. Currently some brewers use the same strain of bottom yeasts to produce both lager and ale. Because this is not the traditional method to produce ales, this product is referred to as bastard ale. The use of only one yeast strain to produce both types of beer eliminates problems related to keeping the strains separate and pure.

Regardless of the type of beer made, a rapid decrease in pH during fermentation will increase its stability and decrease potential problems of contamination. After fermentation, the pH of most lagers drops from approximately 5.2–5.3 to approximately 4.1–4.2, whereas it drops slightly more in ales. These acidic pHs assist in preserving the final product by inhibiting bacterial growth.

F. Aging

At the end of fermentation, the "green beer" is separated from the sediment and transferred to aging vessels. Aging vessels have commonly been wooden (e.g., oak) or glass-lined steel tanks. The beer is aged by storage at 0–2°C for several weeks. Lagers are normally aged for slightly longer periods than ales. Aging allows the beer to develop its final flavor, color, and body characteristics. Clarification occurs to some extent as the yeasts, unstable proteins, and other suspended solids precipitate. Chemical changes occur that create a more mellow, smooth flavor.

G. Finishing

After aging, beer undergoes several processing steps to prepare it for distribution to the consumer. Some of these steps are optional; others can be done by a variety of methods. The choice of the brewer depends on what type of finished product is desired. Most beers will be chill-proofed by the addition of proteases to prevent haze development by residual proteins when the product is held at refrigerated temperatures. Most will also be clarified and filtered to remove remaining solids. The final carbon dioxide level will be adjusted to 0.45–0.52%. The most common carbonation method is to add carbon dioxide back to the product. An alternative way is to add freshly yeasted wort to the beer and allow a natural secondary fermentation. Packaging for beer includes cans, bottles, barrels, and kegs.

To increase the shelf life of canned or bottled beer, it is usually pasteurized. The majority of beer in the past several decades received heat pasteurization (e.g., 60°C for 15–20 min). Beer flavor can be adversely affected if the product is overheated. Thus, there has been and is an interest in using an alternative pasteurization method. "Cold pasteurization" offers some advanatages such as less flavor loss due to heating and better energy efficiency. Cold pasteurization involves the use of chemical agents for preservation or filtration (cold filtration) through membrane filters followed by aseptic packaging. Beer packaged in barrels or kegs is not pasteurized but must be chilled and stored under refrigeration temperatures to maintain maximum quality. [*See* REFRIGERATED FOODS.]

III. Beer Properties

A myriad of beers are produced by different brewers. Properly brewed beers have flavor, color, and body characteristics dictated by the ingredients and yeast strain used, wort composition and conditions for fermentation, aging, and finishing. Among the metabolites produced are ethanol, carbon dioxide, ethyl acetate, other esters, fusel alcohols, diacetyl, 2,3-pentanedione, various sulfur compounds, and amino acids and nucleotides from yeast cells. The type and proportion of these and other compounds in beer contribute the characteristics to a particular brew.

Beers also possess certain factors that aid in preservation of the product. Properly fermented and packaged beer has alcohol produced during fermentation, a relatively low pH, and a low redox potential, which inhibit a variety of spoilage microorganisms. Some compounds extracted from hops are inhibitory to gram-positive bacteria.

Acetic acid bacteria, lactic acid bacteria, coliforms, and wild yeasts often contaminate improperly processed beers. Growth of most contaminating microbes in beer is inhibited by the preservation factors associated with the product. For example, the anaerobic environment during fermentation of packaged beer inhibits the growth of the aerobic acetic acid bacteria. The coliforms are inhibited when the pH level of the beer is 4.3 or lower.

IV. Properties of Brewing Yeasts

A. *Saccharomyces* Characteristics

The genus *Saccharomyces* is composed of ascosporogenous yeasts that produce ovoid, spherical, or elongate cells. *Saccharomyces* strains used in commercial brewing are diploid species. Reproduction by vegetative means is by multilateral budding, whereas sexual reproduction involves the formation of an ascus containing one to four spores. Members of this genus ferment sugars vigorously. Those strains that can yield significant amounts of ethanol by this metabolism are useful commercially.

This genus is a rather diverse group. Some of the growth characteristics for the species *S. cerevisiae* include minimum a_w for growth of 0.90; minimum, optimum, and maximum temperatures for growth of 0–7, 20–30, and 40°C, respectively; and minimum and optimum pH for growth of 2.0–2.4 and 4.0–5.0, respectively.

The taxonomic classification of the yeasts commonly referred to as brewers' yeasts has been debated for many years. According to the most recent taxonomic references, both bottom yeasts and top yeasts are members of the species *S. cerevisiae*. It is estimated that probably more than 1000 strains of *S. cerevisiae* exist. While the differences between strains are usually minor, they may be important to individual brewers. In older literature, the bottom yeasts were widely recognized as *S. carlsbergensis* and later as *S. carlsbergensis* (*uvarum*), or *Saccharomyces uvarum*. These earlier names are still part of the terminology of the brewing industry, and some brewers still believe that differences between strains are sufficient to continue using these names. [*See* TAXONOMIC METHODS.]

B. Factors Affecting Growth of Brewing Yeasts

During brewing, the goal is to maximize the production of desired metabolites of yeast growth while minimizing production of undesirable compounds and yeast biomass. By properly controlling various factors that influence yeast behavior, the brewer can achieve the desired end-product. Factors that influence yeast metabolism include wort composition, oxygen level in the wort, inoculation rate, condition of yeast at pitching, level of microbial contamination, wort pH, and temperature during fermentation. In addition, fermentation can be influenced by the fermenter design and whether a batch or continuous process is used. For example, evidence indicates that continuous fermentation may be faster than batch fermentation and may reduce problems associated with handling the yeast cultures. It may also prove to be more economical in many situations.

Yeasts growing in batch fermentations use the individual sugars in wort in a sequential order. Sucrose is the first used followed by glucose, fructose, maltose, and maltotriose, although not all strains can fully use maltotriose efficiently. Ethanol and flavor compounds are among the compounds produced by carbohydrate metabolism. This metabolism also provides the energy to maintain growth. Yeasts also utilize the 19 amino acids normally present in wort in an orderly fashion. The metabolism of the amino acids is needed to maintain yeast nutrition and to contribute flavor to the product.

C. Strain Development

Over the centuries, brewers have continued to search for yeast strains that will provide the best-quality beer in an efficient manner. In the past, this has often been done by screening for the most suitable strain by trial and error or by forming hybrids of two strains with different desirable qualities and screening for a progeny strain with the desired capabilities. These methods are not only time-consuming but the resulting yeast may have both desirable and undesirable qualities from both parental strains. For example, substrate utilization may be improved but flavor traits may be adversely affected. More recently, the possibility of using molecular methods such as recombinant DNA techniques to improve yeast strains has become an option. The advantage of using the molecular methods is that alterations of single characteristics can be accomplished specifically.

The brewing industry is interested in modifying yeast behavior to accomplish several needs. These needs include increasing the efficiency of fermentation by modifying the uptake and metabolism of wort components; improving the control of fermentation (e.g., timing of late fermentation or yeast floculation properties) and overall organoleptic quality of beer; developing new beers; and improving the value of by-products (e.g., spent yeasts).

Modification of some of these traits using molecular methods has been investigated with some success. While some of these altered strains have been developed experimentally, there are currently few, if any, used in full-scale commercial operations. In an effort to produce "light" beers less expensively, strains have been developed that ferment more of the carbohydrates present in wort not fermentable by the normal strains. This would alleviate the expense of using added enzymes to degrade these carbohydrates.

A second modification that has been accomplished concerns the degradation of α-glucan. This substance causes filtration problems and forms precipitates and hazes in beer. Because genetically developed yeast can solve the problem, the expense of using an added enzyme to do the job is eliminated.

Strains with improved proteolytic capabilities have also been developed. These strains reduce the need to use added proteases to degrade the proteins in beer that are responsible for the haze that can develop when finished beer is refrigerated.

V. Spoilage Problems of Beers

A. Range of Problems

Concern over the transmission of pathogens via beer is rare. In this aspect, beer is unlike most other food and beverage items. The lack of pathogen-related problems is due to the nature of the ingredients, processing methods, and characteristics of the final product.

Spoilage problems, however, can occur. A diverse group of microorganisms can present problems at various stages of brewing (Table I). The type of spoilage and the microorganisms responsible are influenced by the stage of processing and the characteristics related to the stage. Characteristics that affect spoilage include pH, alcohol content, type of ingredients, oxygen level of the product at particular stages, and the level of sanitation within the brewing environment. Wort just prior to fermentation is a rich medium for supporting microbial growth and is extremely susceptible to spoilage.

Table I Microbial Spoilage of Beer

Microorganism	Stage of processing encountered	Problem
Lactobacillus spp.	Any stage	Off-flavors
		Haze
		Ropiness
		Acidification
Pediococcus spp.	Inoculation	Off-flavors
	Fermentation	Haze
	Aging	Ropiness
		Acidification
Acetic acid bacteria	Open fermenters	Off-flavors
	Packaging	Haze
		Surface pellicles
		Ropiness
Hafnia protea	Wort	Off-flavors
	Fermentation	Acidification
Other Enterobacteriaceae	Wort	Off-flavors
Bacillus spp.	Wort	Off-flavors
Zymomonas spp.	Packaging	Off-flavors
Pectinatus spp.	Packaging	Off-flavors
Megasphaera spp.	Bottled beers with pH $>$ 4.1 and $<$3.5% ethanol	Off-flavors
Wild yeasts	All stages	Off-flavors
		Haze
Killer yeasts	Fermentation	Death of desired strain
		Off-flavors
Molds	Raw barley spoilage	Gushing
		Off-flavors

B. Bacterial Spoilage

1. Lactic Acid Bacteria

Species of *Lactobacillus* are among the most frequent and troublesome spoilage microorganisms of beer. They can cause spoilage problems at all stages of processing and even in the finished product. The lactobacilli can tolerate hop substances that are inhibitory to many other gram-positive bacteria. The taxonomy of lactobacilli encountered in beer is confusing and uncertain. One species characterized in the industry as *Lactobacillus pastorianus* may be the most common beer spoiler. It is capable of decreasing the pH, producing diacetyl, and producing a ropy texture in beer. Diacetyl is a chemical that produces a buttery flavor that is undesirable in beer.

Pediococcus damnosus is one of five species of this genus encountered in brewery environments. This lactic acid bacterium like the lactobacilli is resistant to the inhibitory substances in hops. Some strains of this species can tolerate up to 10% (w/v) ethanol.

2. Acetic Acid Bacteria

Species of *Acetobacter* and *Gluconobacter* can spoil beer primarily by converting ethanol to acetic acid. These bacteria are aerobic and are usually only associated with spoilage of beer exposed to air. The increased use of anaerobic conditions in modern breweries has decreased the frequency of problems related to these microorganisms.

3. Enterobacteriaceae

While conditions suitable to support growth of most members of the Enterobacteriaceae are relatively few during the processing of beer, there are opportunities for some species to pose problems. The main concern is spoilage of wort before the decrease in pH and accumulation of ethanol that occurs during fermentation. Most enterics stop growing during early fermentation when the pH of wort drops to <4.4 and when the ethanol content becomes >2% (w/v). One species that is an exception to this behavior pattern is *Hafnia protea*. It has the capability to survive the fermentation changes to a degree and may become associated with the yeast sufficiently that it is passed via pitching to future fermentations. [*See* ENTEROPATHOGENS.]

4. Sporeformers

Although *Clostridium* species may be present on the raw ingredients of beer, these spore-forming microorganisms present no serious spoilage problems in the liquid phase once production commences. Spoilage of spent grains by *Clostridium* has been reported.

Bacillus species, however, have been encountered in spoiled beer at times. Because most, if not all, of the species within this genus are sensitive to hop components, the spoilage bacilli tend to be more of a problem in the stages before the addition of hops to the wort.

5. Anaerobic Gram Negatives

Species of three genera of anaerobic, gram-negative bacteria have become recognized as spoilers. *Zymomonas* was first described in the 1930s and is somewhat unique to British breweries. Occurrence of spoilage problems due to this bacterium elsewhere in the world is rare. This organism produces hydrogen sulfide and acetaldehyde in beer.

Pectinatus and *Megasphaera* species were both described for the first time in the late 1970s. Both are unique to the brewing industry. *Pectinatus* species produce acetic acid, propionic acid, and hydrogen sulfide in beer whereas species of *Megasphaera* produce primarily fatty acids and hydrogen sulfide.

6. Other Bacteria

Species of *Achromobacter, Alcaligenes, Flavobacterium, Acinetobacter,* and *Pseudomonas* are some of the other bacteria that have been associated with infrequent cases of beer spoilage. Many of these miscellaneous spoilage organisms are aerobic, intolerant of acid environments, and die once fermentation starts. Thus, the type of problems they could cause are limited to the mash and early wort stages.

C. Fungal Spoilage

1. Wild Yeasts

Wild yeasts include any yeasts that are present in the brewing process that were not introduced purposely or tolerated for specific purposes in the development of the desired product. The taxonomy of the wild yeasts is confusing. There are probably more than 40 species that could be considered within this group. Problems due to wild yeasts include the production of undesired flavor compounds including esters, acidic, and phenolic substances and fatty acids. They may also ferment the product beyond the desired end-point due to their ability to ferment carbohydrates that the desired brewing yeast cannot utilize. Detection and identification of wild yeasts are important to a brewer if they are to recognize the problem and take corrective action. Identification methods vary from biochemical classification to genetic methods including DNA fingerprinting and gene probes.

2. Killer Yeasts

Killer yeasts can produce a toxic protein, or zymocide, that adversely affects the plasma membrane of other yeast species or strains. Members of several yeast genera are capable of producing this compound including some *Saccharomyces* species. The main problem is the death of the desired fermentative yeast strains caused by the toxic protein. Thus, the product is not properly fermented. It has often been noted that killer yeasts can also produce phenolic off-flavors. Spoilage problems due to killer yeasts can occur in either batch or continuous fermentations but tend to be more common with the continuous process. This may be due to the slightly lower pH encountered in the early stages of continuous fermentation.

3. Molds

Mold contamination of barley is common by a variety of species, making it important to handle barley properly to prevent spoilage prior to malting. Spoiled barley can adversely affect the malting process and can cause a problem known as ''gushing'' in the finished product. Gushing is the sudden release of carbon dioxide when a bottle or can containing beer made from mold-spoiled barley is opened.

VI. Spoilage Control

Controlling spoilage problems in the brewing industry is largely a matter of using the proper sanitation practices. Maintaining a clean, sanitized brewing environment will greatly reduce potential problems related to most of the environmental contaminants. Use of good quality water not only in the beer but also during cleaning and sanitizing is critical. [*See* Foods, Quality Control.]

Using the proper processing practices is also important. Care must be used to maintain the purity of the yeast strain or strains used by a brewer. The brewer should ensure that the raw ingredients are of good quality and that the processing steps proceed as expected. It is also important for a brewer to identify contamination problems early so that corrective action can be taken and damage minimized.

VII. Other Types of Beer Processing

A. African Beers

The native beers of Africa have different ingredients and are brewed somewhat differently than the typical European or American beer. Rather than using barley in the malting step, sorghum, millet, or maize are used, depending on the geographic location within Africa and the customary practice. Brewing in Africa originated in the home for personal or ceremonial uses, and it later expanded into commercial production. Fermentation of these beers is a two-stage fermentation. First, a lactic acid fermentation softens the malt proteins, reduces the pH sufficiently to limit growth of bacterial pathogens, regulates starch conversion to sugar, and contributes to the body of the beer. The second step is a yeast fermentation that is initiated just prior to packaging the product. During distribution, there is an active yeast fermentation.

The finished product is unhopped with a sour, yogurtlike taste and a pinkish brown color and normally contains 2–3% alcohol. It is opaque due to the high concentration of suspended solids and yeast cells. This product is actually thought of as a food more so than a beverage. These beers are highly susceptible to spoilage and have a shelf life that is usually less than 5 days. This limited shelf life is due to the wort not being boiled coupled with the fact that the final product characteristics (e.g., pH, alcohol content) are not sufficient to limit the growth of acid-tolerant spoilage bacteria.

B. Home Brewing

It is possible to brew beer at home that may be of similar quality to commercially produced beers. Many supply stores sell the necessary ingredients and equipment, although normal kitchen utensils can be used for some of the needs. The processing steps that are used in home brewing are basically the same as those used commercially. Naturally, some limitations are met in home brewing that commercial breweries can overcome. For example, in home brewing, excess yeast is usually not removed to the same degree and the storage temperature is usually not controlled as well to limit extended fermentation that can occur after packaging. Several books on the market describe home brewing in detail.

Bibliography

Banwart, G. J. (1990). "Basic Food Microbiology." Van Nostrand Reinhold, New York.

Cantarelli, C., and Lanzarini, G. (eds.) (1989). "Biotechnology Applications in Beverage Production." Elsevier Applied Science, London.

Haggblade, S., and Holzapfel, W. H. (1989). Industrialization of Africa's indigenous beer brewing. *In* "Industrialization of Indigenous Fermented Foods" (K. H. Steinkraus, ed.), pp. 191–283. Marcel Dekker, New York.

Hough, J. S. (1985). "The Biotechnology of Malting and Brewing." Cambridge University Press, Cambridge, England.

Lawrence, D. R. (1988). Spoilage organisms in beer. *In* "Developments in Food Microbiology—3" (R. K. Robinson, ed.), pp. 1–48. Elsevier Applied Science, London.

Spencer, J. F. T., and Spencer, D. M. (eds.) (1990). "Yeast Technology." Springer-Verlag, Berlin.

Stewart, G. G. (1987). Alcoholic beverages. *In* "Food and Beverage Mycology" (L. R. Beuchat, ed.), pp. 307–354. Van Nostrand Reinhold, New York.

Bioassays in Microbiology

David T. Spurr
Agriculture Canada Research Station

Glossary

Bioassay Process of evaluating the active power of a stimulus by examining the effects it produces on living material

International standard Substance with an established form of quality or quantity based on internationally accepted values and used for determining the strength of a test preparation by comparing it with the established standard

Maximum likelihood If $F(x, \theta)$ denotes a frequency function depending on one or more parameters collectively denoted by θ, the likelihood of a sample of independent values $x_1, x_2, \ldots, x_n$ is defined as $L = f(x_1, \theta), f(x_2, \theta), \ldots, f(x_n \theta)$; the method of maximum likelihood is a general method for finding estimates $\hat{\theta}$ of the unknown true paramters θ and does so by choosing those values that maximize L (or log L)

Median effective dose, or ED_{50} Quantity of stimulus that, on the average, produces an effect on 50% of the test population

Median lethal dose, or LD_{50} Quantity of stimulus that, on the average, kills 50% of the test population

BIOLOGICAL ASSAYS are methods for the estimation of the nature, constituents, or potency of a material (or of a process) by the reaction that follows its application to living matter. A bioassay is generally comprised of three components: a stimulus, a subject, and a response. The stimulus may be a physical, chemical, or biological treatment that is applied to some living material, which may be a microorganism, plant, animal, or tissue from a plant or animal. The magnitude of the stimulus is referred to as the dose. The response of the subject to the dose may be some measure of growth or viability. Different amounts of stimuli may be applied, and the relationship between dose and response may be expressed either graphically or mathematically.

I. Introduction

A. What Is a Bioassay

A great deal of microbiological research is concerned with defining the strength of drugs, antibiotics, vitamins, amino acids, sera, or other types of preparations. Although chemical analysis may give precise results, it is not always the preferred method of defining a compound. Bioassays are used to determine the potency of chemical or biological preparations in terms of their effects on living systems. A bioassay may be used when the chemical nature of the stimulus is not known. In addition, bioassays may be more sensitive and performed more rapidly than chemical analysis.

It is not possible to give more than an outline of the topic in this article; therefore, statistical aspects are the concentration here. It is assumed that the reader has a rudimentary knowledge of statistics including regression and analysis of variance and has a working knowledge of microbiology, biochemistry, physical, and analytical chemistry.

B. Uses of Bioassay in Microbiology

Biological assays are a valuable research tool in the field of microbiology. They are often the only practical means of standardizing antibiotics and antibacterial substances. Many medicinal agents, particularly those derived from natural products, may consist of a mixture of biologically active substances. For such agents, we are interested only in the response and not in the chemical composition. In some cases, preparations may be used as a medicine before the chemical structure has been ascertained. In determining the efficacy of penicillin preparations, for example, one must remember that a whole family of penicillins exists, all chemically similar but different enough to cause differences in the degree to which they affect species and strains of bacteria. Choice of subjects becomes an important design consideration when one wants to ascertain the strength of a penicillin preparation that is to be used to control a specific type of infection. With the increasing concern about the environmental cost of the use of chemical insecticides, agricultural scientists are looking for alternative ways of protecting the food supply. Bioassays have been used to monitor the quality of microbial pesticides and determine their potency.

C. Types of Bioassays

Bioassays may be classified by laboratory techniques and culture media used or by the nature of responses, which may be qualitative or quantitative. Direct assays are used to measure the amount of stimulus required to achieve a response, whereas indirect assays measure the degree of response of the subject. The response can be expressed as a continuous variable or can be expressed as to whether it did or did not respond.

II. Qualitative Bioassays

In qualitative bioassays, one is not concerned with the potency of the test preparation. Material is identified by a characteristic reaction produced by a particular type of biological material. Qualitative bioassays such as determining the presence of gram-positive microorganisms in a sample are especially useful in screening studies. Mathematical or statistical problems associated with assays of this type are not discussed in this article.

III. Direct Quantitative Bioassays

In general, direct quantitative bioassays are not statistically complex. They have not been found to be very useful, but an understanding of the statistical principles of such assays is essential to understand other types of bioassays. In this type of assay, the preparation is administered to the subject by continuous infusion or by constant doses at regular intervals until the desired response is attained. It is desirable that the threshold dose or some transformation of the threshold dose, such as log dose, vary according to a normal distribution to make it easier to estimate fiducial limits of the potency of a preparation. Direct assays are not very useful, particularly in the field of microbiology. To carry out a direct bioassay, one must be able to measure the amount of the preparation administered and to identify and measure the response by subjects. Determination of the "cat unit" of digitalis is an example of a direct bioassay that is mentioned in numerous discussions. The potency of digitalis is expressed in terms of the lethal dose to cats. These discussions merely point out the limitiation of a direct assay. One does not know the time lag between infusion and the physiological effects that lead to death. The subject may be able to detoxify the substance being administered, and one may not have any knowledge of the existence of such detoxifying mechanisms, let alone the speed or efficiency of their effects.

A. Relative Potency

Large fluctuations can occur in the results of bioassays, even in the same sample, because biological material is not uniform. Heterogeneity among subjects stemming from differences in nutrition, health, weight, age, etc., may lead to variation in the speed or magnitude of response. Conditions under which the assays are conducted may vary among laboratories or even from day to day within a laboratory.

The relationship between the responses of the test preparations and those of the standard is usually much less variable. Relative potency is estimated as the amount of the standard preparation to that of the test preparation producing the same response, generally the median effective dose (ED_{50}). The reference standard may be some international standard or some convenient working standard. The active principle must be the same for both the standard preparation and the test preparation. Impurities must be biologically inert or at least must not exert any effect on the relative potencies. Internationally accepted biological standards exist for a number of antigens, antibodies, antibiotics, hormones, vitamins, enzymes, and pharmacological substances.

An example of some data from an direct assay is

that used to determine the toxicity of a bacterial toxin by measuring the amount required to cause death of guinea pigs (Table I). The average tolerance to the standard and test preparations was 47.6 and 85.2 μg, respectively. The relative potency (R) of the test preparation is estimated by $R = y_s/y_t = 47.6/85.2 = 0.559$. The analysis of variance (ANOVA) (Table II) is used to measure the precision of the estimate. One now needs some measure of reliability of the estimate of R. The approximate variance of R is

$$\mathrm{Var}(R) = \frac{s^2}{\bar{y}_t^2}\left(\frac{1}{N_1} + \frac{R^2}{N_2}\right) = \frac{137.0}{(85.2)^2}\left(\frac{1}{5} + \frac{0.559^2}{5}\right) = 0.005,$$

and the standard error of R is $s_R = 0.0704$; thus, the 95% fiducial limits are then $R = 0.56 \pm 0.162$.

B. Use of Covariates

By chance, the average weight of the guinea pigs used for the test preparations was greater than that used for the standard (Table I). If a linear relationship exists between the weight and the amount of toxin required to elicit a response, analysis of covariance can be used to remove effects of weight on relative potency and increase the precision of the estimate. Total sums of squares (SS) for both the toxin required (Y) and the animal weights (X) as well as the total sums of products are calculated. These are then divided into between- and within-preparation SS, as given in the first two lines of Table III. The reduction in the error SS for Y due to the regression is $(1431.2)^2/2062.4 = 993.18$ with 1 degree of freedom (df). The original error SS is reduced by this amount to give a new residual term of 102.82 with 7 df. Note that the error mean square is now reduced to $102.82/7 = 14.689$ from 137.00. The next step is to compute the regression slope, $b = 1431.2/2062.4 = 0.69$.

The adjusted means for the two preparations are calculated:

$$\bar{Y}_t - b(\bar{X}_t - \bar{X}.) = 85.2 - 0.69(265.6 - 258.0) = 79.93$$
$$\bar{Y}_s - b(\bar{X}_s - \bar{X}.) = 47.6 - 0.69(250.4 - 258.0) = 52.87$$

The estimate of relative potency is now $R = 52.87/79.63 = 0.6615$.

Using the deviations from regression mean square, 14.689, the approximate variance of the relative potency can be calculated:

$$\mathrm{Var}(R) = \frac{14.689}{(79.93)^2}\left(\frac{1}{5} + \frac{(0.66)^2}{5}\right) = 0.0007.$$

The revised estimate of relative potency is then $R = 0.6615 \pm 0.0257$, and the approximate 95%

Table I Tolerance of Guinea Pigs to Standard and Test Preparations of a Bacterial Toxin

Test preparation			Standard preparation		
Animal No.	Animal wt. (g)	Toxin (μg)	Animal No.	Animal wt. (g)	Toxin (μg)
1	286	104	6	244	48
2	242	70	7	238	42
3	254	72	8	258	50
4	270	88	9	240	38
5	276	92	10	272	60
Mean	265.6	85.2		250.4	47.6

Table II Analysis of Variance of Data in Table I

Source of variation	df	Sums of squares	Mean square
Preparation	1	3534.40	3534.40
Residual	8	1096.00	137.00
Total	9	4630.40	

Table III Analysis of Sums of Squares (SS) and Products (SP)

Source of variation	df	SS(X)	SP(X, Y)	SS(Y)	
Preparation	1	577.6	1428.8	3534.40	
Within preparation	8	2062.4	1431.2	1096.00	
Regression	1				993.18
Deviations from regression	7				102.82
Total	9	2640.0	2860.0	4630.40	

confidence limits are 0.6615 ± 2.365 (0.0257) or 0.6007 and 0.7223.

Exact confidence limits can be calculated using a formula based on Fieller's theorem:

$$\frac{R \pm \frac{t_s}{\bar{y}_t}\sqrt{(1 - g + R^2)/n}}{1 - g},$$

where $g = \frac{t^2 s^2}{n_t \bar{y}_t}$.

The exact 95% confidence limits are 0.6023 and 0.7241.

IV. Indirect Assays—Quantitative Response

In an indirect assay, the degree of response by the subject is measured rather than the amount of stimulus required to cause a response. In this type of bioassay, responses can be measured on a continuous scale, such as diameter of skin lesions, loss in weight, rise in temperature, and survival time. For example, growth rate of larvae of the mealworm (*Tenebrio molitor*) has been used to screen for mycotoxins. The growth rate of larvae of the bertha armyworm (*Mamestra configurata*) was also considered for assay of mycotoxins in stored grain. Bioassays based on cytotoxic action of trichothecene mycotoxins on yeast have been reported to be quite sensitive and useful for routine screening of stored grains and grain products for this class of mycotoxins. Growth rate of algae is sometimes used to measure the degree of chemical pollution. One study has compared statistical models in a study of algal toxicity of several pesticides and tin compounds. The dry weight production of algae has been used for the assay of pentachlorophenol (PCP), a general biocide widely used as a wood preservative. Measures of bacterial growth are often used for standardization of antibiotics. A gradient plate assay was used for assay of tyrothricin using *Streptococcus faecalis*.

A. The Dose–Response Regression

Usually, in assays of this type, a series of doses, z_i, is used so that one or more subjects are exposed to each of the dose levels and the response Y is measured on each subject. The relationship between the dose z and the response Y can be described by a functional relationship $Y = F(z)$. The magnitude of change produced or the speed of onset of the response will increase with dose. Furthermore, the relationship must be monotomic; i.e., if the dose z_1 is greater than z_2, then the response to dose z_1 will be greater than the response to z_2 for all values of z_1 and z_2. The response may be linearly related to dose or some function of dose.

There are two main types of indirect quantitative assays: parallel line assays and slope ratio assays.

B. Parallel Line and Slope Ratio Assays

The parallel line assay is applicable when the response to the stimulus (or some function of the response) is linearly related to the dose, and the lines of the standard and the test preparations are parallel. If the two relationships remain parallel, a common slope is calculated using data derived from both the standard and the test preparations. The relative potency of the two preparations is independent of the dose and is expressed by the horizontal distance between the two parallel lines.

Slope ratio assays are applicable when the intercepts of the standard and the test preparation are not statistically different but the slopes of the regression differ. [*See* STATISTICAL METHODS FOR MICROBIOLOGY.]

C. An Example

Suppose that one has data (Table IV) from an experiment to determine the PCP content of a solution. Six dilutions of standard and of test solution were used with duplicate flasks of an algal suspension at each

Table IV Percent Reduction in Dry Weight of Algae at 96 Hr after Exposure to Different Doses of Pentachlorophenol Preparations from a Standard and a Test Sample

Dose (mg/liter)	x = log(dose)	Standard		Test	
		Flask 1	Flask 2	Flask 1	Flask 2
0.03	−1.523	5.9	3.4	6.3	6.3
0.1	−1.0	15.2	12.6	21.5	15.6
0.3	−0.523	47.2	57.4	37.6	30.0
1.0	0.0	83.5	90.3	47.7	46.8
3.0	0.477	95.8	96.6	83.1	85.7
10.0	1.0	95.8	94.9	92.4	94.9
Total	−1.569 × 4	698.6		567.9	

dose level. The response was the decrease in dry weight production of algae 96 hr after exposure to PCP. Some calculations for the analysis of variance given in Table V are as follows:

Total sums of squares Σy^2 = $\Sigma Y^2 - (\Sigma Y)^2/n$ = $97136.11 - (1266.5)^2/24 = 30301.85$

Preparation and dose SS = $[(5.9 + 3.4)^2 + \ldots + (92.4 + 94.9)^2] - (1266.5)^2/24 = 30166.28$

Preparation SS = $[(698.6)^2 + (567.9)^2]/12 - (1266.5)^2/24 = 711.77$

To calculate the regression equation and associated variances,we need the sums of squares of the independent variable X:

$$\Sigma x^2 = \Sigma X^2 - (\Sigma X)^2/n = 19.282 - (-6.276)^2/24 = 17.64,$$

where X is some transform of the dose Z, and the sums of products:

$$\Sigma xy = \Sigma XY - (\Sigma X)(\Sigma Y)/n = 362.045 - (1266.5)(-6.2745)/24 = 693.16.$$

The regression coefficient is $b = \Sigma xy/\Sigma x^2 = 39.30$, and the regresssion SS = $(693.159)^2/17.64 = 27237.49$.

To estimate the parallelism SS, we find the difference between the regression SS of the regressions from the two preparations and the pooled value: parallelism SS = 15394.61 + 11951.68 − 27237.49 = 108.80.

Relative potency can now be calculated as:

$$m = \log_{10}R = (\bar{y}_t - \bar{y}_s)/b = (47.325 - 58.216)/39.29 = -0.2772 \text{ and } R = 0.528.$$

Using Fieller's theroem, exact upper and lower 95% fiducial limits are calculated using the following formula:

$$M_L, M_U = M \pm \frac{t}{b}\sqrt{\left[(1 - g)\left(\frac{1}{n_t} + \frac{1}{n_s}\right) + \frac{m}{\Sigma x^2}\right]s^2/(1 - g)},$$

where $g = \dfrac{t^2 s^2}{n_t \bar{x}_t^2} = 0.998.$

The limits are

$$M_L, M_U = \frac{-0.2772 \pm 0.0770}{0.998} = -2.01, = -0.355.$$

This corresponds to upper and lower limits of 0.442 and 0.630 for relative potency. Fieller's theorem will be discussed in more detail in Section V.

In the preceding example, calculations were carried out assuming that the two regressions are parallel straight lines. Examination of the ANOVA table will show that there were significant deviations from parallelism and from a straight line. Graphical examination of the data suggests a sigmoid relationship between algal growth suppression and log dose.

V. Indirect Assays—Quantal (All-or-None) Response

A. Formulation of Mathematical Models

The concept of threshold dose can be applied to quantal data. For any subject, there is a threshold level of the intensity of the stimulus below which the response, such as death of the subject, does not

Table V Analysis of Variance for Data in Table IV

Source of variation	df		SS		MS	
Preparations and doses	11		30,166.28			
Preparations		1		711.77		711.77
Regression		1		27,237.49		27,237.49
Parallelism		1		108.80		108.80
Linearity		8		2108.22		263.53
Error	12		135.57		11.30	
Total	23		30,301.85			

occur, and above which it does occur. This is referred to as the tolerance level, and this level will vary among subjects. Because of the many problems associated with the use of direct assays, a more common procedure is to estimate the mean and variance of the threshold dose from the proportion of subjects responding at different doses of a preparation. If the subjects have varying sensitivities to the stimulus, it is reasonable to assume that the sensitivities of individuals would follow some frequency distribution. The frequency distribution of tolerances z is often skewed, but usually a simple transformation, such as $X = \log z$, will convert it to a normal or nearly normal distribution. The transformed scales of a dose z on which tolerances are normally distributed is known as a metametric scale, and the measure of dose is the dose metameter.

B. The Probit Transformation

Assuming that the dose metameter is normally distributed, the relationship between response percent and dose will be a sigmoid curve, as shown in Fig. 1. To make it easier to estimate the ED_{50}, relative potencies, and fiducial limits, probit values are used to transform the sigmoid curve into a straight line. Suppose we view the sigmoid relationship as a cumulative normal curve. The normal equivalent deviate (NED) for any proportional response P between 0 and 1 is defined as the abscissa of the normal curve corresponding to a probability P. To avoid the use of negative numbers, 5.0 can be added to the NED to give a slightly different response, which is now widely known as the probit transformation. Probit values for various proportions of response are given in Table VI. Data in Table VII are used to illustrate the probit and weighted least squares method of estimation of the dose—response relationship. They are taken from an experiment that examined the effects of phomalirazine on the survival of pollen grains. A serial dilution assay of phomalirazine was used with approximately 40 pollen grains at each dose. The steps are as follows:

1. Set up the data as in columns 1, 3, and 4, transform the dose to $X = \log(z)$, and put the dose metameter in column 2.
2. Determine the percentage response $100r/n$ and enter this in column 5.
3. Determine the empirical probits corresponding to the percent response using Table VI, and record the probit values in column 6.

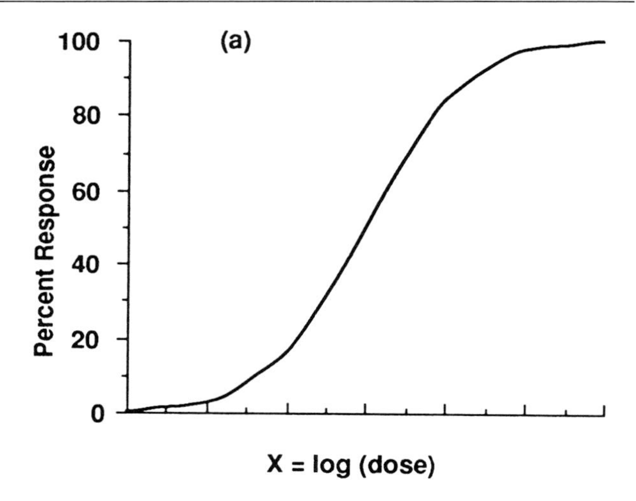

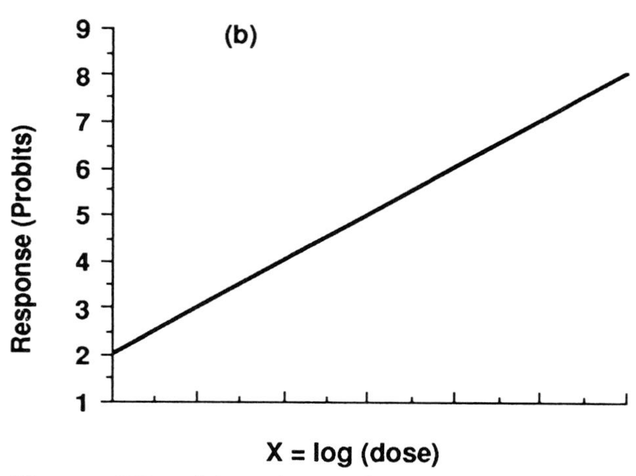

Figure 1 Effect of the probit transformation. (a) The sigmoid relationship between percent response and log dose, and (b) the same relationship as a straight line when percent response is tranformed to probits.

4. Determine the regression of the empirical probit Y on the dose metameter X.
5. Calculate Y' for each of the values of X using the regression equation $Y' = 8.504 + 0.491X$. Record these values in column 7.
6. Working backward from the table of probits (Table VI), calculate the proportion P corresponding to the values obtained for Y', and enter these in column 8.
7. Calculate the entries for columns 9, 10, and 11 as indicated. The sum of column 11 will give the χ^2 based on $k - 2 = 4$ df. The low value of 0.789 indicates that the data gave an exceptionally good fit.

Table VI The Probit Transformation

Response rate	0.00	0.01	0.02	0.03	0.04	0.05	0.06	0.07	0.08	0.09
0.00	—	2.67	2.95	3.12	3.25	3.36	3.45	3.52	3.59	3.66
0.10	3.72	3.77	3.82	3.87	3.92	3.96	4.01	4.05	4.08	4.12
0.20	4.16	4.19	4.23	4.26	4.29	4.33	4.36	4.39	4.42	4.45
0.30	4.48	4.50	4.53	4.56	4.59	4.61	4.64	4.67	4.69	4.72
0.40	4.75	4.77	4.80	4.82	4.85	4.87	4.90	4.92	4.95	4.97
0.50	5.00	5.03	5.05	5.08	5.10	5.13	5.15	5.18	5.20	5.23
0.60	5.25	5.28	5.31	5.33	5.36	5.39	5.41	5.44	5.47	5.50
0.70	5.52	5.55	5.58	5.61	5.64	5.67	5.71	7.74	5.77	5.81
0.80	5.84	5.88	5.92	5.95	5.99	6.04	6.08	6.13	6.18	6.23
0.90	6.28	6.34	6.41	6.48	6.55	6.64	6.75	6.88	7.05	7.33
	0.000	0.001	0.002	0.003	0.004	0.005	0.006	0.007	0.008	0.009
0.97	6.88	6.90	6.91	6.93	6.94	6.96	6.98	7.00	7.01	7.03
0.98	7.05	7.07	7.10	7.12	7.14	7.17	7.20	7.23	7.26	7.29
0.99	7.33	7.37	7.41	7.46	7.51	7.58	7.65	7.75	7.88	8.09

[Abridged from Fisheer, R. A., and Yates, F. (1963). Table IX *in* "Statistical Tables for Biological, Agricultural and Medical Research," 6th ed.; with permission of Oliver and Boyd Ltd., Edinburgh.]

Table VII Example Illustrating the Use of the Probit Method to Estimate the Relationship between the Concentration of Phomalirazine (Z) in the Growth Medium and the Proportion of Pollen Grains Surviving ($p = r/n$)

Dose (z) (1)	log z (X) (2)	n (3)	r (4)	% Response 100 p (5)	Empirical probit (6)	Y' (7)	P (8)
1×10^{-6}	−6.0	39	28	72	5.58	5.56	71.2
3×10^{-7}	−6.5	38	22	58	5.20	5.30	61.8
1×10^{-7}	−7.0	40	21	53	5.05	5.07	52.7
3×10^{-8}	−7.5	40	18	45	4.87	4.81	42.5
1×10^{-8}	−8.0	37	14	38	4.69	4.58	33.6
1×10^{-9}	−9.0	38	6	16	4.01	4.08	18.0

X (2)	nP (9)	r − nP (10)	$\frac{(r-nP)^2}{nP(1-P)}$ (11)	w (12)	nw (13)	nwX (14)
−6.0	27.8	0.2	0.007	0.5635	21.98	−131.8
−6.5	25.3	−1.5	0.251	0.6275	23.84	−155.5
−7.0	21.1	−0.1	0.001	0.6357	25.43	−178.0
−7.5	17.0	1.0	0.102	0.6330	25.32	−190.5
−8.0	12.4	1.4	0.299	0.6147	22.74	−182.0
−9.0	6.8	−0.8	0.129	0.4380	16.64	−149.8
Sum			$\chi^2 = 0.789$		135.96	−987.6

8. To apply tests of significance to the regression equation and to calculate fiducial limits, we must apply weights to relate the results to population size and probit values. Find the weighting coefficients, $W = Z^2/PQ$ from either Finney's (1971) table II or Fisher and Yates' (1963) table XI, where

$$Z = \frac{1}{\sqrt{2\pi}} \exp(-\tfrac{1}{2}Y^2).$$

9. Calculate the weighted sums of squares and cross products.

$$\Sigma(nwx^2 = \Sigma(nwX^2) - (\Sigma nwX)^2/\Sigma nw = 114.14$$
$$\Sigma(nwy^2) = \Sigma(nwY^2) - (\Sigma nwY)^2/\Sigma nw = 27.23$$
$$\Sigma(nwxy) = \Sigma(nwXY) - \Sigma(nwX)\Sigma(nwY)/\Sigma nw = 54.96$$

The new weighted regression becomes $Y = 8.440 + 0.4815X$. The variance of the regression coefficient is $\text{Var}(b) = 1/\Sigma(nwx^2) = 1/114.14 = 0.0009$, and the standard error is 0.094.

One can use the estimates of the relationship between stimulus and response to calculate the dose needed to achieve a specified result, *m*, such as the LD_{50} or LD_{90}. For the LD_{50} (5.0 probits) log dose =

$m = (5.0 - 8.440)/0.4815 = -7.14$. The $LD_{50} = 10^m = 7.18 \times 10^{-8}$, and the variance is

$$\mathrm{Var}(m) = \frac{1}{b^2}\left[\frac{1}{\Sigma nw} + \frac{(m - \bar{X})^2}{\Sigma(nwx^2)}\right]$$

$$= \frac{1}{(0.48)^2}\left[\frac{1}{136.0} + \frac{(0.121)^2}{114.14}\right] = 0.0323.$$

The fiducial limits of log LD_{50} are $m \pm ts_m$ or -7.49 and -6.79.

The lower and upper limits are 3.19×10^{-8} and 1.62×10^{-7}. Two points should be noted here. The preceding method gives only a close approximation to the variance of m, although in this example the approximate formula is sufficiently accurate. A more exact method of computing fiducuial limits is based on Fieller's theorem, which is as follows:

$$m + \frac{g}{1-g}(m - \bar{x}) \pm \frac{Z}{b(1-g)}\sqrt{\frac{1-g}{\Sigma nw} + \frac{(m-\bar{x})}{\Sigma(nwx^2)}},$$

where $g = \dfrac{Z^2_{\alpha/2}}{b^2\Sigma(nwx^2)}$.

The exact limits of log LD_{50} are -7.504 and -6.742, which correspond to lower and upper limits of 3.13×10^{-8} and 1.81×10^{-7}. The procedures outlined above are often sufficiently accurate to describe the relationship between dose–response metameters. Other methods, such as the method of minimum χ^2, are available for estimating the parameters in bioassay models. We wish to estimate the parameters in the best possible manner. Because the true values of the parameters are unknown, it would be meaningless to state that we wish our estimates to be as close as possible to the true values. A useful method for bioassay models is the method of maximum likelihood. In addition to the values for probits and weighting coefficients previously mentioned, working probits are also required for these calculations. These values could be obtained from Finney's (1971) table IV or Fisher and Yates' (1963) table XI or could be calculated with an inexpensive pocket calculator.

C. Natural Response Rate and Natural Immunity

In many experiments, a proportion of the subjects will respond at a zero dose. This is referred to as the natural response rate (NRR), or natural mortality. Also, some subjects may be incapable of responding even at very high doses of the stimulus, and this is referred to as natural immunity rate (NIR) (Fig. 2). After natural mortality has taken its share of the subjects, we wish to express the results of the stimulus in terms of how well the remainder respond. A correction of the proportion responding must be made. One method, the Abbott method, is often used to correct for the proportion (C) that responds in the absence of stimulus. If P^* is the overall proportion that responds, then the corrected proportion is $P = (P^* - C)/(1 - C)$. Similarly, if we let D be the proportion that is incapable of responding, the corrected value considering both the NRR and NIR will be $P = (P^* - C)/(1 - C - D)$. However, Abbott's correction does not account for the reduced number of subjects or the precision of the estimates of C and D. A much better procedure would be the method of maximum likelihood to find estimates of the parameters for NRR and NIR. A number of computer programs (SAS, GENSTAT, GLIM) are available for carrying out maximum likelihood analysis of bioassay data and have provisions for incorporating terms for NRR and NIR in the model.

D. Parallel Line Quantal Assays

As was the case with direct assays and indirect quantitative assays, one often wishes to state the potency of a preparation in relation to some reference standard. Suppose we have quantal bioassay data from two preparations. Separate probit regression lines are obtained for the two preparations. The χ^2 test for parallelism is calculated as the difference between the total χ^2 and the sum of the heterogeneity χ^2 values for the separate lines:

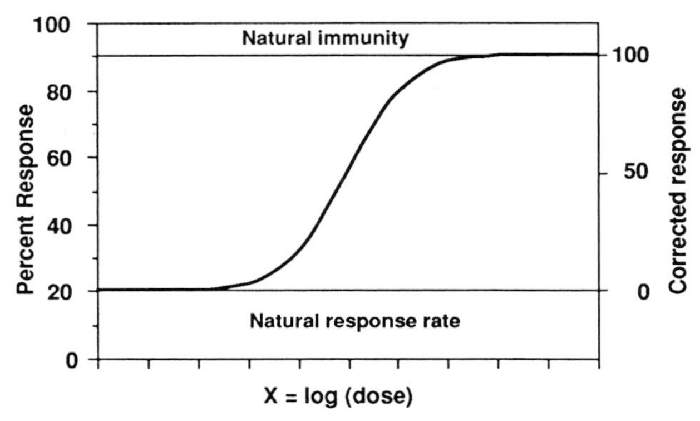

Figure 2 Response rate with and without correction for natural response rate and natural immunity.

$$\chi^2 = \left[\frac{(\Sigma xy)^2}{\Sigma x^2}\right] - \left[\frac{(\Sigma xy_s)^2}{\Sigma x_s^2} + \frac{(\Sigma xy_t)^2}{\Sigma x_t^2}\right]$$

with 1 df. If the χ^2 is not significant, there is no proof of departure from parallelism, so a pooled regression line is computed. The common slope is

$$b = \frac{\Sigma xy_s + \Sigma xy_t}{\Sigma x_s^2 + \Sigma x_t^2}$$

The log of the relative potency is

$$m = \frac{\overline{y}_t - \overline{y}_s}{b} - (\overline{x}_t - \overline{x}_s).$$

Fiducial limits of the relative potency are calculated using Fieller's theorem.

E. Time Response Bioassays

In many quantal response studies, the time that elapses between application of stimulus and the occurrence of a response is an important variable. Assays with successive observations on each individual are carried out to determine the effect of the stimulus over time. Two independent variables are involved: time (t), or its transform, and dose (x), or its transform. Each group of subjects is treated with a particular dose level and observed two or more times. A probit plane of the form $Y_t = A + Bt + Cx$ is fitted to the data. If a control group is available, a probit line of the form $Y_c = A' + B't$ can be fitted to the control data by the method of maximum likelihood to correct NRR and NIR. The statistical analysis must take into account the fact that successive observations on the same group are not independent. Some researchers have performed carcinogenic experiments where animals are exposed to different doses of a carcinogen and kept under observation until they develop a tumor, die, or the experiment is terminated.

F. Wadley's Problem

Wadley's problem refers to a special case of quantal response data, where the exact number of individuals in a group (treated or control) is not known. The data consist only of the number of individuals that respond or the number that do not respond, but not both. The number of subjects that do not respond will be referred to as "survivors." Instead of estimating the NRR, one must use an analogous procedure to estimate the number of individuals at risk, i.e., the effective sample size either as an average value for the whole experiment or varying according to the environment. The plaque neutralization (PN) test is one of the most sensitive and useful tools available for measuring antibody activity. In the PN test, the number of plaques that would develop on any particular antibody-treated plate, if antibodies were absent, is not known and must be estimated from control plates. Because this estimate is subject to sampling error, failure to allow for the sampling variance of the controls would lead to underestimation of the assay variance and could, in some circumstances, cause bias. The PN test has been used to study the relatedness of bluetongue virus isolates. Antibody–antigen interactions were used as a measure of relatedness. Dilutions of different antisera were applied to plates of L-929 cells (human tissue culture cells) containing suspensions of different virus isolates. The volumes of the virus suspension were known, but the number of virus particles (subjects) applied to each plate could only be determined by counting the number of plaques on the control plates.

In a study to investigate the fungicidal activity of metalaxyl, one group of researchers sprayed leaves of *Brassica campestris* plants with a suspension of zoosporangia of *Albugo candida* (the subjects). The stimulus consisted of various doses of the fungicidal seed dressing. The number of pustules that developed on the leaves of the plants was recorded. The number of propagules subjected to any particular dose was unknown and had to be estimated from counts made on plants not receiving any fungicidal treatment.

G. The Logit Transformation

Both the probit and logit transformations have a theoretical basis. Instead of emphasizing the distribution of tolerances among a set of subjects, the quantal response is regarded as the outcome of a process in which the odds in favor of response vary with the dose, and then the fitting of a logistic function to the data would be appropriate. That is, $P = \exp(\alpha + \beta x)/[1 + \exp(\alpha + \beta x)]$ and the proportionate responses P are transformed to $Y = [\ln P/(1 - P)]$. Both the probit and logit transformations give essentially the same answer except for cases where P is very large or very small. Some statisticians favor the logit method because the minimum χ^2 fitting of the logit regression is computationally simpler than the maximum likelihood method used for the probit method; however, with the computer hardware and software that have become available in recent years, this advantage is no longer applicable.

H. The Angular Transformation

Angular transformation is another method based on a tolerance distribution. Some researchers proposed that the weighted regression of $y = \sin^{-1}\sqrt{P}$ on x be used, to make the variance of the transformed variable approximately constant. The log LD_{50} would correspond to the value of x for which $Y = 45°$. A standard maximum likelihood estimation procedure that involves the use of empirical and working angles has been adopted. This analysis is tedious to carry out with hand calculators, especially when extreme doses are used, but will not tax modern computing facilities.

I. Multiple Quantal Regression Models

Assays may consist of one or more stimulus variables plus a number of environmental variables. In some cases, it may be desirable to include two dose variables for one type of stimulus (e.g., dose and volume). Environmental variables may include continuous variables (covariates) such as temperature or relative humidity or may be discrete classes of factors such as replicate, day, method of application, solvent, or some other parameters of the experimental design. Dummy variables are used to describe the design structure of the assay, and the parameters are estimated by a maximum likelihood methods.

VI. Maximum Likelihood Estimation of Most Probable Number

Estimation of the number of bacteria per unit volume in a suspension is not always convenient using the traditional method of counting the number of colonies that grow after incubation. Commercial preparations of nitrogen-fixing bacteria are marketed for inoculating legume seed. The ability of rhizobia to form nodules on the roots of legumes varies with the rhizobial strain and with the species of legume to which it is applied. The standard procedures used for quality control and for registration of commercial preparations consist of inoculating plants with a dilution series of the test material and growing these plants in aseptic test tubes or plastic bags. This method is called the plant infection technique. Two conditions must be met: (1) the organisms must be distributed at random in the bulk suspension from which small samples are to be removed so that the distribution of numbers in replicate samples will be Poisson, and (2) the culture medium and incubation conditions must allow viable growth to occur in every sample containing one or more organisms. An example of data that might be obtained is as follows:

Dilution	1:10	1:100	1:1000	1:10,000	1:100,000
No. with nodules	4	4	2	1	2

Tables are available in the literature for estimating the most probable number (MPN) for data from dilution series of this type. However, the experimenter may choose to use a dilution series for which tables are not available or may obtain results for which tables do not give answers. One method of estimating the MPN is based on the Spearman–Kärber estimator. Using m to denote the natural logarithm of the MPN:

$$m = \exp(\gamma - x_0 - d/2 + d\Sigma p_i),$$

where $\gamma = 0.57722$, the Euler constant, $z_0 = 0.1$, the first dilution, $x_0 = \log_e z_0$, $d = \log_e 10$, the dilution factor, and p_i = the proportion of tubes showing growth. For the data from the preceding example,

$$m = [-0.57722 - \log_e 10^{-1} - \log_e 10/2 + (\log_e 10)(13/4)] = 8.0575.$$

The estimate of MPN is exp(8.0575) = 3157 bacteria/g. The variance of m is

$$\text{Var}(m) = \frac{\log_e 10 \log_e^2}{4} = 0.3990.$$

Lower and upper confidence limits of m are $m \pm (1.96)(0.3990)^{\frac{1}{2}} = 6.8194$ and 9.2956. Limits for the MPN are then 915 and 10,889. This estimate, however, is not unbiased. Other maximum likelihood estimation procedures give unbiased estimates of MPN for any dilution series.

These methods employ the weighting coefficient $w = \exp(2y)/[\exp(e^{y}) - 1]$, and the minimum ($\eta_0$) and maximum ($\eta_1$) working deviates, where $\eta_0 = e^{-Y}$ and $\eta_1 = \exp(e^{Y} - 1) - e^{-Y}$. The working deviate is $\eta = \eta_0 + p(\eta_1 - \eta_0)$. In practice, one would make use of a computer program to arrive at MPN estimates. Calculations given below are used with the intention of helping the reader understand the procedure. Using the previous example and using a first guess of $Y = 0.7$ at the 0.001 dilution (i.e., 2014 bacteria/g), the maximum likelihood estimates are as given in Table VIII. Since a dilution factor of 10 was used and $\log_e = 2.3026$, the other working log log values Y differ by 2.3026 per dilution.

After the first cycle $\bar{\eta} = 0.2949/0.8284 = 0.3568$. One subtracts this amount from the working log log values Y and do the calculations for the second iteration. After three cycles, $\bar{\eta} = 0.819/0.05 = 0.0698$. For the 1:1000 dilution, we now have $\log_e(m/1000) = 0.5189 - 0.0698 = 0.4491$, where $m/1000 = 1.567$ and $m = 1567$ bacteria/g. The variance of $\log_e(m/1000)$ is $1/\Sigma\, nw = 1/(4 \times 0.8192) = 0.3052$ and $s_m = 0.5524$. The lower and upper limits of $\log_e(m/1000)$ are then

$$0.4491 - 1.96 \times 0.5524 = -0.633$$

and

$$0.4491 + 1.96 \times 0.5524 = 1.5318.$$

The fiducial limits of the density of the original sample are 531 and 4627 bacteria/g.

VII. Experimental Design

Results from biological assays are subject to greater variability than chemical or physical assays. However, the precision of biological assays can be improved by appropriate application of the principles of experimental design. In brief, experimental design includes the specification of treatments, experimental units to which they are applied, frequency, time and location of measurement, and specification of the statistical analysis to be carried out. The purpose of experimental design is to remove systematic error, improve precision of estimates of effects, broaden the range of validity about the results, and simplify the experimental procedure, analysis, and presentation of results. A number of the standard texts on experimental design do not address the design of bioassays specifically but do provide the basic principles of design. The first step in planning a bioassay is to choose suitable organisms to use as subjects. If we were to apply the results of an assay of a material containing bacterial toxins to humans, we would need to be sure that the response of the subjects to these toxins is similar to that of humans. Precision requires homogeneous subjects of similar size, weight, and genetic background. The next step is to specify the range and number of doses and the magnitude of differences between successive doses. It is advisable to conduct a preliminary experiment with 5 or 10 subjects per dose to identify the range of doses that produce 0–100% response. From the initial information about the levels of stimulus that give maximum and minimum responses, one can then select dose levels that follow a geometric progression downward from the maximum to the minimum. Fitting a straight line is efficient when the points on the transformed dose axis are equally spaced, hence the suggestion for using a geometric progression. For three doses, one normally does not use less than 2-fold and, more generally, 5-fold or 10-fold dilutions. For most quan-

Table VIII Maximum Likelihood Calculations for Estimation of the Most Probable Number

				1st Cycle			2nd Cycle			3rd Cycle		
Dilution	n	r	p	Y	w	ηw	Y	w	ηw	Y	w	ηw
0.1	4	0	0	5.3	0	0	4.9	0	0	5.1		
0.01	4	0	0	3.0	0	0	2.6	0	0	2.8		
0.001	4	2	0.5	0.7	0.625	0.852	0.3	0.642	0.477	0.5	0.646	0.648
0.0001	4	3	0.75	−1.6	0.182	−0.075	−2.0	0.132	−0.127	−1.7	0.154	−0.104
0.00001	4	2	0.5	−3.9	0.020	−0.484	−4.3	0.014	−0.490	−4.1	0.017	−0.489
0.000001	4	4	1.0	−6.2	0.002	0.002	−6.6	0.001	.001	−6.4	0.002	0.002
Totals					0.829	0.295		0.789	−0.139		0.819	0.057

titative assays, it is advisable to have three or more doses with responses between 10 and 90%. Because precision is poor at extreme doses, very little weight is given to responses at the extreme ends of the response curve. Statistical analysis is easier when each treatment includes the same number of subjects.

Rules by which subjects are allocated to treatments is the next experimental design consideration. To avoid bias, subjects must be allocated to treatments at random. Avoid placing all subjects receiving the same dose in one group, as it confounds differences between doses with differences between groups. The interactions among subjects or environmental differences among groupings such as cages may contribute to bias in the results. Use of replicate sets of cages arranged in completely randomized or randomized complete block design may be advisable. A symmetric dose structure is desirable. In the (k,k)-point assay, k doses of the test preparation and k doses of the standard preparation are used. The two preparations do not necessarily have the same doses, but the difference between successive doses of either preparations should have the same ratio to one another (e.g., 0.1, 0.01, and 0.001 mg/liter for the standard and 50, 5, and 0.5 mg/liter for the test preparation). Other designs such as cross-overs, Latin squares, and any of the incomplete block designs can be and have been adapted to bioassay work. One example is in a study that used a balanced lattice to estimate the potency of vaccines produced from nine strains of *Hemophilus pertussis* relative to that of a standard vaccine. Twelve blocks each consisting of four vaccines were arranged in a 3×3 balanced lattice with a standard preparation added to each block. The 48 ED_{50}'s were subjected to analysis of variance to find the significance of difference among immunizing capacity of the strains. Experimental design includes the specifications for measurements and logistics.

The method of statistical analysis to be used should be considered in specifying the design. Analysis of some of the more complex designs such as incomplete blocks require more knowledge and labor than a simpler design and may not be as precise. The choice of design involves making a compromise between conflicting interests. One must measure the effects of increasing precision of estimation and the quality of the information obtained with complicated designs against the cost of carrying out the experiment and doing the statistical analysis.

Bibliography

Bliss, C. I. (1952). "The Statistics of Bioassay." Academic Press, New York.

Cochran, W. G., and Cox, G. M. (1957). "Experimental Designs," 2nd ed. John Wiley & Sons, New York.

Davis, G. R. F., and Schiefer, B. (1982). *Biochem. Physiol.* **16,** 13–16.

Davis, G. R. F., Smith, J. D., Schiefer, B., and Loew, F. M. (1975). *J. Invert. Pathol.* **26,** 299–303.

Finney, D. J. (1971). "Probit Analysis," 3rd ed. Cambridge University Press, London.

Finney, D. J. (1978). "Statistical Method in Biological Assay." Charles Griffin & Co., London.

Fisher, R. A., and Yates, F. (1963). "Statistical Tables for Biological, Agricultural and Medical Research," 6th ed. Oliver & Boyd, Edinburgh.

Hubert, J. J. (1991). "Bioassay," 3rd ed. Kendall-Hunt, Dubuque, Iowa.

Hughes, P. R., Wood, H. A., Burand, J. P., and Granados, R. R. (1984). *J. Invert. Pathol.* **43,** 343–350.

Kempthorne, O. (1952). "The Design and Anaylsis of Experiments." John Wiley & Sons, New York.

Meynell, G. G., and Meynell, E. (1970). "Theory and Practice of Experimental Bacteriology," 2nd ed. Cambridge University Press, London.

Moore, K. C., and Davis, G. F. R. (1983). *J. Invert. Pathol.* **42,** 413–414.

Sizaret, P. (1988). *Bull. WHO* **66,** 1–6.

Smith, P. D., Brockway, D. L., and Stanch, F. E., Jr. (1987). *Environ. Toxicol. Chem.* **6,** 891–900.

Stone, J. R., Verma, P. R., Dueck, J., and Spurr, D. T. (1987). *Can. J. Plant Pathol.* **9,** 137–145.

Thomas, F. C., Morse, P. M., and Seawright, G. L. (1979). *Arch. Virol.* **62,** 189–199.

Walsh, G. E., Deans, C. H., and McLaughlin, L. L. (1987). *Environ. Toxicol. Chem.* **6,** 767–770.

Biodegradation

Wendy B. Bollag and Jean-Marc Bollag
The Pennsylvania State University

Glossary

Axenic culture Known species of microorganism maintained in the laboratory

Biotransformation Alteration of the structure of a compound by an organism

Cometabolism Biotransformation of a compound by a microorganism that is incapable of using the compound as a source of energy or growth

Detoxification Reduction in the hazardous nature of a compound

Enrichment culture Procedure for isolating microorganisms capable of using particular chemicals for their growth

Microbial infallibility Proposal that any substance can be biodegraded under the appropriate conditions by the appropriate microorganism

Mineralization Complete degradation of a compound to simple inorganic constituents, such as carbon dioxide, water, ammonia, chloride, and sulfate

Recalcitrance Resistance of a compound to biodegradation, resulting in its persistence in the environment

Xenobiotic A synthetic product not formed by natural biosynthetic processes; a foreign substance

BIODEGRADATION can be defined as the decomposition of a substance through the action of biological agents, especially microorganisms. In general, then, biodegradation is the process of decay initiated by microorganisms. In a stricter sense, however, biodegradation has come to signify the complete microbial breakdown, or mineralization, of complex materials into simple inorganic constituents such as carbon dioxide, water, and mineral components. This process is important for several reasons: biodegradation is essential in allowing the recycling of such necessary biological building blocks as carbon, hydrogen, oxygen, nitrogen, and sulfur—without the activity of microorganisms these atoms might be tied up in complex nonbiodegradable substances and, thus, unable to reenter the natural cycles. Furthermore, microbial degradation of dead matter also prevents the accumulation of debris on the earth's surface. Perhaps most importantly, and the focus of this discussion, the activity of microorganisms on certain man-made substances can also, in many cases, result in their removal from the environment, in a reduction in their toxicity, or in both.

This aspect of biodegradation—the decomposition of synthetic compounds, or the lack thereof—has generated a great deal of concern in recent years. Almost daily, reports of overflowing landfills and hazardous pollutants in our air, water, and soil are issued. The "biodegradability" of a substance, that is its susceptibility to decomposition by natural biological processes, has become a criterion by which we evaluate the worthiness of a commercial product. For these reasons, knowledge about the microbial decomposition of anthropogenic chemicals has been actively sought, and, as a result, our understanding of biodegradation, although far from complete, is continually expanding. In this article an overview of biodegradation, in particular that of human-made compounds, will be presented, and possible applications of the biodegradative potential of microorganisms will be discussed.

I. Agents of Biodegradation

Although the transformation of organic molecules can be effected by a variety of abiotic catalysts, the most important degradative agents of these chemicals are microorganisms such as bacteria and fungi. In any given ecosystem, a large number of microbial species may be present and capable of degrading a wide variety of structurally unrelated chemical compounds including xenobiotics, which are chemicals derived from human activity. Moreover, as a result of their remarkable powers of adaptation, microorganisms are able to proliferate in the most diverse and inhospitable environments. Thus, microorganisms appear to possess great biodegradative potential; so great in fact (that the proposal has been made) that under the appropriate conditions, microorganisms can degrade any organic compound. Although this principle of microbial infallibilty has been difficult to prove or disprove, several substances have certainly put this idea to the test (see Section IV).

In general, microorganisms decompose organic substances to generate energy and nutrients for their growth. Thus, a usual consequence of biodegradation of a compound is an increase in the numbers of the microorganisms degrading that substance. In a process termed enrichment culture, this characteristic has been used to identify microorganisms capable of using a particular chemical for their growth. The enrichment technique involves incubating a source of microorganisms, such as sewage, soil, or water, with the substance of interest as the sole source of carbon and energy. Species that can obtain energy from and subsist on the compound (i.e., microorganisms that degrade the chemical) are able to proliferate under these conditions and can be isolated using the appropriate microbiological methods. In general, this enrichment procedure has proven successful in isolating microbial species that biodegrade naturally occurring compounds. More importantly, enrichment techniques have also resulted in the isolation of microorganisms that can mineralize various pesticides and other potentially toxic anthropogenic chemicals. [*See* ORGANIC MATTER, DECOMPOSITION.]

Nevertheless, on occasion the microbial transformation of xenobiotics is observed without the accompanying growth of a degradative microorganism. In some cases, the inability to isolate a microorganism that uses the compound as its sole carbon and energy source may be purely a methodological problem: the active species may require for its growth an accessory factor (or factors) that is not present in the incubation medium. Alternatively, although the microorganism may transform the chemical, it may be unable to convert the compound to products that provide energy and a source of carbon for microbial growth. This phenomenon, known as cometabolism, appears to result from a lack of specificity of certain microbial enzymes. These enzymes can modify the compound, which may be chemically related to the natural substrate of the enzyme, but generate products that cannot be further metabolized by the enzymatic machinery of the microorganism. Thus, the consequences of cometabolism are threefold: (1) the microorganisms responsible for the transformation do not proliferate in the presence of the chemical; (2) as a result of this lack of microbial growth, the rate of transformation of the xenobiotic does not increase over time and is, in fact, generally quite low because of the small starting population of the transforming microorganisms; and (3) a product or products of the transformation reaction may accumulate in the environment. Because these products may be chemically similar to the parent compound, their presence in the habitat may be equally undesirable.

Nonetheless, in a natural ecosystem microbial species exist not in isolation but in the presence of a wide variety of other types of microorganisms. In some instances, the transformation product generated by the cometabolic reaction of one microorganism may serve as a growth substrate for another. Alternatively, a chemical may be completely biodegraded by a series of sequential cometabolic attacks by various microbial species. Thus, cometabolism may be harmful in that this fortuitous transformation can lead to the production of a compound with increased resistance to further degradation or greater toxicity than the parent compound; on the other hand, cometabolism may initiate the biodegradative reactions of a diverse microbial population in an environment to result in the ultimate mineralization of the chemical.

This interaction of various microbial populations in an environment is difficult to investigate in a laboratory setting. Often a microbial species capable of metabolizing a given substance is isolated by enrichment procedures and its degradative capacity studied by the scientist in an axenic culture *in vitro*. Obviously, laboratory conditions may be very different from those encountered by the microorganism in a natural environment. For instance, the mineralizing microbial species may compete poorly with other species for limiting nutrients in the environment; therefore, the microorganism that in a laboratory culture efficiently degrades a compound, may in the natural habitat grow poorly and thus only minimally metabolize the chemical. On the other hand, the interaction between the various populations may be positive, with the degradative microorganism perhaps deriving necessary growth factors or favorable growth conditions from the activity of other microbial species. Thus, the activity of the heterogeneous community, rather than that of the single species, determines a chemical's biodegradation in the natural environment.

Several other factors influence the biodegradation of a compound in the environment. In order for a microorganism to use a substance for growth, other factors required for growth must also be available at suitable concentrations. These factors include nutrients that provide sources of nitrogen, sulfur, phosphorus, calcium, and magnesium, as well as trace metals and an adequate supply of water. In addition, a sufficient supply of a suitable electron acceptor, such as molecular oxygen, nitrate, sulfate, or carbon dioxide, is required. If any one of these factors necessary for microbial growth is scarce or absent, biodegradation of the chemical will be inhibited. Similarly, other factors, such as temperature and pH of the environment, that affect microbial proliferation will influence a molecule's biodegradation. In addition, as microorganisms "feed" on the chemical in question, so too do other organisms feed on the microorganisms; predation may thus influence microbial numbers and the rate of biodegradation. Moreover, in an environment such as soil, the presence of reactive surfaces and particulate matter may interfere with microbial decomposition of a substance. Diffusional limitations may also inhibit substrate-microorganisms interaction, thus slowing the rate of metabolism of a substance. Finally, various inhibitors of microbial growth and metabolism may be present in the habitat and affect mineralization of a compound; indeed, the chemical under investigation may itself be inhibitory and its presence above certain concentrations may preclude its biodegradation. Thus, the interaction of a large number of factors determines the ultimate fate of a particular compound.

II. Biodegradative Enzymes

Microbial enzymes are responsible for the degradation of various substances by microorganisms. As the ultimate result of biodegradation is the production of energy for microbial growth, the enzymes that mediate biodegradative reactions are often those that also initiate essential metabolic pathways. Certain biological transformation mechanisms, however, are unique to microorganisms. For instance, some microbial species are capable of proliferating under both aerobic and anaerobic conditions. Of necessity, this dual capacity requires that these microorganisms possess the enzymatic machinery to use either molecular oxygen or nitrate as final electron acceptor. Alternatively, some species are able to derive the necessary energy for growth by a fermentative process, in which essentially no final electron acceptor is needed. In addition, some microorganisms can produce enzymes that they secrete and that are active extracellularly.

The existence of microbial enzymes that exhibit activity outside the organism has contributed to a recurrent problem experienced by scientists investigating biodegradation: the difficulty in distinguishing between enzyme-catalyzed biotransformation reactions and those that occur as a result of purely physical-chemical effects. Thus, transformed products may be generated in a number of ways, including via: (1) enzyme catalysis occurring in the microorganism, (2) enzyme catalysis occurring extracellularly in the environment, or (3) physico-chemical catalysis. In addition, transformation may be the result of a combination of these mechanisms: for instance, enzyme-generated products may be further transformed by physico-chemical means and vice versa. Further complicating this puzzle is the fact that microorganisms can themselves alter the physico-chemical properties of the environment. For instance, microbial activity can affect the pH or redox conditions of the habitat and thus contribute indirectly to physico-chemical catalysis of transformation. In addition, sterilization, the method most often used for differentiation of biological and non-biological processes, may in fact alter not only the microbial but also the physico-chemical properties of the source of the inoculum under investigation. As an example, most techniques employed for sterilizing soil can also cause changes in the physico-chemical characteristics of the soil itself; thus, the inability of sterilized soil to catalyze a transformation reaction may or may not indicate the involvement of microorganisms and their enzymes. Obviously, the investigator is faced with quite a challenge in attempting to determine whether or not a reaction is enzymatically catalyzed.

Nevertheless, it should be noted that while physico-chemical processes can result in transformation of a chemical, mineralization is almost exclusively a consequence of microbial activity. More importantly, many scientists have succeeded in purifying the microbial enzymes involved in various transformation reactions, thereby abolishing all doubt as to the biological nature of the metabolism.

The ability of microorganisms to transform these human-made compounds raises an interesting question: what is the origin of the enzymes capable of degrading such 'unnatural' chemicals? It is generally acknowledged that xenobiotics that resemble naturally occurring compounds are most likely to be efficiently degraded by microorganisms. Similarly, synthetic compounds with chemical structures that differ greatly from those of natural substances are often degraded poorly. For these reasons, it is generally assumed that xenobiotic-degrading enzymes have evolved by mutation and natural selection from those constitutive or inducible enzymes responsible for general microbial metabolism. The capacity to use synthetic chemicals might thus be gained as a result of hyperproduction of preexisting enzymes due to gene duplication, mutations that alter regulatory control processes, or mutations that generate novel enzymes with new specific activities. Some of the genes encoding pollutant-degrading enzymes can be found on autonomously replicating plasmids that in certain cases are transmissible among different microbial strains. Plasmid exchange within a population could, thus, result in the production of novel microbial strains with a large number of degradative capabilities. In general, then, it appears that the extreme adaptability of microorganisms gives them the capacity to alter their enzymatic machinery to metabolize a wide spectrum of anthropogenic chemicals.

III. Mechanisms of Biodegradation and Biotransformation

In general, xenobiotics are transformed by the same mechanisms that mediate microbial metabolism of naturally occurring compounds. These processes can be categorized as oxidative, reductive, hydrolytic, or synthetic reactions as detailed in the following paragraphs.

Hydroxylation is a common oxidative reaction and is frequently the initial step in the biodegradation pathway of many substances. For instance, the cleavage of aromatic rings and the degradation of aliphatic side chains to result in decarboxylation, deamination, or dealkylation are often preceded by hydroxylation. Moreover, the addition of the hydroxy group often increases the polarity, and thus the water-solubility, of the compound, making it more accessible to biological attack. Enzymes that catalyze this reaction include hydroxylases and mixed function oxidases.

Another important oxidative reaction is *N*-dealkylation which, like hydroxylation, is often a first step in the breakdown of pollutants, in particular alkyl-substituted pesticides. *N*-Dealkylation can be catalyzed by mixed function oxidases. Other oxidative reactions involved in the biodegradation of xenobiotics include decarboxylation, β-oxidation (of fatty acid side chains), hydrolysis of ether linkages, epoxidation, sulfoxidation, and cleavage of aromatic and heterocyclic rings.

Reductive transformations of xenobiotics are less well described; however, as studies on the anaerobic biodegradation of these chemicals are continued, more such mechanisms will undoubtedly be discovered. Nevertheless, several xenobiotics have been demonstrated to undergo reductive metabolism. Examples of reductive reactions are the conversion of a nitro to an amino group, reductive dehalogenation, saturation of double and triple bonds, reduction of aldehydes, reduction of ketones to secondary alchohols, and conversion of sulfoxide to sulfide.

Hydrolytic reactions, in which cleavage of a molecule is effected by the simultaneous addition of water, often initiate the biodegradation of xenobiotics. These reactions are observed frequently in the microbial metabolism of xenobiotics that possess ether, ester, or amide linkages and are catalyzed by esterase, acrylamidase, phosphatase, hydrolase, and lyase enzymes. The result is the hydrolysis of the ether, ester, phosphoester, or amide bond. Another reaction is hydrolytic dehalogenation in which the halogen atom of a pollutant is replaced by a hydroxyl group generated from water.

Synthetic pathways of xenobiotic transformation involve the modification of pollutants via the addition of some chemical group to form conjugated products (conjugation reactions) or the coupling of pollutants to another molecule or molecules to yield dimeric and polymeric compounds (condensation reactions). Although conjugation processes, such as glycoside formation reactions with amino or sulfur groups, occur frequently in plants and animals, only alkylation (in particular methylation) and acylation (especially formylation and acetylation) are commonly observed in microorganisms. The resulting methylated, acetylated, and formylated products are often less toxic than the parent compound; however, in some important instances, the conjugated pollutants may actually prove more toxic to humans and higher organisms, as detailed in Section V.

IV. Biodegradability and Recalcitrance

Through the basic metabolic reactions already described, microorganisms are able to degrade a large number of man-made chemicals. Nevertheless, in some cases microbial degradative capacity can be problematic. The use of pesticides has not only greatly improved the success of modern agriculture but has also allowed the control of disease-carrying insects in developing countries. However, the cost of these chemicals can be quite high, and a need for repeated application may prove prohibitive. For these compounds, then, resistance to microbial decomposition can be a desirable characteristic. Similarly, the deterioration of packaging materials and fabrics, while still in use, would be highly objectionable. Thus, some degree of resistance to biodegradation is a necessary property of various synthetic compounds.

Obviously, resistance to microbial decomposition can also be an undesirable and sometimes a hazardous property of various xenobiotics. The persistence of human-made plastic containers and other waste in the environment is not only aesthetically displeasing but has also created a major concern over the future disposal of such substances in this age of overfull landfills. Other chemicals used in agriculture and industry have been found to contaminate soil and water, and resistance to biodegradation can lead to their accumulation to toxic levels. Often crops grown in such polluted soils concentrate these compounds in their tissues, creating a greater hazard to humans and animals. In addition, the effects of toxic xenobiotics can frequently be magnified; because of their lack of biodegradability, these compounds progress through the food chain. A well-known example of this magnification effect is the profound decimation of the bird of prey population observed years ago as a result of the cumulative toxicity of the insecticide DDT. Thus, it is quite clear that a lack of biodegradability can be a dangerous characteristic of some xenobiotics. [*See* Hazardous Waste Treatment, Microbial Technologies.]

What makes a compound resistant to biodegradation? This question has received a great amount of attention from environmental scientists, but the mechanism of recalcitrance, or resistance to biodegradation, is as yet unclear. It should be noted that there are a number of naturally occurring substances that are quite persistent in the environment. For instance, the organic matter of soil, humus, or a portion at least, appears to be resistant to microbial decomposition under some conditions, and radiocarbon dating has indicated the age of some fossil soils to be between 3000 and 25,000 years.

Several man-made chemicals can persist in the environment. For instance, the pesticides aldrin, DDT, lindane, and parathion are all present in soil more than 15 years after their last application, indicating that these compounds are resistant to microbial attack. In general, xenobiotics with chemical structures similar to those of naturally occurring compounds are more likely to be degraded by microorganisms. On the other hand, synthetic chemicals with dissimilar structures frequently exhibit the phenomenon of recalcitrance; the more dissimilar the molecule the more likely the compound is to lack biodegradability. Although the mechanism of recalcitrance is still only incompletely resolved, certain features are known to predispose a xenobiotic to persist in the environment. For instance, the presence of halogen (e.g., chlorine) or other substituents, such as nitro, sulfonate or methyl groups, tends to diminish biodegradability; two or more such substituents are likely to render the compound still more refractory to microbial decomposition. In addition, the position of the substituent(s) can affect the ability of microorganisms to degrade a given xenobiotic. Nevertheless, although some of the characteristics that predispose a substance to resist microbial attack are known, it is still not possible to predict with any accuracy which xenobiotics will persist in the environment. [*See* Pesticide Biodegradation.]

Several factors can contribute to the phenomenon of recalcitrance of a xenobiotic. Clearly, if no microorganism exists that is capable of transforming the compound, the xenobiotic will not be biodegradable and will persist in the environment. Alternatively, a microbial population able to use the chemical may exist but may not be present at or may be physically restricted from the site of pollutant contamination. Moreover, even if present, the microbial species may lack an essential nutrient or proper conditions for growth. Thus, environmental factors such as temperature, pH, salinity, osmolarity, predator activity, and competition among microbial populations as well as the availability (and identity) of electron acceptors, can influence the biodegradability of a xenobiotic. In this regard, it should be noted that some chemicals that are readily degraded with molecular oxygen as the final electron acceptor may under anaerobic conditions be degraded only poorly or not at all, and vice versa. Microbial growth, enzymatic activity, or both may also be inhibited by the pollutant itself, especially if present at high concentrations, or by another toxin or enzyme inhibitor present in the environment. In addition, the xenobiotic may be inaccessible to the appropriate degradative enzymes because of binding to high molecular weight resistant organic compounds, lack of water solubility, or failure to gain access to intracellular compartments, or there may be steric hindrance at the site acted on enzymatically. Finally, biodegradation of certain chemicals may require a cooperative interaction among two or more microbial species. Thus, if one population is absent or its growth inhibited by any of these factors, recalcitrance of the compound may result. It is quite obvious, therefore, that there are many possible ways in which a synthetic compound may resist microbial attack with the result that at the present time biodegradability or recalcitrance of a xenobiotic are extremely difficult properties to predict.

V. Applications

Although, as discussed earlier, some synthetic chemicals are resistant to biodegradation, microorganisms possess an enormous capacity to transform a wide range of both natural and human-made compounds. This prowess has been recognized and exploited for many years. For instance, since the turn of the century, microorganisms have been used at sewage treatment plants to process waste water. Recently, it has been proposed that microorganisms (or their enzymes) might be used in other systems for bioremediation, the removal of pollutants from a contaminated environment. [*See* BIOREMEDIATION.]

Biodegradative microorganisms may prove particularly useful for the clean-up of toxic chemicals in habitats like soil or ground water. Conventional decontamination procedures involve physically removing the polluted topsoil; these methods, although effective, are cumbersome and cost-inefficient. Furthermore, the pollutant is not degraded but merely moved to a less hazardous site. Quite obviously, complete decomposition of the chemical would be preferable. Indeed, the microbial conversion of toxic compounds to inorganic natural molecules such as carbon dioxide, water, methane, nitrate, and sulfate results ultimately in a *detoxification* of the pollutant. For those synthetic molecules for which a mineralizing microorganism exists then, biodegradation may prove to be an efficient means of detoxifying pollutants in the environment.

Detoxification may also occur without complete mineralization. Often biotransformation reactions can result in a decrease in the toxicity of the parent compound. Nevertheless, occasionally products of these reactions may in fact be more toxic than the original molecule or may persist and accumulate in the habitat. For example, the microbial metabolism of tetrachloroethylene under anaerobic conditions can produce vinyl chloride, which is not only very toxic but is also resistant to further biotransformation. Thus, it is quite clear that the generation of metabolites other than the inorganic products of complete biodegradation should elicit concern until such time as these compounds are proven to be innocuous. However, it should be noted that even pollutants that are eventually mineralized can be converted to intermediates that may transiently accumulate to possibly toxic levels.

In addition to the potential toxicity of microbial metabolites of xenobiotics, there are other limits to the use of microorganisms for decontamination of the environment. First and foremost, a microbial species capable of metabolizing the chemical must exist. As discussed in Section IV, some pollutants are recalcitrant under various environmental conditions. Nevertheless, the principle of microbial infallibility (discussed in Section I) is difficult to prove or disprove, and it may be true that any chemical, either naturally-occurring or synthetic, can be biodegraded under the appropriate conditions. It remains to be seen whether or not degradative microorganisms can be isolated for all xenobiotics using present methods such as the enrichment technique.

Even once a microbial species capable of degrading a particular compound is identified, additional factors can influence the effectiveness of bioremediation. In order for biodegradation to occur, the contaminated site must be inoculated with a sufficient microbial biomass. Transportation of a large number of microorganisms in an active or inducible state may be difficult. Furthermore, the inoculum may not proliferate in the habitat such that there is no guarantee that biodegradation will proceed rapidly and efficiently. As previously discussed, microorganisms require essential nutrients that must be present in or added to the environment. In addition, appropriate conditions for growth must prevail at the polluted site: temperature, pH, availability of electron acceptors, salinity, osmolarity, water saturation, predator activity, and competition with native microbial populations for nutrients can all affect proliferation of the inoculated culture. Obviously, the conditions that the inoculum experiences *in situ,* that is, in a natural environment, are more severe than those obtained in a laboratory setting; and without expansion of the microbial population, biodegradation will occur only slowly and inefficiently.

Another potential hindrance to a rapid rate of biodegradation is the fact that microbial metabolism of some xenobiotics requires the induction of degradative enzymes, occasionally by a chemical other than that present in the contaminated environment. In addition, microorganisms can be physically restricted from entering certain microenvironments with the result that a xenobiotic remains inaccessible to the microbial metabolic machinery. A pollutant may also be inaccessible to intracellular enzymes if the compound is unable to permeate the microorganism because of a lack of uptake by the cell or the adsorption of the chemical to interfering solid substances, such as soil or clay. Thus, the use of microorganisms for the detoxification of polluted habitats has some serious potential limitations.

Some of these limitations may be overcome by using isolated enzymes, rather than the entire microorganism, for the decontamination procedure. Technologies have been developed for purifying large quantities of enzymes that function extracellularly to transform xenobiotics. These catalysts have the advantage that they readily interact with pollutants such that accessibility is seldom a problem. In addition, although the initial production of the enzyme might require induction, once introduced into the contaminated environment the enzyme is active immediately because no further protein synthesis is necessary. Moreover, the activity of isolated enzymes is not dependent on cell growth, therefore, these catalysts might be expected to function under quite severe *in situ* conditions.

Nevertheless, the use of enzymes for decontamination purposes presents some unique obstacles as well. For instance, as discussed earlier, complete mineralization is the preferred biotransformation reaction for detoxification of xenobiotics; however, such degradation necessitates the sequential action of many enzymes on various intermediates derived from metabolism of the parent compound. Obviously, the cost of mass-producing many enzymes for inoculation into a contaminated site would be prohibitive. Furthermore, in the vast milieu of an *in situ* environment, the diffusion of products formed by one enzyme would decrease the availability of the substrate to the next in the enzymatic series. To combat this problem, investigators have successfully enclosed microbial enzymes catalyzing sequential degradative reactions in a permeable sphere and demonstrated the effectiveness of this apparatus in degrading pollutants. Other researchers have suggested that a series of enzymes might be attached to a solid support (e.g., glass beads or soil) in order to effect complete biodegradation of a compound. Nevertheless, the difficulty of mimicking the metabolic machinery of a complex microorganism using these techniques is obvious.

In addition to the potential difficulty of identifying and isolating large quantities of an appropriate enzyme, other factors that may preclude the use of isolated enzymes for decontamination purposes include their possible inactivation by proteases and extremes of pH and temperature and their requirement, in some cases, for cofactors. Current research is seeking solutions to these potential complications of using microbial enzymes for decontamination of the environment. For instance, researchers have successfully attached enzymes to solid supports and demonstrated that not only do the immobilized enzymes still catalyze transformation reactions but also they are more resistant to proteolytic attack and extremes of temperature and pH. Nevertheless, to date there has been limited application of isolated enzymes to field-scale bioremediation efforts and the feasibility of such methods is unknown.

Finally, the incorporation of pollutants into polymers and copolymers has been proposed as a potential means of detoxifying contaminated soils and waste waters. In waste water, the polymers formed as a result of microbial activity precipitate from solution and this precipitate can be easily removed by filtration. Furthermore, in soil pollutants can be oxidatively cross-coupled to humic constituents to form copolymers. The resulting covalent incorporation of the synthetic compounds into humic material has been shown to reduce the toxicity of the pollutants and is thought to decrease their bioavailability as well. Nevertheless, a question remains as to the stability of these humus-bound residues. Clearly, if the compounds are only transiently bound or are subsequently released by the activity of soil microorganisms, they would constitute a future hazard and their binding to soil would present neither a satisfactory nor permanent solution to the pollution problem. However, all research to date indicates that these bound pollutants are quite stable, with only a very limited release into the environment over time. Furthermore, it appears likely that the small amounts of pollutants that are released can be readily mineralized by soil microorganisms to prevent toxic accumulations. Thus, the information currently available suggests that once incorporated into humus, xenobiotics are unlikely to adversely affect the environment.

It should be obvious from the discussion presented herein that microbial biodegradation is a necessary and important process for preserving a clean environment. Nevertheless, it should also be apparent that, although a significant body of knowledge exists about this process, much additional information is needed to fully use the degradative ability of microorganisms. For instance, once the biodegradability of human-made compounds can be predicted with certainty, it should become possible to design and manufacture synthetic materials that are completely biodegradable, thus helping to solve the problem of overflowing landfills. Furthermore, such information concerning the mechanisms and characteristics of biodegradation should enable the efficient and rapid removal of chemical contaminants from our environment. Thus, our expanding knowledge about biodegradation of natural and human-made substances may prove to be a key to a clean and safe future world.

Bibliography

Alexander, M. (1981). *Science* **211,** 132–138.

Alexander, M. (1985). *Environ. Sci. Technol.* **19,** 106–111.

Bollag, J.-M. and Liu, S.-Y. (1990). Biological transformation processes of pesticides. *In* "Pesticides in the Soil Environment," pp. 169–211. Soil Science of America, Madison, Wisconsin.

Chakrabarty, A. M. (ed.). (1982). "Biodegradation and Detoxcification of Environmental Pollutants" CRC Press, Boca Raton, Florida.

Leisinger, T., Cook, A. M., Hütter, R., and Nüesch, J. (ed.). (1981). Microbial degradation of xenobiotics and recalcitrant compounds. Academic Press, London.

Biofilms and Biofouling

Hilary M. Lappin-Scott
University of Exeter

J. William Costerton
University of Calgary

Thomas J. Marrie
Dalhousie University

Glossary

Biofilms Complex association of microorganisms and microbial products attached to a surface

Biofouling Damage caused to a surface by microorganisms attached to the surface

Consortia Spatial grouping of bacterial cells within a biofilm in which different species are physiologically coordinated with each other, often to produce phenomenally efficient chemical transformations

Planktonic Free-floating bacteria living in the aqueous phase and not associated with a biofilm

Sessile Bacteria living within a biofilm

BIOFILMS form when microorganisms attach to surfaces in aquatic ecosystems and produce exopolysaccharides, which aid adhesion to the submerged surfaces. Adhesion to surfaces provides considerable advantages for the bacteria that live within the biofilm, including protection from antimicrobial agents and the many benefits gained from close proximity to other microorganisms (e.g., the exchange of nutrients, metabolites, and genetic material). Although bacteria benefit from this close association with the surface, the combination of growth processes, the production of metabolites, or the physical presence of the biofilm either damages the surface or causes an obstruction so that the efficiency of the surface is reduced. This surface damage is collectively termed biofouling and has been observed on a surprisingly wide range of surfaces, causing problems such as dental decay, metal pipeline corrosion, and colonization of a variety of medical implants.

I. Biofilm Formation

In their quest to develop techniques to isolate and purify bacteria, many of the 18th and 19th century microbiologists removed samples from natural environments and inoculated the samples into liquid growth nutrients. In doing so, the microbiologists removed the bacteria from the surface on which they were growing (including soil particles, food, or body tissue) and dispersed them into nutrients with only the walls of a growth flask available for attachment. Such studies provided the methodology to further our understanding of microbial processes, and these techniques are still taught as the principal method of growing bacteria despite many observations that bacteria do not grow in this manner in natural environments. All aquatic ecosystems that have been studied to date have shown that higher numbers of bacteria reside attached to submerged surfaces than in the aqueous phase itself.

Biofilm studies commenced more than 50–60 years ago with pioneering studies by Cholodny, Henrici, and ZoBell. Some of these early studies involved immersing glass slides into natural environments and observing the biofilms that developed using light microscopy. Other work reported higher bacterial counts on the inside surface of a glass bottle than within the water in the bottle. These early observations of biofilms stimulated investigations into the mechansims of adhesion and studies of the range of aquatic ecosystems and surfaces that were colonized by biofilms.

Biofilm formation commences with the colonization of a surface by bacteria. The bacteria may be attracted by a number of different mechanisms including surface charge, gravity, Brownian motion, and chemoattraction if the surface has nutrients. Once the bacteria have been attracted to the surface by a combination of one or more of the above events, attachment events occur that have been demonstrated to be a two-step process comprised of reversible rather than irreversible binding. The reversible binding is considered to involve weak attractions of Van der Waal forces to hold the bacterium close to the surface before a stronger attachment forms by a combination of both physical and chemical forces. One of these chemical forces is the production of exopolysaccharide-containing material exuded by bacteria. This is termed the *glycocalyx*. The bacteria then grow and divide within the glycocalyx, form microcolonies, and eventually form a confluent biofilm. [*See* GLYCOCALYX, BACTERIAL.]

The attachment of bacteria to a surface is influenced by many factors, including nutrient availability, nutrient concentration, pH, temperature, electrolyte concentration, the flux of materials, and the surface itself. The numbers and types of bacteria that attach depend on many of the physiochemical properties of the surface, including the electrostatic charge, the surface free energy, hydrophobicity, and surface texture/roughness. Surfaces are frequently considered to fall into two categories:

1. The high surface energy materials that are hydrophilic, often carrying a negative charge (e.g., glass, metals, or minerals)
2. The low surface energy materials that are more hydrophobic with a low charge (either positive or negative). The important point is that the charge is low. Such surfaces include organic polymers such as plastics.

The higher the surface energy and activity, the more readily these surfaces adsorb dissolved solutes or nutrients, and the rate of bacterial colonization of a surface is affected by, among other things, the amount of nutrients at the surface. [*See* ADHESION, BACTERIAL.]

The growth of a biofilm has a profound effect on the surface in that before the biofilm forms, the surface is often homogeneous, but bacterial activity during colonization of the surface produces heterogeneous conditions. This is considered to occur as follows: One or more bacteria are attracted to the surface by the events already described. The primary colonizer attaches and by growth and metabolic activity produces a microenvironment around the recently established microcolony, which differs from the rest of the surface (Fig. 1). The primary colonizer therefore alters the surface from a homogeneous to a heterogeneous environment and in doing so may produce conditions that are favorable to attract other bacterial species. In this manner a succession of colonizers can be attracted to the surface and form biofilm until a series of complex communities results with some of the sessile bacteria using the metabolic by-products or supplying essential growth factors for neighboring cells. The glycocalyx holds the sessile bacteria closely together, and this allows consortia to develop within the biofilm.

The biofilm does not increase in size indefinitely as there are factors that bring about the removal of sessile cells or sections of the biofilm. This detachment from biofilms describes the movement or loss of cells from the biofilm into the bulk liquid, which can be the result of erosion, sloughing, or abrasion processes. Erosion results in a continuous loss of cells from the biofilm, the rate of which depends on the biofilm thickness, fluid shear stress and fluid

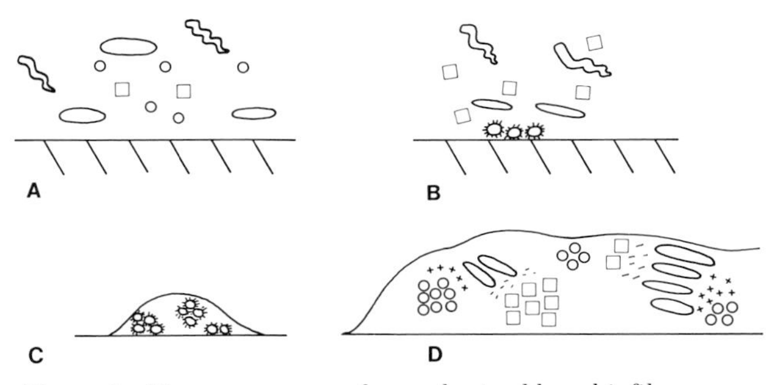

Figure 1 Homogeneous surface colonized by a biofilm produces heterogeneity at the surface. (A) Planktonic bacteria can be attracted to a homogeneous surface. (B) Primary colonizing bacteria attach to surface. (C) Growth of the primary colonizer produces a microenvironment around the microcolony, which differs from the rest of the surface. This can attract and create conditions favorable for other bacteria to colonize the surface. (D) Eventually a mature biofilm develops with many complex events occurring within it. This produces heterogeneous conditions from a homogeneous surface.

velocity. Sloughing involves a large loss of material from the biofilm, so it is a rapid loss. This process has been observed in situations in which a thick biofilm has developed. Abrasion losses occur when an object repeatedly collides with the biofilm.

Biofilms can vary in thickness from a monocell layer to 8–10 cm thick, although they are usually thought of as being about 100 μm thick. Biofilms can be divided into different components including the surface, the base or lower biofilm, the surface biofilms, the aqueous phase, and the gas phase above the liquid. Each component has its own properties in terms of the transfer of nutrients and transport processes.

II. Advantages of the Sessile Growth Mode

In most aquatic ecosystems, more bacteria grow as sessile cells than as planktonic cells. The following advantages to sessile bacteria conferred by this growth mode are summarized in Fig. 2:

1. Protection from antimicrobial agents.
2. Increased availability of nutrients for growth.
3. Increased binding of water molecules, reducing the possibility of desiccation.
4. Establishment of complex consortia, which enhances advantages 1–3.
5. Greater phenotyphic plasticity.
6. Gradation of metabolic activity.
7. Proximity to progeny and other bacteria, facilitating plasmid transfer.

A. Protection from Antimicrobial Agents

In every instance in which a comparative study has been undertaken, the sessile cells were not eradicated by antibacterial agents that successfully controlled their planktonic counterparts. For example, tobramycin killed all the planktonic cells of *Pseudomonas aeruginosa,* yet even when applied at much higher concentrations, did not eradicate a sessile population of the same isolate.

Biofilm formation has been repeatedly demonstrated to protect the sessile cells from most antimicrobial agents including antibiotics, biocides, and host defense mechanisms. The manner of protection was initially considered to be the physical barrier provided by the exopolysaccharide glycocalyx, which did not permit the antimicrobial agents to reach the sessile bacteria. However, data from two noninvasive techniques have helped to create a different concept of biofilms and sessile bacteria. The techniques involved confocal scanning laser microscopy (CSLM) and Fourier transmission infrared spectroscopy (FTIR). CSLM was used by D. E. Caldwell (University of Saskatchewan) and colleagues to monitor the events of biofilm formation. The data from CSLM showed that, rather than being

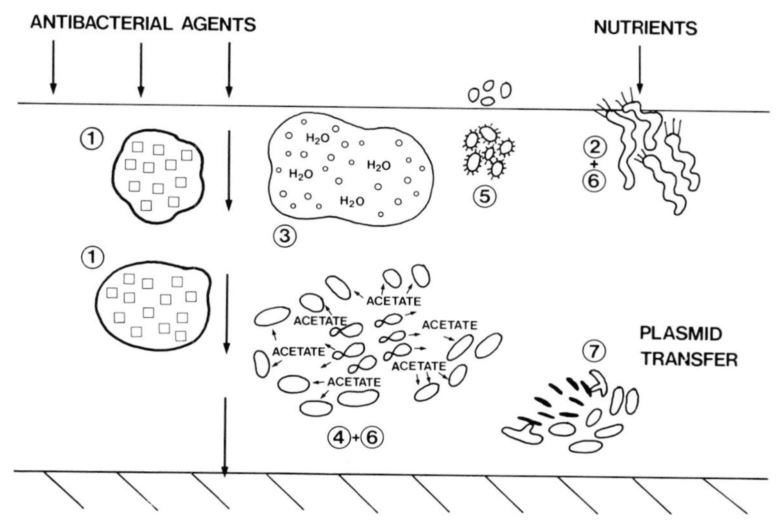

Figure 2 Summary of some of the advantages to living as sessile bacteria and not as planktonic cells. See text for details.

evenly dispersed throughout the biofilm, the sessile bacteria are found in microcolonies. A sudy undertaken by Jana Jass and colleagues at the University of Calgary using FTIR showed that antibiotics can rapidly penetrate through a biofilm and reach the surface below the film. By putting the data from both studies together, it was possible to develop a clearer image of the distribution of sessile cells with the biofilm, which indicates the sessile cells are located in ''clumps'' or microcolonies that the antibacterial agents cannot penetrate effectively (Fig. 3). In this manner, an antibacterial agent applied to the biofilm could be effective against the planktonic population and some of the sessile cells on the outer edges of the microcolonies, but the inner cells could remain viable even after the antibacterial treatment had ceased (Fig. 3).

Other investigators, including Professor M. R. W. Brown and Dr. W. Nicholls, have offered alternative explanations for the different responses of planktonic and sessile bacteria to antibacterial agents. These include the fact that the sessile bacteria are physiologically different from planktonic bacteria (e.g., they have different growth rates), and these differences account for the different susceptibilities of the cells. It should be noted that this subject is not as yet fully resolved and that further investigations are underway to clarify the precise mechanisms that afford sessile cells this protection, but the basic phenomenon of protected cells within biofilms is well-established.

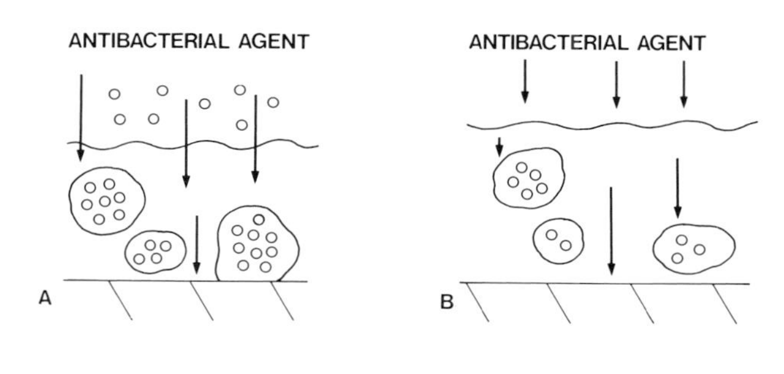

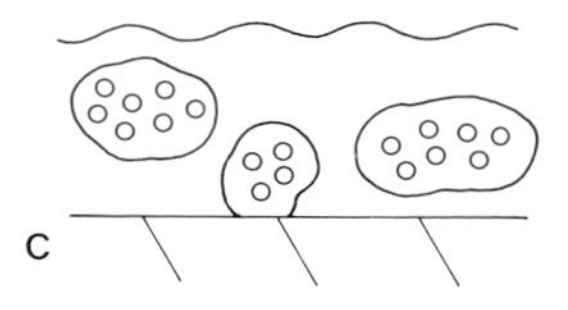

Figure 3 Responses of sessile and planktonic bacteria to antibacterial agents. (A) Planktonic and sessile bacteria are treated with an antibacterial agent. (B) Planktonic cells and outer sessile cells within microcolonies are killed, but inner cells remain viable. (C) Once antibacterial treatment stops, viable sessile cells can grow and divide again. Treatment has only effectively controlled the planktonic population as the sessile cells recover. Some sessile cells detach and become planktonic cells.

B. Concentration of Nutrients in the Biofilm

The favorable environment within the biofilm is considered to include the concentration of available nutrients. The chemical nature of the polyanionic exopolysaccharide matrix surrounding the adherent bacteria allows it to act like an ion exchange column concentrating nutrients and ions, particularly cations from the surrounding fluid. This provides nutrients for the sessile bacteria, which may help them to survive when bulk fluid conditions around the biofilm are not conducive for planktonic growth.

C. Binding Water Molecules

In some ecosystems (e.g., riverbeds, streams, and soils), the available water can decrease or become insufficient to support planktonic growth. However, the highly hydrated glycocalyx, which binds water molecules within the biofilm, protects the sessile cells and reduces the effects of desiccation to which the planktonic cells are subjected.

D. Formation of Consortia within Biofilms

In a mixed culture biofilm, an exchange of metabolites between different bacterial species in the biofilm can occur, as can the removal of toxins by some sessile cells. The physical enclosure of the biofilm allows the recycling of substances produced by the death and lysis of sessile cells, in that the neighboring viable sessile cells can use these cell products as growth substrates. Other interactions that can occur within the biofilm, which may not occur as readily in the planktonic phase, are that, again as a consequence of the close proximity of sessile cells, they may form mixed communities that work together to degrade some compounds to use as a carbon and/or energy source. In a few instances, biofilms form on a surface that is also a substrate (e.g., cellulose). The microbial degradation of cellulose is complex, requiring the combined metabolic capabilities of several different microorganisms, and by forming biofilms, all the microbes required for the degradation can be in close proximity with the cellulose and with each other so that degradation can occur at a rapid rate. [*See* BIODEGRADATION; BIOREMEDIATION.]

E. Phenotypic Plasticity of Bacteria

Bacteria can respond to changes in environmental conditions by altering their phenotypical characteristics (termed *phenotypic plasticity* by M. R. W. Brown). He reported observing changes in bacterial cell envelopes obtained *in vivo* and after culturing in the laboratory. Other reports of changes in cell wall structures relevant to adhesion included changes in the production of exopolysaccharides depending on the availability of carbon. Under nutrient-rich conditions in the laboratory, mucoid strains of some bacteria can become nonmucoid. Some bacteria respond by no longer producing pili, changing their metabolism, or changing the composition of their outer cell membrane.

F. Differences in Metabolic Activity

Work recently undertaken by M. Fletcher at the University of Maryland demonstrated that bacteria attached to a surface are more metabolically active than planktonic bacteria. The metabolic activity was assessed by measuring the assimilation of glucose or leucine by the attached and planktonic bacteria. These data are difficult to interpret as it is not yet fully understood whether the bacteria that have all the nutrients they require are more or less metabolically active than those that do not have their requirements met.

G. Transfer of Plasmid DNA

The close proximity of bacterial cells within the biofilm may allow plasmid transfer to occur more readily than when the bacteria are in the planktonic phase. This is advantageous, as the transferred DNA may carry some useful capabilities that increase the versatility of the recipient. [*See* PLASMIDS.]

III. Biofilms in Medical Systems

Advancing technology has resulted in the emergence of the bionic human. A large number of implantable devices made from a variety of inert materials are used each day in our hospitals, including prosthetic hips made from inert metals and silastic devices used for smaller joints such as the fingers. Indeed, it is estimated that currently between one-quarter and one-half million joint replacements are carried out worldwide each year. Other implantable devices include heart valves, ocular lenses, vascular grafts, and mammary implants, in addition to the screws, pins, and plates commonly used in orthopedic surgery.

Many more devices are used in the short-term treatment of patients in our hospitals. For example, at any one time 10% of all patients in a general hospital will have an indwelling urinary catheter and about 20% of patients in such hospitals will have one or more of their veins accessed using Teflon or polyvinyl chloride, 6–8-cm-long catheters. Some patients will have much longer catheters permanently implanted into their venous system, such as Hickmann catheters and port-a-caths, when repeated access to the venous system is necessary for the administration of chemotherapeutic or antimicrobial agents. Polyvinyl chloride catheters are placed into the peritoneal cavity for dialysis to sustain life in patients who have renal failure, and an endotracheal tube is used to provide a connection to a ventilator so that patients with respiratory failure can have life-sustaining therapy.

The materials from which these devices are made vary but include vitallium (cobalt-chromemolybdenum alloy), titanium, stainless steel, polyethylene, polymethyl methacrylate, silicone rubber, polyethylene terephthalate (Dacron), polytetrafluoroethylene (Teflon), and polyvinyl chloride.

Infection is one of the major limitations in our successful use of implantable and short-term therapeutic devices. The infection rate for implantable devices is low, ranging from 0.5% to 1%; however, the consequences of such infection are drastic, and because the causative bacteria frequently form extensive biofilms on the device surfaces (Fig. 4), the devices must be removed before the infection can be eradicated. This results in major morbidity and, occasionally, mortality.

Devices such as urinary catheters, intrauterine devices, and intravenous catheters (Fig. 5) transverse various surfaces of the body that have a rich flora, and they develop extensive biofilms on these surfaces soon after emplacement. The skin has an especially rich microbial flora (Fig. 6) that tends to be dominated by *Staphylococcus epidermidis*. Because the resident microorganisms attach to the device and gain entry to a sterile area of the body, infection associated with these devices is much more common than that associated with implantable devices. In most instances, infections associated with these short-term therapeutic devices can be easily treated by removal of the device and treatment with appropriate antibiotics. On occasion, however, these infections can lead to serious complications, such as *Staphylococcus aureus* infection of an intravenous catheter, which may lead to bacteria entering the bloodstream and infecting a heart valve (endocarditis).

All the foregoing principles of the formation of bacterial biofilm and the protection from antimicrobial agents apply to these medical devices. A number of strategies have been employed to try to combat this formidable problem. Antibiotics have been bonded to the implantable device with limited success. At the time of insertion of many of these devices, especially prosthetic joints and heart valves, antibiotics are given prophylactically to reduce the

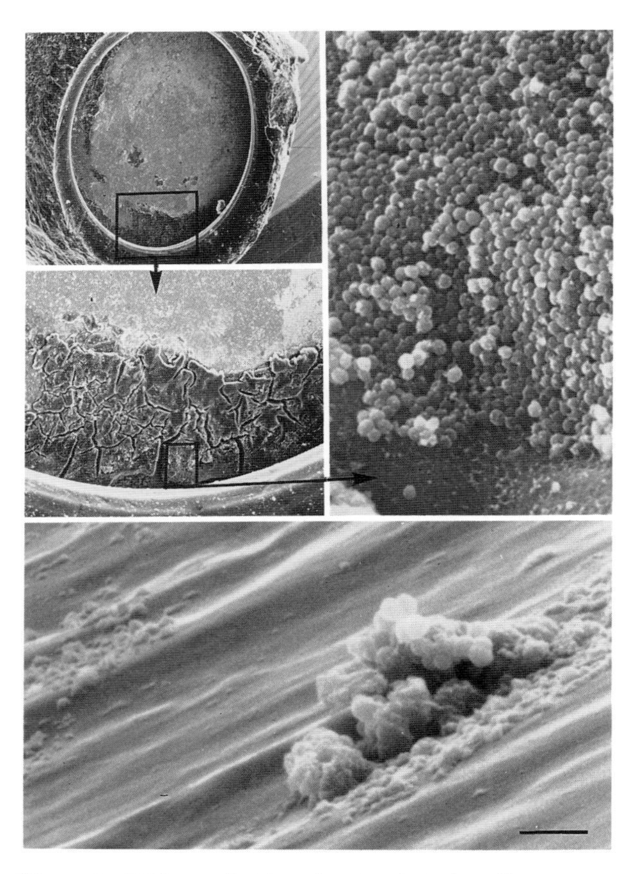

Figure 4 SEM examination of an implanted cardiac pacemaker that had become the focus of a chronic bacterial biofilm infection that was not resolved even by aggressive antiobitic therapy. Upper three panels (follow arrows for increasing magnification) show position and structure of a thick biofilms composed of spherical cells of *Staphylococcus aureus*, and lower panel shows a discrete microcolony of these spherical cells whose enveloping glycocalyx has radically condensed during dehydration for SEM. Bar = 5 μm.

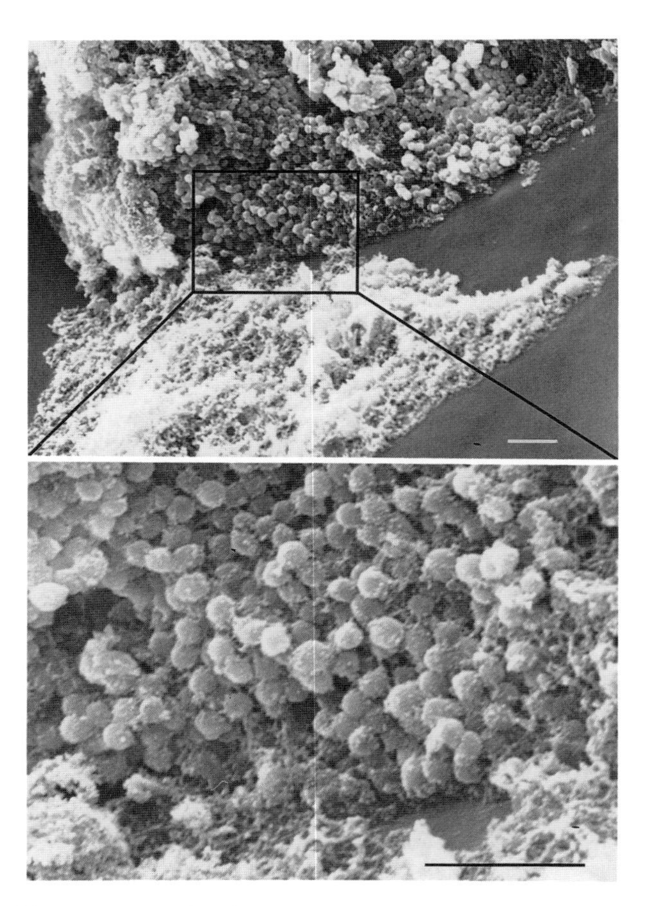

Figure 5 SEM of surface of subcutaneous area of a transcutaneous catheter (Tenckhoff), showing, at low (*upper panel*) and high (*lower panel*) magnifications, development of an extensive biofilm within which spherical bacterial cells are embedded in a glycocalyx matrix whose fibrous residues are seen after dehydration. Bars = 5 μm.

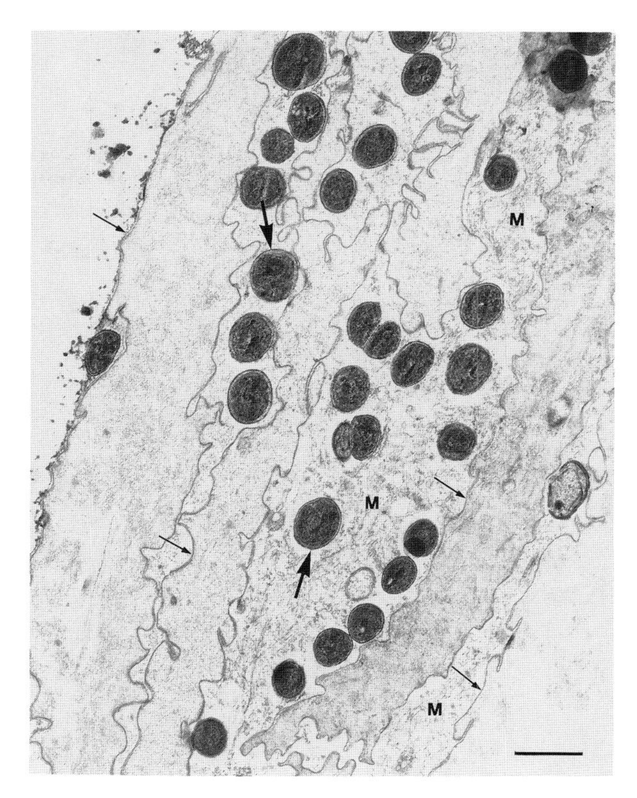

Figure 6 TEM of a ruthenium red–stained preparation of human skin showing presence of large numbers of gram-positive bacterial cells (*large arrows*) embedded in a glycocalyx matrix (M) in spaces between squamous epithelial cells (*small arrows*). These natural microbial inhabitants of the skin are not all killed by surgical preparation, and they rapidly initiate colonization of transcutaneous medical devices. Bar = 1.0 μm.

rate of wound infection and secondary infection of the prosthesis and also to try to eliminate any bacteria that may have contaminated the prosthesis during its insertion. The rationale is that if antibiotics are used at this early stage when a small numer of bacteria are present but before they establish a biofilm, infection can be prevented.

IV. Biofilms in Industrial Systems

The division of biofilm effects into medical and industrial categories is useful but is somewhat anthrocentric, because bacteria employ the same survival strategies wherever they find themselves. When natural waters are diverted to service the heat exchange requirements of a plant or a refinery, bacteria form biofilms on metal surfaces just as they would form these sessile populations on rocks in the original stream. These biofilms have a remarkable ability to insulate against heat exchange (Fig. 7), to reduce fluid flow (Fig. 7C), and to harbor waterborne potential pathogens (Fig. 7A). Bill Characklis' group (Montana State University) described the kinetics of this biofouling process, and it is now clear that this process occurs inevitably in all flowing systems where physical conditions (e.g., temperature) and nutrient conditions are permissive of bacterial growth. This general problem of biofilm formation in industrial water systems is virtually universal, from the gross biofouling of ship bottoms and heat exchangers to the delicate fouling of computer microchips, and it is sufficiently troublesome to cause the expenditure of more than $7 billion U.S. dollars for biocides for its control. Biocide concentrations that kill planktonic bacteria always fail to kill biofilm organisms, and concentrations 50–500 times the planktonic minimal bactericidal concentration (MBC) are usually required to eliminate these adherent populations. Because biocide efficacy is usually assessed by monitoring live planktonic bacteria, biocides are often perceived to be "successful" at MBC concentrations, and only rapid "regrowth" indicates that the biofilm populations have escaped unscathed and have rapidly repopulated the whole system.

It is in the corrosion of metals in industrial systems that we have perhaps discovered the epitome of bacterial biofilm physiology. When bacteria grow as discrete microcolonies within biofilms, each microcolony develops its own microenvironment. Each microenvironment may be acid, if acid metabolites are produced, or rich in specific metal cations, if the glycocalyces of the component cells attract these cations. Local differences in pH, Eh, and ion and metabolite concentrations can readily produce large ionic gradients between adjacent loci on a metallic surface that then become effective anodes and cathodes (Fig. 7B). Consequent metal loss at the anode may be so severe that it can cause "pitting" that will perforate a $\frac{5}{8}$-inch steel pipe in less than 6 months. Cathodic protection is routinely applied in ship and pipeline construction to override these local anodes and cathodes and thus prevent microbial corrosion. When we think

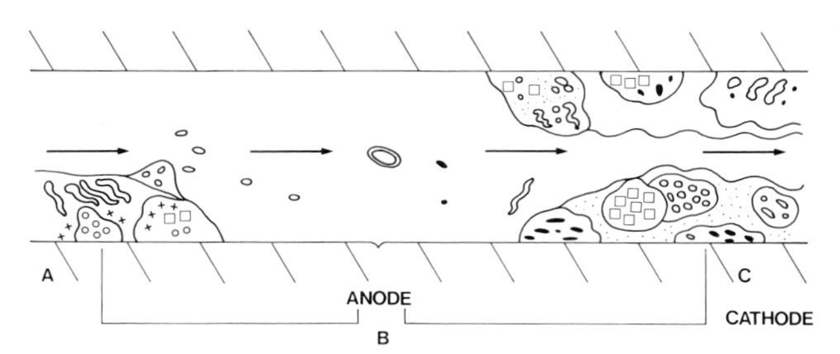

Figure 7 Summary of ways in which biofilms can be harmful in industrial water systems. (A) Biofilm can act as a reservoir of potential pathogens. (B) Corrosion cells can develop within the biofilm when bacterial metabolites and biopolymers with differing charges and composition are being produced by adjacent microcolonies. (C) Biofilm can reduce efficiency of the surface. For example, if the colonized surface is involved in heat exchange, biofilm reduces efficiency of heat transfer because static water in and around the biofilm insulates the surface. The pipe is no longer as efficient in transporting the water because the thick film effectively reduces the volume of water that the pipe can carry.

about bacterial biofilms, we should keep these biological corrosion cells firmly in mind, because they illustrate the remarkable physiological heterogeneity that can develop and be sustained when microcolonies of specific bacteria live in approximate juxtaposition in a highly hydrated, water-permeable matrix with well-developed ion-exchange characteristics. [*See* CORROSION, MICROBIAL.]

It is in biofilms that prokaryotic cells, which seem relatively simple when examined as planktonic cells in single-species cultures, show levels of physiological specialization and sophisticated cooperation that we usually associate with multicellular eukaryotic organisms.

Bibliography

Costerton, J. W., Chen, K.-J., Geesey, G. G., Ladd, T. I., Nickel, J. C., Dasgupta, M., and Marrie, T. J. (1987). *Annu. Rev. Microbiol.* **41,** 435–464.

Costerton, J. W., and Lappin-Scott, H. M. (1989). *ASM News* **55,** 650–654.

Lappin-Scott, H. M., and Costerton, J. W. (1989). *Biofouling* **1,** 323–342.

Marshall, K. C. (1985). Mechanisms of bacterial adhesion at solid-water interfaces. *In* "Bacteral Adhesion. Mechanisms and Physiological Significance" (D. C. Savage and M. Fletcher, eds.), pp. 131–161. Plenum Press, New York, London.

Mittelman, M. W., and Geesey, G. G. (eds.) (1987). "Biological Fouling of Industrial Water Systems: A Problem Solving Approach." Water Micro Associates, San Diego.

Biological Nitrogen Fixation

Gerald H. Elkan
North Carolina State University

Glossary

Biological nitrogen fixation Reduction of atmospheric N_2 (dinitrogen) to a metabolizable active form solely by prokaryotic organisms (eubacteria and archaebacteria)

Cyanobacteria Photosynthetic prokaryotes of some species that can fix nitrogen both free living and symbiotically; found symbiotically with a broad range of lower plant taxa

Frankia Symbiotic, nitrogen-fixing actinomycetes; hosts are found over a broad spectrum of shrubs and trees

***nif* gene** Bacterial gene coding for the nitrogenase enzyme directly (i.e., *nif*HDK), or genes involved in the regulation of nitrogenase

Nitrogenase Highly conserved enzyme complex that catalyzes the conversion of atmospheric dinitrogen to ammonia; nitrogenase consists of two components: (1) dinitrogenase reductase (an iron protein) and (2) dinitrogenase (an iron molybdenum protein)

***nod* genes** Bacterial genes in rhizobia coding for development of the nodulation process on the proper legume host; there are two types: a common nod region, which consists of a structurally and functionally conserved cluster of genes, and host-specific nodulation genes, which cannot complement nodulation defects in other species

Rhizobia Diverse group of nitrogen-fixing bacteria that can establish symbiotic relationships with a limited number of host legumes; the range and effectiveness of the symbiosis depends on a genetically regulated, mutual recognition between bacteria and host

BIOLOGICAL NITROGEN FIXATION (BNF) refers to that property of some taxa of eubacteria or archaebacteria to enzymatically reduce atmospheric N_2 to ammonia. The ammonia produced can then be incorporated by means of other enzymes, into cellular protoplasm. Biological nitrogen fixation has only been observed in procaryotic cells.

I. Significance of Biological Nitrogen Fixation

Nitrogen is a unique, required input in that virtually no naturally available nitrogen is present in mineral form. Yet, above every hectare of soil, at sea level, there are 78 million kg of inert N_2 (dinitrogen) gas. Plants, as eukaryotic autotrophs, need an oxidized or reduced form of N_2 for anabolism. Only a few prokaryotic organisms can "fix" N_2 directly, a technology that humans have learned to mimic using the Haber–Bosch process. World chemical production of NH_3 in 1990 was approximately 80 million mt, three-fourths of which was available for fertilizer.

Thus, humans produced roughly 60 million mt of nitrogenous fertilizer while annual BNF is estimated to be around 175 million mt/yr. As shown in Table I, BNF occurs in every natural environment including the sea. The energy cost for nitrogen fixation is significant. Currently, 1–2% of world fossil energy is used for the production of fertilizer nitrogen, while Hardy (1980) estimates that about 1–2 billion tons of plant carbohydrate (from photosynthesis) may be consumed in BNF.

There are many genetically diverse eubacteria representing 27 families and 80 genera plus 3 thermophilic genera of Archaebacteria. Except from the cyanobacteria (which are listed in Table V), these nitrogen-fixing families and genera are listed in Table II. With the exception of the Azotobacteraceae, there is no genus (or family) whose species are all nitrogen-fixing. The potential amount of nitrogen fixed by these bacteria depends heavily on the ecosystem in which the organisms operate, as will be

Table I Estimated Annual Amount of N_2 Fixed Biologically in Various Systems

System	N_2 fixed/mt/year × 10^6
Legumes	35
Nonlegumes	9
Permanent grassland	45
Forest and woodland	40
Unused land	10
Total land	139
Sea	36
Total	175

Table II Families and Genera of Nitrogen-Fixing Eubacteria Excluding Cyanobacteria

Family genus	Family genus
Acetobacteriaceae	Methanomonadaceae
Acetobacter	*Methylobacter*
Azotobacteraceae	*Methylococcus*
Azomonas	*Methylocystis*
Azotobacter	*Methylomonas*
Azotococcus	*Methylosinus*
Beijerinckia	Pseudomonadaceae
Derxia	*Pseudomonas*
Xanthobacter	Rhizobiaceae
Bacillaceae	*Azorhizobium*
Bacillus	*Bradyrhizobium*
Clostridium	*Rhizobium*
Desulfotomaculum	*Sinorhizobium*
Baggiatoaceae	Rhodospirillaceae
Baggiatoa	*Rhodomicrobium*
Thiothrix	*Rhodopseudomonas*
Vitreoscilla	*Rhodospirillum*
Chlorobiaceae	Spirillaceae
Chlorobium	*Aquaspirillum*
Pelodictyron	*Azospirillum*
Chloroflexaceae	*Campylobacter*
Chloroflexus	*Herbaspirillum*
Chromatiaceae	Streptomycetaceae
Amoebobacter	*Frankia*
Chromatium	Thiobacteriaceae
Ectothiorhodospira	*Thiobacillus*
Thiocapsa	Vibrionaceae
Thiocystis	*Vibrio*
Corynebacteriaceae	Uncertain family
Anthrobacter	*Alcaligenes*
Enterobacteriaceae	*Desulfovibrio*
Citrobacter	
Enterobacter	
Erwinia	
Escherichia	
Klebsiella	

discussed in the following sections. Amounts of nitrogen fixed range from trace amounts for some free-living soil bacteria to 584 kg/ha/yr for the leguminous tropical tree *Leucaena*.

II. The Biological Nitrogen Fixation Process

Dinitrogen, a triple-bonded molecule, is quite stable, requiring much energy to break this bond and reduce the N≡N to ammonia in an exothermic reaction: $3\ H_2 + N_2 \rightarrow 2\ NH_3$. This can be accomplished commercially by the Haber–Bosch process or by prokaryotic organisms using adenosine triphosphate (ATP) as energy to initiate the bond-breaking reaction. The prokaryotes able to fix nitrogen are extremely diverse, representing autotrophs, heterotrophs, aerobes, anaerobes, photosynthetics, single-celled, filamentous, free-living, and symbiotic. Phylogenetically, these organisms are quite heterogeneous and yet the nitrogen-fixing process and enzyme system is similar in all of these organisms and depends on the following: (1) nitrogenase enzyme complex, (2) a high-energy requirement and availability (ATP), (3) anaerobic conditions for nitrogenase activity, and (4) a strong reductant.

Nitrogenase has been purified from all known eubacterial nitrogen-fixing organisms. It consists of two components: (1) dinitrogenase reductase (an iron protein) and (2) dinitrogenase (an iron-molybdenum protein). Both enzymes are needed for activity. Nitrogenase reduces N_2 and H^+ simultaneously using ±75% of the ATP in the reduction of N_2 and ±25% in H^+ reduction. A key characteristic of this enzyme complex is that both components are quickly and irreversibly lost by interacting with free oxygen, regardless of the oxygen requirements of the microbe. The protein usually has a molecular mass of c. 57–72 consisting of two identical subunits coded for by the *nif*H gene. This is a highly *conserved* protein. The iron-molybdenum protein has a molecular mass of about 220 and has four subunits, which are pairs of two different types. The α-subunit is coded for by the *nif*D gene and is 50 kDa in size. The other subunit (β) is coded for by the *nif*K gene and has a mass of 60. These enzymes have been sequenced and demonstrate considerable homology. The nitrogenase complex is big and has been measured as amounting up to 30% of the total cell protein.

Theoretically, as much as 28 moles of ATP are consumed in the reduction of 1 mole of N_2. Depending on the method used, and the nature of the organism in question, *in vitro* studies of energy requirements vary from a minimum of 12–15 moles to 29 moles of ATP/N_2. This energy is not only required for the reduction process but also to maintain the anaerobic requirement needed for the reaction.

Various strategies are used to exclude oxygen from the reaction in nonanaerobic bacteria. Facultative organisms fix nitrogen only under anaerobic conditions. Aerobic organisms exhibit a wide range of methods for protecting the enzyme complex from oxygen. Many grow under microaerophilic conditions accomplished by scavenging free oxygen for metabolism or sharing the ecosystem with other organisms that consume the excess O_2. Some evidence indicates that some free-living aerobes such as *Azotobacter* can change the conformation of nitrogenase protein to a form less oxygen-sensitive. Photosynthetic aerobes such as cyanobacteria can form special cells called heterocysts, which are the site of BNF, and which lack the O_2-evolving mechanism that is part of photosynthesis. The best-known protective mechanisms are found in symbiotic BNF associations (i.e., *Rhizobium*–legume) where the nitrogenase system in the endophyte is protected from excess oxygen by a nodule component—leghemogloblin. Nonleguminous nodules (with *Frankia* or cyanobacteria, etc.) are probably protected by other, as-yet undescribed, oxygen-restrictive mechanisms. There are other protective mechanisms as well, but all of these require diversion of energy from the BNF process itself to maintain a favorable environment for fixation.

In addition to the limitations to BNF caused by free oxygen, generally the presence of combined nitrogen, such as ammonia or nitrate, in the environment will inhibit N_2 fixation. At present, this makes it impractical to mix BNF with chemical N_2 fertilizer for crop production.

III. Free-Living, Nitrogen-Fixing Bacteria

There are many diverse nitrogen-fixing prokaryotes, presently representing 19 families and 48 genera of eubacteria (excluding the cyanobacteria) as well as three thermophilic genera of archaebacteria. Most of these genera fix nitrogen as free-living diazotrophs. These were first described by Winogradsky in 1893 (*Clostridium pasteurianum*), and soon Biejerinck in 1901 (*Azobacter*). Discovery of other free-living, nitrogen-fixing bacteria lagged until the availability of ^{15}N isotopic and acetylene reduction techniques became common. Thus, most of the free-living, nitrogen-fixing bacteria listed in "Bergey's Manual of Systematic Bacteriology" have been described recently using these techniques. It is generally accepted that diazotrophs obtain their carbon and energy supplies from root exudates and lysates, sloughed plant cell debris, and organic residues in soil and water. They are found "completely free living" or in loose associations as a result of root or rhizosphere colonization (the associative bacteria will be discussed separately). The quantity of nitrogen fixed is a matter of some controversy. Russian workers have estimated that *C. pasteurianum* or *Azotobacter* contributes perhaps 0.3 kg N_2/ha/year, compared with about 1000 times that amount in a good leguminous association. Associative organisms such as *Azospirillum* have been estimated to fix from trace amounts to 36 kg/ha. The limitations of the terrestrial BNF system is due not only to the difficulty in obtaining sufficient energy and reductant but also because of the need to divert substrate for respiratory protection of nitrogenase. The cyanobacteria can overcome the environmental constraints faced by most other nitrogen fixers. Being photosynthetically active, these prokaryotes fix CO_2 and are, thus, independent of external energy needs. The families and genera of nitrogen-fixing cyanobacteria and listed in Table III.

Table III Families and Genera of Cyanobacteria with Nitrogen-Fixing Species

Family genus	Family genus
Chroococcaceae	Oscillatoriaceae
Chlorogloea	*Lyngbya*
Chroococcidiopis	*Microcoleus*
Gloeothece	Oscillatoria
Synechococcus	*Phormidium*
Mastigocladaceae	*Plectonema*
Mastigocladus	Pleurocapsaceae
Michrochaetaceae	*Pleurocapsa*
Michrochaete	Rivulariaceae
Nistocaceae	*Calothrix*
Anabaena	*Dichothrix*
Anabaenopsis	*Gleotrichia*
Aphanizomenon	Scytonemataceae
Aulosira	*Scytonema*
Cylindrosperumum	*Tolypothrix*
Nodularia	Stigonemataceae
Nostoc	*Fischerella*
Pseudanabaena	*Hapalosiphon*
Raphidiopsis	*Stigonema*
Richelia	*Westelliopsis*

The free-living cyanobacteria are distributed widely in humid and arid tropical surface soils. Extensive studies, especially under paddy conditions, have been conducted in India, Japan, and the Philippines. Of 308 isolates from Philippine paddy soils, most were identified as *Nostoc* or *Anabaena*. Reports of fixation rates ranged from 3.2 to 10.9 kg/N_2/ha/yr. Reports of 15–20 kg/N_2/ha/yr in rice on the Ivory Coast, 44 kg/N_2/ha/yr in Lake George in Uganda, and 80 kg/N_2/ha/yr in paddy fields in India have been published. Under temperate conditions, BNF by cyanobacteria has been a major problem in eutrophication of lakes.

About 125 strains of free-living cyanobacteria represent families and 31 genera in all taxonomic groups have been shown to fix N_2, but the extent to which fixation occurs and conditions needed for fixation vary greatly. There are basically four types of N_2-fixing cyanobacteria: heterocystous filamentous, nonheterocystous filamentous, unicellular reproducing by binary fission or budding, and unicellular reproducing by multiple fission. Heterocysts are thickened, specialized cells occurring at a regular intervals in some filamentous cyanobacteria. These cells lack the oxygen-evolving component of the photosynthetic apparatus. Heterocysts appear to be the only cells capable of fixing N_2 aerobically as well as anaerobically. The main function of the heterocyst seems to be to fix N_2 serving as a protective mechanism against O_2 and, because there is little or no photosynthesis in the heterocyst, all of the energy translocated is available for N_2 fixation. The other groups of cyanobacteria need to be examined as well because they appear quite active as fixers under anaerobic conditions, especially with paddy rice.

One should not overlook the potential of BNF in unique environments. There are many reports of associations between N_2-fixing bacteria and animals (e.g., termites, ruminants). Because *nif* genes can be transferred and expressed between the enteric bacteria *Klebsiella* and *Escherichia coli*, it may be possible to stimulate N fixation by rumen bacteria. Perhaps ultimately in ruminants, it may be practical to substitute engineered enteric diazotrophs for the plant protein now required by these animals.

IV. Associative Nitrogen-Fixing Bacteria

Within the past two decades, another type of plant–bacteria interaction has been described as "associative." This interaction was shown to result from the adhesion of the bacteria to the root surfaces of wheat, corn, sorghum, and other grasses. The major associative N_2-fixing systems currently described are summarized in Table IV. *In vitro* studies show that many of these bacteria can achieve high rates of N_2 fixation under optimum conditions. However, in the rhizosphere, the ability to survive, grow, and colonize plant roots are preconditions presently limiting the potential for N_2 fixation. The characteristics required for an organism to flourish in the rhizosphere are (1) ability to withstand the changing physical and chemical soil environments, (2) ability to grow well and obtain all needed energy from carbon and mineral supplies in the root zone, and (3) to compete successfully with other rhizosphere organisms for the limited energy and nutrients available. Estimates of *in vivo* fixation are extremely variable and give rise to a recurring question as to whether or not energy substrates in the rhizosphere are sufficient to support growth and N_2 fixation at sufficient

Table IV The Major Associative Diazotrophs

Plant species	Principle microorganism
Rice	
Oryza sativa	*Achromobacter*, Enterobacteriaceae, *Azospirillum brasilense*
Sugarcane	
Saccharum spp.	*Azotobacter*, *Beijerinckia*, *Bacillus*, *Klebsiella*, *Derxia*, *Vihtrio*, *Azospirillum*, Enterobacteriaceae, *Bacillaceae*
Pearl Millet and sorghum	
Pennisetum purpureum	*Azospirillum*, *Bacillus polymyxa*, *Klebsiella*, *Azotobacter*, *Derxia*, *Enterobacter*
Maize	
Zea mays	*Azospirillum lipoferum*, *Azotobacter vinelandii*
Grasses	
Paspalum notatum var. *batatais*	*Azotobacter paspali*
Panicum maximum	*Azospirillum lipoferum*
Cynodon dactylon	*Azospirillum lipoferum*
Digitaria decumbens	*Azospirillum lipoferum*
Pennisetum purpureum	*Azospirillum lipoferum*
Spartina alterniflora Loisel	*Campylobacter*
Wheat	
Triticum spp.	*Bacillus polymyxa*

levels by these associative bacteria. The nitrogenase system is repressed by bound nitrogen. Therefore, in the presence of nitrogen fertilizer BNF by free-living or associative bacteria would be diminished. Thus, a mixed BNF–nitrogen fertilizer system would not work under field conditions unless the organisms could be engineered so that the nitrogenase would be derepressed when bound nitrogen is present. [*See* RHIZOSPHERE.]

V. Symbiotic Nitrogen-Fixing Bacteria

Only three groups of nitrogen-fixing bacteria have evolved mutually beneficial symbiotic associations with higher plants: (1) the filamentous bacteria *Frankia*, forming root nodules with a number of plants such as alder, *Purshia*, Russian olive, etc., (2) heterocystous cyanobacterium with a number of diverse hosts from *Cycads* to *Azolla*, and (3) *Rhizobium* with legumes.

A. *Frankia*

The genera of dicotyledonous plants shown to be nodulated by *Actinorhizae* (*Frankia*) are listed in Table V. To date, these include 178 species in 20 genera representing eight families and seven orders of plants. There is no obvious taxonomic pattern among these hosts. Most of the described species are shrubs or trees and are found in temperate climates, but they have a wide growth range and could be grown in the tropics. In fact, *Purshia tridenta* already is an important rangeland forage crop in Africa, and other *Purshia* species are harvested for firewood. *Casuarina*, a vigorous nitrogen fixer, has been planted in Thailand, where it can be harvested for construction lumber after 5 yr of growth. Almost all the hosts are woody, ranging from small shrubs to medium-sized trees.

Frankia, the nitrogen-fixing endophyte found in nodules of these nonleguminous plants, is a genus of prokaryotic bacteria closely related to *Actinomycetes*. In pure culture, these endophytes behave as microaerophilic, mesophilic, heterotrophic organisms, usually with septate hyphae that develop sporangia. Isolates vary morphologically and nutritionally. Most strains can fix atmospheric nitrogen in pure culture. Nitrogenase genes are highly conserved, and *Frankia* nitrogenase enzymes closely resemble those of other nitrogen-fixing bacteria. *Frankia* were first reliably isolated in 1978. The organism grows slowly, requiring 4–8 wk for visible colonies to be found in culture. *Frankia* are similar to other aerobic *Actinomycetes*, producing a septate, filamentous mycelium that can differentiate into sporangia and vesicles. Cells are routinely gram-positive, but unlike other gram-positive bacteria, *Frankia* has a discontinuous membraneous layer. Molecular methods for taxonomy such as DNA/DNA homology have demonstrated considerable genetic diversity between isolates within this family, but only a limited number of isolates have been analyzed thus far.

Table V Genera of Nitrogen-Fixing Plant Genera with *Frankia* Symbiosis

Genus	No. nodulated species
Alnus	34
Casuarina	25
Ceanothus	31
Cercocarpus	4
Chamaebatia	1
Colletia	3
Coriaria	14
Cowania	1
Discaria	6
Dryas	3
Elaeagnus	17
Hippophae	1
Kentrothamnus	1
Myrica	26
Purshia	2
Rubus	1
Satisca	2
Shepherdia	3
Talguena	1
Trevoa	2

The majority of actinorhizal plants seem to infect plant hosts via a root hair and mediated mechanism, although some infect by direct penetration of the root by means of the intercellular spaces of the epidermis and cortex. As a result of infection, a root meristem is induced as the *Frankia* hyphae penetrate the cells, but the hyphae remain enclosed by a host-produced polysaccharide layer. The *Frankia* grow in the nodule occupying a major part of the host-produced nodular tissue. Root nodule tissue usually comprises 1–5% of the total dry weight of the plant. *Frankia* can form vesicles whose walls adapt to PO_2 in such a way that they can fix N_2 at atmospheric O_2 levels. The mechanism remains obscure.

It appears that the infective abilities of each *Frankia* strain are limited to one or a few plant genera. Currently, four host-specificity groups have been identified: (1) strains infective on *Alnus, Comptonia,* and *Myrica;* (2) strains infective on Casuarinaceae; (3) flexible strains infective on species of Elaeagnaceae and the promiscuous species of *Myrica* and *Gymnostoma;* and (4) strains infective only on species of the family Elaeagnaceae. Strains in specificity groups 1 and 2 infect the host plant via the root hair, while group 4 isolates infect via intercellular penetration. Some strains can use both modes of infection, and these have been closed as "flexible" specificity group 3.

Because these organisms were only isolated and cultured recently, only in the past few years have taxonomic, physiological, and genetic studies been started. To date, not all isolates have been successfully cultured. In contrast to rhizobia, it is possible to routinely obtain N_2 fixation *ex planta* in culture. Even more recently it has been possible to inoculate, with pure cultures, the host plant and obtain effective nodule formation.

B. Cyanobacteria

Cyanobacteria form symbioses with the most diverse hosts of any N_2-fixing systems known. Interestingly, whereas rhizobia and *Frankia* form associations with highly evolved plants, cyanobacteria favor the more primitive plants. Such N_2-fixing symbioses found with lichens, liverworts, pteridophytes (*Azolla*), gymnosperms (i.e., *Cycads*), and angiosperms (i.e., *Gunnera*). The cyanobacterium *Nostoc* forms symbioses with all of the taxa other than the ferns, indicating a potential for genetic manipulation to increase host range. The habitat range is also wide—from tropics to arctic—and includes freshwater, soil, saltwater, and hot springs.

As previously discussed, cyanobacteria are an important source of fixed nitrogen. Whereas the free-living cyanobacteria fix up to 80 kg/N_2/ha, the *Azolla–Anabaena* symbiosis can produce three times that amount. This is based on multiple cropping, but rates in rice paddies have been reliably reported from 1.4 to 10.5 kg/N_2/ha/day as daily averages over the whole growing season.

Nitrogen fixation by cyanobacteria was first reported in 1889 shortly after the rhizobium–legume symbiosis was described. Until about 25 years ago, research with this system was spasmodic, but, partly due to lately recognized importance of these eubacteria in agriculture and the environment, research is accelerating. Cyanobacteria have been classified, often by botanists rather than bacteriologists, mainly using morphological and anatomical criteria. The currently recognized families and genera of cyanobacteria containing nitrogen-fixing species are summarized in Table III. Because morphological characteristics can vary greatly depending on growth conditions, as-yet limited physiological, biochemical, and molecular genetic studies have been conducted. These indicate that there is considerably more diversity in this taxon than is accounted for by the traditional taxonomy. [*See* CYANOBACTERIA, MOLECULAR GENETICS.]

The cyanobacteria possess the requirements for the higher plant type of photosynthesis; i.e., water is the ultimate source of reductant and oxygen is evolved with CO_2 fixation via the Calvin cycle. The photosynthetic pigments are located in the outer cell regions.

Among filamentous, heterocyst-forming cyanobacteria, the ability for BNF appears universal. In nonheterocyst filamentous forms and unicellular forms, the ability to fix nitrogen is much less common. The heterocysts, which are found spaced along the cell filaments, appear colorless. These cells cannot fix CO_2 or evolve O_2 but can generate ATP by photophosphorillation and can fix nitrogen aerobically. This is apparently due to a modified thickened cell wall, thus interfering with oxygen diffusion. In these organisms then, photosynthesis and nitrogen fixation occur in different cells, protecting nitrogenase from excess oxygen stress.

As stated previously, cyanobacteria associate with almost every group of the plant kingdom, forming symbiotic nitrogen-fixing associations. These associations, however, occur with the more primitive plants. Specificity seems less rigid than with rhizobia or *Frankia*. Generally, cyanobacteria can fix nitrogen and grow independently of its plant partner. However, when living symbiotically, photosynthesis is often diminished in favor of increased nitrogen fixation, sufficient for both partners of the symbiosis.

The cyanobacteria usually invade normal morphological structures in the host plant, such as leaf cavities rather than modified structures such as the nodules caused by rhizobia or *Frankia*. Although in some cases, as with the roots of cycads, infection is followed by morphological change.

One reason for the renewed interest in this group of bacteria is the *Azolla–Anabaena* symbiosis. *Azolla* is a free-floating fern commonly found in still waters in temperate and tropical regions often found in rice paddies. There are seven species. *Azolla* is a remarkable plant and, under suitable conditions, can double in weight every 3–5 days. The plant forms a symbiotic relationship with the cyanobacterium *Anabaena azollae*. *Azolla* provides nutrients and a protective leaf cavity for the *Anabaena*, which in turn provides nitrogen for the fern. Under optimum conditions, this symbiosis results in as much (or more) nitrogen fixed than does the legume–*Rhizobium* symbiosis. If inoculated into a paddy and intercropped with rice, this symbiosis can satisfy the N requirements for the rice. In Southeast Asia, where there is a wet and dry season, *Azolla* is recommended for intercropping with paddy rice, but there are problems. *Azolla* is an extremely efficient scavenger of nutrients and will outcompete the rice for phosphate, etc., so careful management is required.

Anabaena azollae symbiotically fixes, perhaps, triple the amount of N_2 fixed by free-living *Anabaena*. The high photosynthetic rate of the *Azolla* no doubt supplies more energy in this symbiosis. The *Anabaena* forms more heterocysts when growing in association with *Azolla*. Similarly, when free-living heterocystic cyanobacteria are grown in a nitrogen-starved environment, they form extra heterocysts. The hypothesis then is that, when sufficient energy and nutrients are present for metabolism, the *Anabaena* can increase N_2 fixation capacity to maximize growth. It has also been reported by several researchers than N_2 fixation in this symbiotic system is not repressed by the presence of bound nitrogen.

Although *Azolla* has been cultivated in China and Vietnam for centuries, this is a new technology for most areas. Studies indicate a great potential (even at a low technological level) for this system as to BNF but also as a green manure supplement and/or animal feed. Little is known, as yet, as to plant-bacterial interactions, but the fact that *Anabaena* can be induced to increase N_2 fixation gives promise that the system can be optimized. The lack of repression of nitrogenase by bound nitrogen of *Anabaena* nitrogenase works in favor of a crop rotation following legumes.

C. The Rhizobium–Legume Symbiosis

The Leguminosae are one of the largest plant families, with worldwide distribution, about 750 genera, and an estimated 16,000–19,000 species. The Leguminosae have traditionally been divided into three distinct subfamilies based on floral differences: Mimosoideae, Caesalpinoideae, and Papilionoideae. Although only about 15% of the total species have been examined for nodulation, these species are representative of all three subfamilies of legumes. Virtually all species within the Mimosoideae and Papilionoideae are nodulated, but about 70% of the species in the subfamily Caesalpinoideae are non-nodulated. It is important to note that *Bradyrhizobium* of the cowpea type has been shown to form an effective symbiosis with a nonlegume, *Parasponia*, a member of the *Ulmaceae* (elm family). Currently, this is the only verified nitrogen-fixing association between a *Rhizobium* and a nonlegume.

The soil-improving properties of legumes were recognized by ancient agriculturalists. For example, Theophrastus (370–285 B.C.) in his "Enquiry into Plants" wrote as follows: "Of the other leguminous plants the bean best reinvigorates the ground;" and in another section, "Beans . . . are not a burdensome crop to the ground; they even seem to manure it." However, it was only in 1888 that Hellriegel and Wilfarth established positively that atmospheric nitrogen was assimilated by root nodules. This was quickly followed by the experiments of Beijerinck (1885), who used pure culture techniques to isolate the root nodule bacteria and proved that they were the causative agents of dinitrogen assimilation. He proposed the name *Bacillus radicicola* for these organisms. The root nodule bacteria were later renamed *Rhizobium* by Frank (1889).

Early researchers considered all rhizobia to be a single species capable of nodulating all legumes. Extensive cross-testing on various legume hosts led to a taxonomic characterization of rhizobia based on bacteria—plant cross-inoculation groups, which were defined as "groups of plants within which the root nodule organisms are mutually interchangeable." The concept of cross-inoculation groupings as taxonomic designators has gradually fallen into disrepute, although some of this philosophy is retained in the current taxonomic scheme.

There is a wide range in the efficiency of the symbiosis. Estimates for the amounts of nitrogen fixed are summarized in Tables VI and VII.

Table VI Nitrogen Fixed by Pulses (kg N/ha)

Plant	Average	Range
Vicia faba (faba beans)	210	45–552
Pisum sativum (peas)	65	52–77
Lupinus spp. (lupines)	176	145–208
Phaseolus aureus (green gram)	202[a]	63–342
Phaseolus aureus (mung)	61	—
Cajanus cajan (pigeon pea)	224	168–280
Vigna sinensis (cowpea)	198	73–354[a]
Canavalia ensiformis (jack bean)	49	—
Cicer arietinum (chickpea)	103	—
Lens culinaris (lentil)	101	88–114
Arachis hypogaea (peanut)	124	72–124
Cyamopsis tetragonolobus (guar)	130	41–220
Calopogonium mucunoides (calapo)	202[a]	370–450

The bacteria of the family Rhizobiaceae as cells without endospores are normally rod-shaped, motile, with one polar or subpolar flagellum or two to six peritrichous flagella, aerobic, and gram-negative, with many carbohydrates utilized. Considerable extracellular slime is usually produced during growth on carbohydrate-containing media. Some strains of rhizobia and *Agrobacterium* show a close relationship in DNA base composition.

Traditionally, rhizobia have been divided into two groups according to growth rate. The term "fast growers" commonly refers to rhizobia associated with alfalfa, clover, bean, and pea because, in culture, these grow much faster (less than one-half the doubling time of slow growers or <6 h) than the "slow growers" exemplified by soybean and cowpea rhizobia (generation time >6 h). Although there is phenotypic and genotypic diversity within these major groupings, and some overlap, numerous studies have demonstrated the validity of this approach.

Table VII Nitrogen Fixed by Tropical and Subtropical Forage and Browse Plants, Green Manure, and Shade Trees (kg N/ha/yr)

Plant	Average	Range
Centrosema pubescens	259	126–395
Desmodium intortum	897	—
Leucaena glauca	277	74–584
Lotonosis bainesii	62	—
Sesbania cannabina	542	—
Stylosanthes sp.	124	34–220
Phaseolus atropurpurea	291	—
Mikanea cordata	120	—
Pueraria phaseoloides	99	—
Enterolobium saman	150	—

The relative fastidiousness of the slow growers has been substantiated by more recent studies. While the major biochemical pathways seem to be similar, evidence suggests that the preferred pathway may be different. 16S RNA analysis of the fast- and slow-growing rhizobia confirmed that these groupings indeed represent different genetic phyla since the similarity coefficient (S_{AB}) of the RNA is 0.53. Thus, with gene analysis, the fast and slow growers, again, fall into widely separate groups. Recent findings using numerical taxonomy, carbohydrate metabolism, antibiotic susceptibilities, serology, DNA hybridization, RNA analysis, and DNA base ratio all demonstrated the validity of the fast- and slow-growing groupings. A summary of some of the differences between these groups is found in Table VIII.

On the basis of the differences between the fast- and slow-growing rhizobia, the traditional rhizobia were divided into two genera. The slow-growing strains were placed in the genus *Bradyrhizobium* comprising two species—*B. japonicum* and *B. elkanii,* which nodulate soybeans. Other bradyrhizobia occur (e.g., the peanut bradyrhizobia) but have not been classified to species of biovar levels. Researchers suggested that until further taxa within the genus are proposed, these should be described with the appropriate host plant given in parentheses [i.e., the peanut rhizobia—*Bradyrhizobium* sp. (*Arachis*)].

The fast-growing rhizobia have been placed in the genus *Rhizobium* comprising of six species—*R. leguminosarum, R. meliloti, R. loti, R. galegae, R. fredii,* and *R. tropici*. Three former species—*R. phaseoli, R. trifolii,* and *R. leguminosarum*—have been combined into the species *R. leguminosarum*. *Rhizobium fredii* is a new species consisting of fast-growing rhizobia that effectively nodulate Chinese soybean cultivars ordinarily nodulated by *B. japonicum*. This species of *Rhizobium* has recently been assigned to a new genus, *Sinorhizobium,* with the type species being *Sinorhizobium fredii*. The current taxonomic scheme is summarized in Table IX.

The taxonomy of the rhizobia is in a state of transition. As more molecular information accumulates, such genetic data will, no doubt, further displace cross-inoculation approaches to classification. The

Table VIII Differences between Fast- and Slow-Growing Rhizobia

Characteristic	Rhizobial type	
	Fast-growing	Slow-growing
Generation time	<6 hr	>6 hr
Carbohydrate substrate	Uses pentoses, hexoses, and mono-, di-, and trisaccharides	Uses pentoses and hexoses solely
Metabolic pathways	EMP—low activity Strain-specific ED—main pathway TCA—fully active PP pathway present	EMP—low activity ED—main pathway TCA—fully active Hexose cycle present
Flagellation type	Peritrichous	Subpolar
Symbiotic gene location	Plasmid and chromosome	Chromosome only
Nitrogen-fixing gene location	*nif*H, D, and K on same operon	*nif*D, K, and H on separate operons
Intrinsic antibiotic resistance	Low	High

ED, Entner–Doudoroff pathway; EMP, Embden–Meyerhoff–Parnas pathway; PP, pentose phosphate pathway; TCA, tricarboxylic acid cycle.

present scheme, however, does function to allow generalized identification of isolates.

1. Nitrogen Fixation

The nitrogenase complex in the rhizobia is a highly conserved enzyme system and, as stated earlier, basically common to the dinitrogen-fixing prokaryotes. Considerable evidence suggests that there are differences in location of nitrogenase genes between fast- and slow-growing rhizobia (see Table VIII). In the fast-growing *Rhizobium,* it has been repeatedly demonstrated that *nif*H, D, and K genes are plasmid-borne. Several investigators have presented evidence that some *nif* functions are chromosomally regulated, and in *Sinorhizobium fredii* the structural *nif* genes are both plasmid and chromosomally located. In *Bradyrhizobium,* plasmids have not been shown to carry *nif* genes. A further difference in the *nif* genes of the two groups is the fact that in *B. japonicum* the *nif*D and K genes and the *nif*H gene are located on separate operons, whereas in the genus *Rhizobium,* the *nif* genes are organized into one transciptional unit.

Table IX Current Taxonomic Classification of the Rhizobia

Recognized genus	Recognized species
Bradyrhizobium	*B. japonicum*
Rhizobium	*R. leguminosarum* *R. meliloti* *R. loti* *R. galagae* *R. tropici* *R. huakuii*
Azorhizobium	*A. caulinodans*
Sinorhizobium	*S. fredii* *S. xinjiangensis*

Analysis and comparison of *nif* DNA in the fast- and slow-growing rhizobia has established the affinity coefficient (S_{AB}) for the nucleotide sequence (*nif*H, *nif*D, and *nif*K) from the two groups. The same analysis was done comparing amino acid sequences of the nitrogenase Fe and MoFe protein polypeptides. A considerable sequence conservation reflects the structural requirements of the nitrogenase proteins for catalytic functions. The S_{AB} *nif* values (based on *nif*H sequences) between fast- and slow-growing rhizobia indicated that these are almost as distant from each other as they are from other gram-negative organisms. The results suggest that *nif* genes evolved in a manner similar to the bacteria that carry them rather than by a more recent lateral distribution of *nif* genes among microorganisms. Again, the phylogenetic difference between fast and slow groups of rhizobia is apparent. Although this general system is common to all of nitrogen-fixing prokaryotes, several concomitant alternative systems have recently been described, but none of these have as yet been shown to be present or active in the rhizobia.

2. Nodule Formation

Nodule initiation and subsequent maturation is an interactive process involving the eukaryotic host legume and the prokaryotic *Rhizobium*. The process is complex, resulting in biochemical and morphological changes in both symbionts and leading to the capacity of reducing atmospheric nitrogen.

Initially, the proper *Rhizobium* species proliferates in the root zone of a temperate leguminous plant and becomes attracted and attached to the root hair. A poorly described chemotactic response attracts the rhizobia to the root surface. At the surface, the bacteria alter the growth of epidermal root hairs so that they are deformed. In some tropical legumes such as peanuts (*Arachis*), root hairs are not the primary invasive sites, but the alternative invasion process is not well described, although infection is reported at the site of lateral root emergence.

The root hair infection process consists of several events leading to nodule formation. These have been summarized as follows: (1) recognition by the rhizobia of the legume, (2) attachment to the root hair, (3) curling of the root hair, (4) root hair infection by the bacteria, (5) formation of an infection thread, (6) nodule initiation, and (7) transformation of the vegetative cells in the nodules to enlarged pleomorphic forms called bacteroids, which fix nitrogen.

Based on morphology, there are two kinds of nodules: determinative and indeterminative. In general, indeterminative nodules are formed by fast-growing rhizobia and are characterized by a defined meristem during nodule growth. Determinative nodules arise from cortex tissue. Legumes nodulated by *Bradyrhizobium* form determinative nodules close to the endodermis, which is near the xylem poles in the root.

The formation of nodules on legumes is the result of a coordinated development involving many plant and bacterial genes. Studies of the nodulation (*nod*) genes of rhizobia have depended on the development of molecular genetic tools. Many of the genes involved in the nodulation process have been located and identified. [*See* Nodule Formation in Legumes.]

Legume roots, grown axenically, do not appear morphologically distinctive from other plant roots so that the ability of these plants to respond to microbial signals and then alter their metabolism to form nodules is not explained by just morphology. It has been conclusively demonstrated that genetic information from both symbionts controls nodulation and the host range of nodulation by a *Rhizobium* species. Metabolically, there are three types of nodules (often termed effective, inefficient, and ineffective). Effective nodules contain a high density of rhizobia actively fixing dinitrogen. Inefficient nodules may contain a similar density of the bacteria, but only a low level of fixed dinitrogen results from the symbiosis. Ineffective nodules result from a symbiosis with bacteria not able to carry out nitrogen fixation. Because the regulatory roles of the plant and bacterial genes in nitrogen fixation have not been generally elucidated, the reasons for differential nitrogen-fixing ability of nodules remains obscure. There is, however, a generally observed legume host–*Rhizobium* interaction that results in optimization of dinitrogen fixation. Such an interaction makes it possible to optimize dinitrogen fixation under field conditions in a cultivar through introduction of an effective *Rhizobium*.

Nodulation genes are defined by their effect on the bacteria's ability to generate the nodulation process on the proper legume host. Because individual rhizobia and bradyrhizobia can each nodulate a limited number of host legumes and plant genes also limit the symbiosis, it follows that a recognition exists between bacteria and host. Thus, there are two types of *nod* genes—a common *nod* region, which consists of a structurally and functionally conserved cluster of genes, and host-specific nodulation genes, which cannot complement nodulation defects in other genera or species. *nod* genes have been studied in varying degrees in different species (usually in the fast-growing rhizobia). In these organisms, four genes have been identified in two transcription units (*nod*D and *nod*ABC). Two additional genes, apparently on the same transcriptional unit, *nod*I and *nod*J, have been identified. Genetic maps of the common *nod* cluster, drawing together the information from many sources, have been published. The *nod*ABC appears to be functionally interchangeable among all rhizobia, and mutations in these genes cause complete nodulation failure. These genes are involved in cell division and root hair deformation. The *nod*IJ genes cause a delay in the appearance of nodules.

The second group of *nod* genes are termed host-specific. These genes are not conserved because alleles from various rhizobia cannot substitute for each other on different hosts. Bacteria carrying mutations in these host-specific genes cause abnormal root hair reactions.

Many genetic *nod* loci have been identified in a variety of rhizobia. The list now includes at least 15 *nod* genes. In many cases, these have been cloned and sequenced and the gene product associated with a step in nodulation.

Although the amino acid sequences of many of the nodulation gene products have been described, the biochemical functions of these genes have not been determined. Possible exceptions are the *nod*D genes, which are positive gene regulators.

The centenary of this first demonstration of biological nitrogen fixation occurred during 1986. During that period, many papers were published expanding the knowledge base, both basic and applied. Currently, the subject of BNF is of great practical importance because of the utilization of fossil fuels in the manufacture of nitrogenous fertilizers. The increased scarcity and higher costs of fossil fuel have made it important to optimize biological nitrogen fixation as an alternative to chemical nitrogen. Additionally, the increasing usage of nitrogen fertilizer has resulted in unacceptable levels of water pollution, which does not occur when the biologically fixed forms of nitrogen are used. With the additional research capabilities resulting from the developing field of biotechnology, it is evident that interest in this field will continue and that we may be reaching a level of accumulated knowledge to allow a full utilization of BNF as an alternative to the Haber–Bosch process.

Bibliography

Dilworth, M. J., and Glenn, A. R. (eds.) (1991). "Biology and Biochemistry of Nitrogen Fixation." Elsevier, Amsterdam. 438 pp.

Elkan, G. H. (ed.) (1987). "Symbiotic Nitrogen Fixation Technology." Marcel Dekker, New York. 440 pp.

Gresshof, P. M. (ed.) (1990). "Molecular Biology of Symbiotic Nitrogen Fixation." CRC Press, Boca Raton, Florida. 269 pp.

Hennecke, H., and Verma, D. P. S. (eds.) (1990). "Advances in Molecular Genetics of Plant–Microbe Interactions." Kluwer Academic Pubs. Dordrecht, The Netherlands. 482 pp.

Sprent, J. I., and Sprent, P. (1990). "Nitrogen Fixing Organisms." Chapman and Hall, London. 256 pp.

Stacy, G., Burris, R., and Evans, H. J. (eds.) (1992). "Biological Nitrogen Fixation." 943 pp. Chapman and Hall, New York.

Biological Warfare

James A. Poupard
Center for the History of Microbiology

Linda Ann Miller
Holy Reedemer Hospital and Medical Center

Glossary

Biological warfare (BW) Use of microorganisms, such as bacteria, fungi, viruses, and rickettsiae, to produce death or disease in humans, animals, or plants. The use of toxins to produce death or disease is often included under the heading of BWR (U.S. Army definition: included in Army report to the Senate Committee on Human Resources, 1977)

Biological weapon(s) Living organisms, whatever their nature, which are intended to cause disease or death in man, animals or plants, and which depend for their effects on their ability to multiply in the person, animal or plant attacked [United Nations definition: included in the Report of the Secretary General entitled Chemical and Bacteriological (biological) Weapons and the Effects of their Possible Use, 1969]

Genetic engineering Methods by which the genomes of plants, animals, and microorganisms are manipulated, includes but is not limited to recombinant DNA technology

Recombinant DNA technology Techniques where by different pieces of DNA are spliced together and inserted into vectors such as bacteria or yeast

Toxin weapon(s) (TW) Any poisonous substance, whatever its origin or method of production, which can be produced by a living organism, or any poisonous isomer, homolog, or derivative of such a substance [U.S. Arms Control and Disarmament Agency (ACDA) definition; proposed on August 20, 1980]

THE MOST GENERAL concept of biological warfare involves the use of any biological agent as a weapon directed against humans, animals, or crops with the intent to kill, injure, or create a sense of havoc against a target population. This agent could be in the form of a viable organism or a metabolic product of that organism, such as a toxin. The accent of this article will focus on the use of viable biological agents because many of the concepts relating to the use of toxins are more associated with chemical warfare. The use of viable organisms or viruses involves complex issues that relate to containment. Once such agents are released, even in relatively small numbers, the focus of release has the potential to enlarge to a wider population due to the ability of the viable agent to proliferate while spreading from one susceptible host to another.

I. Introduction

The roots of biological warfare (BW) can be traced to ancient times and have evolved into more sophisticated forms with the maturation of the science of bacteriology and microbiology. It is important to understand the history of the subject because one often has preconceived notions of BW that are not based on facts or involve concepts more related to chemical warfare rather than biological. Although BW was not employed in the Gulf War, the conflict introduced or altered many of the contemporary issues relating to BW for both the microbiologist and the general public. Many of the contemporary issues relating to BW will potentially deal with third-world conflicts, terrorist groups, or nonconventional warfare. An understanding of these issues becomes important because many of the long-standing international treaties and conventions regarding BW were formulated in an atmosphere of either international conflict or during the Cold War period of international relations. Many of the classic issues have undergone significant alteration by recent events. The issue of BW is intimately bound to such concepts as offensive versus defensive research or to the need for secrecy and national security. It is obvious that BW will continue to be a subject that will demand the attention of contemporary students of microbiology as well as those involved with science policy in the post-Cold War period.

II. Historical Review

A. 300 B.C. to 1925

Many early civilizations employed a crude method of warfare that could be considered BW as early as 300 B.C., when the Greeks polluted the wells and drinking water supplies of their enemies with the corpses of animals. Later the Romans and Persians used these same tactics. All armies and centers of civilization need palatable water to function, and it is clear that well pollution was an effective and calculated method for gaining advantage in warfare. In 1155 at a battle in Tortona, Italy, Barbarossa broadened the scope of BW, using the bodies of dead soldiers as well as animals to pollute wells. Evidence indicates that well poisoning was a common tactic throughout the classical, medieval, and Renaissance periods. In more modern times, this method has been employed as late as 1863 during the U.S. Civil War by General Johnson, who used the bodies of sheep and pigs to pollute drinking water at Vicksburg.

The wide use of catapults and siege machines in medieval warfare introduced a new technology for delivering biological entities. In 1422 at the siege of Carolstein, catapults were used to project diseased bodies over walled fortifications, creating fear and confusion among the people under siege. The use of catapults as weapons was well established by the medieval period, and projecting diseased bodies over walls was an effective strategy employed by besieging armies. The siege of a well-fortified position could last for months or years, and it was necessary for those outside the walls to use whatever means available to cause disease and chaos within the fortification. This technique became commonplace, and numerous classical tapestries and works of art depict diseased bodies or the heads of captured soldiers being catapulted over fortified structures.

In 1763 the history of BW took a significant turn from the crude use of diseased corpses to the introduction of a specific disease, smallpox, as a weapon in the North American Indian wars. It was common knowledge at the time that the Native American population was particularly susceptible to smallpox, and the disease may have been used as a weapon in earlier conflicts between European settlers and Native Americans. In the spring of 1763, Sir Jeffrey Amherst, the British Commander-in-Chief in North America, believed the western frontier, which ran from Pennsylvania to Detroit, was secure, but the situation deteriorated rapidly over the next several months. The Indians in western Pennsylvania were becoming particularly aggressive in the area around Fort Pitt, near what is now Pittsburgh. It became apparent that unless the situation was resolved, western Pennsylvania would be deserted and Fort Pitt isolated. On 23 June 1763, Colonel Henry Bouquet, the ranking officer for the Pennsylvania frontier, wrote to Amherst, describing the difficulties Captain Ecuyer was having holding the besieged Fort Pitt. These difficulties included an outbreak of smallpox among Ecuyer's troops. In his reply to Bouquet, Amherst suggested that smallpox be sent among the Indians to reduce their numbers. This well-documented suggestion is significant because it clearly implies the intentional use of smallpox as a weapon. Bouquet responded to Amherst's suggestion stating that he would use blankets to spread the disease.

Evidence indicates that Amherst and Bouquet were not alone in their plan to use BW against the Indians. While they were deciding on a plan of action, Captain Ecuyer reported in his journal that he had given two blankets and a handkerchief from the garrison smallpox hospital to hostile chiefs with the hope that it would spread the disease. It appears that Ecuyer was acting on his own and did not need persuasion to use whatever means necessary to preserve the Pennsylvania frontier. Evidence also shows that the French used smallpox as a weapon in their conflicts with the native population.

Smallpox also played a role in the American Revolutionary War, but the tactics were defensive rather than offensive: British troops were vaccinated against smallpox, but the rebelling American colonists were not. This protection from disease gave the British an advantage for several years, until Washington ordered vaccination of all American troops. [*See* SMALLPOX.]

It is clear that by the eighteenth century BW had become disease-oriented, even though the causative agents and mechanisms for preventing the spread of diseases were largely unknown. The development of the science of bacteriology in the nineteenth and early twentieth centuries considerably expanded the scope of potential BW agents. In 1915, Germany was accused of using cholera in Italy and plague in St. Petersburg. Evidence shows that Germany used glanders and anthrax to infect horses and cattle, respectively, in Bucharest in 1916 and employed similar tactics to infect 4500 mules in Mesopotamia the next year. Germany issued official denials of these accusations. Although there apparently was no large-scale battlefield use of BW in World War I, numerous allegations of German use of BW were made in the years following the war. Britain accused Germany of dropping plague bombs, and the French claimed the Germans had dropped disease-laden toys and candy in Romania. Germany denied the accusations.

Although chemical warfare was far more important than BW in World War I, the general awareness of the potential of biological weapons led the delegates to the Geneva Convention to include BW agents in the 1925 Protocol for the Prohibition of the Use in War of Asphyxiating, Poisonous or Other Gases, and of Bacteriological Methods of Warfare. The significance of the treaty will be discussed later (see Section III).

B. 1925–1990

The tense political atmosphere of the period following the 1925 Geneva Protocol and the lack of provisions to deter biological weapons research had the effect of undermining the treaty. The Soviet Union opened a BW research facility north of the Caspian Sea in 1929; the United Kingdom and Japan initiated BW research programs in 1934. The Japanese program was particularly ambitious and included experiments on human subjects prior to and during World War II.

Two factors were significant in mobilizing governments to initiate BW research programs: (1) a continuing flow of accusations regarding BW and (2) the commitment of resources for BW research by several national adversaries, thus creating a feeling of insecurity among governments. The presence of BW research laboratories in nations that were traditional or potential adversaries reinforced this insecurity. Thus, despite the Geneva Protocol, it was politically unwise for governments to ignore the threat of BW, and the result was increasingly sophisticated biological weapons.

In 1941, the United States and Canada joined other nations and formed national programs of BW research and development. Camp Detrick (now Fort Detrick) became operational as the center for U.S. BW research in 1943, and in 1947 President Truman withdrew the Geneva Protocol from Senate consideration, citing current issues such as the lack of verification mechanisms that invalidated the underlying principles of the treaty. However, there was no widespread use of BW in a battlefield setting during World War II. BW research, however, continued at an intense pace during and after the war. By the end of the decade, the United States, the United Kingdom, and Canada were conducting collaborative experiments involving the release of microorganisms from ships in the Caribbean. In 1950, the U.S. Navy conducted open-air experiments in Norfolk, Virginia, and the U.S. Army conducted a series of airborne microbial dispersals over San Francisco using *Bacillus globigii, Serratia marcescens,* and inert particles.

Not surprisingly, the intense pace of BW research led to new accusations of BW use, most notably by China and North Korea against the United States during the Korean War. In 1956, the United States changed its policy of "defensive use only" to include possible deployment of biological weapons in situations other than retaliation. During the 1960s, all branches of the U.S. military had active BW programs, and additional open-air dissemination experiments with stimulants were conducted in the New York City subway system. By 1969, however, the U.S. military concluded that BW had little tactical value in battlefield situations, and since it was felt that in an age of nuclear weapons dominated the strategic equation, the United States would be unlikely to need or use BW. Thus, President Nixon announced that the United States would unilaterally renounce BW and eliminate stockpiles of biological weapons. This decision marked a turning point in the history of BW: Once the U.S. government made it clear it did not consider biological weapons a critical weapon system, the door was opened for negotiation of a strong international treaty against BW.

Once military strategists had discounted the value of BW, an attitude of openness and compromise on BW issues took hold, leading to the 1972 Convention on the Prohibition of the Development, Production and Stockpiling of Bacteriological (Biological) and Toxin Weapons and on Their Destruction (see Section III). The parties to the 1972 Convention agreed to destroy or convert to peaceful use all organisms, toxins, equipment, and delivery systems. Following the signing of the 1972 treaty, the U.S. government generated much publicity about its compliance activities, inviting journalists to witness destruction of biological weapons stockpiles.

The problem of treaty verification beleaguered the 1972 Convention. Press reports accusing the Soviet Union of violating the treaty appeared as early as 1975. When an outbreak of anthrax was reported in Sverdlovsk, Soviet Union, in 1979, the United States claimed it was caused by an incident at a nearby Soviet biological defense laboratory that had released anthrax spores into the surrounding community. The Soviet government denied this allegation, claiming the outbreak was caused by contaminated black market meat.

BW continued to be discussed in the public media throughout the 1980s. In 1981, reports describing the American "cover-up" of Japanese BW experiments on prisoners of war began to surface in the public and scientific literature. In 1982, *The Wall Street Journal* published a series of articles on Soviet genetic engineering programs that raised many questions about the scope of Soviet BW activities. The environmental effects of testing biological agents at Dugway Proving Grounds in Utah received considerable press attention in 1988, leading to a debate over the need for such a facility.

The 1980s also were characterized by debate over larger issues relating to BW. A public debate in 1986 considered the possible role of biological weapons in terrorism. Scientific and professional societies, which had avoided the issues for many years, began considering both specific issues, such as Department of Defense support for biological research and more general issues, such as adopting ethical codes or guidelines for their members.

C. 1990 to Present

The last decade of the twentieth century witnessed two significant events that will have long-term effects on developing policies relating to BW. The first event was the Gulf War and the second was the demise of the Soviet Union. The Gulf War raised a serious concern and brought about changed attitudes about BW policy mainly due to the open threat by the Iraqi military to use such agents.

The plans for Operation Desert Storm included provisions for protective equipment and prophylactic administration of antibiotics or approved vaccines as protection against potential biological weapons. Many of the critics of the U.S. Biological Defense Research Program (BDRP) were now asking why the country was not better prepared to protect its troops against biological attack. Fortunately, BW was not used during the Gulf War, but the threat of its use provided several significant lessons. First, although there has been considerable concern that genetic engineering could produce new, specialized biological weapons, most experts predicted that "classical" BW agents, such as anthrax and botulism, would pose the most serious threat to combat troops in Operation Desert Storm. Second, both the war and the United Nations' subsequent disarmament and inspection activities demonstrated the difficulty of verifying the presence and extent of production facilities for BW agents; these difficulties highlight the need for verification protocols for the BW Convention, and verification and treaty compliance certainly will be prominent BW issues in the current decade. In the past, U.S. BW policy was greatly influenced by the perception of Soviet BW capabilities. With the changes occurring in the former Soviet Union, BW policies that focus on a changing world order will need to be formulated.

III. International Treaties

A. The 1925 Geneva Protocol

The 1925 Geneva Protocol was the first international treaty to place restrictions on BW. The Geneva Protocol followed a series of international agreements that were designed to prohibit the use in war of weapons that inflict or prolong unnecessary suffering of combatants or civilians. The St. Petersburg Declaration of 1868 and the International Declarations concerning the Laws and Customs of War, which was signed in Brussels in 1874, condemned the use of weapons that caused useless suffering. Two major international conferences were held at the Hague in 1899 and 1907. These conferences resulted in declarations regarding the humanitarian conduct of war. The conference regulations forbid nations from using poison, treacherously wounding enemies, or using munitions that would cause unnecessary suffering. The so-called Hague Conventions also prohibited the use of projectiles to diffuse asphyxiating or deleterious gases. The Hague Conventions still provide much of the definitive law of war as it exists today.

The Hague Conventions did not specifically mention BW, due in part to the lack of scientific understanding of the cause of infectious diseases at that time. The Conventions have, however, been cited as an initial source of the customary international laws that prohibit unnecessary suffering of combatants and civilians in war. While biological weapons have been defended as humanitarian weapons, on the grounds that many biological weapons are incapacitating but not lethal, there are also biological weapons that cause a slow and painful death. It can be argued, therefore, that the Hague Conventions helped to set the tone of international agreements on laws of war that led to the 1925 Geneva Protocol.

The 1925 Geneva Protocol, formally called the Prohibition of the Use in War of Asphyxiating, Poisonous or Other Gases, and of Bacteriological Methods of Warfare, was opened for signature on 17 June 1925 in Geneva, Switzerland. Over 100 nations have signed and ratified the protocol, including all members of the Warsaw Pact and North Atlantic Treaty Organization (NATO). The 1925 Geneva Protocol was initially designed to prevent the use in war of chemical weapons; however, the protocol was extended to include a prohibition on the use of bacteriological methods of warfare. The Geneva Protocol distinguishes between parties and nonparties by explicitly stating that the terms of the treaty apply only to confrontations in which all combatants are parties and when a given situation constitutes a "war." Additionally, a number of nations ratified the Geneva Protocol with the reservation that they would use biological weapons in retaliation against a biological weapons attack. This resulted in the recognition of the Geneva Protocol as a "no first use" treaty.

B. The 1972 Biological Warfare Convention

International agreements governing BW have been strengthened by the 1972 BW Convention, which is officially called the 1972 Convention on the Prohibition of the Development, Production and Stockpiling of Bacteriological (Biological) and Toxin Weapons and on their Destruction. The Convention was signed simultaneously in 1972 in Washington, London, and Moscow and entered into force in 1975. The preamble to the 1972 BW Convention states the determination of the states parties to the treaty to progress toward general and complete disarmament, including the prohibition and elimination of all types of weapons of mass destruction. This statement places the Convention in the wider setting of international goals of complete disarmament. The 1972 BW Convention is also seen as a first step toward chemical weapons disarmament.

The 1972 BW Convention explicitly builds on the Geneva Protocol by reaffirming the prohibition of the use of BW in war. The preamble, although not legally binding asserts that the goal of the Convention is to completely exclude the possibility of biological agents and toxins being used as weapons and states that such use would be repugnant to the conscience of humankind. The authors of the 1972 Convention, therefore, invoked societal attitudes as justification for the existence of the treaty.

The 1972 BW Convention evolved, in part, from a process of constant reevaluation of the Geneva Protocol. From 1954 to the present, the United Nations has periodically considered the prohibition of chemical and biological weapons. The Eighteen-Nation Conference of the Committee on Disarmament, which in 1978 became the Forty-Nation Committee on Disarmament, began talks in 1968 to ban chemical weapons. At this time, chemical, toxin, and biological weapons were being considered together, in an attempt to develop a comprehensive disarmament agreement. However, difficulties in reaching agreements on chemical warfare led to a series of separate negotiations that covered only BW and toxin weapons. These negotiations resulted in the drafting of the 1972 BW Convention.

The 1972 BW Convention consists of a preamble, followed by 15 articles. Article I forms the basic treaty obligation. Parties agree never in any circumstance to develop, produce, stockpile, or otherwise acquire or retain the following:

1. Microbial or other biological agents, or toxins whatever their origin or method of production, of types and in quantities that have no justification for prophylactic, protective, or other peaceful purposes.

2. Weapons, equipment, or means of delivery designed to use such agents or toxins for hostile purposes or in armed conflict.

Article II requires each party to destroy, or divert to peaceful purposes, all agents, toxins, equipment, and delivery systems that are prohibited in Article I and are under the jurisdiction or control of the party. It also forbids nations from transferring, directly or indirectly, materials specified in Article I and prohibits nations from encouraging, assisting, or inducing any state, group of states, or international organizations from manufacturing or acquiring the material listed in Article I. There is no specific mention of subnational groups, such as terrorist organizations, in the treaty.

Article IV requires each party to the Convention to take any measures to ensure compliance with the terms of the treaty. Article IV has been interpreted by some states as the formulation of civil legislation or regulations to assure adherence to the Convention. This civil legislation could regulate activities by individuals, government agencies, universities, or corporate groups.

Articles V–VII specify procedures for pursuing allegations of noncompliance with the 1972 BW Convention. The United Nations plays an integral part in all of the procedures for investigating allegations of noncompliance. According to Article VI, parties may lodge a complaint with the Security Council of the United Nations if a breach of the treaty is suspected. All parties must cooperate with investigations that may be initiated by the Security Council. Article VII requires all parties to provide assistance or support to any party that the Security Council determines has been exposed to danger as a result of violation of the Convention. Articles VII–IX are general statements for obligations of the parties signing the protocol. Article X gives the parties the right to participate in the fullest possible exchange of equipment, materials, and scientific or technological information of the use of bacteriological (biological) agents and toxins for peaceful purposes. Article XI allows parties to propose amendments to the Convention. The amendments only apply to those states that accept them and enter into force after a majority of the states parties to the Convention have agreed to accept and be governed by the amendment.

Article XII requires that a conference be held 5 years after the entry into force of the BW Convention. Article XIV states that the 1972 BW Convention is of unlimited duration. A state party to the treaty is given the right to withdraw from the treaty if it decides that extraordinary events, related to the subject matter of the Convention, have jeopardized the supreme interests of the country. This article also opens the Convention to all nations for signature. Nations that did not sign the Convention before its entry into force may accede to it at any time.

C. Review Conferences

The 1972 Convention contained a stipulation that a conference be held in Geneva five years after the terms of the Convention entered into force. The purpose of the conference was to review the operation of the Convention and to assure that the purposes of the Convention were being realized. The review was to take into account any new scientific and technological developments that were relevant to the Convention. The first review conference was held in Geneva in 1980. Several points contained in the original Convention were clarified at this conference. The second review conference was held in 1986, and a third was held in 1991. There is general agreement that these conferences serve a definite function in solving contemporary problems that need clarification based on changing events and have made significant contributions in keeping the 1972 Convention relevant to the needs of a changing world situation.

IV. Current Research Programs

Biological weapons research in the United States is under the direction of the BDRP, headquartered at Fort Detrick, Maryland. In accordance with official U.S. policy, the BDRP is solely defensive in nature, with the goal of providing methods of detection for, and protective measures against, biological agents that could be used as weapons against U.S. forces by hostile states or individuals.

Current U.S. policy stems from the 1969 declaration made by President Nixon that confined the U.S. BW program to research on biological defense such as immunization and measures of controlling and preventing the spread of disease. Henry Kissinger further clarified the U.S. BW policy in 1970 by stating that the United States biological program will be confined to research and development for defensive purposes only. This did not preclude research into those offensive aspects of biological agents necessary to determine what defensive measures are required.

The BDRP expanded significantly in the 1980s, in an apparent response to alleged treaty violations and perceived offensive BW capabilities in the Soviet Union. These perceptions were espoused primarily by representatives of the Reagan Administration and the Department of State. At congressional hearings in May 1988, the U.S. government reported that at least 10 nations, including the Soviet Union, Libya, Iran, Cuba, Southern Yemen, Syria, and North Korea, were developing biological weapons. Critics of the U.S. program refuted the need for program expansion.

The BDRP is administered through two separate government organizations—the army and the CIA. Details of the program are described in the April 1989 Environmental Impact Statement published by the Department of the Army, U.S. Army Medical Research and Development Command.

The BDRP is focused in three sites, the U.S. Army Medical Research Institute of Infectious Diseases (USAMRIID) at Fort Detrick, Maryland; Aberdeen Proving Ground in Maryland; and the Dugway Proving Ground in Utah. USAMRIID is designated as the lead laboratory in medical defense against BW threats. Research conducted at the USAMRIID focuses on medical defense such as the development of vaccines and treatments for both natural diseases and potential BW agents. Work on the rapid detection of microorganisms and the diagnosis of infectious diseases are also conducted. The primary mission at the Aberdeen Proving Ground is nonmedical defense against BW threats including detection research, such as the development of sensors and chemiluminescent instruments to detect and identify bacteria and viruses, and development of methods for material and equipment decontamination. The U.S. Army Dugway Proving Ground is a Department of Defense major range and test facility responsible for development, test, evaluation, and operation of chemical warfare equipment, obscurants and smoke munitions, and biological defense equipment. Its principle mission with respect to the BDRP is to perform developmental and operational testing for biological defense material, including the development and testing of sensors, equipment, and clothing needed for defense against a BW attack.

A total of 100 secondary sites have received contracts for biological defense research. Secondary sites include the Swiftwater Lab, operated by the Salk Institute in Swiftwater, Pennsylvania; the Naval Medical Research Institute in California; medical centers; universities; and private biotechnology firms in the United States, Scotland, and Israel.

The second aspect of the BDRP is administered through the CIA. In 1982, Thomas Dashiell of the Office of the Secretary of Defense reported on a classified technology watch program related to BW that was operated by the intelligence community. The program was designed to monitor worldwide developments related to BW that could affect the vulnerability of U.S. and NATO forces to biological attack.

BDRP research focuses on six main areas:

1. Development of vaccines.
2. Development of protective clothing and decontamination methods.
3. Analysis of the mode of action of toxins and the development of antidotes.
4. Development of broad-spectrum antiviral drugs for detecting and diagnosing BW agents and toxins.
6. Utilization of genetic engineering methods to study and prepare defenses against BW and toxins.

The BDRP has often been a center of controversy in the United States. One BDRP facility, the Dugway Proving Ground, was the target of a lawsuit that resulted in the preparation of the environmental impact statement for the facility. A proposal for a high-level containment laboratory (designated P-4) was ultimately changed to new plans for a lower-level (P-3) facility.

The use of genetic engineering techniques in BDRP facilities has also been a focus of controversy. The BDRP position is that genetic engineering will be utilized if deemed necessary. The Department of Defense stated that testing of aerosols of pathogens derived from recombinant DNA methodology is not precluded if a need should arise in the interest of national defense.

Very little is written in the unclassified literature on BW research conducted in countries other than the United States. Great Britain has maintained the Microbiological Research Establishment at Porton Down; however, military research is highly classified in Great Britain and details regarding the research conducted at Porton are unavailable.

Prior to and during the 1991 Gulf War with Iraq, the United States claimed that Iraq had developed BW capabilities. The U.S. military prepared troops for a BW attack by providing protective clothing, masks, and immunizations. Of particular concern was the potential use of *Bacillus anthracis* by Iraq. During the war, Allied forces destroyed a facility the United States claimed was a BW production facility. The Iraqis asserted that the plant produced baby formula. The war ended in February 1991. Biological weapons were not used.

During the 1970s and 1980s, a great deal of U.S. BW policy was based on the assumption of Soviet offensive BW capabilities. Most U.S. accounts of Soviet BW activities were unconfirmed accusations or claims about treaty violations. The Soviet Union was a party to both the 1925 Geneva Protocol and the 1972 BW Convention.

According to Pentagon sources, the Soviet Union operated at least seven top-security BW centers. These centers were reported to be under strict military control. While the former Soviet Union proclaimed that their BW program was purely defensive, the United States has consistently asserted that the Soviet Union was conducting offensive BW research.

V. Contemporary Issues

A. Genetic Engineering

There has been considerable controversy over the potential for genetically engineered organisms to serve as effective BW agents. Recombinant DNA technology has been cited as a method for creating novel, pathogenic microorganisms. Theoretically, organisms could be developed that would possess predictable characteristics, including antibiotic resistance, altered modes of transmission, or altered pathogenic and immunogenic capabilities. This potential for genetic engineering to significantly affect the military usefulness of BW has been contested. It has been suggested that because a large number of genes must work together to endow an organism with pathogenic characteristics, the alteration of a few genes with recombinant DNA technology is unlikely to yield a novel pathogen that is significantly more effective or usable than conventional BW agents.

The question of predictability of the behavior of genetically engineered organisms was addressed at an American Society for Microbiology Symposium held in June 1985. Some symposium participants believed that the use of recombinant DNA increases predictability because the genetic change can be precisely characterized. Other participants, however, felt that the use of recombinant DNA decreases predictability, because it widens the potential range of DNA sources. Other evidence supports the view that genetically engineered organisms do not offer substantial military advantage over conventional BW. Studies have shown that genetically engineered organisms do not survive well in the environment. This fact has been cited as evidence that these organisms would not make effective BW agents. [*See* GENETICALLY ENGINEERED MICROORGANISMS, ENVIRONMENTAL INTRODUCTION.]

Despite the contentions that genetic engineering does not enhance the military usefulness of BW, a significant number of arguments support the contrary. At the 1986 Review Conference of the BW Convention, it was noted that genetic engineering advances since the Convention entered into force may have made biological weapons a more attractive military option.

Several authors have contended that the question of the potential of genetic engineering to enhance the military usefulness of BW is rhetorical, because the 1972 BW Convention prohibits development of such organisms despite their origin or method of production. Nations participating in both the 1980 and 1986 review conferences of the BW Convention accepted the view that the treaty prohibitions apply to genetically engineered BW agents. An amendment to the treaty, specifically mentioning genetically engineered organisms, was deemed to be unnecessary. Additionally, the United States, Great Britain, and the Soviet Union concluded in a 1980 briefing paper that the 1972 BW Convention fully covered all BW agents that could result from genetic manipulation.

While the utility of genetic engineering for enhancing the military usefulness of BW agents has been questioned, the role of genetic engineering for strengthening defensive measures against BW has been clear. Genetic engineering has the potential for improving defenses against BW in two ways: (1) vaccine production and (2) sensitive identification and detection systems. The issues of the new technologies in defensive research have been evident in the U.S. BW program. Since 1982, U.S. Army scientists have used genetic engineering to study and prepare defenses against BW agents. Military research utilizing recombinant DNA and hybridoma technology include the development of vaccines against a variety of bacteria and viruses, methods of rapid detection and identification of BW agents, and basic research on protein structure and gene control. By improving defenses against BW, it is possible that genetic engineering may potentially reduce the risk of using BW.

The primary effect of BW on the government regulations on genetic engineering is the tendency toward more stringent control of the technologies. The fear of genetically engineered BW agents has prompted proposals for government regulation of BW research utilizing genetic engineering research. The Department of Defense had released a statement indicating that all government research was in compliance with the 1972 BW Convention. The government has also prepared an environmental impact statement of research conducted at Fort Detrick.

Government regulations on genetic engineering also affect BW research through limitations on exports of biotechnology information, research products, and equipment. In addition to controls of exports due to competitive concerns of biotechnology companies, a substantial amount of information and equipment related to genetic engineering is prohibited from being exported outside the United States. The Commerce Department maintains a "militarily critical technology" list, which serves as an overall guide to restricted exports. Included on the list are containment and decontamination equipment for large production facilities, high-capacity biological reactors, separators, extractors, dryers, and nozzles capable of disseminating biological agents in a fine mist.

Genetic engineering has altered the concept of BW. A current, comprehensive discussion of BW would include both naturally occurring and potential genetically engineered agents. Many current defenses against BW are developed with genetic engineering techniques. Government regulations on biotechnology have limited BW research, while fears of virulent genetically engineered BW agents have strengthened public support for stronger regulations. Future policies related to BW will need to be addressed in light of their altered status.

B. Mathematical Epidemiology Models

While genetic engineering may potentially alter characteristics of BW agents, mathematical models of epidemiology may provide military planners with techniques for predicting the spread of a released BW agent. One of the hindrances that has prevented BW from being utilized or even seriously considered by military leaders has been the inability to predict the spread of a BW agent once it has been released into the environment. Without the capability to predict the spread of the released organisms, military planners would risk the accidental exposure of their own troops and civilians to their own weapons. The development of advanced epidemiology models may provide the necessary mechanisms for predicting the spread of organisms that would substantially decrease the deterrent factor of unpredictability. [*See* EPIDEMIOLOGIC CONCEPTS.]

C. Low-Level Conflict

Another important factor that has affected the current status of BW is the increase in low-level conflict or the spectrum of violent action below the level of small-scale conventional war, including terrorism and guerrilla warfare. In the 1980s, the low-intensity conflict doctrine, which was espoused by the Reagan administration, was a plan for U.S. aid to anti-Communist forces throughout the world as a way of confronting the Soviet Union without using U.S. combat troops. Despite the significant changes in the world since the inception of the low-intensity conflict doctrine, the probability of increasing numbers of small conflicts still exists for the 1990s and beyond. Although no evidence indicates that the United States would consider violating the 1972 BW Convention and supporting biological warfare, the overall increase in low-level conflicts in the future may help create an environment conducive to the use of BW.

While BW may not be assessed as effective weapons in a full-scale conventional war, limited use of BW agents may be perceived as advantageous in a small-scale conflict. While strong deterrents exist for nuclear weapons, including unavailability and, most formidably, the threat of uncontrolled worldwide "nuclear winter," BW may be perceived as less dangerous. Additionally, the participants of low-level conflicts may not possess the finances for nuclear or conventional weapons. BW agents, like chemical weapons, are relatively inexpensive compared to other weapon systems and may be seen as an attractive alternative to the participants and leaders of low-level conflicts. Low-level conflict, therefore, increases the potential number of forums for the use of BW.

D. Terrorism

A final factor that could significantly affect BW is the worldwide increase in terrorism, or the violent activities of subnational groups. Although there has not been an incident to date of the successful use of BW by a terrorist group, the possibility of such an event has been raised in a number of forums.

The relationship of terrorism and BW can be divided into two possible events. The first is terrorist acts against laboratories conducting BW-related research. The level of security at Fort Detrick is high, the possibility of a terrorist attack has been anticipated, and contingency plans have been made. Complicating the problem of providing security against terrorist attack in the United States is the fact, that, while most BW research projects are conducted within the BW research program of the Army, an increasing number of projects are supported by the government that are conducted outside of the military establishment. These outside laboratories could be the potential targets.

The second type of terrorist event related to BW is the potential use of BW by terrorists. Biological weapons are relatively inexpensive and easy to develop and produce compared to conventional, nuclear, or chemical weapons. BW agents can be concealed and easily transported across borders or within countries. Additionally, terrorists are not hampered by a fear of an uncontrolled spread of the BW agent into innocent civilian populations. To the contrary, innocent civilians are often the intended targets of terrorist activity and the greater chance for spread of the BW agent may be considered to be a positive characteristic.

E. Offensive versus Defensive Biological Warfare Research

The distinctions between "offensive" and "defensive" BW research have been an issue since 1969, when the United States unilaterally pledged to conduct only defensive research. The stated purpose of the U.S. BDRP is to maintain and promote national defense from BW threats. Although neither the Geneva Convention nor the 1972 Convention prohibits any type of research, the only research that nations have admitted to conducting is defensive. The problem is whether or not the two types of research can be differentiated by any observable elements.

Although production of large quantities of a virulent organism and testing of delivery systems has been cited as distinguishing characteristics of an offensive program, a substantial amount of research leading up to these activities, including isolating an organism, then using animal models to determine pathogenicity, could be conducted in the name of defense.

Vaccine research is usually considered "defensive," whereas increasing the virulence of a pathogen and producing large quantities is deemed "offensive." However, a critical component of a strategic plan to use biological weapons would be the production of vaccines to protect the antagonist's own personnel (unless self-annihilation was also a goal). This means that the intent of a vaccine program could be offensive BW use. Furthermore, research that increases the virulence of an organism is not necessarily part of an offensive strategy because one can argue that virulence needs to be studied in order to develop adequate defense.

The key element distinguishing offensive from defensive research is intent. If the intent of the researcher or the goals of the research program is the capability to develop and produce BW, then the research is offensive BW research. If the intent is to have the capability to develop and produce defenses against BW use, then the research is defensive BW research. While it is true that nations may have policies of open disclosures (i.e., no secret research), "intent" is not observable.

Although the terms "offensive BW research" and "defensive BW research" may have some use in describing intent, it is more a philosophical than a practical distinction, one that is based on trust rather than fact.

F. Secrecy in Biological Warfare-Related Research

Neither the Geneva Protocol nor the 1972 BW Convention prohibits any type of research, secret or nonsecret. While the BDRP does not conduct secret or classified research, it is possible that secret BW research is being conducted in the United States outside of the structure of the BDRP. The classified nature of the resource material for this work makes it impossible to effectively determine if secret research is being conducted in the United States or any other nation.

It is not, however, unreasonable to assume that other nations conduct significant secret BW research. Therefore, regardless of the facts, one cannot deny the perception that such research exists in a variety of countries and that this perception will exist for the foreseeable future.

Secrecy has been cited as a cause of decreased quality in BW research. If secret research, whether offensive or defensive, is being conducted in the United States or other nations, it is unclear if the quality of the research is affected by the process of secrecy. If the secret research process consists of a core of highly trained, creative, and motivated individuals sharing information, the quality of the research may not suffer significantly. It must be stated, however, that secrecy by its very nature will limit input from a variety of diverse observers.

Secrecy may increase the potential for violations of the 1972 BW Convention, however, violations would probably occur regardless of the secrecy of the research. Secrecy in research can certainly lead to infractions against arbitrary rules established by individuals outside of the research group. The secret nature of the research may lure a researcher into forbidden areas. Additionally, those outside of the research group, such as policy-makers, may push for prohibited activities if the sense of secrecy prevails. Secrecy also tends to bind those within the secret arena together and tends to enhance their perception of themselves as being above the law and knowing what is "right." As in the case of Oliver North and the Iran–Contra Affair, those within the group may believe fervently that the rules must be broken for a justified purpose and a mechanism of secrecy allows violations to occur without penalty.

The distrust between nations exacerbates the perceived need for secret research. The animosity between the United States and the Soviet Union during the 1980s fueled the beliefs that secret research leading to violations of the 1972 BW Convention was being conducted in the Soviet Union. As the belligerence of the 1980s fades into the New World Order of the 1990s, the questions will not focus on the Soviet Union as much as on the Middle East and third-world countries. There are factions in the United States that believe strongly that other countries are conducting secret research that will lead to violations of the Convention. There is also a tendency to believe that the secrecy in one's own country will not lead to treaty violations, while the same secret measures in an enemy nation will result in activities forbidden by international law.

The importance of the concept of secrecy in BW research is related to the perception of secrecy and arms control agreements. Regardless of the degree of secrecy in research, if an enemy believes that a nation is pursuing secret research, arms control measures are jeopardized. The reduction of secrecy has been suggested as a tool to decrease the potential for BW treaty violations. A trend toward reducing secrecy in BW research was exemplified by the 1986 Review Conference of the 1972 BW Convention, which resulted in agreements to exchange more information and to publish more of the results of BW research. Whether or not these measures have any effect on strengthening the 1972 BW Convention remains to be seen.

Organizations and individuals have urged a renunciation by scientists of all secret research and all security controls over microbiological, toxicological, and pharmacological research. This action has been suggested as a means of strengthening the 1972 BW Convention. The belief that microbiologists should avoid secret research is based on the assumption that (1) secret research is of poor quality due to lack of peer review and (2) secrecy perpetrates treaty violations.

While it may be reasonable to expect microbiologists to avoid secret research, it is not realistic. Secrecy is practiced in almost every type of research including academic, military, and especially industrial. Furthermore, there will always be those, within the military and intelligence structures, who believe that at least some degree of secrecy is required for national security.

Secrecy in BW research is a complex issue. The degree to which it exists is unclear. Individuals are generally opposed to secrecy in BW research while other examples of secrecy in different types of research exist and are accepted. If the assumption is made that some secrecy does exist, then the effect of the secrecy on the quality of research, the need for the secrecy, and the choice of microbiologists to participate in secret BW research remain unanswered questions.

G. Problems Relating to Verification

One of the major weaknesses of the 1972 BW Convention has been the lack of verification protocols. Problems with effectively monitoring compliance include the ease of developing BW agents in laboratories designed for other purposes, and the futility of inspecting all technical facilities of all nations. Measures that have been implemented with the goal of monitoring compliance have included (1) open-inspections, (2) intelligence gathering, (3) monitoring of research, (4) use of sampling stations to detect the presence of biological agents, and (5) international cooperation. The progress achieved with the Chemical Weapons Convention has renewed interest in strengthening mechanisms for verification of compliance with the 1972 BW Convention. While this renewed interest in verification along with the emergence of the Commonwealth of Independent States from the old Soviet Union has brought an optimism to the verification issue, the reticence of countries such as Iraq to cooperate with United Nations inspection teams is a reminder of the complexities of international agreements.

The examples herein are typical of the many issues attached to the concept of BW.

Bibliography

Asinof, R. (1984). Averting genetic warfare. *Environ. Action* **June,** 16–22.

Buckingham, W. A., Jr. (ed.) (1984). "Defense Planning for the 1990s." National Defense University Press, Washington, D.C.

Frisna, M. E. (1990). The offensive–defensive distinction in military biological research. *Hastings Cent. Rep.* **20**(3), 19–22.

Gravett, C. (1990). "Medieval Siege Warfare." Osprey Publishing Ltd., London.

Harris, R., and Paxman, J. (1982). "A Higher Form of Killing." Hill and Wang, New York.

Livingstone, N. C. (1984). Fighting terrorism and "dirty little wars." *In* "Defense Planning for the 1990s" (W. A. Buckingham, Jr., ed., pp. 165–196. National Defense University Press, Washington, D.C.

Livingstone, N. C., and Douglass, J., Jr. (1984). "CBW: The Poor Man's Atomic Bomb." Institute of Foreign Policy Analysis, Tufts University, Medford, Massachusetts.

Marshall, E. (1988). Sverdlovsk: Anthrax capital? *Science* **22 April,** 383–385.

Milewski, E. (1985). Discussion on a proposal to form a RAC working group on biological weapons. *Recombinant DNA Tech. Bull.* **8**(4), 173–175.

Miller, L. A. (1987). The use of philosophical analysis and Delphi survey to clarify subject matter for a future curriculum for microbiologists on the topic of biological weapons. Thesis, University of Pennsylvania, Philadelphia. University Microfilms International, Ann Arbor, Michigan. 8714902.

Murphy, S., Hay, A., and Rose, S. (1984). "No Fire, No Thunder." Monthly Review Press, New York.

Poupard, J. A., Miller, L. A., and Granshaw, L. (1989). The use of smallpox as a biological weapon in the French and Indian War of 1763. *ASM News* **55,** 122–124.

Press, N. (1985). Haber's choice, Hobson's choice, and biological warfare. *Perspect. Biol. Med.* **29**(1), 92–108.

Smith, R. J. (1984). The dark side of biotechnology. *Science* **224,** 1215–1216.

Stockholm International Peace Research Institute (1973). "The Problem of Chemical and Biological Warfare," Vol. II. Humanities Press, New York.

Wright, S. (1985). The military and the new biology. *Bull. Atomic Sci.* **41**(5), 73.

Wright, S., and Sinsheimer, R. L. (1983). Recombinant DNA and biological warfare. *Bull. Atomic Sci.* **39**(9), 20–26.

Bioluminescence, Bacterial

Edward A. Meighen
McGill University

Glossary

Bioluminescence Light emission by living organisms arising by exergonic chemical reactions

Induction Relative increase in luminescence with cellular growth controlled at least in part by compounds synthesized by the bacteria

Luciferases Enzymes that catalyze bioluminescent reactions

Lux Refers to the luminescent system of light-emitting bacteria (e.g., *lux* genes)

BIOLUMINESCENT BACTERIA are found widely distributed in nature in aquatic and terrestrial environments. Their primary habitat is the ocean either as free-living species or associated with other marine organisms in a symbiotic, parasitic, or saprophytic relationship. Although the function and regulation of luminescence in these different modes is not well understood, our knowledge of the biochemistry and molecular biology of the enzymes and genes in the bacterial luminescent system has significantly advanced. Luciferases and other lux-specific enzymes have now been characterized from a number of bacterial species, and the *lux* genes have been cloned and sequenced. Applications of light-emitting systems are becoming of widespread interest, and the ability to transfer the *lux* genes into prokaryotic and eukaryotic cells and organisms to create new luminescent systems has greatly expanded the scope and potential uses of bacterial bioluminescence.

I. Introduction

A. Luminescence

The terms bioluminescence and chemiluminescence refer to the processes of visible light emission that occur via exergonic chemical reactions. These terms should not be confused with fluorescence and phosphorescence, which involve remission of light from the singlet and triplet excited states, respectively, and depend on the prior absorption of light. Bioluminescence differs from chemiluminescence in that the former process occurs in living organisms and is mediated by an enzyme catalyst. Although bioluminescence is considered in many instances as a specialized form of chemiluminescence, in general, the parallel usage of the terms chemiluminescence and bioluminescence in the titles of books, journals, and published articles has supported a clear distinction between them.

Bioluminescence is not the only process that results in light emission from living organisms. The photoluminescent processes of fluorescence and phosphorescence also occur in living cells. It should be noted that in a bioluminescent reaction, the excited molecule is generally considered to be in the singlet state and, thus, its emission spectrum can be equated to that of the analogous fluorescent species.

B. A Bright Past

Light emission by living organisms is widespread and has been a matter of record for over 2500 years. Unfortunately even today, it is often referred to as phosphorescence rather than bioluminescence. The earliest recorded observations were made by the ancient Chinese, which preceded the accounts in Western civilization by the Greeks, including those of the famous naturalist and philosopher Aristotle in the fourth century B.C. Over the subsequent centuries, many luminescent sightings were made, including reports by famous explorers such as Christopher

Columbus and Sir Francis Drake during the period of extensive sea exploration in the fifteenth and sixteenth centuries.

In the nineteenth century, the French physiologist Raphael Dubois first recognized the chemical nature of the components involved in bioluminescence in experiments in which he generated light by mixing hot and cold water extracts from luminous species. From these results, bioluminescence was proposed to involve a heat-stable substrate (luciferin) destroyed or oxidized by the action of a heat-labile enzyme catalyst (luciferase). In addition, mixing of luciferins and luciferases from different luminescent species did not produce light, thus providing evidence for differences in the chemical components. The requirement for oxygen for luminescence was also demonstrated by Dubois in agreement with the classical work over two centuries earlier by the famous English physicist Robert Boyle, who demonstrated the loss of light on placing rotten wood or fish containing luminescent fungi or bacteria in a vacuum. The effects of oxygen limitation on light emission of luninescent bacteria can readily be observed in cultures where air is limited (Fig. 1).

Since the nineteenth century, the structures of the substrates and the luciferases from a number of different luminescent organisms have been identified. Although more than one substrate as well as oxygen is often required for the luminescent reactions in different organisms, the general model proposed by Dubois has not changed.

C. Occurrence of Bioluminescent Organisms

The phenomena of bioluminescence has been observed in the phyla of many terrestrial and aquatic organisms. Included among these species are bacteria, fungi, fish, insects, algae, squid, and clams. Perhaps the most familiar example is the flashing of fireflies as a mating signal. Of these luminescent organisms, the majority are found in the marine or seawater environment. However, despite the widespread distribution among the different phyla includ-

Figure 1 Dependence of luminescene of liquid cultures of bacteria on oxygen. After standing for a few minutes, the culture on the right becomes dim at the bottom as the bubbles of air rise to the top and the oxygen content is depleted.

ing those in the plant and animal kingdoms, there are no luminescent mammalian species, viruses, or higher plants, and, except for certain fish, luminous representatives are not found among the vertebrates. Moreover, in many instances, only one species in a genus may be luminescent while many closely related species do not emit light.

In most cases, the enzymes that catalyze the bioluminescence reactions are called luciferases, and at least one of the component substrates is designated as luciferin. Significant differences exist among the bioluminescence reactions of different organisms including the structures and properties of the luciferases and luciferins. The only common chemical feature of bioluminescence reactions is the requirement for molecular oxygen. Although this result indicates that the luminescent systems have evolved independently, in a few cases, striking similarities in the luciferins of the luminescent systems from different phyla can be noted.

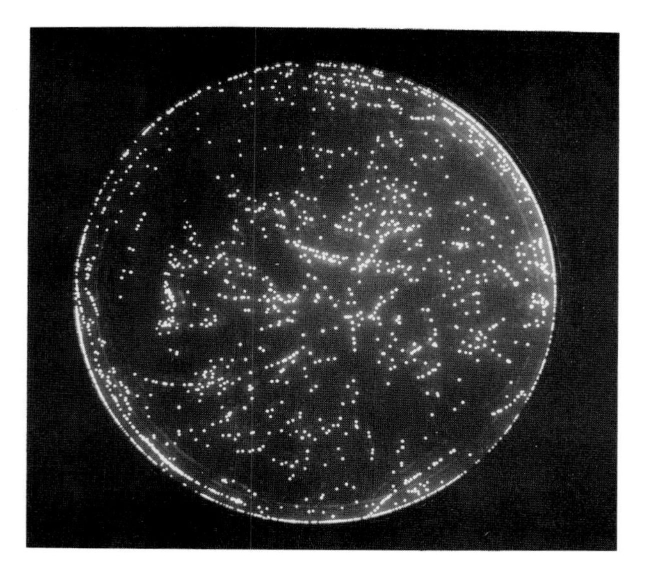

Figure 2 Culture of luminescent marine bacteria grown on solid media.

II. Ecology and Physiology

A. Taxonomy, Habitat, and Distribution

The most widely distributed luminescent organisms are the luminescent bacteria, with most of these species found either free-living or associated with other organisms in the marine environment. The dramatic brilliance of a bright culture of luminescent bacteria growing on solid (Fig. 2) or in liquid (Fig. 1) media illustrates why humankind has been intrigued with light-emitting organisms in the past as well as their potential for use as reporters of metabolic functions. Luminescent bacteria are also found in freshwater and terrestrial environments. The primary habitats of the luminescent bacteria are as saprophytes growing on dead fish or meat, as parasites in crustaceans and insects, as light organ symbionts in fish and squid, as gut symbionts in fish and nematodes, and as free-living species in the ocean. Bioluminescence of many but not all fish arises due to the presence of luminous bacteria. [*See* MARINE HABITATS, BACTERIA.]

Luminous bacteria are primarily classified into three genera with almost all the marine luminescent bacteria being classified in the *Vibrio* and *Photobacterium* genera and the luminescent *Xenorhabdus* species being terrestrial in nature. One luminescent freshwater species, *Vibrio cholerae* (biotype albensis), has also been detected.

The species of luminescent bacteria that have been most closely characterized are *Vibrio harveyi, Vibrio fischeri, Photobacterium phosporeum, Photobacterium leoignathi,* and *Xenorhabdus luminescens*. However, changes in the classification of the luminous bacteria have occurred, with many of the *Vibrio* species having been renamed over the last 20 years. *Vibrio harveyi* was formerly designated both as *Beneckea harveyi* and as *Photobacterium fischeri,* strain MAV. Similarly, *V. fischeri* was previously named *P. fischeri*.

Luminous bacteria are gram-negative motile rods and, in most cases, facultative anaerobes that can grow on a variety of carbon sources. Marine luminescent bacteria also require salt and show optimum growth between 50 and 300 m*M* sodium chloride. These bacteria can be found at temperatures ranging from close to freezing in the Arctic oceans and the lower depths to up to 30°C in shallow and tropical waters.

The *Vibrio* and *Photobacterium* species can be distinguished by the presence of sheathed and unsheathed flagella, respectively. *Vibrio fischeri* species have been found as specific luminous symbionts of the monocentrid fishes. *Vibrio harveyi* species are found primarily in shallow and coastal waters and have not yet been isolated in a specific association with a marine species. Both *P. phosphoreum* and *P. leoignathi* have been isolated as symbionts in specific marine hosts, with the former species generally found at lower depths of the ocean and colder temperatures. Most luminescent marine bacteria that have been studied can also be directly isolated from seawater as well as from the digestive tract of marine fish and as saprophytes growing on dead fish or meat. In this regard, many luminescent strains have been isolated as saprophytes and/or growing on the skin of fish, thus explaining the luminescence of rotting fish observed by Robert Boyle over three centuries ago.

The terrestrial luminescent bacteria, *Xenorhabdus luminescens,* has been isolated primarily as a symbiont of nematodes acting as parasites of caterpillars; however, strains have also been isolated from human wounds. Because the terrestrial as well as the marine luminescent bacteria are believed to be harmless to most organisms including humans, the observation of luminescence in open battlefield wounds reported in the nineteenth century was taken as a positive sign of recovery. It should be emphasized that literally hundreds of different existing strains of luminescent bacteria have not yet been investigated. Moreover, some of these luminescent bacteria are symbionts, in which the conditions for growth in the laboratory are unknown, or obligative symbiotes, which cannot be cultured outside the host. [*See* SYMBIOTIC MICROORGANISMS IN INSECTS.]

B. Functions of Luminescence

The function of light emission by bacteria is not well understood, perhaps because they are found in many types of relationships including symbiotic, saprophytic, parasitic, and free-living associations. In the symbiotic mode in the light organs of fish, the bacteria can receive nutrients and the fish gains a light-emitting source as well as the potential use of additional metabolic pathways supplied by the bacteria. Luminescence by fish has been proposed to serve for communication, defense and/or attraction, and predation. In the angler fish, a glowing bulb dangling in front of the fish can serve to attract prey, in essence turning the fish into a fisherman. The flashlight fish has shutters that can block light emitted by light organs below the eye. Luminescence could be used not only to increase visibility but also to lure prey, divert predators, and possibly even to communicate. In some luminescent fish, light organs on its periphery have been proposed to provide counter-illumination so that its silhouette against the bright ocean surface is less visible to predators lurking below.

The function of luminescence in the parasitic and saphrophytic modes is less obvious, but it may be related as much to the metabolism of the bacteria as to light emission. For example, the production of an extracellular chitinase by marine luminescent bacteria results in the digestion of chitin, thus supporting growth on dead fish and crustaceans. Luminescence would then attract marine organisms to this food source resulting in the propagation and dispersal of the luminescent bacteria. For terrestrial bacteria, the symbiotic relationship enables the nematode to feed on infected caterpillars; luminescence could be used to attract other animals that would eat the luminescent insects and further disperse the luminescent bacteria. However, nonluminescent bacteria exist with similar metabolic capabilities, leaving in doubt the specific role of light emission in these particular habitats.

III. Biochemistry

A. Bioluminescent Reaction

1. Substrates

The substrates for the bioluminescent reaction are reduced flavin mononucleotide ($FMNH_2$), a long-chain fatty aldehyde, and molecular oxygen. The reaction, catalyzed by bacterial luciferase, results in the oxidation of $FMNH_2$ to FMN and fatty aldehyde to the corresponding fatty acid:

$$FMNH_2 + O_2 + R\text{-}CHO \longrightarrow FMN + H_2O + RCOOH + \text{light}$$

The structures of the substrates for the bioluminescent reaction (Fig. 3) are the simplest and the most closely related to compounds in common metabolic pathways of those for any known luminescent reaction. For this reason, the substrates have generally not been referred to as luciferins, although the fatty aldehyde has been considered as the luciferin in some cases because $FMNH_2$ is a common cofactor in many reactions.

Fatty aldehydes from 7 to 16 carbons can function as effective substrates in the reaction with tetradecanal, which is believed to be the natural substrate. Unsaturated or substituted aldehydes of the same chain length also result in high levels of

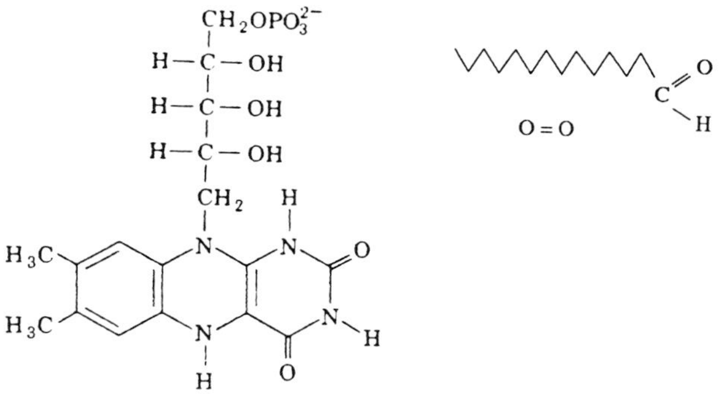

FMNH$_2$

Figure 3 Structure of the substrates for the bacterial bioluminescent reaction.

light, provided that structural changes are not in close proximity to the carbonyl group. The reaction is highly specific for FMNH$_2$.

2. Light Emission

The bioluminescence reaction results in the emission of a blue-green light with a maximum intensity centered at 490 nm and a quantum efficiency between 0.05 and 0.15. The source of energy required for the emission of the blue-green light at 490 nm (60 kcal/mol) arises primarily from the oxidation of aldehyde as well as FMNH$_2$. Differences exist between the luminescence spectrum observed *in vitro* and the color of the emitted light observed *in vivo* for some luminescent species. These differences are believed to be due to proteins found in some luminescent bacteria that act as sensitizers, resulting in the emission of light at a different wavelength. In *Photobacterium* species, the sensitizer is the lumazine protein, which causes a small blue shift in the emitted light toward higher energy or a lower wavelength. In one strain of *V. fischeri,* a yellow fluorescence protein causes a shift in the wavelength of light emission to 540 nm so that the color of the light emitted by the cells is now yellow.

B. Properties of Luciferases

1. Purification and Stability

Luciferases have been purified from the luminescent (lux) systems of a number of different species of bacteria including both marine and terrestrial species. The amounts of luciferase in the luminescent bacteria, particularly from the marine species, is remarkably high and can reach up to 20% of the soluble protein extracted from the cell. The luciferases are devoid of any prosthetic groups, metals, or non-amino acid residues in their structure, are reasonably stable, and can be readily purified by standard chromatographic techniques. There are differences among the thermal stabilities of the luciferases, reflecting to a large degree the native habitat of the bacteria, with *P. phosphoreum* luciferase having the lowest and *X. luminescens* luciferase the highest thermal stability.

2. Primary and Quaternary Structures

All bacterial luciferases are heterodimers ($\alpha\beta$) with molecular masses of approximately 80 kDa (Table I). The luciferase subunits, α and β, have molecular masses of 41 and 37, respectively. The subunits have approximately 30% identity with each other in amino acid sequence because the corresponding genes have arisen by gene duplication. The sequences of the α-subunits from luciferases of different species can differ by as much as 45% with an even greater divergence between the sequences of the β-subunits (up to 55% difference). At present, the sequences of luciferases from over 10 strains representing at least five species from three genera have been determined. Only 30% of the amino acids have been conserved in the β-subunit of all luciferases compared to conservation of 44% of the amino acid residues in the α-subunit, indicating a more rapid divergence of the former subunit.

The kinetic properties of luciferase are primarily dictated by the α-subunit and, consequently, interspecies hybrids formed from the two subunits have properties more closely related to the luciferase contributing the α-subunit. Very little is known about the specific residues at the active site involved in the catalytic mechanism.

Table I Common lux-Specific Enzymes of Bioluminescent Bacteria

Enzyme	Structure	Subunits (mass)	Function
Luciferase	$\alpha\beta$	α (41 kDa) β (37 kDa)	Bioluminescence Reaction
Fatty acid reductase	$r_4s_4t_{2-4}$	r (54 kDa) s (43 kDa) t (33 kDa)	Aldehyde Biosynthesis

r, reductase; s, synthetase; t, transferase.

C. Bioluminescent Assays

1. Standard Assay

The standard assay for the luminescent reaction involves the injection of $FMNH_2$ (50 μM) into an assay solution containing luciferase, and the long-chain aldehyde. Light emission rises rapidly to a maximum and then decays in an exponential fashion (Fig. 4), with the maximum light emission given by I_o and the decay rate by k_L. These parameters depend on the aldehyde (Table II) and the particular bacterial luciferase. Because $FMNH_2$ is oxidized within the first second of the reaction, a stable enzyme–flavin intermediate must be formed. This intermediate is believed to be a flavin peroxy derivative noncovalently bound to the enzyme that can react with the fatty aldehyde, resulting in the formation of the excited state and the emission of light. Because luciferase can only undergo one catalytic cycle in this luminescent assay, integration of the total light emission represented by the area under the curve (I_o/k_L) in Fig. 4 gives the total quanta of light emitted for a single turnover of the enzyme under these conditions.

Table II Dependence of Luciferase Activity on Aldehyde Chain Length

Aldehyde	Relative activity (I_o)[a]	Turnover rate (k_L) (sec^{-1})
Heptanal	0.5	0.05
Octanal	1	0.05
Nonanal	8	0.37
Decanal	8	0.38
Undecanal	1.4	0.073
Dodecanal	1	0.056
Tetradecanal	12	0.9

[a] Maximum light intensity in the standard luminescent assay for *Vibrio harveyi* luciferase at saturating aldehyde concentrations relative to that obtained with dodecanal.

2. Dithionite Assay

An alternative assay is the dithionite assay, in which a solution of the fatty aldehyde is injected into an assay solution containing luciferase and $FMNH_2$ that has been reduced with sodium dithionite. The dependence of light emission on time is similar in the dithionite and standard assays. For these assays, the samples must be rapidly mixed by injection, and suitable apparatus is required to measure and record the luminescence within the first second of mixing.

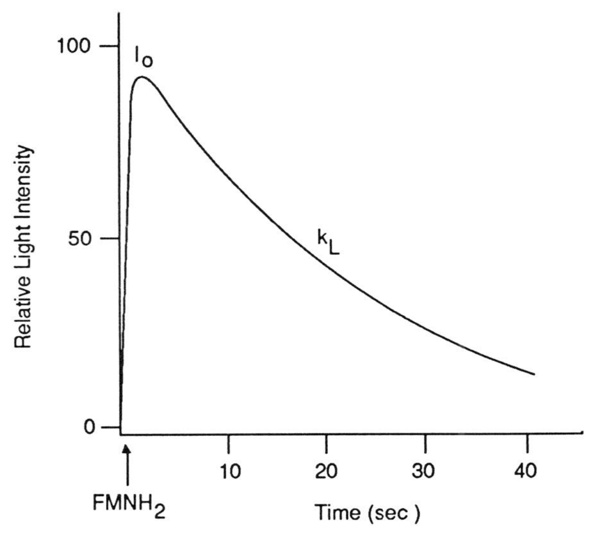

Figure 4 Standard assay for bacterial luciferase. $FMNH_2$ (1.0 ml) was injected into 1.0 ml of phosphate buffer containing 0.001% dodecanal and *Vibrio harveyi* luciferase. The light was detected by a photomultiplier tube and recorded graphically.

3. Coupled Assay

Continuous light emission in luminescent assays only can be obtained by regeneration of $FMNH_2$ by addition of NAD(P)H and an enzyme with NAD(P)H–FMN oxidoreductase activity to solutions containing luciferase, fatty aldehyde, and FMN. Under these conditions, $FMNH_2$ can be continually produced and luciferase can undergo multiple catalytic cycles. Maximum light emission in these NAD(P)H-coupled assays is lower than that in the standard or dithionite assay; however, continuous light emission can be maintained for a number of hours without a large decrease in light intensity. Consequently, using the coupled assay to readily measure luminescence on film, in scintillation counters, or with other equipment without rapidly mixing the components required for the luminescence reaction is possible.

4. Light Standards

Light intensities reported by different investigators are generally given in arbitrary units but are not always referred to as light units. Because the variation in luminescence can extend over an extremely wide range, depending on the assay as well as on the photomultiplier tubes and accessory equipment for light detection, reference to a standard giving light intensity in quanta/sec is important for comparing the relative levels of brightness reported in different experiments. For example, as little as 1 nl of a dense culture of luminescent *Photobacterium* cells can emit too much light to be recorded in a scintillation counter, whereas the level of light emission may be too low to detect with other less-sensitive equipment. Consequently, the terms bright, high response, and relative ease of detection may have significantly different meanings to different investigators.

Two different standards have been used most often to convert relative luminescence into quanta/sec. A primary standard based on the quanta emitted in the chemiluminescent reaction of luminol and H_2O_2 appears to be the best reference point for calibration of light emission. A more convenient light reference with some light-detecting equipment is a standard consisting of a radioactive carbon source.

D. Biosynthesis of Substrates for the Luminescent Reaction

1. Reduced Flavin Mononucleotide

The immediate source of the $FMNH_2$ substrate for the luminescent reaction is as yet unknown. NAD(P)H–FMN oxidoreductases have been purified from luminescent bacteria that catalyze the reduction of free FMN to $FMNH_2$; however, their direct relationship to the bacterial lux system has not been demonstrated. Moreover, transfer of the luciferase genes into nonluminescent bacteria has shown that an adequate supply of $FMNH_2$ is present in other types of bacteria to sustain light emission. Consequently, it is generally accepted that the electron transport chain found in all bacteria is the immediate source of the reducing power ($FMNH_2$). The bioluminescent reaction results in a shunt of the electrons from $FMNH_2$ to oxygen, thus diverting the electron flow away from the more common metabolic pathways.

2. Fatty Aldehyde

The pathway for biosynthesis of the fatty aldehyde substrate for the bioluminescent reaction has been identified and involves enzymes specifically related to the bacterial lux system. The primary reaction involves the reduction of fatty acid to the corresponding aldehyde in an adenosine triphosphate (ATP) and reduced nicotinamide adenine dinucleotide phosphate (NADPH)-dependent reaction:

$$\text{R-COOH} + \text{ATP} + \text{NADPH} \longrightarrow \text{R-CHO} + \text{AMP} + \text{PP}_i + \text{NADP}$$

where PP_i is inorganic pyrophosphate. Aldehyde biosynthesis is catalyzed by a fatty acid reductase enzyme complex (Table I) that contains three different protein components designated as reductase, synthetase, and transferase. Only the reductase and synthetase components are required for the preceding reaction; the synthetase catalyzes the activation of the fatty acid to form a fatty acyl-AMP intermediate with the activated fatty acyl group being transferred to the synthetase. The acyl group is then transferred to the reductase before being reduced with NADPH to the fatty aldehyde. The transferase component generates the fatty acids for the reaction by catalyzing the cleavage of fatty acyl derivatives [(e.g., fatty acyl-carrier protein (ACP)] formed via fatty acid biosynthesis.

Although fatty acid reductase activity has only been measured in extracts of *Photobacterium* species and not in *Vibrio* or *Xenorhabdus* species, the reductase, synthetase, and transferase polypeptides have been identified in these species because each of the subunits can be specifically labeled *in vivo* with radioactive fatty acids or *in vitro* with fatty acid (+ATP) and/or fatty acyl-CoA or -ACP. The conversion of fatty acids to aldehyde has been shown to be highly specific for tetradecanoic acid, consistent with tetradecanal being the natural aldehyde substrate for the luminescent reaction.

The enzyme responsible for fatty acid reductase activity has only been purified from the luminescent bacteria *P. phosphoreum,* even though similar activities have been reported in extracts of other biological systems. The enzyme has a mass of approximately 5×10^5 and is composed of a central core of four reductase subunits that interact in a 1 : 1 stoichiometry with the synthetase subunits. In turn, the transferase subunits interact weakly with the synthetase subunits. Interaction of the transferase with the synthetase affects the activity of both enzyme components. Similarly, the reductase is required for efficient reaction of fatty acyl-AMP with the synthetase subunit. Consequently, formation of the fatty acid reductase complex can control the diversion of fatty acids into aldehyde. The possibility that the aldehyde is then channeled directly to luciferase should be considered because this would prevent the loss of aldehyde via competing metabolic pathways as well as maintain the level of free aldehyde at a low concentration. The fatty acid generated in the luminescent reaction could then be recycled back to fatty aldehyde via the fatty acid reductase reaction.

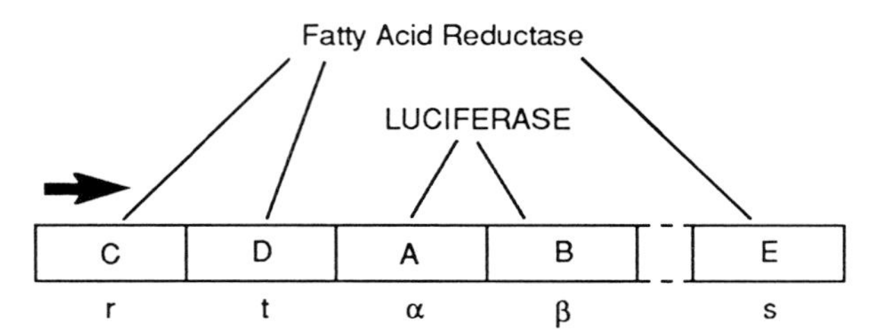

Figure 5 Organization of the common *lux* genes in luminescent bacteria and their corresponding protein products. r, reductase; s, synthetase; t, transferase.

IV. Molecular Biology

A. Cloning of the *lux* Genes

Almost 10 years ago, *lux* genes from luminescent bacteria were first cloned, and the structural genes coding for luciferase and fatty acid reductase were shown to be closely linked. The *lux* genes have been isolated from a number of different luminescent bacteria including marine and terrestrial species. Although the nucleotide sequences of the *lux* genes from different species have diverged to some degree, lux DNA probes from one luminescent bacteria will hybridize to the lux DNA from a different luminescent bacteria. Consequently, the lux DNA can be readily identified in almost any luminescent bacteria and subsequently isolated.

In most cases, the *lux* genes have been characterized by transfer into *Escherichia coli*. Because sufficient $FMNH_2$ and O_2 are present in most if not all bacteria, only aldehyde needs to be added to the *E. coli* cells containing the luciferase genes to generate a luminescent phenotype. However, if the genes for aldehyde biosynthesis are also transferred, then stable luminescent *E. coli* are produced requiring no exogenous additions. The level of luminescence highly depends on the bacteria and, in this case, the particular *E. coli* strain as well as the composition of the recombinant vector.

B. Common *lux* Structural Genes

In all luminescent bacteria, five *lux* structural genes (*luxA–E*) have been detected coding for the α- and β-subunits of luciferase (*luxA* and *luxB*, respectively) and the reductase (*luxC*), synthetase (*luxE*), and transferase (*luxD*) polypeptides of the fatty acid reductase complex. These genes are closely linked and are believed to be part of a single operon in the order *luxCDABE* (Fig. 5).

C. Other *lux* Genes

In addition to the five common *lux* genes, other genes closely linked to the lux system have been found in specific strains or species. As a result, the *lux* gene organization can differ from one luminescent bacteria to the other. For most, but not all, *Photobacterium* species, the *luxF* gene is located between *luxB* and *luxE*, whereas this gene is absent in the *Vibrio* and *Xenorhabdus* lux systems. The *luxF* gene codes for a 24-kDa nonfluorescent flavoprotein with homology in sequence to the luciferase subunits. After *luxE* in marine but not terrestrial bacteria, the *luxG* gene coding for a protein of unknown function has been identified and shown to be transcribed on the same polycistronic messenger RNA as the *lux* structural genes in the *Vibrio* species. An additional gene, *luxH*, is found downstream of *luxG* in *V. harveyi* but not in other luminescent bacteria.

Upstream of *luxC* in *V. fischeri* are regulatory genes (*luxIR*), whereas in other luminescent strains there is an extensive (>400 bp) noncoding region that is rich in AT content. The *luxI* gene is transcribed in the same direction as the *lux* structural genes, whereas *luxR* is immediately upstream of *luxI* on a divergent operon. The gene coding for the lumazine protein, designated as *lumP*, has recently been identified in *P. phosphoreum* 600 bp upstream of *luxC* and transcribed in the opposite direction. Genes located at loci other than that for the *lux* structural genes have also been found in luminescent bacteria, including regulatory genes in *V. harveyi* and a gene coding for a yellow fluorescent protein causing the emission of yellow rather than blue-green light in one specific *V. fischeri* strain.

D. Regulation

The appearance of luminescence during bacterial growth is a spectacular example of enzyme induction (Fig. 6). The development of luminescence lags remarkably behind growth and then can increase >100-fold. The induction of luminescence is very similar to developmental phenomena in higher organisms and apparently is under the control of an environmental sensing mechanism. Induction of luminescence in *V. fischeri* is regulated by a small metabolite, β-ketocaproyl-homoserine lactone, excreted into the media, which upon accumulation turns on expression of the lux system. This compound has been proposed to interact with the *luxR* gene product in *V. fischeri,* causing the stimulation of expression of the operon containing the *lux* structural genes.

An excreted metabolite with a similar structure, β-hydroxylbutyryl-homoserine lactone, has been shown to stimulate expression of the lux system in *V. harveyi,* although its mechanism of action may be different. As other bacteria also excrete compounds that stimulate luminescence in *V. harveyi,* other cellular functions may be under control of this signaling system.

Both inducers of the *Vibrio* lux systems are composed of an amino acid and a fatty acid metabolite and, thus, can serve as a signaling mechanism for the nutritional state of the cell. These molecules are remarkably similar in structure to isocapryl-δ-butyryl lactone, which controls development of sporulation in certain *Streptomyces* species.

A number of factors other than the autoinducers, including both activators and repressors, have been implicated in the control of expression of luminescence including cyclic AMP and its receptor, the lexA protein, and the sigma 32 protein involved in the response to stress. Clearly, multile nutritional and environmental signaling systems are involved in controlling luminescence at the transcriptional and translational levels. Moreover, control of the level of substrates, including oxygen, aldehyde, and $FMNH_2$, as well as the expression of the genes responsible for their synthesis, will clearly change the luminescence intensity. The need for regulatory controls may reflect the necessity for many of these luminescent bacteria to exist in different modes including as symbiotic and free-living species. Because luminescence is expressed only at high cell density for many luminescent bacteria, free-living luminescent bacteria in the ocean do not emit light, thus preventing the potential energy drain. Understanding why luminescent bacteria are found in such a variety of habitats and identification of the factors controlling light emission would greatly advance our knowledge of symbiosis and gene regulation in biological organisms. Few systems could be more ideal.

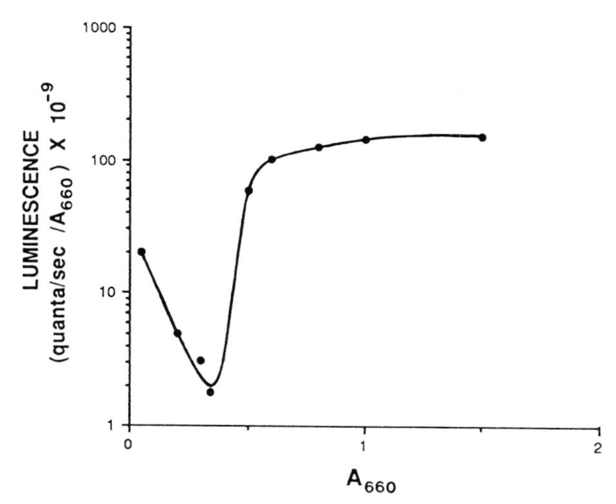

Figure 6 Dependence of specific luminescence on cellular growth (A_{660}) for liquid cultures of *Vibrio harveyi*.

V. Applications

The bacterial luminescent system and light emission is being applied as a sensitive, rapid, and safe assay in an ever-increasing number of biological systems. Recognition of the speed and simplicity of the measurement of light using photomultiplier tubes or film, the high sensitivity of the assay, the linear relationship over an extremely wide range between the concentrations of luciferase or substrates and the light signal, the absence of endogenous activities, and the safety, convenience, and low cost of the assay are some of the primary reasons for the rapid development of luminescent assays. Bacterial luciferases are used *in vitro* to detect metabolites that can be converted into one of the substrates for the luminescent reaction or the *lux* genes used *in vivo* to synthesize luciferase to monitor gene expression and/or metabolic function inside the cell.

A. *In Vitro* Applications

1. Fatty Aldehydes and Precursors

Coupled assays have been developed in which the light-emitting reaction catalyzed by bacterial luciferase is used to monitor the production of oxygen, fatty aldehyde, or $FMNH_2$, or closely related substrate analogs. Extremely low concentrations can be detected with the luminescent response being proportional to concentration. For example, the daily rhythm of release of long-chain aldehyde pheromones from an individual insect can be followed. Fatty alcohols, acids, and esters can also be converted into aldehydes and then analyzed in the bacterial luminescent reaction, providing sensitive assays for these metabolites and/or the enzymes involved in their production. Alternate assays using dark mutants of the luminescent bacteria with defects in aldehyde biosynthesis have also been used to detect fatty acids and aldehydes.

2. $FMNH_2$, NADH, and NADPH

The most widespread *in vitro* application of the luminescent system is the measurement of enzyme reactions in which a product of the reaction is either NADH or NADPH or can be converted into these compounds. NAD(P)H can then be used to reduce FMN in the coupled luminescent assay containing luciferase and a NAD(P)H : FMN oxidoreductase, resulting in a luminescent signal. Consequently, a wide range of enzyme reactions and/or metabolites can be monitored using the coupled luminescent assay. Both soluble and immobilized enzymes (luciferase and NAD(P)H–FMN oxidoreductase) have been used for these assays and have involved analysis of diverse metabolites such as alcohol and estrogens. Moreover, antibodies, antigens, and DNA labeled with enzymes that catalyze reactions leading to the production of NAD(P)H can also readily be monitored.

B. Expression in Prokaryotes and Eukaryotes

The ability to transfer the DNA coding for the luciferases (*luxAB*) as well as the other *lux* structural genes into different prokaryotic and eukaryotic organisms has resulted in the widespread application of the *lux* genes as sensors of gene expression and cellular function. Expression of the luciferase genes inside the cells depends on many factors, including the promoter proficiency and copy number of the recombinant lux DNA, stability and translation efficiency of the lux messenger RNA, and the folding and stability of luciferase in the *in vivo* milieu. Moreover, the level of substrates ($FMNH_2$, O_2, aldehyde) as well as the presence of any inhibitors inside the cell are just as important in determining the level of the luminescence signal. Where applicable, the question of substrate supply can be avoided simply by lysing the cells and measuring the luciferase activity *in vitro*.

1. Prokaryotes

The luciferase genes have been transferred into bacteria from over 25 genera. The recombinant lux DNA can be transferred by transformation, transduction, or transconjugation, with the most efficient method of transfer depending on the particular bacteria. Transformation has generally been used to transfer DNA into gram-positive bacteria and enteric bacteria including *E. coli,* whereas conjugation using biparental and triparental mating has been used for phytopathogenic and marine bacteria. In most cases, only the luciferase genes (*luxAB*) have been transferred, and aldehyde must be added to the cells to give light. Decanal or nonanal are the preferred aldehydes for *in vivo* studies because they more readily cross the cell membrane than longer-chain aldehydes and can give a high luminescent response at saturating concentrations.

In some cases, all five lux structural genes (*luxCDABE*) have been transferred, eliminating the necessity for adding aldehyde because the precursors for aldehyde synthesis (e.g., fatty acyl-ACPs) are readily available in most if not all bacteria. In most cases, the *lux* genes from *V. harveyi* or *V. fischeri* have been used because these lux systems were the first cloned and sequenced. Because the luminescent systems of *V. harveyi* and the terrestrial bacteria *X. luminescens* are the most stable, these *lux* systems may be the most appropriate for future applications.

2. Eukaryotes

Expression of the bacterial luciferase system in eukaryotes has until recently been limited because, in contrast to prokaryotes, a separate promoter is generally required for each gene. However, the *luxA* and *luxB* genes have recently been fused by a number of laboratories to generate a monocistronic luciferase with functional activity. The fused luciferase has now been expressed in yeast, plants, insect cells, and even mammalian cells. Because the generation of activity at higher temperatures (37°C) with the fused luciferase is much more difficult than with the native heterodimer ($\alpha\beta$), the bacterial lux system is much more readily applied at the present time to eukaryotic cells grown at temperatures $<30°C$. It should also be noted that the levels of $FMNH_2$ available in eukaryotic cells are very low and, although *in vivo* luminescence has been obtained on addition of aldehyde in a few cases, significantly higher responses will occur in the *in vitro* assays.

C. *In Vivo* Applications

1. Reporters of Gene Expression

By placing the *lux* genes after different promoters, their strength and regulation under different nutritional and environmental conditions can be monitored by light emission. Such methods have extended from the measurement of the regulation of nitrogen-fixing genes in plant pathogens to the study of the differential expression of promoters during sporulation in gram-positive bacteria.

2. Viability and Distribution

Luminescence provides a simple and powerful detection system for following the dissemination of a particular cell in the environment. Such studies have been particularly useful to study the growth and distribution of plant pathogens in their natural host. Moreover, these bacteria can readily be tested for the effects of different inhibitors on their viability and metabolic function because molecules that affect the cellular metabolism by either disrupting the membrane or inhibiting an essential function will disrupt luminescence. Consequently, tests for susceptibility of a specific bacteria to different antibiotics or biocides can easily and rapidly be monitored by the loss of luminescence. Moreover, the presence of antibiotics, bacteriophages, or other toxic components in different materials can be tested using the relevant bacteria containing the recombinant lux DNA. For example, luminescent *Lactobacillus* has been used to test milk for the presence of components that could prevent fermentation by this bacteria (e.g., antibiotics).

3. Detection of Pathogenic Bacteria

Introduction of the *lux* genes into bacteria results in the luminescent phenotype, which can in some instances be detected in only a few cells after a very short time. With this approach, only a few cells of *Salmonella* could be detected within 1 hr after transduction using viral vectors specific for this species. Consequently, this approach has some potential as an early warning system for contaminating bacteria only limited by the specificity and efficiency of *lux* gene transfer.

Bibliography

Hastings, J. W., Potrikas, C. J., Gupta, S. C., Kurfurst, M., and Makemson, J. C. (1985). *Adv. Microbiol. Physiol.* **26,** 235–291.

Lee, J. (1985). The mechanism of bacterial bioluminescence. *In* "Chemiluminescence and Bioluminescence" (J. G. Burr, ed.), pp. 401–437. Marcel Dekker, New York.

Meighen, E. A. (1988). *Annu. Rev. Microbiol.* **42,** 151–176.

Meighen, E. A. (1991). *Microbiol. Rev.* **55,** 123–142.

Silverman, M., Martin, M., and Engebrecht, J. (1989). Regulation of luminescence in marine bacteria. *In* "Genetics of Bacterial Diversity" (D. Hopwood and K. F. Chater, eds.), pp. 71–85. Academic Press, New York.

Biomonitors of Environmental Contamination

Marylynn V. Yates
University of California, Riverside

Glossary

Bacteriophages Viruses that infect bacterial host cells; they usually consist of a nucleic acid molecule enclosed by a protein coat

Coliform Member of a group of bacteria that has traditionally been used as an indicator of the presence of pathogenic microorganisms in water

Coliphages Viruses that infect *Escherichia coli* host cells

Enteric microorganism Microorganism that replicates in the intestinal tract of warm-blooded animals and is shed in fecal material

Gastroenteritis Clinical syndrome characterized by one or more of the following symptoms: nausea, vomiting, diarrhea, fever, and general malaise

BIOMONITORS, or microbial indicators, are used to indicate the microbiological quality of air, water, and soil. Because it is impossible to test these media for all possible pathogenic microorganisms, it is desirable to find one microorganism (or group of microorganisms) that can be used to indicate whether or not pathogens are present. Although there is no perfect indicator organism, research is ongoing to find a microorganism that is a better indicator of the presence and behavior of some of the newly identified etiologic agents of waterborne disease than traditional indicators.

I. Significance of Microbiological Contamination of the Environment

A. Waterborne Disease Outbreaks

Between 1920 and 1988, 1648 waterborne disease outbreaks were reported in the United States, involving over 446,000 people and resulting in 1083 deaths. These data are summarized in 10-yr increments in Fig. 1. The numbers of reported outbreaks and the numbers of associated cases of illness have risen dramatically since 1971, as compared with the period from 1951 to 1970. The increase in reported numbers of outbreaks may be due to an improved system for reporting implemented in 1971; however, it is still believed that only a fraction of the total number of outbreaks is reported.

Ground water supplies over 100 million Americans with their drinking water; in rural areas, there is an even greater reliance on ground water as it comprises up to 95% of the water used. It has been assumed traditionally that ground water is safe for consumption without treatment because the soil acts as a filter to remove contaminants. As a result, private wells generally do not receive treatment, nor do a large number of public water supply systems. However, the use of contaminated, untreated, or inadequately treated ground water has been the major cause of waterborne disease outbreaks in this country since 1920. In the 1980s, use of untreated or inadequately treated ground water was responsible for 44% of the outbreaks that occurred in the United States (Fig. 2).

When considering outbreaks that have occurred due to the consumption of contaminated, untreated ground water, from 1971 to 1985, sewage was most often identified as the contamination source (Table I). Causative agents of illness were identified in ap-

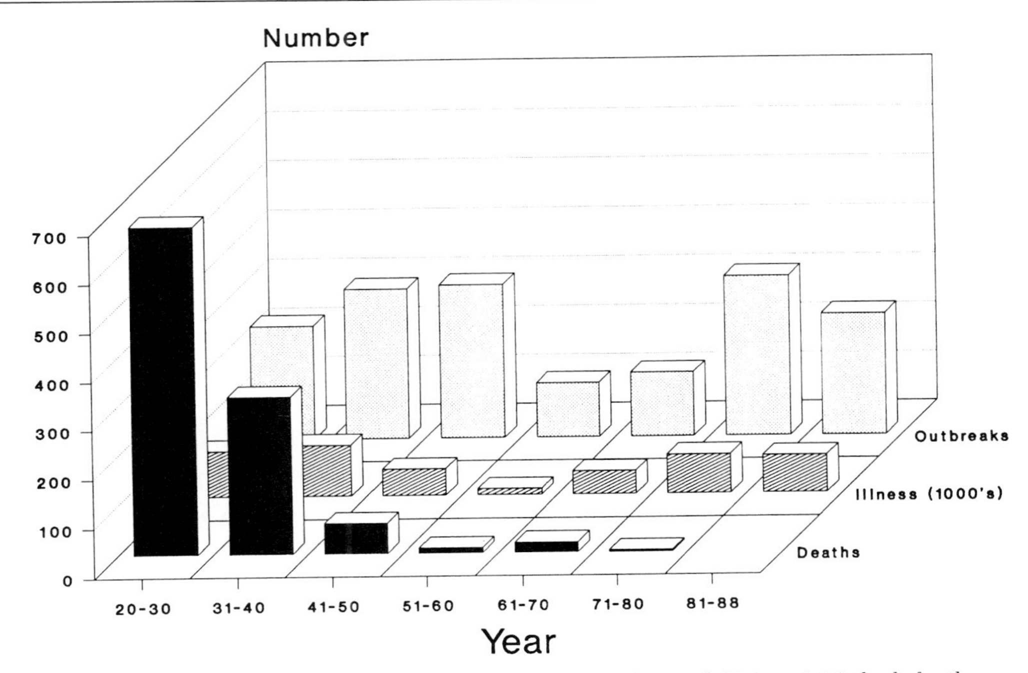

Figure 1 Waterborne disease outbreaks: 1920–1988. [From Craun, G. F. (1990). Methods for the investigation and prevention of waterborne disease outbreaks. EPA/600/1-90/005a. USEPA, Washington, D.C.]

proximately one-half the disease outbreaks during the period from 1971 to 1988 (Table II). The most commonly identified causative agents were *Giardia*, chemicals, and *Shigellae*. In one-half of the outbreaks, no causative agent could be identified, and the illness was listed as gastroenteritis of unknown etiology. However, more recent results suggest that the majority of these outbreaks were caused by enteric viruses and parasites such as *Giardia*. [*See* WATERBORNE DISEASES.]

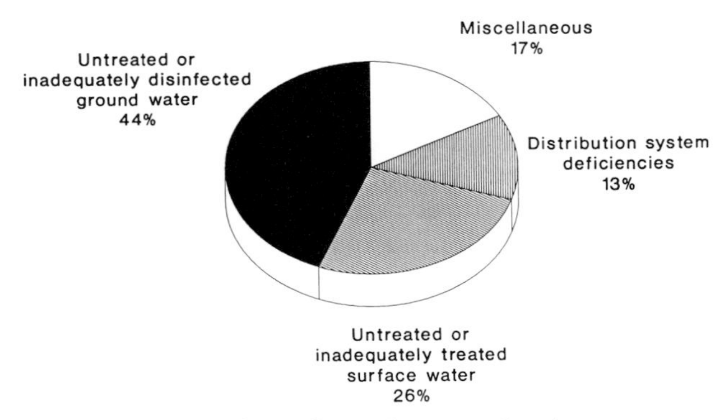

Figure 2 Causes of waterborne disease outbreaks: 1981–1988. [From Craun, G. F. (1990). Methods for the investigation and prevention of waterborne disease outbreaks. EPA/600/1-90/005a. USEPA, Washington, D.C.]

B. Characteristics of Enteric Microorganisms

Microorganisms that infect the gastrointestinal tract of animals are termed enteric microorganisms. They are shed in feces and, thus, are present in domestic sewage. Enteric pathogens can be transmitted by exposure to domestic waste, whether through swimming in contaminated waters, ingesting contaminated water, or eating food that has been irri-

Table I Causes of Waterborne Disease Outbreaks in Untreated Ground-Water Systems: 1971–1985

Cause	Outbreaks (%)
Overflow or seepage of sewage into wells and springs	25
Surface runoff or flooding from contaminated streams	13
Chemical contamination	8
Contamination through limestone or fissured rock	5
Improper system construction	3
Insufficient data to classify	46

[From Craun, G. F. (1990). Methods for the investigation and prevention of waterborne disease outbreaks. EPA/600/1-90/005. USEPA, Washington, D.C.]

Table II Causative Agents of Waterborne Disease Outbreaks: 1971–1988

Disease	Number of outbreaks (%)	Cases of illness (%)
Gastroenteritis, unknown cause	279 (49.47)	64,965 (46.99)
Giardiasis	103 (18.26)	25,834 (18.69)
Chemical poisoning	55 (9.75)	3877 (2.80)
Shigellosis	40 (7.09)	8806 (6.37)
Viral gastroenteritis	26 (4.61)	11,799 (8.53)
Hepatitis A	23 (4.08)	737 (<1)
Salmonellosis	12 (2.13)	2370 (1.71)
Campylobacterosis	12 (2.13)	5233 (3.79)
Typhoid fever	5 (<1)	282 (<1)
Yersiniosis	2 (<1)	103 (<1)
Cryptosporidosis	2 (<1)	13,117 (9.49)
Chronic gastroenteritis	1 (<1)	72 (<1)
Toxigenic *Escherichia coli*	1 (<1)	1000 (<1)
Cholera	1 (<1)	17 (<1)
Dermatitis	1 (<1)	31 (<1)
Amebiasis	1 (<1)	4 (<1)
Total	564 (100)	138,247 (100)

[From Craun, G. F. (1991). *Water Sci. Technol.* **24**, 17–20.]

gated with contaminated water or grown in contaminated soil. Many enteric pathogens can infect an individual without resulting in clinical illness. Thus, although the individual shows no signs of disease, she or he is excreting the organism in her or his fecal material. Others who contact the fecal material can become infected with the pathogen, through what is termed secondary infection. [*See* ENTEROPATHOGENS.]

Bacteria are microscopic organisms, ranging from approximately 0.2 to 10 μm in length. They are distributed ubiquitously in nature and have a wide variety of nutritional requirements. Many types of harmless bacteria colonize the human intestinal tract and are routinely shed in feces. One group of intestinal bacteria, the coliform bacteria, has historically been used as an indication that an environment has been contaminated by human sewage. In addition, pathogenic bacteria, such as *Salmonella* and *Shigella,* are present in the feces of infected individuals. Thus, a wide variety of bacteria is present in domestic wastewater.

Viruses are obligate intracellular parasites; i.e., they are incapable of replication outside of a host organism. They are very small, ranging in size from approximately 20 to 200 nm. Viruses that replicate in the intestinal tract of humans are referred to as human enteric viruses. These viruses are shed in the fecal material of individuals who are infected either purposely (i.e., by vaccination) or inadvertently by consumption of contaminated food or water, swimming in contaminated water, or person-to-person contact with an infected individual. More than 100 different enteric viruses may be excreted in human fecal material; as many as 10^6 plaque-forming units (PFU) of enteroviruses (a subgroup of the enteric viruses) per gram and 10^{10} rotaviruses per gram may be present in the feces of an infected individual. Thus, viruses are present in domestic sewage and, depending on the type of treatment process(es) used, between 50 and 99.999% of the viruses are inactivated during sewage treatment. [*See* ENTEROVIRUSES.]

A third group of microorganisms of concern in domestic sewage is the parasites. In general, parasite cysts (the resting stage of the organism that is found in sewage) are larger than bacteria, although they can range in size from 2 to >60 μm. Parasites are present in the feces of infected persons; however, they may also be excreted by healthy carriers. Cysts are similar to viruses in that they do not reproduce in the environment, but they are capable of surviving in the soil for months or even years, depending on environmental conditions.

Diseases that may be caused by ingestion of enteric microorganisms are shown in Table III.

C. Sources of Microbial Contamination of the Environment

Microorganisms may be introduced into the subsurface environment in a variety of ways. In general, any practice that involves the application of domestic wastewater to the soil has the potential to cause microbiological contamination of ground water. This is due to the fact that the treatment processes to which the wastewater is subjected do not effect complete removal or inactivation of the disease-causing microorganisms present. For examples, see Table IV for expected removals of pathogenic microorganisms after various levels of wastewater treatment. [*See* WASTEWATER TREATMENT, MUNICIPAL.]

Viruses, enteric bacteria, and parasites may be introduced into the subsurface environment in a va-

Table III Pathogens Transmitted by Water

Pathogen	Disease
Bacteria	
Campylobacter jejuni	Gastroenteritis
Enteropathogenic *Escherichia coli*	Gastroenteritis
Legionella pneumophila	Acute respiratory illness
Salmonella	Typhoid, paratyphoid, salmonellosis
Shigella	Bacillary dysentery
Vibrio cholerae	Cholera
Yersinia enterocolitica	Gastroenteritis
Protozoa	
Cryptosporidium	Diarrhea
Entamoeba histolytica	Amoebic dysentery
Giardia lamblia	Diarrhea
Naegleria fowleri	Meningoencephalitis
Enteroviruses	
Adenovirus	Respiratory illness, eye infection, gastroenteritis
Astrovirus	Gastroenteritis
Calicivirus	Gastroenteritis
Coxsackievirus A	Meningitis, respiratory illness
Coxsackievirus B	Myocarditis, meningitis, respiratory illness
Echovirus	Meningitis, diarrhea, fever, respiratory illness
Hepatitis A virus	Infectious hepatitis
Norwalk virus	Diarrhea, vomiting, fever
Poliovirus	Meningitis, paralysis
Rotavirus	Diarrhea, vomiting

riety of ways. Viruses have been isolated from the ground water beneath cropland being irrigated with sewage effluent. Viruses have also been detected in the groundwater at several sites practicing land treatment of wastewater. The burial of disposable diapers in sanitary landfills is a means by which pathogenic microorganisms in untreated human waste may be introduced into the subsurface. Viruses have been detected as far as 408 m downgradient of a landfill site in New York. Land application of treated sewage effluent for the purpose of groundwater recharge has also resulted in the introduction of viruses to the underlying groundwater.

Septic tank effluent may be the most significant source of pathogenic bacteria and viruses in the subsurface environment. Septic tanks are the source of approximately 1 trillion gallons of waste disposed to the subsurface every year and are frequently reported as sources of ground-water contamination.

Another source of microorganisms to the subsurface is municipal sludge. Land application of municipal sludge is becoming a more common practice as alternatives are sought for the disposal of the ever-increasing amounts of sludge produced in this country. The sludge that is produced during the process of treating domestic sewage contains high levels of nitrogen and other nutrients that are required by plant materials. However, it also contains pathogenic microorganisms at concentrations sufficient to cause disease in exposed individuals (see Table V).

Several studies conducted in the late 1970s suggested that viruses are tightly bound to sewage sol-

Table IV Pathogen Removal in Treated Sewage

	Enteric viruses	*Salmonella*	*Giardia*
Infective dose (particles)	1	$>10^3$	25–100
Amount in feces	10^6–10^{10}/g	10^{10}/g	9×10^6/stool
Concentration in raw sewage (no./liter)	10^5	5000–80,000	9000–200,000
Removal during:			
Primary treatment			
% Removal	50–98.3	95.8–99.8	27–64
Number remaining	1700–50,000	160–3360	72,000–146,000
Secondary treatment			
% Removal	53–99.92	98.65–99.996	45–96.7
Number remaining	85–47,500	3–1075	6480–109,500
Tertiary treatment			
% Removal	83–99.9999998	99.99–99.999999995	98.5–99.99995
Number remaining	0.0002–17	0.000004–7	0.099–2951

[From Stewart, M. (1990). Metropolitan Water District of Southern California. Internal report.]

Table V Concentrations of Microorganisms in Digested Sludges

Organism	Type of stabilization	
	Anaerobic	Aerobic
	(No./g dry weight)	
Enteroviruses	0.2–210	0–260
Rotaviruses	14–485	ND
Salmonella	$3–10^3$	3
Total coliforms	$10^2–10^6$	$10^5–10^6$
Fecal coliforms	$10^2–10^6$	$10^5–10^6$
Shigella sp.	20	ND
Yersinia enterocolitica	10^5	ND
Ascaris		4[a]
Trichuris		1.3[a]
Toxocara		0.4[a]

[a] Average of all types of digested sludge, percent viable.
ND, no data.
[Adapted from Gerba, C. P. (1988). Development of qualitative pathogen risk assessment methodology for municipal sludge land filling. EPA/600/6-88/006. USEPA, Washington, D.C.]

ids and are not easily released into the soil. However, in a more recent study, viruses were detected in a 3-m-deep well at a site where anaerobically digested sludge was applied to a sandy soil 11 wk after sludge application.

D. Risks Associated with Waterborne Infectious Disease

As discussed previously, microorganisms (not chemicals) are the major cause of illness associated with the consumption of contaminated food and water. On a worldwide basis, in 1980, approximately 25,000 people died each day from consumption of contaminated water, and it has been estimated that 80% of all diseases in the world may be related to contaminated water. Food and waterborne disease is one of the major causes of diarrhea, resulting in an estimated 1 billion cases every year in children under the age of 5 yr. In underdeveloped or developing countries, acute gastroenteritis is the leading cause of death of children under the age of 4 yr. The importance of waterborne disease in the United States is much less than that in developing countries; however, waterborne disease continues to occur, resulting in millions of dollars of lost productivity in the workplace.

Based on recent reported waterborne outbreak data, the risk of acquiring an illness from contaminated water in the United States has been estimated to be approximately 4×10^{-5} per year or 2.8×10^{-3} during a lifetime. This risk estimate is probably low due to the fact that many waterborne outbreaks are not reported. There are several reasons for this, including the fact that many waterborne pathogens cause gastroenteritis, which is generally not severe enough to require medical attention, so unless a large number of people are involved, the outbreak goes unrecognized. Another reason is that reporting of waterborne disease outbreaks is not required in the United States. In addition, it is difficult to assess the number of cases of illness associated with waterborne pathogens because of secondary infection to individuals who were not directly exposed to the contaminated water.

The impacts of very low levels of pathogens in drinking water are very difficult to document. However, the fact that only one or two virus or parasite particles may be required to cause infection necessitates an attempt to quantitate this risk. The probability of infection resulting from the ingestion of one organism has been estimated for several different pathogens using data from dose–response curves derived from human feeding studies and assuming an ingestion rate of 2 liters/day (Table VI). It is important to remember that infection will not necessarily result in illness in all cases. The estimated annual risk of infection, illness, and mortality from

Table VI Probability of Infection for Enteric Microorganisms

Microorganism	Probability of infection from exposure to one organism
Campylobacter	7×10^{-3}
Salmonella	2.3×10^{-3}
Salmonella typhi	3.8×10^{-5}
Shigella	1.0×10^{-3}
Vibrio cholera classical	7×10^{-6}
Vibrio cholera El Tor	1.5×10^{-5}
Poliovirus 1	1.49×10^{-2}
Poliovirus 3	3.1×10^{-2}
Echovirus 12	1.7×10^{-2}
Rotavirus	3.1×10^{-1}
Entamoeba coli	9.1×10^{-2}
Entamoeba histolytica	2.8×10^{-1}
Giardia lamblia	1.98×10^{-2}

[From Rose, J. B., and Gerba, C. P. (1991). *Water Sci. Technol.* **24**, 29–34.]

echovirus 6 (assuming 2 liter/day ingestion) is shown in Fig. 3. The risk of mortality from ingesting hepatitis A virus in the same volume of water would be much higher (Fig. 4), due to the relatively higher mortality rate for hepatitis A (0.6% compared to 0.29%).

II. Indicators of Microbial Contamination of the Environment

More than 100 different pathogens may be excreted in the fecal material of an infected individual. Obviously, it would not be technically or economically feasible to test all drinking water or wastewater for the presence of all pathogens that could potentially be present. Therefore, in 1914, the United States Public Health Service adapted the coliform group of bacteria to serve as an indicator of the fecal contamination of drinking water. Since that time, all public water supplies have been routinely tested for the presence of coliform bacteria. The detection of coliforms in drinking water is used as an indication that the water has been contaminated by fecal material, and therefore may contain pathogenic microorganisms. Drinking water that is found to contain coliforms must be treated to inactivate any pathogens that may be present in the water and its distribution must be discontinued until the source of the problem has been ascertained and remedied. The absence of coliform organisms in water has been interpreted to mean that there has been no fecal contamination of the water supply and that pathogenic microorganisms are not present. The low rates of waterborne disease outbreaks (relative to other countries) that occur in the United States are due primarily to the implementation of the total coliform standard to assess the sanitary quality of water. However, as discussed previously, a large number of waterborne disease outbreaks continues to occur every year in the United States. [*See* WATER, DRINKING.]

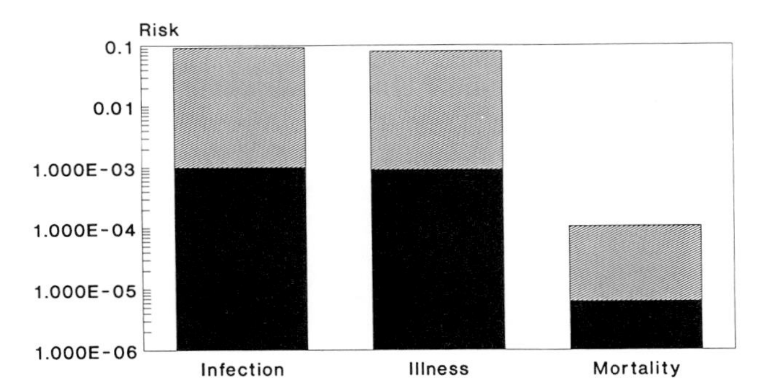

Figure 3 Estimated annual risk by exposure to echovirus 6. ■, annual; ▨, lifetime; concentration, 1 viral infectious unit per 1000 liters. [From Bull, R. J., Gerba, C., and Trussell, R. R. *Crit. Rev. Environ. Contr.* **20,** 77–113.]

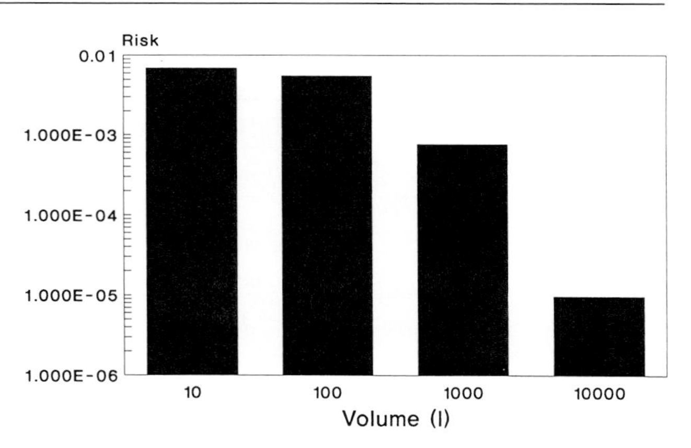

Figure 4 Estimated annual risk of mortality by exposure to hepatitis A. Indicated volume contains 1 infectious unit. [From Bull, R. J., Gerba, C., and Trussell, R. R. (1990). *Crit. Rev. Environ. Contr.* **20,** 77–113.

A. Concepts of an Indicator Organism

The lack of a consistent correlation between the presence of the traditional indicators of water quality (i.e., free chlorine residuals, coliforms, and turbidity) and waterborne pathogens (especially viruses and parasites) has spurred efforts to identify a new indicator (or group of indicators) that would better reflect the virological and parasitological quality of water. Criteria for an ideal indicator microorganism have been discussed by many researchers over the past several years. A compilation of the most important of these criteria is given here. An ideal indicator microorganism should:

1. Always be present when pathogens are present and always be absent when pathogens are absent.
2. Be present at a density that has some constant, direct relationship to the density of the pathogen.
3. Be present and detectable in all types of contaminated media.

4. Be unable to reproduce in contaminated media.
5. Be nonpathogenic to humans and easily identifiable.
6. Survive in the environment at least as long as the pathogens.
7. Be detectable by simple, rapid, and economical methods.

Obviously, there is no one microorganism that can meet all of these criteria. The microorganism used as an indicator should also be chosen on the basis of the particular application of the information. Potential applications of an indicator include being an indicator of:

1. Fecal contamination.
2. The presence of domestic sewage.
3. The presence of pathogens.
4. The efficiency of a particular water or waste treatment process.
5. The environmental fate of a pathogen of interest.
6. The movement of particles suspended in water during subsurface transport.

B. Traditional Indicators

1. Total Coliforms

The total coliform group of bacteria is defined as all aerobic and facultative anaerobic, gram-negative, non-spore-forming, rod-shaped bacteria that ferment lactose with the production of gas within 48 hr at 35°C. The total coliform group consists of members of the genera *Escherichia, Citrobacter, Klebsiella,* and *Enterobacter.* Coliforms are shed in the feces of all humans at an average concentration of approximately 10^7 organisms/g or 10^9/day. Thus, total coliforms can be, and are, used as indicators of the presence of fecal contamination. However, some members of the total coliform group are not specific to fecal material and may be found in soil or on vegetation. Therefore, the presence of total coliforms is not definitive proof that the environment has been contaminated by fecal material. Another negative aspect of using coliforms as indicators is that some coliforms are capable of multiplying in environmental waters.

2. Fecal Coliforms

The fecal coliforms are a subgroup of the total coliforms that are capable of fermenting lactose at 44.5°C. Approximately 95% of the coliforms present in human fecal material are members of the fecal coliform subgroup, which consists of the genus *Escherichia* and certain species of *Klebsiella.* Because fecal coliforms are almost exclusively found in the waste of warm-blooded animals, this group more accurately reflects the presence of fecal contamination from animal waste in water than does the total coliform group. The potential for regrowth or multiplication of fecal coliforms is less than that for the total coliform group.

3. Fecal Streptococci

Fecal streptococci have also been used as indicators of fecal contamination. *Streptococcus faecalis* is found in the feces of warm-blooded animals, including humans, whereas other streptococci such as *Streptococcus bovis* and *Streptococcus equinus* are specific indicators of nonhuman fecal contamination. *Streptococcus bovis* and *S. equinus* die off very rapidly in the environment; therefore, their presence is indicative of very recent fecal contamination by cattle and horses, respectively.

The ratio of the fecal coliform : fecal streptococci (FC : FS) concentrations in a sample has been used to indicate whether contamination is derived from a human or animal source. An FC : FS ratio of <0.7 is indicative of animal sources, whereas an FC : FS ratio of >4.0 is indicative of human sources. This relationship is only valid for very recent contamination (24 hr) because fecal coliforms may not survive as well as fecal streptococci, depending on the environmental conditions.

4. Anaerobic Bacteria

Many genera of anaerobic bacteria, including *Clostridium, Bacteroides,* and *Lactobacillus,* are present in human and animal fecal material. *Clostridium perfringens* is a spore-forming bacterium and, as such, persists for a relatively long time in the environment (compared to the coliforms). This organism has been suggested for use as an indicator of past pollution because of its persistence in the environment. For the same reason, *Clostridium* spores have also been suggested as tracers of pathogen movement in the subsurface. However, the extreme resistance of bacterial spores to environmental stresses may render them inappropriate as indicators of the presence of pathogens, which are less resistant. Few data are available on the relative densities of *Clostridium* sp. and other indicators and pathogens in environmental samples.

Other potential indicators of fecal contamination

are several species of *Bacteroides*. These anaerobic bacteria are present in fecal material at higher concentrations than the coliforms or fecal streptococci, and their potential for regrowth in the environment is lower than these indicators. However, the need to maintain strict anaerobic conditions during the isolation and identification of these bacteria has resulted in little interest in their widespread use as indicators.

C. New Indicators

Increasing amounts of evidence collected during the past 15–20 years suggest that the coliform group may not be an adequate indicator of the presence of pathogenic viruses and possibly protozoan parasites in water. For example, in a study of the removal of viruses and indicator bacteria at seven drinking water treatment plants in Canada, none of the bacterial indicators were correlated with the concentration of viruses in the finished water (Table VII).

One group of microorganisms that has been proposed as an indicator of the presence of pathogenic viruses is the bacteriophages, which are viruses that infect bacteria. Bacteriophages have many of the same characteristics as human enteric viruses but are much easier, less expensive, and less time-consuming to detect in environmental samples. [*See* BACTERIOPHAGES.]

The use of bacteriophages, and more specifically coliphages (viruses that infect *Escherichia coli*), as indicators of the efficiency of water and wastewater treatment processes in inactivating pathogens has received some study. As shown in Table VIII, viruses are generally more resistant to inactivation by disinfectants than are indicator bacteria. Parasites, in general, are more resistant to inactivation than either viruses or bacteria. Studies performed using MS2 coliphage have shown that it is usually one of the most resistant of the organisms tested, which has generated interest in its potential for use as an indicator for pathogenic viruses.

A comparative study of microbial inactivation was performed at a drinking water treatment facility in Mexico. In this study, of the 33 finished water samples that met all U.S. standards for drinking water (i.e., free chlorine residual, turbidity, and total coliforms), 8 (24%) were found to contain one or more enteric viruses. The presence of enteroviruses was not significantly correlated with any of the bacterial indicators or coliphages. However, the presence of rotavirus was significantly correlated ($p \leq 0.05$) with the concentration of free chlorine (negative correlation) and with the concentration of coliphages in the finished water. This study demonstrated that perhaps no single organism may be used as an indicator of the inactivation efficiency for all pathogens.

Very little, if any, information exists on the effectiveness of various treatment processes against Norwalk virus and many of the other viral agents of waterborne gastroenteritis. More research is necessary to determine the resistance of these pathogens to various disinfection processes. The organism chosen as an indicator must be at least as resistant to treatment as the etiologic agents of the majority of the waterborne disease outbreaks.

Bacteriophages have also been used as models for pathogens in studies on the fate and transport of microorganisms through soil and ground-water systems. Chemicals, which are generally used as tracers of water and contaminant movement, are inappropriate tracers of microbial transport for several reasons. One important reason is size: Microorganisms are so much larger than chemical molecules

Table VII Correlation Analysis of Virus Density with Bacterial Data from Seven Drinking Water Treatment Plants in Canada

Water sample (No. samples)	Standard plate count		Total coliforms	Fecal coliforms	Streptococci	*Staphylococcus aureus*	*Pseudomonas aeruginosa*
	20°C	35°C					
Raw (144)	0.56	0.62	0.68	0.59	0.45	0.19	0.41
Chlorinated (15)	0.34	0.25	0.07	0.04	NA	NA	0.05
Sedimented (112)	0.19	0.10	0.03	0.001	−0.08	−0.06	0.06
Filtered (111)	0.09	0.02	0.09	0.12	−0.06	NA	0.20
Ozonated (42)	0.11	0.07	−0.06	NA	−0.04	NA	−0.05
Finished (145)	0.06	−0.02	−0.04	NA	NA	0.02	−0.02

NA, not available.
[From Payment, P., Trudel, M., and Plante, R. (1985). *Appl. Environ. Microbiol.* **49,** 1418–1427.]

Table VIII Inactivation of Indicator and Pathogenic Microorganisms by Various Disinfectants

Microorganism	Free chlorine (4–5°C, pH 6–8) C · T	Inactivation (%)	Chloramine (5°C, pH 7–9) C · T	Inactivation (%)	Chlorine dioxide (5°C, pH 6–7) C · T	Inactivation (%)	Ozone (4–5°C, pH 6–8) C · T	Inactivation (%)	UV (various temperatures and pH's) (mW-sec/cm)	Inactivation (%)
Escherichia coli	2.5	99.9	113	99	0.48	99	0.006–0.02	99	6.5	99.9
Poliovirus 1	1.1–2.5	99	1420	99	0.2–6.7	99	0.2	99	21	99.9
Rotavirus SA11	0.03	99.9	4034	99	0.2–0.3	99	0.019–0.064	99.9	25	99.9
Human rotavirus	0.03	99.9	—	—	—	—	—	—	0.006–0.036	99.9
Hepatitis A virus	1.8	99.99	592	99.99	1.7	99	—	—	—	—
MS-2 coliphage	0.25	99.99	2100	99.99	5.1	99	—	—	—	—
Giardia lamblia	90–170	90	—	—	—	—	0.53	99	100	99.9
Giardia muris	>150	90	1400	99	10.7	99	1.94	99	—	—
Cryptosporidium	> >18 hr	<95	—	—	—	—	–	—	—	—

C · T, product of the disinfectant concentration in mg liter^{-1} and contact time in min for 99% inactivation.
[Adapted from Bull *et al.* (1990). *Crit. Rev. Environ. Contr.* **20,** 77–113; and Sobsey, M. (1989). *Water Sci. Technol.* **21,** 179–195.]

that they are transported only through large pores. Bacteriophages are the same size and possess many of the same adsorptive properties as pathogenic viruses and, thus, are attractive as models or surrogates. Comparative studies have shown that some bacteriophages are transported at least as far and survive for at least as long as hepatitis A virus and other enteric viruses, suggesting that bacteriophages may be acceptable model organisms. However, until data on the transport and fate behavior of Norwalk virus and other pathogenic human viruses is obtained, determination of which organism is the best model to use for this purpose is impossible.

III. Regulations Pertaining to Microbial Contamination of the Environment

The fact that microorganisms are responsible for numerous waterborne disease outbreaks led the U.S. Environmental Protection Agency (EPA) to propose Maximum Contaminant Level Goals for viruses and *Giardia,* a protozoan parasite, in 1985. These standards are in addition to the standard for the indicator microorganism, total coliforms. Rather than require public water systems to monitor the water for the presence of these pathogenic microorganisms, the EPA proposed treatment technique regulations to ensure that the levels of these pathogens would be reduced to nonharmful numbers in treated drinking water. In June 1989, surface water treatment requirements that require a minimum of 99.9% removal of *Giardia* and 99.99% removal of viruses were finalized.

The draft ground-water treatment requirements were published in March 1990. In the case of ground water, it will probably be possible to obtain a variance from the disinfection requirement. One of the variance criteria requires that an assessment of the source water be performed to determine its vulnerability to viral and bacterial fecal contamination. This may consist of a sanitary survey that will determine the locations of all potential sources of microbial contamination (e.g., septic tanks, landfills, sewer lines, wastewater disposal sites) and a calculation of whether or not they could detrimentally affect the microbiological quality of the drinking water.

The EPA has recently proposed new standards for the disposal of sewage sludge. In the proposed rule, sludge is classified based on pathogen and indicator organism reduction requirements and/or specified maximum concentrations of pathogens and indicator organisms (fecal coliforms and fecal streptococci). The crops that can be grown on the land and human and animal access limitations to sludge-applied land depend on the class of sludge applied, with the strictest controls on land receiving the least treated sludge.

In addition to federal standards, several states have laws and regulations designed to minimize the potential for pathogen contamination of drinking water. For example, most states have prescribed minimum setback distances between septic tanks and drinking water wells. Setback distances range from 7.6 m in South Carolina to 122 m in Rhode Island and Massachusetts.

Another practice regulated by many states is the reuse of treated sewage effluent. As stated previously, even tertiary treated sewage effluent may contain concentrations of pathogens high enough to pose a threat to human health. The state of California has recently (1990) proposed regulations pertaining to the use of effluent for irrigation and other purposes. In a manner similar to the EPA's sludge standards, the California State Department of Health Services proposed different classifications for effluent. The class is determined by the level of treatment to remove pathogens and indicator organisms the effluent has received. The most highly treated effluent must be treated using processes that have been shown to reduce levels of viruses, parasites, and helminths to acceptable levels. This effluent has relatively fewer restrictions on its use than does the next class. The lowest classes must be treated using processes that reduce concentrations of indicator microorganisms (fecal coliforms and fecal streptococci), rather than specifying pathogen reduction levels. The least treated effluent has the most severe restrictions placed on its use.

Bibliography

Berg, G. (1978). "Indicators of Viruses in Water and Food." Ann Arbor Science, Ann Arbor, Michigan.

Craun, G. F. (1990). Methods for the investigation and prevention of waterborne disease outbreaks. EPA/600/1-90/005a. USEPA, Washington, D.C.

Goyal, S. M., Gerba, C. P., and Bitton, G. (1987). "Phage Ecology." John Wiley & Sons, New York.

IAWPRC Study Group on Health Related Water Microbiology (1991). *Water Res.* **25,** 529–545.

Snowdon, J. A., and Cliver, D. O. (1989). *CRC Crit. Rev. Environ. Contr.* **19,** 231–249.

Sobsey, M. D. (1989). *Water Sci. Technol.* **21,** 179–195.

Biopharmaceutins

Jan S. Tkacz
Merck Research Laboratories

Glossary

Antibiotic Substance that is produced as a secondary metabolite by a microorganism and that has the capacity in low concentration to halt the growth of or to kill another microorganism

Fermentation Culturing of a microorganism under conditions that support the biosynthesis and accumulation of a desired metabolic product

Pharmacophore Portion of a molecule that is responsible for the pharmacological activity displayed by the compound

Secondary metabolite Relatively low-molecular weight substance synthesized by one species of organism or several species from a limited number of genera that is not part of the pool of materials essential for the metabolism and growth of that (those) organism(s)

BIOPHARMACEUTINS are secondary metabolites of microbial origin that can elicit a pharmacological effect and that consequently may be useful as a therapeutic agents in an area of human medicine other than infectious disease.

I. Introduction to Secondary Metabolites

Many microorganisms, especially the actinomycetes and filamentous fungi, produce metabolites that are not essential for their growth—nonessential at least under the culture conditions used in the laboratory. Molecules of this type fall into a wide variety of structural groups, and the genetic capability to make one compound is usually possessed by just one or a few species. In view of their nonessential character and restricted phylogenic distribution, these compounds have been termed secondary metabolites, but it is perhaps more appropriate to think of them as specialized metabolites. An elucidation of their natural roles has not been a prerequisite to their discovery because many exhibit bioactivity, i.e, a capacity to elicit a biological response in another organism. A toxin such as aflatoxin from *Aspergillus flavus* provides an example, but undoubtedly the best-known bioactive secondary metabolites from microorganisms are the antibiotics, which are among the most widely prescribed drugs in use today. Microbial metabolites active in other ways have also been found, and an up-to-date catalog of secondary metabolites with beneficial applications would extend beyond the antibiotics to include examples such as bialaphos, a product of *Streptomyces hygroscopicus* that is employed as an herbicide, and avermectin from *Streptomyces avermitilis,* which is utilized in agriculture and veterinary medicine as a potent parasiticide effective against helminths and arthropods.

The bioactivity of a secondary metabolite is often a consequence of its ability to interact in a highly specific fashion with a particular target molecule either on the surface of or within the susceptible cell. In some cases, the target molecule is an enzyme, and the interaction results in inhibition of the biochemical reaction that the enzyme normally catalyzes. In other instances, the metabolite may block a site on a receptor molecule, making the receptor inaccessible to its physiological effector. For still other examples, the target molecule is not a protein, and the binding causes a physical chemical change in the target that renders it nonfunctional. The biochemical research that led to these general insights also set the stage for a fundamental change in the way the search for novel secondary metabolites could be conducted. Rather than first observing the metabolite's activity in an organism and then elucidating the mechanism of its action at the molecular level, researchers realized that they could begin by seeking a metabolite capable of interacting with a cellular component that they purposefully choose as a target. Once the compound was in hand, its impact upon the organism could be assessed. Fortunately, this turnabout has come at a time when the understanding of many human diseases has similarly advanced to the molecular level, thereby providing a rational basis for the selection of biochemical targets for drug intervention. Therefore, it is now realistic to view microorganisms in the same way that medicinal herbs were regarded at the start of the twentieth century, namely, as potential sources of therapeutically useful drugs.

In fact, four classes of prescription drugs are already available that are made by microorganisms but that are not used as antibiotics. Although only one of these classes, the cholesterol-lowering agents, was discovered using the targeted approach just described, the method has identified other potentially useful microbial compounds that will be evaluated in clinical trials in the near future. In anticipation that the number of secondary metabolites found to exhibit pharmacological effects will continue to increase, the word biopharmaceutin has recently been suggested as a general designation for them; it provides a descriptive term referring to microbial products that have applicability in human medicine but are not antibiotics for fighting infection.

The next three sections provide brief summaries of three important groups of biopharmaceutins focusing on the present understanding of how each exerts its therapeutic effect. The final section returns to the general topic of secondary metabolites and examines the impact that recombinant DNA technology may have in this field. [*See* ANTIBIOTIC RESISTANCE; ANTIFUNGAL AGENTS; INSECTICIDES, MICROBIAL; MYCOTOXICOSES.]

II. Ergot Alkaloids

The ergot alkaloids appear to have been the first microbial secondary metabolites exploited for their medicinal effects. A text from the second half of the sixteenth century refers to the induction of childbirth with ergot fungi, and there are indications that the ancient Chinese were aware of the effect of ergot upon the uterus. However, throughout most of human history, ergot alkaloids constituted a scourge rather than a blessing. Intoxication with these alkaloids was widespread in Europe during the Middle Ages, when cereals rather than potatoes were the staples of the diet and when rye, among the cereal crops the most susceptible to infestation by the ergot fungi, was a major source of the flour in bread. In the New World, an outbreak of ergot poisoning in 1692 among the inhabitants of Salem, Massachusetts, appears to have formed the basis of the societal hysteria that culminated in the infamous witchcraft trials. It was not until the beginning of the twentieth century when the active compounds were obtained in pure form and the link between dosage and clinical effect was established that the use of ergot alkaloids in medicine gained a firm footing. In modern practice, only 2 of the 40 ergot alkaloids produced by fungi are employed: one to treat migraine headaches and the other to decrease the loss of blood from the uterus just after birth.

The pain of a migraine headache is brought about by vascular changes in branches of the cranial arteries. The initial phase of vasoconstriction is followed by dilation of the extracranial branches of these arteries. The exact nature and sequence of physiological events leading to this syndrome are not entirely clear, but increased amounts of the body's own vasodilating peptides are found in tissue fluids during migraine attacks. The ergot alkaloids can induce a prolonged vasoconstriction of dilated cranial arteries and thereby provide relief. Ergotamine (Fig. 1) is the ergot alkaloid of choice for this purpose and is available under a number of tradenames (Cafergot, Bellergal, Ergomar, Ergostat, and Wigraine).

During pregnancy, the uterus is highly sensitive to ergot alkaloids. Although these compounds are no longer employed to initiate or accelerate birth, the increase in uterine contraction that results from their administration just after birth greatly decreases the blood loss that would otherwise occur. For a number of pharmacological reasons, the most appropriate ergot alkaloid for this purpose is ergonovine (Fig. 1; tradename: Ergotrate Maleate) or a compound chemically prepared from it called methylergonovine.

The basis for the activity of ergot alkaloids in each of these cases appears to be their ability to interact with the cellular receptors of natural effector molecules such as serotonin, adrenalin, and noradrenalin. Vasoconstriction, especially in the arms and legs, can also be seen as the underlying cause of the gangrenous form of ergot intoxication known as sacred fire or St. Anthony's fire in the Middle Ages.

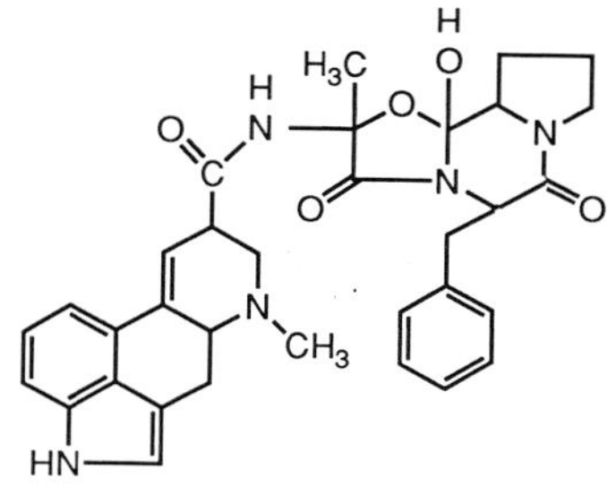

Ergotamine

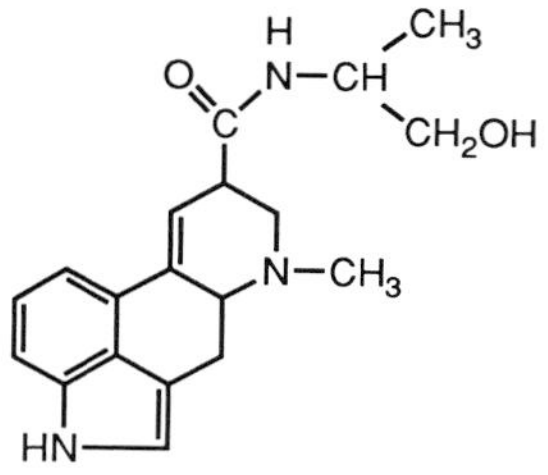

Ergonovine

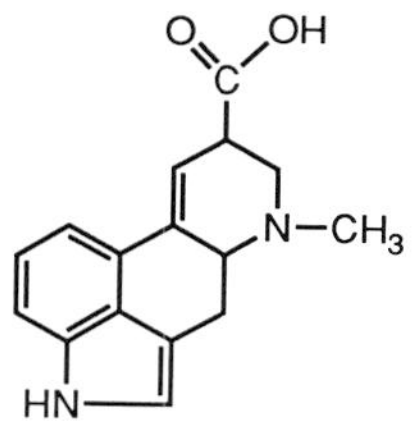

Lysergic Acid

Figure 1 Structures of the medically important ergot alkaloids ergotamine and ergonovine. Lysergic acid is a microbial product that is modified chemically to prepare ergonovine on a commercial scale.

Prolonged poisoning resulted in tissue necrosis in the extremities with ensuing desiccation or putrefaction.

The classical ergot fungi are members of the genus *Claviceps*. Ergot alkaloids are produced by a number of species in this genus including *C. purpurea, C. paspali, C. gigantea, C. fusiformis,* and *C. microcephala*. They have also been found in *Aspergillus fumigatus, Balansia* sp., *Penicillium chermesinum, Phycomyces nitens,* and *Rhizopus arrhizus*. The *Claviceps* fungi have complicated life cycles, and during growth on rye they form specialized structures known as sclerotia. These replace rye grains, which they approximate in shape but exceed in size. The synthesis of ergot alkaloids accompanies the formation of these sclerotia, and for this reason the first commercial process for production of the compounds involved the deliberate infection of rye plants in the field. The yield from this method depends, of course, on climatic conditions, and, consequently, attempts have been made to develop a fermentation process for production. Despite much effort, a cost-effective procedure has not been found for the medically important compounds. However, ergonovine can easily be prepared by chemical modification of lysergic acid (Fig. 1), and two commercially viable fermentation processes yield this precursor. [*See* MYCOTOXICOSES.]

III. Cyclosporin A

For many years, the success of organ transplant surgery was limited because the implanted organ was subject to rejection by the recipient's immune system. Over the past decade, however, the use of cyclosporin A to suppress the immune response has resulted in a dramatic improvement in the success rate for kidney, heart, and liver transplants. Cyclosporin A (tradename: Sandimmune) is produced by the filamentous fungi *Cylindrocapon lucidum* and *Beauveria nivea*, and its structure is shown in Fig. 2. It is a cyclic polypeptide containing 11 amino acid residues, most of which are not typically found in proteins. Closely related compounds (cyclosporins B–I and K–Z) have been found either as minor fermentation products of these two fungi or in other fungi (*Tolypocladium geodes, Neocosmospora vasinfecta,* and *Fusarium solani*). Among the cyclosporins that have been tested for biological activity, none better than cyclosporin A has been found.

Cyclosporin A is the first drug to act selectively on a component of the immune system, namely, a class of white blood cells or lymphocytes known as T cells. It does not affect phagocytes or the elements of bone marrow involved in the formation of blood cells. The initiation of an immune response requires the activation of T cells, and cyclosporin A thwarts this activation process. Although the mechanism by which it does so is still the subject of ongoing and energetic research, several parts of the puzzle are known. In resting T cells, the genes for several cell growth factors known as cytokines are not transcribed. One of these cytokines is interleukin-2 (IL2). Presentation of an antigen at the surface of the T cell is the signal for transcription of the IL2 gene and the production of the IL2 protein to begin. Cyclosporin A blocks the appearance of IL2 during this activation process. Transcription of the IL2 gene requires a nuclear protein called NF-AT (nuclear factor of activated T cells). The functional form of NF-AT is found exclusively in the nuclei of activated T cells and is composed of two parts or subunits. One subunit can be found in resting T cells, but in the cytoplasm not in the nucleus. The other

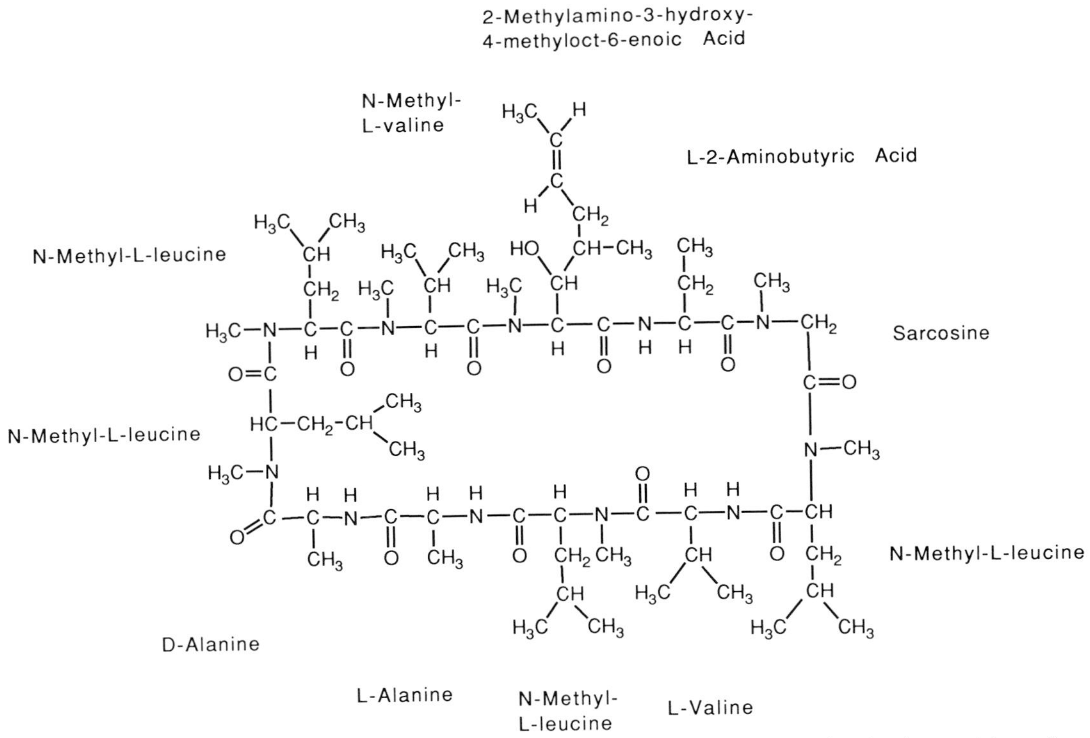

Figure 2 Structure of cyclosporin A. It is a cyclic peptide containing 11 amino acid units: four residues of N-methyl-L-leucine, one residue of L-alanine and another of D-alanine, and one residue each of N-methyl-L-valine, L-2-aminobutyric acid, sarcosine, L-valine, and 2-methylamino-3-hydroxy-4-methyloct-6-enoic acid.

subunit is absent from resting T cells and is synthesized in response to the presentation of antigen at the cell surface. During the early stages of cell activation, the cytoplasmic subunit must enter the nucleus, where it combines with the other subunit to form functional NF-AT, which can then activate transcription of the IL2 gene. Cyclosporin A blocks this transcription by interfering with the import of the cytoplasmic NF-AT subunit into the nucleus. It is also known that cyclosporin A binds to a cellular protein named cyclophilin and that the complex, in turn, interacts with and inhibits calcineurin, a calcium-activated enzyme with the ability to remove phosphate groups from some phosphorylated proteins. It is possible that the cytoplasmic subunit of NF-AT is a phosphorylated protein and that it must be dephosphorylated before it can enter the nucleus. Future research may verify this possibility or may uncover yet other biochemical events in the signaling mechanism, which is set into motion when antigen is presented at the surface of the T cell.

IV. Mevinolin and Related Compounds

Heart disease resulting from atherosclerosis is a leading cause of death in the industrialized world. Atherosclerosis involves the abnormal thickening of arterial walls initiated by the localized accumulation within the inner layer of the blood vessel of phagocytic white blood cells filled with cholesterol, cholesterol esters, and other lipids. As the deposits or plaques become fibrotic and calcified, elasticity of the vessel is reduced, the lumen narrows, and the blood supply to the organ or tissue served by the artery is decreased. Reduction of blood flow to the heart can become life threatening.

Because there are no effective treatments to cause regression of atherosclerotic lesions, emphasis has been placed on prevention of the disease. Foremost among the factors that place an individual at risk for atherosclerosis is chronic elevation of serum cholesterol levels, i.e., hypercholesterolemia. Cholesterol is an essential component of the body acting not only as the precursor of bile acids and steroid hormones but also as a structural element in all cellular membranes. Serum cholesterol levels are a reflection of both the dietary intake of cholesterol and the body's own synthesis of the compound from mevalonic acid. Consequently, lowering the cholesterol in the diet is one way to reduce serum cholesterol, and inhibiting the synthesis of cholesterol should be another.

In the biochemical pathway leading to the 27-carbon cholesterol molecule, 5-carbon building blocks derived from mevalonic acid are sequentially linked to form 15-carbon intermediates, and two of these are condensed to make the 30-carbon backbone that is modified to produce cholesterol. The rate-limiting step in this pathway is the synthesis of mevalonic acid from an activated form of the 6-carbon compound, 3-hydroxy-3-methylglutaric acid. This activated form is known as 3-hydroxy-3-methylglutaryl coenzyme A (HMG-CoA), and the enzyme that mediates its conversion to mevalonic acid is HMG-CoA reductase (Fig. 3). Two structurally related microbial metabolites are specific and potent inhibitors of HMG-CoA reductase (Fig. 4). One called compactin or mevastatin was discovered by Japanese researchers, and the other known as mevinolin or lovastatin was found in the United States. Studies on the metabolic transformations of compactin in dogs led to the identification of pravastatin, which is a hydroxylated and open-acid form of the parent compound (Fig. 4). Today mevinolin and pravastatin, under the tradenames Mevacor and Pravachol, respectively, are in clinical use to lower serum cholesterol levels in humans.

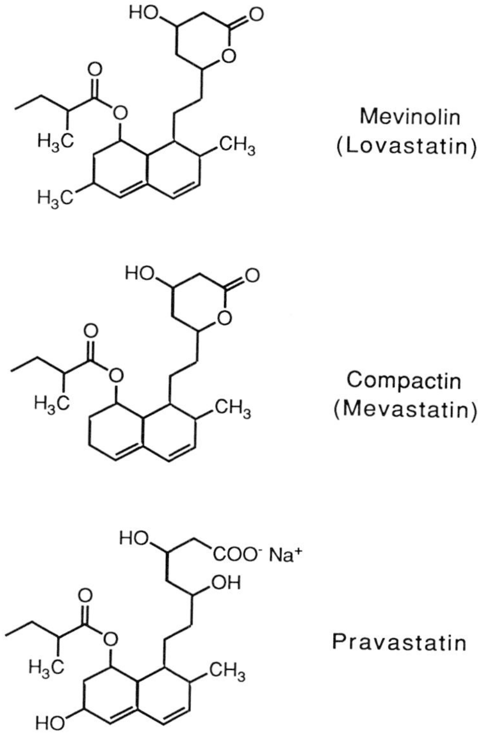

Mevinolin (Lovastatin)

Compactin (Mevastatin)

Pravastatin

Figure 4 Structures of mevinolin, compactin, and pravastatin, three inhibitors of 3-hydroxy-3-methylglutaryl coenzyme A reductase.

Mevinolin and compactin are actually prodrugs or compounds that are converted into drugs in the body. The conversion involves a ring fission to produce the open-acid structure found in pravastatin. A comparison of this structure (Fig. 4) with either the substrate or the product shown in

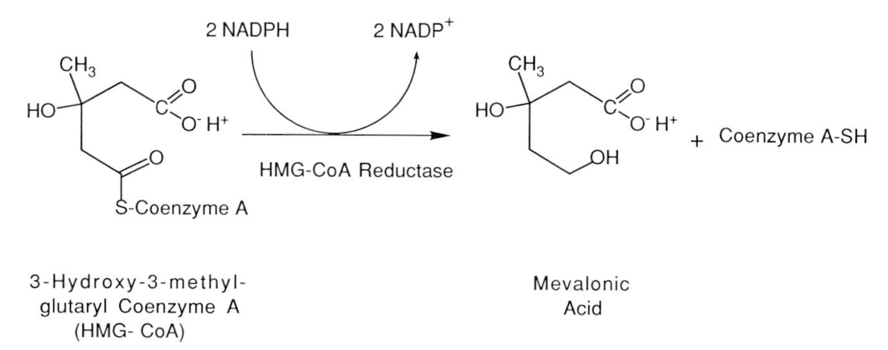

Figure 3 Formation of mevalonic acid from 3-hydroxy-3-methylglutaryl coenzyme A (HMG-CoA) catalyzed by the enzyme HMG-CoA reductase.

Fig. 3 reveals a structural similarity that implicates this portion of the three inhibitor molecules as the pharmacophore and raises the possibility that binding of the inhibitors to the enzyme competes with binding of the substrate. The latter expectation has been verified experimentally; in fact, the active form of mevinolin binds to the enzyme nearly 10^4 times more tightly than the substrate does. Thus, these inhibitors prevent HMG-CoA reductase from converting its substrate to mevalonic acid by obstructing the active site of the enzyme.

With a reduction of cholesterol synthesis, the demand for cholesterol in the liver remains; cells within the organ meet this demand by increasing the number of outer-membrane receptors, which transport cholesterol-containing structures called low-density lipoproteins from the circulating blood to the interior of the cells. The action of these transport molecules brings about the desired effect of lowering the cholesterol level in the circulation.

All the HMG-CoA reductase inhibitors are produced by filamentous fungi. Compactin was originally found in *Penicillium brevi-compactum* and *Penicillium citrinum* but is also made by species of *Eupenicillium, Hypomyces, Paecilomyces,* and *Trichoderma*. Mevinolin is made by species of *Monascus, Phoma, Doratomyces,* and *Gymnoascus,* but the commercial fermentation process for its production employs *Aspergillus terreus,* the organism from which mevinolin was originally obtained.

V. Outlook

In addition to the examples already discussed, several microbial products are in clinical use as cancer treatments, *viz.*, adriamycin, bleomycin (Blenoxane), dactinomycin (Cosmegen), daunorubicin (Cerubidine), mithramycin (Mithracin), and mitomycin-C (Mutamycin). These drugs are not considered in further detail because they do not possess the high degree of pharmacological selectivity that characterizes the other biopharmaceutins. They are toxic compounds whose utility as antineoplastic agents stems from the difference in sensitivity between rapidly dividing and nondividing cells.

With the success of cyclosporin A and mevinolin, it appears that microbial metabolites have entered a new era. Pharmaceutical companies are now keenly aware of the value of microorganisms as a source of new drug candidates. For example, screening to find molecules with the same action as cyclosporin A has led to the discovery of a product known as FK506. Unlike cyclosporin A, FK506 is produced by *Streptomyces* species and belongs to a structural class of molecules (macrolides) quite different from cyclic peptides. Nevertheless, the two compounds have remarkably similar modes of action. In ongoing clinical trials, FK506 appears to be at least as potent as cyclosporin A, and the information already gained suggests that its use may expand the scope of transplantation surgery to include pancreatic islets and intestines. And there are indications that still other biopharmaceutins are moving along industrial research pipelines. Bioactive microbial products have an additional value in pharmaceutical research; the insight their pharmacophores provide can be used in medicinal chemistry programs to synthesize molecules with improved phramacological qualities such as oral bioavailability, increased serum half-life, and metabolic stability.

Although a secondary metabolite is not required for the growth of the organism that makes it, to conclude that it is of no value to the organism would be an error. Laboratory culture conditions are certainly not an appropriate circumstance in which to evaluate the importance of such a product. Furthermore, unless the metabolite has a significant role to play, it is difficult to imagine that natural selection would allow the producing organism to preserve the considerable amount of genetic information required to encode the enzymatic and regulatory machinery involved in its synthesis. The genes involved in secondary metabolism are often clustered in the genome; this is the case particularly in *Streptomyces* species. For two important classes of secondary metabolites (the peptides and the macrolides), these genes encode large multifunctional enzymes or enzyme complexes having the capacity to perform virtually all the reactions needed to form the product. An extreme illustration is provided by the multifunctional enzyme that makes cyclosporin A; it is a single polypeptide with a molecular weight of 800,000 and catalyzes 40 individual reactions! Most of the multifunctional enzymes that have been studied are smaller and perform fewer catalytic steps. A certain type of reaction may be required several times in the synthesis of a product, and the enzyme accomplishes this not by using the same active site repetitively but by having several similar active sites, each performing its reaction in the proper sequence. In other words, these large proteins appear to function as self-contained molecular assembly lines. Through the application of recombinant DNA technology to the genomic clusters encoding the catalytic sites, it should soon become possible to reprogram these assembly lines to manufacture altered products. This could be accomplished by modifying the genetic information for one catalytic site or replacing it with information for another site that catalyzes a chemically analogous but distinct reaction. The likelihood for the success of this approach will probably be greatest, at least initially, in those organisms where the genetic information is arranged in the DNA molecule in the same order as the sequence of biochemical steps performed by the enzyme. Recombining the information developed during the evolutionary divergence of microorganisms in this way is an intriguing prospect for the future development of useful secondary metabolites.

Bibliography

Alberts, A. W. (1988). *Am. J. Cardiol.* **62,** 10J–15J and accompanying papers.

Borel, J. F., Di Padova, F., Mason, J., Quesniaux, V., Ryffel, B., and Wenger, R. (1990). *Pharmacol. Rev.* **41,** 239–242 and accompanying papers.

Brown, M. S., Kovanen, P. T., and Goldstein, J. L. (1981). *Science* **212,** 628–635.

Campbell, I. M. (1984). *Adv. Microbiol. Physiol.* **25,** 1–60.

Caporeal, L. R. (1976). *Science* **192,** 21–26.

DeFranco, A. L. (1991). *Nature (London)* **352,** 754–755.

Donadio, S., Staver, M. J., McAlpine, J. B., Swanson, S. J., and Katz, L. (1991). *Science* **252,** 675–679.

Endo, A. (1988). *Klin. Wochenschr.* **66,** 421–427.

Esser, K., and Düvell, A. (1984). *Process Biochem.* **19,** 142–149.

Kleinkauf, H., and von Döhren, H. (1990). *FEBS Lett.* **268,** 405–407.

Monaghan, R. L., and Tkacz, J. S. (1990). *Annu. Rev. Microbiol.* **44,** 271–301.

Robbers, J. E. (1984). *Adv. Biotech. Proc.* **3,** 197–239.

Shevach, E. M. (1985). *Annu. Rev. Immunol.* **3,** 397–423.

Vining, L. C. (1990). *Annu. Rev. Microbiol.* **44,** 395–427.

Biopolymers

Ian W. Sutherland
Edinburgh University

Michael I. Tait
Aberdeen University

Glossary

Capsule Exopolysaccharide attached, possibly covalently to the cell surface

Exopolysaccharide (EPS) Extracellular polysaccharide excreted outside the microbial cell

Heteropolysaccharide Polysaccharide composed of more than one type of monosaccharide subunit

Homopolysaccharide Polysaccharide composed of a single type of monosaccharide

Polyhydroxyalkanoate Heteropolymers of hydroxy acids found as intracellular storage products in prokaryotic cells

Polyhydroxybutyric acid Intracellular homopolymer composed of β-hydroxybutyric acid

Slime Amorphous polysaccharide found outside and not attached to the microbial cell

BIOPOLYMERS are microbial products—either exopolysaccharides or polyhydroxyalkanoates—that can be manufactured commercially from renewable resources, are biodegradable, and provide alternatives to traditional plant and algal gums or to plastics made from hydrocarbons. Some are already well-established biotechnological products with a sizable market. Others have potentially useful physical or chemical properties or are involved in processes such as water or sewage purification.

I. Structure of Biopolymers

A. Exopolysaccharides

The number of microorganisms capable of producing exopolysaccharides is very large indeed, but the number of structures of these polymers that have been accurately determined is relatively small. Microbial exopolysaccharides can be divided into homopolysaccharides and heteropolysaccharides. Most of the homopolysaccharides are neutral glucans, whereas most heteropolysaccharides are polyanionic because of the presence of uronic acids or pyruvate ketals. Three types of homopolysaccharide structure have been found. Several are linear neutral polymers composed of a single linkage type. (Microorganisms do not appear to yield the "mixed linkage" type of glucan found in cereal plants such as oats and barley.) The second group, homopolysaccharides exemplified by scleroglucan, possess tetrasaccharide repeating units caused by the 1,6-α-D-glucosyl side chains present on every third main-chain residue. Finally, branched homopolysaccharide structures are found in dextrans. A small number of polyanionic homopolymers are also known.

Almost all microbial heteropolysaccharides are composed of repeating units varying in size from disaccharides to octasaccharides. These frequently contain 1 mole of a uronic acid. This is commonly D-glucuronic acid, but some heteropolysaccharides contain D-galacturonic acid, and D-mannuronic acid is found in bacterial alginates and a few other polysaccharides. Very occasionally, two uronic acids are present. The uniformity of the repeat units is based on chemical studies, and some irregularities may possibly be found, especially in polymers composed of larger and more complex repeat units. The heteropolymers commonly possess short side

Encyclopedia of Microbiology, Volume 1

chains, which may vary from one to four sugars in length. In a few polymers of more complex structure, the side chains are also branched. Bacterial alginates are exceptional as they are heteropolysaccharides composed of D-mannuronic and L-guluronic acids in an irregular linear structure.

1. Homopolysaccharides

a. β-D-Glucans

i. Cellulose Cellulose, although normally a polymer isolated from the walls of eukaryotic plants, slime molds, and algae, is also produced as an exopolysaccharide by *Acetobacter xylinum* and possibly other, mainly gram-negative bacterial species. In these bacteria, cellulose is excreted into the medium, where it rapidly aggregates as microfibrils. Some cellulose from bacteria is produced commercially as a source of pure polymer, free from lignin and other related material. It has a number of uses including the manufacture of wound dressings for patients with burns or other extensive loss of tissue. [*See* CELLULASES.]

ii. Curdlan A number of bacterial strains, including *Agrobacterium* and *Rhizobium* species, each produce several exopolysaccharides. One of these, curdlan, is a neutral gel-forming 1,3-β-D-glucan of relatively low molecular weight (~74,000) that is insoluble in cold water (Fig. 1).

iii. Scleroglucan Scleroglucan and related β-D-glucans are produced by several fungal species including *Sclerotium rolfsii* and the wood-rotting Basidiomycete *Schizophyllum commune*. Despite their close structural similarity to curdlan, they are soluble polysaccharides yielding highly viscous aqueous solutions in which the polymer molecules adopt a triple helical conformation. The main chain is composed of 1,3-β-D-linked glucose residues, attached to which are 1,6-β-D-glucosyl residues (Fig. 1). The side chains are probably attached regularly on every third glucose in the main chain; the molecular weight is about 1.3×10^5. Some of the glucans, such as those from *Sclerotium glucanicum*, are of lower molecular weight (~18,000), with glucosyl substituents on every fourth or every sixth main-chain unit. The side chains may not always be regularly distributed on the main chain; distribution may possibly be random.

b. α-D-Glucans

i. Dextrans Dextrans are exopolysaccharides composed predominantly of (1 → 6)α-linked D-glucosyl residues. In some dextrans, there may be almost no other type of linkage; alternatively, up to 50% of the glucose residues may be linked 1,2; 1,3; or 1,4. Industrial dextran production is mainly from a strain of *Leuconostoc mesenteroides*, which yields a polysaccharide with about 95% 1,6 linkages and 5% 1,3 linkages and has a molecular weight of about $4\text{–}5 \times 10^7$ (Fig. 2). For many purposes, the molecular weight is reduced by mild acid hydrolysis.

ii. Elsinan A further α-D-glucan, obtained from culture filtrates of *Elsinoe leucospila*, has some structural similarity to pullulan, as well as to the polysaccharide nigeran from *Aspergillus niger*. It is a linear polysaccharide composed mainly of maltotriose units linked 1,3α- to each other but also containing a small number of similarly linked maltotetraose units. The structural units are similar to pullulan (which also contains a small number of maltotetraose units), but the linkage is 1,3 instead of 1,6. Elsinan is readily soluble in water, giving highly viscous solutions. At higher concentrations, gels are formed. Nigeran also contains 1,3α-D-glucosyl linkages; these alternate on a regular basis with 1,4 linkages.

iii. Pullulan The exopolysaccharide pullulan from the dimorphic fungus *Aureobasidium pullulans* is composed of maltotriosyl repeat units linked

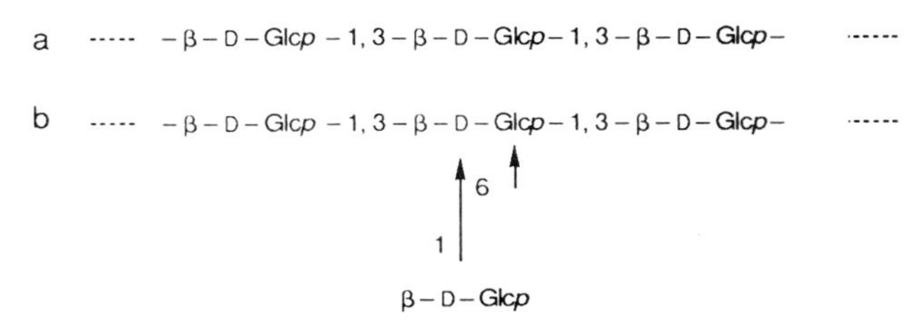

Figure 1 The structure of curdlan (a) and of scleroglucan (b).

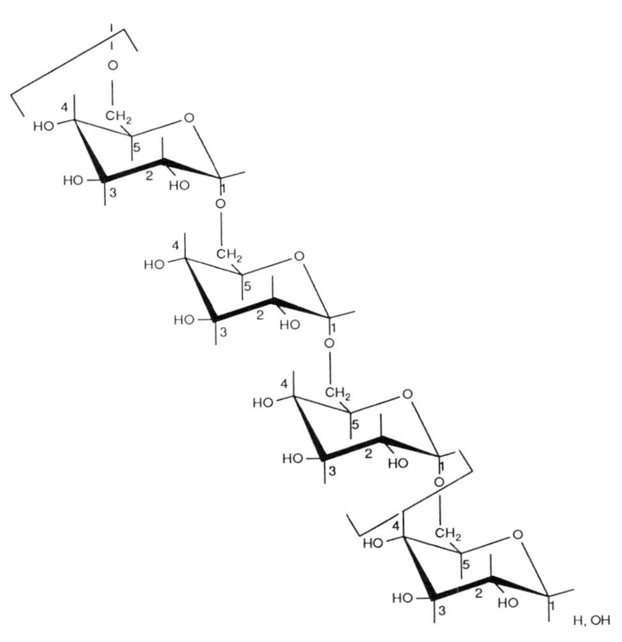

Figure 2 Typical dextran structure.

1,6-α- to form a polymer of molecular weight 10^5–10^6. Similar products are formed by other fungal species including *Tremella mesenterica* and *Cyttaria harioti*. This polymer has potential uses because of its capacity for forming films capable of excluding oxygen.

iv. Sialic Acids Strains of *Escherichia coli* and *Neisseria meningitidis* secrete sialic acid–containing exopolysaccharides. The products from *E. coli* K1 and *N. meningitidis* group B are (2 → 8)-α-linked. Group C meningococcal sialic acid is a homopolysaccharide in which the residues are (2 → 9)-linked. Another sialic acid (from different strains of *E. coli*) is nonacetylated and is composed of alternate 2,8-α- and 2,9-α- linkages. In addition to these homopolymers, some strains of *N. meningitidis* synthesize heteropolysaccharides containing both sialic acid and D-glucose. Several of the bacterial sialic acids possess hydrophobic 1,2 diacylglycerol groups at the polysaccharide chain termini, causing the polymers to aggregate in the form of micelles.

2. Heteropolysaccharides

a. Microalgal Exopolymers

Whereas the commercially important polysaccharides synthesized by macroalgae (i.e., agars, carrageenans, and alginates) have been studied extensively, microalgal exopolymers have received comparatively little attention. Information on the structure of these is sparse, and most of the data that are available refers to compositional studies of what may be heterogeneous mixtures of polymers. Consequently, this caveat applies to the results in this section.

b. Red Algal Exopolymers

The exopolymer synthesized by red algae belonging to the genus *Porphyridium* has properties similar to xanthan and has been suggested as an industrial alternative to the bacterial polymer. *Porphyridium* exopolymer appears to be a sulfated anionic proteoglycan composed of polysaccharide covalently linked to a polypeptide moiety comprising 1–5% of the total dry weight. The major glycose and sugar acid components are D-xylose, D-glucose, D- and L-galactose, L-rhamnose, and D-glucuronic acid (or 2-*O*-methyl-D-glucuronic acid). The relative proportions of each of these vary with growth conditions and between species. The polymer contains 2–10% ester sulfate. Reported molecular weights are in the range of $0.5–10 \times 10^6$. No structural details have been reported.

Members of another genus of red algae, *Rhodella*, synthesize an exopolymer with a similar composition (i.e., containing xylose, glucose, galactose, rhamnose, glucuronic acid, and protein, the amounts of each again varying with growth conditions and between species). Arabinose and mannose have also been detected in *Rhodella* exopolymers; the sulfur content is 1.5–10% of the polymer dry weight. The galactosyl residues are 3- and 4-linked, and some are present as 3,6-anhydrogalactose. These characteristics are shared by the macroalgal galactans (i.e., agars and carrageenans). However, the evidence available at present indicates that *Rhodella* exopolymer is *not* a heterogeneous mixture of polymers, one of which is a galactan similar to agar or carrageenan. The xylose residues in this polymer are also linked at the 3- and 4- positions.

c. Green Algal Exopolymers

Less information is available on the exopolymers synthesized by green algae than on the red algal polymers. *Chlamydomonas* spp. exopolysaccharides have been studied because of their assumed importance in the use of these organisms for soil conditioning. These polymers contain glucuronic acid and different glycosyl residues. The identity of the latter varies between species: *C. humicola* exopolysaccharide contains glucuronic acid, glucose, and xylose in an approximate molar ratio of 1 : 1 : 3, whereas *C. peterfi* and *C. sajao* exopolysaccharides contain glucuronic acid and galactose in the molar ratio 1:2.

The exopolymer synthesized by *Chlorella stigmatophora* also has a high uronic acid content (30%), whereas that from *C. salina* has a uronic acid content of only 6%. This difference is reflected in the different heavy metal-binding capacities of these two polymers. Both polymers contain 8–9% sulfate.

Studies of other green algal exopolymers indicate that these are uncharged. For example, *Botryococcus braunii* exopolysaccharide is composed of galactose, fucose, 3-*O*-methyl fucose, rhamnose, 3-*O*-methyl rhamnose, and glucose; and *Collinsiella cava* exopolysaccharide contains arabinose, rhamnose, and glucose.

d. Cyanobacterial Exopolymers

In common with the algal exopolymers, those synthesized by cyanobacteria have yet to be structurally defined, although some details of their composition are known. All those studied so far are heteropolymers containing both sugars and uronic acids.

The exopolymers synthesized by *Phormidium* sp. and *Gloethece* sp. differ from other bacterial exopolysaccharides in that they appear to be sulfated. *Phormidium* exopolymer, also known as emulcyan, has attracted interest because of its emulsifying properties. It has a molecular weight of 1.2×10^6 and contains rhamnose, mannose, galactose, an unidentified uronic acid (not galacturonic or glucuronic or mannuronic), protein, and sulfate.

The exopolysaccharide synthesized by *Macrocystis flos-aquae* has a composition similar to that of pectin, containing 83% galacturonic acid, 5.5% rhamnose, 5% mannose, 3% xylose, 2% glucose, and 1.5% galactose. Details of whether its structure resembles that of pectin are not yet available. *Anabaena flos-aquae* synthesizes two exopolymers, a xyloglucan containing glucose and xylose in the molar ratio 8:1 and a polymer containing uronic acid, glucose, xylose, and ribose in the molar ratio 10:6:1:1. *Nostoc calcicola* is reported to form a complex proteoglycan containing galacturonic and glucuronic acids, glucose xylose, mannose, galactose, and fucose as its major components. The polysaccharide component comprised more than 70% of the polymer and could not be separated from the polypeptide component.

e. Bacterial Alginates

Commercially available alginates are currently isolated from marine algae such as *Laminaria* and *Macrocystis* species. Bacterial alginates, linear polymers of irregular structure also composed of D-mannuronic acid and L-guluronic acid, have been proposed as substitutes for the algal products. Structural investigations on alginates from the range of bacterial species now known to synthesize this type of polysaccharide have revealed a range of different polymer types—all with similar chemical composition. All the bacterial alginates are composed of the same two uronic acids as algal alginate, but in addition, a number of them are highly acetylated, the acetyl groups being carried solely on D-mannuronosyl residues in the polymers. As can be seen from the details presented in Table I, bacterial alginates are a family of exopolysaccharides that, despite their similar composition, vary considerably in their structure. The material from *Azotobacter vinelandii* bears the closest resemblance to algal alginate, although unlike the latter, it contains *O*-acetyl groups. Both are composed of three types of structure: poly-D-mannuronic acid sequences; poly-L-guluronic acid sequences; and mixed sequences (Fig. 3). These are sometimes termed *block structures*. *Azotobacter chroococcum* synthesizes two polysaccharides, one of which is an acetylated alginate with high D-mannuronic acid content. Most of the *Pseudomonas* species that synthesize alginate also yield exopolysaccharides with relatively high mannuronic acid content. The interesting feature of the *Pseudomonas* products is that, unlike the algal or *Azotobacter* alginate, there are no contiguous se-

Table I Bacterial Strains Producing Alginatelike Polymers

Strain	Ratio of mannuronic acid to guluronic acid
Azotobacter vinelandii	Wide range
Azotobacter chroococcum	0.9:0.1
Pseudomonas aeruginosa	From 1:0 to ~0.6:0.4
Pseudomonas cepacia	
Pseudomonas fluorescens	~0.6:0.4
Pseudomonas maltophila	
Pseudomonas mendocina	
Pseudomonas phaseolicola	0.95:0.05
Pseudomonas pisi	0.83:0.17
Pseudomonas putida	~0.6:0.4

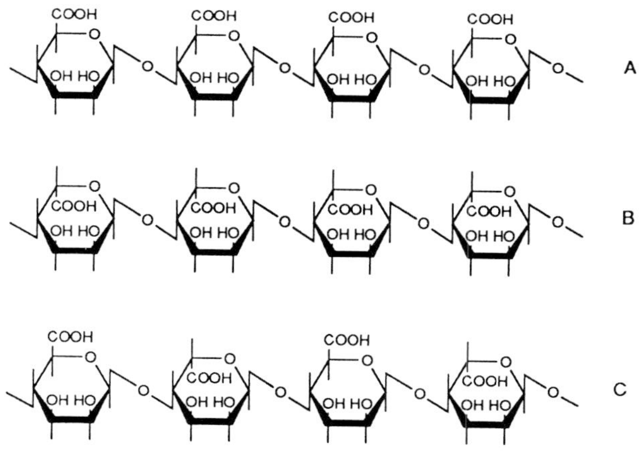

Figure 3 Bacterial alginate. The three sequences of uronic acids that may be found are indicated. Some of the D-mannuronosyl residues carry *O*-acetyl groups (A) Polymannuronic acid; (B) polyguluronic acid; (C) mixed block structures.

quences of L-guluronic acid residues. Whereas the properties of *A. vinelandii* polysaccharide can therefore be expected to have much in common with the commercial algal products, the others possess different properties. Alginates from the plant pathogenic *Pseudomonas* species are also generally of lower molecular weight, relatively low acetyl content, and higher polydispersity. In the highly acetylated polysaccharides from either *A. vinelandii* or *P. aeruginosa,* some D-mannuronosyl residues may carry *O*-acetyl groups on the C_2 and C_3 positions. Between 3 and 11% of the mannuronosyl residues carry acetylation on both carbon atoms.

f. Emulsan and Related Polysaccharides

Acinetobacter calcoaceticus is a capsulate gram-negative bacterium, the exopolysaccharides from which are potentially useful emulsifying agents. Structural studies have shown that two of these polymers differ considerably in their chemical structure, despite their common functional properties. Strain BD4 polysaccharide is a heptasaccharide repeat unit composed of rhamnose, mannose, glucose, and glucuronic acid. Emulsan from strain RAG-1 contains D-galactosamine, an aminouronic acid, and an amino sugar. It also contains 15% fatty acyl *O*-esters.

g. Gellan and Related Polymers

The discovery of gellan from *Auromonas elodea* was followed by the observation that it was one of a series of structurally related polysaccharides. In its native form, gellan carries both *O*-acetyl and glyceryl substituents on a linear polymer composed of tetrasaccharide repeat units. Other polysaccharides (termed *welan and rhamsan*) with the same main-chain sequence were later obtained from other bacterial isolates. These carried differing side chains—a rhamnose-containing or a glucose-containing (gentibiosyl) disaccharide or, in one polysaccharide, either L-rhamnose or L-mannose. This latter polysaccharide is thus highly unusual in containing L-mannose and a variable side chain. The ratio of L-rhamnose to L-mannose is approximately 2:1. Two further polymers were even more unusual in that in both, one of the main-chain sugars varied. It could again be either L-mannose or L-rhamnose. One of the exopolysaccharides (S88) also contained 5% acetate. The structures proposed for two of this series of polysaccharides are shown in Fig. 4.

h. Hyaluronic Acid and Heparin

Although no microbial strains produce heparin, a strain of *E. coli* serotype K5 does form a capsular polysaccharide in which the disaccharide repeat unit is essentially a form of desulfatoheparin. The polymer is composed of a repeating unit of 4-β-D-glucuronosyl-1,4-α-*N*-acetyl-D-glucosamine. Desulfatoheparin from eukaryotic sources is normally composed of alternate disaccharides containing D-glucuronosyl amino sugar and L-iduronosyl-1,4-α-*N*-acetyl-D-glucosamine. The bacterial product resembles type II glycosaminoglycuronan chains, which are synthesized in eukaryotes in the Golgi complex and then polymerized onto core proteins.

After polymerization they may be modified through the action of uronosyl epimerase and sulfatotransferases. These enzymes introduce the L-iduronosyl and sulfate residues, respectively, both of which are absent from the bacterial product. Because of its structural similarity to heparin, this bacterial polysaccharide may prove useful in medical

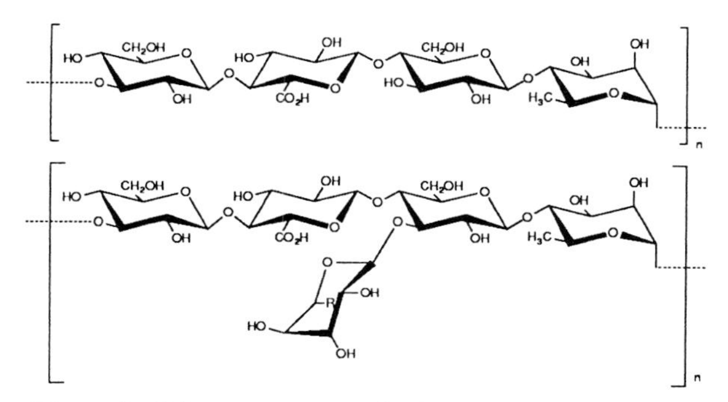

Figure 4 Primary structure of gellan gum (top) and welan gum (bottom). R, CH_3 CH_2OH

research and in determination of the specificity of heparinases and related enzymes. Another interesting polymer of this general type from *E. coli* K4 possesses a chondroitin backbone to which β-D-fructofuranosyl residues are attached at the C_3 position of the D-glucuronic acid. After the fructosyl residues have been removed by mild acid treatment, the polysaccharide is a substrate for both hyaluronidase and chondroitinase.

Group A and group C streptococci produce hyaluronic acid, apparently identical in chemical structure to that obtained from eukaryotic material. The polysaccharide is composed of repeating units of 1,4-β-linked disaccharides of D-glucuronosyl-1,3-β-*N*-acetyl-D-glucosamine. The product from one group C streptococcal strain forms a high-molecular-weight cell-bound product and a soluble exopolysaccharide of average molecular weight 2×10^6. Bacterial production of hyaluronic acid has been commercialized to produce high-quality (and high-cost) polysaccharide for use in surgery as well as in cosmetic preparations.

i. *Rhizobium* Heteroglycans

Some species of *Rhizobium* (e.g., *R. trifolii* and *R. leguminosarum*) form a neutral capsular heteropolysaccharide composed of D-glucose, D-galactose, and D-mannose in the molar ratio 1:3:2, forming a hexasaccharide repeat unit (Fig. 5). This polymer is produced, as well as soluble polyanionic material, the relative proportions of the two being dependent on the strain and on the growth conditions used. The main chain of the gelling polysaccharide comprises a sequence of α-D-glucosyl-1,3-α-D-mannosyl-1,3-β-D-galactose. The glucose residue carries two side chains, one of a single D-galactosyl unit and the other a disaccharide composed of D-galactose. Like curdlan, it is insoluble in water at room temperature

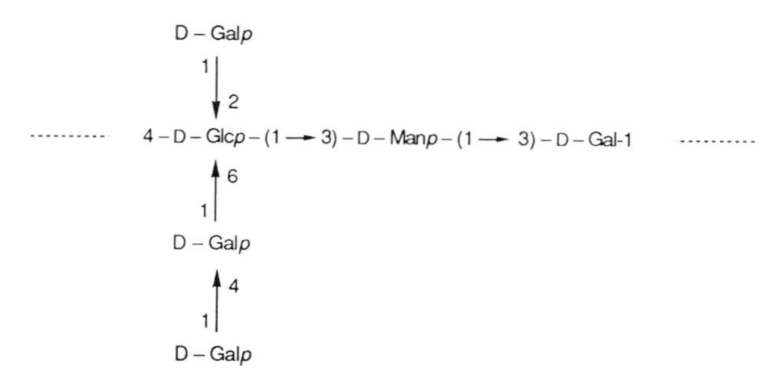

Figure 5 The neutral gel-forming polysaccharide from *Rhizobium meliloti*.

but can be dissolved on heating and gels on subsequent cooling.

j. Succinoglycan

The name succinoglycan has been applied to a group of biopolymers from bacterial strains that also yield curdlan. As with other examples of microorganisms that can synthesize more than one polysaccharide, strain selection and choice of growth conditions can ensure high yields of one specific polymer. Succinoglycans differ from curdlan in being water-soluble and are composed of octasaccharide repeat units, thus conforming to a pattern found in a number of other exopolysaccharides from *Rhizobium* species. There is close structural similarity to the other *Rhizobium* polysaccharides, which all possess highly conserved structures composed of octasaccharide repeat units. Succinoglycans contain D-glucose and D-galactose in the molar ratio 7:1. Attached to the sugars are three different acyl groups: acetate and succinate esters and pyruvate ketals. Pyruvate is normally present in stoichimetric amounts, whereas the molar ratios of acetate and succinate are commonly of the order of 0.2 and 0.4–0.5, respectively. The monosaccharide components are, of course, neutral, but the pyruvate ketal situated on the side-chain terminus (D-glucose) confers anionic properties to the polysaccharide. Despite its 1,3; 1,4; and 1,6-linked β-D-glucosyl residues, succinoglycan is resistant to the action of the endo-β-glucanases that are available commercially.

k. Xanthan

In tonnage terms (>20,000 tonnes per annum), one of the major commercial biopolymers produced is xanthan, the exopolysaccharide from *Xanthomonas campestris* pv. *campestris*. Definitive studies on the structure of xanthan showed that it is composed of pentasaccharide repeat units, which are of particular interest as they represent trisaccharide side chains attached to a cellulosic backbone (Fig. 6). Alternate glucose residues in the backbone carry the side chains composed of D-mannose and D-glucuronic acid. Most commercial xanthan preparations are fully acetylated on the internal D-mannose residue but only carry pyruvate ketals on about 30% of the side-chain terminal mannose residues. Recently, xanthan with a higher pyruvate content has become available commercially, and strains totally lacking pyruvate have also been developed. Through the use of mutants, different *X. campestris* pathovars, and different nutrient conditions, it is possible to obtain

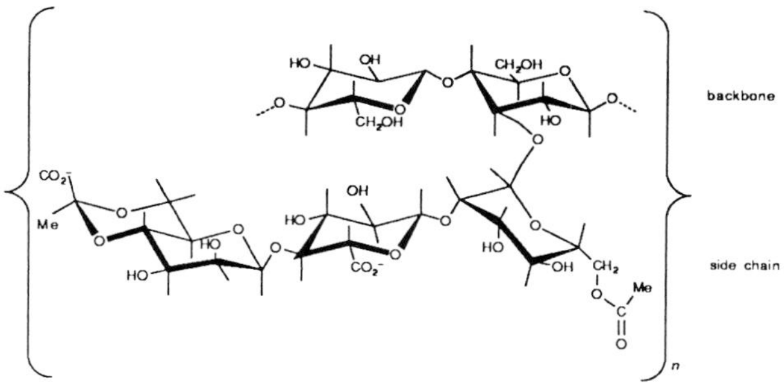

Figure 6 Primary structure of xanthan.

a range of different polysaccharides that all conform to the general structure of xanthan but differ in the completeness of carbohydrate side chains and of acyl groups. One group of pathovars yields xanthan in which the content of pyruvate is very low but in which the internal mannosyl residue in the side chain carries 2 moles of acetate. Mutant polymers lacking the terminal D-mannose (and pyruvate) or lacking the terminal disaccharide are also available. The latter naturally differs from xanthan in being a neutral polymer, although it still possesses useful rheological properties. The various types of repeating unit to be found in different xanthan preparations are illustrated in Fig. 7.

As well as xanthan from wild-type or mutant strains of *X. campestris* pv. *campestris* and related strains, several other polysaccharides have been shown to share part of the structure and also possess exploitable properties. The ability of mutant strains to produce "polytrimer" and "polytetramer" (i.e., xanthan molecules lacking the side-chain disaccharide and monosaccharide termini, respectively) has been mentioned. In addition to these variants,

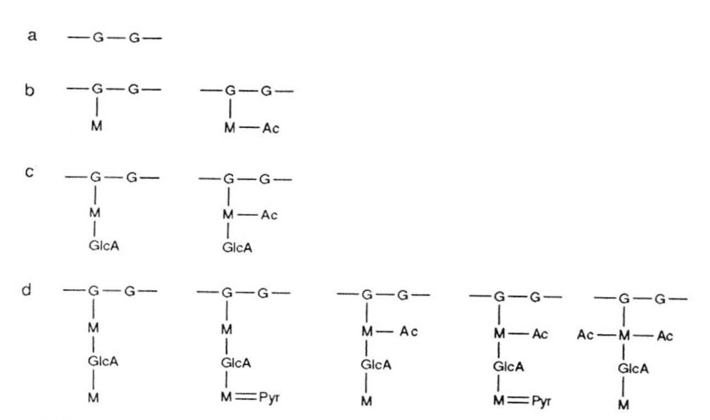

Figure 7 Various repeat units occurring in preparations of xanthan from wild-type (d) and mutant (a–c) bacteria. M, D-mannose; G, D-glucose; Ac, acetyl; Pyr, pyruvate.

several strains of *Acetobacter xylinum* yield apparently xanthan-like polysaccharides. As bacterial cellulose is a normal product of *A. xylinum*, the ability to modify cellulose by the addition of side chains is perhaps not unexpected. One product has been characterized as having a cellulosic main chain together with a pentasaccharide side chain on alternate main-chain sugars. This polymer closely resembled xanthan, the terminal β-D-mannosyl residue being replaced by a L-rhamnosyl-gentibiosyl sequence. As is the case with xanthan, this polysaccharide is acetylated. Another strain of the same bacterial species is thought to synthesize a polysaccharide that differs only in the conformation of one of the side-chain linkages. These xanthan-like polymers are produced in addition to cellulose in some strains or replacing it in others. They form highly viscous aqueous solutions but may not be as potentially valuable as xanthan and have not yet been produced commercially. Another polysaccharide of the same composition but apparently different structure has also been found in a cellulose-negative *A. xylinum* strain.

I. XM6

A polysaccharide with interesting gelation properties is produced by an *Enterobacter* strain (XM6); this polymer proved to resemble very closely two polysaccharides from *Klebsiella* aerogenes serotype 54 strains. Each polysaccharide was composed of the same tetrasaccharide repeat unit (Fig. 8). However, the *Klebsiella* two exopolysaccharides carried *O*-acetyl groups on every or on alternate L-fucose residues, respectively, whereas the XM6 exopolysaccharide is devoid of acyl substituents.

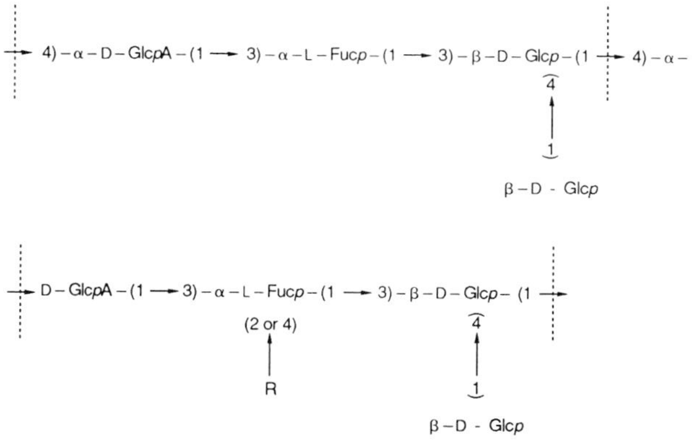

→ 4)-α-D-GlcpA-(1 → 3)-α-L-Fucp-(1 → 3)-β-D-Glcp-(1 → 4)-α-
4 ↑ 1 β-D-Glcp

→ D-GlcpA-(1 → 3)-α-L-Fucp-(1 → 3)-β-D-Glcp-(1 →
(2 or 4) ↑ R; 4 ↑ 1 β-D-Glcp

Figure 8 Chemical repeat unit of *Klebsiella* K54. R group is mono-*O*-acetyl, which occurs on alternative fucose units. On deacetylation, the repeat degenerates to a tetrasaccharide repeat identical to that of XM6.

m. Galactoglucans

i. **Agrobacterium, Rhizobium,** ***and*** **Zoogloea *Polysaccharides and Related Polymers*** Floc-forming bacteria play an important role in the purification of activated sludge or water with a high content of organic material. In the purification process, the exopolysaccharides synthesized by *Zoogloea ramigera* and related species function in flocculation of particulate material. Some of the polysaccharides are claimed to have an unusually high affinity for metallic ions and for amino acids. The polymer from one strain contains D-glucose, D-galactose, and pyruvate. It has been suggested that the main chain of this polymer contains 1,4-β-linked D-glucose residues, but there are still conflicting views about its exact structure.

Several other bacterial polysaccharides composed solely of the neutral sugars D-glucose and D-galactose have been reported. The structures of most of these polymers have still not been elucidated, although several appear to contain equimolar amounts of glucose and galactose, together with acyl substituents (Table II). However, one interesting polymer has been isolated from strains of *Pseudomonas marginalis* and shown to be composed of disaccharide repeat units to which pyruvate and succinate groups are attached.

Exopolysaccharides represent a very varied series of chemical structures, although some reveal considerable conservation of structure. These may be from either closely related bacterial species or from groups with lower affinity to each other. Those polysaccharides that have been extensively studied are composed of a relatively limited range of monosaccharides and acyl substituents. The structural studies on some of the polysaccharides, especially those used industrially, provide opportunities for the correlation of carbohydrate structure to physical properties of potential industrial interest.

B. Structure of Polyhydroxyalkanoates

Polyhydroxybutyric acid (PHB) was identified many years ago as a bacterial storage polymer of relatively widespread occurrence in prokaryotic cells. Although most attention has been paid to bacterial PHB, the observation that copolymers with higher homologues such as hydroxyvaleric acid could also be formed extended considerably the potential for these bacterial ''plastics.'' The random copolymers can be obtained through growth of *Alcaligenes eutrophicus* with glucose and propionic acid as substrates. The molecular weight of polyhydroxyalkanoates appears to depend on the bacterial species from which they are isolated and can range from 50,000–60,000 (*Pseudomonas extorquens*) to 250,000–300,000 (*Methylobacterium*). The melting point of copolymers decreases as the content of hydroxyvaleric acid increases. The molecule also becomes more flexible and tougher, thus extending the range of potential applications. All polyhydroxyalkanoates are linear polymers.

II. Synthesis

A. Synthesis of Exopolysaccharides

Two distinct mechanisms of synthesis are known for exopolysaccharides. Dextrans are produced by an extracellular process involving enzymes that are either secreted from the bacteria or are loosely associated with the cell surface of gram-positive bacteria such as *Leuconostoc mesenteroides*. These enzymes are induced by the presence of sucrose in the culture medium, are absent from glucose-grown cultures and are lipoproteins which can be purified from the culture supernatant. A specific substrate, sucrose, is needed together with an acceptor in the

Table II Galactogluco-Polysaccharides

Source	Glucose	Galactose	Pyruvate	Acetate	Succinate
Achromobacter sp.	1	1*	0.81–0.99	—	—
Agrobacterium radiobacter	0.9	1	0.83	—	—
Pseudomonas marginalis	1	1[a]	1	—	1
Rhizobium meliloti	1	1	0.9	—	—
Zoogloea ramigera	2	1	+	(Ill-defined)	
	11	3	1.5		

[a] pyruvylated sugar.

form of low-molecular-weight dextrans. The enzyme *dextransucrase* cleaves the substrate molecule and adds the glucosyl moiety to the acceptor. The fructose component of the substrate is utilised by the bacteria. In commercial dextran production the presence of iron and manganese are important in a complex medium based on corn steep liquor. Fermentation is carried out at 25°C. The fermentation broth becomes increasingly viscous as dextran is formed and vigorous stirring is required to ensure adequate aeration.

Other exopolysaccharides, whether heteropolymers or homopolymers, are synthesized intracellularly by a more complex process in which sugar nucleotides (nucleoside diphosphate sugars) are the activated form of the monosaccharides. The sugar nucleotides also provide the microbial cell with a means of interconversion of various monosaccharides through epimerization, dehydrogenation, and decarboxylation reactions. The monosaccharides are sequentially transferred by highly specific sugar transferases to an isoprenoid lipid acceptor molecule (bactoprenol, C_{55}-isoprenoid lipid) located in the cytoplasmic membrane of prokaryotes. In this way the repeating unit of heteropolysaccharides is synthesized, and at this stage acyl groups are also added. The repeating units are then polymerized, and the polysaccharides are excreted into the extracellular environment (Fig. 9). The products are usually of fairly discrete molecular weight, although the method by which this is controlled is unknown.

One modification to this biosynthetic mechanism is found in the production of bacterial alginates. The primary product is thought to be an acetylated homopolymer, poly-D-mannuronic acid. This is modified in a postpolymerization reaction by the extracellular enzyme *polymannuronic acid epimerase*, which converts some of the nonacetylated D-mannuronosyl residues to L-guluronic acid.

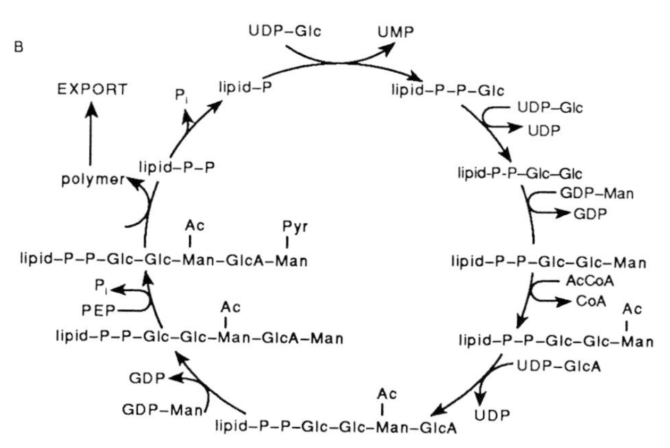

Figure 9 (A) Structure of isoprenoid alcohol used as a carrier lipid in polysaccharide biosynthesis. (B) A scheme for xanthan biosynthesis.

B. Synthesis of Poly-β-hydroxybutyric Acid

Synthesis of PHB is achieved by growth of *Alcaligenes eutrophus* in fed-batch culture in a medium with glucose as carbon source. Phosphate content is limited to limit the amount of cellular material produced, whereas other nutrients are in excess. After phosphate depletion has been reached, glucose is added to the cultures and PHB accumulation up to concentrations of 75% cell dry weight can be obtained. Copolymers are produced by simultaneous addition of glucose and propionic acid in carefully controlled ratios. Excess propionic acid is toxic to the bacteria.

Biosynthesis of PHB involves conversion from the precursor acetyl CoA by the enzymes *3-thioketolase* (acetyl CoA acetyltransferase), *acetoacetyl CoA reductase* (hydroxybutyryl CoA reductase), and PHB synthetase (Fig. 10). These enzymes from various PHB-synthesizing bacterial species vary in their cofactor and substrate specificity and other properties. The acetoacetyl CoA reductase from *Zoogloea* is NADP-specific. In *Rhodospirillim rubrum*, the conversion of L-(+)-3-hydroxybutyryl CoA to D-(−)-3-hydroxybutyryl CoA via crotonyl CoA is catalyzed by two stereospecific enoyl-CoA hydratases. Polymer degradation in several species also involves alternative mechanisms. [*See* ACETOGENESIS AND ACETOGENIC BACTERIA.]

III. Regulation

A. Exopolysaccharides

Several recent studies have indicated that the genes controlling enzymes specifically associated with exopolysaccharide synthesis are grouped together

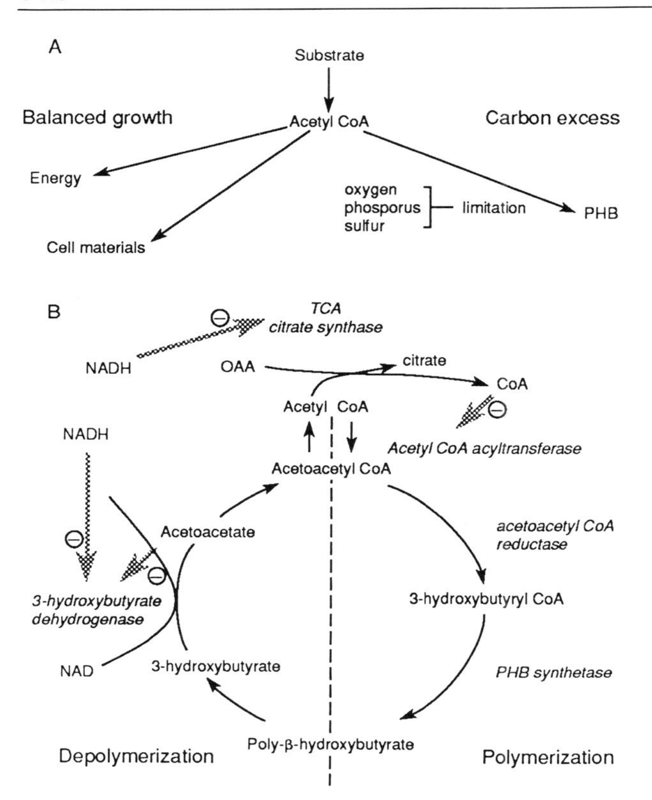

Figure 10 (A) Products of acetyl CoA under balanced growth conditions and carbon excess. (B) Biosynthesis and breakdown of PHB. Under conditions of carbon (energy) excess, NADH levels inhibit both depolymerization of PHB and tricarboxylic acid cycle (TCA). The resulting reduced levels of free CoA means that acetyl COA acyltransferase can catalyze the condensation of acetyl CoA to acetoacetyl CoA and that PHB can be synthesized. In energy (carbon) deficiency, the control is reversed.

regions of about 17 kb. The organization within such regions varies. In *X. campestris,* the genes for enzymes effecting addition of monosaccharides and acyl groups are closely associated with those for the polymerase and related functions. In *E. coli,* one chromosomal segment controls enzymes involved in synthesis of those sugar nucleotides specifically required for polysaccharide formation, along with the enzymes for monosaccharide transfer and the polymerase. This region is flanked by others controlling polymer translocation and postpolymerization reactions, respectively (Fig. 11). The central region alone is responsible for the specific polysaccharide structure and could be replaced by equivalent segments from other *E. coli* strains. Defects in the

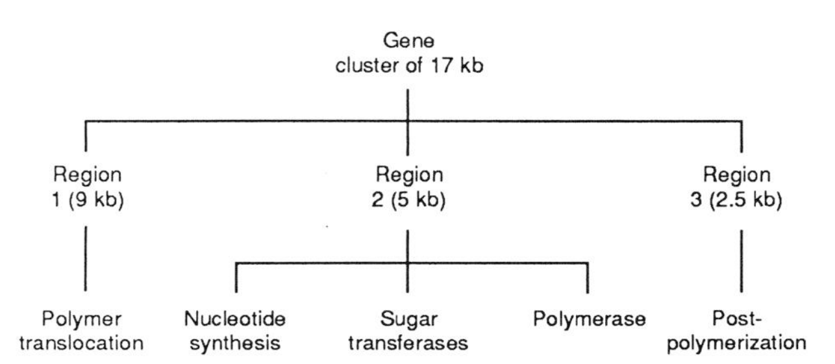

Figure 11 *Escherichia coli*—polysaccharide gene cassettes.

mechanism for polymer translocation caused polysaccharide to accumulate intracellularly or in the periplasm with lethal results. In some bacteria, exopolysaccharide synthesis is controlled through megaplasmids rather than chromosomally. In many bacteria, complex positive and negative regulatory elements such as the *lon* protein have been identified. Alginate synthesis in *P. aeruginosa* is responsive to various environmental stimuli.

B. Polyhydroxyalkanoates

The enzyme 3-ketothiolase controls PHB synthesis in *A. eutrophicus* with CoA as the metabolic effector. The three enzymes responsible for PHB synthesis have been sequenced, and it has been demonstrated that transcription occurs in the sequence polymerase-thiolase-reductase. The polymerase gene encodes a protein of a molecular mass of 63.9 kDa. The genes for PHB biosynthesis can be transferred to unrelated bacteria not normally capable of PHB synthesis and lead them to produce the biopolymer in relatively large amounts.

Bibliography

Anderson, A. J., and Dawes, E. A. (1990). *Microbiol. Rev.* **54,** 450–472.

Bertocchi, C., Navarini, L., Cesaro, A., and Anastasio, M. (1990). *Carbohydr. Polymers* **12,** 127–153; 169–187.

Boulnois, G. J., and Jann, K. (1989). *Mol. Microbiol.* **3,** 1819–1823.

Byrom, D. (1991). "Biomaterials." Stocton Press, New York.

Crescenzi, V., Dea, I. C. M., Paoletti, S., Stivala, S. S., and Sutherland, I. W. (1989). "Biomedical and Biotechnological Advances in Industrial Polysaccharides." Gordon and Breach, New York.

Dawes, E. A. (1990). "Novel Biodegradable Polymers." Kluwer Academic Publishers, Dordrecht.

Huang, S., Lee, H. C., and Mayer, R. M. (1979). The purification and properties of dextransucrase from *Streptococcus sanguis*. *Carbohydr. Res.* **74,** 287–300.

Lawford, G. R., Kligerman, A., and Williams, T. (1979). Dextran biosynthesis and dextransucrase production by continuous culture of *Leuconostoc mesenteroides*. *Biotechnol. Bioeng.* **21,** 1121–1131.

Linton, J. D. (1990). *FEMS Microbiol. Rev.* **75,** 1–18.

Sutherland, I. W. (1990). "Biotechnology of Microbial Exopolysaccharides." Cambridge University Press, Cambridge.

Tait, M. I., Sutherland, I. W., and Clarke-Sturman, A. J. (1986). *J. Gen. Microbiol.* **132,** 1483–1492.

Yalpani, M. (1987). "Industrial Polysaccharides." Elsevier, Amsterdam.

Bioreactor Monitoring and Control

Gregory Stephanopoulos and Robert D. Kiss
Massachusetts Institute of Technology

Glossary

Closed-loop (feedback) Control architecture in which the measured value of the controlled variable is returned, or fed back, to a controller mechanism acting to maintain the controlled variable at its set point or along its desired trajectory

Controlled variable Variable that is maintained or controlled at some desired value by a control mechanism

Manipulated variable Variable that is adjusted freely by a control mechanism to regulate the controlled variable

Open-loop Without feedback; the system has no means of controlling deviations of the process output from its desired value or trajectory

Optimal set point Provides the best (maximum or minimum) value of a given performance criterion

Proportional–integral–derivative Three-mode control algorithm designed to make the difference between the actual value and set point of the controlled variable as small as possible

Robustness Characteristic describing a system that performs well even in the presence of measurement noise and discrepancies in the system model

Set point Desired value of a controlled variable

Suboptimal control Control of inferior performance relative to the optimal control, but usually easier to implement than the optimal control

BIOREACTOR MONITORING AND CONTROL provide the means for achieving consistent bioreactor operation and improving process performance. The monitoring of a bioreactor, or fermentor, typically requires the application of a variety of process sensors to record the time course of the physical and biological parameters of interest. Given such information about the conditions within the bioreactor, and the ability to interpret the meaning of this information through an appropriate model or empirical relationship, a control strategy can be implemented to maintain the bioreactor state within desired limits or to vary its operation along a predetermined time trajectory necessary to meet the process objectives.

I. Introduction

Bioreactors provide the physical environment in which biochemical transformations take place, mediated by the catalytic activity of enzymes or intact, viable microbial or other cells. Such systems usually involve complicated networks of biochemical reactions affected by the conditions of a rather complex, multiphase bioreactor environment. Therefore, the control of biochemical events and bioreactor performance requires strict environmental control within the bioreactor, which, in turn, depends critically on information about the state of the bioreactor. Advances in bioreactor monitoring provide a more complete picture of the biochemical process, allowing the development of more effective control strategies. The end result of such improvements in bioreactor monitoring is the ability to achieve much greater control of the cellular environment, leading to better bioprocess performance. [*See* BIOREACTORS.]

Products of bioprocesses are the direct outcome of cellular metabolism or enzymatic reactions, uniquely determined by the set of environmental conditions, such as temperature, pH, substrate, or dissolved oxygen (DO) concentration, imposed on the bioreactor by the monitoring and control system being employed. The monitoring and control system can regulate the environmental variables in a static or dynamic manner. In static regulation, data from preliminary experiments are used to determine the optimal set points for the bioreactor environmental conditions. Control is then realized by altering or manipulating certain process variables such as heating/cooling rate, acid/base addition rate, nutrient feed rate, and aeration rate to maintain the environmental conditions at their desired set points. In this way, the available operating parameters are manipulated to achieve and to maintain an optimal static environment. In some cases, further process improvements are possible by varying in time some of the environmental conditions. The monitoring and control system then operates to force the selected environmental conditions along the desired time-varying profiles. Such profiles may be determined *a priori* through the use of mathematical modeling and optimization or may be determined in real-time during the course of the bioprocess. In either case, reliable process monitoring is crucial to achieve the desired controls for this dynamic optimization strategy.

In this article, an overview is offered of established practices and research efforts in the respective areas of bioreactor monitoring and bioreactor control. It should be noted that despite the external simplicity and apparent high success rates of common monitoring and control strategies, there is still considerable room for improvement in bioreactor operation, as evidenced, for example, by the great variability in final fermentation yields. Many factors contribute to process variabilities, such as inoculum quality, aeration interruption, inhibitory levels of substances, or the presence of compounds that shift the metabolism toward undesirable products. Correction and/or detection of process variabilities requires detection of such events through improved monitoring and identification of proper control strategies through improved mechanistic understanding and effective modeling. This constitutes a continuing effort in bioreactor process optimization.

II. Bioreactor Monitoring

A. Direct Measurements

Many aspects of a fermentation environment can be monitored, using a variety of sensors. Because this article is mainly concerned with the application of monitoring techniques in conjunction with control strategies, discussion focuses on measurements that can be made automatically during the course of a fermentation. Fermentation sensors have typically been categorized as either physical or chemical environment sensors. Measurements can be further distinguished as those that can be obtained continuously and those that can be obtained only intermittently. The reason for this further classification is the possible use of measurements in real-time, or on-line, for control purposes. If a measurement crucial to the calculation of control parameters can only be made once every 30 min (intermittently), then control actions based on actual measurements can only be adjusted once every 30 min, unless an estimate of the unknown quantity can be obtained in the interim.

1. Continuous Measurements

A continuous measurement is one that can be obtained "instantaneously" in comparison to the typical fermentation time scale. These measurements have the form of a continuous signal (usually a voltage or current). The signal must be calibrated to the actual value of the measured variable by the controlling computer or data-acquisition module, which typically acquires the signals at frequencies on the order of seconds to minutes.

Continuously measured physical environmental variables are temperature, pressure, agitation rate, aeration rate, power input rate, liquid feed rate, liquid volume, foam level, and broth turbidity. Temperature is usually monitored by thermistors or thermocouples. Diaphragm gauges and pressure transducers are used to detect pressure. Agitation rate, or impeller speed, is usually measured via a generator-type tachometer. Aeration rate, or air flow rate, can be measured and acquired by a computer through the use of a thermal mass flowmeter. Shaft power input, useful in estimating oxygen transfer capability, can be measured with either torsion dynamometry or a strain gauge, the latter being more accurate but considerably more expensive. Liquid feed rates, typically small relative to fermentor volume, are usually measured by placing additional vessels on a load cell system that continuously weighs the contents; thus, this measurement is much more accurate in the long term (minutes) rather than the short term (seconds). Fermentor liquid volume or, more correctly, liquid mass is also continuously estimated by placing the fermentor on a load cell system. Foam is detected by either conductance or capacitance probes. Finally, broth turbidity, while not frequently found in practice, can be measured continuously as an indication of biomass density. One measurement method is a continuous (and aseptic) withdrawal and pumping of a small sample through a flow cell turbidimeter or spectrophotometer. Recently, steam sterilizable, spectrophotometrically based biomass sensors, which fit into standard fermentor ports, have become available.

The major continuously measured chemical environmental variables are pH, redox potential, DO, dissolved carbon dioxide, and culture fluorescence. The pH is easily measured with a combination glass electrode, which is steam sterilizable. The redox potential measurement is typically made with a combined platinum electrode, also steam sterilizable. This measurement can be useful when the DO level is below the minimum sensitivity of the DO probe, although the interpretation of redox is difficult. DO is measured via steam sterilizable polarographic or galvanic electrodes. Steam-sterilizable electrochemical probes for dissolved carbon dioxide have been introduced only recently. Measurement of culture fluorescence, made by directing ultraviolet light into the culture and measuring the fluorescence emitted with a photodiode or photomultiplier, can provide information on the biochemical or metabolic state of the cell population. Such fluorescence measurements have been shown to be useful as an indicator of intracellular reduced nicotinamide adenine dinucleotide and reduced nicotinamide adenine dinucleotide phosphate (measures of the cell's energy and reductive power) and as an indicator of substrate exhaustion. Several devices, called biosensors, for continuous, on-line assay of specific liquid-phase components, have been undergoing research and development in the last 5–10 yr. One type of biosensor is based on coupling the action of enzyme-catalyzed reactions with electrochemical detectors. In another type, called enzyme thermistors, the heat released by the enzyme-catalyzed reaction is detected by a nearby calorimeter. Not currently available, but in the research stage, are enzyme transistors in which products of an enzymatic reaction cause changes in the electronic properties of solid-state devices, which can easily be measured. Specific sensors are under development for the measurement of small biomolecules and metabolically important ions such as NH_4^+, Mg^{2+}, Ca^{2+}, and PO_4^{3-}. These different biosensors, while quite promising, still require much more work in addressing problems with sterilization damage, stability, and calibration drift.

2. Intermittent Measurements

An intermittent measurement is one that can be obtained at a frequency of the same time scale as a typical fermentation. The reason for the reduced measurement frequency is typically due to the time required to obtain or transport a sample to an analyzer, the time required to clean or regenerate an analyzer, or the time required to incubate a reaction mixture or carry out a separation procedure. All of the available measurements in this category are of the chemical type.

Chemical environmental variables that are measured intermittently are exhaust oxygen and carbon dioxide concentrations and liquid-phase concentrations of substrates and products. Exhaust gas composition can be measured by a variety of techniques, although mass spectrometry (MS) has gained increased popularity. Availability of less expensive spectrometers is making MS more accessible for research applications, and reliable and robust instruments have made MS more practical for industrial application. Although a given spectrometer can analyze a gas sample in <1 min, typically several fermentors are multiplexed to one spectrometer and, thus, the sampling frequency for each fermentor is reduced. Mass spectrometers are most often used for the determination of the oxygen uptake rate (OUR), carbon dioxide evolution rate (CER), and the respiratory quotient (RQ; CER/OUR), all useful indicators of metabolic activity.

Measuring liquid-phase concentrations almost always involves obtaining a cell-free sample aseptically from the bioreactor. This is typically accomplished by automatically transferring a culture stream to some type of cross-flow filter or membrane device for cell removal. For volatile medium components such as ethanol, continuous flow of a carrier gas through a length of tubing, permeable to the medium compound of interest, sweeps penetrating compounds off to a gas analysis device. In either case, the cell-free sample is sent to an analytical device where a chromatographic separation or quantitative reaction may be performed. On-line high-pressure liquid chromatography has been used, at the research scale, to measure substrates such as sugars and alcohols and products such as alcohols and organic acids. On-line gas chromatography has been used to quantitate volatiles such as alcohols and solvents. Another useful strategy for on-line chemical analysis is selective conversion of the broth component of interest to a readily measurable product, typically by enzymatic methods. Several immobilized enzymatic probes have been developed for measuring compounds such as glucose, L-lysine, and L-glutamine. These probes rely on the participation, and subsequent detection, of DO in the enzymatic reaction. Flow injection analysis (FIA) is another immobilized enzyme technique that relies on spectrophotometric detection of reaction products. FIA systems have been developed for quantitation of several compounds, including glucose, ammonia, ethanol, and phosphate. Most enzyme-based sensors for liquid-phase substrates and products are applicable only in limited concentration ranges and, thus, some sort of automatic dilution apparatus needs to be incorporated into the on-line analysis system.

B. Indirect Measurements

In many cases, fermentation variables of interest, particularly biomass, substrate, and product concentrations, cannot be measured on-line. This is particularly troublesome if a desired control strategy requires such an unmeasurable quantity as input. In these cases, it may be possible to relate variables that are measurable to the desired quantity or quantities by the use of an independently determined experimental correlation or through an algorithmic sensor. The variable of interest is then said to be indirectly measured, or inferred, from the measurement of a related variable.

1. State Variable and Parameter Correlations

During preliminary characterization of a fermentation process, correlations between process variables are often established. If, for example, a linear relationship exists between variable A and variable B, then an on-line measurement of variable A can be used to obtain an on-line estimate of variable B. A simple example of such a correlation is the linear (under certain conditions) relationship between optical density and biomass concentration. An on-line measurement of optical density can, thus, be used as an on-line estimate of biomass concentration. There are also several useful correlations between respiratory rates and other process variables. The rate of heat production, which can be measured by calorimetry, has been shown to be proportional to oxygen uptake rate for a number of different fermentations. Thus, if off-gas analysis is not available, OUR may be estimated from a heat of fermentation measurement, if available. In baker's yeast fermentation, a correlation between the rate of ethanol production and the RQ has been experimentally determined; thus, RQ can be utilized as an indicator of ethanol formation. Furthermore, if the RQ measurement is used in conjunction with a DO measurement, it provides an indication of whether the metabolic shift to ethanol production is a result of oxygen starvation or overfeeding. In amino acid fermentations, the RQ may, under some conditions, be directly related to the yield of amino acid from substrate, and such a correlation could be used to infer yield on-line. In fermentations without acidic or basic product formation, pH neutralization information, such as the rate of base addition required to maintain constant pH, can provide an estimate of the rate of biomass formation. Finally, correlations may exist between specific growth rates or specific product formation rates and substrate or product concentrations. If a control strategy aims to regulate such specific rates, an on-line measurement of the appropriate substrate or product concentration may be useful for the on-line estimation of the specific rate.

Most of the preceding correlation-based, indirect measurements depend on conducting preliminary experiments to establish the desired relationship between variables. Furthermore, these correlations often have limited regions of applicability or are only valid for a particular fermentation. The desire for more general indirect measurement or estimation schemes has led to the development of model-based and material balance-based algorithmic sensors, discussed next.

2. Algorithmic Sensors

The indirect measurements obtained using the sensors described in this section are a direct result of the introduction of computers to the biochemical engineering environment. Although the actual measurements available on-line during a given fermentation may be limited, the introduction of computers has allowed the estimation of the values of unmeasurable process variables and parameters by combining mass and energy balances and process mathematical models in concert with the available measurements. The initial goal in developing such algorithmic sensors was the on-line estimation of biomass concentration and specific growth rate, but the methodology has since been applied in the on-line estimation of unmeasurable quantities such as substrate and product concentrations.

One of the first examples of an algorithmic sensor was the application of an on-line material balance to the on-line determination of the biomass growth rate. To this end, the biomass growth and product formation processes were represented by a lumped stoichiometric equation, as shown here:

$$a\ C_xH_yO_z + b\ O_2 + c\ NH_3 \rightarrow C_\alpha H_\beta O_\gamma N_\delta + d\ H_2O + e\ CO_2 + f\ C_{\alpha'}H_{\beta'}O_{\gamma'}N_{\delta'}, \quad (1)$$

where $C_xH_yO_z$, $C_\alpha H_\beta O_\gamma N_\delta$, *and* $C_{\alpha'}H_{\beta'}O_{\gamma'}N_{\delta'}$, represent the empirical formulae for the carbon and energy source, cell mass, and product, respectively. There are seven unknowns in Equation 1 (assuming empirical formulae are given for substrate, cell mass, and product): the stoichiometric rate (mmol/hr) coefficients *a–f,* and the extent to which the reaction has occurred. There are four elemental balance equations (carbon, hydrogen, oxygen, and nitrogen) that can be applied toward solving for the unknowns. Therefore, to determine all seven unknowns, a minimum of three measurements must be available. For example, if measurements of OUR, CER, and ammonia addition are available, it is then possible to estimate substrate uptake and biomass and product formation rates. This methodology has been utilized in several cases (including fed-batch baker's yeast fermentation), with varying degrees of success, to estimate biomass concentration on-line. More complicated on-line stoichiometric balance approaches, requiring additional on-line measurements for closure, have been proposed for the elucidation of fundamental biochemical structures, such as intracellular reaction fluxes, although such approaches have yet to be applied in practice. Several possible pitfalls exist in such direct material balancing approaches. The cell composition may vary during the course of the fermentation; side-product formation can invalidate the stoichiometric equations; and errors in process variable estimates, once made, tend to propagate and grow with time.

To overcome some of the limitations of the material balance approaches, more advanced methods have been developed for the estimation of key fermentation parameters such as specific metabolic rates and yields. These advanced techniques take into account not only known relationships among process variables, but also consider measurement noise effects, process noise effects (inaccuracies in the process model), and error propagation. Most applications of this type of estimator have involved the use of the extended Kalman filter for generating on-line estimates of fermentation variables such as biomass, substrate, and product concentrations, specific growth rate, and substrate and product yields. One application even used a type of extended Kalman filter as a means of detecting contamination of a fermentation in real-time.

Some of the most recent research on bioreactor state estimation algorithms has utilized neural network approaches. These methods can offer an interesting solution to the estimation problem because they do not require any *a priori* knowledge about the structure of the relationships that exist among fermentation variables. A neural network model uses representative examples of fermentation data for the training of the network, after which period the network learns to generalize the dynamic behavior of a given fermentation, including the ability to reject certain levels of measurement noise. Neural networks offer the ability to learn complex relationships without requiring knowledge of a detailed physical or biological model. Simulation work has shown the technique capable of estimating biomass and substrate concentrations on-line. There have been no applications of such technology to any practical fermentation systems yet, but much more is expected from such approaches in the future, including applications to bioreactor process control and fault detection and diagnosis.

All of the preceding approaches to estimation of unmeasurable bioreactor variables offer valuable methods of circumventing the lack of accurate, reliable on-line sensors for all variables of interest. Of course, the best estimate of the desired quantity would come from a specific sensor. Current efforts continue to address this need.

III. Bioreactor Control Strategies

A. Direct Environmental Controls

The objective of direct environmental controls is to maintain some of the bioreactor's environmental variables at specific set points. These set points usually have been determined by small-scale experiments and are deemed optimal for a particular objective. Depending on the frequency of use, application, and their dependence on the manipulated variables, these controls are further classified into two distinct categories: conventional controls and other environmental controls.

1. Conventional Controls

The control of temperature, pH, DO concentration, and foam level are examples of conventional controls. Bioreactor vessel pressure is routinely monitored but usually not automatically controlled. These conventional controls are present in almost all experimental bioreactors, most pilot-plant bioreactors, and some production fermentors. In the control of temperature, the temperature itself is obviously the measured and controlled variable and the cooling and heating fluid (typically water) flow rates are the manipulated variables used to maintain the temperature at its set point. Acid and base addition are the manipulated variables for pH control, whereas aeration and/or agitation rates are the manipulated variables for the control of DO. In a foam control system, a foam detector activates addition of a chemical antifoam (some types of foam control systems, such as the centrifugal defoamer, do not need a detector). The control laws employed for these conventional controls are usually simple, ranging from on/off to proportional–integral–derivative algorithms. An important characteristic of these types of controls is that effective control action requires very little knowledge of the biological process. In fact, these variables are often controlled at set points via individual closed-loop feedback controls in which control action is taken very frequently, on the order of seconds, based on continuous measurements.

There are very few problems with these controllers, most of them stemming from sensor malfunction and nonrepresentative readings resulting from reactor heterogeneities. Due to the sensitivity of most microbial bioprocesses to fluctuations in temperature, pH, and DO, these controls must be reliable with respect to errant sensor inputs. Thus, controllers that incorporate sensor redundancies as well as software for measurement error detection are justifiable. Strong coupling among the control variables is not usually encountered in most fermentation processes. However, one can envision several possibilities of interaction among the control loops of this and the next category. Examples include the simultaneous control of the DO and substrate level in an animal cell culture (the interaction occurring through the strong effect of the agitation rate, used for DO control, on cell growth rate and subsequent substrate consumption) or the simultaneous control of substrate and biomass concentrations through the manipulation of feed concentration and flow rate in a continuous bioreactor.

2. Other Environmental Controls

These controls aim to regulate some of the remaining variables of the abiotic, or nonliving, environment of a bioreactor, such as substrate, product, and precursor concentration. Few of these controls have found their way to applications in bioreactor sizes of any industrial significance. However, it has been established that significant enhancement of bioreactor productivity can be achieved by controlling, for example, a substrate at a level that optimally balances growth and undesirable side reactions, or an expensive precursor at a level that ensures high yields and satisfactory rates, or an inhibitory substrate at a level that minimizes inhibition while maintaining an acceptable production rate. Nevertheless, the rate of implementation of these controls in industry is very low, due, in part, to lack of robustness in the examples demonstrated to date and to the usual inertia that often impedes the transfer of such technologies to industrial practice.

Examples of research work in this area include the control of an unstable steady state deemed optimal for the continuous biomass production of a methylotrophic bacterium, the control of the glucose (product) concentration against lactose (substrate) concentration disturbances in the feed in a continuous lactose enzyme reactor subject to enzyme deactivation, and substrate (glucose and sucrose) control in a continuous immobilized yeast reactor for ethanol production. Some simulation studies have examined the following: the possibilities of manipulating dilution rate and substrate feed for the simultaneous control of biomass and substrate concentration in a quasi-steady-state fed-batch fermentor, single-variable biomass control and noninteracting multivariable substrate and biomass control in a continuous bioreactor, biomass control in a continuous mixed-culture fermentor, and biomass control at the point of maximum productivity in a continuous bioreactor. Other applications include control in wastewater treatment plants and dairy product processing plants.

There are several key problems in implementing these types of control strategies: (1) the lack of reliable sensors for on-line measurement of the controlled variable, (2) the absence of reliable models for predicting the effect of the manipulated variable on the controlled variable, and (3) the possibility that the optimal set points may vary with time. Such variations in the optimal set points are caused by changes in the catalytic activity brought about by enzyme deactivation in enzyme reactors and by culture adaptation or degeneration in fermentation systems. As an example of the preceding points, consider the microaerobic yeast fermentation in continuous culture. In that system, it has been observed that the specific ethanol productivity attained a maximum at a very low DO concentration (in the parts per billion range). This maximum was at least 40% higher than the specific productivity observed under strictly anaerobic conditions, indicating the clear incentive for controlling DO at the optimal level. However, no DO probes currently exist that function well in the aforementioned low range; furthermore, how aeration and agitation rates should be manipulated to establish optimal operation is unclear, and the optimal point is expected to vary with time. Therefore, controlling the DO in an optimal fashion would clearly be difficult, if not impossible.

Such problems are not specific to this particular system, as similar problems can be found in the majority of industrially important bioprocesses. In amino acid fermentations, shifts in the DO level may favor the production of one amino acid over another and the optimal DO level may be quite low and out of the range of commercially available probes.

Attempts to resolve some of the problems and issues surrounding these types of controls include, as discussed in detail earlier, the development of improved and specific sensors for more reliable online monitoring of the controlled variables. In addition, the use of the previously mentioned on-line material balances has been employed to estimate the controlled variable when it cannot be measured accurately or at sufficient frequency. Also, the development of correlations between manipulated and controlled variables, both through empirical and mechanistic means, continues to offer aid in the construction of reliable control laws.

While such empirical correlations are important and useful in the formulation of the control laws, there is a distinct need for better mechanistic understanding of the underlying biological processes responsible for the observed system behavior. In the microaerobic yeast fermentation example mentioned earlier, experimental and modeling studies indicated that the specific ethanol productivity maximum occurs because at very low DO levels ethanol formation is limited by adenosine diphosphate (ADP) availability in the decarboxylation pathway, while glucose transport limits the metabolic rates at relatively higher oxygen levels. Consequently, if a decrease in ethanol production is accompanied by a similar decrease in glycerol production, it indicates that glucose transport is limiting ethanol production and the control should aim at decreasing the DO level. If, however, glycerol production remains high after ethanol production drops, it indicates ADP limitation and, thus, the control should aim at increasing the DO level. This example clearly demonstrates how a basic understanding of the underlying biochemical mechanisms can be translated into a simple control action that utilizes the measurement of product concentrations (ethanol and glycerol) to maintain the optimal operation of the fermentor despite culture adaptation with time (varying optimal set point) and unavailability of probes to measure very low DO concentrations.

The challenges in implementing these types of controls are the need for sensitive and reliable sensors and/or robust estimation methodologies for bioreactor monitoring, as discussed earlier, and reliable cause–effect relationships for formulating effective control laws.

B. Metabolic Activity Controls

The stated purpose of the controls in the previous section is to maintain the bioreactor's environment at a desired state, which reflects the operator's assessment of optimality. These controls, however, are not the only answer to the pursuit of bioreactor performance improvement. Because the optimal set points may indeed vary with time, performance may be more effectively improved by attempting to maintain the metabolic activity of the culture, rather than the abiotic environmental variables, at an optimal state. This type of strategy obviously requires some measurement or evaluation of the culture's metabolic state as input for determining the appropriate control action.

A well-known example of this type of control strategy is found in the fed-batch baker's yeast fermentation. In this system, overfeeding of sugars results in ethanol formation and poor biomass yields, even under aerobic conditions (Crabtree effect). Conversely, underfeeding results in cell starvation and specific growth rate deterioration, leading to low volumetric productivities. To achieve optimal results, the sugar feed rate is regulated so that the fermentor sugar concentration remains ideal for growth. Two types of feedback controls have been investigated for the regulation of the nutrient feed rate. In the first type, an indirect control parameter, which can be monitored on-line, serves as an indicator of metabolic activity. Since carbon dioxide evolution accompanies ethanol formation, the RQ and CER have been applied as the indirect indicators of metabolic activity toward ethanol production. In other cases, the DO level and culture pH have been used for this type of control. In the second type of controls, the formation of the undesired product is detected directly by measuring its concentration in the broth or exit gas. The production rate of the undesired product is then related to metabolic activity, which is then used as the controlled variable in the control loop.

Another example of metabolic activity control is the production of L-lysine by nutritional, or auxotrophic, mutants in fed-batch culture. In this system, the need to supply the required amino acid growth factor, L-threonine, must be balanced against the possibility of building up excess levels of L-threonine, which can inhibit L-lysine formation. By balancing these two opposing effects of L-threonine, the metabolic activity of the culture can be controlled at desired levels based on the desire to maximize either fermentation yield (g product/g substrate) or volumetric productivity (g product/liter/hr). The culture respiratory behavior, namely the process trends in RQ, OUR, and CER, was demonstrated to be an effective indirect indicator of the culture's metabolic activity toward efficient overproduction of L-lysine. Feed rate to the culture was thus manipulated based on the on-line measurements of the respiratory trends. This type of approach demonstrated the utility of maximization and maintenance of culture metabolic activity through the use of empirically obtained relationships between available on-line measurements (respiration) and fermentation performance (yield and productivity) variables.

The previously mentioned techniques for metabolic activity indication have been met with varying degrees of success. The RQ has been extensively used (mainly in cell mass production processes) because of the availability of gas analyzers for oxygen and carbon dioxide and, in the case of baker's yeast fermentation, the relatively reliable representation of the cell's metabolic state with regard to ethanol formation. In some of the initial yeast applications, the formation of the control law for the sugar feed rate required some glucose to ethanol yield parameters, which were indirectly determined and variable in nature. This reduced the applicability of the method. In subsequent applications, required parameters were determined on-line through gas analysis and material balancing, thus contributing to the robustness of the control structure. Such application of on-line material balancing allowed the calculation of the anticipated substrate demand, and the feed could then be manipulated to meet that demand (feed on-demand control). It should be noted that, in some cases, the RQ may, in itself, be too sensitive a measurement of metabolic activity to be of practical use. Such a case could arise if significant side-product formation exists, in which case the measured RQ reflects an aggregate of the contributions of the desired product pathway's respiratory characteristics and the side-product pathway's respiratory characteristics. While this is, in fact, the basis of the utility of RQ in controlling the production of yeast cell mass, it can lessen the utility of RQ-based control in the production of excreted compounds such as amino acids.

Another indirect physiological indicator is the DO level, which has been used in the fermentation of methanol-utilizing microorganisms. However, in this case, the use of the DO measurement is of marginal value due to the ambiguity of the biological interpretation of the measurement. For example, an increasing DO concentration could be interpreted either as depletion of the substrate, suggesting a feed increase, or as damaged metabolic capability, suggesting a feed decrease. One can conclude that, if the control objective is to prevent the accumulation of growth-inhibiting substances (i.e., methanol), the direct measurement of the substance in question gives the best results. If, however, the objective is to balance two opposing factors in bioreactor productivity, a sensitive indicator of metabolic activity, such as RQ, is the preferred control variable. Direct control of the substance that determines the shift in metabolism (glucose in the yeast fermentation, L-threonine in the L-lysine fermentation, methanol in the methylotrophic fermentation) is, in most cases to date, either unrealistic or unreliable. This point is also well illustrated in the baker's yeast fermentation, where the use of complex carbohydrate mixtures precludes the measurement of a representative sugar concentration. In addition, even when a simple sugar is used, there may still be a range of concentrations over which the shift in metabolism takes place, possibly dependent on other environmental factors that are not monitored or are even unknown.

The majority of the controls in this category are based on measurements attainable with presently available sensors. The main problem is to relate these measurements to the particular aspect of metabolism that determines the control law. Most of the examples presented utilized correlations derived by combining phenomenological relationships and scientific judgment. Better understanding of the fundamental mechanisms governing the underlying biochemical processes will certainly improve these correlations and contribute to improved quality of these control strategies.

C. Specific Growth Rate Controls

These controls have been applied mainly in fed-batch fermentations for antibiotic and enzyme production. There are two distinct phases in these (as well as other) processes: a growth phase and a product formation phase. During the growth phase, the specific growth rate is controlled at its maximum value to accumulate biomass in the shortest time possible. There is an optimal biomass density beyond which adequate aeration cannot be achieved and subsequent specific productivities are reduced. At that point, product formation is initiated by shifting the specific growth rate to a lower (maintenance) value and adding product-inducing agents, if required.

Clearly, the two main problems in these types of controls are the determination of the specific growth rate and efficient means of affecting changes in growth. Directly measuring the specific growth rate is not presently possible, so indirect correlations, often purely empirical, are routinely used. For example, a linear relationship was observed between the specific growth rate and CER in a fed-batch cellulase production study; thus, CER was used as the measured variable. In another case, a study of β-galactosidase production, CER and RQ were used as indicators of the specific growth rate. These indicators determined the need for intermittent addition of carbon source, as well as the shift between two different sets of conditions for temperature, pH, aeration, and agitation. In other cases, the specific growth rate has been estimated via on-line material balancing (based on off-gas and ammonia addition measurements) used in conjunction with advanced estimation-filtering techniques. This type of strategy has been shown to be useful in certain cases but has failed in the estimation of biomass in a pencillin fermentation, due to the presence of a large amount of residual complex nutrient. The use of on-line filtration technique for the direct measurement of biomass (and consequent calculation of specific growth rate) and feedback control, however, yielded significant fermentation improvements in this case.

The most common means for effecting changes in the specific growth rate is by adjusting the nutrient feed rate, although a combined manipulation of aeration and agitation is also occasionally used. As in the biomass production example of the previous section, the substrate feed must be carefully controlled in many antibiotic or extracellular enzyme processes to avoid repressive effects resulting from overfeeding. Similarly, underfeeding causes starvation and may induce severe and irreversible damage to the product formation mechanisms. These types of controls have found diverse application, yielding impressive performance improvements in some cases. As improvements in the ability to monitor specific growth rate become available, the introduction of specific growth rate controls to increased numbers and types of fermentations is expected.

D. Optimal Controls

Fermentations are inherently dynamic systems with distinct phases of growth and production. Controlling the transition from one phase to the other is but one way to improve the process. Considering the different criteria that can be optimized, the various parameters, besides the feed rate, that can be controlled, and the dynamic nature of the process, bioreactors are natural candidates for application of optimal control theory.

In an optimal control problem, the given system is described by a mathematical model incorporating the factors, including time-varying input functions, governing the system dynamics. There is also an objective function whose value is determined by the system behavior over the entire course of the process and is, in some sense, a measure of the quality of that behavior. The optimal control problem is that of selecting the input function so as to optimize (maximize or minimize) the objective function. For example, in a fermentation system, the model may describe the dynamics of biomass, substrate, and product concentrations and include a time-varying feed rate as an input function. The objective function may be the maximization of product concentration or volumetric productivity (g product produced/liter/hr), and the optimal control solution determines the feed rate profile achieving the desired maximization.

Numerous studies have detailed the subject of optimal control of batch, fed-batch, and continuous bioreactors. Specific examples include the optimal temperature profile of a batch penicillin fermentation, the optimal pH and temperature profile of a gluconic acid fermentation, the optimal substrate profile in a turimycin fermentation, the optimal feed rate profile in a fed-batch L-lysine fermentation, the optimal scheduling of repeated fed-batch fermentation for biomass production, the optimal feed rate profile in a penicillin fermentation, which also considered the effects of cell age, and the optimal flow rate profile in a continuous enzymatic reactor subject to enzyme deactivation. Maximization of productivity has been chosen as the objective function in most applications. Productivity, however, is not the only criterion, and, actually, in other cases, a different index should be optimized via the optimal control. For fermentations intensive in raw material use (amino acids, citric acid), the yield (product produced per substrate consumed) may be critical; for fermentations producing small quantities of high added-value products, the maximization of final product concentration may be the main objective both to maximize production and facilitate downstream processing steps; for processes with large separation-purification capacities, minimization of fermentation time may be the desired goal. Of course, different criteria may be combined to form an objective function that reflects tradeoffs between several different fermentation performance measures, such as yield and productivity. However, overall, composite performance indices, which take into consideration all the diverse factors in a fermentation, are very difficult to formulate, and the identification of a single, most important criterion is usually a critical step in developing a bioreactor optimization. Also, different performance indices yield different control strategies, and the index of a given fermentation may change with market conditions or company priorities. Finally, the implications of a certain bioreactor operation on downstream processing steps should also be considered in the overall process optimization.

The most important problem with all applications of dynamic bioreactor optimization is the quality of the models used. In the majority of the cases studied to date, the models employed were derived under static, or steady-state, conditions, did not consider any history or time delay effects, and, in general, exhibited poor dynamic performance. A few studies have used more complicated, "structured" models based on some simplified picture of the critical metabolic processes, which gave better dynamic performance. Although an improvement when compared to the previous unstructured models, not all aspects of growth and product formation could be predicted with such models, and, more importantly, along with the introduction of structure came the addition of more, empirical, nonobservable parameters. The use of models that summarize past history effects on present culture performance, called delay kernel models, have been proposed with good results. These models, determined from simple dynamic experiments, capture in a simple set of equations the biological essence of the more complicated structured models. The many dynamic bioreactor optimizations studied have continually pointed out that easily observable, reliable dynamic models are essential for the formulation of a robust control structure.

In addition to the need for reliable models, another weakness of optimal control theory applied to bioreactors is the fact that most of these optimizations have been formulated as open-loop structures. For example, in the case where the optimization is used to solve for the optimal feed profile, said profile is solely a function of time. As a result, the obtained control cannot compensate for model uncertainties, measurement errors, and other process noises unaccounted for in the derivation of the control policy. Even if the process model used describes the system dynamics well, the model requires initial conditions to be specified, and there is enough variability from run to run in the quality of the inoculum, nature of the fermentation medium, and actual bioreactor operation to raise serious doubts about the suitability of open-loop controls for these systems. Therefore, either the optimization formula must be formulated with a closed-loop feedback structure or a solution must be pursued that links control implementation to actual bioreactor measurements.

There are several reasons why few implementations of optimal control algorithms exist. First, many biological systems are complicated by constraints on certain variables that make the determination of the optimal profiles computationally difficult. For example, determining the optimal feed rate profile for a fed-batch system in which there exists a biomass maximum beyond which aeration is impaired is very difficult. Even if it is possible to calculate the optimal profiles, optimal control strategies are rarely applied in practice, and, thus, it is not yet conclusive as to how much more can be gained from the implementation of such policies. A number of reasons preclude the use of optimal control strategy implementations: The improvement from the use of sophisticated optimal control may not be substantial, the optimal control policy may be difficult to implement due to observability problems, the model used may not be totally reliable, or the model parameters may not be accurately estimated. In all these cases, the use of a suboptimal policy may be justified and, in fact, preferred. Thus, the possible gain from optimization must be balanced against the costs associated with the added process complexity. In some cases, a simple control strategy may be more appropriate.

IV. Monitoring and Control Integration for Advanced Bioreactor Operations

The majority of the control structures presented in this article have been applied to small-scale experimental reactors. A few are profitably employed at the present time in pilot plants, but, for the most part, only the basic controls described in Section III.A.1 are being practices in production-scale bioreactors. On the other hand, computers are being introduced at an increasing rate to all stages of industrial reasearch, development, and production. Most of the installed computational capacity is still grossly underutilized, mostly occupied with data logging, report generation, and simple computational tasks such as the calculation of respiratory data, such as RQ, from off-gas measurements.

The time is right for a general upgrade of the quality of information that is presently being generated during a biological process such as a fermentation. Advances in sensors, estimation and fault detection techniques, and process modeling of biological systems are ready to lead the way in the development of advanced bioreactor operational strategies. As discussed previously, methodologies and sensors are presently available for the on-line determination of useful biological parameters, such as specific rates and yields, along with the basic bioreactor state variables, such as biomass, substrate, and product concentration. Significant progress on the modeling front, such as the work with structured models and delay kernel models, discussed previously, also make advanced control strategies serious contenders for fermentation improvement. In addition, control theory and the practical applications of control strategies are well established in many industries and much work, as discussed, has been conducted on applying control principles to bioprocesses. The tools are thus present; the difficulty is to appropriately combine and integrate the aspects of monitoring and control in developing an advanced bioreactor control system.

The preceding elements, namely reliable model and state measurement-estimation capability, are the two key components of the advanced bioreactor control structure depicted in Fig. 1. This structure can be constructed with presently available technology. It has the attractive features of adaptability, regulation, and optimization. The structure is adaptive because it contains an adaptation mechanism that automatically adjusts its controller parameters so as to compensate for variations in the characteristics of the process and, thus, tunes the controllers on-line to make them most efficient. The structure regulates the process through typical feedback controls designed to drive the process to the desired state set points, which are determined by a model-based optimal control strategy formulated to exploit the availability of on-line measurements and estimates. The state estimator may employ any of the techniques discussed previously. The optimal control strategy determines the new system set points, compares those values with the current set of measurements and estimates, determines the difference between the two (the error term), and passes that information to the feedback controllers, which will attempt to drive the process variables toward the desired optimal set points. The structure also contains a fault detection block, which performs tasks such as testing the consistency of measurements and identifying process faults such as sensor malfunctions or general process irregularities. This fault detector could alert the state estimator to disregard faulty measurements in performing the estimation

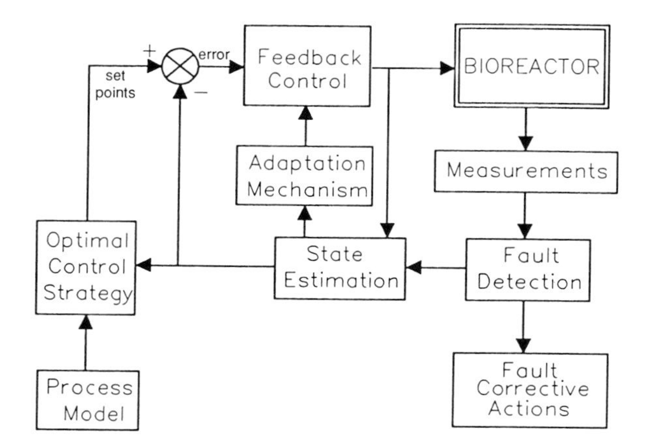

Figure 1 Block diagram for the proposed advanced bioreactor monitoring and control system.

tasks and could either automatically activate the proper controls in response to such process problems or alert an operator to take the necessary actions. Finally, the control structure of Fig. 1 requires little additional hardware beyond what is usually available in typical bioreactor operations. The main challenge is to assemble the various components and demonstrate the integrated software system in a specific application.

The single most important factor that will determine the rate at which advanced control structures are introduced in bioreactor operation is the anticipated improvement in overall process economics or the magnitude of the process problem that such a system would be called upon to address. In this regard, of critical importance is the identification of real problems requiring the added sophistication or processes by which the anticipated performance increases can make the difference between approval and rejection. Some examples where the advanced approach may offer real benefits include continuous, large-volume organic acid fermentations, where significant yield increases can be achieved as long as the biocatalytic activity of the cells is adequately monitored and regenerated as required; high productivity continuous ethanol processes, which are presently impaired by oscillations that conventional controls have failed to regulate properly; and single-cell protein processes, which can be improved by more efficiently balancing the opposing effects of growth and inhibition. Additionally, many of the cases mentioned earlier can provide further examples of potential advanced control applications.

One can conclude that correct problem definition will determine the introduction of more advanced control applications to bioreactor operation. Once need has been established, there is little doubt that existing technologies can meet the challenges through advanced bioreactor operations.

Bibliography

Bailey, J. E., and Ollis, D. F. (1986). "Biochemical Engineering Fundamentals," 2nd ed. McGraw-Hill, New York.

Cass, A. E. G. (ed.) (1990). "Biosensors: A Practical Approach." Oxford University Press, New York.

Moser, A. (1988). "Bioprocess Technology: Kinetics and Reactors." Springer-Verlag, New York.

San, K. Y., and Stephanopoulos, G. N. (1984). *Biotechnol. Bioeng.* **26,** 1176–1218.

Schügerl, K. (ed.) (1991). Measuring, modeling and control. *In* "Biotechnology—A Multi-Volume Comprehensive Treatise," Vol. 3 (H. J. Rehm and G. Reed, eds.). Verlagsgesellschaft, Weinheim.

Stephanopoulos, G. (1984). "Chemical Process Control: An Introduction to Theory and Practice." Prentice-Hall, Englewood Cliffs, New Jersey.

Turner, A. P. F., Karube, I., and Wilson, G. S. (eds.) (1987). "Biosensors: Fundamentals and Applications." Oxford University Press, New York.

Wang, N. S., and Stephanopoulos, G. N. (1984). *CRC Crit. Rev. Biotechnol.* **2,** 1–103.

Bioreactors

Larry E. Erickson
Kansas State University

Glossary

Airlift reactor Column with defined volumes for upflow and downflow of the culture broth; vertical circulation occurs because air is bubbled into the airflow volume

Batch bioreactor Culture broth is fed into the reactor at the start of the process; air may flow continuously

Bubble reactor Aerated column without mechanical agitation

Fed batch Liquid media is fed to the reactor continuously; the broth accumulates in the reactor as there is no outflow of liquid

Heterotrophs Microorganisms growing on an organic compound that provides carbon and energy

Photoautotrophs Microorganisms that use light for energy and carbon dioxide for their carbon source

BIOREACTORS are vessels or tanks in which whole cells or cell-free enzymes transform raw materials into biochemical products and/or less undesirable by-products. The microbial cell itself is a miniature bioreactor; other examples include shake flasks, Petri dishes, and industrial fermentors. Diagnostic products based on enzymatic reactions, farm silos for silage fermentations, bread pans with fermenting yeast, and the soil in a Kansas wheat field may also be viewed as bioreactors. While the bioreactor may be simple or highly instrumented, the important consideration is the ability to produce the desired product or result. The bioreactor is designed and operated to provide the environment for product formation selected by the scientist, baker, or winemaker. It is the heart of many biotechnological systems that are used for agricultural, environmental, industrial, and medical applications.

I. Introduction

The importance of the bioreactor is recorded in early history. The Babylonians apparently made beer before 5000 B.C. Wine was produced in wineskins, which were carefully selected for their ability to produce a beverage that met the approval of the King and other members of his sensory analysis taste panel. Food and beverage product quality depended on art and craftsmanship rather than on science and engineering during the early years of bioreactor selection and utilization. Early recorded history shows that some understood the importance of the reactants and the environmental or operating conditions of the reactor. This allowed leavened bread and cheese to be produced in Egypt more than 3000 years ago.

The process of cooking food to render it microbiologically safe for human consumption as well as to improve its sensory qualities is also an ancient tradition. The process of thermal inactivation of microorganisms through the canning of food to allow safe storage was an important early achievement in bioreactor design and operation.

As humans learned to live in cities, waste management including wastewater treatment emerged as a necessity for control of disease. One of the first process engineering achievements was the biological treatment of wastes in bioreactors designed and built by humans for that purpose. Because a significant fraction of the population of a city could die from disease spread by unsanitary conditions, these early bioreactors represented important advancements.

After microorganisms were discovered, microbiologists and engineers increased their understanding of the biochemical transformations in bioreactors. Simple anaerobic fermentations for the production of ethyl alcohol, acetone, and butanol were developed. Aerobic and anaerobic treatment of wastewater became widely used. Sanitary engineering became a part of civil engineering education.

In the 1940s, the field of biochemical engineering emerged because of developments in the pharmaceutical industry that required large-scale bioreactors for the production of streptomycin and penicillin. Progress in bioreactor design and control resulted from research on oxygen transfer, air and media sterilization, and pH control. The central concern of the early biochemical engineers was the development of bioreactors that could achieve and maintain the chemical and physical environment for the organism that the biochemist/microbiologist recommended. The ability to scale-up from laboratory bioreactors to large fermentors required the development of instrumentation such as the sterilizable oxygen electrode. Early courses in biochemical engineering were concerned with the analysis, design, operation, and control of bioreactors. While the field of biochemical engineering is less than 50 years old and some of the pioneers are still available to provide a first-person account of those exciting days, great progress has been made in bioreactor engineering.

II. Classification of Bioreactors

Several methods have been used to classify bioreactors. These include the feeding of media and gases and the withdrawal of products; the mode of operation may be batch, fed batch, or continuous. The classification may be based on the electron acceptor; the design may be for aerobic, anaerobic, or microaerobic conditions. In aerobic processes, the methods of providing oxygen have resulted in mechanically agitated bioreactors, airlift columns, bubble columns, and membrane reactors. The sterility requirements of pure culture processes with developed strains differ from those of environmental mixed culture processes, which are based on natural selection. There are bioreactors in which the vessel is made by humans and natural bioreactors such as the microbial cell, the flowing river, and the field of native grass. In this article, the clasification of bioreactors will be based on the physical form of the reactants and products.

A. Gas Phase Reactants or Products

Oxygen and carbon dioxide are the most common gas phase reactants and products. Others include hydrogen, hydrogen sulfide, carbon monoxide, and methane. Oxygen is a reactant in aerobic heterotrophic growth processes, whereas it is a product in photoautotrophic growth. Generally, the concentration of the reactants and products in the liquid phase in the microenvironment of the cell influences the kinetics of the cellular reactions. Mass transfer to and from the gas phase affects bioreactor performance in most processes with gas phase reactants or products. The anaerobic reactor is designed to exclude oxygen. In some cases, inert gases are bubbled into the anaerobic reactor to provide gas–liquid interfacial area to remove the product gases.

Because the solubility of oxygen in water is very low, the dissolved oxygen in the broth is rapidly depleted if oxygen transfer from the gas to the liquid phase is disrupted in aerobic processes. The distribution of dissolved oxygen throughout the reactor volume and the transient variation affect reactor performance. When mold pellets or biofilms are present, the diffusion of oxygen into the interior should be considered. A significant fraction of the bioreactor literature is devoted to oxygen transfer and the methods recommended for the design and operation of aerobic bioreactors. The phase equilibrium relationship is based on thermodynamic data while the rate of oxygen transfer depends on the gas–liquid interfacial area and the concentration driving force. Mechanical agitation increases the gas–liquid interfacial area. Aeration provides the supply of oxygen, and it affects the gas–liquid interfacial area.

Oxygen has been supplied by permeation through membranes in cultures in which bubbles may damage shear-sensitive cells. The membrane area and concentration driving force determine the oxygen transfer rate in these bioreactors.

Most large-scale bioreactors have either oxygen or carbon dioxide among the reactants or products. In many anaerobic fermentations the formation of carbon dioxide results in bubbling, and often no additional mixing is required for either mass transfer or suspension of the microbial cells. Methane is produced through anaerobic digestion of waste products. It is also a product of microbial action in landfills, bogs, and the stomach of the cow.

B. Liquid Phase Reactants or Products

Many bioreactors have liquid phase reactants and products. Ethanol, acetone, butanol, and lactic acid are examples of liquid products that can be produced by fermentation. The kinetics of biochemical reactions depend on the liquid phase concentrations of the reactants and, in some cases, the products. The Monod kinetic model and the Michaelis-Menten kinetic model show that many biochemical reactions have first-order dependence on reactant (substrate) concentration at low concentrations and zero-order dependence at higher concentrations. Rates are directly proportional to concentration below 10 mg/liter for many reactants under natural environmental conditions. At very high concentrations, inhibition may be observed.

Hydrocarbons that are relatively insoluble in the water phase, such as hexadecane, may also be reactants or substrates for biochemical reactions. Microbial growth on hydrocarbons has been observed to occur at the liquid–liquid interface as well as in the water phase. The oxygen requirements are greater when hydrocarbon substrates are used in place of carbohydrates. At one time, there was great interest in the production of microbial protein from petroleum hydrocarbons. The commercialization of the technology has been most extensive in the Soviet Union. The airlift bioreactor is uniquely suited for this four-phase process because of the tendency of the hydrocarbon phase to migrate to the top of the fermentor. The hydrocarbons are found suspended as drops in the water phase, adsorbed to cells, and at the gas–liquid interface. The cells are found adsorbed to hydrocarbon drops, suspended in the water phase, and at the gas–liquid interface. In the airlift fermentor, the vertical circulation mixes the hydrocarbons and cells that have migrated to the top of the fermentor with the broth that enters the downflow side of the column.

One of the oldest and most widely practiced fermentations is the microbial production of ethanol and alcoholic beverages such as beer and wine. Because ethanol inhibits the fermentation at high concentrations, the process of inhibition has been extensively studied for this fermentation. Ethanol affects the cell membrane and the activities of enzymes. This inhibition limits the concentration of ethanol that can be obtained in a fermentor. Because ethanol is also produced for use as a motor fuel, there is still considerable research on ethanol production. Because the cost of the substrate is a major expense, inexpensive raw materials such as wastes containing cellulose have been investigated.

C. Solid Phase Reactants or Products

There are many examples of bioreactors with solid phase reactants. The cow may be viewed as a mobile bioreactor system that converts solid substrates to methane, carbon dioxide, milk, and body protein. While the cow is a commercial success, many efforts to transform low-cost cellulosic solid wastes to commercial products in human-made bioreactors have not achieved the same level of success.

Solid substrates such as soybean meal are commonly fed into commercial fermentations. Through the action of enzymes in the fermentation broth, the biopolymers are hydrolized and more soluble reactants are obtained.

Many food fermentations involve the preservation of solid or semisolid foods such as in the conversion of cabbage to sauerkraut and meats to sausage products. Cereals, legumes, vegetables, tubers, fruits, meats, and fish products have been fermented. Some fermented milk processes result in solid products such as cheeses and yogurts.

Other examples include the composting of yard wastes, leaching metals from ores silage production, biodegradation of crop residues in soil, microbial action in landfills, and the remediation of contaminated soil.

In many of these fermentations, mixing is difficult or expensive. Transport of essential reactants may depend on diffusion; the concentrations of reactants and products vary with position. Rates may be limited by the transport of essential reactants to the microorganisms.

Most compounds that are present as solids in bioreactors are somewhat soluble in the water phase. For reactants that are relatively insoluble, biochemical reaction rates may be directly proportional to the available interfacial area. The surface of the solid may be the location of the biochemical transformation. An example of microorganisms growing on the surface of a solid substrate is mold on bread. To design bioreactors for solid substrates and solid products, the solubility and the transport processes should be addressed as well as the kinetics of the process.

D. Microorganisms in Bioreactors

The rate of reaction in bioreactors is often directly proportional to the concentration of microbial biomass. In biological waste treatment, the influent concentration of the organic substrate (waste) is relatively low, and the quantity of microbial biomass that can be produced from the waste is limited. The economy of the operation and the rate of biodegradation are enhanced by retaining the biomass in the bioreactor. In the activated sludge process, this is done by allowing the biomass to floculate and settle; it is then recycled. The trickling filter retains biomass by allowing growth on the surfaces of the packing within the bioreactor.

A variety of immobilized cell reactors and immobilized enzyme reactors have been designed and operated because of the economy associated with reuse of cells and enzymes. In the anaerobic production of ethanol, lactic acid, and other fermentation products, the product yield is greatest when the organisms are not growing and all of the substrate is being converted to products. Continuous processes can be designed in which most of the cells are retained and the limiting maximum product yield is approached. Ultrafiltration membrane bioreactors have been used to retain cells, enzymes, and insoluble substrates.

In nature, cells are retained when biofilms form along flow pathways. The biofilms allow microorganisms to grow and survive in environments where washout would be expected. The excellent quality of groundwater is the result of microbial biodegradation and purification under conditions where microbial survival is enhanced by biofilm formation and cell retention on soil and rock surfaces. The ability of microorganisms to survive even after their food supply appears to be depleted is well established; this accounts for our ability to find microorganisms almost everywhere in nature. When spills occur, organic substances will often be degraded by microorganisms if the nutritional environment is balanced. Nitrogen, phosphorous, and other inorganic nutrients often must be added. [*See* BIOFILMS AND BIOFOULING; BIODEGRADATION.]

The concentration of cells adsorbed to the surface and the concentration in the water phase depends on an adsorption phase equilibrium relationship and the operating conditions. In many environmental applications, most of the cells are adsorbed to surfaces. However, in large-scale fermentors with high cell concentrations and rich media feeds, only a small fraction of the cells are found on surfaces.

E. Photobioreactors

Light is the energy source that drives photoautotrophic growth processes. Because light is absorbed by the growing culture, the intensity falls rapidly as the distance from the surface increases. Photobioreactors are designed to produce the quantity of product that is desired. Heat transfer is an important design aspect because any absorbed light energy that is not converted to chemical energy must be dissipated as heat.

III. Principles of Bioreactor Analysis and Design

The basic principles of bioreactor analysis and design are similar to those for chemical reactors; however, many biochemical processes have very complex biochemistry. The chemical balance equations or stoichiometry of the process must be known or investigated. The yield of microbial biomass and products depends on the genetics of the strain and the operating conditions. The consistency of data from experimental measurements can be evaluated using mass balances such as the carbon balance and the available electron balance.

Microorganisms obey the laws of chemical thermodynamics; some heat is produced in heterotrophic growth processes. The free energy change is negative for the complete system of biochemical reactions associated with heterotrophic growth and product formation. Thus, the chemical energy available for growth and product formation decreases as a result of microbial assimilation of the reactants.

The rate of growth and product formation depends on the number of microorganisms and the concentrations of the nutrients. The kinetics of growth and product formation are often written in terms of the concentration of one rate-limiting substrate; however, in some cases, more than one nutrient may be rate limiting. The kinetics must be known for rational design of the bioreactor.

Heat is evolved in microbial bioreactors. For aerobic processes, the quantity of heat generated (heat of fermentation) is directly proportional to the oxygen utilized. Thus, the heat transfer and oxygen transfer requirements are linked by the energy regularity of approximately 450 kJ of heat evolved per mole of oxygen utilized by the microorganisms.

Transport phenomena is widely applied in bioreactor analysis and design. Many fermentation processes are designed to be transport limited. For example, the oxygen transfer rate may limit the rate of an aerobic process. Bioreactor design depends on the type of organism as well as the nutritional and environmental requirements. For example, in very viscous mycelial fermentations, mechanical agitation is often selected to provide the interfacial area for oxygen transfer. Likewise, animal cells that grow only on surfaces must be cultured in special bioreactors, which provide the necessary surface area and nutritional environment. In other cases, animals are selected as the bioreactors, because the desired biochemical transformations can best be achieved by competitively utilizing animals; cost and quality control are both important when food and pharmaceutical products are produced.

IV. Sensors, Instrumentation, and Control

The ability to measure the physical and chemical environment in the fermentor is essential for control of the process. In the last 50 years, there has been significant progress in the development of sensors and computer control. Physical variables that can be measured include temperature, pressure, power input to mechanical agitators, rheological properties of the broth, gas and liquid flow rates, and interfacial tension. The chemical environment is characterized by means of electrodes for hydrogen ion concentration (pH), redox potential, carbon dioxide partial pressure, and oxygen partial pressure. Gas phase concentrations are measured with the mass spectrometer. Broth concentrations are measured with gas and liquid chromatography; mass spectrometers can be used as detectors with either gas or liquid chromatography. Enzyme thermistors have been developed to measure the concentration of a variety of specific biochemicals. Microbial mass is commonly measured with the spectrophotometer (optical density) and cell numbers through plate counts and direct microscopic observation. Instruments are available to measure components of cells such as reduced pyridine nucleotides and cell nitrogen. However, one of the greatest needs is a better online sensor for biomass.

The basic objective of bioreactor design is to create and maintain the environment that is needed to enable the cells to make the desired biochemical transformations. Advances in instrumentation and control allow this to be done reliably. [*See* BIOREACTOR MONITORING AND CONTROL.]

V. Metabolic Engineering

Genetic modification has allowed many products to be produced economically. With the use of recombinant DNA technology and metabolic engineering, improved cellular activities may be obtained through manipulation of enzymatic, regulatory, and transport functions of the microorganism. The cellular modifications of metabolic engineering are carried out in bioreactors. Successful manipulation requires an understanding of the genetics, biochemistry, and physiology of the cell. Knowledge of the biochemical pathways involved, their regulation, and their kinetics is essential.

Living systems are bioreactors. Through metabolic engineering, man can modify these living bioreactors and alter their performance. Metabolic engineering is a field of reaction engineering that utilizes the concepts that provide the foundation for reactor design including kinetics, thermodynamics, physical chemistry, process control, stability, catalysis, and transport phenomena. These concepts must be combined with an understanding of the biochemistry of the living system. Through metabolic engineering, improved versions of living bioreactors are designed and synthesized.

VI. Stability and Sterilization

While beneficial genetic modification has led to many industrially successful products, contamination and genetic mutations during production operations have resulted in many batches of useless broth. Batch processes are common in bioreactors because of the need to maintain the desired genetic properties of a strain during storage and propagation. Continuous operation is selected for mixed culture processes such as wastewater treatment, where there is natural selection of effective organisms.

Bioreactors that are to operate with pure cultures or mixed cultures from selected strains must be free of contamination; i.e., the reactor and associated instrumentation must be sterilizable. The vessels that are to be used for propagation of the inoculum for the large-scale vessel must be sterilizable as well. Methods to sterilize large vessels, instrumentation, and connecting pipes are well developed; however, there is a continuing need to implement a wide variety of good manufacturing practice principles to avoid contamination problems.

Steam sterilization has been widely applied to reduce the number of viable microorganisms in food and in fermentation media. As temperature increases, the rates of biochemical reactions increase exponentially until the temperature affects the stability of the enzyme or the viability of the cell. The Arrhenius activation energies, which have been reported for enzymatic reactions and rates of cell growth, are mostly in the range of 20–80 kJ/gmol, whereas activation energies for the thermal inactivation of microorganisms range from 200 to 400 kJ/gmol. Many of the preceding principles also apply to the thermal inactivation of microorganisms in bioreactors. When solids are present in foods or fermentation media, heat transfer to the interior of the solid is by conduction. This must be considered in the design of the process because of the increase in the required sterilization time.

VII. Conclusions

Bioreactors are widely used for a variety of purposes. The knowledge base for their application has increased significantly because of the advances in chemical, biochemical, and environmental engineering during the last 50 years.

Bibliography

Bailey, J. E., and Ollis, D. F. (1986). "Biochemical Engineering Fundamentals," 2nd ed. McGraw-Hill, New York.

Carberry, J. J., and Varma, A. (eds.) (1987). "Chemical Reaction and Reactor Engineering." Marcel Dekker, New York.

Characklis, W. G., and Marshall, K. C. (eds.) (1990). "Biofilms." Wiley Interscience, New York.

Chisti, M. Y. (1989). "Airlift Bioreactors." Elsevier, New York.

Erickson, L. E., and Fung, D. Y. C. (eds.) (1988). "Handbook on Anaerobic Fermentations." Marcel Dekker, New York.

Fan, L. T., Gharpuray, M. M., and Lee, Y. H. (1987). "Cellulose Hydrolysis." Springer-Verlag, Heidelberg.

Lubiniecki, A. S. (ed.) (1990). "Large-Scale Mammalian Cell Culture Technology." Marcel Dekker, New York.

Moo-Young, M. (ed.) (1988). "Bioreactor Immobilized Enzymes and Cells: Fundamentals and Applications." Elsevier, New York.

Twork, J. V., and Yacynych, A. M. (eds.) (1990). "Sensors in Bioprocess Control." Marcel Dekker, New York.

Van't Riet, K., and Tramper, J. (1991). "Basic Bioreactor Design." Marcel Dekker, New York.

Bioremediation

J. M. Thomas and C. H. Ward
Rice University and National Center for Ground-Water Research

R. L. Raymond
Du Pont Environmental Remediation Services

J. T. Wilson
R. S. Kerr Environmental Research Laboratory and U.S. Environmental Protection Agency

R. C. Loehr
University of Texas at Austin

Glossary

Anoxic Without oxygen

Biodegradation Metabolism or transformation of a substance by microorganisms

Bioventing Accelerated biodegradation of vapor-phase contaminants in contaminated subsurface materials by forcing and/or drawing air through the subsurface.

Capillary fringe Region between the unsaturated and saturated zones in which the soil pores are saturated, the pressure is less than atmospheric, and contaminants may be held by capillary forces

Cometabolism Biodegradation of an organic compound by a microorganism that cannot use the compound for growth

Composting Accelerated biodegradation of contaminants at high temperatures by aerating and adding bulking agents and possibly nutrients to waste in a compost pile

Immobilization Removal of contaminants from a waste onto soil by chemical and/or physical processes

Land treatment Accelerated biodegradation of contaminants by aerating and possibly adding nutrients to contaminated surface soil or contaminated material that has been applied to surface soil

Liquid/solids treatment Accelerated biodegradation of contaminants by aerating and possibly adding nutrients to waste or contaminated soil that has been made into a slurry

Oxic Containing oxygen

Plume Dissolved contaminants that move with groundwater

BIOREMEDIATION is defined in this article as the process by which microorganisms are stimulated to rapidly degrade hazardous organic contaminants to environmentally safe levels in soils, subsurface materials, water, sludges, and residues. Stimulation is achieved by the addition of nutrients and a terminal electron acceptor, usually oxygen, because most biological reactions occur faster under aerobic than anaerobic conditions. Under anaerobic conditions, nitrate has been used as the terminal electron acceptor. The microorganisms use the contaminants as a food source and convert the contaminants into biomass and harmless by-products of metabolism such as CO_2 and inorganic salts. Usually bioremediation is used to degrade contaminants that are sorbed to surfaces or dissolved in water rather than to degrade pure chemical(s). As a result, the process is used in conjunction with other techniques in remediation of contaminated sites. The contaminants can be biodegraded *in situ* or removed and placed in a bioreactor, which can be placed off or at the site where the contamination occurred.

I. Introduction

A. Concepts

Heterotrophic microorganisms are the principal users of organic matter in the biosphere and are key in cycling carbon from the organic to the inorganic state. Provided that sufficient inorganic nutrients and a terminal electron acceptor for metabolism are present, all naturally occurring organic materials can be biodegraded eventually. However, even simple organic compounds (e.g., acetate) may persist under conditions that do not favor microbial activity. These conditions include extremes in temperature or pH, the presence of toxicants or antimicrobial agents, the inhibition or exclusion of microbial enzymes, and the lack of water, nutrients, and an electron acceptor. Critical environmental factors for microbial activity and their optimum levels are shown in Table I. Complex organic materials such as cellulose and lignin often are degraded by groups or consortia of microorganisms rather than just one species. [*See* HETEROTROPHIC MICROORGANISMS.]

With the advent of petroleum refining and manufacture of synthetic chemicals, many potentially hazardous organic compounds have been introduced into the air, water, and soil. One method for removing these undesirable compounds from the environment is bioremediation, an extension of carbon cycling. Given the right organism(s), time, nutrients, an electron acceptor, and nonextreme environmental conditions, many synthetic organic compounds may biodegrade. The ultimate goal in bioremediation is conversion of undesirable organic compounds into innocuous materials, usually CO_2, water, inorganic salts, and biomass. However, undesirable metabolic by-products may accumulate when biodegradation of compounds is incomplete. Biodegradation of naturally occurring and synthetic organic compounds requires or is faster when several species of microorganisms are present. Bioremediation normally is achieved by stimulating the indigenous microflora present in or associated with the material to be treated. In instances where the indigenous microflora fails to degrade the target compounds or has been decimated by the presence of toxicants, microorganisms with specialized metabolic capabilities may be added. [*See* BIODEGRADATION.]

The technical basis for modern bioremediation technology has a very long history (e.g., composting of organic wastes into mulch and soil conditioners). Bioremediation technology has grown to include the biological treatment of sewage and wastewater, food processing wastes, agricultural wastes, and more recently, hazardous waste. In this article, bioremediation is used in a limited sense and is defined as the biological treatment of hazardous organic contaminants. The following sections will describe various bioremedial techniques available for treatment of hazardous organic wastes. These include techniques for remediation of (1) contaminated surface soils and sludges [i.e., land treatment, and slurry phase processes (bioreactors)], (2) contaminated unsaturated subsurface material (i.e., bioventing), and (3) contaminated saturated subsurface material in which contaminant plumes and sources are treated. The last section of this article discusses the use of microorganisms with specialized metabolic capabilities in bioremediation. Although several papers have been published in the peer-reviewed literature concerning the use of microorganisms in remediation of inorganic wastes (e.g., metals), this type of application will not be discussed in this article. [*See* HAZARDOUS WASTE TREATMENT, MICROBIAL TECHNOLOGIES.]

Table I Critical Environmental Factors for Microbial Activity

Environmental factor	Optimum levels
Available soil water	25 to 85% of water-holding capacity
Oxygen	Aerobic metabolism: greater than 0.2 mg/liter dissolved oxygen, minimum air-filled pore space of 10%; Anaerobic metabolism: O_2 concentrations less than 1%
Redox potential	Aerobes and facultative anaerobes: greater than 50 mV; Anaerobes: less than 50 mV
pH	5.5 to 8.5
Nutrients	Sufficient nitrogen, phosphorus, and other nutrients so not limiting to microbial growth (suggested C:N:P ratio of 120:10:1)
Temperature	4° to 45°C

(From U. S. EPA report 600/9-89/073)

B. Enhanced Bioremediation of Hazardous Wastes— A Historical Perspective

1. Federal Regulations and Bioremediation Technology

The history of enhanced bioremediation for hazardous waste began as early as the 1940s and 1950s with the use of land treatment (a biological treatment process) as a management alternative for selected industrial wastes (i.e., those emanating from the petroleum industry). At that time, no specific regulations were directed toward such land management practices. In 1972, the Federal Water Pollution Control Act Amendments (Public Law 92-500) were passed, which required the investigation of waste management practices to minimize or eliminate surface water pollution; however, groundwater quality was not emphasized strongly. Land treatment of municipal wastewaters was one of the waste management practices emphasized. In 1974, this resulted in increased funding of the multidisciplinary research program at the U.S. Environmental Protection Agency (EPA) Robert S. Kerr Environmental Research Laboratory in Ada, Oklahoma (RSKERL-Ada), to determine the feasibility of using land treatment as technology for treating municipal wastewaters. It soon became apparent that industrial wastewater treatment practices implemented in response to requirements of PL 92-500 had resulted in a significant increase in production of waste residuals for disposal. Land treatment was one technology used for management of such residuals. Also in 1974, Congress passed the Safe Drinking Water Act, which emphasized the protection of groundwater. This resulted in increased funding for the agency's groundwater research program being implemented at the RSKERL-Ada.

The Resource Conservation and Recovery Act (RCRA) was passed in 1976. This legislation imposed the first federal regulations for management of hazardous waste and residues and mandated the EPA to list wastes and constituents considered hazardous, as well as options for hazardous waste disposal. In particular, RCRA imposed restrictions on the design and operation of hazardous waste land disposal facilities, including land treatment units. In 1980, the EPA designated RSKERL-Ada to implement and direct the agency's hazardous waste land treatment research program to support development of regulations and guidance related to land treatment of hazardous waste. At that time, researchers at RSKERL-Ada and at the EPA-sponsored National Center for Ground Water Research were investigating the potential for bioremediation in the subsurface. As a result of the growing concern over human health and the environment, including groundwater contamination, the Hazardous and Solid Waste Amendments (HSWA) to RCRA were enacted in 1984. These amendments restricted land disposal of hazardous waste. Because land treatment was considered in RCRA as a land disposal option, use of land treatment as a management alternative for untreated hazardous waste also was restricted.

2. Early Demonstrations of Hazardous Waste Bioremediation

Milestones in the use of enhanced bioremediation as a technology for treatment of hazardous organic wastes date from the early 1970s (Table II). The first large-scale demonstration of bioremediation of petroleum wastes was the land treatment facility at the Shell Oil refinery in Deer Park, Texas, in 1972 (reported in an EPA document by C. B. Kincannon). Although land treatment of industrial wastes had been used before 1972, the Shell Oil demonstration was the first managed system for hazardous waste that emphasized pollution control. Other important historial landmarks in bioremediation can be attributed to R. L. Raymond and co-workers J. O. Hudson and V. W. Jamison at Sun Research and Development Company, Philadelphia, Pennsylvania. In 1974, the first patent on bioremediation of gasoline-contaminated groundwater, "Reclamation of Hydrocarbon Contaminated Ground Waters," was obtained by Raymond. The process involved stimulating the indigenous groundwater microflora with oxygen and nutrients to degrade hydrocarbons, especially gasoline. In 1975, Raymond and co-workers authored a document for the American Petroleum Institute (API), "Beneficial Stimulation of Bacterial Activity in Ground Waters Containing Petroleum Products," which described a successful demonstration of the patented technique. This study was later published in 1976 in an AIChE symposium series. In 1976, the first field study that demonstrated that hydrocarbon concentrations were decreased and that populations of hydrocarbon-degrading microorganisms increased after the application of oil to

Table II Milestones in the Use of Bioremediation as a Technology

Year	Milestones
1972	PL 92-500 requires investigation of technologies that reduce or eliminate surface water pollution
1972	First field demonstration of land treatment hazardous waste that emphasizes pollution control
1974	Task forces formed to investigate land treatment as a technology
1974	Safe Drinking Water Act
1974	Patent for bioremediation of gasoline-contaminated ground water obtained by R. L. Raymond
1975,1976,1978	R. L. Raymond's patent demonstrated in the field
1976	RCRA regulates hazardous waste disposal
1979	J. T. Dibble and R. Bartha publish a paper on remediation of kerosene-contaminated soil using land treatment
1984	HSWA amendments to RCRA restrict land disposal

soil was conducted by these co-workers. Raymond and co-workers authored a second document for API in 1978, "Field Application of Subsurface Biodegradation of Gasoline in a Sand Formation," which described another demonstration of Raymond's patent for bioremediating subsurface materials contaminated with gasoline. In 1979, J. T. Dibble and R. Bartha published a paper in *Soil Science* that described the land treatment of soil contaminated with kerosene.

II. Natural Bioremediation

In most cases of environmental contamination, biodegradation begins after the indigenous microflora adapts to degrade the contaminants and/or to the conditions imposed by the contamination. Adaptation is defined as an increase in the rate of biodegradation of a compound to which the microorganisms are exposed. Explanations for adaptation suggested by B. A. Wiggins and co-workers published in *Applied and Environmental Microbiology* in 1987 include (1) an increase in the number of contaminant-degrading microorganisms, which may depend on (a) removal of or adjustment to toxins, (b) the addition of nutrients, and (c) the reduction or elimination of predators of the contaminant-degrading organisms, (2) a mutation or transfer of genetic information, which enables an organism to degrade the contaminant that it previously could not degrade, (3) induction of the enzyme(s) responsible for contaminant degradation, and (4) depletion of a preferentially metabolized substrate before metabolism of the target substrate. The time required for adaptation to occur is site- and contaminant-specific. The effect of adaptation on the reduction of several polycyclic aromatic hydrocarbons in soil is shown in Table III. As a result of adaptation, many contamination events go unnoticed because nature has taken its course. However, the rate of natural bioremediation may not be fast enough to prevent environmental and health hazards.

One of the best characterized sites where natural bioremediation has been shown to retard the movement of contaminants in the subsurface is an abandoned wood-preserving facility in the Gulf Coast region in Texas. An investigation of the site pub-

Table III Adaptation of Soil Microflora to Compounds Present in Complex Fossil Fuel Waste

	Unexposed soil		Exposed soil	
Compound	Initial concentration (mg/kg dry weight)	Reduction in 40 days (%)	Concentration after second waste addition[a] (mg/kg dry weight)	Reduction in 22 days (%)
Naphthalene	38	90	38	100
Phenanthrene	30	70	30	83
Anthracene	38	58	38	99

[a] Exposed soil was reamended with fossil fuel waste after 168 days incubation at initial level.
(Adapted from U.S. EPA report 600/9-89/073).

lished in 1985 in *Environmental Toxicology and Chemistry* by J. T. Wilson and co-workers at the RSKERL-Ada and Rice University indicated that microorganisms in aquifer samples from the margin of the plume of ground water contamination rapidly degraded naphthalene, 1-methylnaphthalene, 2-methylnaphthalene, dibenzofuran, acenaphthene, and fluorene, which were the predominant compounds detected in the plume. Biodegradation of these compounds was not detected in aquifer samples collected outside of the contaminant plume, suggesting that the microflora exposed to these chemicals had adapted to biodegrade them. Measurements of dissolved oxygen in the field indicated that ground water beyond the edge of the contaminant plume was oxic whereas the contaminant plume was anoxic, suggesting that contaminant biodegradation caused removal of oxygen. It was postulated that the most active site of microbial activity was at the edge of the plume where oxygen-containing groundwater was mixed with the contaminant plume, thus retarding contaminant migration. Field measurements and computer model simulations by R. C. Borden at Rice University actually showed that the plume was narrower and shorter than would be expected if only dilution caused by dispersion, sorption to the subsurface matrix, and nonbiological degradation were influencing contaminant migration.

When the rate of natural bioremediation is not fast enough to prevent health risks and spread of contamination, which is true in most cases, intervention using one or several remedial techniques is required. Bioremediation may be only one in a series of techniques used.

III. Surface Soil and Sludges

A. Solid Phase Processes

1. Land Treatment

a. Technology

Land treatment is the process by which soil contaminated with hazardous or nonhazardous waste is treated in place or excavated and then treated or by which wastes are applied and incorporated into surface soil. The process relies on the assimilative capacity of soil to degrade and immobilize constituents in the waste. The soil–waste mixture is managed in a manner that (1) enhances immobilization of waste by soil, (2) stimulates degradation of waste by indigenous microflora or added microrganisms, (3) minimizes volatilization and leaching of waste out of the treatment area, and (4) controls surface water runoff. The mechanisms of immobilization and degradation include sorption, hydrolysis, photolysis, chemical degradation, and biodegradation; volatilization also contributes to removal of contamination. Factors that affect biodegradation in land treatment include the type and concentration of the waste, presence of waste-degrading organisms, pH, temperature, and the availability of oxygen, water, and nutrients. The effect of additional carbon and energy sources (manure) and adjusting the pH from 6.1 to 7.5 on degradation of several polycyclic aromatic hydrocarbons in soil is shown in Table IV. Usually, the indigenous soil microflora is stimulated to degrade the wastes; however, organisms with specialized metabolic capabilities may be added when the soil has been sterilized because of the presence of toxicants. In this case, the toxicants must be removed or detoxified before the land treatment process can be successful.

There are two types of land treatment processes: surface soil land treatment and on-site land treatment. Both types involve the addition of nutrients, tilling the soil–waste mixture to increase nutrient and oxygen availability to the microflora, and adjusting the pH and moisture content when needed. The difference between the two processes is that surface soil land treatment involves the treatment of contaminated surface soil in place or the addition and incorporation of waste to surface soil. The

Table IV Effect of Manure and pH Amendments on Polycyclic Aromatic Hydrocarbon Degradation in a Complex Waste Incorporated into Soil

	Half-life in waste/soil mixture[a] (days)	
Compound	Without amendments	With amendments
Acenaphthene	96	45
Fluorene	64	39
Phenanthrene	69	23
Anthracene	28	17

[a] Amount of time required for half of the compound to degrade. (Adapted from U.S. EPA report 600/9-89/073.)

waste is incorporated into the upper 6 to 12 inches of top soil. The actual treatment zone in which degradation and immobilization occur may be several feet in depth. However, most degradation and immobilization will occur in the upper 6 to 12 inches.

On-site treatment involves construction of a lined and walled holding facility in which the waste is mixed with clean soil or in which a soil-contaminant mixture is placed. The facility has drainage and leachate collection systems and is usually constructed above ground. To minimize costs and liability from transportation, on-site treatment facilities are usually constructed at the contamination site or where the waste is generated.

Land treatment has been used to treat wastes from coal gasification/liquefaction, food processing, leather tanning, paper and pulp production, petroleum refining, and wood preserving, as well as municipal wastewater and sludge. The process has been used in a wide range of hydrogeologic conditions and in the major climatic regions of the United States, Europe, and Canada. Land treatment usually requires minimal management compared with other bioremedial techniques but does require aeration of the soil and large amounts of land. The average cost of land treatment relative to other bioremedial methods is shown in Table V.

Restrictions on the application of hazardous wastes to land became effective after the HSWA Amendments to the RCRA were enacted in 1984. The amendments required that before disposal on land, (1) certain wastes be treated by the "best demonstrated available technology" to meet treatment standards established by EPA or (2) it must be demonstrated that the waste does not migrate out of the treatment zone during the time that the waste is considered hazardous. These restrictions applied to land treatment, which is considered land disposal under RCRA, even though the technology involves some treatment. Exemptions to pretreatment of these wastes could be obtained if it could be demonstrated that the hazardous constituents in the waste do not migrate out of the treatment zone into groundwater or the atmosphere (no migration variance). The petroleum industry, which relies on land treatment for most of its wastes, was affected greatly by the standards issued by EPA in 1988 that restricted the application of petroleum wastes on land. Because the industry was not prepared to meet these standards, a 2-year national variance was granted by EPA, during which time adequate technologies to meet the standards could be developed and "no migration" in their land treatment facilities could be demonstrated. Because "no migration" is almost impossible to achieve for hazardous wastes, industry must pretreat its hazardous waste before land application or use other methods for waste management. [*See* PETROLEUM MICROBIOLOGY.]

Table V Relative Costs of Bioremediation Methods

Method	Cost ($/yd^3)
Land treatment	10 to 40
Composting	100
Liquid–solids	100 to 150
Subsurface[a]	
hydrogen peroxide	490
nitrate	200
bioventing	60

[a] Costs for different types/sources of electron acceptors at an aviation fuel spill at Traverse City, Michigan.

b. Demonstration

Remediation of kerosene-contaminated soil was demonstrated in a field experiment published in 1979 in *Soil Science* by J. T. Dibble and R. Bartha. About 1.9×10^6 liters of kerosene was spilled on 1.5 ha of agricultural land when an underground pipeline ruptured. Initial remedial action involved excavating 200 m^3 surface soil. The remaining contaminated soil was rehabilitated by liming, fertilizing, and tilling. The soil was classified as a sandy loam and had good drainage properties. The field was leveled, limed, and then subsoiled to a depth of 46 cm. The field was tilled monthly for 7 months and then subsoiled again to facilitate aeration. Fertilizer (N:P:K) was applied twice. Soil cores were collected during the treatment period to determine the concentration of residual kerosene, which was reported as (wt/wt) oil in dry soil.

During the 21-month treatment period, the kerosene concentration decreased from 0.87% to trace levels in the upper 30 cm of soil; about 0.25% remained in the lower 30 to 45 cm. Initially, the concentration of kerosene was higher in the upper than the lower treatment zone, a trend that reversed after 6 months. The rate of kerosene removal was directly related to temperature, with the highest rates determined during July and August. Gas chromatographic analysis of soil samples at the start of treatment indicated the presence of an unresolved mixture of hydrocarbons and *n*-alkanes with 10 to 17 carbons (C_{10} to C_{17}). During the first 7 months, the *n*-alkanes and most of the hydrocarbon mixture disappeared. Of the *n*-alkanes, the *n*-C_{13} to *n*-C_{17} compounds disappeared faster than those with a lower molecular weight. This pattern of *n*-alkane utilization suggested that biodegradation was an important mechanism in oil removal from this soil because the pattern followed the order of preferential metabolism by microorganisms.

2. Composting

a. Technology

The concept of composting, which has been used for centuries to convert organic wastes into mulch and soil conditioners, has been applied to hazardous waste treatment. Composting exploits the process by which biomass, inorganic by-products, and energy in the form of heat are produced when microorganisms degrade organic matter aerobically. The process requires that the material be biodegradable, have an adequate water content, be present in sufficient quantities to retain heat, and be porous enough to allow gas exchange. The amount of free water should be kept to a minimum to allow adequate circulation of oxygen. For substrates that contain little nitrogen (e.g., leaf litter), nitrogen additions may be required. Temperatures as high as 60°C may be achieved. As a result of composting, the waste is (1) converted to a less complex but stable material (stabilization), (2) decreased in water content, and (3) decreased in mass. Microorganisms involved in the composting process are associated with the organic material. Traditionally, organic matter with a high carbohydrate content such as leaves, grass, and food wastes has been composted. [*See* ORGANIC MATTER, DECOMPOSITION.]

Typical bioremediation composting systems include the windrow, the Beltsville, and the in-vessel (Table VI). The windrow system is an open system in which the pile is made into one or several rows and aeration is achieved by periodically turning the compost pile. This type of composting often is used to treat wastewater sludge. Like the windrow system, the Beltsville system (static pile) also is uncovered; however, aeration is achieved via an air distribution system located under the pile. The aeration system draws air from the atmosphere through the pile and into an air pollution control system. The in-vessel system involves adding the waste to a closed or open vessel that is equipped with a temperature-controlled aeration system in which the material is mechanically mixed and/or through which air is forced using blowers.

Table VI Characteristics of Different Types of Composting Systems

Type	Open/closed	Aeration	Examples of wastes treated
Windrow	Open	Turning pile	Wastewater sludge, food processing wastes, manures
Beltsville	Open	Distribution system	Leaves, wastewater sludge, other organic wastes
In-vessel	Open/closed	Distribution system	Wastewater sludge, food processing waste

Materials that are amenable to the bioremediation composting process include sewage sludge, soils contaminated with diesel fuel and similar petroleum products, and wastes from brewing processes, antibiotic fermentations, food processing, mineral oil, munitions residue, and agriculture. Bulking agents may be added to increase the porosity and facilitate aeration. Materials used as bulking agents include fibrous plant material, wood chips, and bark. For compounds that are difficult to biodegrade, such as those found in munitions residue, the waste is mixed with a feedstock of highly biodegradable material. The feedstock serves as the carbon source for the microorganisms while the waste is biodegraded fortuitously (cometabolism). The ratio of waste to feedstock, or loading rate, is important because a high ratio may retard or inhibit microbial activity. Materials used as feedstock should be inexpensive; examples include food processing wastes, manure, and plant material.

Composting presents several environmental and regulatory concerns such as (1) creation of a larger waste volume caused by the addition of bulking agents, (2) containment of the composting material during treatment, (3) control of leachate and volatiles, and (4) disposal of the resulting residue. Composting may serve as one step in the remediation and ultimate disposal of nonhazardous and hazardous wastes. The average cost of composting relative to other bioremedial techniques is shown in Table V.

b. Demonstration

A field demonstration of aerated static (unturned) pile composting was conducted by Roy F. Weston, Inc., for munitions processing wastes that included trinitrotoluene (TNT), hexahydro-1,3,5-trinitro-1,3,5-triazine (RDX), and octahydro-1,3,5,7-tetranitro-1,3,5,7-tetraazocine (HMX). The wastes had been disposed of in unlined lagoons, resulting in contamination of the underlying sediments and groundwater. Two static compost piles were constructed that contained (percent by weight) contaminated sediment, 24%; alfalfa, 10%; used livestock bedding, 25%; and horse feed, 41%. The piles, about 10 y^3 in volume, were housed on concrete pads under roofs to minimize leachate formation and migration. Both piles were aerated mechanically; however, one was maintained at 35°C (mesophilic) and the other at 55°C (thermophilic). A small amount of fertilizer containing nitrogen, phosphorus, and potassium was added. The piles were composted for 153 days, during which concentrations of the contaminants, intermediates of biodegradation, and moisture content were monitored. The concentrations of total explosives in the thermophillic and mesophillic piles declined from 17,870 to 74 mg/kg and 16,460 to 326 mg/kg, respectively, after 153 days of treatment. Calculated half-lives (days) for TNT, RDX, and HMX under thermophilic conditions were 11.9, 17.3, and 22.8, respectively, and under mesophyllic conditions were 21.9, 30.1, and 42.0, respectively. Results indicated that significant transformation of TNT, RDX, and HMX occurred during the composting process.

B. Liquid–Solids Processes (Slurries)

1. Technology

Liquid–solids contact, or slurry-phase bioremediation, treats hazardous waste in a closed reactor, or open pit or lagoon (surface impoundments). The process is similar to conventional biological suspended growth treatment (e.g., activated sludge). When treatment takes place in a closed reactor, conditions that affect microbial activity (e.g., temperature and pH) and nutrient and electron acceptor availability can be controlled. The waste to be treated is suspended in a slurry and mixed to maximize mass transfer between microorganisms and electron acceptor, inorganic nutrients, organic nutrients, and the waste to be treated. Because bioremediation usually occurs faster under aerobic than anaerobic conditions, the electron acceptor used most often is oxygen. Anaerobic or anaerobic/aerobic cycling treatment also is a possibility. Under aerobic conditions, oxygen is supplied by spargers, compressors, or floating or submerged aerators. In addition to supplying oxygen, the process of aeration also serves to mix the slurry; however, mechanical mixing is common. As a result of mixing and aeration, volatilization of compounds in the waste may occur. If control is needed, the reactors are equipped with monitoring and/or trapping systems. Single-batch reactors and sequenced-batch reactors into which the waste is fed continuously or semi-continuously may be used. The reactors may be mobile aboveground tanks or lined *in situ* lagoons or pits. [*See* BIOREACTOR MONITORING AND CONTROL.]

In addition to inorganic and organic nutrients, neutralizing agents may be added to correct adverse conditions (e.g., pH extremes or the presence of toxic compounds), which may retard or inhibit microbial activity. Surfactants and dispersants may be added to increase the surface area of the slurry, rendering it more available for microbial degradation.

After the waste has been biodegraded to acceptable levels in a closed reactor, the solids are allowed to settle. The liquid phase is removed and discharged in an environmentally appropriate manner, depending on its content. The solid phase can be treated further using land treatment or disposed of appropriately. In the case of a waste pit or lagoon, the liquid phase may be removed and the solids left behind when the waste has been bioremediated to environmentally acceptable levels.

Materials most often treated using this technology are wood preserving and oil refinery wastes containing phenols, benzene, toluene, the xylene isomers, ethylbenzene, naphthalene, and the polycyclic aromatic hydrocarbons. Materials that may not be treatable by slurry phase processes without dilution include concentrated sludges and sludges with a high content of oil and grease (>30%), which adversely affects mixing. Asphaltic and tar-like materials also are difficult to treat. The average cost for liquid–solid contact relative to other bioremedial processes in shown in Table V.

2. Demonstration

A pilot-scale demonstration of bioremediation of petroleum sludge using slurry phase processes was conducted at a Gulf Coast refinery by Remediation

Technologies, Inc., in 1989. The sludge used in the demonstration was highly weathered and had been subjected to biological, chemical, and physical degradation before treatment. The sludge contained benzene, toluene, ethylbenzene, the xylene isomers, and a predominance of two- and three-ringed polycyclic aromatic hydrocarbons.

A single batch of sludge was treated in a 3.8×10^3-m^3 batch bioreactor for 56 days; the loading rate was 10% of the total mass treated in the bioreactor. The volume of the slurry was 3.6×10^6 liters (947,000 gal). The slurry was mixed and aerated mechanically and seeded with the microorganisms contained in 8.3×10^4 liters of waste-activated sludge from a refinery wastewater treatment system. Calcium carbonate (for pH control) and nutrients were added during the treatment. Nitrogen, as NH_4NO_3, and phosphorus, as P_2O_5 or H_3PO_4, were added periodically to approximate a carbon:nitrogen:phosphorus ratio of 100:2:0.2. The slurry was monitored routinely for temperature, pH, dissolved oxygen concentrations, the rate of dissolved oxygen uptake, microbial numbers, the reactor liquid level, and amperages required by the mixing/aeration equipment. During treatment, conditions within the reactor were maintained optimal for microbial activity. Chemical analyses were conducted periodically for solids, soluble chemical oxygen demand, and concentrations of individual hydrocarbons.

Results indicated that there was a net solids reduction of 10%. Volatile hydrocarbons were below detection limits after 1 day of treatment, primarily by volatilization. The overall removal of the semivolatile hydrocarbons was greater than 90%, most of this occurring during the first 2 weeks of treatment. Of the 425,000 kg of sludge treated, 910 kg of nonmethane hydrocarbons was volatilized. Oil and grease concentrations were reduced by 50% between 80 and 90 days. A concentration of about 10^7 to 10^8 microorganisms/ml of slurry was detected during the treatment process, indicating that a healthy microbial population was present.

IV. Subsurface Materials

Contamination of the subsurface usually is discovered as plumes dissolved in groundwater or fumes in soil air that work their way into basements and engineering excavations (Fig. 1). The contaminated ground water and soil air usually emanate from a location that contains the contaminants as a discrete insoluble phase. Examples include fuels and solvents spilled from aboveground and underground storage tanks, pipeline spills, and tars and waste oils that were intentionally disposed to the subsurface after creosoting of wood or the manufacture of coke or illuminating gas. The greatest mass of contaminants in the subsurface is usually in the form of undissolved chemical products. As a result, the source areas continue to contaminate ground water and soil air for a long time.

If the undissolved material is more dense than water (DNAPL—dense nonaqueous phase liquid), such as the chlorinated aliphatic solvents, it moves under the influence of gravity, is retarded by sorption and capillary action in the flow path, and may continue to move toward the bottom of the aquifer (Fig. 1). If the insoluble material is less dense than water (LNAPL—light nonaqueous phase liquid), such as gasoline, it moves under the influence of gravity and concentrates at the capillary fringe just above the water table.

For both DNAPLs and LNAPLs, changes in the elevation of the water table tends to smear the contamination over greater vertical distances. If possible, the undissolved phase should be pumped out before *in situ* bioremediation is initiated. The quantity of contamination held behind by capillary attraction is termed the "residual saturation." The stoichiometric demand of the contaminants at residual saturation determines the quantities of nutrients and electron acceptors that must be delivered to the subsurface. Fluids must be circulated through the subsurface to bring nutrients and electron acceptors into zones of contamination.

In this section, methods used to bioremediate contaminated unsaturated and saturated subsurface material are discussed. In the unsaturated subsurface, bioventing is used to remediate many petroleum-derived contaminants that are trapped in or sorbed to the solid matrix. Problems in the saturated zone include plume and source contamination, which are remediated using different approaches and technologies.

A. Unsaturated Subsurface Material

1. Bioventing

a. Technology

Soil venting, also called "soil vacuum extraction," is widely practiced for removal of oily phase contaminants above the water table. A well is screened near the point of contamination but above the water

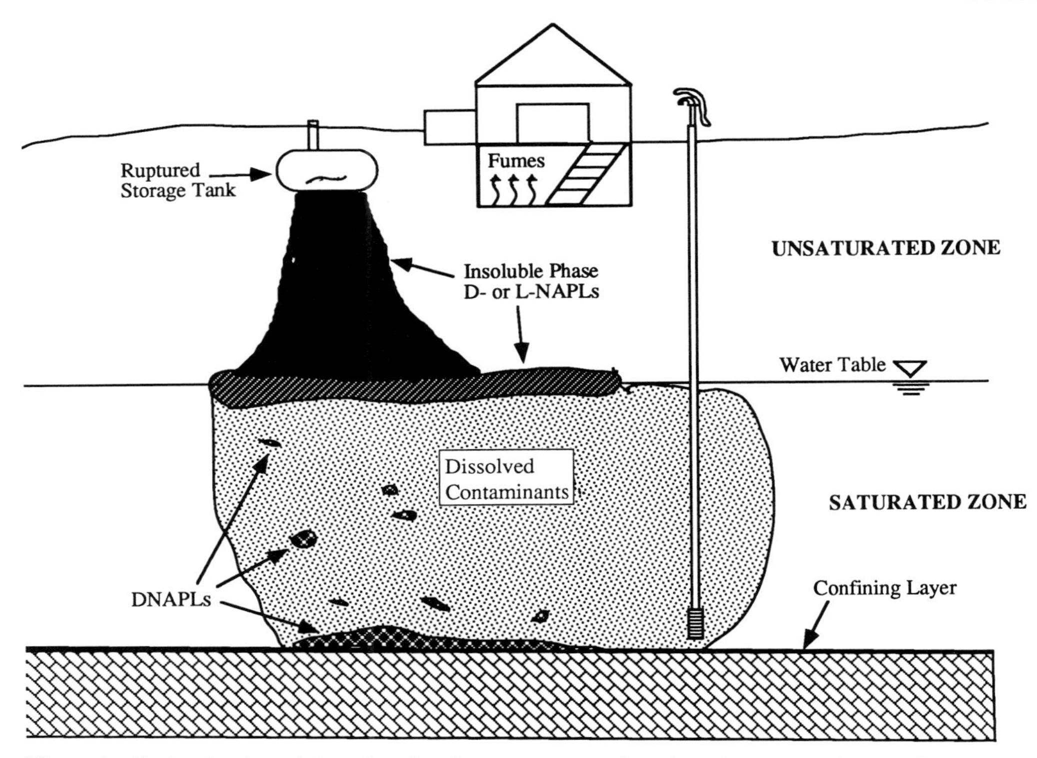

Figure 1 Contamination of the subsurface by nonaqueous phase liquids (NAPLs) that are denser than water (DNAPLs) or less dense than water (LNAPLS).

table (Fig. 2). A vacuum is placed on the well to extract soil air that contains volatile emissions from the wastes. The emissions are disposed to the atmosphere, adsorbed to activated carbon, or destroyed in an internal combustion engine or catalytic convertor. Before the application of vacuum, the soil air in contact with contaminants is usually anoxic, and aerobic metabolism of the waste is precluded. The extraction well pulls oxygenated air through the contaminated subsurface materials. After a period of acclimation, the concentration of oxygen in the extracted air usually decreases, and the concentration of carbon dioxide increases, indicating that biodegradation of the oily contaminant is contributing to the remediation.

The potential contribution of biodegradation during the soil venting process is shown in the following example. The concentration of oxygen in 1 liter of air will support metabolism of approximately 60 mg of hydrocarbon vapors. One liter of air in equilibrium with liquid octane (representative of gasoline) has 25 mg octane vapors at 10°C and 45 mg at 20°C. Therefore, air can support more biodegradation of octane than it can carry away as a vapor. The advantage of biodegradation over stripping is even greater for heavier hydrocarbons (e.g., heating oil or diesel fuel), which have vapor pressures lower than that of octane. For example, 1 liter of air in equilibrium with hexadecane (representative of diesel fuel) has less than 0.1 mg hexadecane vapors at 10°C.

Robert Hinchee with the Battelle Memorial Institute in Columbus, Ohio, coined the word "bioventing" to describe biological destruction of volatile contaminants during soil venting. The oxygen can be consumed by microbes in direct contact with such wastes or by microbes that consume the volatile emissions swept away from the waste in the flowing soil air.

Bioventing is carried out by organisms indigenous to the subsurface. There is no evidence that inoculation of the unsaturated zone is either feasible or beneficial. When microbes have ready access to hydrocarbons and the oxygen content of the soil air is greater than 1%, the rate of hydrocarbon oxidation is not dependent on hydrocarbon concentration (zero-order process). However, the rate of hydrocarbon oxidation may be limited by nutrients in recent contamination events or biomass turnover by

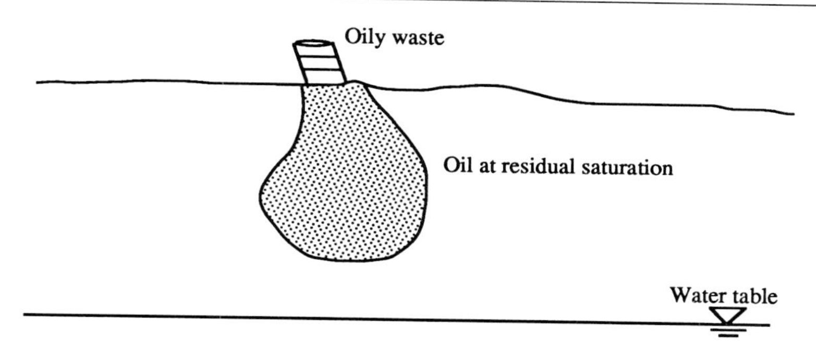

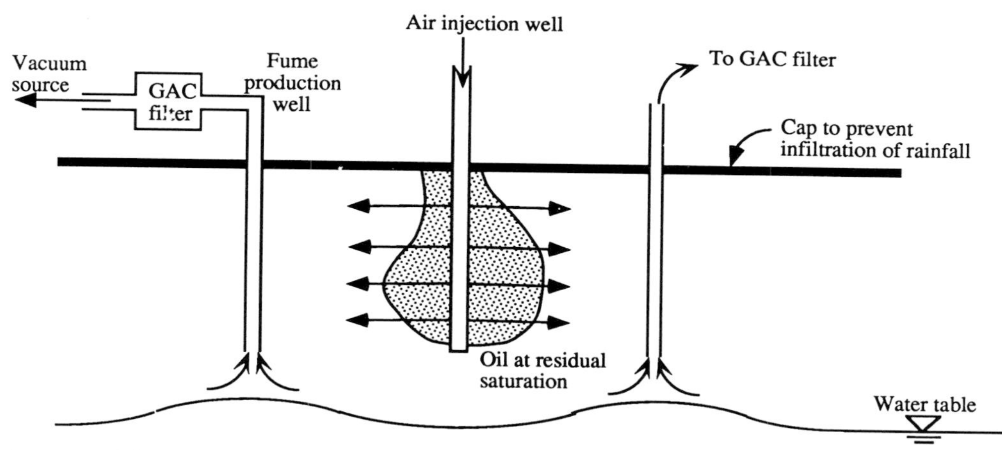

Figure 2 Schematic diagram of soil venting. [Adapted from J. T. Wilson and C. H. Ward (1987). *Dev. Indus. Microbiol.* **27**, 112.]

predators in older contamination events. The oxidation of vapors in the soil air at a distance from the contamination may be limited by hydrocarbon concentration (first-order process). Table VII compares the rate of hydrocarbon oxidation in contaminated subsurface material from a variety of locations. The rates vary from about 2 to 20 mg hydrocarbon/kg aquifer material/day.

The rates of hydrocarbon oxidation are determined by soil temperature and the availability of mineral nutrients. They may also be influenced by predation by protozoa. Laboratory studies with soil microcosms found that the rate of oxidation of hydrocarbon vapors in soils increased from two- to threefold with each 10°C rise in temperature. A field study found a sharp reduction in the rate of hydrocarbon oxidation in the winter months. The influence of mineral nutrients is less predictable and may be related to the native fertility of the subsurface material. A site in Michigan only attained high rates after the material was fertilized with nitrogen, phosphorus, and potassium. The fertilizer increased the rate of oxidation of hydrocarbon vapors by fourfold. However, hydrocarbon oxidation at a site in Florida was not stimulated by the addition of fertilizer. Protozoa are present in high numbers at the bioventing site in Michigan and may limit the rate of hydrocarbon oxidation that can be attained at that site by cropping the biomass.

At the U.S. Coast Guard Air Station in Traverse City, Michigan, bioventing of a spill of aviation gasoline has been compared with the use of hydrogen peroxide and nitrate to stimulate biodegradation of JP-4 jet fuel below the water table. Table VII contrasts the rates of hydrocarbon oxidation in the three systems. The rate of treatment attained during bioventing was equivalent to the rate achieved by biological activity below the water table.

Table VII Relative Rates of Hydrocarbon Oxidation at Fuel Spill Sites

	Basis of estimate[a] (mg hydrocarbon/kg dry aquifer material/day)		
Electron acceptor	Acceptor depleted	Hydrocarbon consumed	CO_2 produced
Bioventing[b]			
Michigan	22	18	
Maryland	2.9		1.5
Florida	2 to 20		
	6.0 and 7.8		4.8 and 5.2
Nevada	4.4 to 5.9		0.1 to 0.16
Alaska	1.4 to 13.2		0.15 to 0.64
Hydrogen peroxide[c]			
Michigan	4.6	20	
Nitrate[c]			
Michigan	20	14	

[a] The amount of hydrocarbon degraded per day is determined directly or estimated from stoichiometry based on the depletion of electron acceptor or production of CO_2 in samples of groundwater or soil air.
[b] Above the water table.
[c] Below the water table.

The rate of treatment of a spill with conventional vacuum extraction depends on the vapor pressure of the contaminants and on mass-transfer limitations on their volatilization to the soil air. Before venting, the soil air is at or near equilibrium with the contaminants, and vapor concentrations are high. As extraction proceeds, the vapor concentrations usually drop several orders of magnitude. The contaminants tend to volatilize at a constant rate, and the concentration of vapors in the soil air is controlled by the extent to which the vapors are diluted in air being drawn into the subsurface.

Recent practice in bioventing is to draw air through the subsurface at the minimum flow that will maintain aerobic conditions (at least 5% oxygen) throughout the spill. Low flow rates maximize the residence time of air in the subsurface, allowing adequate time for biotransformation. If there is an adequate length of uncontaminated material between the spill and the atmosphere, air can be injected rather than extracted. At low flow rates the fumes volatilized from the spill will be biologically degraded before the air returns to the atmosphere. Current designs usually relate residence time to the rate of hydrocarbon consumption. Hydrocarbon oxidation at 5 mg/kg/day (Table VII) will deplete oxygen in soil air in 1.0 to 1.5 days.

Table VIII presents general estimates of the volume of subsurface material occupied by air, water, and a waste oil with a viscosity near that of gasoline, under flooded and drained conditions. The proportions are then used to estimate the number of volumes of fluid that must be drawn through the contaminated subsurface material to remediate it. The concentration of hydrogen peroxide selected for comparison is near the maximum that can be tolerated on a continuous basis by hydrocarbon oxidizing organisms. The concentration of nitrate selected is the drinking water standard for nitrate in the United States. Air by itself is roughly equivalent to water amended with high concentrations of hydrogen peroxide or nitrate. The average cost for bioventing as opposed to the use of nitrate or hydrogen peroxide as the source of the electron acceptor at the Traverse City site is shown in Table V.

Table VIII Volumes of Water or Air Required to Completely Renovate Subsurface Material That Originally Contained Hydrocarbons at Residual Saturation[a]

	Proportion of the total occupied by			Air or water pore volumes required to meet the electron acceptor demand				
	Hydrocarbon when drained	Air when drained	Water when flooded	Air during bioventing	Air	Pure oxygen	Hydrogen peroxide	Nitrate
Stone to coarse gravel	0.005	0.4	0.4	250	5000	1250	200	250
Gravel to coarse sand	0.008	0.3	0.4	530	8000	2000	320	400
Coarse sand to medium sand	0.015	0.2	0.4	1500	15000	3800	600	750
Medium to fine sand	0.025	0.2	0.4	2500	25000	6300	1000	1250
Fine sand to silt	0.040	0.2	0.5	4000	32000	8000	1300	1600

[a] Assumes hydrogen peroxide provided at 500 mg/liter, nitrate provided at 10 mg/liter as N, and that 30% of the hydrocarbons can be metabolized with nitrate.

B. Saturated Subsurface Material

1. Plumes

a. Technology

A plume refers to the circumstance where contaminants occur as solutes in ground water. Because there is very little disperson along flow paths in aquifers, it is difficult to mix remedial amendments (e.g., electron acceptors or nutrients) into contaminated ground water *in situ*. If clean water containing amendments is injected into a plume, it simply displaces the contaminated water, giving the false appearance that the aquifer has been remediated. The bioremediation of a plume *in situ* requires that the water be pumped from the aquifer, amended with the appropriate requisites for growth, and returned to the aquifer (Fig. 3). As of this writing, the regulatory philosophy of most states in the United States and much of Europe prohibits injection of contaminated groundwater into an aquifer. As a practical consequence, bioremediation of contaminated ground water is conventionally conducted by pumping and then treating in an aboveground bioreactor.

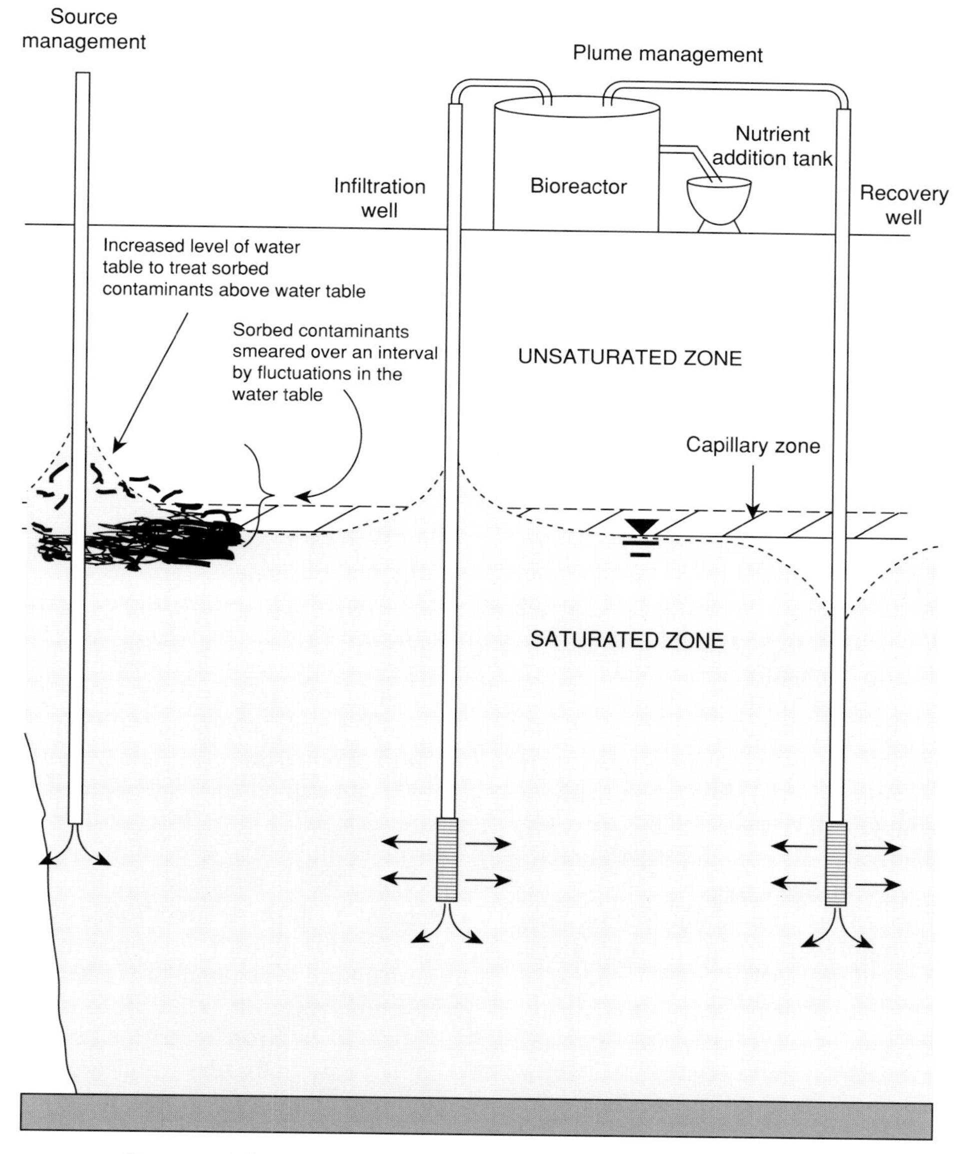

Figure 3 Schematic diagram of plume source management techniques.

Special designs are often required for bioreactors to treat ground water. The common contaminants of ground water are volatile and are often stripped rather than degraded in conventional wastewater biological treatment units. Further, the effluent concentrations required to meet clean-up standards (as little as 1 μg/liter) are often below the concentration required for the production of biomass.

Plumes of naturally occurring organics in groundwater are almost invariably anoxic. Some contain appreciable concentrations of reduced iron and manganese. If oxygen or hydrogen peroxide are used to treat the ground water, precipitation of hydrated iron and manganese hydroxides can quickly foul the well system. Many plumes contain high concentrations of biogenic methane. Frequently, there is more oxygen demand associated with methane than with organics of regulatory interest. If methane is not considered in the design of a bioreactor, the total oxygen demand of contaminated groundwater seriously can be underestimated.

b. Demonstration

A pilot-scale demonstration of bioremediation of a plume contaminated with several chlorinated aliphatic solvents was conducted by Stanford University in a shallow semi-confined aquifer on Moffett Naval Air Station in California. Bioremediation depended on the activity of methane oxidizing bacteria (methanotrophs), which normally oxidize methane into carbon dioxide via the enzyme, methane mono-oxygenase. Because of the nonspecificity of the methane mono-oxygenase, several chlorinated aliphatic solvents also were biodegraded. After artifically creating a plume of vinyl chloride, trichloroethylene, *cis*-dichloroethylene, and *trans*-dichloroethylene at concentrations of 0.034, 0.046, 0.100, and 0.052 mg/liter, respectively, naturally occurring methanotrophic bacteria in the aquifer were stimulated by amending ground water with methane and oxygen to achieve final concentrations of 7 and 21 mg/liter, respectively. After adaptation to methane degradation, 95% of the vinyl chloride was destroyed by one application of methane and oxygen. Removal of trichloroethylene, *cis*-dichloroethylene, and *trans*-dichloroethylene were 20, 40, and 85%, respectively. The removal of trichloroethylene and *trans*-dichloroethylene was not adequate for *in situ* treatment of these compounds. Higher concentrations of methane do not increase the extent of transformation, apparently because of competitive inhibition. [*See* METHANOGENESIS.]

Practical bioremediation of ground water contaminated with chlorinated solvents is yet to be obtained; however, research is progressing on several fronts. Devices are being designed that recirculate groundwater in a plume, allowing reamendment with methane and oxygen an arbitrary number of times until acceptable concentrations of the chlorinated compounds are reached. Aboveground reactors are being developed that treat groundwater directly or that treat air that has been used to strip the chlorinated compounds from water.

2. Sources

a. Technology

Water has been supplied to contaminants trapped below the water table through infiltration wells and infiltration galleries (Fig. 3). In the former case, the water is inserted into the subsurface over a selected depth interval through the screens of infiltration wells. The water moves horizontally away from the infiltration wells and joins the regional flow in the aquifer. Water can also be introduced through an infiltration gallery. Material is excavated and replaced with gravel or cobbles that allow uniform distribution of water within the gallery. Water is delivered to the gallery at a rate that floods all the native material between the gallery and the oily waste. Flow from an infiltration gallery is down into the aquifer.

Water in infiltration wells has been oxygenated by bubbling air through the water column within the well, which achieves an oxygen concentration of about 10 mg/liter. The water also can be equilibrated with pure oxygen, which achieves about 40 mg/liter. Hydrocarbon-degrading bacteria generally are believed to tolerate between 500 and 1000 mg/liter of hydrogen peroxide, which decomposes to produce 250 to 500 mg/liter oxygen. Because hydrogen peroxide can deliver the highest amount of oxygen, the oxidant is used more frequently than bubbling with air or equilibrating the infiltration water with pure oxygen. Water has been amended also with nitrate. Regulators in the United States are reluctant to permit injection of nitrate at concentrations greater than the drinking water standard (10 mg/liter as nitrogen). This concentration of nitrate is equivalent to 28 mg/liter O_2 in electron accepting capacity.

Conventionally, the water delivered to remediate contamination below the water table is supplemented with phosphate and ammonia–nitrogen. To reclaim the investment in mineral nutrients, the infiltrated water may be captured in a recovery well, reamended with electron acceptor, and recycled to the infiltration system. If the recycled water contains significant concentrations of organic compounds, biofilms will develop in the infiltration systems. Catalase (enzyme that degrades hydrogen peroxide to one-half O_2 and H_2O) in the biomass developing in the infiltration system can decompose hydrogen peroxide prematurely, resulting in loss of the oxygen produced to the atmosphere. The average cost of using hydrogen peroxide as opposed to nitrate as an electron acceptor or bioventing at the Traverse City site is shown in Table V.

Aromatic compounds, and the alkylbenzenes in particular, are preferentially degraded when contaminant sources are perfused with solutions of oxygen and nitrate. A spill can be cleansed of alkylbenzenes, while significant quantities of alkanes remain. As a consequence, it is easier to meet a benzene drinking water standard in monitoring wells than a total petroleum–hydrocarbon standard for aquifer material.

b. Demonstration

A pilot-scale field demonstration of bioremediation of a contaminant source in the subsurface was conducted at the site of an aviation fuel spill in Traverse City, Michigan. The objectives were to (1) quantitatively demonstrate the method as a basis for process design, (2) investigate the use of hydrogen peroxide as a source of oxygen in subsurface bioremediation, and (3) assess the use of the model, BIOPLUME II, in subsurface bioremediation. The demonstration was conducted in a 30 × 100-ft plot, one end of which was located upgradient of the contamination source and the other end which was located in the plume resulting from the source. The mass of hydrocarbon located in the test plot was estimated using the results from cores collected before the demonstration was initiated. The estimate was used in the BIOPLUME II model to design the well system and the amount of hydrogen peroxide required for bioremediation of the test plot. Infiltration wells were installed upgradient of the source, and multilevel sampling wells were installed at 7, 31, 50, 62, 83, and 108 ft from the infiltration point. Core material and ground water were collected during the demonstration for determinations of benzene, toluene, ethylbenzene and the xylene isomers (BTEX). Groundwater samples were collected for measurements of dissolved oxygen, ammonia, phosphate, pH, and conductivity.

Hydrogen peroxide, ammonia, and phosphate were added to the subsurface through infiltration wells. The initial concentration of hydrogen peroxide added was 50 mg/liter, which was increased incrementally to 750 mg/liter during a 10-month period. The concentration of phosphorus and ammonia in the infiltration water was 75 and 100 mg/liter, respectively.

The estimate of total hydrocarbon mass in the test plot used in designing the well system and determining the amount of hydrogen peroxide required for bioremediation was 5590 mg/kg. The demonstration was conducted for 30 months, during which time the concentrations of BTEX, oxygen, ammonia, and phosphate in ground water were decreased. BTEX was less than or close to the detection limit at all depths in the (1) 7-ft wells after about 10 months, (2) 31-ft well after about 12 months, and (3) 50-ft wells after about 15 months. Wells located 62, 83, and 108 ft from the infiltration point contained BTEX after 27 months of treatment, mainly at depths 16.5 to 18.0 ft below the soil surface where most of the contamination was located. At a depth of 16.5 ft, ground water from the 62- and 83-ft wells contained about 40 and 200 to 1000 μg/liter BTEX, respectively. Ground water from the 108-ft well at a depth of 18.0 ft contained about 400 μg/liter BTEX. The concentrations of aviation gas in core samples collected after 30 months of treatment at 7, 31, 62, and 108 ft from the infiltration wells were <1.0, 68, 39, and 319 mg/kg, respectively. Ground water within a 50-ft radius of the infiltration wells met the U.S. federal drinking water standard for alkylbenzenes.

V. Need for Inocula

The addition or inoculation of organisms with specialized metabolic capabilities (seed microorganisms) to initiate or enhance the degradation of organic wastes is not a new concept. The practice is used commonly in wastewater treatment facilities, where adapted populations are obtained from existing processes to seed new processes. Seed microorganisms are used frequently in slurry phase processes that take place in contained bioreactors and are controlled easily. Although the indigenous microflora is stimulated to degrade target compounds in most *in situ* bioremedial operations, seed microorganisms are sometimes used in surface operations such as land treatment and surface impoundments when the population of contaminant-degrading microorganisms is low or absent. In addition, microorganisms have been injected into the subsurface. For seed microorganisms to be effective in *in situ* bioremediation, the organisms must come into contact with the contaminant, compete with the indigenous microflora for nutrients and electron acceptors, retain their specialized metabolic capabilities, and survive under stressful conditions such as predation, the presence of toxins, and extremes in temperature and pH.

The injection of microorganisms into the subsurface logistically presents a more difficult problem than the addition of a seed to surface operations. The organisms must be transported through the subsurface matrix to the zone of contamination, colonize the matrix, increase in numbers, and maintain their biodegradative capabilities. Microbial transport is a multifaceted phenomenon that involves the properties of both the organism and the subsurface matrix. Physical and biological properties of the organism that may affect transport are the shape, size, condition, motility, and stickiness of the cell. Important characteristics of the subsurface that may affect transport include the porosity, hydraulic conductivity, and compostion of the formation. Provided that the average pore space is large enough to allow transport of the organisms, passage of the organism through the subsurface will be affected by complex interactions between solid surfaces and the physical and biological characteristics of the cell. In subsurface matrices without preferential flow paths, injected microorganisms move only short distances (centimeters). Hence, "seeding" of aquifers in most instances is not feasible.

Selective enrichment and genetic engineering techniques are used to obtain microorganisms with specialized metabolic capabilities. Selective enrichment usually involves exposing a sample of soil or water to successively higher and higher concentrations of the target contaminant in the presence of ample inorganic nutrients and an electron acceptor, which selects for organisms capable of degrading the compound of interest. These organisms are added frequently to surface and slurry phase operations. Genetic engineering has been used to manipulate many microbial strains; however, genetically engineered organisms cannot be released into the environment without prior government approval. [*See* Genetically Modified Organisms: Guidelines and Regulations for Research.]

In surface operations such as land treatment and surface impoundments, seed microorganisms have been found to decrease the lag time before biodegradation is initiated; after some time, however, the enhancement afforded by the seed is negligible. In addition, seed microorganisms have been beneficial when added to matrices that were devoid of microorganisms. Transport of microorganisms with specialized metabolic capabilities and their role in contaminant degradation in the subsurface has not been demonstrated. R. M. Atlas, a noted petroleum hydrocarbon microbiologist, has stated that microorganisms that degrade hydrocarbons can be found in most environments and that seeding for hydrocarbon biodegradation is unnecessary. However, compounds that are more resistant to biodegradation, such as the heavily chlorinated biphenyls, dioxins, phenols, and even the polycyclic aromatic hydrocarbons with more than three rings, may require or benefit from inoculation techniques. [*See* Petroleum Microbiology.]

VI. Summary

Bioremediation is the transformation of organic wastes by microorganisms into biomass, carbon dioxide, water, and inorganic salts, depending on the structure of the compounds in the waste. However, transformation of organic waste can be incomplete and result in products that are still environmentally significant. Surface operations include land treatment, composting, and slurry phase processes. Contaminated surface soils and some wastes can be treated using land treatment, which takes advantage of the soils assimulative capacity to biodegrade and immobilize the waste. Composting is the process by which a waste is mixed with a readily biodegradable material to facilitate biodegradation of the waste. Slurry phase processes involves forming a slurry out of the waste or contaminated soil and treating the slurry in a surface impoundment or contained bioreactor. These processes may be one or the final step in waste disposal. The need for further treatment will depend on the nature of the waste after bioremedial action.

In the subsurface environment, bioremediation is usually the last step in the remedial process. Before bioremediation is initiated, physical methods such as well systems are used to remove the recoverable phase of contamination. Bioremediation is used to treat the residual contamination that may be dissolved, entrained in pore spaces, and sorbed to the matrix. In some instances, however, bioremediation may not degrade the contaminants to acceptable regulatory standards, and alternate methods (e.g., pump and treat at the surface) may be required. Treatment at the surface may include air stripping and sorption by activated carbon or treatment of vapor phase contaminants in aboveground soil bioreactors similar to inplace treatment with bioventing.

A. Disclaimer

Although the research described in this article has been supported by the U.S. Environmental Protection Agency through Assistance Agreement No. CR-812808 to Rice University to support the National Center for Ground Water Research, it has not been subjected to agency review and therefore does not necessarily reflect the views of the agency and no official endorsement should be inferred.

Acknowledgments

We thank J. E. Matthews, M. R. Scalf, J. W. Keeley, and D. H. Kampbell from the R. S. Kerr Environmental Research Laboratory and F. B. Closmann, L. A. Rogers, and H. F. Stroo from Remediation Technologies, Inc., for their assistance and helpful discussions concerning sections of this article. Financial support from the Brown Family Fund, Houston, Texas, is gratefully acknowledged.

Bibliography

Atlas, R. M. (1977). *CRC Crit. Rev. Microbiol.* **5,** 371–386.

Bartha, R. (1986). *Microb. Ecol.* **12,** 155–172.

Finstein, M. S., Miller, F. C., and Strom, P. F. (1986). Waste treatment composting as a controlled system. *In* "Microbial Degradations" (Biotechnology, Volume 8) (W. Schonborb, ed.), pp. 363–398. German Chemical Society, Weinheim, FRE.

Ghiorse, W. C., and Wilson, J. T. (1988). *Adv. Appl. Microbiol.* **33,** 107–172.

Lee, M. D., Thomas, J. M., Borden, R. C., Wilson, J. T., Bedient, P. B., and Ward, C. H. (1987). *CRC Crit. Rev. Environ. Control* **18,** 29–89.

Loehr, R. C., Jewell, W. J., Novak, J. D., Clarkson, W. W., and Friedman, G. S. (1979). "Land Application of Wastes," vols. 1 and 2. Van Nostrand Reinhold Company, Van Nostrand Reinhold Environmental Engineering Series, NY.

Loehr, R. C., and Malina, J. F. (1986). "Land Treatment—A Hazardous Waste Management Alternative." Center for Research in Water Resources, The University of Texas at Austin, TX.

Sims, J. L., Sims, R. C., and Matthews, J. E. (1989). "Bioremediation of Contaminated Surface Soils." U.S. Environmental Protection Agency Document EPA-600/9-89/073, Ada, OK.

Stroo, H. F. (1992). *J. Environ. Qual.* **21,** in press.

Thomas, J. M., and Ward, C. H. (1989). *Environ. Sci. Technol.* **23,** 760–766.

Biotechnology Industry: A Perspective

Vedpal S. Malik
Philip Morris Research Center
Erik P. Lillehoj
Cambridge Biotech Corporation

Glossary

Biotechnology Applied biological science concerned with the application of data to problems (i.e., bioengineering or recombinant DNA technology)

Xenobiotics Human-made chemical structures

Even in the case of the Earth, if seeds are sown in the same field and at the same time, they will grow in different forms depending on their nature.

Manusmriti 9:38
400 B.C.

BIOTECHNOLOGY is the exploitation of living systems and organisms for the benefit of humans. Although there has been a recent technological revolution in biology, the practice of biotechnology is one of the oldest industries in the world. However, the new term "biotechnology" has emerged to include processes that exploit versatile metabolic machinery or components of living organisms to produce valuable metabolites from renewable resources. At this time in human history, when the supply of many valuable raw materials is diminishing, use of microorganisms for producing molecules of complex chemical structure is increasing. The use of countless microbes is being exploited at many new frontiers. Chemical synthesis, which is energy-intensive and pollution-producing, may eventually be replaced by low-energy consumption and pollution-free microbial synthesis. In this way, biological machineries may transform waste and toxic pollutants into complex chemical structures that are difficult to assemble chemically (Table I).

I. The Industrial Organism

A. Source of Products

During World War I, Neuberg in Germany used large-scale yeast fermentations to produce glycerol. On the British side, Weismann used *Clostridium* to produce acetone-butanol. However, these fermentations were competed out by chemical means of solvent production, although they did lay the foundation for industrial production of citric acid by *Aspergillus niger*. The commercial production of penicillin in the 1940s by submerged fermentation was of great benefit to the Allied war effort and remains the technological triumph for producing microbial products of human interest (see Table II). Various landmarks in the history of commercially valuable antibiotics are listed in Table III.

As in the past, naturally occurring organisms con-

Table I Some Microbial Products of Economic Value

Product category	Example	End use
Amino acids	Cysteine, lysine, glutamate, methionine, proline, threonine, phenylalanine, tryptophan	Food enrichments, flavorants
Biopolymers	Polyamylose, poly-β-hydroxybutyrate	Biodegradable wrapping
	Alginate, cellulose, curdlan, levan, dextran, xanthan	Food thickening
Microbial cells	Yeast, lactobacilli, streptococci	Brewing, baking, wine, and cheesing making
Steroids	Cortisone	Arthritis
Ergot alkaloids	Ergotamine, agroclavine	Therapeutics
Microbial proteins	*Methylophilus methylotrophus*, yeast, algae	Single-cell proteins
Vitamins	B_2, B_{12}, D, C, nicotinic acid	Food
Peptides, proteins	Aspartame, thaumatin	Sweetener
Carotenoids	β-Carotene	Colorant
Organic acids, etc.	Geraniol, isobutylene, linalool, nerol, nucleotides, acetic acid, benzoic acid, citric acid, fumaric acid	Flavorants
	Gibberellins	Plant growth promoters
Organic feed stocks	Ethanol, butanol, acetic acid, propylene, glycerol, 2,3-butanediol, methyl ethyl ketone	Solvents, fuels
Secondary metabolites	Cyclosporin, FK-506	Immunomodulators
Enzyme inhibitors	Antipapain, chymostatin, leupeptin, pepstatin	Protease inhibitors
	L-phenylacetylcarbinol	Precursor of L-ephedrine

tinue to be a source of many valuable products. For example, more than 6000 antibiotics are produced by microorganisms and more than 150 have been commercialized. Antiparasitic agents such as avermectins are microbially produced. Microbial products such as levastatin are used to lower blood pressure, and cyclosporin and FK-506 are stimulators of host defense against pathogens. Taxol is a plant product with antitumor potential.

The antibiotic market is worth over $10 billion annually. Ranitidine, captopril, and infedipine each sold over $1 billion in 1990. However, molecular biology is now impacting on drug discovery. This new technology is providing new assay tools, humanized antibodies, cloned and expressed human drug receptors, and transgenic animals. For example, human 5-HT_{1D} and 5-HT_2 receptors have been expressed in nonhuman, nonneuronal cell lines (mouse fibroblasts) that do not express monoamine receptors. This may allow targeting of drugs to single human proteins. Subtypes of receptors for the same neurotransmitters may belong to a distinct class. Cloning and sequencing can reveal differences between receptor subtypes and identify new targets for drug action in the intracellular signaling system. The power of the detection system is the key to finding specific compounds from naturally occurring organisms.

In general, the productivity of organisms directly isolated from nature (wild type) is unsatisfactory. Mutagenesis and/or breeding is used to enhance the output of the targeted product. In the past, the hit-or-miss methods of classical genetics had limitations in improving the biotechnological processes. Recombinant DNA technology now provides new ways of modifying to order the genomes of organisms in a directed manner. Genetic incompatibility of species, a restriction in classical genetics, has been overcome by new technology. Because of recent advances in methodology, genetic manipulations of the molecular biology of many organisms is rapidly becoming possible. As a result, naturally occurring organisms can be endowed with additional genetic information from other sexually incompatible organisms. Thus, construction of novel organisms that do not exist in nature with new metabolic capability is now routine. Molecular biology is finally contributing toward the development of new commercial products and the new biological methodologies have spawned a new industry. [*See* RECOMBINANT DNA, BASIC PROCEDURES.]

This is a special period of rapid progress in bio-

Table II A Few Economically Important Antibiotics

Antibiotic	Chemical form	Producing organism	Producing company	Activity spectrum
Actinomycin	Peptide	*S. antibioticus*	Bayer AG, Merck	Antitumor
Adriamycin	Anthracycline	*S. peuceticus* var. *caesium*	Farmitalia, Kyowa Hakko Kogyo, Rhone-Poulenc	Antitumor
Avermectins	Macrolide	*S. avermitilus*	Merck	Antihelmintic
Butiriosin	Aminoglucoside	*Bacillus circulans*	Kyowa Hakko Kogyo	Antibacterial
Candicidin B	Polyene	*S. griseus*	Penick	Antifungal
Cephalosporin C	β-Lactam	*Acremonium chrysogenum*	Antibiotics SA, Brystol-Myers, CIBA-Geigy, Eli Lilly, Farmitalia, Fujisawa, Glaxo Hoechst AG, Merck, Novo, Takeda	Antibacterial
Cephamycin C	β-Lactam	*S. lactamdurans*	Merck	Antibacterial
Clavulanic acid	β-Lactam	*S. clavuligerus*	Beecham Pharmaceutical	Antibacterial
Cyclosporin A	Peptide	*Tolypoeladium inflatum*	Biochemie GmbH, Sandoz	Immunosuppressor
Daunorubicin	Anthracycline	*S. coeruleorubidus*	Rhone-Poulenc, Farmitalia	Antitumor
Erythromycin	Macrolide	*S. erythreus*	Abbott, Eli Lilly, Pfizer, Roussell-Uclef, Upjohn	Antibacterial
Gentamicins	Aminoglycoside	*Micromonospora purpurea*	Schering	Antibacterial
Monensin	Polyether	*S. cinnamonensis*	Eli Lilly	Coccidiostat
Penicillin	β-Lactam	*Penicillium chrysoginum*	Bristol-Myers, Gist-Brocades, Merck, Pfizer	Antibacterial
Rifamycin	Ansamycin	*Nocardia mediterranei*	CIBA Geigy, Dow, Lepetit	Antibacterial, antiviral
Spectinomycin	Aminocyclitol	*S. spectabilis*	Upjohn	Antibacterial
Streptomycin	Aminoglycoside	*S. griseus*	Gist-Brocades, Merck	Antibacterial
Tetracycline	Polyketide	*S. aureofaciens*	American Cyanamid	Antibacterial
Thienamycin	β-Lactam	*S. cattleya*	Merck	Antibacterial

technology and many useful products have already been marketed, employing recombinant DNA and hybridomas. As shown in Table IV, tremendous growth in the U.S. biopharmaceutical market is expected. The distribution of these projected revenues between the various subdisciplines of biotechnology are listed in Table V. Hundreds of dynamic small companies have been formed to tap the excitement of new technological advances; some of these companies are already well established with significant marketing resources and an outstanding record for delivering products of the new technology. Already delivered are products such as medicinal drugs (antibiotics, vitamins, hormones, immunosuppressants, and antitumor and antiparasitic agents), diagnostic agents, enzymes, fine chemicals, pesticides, biopolymers, vaccines, and amino acids.

Synthesizing a new compound on a laboratory scale is only the first step in a long trek toward producing and marketing a new product. To realize its full potential, production must be optimized and scaled up, clinical trials performed to seek approval of appropriate governmental regulatory agencies, and finally the product packaged and marketed. All of these steps require several years and enormous financial costs to complete. Thus, for newly emerging small biotechnology companies to be successful, they must select their niche carefully. However, the rewards are enormous. The next century will be dominated by production of safe bioproducts for the benefit of life and the environment and for the survival of human society.

B. The Microbial World: Source of Novel Genes

Because of the small genome size of microorganisms, it is in general easier to isolate a particular gene of interest from them than from higher life forms. Unlike eukaryotic genes, bacterial genes usually do not possess introns and can be easily expressed in other organisms. Furthermore, many

Table III Chemicals Celebrate a Century of Combat against Bacteria.

Ever since doctors found that bacteria cause disease, chemists have sought the "magic bullets" to kill them.

1865. Joseph Lister, surgeon at the Glasgow Royal Infirmary, uses aqueous phenol to disinfect instruments, wounds, incisions, and air. By 1869 mortality on his wards drops to 15% from 45%.

1876. Prussian country doctor Robert Koch isolates bacteria from animals with anthrax, cultures them, and shows that they induce the disease in other animals. His classic method of proof finds wide use of linking microorganisms to disease and he receives the Nobel Prize in 1905.

1877. Louis Pasteur, physiological chemist, at the Ecole Normale Superieure, Paris, finds that invading atmospheric bacteria kill anthrax germs in cultures, that anthrax germs cannot survive in soil, and that animals infected by other germs are not infected by anthrax. This antagonism among microorganisms is later called antibiosis.

1898. Bacteriologist Rudolph Emmerich of the University of Munich isolates bacteriocidal pyocyanase from *Psuedomonas aeruginosa*. Though moderately successful in patients, pyocyanase is abandoned as unreliable and toxic by 1913.

1910. Paul Ehrlich, director of the Royal Institute for Experimental Therapy, Frankfurt, tries 605 arsenic compounds against microbes and finds that No. 606, arsphenamine, is effective against *Treponema pallidum*, the syphilis spirochete. Ehrlich, who had won the Nobel Prize in 1908 for work on immunology, was inspired by preferential staining of certain tissues with methylene blue. He theorized that some chemicals, which he called "magic bullets," might bind to and kill bacteria without harming human cells.

1928. Pathologist Alexander Fleming at St. Mary's Hospital, London, finds that *Penicillium notatum* molds contaminating his *Staphylococcus* cultures kill them. He names the inhibiting substance penicillin and realizes its importance, but lack of funds prevents him from isolating it for tests.

1932. Pathologist Gerhard Domagk of Farbenfabriken Bayer's laboratory in Wuppertal-Elberfeld, Germany, tries thousands of dyes against bacteria and discovers that 4-(2,4-diaminophenyl)-azobenzenesulfonamide cures streptococcal infections in mice. Others will find that animals metabolize the dye to sulfanilamide. Domagk wins the 1939 Nobel Prize for the sulfa drugs, but the German government forbids him to accept. Domagk gets his medal in 1947.

1938. Pathologist Howard W. Florey and biochemist Ernst B. Chain of Oxford University evaluate Fleming's penicillin, made confident of success by the known effectiveness of the sulfa drugs. They come to the United States in 1941 at the outbreak of war to arrange commercial production. Fleming, Florey, and Chain share a Nobel Prize in 1945.

1939. Bacteriologist Rene Dubos of Rockefeller Institute for Medical Research, New York City, isolates gramicidin for

Continues

Continued

soil bacterium *Bacillus brevis*. This becomes the first polypeptide antibiotic and the first antibiotic to be produced commercially, though it proves too toxic for any but external use.

1940. Stimulated by Dubos' success, Selman A. Waksman, microbiologist at Rutgers University, sifts through soil actinomycetes and isolates actinomycin. In 1942 he reports streptothricin and in 1943 streptomycin. Waksman coins the word antibiotic. Streptomycin, the first of the aminoglycosides, proves effective against gram-negative and anaerobic bacteria, whereas penicillins known to date act against only gram-positive aerobes. Waksman wins a Nobel Prize for streptomycin in 1952.

1943. Chemists William B. Stillman, A. B. Scott, and J. M. Clampit at Norwich Pharmacal Co. report nitrofurazone as the first of the nitrofuran antibacterial drugs.

1947. John Ehrlich isolates chloramphenicol from *Streptomyces venezuelae*, which is both a rare example of a nitrochloro compound in nature and a broad-spectrum antibiotic.

1948. Botanist Benjamin M. Duggar retires from the University of Wisconsin, goes to work for Lederle Laboratories, and isolates aureomycin from *S. aureofaciens*, thereby discovering the first of the tetracyclines.

1952. James McGuire of Eli Lilly isolates erythromycin from *S. erythreus*.

1957. P. Sensi of Lepetit isolates rifamycins, the first of the ansamacrolides from *Nocardia mediterranei* in a project code-named "Rififi" after the then-popular movie and modifies one of them chemically to rifampin.

1962. Chemists George Y. Lesher and Monte D. Gruett of Sterling Drug Co. report nalidixic acid as the first of the quinolone antibacterial drugs. Donald J. Mason, Alma Dietz, and Clarence DeBoer of Upjohn isolate lincomycin, the first of the lincosaminides, from *S. lincolnensis*.

1975. Workers at Merck isolate thienamycin from *S. cattleya*, the first of the carbapenems, and develop it into Primaxin, the widest spectrum antibacterial known to date.

1980. Scientists at Takeda isolate the first of the monobactams after a massive screening of soil molds.

1986. The U.S. antibacterial drug market tops $2.5 billion, heading toward $3 billion in 1990. Medicinal chemists pursue such compounds as monobactams, carbapenams, and quinolones in vigorous efforts to outflank bacterial resistance and meet pressures to cut treatment cost.

[Reprinted, with permission, from Stinson, S. C. *Chem. Eng. News*, September 29 (1986). **64(39),** 36. Copyright 1986, American Chemical Society.]

organisms can be easily selected for certain biochemical activities—for example, degradation of a toxic compound or growth on a particular medium. In this way, many microorganisms have been selected from nature for a targeted function and the

Table IV Total U.S. Biopharmaceuticals Market: Revenue and Growth Forecasts, 1985–1995

	Revenues ($ millions)	Revenue growth (% growth)
1985	0.0	—
1986	26.2	—
1987	253.0	865.5
1988	299.1	18.2
1989	557.5	86.4
1990	1499.6	169.0
1991	2129.6	42.0
1992	3033.2	42.4
1993	3607.5	18.9
1994	4238.2	17.5
1995	4855.5	14.6

[Source: Market Intelligence Research Co. From Roth, S. (1991). *Gen. Eng. News* **11**(2), 5.]

respective gene(s) expressed in other living systems to obtain a desired result.

Scientists at one company incorporated a chitinase gene of *Serratia marcescens* into tobacco, potato, lettuce, and sugar beet. Chitinase breaks down the chitin of the cell wall of fungi, and the engineered plants were resistant to infection by the soil-borne phytopathogen *Rhizoctonia solani*. These researchers have chemically synthesized a gene encoding an antifreeze protein similar to that of the winter flounder fish. This gene was expressed in vegetable plants and yeast. Such engineered plants may better withstand freeze–thaw cycles without loss of flavor or texture. This antifreeze protein may also be added to ice cream and ice milk to prevent grainy texture or ice crystals. The nucleotide sequence encoding the bacterial acetolactate synthase protein that is resistant to sulfonylurea, triazolopyrimidine sulfonamide, and imidazolinone herbicides have been expressed in several plants to confer respective resistance phenotypes.

Table V Biopharmaceutical World Markets ($ Million)

	1995	2000
Cancer	$730	$1900
AIDS	100	500
Cardiovascular disease	680	1500
Monoclonal therapies	200	490
Growth factors	670	1500

[Source: Consulting Resources Corporation, Lexington, Massachusetts. From Shamel, R. E., and Chow, J. (1990). *Gen. Eng. News* **10**(3), 4.]

II. Secondary Metabolites: The Billion Dollar Molecules

Microbial synthesis of bioactive chemotherapeutics is an important venue of classical fermentation technolgy (Fig. 1). These molecules have a wide application in veterinary medicine, agriculture, and human health with antiviral, antibacterial, antifungal, anticoccidial, antiprotozoal, antihelmintic, antitumor, pesticidal, and insecticidal activity.

Most of the microbially produced molecules are modified chemically to produce analogs with superior biological activity. In this way, many novel analogs of penicillins, cephalosporins, aminoglycosides, chloramphenicols, etc., have been prepared. By classical mutagenesis, the yields of antibiotics have been increased to a level where their production is profitable. Penicillins (60 g/liter), cephalosporins (40 g/liter), streptomycin (40 g/liter), erythromycin (6 g/liter), tetracycline (30 g /liter), rifampin (10 g/liter), and lincomycin (10 g/liter) are produced in many industrial fermentations at the indicated levels.

Recombinant DNA technology may now be used to create hybrid antibiotics. This may be achieved by fusion of pathways that generate chemically diverse as well as similar structures. Fusion of pathways of various macrolide-producing streptomyces has already begun and could eventually yield a new clinically useful antibiotic. The genes for the enzyme expandase from the cephalosporin-producing organism *Cephalosporium acremonium* may be introduced into the penicillin producer *Penicillium chrysogenum*. This could allow production of a molecule with an expanded ring structure but with side chains similar to penicillin. This new biosynthesized analog may be hydrolyzed by existing enzyme technologies to generate 7-aminocephalosporanic acid (7-ACA). The last steps involved in the biosynthesis of cephalosporin may be blocked by inserting deletions of corresponding genes. This could lead to economy of the pathway while yielding a product that could be converted to 7-ACA. Japanese biotechnologists have already introduced hydrolytic enzymes in *Acremonium chrysogenum* to synthesize 7-ACA by fermentation. Genes for efficient utilization of soy, carbon and nitrogen sources, may also be trans-

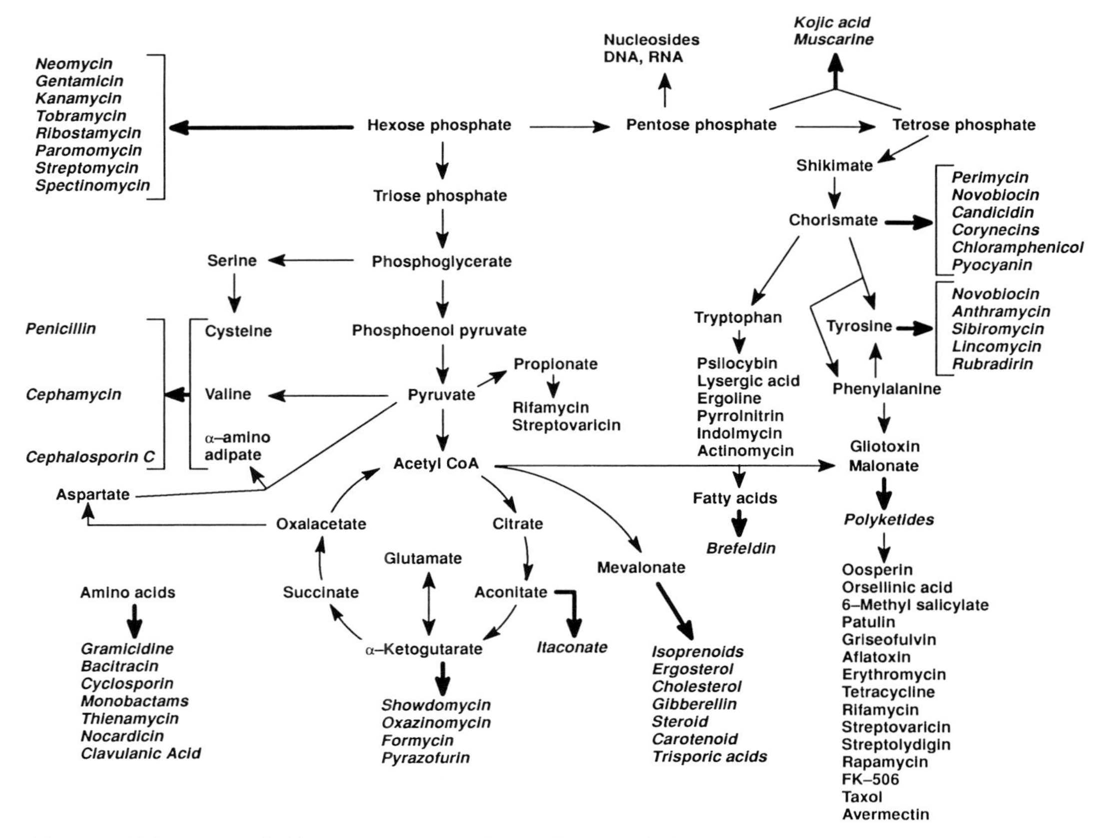

Figure 1 Primary metabolites as precursors of secondary metabolites. Heavy arrows emphasize channeling of primary metabolites into secondary metabolites. [Reproduced from Malik, V. S. (1982). *Adv. Appl. Microbiol.* **28**, 27–115.]

planted into organisms that were originally isolated from soil.

III. Engineering Proteins: The Next Revolution

Many uses of enzymes for producing food ingredients, amino acids, laundry detergents, pollution control, and asymmetric chemical syntheses have been well recognized (see Table VI). It is believed that worldwide use of enzymes will increase. This is because enzymes are often products of single genes, and modern methods of recombinant DNA technology can be used to enhance their yields and modify their specificities in suitable microbial hosts. To reduce the expense of producing and using enzymes, bacterial strain improvement, enzyme and cell immobilization, and stability enhancement are being explored. First, the chemistry of a protein should be understood before its performance can be improved. The long-term goal is the rational design of efficient, biologically active enzymes. Protein engineering is based on genetic engineering, X-ray crystallography, protein structural modification, synthesis, analysis interpretation, scale-up, and purification. Large corporations have in-house efforts directed toward these goals. The rational design of peptide-based HIV proteinase inhibitors using the three-dimensional structure of the enzyme has already been achieved.

Site-directed mutagenesis enables precise amino acid replacement anywhere in the protein structure. These changes are first introduced at the nucleotide level in the coding sequence, the modified gene inserted into an appropriate host for protein expression, and the modified protein isolated. Proteins whose three-dimensional X-ray structures have been determined are good candidates for such genetic modification, although trying to make mean-

Table VI Some Industrial Enzymes: Their Sources and Uses

Enzyme	Source	Uses
Proteases	*Bacillus subtilis, B. licheniformis*	Detergents, meat tenderizer, chill-proofing beer, cheese, and flavor production
Rennin	Calf and lamb stomachs, microbes	Cheese production
Papain	Papaya	Meat, beer, leather, textiles, pharmaceuticals, tenderizer, digestive aid, dental hygiene, clarification of beer haze
Bromelin	Pineapple cannery residues	Meat, beer, pharmaceuticals
Ficin	Figs	Meat, beer, pharmaceuticals, leather
Pepsin	Hog stomachs	Cereals, pharmaceuticals, feeds
Trypsin	Hog and calf pancreases	Meat, pharmaceuticals
Amylases	*Aspergillis niger, A. oryzae, Bacillus licheniformis, B. subtilis*, recombinant organisms, barley malt	Hydrolyze starch for ethanol production, detergents, baked goods, milk, cheese, beer, fruit juices, digestive aid, dental hygiene
Invertases	*Saccharomyces cerevisiae*	Produce invert sugar, confectionery, distilled beverages
α-D-Galactosidase	*Mortierella vinaceae*	Raffinose hydrolysis
β-D-Galactosidase	*Aspergillus niger*	Lactose hydrolysis
Glucose isomerases	*Streptomyces olivaceus, E. coli, Arthrobacter globiformis, Actomoplanes missouriensis*	Convert glucose to fructose, production of high-fructose corn syrup for soft drinks, also other beverages and foods
11-β-Hydroxylase	*Curvularia lunata*	Steroid conversions
Aminoacylase	*Aspergillus oryzae*	Optical resolution of α-amino acids
Aspartase	*Escherichia coli*	L-aspartic acid from fumaric acid
Fumarase	*Brevibacterium ammoniagenes*	L-malic acid from fumaric acid
Hydantoinase	*Bacillus brevis*	D-phenylalanide from D,L-phenyl hydantoin
Nitrilase	*Corynebacterium* sp.	Acrylamide from acrylonitrile L-phenylalanine
Pectinases	Fungi, tomatoes	Hydrolyze pectic substances, clearing of fruit juices, wine, coffee, cocoa
Glucose oxidases	Microorganisms	Oxidize glucose to gluconic acid, eggs, preservation of flavor and color in eggs, food and fruit juices, dental hygiene
Pullulanase	Microorganisms	Production of beer, maltose and glucose
Cellulase	Microorganisms	Ethanol production, digestive aid
Urokinase	Urine	Treatment of thrombosis
Asparaginase	*E. coli*	Antitumor agent
Lipases	Hog kidneys and calf pancreases and glands, recombinant organisms	Hydrolyze fats and fatty acid esters, detergents, chocolate, cheese, feeds, digestive aids
Catalases and lipoxidase	Livers	Decompose hydrogen peroxide, milk sterilization, production of cheeses, bleaching agent in baking
Penicillin acylase	*E. coli*	Production of 6-amino-penicillanic acid for synthesis of various β-lactam antibiotics

[Adapted from Malik (1989). *Adv. Appl. Microbiol.* **34,** 263–306.]

ingful amino acid sequence changes is still very much guesswork.

The protein-digesting enzyme subtilisin is commercially used in laundry detergents where it is exposed to chlorine, oxidants, heat, and extremes of pH. Some scientists have prepared more than 80 mutants of subtilisins. They have observed that replacing methionine at position 222 with more oxidation-resistant amino acids (alanine, serine, or threonine) stabilized the enzyme but decreased its activity. One company has engineered a bleach-resistant detergent protease "Durazym" in which methionine is substituted by other amino acids. This enzyme is active in detergents to remove protein-based stains. This company is also designing additional enzymes for various purposes. Use of other enzymes (e.g., heat stable alkaline proteases, amylases, lipases) in automatic dishwashing detergents is increasing as a result of demand for safer products.

Similarly, cysteine residues have been introduced to form new disulfide cross-links in lysozyme. Such internal cross-links have been introduced with the goal of protecting the enzyme from denaturation-induced inactivation without destroying catalytic function. A particular methionine susceptible to oxidation has been substituted with valine in α_1-antitrypsin by scientists. This engineered protein gives protection to lungs against oxidation and the action of the enzyme elastin. Metal recognition can be designed into proteins for use in protein purification. Other examples of enzyme engineering include the shifting of trypsin specificity from arginine and lysine amino acids to lysine alone. The crucial serine residue in the active site of alkaline phosphatase has been substituted with cysteine without loss of catalysis. Substrate preference of thermophilic xylose isomerase has been altered for D-xylose to D-glucose. *Bacillus stearothermophilus* lactate dehydrogenase has been mutated to a potent malate dehydrogenase and a broad specificity hydroxyacid dehydrogenase for synthesis of chiral intermediates for pharmaceutical use. Chymosin, an aspartic proteinase found in the fourth stomach of unweaned calves, has been mutated to obtain pH optima shift and a substrate-specificity pocket variant. Engineering of the lactococcal proteinase specificity for use in cheese manufacturing has also been envisioned. Proteins that have commercial value are going to be engineered in the near future to generate a second generation of products. Hormones and biological response modifiers may be an important fraction of these new products. Taylor-made superactive insulins and receptor-specific variants of human growth hormone have been produced. Replacement of finger and growth factor domains of tissue plasminogen activator with plasminogen kringle I has also been reported.

IV. Bioremediation and Pollution Control: An Odyssey

Manmade chemical structures (xenobiotics) to which microbes have not been exposed during the course of evolution are generally resistant to biodegradation. These recalcitrant molecules can remain in the environment virtually unchanged for eons. However, synergistic communities of microorganisms can be enriched to metabolize such pollutants. First, organisms can be selected that modify the molecule by stoichiometric conversion to a single substrate but do not utilize it as a carbon source. Then, several key reactions can be pooled into a single organism that best survives the environment into which the pollutant appears. The U.S. Environmental Protection Agency (EPA) list of priority pollutants includes pesticides, halogenated aliphatics, aromatics, nitroaromatics, chloroaromatics, polychlorinated biphenyls, phthalate esters, polycyclic aromatic hydrocarbons, and nitrosamines. The U.S. market for bioremediation of such hazardous wastes is now on the order of $2–3 billion/yr. Use of genetically engineered and cultivated organisms to accelerate biodegradation would contribute to the diffusion of the toxic chemical waste time bomb. [*See* BIODEGRADATION; BIOREMEDIATION.]

It may be possible to dissolve radioactive uranium oxides from low-level nuclear wastes and subsequently change the valence of the uranium-containing compounds. This would cause them to precipitate into an immobile and nonradioactive form.

Scientists have shown that microbes that possess benzoate dioxygenase and 1,2-dihydrodihydroxybenzoate dehydrogenase and catechol 1,2-oxygenase do not grow on benzoate or monohalobenzoate and can accumulate muconic acid from selected aromatic substrates (U.S. patent 5,026,648). This sort of strategy may be used for converting toxic pollutants into useful intermediates of the metabolism.

V. Living Organisms: Factories for Producing Foreign Proteins

Recombinant DNA technology allows the isolation of complementary DNAs (cDNAs) that determine the synthesis of valuable proteins that exist in the human body in only minute quantities. These cDNAs are being expressed in a variety of cells to mass produce their protein products commercially (see Table VII). Human insulin and human growth hormone are already being produced in *Escherichia coli* and sold commercially. One company has produced a recombinant granulocyte colony-stimulating factor to fight infections associated with some anticancer drug therapies. Tissue plasminogen activator, produced by transplanting the human gene in bacteria, is a commercial product worth $300 million. Another company's γ-interferon is approved for treating chronic granulomatous dis-

Table VII Current Clinically Useful Polypeptides Produced via Biotechnbology

Disorders and products	Manufacturers
AIDS	
rCD4	Biogen, Genentech, SmithKline Beecham
Vaccine	MicroGene Systems, Oncogene, Repligen
Cancer	
Monoclonal antibodies	Bristol-Myers/Oncogen, Chiron, Cytogen, Genetics Institute, Idec Pharmaceuticals, ImmunoGen
Fibroblast growth factor	California Biotechnology
Human platelet factor VIII	Behringwerke, Genetics Institute
Interferon-α	Biogen, Boehringer Ingelheim, Burroughs Wellcome, Genentech, Hoffmann–La Roche, Immunex, Schering-Plough
Interferon-β	Cetus
Interferon-γ	Amgen, Bioferon, Biogen, Genentech
Interleukin (various)	Amgen, Biogen, Cetus, Genentech, Hoechst, Hoffmann–La Roche, Immunex, Schering-Plough, Syntex
Tumor necrosis factor	BASF, Biogen, Cetus, Chiron, Genentech, Novabiochem
Cardiovascular	
Saruplase	Gruenetal
Tissue plasminogen activator	Behringwerke, Boehringer Ingelheim, Boehringer Mannheim, Genentech, Genetics Institute, Genzyme, SmithKline Beecham, Tomae
Arterial natriuretic peptide	California Biotechnology, Genentech
Superoxide dismutase	Biotechnology General, Chiron
Alteplase	Genentech
Common Cold	
ICAM (intracellular adhesion molecule)	Boehringer Ingelheim
Hormone Disorders	
Human insulin	Eli Lilly, Hoechst, Novo Nordisk
Human somatotropin	Biotechnology General, Carlbiochem, Celltech, Eli Lilly, Genentech, Novo Nordisk, Serono
Nervous System	
Ciliary neurotropic factor	Regeneron Pharmaceuticals
Neurotrophin-3	Regeneron Pharmaceuticals
Nerve growth factors	Athena Neuroscience, Cephalon, CNS Research
Transplant Rejection	
OKT-3	Centocor, OrthoPharm
Wound Healing	
Bone growth factor	Chiron, Genetics Institute
Epidermal growth factor	Amgen, Chiron, Intergen, Johnson & Johnson
Fibroblast growth factors	California Biotechnology, Synergen
Platelet-derived growth factor	Chiron, Genetics Institute, Intergen
Transforming growth factor	Bristol-Myers/Oncogen, Collagen, Immunodiagnostic Systems

[From the PMA 1990 Annual Survey Report.]

ease, a genetic deficiency of the immune system that results in life-threatening infections. Erythropoietin is being sold to increase red blood cell production and fight anemia. Several interferons and lymphokines produced by microbial means are on the market and many more are in clinical trials to test their efficacy.

Once a gene has been isolated, its protein product must be produced and characterized. The choice of host for expression of the foreign gene often depends on the initial purpose for isolating the gene. Although bacterial host cells can be exploited for high-level expression of inserted genes, they do not possess the ability to correctly modify certain eukaryotic protein—for example, glycosylation of glycoproteins. Furthermore, foreign genes expressed in bacteria often accumulate as aggregated and denatured inclusion bodies that require highly concentrated solutions of urea or guanidine for their solubilization. In many cases, only a fraction of the expressed protein can be renatured into a fully active form. Eukaryotic host cells, grown in tissue culture, circumvent these problems but present a new set of handicaps. Many such hosts require serum supplementation into their growth medium and the high concentration of unwanted serum proteins may complicate down-stream purification procedures. A new generation of serum-free media are now being commercially introduced to facilitate isolation of recombinant proteins expressed in eukaryotic cells.

The development of DNA microinjection techniques has spawned the emergence of transgenic animals carrying foreign genes and propagating them to their offspring. In this manner, gene expression can be studied *in vivo*. For example, human growth hormone has been successfully transferred to mice, rabbits, sheep, and pigs, with the hope of someday providing a genetic therapy for human dwarfism. Transgenic cows carrying bovine somatotropin have increased milk production by 10–25%, although the social benefits of this effect are still debated. One company has produced genetically engineered pigs that synthesize human hemoglobin. Other transgenic animals carrying human genes may provide an abundant source of recombinant therapeutic proteins in the future.

Tissue-specific expression of several therapeutic proteins has been achieved in dairy cattle for high-volume production. These technological advances may allow cheap production of foreign proteins that conventional expression systems cannot provide. Researchers at the Institute of Animal Physiology and Genetic Research in Edinburgh, Scotland, were the first to produce a transgenic sheep that secreted in their milk low levels of human antihemophilic factor IX and human α_1-antitrypsin. In the latter case, up to 35 g of recombinant human α_1-antitrypsin are secreted.

A biotechnology company and Tufts University scientists have engineered transgenic goats to secrete tissue plasminogen activator in their milk. Elsewhere, plans are underway to produce in bovine milk a human milk protein with antibacterial and iron transport properties for infant formula. Proteins difficult to produce by other means may be produced in this way.

VI. Engineering Vaccines and Antibodies

Unlike traditional polyvalent vaccines, genetically engineered vaccines contain only one or a few major antigens of the pathogenic agent that are capable of producing a neutralizing immune response upon infection. Such subunit vaccines may not be as efficacious as those prepared from the whole organism, but they are generally more stable and safer to use. Diseases that affect a large proportion of the world's population have received high priority for production of recombinant vaccines. For example, in China, >85% of the population has been exposed to hepatitis B. The gene for the surface antigen HbsAg of the hepatitis B virus was cloned and, when expressed in yeast and used as an immunogen, generated a protective immune response. Human diseases such as measles, polio, tuberculosis, leprosy, and malaria may also someday be controlled or even eradicated by use of genetically engineered vaccines.

The great diagnostic and therapeutic potential of murine monoclonal antibodies is limited in humans by their species-associated immunogenicity. Genetic engineering approaches have been used to humanize these antibodies, that is by replacing the mouse antigen-combining regions (VH and VL) with human constant heavy and light regions. Such chimeric antibody molecules are not only less immunogenic in humans, but also retain the antigen specificity of the original monoclonal antibody. Another approach makes use of antibody fragments that retain antigen-binding properties. $F(ab')_2$, Fab, and Fv fragments of monoclonal antibodies have reduced (albeit not abolished) immunogenicity. However, their greater potential lies in their ability to penetrate target tissues better than intact antibody molecules. For this reason, antibody fragments are more effective diagnostic agents for applications such as radioimaging, where radioisotopes are conjugated to them. These are also more efficient immunotherapeutic molecules when conjugated to radioisotopes, natural toxins, or synthetic cytotoxic drugs. Eventually, it is hoped that the minimum amino acid sequence of an antibody capable on its own of binding to an antigen will be found and applied to these goals. Single-chain antibodies consisting of VH and VL regions covalently joined by a peptide linker have been shown to possess the same antigen-binding characteristics as the antibody from which it was derived. Carried to the extreme, short synthetic peptides whose sequences mimic those of the antigen-combining site of antibodies may soon find clinical application in this regard.

VII. Biopesticides: An Alternative to Environmental Pollutants

Mycogen's genetically engineered pesticides (MVP and M-TRAK) have been approved by the EPA for sale in the United States. To produce these biopesticides, Mycogen inserted the *Bacillus thuringiensis* (Bt) insecticidal protein gene into *Pseudomonas* (U.S. patent 5,002,765). The recombinant bacteria are fermented and then killed for application to crops. These killed cells encapsulating insecticidal proteins are deadly meal for caterpillar insects that attack cabbage, broccoli, lettuce, and other vegetable crops and against the Colorado potato beetle, which damages potato, tomato, and eggplant. Scientists have expressed the Bt toxin gene in the organisms that colonize plant roots.

VIII. Plant Biotechnology: Triumph of Genetic Engineering

Genetic engineering of plants is used to design crops to the specifications of the food industry. This new biology does not replace, but rather augments, conventional plant breeding techniques. Plant varieties with good agronomic characteristics can be further improved for better processing, solid content, flavor, color, texture, acidity, sweetness, size, and shape of edible parts. Other characteristics that can be modified include suitability for mechanical harvesting, self-storage, and resistance to disease, pests, insects, and frost. Genetic modification of plants offers possibilities for producing many specialty chemicals and pharmaceuticals in crop plants. Many monocotyledon and dicotyledon crop plants are amenable to improvement by recombinant DNA methods. Genetically engineered seeds of soybean, cotton, rice, corn, rape, sugar beet, tomato, alfalfa, wheat, and potato will be commercially sold by the end of this century.

Engineering herbicide tolerance into crops has been achieved by (1) altering the level and sensitivity of the target enzyme for the herbicide in the plant or (2) incorporating the gene that detoxifies the herbicide. Scientists have produced glyphosate (Roundup)-resistant plants by introducing a genetic construction for the overproduction of 5-enolpyruvylshikimate-5-phosphate (EPSP) synthase or of glyphosate-tolerant mutants of EPSP enzymes. Expression of the mutant acetolactate synthase (ALS) genes have produced sulfonylurea-tolerant plants. Bacterial genes that acetylate or hydrolyze gluphosinate and bromoxynil have been introduced to obtain resistant cultivars as well. Expression of the insect control protein gene of *B. thuringiensis* into tomato and cotton plants has made them resistant to several insect pests. Tobacco plants engineered to express a cowpea proteinase inhibitor are partially resistant to the tobacco budworm and further work may prove the benefit of such an approach.

Notable resistance to plant virus infections (alfalfa mosaic virus, cucumber mosaic virus, potato virus X, and potato virus Y) has been produced by expression of the coat protein gene of the respective virus in plants. Such an approach of constructing virus-resistant plants has serious possibilities for agriculture in the third world, where losses due to virus epidemics are immense. [*See* PLANT DISEASE RESISTANCE: NATURAL MECHANISMS AND ENGINEERED RESISTANCE.]

IX. Flavors and Fragrances: Nature's Identical Materials

One corporation intends to supply the flavor and fragrance industry with high-quality plant-derived flavors. They have used the technique of somoclonal variation to produce tobacco cultivars with high levels of the musky fragrance sclareol (U.S. patent 5,012,040). Sclareol and abienol are used to create ambrox, a fragrance needed in many products. Two researchers in Holzminden, Germany, have described microbial preparation of vanillin from eugenol and isoeugenol. Biochemically produced aroma chemicals already available are aldehydes and ketones, (acetaldehyde, diacetyl), acids (acetic, butyric, caproic, caprylic, isobutyric, isovaleric, 2-methylbutyric), esters (ethyl and butyl acetates, ethyl butyrate, caproate, isobutyrate, isovalerate, 2-methylbutyrate, menthylacetate), and lactones (λ-decalactone).

Specific activity of isolated enzymes or whole cells can convert precursors to valuable aroma chemicals—for example, oxidation of valence to nootkatone in the presence of citrus cell cultures and reduction of citronellal to citronellol by certain yeasts; fungal hydroxylation of patchoulol, a sesquiterpene in the patchouli essential oil to 10-hydroxypatchoulol followed by chemical oxidation/decarboxylation to the fragrance compound norpatchoulenol; the microbial oxidation of l-carvone, a spearmint flavor compound starting with α- or β-pinene. Long-chain, α/ω dibasic acids for synthesizing macrocyclic musks are already commercial products. *Streptococcus diacetylactis* produces diacetyl(2,3-butanedione), a flavor component of cultured dairy products. *Yarrowia lipolytica* hydrolyzes and then β-oxidizes castor oil to form γ-hydroxydecanoic acid. The latter is lactonized to γ-decalactone with a fruity peachlike aroma.

The role of lactococcal peptidases in the production of flavor peptides during cheese manufacturing is known. A supplier of industrial enzymes in Denmark has reported that natural geranyl acetate of high purity is produced by catalytic esterification of natural geraniol and acetic acid using specific lipases. Microbial esterases selectively hydrolyze the L-isomer in a racemic mixture of D,L-menthyl acetate, allowing separation of L-menthol from synthetic racemic mixtures. *Corynebacterium glutamicum* produces 3 g/liter of tetramethylpyrazine in glutamic acid fermentation. Pyrazines are powerful aroma compounds that occur naturally in food as a result of browning or heating. Biotechnology will expand the opportunities for producing difficult flavors and fragrances to synthesize chemicals. [*See* FOOD BIOTECHNOLOGY.]

X. The Future: New Products for a New Century

Where there is no vision, the people perish.

Prov. 29:18

Biotechnology has been rejuvenated by the technological advances in new biology. The last decade of the twentieth century will change promises into products since there are many products for human health in clinical trials. Many experiments are in progress to produce more compounds to benefit agriculture and animals.

Fusion of various metabolic pathways could yield molecules of commercial importance. The challenge would lie in producing molecules such as taxol and steroid hormones by microbial fermentations.

Sequencing of the human genome is already gaining momentum. Knowledge of the complete base sequence of the human genome will provide a blueprint of human life. DNA sequences that code for many receptors and other molecules that direct human metabolism, signal transduction, the nervous system, immunity, and aging will be eventually fathomed. This knowledge may be used for producing unlimited quantities of certain materials that exist in the human body only in minute quantities. Some of these may be of pharmaceutical value. For example, the availability of all genes involved in steroid metabolism may lead to better understanding of human reproduction and improvement of birth control. It may be possible to transplant human genes into yeast to produce therapeutic steroids identical to peutic steroids identical to those synthesized by humans. Furthermore, many molecules that are involved in stimulation of the immune system and pain suppression may be commercially produced.

Knowledge and availability of many receptors involved in modulation of living activities has already heralded a new era of innovation in the pharmaceutical industry. Many microbially produced ligands that could interact with these receptors may come out of screening programs. The future outlook is bright and the immense power of the new technologies is adding speed to solution of tasks that were unthinkable only a decade ago. The limiting factor is the human imagination and ingenuity.

Bibliography

Adams, M. D., Kelley, J. M., Gocayne, J. D., Dubnick, M., Polymeropoulos, Xiao, H., Merril, C. R., Wu, A., Olde, B.,

Moreno, R. F., Kerlavage, A. R., McCombie, W. R., and Venter, G. J. (1991). *Science* **252,** 1651.
Arnold, F. H., and Haymore, B. L. (1991). *Science* **252,** 1796–1797.
Bailey, J. E. (1991). *Science* **252,** 1668.
Berg, P. (1991). *Bio/technology* **9,** 342–343.
Berka, R. M., Kodama, K. H., Rey, M. W., Wilson, L. J., and Ward, M. (1991). *Biochem. Soc. Trans.* **19,** 681–685.
Bird, R. E., Hardman, K. D., Jacobson, J. W., Johnson, S., Kaufman, B. M., Lee, S.-M., Lee, T., Pope, S. H., Riordan, G. S., and Whitlow, M. (1988). *Science* **242,** 423–426.
Bloch, W. (1991). *Biochemistry* **30,** 2735–2747.
Borrebaeck, C. A. K., and Larrick, J. W. (1990). "Therapeutic Monoclonal Antibodies." Stockton Press, New York.
Brann, D. W., and Mahesh, V. B. (1991). *FASEB J.* **5,** 2691–2698.
Burke, G. T., Hu, S. Q., Ohta, N., Schwartz, G. P., Zong, L., and Katsoyannis, P. G. (1990). *Biochem. Biophys. Res. Commun.* **173,** 892.
Cenningham, B. C., and Wells, J. A. (1991). *Proc. Natl. Acad. Sci. USA* **88,** 3407–3411.
Chaplin, M. F., and Bucke, C. (1990). "Enzyme Technology." Cambridge University Press, Cambridge. U.K.
Clarke, A. R., Colebrook, S., Cortes, A., Emery, D. C., Halsall, D. J., Hart, K. W., Jackson, R. M., Wilks, H. M., and Holbrook, J. J. (1991). *Biochem. Soc. Trans.* **19,** 577–581.
Coombs, J. C., and Alston, Y. R. (1990). "The Biotechnology Directory." Stockton Press, New York.
Copsey, D., and Delnatte, S. Y. J. "Genetically Engineered Human Therapeutic Drugs." Stockton Press, New York.
Crease, R. (1989). *Science* **246,** 883–884.
Crueger, W., and Crueger, A. (1982). "Biotechnology." Science Technology, Inc., Madison, Wisconsin.
Culver, K., Cornetta, K., Morgan, R., Morecki, S., Aebersold, P., Kasid, A., Lotze, M., Rosenberg, S. A., Anderson, W. F., and Blaese, R. M. (1991). *Proc. Natl. Acad. Sci. USA* **88,** 3155–3159.
Darnell, J., Lodish, H., and Baltimore, D. (1991). "Molecular Cell Biology." W. H. Freeman and Co., New York.
Dibner, M. D. (1990). "Biotechnology Guide USA." Stockton Press, New York.
Erlich, H. A., Gelfand, D., and Sninsky, J. J. (1991). *Science* **252,** 1643.
Farbood, M. I. (1991). *Biochem. Soc. Trans.* **19,** 690–695.
Finberg, R. W., Wahl, S. M., Allen, J. B., Soman, G., Strom, T. B., Murphy, F. R., and Nichols, J. C. (1991). *Science* **252,** 1703.
Fraser, T. H., and Bruce, B. J. (1978). *Proc. Natl. Acad. Sci. USA* **75,** 5936–5940.
Gadani, F. (1990). *Arch. Virol.* **115,** 1.
Gill, G. S., Zawoski, P. G., Marotti, K. R., and Rehberg, E. F. (1990). *Bio/technology* **8,** 956–958.
Goodenough, P. W., and Jenkins, J. A. (1991). *Biochem. Soc. Trans.* **19,** 655–662.
Greenshields, R. N. "Resources and Applications in Biotechnology. The New Wave." Stockton Press, New York.
Habby, G. L. (1985). "Penicillin: Meeting the Challenge." Yale University Press, New Haven, Connecticut.
Hale, G., *et al.* (1988). *Lancet* **2,** 1394–1399.
Hall, M. J. (1989). *Bio/technology* **7,** 427–430.
Heitman, J., Movva, N. R., Heistand, P. C., and Hall, M. N. (1991). *Proc. Natl. Acad. Sci. USA* **88,** 1948–1952.
Huston, J. S. (1988). *Proc. Natl. Acad. Sci. USA* **85,** 5879–5883.
Isogai, T., Fukagawa, M., and Aramori, I. (1991). *Bio/technology* **9,** 188–191.
Jaenisch, R. (1988). *Science* **240,** 1468.
Joshi, R. L., Joshi, V., and Ow, D. W. (1990). *EMBO J.* **9,** 2663.
King, K., Dohlman, H. G., Thorner, J., Caron, M. G., and Lefkowitz, R. J. (1990). *Science* **250,** 121.
Kok, J. (1991). *Biochem. Soc. Trans.* **19,** 670–674.
Landisch, M. R. (1991). *Enzyme Microbiol. Technol.* **13,** 280.
Landt, O., Grunert, H., and Hahn, U. (1990). *Gene* **96,** 125–128.
Lillehoj, E. P., and Malik, V. S. (1989). *Adv. Biochem. Engin. Biotechnol.* **40,** 19–71.
Lillehoj, E. P., and Malik, V. S. (1991). *Adv. Appl. Microbiol.* **36,** 280–338.
Mahato, S. B. (1991). *Inform* **2,** 214–217.
Malik, V. S. (1981). *Adv. Appl. Microbiol.* **27,** 1–84.
Malik, V. S. (1982). *Adv. Appl. Microbiol.* **28,** 27–115.
Malik, V. S. (1986). Genetics of secondary metabolism. *In* "Biotechnology," Vol. 4 (H. J. Rehm and G. Reed, eds.), pp. G. Reed, eds.), pp. 39–68. Springer-Verlag, New York.
Malik, V. S. (1989). *Adv. Appl. Microbiol.* **34,** 263–306.
Malik, V. S. (1989). *Adv. Gen.* **20,** 37–114.
Malik, V. S., and Lillehoj, E. P. (1992) (in press).
McKusick, V. A. (1991). *FASEB J.* **5,** 12–20.
Meng, M., Lee, C., Bagdasarian, M., and Zeikus, J. C. (1991). *Proc. Natl. Acad. Sci. USA* **88,** 4015–4019.
Morihara, K., and Veno, Y. (1991). *Biotechnol. Bioeng.* **37,** 693–695.
Morrow, K. J. (1990). *Gen. Eng. News* **10,** 9.
Moss, B. (1991). *Science* **252,** 1662.
Moyler, D. (1991). *Chem. Ind.* **7 January 1991,** 11–14.
Mulholland, F. (1991). *Biochem. Soc. Trans.* **19,** 685–690.
Netzer, W. J. (1991). *Bio/technology* **8,** 618–622.
Nisbet, L. J. (1982). *J. Chem. Technol. Biotechnol.* **32,** 251–270.
Okafor, N. (1985). *Proc. Biochem.* **February 1985,** 23–25.
Pitts, J. E., Quinn, D., Uusitalo, J., and Penttila, M. (1991). *Biochem. Soc. Trans.* **19,** 663–665.
Potera, C. (1990). *Gen. Eng. News* **10**(6), 3.
Pratt, A. J. (1989). *Chem. Britain* **March,** 282–286.
Primrose, S. B. (1987). "Modern Biotechnology." Blackwell Scientific Publications, Oxford.
Rehm, H. J., and Reed, G. (1992). "Biotechnology," 12-vol. reference work VCH. Weinheim, Cambridge.
Roberts, N. A., Martin, J. A., Kinchington, D., Broadhurst, A. V., Craig, J. C., Duncan, I. B., Galpin, S. A., Handa, B. K., Kay, J., Krohn, A., Lambert, R. W., Merrett, J. H., Mills, J. S., Parkes, K. E. B., Redshaw, S., Ritchie, A. J., Taylor, D. L., Thomas, G. J., and Machin, P. J. (1990). *Science* **248,** 358–361.
Roth, S. (1991). *Gen. Eng. News* **11**(2), 5.
Safer, P., Ahern, T. J., Angus, L. B., Barone, K. B., Brenner, M. J., Horgan, P. G., Morris, G. E., Stoudemire, J. B., Timony, G. A., and Larsen, G. R. (1991). *J. Biol. Chem.* **266,** 3715.
Schimmel, P. (1990). *Biochemistry* **29,** 9495–9502.
Seetharam, R., and Sharma, S. K. (1991). "Purification and Analysis of Recombinant Proteins." Marcel Dekker, New York.
Shamel, R. E., and Chow, J. (1990). *Gen. Eng. News* **10**(3), 4.
Siegel, B., and Brierley, R. (1989). *Biotechnol. Bioengin.* **34,** 403–404.

Bread

Richard A. Ledford
Cornell University
Martha W. Ledford

Glossary

Gluten Protein in moistened and worked dough that imparts elastic properties

Saccharomyces cereviseae Yeast species important in the fermentation of bread and other industrial fermentations

Triticum aestivum Species of wheat used for most breadmaking

BREAD has been called the "Wings of Life." This description aptly denotes the place of bread in human nutrition. This article gives some of the vast history of bread, indicates the major ingredients used in breadmaking and discusses their functions, describes the basic operations involved in making bread, characterizes several types of breads, and indicates briefly the role of bread in the diet. Emphasis is placed on the microbiological aspects of breadmaking.

I. Brief History

An important discovery made very early in history was that cereal grains could be sun-dried and stored for long periods of time, months and years, without spoiling. The very earliest breads probably made use of unmilled grain mixed with water to give a coarse mash, and they were most likely sun-dried. Next, someone found that milling the grain into a crude powdery flour and mixing it with water followed by baking improved bread. These early breads were all unleavened and must have resembled poorly made pie crust. The Egyptians are credited with discovering bread fermentation, a discovery that revolutionized breadmaking. Although not known, this important discovery was probably by accident. Dough was probably mistakenly allowed to sit for several hours or longer and, rather than discard it, the aged dough was baked producing a spongy light loaf. News of this great discovery spread to other countries, especially to Italy where the fermentation process was improved considerably. The Romans later removed yeast from the surface of fermenting wine and used it to leaven bread. This method of yeast preparation, called barm, gained acceptance and was practiced in other countries including Britain. *Saccharomyces cerevisae* was later identified as the species of yeast important in bread fermentation, and today a few manufacturers supply preparations of this yeast to bakeries. Recent developments in yeast strain selection and development have led to new yeast products for use in baking.

II. Major Ingredients

The major ingredients used in breadmaking include wheat and flour, leavening agents, milk, eggs, shortenings, sugars, salts, flavors, and spices. The use of water, a vital component of doughs, to partially dissolve other ingredients of a dough is essential. Although the ingredients used in breadmaking vary to a small extent, the resulting baked products may be very different. The amount of salt, the type of flour, and oven conditions are just some of the determining factors. However, the type of yeast and fermentation conditions are the most significant controlling factors in the production of breads.

The wheat kernel is composed of several major parts significant in breadmaking: germ, endosperm, and bran (Fig. 1). The germ is the seed part and

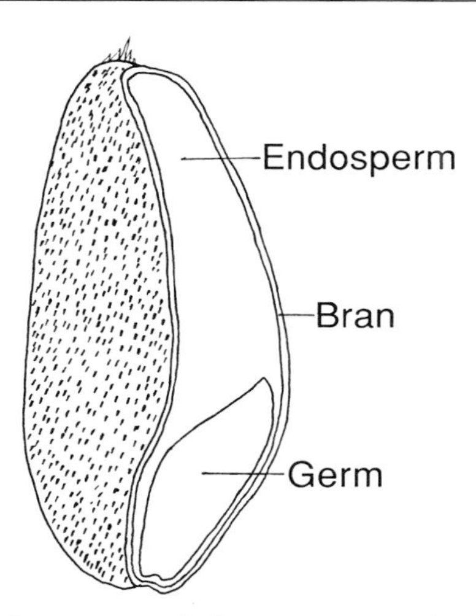

Figure 1 Components of wheat important in breadmaking.

consists of fatty substances. The endosperm is an important component in breadmaking because it contains the fine starch particles and gluten-forming proteins. Bran is found at the surface of wheat kernels, called wheat berries, and it consists of several hard layers that protect the endosperm.

There are several types of wheat including hard red spring, durum, hard red winter, soft red winter, and white. "Hard" and "soft" describe the consistency of the kernel and flour; both are used in different applications in baking. Bread and rolls are made from hard flours, whereas soft flours are used for making delicate items such as cakes, pastries, and cookies. "Winter" and "spring" indicate the planting seasons of the wheat. Thus, winter wheat is grown in relatively warm climates whereas spring wheat is cultivated in colder areas. Approximately 70% of the wheat-growing acreage in the United States is used for the hard type, which has a high percentage of good-quality protein. Milling of the hard types is generally easy and produces good yields of flour. *Triticum aestivum* is the species of wheat that contains the hard red spring, hard red winter, soft red winter, and white wheats (winter and spring).

Wheat is washed, cleaned, and tempered by adding moisture to a standard level before it is milled. Removal of bran and flaking out of germ particles is facilitated by the tempering process. Soaking wheat with water causes the bran to become tough and rubbery while the internal components remain soft and friable. Milling consists of conveying wheat through corrugated rollers, which causes cracking and the release of some endosperm and flour, called break flour. This breaking separates wheat into bran, break flour, and a component termed middlings, which is conveyed through smooth rollers to separate additional germ in flake form. This operation is repeated several times to remove fatty portions and bran particles, termed shorts, which are used for special foods and animal feeds. The flour that remains after impurities are removed is called straight flour and, by sifting, it is further separated into patent flour, the best portion, and two other fractions referred to as first and second clear. The sifting operation utilizes several mesh sizes to separate the fine and coarse particles. In general, the finer portions are derived from the center of the endosperm and contain high-quality gluten. Coarser particles are from areas closer to the bran and are of darker color. Coarseness is characteristic of clear flour, which is often used for rye bread. On the other hand, patent bread flour contains fine starch particles. The average composition of patent bread flour is moisture (12%), protein (12%), carbohydrates (starch, sugar, dextrins; 75%), fat (1%), and mineral matter (0.5%).

Whole wheat flour is prepared by cleaning, grinding, and sifting the entire wheat berry. The higher fat content of this flour contributes to a shorter storage life because of the possible development of a rancid flavor. Rye flour is milled from rye. Pumpernickel flour is milled from the whole rye berry. Bran flour has more bran flakes, and gluten flour has more of the starch removed and contains approximately 47% protein and 53% carbohydrate.

Freshly milled flour is referred to as green flour. Natural aging of flour causes disappearance of a slightly yellowish color and changes in the proteins. Scientists have found that bleaching with compounds such as chlorine dioxide accelerates the aging process. Cake flour is treated with chlorine to soften the gluten and bleach out the color. Bread flours may be subjected to bromates to improve the bread texture.

III. Functions of the Major Ingredients

Primary components of flour significant in breadmaking are gluten and enzymes. Gluten is a mixture of wheat proteins, which, when mixed with water in the proper amounts, forms an elastic dough capable of retaining carbon dioxide produced by fermentation. Gluten is composed of almost equal amounts of two proteins, glutenin and gliadin. Glutenin provides strength to the dough to retain carbon dioxide and imparts the spongy structure to bread. Gliadin contributes the elastic characteristics to the dough. Amylases and proteases are the main enzymes of flour. Amylases degrade some of the starch to dextrins and maltose, which is important for the fermentation of doughs, especially those in which no sugar is added. Protease changes some of the proteins of doughs into a soluble form, which contributes to the elasticity of doughs.

Saccharomyces, the most important genus of yeasts involved in breadmaking, has been described as "the oldest plant cultivated by man." Wall carvings dating back to 2300 B.C. depict the use of yeast in breadmaking. Yeasts are used to produce gas, essentially carbon dioxide, and flavor and aroma in breads. Active enzymes present in yeasts cause changes mostly in carbohydrates during the fermentation of the dough forming carbon dioxide and alcohol, which evaporates during baking.

In nature, yeasts are found widely, including on the surfaces of fruits and cereal seeds. They are also present in the soil and in waters. Yeasts are easily spread by wind currents, insects, and animals and, thus, are ubiquitous in nature.

The industrial importance of yeasts is due to their fermentative and oxidative respiration reactions. The Embden-Meyerhof-Parnas (EMP) pathway and the hexose monophosphate shunt (HMS) are the fermentative mechanisms present in yeasts. The EMP pathway, in the absence of oxygen, accounts for >95% of the glycolytic metabolism of bakers' yeast whereas the HMS, under aerobic conditions, accounts for 6–30% of yeast glycolysis. Air stimulates oxidative respiration and growth of yeast, and it is used in the large-scale production of yeasts for fermentation, including that in breadmaking.

Not only are active yeasts used in fermentation, but inactive yeasts, referred to as dried yeasts, are sources of nutritional and flavor compounds. Bakers' yeast is commercially available in several active forms: compressed, active dry, and instant active dry. These forms differ in the level of activity and stability in storage. The compressed form is produced with minimal processing after yeast culturing and therefore is the most active and consistent form in quality. The baking industry has widely used this form over the years.

An important criterion for commercial yeast is acceptable resistance to drying. As with any microorganism, there is appreciable loss in activity during drying; however, some yeast strains are more resistant than others. Manufacturers of commercial yeast, with better strains and improved drying technology, have developed highly active dry yeast products for different baking applications. These more recently developed yeast products are known as instant or quick-rising yeast. The improved drying technology produces yeast particles that are very porous, a property that imparts the ability to rehydrate rapidly. While rapid rehydration is desirable, the yeast is more exposed to oxygen because of the porous structure, thus causing some loss of activity during storage. Therefore, these instant-rising yeast preparations are marketed in packages that are hermetically sealed with the inclusion of inert gas or vacuum. In terms of the important consideration of activity, compressed yeast, among the various earlier available forms of commercial yeasts, is considered to have the preferred level of activity. Instant-rising yeast has a level of activity that is very close to that of compressed yeast. It also has the desirable property of being dry, which facilitates storage and shipping.

In recent years, improvements of industrial strains of yeast have included the following methods: selection and isolation of single cells, selection and isolation after mutagenesis, and the production of new hybrids. Advances in yeast genetics have led to the establishment of breeding collections of known desirable characteristics. Also, yeast gene banks have been established with the DNA of appropriate strains. These have made possible the construction of industrially desirable strains.

Isolation efforts were advanced by the development of pure culture techniques introduced first in the brewing industry. These methods were extended to the baking industry and enabled the isolation and identification of single cells from yeast populations. The objective was to isolate strains that had desirable characteristics for application in baking. For many years, isolated yeast strains were used in the baking industry. However, improvements were needed and advances in the use of mutagens and genetic techniques were applied to bakers' yeast. [*See* ISOLATION.]

Although mutagenesis of bakers' yeast has been studied in some countries, notably the Soviet Union and Poland, this approach for improving commercial strains has not been widely used in the United States. Agents for mutagenizing yeast include ultraviolet (UV) light, *N*-methyl-*N*′-nitro-*N*-nitrosoguanidone, ethyl methane sulfonate, nitrous acid, *N*-nitrosourea, and diethylystilbestrol. Among these agents, UV light appears to have been used most. Nucleic acids absorb UV light, causing changes in the bonds of purines and pyrimidines. Covalent bond formations between adjacent pyrimidines in the same DNA molecule is a result of UV exposure to yeast cells. This interferes with base pairing and explains the mutagenic effects of UV light, which has the potential of improving yeast characteristics for application in baking.

Hybridization techniques have been used by the industry to produce yeast strains with more desirable characteristics. Techniques involving mass-matings involve mixing large numbers of haploid yeast cells of opposite mating types. Subculturing a mating mixture several times encourages hybrid cells to outgrow haploid cells. Micromanipulators are used to isolate single yeast cells, which, after propagation in a suitable medium, are screened for desirable baking characteristics.

Techniques involving protoplast fusion were used in the more recent past to further contribute to the improvement of yeast strains. The development of this approach was enhanced by the isolation of enzymes to digest the rigid polysaccharide cell wall material from yeast cells. Protoplast fusion has been employed to develop yeast strains that have improved production of carbon dioxide, resulting in better leavening characteristics. In addition, tolerance to higher levels of osmotic pressure, caused by increased amounts of sugar in sweet dough, has been constructed in yeast strains through protoplast fusion.

During recent years, advances in studies of genetic recombinations have been evident in yeast research. The application of recombinant DNA techniques has contributed to our understanding of the mechanism of expression of a number of genes in yeast cells. Genetic improvements of commercial yeast strains are now feasible through the application of these modern methods of biotechnology and are particularly useful in changing small segments of the yeast genome that involve one or two genes. Scientists have yet to gain a complete understanding of the yeast genetics that are important in desirable baking characteristics. [*See* RECOMBINANT DNA, BASIC PROCEDURES.]

One characteristic that scientists are interested in improving in yeast to make it more attractive for baking is rapid maltose-fermenting ability. Maltose is produced by enzymes, naturally present in flour and dough, and is the predominant fermentable sugar, especially in dough formulations with little or no added sugar (lean doughs). Yeast usually requires an induction period before it utilizes maltose, and this lag results in a prolonged period of fermentation. Using genetic approaches, scientists have been successful in constructing strains that have constitutive enzymes that enable them to ferment maltose.

Research on bakers' yeast has sought to improve the tolerance of yeast to several physical conditions including high osmotic pressures and freezing. Dough formulations with added sugar (>4%), as mentioned earlier, usually will have diminished leavening activity because of the inhibition of yeast growth, in part because of the increased osmotic pressure conditions of the dough. Greater amounts of yeast are needed to compensate for less yeast activity in these types of dough, and such an increase is expensive. Research results indicate that osmotolerance is achieved when a yeast strain has the ability to accumulate glycerol inside the cell and produce a high amount of trehalose and a low amount of invertase. Although progress is being made, the genetics of osmotolerance in yeast is not completely known. Over recent years, increasing numbers of supermarkets have developed in-store bakeries, which, for the most part, are bake-off operations. Frozen dough is often purchased for these bake-off operations to provide significant savings in time, labor, and maintenance. This development has stimulated the interest of frozen dough manufacturers to improve the performance and quality of their products. While, in general, freezing injures yeast cells, some conditions of freezing have been found to be more detrimental than others. Generally, research has indicated that the best results are obtained when doughs are frozen to −10°C rather than to a higher temperature, with carefully controlled cooling and thawing rates. Studies on strain improvement using techniques such as hybridization as well as research on optimizing rates of freezing and thawing of dough are contributing to improved functionality of frozen doughs. Indeed, the prospects for improving yeast strains through the application of recombinant DNA technology to enhance a variety of characteristics, including thermotolerance, is exciting.

A number of enzymes, mostly produced by bacteria and fungi, are commercially available for use by the baking industry. These include α- and β-amylases and proteases, the most important in improving baked products, as well as amyloglucosidase, pentosanase, glucanase, and phytase.

Amylases, also natural constituents of wheat

flour, catalyze the conversion of starch in the flour to maltose, which is used by the yeast in the fermentation. Alpha- and β-amylases produce dextrins and maltose, respectively. Staling is also retarded by amylases. The proper amount of amylases is needed to ensure sufficient formation of maltose, the main fermentable sugar. Too much α-amylase in the flour causes texture defects because of excessive dextrin formation. Too little amylase activity is likely to result in the defect of low loaf volume. To avoid these possible defects, the flour miller adjusts the α-amylase in the flour, and the baker may even standardize it further using commercially available enzyme. Staling of baked goods, a major problem in the United States, is caused by a shift in the distribution of water that accompanies a change in the starch molecules from a soluble to an insoluble form. By cleaving a few bonds in the starch molecules, α-amylase can retard staling for several days. Protease additions to dough containing high amounts of gluten improve bread quality as well as assist in the handling of these types of dough during the baking process. Peptide bonds of gluten are hydrolyzed by proteases, resulting in reduced elasticity and improved extensibility of the dough. The use of machines in the baking process of some products is facilitated by the application of proteases, which, by liberating amino acids, promote the Maillard reaction with sugars and yeast growth. Flavor and crust color are improved by the action of proteases.

Polysaccharides, consisting of the pentose sugars D-xylose and L-arabinose, are present in wheat flours in significant concentrations. Beneficial effects on bread quality are produced when these pentosans are hydrolyzed. The addition of pentosanases improves dough handling as well as the loaf volumes of the finished products. The improvements are also evident when pentosanases are used in the manufacturing of reduced-calorie, high-fiber breads.

Salt is an essential ingredient in numerous food fermentations. In bakery products, it serves several functions. Salt contributes to flavor itself and accentuates ingredients. In most food fermentations, salt controls the growth of microorganisms, desirable and undesirable, and, thus, the rate of fermentation and spoilage. In bakery products, salt has a strengthening effect on the gluten of the dough, and it influences the crust color of yeast-leavened products.

Milk in various forms (fluid, dried, buttermilk) is frequently used in bakery products. The major protein of milk, casein, increases the absorptive and water-retention properties of dough. Milk adds firmness to dough, especially when a softer type of flour is used.

IV. Basic Operations in Breadmaking

The basic operations of breadmaking include the mixing of ingredients, dough fermentation, dividing and rounding, proofing, and baking. Numerous variations of these operations exist in baking technology as practiced in different bakeries and in homes.

Mixing of ingredients, the first major operation, is carefully done not only to uniformly blend the ingredients that make up the dough but also to develop the physical structure of gluten into a continuous network of thin, hydrated protein films. Two main methods are used: the straight dough and the sponge and dough methods. The straight dough method, used mostly in small bakeries, involves mixing all of the ingredients in a single batch. The mixing is continued until the dough has a smooth appearance and the optimum elastic characteristics. The sponge and dough method, widely used in the baking industry, involves mixing more than one-half of the total dough flour, yeast, yeast food, some of the other ingredients, and enough water to produce a fairly stiff dough. A major part of the fermentation time (a period of about 4 hr, depending on the amount of the total flour used) takes place in this sponge, after which the remainder of the dough ingredients are added. After mixing to the optimum condition of the dough, a second fermentation takes place for a period of up to 1 hr. Mixing in the proper amount, sufficient to attain the optimum rheological properties of the dough, is crucial to the success of the baking process. When the dough has reached the peak of development, its appearance will be silky and dry, with no wetness or roughness. At this point, the dough is capable of being stretched into a thin film without breaking.

Fermentation of the mixed dough for the proper length of time is the next operation during which the yeast uses the available sugars to produce primarily carbon dioxide and alcohol. During the fermentation, dough volume increases several times the original size, and the gluten becomes more elastic due to several factors, including the action of proteolytic enzymes and the formation of various end-products of the fermentation. As the gluten becomes more elastic, it can better retain carbon dioxide in the dough. Ideally, gas retention capacity and gas production activity peak at the same time, a condition that contributes to the highest quality in the finished product. During the fermentation, the dough should be in the optimum environment with regard to temperature (approximately 27°C) and relative humidity (75%).

Dividing and rounding the fermented dough, the next operation, involves cutting the dough into pieces of the proper size, rounding them to produce a surface that will retain carbon dioxide, degassing, and molding the pieces into loaf shapes for placement in pans and proofing. In most commercial bakeries, these operations are done by machines, and it is important that they be done as rapidly as possible to ensure uniformity of weights of the different pieces. Removal of carbon dioxide enables the baker to obtain more uniform pieces in terms of grain and texture in the finished product. The dividing operation produces dough pieces that have lost much of the carbon dioxide. The objective of the rounding step is to produce a continuous surface skin that will retain carbon dioxide, which continues to be produced by the yeast. At this time, application of dusting flour assists in eliminating the dough's stickiness.

Next, intermediate proofing is a brief period (approximately 10 min) of fermentation during which the dough regains part of the gas lost earlier and becomes more elastic and pliable. The temperature is controlled (27–29°C) as well as the humidity (75%) during this period. Molding, the next operation, involves transforming the round dough pieces into thin, oval dough sheets, under pressure rollers in bakeries, producing the greatest degree of carbon dioxide expulsion and redistribution of gas cells to create fine, uniform grains in the finished loaf. Next, molding involves shaping the dough into sizes proper for baking pans.

The objective of the final proofing operation is to produce dough that has the optimum amount of carbon dioxide as well as mellowness and extensibility. This is necessary because of the physical treatment the dough receives during earlier operations in commercial bakeries. The final proofing, often termed pan proofing, allows the dough to recover its carbon dioxide content. Conditions of temperature (approximately 35°C), humidity (85%), and time (approximately 1 hr) are carefully controlled during this proofing period. Generally, however, the proofing time is determined by the height of the dough rise within the pan, which is usually determined by each commercial bakery.

The final operation is the important baking process. In commercial bakeries, pans with dough are conveyed through a series of four or more oven zones that vary in temperature from approximately 200° to 230°C. The baking time varies with the loaf size but is generally in the range of 20–22 min.

During the baking process, the biological activities of the dough that continue through proofing are inactivated. Early in baking, a thin surface film is produced. It continues to expand until the rising temperature and reduction in available moisture of the dough inhibit further expansion. A physical phenomena that occurs during the baking process is termed oven spring. It causes the sudden expansion of the dough volume by approximately one-third. The heat of baking increases the pressure and decreases the solubility of carbon dioxide, which contributes to the expansion. In addition, heat vaporizes the dough liquids with low boiling points, primarily alcohol, which causes further increases in the gas pressure increasing the expansion. These are the main forces responsible for oven spring. In addition to these physical forces, yeast activity continues to produce carbon dioxide until the thermal death point (approximately 60°C) is reached. Significant changes in the physical condition of gluten and starch occur during the baking process. Moisture migrates from the gluten phase to the starch phase, which then promotes starch gelatinization or swelling. This gelatinization accelerates the activity of amylases and continues until heat inactivates the enzymes. Substances such as caramelized sugars and various carbonyl compounds that impart the desirable flavor characteristics are formed during the baking process.

The preceding description of the basic operations in breadmaking reflects those primarily practiced by commercial bakeries. Although numerous variations are used in the industry, the basic principals of different operations are essentially the same.

V. Types of Bread

A. Unleavened Bread

The early history of bread refers to flat cakes made by the Hebrews, Chinese, and Egyptians. Made basically from solids, mostly flour, and liquids, predominantly water, unleavened breads are still important today, mostly in specialized uses. Dried fruits are frequently included in this type of bread as are nuts and seeds. A wide variety of these products are made chiefly in homes.

B. White Bread

White breads, the primary types made by the technology described earlier, are most often produced from finely ground flour. In the United States today, approximately 95% of all baking is done commercially, with white, enriched, sliced, and wrapped bread being the most commonly produced item. The National Research Council in the United States recommended some years ago that bread made from refined flour be enriched with thiamine, niacin, riboflavin, and iron to the levels that these nutrients are lost during processing the wheat. This enrichment program is a factor in bread being an important food in improved human nutrition in this country.

C. Rye Bread

Commercial bakeries produce rye bread in numerous varieties to meet the needs of different consumers. Variations in flavor, shape, and color of the crumb are among the main differences. Not only does the baker of rye bread use the basic ingredients of wheat and rye flours, water, yeast, and salt but additional optional ingredients including potato flour, molasses, sugar, caramel color, buttermilk, and others may be used to produce rye bread with the greatest appeal to consumers of a particular region.

Rye bread production technology is essentially the same as that for white bread. A basic difference between rye and wheat flours is that rye proteins do not form the type of gluten on mixing with water that wheat proteins do. This is a factor in less carbon dioxide retention in rye dough; therefore, there is a dependency on wheat flour as an ingredient to obtain an aerated loaf.

The sourdough method is usually used in the production of rye bread. Leavening, acidification, and flavor development are the primary functions of sourdough methodology. Lactic acid bacteria produce the acidification, with lactic and acetic acids being the main acids produced. Yeast indigenous in the sour dough produces the carbon dioxide for leavening; bakers' yeast is frequently supplemented in the final dough. *Lactobacillus brevis, Lactobacillus plantarum,* and *Lactobacillus fermentum* are the primary bacteria associated with sour doughs. *Lactobacillus brevis,* thought to be the principal producer of the typical rye flavor, is heterofermentative in that it produces lactic and acetic acids as well as carbon dioxide as the main products.

D. French and Italian Breads

Similar production methods are used for both French and Italian bread varieties. The flour used generally is a strong protein flour with a protein level of 13–15%. Time allowed for mixing of both the sponge and dough is limited. Normally, sponge fermentation is for 4–5 hr. Handling of the mixed, fermented dough is generally the same as that of white bread. The shape of the product is usually that of an elongated loaf. The final proofing operation is approximately 45 min at 35°C and 85% humidity. Baking is normally directly on the oven hearth, producing a loaf that is light with a thin, crisp crust.

VI. The Role of Bread in the Diet

Bread is a relatively low-cost source of the essential nutritional components in a well-balanced diet. Nutritional programs that have been conducted with adequate nutrition at reasonable cost as their goal have emphasized the consumption of bread. In such surveys, it is interesting to note that less emphasis is placed on types of bread such as white or brown than on the quantity. Some or all of the needed valuable nutrients are present in bread: water or other liquids, carbohydrates, proteins, lipids, and certain minerals and vitamins.

Water is a primary ingredient in breadmaking; therefore, it is important that the baker understand the chemical and biological composition of dough waters, because success or failure of the finished bread product depends in large part on water quality. It is known that hard waters retard fermentation, whereas excessively soft waters have been found to soften the gluten and yield a soft sticky dough.

Bread enriched with thiamine, niacin, riboflavin, and iron supplies a significant amount of nutrients needed for good health. It is, indeed, the "Wings of Life."

Bibliography

Dziezak, J. D. (1991). *Food Technol.* **45,** 78–85.
Dziezak, J. D. (1987). *Food Technol.* **41,** 104–121.
Jordan, J. (1976). "Wings of Life." The Crossing Press, Trumansburg, New York.
Lamlin, J. (1988). *Country J.* **15,** 35–38.

Nagodawithana, T. W., and Trivedi, N. B. (1990). Yeast selection for baking. *In* "Yeast Strain Selection" (C. J. Panchol, ed.), pp. 139–184. Marcel Dekker, New York.

Pomeranz, Y., and Shallenberger, J. R. (1971). "Bread Science and Technology." The Avi Publishing Co., Westport, Connecticut.

Pyler, E. J. (1973). "Baking Science and Technology," Vols. 1 and 2. Siebel Publishing, Chicago.

Sheppard, R., and Newton, E. (1957). "The Story of Bread." Routledge and Kegan Paul, London.

Trivedi, N. B., Cooper, E. J., and Bruinsma, B. L. (1984). *Food Technol.* **38,** 51–57.

Careers in Microbiology

Alice G. Reinarz
University of Texas at Austin

Glossary

Biochemistry and biophysics Chemical and physical techniques used to study structure and function of biomacromolecules; study mechanism and regulation of molecular biosynthesis, energy conversions, and assembly of cellular structures

Genetics Study of transmission of heritable information in organisms; includes topics such as organization and function of DNA, recombination, and regulation of gene expression

Microbial ecology Study of the interactions among microorganisms and impact of microorganisms on their environment; includes coevolution of species in natural ecosystems; study of distribution and abundance of microorganisms in nature

Microbiology Study of organisms so small that visualization requires a microscope; uses microorganisms as models for basic science and studies their clinical significance in human, animal, and plant disease, with applications in physiology, genetics, structural studies, development, and population growth; models include bacteria, fungi, viruses, and cells of the immune response; industrial applications involve making products of value and bioremediation

Molecular biology Application of technologies to study and manipulate biomacromolecules; utilized to study biological phenomena at the molecular level; techniques with which genes can be purified, sequenced, changed, and introduced into cells; provides an integrated experimental approach to problems in genetics, biochemistry, and prokaryotic and eukaryotic cell biology

WITH EXCITING NEW PROGRESS in the science of microbiology, many varied career options have developed. It is expected that there will be too few scientists for positions in the future; hence, jobs for well-trained microbiologists should be widely available. Bachelor's-, Master's-, and Doctoral-level education in microbiology is obtainable at institutions across the United States. Career options for professional microbiologists are found in academic, industrial, and governmental settings. Many microbiologists also choose to work in the health professions. Salary expectations vary with degree of training, location of the job, and position. Professional information is available from organizations such as the American Society for Microbiology and from many publications.

I. Why Choose Microbiology?

In seeking a career, one important consideration is the stimulation and satisfaction that is likely to derive from the major work interest. For scientists, challenges lie in exploring the unknown, participating in developments that may ultimately benefit humanity, and being part of an enterprise that is truly a worldwide collaboration. Microorganisms are recognized as some of modern biology's most useful models, and microbiologists now work with powerful instruments and techniques, some of which were developed only recently. There is a sense of excitement and enthusiasm among microbiologists as the science enters an unprecedented age of discovery. In fact, microbiologists are encouraged to recruit actively bright people into the field. [*See* HISTORY OF MICROBIOLOGY.]

An element in choosing a discipline is availability of jobs. The well-documented lack of students selecting education in science and engineering is one of the factors suggesting ample job opportunities in the Natural Sciences including microbiology. Despite strong interest in science and engineering expressed in high school, many students are lost to other areas of study. Only a small fraction finish science degrees. Tracking persistence of interest in science and engineering from high school through Ph.D. degrees shows a dramatic decline. University studies on choice of major field of study have demonstrated that a high proportion of college freshmen, after first declaring interest in natural science, may switch to other areas. Students first choosing business administration or communications switch with less frequency, however.

Exit of students from the science professional pool comes at a time when the job opportunities for scientists are increasing. The result is the predicted shortfall of degree holders in science and engineering. The estimate is a shortfall of 450,000 scientists, engineers, and mathematicians (at the Bachelor's level) by the year 2000. In fact, employment projections developed by the Bureau of Labor Statistics suggest that the number of jobs in private industry for scientists and engineers is expected to increase significantly. For more information on education trends and job availability, the National Science Foundation (for address, see Appendix A) has many useful publications.

Because of the concern that too few people, and particularly underrepresented groups such as women, ethnic minorities, and disabled persons, are choosing science careers, many programs have been designed to attract students. Some programs provide grants to groups, schools, and colleges to reach out to students (K–12) and encourage their study and career interest in science. Summer institutes for science teachers and talented high school students are supported at some colleges. Both pre- and postdoctoral fellowships are available to underrepresented groups. (For names and addresses of representative programs, see Appendix A.) For example, colleges and universities have listings of scholarships to enable minority students to finance their education. The admissions and financial aids offices of various institutions can supply information on opportunities. Relatively more funds are available to graduate than to undergraduate students. Corporations are involved in sponsoring students in university programs, and those particular students are often recruited to industrial jobs after finishing their education. Many universities have special student services directed toward the retention and success of qualified students from underrepresented groups. Information is available from institutions through the Dean of Students or Office of Student Affairs.

Because of its pivotal position among the cellular and molecular sciences, microbiology provides training for many subdisciplines. Microbiologists have backgrounds appropriate to pursue biochemistry, molecular biology, genetics, and cell biology. In addition to the well-recognized professional opportunities for microbiologists, forecasts of the impact of new technology on emerging occupations for the future suggest that many positions will become available in environmental research/applications and with biotechnology firms.

II. Educational Background for Microbiology

Education in microbiology is available throughout the United States. A list of the 374 institutions that grant degrees in microbiology or in an area of the life sciences with emphasis on microbiology is available for a small charge from the Publication Sales Office of the American Society for Microbiology (for address, see Appendix A). It is possible to pursue training in microbiology in virtually all regions of the United States.

To become a microbiologist, choices of appropriate classes should begin in high school so that the student will have a background in sciences (biology, chemistry, physics) and mathematics. In addition, the high school student is well advised to develop skills in logical thinking and in oral and written presentation of arguments. Experience in public speaking, computer skills, and management of group activities will be helpful in the workplace. The high school student should plan to take the Scholastic Aptitude Test (SAT) or American College Test (ACT) and, for some schools, College Board Achievement Tests in selected subjects. This must be done in ample time to apply to the college or university of choice. The school's Admissions Office will send free information regarding the application procedure. These inquiries should begin at least by the junior year in high school. Early attention to test-taking and investigating various colleges is critical to finding scholarship support.

The length of time that one is willing to commit to education and training is a factor in career options. For students who effectively devote full-time effort to college, the Bachelor of Arts (B.A.) or Bachelor of Science (B.S.) will require approximately 4 years. The Master of Arts (M.A.) or Master of Science (M.S.) typically requires 2 additional years. Some university programs do not require a Master's degree if the student chooses to work directly for the Doctorate. There is greater variability at this level, but the B.A. or B.S. degree holder can expect to spend 4–7 years, or sometimes more, in gaining the Doctorate of Philosophy (Ph.D.). A main requirement for the Ph.D. is a substantial, independent research project. Other advanced degrees that may be sought include Doctorates in Medicine (M.D.), Science (D.Sc.), Public Health (D.P.H.), and Veterinary Medicine (D.V.M.). Following the doctoral degree award, many scientists choose to work with a leader in their field of choice for specialized training. This "postdoctoral" work generally lasts 1–4 years in the laboratory of the mentor.

Students planning graduate study should be aware that there is considerable competition among schools for the best candidates. Various levels of support are available. Descriptions of graduate programs are found in "Peterson's Graduate Programs in the Biological and Agricultural Sciences." Having chosen several possible institutions, the student can write the Graduate Adviser or Graduate Studies Committee Chairman for specific program and support information. Particularly at the graduate level, women, minority, and disabled students should seek information on unique opportunities through the National Science Foundation and American Association for the Advancement of Science (for addresses, see Appendix A).

For occupations such as directing a research or clinical lab, the Doctorate degree will be required. The lab director will be responsible for the productivity and effectiveness of the group. The doctoral degree holder will have the greatest latitude in choice of employment options. The Master's degree is generally held by individuals to whom considerable responsibility is given in the daily operation of a laboratory. These individuals have some role in charting the direction and emphasis of their work. Holders of Bachelor's-level degrees have the least freedom of choice in their daily work activity but also have less of the stressful responsibility that accompanies the lab operation.

Students are often worried about choosing whether to pursue a B.A. or B.S. degree. In most universities, the programs have many courses in common. The more traditional B.A. degree usually requires four semesters of foreign language. The B.S. degree may require one less semester of foreign language but may have additional calculus, computer science, and/or laboratory component plus some additional microbiology course requirements that the B.A. does not have. The B.S. degree is popular at present among students who plan to pursue graduate work. Choosing a good variety of microbiology classes and attaining a strong background in chemistry, physics, and math is probably more significant than deciding between the B.A. and B.S., however.

For students planning to enter graduate school, experience gained by working one or two semesters in a research lab is highly recommended by faculty advisers. These opportunities are available through organized (often "honors") courses by faculty mentor invitation. Additionally, many universities and medical schools sponsor summer programs for enrichment. These summer experiences are valuable to students, and the institutions accept only outstanding candidates. Some programs are designed for excellent minority students. Modest stipends to students are provided.

More opportunities are becoming available for research technicians. The large number of public community and junior colleges across the United States have made it relatively convenient and inexpensive for students to finish a 2-year program with an Associate's degree. The technician has specific training and specialized work experience in the routine tasks of the research or clinical laboratory.

III. Career Options for Microbiologists

With degree(s) in hand, the microbiologist has many job choices. Location may be an issue. Most scientists are employed on the East or West Coast, but a growing number of positions are becoming available in metropolitan areas and university centers across the United States. Because there are simply more jobs at B.A. or B.S. than at the Ph.D. level, Bachelor's degree holders will have a wider choice of location. Advanced degree positions tend to be clustered, but not exclusively, in large cities or in smaller communities that have a university or research consortium.

A. Academic

For the Ph.D. microbiologist who works in an academic institution, commitment to both teaching and research is required. At universities with graduate programs, the faculty member will be expected to direct, and fund through grants, a research laboratory for the pursuit of research and the training of graduate students. In addition, organized classes for undergraduate and graduate students are taught. In the institutions that offer only Bachelor's degrees, the faculty member will teach lecture and laboratory courses and will very often carry out a research program with undergraduate students in the lab. In colleges and universities, the promotional track begins at assistant professor. The assistant professor typically holds this position for approximately 6 years and is, at that juncture, promoted or dismissed. The next step, associate professor, has "tenure," which provides considerable freedom in that the tenured faculty member's position cannot be terminated except under the most unusual of circumstances. The next level, professor, carries the highest prestige and salary.

In addition to universities, research and teaching opportunities for microbiologists are found in medical, dental, and veterinary schools. Faculty members (assistant professor, associate professor, and professor) require a Ph.D. or M.D. The faculty's teaching commitment generally involves less classroom contact at health education institutions as compared to universities. The research missions of health science centers typically relate to clinical topics as contrasted to the basic science research focus of universities.

In a university or health science center, research labs will employ Master's- and Bachelor's-level personnel as well. These positions—research technician, research assistant, research associate—vary a good deal in responsibility and salary commensurate with the background and experience of the individual. The research lab director, through grants, also supports graduate students and postdoctoral research associates. The teaching labs are typically staffed by graduate student teaching assistants under direction of a faculty member, and these positions are funded by the institution.

Some microbiologists in academic institutions are also engaged by industrial or governmental units as consultants. The particular expertise of the academician is sought. The duration and extent of the consultantship depends on the situation, and these interactions must be approved by the university's administration. Consulting provides an additional income and collaboration source for the faculty member. In addition, many individuals who are full-time consulting microbiologists may have gained their expertise in industry or in academia.

Appointments for faculty at community and junior colleges involve teaching commitments exclusively. These colleges, all across the United States, provide introductory and survey classes in biological sciences and some more specialized classes including microbiology lecture and lab. A frequently cited advantage of this occupation is the opportunity to interact with students. A Master's or Ph.D. degree is required.

B. Industry and Research Institutes

Two areas showing particularly rapid growth in opportunities for microbiologists are industry (particularly the chemical industry) and environmental science. Many new positions are developing in areas of applied microbiology including water and wastewater technologies, handling of hazardous wastes, and quality assurance for product industries.

Many corporations in the United States and abroad employ microbiologists. These include companies in the food, pharmaceutical (drugs, antibiotics, vaccines), environmental, agricultural, fermentation, and chemical industries. Corporate positions can be found in research and development, production, quality control, management, and marketing. There are positions at all levels (Bachelor's, Master's, and Ph.D.) with salaries and job flexibility increasing with higher degree of training and experience. In many companies, advancement can occur along the technical or scientific pathway or in a direction of more supervisory or management responsibility. Directors and corporate officers have the greatest responsibility, prestige, and salary. Research done in industrial labs is most often related to the focus of the company and is profit-driven. This includes searching (and patenting) new products, processes, and safety/reliability testing.

Some entrepenuerial microbiologists establish their own businesses with venture capital or with assistance from the federal government's Small Business Association. These companies generally start with only a small number of employees and a limited market. Because of the high rate of failure for all small businesses, this would have to be considered a risky career choice but one that may ultimately provide substantial satisfaction and possibly much greater financial reward than that of salaried microbiologists.

Privately endowed or university-affiliated research institutes employ microbiologists at all degree levels. Doctoral degree holders follow the promotion pathway of assistant professor to associate professor to professor. Some may serve as consultants to industry. In general, no teaching is required, although some institutes are introducing graduate programs. The lab director must seek funds through grants. At some institutes, the research tends to focus in one area, whereas at others research topics are considerably diverse.

Many private foundations support research through grants to a university or research institute. Regional foundation libraries maintain a reference collection detailing funding opportunities through corporations and foundations.

Industrial or research institute microbiologists may have cross-appointments at a university. They may teach classes in their specialty at the undergraduate or graduate level. Often, their title will bear the term ''adjunct'' (such as adjunct associate professor) to show their primary employment outside the university.

C. Federal, State, and Local Government

Government agencies at all levels need microbiologists. For example, the United States government supports research laboratories and regulatory bureaus at the Federal level. The employees of these facilities are civil service workers and move through the government service (GS) hierarchy. Greater education and experience credentials enable the person to have a higher GS rank with the most responsibilities and highest salaries. Federal agencies employing microbiologists include the Food and Drug Administration, Department of Agriculture, Department of Energy, Department of Defense, and Environmental Protection Agency. Additionally, the Department of Health and Human Services (HHS) has major programs for research at the many branches of the National Institutes of Health. HHS also supports science education and research through the National Science Foundation. Among the many large laboratory operations of the Federal government are the NIH labs in Bethesda, Maryland, and the Centers for Disease Control, headquartered in Atlanta, Georgia. Smaller regional labs are scattered across the United States.

At the National Institutes of Health campus, thousands of postdoctoral-level scientists conduct research and have clinical opportunities. The number of Ph.D. and M.D. holders is approximately equal. Some of these persons serve also as consultants to industry. After several years of work, some individuals become ''tenured'' and are invited to remain in permanent positions. Information about postdoctoral research fellowship opportunities is available from the Office of Education, Building 10, National Institutes of Health, Room 1C-129, 9000 Rockville Pike, Bethesda, Maryland 20892. These labs also employ thousands of technical personnel at the Bachelor's and Master's degree level.

State and local governments also have both laboratory and regulatory functions. Microbiologists in these positions might be involved in routine monitoring of water, food, or milk. Public health facilities associated with hospitals and clinics employ microbiologists. Diagnosis and reporting of some infectious diseases are handled in state laboratories. These occupations are generally organized in grades (such as Virologist I through Virologist V) with advancement and salary dictated by training and experience.

D. Clinical Microbiologists

Many microbiologists choose careers in health-related professions. For example, clinical laboratory directors are often M.D.- or Ph.D.-level microbiologists. The director is responsible for the accuracy of tests performed in the lab and also for interactions with the physicians who require information about their patients' tests. The lab may be affiliated with a hospital or clinic or it may be a private organization. Naturally, this type of career requires considerable management skills as well as expertise in infectious diseases, instrumentation, and a myriad of diagnostic and immunological tests.

E. Medical Technologists and Medical Laboratory Technicians

Some students seek a B.S. degree in medical technology or a B.A. in microbiology with the intent of working in a clinical or diagnostic lab. In addition to the degree, professionals in this area take a test for certification by one or more agencies. For example, medical technologists (MTs) and microbiologists who pass the tests are registered by the American Society of Clinical Pathologists and are designated MT(ASCP) and M(ASCP), respectively. The American Academy for Microbiology maintains a national registry. By examination in specialty areas, earned designations are registered microbiologist [RM(AAM)] or specialist microbiologist [SM(AAM)].

MTs collect specimens and perform a variety of tests on patient samples. Labs employing MTs are located in large cities and small towns throughout the country, often housed in hospitals or clinics. This work demands careful, conscientious effort because the diagnosis and treatment regimen to be chosen for the patient often depends on the lab tests.

Reliability and accuracy are critical to the medical laboratory technician (MLT) also. Although the MLT may do the more routine tests of the clinical lab, extreme care and excellent record-keeping are required. Education in a 2-year program, often a junior or community college, with specialized experience on the job is required for an MLT.

F. Microbiologists in Medical, Dental, or Veterinary Professions

Many individuals pursue microbiology, including some who receive advanced degrees, and then go to medical, dental, or veterinary school. For example, after graduate work in infectious diseases or immunology (M.A. or Ph.D.), a person may complete medical school (M.D.) and then choose a specialty such as internal medicine, pediatrics, oncology, or pathology. Also, many medical schools offer a M.D.–Ph.D. training program. Although it requires additional time for completion, the dual-degree recipient has insight and experience both in working with patients and with clinical research.

New opportunities for microbiologists or MTs (B.A. or B.S.) are found in reference labs for veterinary laboratory diagnosticians. This is only one example of emerging professions for individuals with good laboratory skills.

G. Other Options

With training in microbiology, other opportunities exist outside academic, industrial, government, or health-related jobs. For example, with a degree in microbiology and a law degree, one might be an expert in the patent laws concerning new developments in molecular and cellular science for the biotechnology industry. For microbiologists with appropriate talents and interest, scientific writing and illustration are possible careers.

Microbiologists may also choose to be science teachers or administrators in secondary schools. With the realization that an adequate supply of scientists depends on training and encouragement of students at elementary, middle, and high school levels, more incentives to enter the teaching profession have been suggested. More competitive fellowships, active recruitment of women and minority group members, and greater access to summer in-service institutes are being suggested to draw talented people into precollege science teaching.

Another option to consider is to finish Bachelor's-level training in microbiology and then seek an Master's in Business Administration (M.B.A.) degree. This background provides managerial opportunities, particularly in industry, as area leaders or laboratory supervisors. In companies that market medical diagnostics or pharmaceuticals, positions are available in technical sales or in the development of new products for the marketplace.

IV. Salary Expectations

Libraries and career information centers at colleges and universities have resources that give representative salary data for various occupations. Professional organizations (see Appendix A) will send, on request, brochures or pamphlets describing their discipline, job opportunities, and outlining salary expectations. It is not appropriate to list salary figures here because these data lose their reliability so quickly. If one is interested, current information should be sought.

As a rule, when one compares salary data, the most significant element is education. In all sectors, beginning wages are least for those with B.A./B.S. degrees, higher for M.A./M.S. holders, and highest for Ph.D. holders. *Average* salaries at all degree levels increase in the following progression: lowest in academic careers (some difference between public and private schools), next higher in government agencies, next in hospitals or clinics, and highest in industry. At all levels and in all sectors, tremendous variations can occur among individuals depending on their experience, expertise, and relative success in their own career. In addition to salary, other components such as opportunity for travel, to consult, or to write independently are significant. Working hours, retirement, and fringe benefits are important variables.

Salaries paid to microbiologists are adequate for a reasonable standard of living but, clearly, there are other professions that offer more money. Although salary is one tangible reward for work, other factors should be considered. For example, microbiologists can expect good working conditions and anticipate stimulation by personal interactions with co-workers and others in the field. Depending on degree and occupation, freedom to pursue one's interests is often cited as an important factor in career choice among scientists. Additionally, satisfaction can be derived from making contributions: researchers to the body of knowledge, clinical microbiologists to patients, academicians to students, industrial microbiologists to products and services.

V. Resources

A. Professional Affiliations

While an undergraduate, a student can participate in campus chapters of preprofessional organizations. The student learns about career options and also has interactions with peers who share the same interests and with faculty in the discipline. The faculty (particularly the organization's adviser) will be available if problems arise and will also be role models for the student.

Microbiologists (at all degree levels) and graduate students in the discipline are eligible for membership in the American Society for Microbiology (ASM). Members receive the monthly publication *ASM News* and member rates for meetings and primary research journals (e.g., *Journal of Bacteriology, Molecular and Cell Biology, Journal of Virology, Infection and Immunity*). In addition, ASM members receive schedules of all meetings and continuing education workshops involving microbiology and related topics. The 35,000-member organization promotes exchange of information among professionals and is an effective voice to address legislation (national and state) and public opinion that relates to microbiology. An application for membership can be obtained from the Membership Committee, American Society for Microbiology, 1325 Massachusetts Avenue, NW, Washington, DC 20005 [phone: (202) 737-3600]. For the microbiologist seeking employment, the reasonably priced placement service with interview opportunities held in conjunction with national ASM meetings is a valuable service.

Other professional organizations that may also serve the interests of microbiologists are listed in Appendix B.

B. Sources of Useful Publications

Both precollege and college teachers and students may want to order a booklet of information, "Your Career in Microbiology: Unlocking the Secrets of Life" and two posters on careers in microbiology from the ASM. These items are designed to attract high school and undergraduate students to microbiology. Write to the Office of Education and Training at the ASM.

An inexpensive booklet, "On Being A Scientist," is published through the Committee on the Conduct of Science by the National Academy of Sciences. Discussing the methodology and philosophy of science and some difficult issues such as scientific fraud and misappropriation of credit, this publication is appropriate for microbiologists at the level of undergraduate student through degreed professionals. It can be ordered from the National Academy Press, 2101 Constitution Avenue, NW, Washington, DC 20418 (phone: 1-800-624-6242).

Public, school, and university libraries have many books that provide interesting reading about microbiology, its practitioners, and its history and milestones. Introductory level microbiology textbooks provide an overview of the science and are available in college libraries. Biographies of Louis Pasteur and Antony Van Leeuwenhoek are particularly appropriate for younger readers. Reading lists relating to developments in microbiology, new techniques, and the ethics and regulation of modern technologies are incorporated in ASM brochures and posters. Appropriate choices can be recommended by librarians.

A multitude of free or very inexpensive materials (booklets, directories, newsletters, books, and teaching aids) relating to science and science education are available. Items can be chosen at levels that are appropriate for youngsters. Lists of publications of the National Science Foundation and the American Association for the Advancement of Science are available free of charge (for addresses, see Appendix A).

Many science-oriented periodicals are available by personal subscription or through libraries. These magazines are useful in maintaining a current overview of science and technology. They are appropriate for readers at varied levels of interest and sophistication. Examples, but by no means an all-inclusive list, are *Science, Scientific American, Discover,* and *Omni.*

Appendix A Programs to Stimulate Student Interest in Science[a]

American Association for the Advancement of Science
"Publications"
Directorate for Education and Human Resources Programs
1333 H Street, NW
Washington, D.C. 20005

Carnegie Corporation of New York
"List of Grants and Appropriations"
437 Madison Avenue
New York, New York 10022

Ford Foundation
Ford Foundation Predoctoral and Dissertation Fellowships for Minorities
"Program Announcement of National Research Council"
2101 Constitution Avenue, NW
Washington, D.C. 20418

Howard Hughes Medical Institute
"Grants Program Policies and Awards"
6701 Rockledge Drive
Bethesda, Maryland 20817

National Science Foundation
"Publications of the National Science Foundation" and "Guide to Programs"
1800 G Street, NW
Washington, D.C. 20550

[a] Some grants and fellowships are targeted to groups underrepresented in science.

Appendix B Professional Organizations That May Serve the Interests of Microbiologists

American Association of Immunologists[a]
9650 Rockville Pike
Bethesda, Maryland 20814

American Chemical Society
1155 16th Street, NW
Washington, D.C. 20036

American Genetics Association
818 18th Street, NW
Washington, D.C. 20006

American Institute of Biological Sciences
730 11th Street, NW
Washington, D.C. 20001

American Society for Biochemistry and Molecular Biology[a]
9650 Rockville Pike
Bethesda, Maryland 20814

American Society for Cell Biology[a]
9650 Rockville Pike
Bethesda, Maryland 20814

American Society for Medical Technology
2021 L Street, NW
Washington, D.C. 20036

American Society for Microbiology
1325 Massachusetts Avenue, NW
Washington, D.C. 20005

Genetics Society of America
PO Box 6018
Rockville, Maryland 20850

Society for Industrial Microbiology
PO Box 12534
Arlington, Virginia 22209

[a] Organization is part of the Federation of American Societies for Experimental Biology.

Bibliography

"Directory of Research Grants" (1991). The Oryx Press, Phoenix, Arizona.

National Research Council (U.S.) (1990). "Fulfilling the Promise: Biology Education in the Nation's Schools." National Academy Press, Washington, D.C.

National Science Foundation. (1989). "Profiles—Biological Sciences: Human Resources and Funding." NSF 89-318. National Science Foundation, Washington, D.C.

National Science Foundation (1990). "Women and Minorities in Science and Engineering." NSF 90-301. National Science Foundation, Washington, D.C.

"Peterson's Graduate Programs in the Biological and Agricultural Sciences (1991). Peterson's Guides, Princeton, New Jersey.

Shakhashiri, B. Z. (1990). *ASM News* **56**(1), 17–21.

U.S. Department of Labor (1990–1991). "The Occupation Outlook Handbook." Bulletin 2350. U.S. Department of Labor, Bureau of Labor Statistics, Washington, D.C.

Wright, J. W. (1990). "The American Almanac of Jobs and Salaries." Avon, New York.

Catabolic Plasmids in the Environment

William H. Wallace and Gary S. Sayler
The University of Tennessee

Glossary

Biodegradation Process of biochemical metabolism of a compound as opposed to chemical or physical degradation

Cointegrate Segment of DNA containing two replicons; often used to described the fusing of two plasmids

Constitutive expression Genes that are expressed at all times usually at low level

Degradative pathways Specific enzymatic biochemical reactions for sequential metabolism of a particular substrate

Enzymes Protein molecules acting as catalysts to reduce energy of activation for a specific chemical reaction

Gene Specific segment of a DNA molecule in which the nucleotide base sequence determines (1) the synthesis of a rRNA or tRNA molecule, (2) a functionally active site on DNA, or (3) the amino acid sequence of a polypeptide chain

Genetic regulation Cellular methods by which expression of genes are turned on (induced) and/or off (repressed)

Incompatibility If two plasmids are subject to common regulation and if both are present in the same cell, one or the other will be lost following several generations of bacterial growth

Inducible expression Transcription activation of specific genes as a result of the presence of an inducer interacting with a regulator factor

Molecular breeding Enrichment technique where a mixed population of organisms with and without degradative plasmids are mixed together and allowed to grow for long periods of time in the presence of a pollutant; in theory this technique provides the genes necessary for evolution of a degradative pathway

Operon Genetic unit of coordinated transcription

Plasmid Autonomously replicating circular DNA, conserved in an extrachromosomal state through successive bacterial cell divisions

Replicon Region of DNA that contains the origin for initiation of replication

Restriction enzymes Class of enzymes that catalyze double-strand breaks in DNA molecules to yield restriction fragments; site of cleavage is specific for each enzyme

Retrotransfer Transfer of genes recipient to donor bacterium mediated by some plasmids; as opposed to the classical direction of gene transfer from donor to recipient

Sole source of carbon Organic compound that a cell can metabolize into the biosynthetic and energy materials necessary for all cellular processes and energy needs of the cell

Transposon Nonreplicating segment of DNA that is capable of inserting itself into other DNA molecules via an event called transposition

PLASMIDS are found in virtually every bacterial species and are probably present in most individual bacterial cells. Plasmids account for only a small portion of the bacterial cell's genome, ranging from 1 to 3% and in a few rare cases up to 15%. However, this small portion of the cell's genetic information codes for important cellular functions—often the

Encyclopedia of Microbiology, Volume 1

deciding factor in cell survival. Plasmids are usually defined by the functions it encodes. Resistance (R) plasmids code for resistance to particular antibiotics, fertility (F) plasmids contain transfer genes for mobilization to other bacteria, and catabolic plasmids encode for enzymes that degrade or transform various compounds. Many of the man-made compounds only recently released into the environment are significantly degraded by microbes in which plasmids are responsible for either total or partial degradation of these compounds (Table I).

Bacterial plasmids including catabolic plasmids are classified into incompatibility groups to facilitate identification of common mechanisms for plasmid replication and segregation. This article will discuss and provide a comparative analysis of the topics outlined above for catabolic plasmids of the following incompatibility groups: IncP1, IncP2, IncP3, IncP7, and IncP9). [*See* PLASMIDS.]

I. Introduction: General Occurrence and Distribution of Catabolic Plasmids

Catabolic plasmids have been isolated from bacterial hosts living in water, sediments, and soil. Recent studies with subsurface sediments have identified plasmids that show hybridization homology to the TOL plasmid (pWWO), suggesting possible catabolic potential in deep subsurface terrestrial environments (0.2–260 m). Most researchers believe the presence of a particular compound, even those that are toxic to cells at high concentrations, will eventually select bacteria from the indigenous population, which are resistant to, and in addition may often survive and degrade the contaminating substance. In many cases, this rapid selection is due to catabolic plasmids, which have the advantage over chromosomal genes in gene dosage and the ability to be rapidly spread through the bacterial community, often without regard to genus.

The number of bacterial genera in which catabolic plasmids have been isolated has increased over the last several years. Early isolation of catabolic plasmids were restricted to predominantly *Pseudomonas* species known for their ability to degrade a plethora of compounds. The genus *Pseudomonas* continues to predominate in catabolism of compounds. However, as more and more environments are examined, the list of bacterial species with catabolic plasmids has increased. Table I lists the catabolic plasmids discussed in this report and bacterial species from which these plasmids were isolated.

II. Plasmid Occurrence for Representative Incompatibility Groups

A. Distribution in Microbial Populations

Degradation studies of various compounds have led to the isolation of numerous bacteria with enormous degradative potential. Most of these environmental bacteria contain plasmids. Often these plasmids are responsible for catabolism of various compounds (Table I). Studies have shown as many as 50% of the bacterial isolates from contaminated sites may contain plasmids. In some cases, the incidence of plasmids in bacterial strains has been positively correlated with the presence of pollution. In addition, plasmids from contaminated sites are often larger than plasmids isolated from microbes at clean sites. Bacteria present at contaminated areas often contain plasmids which range from 40 to greater than 400 kilobase pairs. This finding is important considering that most degradative plasmids are larger than 50 kilobase pairs. Investigators have also determined that bacteria from polluted sites often contain a greater number of plasmids per organism than plasmid bearing strains from unpolluted sites. These general trends appear to be consistent for environments as diverse as, running streams, sediment, deep subsurface, and shallow aquifers examined to date.

A preliminary study examining bacterial isolates from the deep subsurface for the presence of genes homologous to the TOL plasmid have demonstrated the presence of TOL related DNA sequences, which appear to be plasmid associated. However, further characterization of the naturally occurring plasmids from polluted and nonpolluted sites will be required to determine if these plasmids are degradative and if the degradative genes are actually being actively transcribed. [*See* BIODEGRADATION.]

B. Ancestry and Homology with Other Catabolic Plasmids and Noncatabolic Plasmids

All plasmids probably have common mechanisms by which they evolved, exchanging pieces of DNA with

Table I Pollutant Substrates Metabolized by Catabolic Plasmids

Substrate	Plasmid	Incompatibility group	Size range (kb)	Transmissibility	Host bacteria	Environmental source
4-Chlorobiphenyl	pSS50	IncP1	53	+	*Alcaligenes* and *Acinetobacter*	sediment
4-Chlorobiphenyl	pKF1	IncP1	78	ND	*Acinetobacter* and *Arthrobacter*	soil
2,4-Dichloro-phenoxyacetate	pJP4	IncP1	78	+	*Alcaligenes eutrophus*	soil
Camphor	CAM	IncP2	>200	+	*Pseudomonas sp.*	soil
E-Caprolactam	pBS271	IncP2	500	+	*Pseudomonas sp.*	soil and water
E-Aminocaproic acid	pBS271	IncP2	500	+	*Pseudomonas sp.*	soil
2-Fluoroacetate	pU01	IncP2	65	+	*Morexella sp.*	soil
2-Chloroacetate	pU01	IncP2	65	+	*Morexella sp.*	soil
2-Chloropropionic acid	pU202	IncP2	230	−	*Pseudomonas sp.*	soil and water
3-Chlorobenzoate	pAC25	IncP2	117	+	*P. putida*	soil and water
Octane	OCT	IncP2	>200	+	*P. olevorans*	soil
Parathion	pPDL2	IncP2	43	+	*Flavobacterium sp.*	soil
Phenol	pPGH1	IncP2	200	+	*P. putida*	soil
Phenylacetate	pWW17	IncP2	270	+	*Pseudomonas*	soil and water
2,4-Dichloro-phenoxyacetate	pJP2	IncP3	55	+	*A. paradoxus*	soil
Naphthalene	pBS4	IncP7	173	+	*P. fluorescens*	soil
p-Cresol	pND50	IncP9	ND[a]	+	*P. putida*	soil
Naphthalene	NAH7	IncP9	83	+	*P. putida*	soil
Phenanthrene	pKA2	IncP9	31	−	*Beijerinckia sp.*	soil
Nicotine	Nic	IncP9	ND	+	*P. convexa*	soil
Salicylate	Sal1	IncP9	84	+	*P. putida*	soil
Styrene	pEG	IncP9	37	+	*P. fluorescens*	soil
Toluene	pWWO	IncP9	117	+	*P. putida*	soil
p- and *m*-xylene	pWWO	IncP9	117	+	*P. putida*	soil and water
Alkylbenzene sulfonates	ASL	ND	91	+	*P. testosteroni*	soil
Trimethylbenzene (1,2,4-)	pGB	ND	85	−	*P. putida*	soil and water
Dichlorobenzene (1,2-)	pP51	ND	110	ND	*Pseudomonas sp.*	soil and water
Trichlorobenzene (1,2,4-)	pP51	ND	110	ND	*Pseudomonas sp.*	soil and water
Dichlorobenzoate (3,5-)	pWWO-NAH7	Genetic[b] construction	—	−	—	—
Salicylate (3-,4-, & 5-)	pWWO-NAH7	Genetic construction	—	−	—	—

[a] ND, not determined.
[b] Genes from pWWO and NAH7 placed on the broad host range plasmid pKT231.

the host chromosome and other plasmids. These DNA rearrangements include deletions, insertions, and inversions often occurring several times until a stable molecule results. The genetic exchange hypothesis is supported by reports in the literature of large plasmids disassociating into several small plasmids, i.e., octane (OCT) plasmid. There is evidence that some degradative pathways are partially encoded by both plasmids and the chromosome and often with some overlap for the same or very similar enzymes, i.e., pJP4 and pWWO. This type of cooperation and duplication in degradative functions may suggest genetic exchange between plasmids and the chromosome. Isolation of R prime plasmids (plasmids containing large fragments of chromosomal genes) lends credibility to genetic exchange between plasmids and chromosome. Examples of R prime plasmids include pyocin determinants in *Pseudomonas aeruginosa*, carbohydrate metabolism in *Pseudomonas putida,* and tryptophan biosynthetic pathway in *Escherichia coli*. Furthermore, other evidence has shown that degradative gene clusters on plasmids are often transposable, i.e., TOL, pJP4, NAH7, and suggested in pSS50. DNA exchange and recombination has been suggested not only for structural genes but also for regulatory genes of catabolic pathways.

All catabolic and noncatabolic plasmids encode genes responsible for plasmid replication and maintenance in the cell. These coding regions are often referred to as "backbone" regions and are conserved in many types of plasmids. The backbone regions of the catabolic plasmids pSS50, pBR60, pJP4, and R751, belonging to the incompatibility group IncP1, have been shown to be conserved. DNA homology for plasmid replication and conjugal transfer functions have been demonstrated. The genetic regions responsible for catabolic functions in pSS50 have not been extensively studied to date and data are not available for comparison with other chlorinated biphenyl-degrading plasmids.

The plasmid pKF1, was the first identified plasmid found to mediate the partial degradation of chlorobiphenyls. This plasmid was shown to replicate in both gram-positive and gram-negative bacteria. Isolation of the 2,4,5-trichlorophenoxyacetic acid, (2,4,5-T) degrading plasmids from strain *P. cepacia* AC1100 was accomplished by specialized enrichment techniques known as molecular breeding. However, these plasmids are poorly characterized and will not be discussed in detail.

The catabolic regions of plasmid pJP4 have been compared to regions of another chlorocatechol-degrading plasmid, pAC27. The DNA sequences for the enzymes pyrocatechase and cycloisomerase in plasmids pJP4 and pAC27 demonstrated >50% DNA homology. In addition to homology with structural genes, the promoter regions for catabolic genes of plasmids pAC27 and pJP4 were found to be very similar. DNA homology was also demonstrated for genes that encode resistance to mercury ions. Mercury resistance appears to be a common feature of many IncP plasmids, often associated with the transposon Tn501.

Campher (CAM) and OCT plasmids are both very large, >200 kbp. Their size makes physical and genetic analysis difficult. However, limited genetic studies of the *alk* genes on the OCT plasmid revealed homology to *xylD* and *xylS* from the TOL plasmid and *NAH7* plasmid. The G+C content of the *alk* genes (47%) is very different from the host bacteria *P. putida* with a G+C content of its chromosome of 62.5%. In addition, the codon usage of *P. putida* is different from the *alk* genes, suggesting that the *alk* genes probably originated from an unrelated organism.

Plasmid pJP2, incompatibility group IncP3, shows significant homology with IncP1 plasmid pJP4. Both of these plasmids encode for enzymes that degrade 2,4-dichlorophenoxyacetic acid (2,4-D). An 25-kb DNA fragment has been identified that harbors the catabolic genes in both of these plasmids. This DNA fragment is homologous in both pJP2 and pJP4. Restriction mapping of the regions adjacent to the catabolic genes suggests that the catabolic gene cluster is contained on a large transposon. If in fact the catabolic genes are located on a transposon it would explain why these same genes are found on other plasmids.

The naphthalene-degrading plasmids pBS211, pBS213, and pBS243 belong to IncP7. Studies on key enzymes responsible for the degradation of naphthalene indicate that these plasmids are similar to the IncP9 naphthalene plasmids and oxidize naphthalene through salicylate to catechol and further degrade catechol via the meta cleavage pathway. To our knowledge, further biochemical or genetic characterization has not been performed on the naphthalene IncP7-degrading plasmids. However, because the IncP7 plasmids express the key enzymes naphthalene oxygenase, salicylate hydroxylase, and catechol-2-3-dioxygenase under conditions similar to those of the IncP9 NAH7 plasmid, these plasmids are suspected to be biochemically and genetically similar.

Plasmids NAH7 and pWWO belonging to IncP9 group are the best studied of all catabolic plasmids. The gene encoding the 2,3-dioxygenase on pWWO (*xylE*) and the 2,3-dioxygenase gene (*nahH*) of plasmid NAH7 have been sequenced and shown to demonstrate 80% similarity in DNA and protein sequence, suggesting a common ancestor. The gene order of arrangement for enzymatic reactions and DNA homology between pWWO (TOL) and NAH7 plasmids would also suggest that the catabolic regions of these plasmids were derived from a common ancestral sequence. This DNA homology also exists between plasmid transfer and replication functions in addition to catabolic genes, suggesting a common evolutionary origin. Naphthalene-degrading plasmids NAH7, SAL1, pWW60-1, and pNAH484 appear to be conserved, at least for the enzymes responsible for the conversion of naphthalene to catechol.

III. Types of Substrates Metabolized

A. Naturally Occurring Substrates

Naturally occurring hydrophobic hydrocarbons are often toxic to microbes. Degradation of these hydrocarbons, often via plasmid-encoded enzymes, reduces the toxic concentrations, allowing the host cell to survive. Table II lists some of the diverse groups of aromatics (cresol), terpenes (camphor), and polynuclear aromatic hydrocarbons (naphthalene), that are catabolized by catabolic plasmids. Many scientists feel that the long-term presence of these natural compounds have selected microbes that can readily adapt to catabolize similar synthetic compounds.

B. Synthetic Substrates

Industrial nations have introduced large amounts of xenobiotics into the environment in the last 75 years. These compounds are often toxic to plant and animal life. Fortunately, many of these compounds are readily degraded by microbes, however, some are recalcitrant. This report will not attempt to address all possible substrates that can be biodegraded but will focus on compounds that are substrates for catabolic plasmids listed in Table II.

IV. Biochemical Characterization

A. Type of Enzymes Involved in Plasmid-Encoded Degradative Pathways

The biochemistry of 4-chlorobiphenyl degradation has not been elucidated for plasmid pSS50. There-

Table II Catabolic Plasmids, Incompatibility Groups, and Other Relevant Information

Reference plasmid	Incompatibility group	Compounds degraded[a]		Original host bacteria	Environmental source
		Natural	Human-made		
pSS50	IncP1		4CB	*Alcaligenes* and *Acinetobacter*	Sediment
pJP4	IncP1		2,4-D, 3CBA, and MCPA	*Alcaligenes*	Soil and water
OCT	IncP2		octane, hexane, and decane	*P. putida*	Soil
CAM	IncP2	Camphor	camphor	*P. putida*	Soil
pJP2	IncP3		2,4-D and MCPA	*Alcaligenes*	Soil and water
Naph	IncP7	Naphthalene		*Pseudomonas*	Soil
pWWo	IncP9		Toluene and p- and m-xylene	*P. putida*	Soil
NAH7	IncP9	Naphthalene		*P. putida*	Soil
Cresol	IncP9	Cresol	Cresol	*P. putida*	Soil

[a] Abbreviations: 4CB, 4-chlorobiphenyl; 2,4-D, 2,4-dichlorophenoxyacetic acid; 3CBA, 3-chlorobenzoic acid; MCPA, 4-chloro-2-methylphenoxyacetic acid.

fore, the discussion of pathway biochemistry for IncP1 plasmids will be limited to plasmid pJP4. A schematic diagram of enzymatic reactions encoded by plasmid pJP4 are given in Fig. 1. Plasmid pJP4 encodes for enzymes that degrade 3-chlorobenzoate (3CB) and the widely used pesticides 2,4-D and 2-methyl-4-chlorophenoxyacetic acid. Strain *Alcaligenes eutrophus* containing pJP4 synthesizes two separated sets of enzymes for the ortho cleavage of catechol and halocatechols. Genes for the catabolism of catechol are found presumably on the chromosome, and the genes for 3CB and 2,4-D degradation are encoded on plasmid pJP4 with the exception of the maleylacetate reductase gene, which is also probably chromosomally encoded. Transposon mutagenesis of pJP4 has demonstrated that a single insertion into plasmid genes resulted in the loss of the ability to degrade both 2,4-D and 3CB, suggesting that both pathways converge and the insertion was in a gene common to both pathways. It is suspected that the convergence of the pathway is after the formation of 3-chlorocatechol, which is an intermediate in 2,4-D degradation. The enzymatic reactions for 2,4-D metabolism involve first the cleavage of the ether linkage to yield 2,4-dichlorophenol and glyoxylic acid. The 2,4-dichlorophenol is hydroxylated in the ortho position to produce 3,5-dichlorocatechol. Ring fission occurs to form *cis,cis*-2,4-dichloromuconic acid. The next step is a cycloisomerization with the spontaneous loss of chloride to yield the hypothetical intermediate 2-chloro-4-carboxymethylene but-2-enolide. The next reaction hydrolizes the ring to form chloromaleylacetic acid, which is dechlorinated to yield succinate.

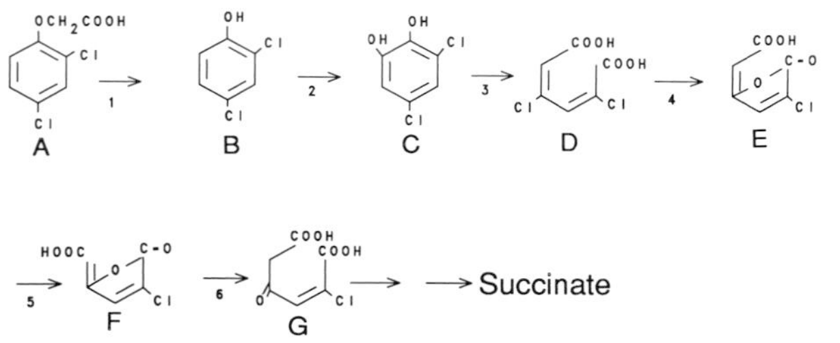

Growth on alkanes in *Pseudomonas olevorans* depends on the presence of the OCT plasmid. The OCT plasmid encodes genes that convert octane to an octanaldehyde. The aldehyde is then converted to a fatty acid and undergoes β-oxidation. Further metabolism of the aldehyde is by chromosomally encoded enzymes. The *P. putida* host strain also contains a chromosomal gene for a NAD^+-dependent alcohol dehydrogenase, whereas the plasmid-encoded alcohol dehydrogenase is $NAD(P)^+$-dependent. In fact, all plasmid-encoded enzymatic reactions for octane degradation except alkane hydroxylase are duplicated on the chromosome. Figure 2 depicts the biochemical steps for conversion of octane or octanoate. The OCT plasmid encodes for alkane hydroxylase, alcohol dehydrogenase, and aldehyde hydroxylase. The first enzymatic reaction for alkane degradation is alkane dehydrogenase, which is comprised of proteins Alk B, Alk G, and Alk T. Alk B is a membrane-bound monooxygenase; its amino acid sequence has been determined and shown to have eight potential membrane-traversing regions and several hydrophilic loops. The Alk G (rubredoxin) protein has a molecular mass of 19 kDa is capable of binding two

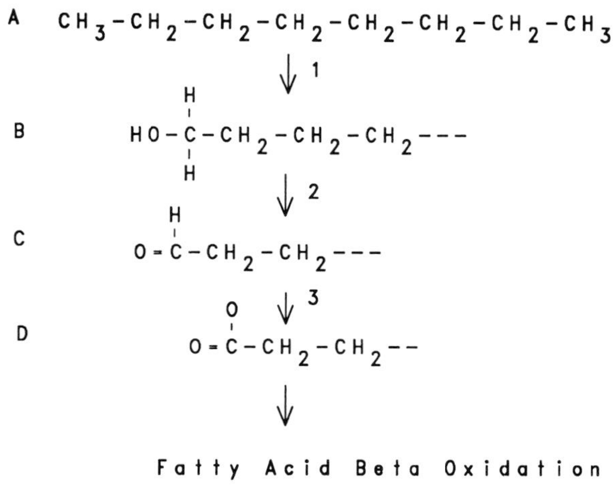

Figure 1 Proposed degradative pathway for 2,4-D degradation encoded for by plasmid pJP4. Chemical compounds: (A) 2,4-D, (B) 2,4-dichlorophenol, (C) 3,5-dichlorocatechol, (D) 2,4-dichloromuconate, (E) *Trans*-2-chlorodiene lactone, (F) *cis*-2-chlorodiene lactone, (G) 2-chloromaleylacetate. Enzymatic reactions and genes: (1) 2,4-D monooxygenase, *tfdA*, (2) 2,4-dichlorophenol hydroxylase, *tfdB*, (3) chlorocatechol-1,2-dioxygenase, *tfdC*, (4) chloromuconate cycloisomerase, *tfdD*, (5) chlorodiene lactone isomerase, *tfdF*, (6) chlorodienelactone hydrolase, *tfdE*.

A $CH_3-CH_2-CH_2-CH_2-CH_2-CH_2-CH_2-CH_3$

↓ 1

B $HO-CH_2-CH_2-CH_2---$

↓ 2

C $O=CH-CH_2-CH_2---$

↓ 3

D $O=C(OH)-CH_2-CH_2--$

↓

Fatty Acid Beta Oxidation

Figure 2 Proposed pathway for catabolism of octane by the OCT plasmid. Chemical compounds: (A) octane, (B) octanol, (C) octanal, (D) octanoate. Enzymatic reactions and genes: (1) alkane hydroylase, *alkB, alkG, and alkT*, (2) alcohol dehydrogenase, *alkJ*, (3) aldehyde dehydrogenase, *alkH*. [From Shapiro, J. A., Benson, S., and Fennewald, M. (1980). Genetics of Plasmid-determined Hydrocarbon Oxidation. *In* "Plasmids and Transposons." Academic Press, San Diego.]

molecules of iron per molecule, and is the electron carrier for the system. The Alk T (rubredoxin reductase) protein has been shown to be soluble, and its amino acid composition has been determined. Regions where FAD and NAD bind to the protein have been identified. A model suggesting the protein interactions for alkane hydroxylation has been proposed, although these proteins are not found in equal molar amounts in the cell. Both the alcohol dehydrogenase Alk J and aldehyde dehydrogenase Alk H have been sequenced and shown to be 60.9- and 52.7-kDa proteins, respectively. Regulatory proteins have been identified, but as yet their functions remain unclear.

The CAM plasmid, also belonging to the IncP2 group, harbors genes for the conversion of camphor to isobutyrate. Although the CAM plasmid contains some genes to further metabolize isobutyrate, the pathway is incomplete. However, *P. putida*, the original host of the CAM plasmid, contains chromosomal genes for metabolism of isobutyrate. Eleven enzymatic steps, schematically depicted in Fig. 3, are required to convert camphor to isobutyrate and acetate. The first reaction involves a two-electron reduction of molecular oxygen and concomitant hydroxylation of the camphor molecule at the 5-exo position. The 5-exo-hydroxylation of camphor is catalyzed by a soluble cytochrome P-450-containing hydroxylase complex. The camphor hydroxylase complex is composed of three proteins: putidaredoxin reductase, putidaredoxin, and cytochrome P-450. The putidaredoxin reductase is a 48-kDa protein containing FAD as a prosthetic group.

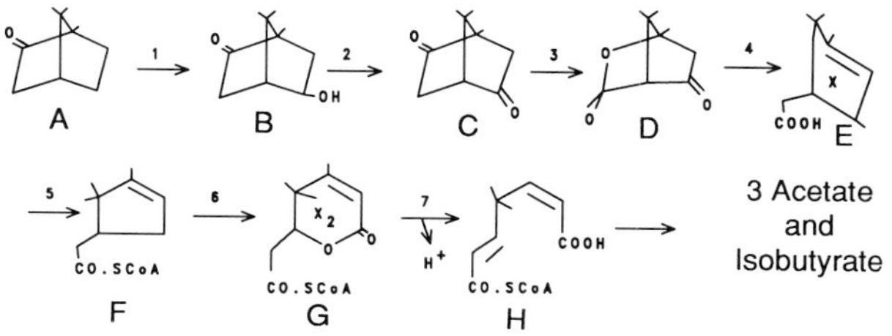

A B C D E F G H 3 Acetate and Isobutyrate

Figure 3 Proposed pathway for degradation of camphor in *P. putida* containing the CAM plasmid. Chemical compounds: (A) camphor, (B) 5-exo-alcohol, (C) 2,5-diketone, (D) 1,2-campholide, (E) Me_3-cyclo-pentenyl-Ac, (F) Me_3-CoA ester, (G) Me_3-primelyl-CoA δ lactone, (H) Me_3-primelyl-CoA. Enzymatic reactions and genes: (1) 5-exo-hydroxylase, *camABC*, (2) F dehydrogenase, *camD*. The remaining reactions are genetically and biochemically poorly characterized. [Adapted from Koga, H., Rauchfuss, B. and Gunsalus, I. C. (1985). *Biochem. Biophys. Res. Comm.* **130**, 412–417.

The putidaredoxin protein is important in the transfer of electrons into the P-450 heme. The active site for hydroxylation of camphor is located at the center of the cytochrome P-450_{cam} molecule. The P-450_{cam} protein was the first of the ubiquitous P-450 enzymes to be purified and studied in detail. The remaining reactions have not been studied nearly to the extent of the first reaction and, thus, will not be discussed.

Incompatibility group IncP3 is represented by plasmid pJP2. Plasmid pJP2 has only had limited biochemical characterization. The pathway for 2,4-D degradation is suspected to form the intermediate 3,5-dichlorocatechol rather than a monochlorocatechol, as shown with plasmid pJP4.

Incompatibility group IncP7 is represented by plasmids that degrade naphthalene. Plasmids pBS211, pBS213, and pBS243 have been examined for the presence of key enzymes responsible for the degradation on naphthalene. The naphthalene-degrading enzymes appear to be similar to those in the well-characterized NAH7 plasmid. The enzymatic activities were induced by salicylate, as is the case with the enzymes of NAH7 plasmid. Therefore, the presence of key enzymes and their inducibility with salicylate would suggest that the IncP7 plasmids are similar to the IncP9 naphthalene-degrading plasmids at least in these two facets.

The biochemistry for toluene degradation by the IncP9 TOL plasmid has been extensively studied, and the enzymatic steps are provided in Fig. 4. The initial reactions in the catabolism of toluene are successive single-step oxidations of the methyl group. Oxidation reactions catalyzed by dihydroxycyclohexadiene carboxylate dehydrogenase and catechol 2,3-oxygenase lead to ring fission products. The 2-hydroxymuconic semialdehyde is then converted to pyruvate and acetaldehyde.

Another well-characterized IncP9 plasmid is NAH7. The NAH7 plasmid allows host bacteria to utilize naphthalene as a sole source of carbon and energy. The NAH7 plasmid encodes for genes responsible for the 14-step pathway for degradation of naphthalene to pyruvate and acetaldehyde through the intermediate salicylate. Each of these enzymatic reactions are shown in Fig. 5. The first reaction involves the enzymatic incorporation of molecular oxygen into the aromatic ring to form *cis*-naphthalene dihydrodiol. This reaction is catalyzed by the multicomponent enzyme system naphthalene dioxygenase. Further oxidation is accomplished via a dehydrogenase to form 1,2-dihydroxynaphthalene. A series of four enzymatic reactions

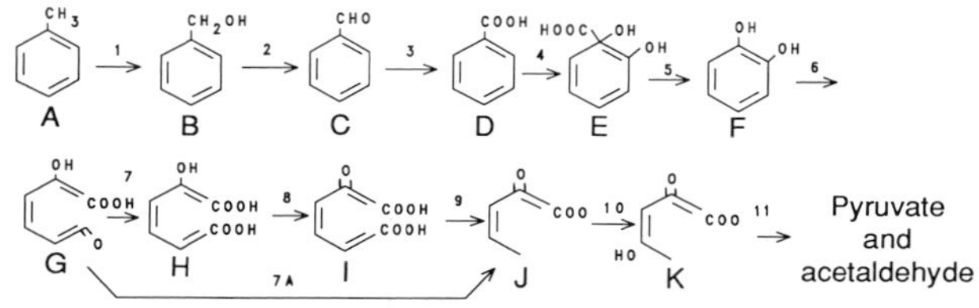

Figure 4 Degradative pathway for toluene degradation encoded for by the pWWO plasmid. Chemical compounds: (A) toluene, (B) benzyl alcohol, (C) benzaldehyde, (D) benzoate, (E) 1,2-dihydroxycyclo-3,5-hexadiene carboxylate, (F) catechol, (G) 2-hydroxymuconic semialdehyde, (H) 2-hydroxy-2,4-hexadiene-1,6-dioate, (I) 2-oxo-3-hexene-1,6-dioate, (J) 2-oxo-4-pentenoate, (K) 4-hydroxy-2-oxovalerate. Enzymatic reactions and genes: (1) xylene oxygenase, *xylA*, (2) benzyl alcohol dehydrogenase, *xylB*, (3) benzaldehyde dehydrogenase, *xylC*, (4) toluate oxygenase, *xylD*, (5) dihydroxycyclohexadiene carboxylate dehydrogenase, *xylL*, (6) catechol 2,3-dioxygenase, *xylE*, (7A) 2-hydroxymuconic semialdehyde hydrolase, *xylF*, (7) 2-hydroxymuconic semialdehyde dehydrogenase, *xylG*, (8) 4-oxalocrotonate isomerase, *xylH*, (9) 4-oxalocrotonate decarboxylase, *xylI*, (10) 2-oxopent-4-enoate hydratase, *xylJ*, (11) 2-oxyohydroxypent-4-enodate aldolase, *xylK*. [Modified from Burlage, R. S., Hooper, S. W., and Sayler, G. S. (1989). *Minireview Appl. Environ. Microbiol.* **55,** 1323–1328.]

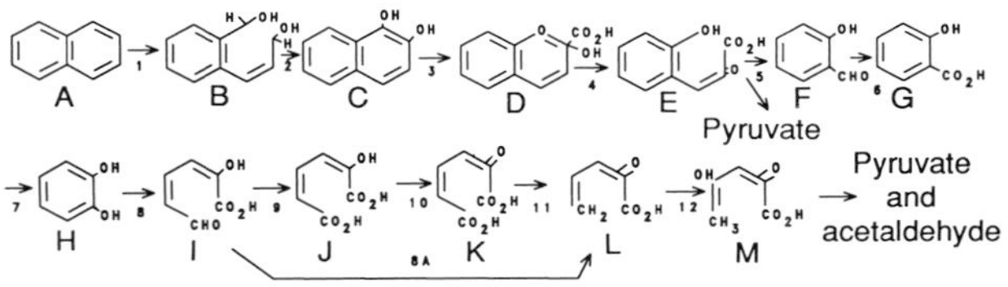

Figure 5 Naphthalene degradative pathway encoded for by the NAH7 plasmid. Chemical compounds: (A) naphthalene, (B) *cis*-dihydrodiol naphthalene, (C) 1,2-dihydroxy naphthalene, (D) 2-hydroxychromene-2-carboxylate, (E) 2-hydroxybenzalpyruvate, (F) salicyaldehyde, (G) salicylate, (H) catechol, (I) 2-hydroxymuconic semialdehyde, (J) 2-hydroxymuconate, (K) 4-oxal-crotonate, (L) 2-oxo-4-pentenoic acid, (M) 2-oxo-4-OH pentanoic acid. Enzymatic reactions and genes: (1) naphthalene dioxygenase, *nahA*, (2) *cis*-dihydrodiol naphthalene dehydrogenase, *nahB*, (3) 1,2-dihydroxynaphthalene dioxygease, *nahC*, (4) 2-hydroxychromene-2-carboxylate isomerase, *nahD*, (5) 2-hydroxybenzalpyruvate aldolase, *nahE*, (6) salicyaldehyde dehydrogenase, *nahF*, (7) salicylate 1-monoxygenase, *nahG*, (8A) 2-hydroxymuconic semialdehyde hydrolase, *nahN*, (8) catechol 2,3-dioxygenase, *nahH*, (9) hydroxymuconic semialdehyde dehydrogenase, *nahI*, (10) 4-oxalocrotonate tautomerase, *nahJ*, (11) 4-oxalocrotonate decarboxylase, *nahK*, (12) 2-oxopent-4-enoate hydratase, *nahL*. [Modified from Yen, K.-M., and Serder, C. M. *Crit. Rev. Microbiol.* **15,** 247–267.]

are required to produce salicylate. Salicylate is then converted to pyruvate and acetaldehyde in seven additional reactions.

Studies indicate that plasmid pND50 encodes for enzymes that degrade cresol to *p*-hydroxybenzoate, and in the following reaction protocatechuate is formed. The protocatechuate is then metabolized via the ortho cleavage pathway to succinate and acetyl-CoA.

V. Genetics of Pathways

A. Regulation (Repression and Induction)

Bacterial species that inhabit soil, sediments, and water are often faced with very limited and/or intermittent amounts of compounds that can be used as an energy source and metabolized into biosynthetic molecules. Therefore, one would expect a very tightly regulated system that could be rapidly adjusted to utilize a wide variety of compounds that may be present at very low concentrations. In this section, we will examine the regulation of some of these regulatory systems for catabolic plasmids.

Genetic regulation has not been elucidated in bacterial strains containing the IncP1 plasmid pSS50. IncP1 plasmid pJP4 genes, which encodes-enzymes for metabolism of 2,4-D are induced in the presence of 2,4-D and 3-CB. A negative control regulator gene, *tfdR*, has recently been identified. Bacterial strains deficient in *tfdR* express enzymes for 2,4-D degradation constitutively in the absence of inducer; however, these same organisms regained normal regulation when the *tfdR* gene was present in *trans*.

The two enzymes alkane hydroxylase and alcohol dehydrogenase encoded on the IncP2 OCT plasmid are coordinately induced by a wide range of straight-chain compounds. Compounds with a chain length of C_6–C_{10} that do not contain a terminal or subterminal hydroxyl group at both ends can function as inducers but not necessarily as growth substrates. Genes for alkane oxidation on the OCT plasmid are induced in *P. aeruginosa* strains by alkane growth substrates, aliphatic diols, straight-chain diethers, and dicyclopropyl compounds. Induction of OCT genes is specific for chain length and is affected by the degree of oxidation of the carbon chain. The enzymes for octane degradation are repressed by octanol. However, one must be careful in suggesting generalities about alkane enzyme induction. Enzyme induction and metabolism of different chain length appear to vary from strain to strain. Experiments have shown that some strains of *Pseudomonas* will metabolize alkanes as short as hexane and as long as heptadecane, yet other strains are much more limited in degradation of long-chain alkanes. Studies with *P. putida* have shown that only the first two enzymes (alkane hydroxylase and alcohol dehydrogenase) are inducible and that the remaining enzymes are constitutive. This is apparently because the first and probably the second enzymes are plasmid-encoded and subsequent enzymes are encoded for by chromosomal genes. While the logic for the locations of these pathway genes is unclear, the first two enzymatic reactions may serve to funnel the alkanes into a common cellular pathway. Gene regulation of degradative pathways encoded by the IncP2 CAM plasmid, plasmid pJP2 (IncP3), and the naphthalene-degrading plasmids belonging to the IncP7 group have not been genetically characterized to date.

The genetic regulation of the pWWO (TOL) plasmid has been intensely investigated and its regulation is well understood. Induction of upper-pathway enzymes can be by toluene or *m*-methylbenzyl alcohol, the lower pathway by *m*-toluic acid or benzoate. The genes for toluene degradation are organized into two operons, the first operon (*xylCAB*) encodes for enzymes that metabolize toluene to benzoate. The second operon (*xylDLEGF*) encodes for enzymes that degrade benzoate to acetaldehyde and pyruvate. Both of these operons are positively regulated by two regulatory genes, *xylS* and *xylR*. When cells are presented with toluene, the *xylR* gene product, which is transcribed constitutively, activates the transcription of the first operon and the regulatory gene *xylS*. The *sylS* gene product then activates the transcription of the second operon. The *xylS* gene can act independently of the *xylR* gene in the presence of benzoate and activate the *xylDLEGF* operon. Thus, this positively regulatory system for toluene or benzoate degradation represents a very efficient mechanism of controlling metabolic pathways. Studies have also suggested a possible chromosomally encoded gene that, in the presence of unsubstituted compounds like benzoate or benzyl alcohol, can induce the transcription of the *xylDLEGF* genes. It has been postulated that this chromosomal gene may encode for a positive-regulatory protein required for expression of a chromosomally encoded benzoate dioxygenase gene. Interestingly, numerous studies have investigated the induction of TOL genes; however, to our knowledge, no information is available concerning repression of these pathways.

The catabolic genes for naphthalene degradation on the NAH7 plasmid are induced by salicylate or possibly salicylaldehyde. Researchers have demonstrated that constitutive expression of the early naphthalene genes, which converts naphthalene into salicylate, is at a low level. The newly formed salicylate can act as inducer of the entire pathway to completely induce the naphthalene-degrading genes (Fig. 6). Transposon mutagenesis of the regions of the NAH7 operon has identified a regulatory region involved in the induction of the naphthalene genes. This regulatory region (*nahR*) encodes for a 36-kDa protein. Although the mechanism of how the *nahR* gene and salicylate molecule interact to exert positive regulation on the naphthalene genes is not certain, several investigations have demonstrated that the *nahR* protein can bind to the promoter regions of both operons *nah1* and *nah2* in the presence or absence of salicylate. It is proposed that salicylate interacting with the NahR protein, perhaps altering NahR's interaction with the DNA activates transcription of the *nah* genes. The naphthalene-degrading genes are not repressed by compounds normally and preferentially utilized by bacterial cells such as succinate, glucose, or complex-rich media. [*See* CATABOLITE REPRESSION.]

B. Gene Mapping

Gene mapping determines the number of genes in the pathway and the order of genes. Figure 7 depicts schematic drawings of gene order for the following catabolic plasmids: pSS50, pJP4, OCT, CAM, pWWO, and NAH7.

Restriction analysis of plasmid pSS50 has revealed a restriction-rich region located between *oriV* and *trfA* genes. This restriction-rich region contains

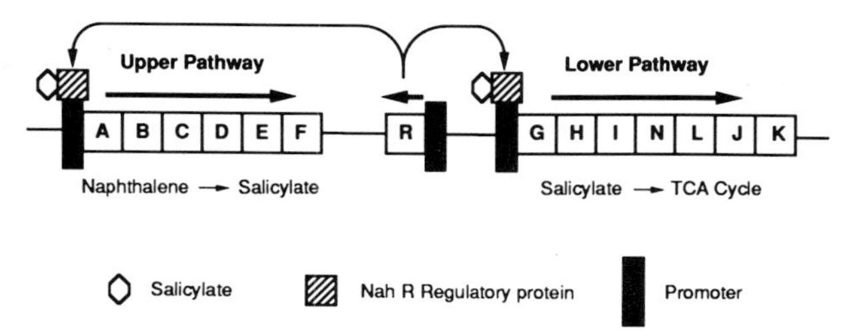

Figure 6 NAH7 catabolic gene organization and regulation. Schematic diagram of the *nah* genes and regulatory elements encoded on the NAH7 plasmid.

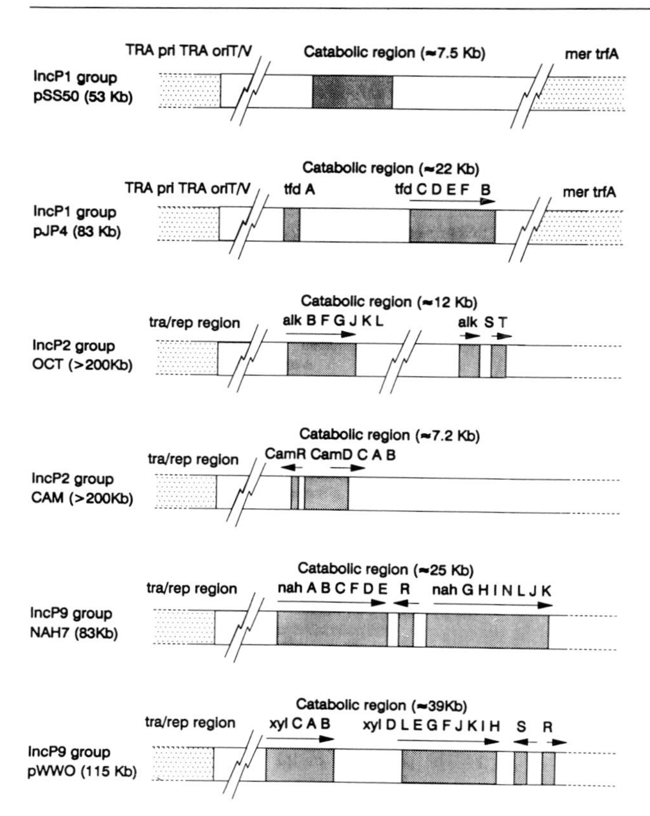

Figure 7 Gene order for several catabolic plasmids. Arrows over genes indicate proposed direction of transcription. Drawings are only to scale for catabolic regions, where 1 cm equals approximately 4 kbp. Letters represent pathway genes. Abbreviations: mer, mercury resistance gene; oriT, origin of transfer; oriV, origin of vegetative replication; pri, primase gene; TRA, transfer genes; tra/rep, transfer and replication genes.

a 2.2-kb repeat sequence and flanking regions suspected to encode for the catabolic genes. The original plasmid is known to undergo deletions resulting in a smaller plasmid with reduced catabolic potential. The deletion of DNA is correlated with the simultaneous loss in the ability to mineralize 4-chlorobiphenyl. The remaining smaller plasmid retains the ability to degrade 4-chlorobenzoate, an intermediate in 4-chlorobiphenyl degradation. Studies in this and other laboratories suggest that the degradative genes on pSS50 are located on a transposable element that freely exchanges with the chromosome.

Plasmid pJP4 is approximately 80 kb with only a small portion of this genetic material encoding for 2,4-D degradation. The *tfdA* gene has been localized to a 1.4-kb DNA fragment and the complete sequence is known. Using this 1.4-kb DNA sequence, a 32-kDa protein was synthesized *in vivo* and identified as 2,4-D monooxygenase. The *tfdA* gene was found to be located approximately 13 kb from a 6.3-kb fragment that contains the *tfdCDEF* structural genes. To our knowledge, only the promoter region for *tfdCDEF* and the structural gene for *tfdC* have been sequenced to date. A negative control regulatory gene, *tfdR*, has been identified on plasmid pJP4, but its mechanism for regulation of 2,4-D degradation is not understood. Recently, a second gene on pJP4 that encodes for 2,4-D monooxygenase was identified and designated *tfdAII*. The *tfdAII* gene is located adjacent to the *tfdC* gene. The reason the *tfdA* gene is duplicated remains unclear.

The degradative genes encoded on the OCT plasmid have been localized into two operons. The operon *alkBAC* encodes for seven proteins and is transcribed as a single messenger RNA molecule, 7.3 kb in length. The *alkB* locus encodes for the alkane hydroxylase located in the cytoplasmic membrane of cells containing the OCT plasmid. The *alkA* locus encodes for rubredoxin 1 and 2, one of which comprises a soluble component of alkane hydroxylase. An *alkC* locus encodes for four proteins of aldehyde dehydrogenase and alkanol dehydrogenase and two other proteins whose functions are unknown at this time. A second genetic locus, *alkST*, has been identified on the OCT plasmid. This DNA region encodes for the *alkT* gene, which is the rubredoxin reductase protein and a regulatory gene *alkS*. The *alkS* gene product, a 99-kDa protein, acts as a positive regulator and is required for induction of the alkane-degrading genes.

The CAM plasmid has received limited characterization due to its large size; only the genes for the first two biochemical reactions have been located. Genetic analysis of the CAM plasmid has located the *camDCAB* structural genes and *camR* regulatory gene on a 7.2-kb DNA fragment. A detailed restriction map has been constructed and the genes have been cloned and partially sequenced. The *camDCAB* genes form an operon that is negatively regulated by the *camR* gene located just upstream and transcribed in the opposite direction of the *camR* gene.

The IncP3 plasmid pJP2, which degrades 2,4-D has a smaller size (57 kb) than pJP4 (80 kb) and demonstrates different restriction patterns from plasmid pJP4. However, studies have shown that restriction analysis can be misleading, and often there is DNA conservation between genes only at specific regions. These specific regions may encode for essential regions for enzyme function while alterations in the sequences outside of these areas will not affect the physical properties of the enzyme. Whether or not this is the case with enzymes for 2,4-D degradation in pJP4 and pJP2 is not known. However, DNA hybridization studies indicate homology does exist between these two plasmids, probably between the degradative genes. Cloning studies have identified a 20-kb DNA fragment in plasmid pJP2, which encodes for 2,4-D catabolism. Restriction analysis and DNA hybridization studies suggest that the 2,4-D genes are more tightly clustered on plasmid pJP2 than on plasmid pJP4.

The IncP7 group naphthalene-degrading plasmids has not been genetically analyzed to date.

The IncP9 pWWO plasmid (TOL) is approximately 117 kbp in size and is self-transmissible. An interesting feature of the TOL and TOL-like plasmids is that the genes for degradation enzymes are located on a 56-kbp transposon. The catabolic genes are clustered into two operons. The first operon, *xylCAB,* encodes for enzymes that convert xylenes and toluene to the aromatic carboxylic acids *m*-toluate and benzoate. The second operon, *xylDLEGF,* which encodes for the conversion of aromatic carboxylic acids to anaplerotic compounds for the tricarboxylic acid cycle, is located approximately 14 kb from the first operon. Two regulatory operons, *xylR* and *xylS,* are adjacent to the *xylDLEGF* operon. The *xylDLEGF* operon actually contains the additional genes *xylD, L, E, G, F, J, I,* and *H* in this order.

The NAH7 plasmid is similar to the TOL plasmid in many aspects. The *nah* genes are clustered into two separate operons located within a 30-kb region on the NAH7 plasmid. The probable gene order of the *nah1* operon is *Nah A, B, C, F, D,* and *E* encoding genes that convert naphthalene to salicylate, designated the upper pathway. Oxidation of salicylate to acetylaldehyde and pyruvate is encoded by the genes of the second *nah2* operon. The proposed gene order is *nahG, H, I, NL, J,* and *K.* Coordinate regulation of the *nah* genes is by *nahR,* a gene located between the two gene clusters, and exerts positive control on both *nah1* and *nah2* operons.

To our knowledge, the cresol plasmids have not been genetically characterized.

C. Noncatabolic Regions

Because very little genetic information is available on the noncatabolic regions of degradative plasmids, we will not address each plasmid individually as in earlier sections; instead, in this section we will discuss all catabolic plasmids in general terms. The plasmid backbone sequences necessary for plasmid maintenance and transfer, may be similar because of the similarity of their basic functions, would be similar from plasmid to plasmid, suggesting that the DNA sequences encoding these functions may also be similar. One phenomenon, observed for all catabolic and noncatabolic broad host-range plasmids, is the paucity of restriction sites in regions encoding plasmid maintenance and transfer. Scientists hypothesize that selection pressure would favor few restriction sites in these essential regions of broad host-range plasmids to protect against degradation as they move from host to host.

Most degradative plasmids are self-transmissible; in fact, all the plasmids outlined above are self-transmissible. The process by which bacteria transfer their plasmids is known as conjugation. The mechanism of conjugation has been examined in detail in fertility plasmids and shown to require numerous genes known as *tra* genes. Approximately 33 kb of DNA has been shown to be required for the conjugation process encoding for processes such as pili formation, stabilization of mating pairs, and DNA nicking. [*See* CONJUGATION, GENETICS.]

Other housekeeping functions such as plasmid replication and copy number control would be expected to be similar for all the IncP catabolic plasmids discussed in this article.

VI. Engineered Plasmids

A. Molecular Construction

Genetic manipulation and construction of genetically engineered plasmids, which degrade contaminating compounds, are still in their inception. However, the few examples of genetically engineered catabolic plasmids have been with the best understood and characterized plasmids. In 1973, a cointergret was formed with the incompatible CAM and OCT plasmid by treating the plasmids with ultraviolet irradiation and selecting for a fused cointergrate. The result was a CAM–OCT plasmid that allowed the host bacteria to degrade camphor and alkanes. The advantage of multiple degradative pathways in one cell can easily be recognized as providing the cell with alternative carbon sources. In addition, most contaminate sites are rarely contaminate with one specific substance; therefore, multiple degradative pathways would increase the probability of cell survival in complex environments.

An example of genetic manipulation to expand a strain's degradative potential was with *Pseudomonas* sp. B13. *Pseudomonas* sp. B13 could only degrade 3-CB due to a narrow substrate specificity of the first enzyme of the pathway, benzoate 1,2-dioxygenase. The broad substrate-specific enzymes of the TOL plasmid, toluate 1,2-dioxygenase, 1,2-dihydroxycyclohexadiene carboxylase dehydrogenase, and regulatory genes for these enzymes were cloned onto a plasmid, and this plasmid was moved into *Pseudomonas* sp. B13. The presence of these enzymes allowed growth on 4-chlorobenzoate and 3,5-dichlorobenzoate. The cloned *nahG* gene from the NAH7 plasmid was also introduced into *Pseudomonas sp*. B13, thus allowing catabolism of 3-, 4-, and 5-chlorosalicylates.

B. Stability and Maintenance

Broad host-range vectors (plasmids that contain genes necessary for replication and maintenance and do not depend on host-encoded functions) can be mobilized into a variety of bacteria where they will be replicated and maintained. The level of knowledge in molecular biology at the present time allows selection of a plasmid that will be maintained in a particular host bacterium. However, in some cases the cloned DNA fragment may be unstable and lost in a particular host. Instability of cloned DNA fragments can often be circumvented if the researcher considers properties of the host bacteria. Molecular biology has permitted construction of bacterial strains with specific properties that improve the stability of foreign (cloned) DNA. One such property is recombination-deficient host strains (*rec*), which cannot exchange homologous DNA fragments between plasmids and the chromosome. The inability of a strain to perform homologous recombination is very important in maintaining a plasmid DNA, which could otherwise be lost (recombined) with the chromosome or a second plasmid. Bacteria have a restriction modification system that allows recognition and destruction of foreign DNA while safeguarding their own DNA. However, this system can be a problem when attempting to express foreign genes in bacterial host. Construction of a restriction-deficient host also improves stability and maintenance of foreign DNA on a plasmid vector. Both restriction and recombination-deficient bacterial hosts are available in several *Pseudomonas* species, which are useful for expression of degradative genes. Plasmid vectors that provide low copy or high copy numbers per cell can be chosen. Typical plasmid copy numbers range from 1 or 2 to 50 plasmid copies per cell. The copy number ultimately controls gene dosage and can influence degradation rates if the recombinant plasmid contains degradative genes. However, some genes may have toxic effects if expressed in high concentrations (i.e., overproduction of membrane proteins); therefore, a researcher must use caution when selecting plasmid copy number.

VII. Environmental Constraints

A. Host Range, Transmission, and Stability within and outside Genus

To increase the role of degradative plasmids in pollution control and catabolism of contaminating compounds, host range, plasmid transfer, stability, and expression of catabolic genes become an important consideration. There is limited information on transfer of catabolic plasmids in nature. However, because characterization of numerous catabolic plasmids in the laboratory has demonstrated the presence of transfer genes on the majority of catabolic plasmids, conjugation is probably an important mode of gene transfer in naturally occurring bacterial populations. Recent studies with soil bacteria have demonstrated intrageneric and intergeneric transmission and expression of plasmids. These same studies have shown that nonmobilizable plasmids in the presence of mobilizable plasmids can also move into new host bacteria. A phenomenon known as retrotransfer (recipient to donor) has been demonstrated for IncP1 plasmids. Retrotransfer often involves the intergration of chromosomal genes into a plasmid vector, thus, increasing the variety of genes which can be exchanged between bacteria.

Studies have demonstrated that the TOL plasmid is derepressed for transfer in its natural host *P. putida* mt-2. *P. putida* containing the TOL plasmid expressed thick flexible conjugative pili constitutively. It was also determined that the TOL plasmid is transferred much more effectively on solid surfaces rather than liquid media. Studies to investigate degradative plasmid transfer in environmental settings are still in their infancy. IncP1 plasmid pSS50 was originally isolated from *Acinetobacter* and *Alcaligenes* spp. Plasmid pSS70 is approximately 77 kb, but it is unstable under laboratory conditions and spontaneously undergoes deletions of 23 kb with the simultaneous loss in the ability to mineralize 4-chlorobiphenyl. The plasmid derivative of pSS50 does retain the ability to dechlorinate 4-chlorobenzoic acid. Plasmid pSS50 or its derivatives have been mobilized into *P. putida* and *E. coli*. Restriction analysis and DNA hybridization studies indicate that the complete plasmid is maintained in *E. coli* but not in *P. putida*. *Pseudomonas putida* transconjugants retain the plasmid backbone regions but consistently lose the DNA segment thought to encode for catabolic abilities. Attempts to transfer the plasmid into *P. aeruginosa* strains were unsuccessful.

Plasmid pJP4 was originally isolated from *Alcaligenes eutrophus*. The presence of pJP4 allowed *Alcaligenes paradoxus*, *P. putida*, *Pseudomonas oxalacticus*, and *Pseudomonas cepacia* to utilize 2,4-D as a sole source of carbon and energy. However, when pJP4 was introduced into *E. coli*, *Rhodopseudomonas sphaeroides*, *Rhizobium* sp. *Pseudomonas fluorescens*, *Agrobacterium tumefaciens*, and *Acinetobacter calcoaceticus*, the resulting strains could not utilize 2,4-D as their sole source of carbon and energy. Studies indicate that the enzymes encoded by pJP4 were expressed in these hosts; however, these strains probably do not encode for enzymes that further degrade the product (chloromaleylacetic acid) from the enzymatic reactions encoded by pJP4. Another possible constraint in expressing degradative genes in various bacterial hosts is the intolerance to possible toxic intermediates, which may affect some bacteria more than others. Reports on pJP4 did not include information about plasmid stability in various hosts; therefore, we assume that plasmid pJP4 was stable under these conditions.

The IncP2 group of plasmids have a limited host range restricted to the genus *Pseudomonas*. The CAM plasmid has been transferred to many of the *Pseudomonas* species, including *P. aeruginosa* and *P. fluorescens*. In these two species the CAM plasmid was maintained and the genes for camphor metabolism were expressed. The OCT plasmid can also be transferred to various fluorescent pseudomonads with stable expression of the plasmid-encoded OCT genes. The CAM–OCT plasmid resulting from fusion of the two plasmids is stable and has a host range similar to that of the two parental plasmids.

To our knowledge, little or no work concerning plasmid transfer and stability has been done with either pJP2 (IncP3) or naphthalene-degrading plasmids (IncP7).

The IncP9 plasmid pWWO can be transferred to *E. coli*. The replicative and transfer functions are expressed in this host; however, the TOL phenotype is not expressed. Examination of specific enzyme activities encoded by TOL plasmid genes had very limited activity. Evidently, this limited activity was not enough to allow growth of *E. coli* cells on toluene as the sole source of carbon and energy. The pWWO (TOL) plasmid can be transferred to *P. aeruginosa* where it is replicated, conjugative and expresses TOL catabolic genes. Genes from the TOL plasmid have been expressed in *E. coli*, when the structural genes were fused with an efficient *E. coli* promoter.

A 10.5-kb fragment of the NAH7 plasmid expressing naphthalene dioxygenase was inserted into plasmid vector pBR322. This plasmid was transferred into *E. coli* HB101 and shown to express naphthalene dioxygenase in the biosynthesis of indigo. Identifying a suitable bacterial host for gene expression is sometimes difficult, and it is often recommended to work with plasmids and plasmid derivatives in the original host. Studies have shown *P. aeruginosa*, the best genetically and biochemically characterized *pseudomonad*, to cause deletions within cloned *E. coli* DNA containing tryptophan operon fragments.

The *E. coli* cloning system, frequently chosen as a host bacteria, has many advantages. The genetics and physiology of the *E. coli* cell has been well characterized and numerous strain and vector constructions make cloning in *E. coli* attractive. However, there are disadvantages to the *E. coli* system. *Pseudomonas* gene expression is inefficient in *E. coli* mainly because *Pseudomonas*, promoters are not recognized efficiently by *E. coli* RNA polymerase. Poor promoter recognition may be due to the absence ancillary factors, such as sigma factors, which assist RNA polymerase in the recognition of specific DNA promoters. Other factors such as nucleotide mismatches with *E. coli* promoter sequences and differences in codon usage between *Pseudomonas* and *E. coli* are disadvantages for expression of biodegradative genes in the *E. coli* system.

VIII. Summary and Conclusions

Since the discovery of the CAM catabolic plasmid in 1973, the field of biodegradation has progressed steadily. Due to the hard work, dedication, and years of diligent labor, an understanding of the biochemistry and genetics of biodegradation is impacting the scientific fields of biotechnology and biology. Although the genetics and physiology of bacteria isolated from the environment may never reach the levels of the enteric *E. coli* or *Salmonella* bacteria, the progress made with broad host-range vectors, transposons, promoter probes, vectors designed to express plasmid-encoded genes, etc., will allow the rapid and productive investigation of many catabolic gene functions. Future work will provide ideas for and insight into the evolution of catabolic plasmids and the construction of plasmids that will enhance natural degradation rates and, in some instances, convert unwanted toxic compounds into valuable compounds.

Acknowledgment

Grant support was provided by U.S. Air Force contract #F49620-89-C-0023.

Bibliography

Burlage, R. S., Hooper, S. W., and Sayler, G. S. (1989). *Appl. Environ. Microbiol.* **55,** 1323–1328.

Burlage, R. S., Bemis, L. A., Sayler, G. S., and Larimer, F. (1990). *J. Bacteriol.* **172,** 6818–6825.

Don, R. H., and Pemberton, J. M. (1981). *J. Bacteriol.* **145,** 681–686.

Koga, H., Yamaguchi, E., Matsunaga, K., Aramaki, H., and Horiuchi, T. (1989). *J. Biochem.* **106,** 831–836.

Owen, D. J., Eggink, G., Hauer, B., Kok, M., McBeth, D. L., Yang, Y. L., and Shapiro, J. A. (1984). *Mol. Gen. Genet.* **197,** 373–383.

Pemberton, J. M. (1983). *Int. Rev. Cytol.* **84,** 155–183.

Reineke, W., and Knackmuss, H. J. (1988). *Annu. Rev. Microbiol.* **42,** 263–287.

Schell, M. A. (1990). Regulation of the naphthalene degradation genes of plasmid NAH7: Example of a generalized positive control system in *Pseudomonas* and related bacteria. *In* "*Pseudomonas* Biotransformations, Pathogenesis, and Evolving Biotechnology" (S. Silver, A. M. Chakrabarty, B. Iglewski, and S. Kaplan, eds.), pp. 165–176. *Amer. Soc. Microbiol.*, Washington, D.C.

Shields, M. S., Hooper, S. W., and Sayler, G. S. (1985). *J. Bacteriol.* **163,** 882–889.

Witholt, B., Sijtsema, J., Kok, M., and Eggink, G. (1990). Oxidation of alkanes by *Pseudomonas olevorans*. *In* "*Pseudomonas* Biotransformations, Pathogenesis, and Evolving Biotechnology" (S. Silver, A. M. Chakrabarty, B. Iglewski, and S. Kaplan, eds.), pp. 141–150. ASM, Washington, D.C.

Catabolite Repression

Milton H. Saier, Jr., and Matthew J. Fagan
University of California, San Diego

Glossary

α-Factors Sigma subunits of bacterial RNA polymerases that function in sequence recognition and confer promoter specificity for transcriptional initiation

Amphibolic pathways Metabolic pathways that serve both anabolic (biosynthetic) and catabolic (degradative) functions

Antisense RNA RNA synthesized from the complimentary (non-mRNA encoding) DNA strand

Autogenous regulation Self-regulation: transcriptional regulation by components encoded within the regulated operon

Catabolic operon Operon (transcriptional unit) that encodes catabolic (degradative) enzymes and transport proteins

Continuous culture Continuous growth of a bacterial culture, usually in a chemostat

Enteric bacteria Gram-negative intestinal eubacteria and their close relatives (i.e., *E. coli* and *S. typhimurium*)

Fermentative growth Growth resulting from the utilization of a carbon source in the absence of oxygen. Fermentative carbon utilization is always incomplete resulting in the release of organic acids.

Intracellular homeostasis Capacity to maintain constant conditions in the cytoplasm of a cell

Km Michaelis-Menten kinetic constant, reflecting the apparent affinity of an enzyme or transport system for its substrate

Methyl α-glucoside Glucose analog, which is nonmetabolizable in *E. coli*; it is used to study sugar transport and accumulation via the glucose permease separate from events involved in sugar metabolism

Open (preinitiation) complex High affinity complex of the DNA with RNA polymerase in which the two DNA strands are separated in the region of RNA polymerase binding

Operator Control region of an operon which binds the repressor and/or activator proteins

Oxidative growth Growth resulting from the utilization of a carbon source in the presence of oxygen; carbon utilization can be complete, resulting in oxidation to carbon dioxide

Pleiotropic Multifaceted, that is, having global effects on many target systems

Promoters Specific DNA sequences located upstream from the start codon of the first cistron in an operon that serve as binding sites for RNA polymerase and the cyclic AMP receptor protein

***ptsH* or *ptsI* mutants** Mutants defective for HPr or Enzyme I of the phosphoenolpyruvate : sugar phosphotransferase system (PTS); leaky mutants retain residual activity of one of these proteins.

Second messengers Cytoplasmic molecules that mediate the activities of hormones or other primary stimulants. In microorganisms, these stimulants may be indicators of energy availability or metabolic state

CARBON CATABOLITE REPRESSION is the phenomena whereby the presence of a rapidly metabolizable carbon source in the growth medium of a microorganism inhibits, at the transcriptional level, synthesis of carbon catabolic enzyme systems and related proteins. The carbon catabolic enzyme systems include the solute-specific permease proteins as well as the enzymes which initiate metabolism of the carbon sources. Peripheral functions related to carbon utilization include motility, adhesion, bioluminescence, pigment formation, differentiation, and intracellular macromolecular degradation. Enzymes comprising central pathways of carbon utilization and synthesis have also been shown to be responsive to carbon catabolite repression. The repressive mechanisms in bacteria and yeast are multifaceted as several such mechanisms exist. The only well-understood mechanism, however, is the cyclic AMP/cyclic AMP receptor protein system of *Escherichia coli*. In eukaryotes, mitochondrial functions can also be pleiotropically subject to catabolite repression. The unrelated phenomena of nitrogen, sulfur, and phosphorus mediated repression are not considered in this article.

I. Historical Perspectives and Introduction

Catabolite repression was initially defined as the phenomenon whereby the presence of a rapidly metabolizable carbon source in the growth medium of a microorganism inhibits the synthesis of enzymes involved in the metabolism of other carbon-containing compounds. The phenomenon was discovered in 1901 and was then studied for over half a century from a physiological standpoint. Because glucose was often the most effective carbohydrate causing repression of the synthesis of the target catabolic enzymes, the phenomenon became known as "the glucose effect." The early analyses of the glucose effect led to the postulate that it occurs whenever growth conditions are such that catabolism (degradation) exceeds anabolism (biosynthesis). This postulate in turn led to a second hypothesis, namely that it was the accumulation of one or more cytoplasmic catabolite(s), derived from the repressing carbohydrate, that gave rise to the glucose effect. This concept, now known to be at least partly in error, caused the term "catabolite repression" to be coined. We now know that many proteins in addition to carbon catabolic enzymes are subject to catabolite repression.

The glucose effect consists of at least three kinetically and mechanistically distinguishable phenomena, termed "permanent repression," "transient repression," and "inducer exclusion." Permanent repression, which occurs without release as long as the repressing carbon source is present in the medium, is sometimes equated with catabolite repression. Transient repression is an exceptionally severe form of repression that occurs for a limited period of time following addition of glucose, a nonmetabolizable glucose analog, or another carbohydrate to a bacterial culture actively growing on a different carbon source. Inducer exclusion is defined as the exclusion of the inducer of a catabolic operon from the cell upon addition of the repressing carbohydrate to the culture medium. All three of these processes depress the rate of initiation of transcription of the target genes encoding the repressible catabolic enzymes and therefore should be considered together as different aspects of the same topic. While at least some mechanisms of inducer exclusion and permanent repression are well understood, that of transient repression remains an enigma.

In the mid-1960s, inducer exclusion was first clearly demonstrated, and bacterial cyclic adenosine-3′,5′-monophosphate (cAMP) was first discovered. It was shown that exogenous glucose both inhibited the synthesis of cAMP in intact *E. coli* cells and stimulated the efflux of cAMP from the cell cytoplasm. Cyclic AMP efflux, catalyzed by the cyclic nucleotide transport system, was later shown to be driven by the membrane potential as an energy source, whereas inhibition of cAMP synthesis, catalyzed by the enzyme adenylate cyclase, involved the proteins of the phosphoenolpyruvate (PEP)–sugar phosphotransferase system (PTS). Both catabolite and transient repression were shown to be promptly reversed by addition of cAMP to the culture medium, and after *extended* growth in the presence of cAMP plus an inducer, the phenomenon of inducer exclusion was also abolished. These observations led to the postulate that cAMP and its receptor protein (CRP), also known as the catabolite gene activator protein (CAP), mediated catabolite repression. It is, however, clear that in Gram-negative bacteria such as *E. coli*, cAMP-independent mechanisms of catabolite repression are operative and that cAMP does not mediate catabolite repression in the gram-positive bacterium *Bacillus subtilis*. A different mechanism of catabolite repression is also operative in eukaryotes such as the well-studied yeast *Saccharomyces cerevisiae*. In both *B. subtilis* and *S. cerevisiae*, adenosine triphosphate (ATP)-dependent protein kinases may play a role in catabolite repression.

As summarized in Table I, the syntheses of many bacterial proteins and organelles, most of which are directly or indirectly involved in carbon and energy utilization, are subject to catabolite repression. These include enzymes that function in the extracellular degradation of carbon-containing macromolecules (carbohydrases, lipases, and some proteases), the chemoreceptors, permeases and enzymes that initiate the catabolism of exogenously supplied carbon compounds, converting them to common intermediates of the central metabolic pathways, and enzymes that comprise the central metabolic pathways involved in carbon and energy utilization or generation. These pathways include *amphibolic pathways* (i.e., the Krebs cycle, the electron transfer chain, and the glyoxylate shunt), *catabolic pathways* (those involved in fatty acid, amino acid, and nucleotide degradation), and *anabolic pathways* (gluconeogenesis, glycogen biosynthesis, and fatty acid biosynthesis). In yeast, essentially all mitochondrial functions are subject to glucose repression.

A number of bacterial surface organelles such as flagellae (organelles of motility) and fimbriae (organelles of adhesion) are subject to catabolite repression. In enteric bacteria, these organelles, like almost all of the carbon catabolic enzymes, are regulated by cAMP and its receptor protein, CRP. Functions involving the emission of bioluminescent light and the capture of or protection from light (i.e., pigment-containing proteins as well as related electron carriers) are also subject to repression. Additionally, in both nonsporulating and sporulating bacteria, proteins that allow toleration of extreme environmental conditions and of starvation conditions are made in the stationary phase of growth, and these proteins are similarly subject to catabolite repression. Finally, several functions that have no clear relationship to carbon utilization have been reported to be under cAMP and catabolite-repressive control (Table I). Different mechanisms of catabolite repression may be operative in different organisms, and in any one organism several mechanisms of catabolite repression may exist. [*See* FLAGELLA; ADHESION, BACTERIAL.]

Table I Some Bacterial Functions Subject to Catabolite Repression

Organelle or protein	Representative organism (genus)	Function
1. Extracellular degradative enzymes	*Serratia* *Bacillus*	Carbon and energy metabolism; carbohydrate biosynthesis
2. Carbohydrate chemoreceptors, permeases, and catabolic enzymes	*Escherichia* *Bacillus*	
3. Central enzymes of carbon metabolism	*Escherichia* *Bacillus*	
4. Glycogen biosynthetic enzymes	*Escherichia*	Carbon storage
5. Flagellae	*Escherichia*	Motility
6. Fimbriae	*Salmonella*	Adhesion
7. Luciferase repressor	*Vibrio*	Bioluminescence
8. Pigment proteins	*Serratia* *Rhodobacter*	Photoprotection Photosynthesis
9. Scavenging proteins and stress resistance proteins	*Escherichia*	Starvation survival
10. Sporulation-inducing proteins	*Bacillus* *Myxobacteria* *Streptomyces*	Sporulation (resting state)
11. Chloramphenicol acetyl transferase	*Escherichia*	Drug resistance
12. Proteins catalyzing phage repressor degradation	*Salmonella* *Escherichia*	Phage lysogeny
13. Outer membrane proteins	*Escherichia*	Outer membrane permeability

II. The cAMP System in *Escherichia coli*

Cyclic AMP serves as one of several second messengers of hormone action in animal cells, and in enteric bacteria it is a primary agent mediating control of carbon utilization. In all living cells that contain it, cAMP is made from ATP in a reaction catalyzed by the enzyme adenylate cyclase, is degraded to 5′-AMP in a hydrolytic reaction catalyzed by cAMP phosphodiesterase, and is excreted from the cell by an energy-dependent cAMP transport system. In enteric bacteria, adenylate cyclase is encoded by the *cya* gene (85 min on the *E. coli* chromosome; 83 min on the *Salmonella typhimurium* chromosome), the cAMP phosphodiesterase is encoded by the *cpd* gene (64 min on the *S. typhimurium* chromosome; not mapped in *E. coli*), and the gene encoding the cAMP transport system has not been identified in either organism. The action of cAMP in mediating repression in response to carbon availability is effected by the CRP encoded by the *crp* gene (74 min on the *E. coli* chromosome; 72 min on the *S. typhimurium* chromosome). Mutants that lack either adenylate cyclase (*cya* mutants) or the cAMP receptor protein (*crp* mutants) are negative for the utilization of almost all carbon sources including sugars, glycolytic intermediates, and Krebs cycle intermediates.

In contrast to the wild-type strains, the catabolic enzymes and permeases responsible for the utilization of these compounds are generally not inducible in the *cya* and *crp* mutants. This fact reflects the involvement of the cAMP–CRP complex in positive regulation of the transcriptional expression of the corresponding structural genes (see later). Of the sugars, *cya* and *crp* mutants of *E. coli* and *S. typhimurium* utilize only glucose, fructose, and galactose, all at reduced rates. Interestingly, the genetic defects manifest themselves in different ways. For example, synthesis of the glucose transport system or permease (glucose Enzyme II of the PEP-dependent PTS) is normally inducible by growth in the presence of glucose in the wild-type strain, but not in *cya* and *crp* mutants. In these mutants, the glucose permease exhibits basal activity, which is apparently sufficient for growth. By contrast, the fructose permease is expressed at reduced levels in *cya* and *crp* mutants in both the uninduced and the induced states, but the degree of induction is virtually the same in the mutants as in the wild-type strain. Thus, the detailed molecular mechanism of action of the cAMP–CRP complex must be different in these two cases because the glucose-specific induction mechanism is dependent on the cAMP–CRP complex while the fructose-specific induction mechanism is not. The situation for galactose will be discussed in Section VII.

While the cAMP–CRP complex clearly controls synthesis of the constituents of the PTS, as noted above for the glucose and fructose permease proteins, a reciprocal relationship also exists. Thus, the PTS controls the synthesis of cAMP. Consequently, to understand PTS-mediated regulation of adenylate cyclase, it is important to know the constituents of the PTS. This system consists of two general energy-coupling proteins that are always present in substantial levels. These proteins are termed Enzyme I and HPr. There are also two homologous proteins encoded within the fructose regulon, and they are termed Enzyme I^{fru} and FPr. Enzyme I^{fru} is normally expressed at low, cryptic levels, whereas FPr is inducible by growth in the presence of fructose; their roles in the regulation of cellular physiology have not been defined, but mutations altering their expression or activities alter physiological processes, suggesting that they possess regulatory functions. In addition to these energy-coupling proteins, sugar-specific permeases, often designated the Enzymes II or Enzyme II–III pairs, are constituents of the PTS.

Recent molecular analyses of these permease proteins have revealed that they generally consist of three domains or proteins that serve a uniform function. The first, designated IIA, bears the first permease-specific phosphorylation site; the second, designated IIB, bears the second permease-specific phosphorylation site; and the third, designated IIC, is the transmembrane segment that bears the sugar-binding site and the transmembrane channel. Thus, the proteins of the PTS comprise a phosphoryl transfer chain that sequentially transfers phosphate from the phosphoryl donor, PEP, to several protein intermediates before transferring it to sugar.

III. Catabolite Repression and the Regulation of Intracellular cAMP

The intracellular concentration of cAMP in a bacterial cell can theoretically be regulated by modulating the rates of cAMP synthesis, degradation, and/or transmembrane transport. The cyclic nucleotide phosphodiesterase is clearly active in intact gram-negative enteric bacterial cells because genetic loss of this enzyme results in a marked enhancement of the internal levels and net production of cAMP. However, the activity of this enzyme does not appear to respond to changes in environmental conditions in a fashion that can account for catabolite repression. By contrast, the activities of both adenylate cyclase and the cAMP transport system have been shown to depend on exogenous carbon and energy sources available to the cell, and these regulatory interactions appear to be of physiological importance.

IV. Phosphotransferase System-Mediated Control of Adenylate Cyclase

Several studies have shown that adenylate cyclase of enteric bacteria is subject to regulation by a mechanism that depends on the activities of the proteins of the PTS. The effects of mutations in genes coding for the proteins of the PTS on adenylate cyclase regulation have been studied in some detail, and several conclusions have been drawn from the results reported:

1. Net synthesis of cAMP ceases shortly after wild-type cells reach the stationary growth phase. This effect is due to the loss of *in vivo* adenylate cyclase activity and can be reversed only by growth under conditions that allow active protein synthesis.
2. The net production of cAMP is not altered by a mild deficiency in Enzyme I, although complete loss of Enzyme I depresses cAMP synthesis. Adenylate cyclase activity generally depends on the carbon source and growth conditions employed, being greatest when examined in washed cells that had been grown on a carbohydrate that can be rapidly metabolized.
3. Sugar substrates of the PTS inhibit cAMP production, and leaky *ptsH* and *ptsI* mutations, which, respectively, result in the genetic reduction of active HPr and Enzyme I, enhance sensitivity to this effect. Inhibition of adenylate cyclase is a permanent effect when the sugar is added to freshly washed cell suspensions.
4. Mutations to carbohydrate repression resistance, due to mutations in the *crr* gene, which encodes the glucose Enzyme III (III^{glc}) of the PTS, decrease the amount of cAMP produced to a level substantially below that found in the wild-type strain. *crr* mutations depress cAMP synthesis in most strains of *E. coli* and *S. typhimurium* tested, but cAMP synthesis is then insensitive to regulation by sugar substrates of the PTS.
5. A variety of sugars that are not substrates of the PTS also depress adenylate cyclase activity if the corresponding catabolic enzymes are present at high levels. However, different mechanisms, not dependent on the PTS, appear to be operative.
6. In mutant strains of *S. typhimurium,* which contain very low Enzyme I activity ($<0.1\%$ of wild-type activity), exceptionally low concentrations of methyl α-glucoside ($0.05\ \mu M$) inhibit net cAMP synthesis. Because the apparent K_m of the PTS for methyl α-glucoside uptake in wild-type *S. typhimurium* cells is >1000 times this value, this result argues against a regulatory mechanism involving direct interaction of a sugar receptor (i.e., an Enzyme II) with adenylate cyclase and provides definitive evidence that the PTS proteins function catalytically to regulate adenylate cyclase.

A mechanism that accounts for these results has been proposed and is shown in Fig. 1. The left portion of the figure shows the phosphate transfer chain of the PTS. The phosphoryl group of PEP can be sequentially transferred to Enzyme I, then to HPr, and finally to sugar in a reaction catalyzed by the phosphorylated Enzyme II complex. III^{Glc} is the central Regulatory Protein, originally termed RPr, which can be phosphorylated as a result of transfer of the phosphoryl moiety from phospho-HPr. Consequently, this protein is assumed to exist in the cell in two alternative states: an underivatized protein and a phosphorylated protein. It is further assumed that adenylate cyclase and the carbohydrate uptake systems, which are sensitive to PTS-mediated regulation, possess allosteric regulatory sites, by virtue of which their catalytic activities are sensitive to regulation. The allosteric effector molecules are the derivatized and free forms of III^{Glc}. Thus, the target carbohydrate permeases and catabolic enzymes may normally exist in an active configuration, but binding of free III^{Glc} to the allosteric site would alter the conformation of the target proteins such that they would function with reduced efficiency. Uptake activity would therefore be inhibited. In contrast, adenylate cyclase is believed to normally exist

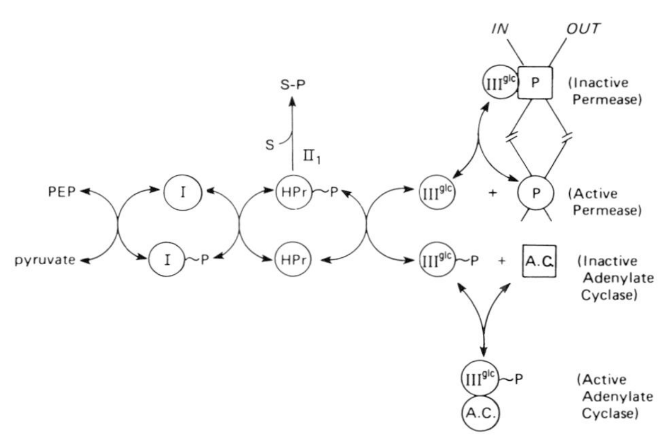

Figure 1 Proposed mechanism for regulation of non-PTS carbohydrate uptake systems and adenylate cyclase by the PTS in *E. coli* and *S. typhimurium*. Enzyme III^{Glc} is the same as RPr (regulatory protein) in the originally proposed model. I, Enzyme I; II, an Enzyme II specific for a particular sugar (S); P, a non-PTS permease or glycerol kinase; A. C., adenylate cyclase. [Reproduced, with permission, from Saier, M. H., Jr. (1989). Protein phosphorylation and allosteric control of inducer exclusion and catabolite repression by the bacterial phosphoenolpyruvate : sugar phosphotransferase system. *Microbiol. Rev.* **53,** 109–120.]

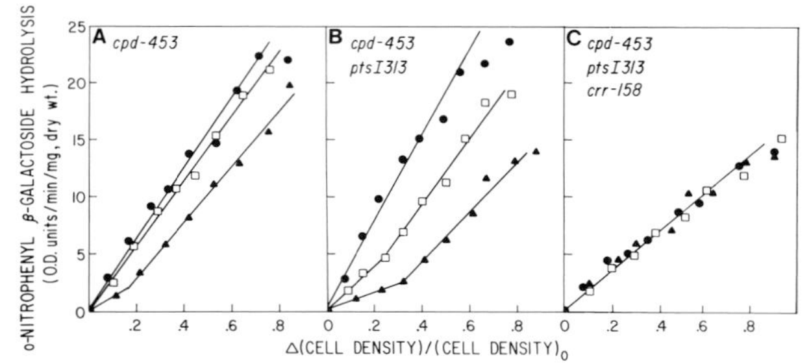

Figure 2 Transient and permanent repression of β-galactosidase synthesis in a series of mutant strains of *E. coli*. Cells were grown continuously in a salts medium containing 1% sodium pyruvate. With cells in the exponential growth phase at 34°C, isopropyl β-thiogalactoside (0.5 m*M*) was added alone (●) or together with 5 m*M* methyl α-glucoside (□) or 5 m*M* glucose (△). Subsequently, aliquots were periodically removed for measurement of cell density and β-galactosidase activity. (A) Strain *cpd-453*, a cAMP phosphodiesterase-negative strain. (B) *cpd-453 ptsI313*, a mutant strain that produces a temperature-sensitive Enzyme I with low activity after growth at 34°C. (C) *cpd-453 ptsI313 crr-158*, a carbohydrate repression-resistant mutant derived from strain *cpd-453 ptsI313*. [Reproduced, with permission, from Leonard, J. E., Lee, C. A., Apperson, A. J., Dills, S. S., and Saier, M. H., Jr. (1981). The role of membranes in the transport of small molecules. *In* "Organization of Prokaryotic Cell Membranes," Vol. I (B. K. Ghosh, ed.), 1–52. CRC Press, Boca Raton, Florida. Copyright CRC Press, Inc., Boca Raton, Florida.]

in a relatively inactive state that exhibits a low rate of cAMP synthesis. This enzyme can be activated when the phosphorylated form of III^{Glc} binds to its allosteric regulatory site. Thus, according to the model illustrated in Fig. 1, the non-PTS sugar-uptake systems are subject to negative control by free III^{Glc}, giving rise to inducer exclusion, whereas adenylate cyclase is subject to positive control by phospho-III^{Glc}, giving rise to catabolite repression. [*See* PEP: CARBOHYDRATE PHOSPHOTRANSFERASE SYSTEM.]

V. Transient versus Permanent Repression

The induction of β-galactosidase synthesis in *E. coli* has been studied employing conditions in which the phenomenon of inducer exclusion is unimportant, and under these conditions, a reasonably good correlation is observed between repression of β-galactosidase synthesis and inhibition of cAMP production *in vivo*. Repression of β-galactosidase synthesis by glucose and methyl α-glucoside is shown as a function of time in Fig. 2, for a series of *E. coli* mutants, all of which lacked cAMP phosphodiesterase. In the parental strain (*cpd-453*), glucose weakly repressed β-galactosidase synthesis, while the nonmetabolizable glucose analogue, methyl α-glucoside, was almost without effect. Reduction of the Enzyme I content of the cell, due to the *ptsI313* mutation, enhanced both the intensity and duration of transient repression. Sensitivity to permanent repression was also increased. The *crr-158* mutation in the *cpd-453 ptsI313* genetic background depressed the rate of β-galactosidase synthesis but abolished both transient and permanent repression by glucose. Under all of these conditions and in all three strains, relative rates of β-galactosidase synthesis correlated approximately with intracellular cAMP in both the presence and absence of sugars.

The results summarized above provide information about the repression of β-galactosidase synthesis by a sugar substrate of the PTS. However, they do not bear on the influence of Enzyme I on transient or permanent repression caused by carbohydrates that are *not* substrates of the PTS. This problem has been examined employing *E. coli* strains that synthesized the glycerol catabolic enzymes as well as the hexose phosphate permease, which transports glucose-6-phosphate constitutively. One such strain carried a deletion mutation within the structural gene for Enzyme I of the PTS. Both transient and permanent repression of β-galactosidase synthesis by glycerol and glucose-6-phosphate were demonstrated in the parental and Enzyme I-negative strains. However, the absence of Enzyme I had several effects on β-galactosidase synthesis. First, the rate of β-galactosidase synthesis in the absence of glycerol or glucose-6-phosphate was reduced, and this rate could be restored by the addition of cAMP. Second, both the duration and intensity of transient repression were enhanced by the *ptsI* mutation when either glycerol or glucose-6-phosphate served as the repressing carbohydrate. Third, the intensity of permanent repression was also enhanced by the loss of Enzyme I function. These results appear qualitatively similar to those depicted in Fig. 2, in agreement with the suggestion that Enzyme I exerts a primary effect by maintaining adenylate cyclase in an activated state. However, the results do not lead to a mechanistic understanding of transient repression.

As a result of studies on the regulation of hexitol uptake, a mechanism has been proposed to explain the phenomenon of transient repression. Transient repression of enzyme synthesis, presumably due to lowered internal cAMP, may occur as a consequence of the transient accumulation of cytoplasmic carbohydrate phosphates as follows:

1. Addition of a new carbon source to the growth medium of bacteria, which can immediately transport and metabolize it, results in increased cellular concentrations of carbohydrate phosphates.
2. Adenylate cyclase is subject to inhibition by intracellular organic phosphate esters so that the activity of this enzyme is depressed, and internal cAMP concentrations fall.
3. Due to decreased internal cAMP, synthesis of the permeases specific for the repressing carbohydrates is partially inhibited.
4. In response to decreased sugar permease activity, cellular sugar-phosphate pools are reduced to a new steady-state level, and inhibition of adenylate cyclase and enzyme synthesis is relieved.

While the proposed mechanism appears reasonable, much experimental evidence will be required to establish or refute its validity. Carbohydrate permeases appear to be sensitive to inhibition by several distinct mechanisms. A multiplicity of regulatory mechanisms for adenylate cyclase have been postulated but not yet demonstrated, and physiologically relevant inhibition of adenylate cyclase by sugar phosphates has not been reported. Moreover, relief from transient repression is usually observed to occur suddenly after a specific time interval following addition of the repressing carbohydrate. Gradual relief from transient repression would be expected from the proposed model. Thus, we are still far from a molecular understanding of the phenomenon of transient repression.

VI. Physiological Significance of cAMP Transport

In 1965, it was first demonstrated that addition of glucose to a resting suspension of *E. coli* cells in carbohydrate-free medium inhibited net cAMP synthesis and stimulated efflux of the cyclic nucleotide from the cell. The mechanistic basis for the former observation is discussed in the preceding two sections, and the explanation for stimulation of cAMP excretion relates to the energy dependency of the process. The physiological significance of cAMP excretion to the regulation of the intracellular cAMP concentration and the synthesis of catabolic enzymes has been demonstrated. Internal and extracellular concentrations of cAMP in glucose-limited continuous cultures of *E. coli* as a function of growth rate have been determined, and the results showed that in accord with expectation for an energy-dependent cAMP transport system, cAMP efflux is most rapid when glucose is rapidly metabolized.

Studies on the induction of biodegradative threonine dehydratase in a cAMP phosphodiesterase-negative strain of *E. coli* revealed that induction of the enzyme required anaerobic conditions, and internal cAMP concentrations were measured during the aerobic to anaerobic transition. Intracellular cAMP was approximately constant (5–10 μM) during exponential growth under both aerobic and anaerobic conditions, but when the aerobically growing culture was depleted of oxygen, the internal cAMP concentration increased 30-fold within 10 min and subsequently declined to normal. During the transition period when internal cAMP levels increased, no increase in the rate of cAMP synthesis was observed. Since a cAMP phosphodiesterase-negative strain was employed in these studies, the "spike" of internal cAMP could only be attributable to a decrease in the rate of cAMP efflux, which, in turn, was presumably due to a lowering of the cellular energy level.

In response to the elevated internal cAMP concentration, threonine dehydratase was synthesized. Increased cellular cAMP evidently served as a signal for the initiation of threonine dehydratase synthesis, but threonine dehydratase synthesis allowed energy generation by the nonoxidative degradation of threonine. Threonine metabolism thus restored the cellular energy level, increased the rate of cAMP efflux, and caused the intracellular cAMP concentration to drop to normal.

VII. Mechanism of Action of the cAMP–CRP Complex

Transcription of the galactose (*gal*) operon is not observed *in vitro* unless cAMP and the CRP are present. This CRP-activated transcription is initiated at one of two *gal* promoters, P_1 , and the cAMP–CRP complex is required for formation of an open (preinitiation) complex by RNA polymerase at this promoter. It is thought that the cAMP–CRP complex binds and promotes or stabilizes RNA polymerase binding. Interestingly, the second *gal* promoter, P_2 , is *negatively* regulated by the cAMP–CRP complex, and consequently, in the absence of this complex, RNA polymerase transcribes the operon at a rate that is suboptimal but sufficient to allow reasonably rapid galactose utilization. The *gal* operon thus provides an example of genetic regulation involving both positive and negative control by the cAMP–CRP complex. While most operons subject to cAMP action, and all such operons subject to catabolite repression, are positively regulated by the cAMP–CRP complex, others, such as those that encode the five enzymes that interconvert glutamate and glutamine as well as some outer membrane proteins, are negatively regulated.

Interestingly, the cAMP–CRP complex negatively regulates transcription of the *cya* and *crp* genes and therefore inhibits both adenylate cyclase and CRP synthesis. Thus, expression of both the *cya* and *crp* genes is autogenously regulated. It has recently been found that the negative autoregulation of *crp* may be indirect and due to the activation by the cAMP–CRP complex of a divergent promoter located within the *crp* promoter region. The divergent, antisense RNA produced from this promoter may inhibit *crp* transcription both *in vitro* and *in vivo*. This situation is probably not characteristic of operons negatively regulated by the cAMP–CRP complex. Thus, the latter operons are probably directly repressed by altering RNA polymerase binding to the promoters. In the case of *crp* gene regulation, the cAMP–CRP complex *activates* a gene encoding a repressing antisense RNA. The antisense RNA is believed to pair with bases 2–11 in the nascent messenger RNA (mRNA), yielding a stemlike structure that may function as a transcriptional terminator.

CRP is a homodimeric protein with a subunit molecular weight of 22,500 (201 amino acyl residues). It consists of two domains, an N-terminal basic domain with a characteristic helix-turn-helix DNA-binding motif, and a much larger C-terminal domain that binds cAMP. Cyclic AMP-binding causes CRP to undergo a conformational change, which allows binding of the protein to CRP-specific DNA-binding sites. The details of this conformational change have been elucidated. The sequence of one such DNA binding site in the *lac* operon, an imperfect inverted repeat, is GTGAGTT/AGCTCAC. Most operons have one or two CRP-binding sites, but as many as five sites corresponding to the CRP consensus sequence have been reported, for example, for the mannitol and glucitol operons of *E. coli*. The significance of multiple CRP binding sites is not known.

Examining different operons under cAMP control, one does not find a constant distance between the CRP-binding site and the transcriptional start site. However, for any one promoter, this distance may be critical, for the analogous orientation of the bound CRP to the bound RNA polymerase may be a deciding factor. CRP probably interacts directly with RNA polymerase during transcriptional activation and may additionally induce a bend in the DNA. These interactions presumably give rise to a complex of the DNA with the two proteins, which together are capable of initiating transcription. In some cases, the transcriptional initiation complex includes additional proteins.

VIII. Catabolite Repression in the Absence of cAMP

Several studies have indicated that synthesis of carbohydrate catabolic enzymes may be influenced by metabolites other than inducer and cAMP. The repression of β-galactosidase synthesis can occur in *cya* or *crp* deletion mutants of *E. coli,* which are incapable of cAMP synthesis or action. β-galactosidase synthesis in these mutants was found to be subject to catabolite repression, particularly when the nitrogen source was limiting. This observation established that cAMP–CRP independent mechanisms of catabolite repression exist in *E. coli.*

The genes encoding several enzymes in *E. coli* appear to be subject to carbon catabolite repression but are apparently not affected by *cya* and *crp* mutations. Among these enzymes are putrescine aminotransferase and pyrroline dehydrogenase. Their syntheses in wild-type *E. coli* under repressing conditions are not restored by addition of cAMP, although in parallel experiments, β-galactosidase synthesis was enhanced.

Several genes that confer resistance to stress or starvation conditions are also known to be subject to catabolite repression by a cAMP-independent mechanism. These genes, designated *pex,* are induced in response to starvation for carbon, nitrogen, or phosphorous, or to stress conditions such as heat, oxidative conditions, acidic conditions, or osmotic stress. Some of these genes are transcribed by RNA polymerase containing minor, stationary phase-specific sigma factors such as the starvation-induced σ^{32}, thus accounting for their specific induction during the stationary growth phase. A few of the protein products of these *pex* genes have been found to function in protein stabilization and folding. These "chaperone" proteins represent just one class of proteins that confer resistance to conditions of starvation and stress. It seems that several cAMP-independent mechanisms of catabolite repression must be operative in *E. coli* and that a large number of genes are subject to these kinds of control.

It has been reported that a guanosine tetraphosphate compound called ppGpp (also called magic spot) improves transcription of the lactose (*lac*) operon in an *in vitro*-coupled transcription–translation system. It also appears to regulate other catabolite-repressible genes such as those encoding the enzymes of glycogen biosynthesis. This molecule might therefore mediate a second mechanism of carbon catabolite repression, but more work will be required to establish this possibility.

It is interesting to note that indole-3-acetate has the ability to activate transcription of some operons that are under cAMP control. In this case, the operon examined, the arabinose utilization (*ara*) operon in *E. coli,* could be activated in the absence of CRP. While the *ara* operon appears to be regulated independently by indole acetate and the cAMP–CRP complex, it is possible that other operons are regulated only by the former compound. If this is the case, indole acetate could mediate cAMP-independent catabolite repression provided that the cytoplasmic levels of this compound are regulated by carbon availability.

IX. Catabolite Repression in the Gram-Positive Bacterium *Bacillus subtilis*

Many genes in the gram-positive bacterium *B. subtilis* are subject to catabolite repression by a cAMP-independent mechanism. These genes include those involved in carbon and energy metabolism as well as those that are activated in response to sporulation-inducing conditions. In all cases examined, catabolite repression occurs at the level of transcriptional initiation as in *E. coli.* Catabolite repression of one gene, *amyE,* encoding the enzyme α-amylase, is due to the presence of an operatorlike sequence in the DNA located just downstream from the *amyE* promoter and overlapping the transcriptional start site. This region exhibits twofold symmetry, is homologous to other catabolite-repressed *B. subtilis* promoter sequences, and presumably functions to bind an unidentified transcriptional regulatory protein, which mediates catabolite repression.

At least two mechanisms of catabolite repression have been shown to occur in *B. subtilis.* One depends on the generation of glycolytic metabolites such as fructose-1,6-bisphosphate while the other does not. The gluconate (*gnt*) operon in *B. subtilis* is regulated by the former mechanism. Some evidence suggests that this mechanism involves a fructose-1,6-bisphosphate-activated, ATP-dependent protein kinase, which phosphorylates HPr, the small phosphocarrier protein of the PTS on a serine residue. If HPr cannot be phosphorylated by the kinase, the target genes are resistant to catabolite repression. These observations suggest that the metabolite-dependent mechanism of catabolite repression operative in *B. subtilis* is different from both the CRP-dependent mechanism and the σ^{32}-dependent mechanism of catabolite repression in *E. coli.* However, the details of the process have yet to be elucidated.

Sporulation in *B. subtilis* is subject to catabolite

repression by an apparently distinct group of proteins that, however, are apparently activated by a phosphorylation cascade mechanism (see Fig. 3). Two sporulation-specific, ATP-dependent protein kinases, one embedded in the membrane, the other cytoplasmic, phosphorylate a protein called SpoOF. One of these two kinases may be activated by the starvation conditions that relieve catabolite repression and thereby allow initiation of sporulation. SpoOF, a "response regulator," then transfers its phosphoryl group to a phosphotransfer protein, SpoOB, which in turn transfers the phosphoryl group to a second response regulator termed SpoOA. SpoOB is regulated by a guanosine triphosphate-binding protein that may be responsive to cell-cycle signals (Fig. 3). SpoOA in its phosphorylated form then binds to a DNA sequence, TGNCGAA (N = any nucleotide), to activate or repress transcription of the adjacent gene. This scheme, possibly incomplete, but illustrated in Fig. 3, provides for many levels of control so that sporulation, known to be induced by a variety of starvation conditions, is not induced capriciously.

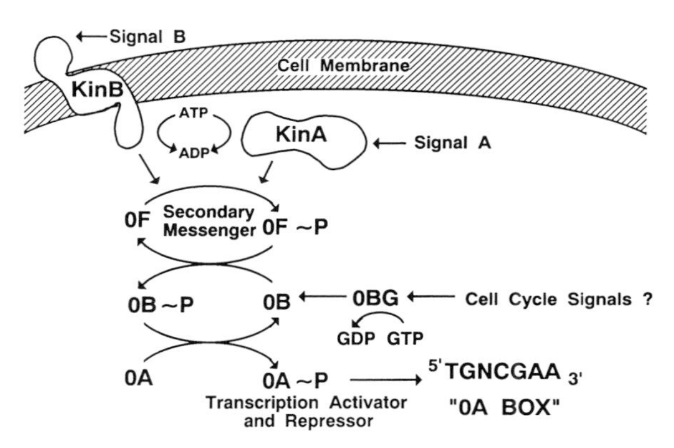

Figure 3 Proposed scheme for the initiation of sporulation in the Gram-positive bacterium *B. subtilis*. Protein kinases A and B (KinA and KinB) phosphorylate the SpoOF protein (OF) in response to unidentified stimuli, Signals A and B, respectively. Phosphorylated SpoOF~P then phosphorylates SpoOB (OB), which is regulated by a GTP-binding and hydrolyzing protein, OBG, which in turn may be responsive to cell-cycle signals. SpoOB then phosphorylates SpoOA (OA), which functions as a transcription factor that binds to a DNA sequence (the so-called "OA BOX"), as indicated. [Reproduced, with permission, courtesy of James A. Hoch, Scripps Clinic and Research Foundation, La Jolla, California.]

X. Catabolite Repression in Yeast

Glucose (carbon) catabolite repression in the yeast *S. cerevisiae* is a complex phenomenon mediated by several different transcriptional activators/ repressors and other accessory proteins in conjunction with protein kinases, some of which are cAMP-dependent and some of which are cAMP-independent. These transcriptional regulators mediate their effects by binding to typical eukaryotic Upstream Activating Sequences (UASs), which are located at varying distances upstream from the initiation codon. The physiological significance of catabolite repression in eukaryotes, as in prokaryotes, is that it allows the organism to rapidly and efficiently alter the expression of key metabolic genes in response to changes in carbon source availability in the environment. In yeast, in an environment suitable for fermentative growth, the presence of glucose triggers the transcriptional repression of mitochondrial and nuclear-encoded genes required for oxidative growth and gluconeogenesis. Conversely, in a nonfermentative environment, the absence of glucose releases these genes from repression and allows their expression. In addition to allowing alternate expression of genes for oxidation and fermentation, yeast also possess catabolite-controlled mechanisms that allow creation of a hierarchy of preferred carbon sources (i.e., galactose, sucrose, lactose, maltose, ethanol), with glucose at the top of the hierarchy.

Well-studied target genes in *S. cerevisiae,* exemplifying switching in response to fermentative versus oxidative conditions, are the glucose-repressible *CYC1* and *COX6* genes, which encode mitochondrial iso-1-cytochrome *c* and cytochrome *c* oxidase subunit VI, respectively. These genes as well as other oxidative, tricarboxylic acid (TCA) cycle, and gluconeogenic genes are strongly repressed in the presence of glucose. Under the same conditions, genes encoding glycolytic enzymes are induced. The transcriptional regulatory proteins that bind the UASs preceding the *CYC1* and *COX6* genes belong to a class of trimeric DNA-binding proteins, the HAP2, HAP3, and HAP4 proteins. An additional gene product, HAP1, binds to a separate UAS preceding the *CYC1* gene.

How do the HAP proteins regulate transcription, and what is the nature of the signal initiated by the presence of glucose? These two questions have recently been addressed at a mechanistic level, and most of the evidence points to a protein phosphorylation cascade involving the product of the SNF1 (CAT1) gene, a serine–threonine protein kinase that shows homology with mammalian protein kinases. Other gene products are also necessary for the reversible repression phenomenon. The following mechanism has been proposed: Glucose is sensed, possibly by a protein that acts as a positive effector of the SNF1 protein kinase. This kinase then participates in a signal transduction pathway that phosphorylates some components of the transcriptional activation–repression complex, possibly one of the HAP proteins. Reversible phosphorylation of such a protein controls activation or repression of the target genes.

The involvement of the SNF1 protein kinase in catabolite repression of the *CYC1* and *COX6* genes is not exclusive. This same kinase allows release from repression of a number of genes that are probably regulated by different sets of DNA-binding proteins. Thus, it appears that the SNF1 protein kinase is a global regulatory protein that transmits a pleiotropic signal controlling yeast carbon catabolite repression through protein phosphorylation.

In addition to the HAP-mediated control of fermentative versus oxidative genes, a distinct DNA-binding protein, RAP1, binds to the UASs that precede many glycolytic genes. Included in the list of genes known to contain RAP1-binding sites upstream of their promoters are the genes encoding enolase (*ENO1* and *ENO2*), phosphoglycerate kinase (*PGK*), triosephosphate isomerase (*TPI*), alcohol dehydrogenase 1 (*ADH1*), and pyruvate kinase (*PYK1*). Protein phosphorylation seems to be involved in RAP1 binding because only phosphorylated forms of the protein are capable of binding the DNA. Glucose presumably enters the cell, and it or some metabolite of glucose may directly, or via a phosphorylation cascade, signal the phosphorylation of RAP1. Phosphorylation of RAP1 results in DNA binding and consequent activation or repression of metabolic genes.

SNF1 is a cAMP-independent protein kinase that apparently mediates glucose catabolite repression. However, in addition to the SNF1-mediated pathways, yeast also have cAMP-dependent pathways that mediate catabolite repression. One gene regulated by a cAMP-dependent protein kinase encodes alcohol dehydrogenase II (*ADHII*). Expression of the *ADHII* gene is repressed under fermentative conditions by the action of a regulatory protein called alcohol dehydrogenase regulatory protein (ADR1). ADR1 is a DNA-binding protein that uses a zinc-finger binding motif to bind the DNA. The levels of ADR1 in the cell and its DNA binding are not affected by the carbon source employed for growth. Rather, the phosphorylation state of the protein is indirectly controlled by the carbon source, and its phosphorylation regulates transcriptional initiation of the *ADHII* structural gene. Glucose or some metabolite of glucose is sensed, triggering a transient rise in cAMP levels, in turn activating a cAMP-dependent protein kinase. This kinase phosphorylates ADR1, which then functions to repress transcription of the *ADHII* structural gene.

The galactose (*GAL*)/melibiose (*MEL*) regulon of *S. cerevisiae* is also subject to catabolite-controlled transcriptional regulation. The mediator of this control is the 881-amino acid DNA-binding protein encoded by the *GAL4* gene. Several structural genes are directly controlled by GAL4. GAL4 is a multidomain protein that binds the DNA via an N-terminal zinc-finger binding domain but activates transcription in a process dependent on a C-terminal domain. Another protein, GAL80, is a negative regulator that functions antagonistically to GAL4 by binding to it, preventing GAL4 from activating transcription.

GAL4/GAL80-mediated transcriptional regulation may occur as follows. In the absence of glucose and galactose, GAL4 binds to the DNA, and GAL80 binds to GAL4, preventing GAL4 from activating transcription. In this uninduced state, GAL4 is unphosphorylated. The presence of galactose, or a metabolite of galactose, triggers the release of GAL80 from GAL4, and GAL4 becomes phosphorylated. This corresponds to the activated state. The presence of glucose triggers the release of GAL4 from the DNA, which gives rise to the repressed state.

Other mechanisms of carbon catabolite repression are less well characterized and may involve two distinct forms of RNA polymerase found under different growth conditions as well as differential translational efficiency of specific mRNAs. In cells grown without glucose, the structural genes allowing catabolism of alternative sugars (i.e., sucrose, galactose, etc.), gluconeogenesis, ethanol utilization, and the functioning of the TCA cycle and electron transport are all derepressed. This derepression is correlated with a change in the mRNA pools within the cell. The mRNAs isolated from derepressed cells are more efficiently translated than those from repressed cells. This fact has led to the hypothesis that the mRNAs synthesized under these two conditions are somehow different and that those made under derepressing conditions have higher affinity for the translational machinery than the mRNAs made under repressing conditions. This possibility affords the cells an additional means to rapidly adjust to new environmental conditions, resulting in higher levels of protein synthesis.

Two forms of mitochondrial RNA polymerase may be present in repressed versus nonrepressed cells. A core enzyme form is apparently synthesized under repressing conditions while a holoenzyme form is probably made under derepressing conditions. As a consequence, mitochondrial gene expression is enhanced sixfold under nonrepressing conditions.

Catabolite repression in other eukaryotes is not well understood mechanistically. However, it seems likely that similar mechanisms of catabolite repression will apply to many carbohydrate catabolic and anabolic enzymes in a variety of eukaryotic organisms. The bewildering complexity of the regulatory interactions in yeast suggest that bacterial model systems should be utilized as a basis for understanding gene expression in eukaryotes.

XI. Major Unexplained Questions Related to Catabolite Repression

The concept of intracellular homeostasis is often the basis upon which current studies of biochemical and physiological regulation are founded. However, in the early days, when the notion of genetic regulation was first being propounded, it was of interest to know how a cell could switch from one condition to another with a minimal amount of energy expenditure. Studies on the *lac* operon of *E. coli* fostered much interest, but molecular biology is little different from other sciences in being faddish, and interest shifted to other "hot" topics, leaving many unanswered questions to gather dust. The scientific world remains more or less ignorant of the fact that the "prototype" gene is still far from understood and that it may not be representative of all superficially similarly regulated systems. For example, the mechanisms of cAMP-independent catabolite repression in *E. coli, B. subtilis,* and *S. cerevisiae* discussed in Sections VIII–X are still not understood. It does appear, however, that in eubacteria alone, at least five mechanisms of catabolite repression are operative, one involving cAMP and the cAMP receptor protein, a second involving σ^{32}, a third involving other second messengers, a fourth involving a metabolite-activated protein kinase, and a fifth involving a sensor kinase-response regulator pair.

A second example concerns the hyperactivation of the *lac* operon in *E. coli* cells approaching stationary phase, reported over a decade ago. This phenomenon of "stationary phase induction" is still being studied; the hyperinduction of β-galactosidase synthesis seems to be linked to the depletion of dissolved oxygen in the culture medium. It commences concurrently with diminishing mean cell size, but the mechanism is not understood. It is possible that σ^{32} plays an important role in this process.

As a third example, the strength of carbon catabolite repression in *E. coli* is influenced by the availability of a nitrogen source. If nitrogen is limiting for growth, the intensity of catabolite repression increases, sometimes as much as 1000-fold. The mechanistic interactions giving rise to this increased sensitivity are not understood. Possibly, the availability of organic nitrogen regulates the cytoplasmic accumulation of intermediates of carbon metabolism. Thus, it is clear that our understanding of catabolite repression in bacteria as in eukaryotes is still in its infancy. Much more work will be required before we have a detailed understanding of this multifaceted phenomenon.

Acknowledgments

We thank J. Reizer, C. Vermuelen, and J. Deutscher for helpful discussions and A. Reizer for assistance in the preparation of the manuscript. Work in the authors' laboratory was supported by Public Health Service grants 5R01AI 21702 and 2R01AI 14176 from the National Institute of Allergy and Infectious Diseases.

Bibliography

Hopwood, D. A., and K. F. Chater (eds.) (1989). "Genetics of Bacterial Diversity." Academic Press, London.

Krulwich, T. A. (ed.) (1990). "Bacterial Energetics," Vol. XII. Academic Press, San Diego, California.

Saier, M. H., Jr. (1985). "Mechanisms and Regulation of Carbohydrate Transport in Bacteria." Academic Press, Orlando, Florida.

Saier, M. H., Jr. (1989). *Microbiol. Rev.* **53,** 109–120.

Wills, C. (1990). *Crit. Rev. Biochem. Mol. Biol.* **25,** 245–280.

Caulobacter Differentiation

Austin Newton and Noriko Ohta
Princeton University

Glossary

Chemoheterotroph Organism that obtains energy and carbon from organic chemicals

Chemotaxis Movement of an organism toward or away from a chemical

Cosmid Hybrid genetic cloning vector containing plasmid sequences and *cos* site of bacteriophage lambda

Dichotomous replication Single chromosome replicating at multiple forks

Holdfast Adhesive material located at the pole of *Caulobacter* responsible for cellular attachment

Integration host factor Histonelike DNA-binding protein that can wrap DNA in prokaryotes into higher-order structure

Pili Short filamentous structures on bacterial cells that aid in adherence to surfaces

Prostheca Cellular appendage that is derived at least in part from the cell envelope

Rosettes Clusters of stalked or dividing cells attached at a common point by holdfasts at the tips of their stalks

***CAULOBACTER* DIFFERENTIATION** occurs during an asymmetric cell division cycle that produces two different cell types: a nonmotile stalked cell and a flagellated swarmer cell. The stalked cell of this gram-negative bacterium divides repeatedly like a stem cell to generate another new swarmer cell at the pole opposite to the stalk. The swarmer cell cannot divide until it loses its flagellum and develops into a stalked cell. The developmental events required to form the new swarmer cell and for differentiation of the swarmer cell into a stalked cell are under tight temporal and spatial control: They are stage-specific in the cell cycle and occur only at one pole of the cell. Differentiation is part of the normal proliferative cell cycle and does not occur in response to starvation or other changes in cultural conditions, as observed in myxobacteria, *Anabaena,* and the sporulating *Bacilli.* The attraction of *Caulobacter* as a model system for the study of differentiation is the relative simplicity of its life cycle, the ease with which developmental structures can be isolated and studied, and the availability of an array of genetic, molecular, and biochemical techniques. These features make this bacterium a fertile organism for elucidating the underlying mechanisms that regulate cell differentiation.

I. Distribution and Growth of *Caulobacter*

A. Distribution and Characteristics

Caulobacter, a chemoheterotrophic aerobic bacterium, is widely distributed in nature. It is a scavenger found in ponds, soil, saltwater, and even in running tap water, where they were first observed. The genus *Caulobacter* contains many species. The variety of Caulobacters is illustrated by the 19 species proposed some 10 years ago, mainly on the basis of cellular morphology (Table I), although this classification has recently been simplified. All members of this genus are characterized by a cellular stalk, or prostheca, that is an extension of the cell envelope at one pole of the cell. It is composed of an outer membrane, peptidoglycan, and inner membrane and is punctuated by cross bands. The predivisional cell, as seen in a metal-shadowed preparation by electron microscopy (Fig. 1A), carries a stalk at one pole and a flagellum at the opposite pole that is destined to become the swarmer cell at division (Fig. 1B). These two appendages are convenient ex-

Table I Species Proposed for *Caulobacter*

Species	No. of described isolates	Morphology[a]	Nonmorphological traits		
			Pigmentation	Requirement(s)	Other distinction[b]
C. vibrioides	30	Vibrioid	None, yellow	Vitamin B_2, often also others	
C. henricii	10	Vibrioid	Yellow, orange	Vitamin B_2	
C. intermedius	1	Vibrioid	None	Biotin and others	
C. crescentus	7	Vibrioid	None	None	
C. robiginosus	1	Vibrioid	Red-brown	Amino Acids	Sucrose +, lactose −
C. rutilis	2	Vibrioid	Red-brown	Amino Acids	Sucrose −, lactose +
C. subvibrioides	21	Subvibioid	None, dark orange	Unidentified (not B vitamins)	
C. fusiformis	3	Fusiform	Dark yellow, dark orange	Unidentified	
C. rossii	1	Fusiform	None	Unidentified	
C. kusnezovii	9	Fusiform	Red-brown	Unidentified	
C. leidyi	1	Fusiform	None	None	
C. bacteroides	66	Bacteroid	None, yellow, orange	None or unidentified	
C. metschnikovii	2	Bacteroid	Yellow-orange	Not reported	
C. fulvus	2	Bacteroid	Yellow-brown, red-brown	Unidentified	
C. flexibilis	2	Bacteroid, curved	None	Unidentified	
C. glutinosus	9	Not described	None	Not reported	Colonies adherent
C. halobacteroides	1	Bacteroid	None	Unidentified	Marine, amino acids +
C. maris	1	Bacteroid	None	Unidentified	Marine, amino acids −
C. variabilis	3	Bacteroid, stalk polar, or subpolar	None, red-orange	Unidentified	

[a] Vibrioid = cell tapered, long axis distinctly curved; subvibrioid = cell tapered, long axis not strongly curved; fusiform = cell tapered, long axis not curved; bacteroid = poles rounded, long axis not curved.
[b] +, utilized as carbon source; −, not utilized.
[From Poindexter, J. S. (1981). The caulobacters: Ubiquitous unusual bacteria. *Microbiol. Rev.* **45**, 123–179. American Society for Microbiology, Washington, D.C.]

perimental markers to distinguish between the old stalked cell and the new swarmer cell.

Another characteristic of *Caulobacter* is the presence of an adhesive holdfast at one cell pole. In *Caulobacter crescentus,* the most widely studied species, this material is located at the tip of the stalk or at the base of the polar flagellum, where it mediates the attachment of cells to solid surfaces in the medium, to other microorganisms, or to the holdfasts of nearby stalked cells to form rosettes. The highly motile swarmer cells frequently collide with solid surfaces where they attach, and most attached stalked cells observed in culture probably develop from attached swarmer cells. The swarmer cell does not require attachment in order for development to occur, however (see Fig. 2).

Attention has been drawn to the similarity of *Caulobacter* and another monoflagellated genus, *Pseudomonads.* Both lack the Entner–Doudoroff pathway for carbohydrate metabolism and contain DNA with an extremely high guanine plus cytosine content, which in *Caulobacter* ranges from 62 to 67%. Marine and terrestrial *Caulobacter* species are, however, a large and extremely diverse group of organisms, as indicated by an analysis of the 16S RNA sequences by the method of Woese. *Caulobacter* isolates, many of which are similar to *C. crescentus* by this measure, are related to the α-

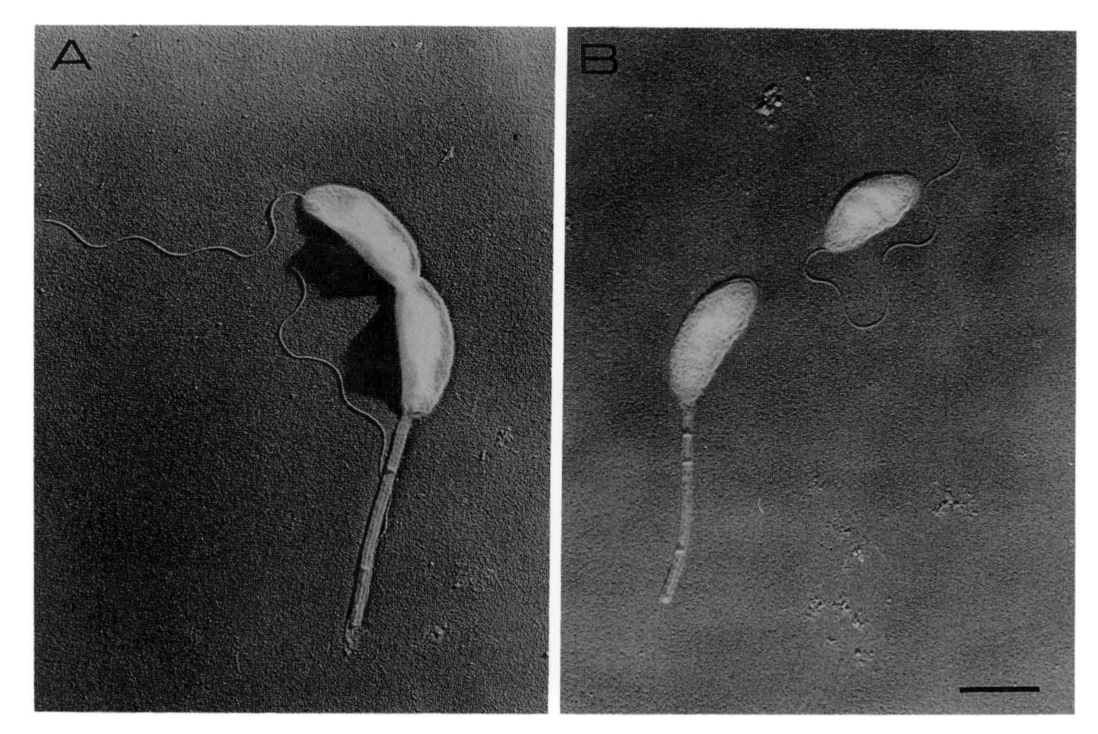

Figure 1 Electron micrograph of *Caulobacter crescentus* cells. (A) Dividing cell with polar flagellum and stalk. (B) Newly divided stalked cell and swarmer cell. Scale bar = 1 μm. Electron micrograph courtesy of J. Poindexter. [From Newton, A. (1989). Differentiation in *Caulobacter:* Flagellum development, motility and chemotaxis. *In* "Genetics of Bacterial Diversity" (D. A. Hopwood and K. F. Chater, eds.), pp. 199–220. Academic Press, San Diego.]

purple bacteria, but they are such a diverse genus that their relationship to any other single group of bacteria is not apparent. In fact, all species do not fall within any one subgroup of the α-bacteria.

B. Growth

Caulobacter species show various requirements for growth in a salts medium (Table I), but *C. crescentus* is a prototroph that is normally grown in the laboratory at temperatures ranging from 24° to 37°C, with optimal growth occurring around 30°C. The cells thrive on media containing low concentrations of salts and organic compounds and grow poorly on medium designed for *Escherichia coli*. When cultured in a nutrient broth, the cells divide with a doubling time of ca. 90 min, and in a defined salts medium with glucose as the carbon source they display a doubling time of ca. 180 min. The ability of *C. crescentus* to form colonies on agar surfaces and to grow well in liquid culture has been critical to the application of genetic analysis and large-scale biochemical studies of the organism. With few exceptions, studies of the cell cycle and differentiation have been carried out with the prototrophic *C. crescentus* strain CB15 (ATCC19089). The timing of cell cycle events referred to in this chapter is for cells growing in a defined minimal salts–glucose medium.

II. Cell Cycle and Developmental Events

A. Asymmetric Cell Division

Asymmetric cell division is a central mechanism in generating cellular diversity, and it is this feature of the *Caulobacter* life cycle that makes this bacterium such an important model system for the study of cell differentiation. The stalked cell and the new motile swarmer cell produced at division are morphologically and functionally distinct from one another; each of the two cells inherits its own developmental program that is expressed in an invariant

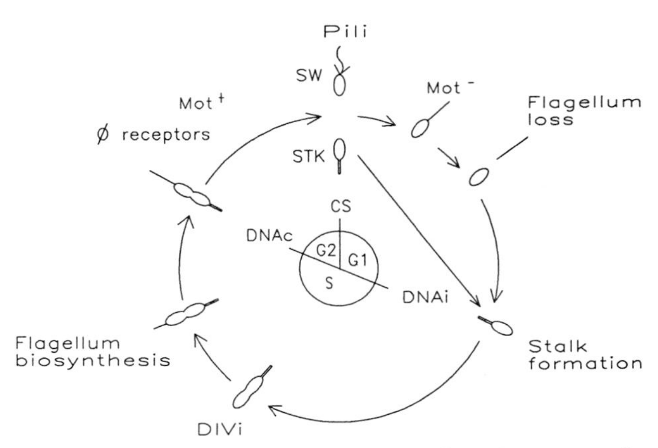

Figure 2 *Caulobacter crescentus* cell cycle and sequence of developmental events. Two distinct daughter cells, swarmer (SW) and stalked cell (STK), are produced at division. Developmental events indicated include pili assembly (Pili), loss of motility (Mot^-), initiation of division (DIVi), activation of polar DNA bacteriophage receptors (ϕ receptors), gain of motility (Mot^+), and cell separation (CS). Cell cycle periods G1, S, and G2 are defined in the text. [From Newton, A., and Ohta, N. (1990). Regulation of the cell division cycle and differentiation in bacteria. *In* "Annual Reviews of Microbiology" (L. N. Ornston, A. Balows, E. P. Greenberg, eds.), pp. 689–719. Annual Reviews, Palo Alto, California.]

sequence of morphological events during the cell cycle (Fig. 2).

B. Sequence of Developmental Events

Cell division in *C. crescentus* occurs over approximately one-half of the stalked cell cycle by a progressive pinching of the cell envelope, and the first visual indication of the incipient swarmer cell is the appearance of the transverse division site. This site is placed closer to the stalk-distal pole of the cell and it defines the part of the dividing stalked cell that will become the swarmer cell at division. Surface structures are assembled at this new cell pole in a defined sequence, beginning with biosynthesis of the flagellum, which continues through the remainder of the cell cycle. After the initiation of flagellum biosynthesis specialized DNA bacteriophage receptor sites and a holdfast are formed at the base of the flagellum, and the cell gains motility immediately before cell separation when the flagellum begins to rotate. The pili are assembled at the flagellated pole of the motile swarmer cell immediately after cell division. This sequence of events and the approximate times at which they occur are diagrammed on the cell cycle in Fig. 2.

The swarmer cell loses motility after 15–20 min, retracts the pili, loses the bacteriophage receptor sites, and after about 60 min forms the cellular stalk at the point of holdfast and flagellum attachment. The mature stalked cell then undergoes cell division as described previously for the old stalked cell (Fig. 2).

Developmental events in *Caulobacter* are reproducibly and precisely timed in the cell cycle under laboratory conditions, but in nature where nutrient conditions are often not optimal there appears to be some adjustment in the relative time required for certain events, even though the order of events does not change. In slowly growing cultures, for example, the number of swarmer cells relative to stalk and dividing cells increases, which indicates that the swarmer cell takes longer to develop into a stalked cell when nutrients are limiting. Because motile swarmer cells are chemotactic, they are equipped to act as ideal food-seeking modules that would develop into stalked cells and go on to divide only when optimal nutritional conditions are located. Also, evidence indicates that the stalk functions in the active uptake of amino acids and perhaps other nutrients from the medium.

C. Initiation of DNA Synthesis and Chromosome Replication

The most striking biochemical difference between stalked and swarmer cells is illustrated by the patterns of chromosome replication. The stalked cell cycle is made up of an S phase (DNA synthetic period) and G2 phase (postsynthetic gap period), while the swarmer cell cycle goes through an initial G1 phase (presynthetic gap period) and an S phase and G2 phase. Thus, after cell division the stalked cell initiates DNA synthesis immediately, but the swarmer cell enters a presynthetic gap period of almost 60 min, the time required for loss of motility, stalk formation, and preparation for the initiation of DNA synthesis (Fig. 2). Clearly, an understanding of how DNA initiation is regulated will provide valuable insights into how the different genetic programs are established and executed in the sibling stalked and swarmer cells.

The G1, S, and G2 designations proposed to describe phases of the *Caulobacter* cell cycle are borrowed from eukaryotic cell cycle terminology. This seems justified despite the fact that *Caulobacter*, like other prokaryotes, does not undergo an M, or mitotic, phase in its division cycle. In *E. coli* and other rapidly growing bacteria, "C" has been used to designate the time required for chromosome replication and "D" the time between the termination of chromosome replication and cell division. These designations reflect the fact that at rapid growth rates chromosome replication is dichotomous and DNA synthesis occurs continuously in the cell cycle with no apparent gaps. In *C. crescentus*, by contrast, dichotomous or multifork replication does not occur, and the pre- and postsynthetic gaps are observed regardless of how fast the cells divide.

III. Experimental Procedures and Approaches

A. Synchronous Cultures

An enormous advantage of working with *C. crescentus* for studies of both differentiation and cell cycle regulation is the ease with which synchronized cells can be prepared. Several methods have been developed, but the most ingenious is based on the "baby machine" or mitotic release method devised originally to synchronize cultured animal cells. In this procedure, the *Caulobacter* cells are grown in glass Petri plates. Stalked cells attached to the bottom of the plate divide and release motile swarmer cells into fresh medium, which can then be collected from the plate. Figure 3 shows the use of a synchronous culture of swarmer cells to determine the timing of DNA bacteriophage receptor formation. The receptor, which is detected by the ability of the cells to adsorb bacteriophage particles, appears just before cell division. The receptors are then lost after cell division as the swarmer cell matures into a stalked cell. Because these experiments monitor the entire cell cycle beginning with the bacteriophage-sensitive swarmer cells, loss of the receptors is observed first in the culture. The timing of other morphological events, including loss of motility, stalk formation, and pili formation, has been determined using synchronous cultures in an almost identical fashion. [*See* BACTERIOPHAGES.]

B. Genetics

Combinations of classical genetics and modern molecular biology have now made it possible to manip-

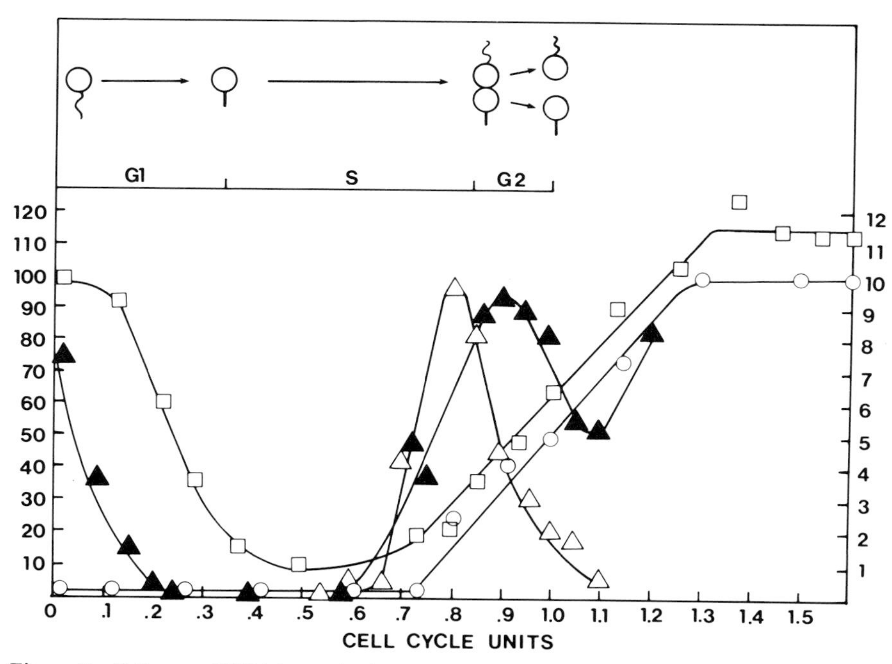

Figure 3 Patterns of DNA bacteriophage ϕLC72 receptor formation and hook protein and 25-kDa flagellin synthesis in a synchronous culture of *Caulobacter* cells. The rates of protein synthesis were determined by radioimmunoassays, and the time of cell division at 180 min is normalized to 1.0. ○, % cell division; □, % phage adsorption; △, hook units; ▲, flagellin units. [From Huguenel, E. D., and Newton, A. (1982). Localization of surface structures during procaryotic differentiation: Role of cell division in *Caulobacter crescentus*. *Differentiation* **21,** 71–78. Springer-Verlag, New York.]

ulate and, in effect, redesign the genomes of many organisms, and *Caulobacter* is no exception. The genetic analysis of developmental regulation in these bacteria is facilitated by a variety of procedures, including generalized transduction, conjugation, mutagenesis by Tn*5* transposons, complementation using stably replicating plasmids, gene replacement, and molecular cloning using cosmid vectors.

Many mutations in nutritional, motility, and cell division genes have been isolated and mapped. A major advance in genetic mapping and the study of gene organization in *Caulobacter* was the correlation of the genetic linkage map with a circular physical map of *DraI, AseI,* and *SpeI* DNA restriction fragments using pulsed-field gel electrophoresis (PFGE; Fig. 4). This elegant technique has substantially reduced the time previously required to map a genetic marker; a 5.6-kb Tn*5* transposon inserted in or linked to any gene can be rapidly located by the resulting change in mobility of a restriction fragment whose position on the physical map is already known. Analysis by PFGE has also permitted the size of the *C. crescentus* genome to be estimated at approximately 4000 kb contained in one circular chromosome that replicates bidirectionally.

IV. Flagellum Biosynthesis

A. Structure

Most studies of *Caulobacter* differentiation have focused on flagellum formation. This organelle is the most prominent of the polar surface structures, and it displays the key properties of temporal and spatial

Figure 4 Correlation of genetic and circular physical map of the *Caulobacter crescentus* chromosome. The concentric circles are restriction maps for *DraI*, *AseI*, and *SpeI*, respectively, with the *SpeI* map on the inside. The sizes of restriction fragment are indicated in kilobases. The positions of representative genetic markers are on the outer circle. [From Dingwall, A., Shapiro, L., and Ely, B. (1990). *METHODS: A Companion to Methods in Enzymology* **1,** 160–168.]

control exhibited by these cells during development. Because of its location on the cell surface, it can be easily isolated for biochemical studies and its function in cell motility permits a straightforward mutational analysis by the isolation of nonmotile mutants. The structural components of the flagellum can be seen in electron micrographs of negatively stained preparations like the one shown in Fig. 5A. It is composed of three essential parts: The basal body or motor is embedded in the cell envelope and acts as a torque generator responsible for flagellum rotation; the hook acts as a universal joint; and the flagellar filament, which is coupled to the basal body by the hook, rotates to propel the cell through the medium. One unusual feature of this structure is the presence of the three distinct, but sequence-related 29-, 27-, and 25-kDa flagellins in the flagellar filament (Fig. 5B). The *E. coli* filament by contrast is assembled from a single 50-kDa monomeric flagellin species. [*See* FLAGELLA.]

B. Periodic Synthesis of Flagellar Proteins

Because the hook and flagellar filament are shed during development (Fig. 2), they can be recovered from the culture medium for purification. Antibodies prepared against the 70-kDa hook protein and the major 25-kDa flagellin (see Fig. 5B) were used initially in experiments to measure the rates of flagellar

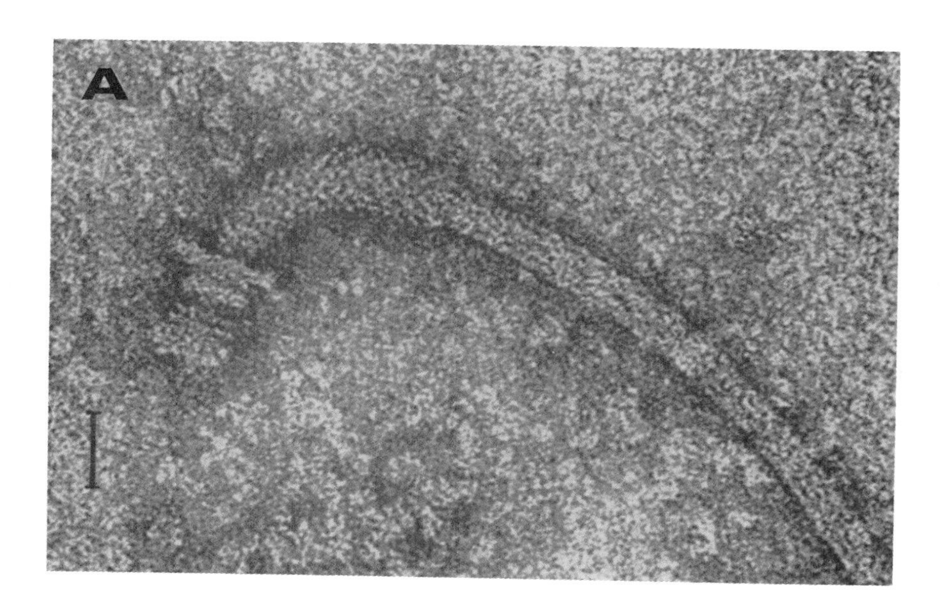

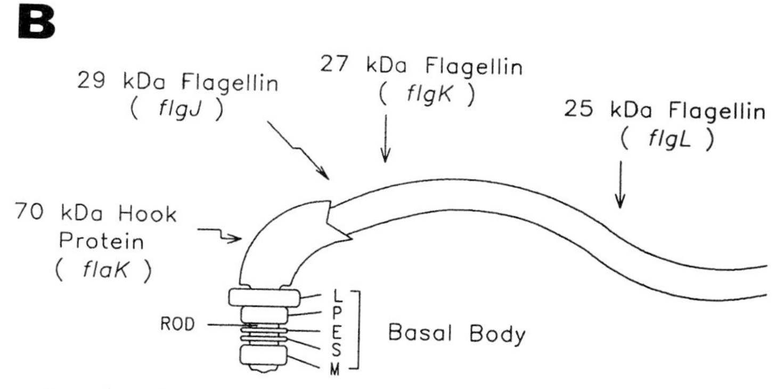

Figure 5 Structure of *Caulobacter crescentus* flagellum. (A) Electron micrograph of a negatively stained *C. crescentus* flagellum. Bar = 30 nm. [From Johnson, R. C., Walsh, M. P., Ely, B., and Shapiro, L. (1979). *J. Bacteriol.* **138,** 984–989. American Society for Microbiology.] (B) Schematic drawing of the flagellum indicating the major protein components.

protein synthesis in synchronous cell cultures, as illustrated in Fig. 3. The results of these and similar experiments showed that the hook and flagellin proteins are synthesized periodically in the cell cycle at a time that coincides with flagellum assembly. A fundamental conclusion about developmental regulation from this finding was that the stage-specific pattern of flagellum biosynthesis, and more generally polar morphogenesis in *Caulobacter,* is controlled at the level of gene expression. The results also raised other questions about flagellum biosynthesis, *viz.:* Is periodic gene expression a general feature of flagellar (*fla*) gene expression? Are the *fla* genes regulated at the level of flagellar (*fla*) gene expression? Are the *fla* genes regulated at the level of transcription? And if so, how is the time and order of *fla* gene expression determined in the cell cycle? The answers to many of these questions have come from the application of classical genetics and modern molecular approaches.

C. Genetic Regulation of Flagellar Gene Expression

The *Caulobacter* flagellum is built from the inside of the cell to the outside of the cell. Components of the basal body are laid down first in the cell envelope, followed by assembly of the hook and flagellar filament in that order. Well over 50 genes are required for flagellum biosynthesis (*fla* genes) and, in addition, at least 8 genes are required for chemotaxis (*che* genes) and 3 genes are required for motility (*mot* genes). Thus, between 1 and 2% of the information content of the *Caulobacter* genome is devoted to flagellum formation and function, which gives some indication of its importance for bacterial survival. [*See* CHEMOTAXIS; MOTILITY.]

Many of the *fla* genes have now been cloned and their organization in operons has been defined by genetic complementation, DNA sequencing, and nuclease S1 mapping. The patterns of transcription have been determined for most of these cloned genes. Remarkably, all of the *fla* genes examined are transcribed periodically in the cell cycle in a sequence that corresponds generally to the order in which their respective gene products are assembled into the flagellum. Thus, the basal body genes are transcribed early in the cell cycle, followed in order by expression of genes encoding the 70-kDa hook protein, the 29-, the 27-, and finally the 25-kDa flagellin. Examination of the flagellar structure in Fig. 5B shows that this sequence of gene expression reflects precisely the order in which the proteins are assembled and arranged in the flagellum itself.

Results of an epistasis analysis of the *fla* mutations indicate how the ordered pattern of gene expression is controlled and coordinated with flagellum assembly. The *fla* genes are organized in a complex, multitiered, regulatory hierarchy in which genes at each level of the hierarchy are required for transcription of genes at lower levels of the hierarchy. Because the regulatory interactions of many *fla* genes have not been precisely defined, an abbreviated hierarchy for 13 of the *fla* genes contained in the six transcription units that have been most intensively investigated is shown in Fig. 6. As a specific example of their interactions, expression of basal body genes in the *flaO* operon is required for expression of the 70-kDa hook protein gene, and the expression of the hook protein gene is required in turn for expression of the 27- and 25-kDa flagellin genes. The *fla* gene hierarchy thus mediates a transcriptional cascade that ultimately regulates the timing of protein synthesis and flagellum assembly. Negative autoregulation has also been observed in the *flaO, flaN,* and *flbG* operons, and this may be one mechanism for turning off transcription at the end of the synthetic periods.

D. Transcriptional Specificity

The transcription start sites of at least 12 *fla* genes have been mapped and the 5′ nucleotide sequences determined. The most striking feature to emerge from an examination of these sequences is the presence of characteristic sequence elements at each level of the hierarchy, including those similar to σ^{28}, σ^{32}, and σ^{54} promoters (refer to Fig. 6). In other bacteria, these promoters are recognized by minor sigma factors that are dedicated to the transcription of genes with specialized functions. Most of the *fla* genes in *Salmonella* and *E. coli* are transcribed from σ^{28} promoters, while heat-shock genes contain σ^{32} promoters, and genes involved in nitrogen metabolism, as well as many specialized functions in a variety of bacteria, have σ^{54} promoters. *Caulobacter* has apparently appropriated these and other regulatory elements to achieve the complex regulatory pattern observed for *fla* gene expression.

Extensive genetic and biochemical experiments support the conclusion that genes at the bottom of

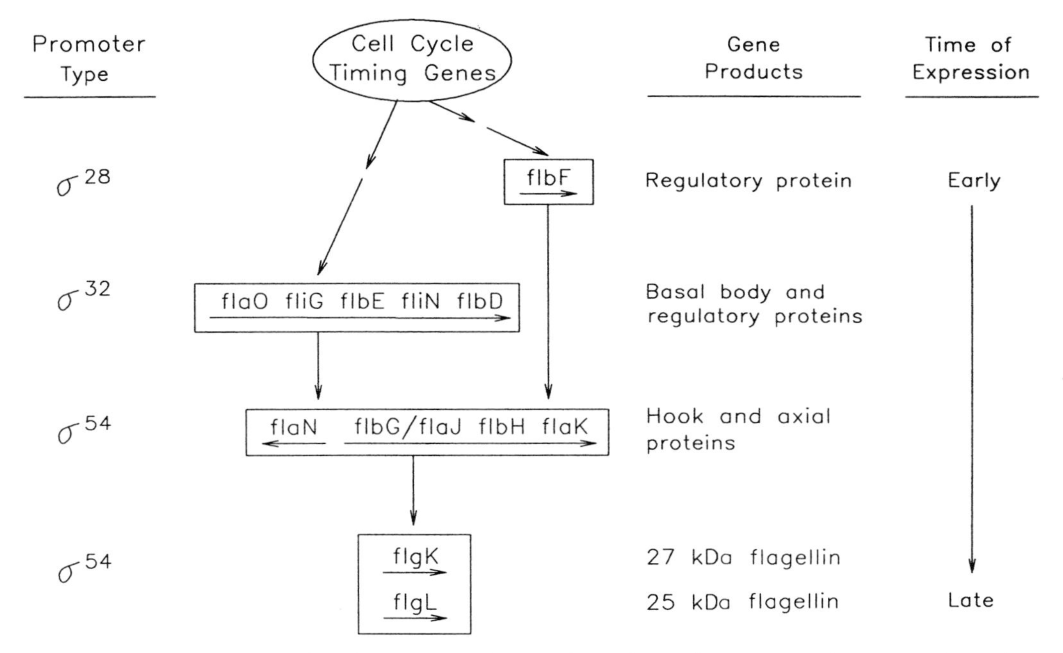

Figure 6 Organization of selected *fla* genes in the regulatory hierarchy. Horizontal arrows indicate the direction of transcription in the six transcription units shown and the vertical arrows connecting the boxes indicate positive transcriptional control. Transcription units at the top of the hierarchy are expressed early in the cell cycle and those at the bottom are expressed late, as indicated on the right of the figure.

the *Caulobacter fla* gene hierarchy are transcribed by a σ^{54} RNA polymerase. The assignment of the σ^{28} promoter to *flbF* and the σ^{32} promoter to *flaO* in Fig. 6 is based largely on homologies revealed by "sequence gazing," but preliminary mutagenesis studies support these assignments. Interestingly, few of the genes examined in *Caulobacter,* thus far, appear to contain the canonical -35 (TTTGACA), -10 (TATAAT) promoter recognized by the major *E. coli* σ^{70}.

Regulation of the *fla* gene hierarchy also appears to depend on the transcriptional activation by a set of flagellar-specific, *trans*-acting factors as well as a diverse set of σ factors. Expression of the four σ^{54}-dependent genes at the bottom of the hierarchy requires the FlbD protein, which belongs to a family of response regulators known to be required for transcription by σ^{54} RNA polymerases. FlbD appears to activate the *Caulobacter* promoters by its interaction with a conserved 19-bp DNA sequence element *ftr* (flagellar transcription regulation) that is located approximately 100 bp from the transcription start site of each gene. A third sequence element required for transcription from the σ^{54} promoters is the integration host factor (IHF) binding site that is located between the promoter and the *ftr* sequence. The analyses of genes that require IHF for expression in other systems suggests that this versatile protein may activate transcription by bringing FlbD bound at an *ftr* sequence in contact with σ^{54} RNA polymerase bound at the promoter by bending the DNA between the two distant regulatory sites.

The exact roles of these *cis*-elements and *trans*-acting proteins must be dissected in order to determine how expression of the *Caulobacter fla* gene hierarchy is fine-tuned. It is already apparent, however, that the exquisitely ordered sequence of *fla* gene expression during the cell cycle (Fig. 6) can be understood in broad outline by a cascade in which a specific transcriptional factor(s) encoded by genes at each level of the hierarchy activates genes at the next lower level.

E. Initiation of the Cascade

If the organization of genes in the regulatory hierarchy provides a genetic mechanism for ordering the sequence of gene expression, it does not explain how the transcriptional cascade is started. The simplest solution to this particular regulatory problem is one or more "master timing genes" at the top of the hierarchy that initiates the cascade. The action of this master gene can be likened to pushing over the first in a long row of dominoes. Its identity has not been uncovered, but many developmental events in *Caulobacter* are tied to steps in the cell cycle, as discussed in the last section of this chapter, and a convenient trigger of the *fla* gene cascade would be an early event in the cell cycle. Previous work has shown that DNA synthesis is required for expression of the hook protein gene and the 25- and 27-kDa flagellin genes, and chromosome replication is one candidate for this cell cycle clock.

V. Control of Spatial Localization

The requirement for spatial information in the developmental program of *Caulobacter* is evident in the polar placement of the differentiated structures seen on the predivisional cell in Fig. 1A. The genetic blueprint must specify the topological location for assembly of the structures on the cell surface as well as the time of subunit synthesis in the cell cycle. One of the first systematic studies of this problem was carried out on the chemotactic proteins, including the methyltransferase, the methylesterase, and the methyl receptor proteins (MCPs). All of these proteins are synthesized periodically in the cell cycle just before cell division in a pattern very similar to that observed in Fig. 3 for the hook and flagellin proteins. Other experiments showed that the entire chemotactic apparatus is segregated exclusively to the swarmer cell at division.

Using sophisticated fractionation procedures that allowed separation of membrane vesicles derived from the swarmer and the stalked parts of the predivisional cell, researchers showed that newly synthesized MCPs were localized in the incipient swarmer cell, preferentially to the region of the cell envelope adjacent to the flagellum. Other experiments have shown that the 27- and 25-kDa flagellins are targeted to the same membrane domain before they are assembled into the flagellar filament.

The general export pathway in *E. coli* is responsible for localization of proteins to the periplasm and outer membrane compartments, and using this system as a paradigm it is reasonable to suppose that spatially localized *Caulobacter* proteins contain an analogous set of signal sequences that target them to a spatially restricted membrane domain. Experiments are now being undertaken to determine if the flagellar and chemotactic proteins do in fact contain such specialized signal sequences.

Another level of control over protein localization has been suggested by recent experiments using fusions of *fla* gene promoters to various reporter genes. The observation in several cases that the transcriptional activity was segregated preferentially to swarmer cells at division raised the possibility that transcription of at least some *fla* genes occurs within the swarmer cell compartment of the predivisional cell. There is as yet no direct evidence for this conclusion, but examples of genes whose transcription is restricted to either the mother cell or the forespore compartment of *Bacillus subtilis* during sporulation have been reported.

What specifies the cell pole as the target for assembly of developmental structures in *Caulobacter?* A signal recognition protein or complex has not been identified, but we have proposed that a site for polar localization may be part of an organizational center that is laid down at the two new cell poles at the time of cell division. This idea is considered more fully in the following section.

VI. Cell Division and Regulation of Differentiation

Temperature-sensitive (ts) cell division cycle (*cdc*) mutations have been used to define the organization of cell cycle steps in *Caulobacter* and to define the role of these genes in differentiation. A large number of *cdc* mutants that fail to divide normally at the restrictive temperature form long filamentous cells like those shown in Fig. 7. These ts strains, which have defects in genes required for either DNA synthesis (*dna*) or cell division (*div*), have been used to define a DNA synthetic pathway, DNAi → DNAe → DNAc, where DNAi, DNAe, and DNAc represent DNA initiation, chain elongation, and completion, respectively. The *div* mutants were placed on the cell division pathway, DIVi → DIVp → CS, as a function of how far they progressed in cell division at the restrictive temperature: DIVi (division initiation) mutants formed unpinched cellular filaments blocked at the initiation of cell division; DIVp (division progression) mutants formed partially pinched filaments; and CS (cell separation) mutants formed highly pinched filaments blocked at the very last stage of cell division.

Close inspection of the mutant cells by electron microscopy led to the striking observation that the extent of differentiation at the flagellated cell pole depended on the stage of cell division completed.

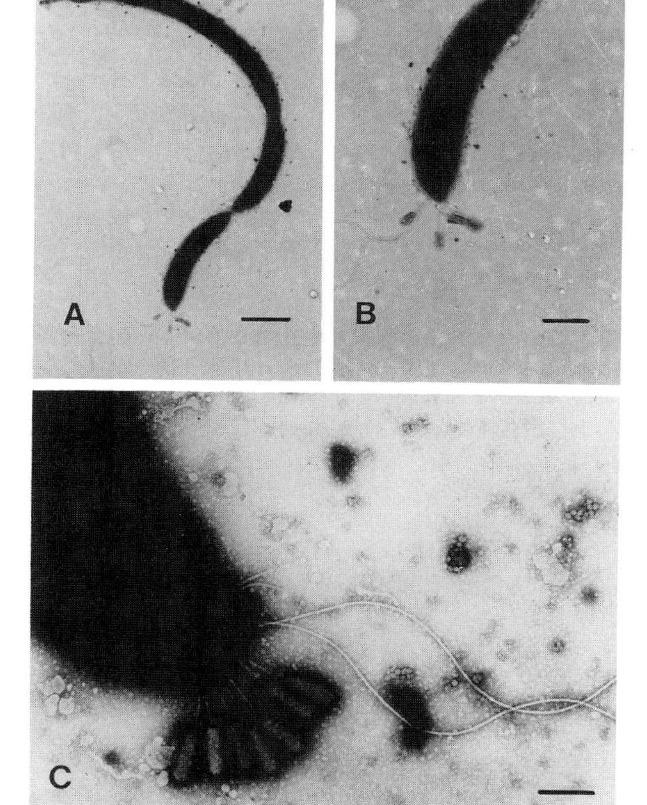

Figure 7 Distribution of DNA bacteriophage and flagella on negatively stained, filamentous cells of a DIVp mutant grown at the restrictive temperature. (A, B) Electron micrographs of the same cell with a higher magnification shown in B. (C) Flagellated pole of another filamentous cell in the same culture. Bar = 1 μm (A), 0.5 μm (B), and 0.2 μm (C). [From Huguenel, E. D., and Newton, A. (1982). Localization of surface structures during procaryotic differentiation: Role of cell division in *Caulobacter crescentus*. *Differentiation* 21, 71–78. Springer-Verlag, New York.]

DNA replication mutants blocked at DNAi or DNAe did not assemble detectable polar structures, including the bacteriophage receptors and flagellum. Mutants blocked in the division pathway at DIVi or DIVp, on the other hand, assembled the bacteriophage receptors and multiple flagella at the stalk-distal pole; these flagella are inactive, however, and the filamentous cells are not motile. The localization of the adsorbed bacteriophage particles and the multiple flagella at the stalk-distal cell pole is dramatically illustrated by the electron photomicrographs of a DIVp mutant in Fig. 7. It is also apparent that the DIVp mutant fails to form a new stalk at the flagellated cell pole. Filamentous mutant cells blocked at CS, the last step in the division pathway, are motile, and they also undergo stalk formation. They do not, however, assemble pili.

In summary, these observations indicated that DNA replication is required for flagellum biosynthesis and bacteriophage receptor formation, completion of DIVp is required for turning on motility and stalk formation, and completion of cell separation is required for pili formation. This sequence of developmental events corresponds exactly to that observed during the normal cell cycle as it is illustrated in Fig. 2. The strong implication of these results is that differentiation in *Caulobacter* is coupled in some way to steps in the underlying DNA synthetic and cell division pathways. [*See* DNA Replication.]

The same conclusion is supported by a detailed suppressor analysis that has provided evidence for an interaction of cell division and developmental gene products. One model suggested by this study is that the cell division proteins are localized at the division site and that after cell separation they then remain at the new cell poles where they function as specialized organizational centers responsible for targeting of proteins for assembly of the differentiated surface structures. Although highly speculative, this model provides a provocative framework for future research. We can anticipate that these and related studies will detail the molecular and biochemical mechanisms that generate cellular polarity and drive the developmental programs responsible for the characteristic temporal and spatial patterns of differentiation in *Caulobacter*.

Acknowledgments

We are grateful to John Smit for unpublished information on the classification of caulobacters prior to publication, to Bert Ely for information on the growth conditions, and to colleagues in our laboratory for their comments on the manuscript. Work from this laboratory has been supported by Public Health Service grants from the National Institutes of Health and Grant MV-386 from the American Cancer Society.

Bibliography

Ely, B. (1991). Genetics of *Caulobacter crescentus*. *In* "Methods in Enzymology" (J. Miller, ed.), pp. 372–384. Academic Press, San Diego.

Gober, J. W., and Shapiro, L. (1991). *BioEssays* **13,** 277–283.

Newton, A. (1987). *Microbiol. Sci.* **4,** 338–341.
Newton, A. (1989). Differentiation in *Caulobacter:* Flagellum development, motility and chemotaxis. *In* "Genetics of Bacterial Diversity" (D. A. Hopwood and K. F. Chater, eds.), pp. 199–220. Academic Press, San Diego.
Newton, A., and Ohta, N. (1990). Regulation of the cell division cycle and differentiation in bacteria. *In* "Annual Reviews of Microbiology" (L. N. Ornston, A. Balows, and E. P. Greenberg, eds.), pp. 689–719. Annual Reviews Inc.
Poindexter, J. S. (1989). Bacteria that divide by binary transverse fission, genus *Caulobacter*. *In* "Bergey's Manual" (J. T. Staley, ed.) pp. 1924–1939. Williams and Wilkins, Baltimore.

Cell Membrane: Structure and Function

Philip L. Yeagle
University of Buffalo

Glossary

Entropy Measure of the randomness of a system; tending toward chaos

Flip-flop Transmembrane movement of lipids, flipping across a lipid bilayer

Hydrophobic Molecules or portions thereof that cannot participate in hydrogen-bonding or other polar interactions with liquid water

Integral membrane protein Membrane protein with substantial portions of its mass buried in the hydrophobic interior of the membrane

Lamellar Lipid bilayer form, with the hydrocarbon chains of the phospholipids opposing each other in two leaflets

Peripheral protein Membrane protein associated with the surface of the membrane, readily removed from the membrane in many cases, and little or none of its mass incorporated into the membrane

CELL MEMBRANES constitute the boundary of the cytoplasm of all cells and thus the outer perimeter of intracellular functions. All cells (and enveloped viruses) contain as their most fundamental element of structure a plasma membrane, or its equivalent, surrounding the cell (or virion). The early steps in the evolution of prokaryotic cells probably required the distinction between the "inside" of the protocell and the "outside" of the protocell provided by a membrane. Because the lipid bilayer—the fundamental structural element of a membrane—forms spontaneously and because the lipids of that bilayer are not a gene product, the lipid bilayer may be the earliest and most fundamental structure of life. Prokaryotic cells contain no intracellular membranes. In contrast, the intracellular volume of eukaryotic cells is largely filled with organelle membranes, in which the labor of the cell is divided into specialized functions. Cell membranes compartmentalize these specialized functions as a necessary contribution to their individual operation. Cell membranes contain much of the machinery for cellular function. Many of the synthetic, developmental, and metabolic activities of the cell occur on or around the membranes of the cell. Cell membranes are made of two major components: proteins (gene product) and lipids (product of enzymes). Remarkably, the structure of these membranes is stabilized by noncovalent forces, collectively referred to as the hydrophobic effect.

I. Elements of Membrane Structure

The membranes of cells constitute the obvious components of cellular structure as observed under an electron microscope. Outside of the plasma membrane many prokaryotes have an outer membrane or cell wall. The outer membrane is semipermeable in that proteins such as porins offer transit to relatively large molecules (<600 daltons in many cases though in some cases larger) through nonspecific pores. In addition to the plasma membrane surrounding the cell, eukaryotic cells contain many organelles within the cell, each constituted of, and bounded by, a membrane (see Fig. 1). For example, the nucleus is

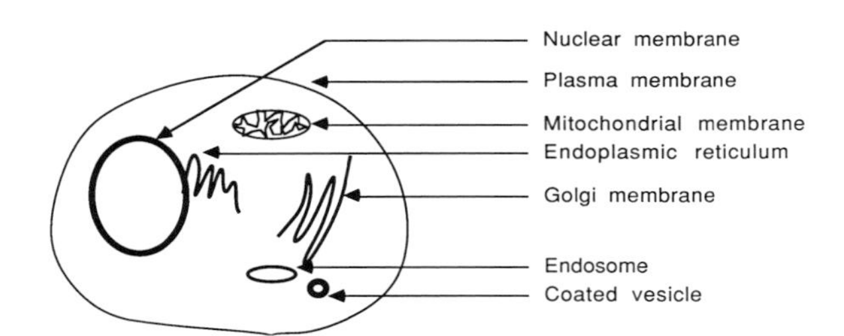

Figure 1 Schematic representation of some of the membrane systems within a mammalian cell.

bounded by a double membrane, which not only compartmentalizes the genetic material but also is involved in regulation of gene expression. The membranes of the endoplasmic reticulum provide the biosynthetic machinery for much of lipid and membrane protein biosynthesis. On the inner mitochondrial membrane, the major production of adenosine triphosphate (ATP) for cellular use occurs. Thus, the functions of cellular membranes are complex and crucial to cell viability; however, the most fundamental factors leading to the structures of these complex membranes are relatively simple. [*See* CELL WALLS OF BACTERIA.]

A. The Hydrophobic Effect

The membranes of cells are built of two major components, lipid and protein, and are stabilized in the membraneous structures they form by a third component, water. The hydrophobic effect describes the noncovalent interactions of these three components that determine the structure of biological membranes.

The structure of liquid water and the relative ability of other chemical structures to incorporate into that water structure form the concept known as the hydrophobic effect. Water molecules are linked by intermolecular hydrogen bonds into a transient lattice structure in the liquid state. Although the details of the liquid state are not well understood, the formation of these hydrogen bonds is known to provide many of the unique properties of liquid water. For molecules other than water to dissolve in water, the chemical structures of those molecules must participate in the water structure. The most simple requirement for incorporation in the water structure is a capability on the part of an intruding molecule to donate and/or accept a hydrogen bond to/from water molecules. Thus, a molecule such as urea will readily dissolve in water due to its ability to both donate and accept hydrogen bonds involving water molecules. On the other hand, a molecule such as *n*-hexane can neither donate nor accept hydrogen bonds from the water and, thus, cannot readily incorporate into the water structure. This simple description is the basis for the statement "oil and water do not mix."

There are various ways to describe what the water structure does in response to the intrusion of a moleculelike hexane. Conceptually, the most simple is the development of a cage-, or clathrate-, type structure of the water around the intruding molecule. All hydrogen bonds of the water must form, but they must form in a much more organized fashion to surround the intruding molecule. Organization is in general energetically costly. The cost is referred to as entropy. An energetically favorable entropy change is toward chaos. The formation of water cages around hydrocarbons is a more organized state and, thus, is energetically unfavorable. This is the hydrophobic effect.

The hydrophobic effect determines the structure of cell membranes. The hydrophobic effect also leads to the sequestration of chemical structures that cannot interact with the water structure into aggregates that occupy space not in contact with water. These structures are referred to as hydrophobic.

B. The Lipid Bilayer

In regard to their ability to interact with the structure of water, the lipids of biological membranes are characterized by chemical structures of two natures. One part of their chemical structure is like the preceding ethanol example: part of the molecule can participate in the water structure and, if by itself, readily dissolve in water. The other part of the chemical structure of these lipids is hydrophobic; therefore, energetics requires that part of the chemical structure to be sequestered, out of contact with water.

Figure 2 shows a structure for a lipid found in biological membranes. The key elements of this structure are common to most of the lipids found in biological membranes.

The hydrophobic portion of the phospholipid is dominated by hydrocarbon chains, often (but not always) consisting of fatty acids esterified to the glycerol. This region must be protected from interaction with the water structure by the aggregate structure that the phospholipids form. Table I lists

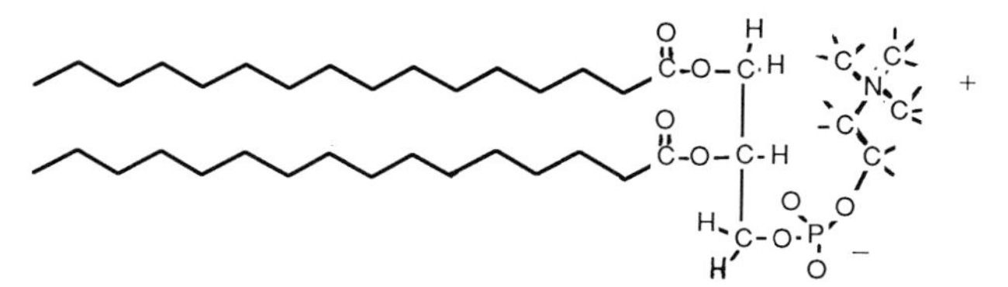

Figure 2 Chemical structure of phosphatidylcholine. This phospholipid has a zwitterionic headgroup with both a positive and a negative charge.

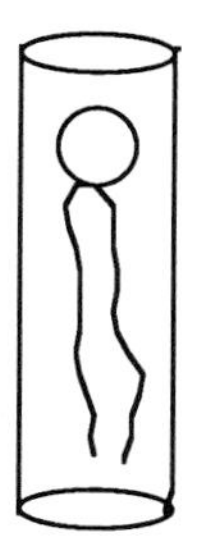

Figure 3 Schematic representation of the structure of a lipid, approximated as a cylinder.

some of the fatty acids commonly found esterified to the phospholipids.

The third position of the glycerol is esterified to a phosphate and thence to the phospholipid headgroup. This is the hydrophilic or polar portion of the molecule. For phospholipids, the headgroups are commonly alcohols, such as ethanolamine, choline, serine, inositol, and glycerol. The phospholipid is then named according to its headgroup (e.g., phosphatidylcholine from choline, phosphatidylglycerol from glycerol).

When a molecule with the structure of a phospholipid finds itself in an aqueous environment, what kind of a macromolecular structure is likely to be formed? The phospholipid is considered to be amphipathic; one portion is hydrophobic and the other portion is hydrophilic or polar. The hydrophobic effect requires that the hydrophobic portion of the phospholipid, the phospholipid hydrocarbon chains, be sequestered away from the water. In the case of phosphatidylcholine, the cross-sectional area of the phospholipid headgroup (measured perpendicular to the long axis of the molecule) is similar to the cross-sectional area of the phospholipid hydrocarbon chains. Therefore, the overall shape of the molecule can be approximated by a cylinder (Fig. 3). To satisfy the hydrophobic effect, one can readily see that the cylinders will stack with the hydrocarbon chains together, as shown schematically in Fig. 4. Thus, the lipid bilayer is formed, which is the fundamental element of membrane structure common to all biological membranes (Fig. 5).

Table I Fatty Acids

Code	Name
14:0	Myristic acid
16:0	Palmitic acid
18:0	Stearic acid
18:1	Oleic acid
18:2	Linoleic acid
20:4	Arachidonic acid
22:6	Docosahexenoic acid

C. Model Membranes

The phospholipid bilayer forms spontaneously when phospholipids are introduced into an aqueous environment. These bilayers form the basis for several model membrane systems. These model membrane systems in general mimic some of the properties of biological membranes because both the model systems and the biological membranes contain a lipid bilayer as a common structural element.

The most simple form of these model membrane systems is the multilamellar liposome. This structure forms spontaneously upon hydration of a film of phospholipid dried out of organic solvent onto the side of a flask. The liposome consists of a concentric set of bilayers, which in cross-section resemble an onion. These systems have been used to study the physical properties of lipids in bilayers.

If this material is then sonicated (subjected to intense ultrasonic irradiation), small unilamellar vesicles are formed. These consist of a shell of a single phospholipid bilayer surrounding a small aqueous space. Such vesicles are about 210–250 Å in diame-

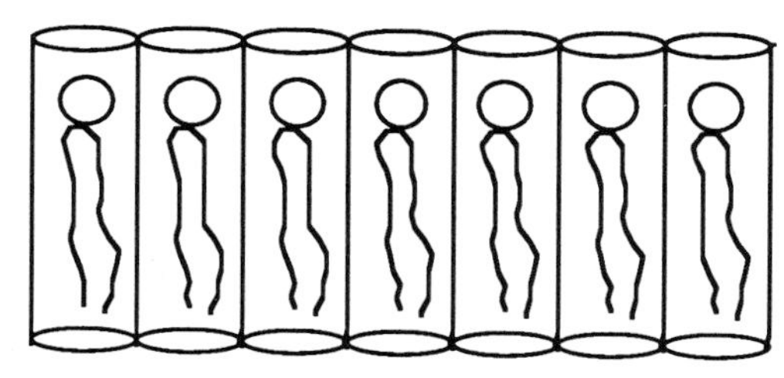

Figure 4 Schematic representation of how the cylindrically shaped lipids pack together.

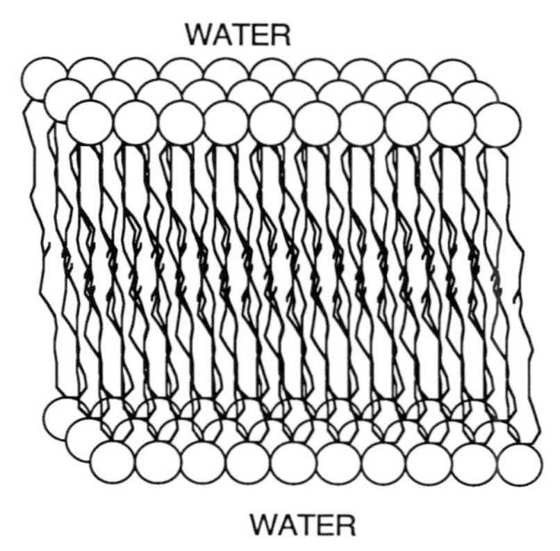

Figure 5 Schematic representation of the lipid bilayer, common to all mammalian cell membranes. The polar headgroups of the lipids face the water, and the hydrocarbon chains of the lipids are sequestered away from the water in the middle of the bilayer.

ter, depending on the phospholipid composition. They can be fractionated into a remarkably homogeneous preparation suitable for physical chemistry measurements. They are too small, however, to effectively trap any macromolecules in their interior.

Several other techniques have been developed to form large unilamellar vesicles. These vesicles are suitable for trapping materials in their interior. They have been used for drug delivery systems. These vesicles are typically several hundred nanometers in diameter, but some specialized techniques are available to make them much larger.

These model membranes have been, and remain, important subjects of study for increasing our understanding of the properties of lipid bilayers.

II. Structure and Properties of Membrane Lipids

A. Phospholipids

The structures of phospholipids were outlined in the preceding section. All the phospholipids are built on the basic structure of a phosphate esterified to a headgroup and to a glycerol. With the variety of headgroups, hydrocarbon chains, and the attachment of hydrocarbon chains to the glycerol backbone, literally thousands of different species of phospholipids are known.

The most common phospholipid structures exhibit an unsaturated, branched or alicyclic hydrocarbon chain at position 2 of the glycerol and a saturated hydrocarbon chain at position 1 of the glycerol. This is a direct result of the biosynthesis pathways for the phospholipids. Most phospholipids form bilayers spontaneously as already described.

Three classes of digestive enzymes, the phospholipases, can act on these phospholipids. Phospholipase A_2 hydrolyzes the ester bond at position 2 of the glycerol, releasing as products a fatty acid (such as arachidonic acid as a precurser to prostaglandin biosynthesis) and a lysophospholipid. Phospholipase C removes the phosphorylated alcohol from the glycerol. In the case of phosphorylated inositol, this results in the production of the second messengers inositol-(di,tri)-phosphate and diacylglycerol. Phospholipase D catalyzes the hydrolysis of the terminal phosphoester bond of phospholipids and can bring about transphosphatidylation. The transphosphatidylation activity of phospholipase D is involved in some of the pathways of phospholipid biosynthesis.

B. Sphingolipids

The chemical structure of sphingolipids is most strongly distinguished from the phospholipids by means of the backbone. The sphingosine base, or *trans*-D-erytho-1,3-dihydroxy-2-amino-4-octadecene, is the backbone of sphingolipids, whereas the glycerol is the backbone of phospholipids. The sphingosine base contributes its long hydrocarbon chain as one of the two hydrocarbon chains in sphingolipids. A fatty acid is amide-linked to the nitrogen of the sphingosine base at carbon 2 position, contributing its hydrocarbon chain as the second long hydrocarbon chain in sphingolipids.

Sphingomyelin, a member of this family found in abundance in plasma membranes, is constructed with a phosphocholine esterified to the primary hydroxyl of the sphingosine base and a fatty acid linked by an amide bond through the amine. In natural sphingomyelins, these fatty acids are often saturated.

C. Glycolipids

Glycolipids, in general, have carbohydrates as headgroups. This may be either a simple sugar attached to a diacylglycerol, such as monoglactosyl diglyceride, or a complex carbohydrate attached to a ceramide, such as a ganglioside. In the case of the latter, sialic acid is part of the structure, lending net negative charge to the glycolipid. The latter apparently can serve as binding sites on the cell surface for interesting "ligands" such as viruses.

D. Sterols

Cholesterol (Fig. 6) is the best-known member of the lipid family of sterols. Cholesterol consists of a planar ring at the heart of an amphipathic structure. The hydroxyl provides the polar headgroup of this lipid, whereas the remainder of the molecule is hydrophobic. Cholesterol orients in bilayers, with its sterol nucleus parallel to the long axis of the phospholipid hydrocarbon chains. Cholesterol is the sterol found in mammalian cells. (In yeast and plants, sterols of different structure are found.) The sterols are predominantly found in the plasma membranes of these cells, in which their role is essential and structurally specific. For example, the yeast sterol, ergosterol, cannot substitute for cholesterol in mammalian cells for normal cellular function and viability. The molecular basis appears to be the ability of the specific sterol to modulate the activity of crucial membrane enzymes. Cholesterol at high levels appears to "stiffen" the membrane by increasing the motional order of the lipid hydrocarbon chains in the bilayer of the membrane.

E. Lipid Phase Behavior

The lipids of biological membranes can form a variety of phase structures. In forming these structures, lipids reveal some of their important properties. To understand the phase behavior of membranes, it is important to recognize that the lipid bilayer of these membranes is an ordered structure; i.e., the hydrocarbon chains are constrained to be, on average, approximately parallel to each other and the individual lipid molecules are not likely to move in and out of membranes readily due to the hydrophobic effect. The most commonly found state in biological membranes is the liquid crystal state, in which the lipids are capable of lateral diffusion in two dimensions in the plane of the bilayer, but their movement is constrained in the third dimension.

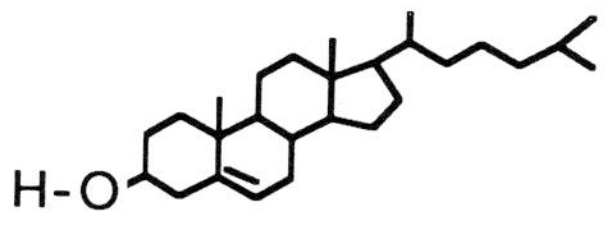

Upon cooling, the liquid crystalline phase can undergo a phase transition to a solid gel state. This transition can be readily detected by calorimetry. In the gel state, the lipid hydrocarbon chains stretch out in the all-trans configuration (about the individual carbon–carbon bonds), increasing the bilayer thickness. Lateral diffusion is dramatically reduced in the gel state. Observations of the transition from the liquid crystal to the gel phase in certain cellular membranes was important in establishing that the cell membranes consisted of a lipid bilayer as a fundamental structural element.

Some membrane lipids, such as phosphatidylethanolamine, are capable of a phase transition to the hexagonal II phase. Intermediates in this transition process may be involved in membrane fusion. Nonlamellar phases, such as the hexagonal II phase, disrupt biology and *in vivo* would represent membrane degeneration.

Cholesterol

Figure 6 The chemical structure of cholesterol. The lines connect carbon atoms, with the exception of the oxygen and hydrogen explicitly noted. The hydrogen atoms on the carbons are not indicated.

F. Lipid Dynamics

Motional order is a concept related to that of entropy, i.e., the degree of ordering of the system. The more degrees of motional freedom experienced by a system, or part of a system, the less ordered is that system. The degree of ordering of a particular segment of the lipid hydrocarbon chain depends on the part of the chain under consideration. Motional order is highest near the glycerol backbone of the lipid and lowest in the center of the bilayer. Thus, the interior of a lipid bilayer or a biological membrane is quite different than liquid hydrocarbon, which the interior of the membrane resembles in terms of hydrophobicity. The membrane bilayer is a highly anisotropic medium. Other molecules introduced into a bilayer are, in turn, oriented by the bilayer (see Fig. 7), which may affect the behavior of the foreign molecule and the properties of the membrane itself.

Biological membranes are permeable to water. This is readily understood in terms of the dynamics of the bilayer structure. In addition to lateral diffusion, the lipid hydrocarbon chains are capable of isomerization around the carbon–carbon single bonds. This can lead to localized, transient defects in the bilayer structure of a size sufficient to accommodate water molecules. Thus, even though the hydrophobic effect states that water cannot, on aver-

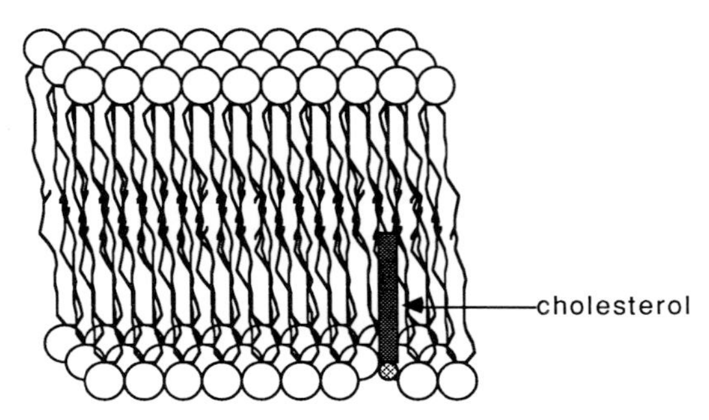

Figure 7 Orientation of cholesterol by the phospholipids in the membrane.

age, exhibit high occupancy of sites within the hydrophobic region of the bilayer, water can rapidly move through the lipid bilayer by means of these defects in bilayer structure. If the bilayer undergoes a phase transition to the gel state, the packing of the lipids in the gel state largely eliminates these defects, and water permeability is reduced.

Likewise, lateral diffusion is dramatically reduced in the gel state. Other perturbations, such as an increase in cholesterol, do not strongly affect lateral diffusion. Lateral diffusion coefficients in the range of 10^{-8} cm^2 sec^{-1} are commonly found for lipids in model and biological membranes.

It is apparent that the lipid bilayer of a biological membrane is normally in the liquid crystal state. On the one hand, the gel state would inhibit motional freedom important to membrane protein function; on the other, the hexagonal II state would destroy the permeability barrier of the membrane.

In general, the lipid composition on one side of a bilayer leaflet in a cell membrane is not identical to the lipid composition on the other side of the bilayer leaflet. An example is found in the mammalian erythrocyte membrane, in which the outer monolayer of the plasma membrane is enriched in phosphatidylcholine and sphingomyelin and the inner monolayer is enriched in phosphatidylethanolamine and phosphatidylserine. How is this transmembrane asymmetry of lipids created? How is this maintained? And what role does it play in cell biology?

Flip-flop is the term applied to transmembrane movement of phospholipids. This is thermodynamically unfavorable and does not occur readily in most model membrane systems. In biological membranes, flip-flop does occur and is essential to cell membrane structure and synthesis (synthesis of new lipid occurs on the cytoplasmic face of the endoplasmic reticulum, and flip-flop must occur to move newly synthesized lipid to the lumenal surface of the same membrane). It has been suggested that membrane protein components facilitate the transmembrane movement of lipids. Furthermore, in the erythrocyte, the maintenance of the transmembrane lipid asymmetry is also promoted by proteins in an energy-dependent manner (ATP-dependent).

III. Structure and Properties of Membrane Proteins

Membrane proteins confer on membranes many of the properties of cellular function. As in other areas in biology, proteins provide specificity of interactions and catalysis of biochemical reactions. In addition, the basic structure of the membrane provides protein-mediated functions that are impossible to achieve in solution. For example, the establishment and utilization of a transmembrane [H^+] gradient for synthesis of ATP is uniquely possible through cell membranes and their proteins. It is topologically impossible to achieve without the structure of the membrane.

A. Classification of Membrane Proteins

To understand membrane protein structure, it is useful to classify membrane proteins according to their disposition in the membrane. The most general classifications are those of peripheral and integral membrane proteins. Peripheral membrane proteins are associated with the membrane but do not have any of their mass buried in the membrane. The structures of proteins in this class are similar to those of typical water-soluble proteins. The class of peripheral membrane proteins is divided into two subclasses. One, called associated membrane proteins, refers to globular proteins that bind to the surface of a membrane, either to the lipid bilayer or to other protein components in the membrane [integral membrane proteins (see following)]. The second subclass is that of skeletal membrane proteins. These proteins form the membrane skeleton that lies just beneath the plasma membrane of cells. This membrane skeleton is attached to the membrane by associated proteins and often connects to the cytoskeleton of the cell. Figure 8 shows schematically these two subclasses of peripheral membrane proteins.

The other major class of membrane proteins is the class of integral membrane proteins. These proteins

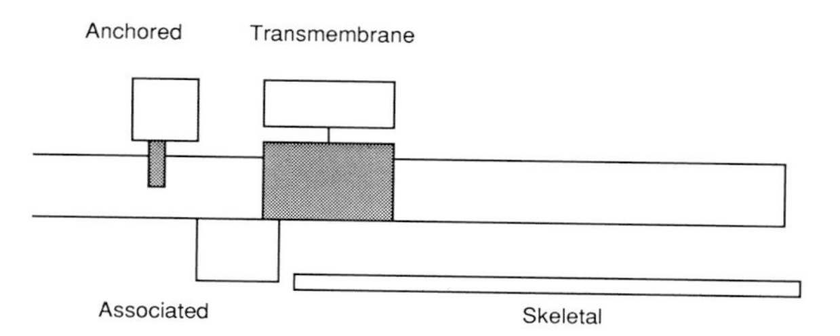

Figure 8 Classification of membrane proteins.

bury substantial portions of their total mass in the hydrophobic region of the lipid bilayer of the membrane. These proteins cannot be readily removed from the membrane, except in the presence of detergents that can solubilize the individual lipid and protein components of the membrane. While these proteins are constrained to the nearly two-dimensional world of the membrane, integral membrane proteins can diffuse in the plane of the membrane, as do the lipids. Lateral diffusion of proteins is slower in general than for lipids. Lateral diffusion of integral membrane proteins can be restricted by interactions between the integral membrane protein and the membrane skeleton.

This class of membrane proteins can be divided into two subclasses. One is referred to as transmembrane proteins (Fig. 8). These integral membrane proteins not only bury much of their mass within the lipid bilayer but also expose significant portions of their structure to both sides of the membrane. Thus, while the disposition of integral membrane proteins is always absolutely asymmetric in a membrane, a portion of the transmembrane protein is exposed on one side of the membrane and a different portion of the protein is exposed on the other side of the membrane.

The other subclass is referred to as anchored membrane proteins. As shown in Fig. 8, anchored membrane proteins have a component that penetrates the bilayer and attaches the protein to the membrane but does not completely traverse the membrane. The most common members of this class are anchored to the membrane by covalent attachment to a lipid, such as a phosphatidylinositol derivative. This may allow for an interesting mechanism for release of the protein from the membrane by enzymatic cleavage of the phospholipid (such as with a phospholipase). The result is then a soluble protein with often a different intrinsic activity from the membrane-bound form.

B. Membrane Protein Structure

The structure of peripheral membrane proteins is analogous to the structure of typical water-soluble proteins in many aspects. However, the integral membrane proteins exhibit distinctly different structures. This is necessitated by the inclusion of a portion of the mass of the protein in the hydrophobic interior of the lipid bilayer. The portion of the protein in the membrane must be compatible with the bilayer interior, as governed by the hydrophobic effect. Many membrane proteins have utilized an interesting topological solution of this problem. The amino acid sequence of these proteins contains concentrations of the hydrophobic amino acids in linear arrays. These linear sequences of hydrophobic amino acids are long enough (19–23 hydrophobic amino acids) to span the hydrophobic interior of the membrane in the form of an α-helix. These hydrophobic transmembrane α-helices constitute therefore a fundamental element of the structure of integral membrane proteins. Interaction energies among these transmembrane helices lead to energetically favorable self-association within the membrane. The result can be a structure more stable thermodynamically than is observed for many water-soluble proteins.

The intramembraneous, hydrophobic portion of a membrane protein often acts as a domain within the structure of the protein. Therefore, the membrane protein as a whole is characterized by domain structure. In these terms, the integral membrane proteins can be subdivided further on the basis of their domain structure, as represented schematically in Fig. 9.

The one-domain protein buries most of its mass within the lipid bilayer. An example of such a protein would be the light activated proton pump, bacteriorhodopsin. A two-domain membrane protein has one intramembraneous domain (consisting of one or more transmembrane stretches of the polypeptide) and one extramembraneous domain with a structural characteristic similar to that of water-soluble proteins. Human erythrocyte glycophorin is a simple example of a two-domain protein. The three-domain protein has two extramembraneous domains separated by a extramembraneous domain. An example of this structure is found in some bacterial H^+ pump proteins. The last example is that of the viral fusion proteins, which are similar to two-domain proteins except that they have an additional hydrophobic sequence, not involved in anchoring

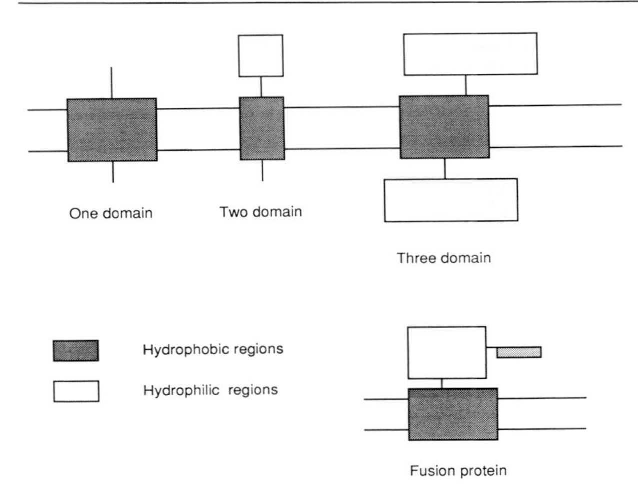

Figure 9 Subclasses of integral membrane proteins.

the protein to the viral membrane but, rather, involved in the fusion activity exhibited by this class of proteins.

An additional level of organization in membrane protein structure analogous to quaternary structure of water-soluble proteins has been observed. Membrane proteins often form dimers or higher oligomers in membranes. It has been suggested that formation and disruption of oligomers in the membrane can modulate the activity of membrane proteins.

IV. Membrane Transport

Biological membranes provide a permeability barrier separating the inside of the cell from the outside and the organelles within the cell from one another. Therefore, to move solutes from the outside to the inside and from one intracellular compartment to another, transport mechanisms must exist to move materials across membranes. Often through transport, exquisite control of biological processes is achieved. Transport occurs through both energy-independent and energy-dependent pathways.

Figure 10 Schematic representation of the integration of membrane proteins into a membrane bilayer of a plasma membrane. This representation also includes the carbohydrate portion of the glycolipids and glycoproteins exposed to the extracellular surface.

A. Passive Diffusion

Passive diffusion does not require energy—it occurs spontaneously. It can occur by flow of material down a concentration gradient or even when no net transport is observed. Net passive diffusion is driven by the energy arising from the difference in chemical potential of a material (e.g., manifest in a concentration gradient) across a membrane. For example, cells normally experience higher sodium on the outside than on the inside. Even in the absence of specific transport proteins, sodium can ''leak'' back into the cell by passive diffusion.

How does passive diffusion occur? The mechanism is the same as that described earlier for water. Thus, the rate of transport depends on how effectively the molecule to be transported ''fits'' into the defect in the membrane that provides the pathway for transit of the membrane as well as on the charge of the molecule. Sucrose transits the membrane much more slowly than does water, because the sucrose is substantially larger than the water and does not fit into the defect as well. Sodium transits a lipid bilayer more slowly than glucose because of its charge (and strong hydration).

B. Facilitated Diffusion

Molecules such as glycerol may be too large to diffuse across the membrane rapidly enough by the preceding mechanism for the metabolic requirements of bacteria under some conditions. Many cells have developed another, more efficient means for transmembrane diffusion of molecules such as glycerol. Some membrane proteins are capable of facilitating the diffusion of solutes across a membrane. In this capacity, these transmembrane proteins act like channels or pores in the membrane that are relatively specific for the molecules of interest. The lining of this channel may be relatively hydrophilic to accommodate the species in transit much better than the hydrophobic interior of the lipid bilayer. An example is the glycerol facilitator of *E. coli*. This is a transmembrane protein that facilitates the movement of glycerol across the membrane at rates much faster than those observed across lipid bilayers. Net facilitated transport only occurs, however, in the presence of a transmembrane gradient of free glycerol. Therefore, this facilitated diffusion cannot be used to build up a transmembrane gradient of glycerol. The porins of the outer membrane of gram-negative bacteria provide another example of facilitated diffusion.

C. Active Transport

In many situations within a cell and across the plasma membrane of a cell, cell viability depends on the maintenance of a transmembrane gradient of some chemical species. For example, most cells depend on a sodium gradient (from outside to inside) and a potassium gradient (from inside to outside) across the plasma membrane. This gradient is involved in the conduction of nerve impulses, in the transport of essential nutrients, and in signal transduction. The establishment of such a gradient requires pumping ions against a concentration gradient, which can only be done by coupling the transport function (of the ion) to an energy source such as ATP hydrolysis. Therefore, integral membrane proteins are found in the plasma membrane of cells that are capable of utilizing the energy in the hydrolysis of ATP to pump sodium ions from a lower concentration to a higher concentration. In this case, the enzyme is referred to as the Na^+/K^+ ATPase. This name describes the ATP hydrolyzing function as well as the absolute requirement for sodium and potassium for the reactions catalyzed by this enzyme. For the coupling to occur between the energy source (ATP hydrolysis) and the energy-requiring function (cation transport), the consumption of the energy source is allowed to procede only in conjunction with transmembrane translocation of Na^+ and K^+. This is the secret of active transport. [*See* ATPASES AND ION CURRENTS; ION TRANSPORT.]

Many other examples of active transport can be found in cells. For example, the uptake of maltose in *E. coli* occurs by an active transport mechanism. Hydrolysis of ATP releases energy that is used to pump maltose up a concentration gradient of as much as five orders of magnitude. Another example can be found in some unicellular organisms that pump ions directionally—out one end of the cell, while ions flow back in the other end. This sets up ion currents around the exterior of the cell.

V. Receptors

A receptor transmits information from the exterior to the interior of the cell. Information usually comes in the form of an agonist, such as a hormone, to the cell surface. The receptor contains a high-affinity binding site for the agonist. The cell surface receptor is a transmembrane protein. Information that the receptor-binding site is occupied by the agonist is transmitted to the cytoplasmic surface of the plasma membrane by the receptor. The signal is "transduced" into enzyme activation and/or second messengers inside the cell that result in a cellular response. Some examples of receptors will now be described, providing insight into the capabilities of these signal transduction systems.

A. Receptor Function Involving G Proteins

Examples of receptors utilizing G proteins include the β_2-adrenergic receptor (catecholamines), photoreceptor, α_1-adrenergic receptor (catecholamines), muscarinic cholinergic receptors, vasopressin receptor, and angiotensin receptor. With respect to this group of receptors, transmembrane signaling involves three different kinds of proteins. The first kind of protein is the receptor, an integral membrane protein. This receptor may be in the plasma membrane and, thus, located on the cell surface, like the β-adrenergic receptor. Alternatively, this protein may be located in an intracellular membrane, as in the visual transduction system of retinal rod cells. The receptor is the recognition component of the signal transduction system. It is specific for the particular signal to be recognized. Signals include catecholamines, serotonin, purines, peptides, and light.

The second kind of protein is the coupling factor, often referred to in these systems as G protein. This protein may be membrane-bound or soluble, or it may shuttle between the two forms. The G protein binds guanosine triphosphate (GTP) and has hydrolytic activity, turning over GTP to guanosine diphosphate.

The G protein, when activated by the receptor-mediated response, in turn activates the third kind of protein in this signal transduction system, a key enzyme that produces the intracellular second messenger. In this general way, a signal is transduced to an intracellular biochemical event with serious functional consequences for the cell in question.

The β-adrenergic receptor contains a plasma membrane-bound receptor protein. When the specific agonist binds to this receptor, the receptor couples to the G protein, which in turn can activate the adenylate cyclase and raise intracellular cyclic adenosine monophosphate (cAMP) levels. The increase in cAMP can lead to activation of cAMP-dependent protein kinase, which in turn can lead to protein phosphorylation and specific metabolic consequences.

The visual pigment of vertebrate retinal rod cells, rhodopsin, also functions as a receptor. Light-activated rhodopsin is capable of activating G protein, sometimes called transducin in this system, which in turn can activate phosphodiesterase. In the outer rod segment, this leads to a reduction in cyclic guanosine monophosphate, which may result, in one current hypothesis, in the closing of Na^+ channels in the plasma membrane.

B. Receptor-Mediated Endocytosis

The receptor for low-density lipoproteins (LDL receptor) provides an interesting example of receptor-mediated endocytosis. The LDL receptor is a transmembrane protein expressed on the surface of, for example, liver cells. It binds the protein component of the LDL. The occupied receptor then migrates to a coated pit on the cell surface (coated with clathrin on the cytoplasmic face of the pit). Endocytosis proceeds through a membrane fusion event and a coated vesicle is formed containing the occupied receptor. Subsequent intracellular fusions with endosomes and finally lysosomes leads to the ultimate catabolism of the LDL, but not necessarily the receptor. The receptor can be recycled, through transport vesicles, back to the plasma membrane of the cell. The catabolic products of the LDL (especially cholesterol) will regulate the rate of synthesis of cholesterol by the cell and the expression of the LDL receptor on the cell surface. Other examples of receptor-mediated endocytosis include the transferrin receptor for ion transport.

C. Acetylcholine Receptor

The nicotinic acetylcholine receptor offers yet another example of signal transduction. This is a transmembrane protein in the plasma membrane of nerve cells. It recognizes acetylcholine and changes conformation upon binding of this agonist. The protein contains an ion channel that is gated by the binding of the acetylcholine. This receptor offers a major pathway of signal transduction for rapid synaptic responses. The pathway in which this receptor functions is initiated by release of neurotransmitter at the presynaptic membrane through fusion of the secretory vesicles with the plasma membrane. Diffusion of the neurotransmitter occurs to the postsynaptic membrane where it binds to the acetylcholine receptor. This opens the sodium channel in that receptor, allowing the subsequent events of nerve conduction to occur.

VI. Membrane Fusion

Fusion is the term applied to two types of membrane events. One is the joining of two membranes to make one membrane, such as the fusion of a virus with a cell plasma membrane. The other is creation of a new vesicle from an existing membrane such as in endocytosis. The events leading up to the fusion of two lipid bilayers have been suggested to include the following events: (1) aggregation or adhesion of the membranes that will fuse; (2) close approach of the lipid bilayers of the membranes, leading to removal of some of the water separating the membranes (partial dehydration); (3) destabilization of the bilayer at the point of fusion (two bilayers closely opposed will not spontaneously fuse by themselves); and (4) mixing of the bilayers and ultimate separation from the point of fusion into the new membrane structure(s). In this section, some of the membrane fusion systems that have been intensively studied are described.

A. Enveloped Virus Fusion

Many enveloped viruses are effective at promoting membrane fusion. Among the most studied are the Sendai virus, which fuses at the plasma membrane of the cell, and influenza, which may fuse in endocytic vesicles (endosomes) after pH reduction by the vesicle H^+ ATPases. These viruses possess an outer limiting membrane containing glycoproteins responsible for recognition of the target cell and mediation of the fusion event between the viral envelope and the target cell plasma membrane. Following the fusion event in a successful infection, the virus then undergoes disassembly (uncoating), whereupon viral replication can begin.

Current hypotheses for the mechanism of enveloped virus fusion postulate a role for the envelope glycoproteins. Among these glycoproteins is a class of "fusion" proteins, which appear to be required in the fusion process. For example, in the Sendai virus and in RSV, this protein has been identified as the F, or fusion, protein. Similar glycoproteins for herpes simplex virus (HSV I) fusion have been identified. Many investigators have suggested that these fusion proteins cause membrane bilayer destabilization, which is promoted by a "fusogenic" sequence common to many enveloped viruses.

The fusion proteins of many enveloped viruses must be proteolytically cleaved to generate a fusion-competent virus. After cleavage, new, highly hydrophobic N termini are generated on the fusion proteins. For example, in Sendai, the fusion protein is made in the F_0 form, which can be cleaved with trypsin to produce F_1 and F_2 . F_1 and F_2 are linked by disulfide bonds. The amino acid sequence of these new N termini contain regions of homology, particularly within the first 20 or so amino acids, which are highly hydrophobic. The sequence N-Phe-*X*-Gly in particular, where *X* is any hydrophobic amino acid, is common to many of the above-mentioned viruses. Exposure of the fusogenic sequence to the target membrane is apparently required for enveloped virus fusion.

In summary at this point, the available data suggest that viral fusion proceeds by a mechanism echoing the general fusion scheme previously listed.

1. Aggregation of the virus to the target membrane occurs through binding of envelope glycoproteins to cell surface receptors.
2. The close approach of the viral and target membranes may occur after a rearrangement of viral proteins or by clustering and lateral displacement of cellular membrane proteins.
3. The fusogenic sequence of the fusion protein may destabilize a site in the target membrane, thereby facilitating the membrane fusion event

B. Intracellular Vesicular Fusion

Another important class of membrane fusions are those involving intracellular membranes. These include the fusion of transport vesicles from the endoplasmic reticulum moving to the Golgi and fusing with the Golgi membrane. Likewise, within the Golgi, vesicles move materials from one part of this organelle to another. These fusions have been shown to be the product of a complicated series of factors, both in the transport vesicles and in the target membrane. As opposed to some of the fusions already described, this class of membrane fusion must be tightly controlled to achieve proper transport and to maintain organelle integrity. Key factors that have been identified include (1) GTP-binding proteins, which function in the interaction between vesicles and target membranes and require GTP turnover to facilitate the interaction between the transport vesicles and the target, and (2) *N*-ethylmaleimide-sensitive (NSF) protein, which is part of the apparatus that sets up the membrane fusion event, although it is probably not involved directly in the fusion. The latter functions in an energy-dependent fashion, with the binding of ATP and the hydrolysis of ATP being involved in the cycle of NSF protein on and off the target membrane. Thus, GTP hydrolysis and ATP hydrolysis are involved, which may provide the energy required for the regulation of these fusions that is so important to cell viability. As yet, the details of this pathway are not yet fully described.

C. Secretion

Secretion also involves a fusion event. In fusion of secretory vesicles to the plasma membrane, calcium appears to be an important trigger. This may not be the same fusion that is observed in model membrane systems with negatively charged lipids and calcium, because the levels of calcium required are much higher in the model membrane fusion than in secretion. However, some cytoplasmic proteins, such a synexin, have demonstrated an ability to lower considerably the threshold at which calcium will promote a fusion event in a model system. Much is yet to be learned in this interesting fusion system.

Bibliography

Aloia, R. C., Curtain, C. C., and Gordon, L. M. (1988). "Methods for Studying Membrane Fluidity." A. R. Liss, New York.

Duzgunes, N., and Bronner, F. (1988). "Membrane Fusion in Fertilization, Cellular Transport, and Viral Infection." Academic Press, Orlando, Florida.

Esfahani, M., and Swaney, J. B. (1990). "Advances in Cholesterol Research." Telford Press, Caldwell, New Jersey.

Nicolau, C., and Chapman, D. (1990). "Horizons in Membrane

Biotechnology.'' *Progress in Clinical and Biological Research*, Vol. 343. Wiley-Liss, New York, New York.
Tanford, C. (1980). "The Hydrophobic Effect." Wiley-Interscience, New York, New York.
Vance, D. E., and Vance, J. E. (1985). ''Biochemistry of Lipids and Membranes.'' Benjamin/Cummings Publishing, Menlo Park, California.
Yeagle, P. L. (1987). ''The Membranes of Cells.'' Academic Press, Orlando, Florida.
Yeagle, P. L. (1988). ''Biology of Cholesterol.'' CRC Press, Boca Raton, Florida.
Yeagle, P. L. (1991). ''The Structure of Biological Membranes.'' CRC Press, Boca Raton, Florida.

Cellulases

Pierre Béguin and Jean-Paul Aubert
Institut Pasteur, Paris

Glossary

β-Glucosidase β-D-glucoside glucohydrolase (E.C. 3.2.1.21); cleaves β-glucosidic bonds in cellobiose and other β-glucosides

Carboxymethyl cellulose Amorphous derivative of cellulose; soluble forms with a degree of substitution of about 0.7 are commonly used as substrates for endoglucanases

Cellobiohydrolase 1,4-β-D-glucan cellobiohydrolase (E.C. 3.2.1.91); cleaves cellulose chains from the nonreducing end, liberating a cellobiose molecule at each step

Cellobiose β-1,4-D-glucopyranosyl-glucopyranose; a dimer of glucose

Endoglucanase 1,4-β-D-glucan glucanohydrolase (E.C. 3.2.1.4); cleaves cellulose chains randomly at multiple internal sites within the molecules

CELLULASES constitute a family of enzymes that hydrolyze β-1,4-glucosidic bonds in native cellulose and derived substrates. Their action on nonsubstituted cellulose leads to the formation of cellobiose, which can be further hydrolyzed by β-glucosidases. By hydrolyzing cellulose, the most abundant form of organic carbon synthesized by plants, cellulases play a key role in the carbon cycle of the biosphere. By the same argument, cellulases will be required if biological conversion of plant biomass into fuels and basic chemicals is to become economically feasible.

I. Structure of Lignocellulose

Cellulose is the major carbohydrate polymer synthesized by plants. Cellulose fibrils have a high tensile strength, which has been used in the textile industry since the dawn of history and which endows secondary plant cell walls (Fig. 1) with the capacity to withstand mechanical stress, including osmotic pressure. Cellulose is a linear polymer made of glucose subunits linked together by β-1,4 bonds. Stereochemically, however, the basic repeating unit is cellobiose (Fig. 2).

Cellulose is totally insoluble in water. In most forms of native cellulose, the carbohydrate chains, which contain between 100 and 14,000 glucose residues, form bundles, or microfibrils, in which the molecules are oriented in parallel and held together by hydrogen bonds. Such microfibrils consist of highly ordered, crystalline domains interspersed by more disordered, amorphous regions. The degree of crystallinity can vary between 0%, as in the case of acid-swollen cellulose or chemically derivatized, soluble cellulose such as carboxymethylcellulose (CMC), and nearly 100%, in the case of cellulose isolated from the cell walls of the alga *Valonia*. Cellulose from cotton is about 70% crystalline, and most commercial celluloses are between 30 and 70% crystalline.

In most cases, cellulose fibrils present in plant cell walls are embedded in a matrix of hemicellulose and lignin (Fig. 1). Hemicellulose designates a set of complex carbohydrate polymers, with xylans and mannans as the main components. The xylan backbone carries a variety of side chains including acetyl-, arabinofuranosyl-, and methylglucuronyl groups. Phenolic components including ferulic (4-hydroxycinnamic) and *p*-coumaric acids are covalently bound to the side chains, and some are thought to be involved in cross-links between hemicellulose and lignin mediated by ether linkages.

Lignin is a phenylpropanoid polymer produced

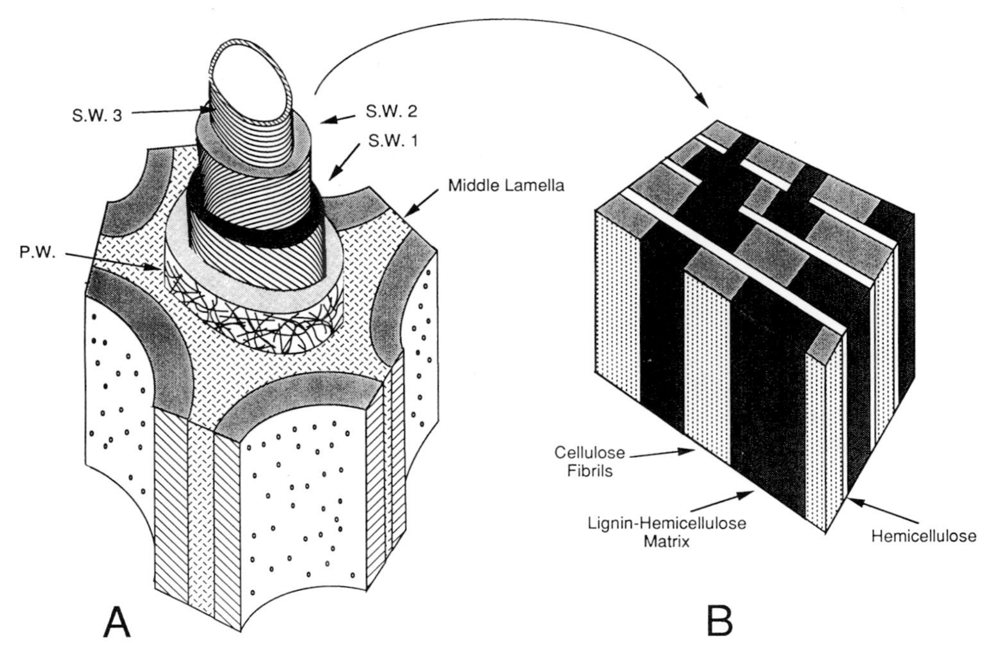

Figure 1 Schematic illustration of the molecular architecture of woody tissue, showing a cutaway view of the cell wall layers (A) and the probable relationship of lignin and hemicellulose to the cellulose fibrils in the secondary walls (B). The diameter of each cell is approximately 25 μm. P.W., primary wall; S.W. 1–S.W. 3, secondary cell walls. [Adapted with permission from Kirk, T.K. (1983). Degradation and conversion of lignocelluloses. *In* "The Filamentous Fungi," Vol 4. (J.E. Smith, D.R. Berry, and B. Kristiansen, eds.), p. 271. Edward Arnold, London.]

by the free-radical condensation of aromatic alcohols. The resulting compound is an amorphous, highly branched random polymer, which is highly resistant to biodegradation and protects cellulose fibers against enzymatic attack by microbes.

II. Diversity, Ecology, and Physiology of Cellulolytic Microorganisms

Fungi and bacteria are the main natural agents of cellulose degradation. However, protozoa living in the hindgut of lower termites have long been known to digest cellulose, and more recent work suggests that, in higher termites, cellulase activity is produced by the midgut tissue and the salivary glands of the termite itself. In addition, plants synthesize cellulases, which play a role in morphogenesis and developmental processes (e.g., in the ripening of fruits, including the avocado). This topic will not be further addressed in this article.

The fungi and bacteria that can degrade crystalline cellulose and use it as a source of carbon and energy include a variety of aerobes and anaerobes, mesophiles, and thermophiles. They are abundant in nature and play an important role in the carbon cycle by recycling CO_2 fixed by photosynthesis. These

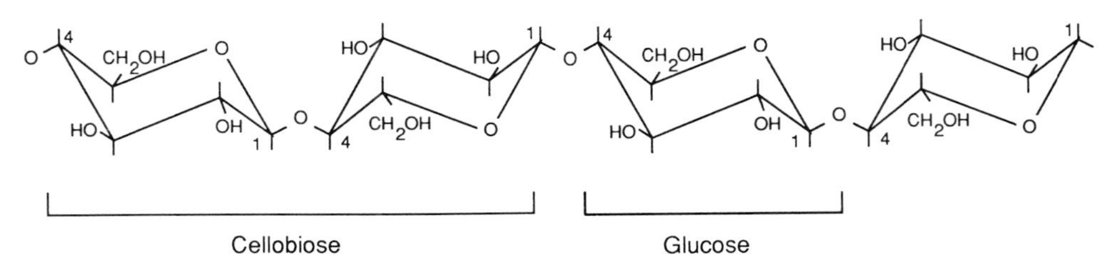

Figure 2 Chemical structure of cellulose.

organisms occupy various ecological niches where plant residues accumulate. As a general rule, fungi and bacteria found in such natural habitats constitute mixed populations, including several species of both cellulolytic and noncellulolytic microorganisms that interact synergistically. These interactions lead to efficient cellulose degradation with formation of CO_2 and H_2O in aerobiosis or CO_2, CH_4, and H_2O in anaerobiosis.

The soil surface is the most important aerobic environment where dead plant material (wood, leaves, straw, etc.) accumulates. The most recalcitrant substrate is wood, which is highly lignified. White-rot fungi, such as the well-studied basidiomycete *Phanerochaete chrysosporium,* can degrade both lignin and cellulose and play a major role in wood decay. Other aerobic fungi are devoid of ligninase activity but are efficient cellulose degraders. This is the case, for example, of *Trichoderma reesei,* whose cellulase system has been extensively studied. Aerobic cellulolytic bacteria have long been known to be present in soil. Cellulolytic species found in soil include *Cellulomonas, Cytophaga, Pseudomonas,* bacilli, and a number of actinomycetes. Some cellulolytic actinomycetes are thermophilic, such as *Thermomonospora* and *Microbispora*; furthermore, several actinomycetes including nocardiae and rhodococci can attack lignocellulose.

Cellulose-rich material is also found in anaerobic habitats, including the rumen and intestinal tracts of animals, sewage sludge digestors, composts, freshwater and seawater muds, and sediments. Cellulolytic and noncellulolytic microorganisms living in such environments have in general a fermentative metabolism, and they are frequently associated with methanogens.

In the rumen system, anaerobic fungi such as *Neocallimastix frontalis* are associated with cellulolytic bacteria, in particular *Ruminococcus flavefaciens* and *Ruminococcus albus*. Fermentation products formed are acetate, lactate, formate, succinate, ethanol, CO_2, and H_2. These products are partly used as food by the animal. Residual acetate, formate, CO_2, and H_2 are converted into CH_4 by methanogens. [*See* RUMEN MICROBIOLOGY.]

In other anoxic environments, cellulose degradation seems to be mainly due to bacteria—for example, *Acetivibrio cellulolyticus* and various *Clostridia,* either mesophiles such as *C. cellulolyticum* and *C. cellulovorans* or thermophiles such as *C. thermocellum* and *C. stercorarium*. The main fermentation products formed by these bacteria are lactate, acetate, ethanol, CO_2, and H_2. In addition, cellobiose, glucose, and some cellodextrins of low molecular weight, resulting from cellulose hydrolysis, are produced in excess. These sugars are fermented by other noncellulolytic saprophytic bacteria, such as *Clostridium thermohydrosulfuricum* and *Thermoanaerobacter ethanolicus,* with formation of additional amounts of lactate, acetate, ethanol, CO_2, and H_2. The removal of excess sugars derived from cellulose hydrolysis increases the efficiency of the various cellulase systems. Lactate and ethanol can be further used as sources of energy by other bacteria producing acetate, CO_2, and H_2.

It is well known that accumulation of H_2 inhibits most fermentative processes. In natural anoxic environments, this inhibition is relieved by hydrogen-consuming bacteria. Most of these bacteria belong to one of two groups: acetogenic bacteria, such as *Clostridium thermoaceticum,* which produces acetate from CO_2 and H_2, and methanogens, such as *Methanobacterium thermoautotrophicum* and *Methanosarcina barkeri,* which convert acetate, CO_2, and H_2 into CH_4. Thus, the action of a complex mixed population of bacteria finally transforms cellulose into CO_2, CH_4, and H_2O. Once CH_4 reaches aerobic zones, it can be oxidized into CO_2 by methanotrophic bacteria. [*See* ACETOGENESIS AND ACETOGENIC BACTERIA; METHANOGENESIS.]

Efficient cellulose degraders, able to grow at the expense of crystalline cellulose, produce a cellulase system constituted of several enzymes, each with a different specificity. However, many other bacteria, unable to use cellulose as a carbon source, synthesize enzymes endowed with cellulase activity. For example, enzymes that can hydrolyze CMC have been characterized in *Bacillus subtilis* and related species and in phytopathogens. The significance of these systems is not yet entirely elucidated. In phytopathogens, a restricted cellulolytic action may be sufficient and/or required for penetration of the bacteria into the plant tissue. In saprophytic bacteria, limited cellulase activity could be useful for the utilization of soluble cellodextrins resulting from the action of the "true" cellulolytic organisms with whom they are associated.

Due to the extreme insolubility of lignocellulose and cellulose, enzymes that can digest these compounds must be either secreted into the medium, as in fungi, or bound to the cell surface, as in several cellulolytic bacteria. In addition, the compact structure of wood restricts the accessibility of the enzymes to their substrates. This is probably why fungi, rather than bacteria, are the main degraders of wood. Indeed, the hyphae produced by these organisms give them the ability to penetrate cracks and to progress inside the wood as branched tiny threads, which deliver *in situ* a mixture of different depolymerizing enzymes. No organism known to date can grow only on lignin. Thus, it is thought that a white-rot fungus growing inside of wood first degrades exposed cellulose and hemicellulose sectors. This supplies soluble sugars as nutrients. Once the fungus starts to starve, its metabolism shifts toward lignin degradation, thereby exposing new cellulose and hemicellulose sectors, and the process continues. Several bacteria, in particular *Cellulomonas* and *Clostridia*, have been shown to bind to cellulose fibers. This property confers to these microorganisms an obvious selective advantage over other saprophytic bacteria living in the vicinity, because they have a direct access to the soluble sugars released by cellulose hydrolysis. [*See* TIMBER AND FOREST PRODUCTS.]

III. Biochemistry of Cellulases

Although the hydrolysis of a linear homopolymer such as cellulose may seem straightforward at first glance, the conversion of cellulosic material into glucose is a process of considerable complexity. First, as already mentioned, natural cellulose rarely if ever occurs in a pure form. The matrix of hemicellulose and lignin embedding cellulose fibers severely restricts the access of cellulolytic enzymes to the substrate. Thus, most cellulolytic organisms also produce hemicellulases (xylanases, acetyl xylan esterases, mannanases, etc.). Hydrolysis of hemicellulose appears also to favor delignification, possibly through the removal of lignin bound by ether linkages to aromatic components of the hemicellulose side chains. The degradation of lignin has been most extensively studied in the white-rot fungus *P. chrysosporium*. The process involves the generation of aryl cation radicals catalyzed by lignin peroxidase and Mn-peroxidase, followed by spontaneous, nonenzymic decay into a variety of final products. Several actinomycetes also attack lignocellulose. Lignin degradation by actinomycetes involves a much less extensive mineralization to CO_2 than in the case of fungi. Lignin solubilization possibly proceeds through a set of reactions involving limited oxidation of lignin and cleavage of cross-links between lignin and hemicellulose.

Second, cellulose occurs in multiple forms differing in crystallinity, degree of polymerization, and pore size distribution. This is also true of chemically pure, commercial cellulose preparations, which are heterogeneous with respect to physical parameters. Furthermore, the physicochemical parameters of the substrate change during the course of degradation. As a consequence, enzymology is difficult to study rigorously, except using low-molecular weight, soluble substrates such as cellodextrins and their chromogenic derivatives.

Third, all microorganisms that are known to degrade crystalline cellulose produce a number of cellulases with different specificities and modes of action. The activity of each individual component can usually be assayed on soluble model substrates, but hydrolysis of the native substrate requires the synergistic interaction of several components. The biochemical complexity of cellulase systems arises both from the presence of different genes encoding different polypeptides and from posttranslational modifications (glycosylation, proteolysis).

A. Cellulolytic Systems

Cellulase systems are probably as manifold as the microorganisms that produce them; however, a few systems have been investigated extensively and currently serve as paradigms for the interpretation of other, less thoroughly characterized cellulase systems. Two such systems will be discussed here. In the first, cellulolytic enzymes are found as individual proteins in the extracellular medium; in the second, the enzymes are physically associated to form high-molecular weight complexes maximizing synergistic interactions.

1. Unassociated or Transiently Associated Cellulases: The *Trichoderma reesei* Cellulase System

Trichoderma reesei is a filamentous fungus belonging to the *Fungi imperfecti*. The main features of the *T. reesei* cellulase system are indicated in Fig. 3. The three major types of enzyme it produces are endoglucanases, cellobiohydrolases, and β-glucosidases. Endoglucanases act almost exclusively on the noncrystalline regions of the substrate. Half a dozen of them have been purified, and two distinct

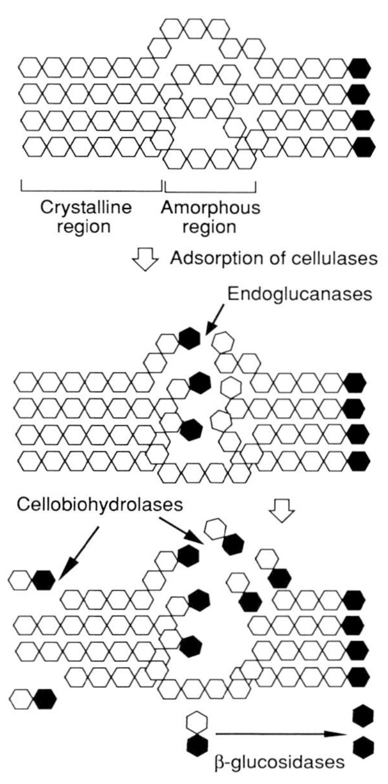

Figure 3 Schematic representation of the synergistic action of *Trichoderma reesei* cellulases. Glucose residues are represented by hexagons; glucose residues carrying reducing ends are shown in black.

endoglucanase genes have been cloned. Additional enzymes may be encoded by other genes that have not been cloned or result from alternative posttranslational modifications of the same gene product(s). In contrast to most endoglucanases, cellobiohydrolases hydrolyze soluble CMC quite poorly, most likely because hydrolysis cannot proceed beyond substituted residues. Two species with different enzymatic and immunological properties have been isolated, and the corresponding genes have been cloned and sequenced. As cellobiohydrolases are capable of hydrolyzing crystalline cellulose, they are regarded as key enzymes for the degradation of the native substrate. Extracellular β-glucosidases hydrolyze cellobiose and other low-molecular weight, soluble cellodextrins to yield glucose. Furthermore, a plasma membrane-bound β-glucosidase is thought to be involved in transglycosylation reactions leading to the formation of inducers (see Section IV).

The synergy between the three types of enzymes in the degradation of native cellulose was first explained in the following terms. Endoglucanase cleavages occurring in the amorphous regions of cellulose fibers generate nonreducing ends acting as start sites for cellobiohydrolases, which proceed with the degradation of the crystalline regions of the substrate. β-glucosidases prevent the buildup of cellobiose, which is a competitive inhibitor of cellobiohydrolases. The model fits well with the known properties of individual enzymes but does not account for some observations whose explanation is still debated. For example, two structurally and enzymatically different cellobiohydrolases, displaying synergism with each other, are each found in *T. reesei* and *Penicillium pinophilum*. Furthermore, *Trichoderma koningii* cellobiohydrolase is not synergistic with all endoglucanases of the same species, and interspecific synergism with endoglucanases from other species works only in some cases. Specific, physical interactions may be required for the synergy between some cellulolytic components. Although such components generally occur as independent enzymes in the extracellular medium and can be purified individually, there may be transient complex formation *in vivo*, as is the case for *T. reesei* cellobiohydrolases I and II.

Many of the features of cellulose degradation by *T. reesei* have been demonstrated for other aerobic fungi. In addition, several cellulolytic bacteria (e.g., *Cellulomonas fimi, Microbispora bispora, Streptomyces flavogriseus,* and *Clostridium stercorarium*) also contain both endoglucanases and β-1,4-glucanases with an exo mode of action. There is evidence of endo–exo synergy in the case of *C. stercorarium* and *M. bispora*. Exo–exo synergy has been demonstrated in the case of *M. bispora,* the effect being most apparent for the hydrolysis of cotton.

2. Permanently Associated Cellulases: The Cellulosome Concept

The concept of physical association between cellulolytic components has become central to the understanding of cellulase systems from several anaerobic microorganisms. These organisms produce cellulase systems with a very high specific activity, in which most individual components are associated in high-molecular weight complexes termed cellulosomes. The most extensively studied example is the thermophilic, anaerobic bacterium *Clostridium thermocellum*.

The cellulase system of *C. thermocellum* is composed of both a low-molecular weight fraction, with $M_r < 100{,}000$, corresponding to unassociated endoglucanases, and a high-molecular weight, multienzyme complex, or cellulosome, of about 2–4 MDa, which is responsible for the hydrolysis of crystalline cellulose. At least 14 different cellulosome components, of M_r ranging from 40 to 250 kDa, can be identified by sodium dodecyl sulfate–polyacrylamide gel electrophoresis. The catalytic subunits include several endoglucanases and xylanases and possibly a cellobiohydrolase.

The surface of *C. thermocellum* cells is studded with protuberances containing cellulosome aggregates of M_r of up to 100 MDa, which are responsible for the adhesion of the bacteria to the substrate and for its degradation. Attachment of the (multi)cellulosomes to the cells is not permanent; the complexes are released from the cells and proceed independently with the degradation of cellulose fibers.

The cellulosome binds strongly to cellulose but can be eluted at low ionic strength, which suggests hydrophobic interactions. Full dissociation of the complex requires particularly harsh chaotropic conditions, making the isolation and characterization of individual components in an active form particularly difficult. Cloning and expression of cellulase genes in *Escherichia coli* was therefore a useful approach for the characterization of individual enzymes. Fifteen endoglucanase genes, two β-glucosidase genes, and two xylanase genes have been cloned from the *C. thermocellum* type strain NCIB 10682. The cellulosomal localization of several cloned gene products was demonstrated by immunoblotting.

Current ideas about the structure and mode of action of the cellulosome are based on electron micrographs showing clusters of subunits arranged parallel to the major axis of the cellulosome and from the purification and characterization of two cellulosome components. One, termed S_L, is a 250-kDa glycoprotein with no catalytic activity. The other, termed S_S, is an 82-kDa protein with activity against CMC but not against crystalline cellulose. Hydrolysis of crystalline cellulose requires the presence of both components. In addition, binding of the catalytic subunit S_S to cellulose is promoted by the S_L-subunit. A tentative model of the cellulosome based on these observations is shown in Fig. 4.

Catalytic subunits are assumed to be anchored to a large, multidomain scaffolding subunit (possibly

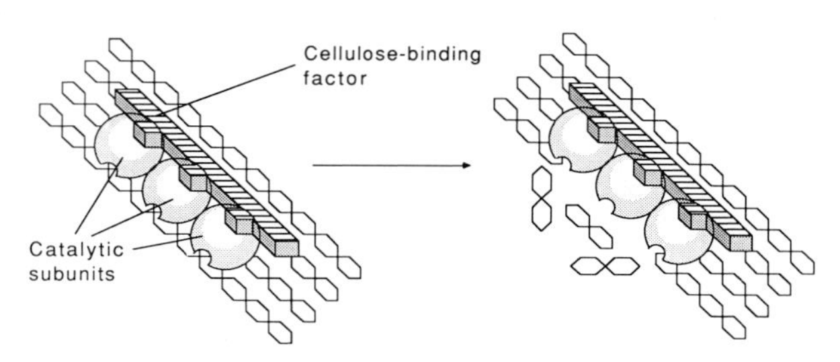

Figure 4 Tentative model for the hydrolysis of cellulose by the cellulosome of *Clostridium thermocellum*. Binding and alignment of the catalytic subunits at regular intervals along the cellulose molecule are assumed to be mediated by a cellulose-binding factor. Simultaneous hydrolysis at multiple sites generates small fragments which no longer bind to the crystal lattice.

the S_L-subunit), which also serves as a cellulose-binding factor. Catalytic subunits aligned at closely spaced intervals along the cellulose fiber would cleave a single cellulose molecule at multiple sites quasi-simultaneously. The hydrolysis products would be subsequently hydrolyzed to cellobiose without leaving the complex. The model fits with the hypothesis that crystalline cellulose hydrolysis requires the formation of low-molecular weight products, which are not held in place within the crystal lattice.

Features reminiscent of the (multi)cellulosome organization of *C. thermocellum* cellulases have been observed in several cellulolytic bacteria. True cellulosomes, albeit with a simpler subunit composition, are produced by *Clostridium cellulovorans, Acetivibrio cellulolyticus,* and *Bacteroides cellulosolvens*. High-molecular weight cellulolytic complexes are not only found in bacteria: The anaerobic rumen fungus *N. frontalis* produces a very potent cellulolytic complex of about 1–2 MDa, which has a strong affinity for cellulose.

Multicellulosomelike protuberances similar to those of *C. thermocellum* are displayed by a number of cellulolytic bacteria, including *Clostridium cellobioparum, Ruminococcus albus,* and *Cellulomonas* sp. In most of these organisms, the cellulase system also contains a high-molecular weight glycoprotein immunologically related to S_L. It is not clear how similar these cellulase systems actually are to the cellulosome of *C. thermocellum*. There may be significant differences between the stabilities of the complexes and the strengths of their attachment to the cell surface. For example, electron micrographs of *Fibrobacter* (formerly *Bacteroides*) *succinogenes* cultures show that the bacteria adhere closely to cellulose fibers and that degradation of the substrate occurs mostly at the contact surface immediately underneath the cells.

B. Structure and Function of Cellulolytic Enzymes

The introduction of recombinant DNA technology has considerably accelerated the analysis of the structure and mechanism of cellulolytic enzymes. Structural paradigms have become apparent, and the function of putative catalytic site residues can be assessed by oligonucleotide-directed mutagenesis. Furthermore, at least three cellulases have been crystallized, and the three-dimensional structures of two of them have been solved by X-ray diffraction. These enzymes will therefore become good systems for detailed investigations whose results could be extrapolated with reasonable confidence to homologous cellulases.

1. Cloning of Cellulase Genes

The construction of gene libraries from cellulolytic microorganisms is reasonably straightforward using generally available techniques. Libraries from eukaryotes (fungi) are usually constructed from complementary DNA (cDNA) synthesized from messenger RNA (mRNA). In the case of bacteria, libraries are constructed directly from genomic DNA.

The easiest tests for screening cellulase genes are based on the detection of the activity of the encoded protein expressed in a foreign, noncellulolytic host such as *E. coli*. The most widely used plate test is based on the detection of zones of CMC hydrolysis around endoglucanase-positive clones upon staining with Congo red.

Tests based on the hydrolysis of CMC are usually specific for the detection of endoglucanase activity. A second, less specific test is the hydrolysis of the fluorogenic compound methylumbelliferyl-β-D-cellobioside, which is a substrate for several cellobiohydrolases and also for a variety of endoglucanases, β-glucosidases, and xylanases.

Nonenzymatic screening has been used to identify certain genes. The cloning of several *C. fimi* cellulase genes was based on the detection of proteins cross-reacting with antiserum raised against unfractionated *C. fimi* cellulases. Genes encoding *T. reesei* cellulases were cloned after differential hybridization with cDNA synthesized from mRNA isolated under cellulase-inducing or noninducing conditions, and several fungal genes homologous to the *T. reesei* gene encoding cellobiohydrolase I were cloned using the latter gene as a hybridization probe.

2. Modular Structure of Cellulases

The sequences of over 50 cellulases have been deduced from the sequences of the corresponding genes. When compared with each other, the most striking feature of cellulases is that many of them are composed of similar basic modules, which are arranged in variegated combinations in different enzymes. In several cases, the study of truncated proteins generated by proteolytic cleavage or genetic deletion has demonstrated that such modules correspond to individual protein domains. Various types of domains have been identified.

a. Catalytic Domains

Catalytic domains range in size between 300 and 500 residues. Sequence comparisons based on the analysis of hydrophobic clusters indicates that the overwhelming majority of cellulase active domains can be classified into six broad families, designated A–F. From the total lack of similarity among different families, it was concluded that each of the families corresponds to a different type of protein structure. The tertiary structure of the two cellulases determined to date is consistent with this suggestion. *Trichoderma reesei* cellobiohydrolase II (family B) has a 7-stranded β-barrel structure, whereas *C. thermocellum* endoglucanase CelD (family E) has a core built from 12 α-helices and a lateral domain consisting of β-sheets.

Despite probably sharing the same gross structural organization, enzymes from the same family can display significantly different substrate specificities and modes of action. For example, several bacterial endoglucanases are homologous to *T. reesei* cellobiohydrolase II. This can explained by modeling the structure of the bacterial enzymes according to the known three-dimensional structure of the cellobiohydrolase. The active site of *T. reesei* cellobiohydrolase II is shaped like a tunnel through which the substrate chain must be threaded from the nonreducing end, whereas the active sites of the endoglucanases are predicted to have a much more open structure, shaped like a horseshoe, allowing the protein to bind anywhere along the cellulose molecule. Significant variability among enzymes of the same family is also observed in their patterns of specificity toward different substrates (cellodextrins, CMC, chromogenic cellobiosides, xylan, lichenan, barley β-glucan).

From the point of view of evolution, a surprising degree of similarity exists among cellulases from widely divergent organisms. A striking example is avocado endoglucanase, which shares >35% amino acid identities with a set of bacterial endoglucanases. This suggests that horizontal transfer of cellulase genes must have occurred on a wide basis. The best evidence for such a horizontal transfer is provided by the nucleotide sequence of an endoglucanase gene from *Cellulomonas uda*. The gene is largely homologous to the *celZ* gene of *Erwinia chrysanthemi*. Furthermore, the gene has a G/C content, which is significantly lower than that of the bulk of *C. uda* DNA, suggesting that it was acquired relatively recently from an organism with a lower G/C content.

b. Cellulose-Binding Domains

Many cellulases contain domains that are physically and functionally independent of the catalytic domains and whose role is to anchor the enzymes to the substrate. All known *T. reesei* cellulase sequences contain a conserved segment of about 30 residues, which is located either at the NH_2- or at the COOH end of the polypeptide and which is separated from the catalytic domain by a region rich in proline and hydroxy amino acids. Mild papain treatment of *T. reesei* cellobiohydrolases I and II resulted in the formation of core enzymes retaining full activity against chromogenic cellobiosides but loosing 85–100% (for cellobiohydrolase I) or 40–60% (for cellobiohydrolase II) of their activity against crystalline cellulose. Binding to cellulose was reduced by 65% for both enzymes, indicating that papain treatment cleaves the cellulose-binding domain. Analysis of the cellulose-binding domain of *T. reesei* cellobiohydrolase I by nuclear magnetic resonance spectroscopy indicates that it is wedge-shaped, with overall dimensions of about 30 × 18 × 10 Å. One face of the wedge is flat and contains three tyrosyl residues involved in binding to the substrate.

Using the same approach, a cellulose-binding domain of about 100 residues was characterized in cellulolytic enzymes from *C. fimi*. Very similar domains have since been found in endoglucanases from *M. bispora* and *Butyrivibrio fibrisolvens* and in endoglucanases, xylanases, and an arabinofuranosidase from *Pseudomonas fluorescens* var. *cellulosa*. Removal of the cellulose-binding domain from the enzymes results in loss of affinity for cellulose but in little, if any, change in activity. Conversely, genetic fusion between the sequence encoding the cellulose-binding domain and enzymes such as alkaline phosphatase or *Agrobacterium* sp. β-glucosidase results in the synthesis of chimeric proteins retaining their original enzymatic activity but acquiring the ability to bind to cellulose. These properties suggest interesting applications of the cellulose-binding domain as a tag for affinity purification of recombinant proteins.

c. Other Domains

A large number of cellulases contain segments up to several hundred residues with no known function and often little similarity to the sequence of other cellulases. It is possible that several of these segments are new types of cellulose-binding domains. In addition, a domain of about 65 residues, containing two homologous segments of 23 amino acids each, is highly conserved in most endoglucanases and in a xylanase of *C. thermocellum*, as well as in endoglucanases from *C. cellulolyticum* and *C. cellulovorans*. Deletion experiments indicate that the conserved domain is not required for activity nor for cellulose binding. Furthermore, this conserved domain is generally present in enzymes that are found in cellulosomes, suggesting that it may anchor the catalytic subunits to the cellulosome.

IV. Regulation of Cellulase Synthesis

Despite an impressive number of reports on the subject, the mechanism of regulation of cellulase synthesis is still unclear. In most cellulose degraders, readily utilized carbon sources repress cellulase synthesis by a mechanism thought to be similar to catabolite repression. In addition, cellulase synthesis is induced in many organisms by the presence of cellulose. In *T. reesei*, induction requires the presence of low constitutive levels of cellulases generating cellobiose. The actual inducer(s), possibly sophorose, appears to be formed from cellobiose by transglycosylation carried out by β-glucosidase. [*See* Catabolite Repression.]

Transcriptional regulation has been studied in some bacteria. In *Thermomonospora fusca*, catabolite repression might be mediated through cyclic adenosine monophosphate (cAMP), although large changes (up to 200-fold) in cellulase synthesis rate accompany relatively small changes (10-fold) in cAMP level. In addition, three putative closely linked promoters have been located by S1 mapping of the *T. fusca celE* gene mRNA and an activator protein, binding DNA downstream of *celE* promoters, has been identified. In both *C. fimi* and *C. thermocellum,* transcription of individual cellulase genes follows differential regulatory patterns, and some genes are expressed from dual promoters submitted to differential regulation. The identification of regulatory genes is not easy, because *in vivo* genetic analysis is not possible in these bacteria. In contrast, some clues may come from studying the regulation of the endoglucanase genes *celZ* and *celY* of *E. chrysanthemi*. Although not a true cellulolytic organism, *E. chrysanthemi* has the advantage of belonging to the enterobacteria, for which many tools for genetic analysis are available.

It is worth noting that, even if they are not characterized at the molecular level, a large number of cellulase negative, constitutive, or hyperproducing mutants have been isolated from several wild-type bacterial and fungal species. Most effort has been concentrated on *T. reesei,* essentially for improving cellulose-based bioconversion processes by empirical means (see Section V). However, transformation systems for *T. reesei* have recently become available, and a wide range of manipulations can now be carried out in this species, allowing the study of its cellulase system and its regulation.

V. Applications of Cellulases in Biotechnology

A. Cellulose as a Renewable Carbon Source for the Production of Fuels and Chemicals

From the days of the first oil crisis, considerable efforts have been devoted to the conversion of biomass to petroleum substitutes, which could serve either as fuel or as raw materials for the chemical industry. The interest of biotechnologists was drawn by the vast supply of cellulose, which could potentially be used for the production of glucose, which can in turn be fermented into alcohols, acetone, and volatile fatty acids. Excitement abated significantly in the course of the 1980s due to the drop of oil price and to the realization that long-term basic and applied research would be required before commercial exploitation of most natural cellulosic materials could become economically feasible. More recently, however, concern about global climatic warming due to CO_2 emissions from the burning of fossil fuels has stimulated a renewed interest in the use of biomass for the production of energy. Switching from fossil to biomass-derived fuels would indeed reduce the net output of CO_2 from energy production to close to zero. Thus, in the long term, the prospect of seeing cellulosic biomass contribute significantly to the production of fossil fuel substitutes is still alive, but much fundamental and applied research is still needed. [*See* PETROLEUM MICROBIOLOGY.]

1. Pretreatment of Cellulosic Materials

Most of the cheaply available supply of cellulose is in the form of lignocellulose. Although a balance exists in nature between synthesis and degradation of lignocellulose, degradation is too slow for industrial applications. Hence, many efforts have been devoted to the development of chemical and physical pretreatments that enhance the digestibility of natural lignocellulosics. Rather than reducing the degree of crystallinity of cellulose, the critical factor appears to be the increase in surface area available for enzymatic attack. Increasing the accessibility of the substrate to cellulolytic enzymes may be brought about by mechanical disruption and/or partial removal of the lignin and hemicellulose matrix. The hydrolysis of hemicellulose can be acid-catalyzed at high temperature, either in the presence of added acid or due to the liberation of acetic acid from the acetyl groups of xylan. As hydrolysis of hemicellulose proceeds, associated lignin forms clumps that separate from the cellulose fibers. Various pretreatment procedures have been developed, including alkaline swelling, ball milling, mild acid hydrolysis, delignification with oxidizing agents, steam explosion, and high-pressure, high-temperature treatment with water or dilute methanol. Several points must be considered in the assessment of a pretreatment process: (1) efficiency, i.e., suitability of the pretreated product as a substrate for enzymatic hydrolysis; (2) price, including cost of environmentally sound operation (treatment of effluents); and (3) yield, which is influenced by factors such as recovery of carbohydrates solubilized during pretreatment and losses due to the formation of by-products (e.g., furfural, hydroxymethylfurfural). Another problem to be avoided is the inhibitory effect of by-products on the fermenting microorganisms used to convert glucose into the desired final products. Pretreatment conditions optimizing these parameters must be determined empirically for each kind of substrate.

2. Mass Production of Cellulases

As is often the case for products derived from industrial microorganisms, improvements in the production of cellulase by cellulolytic microorganisms have been achieved largely by empirical means combining cycles of random mutagenesis and selection with suitable culture conditions. The most extensive efforts have addressed the improvement of cellulase production by *T. reesei*. Wild-type *T. reesei* is already a remarkably potent cellulase producer; therefore, it was an organism of choice for the development of mass production of cellulolytic enzymes. Preliminary screening was greatly facilitated by the development of plate tests, in which spreading of individual mycelial colonies was prevented physically or chemically (growth retardants), and the production of cellulases could be assessed semiquantitatively by visualizing the clearing of cellulose powder incorporated in the medium around each colony. Mutants with reduced sensitivity to catabolite repression were isolated by including in the medium an easily metabolized carbon source and 2-deoxyglucose as a selecting agent.

The appropriate choice of culture conditions is another critical factor, even for strains selected for reduced catabolite repression. By careful design of fed-batch cultivation, lactose can be converted to cellulase by *T. reesei* strain CL-847 with a protein yield on lactose close to 40% and with final extracellular cellulase concentration in excess of 30 g/liter.

Trichoderma reesei can secrete more protein to the extracellular medium than any other species used for the production of proteins from cloned genes. This has led to a line of research other than the development of commercial cellulose degradation. Efforts are currently being directed at harnessing the potential of the organism and its cellulase genes to develop highly efficient host–vector systems for the secretion of foreign proteins.

B. Modification of Lignocellulosic Materials

Before complete hydrolysis of cellulose into monomeric sugars becomes economically feasible, a number of other applications of cellulases or hemicellulases are likely to be developed, which involve the modification, without complete hydrolysis, of various lignocellulosic materials.

In the food-processing area, cellulases are already being used to soften fruit pulp and increase the recovery of oil and fruit juices. Pretreatment of animal feed has also been the topic of numerous studies. The situation is a rather complex one. Many factors must be taken into account, such as the physiology of ruminants, including, for example, the influence of fiber length on the residence time of cellulosic material in the rumen. Certainly total saccharification is not a prerequisite, nor even desirable. One aspect that has received increased attention is the pretreatment of silage. Partial attack of cellulose and hemicellulose would release soluble carbohydrates, both from the hydrolyzed substrate and from intracellular cell sap, thereby speeding up the onset of lactic acid fermentation and preventing butyric fermentation by clostridia. Recombinant lactic acid bacteria containing cloned cellulase genes have been constructed. The economic viability of silage pretreatment with cellulases remains to be assessed, however.

Considerable interest is currently focused on the use of hemicellulose-degrading enzymes for the bleaching of paper pulp. The action of chlorine bleach on lignin results in the release of polychlorinated phenols in paper mill effluents, which is increasingly considered unacceptable. The action of xylanases significantly reduces the amount of chlorine required for bleaching, due to the disruption of the lignin–hemicellulose matrix, which causes partial delignification. Beside price considerations, research efforts are aimed at obtaining hemicellulase preparations devoid of any cellulase, particularly endoglucanase activity, which reduces the viscosity of the pulp and is therefore strongly detrimental to the tensile strength of the paper. Furthermore, because residual hemicellulose contributes to the finishing of paper, it could be advantageous to use enzymes such as feruloyl esterases, which cleave the side chains linking hemicellulose to lignin without destroying the backbone of hemicellulose molecules.

Cellulases have recently found a somewhat unexpected outlet as an additive to laundry detergents. Repeated washings cause cotton fabrics to assume a dull, feltlike aspect and a rough texture, due to the appearance of microfibrils arising from the wear and tear of cotton fibers. Limited action by cellulases added to laundry detergent removes the microfibrils and contributes to restoring the original aspect and texture of the fabric.

The potential uses of cellulases are therefore

many and divergent. They are valuable in various specialized niche applications, even if their use for the bulk commercial conversion of wood remains dependent on further technological advances.

Bibliography

Aubert, J.-P., Béguin, P., and Millet, J. (eds.) (1988). "Biochemistry and Genetics of Cellulose Degradation." FEMS Symposium No. 43. Academic Press, London.

Béguin, P. (1990). *Annu. Rev. Microbiol.* **44,** 219–248.

Coughlan, M. P. (ed.) (1989). "Enzyme Systems for Lignocellulose Degradation." Elsevier Applied Science, London and New York.

Eriksson, K. E., Blanchette, R. A., and Ander, P. (1990). "Microbial and Enzymatic Degradation of Wood and Wood Components." Springer-Verlag, Berlin.

Gilkes, N. R., Warren, R. A. J., Miller, R. C., Jr., and Kilburn, D. G. (1988). *J. Biol. Chem.* **263,** 10401–10407.

Joliff, G., Béguin, P., Juy, M., Millet, J., Ryter, A., Poljak, R., and Aubert, J.-P. (1986). *Bio/Technology* **4,** 896–900.

Kubicek, C. P., Eveleigh, D. E., Esterbauer, H., Steiner, W., and Kubicek-Pranz, E. M. (eds.) (1990). "*Trichoderma reesei* Cellulases. Biochemistry, Genetics, Physiology and Application." The Royal Chemical Society, Cambridge.

Lamed, R., and Bayer, E. A. (1988). *Adv. Appl. Microbiol.* **33,** 1–46.

Ljungdahl, L. G., and Eriksson, K.-E. (1985). *Adv. Microbial Ecol.* **8,** 237–299.

Rouvinen, J., Bergfors, T., Teeri, T., Knowles, J. K. C., and Jones, T. A. (1990). *Science* **249,** 380–386.

Tomme, P., van Tilbeurgh, H., Pettersson, G., Vandekerckhove, J., Knowles, J., Teeri, T., and Claeyssens, M. (1988). *Eur. J. Biochem.* **170,** 575–581.

Wu, J. H. D., Orme-Johnson, W. H., and Demain, A. L. (1988) *Biochemistry* **27,** 1703–1709.

Cell Walls of Bacteria

R. J. Doyle
University of Louisville

Glossary

Autolysin Enzyme produced by a bacterium capable of hydrolyzing specific bonds in its own peptidoglycan

Cell wall Peptidoglycan layer surrounding the cytoplasmic membrane of eubacteria; the wall imparts shape to eubacteria; the cell wall resembles a polyelectrolyte mat in physical properties

Cell wall turnover Loss of cell wall materials due to autolysins; products of cell wall turnover may or may not be reutilized by the bacterium during growth

Lysozyme Enzyme capable of hydrolyzing certain glycosidic linkages of peptidoglycans; found in almost all life forms, ranging from bacterial viruses to human cells; the enzyme is also frequently found in serums

Peptidoglycan Complex composed of alternating muramic acid and glucosamine, the muramic acid of which may be substituted with peptides; also known as murein

Protoplast Structure free of all cell wall and outer membrane materials, consisting of cytoplasmic membrane and intracellular components; in the absence of osmotic protection, protoplasts will burst, or lyse

Spheroplast Protoplast surrounded by a damaged or defective cell wall; a spheroplast is osmotically sensitive

Teichoic acid Polyelectrolyte, usually composed of poly(glycerol phosphate) or poly(ribitol phosphate) covalently linked to the peptidoglycan of gram-positive bacteria; the glycerol or ribitol residues may be substituted by D-Ala or by carbohydrates; a few teichoic acids may possess carbohydrates in the polymer chain

Teichuronic acid A uronic acid—containing polysaccharide covalently attached to peptidoglycan of some gram-positive bacteria

NEARLY 40 YEARS AGO, Salton found that burst or disrupted bacteria yielded a cytoplasm-free fraction that retained the characteristic shape of the intact cell. The shape-determining structure was found to consist of small peptides and two amino sugars. The peptides were determined to be covalently attached to a unique amino sugar, muramic acid. Muramic acid is glucosamine to which a lactyl group is attached at the 3-O position. The other amino sugar, glucosamine, was later found to alternate with muramic acid to form a glycan chain. When peptides originating from one glycan strand are cross-linked with peptides of another glycan strand, a mat or matrix of peptidoglycan is formed.

In the 1960s and 1970s, it was found that peptidoglycan could have other polymers attached at various places. These polymers are now known to be polysaccharide, teichoic acid, teichuronic acid, and lipoprotein. Usually only one of these polymers is found attached to peptidoglycan of a given species. Figure 1 shows a segment of a typical peptidoglycan structure. The amino function of the amino sugars is frequently acetylated, forming *N*-acetylglucosamine and *N*-acetylmuramic acid. In a few species, the 6-O position of muramic acid is esterified with an acetyl group. The peptides of the insoluble wall usually are limited to four to six amino acids. The most common amino acids are L-alanine, D-glutamic acid, L-lysine [or meso-diaminopimelic acid (DAP), a lysine containing a terminal carboxylate function at the ϵ terminus], and D-alanine. In fact, D-Ala, D-Glu, meso-DAP, and muramic acid are found only in bacteria.

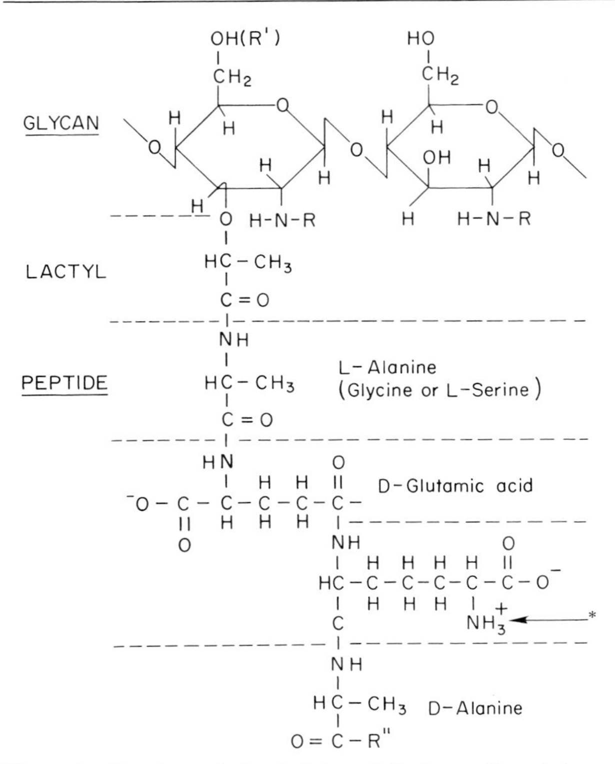

Figure 1 Structure of a bacterial peptidoglycan. R, acetyl or hydrogen; R′, teichoic acid of teichuronic acid; R″, amino acid or peptides. *, Diaminopimelic acid (or L-lysine, L-ornithine, L-diaminobutyric acid), function involved in cross-linking.

The specific inhibition of peptidoglycan biosynthesis constitutes the basis for the mechanism of action of several antibiotics, such as penicillin and vancomycin. In this article, only the cell walls of eubacteria are considered, but it should be kept in mind that other kinds of bacteria also possess shape-determining wall structures. For example, almost all of the archaebacteria have peptidoglycanlike structures in their cell walls. Archaebacterial walls, however, are not composed of muramic acid or D-amino acid-containing structures. Some archaebacteria contain a "pseudomurein" cell wall, which is analogous to the peptidoglycan of eubacteria. Pseudomurein contains L-talosaminuronic acid instead of muramic acid. The L-talosaminuronic acid is β-1,3-linked with *N*-acetylglucosamine to form the glycan backbone, whereas the amino acids of pseudomurein are always in the L-configuration. Pseudomurein occurs mainly in the order Methanobacteriales.

I. Cell Walls of Gram-Positive and Gram-Negative Bacteria

Bacteria can be classified as either gram-positive or gram-negative, based on stainability with gentian violet (crystal violet). Over 100 years ago, Christian Gram observed that when bacteria were stained with gentian violet in the presence of iodine or iodides and ultimately washed with a nonaqueous solvent, such as acetone or ethanol, some of the bacteria retained the stain. These bacteria are referred to as gram-positive. When the dye is readily removed by nonaqueous solvents, the bacteria are called gram-negative. Only recently has the mechanism of the Gram stain been somewhat understood. It now seems clear that the crystal violet–I_2–cell complex cannot be decolorized because the peptidoglycan layer forms a physical trap. Many researchers have shown that the peptidoglycan layer of the gram-positive bacterium is 20–40 nm in width around the cell periphery. In gram-positive rods, the width of the peptidoglycan layer is surprisingly uniform. In gram-negative bacteria, the peptidoglycan layer is much narrower, ranging from 2 to 7 nm. The gram-negative bacterium, however, has another layer external to the peptidoglycan. This layer is referred to as the outer membrane. Some researchers refer to the outer membrane as a part of the cell wall, but in reality it is a truly membranous structure that does not impart any shape-determining characteristics. In this article, the cell wall is considered to be peptidoglycan and peptidoglycan-associated polymers. The function of the outer membrane seems to be a passive permeability barrier. The relatively thin diameter of the peptidoglycan of the gram-negative cell cannot retain the Gram stain upon washing with nonaqueous solvents. In fungi, walls are also quite thick and the cells stain gram-positive. [*See* Cell Membrane: Structure and Function.]

In the Gram stain, it must be considered that teichoic acids and other peptidoglycan-associated molecules contribute to stainability; however, the extent of these contributions is unclear. When teichoic acids or teichuronic acids are removed from wall preparations, the walls seem more compact, suggesting that peptidoglycan-associated polymers contribute significantly to cell wall volume.

Table I and Figs. 2 and 3 review some of the major characteristics of the cell walls of gram-positive and gram-negative bacteria. The cell wall of the gram-positive cell is shown to be exposed to the environ-

Table I Summary of Characteristics of Gram-Positive and Gram-Negative Cell Walls

	Gram-positive	Gram-negative
Peptidoglycan	Up to 40% of dry weight of cell Relatively thick Extent of cross-linking varies from 10 to >90% Determinant of cellular shape	Low contribution to cellular dry weight Relatively thin, ranging from 2 to 6 nm Extent of cross-linking similar to gram-positive Determinant of cellular shape
Lysozyme	Resistance or susceptibility to lysozyme depends on O,N-acetylation	Same as gram-positive
Location	Usually the external layer	Usually protected by outer membrane
Auxiliary polymers		
Teichoic acid	Ribitol- or glycerol phosphate polymers usually present	Absent
Teichuronic acid	Uronic acid-containing polysaccharides are present in many species when cultured in limiting phosphate	Absent
Protein	Only a few reports of protein covalently linked to peptidoglycan exist, although proteins can apparently be embedded in the wall matrix	A small lipoprotein (Braun's lipoprotein) is found covalently linked to peptidoglycan in many gram-negative bacteria
Polysaccharide	Nonteichuronic acid polysaccharides occur as group-specific antigens in some gram-positive cells	Only a few reports describing a polysaccharide–peptidoglycan complex in gram-negative bacteria exist
Lipopolysaccharides	Most gram-positive bacteria do not possess lipopolysaccharides	Are components of outer membrane and are not bound to peptidoglycan

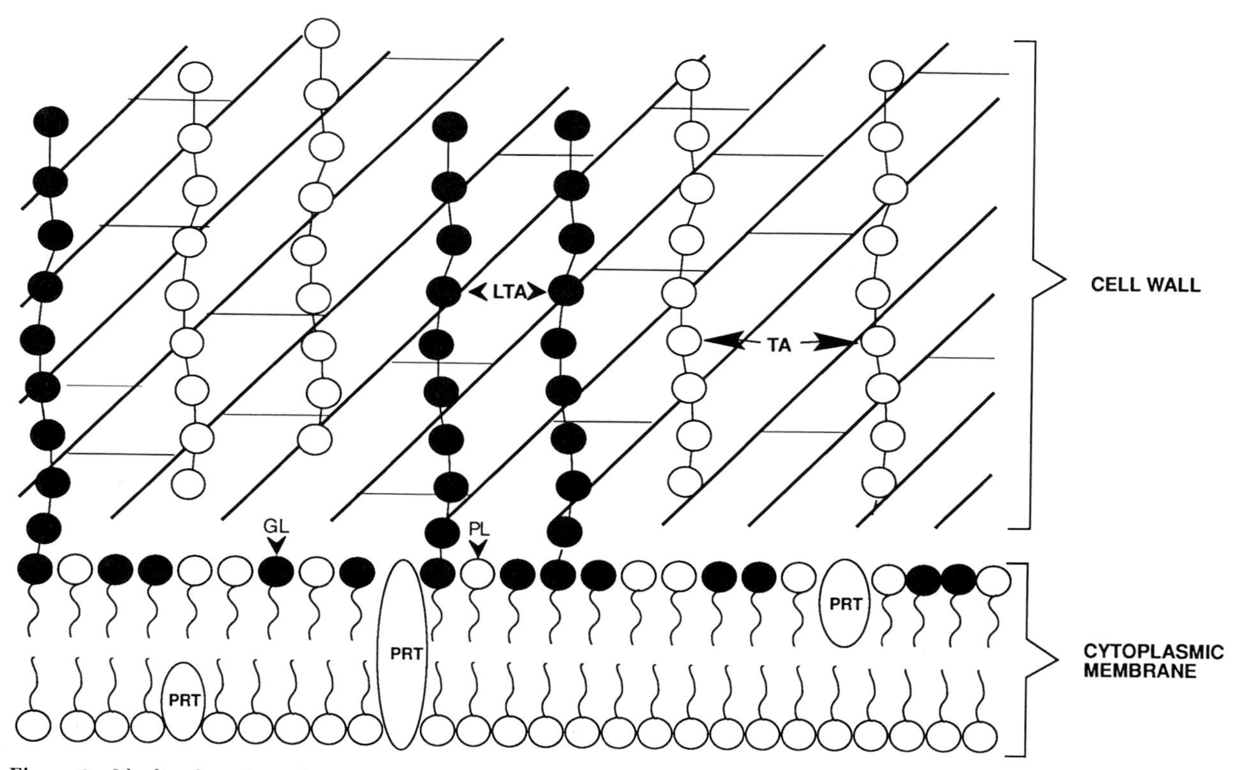

Figure 2 Idealized section of a surface of a gram-positive bacterium. Lipoteichoic acid (LTA) originates in the cytoplasmic membranes and may penetrate the wall. The wall glycan is shown by thick, slanted lines, whereas the cross-linking peptide(s) is shown by the thin, horizonal lines. The wall may contain covalently bound teichoic acid (TA) or teichuronic acid. The wall may also contain protein molecules intercalated into the peptidoglycan matrix. GL, glycolipid; PL, phospholipid; PRT, protein.

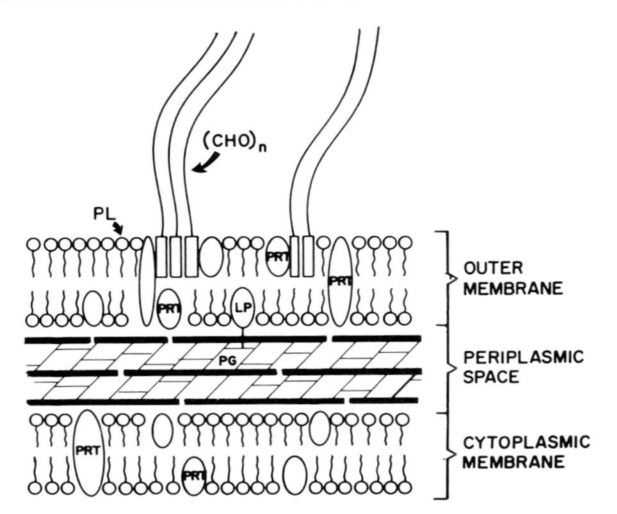

Figure 3 Idealized representation of the gram-negative cell wall. The peptidoglycan (PG) is identified as the insoluble matrix between the cytoplasmic membrane and the outer membrane. The wall of the gram-negative bacterium is much thinner than that of the gram-positive bacterium. Lipoprotein (LP) may be covalently attached to the PG. Lipopolysaccharide is an outer membrane component, the lipid end of which is associated with the surface leaflet of the outer membrane and the polysacccharide $(CHO)_n$ portion exposed to solvent environment. PL, phospholipid; PRT, protein.

ment, whereas the wall (peptidoglycan) layer of the gram-negative cell is protected by the outer membrane. Lysozyme will readily digest the walls of some gram-positive bacteria, but in intact gram-negative cells, the peptidoglycan is frequently resistant to lysozyme because the enzyme cannot penetrate the outer membrane.

Gram-positive cells possess a wall constituting up to 40% of the dry weight of the bacterium. Wall synthesis requires considerable energy in the form of high-energy cofactors such as adenosine triphosphate (ATP) or phosphoenolpyruvate. Surprisingly, in some gram-positive bacteria, 50% of the wall is lost during a single generation due to cell wall turnover (the need for wall turnover is discussed in a later section). Gram-negative bacteria possess walls that contribute very little to the cellular dry weight. [*See* GRAM-POSITIVE COCCI.]

II. Ultrastructure of Bacterial Cell Walls

Cell walls are easily identified in the electron microscope. Usually, a wall preparation is stained with metal ions, uranyl acetate, or phosphotungstic acid. Cocci or bacilli, cell side walls, septa, and cell poles can be distinguished readily. Figure 4 shows a sample of cell walls of *Bacillus subtilis* (a gram-positive rod) stained with uranyl acetate. The preparation, obtained by bursting intact cells with high-intensity sound waves, contained side walls of cell cylinders, completed septa, and cell poles (separated septa). The cytoplasmic contents were solubilized upon breakage of the cells. Residual cytoplasmic constituents were removed by hot detergents.

Cell walls can also be embedded in plastics and subjected to cleavage by the microtome. Walls may be stained pre- or postfixation and embedding. When cross-sections are examined by electron microscopy, the walls exhibit dense areas of staining. Figure 5 shows a cross-section of *B. subtilis,* in which the wall is easily discernible. The wall of this bacterium is about 30 nm thick and is not covered by external structures. In contrast, the wall of *Escherichia coli,* a gram-negative bacterium, is shown to be very narrow in diameter and is surrounded by an outer membrane (Fig. 6).

Figure 4 Cell wall preparation of *Bacillus subtilis*.

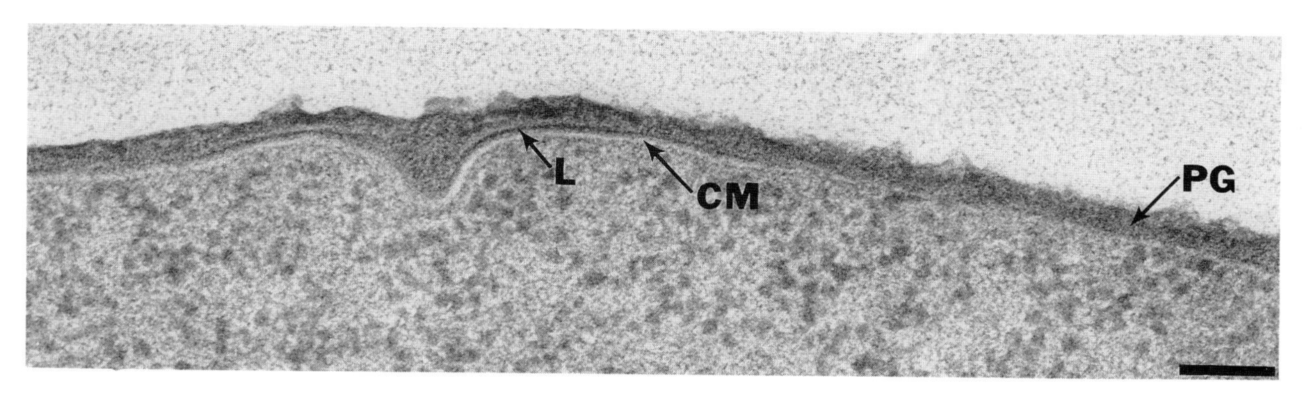

Figure 5 Electron micrograph of a bacterial cell envelope showing the peptidoglycan (PG) structure. The gram-positive *Bacillus subtilis* cell wall contains PG and teichoic acid. Lipoteichoic acid (L), originating at the cytoplasmic membrane (CM), may be found in the PG matrix or in the space between the PG and CM. Bar = 0.1 μm. [Micrograph courtesy of J. A. Hobot, University of Wales.]

III. Functions of Bacterial Walls

The foregoing paragraphs have alluded to some of the functions of cell walls of bacteria. The wall was considered to serve as a matlike structure maintaining cellular integrity under various osmotic conditions. In addition, the importance of the wall as a shape determinant has been confirmed by various experiments. In recent years, it has become clear that the wall has other functions as well (Table II). For example, intact peptidoglycan is needed for flagellar motion, because the rotation of flagella seems to require a "rigid" structure or fulcrum to rest against. Furthermore, in many bacteria, the wall is the receptor site for bacteriophages. The best-known example is *B. subtilis* in which several bacteriophages bind to composite peptidoglycan–teichoic acid structures.

In some bacteria, adhesin molecules are embedded in the matrix of the cell wall, although apparently not covalently linked to wall components. These adhesins function to bind bacteria to surfaces, where nutrients may be in greater abundance. Adherent bacteria are frequently more resistant to antibiotics, antibodies, and other immune factors and to antiseptics than are suspended or nonadherent bacteria. Wall-associated adhesins are more common in gram-positive bacteria than in gram-negative bacteria. Pyogenic cocci, including *Streptococcus pyogenes* and *Staphylococcus aureus,* are classic examples of bacteria that possess adhesins associated with the peptidoglycan matrix. [*See* ADHESION, BACTERIAL.]

Streptococcus pyogenes possesses a wall-associated polypeptide called M protein. The M protein is the type antigen of pyogenic streptococci.

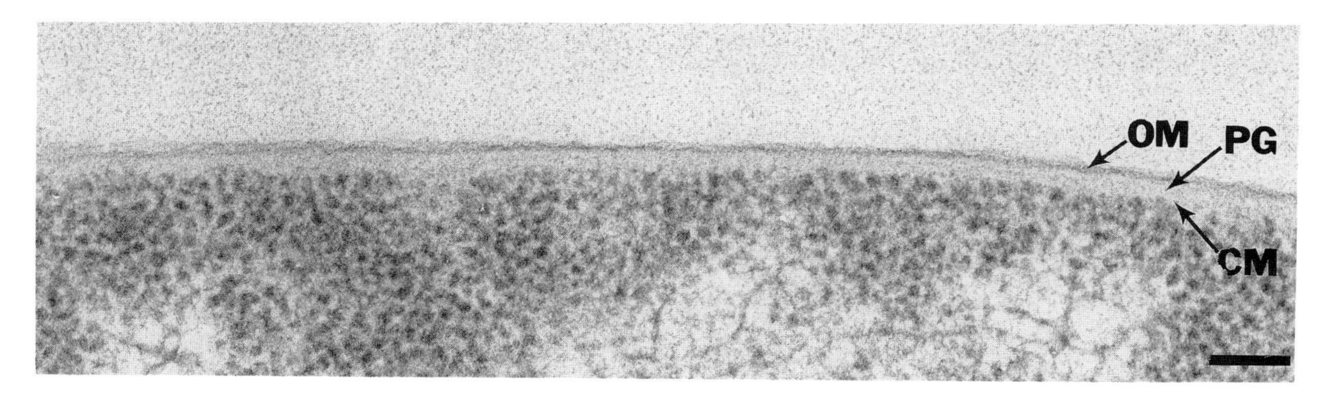

Figure 6 Cross-sectional representation of the surface of the gram-negative *Escherichia coli* showing peptidoglycan (PG), cytoplasmic membrane (CM), and outer membrane (OM). Bar = 0.1 μm. [Micrograph courtesy of J. A. Hobot, University of Wales.]

Table II Functions and Properties Ascribed to Bacterial Cell Walls

Function or property	Comments
Osmotic protection of protoplast	The high internal colligative pressure of bacteria will result in cellular lysis when the wall is damaged or removed
Determinant of cellular morphology	The walls of bacteria are responsible for their characteristic shapes; isolated cell walls retain their original shapes
Fulcrum for flagellar motion	Peptidoglycan provides a fulcrum for rotation of flagella
Receptor for bacterial viruses	Many bacteriophages bind to the cell wall as the initial step in the infection process
Mitotic apparatus	The segregation of growing chromosomes in some bacteria depends on the presence of an intact cell wall
Adhesion	The wall or wall-associated molecules may act as adhesins, providing a mechanism for adhesion to environmental substrata
Nutrient-scavenging organelle	Wall components may be able to scavenge micronutrients, such as metal ions, from the environment
Target for antibiotic action	Some antibiotics specifically interfere with wall synthesis, resulting in growth inhibition or cellular lysis
Protection from environmental challenge	Some toxins or enzymes may not be able to gain access to the cytoplasm because they cannot traverse the wall matrix
Modulation of immune response	Wall materials may induce autoimmune disorders or cause enhanced mitogenicity of lymphocytes; walls also act as adjuvants

The protein has several domains, one domain of which appears to be intercalated within the peptidoglycan matrix but is not covalently bound to the wall. The wall-associated domains are surprisingly homologous in sequence from strain to strain. The wall-associated domain(s) of M proteins is(are) near the carboxy terminus of the protein and seem(s) to be protected from proteolysis. A surface-exposed domain near the N terminus is capable of binding glycerol phosphate residues of lipoteichoic acids. Once lipoteichoic acids are bound by M protein, the hydrophobic ends are free to bind with cell surfaces of mucosal cells, thereby enabling the bacterium to adhere. This complicated type of adhesion mechanism may be inhibited by antibodies against poly(glycerolphosphate), M protein, or lipoteichoic acid. Another likely way to interfere with streptococcal adhesion is to inhibit the association between M protein and peptidoglycan.

Some of the mutans streptococci, including *Streptococcus sobrinus* and *Streptococcus cricetus* possess peptidoglycan-associated lectins. The lectins are specific for oligomers of isomaltose. Chaotropes do not readily remove the lectins from the streptococcal cell walls, although the lectins can be removed by enzymes that degrade peptidoglycan. To date, the nature of the association between the peptidoglycan and the glucan-specific lectin remains to be defined. This type of lectin is probably involved in the sucrose-dependent colonization of oral streptococci on enamel surfaces. Similarly, *Staphylococcus saprophyticus* appears to express a surface lectin, a domain of which is tethered to peptidoglycan by noncovalent interactions. In studies on initiation of infections, much emphasis is being placed on the intervention of adhesion. The role of the cell wall in stabilizing adhesin molecules is only beginning to be understood. Wall-specific antibiotics may be partially effective due to their effects on modifying the ability of cell walls to complex with certain domains of lectins.

Researchers have shown that the cell wall can bind Mg^{2+} and other metal ions. In addition, cell walls of *B. subtilis* can weakly complex with amino acids such as lysine and arginine. These observations suggest that walls may function to bind potential nutrients from the environment. If walls indeed sequester nutrients, there would be a requirement that the association between the wall ligands and the nutrients be relatively weak so as to permit availability of the nutrients to the cytoplasmic membrane.

Cell wall materials of pathogens frequently provoke immune responses. One researcher has shown that walls tend to persist in joints for several months. These walls may give rise to arthritic conditions. Walls are antigenic, and they are also adjuvants. Freund's complete adjuvant, for example, contains mycobacterial walls. Recent studies have shown that muramyl peptides have adjuvant activities. Walls also stimulate mitosis in B-lymphocytes. These are the lymphocytes that produce antibodies.

Segregation of the replicating genome(s) of bacteria has been ascribed to the cell wall of at least some bacteria. The now-revered replicon model stated that DNA replication began from a chromosomal origin and that membrane growth between replicated origins caused nuclear segregation. The replicon hypothesis has now been modified to include the cell wall. The renovated replicon hypothesis states that the origin and terminus of replication are bound to structures or sites in the cell wall. These sites are likely to contain DNA–membrane–wall complexes. Furthermore, the model requires that side walls expand from many ("diffuse") sites. Thus, DNA segregation can be realized because of side-wall extension. The wall then is assumed to carry the bidirectionally replicating chromosome chains away from each other. Details of the renovated replicon hypothesis are outlined in Fig. 7. This new model provides for the precise partitioning of chromosomes and for the accurate division of the cytoplasm. Several lines of evidence support a role for the wall as a type of primitive mitotic apparatus. First of all, isolated wall preparations of *B. subtilis* contain segments of DNA rich in markers for the origin and terminus of replication. Second, walls partially protect DNA from shearing and from nuclease digestion. Third, the walls of *B. subtilis* are known to enlarge by diffuse intercalation of peptidoglycan (see Section VI). The use of cell wall in segregating chromosomes would circumvent some of the problems associated with membrane fluidity, a factor not considered in the original replicon hypothesis. The renovated replicon hypothesis is attractive for bacilli but may not be valid for organisms known to expand their surfaces via discrete zones, such as streptococci. [*See* DNA REPLICATION.]

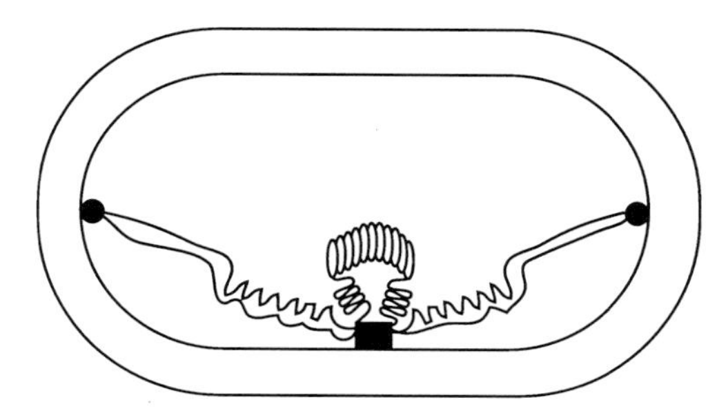

Figure 7 The renovated replicon model. A symmetrical structure (replisome) is formed during chromosome replication. This allows the bacterial rod to find its middle. The structure consists of the replicating chromosome and its associated proteins and is held to the poles of the cell at unique sites that bind the two origin regions of the DNA. The terminus, together with the replisome, is connected to the membrane, but in a nonrigid fashion. As replication takes place, tension develops in the DNA strands extending from both poles to the replisome. The stresses then center the replisome together with the terminus, and this in turn fixes the site of the next cross-wall. The origin binding sites are thought to be probably at the tips of the poles. [From Koch, A. L., and Doyle, R. J. (1986). *FEMS Microbiol. Lett.* **32,** 247–254.]

IV. Preparation of Cell Walls

The strategy for the isolation of bacterial cell walls is generally simple. Usually, dense cell suspensions (up to 50 mg/ml dry weight) in distilled water or buffer are subjected to mechanical disruptions. The disrupted cell walls are then isolated by differential centrifugation. Some of the common methods employed for breaking bacterial cells are outlined in Table III. All of the methods depend on the forced breakage of cells in the suspension. Some of the methods yield poor recoveries. For example, freezing and thawing or the grinding of thick suspensions in the presence of alumina provide only small percentages of cell breakage. Sonic oscillations, the French press, or rapid shaking in the presence of beads gives rise to higher recoveries of cell walls.

When gram-positive bacteria are broken and centrifuged, several phases can be observed in the centrifuge tube. The soluble phase contains cytoplasmic contents. An insoluble top phase, usually brownish, consists largely of membrane. This phase can be carefully scraped away from the underlying wall phase. The wall phase, usually grayish white, is scraped off and suspended once again. The very bottom phase is unbroken cells and is discarded.

Table III Common Methods of Bursting Bacteria for Cell Wall Preparation

Method	Comment
French pressure cells, Ribi fractionator	Suspensions of bacteria, when forced through small openings, undergo rapid pressure changes and burst
Sonic oscillations	High-intensity sound waves cause local cavitations around cells and result in cellular breakage
Ballotini beads (or glass or plastic beads) (Braun homogenizer)	Beads are made to collide with the bacteria, leading to disruption of the cells
Freezing–thawing	Cycles of freezing and thawing may burst sensitive cells
Waring blender	Blenders may cause some breakage, the degree of which may be increased with beads
Mortar and pestle	Bacterial suspensions ground together with alumina may be disrupted

The centrifugation–scraping–resuspension process is repeated several times; each time the wall phase becomes less contaminated with membrane and cytoplasmic constituents.

For gram-positive bacteria, up to 100 mg may be isolated in 1 day by the foregoing procedures employing a single centrifuge. For gram-negative bacteria, yields are much lower. Samples of 1–2 mg may be obtained when starting with 2–3 g dry weight of cells. There are several criteria for wall "purity." One criterion is to note the absence of unbroken cells or membrane fragments in electron micrographs. Another criterion is to note the absence of cytoplasmic enzymes. Unless treated with chaotropic agents, such as hot sodium dodecyl sulfate, wall preparations usually contain small amounts of protein, RNA, and DNA. These latter components may not be contaminants but, rather, actually may adhere to sites on the wall via stereospecific interactions. Wall preparations of *B. subtilis* have been shown to contain unique membrane constituents and DNA enriched for markers near the origin and terminus of replication. Furthermore, most wall preparations contain adherent autolysins. The final "purification" of wall involves chaotrope extraction and extensive washing in water or buffers. These wall preparations are completely free of membrane and nucleic acid but may possess proteins embedded in the wall matrix. Extraction of whole cells with detergents or protein solvents is usually not successful in yielding intact cell walls. Breakage of the cells makes it much easier to obtain a wall devoid of cytoplasmic contaminants. Walls obtained by mechanical breakage typically include fragments with cell poles, septa, cell side walls, or large, virtually intact walls. For *B. subtilis,* controlled autolysis of wall preparations leads to an enrichment of cell poles. To date, however, no interesting wall chemistry has been reported showing a unique composition of the cell poles.

V. Biosynthesis of Cell Walls

The insoluble cell wall material is synthesized by a complex series of reactions. The bacterium employs a clever strategy for wall synthesis. Precursors of peptidoglycan are synthesized in the cytoplasm. These precursors are then added to in the cytoplasmic membrane and ultimately escorted through the membrane to the cell exterior, where they are added to preexisting wall and finally cross-linked. These steps in peptidoglycan synthesis can frequently be specifically inhibited by antibiotics such as penicillin. Wall antibiotics give rise to cells with defective peptidoglycan (recent papers suggest that teichoic acid may also be a target for wall antibiotics), resulting in growth inhibition or lysis. The most effective antibiotics have historically been those that inhibit peptidoglycan synthesis (Table IV).

Figure 8 outlines the major features of peptidoglycan biosynthesis. In the first series of biosynthetic reactions, a uridine diphosphate–muramyl (or –*N*-acetylmuramyl) (UDP–MurNAc) pentapeptide is formed. The pentapeptide is covalently bound to the lactyl group of the muramic acid. Almost all of the reactions leading up to the UDP–MurNAc pentapeptide require an ATP or ATP equivalent, so in wall synthesis ribosomes are not involved. The peptides of peptidoglycan are synthesized by specific enzymes and the hydrolysis of an ATP for each amino acid added. Protein synthesis inhibitors such as tetracyclines or erythromycin do not inhibit the synthesis of peptides in cell walls.

Table IV Antibiotics Interfering with the Biosynthesis of Peptidoglycan

Antibiotic	Site of action
Phosphonomycin	Reactive analog of phosphoenolpyruvate; prevents synthesis of UDP–MurNAc from UDP–GlcNAc
Cycloserine	Structural analog of D-alanine; interferes with synthesis of D-Ala and D-Ala-D-Ala
Bacitracin	Polypeptide that inhibits a membrane-associated phosphatase; specifically inhibits the dephosphorylation of bactoprenolpyrophosphate
Vancomycin and ristocetin	Complicated effects on wall synthesis; the primary target seems to be the wall–membrane interface, where newly synthesized wall materials are added to preexisting wall
Penicillin (and penicillin analogs and derivatives)	There may be several effects of penicillin on wall synthesis; the best-known effect is the inhibition of peptidoglycan cross-linking

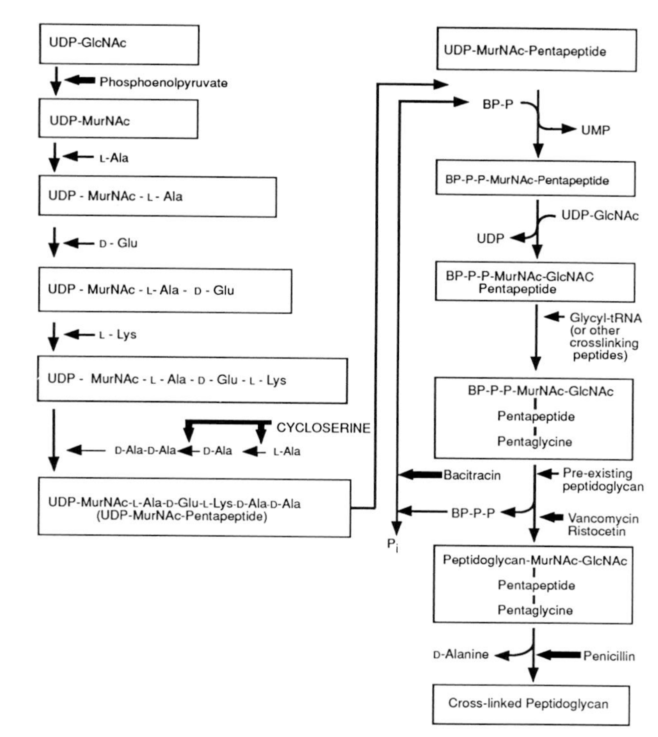

Figure 8 Major features of peptidoglycan synthesis in bacteria. Sites showing where the synthesis can be interrupted by antibiotics are shown by thickened arrows. BP, bactoprenol phosphate; P_i, inorganic phosphate; tRNA, transfer RNA; UDP and UMP, uridine di- and monophosphate, respectively.

Interestingly, the synthesis of the pentapeptide involves the sequential addition of L-Ala, D-Glu, L-Lys (or frequently, depending on the bacterium, some optical isomer of diaminopimelic acid) followed by the addition of a dipeptide, D-Ala-D-Ala. The D-Ala-D-Ala is synthesized from L-Ala in two reactions involving an alanine racemase and a D-Ala-D-Ala ligase. Cycloserine inhibits both the racemase and the ligase. For reasons discussed later, the synthesis of a pentapeptide, rather than a tetrapeptide, favors wall cross-linking at a later stage.

The UDP–MurNAc peptapeptide is then transferred from the relatively hydrophilic cytoplasm to the hydrophobic membrane by a "carrier" molecule, called bactoprenol phosphate (BP). The BP is an undecaprenol phosphate similar to the farnesyl carrier in other living systems. The reaction involves UDP–MurNAc pentapeptide with BP to yield membrane-soluble BP-P–MurNAc pentapeptide plus uridine monophosphate (UMP). The reaction requires an enzyme appropriately called a translocase. The first glycosidic bond is synthesized by the interaction of UDP–GlcNAc with the BP-P–MurNAc pentapeptide to yield BP-P–MurNAc–pentapeptide–GlcNAc plus UMP. The final reaction in the membrane involves the addition of amino acids to the ammonium group on the L-Lys or DAP (third amino acid of the pentapeptide). In *S. aureus*, the newly added amino acids consist of a $(gly)_5$ transfer RNA complex. In some bacteria, a single amino acid may serve us the cross-linking agent, whereas in others, the ammonium group of the L-Lys (or DAP) may form a peptide bond with a carboxylate or a D-Ala. The cross-linking step requires a transpeptidase capable of cleaving the terminal D-Ala and condensing the carboxylate function of the remaining D-Ala with the ammonium function of the cross-linking amino acid. The transpeptidation reaction does not require ATP or other high-energy metabolic intermediate.

The cross-linking reactions are sensitive to penicillins and other β-lactam antibiotics. In addition, cephalosporins act at the same loci as penicillin. Cells grown in the presence of penicillin frequently form protoplasts or spheroplasts when provided suitable osmotic protection (see Fig. 9). Under certain conditions, cell wall antibiotics may kill cells in the absence of any detectable wall defects. In addition to the transpeptidases, other cell wall-metabolizing enzymes may be inhibited by penicillin. These proteins are collectively called penicillin-binding proteins (PBPs). A PBP may be detected when it binds a radioactive penicillin.

Factors involved in the regulation of the extent of peptidoglycan cross-linking or glycan chain length are not completely understood. In some bacteria, the extent of cross-linking is very low. For example, the peptidoglycan is about 15% cross-linked in *Bacillus thuringiensis,* whereas in some *Legionella pneumophila* strains cross-linking of >80% is known. Furthermore, gram-negative and gram-positive bacteria may be cross-linked to the same extent, although the latter supports more turgor. Glycan chain lengths of up to several hundred disaccharides have been reported, although reports of 25–35 disaccharides are the most common.

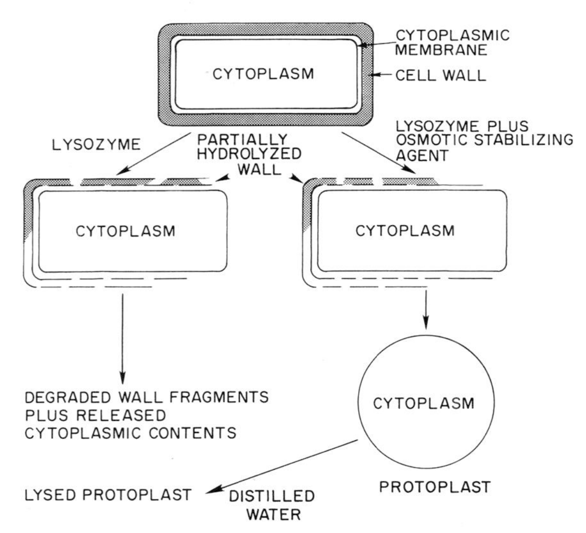

Figure 9 How the cell wall protects a bacterium from osmotic lysis.

Protease-deficient (highly autolytic) mutants of *B. subtilis* have short glycan strands, whereas isogenic hyper-protease-producing strains possess very long glycan strands.

In the biosynthesis of peptidoglycan, the bacterium expends a great deal of energy, but the costs are minimal when considering the need of a structure that prevents lysis. The bacterium cleverly engineers the assembly of the extracellular wall materials. The assembly and dynamics of cell walls are considered in the next section.

VI. Cell Wall Assembly and Determinants of Cell Shapes of Bacteria

Different kinds of bacteria seem to have different means to assemble their macromolecular cell walls. The common denominator for the assembly of all cell walls in bacteria is that the wall material is synthesized as short oligo-peptidoglycan precursors and then inserted into a preexisting wall (see Section V). When newly inserted wall is first added to preexisting (''acceptor'') peptidoglycan, it does not add to the mechanical strength of the wall. Only when the peptidoglycan oligomers are cross-linked to yield a three-dimensional network or mat does the newly assembled wall assume the ability to withstand turgor or stress.

Wall assembly in streptococci occurs primarily at discrete, easily identifiable zones. The zonal growth is coupled with a wall-thickening process by which new wall is added to old wall during surface expansion. Main features of the current understanding of streptococcal wall assembly are outlined in Fig. 10. The unit cell (smallest cell in a population) contains a circumferential wall band around the middle of the cell. The wall band is split by autolysin(s) and new wall added between the split sites. As more and more wall is added, the split bands (now two new bands) migrate (pushed) away from each other, generating new surface area. The separation of the old from the new wall is accompanied by wall thickening (mature wall is thicker than newly assembly wall) and formation of a cross wall. When the walls are completed, the cells separate (rapidly growing cells do not separate but, rather, form chains of streptococci) giving rise to two unit cells. A cell wall of a unit cell is therefore 50% new wall and 50% ''old''

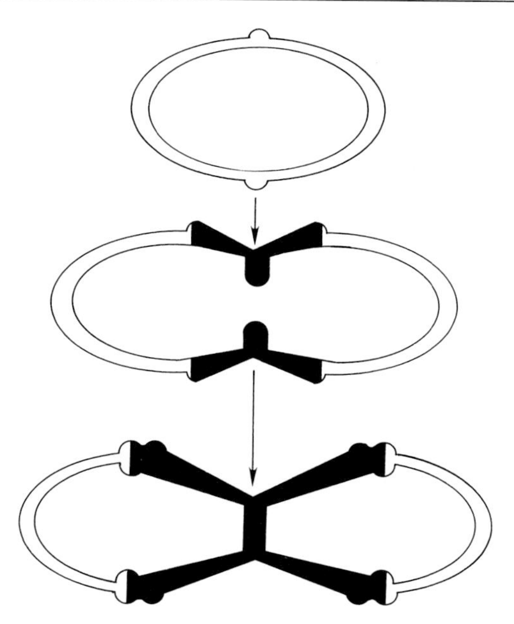

Figure 10 Cell wall growth in a streptococcus. □, old wall; ■, new wall. [Major features of the model were initially published by Higgins, M. L., and Shockman, G. D. (1976). *J. Bacteriol.* **127**, 1346–1358.]

wall. During wall assembly in streptococci, there is no excision or turnover of wall materials. As discussed later, turnover or excision of old wall is required for normal growth processes of gram-positive bacilli.

The characteristic streptococcal shapes arise partially as a result of internal pressure. In other words, the wall expands as a result of cellular turgor pressure. During growth, a bacterium will take up nutrients, metabolize them, and synthesize new bacterial components. Unless surface expansion accompanies the uptake of nutrients, the cell would ultimately develop a high enough internal pressure to burst the cell. The change in surface as a result of turgor pressure is the basic building block for the surface stress theory (SST). A researcher observed that soap bubbles could be made to assume streptococcal shapes and that two soap bubbles could coalesce to form shapes approximating those of dividing streptococci. Based on the following assumptions, turgor was shown to be the dominant factor in streptococcal morphology: (1) surface expansion occurs only in zones defined by the splitting of the equatorial wall band; (2) wall material, when newly assembled, assumes tension once it is cross-linked; (3) septal or unsplit wall is at least twice as thick as adjacent wall; and (4) turgor pressure does not change as wall is assembled or, stated differently, growing wall responds to increased turgor in such a way as to minimize its surface tension. In streptococci, the addition of new wall must therefore occur at the leading edge of the septum. Only when the septum is split (and therefore externalized) does the wall assume tension. This tension causes the characteristic streptococcal morphology. Interestingly enough, thicknesses of walls much greater or less than the 2 : 1 ratio observed for septa—split and externalized wall—would give rise to unstable morphologies. The preceding work constitutes one of the greatest advances in understanding cellular morphologies since bacterial forms were first observed by Leeuwenhoek over three centuries ago.

Rod-shaped bacteria have different strategies than the streptococci for surface assembly. In the gram-positive rod, wall material is inserted at many sites in the cell cylinder. The wall material is then cross-linked, whereupon it assumes tension due to cellular turgor. Addition of more wall pushes older wall from the surface where it stretches due to the additional tension. The continued addition of new wall, accompanied by wall stretching, results in surface expansion (Fig. 11 outlines the main features of the inside-to-outside and split-and-stretch model for gram-positive rods). The outermost wall material (oldest wall) is the most highly stretched. The external tension may be relieved by the clipping of peptidoglycan bonds by autolysins.

Frequently, the hydrolysis of the surface peptidoglycan by autolysins results in the release of wall fragments into the growth medium. This is a process called turnover. As far as is known, turnover occurs in all gram-positive rods, although autolysin-

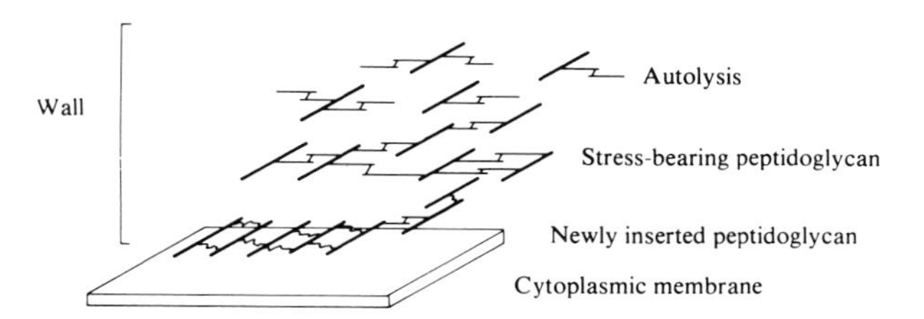

Figure 11 Cell wall growth in a gram-positive rod. Newly added wall "pushes" older wall away from the cytoplasmic membrane, resulting in peptidoglycan stretching. The oldest, peripheral wall may be cleaved by autolysins. [From Koch, A. L. *et al.* (1982). *J. Gen. Microbiol.* **128**, 927–945.]

deficient cells turn over their walls at a much reduced rate. Figure 12 supports the view that wall assembly in gram-positive rods occurs via random addition of precursors in the cell cylinder of *B. subtilis*. Because *B. subtilis* possesses a peptidoglycan-bound glucosylated teichoic acid, the lectin concanavalin A (con A) can be used to probe the cell surface of the bacterium. Furthermore, a temperature-sensitive mutant (phosphoglucomutase-deficient) of *B. subtilis* in teichoic acid glucosylation should yield cells incapable of binding the lectin when the cells are grown at the nonpermissive temperature. By switching cultures from permissive to nonpermissive and back to permissive, it should be possible to follow the sites of wall insertion and wall turnover by use of fluorescein-labeled con A (Fl-con A). In Fig. 12, it is seen that when the lectin was mixed with cells grown at a permissive temperature, fluorescence was distributed evenly on the cell surface, with the exception of the septum, where wall thickness is at least double. When the mutant was transferred to a nonpermissive temperature and incubated for 1.8 generations, con A-reactive material from the cell cylinder was uniformly lost, but the cell poles retained the ability to bind the lectin. One researcher has suggested that the cell poles are less susceptible to turnover because they are not formed under conditions of high stress. When the mutant was grown several generations at the nonpermissive temperature and shifted to permissive conditions, con A-reactive sites appeared on cell cylinders after about 0.8 generations (Fig. 13). The results for the Fl-con A experiments show that old wall disappeared randomly from cell surface side walls. Furthermore, new wall apparently was added from many sites. These findings support the concept that surface extension in gram-positive rods is by an inside-to-outside growth mechanism, accompanied by the splitting of older wall during growth.

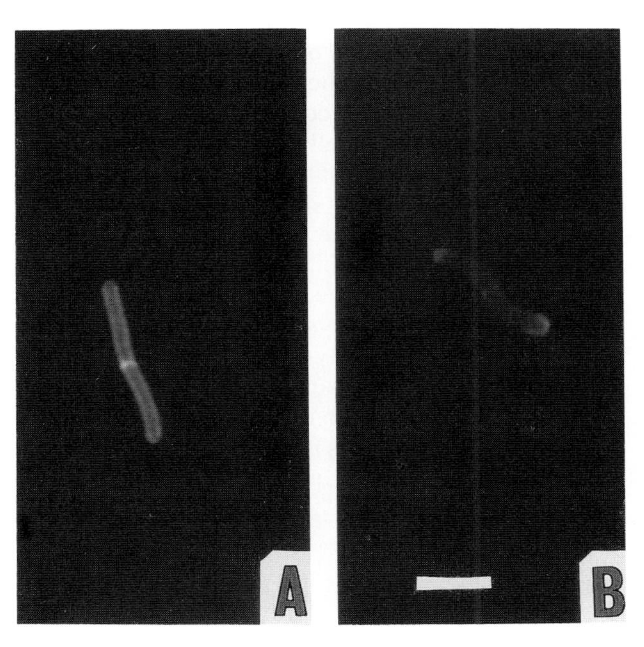

Figure 12 Interaction of fluorescein-conjugated concanavalin A and *Bacillus subtilis gta* C33. Cells were grown for 10 generations at 35°C and then shifted to 45°C. (A) 0 generations after the shift. (B) 1.8 generations after the shift. Phosphoglucomutase remains active at 45°C for approximately 1 generation of growth after the temperature shift. Septal and polar fluorescence represents cell wall formed before enzyme inactivation. Loss of fluorescence within the cell cylinder is indicative of cell wall turnover in that region. Bar = 5 μm. Cells were grown in Penassay broth. [From Doyle, R. J. *et al.* (1981). *Curr. Microbiol.* **5**, 19–22.]

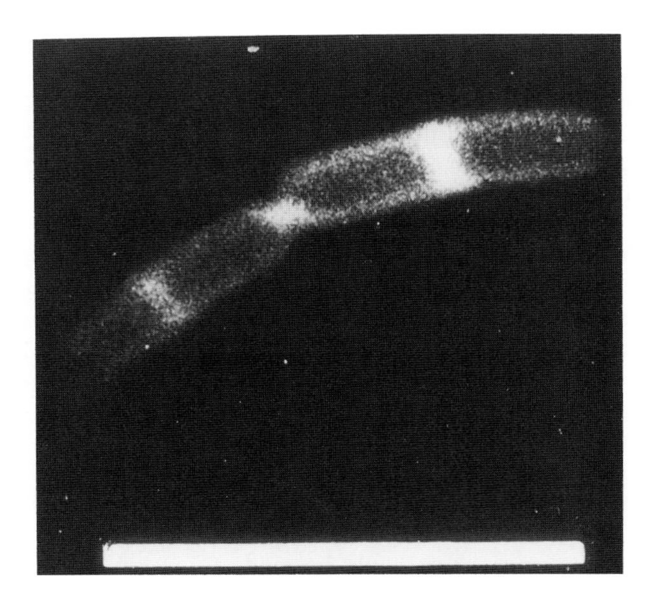

Figure 13 Interaction between fluorescein-labeled concanavalin A and *B. subtilis* C33 after a shift from nonpermissive to permissive conditions. Cells were prepared for photography 0.8 generations after the temperature shift. Note that the cell poles are not as fluorescent as the cell cylinders or the septa. Growth rate in the minimal medium was 40 min/generation at the permissive temperature. Cells grown at the nonpermissive temperature could not bind the lectin. Bar = 15 μm. [From Mobley, H. L. T. *et al.* (1984). *J. Bacteriol.* **158**, 169–179.]

The foregoing discussion illustrates one kind of experiment designed to study the insertion and fate of cell wall in gram-positive bacilli. Other kinds of approaches have also been employed, but usually the same conclusions have been reached. In another kind of experiment employing an autolysin-deficient mutant (the phenotype shows poor cell separation and low cell wall turnover), cells were labeled with tritiated *N*-acetylglucosamine, washed, and suspended in fresh growth media. At intervals, samples were removed, fixed to slides, and coated with a photographic emulsion. Results (Fig. 14) showed the random loss of label from cell cylinders and retention of label at poles or preformed septa. These results are in accord with the inside-to-outside growth model. Wall assembly in gram-positive rods is clearly different from that of gram-positive streptococci. The implications for the various means of wall assembly are less clear, especially in terms of DNA partitioning and in susceptibility to peptidoglycan synthesis inhibitors.

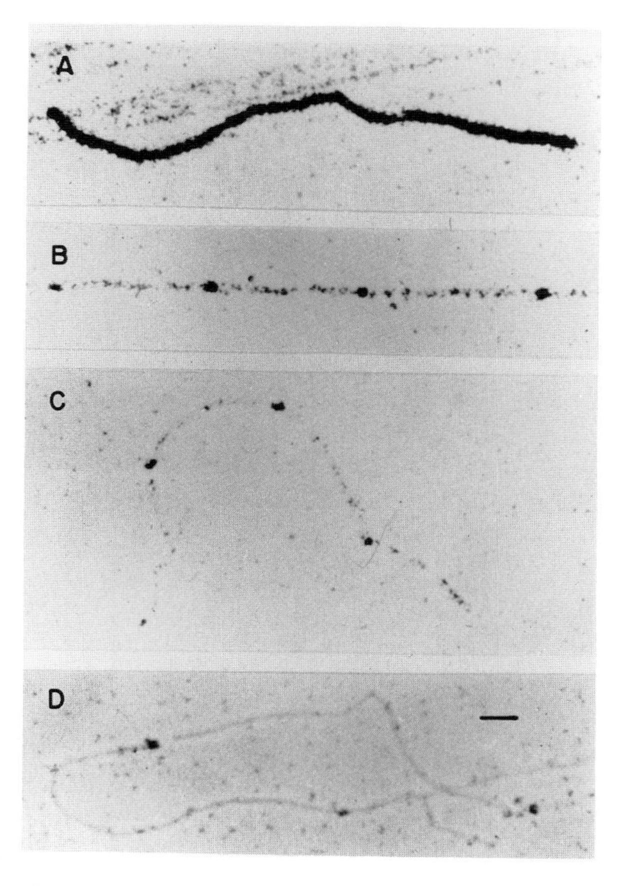

Figure 14 Autoradiography of *Bacillus subtilis* BUL96 labeled with [^{3}H]GlcNAc. Cells labeled for two generations in a rich medium (24 min/generation) were filtered, washed, and transferred to fresh medium. Samples were taken at intervals, fixed to slides, coated with nuclear track emulsion, maintained at refrigerated temperatures, and developed after 7 days. Interval (generations) after the transfer to fresh medium: 0 (A), 1.4 (B), 1.9 (C), and 2.9 (D). Bar = 5 μm. [From Mobley, H. L. T. *et al.* (1984). *J. Bacteriol.* **158**, 169–179.]

The peptidoglycan layer of gram-negative cells is much thinner than that of the gram-positive organism. Over the past several years, there has been a debate concerning the number of layers of glycan strands in the gram-negative rod. A view for many years was that the peptidoglycan consisted of a single layer and that the peptidoglycan was resistant to turnover. Researchers have convincingly shown that the peptidoglycan of *E. coli* turns over, but the turnover products are reutilized during growth. Other researchers have presented some compelling results to suggest that the peptidoglycan of *E. coli* is at least three layers thick in glycan equivalents. These workers have also suggested that the sacculus of *E. coli* enlarges by an inside-to-outside growth mechanism. *E. coli,* however, does not form septa similar to those of gram-positive bacilli, so how its cell poles are formed or whether or not its poles turn over is not clear. In general, one of the unsolved problems in bacterial wall assembly is to distinguish between side walls and septa (or poles) in gram-positive (or gram-negative) bacilli.

Because cell walls are constructed outside the protoplast, the bacterium is always safe from osmotic lysis. For a rod-shaped bacterium, the inside-to-outside growth, coupled with cell wall turnover, is a convenient means to enlarge its surface area during division. The gram-positive streptococcus, in contrast, makes use of zonal growth to extend its surface. Both kinds of bacteria apparently rely on surface tension to direct their final morphologies.

The participation of autolysins in wall assembly and metabolism has now been well established. It is less clear, however, about the mechanism(s) of autolysin regulation. Unregulated autolytic activity will lead to cellular lysis. Autolysin must be active at certain times in a cell cycle to promote orderly growth. For example, the splitting of wall bands may be taken as one of the initial events in cell growth of streptococci. Similarly, the coupling between cell wall turnover and normal growth processes in gram-positive rods assumes that autolysin is active during surface extension. In some bacteria, including some members of genera *Bacillus, Listeria, Neisseria, Moraxella, Staphylococcus,* and others, dissipation of protonic potential results in cellular autolysis. It now seems likely that during metabolism extruded protons are bound to the wall matrix, creating a local environment of low pH. The low pH can be maintained in the wall matrix so long as the cell is metabolizing. Autolysin on the periphery of the wall may be active because ions from the medium may titrate the hydrogen ions. This is a means for wall regulation, which does not require the synthesis of separate regulatory molecules. Streptococcal autolysins can be inhibited *in vitro* by lipoteichoic acids and phospholipids. This implies that the inhibitory molecules are involved in the regulation of autolysins. In gram-negative rods, very little is known about the regulation of autolysins. In *E. coli,* there are several autolysins, but none seem to be influenced by the energized membrane.

By promoting autolysin activities, it may be possible to modify certain disease processes. For example, autolysin may degrade walls (or whole cells) of streptococci in joints, leading to elimination of sites of inflammation. Similarly, activated autolysin may be able to reduce microorganism levels in biofouled pipes and other surfaces.

VII. Biophysical Properties of Cell Walls

Bacterial cell walls are insoluble in most aqueous and nonaqueous solvents. Strong chaotropes, such as concentrated urea or lithium thiocyanate, do not solubilize the walls. The physical strength of cell walls depends on the cross-linking of peptidoglycan, the linear β-1,4-linked glycan, and the extensive network of ionic and hydrogen bonds formed between charged species. Walls can be titrated, yielding typical acid-base titration curves dictated largely by ammonium, carboxylate, and, in the case of gram-positive cells, phosphate groups. The wall is porous, permitting the passage of nutrients and low-molecular weight proteins.

Some components of cell walls have typical polyelectrolyte properties. The teichoic acid of *B. subtilis* exists as an extended, rigid-rod structure in water and dilute buffers; however, in salts the teichoic acid assumes a random coil configuration. Walls exhibit expanded volumes in the absence of salts, showing that walls are not truly rigid structures. In fact, the concept that walls are rigid has been a textbook error propagated for several decades. Turgor pressure from a protoplast also causes walls to be stretched. Stretched bonds in peptidoglycan are more easily hydrolyzed by autolysins or lysozyme.

In some bacteria, there is an asymmetry of surface charge. The outside wall surface of *B. subtilis* is serrated and capable of complexing with cationized ferritin (Fig. 15). The serrated surface probably results from the actions of autolysins and cellular turgor. Wall on the outer surface is stretched more than wall on the surface near the cytoplasmic membrane.

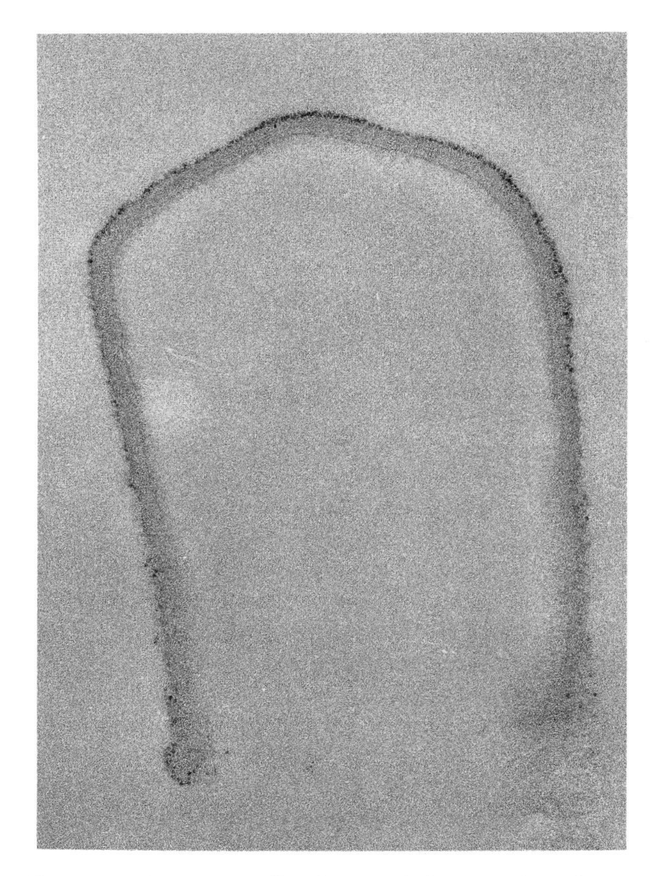

Figure 15 Asymmetric distribution of charge on the cell wall of *Bacillus subtilis*. Cationized ferritin binds strongly to the outer wall surface, but not at all to the inner face. [Micrograph courtesy of T. J. Beveridge, University of Guelph.]

The prospect that wall biosynthesis may be interrupted continues to form the basis for development of antibiotics. Other important studies include means to stimulate autolysins. Finally, because the wall is frequently essential for adhesin activity, great efforts are being made to study how cell walls influence microbial ecology.

Bibliography

Doyle, R. J., and Koch, A. L. (1987). *CRC Crit. Rev. Microbiol.* **15,** 169–222.

Doyle, R. J., and Sonnenfeld, E. S. (1989). *Intl. Rev. Cytol.* **118,** 33–92.

Doyle, R. J., Chaloupka, J., and Vinter, V. (1988). *Microbiol. Rev.* **52,** 554–567.

Koch, A. L. (1990). *Am. Sci.* **78,** 327–341.

Koch, A. L., and Doyle, R. J. (1985). *J. Theoret. Biol.* **117,** 137–157.

Koch, A. L., Higgins, M. L., and Doyle, R. J. (1981). *J. Gen. Microbiol.* **123,** 151–161.

Krell, P. J., and Beveridge, T. J. (1987). *Intl. Rev. Cytol.* **17,** 15–87.

Rogers, H. J., Perkins, H. R., and Ward, J. B. (1980). "Microbial Cell Walls and Membranes." Chapman and Hall, London.

Weidel, W., and Pelzer, H. (1964). *Adv. Enzymol.* **26,** 193–232.

Centrifugation and Filtration: Detection of Bacteria in Blood

Mathias Bernhardt and Ronald F. Schell
University of Wisconsin, Madison

Glossary

Aminoglycosides Class of antimicrobials that inhibit protein synthesis; used widely against gram-negative bacteria

Bacteremia Presence of bacteria in blood

Ficoll–Hypaque Mixture of Ficoll, a polymer of sucrose, and Hypaque-M (diatrizoate meglumine), a contrast material used in angiography, which is used to separate bacteria from erythrocytes

Leukocytosis Increase in white blood cells

Septicemia Presence of bacteria in blood along with signs of clinical infection

Tachycardia Rapid heartbeat

BACTEREMIA is the presence of bacteria in the blood. The presence of bacteria in itself is no cause for alarm, because this is a daily occurrence in our lives. For instance, every time we brush our teeth or eat, microscopic abrasions in the gums caused by the toothbrush or the morning toast allow for the entry of oral flora. However, these bacteria are rapidly eliminated from the blood stream via phagocytosis. Septicemia, on the other hand, is the presence of bacteria in the blood along with signs of clinical infection such as fever, chills, tachycardia, hypotension, shock, or leukocytosis. In this scenario, the bacteria somehow evaded the body's defense mechanisms and are able to colonize a normally sterile site. Septicemia is potentially life threatening. In the United States alone, approximately 200,000 cases of septicemia occur annually with a mortality rate of 40–50%.

I. Overview of Blood Culturing

A. Background of Blood Culturing

The basic procedure of performing a blood culture has changed little since its inception as a diagnostic tool. A blood sample is obtained from the patient via venipuncture, inoculated to the appropriate broth medium (a blood culture bottle), incubated, and repeatedly checked for growth of the organism(s). Once the organism is isolated on solid medium, drug susceptibilities are performed and the proper choice of antibiotics can be administered to the patient. While relatively simple, isolation of the causative organism can take anywhere from 24 hr to several days. Given the high mortality rate, rapid detection and isolation of the infecting microorganism(s) are paramount for the administration of effective antimicrobial therapy and survival of the patient. [*See* ISOLATION.]

B. Current Technology

To better aid in the rapid detection of bacteremia, several rapid detection blood culturing devices have been developed. These include the BACTEC system, the Oxoid Signal blood bottle, the Roche Septi-Chek system and the Isolator. With the exception of the Isolator, these methods rely on the incubation of blood in a liquid medium but differ in the detection of bacterial growth. The BACTEC system uses infrared light and a computer to detect CO_2 production in the blood bottles. In the Oxoid bottle, CO_2 produced by the growing bacteria pushes the liquid medium containing the bacteria into a reservoir attached to the top of the bottle. The Septi-Chek system detects actual colonies on a slide paddle compromised of chocolate, McConkey, and malt agars attached to the bottle. This paddle is coated with the blood–broth mixture immediately following inoculation of the bottle. The Isolator uses lysis of red blood cells and centrifugation for the recovery of bacteria. After centrifugation, the supernatant that contains the lysed blood cells is removed and the concentrate (containing the bacteria) is streaked out on several plates of different media. The plates are incubated and checked for the growth of isolated colonies.

Still, several days may be required for detection and isolation of bacteria using these new systems. In addition, although the BACTEC system, Oxoid Signal, Roche Septi-Chek, and Isolator provide enhanced detection of bacteria, some drawbacks have been documented such as contamination and different recovery rates for members of the family Enterobacteriaceae, staphylococci, yeasts, streptococci, and anaerobes.

II. Lysis and Filtration Technology: A Brief History

A. Lysis

The use of lysis technology in the detection of bacteremia is usually associated with the use of filtration technology. The original concept was to devise a detection method for bacteremia that would not necessarily involve the use of a liquid medium. Blood bottles, though widely accepted for the detection of microorganisms in blood, do pose disadvantages. The most critical is that blood bottles do not provide an optimum growth environment for microorganisms that may be present in the blood. This is because inhibitory agents such as complement; antibodies; sodium polyanethol sulfonate (SPS), an anticoagulant; and antimicrobials that may have been administered prior to culture are not removed. This in turn leads to inhibited growth of microorganisms and, hence, longer detection times.

The challenge was to devise a system in which the bacteria could be cultured in absence of these inhibitory agents. The simplest way to do this was to pass the blood through a membrane filter where the bacteria would be trapped. Prior to this step, however, the blood cells had to be lysed. This was accomplished by several investigators through the use of lysing agents such as Triton X-100 and streptokinase–streptodornase. After filtration, the filters were either immersed in a broth or, more commonly, placed on an agar plate with the filtrate side up. Nutrients would diffuse through the filter, allowing the bacteria to grow as isolated colonies. Still, the procedure has been considered slow, cumbersome, and impractical for clinical use. Furthermore, problems such as filter clogging and decreased recovery due to lysis of bacteria in addition to blood cells were also reported.

One system on the market today uses lysis centrifugation technology, which does not involve any filtration. This is the Isolator developed by Gordon Dorn. The Isolator uses a chemical cocktail that lyses blood cells in addition to preventing coagulation of the blood sample. The cocktail consists of Saponin, which lyses the red and white cells; SPS and ethylenediaminetetraacetic acid (EDTA), both of which are anticoagulants; and polypropylene glycerol, which inhibits the natural foaming tendency of Saponin. Some studies have shown that both the EDTA and Saponin may be toxic toward *Streptococcus pneumoniae* and *Pseudomonas aeruginosa,* respectively. Another disadvantage is that while the Isolator may inhibit the action of aminoglycosides via SPS, it does not remove them entirely from the system.

B. Filtration

The use of membrane filter procedures for the isolation of microorganisms from blood has long been established. As already mentioned, these procedures were not very practical for clinical use. The early trials incorporated lysis technology as a vital component. However, filtration for the isolation of bacteria does not have to involve lysis. In recent years, a centrifugation–filtration method has been developed that employs the use of a density mixture for the separation of red blood cells from the microorganisms.

III. Detection of Bacteria by Centrifugation and Filtration

A. Background

As it stands today, the culture of blood in a liquid medium is still the most common means of diagnosing bacteremia. However, this is a slow process. In addition, growth conditions in blood bottles are not necessarily conducive to rapid uninhibited growth. A detection system was needed that would allow the growth of bacteria on a filter membrane without use of lysing agents. The potential advantages of such a system include faster isolation of bacteremic agents, collection of leukocytes that may harbor phagocytized bacteria, no subculturing, removal of inhibitory agents (antibiotics) from blood, detection and faster identification of the causative organism, and detection of mixed infections or contaminants. Antibiotic susceptibility tests can be accelerated due to growth of distinct isolated colonies that do not need further purification. In addition, the efficacy of treatment (via the proper course of antibiotic therapy) can be determined by quantitation of bacteria on the filters.

B. Centrifugation and Filtration Procedure

The centrifugation and filtration procedure is shown in Fig. 1. Fresh human whole blood is first obtained from a healthy volunteer who has not received antibiotics during the preceding 2 wk. This blood sample is then seeded with a known quantity of a single species of bacteria and mixed. Concurrently, sterile glass tubes are loaded with 8 ml of Ficoll–Hypaque (density mixture: $D = 1.149 \pm 0.002$ g/ml). Then, 5 ml aliquots of the seeded blood are added to the tubes containing the Ficoll–Hypaque. The tubes are stoppered, inverted five times to thoroughly mix the Ficoll–Hypaque with the seeded blood, and centrifuged (386 g) for 30 min at room temperature.

During centrifugation, the red blood cells, which are denser than the bacteria, are driven to the bottom of the tube. This is further enhanced by the Ficoll, a polymer of sucrose. Ficoll acts by adhering to the red blood cells. This facilitates the sedimentation of the red blood cells during centrifugation. As the red blood cells are tumbling through the mixture, the bacteria are "washed free" from the erythrocytes and are retained throughout the Ficoll–Hypaque because they are less dense.

Centrifugation of the blood–density mixture results in the red blood cells being pelleted at the bottom of the tubes. Mononuclear and polymorphonuclear cells are distributed predominantly in the upper portion of the Ficoll–Hypaque. After centrifugation, the entire mixture, except the eythrocytes, is removed and filtered through a 0.22-μm-pore size filter under negative pressure with a single-place sterility test manifold attached to a vacuum pump.

The filter is then removed from the filtration apparatus, placed with the filtrate side up on a chocolate agar plate, and incubated at 35°C in a humidified atmosphere containing 5% CO_2. Isolated colonies of bacteria are detected within 18 hr.

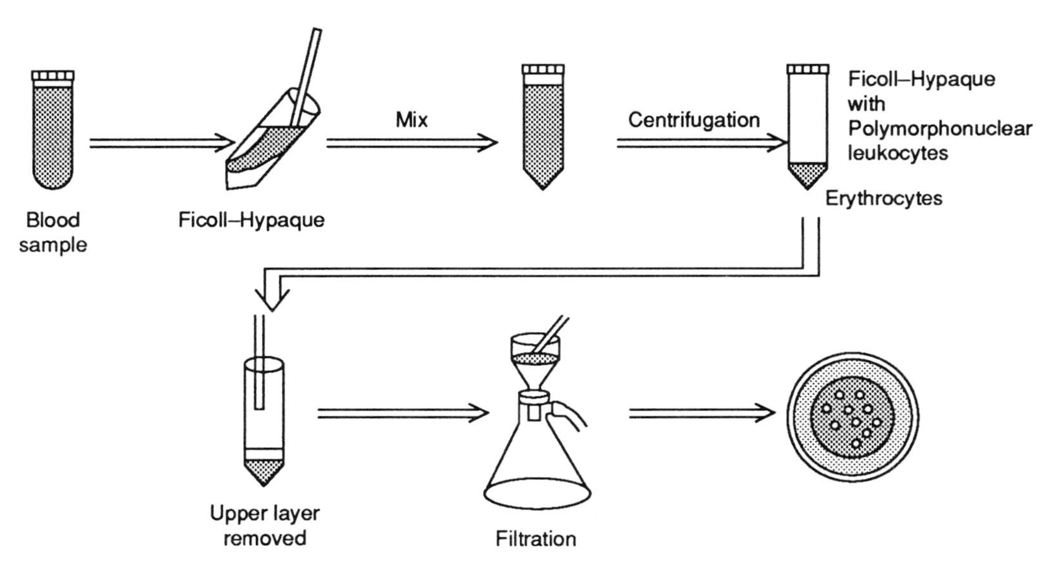

Figure 1 Ficoll–Hypaque centrifugation and filtration procedure.

C. Data

The recovery of bacteria from seeded whole blood is shown in Table I. In general, fewer microorganisms are recovered by the filters compared with the original inoculum. However, when 10 ml of Ficoll–Hypaque is used, improved recovery is observed (Table II). This is because the extra 2 ml of Ficoll–Hypaque provide an additional travel distance for the bacteria to be separated from the blood cells.

The centrifugation filtration procedure compares well to current blood culturing systems on the market today, namely, the Isolator and the Septi-Chek. The centrifugation and filtration system has also shown itself to be very effective in the removal of antibiotics when used in conjunction with a nonionic polymeric adsorbent resin.

Table II Microorganisms (mean ± SD) Recovered from Blood by Centrifugation and Filtration[a]

Organism	Filtration	Inoculum	% Recovery
Staphylococcus aureus	45.5 ± 4.0	63.0 ± 6.0	72.0
Escherichia coli	18.0 ± 1.0	20.0 ± 2.0	90.0
Haemophilus influenzae	4.0 ± 1.0	7.0 ± 1.0	57.1
Klebsiella pneumoniae	49.0 ± 3.0	54.2 ± 7.0	90.4
Proteus mirabilis	5.0 ± 2.0	7.0 ± 1.0	71.4
Listeria monocytogenes	36.0 ± 4.0	47.0 ± 5.0	76.0

[a] Ten ml of Ficoll–Hypaque were used.

D. Concerns

Overall, the centrifugation and filtration procedure recovers less microorganisms than are present in the inoculum (Table I). Some loss of microorganisms is expected, namely, due to retention of small quantities of seeded blood in the pipette used to transfer blood to the gradient. In addition, the tubes retain a small portion of the Ficoll–Hypaque containing seeded blood. It is also possible that some bacteria adhered to the sides of the glass tubes and were missed by filtration. Some may argue that a 20 or 35% recovery of bacteria is poor; however, the sensitivity of the conventional blood culturing bottle is less. When 23 blood culture media were inoculated with 7–15 microorganisms per bottle, only 1 of the 23 different types of blood culture media supported the growth of all the bacteria tested. Furthermore, and most importantly, blood is normally a sterile body fluid. Any number of bacteria isolated from blood is significant, be it 1 colony-forming unit (CFU)/ml or 50 CFU/ml blood.

Table I Microorganisms (mean ± SD) Recovered from Blood by Centrifugation and Filtration[a]

Organism	Filtration	Inoculum	% Recovery
Staphylococcus aureus	15.3 ± 8.7	37.5 ± 4.9	40.8
Enterococcus faecalis	28.0 ± 4.5	32.0 ± 8.4	87.5
Streptococcus mitis	22.6 ± 6.6	65.0 ± 8.4	34.7
Streptococcus mutans	141.3 ± 10.0	172.6 ± 20.5	81.8
Streptococcus pneumoniae	52.3 ± 6.3	56.0 ± 1.4	93.3
Streptococcus salivarius	12.3 ± 8.0	15.5 ± 2.1	79.3
Streptococcus sanguis	40.0 ± 6.5	49.0 ± 4.2	81.6
Escherichia coli	38.6 ± 3.5	79.0 ± 9.8	48.8
Haemophilus influenzae	19.0 ± 6.2	46.3 ± 16.6	41.0
Klebsiella pneumoniae	35.3 ± 5.1	59.5 ± 7.7	59.3
Neisseria meningitidis	93.0 ± 7.8	131.0 ± 13.4	70.9
Pseudomonas aeruginosa	171.3 ± 12.0	159.5 ± 9.2	107.3

[a] Eight ml of Ficoll–Hypaque were used.

The drawbacks of this system are minor compared to the advantages. Also, this system is still in its infancy and has not been refined yet. As it stands, centrifugation and filtration does not require lysing agents, multiple filters, dilutions, sophisticated equipment, or excessive centrifugation speeds. Most importantly, it offers faster isolation of bacteremic agents, which translates into better patient care.

IV. Future Directions

A. Refinement

The drawbacks mentioned earlier can be eliminated once the centrifugation and filtration technique becomes a closed system. This would be accomplished by manufacturing a double-ended blood collection tube that contains the Ficoll–Hypaque along with an anticoagulant such as SPS. Blood would be drawn through one end of the tube. The other end of the tube would have a small rubber cone protruding into the tube. During centrifugation, the erythrocytes pellet around the cone but leave the tip exposed. The cone would then be punctured from the bottom of the tube during the filtration procedure. Ficoll–Hypaque-containing bacteria would be drawn directly onto the filters. All this could be accomplished without ever opening the tube. Using this approach, loss of microorganisms should be minimal because there are no pipetting steps as in the current procedure.

Should there be any back-flushing of Ficoll–Hypaque into the vein during blood collection, this should cause no alarm. Ficoll is a polymer of sucrose, and Hypaque is used commonly as a contrast medium in procedures such as angiograms.

B. Clinical Trials

Comparing a new technique such as centrifugation and filtration against established blood culturing devices in a laboratory setting is valuable in determining its usefulness. However, it still needs to be evaluated in a clinical setting to determine its ultimate effectiveness.

Bibliography

Bernhardt, M., Pennell, D. R., Almer, L. S., and Schell, R. F. (1991). *J. Clin. Microbiol.* **29,** 422–425.

Dorn, G. L., Haynes, J. R., and Burson, G. (1976). *J. Clin. Microbiol.* **3,** 251–257.

Herlich, M. B., Schell, R. F., Francisco, M., and LeFrock, J. L. (1982). *J. Clin. Microbiol.* **16,** 99–102.

Sullivan, N. M., Sutter, V. L., and Finegold, S. M. (1975). *J. Clin. Microbiol.* **1,** 30–36.

Sullivan, N. M., Sutter, V. L., and Finegold, S. M. (1975). *J. Clin. Microbiol.* **1,** 37–43.

Chemotaxis

Michael D. Manson
Texas A&M University

Jeffrey E. Segall
Albert Einstein College of Medicine of Yeshiva University

Glossary

Adaptation Process that terminates a cell's response to a stimulus and restores signal production to the prestimulus level

Biased random walk Motility pattern that allows bacteria to migrate in chemical gradients; tumbles (abrupt, random course changes) are suppressed for cells swimming in the favorable direction

Cellular slime mold Free-living amoebae that remain as individual cells upon aggregation, forming multicellular structures including pseudoplasmodia and fruiting bodies

Chemoreceptor Cell-surface protein that binds specific extracellular chemicals (chemoeffectors) and as a result produces an intracellular signal that modifies cell motility

Ectoplasm Peripheral cytoplasm of an amoeboid cell that underlies the cell membrane and is more rigid and gellike in structure than endoplasm; also called the cell cortex

Endoplasm Central cytoplasm of an amoeboid cell; of varying viscosity, but fluid

Hyaline Clear, transparent; used in describing the peripheral cytoplasm of amoebae

Pseudopod Membrane-bound cytoplasmic protrusion extended from the cell body to initiate amoeboid movement

Temporal sensing of gradients Ability of bacterial cells to detect spatial chemical gradients by comparing concentrations in time as they swim through their environment

CHEMOTAXIS is the oriented movement of organisms in response to a gradient in the external concentration of a chemical compound. Strictly speaking, these responses are classified as either taxes or kineses. Taxis occurs if an organism steers or directs its changes in orientation to move up or down the gradient, whereas kinesis occurs if the direction of change is random with respect to the gradient. By this definition, bacteria carry out chemokinesis. However, the ability of bacteria to accumulate in specific areas of a gradient has historically been described as chemotaxis, and we shall continue that tradition.

Positive chemotaxis leads to accumulation of cells at higher concentrations in the gradient, and negative chemotaxis leads to accumulation of cells at lower concentrations in the gradient. Chemotactic responses are seen in many, if not all, motile microorganisms, including eubacteria, archaebacteria, amoebae, ciliates, and flagellates. This article will focus on two systems that have provided insight into the molecular mechanisms involved in chemotaxis: the enteric eubacterium *Escherichia coli* and the eukaryotic amoebae of the cellular slime mold *Dictyostelium discoideum*.

I. Bacterial Chemotaxis

A. Significance of Bacterial Chemotaxis

Bacterial chemotaxis was first reported in the 1880s by Wilhelm Pfeffer, who observed the accumulation of bacteria around sources of oxygen or nutrients. The modern era of research in bacterial chemotaxis was initiated in the late 1960s by Julius Adler, who saw that the powerful tools of bacterial genetics and the extensive knowledge available about bacterial physiology could be applied to the analysis of a simple sensory system. Bacterial chemotaxis will probably be the first behavior to be understood at the molecular level—from the reception of an environmental signal to the control of the motor output.

Bacterial chemotaxis also has ecological and clinical relevance. It has been implicated in the aggregation of soil-dwelling rhizobia around incipient root hairs, and the crown-gall bacterium *Agrobacterium tumifaciens* may be drawn to plant wounds via chemotactic responses to plant exudates. Chemotaxis is a bacterial virulence factor in animals as well as plants. *Vibrio cholera* and *Salmonella* sp. penetrate the mucosal lining of the gut in response to gradients of chemicals released by the underlying cells of the intestinal wall, and the pathogenicity of strains is correlated with their chemotactic ability. Chemotaxis is likely to be a significant component of many bacterial diseases.

About 50 genes (3–5% of a typical bacterial genome) encode proteins involved in motility and chemotaxis, implying that these properties confer substantial selective advantage. The basis of this advantage has been characterized in relatively few instances, however. Even for the enteric bacteria *E. coli* and *Salmonella typhimurium*, in which the mechanisms of chemotaxis have been most thoroughly studied, the role of chemotaxis in the normal life cycle is not well understood.

The focus here is on how bacteria are attracted to or repelled by chemicals. However, some bacteria also migrate in gradients of light (phototaxis), temperature (thermotaxis), and osmotic strength (osmotaxis) or in magnetic fields (magnetotaxis) and electric fields (galvanotaxis). A number of species respond to several of these modalities.

B. Mechanism of Gradient Sensing by Bacteria

Rod-shaped bacteria, such as *E. coli* and *S. typhimurium*, are only a few micrometers in length. In typical chemical gradients, concentration does not vary significantly over this distance. As a result, cells cannot make useful "head-to-tail" comparisons of concentration to determine the gradient direction. The cells surmount this problem by measuring the change in concentration of a chemical as they swim; they use a temporal sampling mechanism to sense a spatial gradient. The cells respond by increasing the time spent swimming in the favorable direction (up an attractant gradient or down a repellent gradient) to bias their three-dimensional random walk (Fig. 1). The

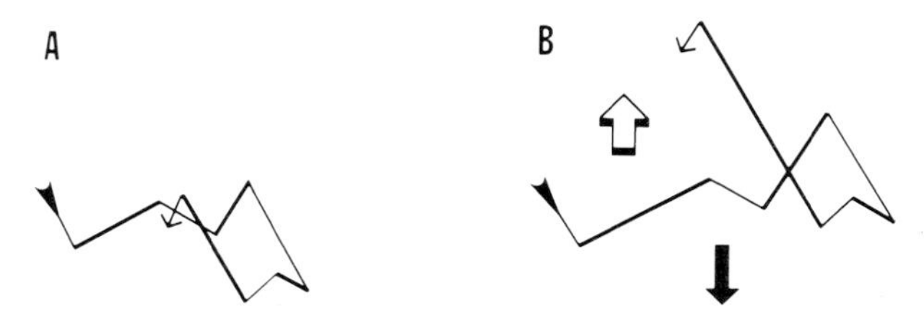

Figure 1 Idealized behavior of an *Escherichia coli* or *Salmonella typhimurium* cell swimming in the absence (A) and presence (B) of spatial gradients of chemotactically active compounds. The cell in A begins at the solid arrow and undergoes nine runs and eight tumbles (tumbles are the abrupt changes in direction). The runs are of variable length, and the net motion is a three-dimensional random walk. The cell in B is swimming in the presence of a gradient of attractant (open arrow pointing in the direction of higher concentration) or repellent (closed arrow pointing in the direction of higher concentration). The swimming path is identical to that in A except that the runs in the "favorable" direction (up the attractant or down the repellent gradient) have been lengthened by the amount cos θ, where θ is the angle with the gradient vector. The result is a biased three-dimensional random walk in which the step length in the favorable direction is increased.

maximum velocity of migration in a chemical gradient is about 20% of the swimming speed. The biased random walk enables cells to accumulate in capillary tubes containing attractants. It also causes a bacterial colony to form swarm rings in soft (0.3–0.4%) agar as cells move out from a point of inoculation in gradients they create by metabolizing an attractant in the medium.

In peritrichously flagellated bacteria such as *E. coli*, the random walk is generated by periods of smooth-swimming (runs) interrupted by abrupt and random changes in direction (tumbles). Runs in a favorable direction are extended because tumbles are suppressed. In bipolarly flagellated bacteria that reverse 180° rather than tumble, reversals are suppressed when the cells are swimming in the favorable direction. In bacteria that change direction via Brownian motion during a pause (a brief cessation of swimming), pauses are suppressed. [*See* MOTILITY.]

Rather than tracking the movements of free-swimming bacteria, observations of chemotaxis can be made with tethered cells. These cells have had their flagellar filaments mechanically sheared down to stubs and then attached to a surface coated with anti-filament antibody. Because the filaments cannot move, rotation of the flagellar motor causes the cell bodies to turn, typically at a few revolutions per second. The cells alternate between clockwise (CW) and counter-clockwise (CCW) rotation, with CCW flagellar rotation corresponding to running and CW rotation to tumbling. If the cells are tethered in a flow chamber, attractants or repellents can be introduced in step changes or in ramps, and the change in the ratio of CW–CCW rotation recorded. With either free-swimming or tethered bacteria, video techniques can be used to record the responses of cells over an entire microscope field, and data can be analyzed for individual cells or for populations of cells.

Cells respond to addition of attractants by running and to addition of repellents by tumbling. The length of the response to a compound depends on the magnitude of the concentration change. More subtle shifts in the CW–CCW ratio are observed when tethered cells are exposed to gradual, continuous increases or decreases in attractant or repellent (ramp stimuli). The ability to measure changes in concentration with time implies that the cells can monitor the present concentration and store information about what the concentration was in the recent past. This chemical memory is provided by the process of adaptation.

C. Bacterial Chemoreceptors

Attractants tend to be nutrients, and repellents tend to be compounds associated with crowded or otherwise suboptimal environments. Bacterial chemoreceptors are proteins at the cell surface that bind small molecules in a stereospecific way. Many attractants and repellents are recognized as such because they bind to specific receptors rather than because of any effects they have on the metabolism of the cells.

The first bacterial chemoreceptor to be characterized was the galactose–glucose-binding protein (GBP) of *E. coli*. Binding proteins for maltose (MBP), ribose (RBP), and dipeptides (DBP) are also chemoreceptors. These soluble proteins are located in the periplasmic space, the compartment between the outer and inner membranes of gram-negative bacteria. They serve as components of high-affinity transport systems for their substrates, which bind in a cleft between the N-terminal and C-terminal domains of the polypeptide to stabilize a closed conformation of the protein. It is assumed that when the proteins are in the closed conformation they have an increased affinity for membrane-associated chemotaxis and transport components. Some mutant binding proteins lose chemotactic function while retaining transport function, demonstrating that transport is not a prerequisite for chemosensing. The dual function of these four binding proteins is exceptional, since most binding proteins found in *E. coli* function only in transport.

Transport is needed, however, to elicit chemotactic responses to sugars that are substrates of the phosphotransferase system (PTS), although further metabolism of the sugar is not required. Each membrane-bound, sugar-specific Enzyme II phosphorylates its substrate, and transport-induced changes in the physphorylation of PTS proteins may link the transport and chemosensory pathways.

Sensing of oxygen is directly coupled to metabolism and is mediated by cytochrome oxidase. The oxygen response is really a tactic response to changes in the protonmotive force (pmf), which is the work per unit charge needed to move a proton from the outside to the inside of the cell. In a metabolizing cell, the pmf has a negative sign, indicating that protons passing through the cytoplasmic membrane into the cell release energy that can be utilized for other processes, such as flagellar rotation, nutrient transport, and adenosine triphosphate (ATP) synthesis. Increases in the absolute value of the pmf are sensed as attractants and decreases are sensed as repellents.

The methyl-accepting chemotactic signal transducers are receptors that play a central role in chemoefffector recognition and signal generation. *Escherichia coli* has four transducers: Tsr (taxis toward serine and away from some repellents); Tar (taxis toward aspartate and away from some repellents); Tap (taxis toward peptides); and Trg (taxis toward ribose and galactose). Transducers are membrane-associated polypeptides of 510–557 residues. Their positively charged N-terminus is in the cytoplasm and is followed by a stretch of hydrophobic residues that comprise the first transmembrane region. Next comes the periplasmic receptor domain, which contains about 160 amino acids. The second transmembrane region brings the polypeptide back through the membrane to the cytoplasmic signaling domain, which contains 300–350 residues.

Transducers can be primary or secondary chemoreceptors for attractants. Tsr functions only as a primary receptor, binding L-serine with high affinity and other amino acids with substantially lower affinity. Tar is a primary chemoreceptor for aspartate and, with lower affinity, for some other amino acids. Tar from *E. coli* (but not from *S. typhimurium*) also serves as a secondary receptor, interacting with MBP to initiate the response to maltose. Trg and Tap act only as secondary receptors. Trg interacts with RBP and GBP to initiate responses to ribose and galactose, and Tap interacts with DBP to mediate taxis toward dipeptides.

Tsr mediates repellent responses to L-leucine, indole, and low external pH, and it senses reductions in cytoplasmic pH as well. Tar mediates the repellent responses to cobalt and nickel ions. The properties of chimeric transducers with the periplasmic domain of Tsr and the cytoplasmic domain of Tar, or vice versa, indicate that most repellents interact with the periplasmic domain. However, changes in internal pH are sensed because they alter the extent of protonation of a histidine residue in the cytoplasmic domain and, thus, presumably the conformation of the domain.

The transducers are not present in equal numbers. Tsr is the most abundant, with an estimated 2000–2500 molecules per cell, and Tar is present in 1000–1500 copies per cell. Thus, Tsr and Tar are the major transducers. Trg and Tap occur in several hundred copies per cell and are designated as minor transducers. Binding proteins, which are present at 10,000–100,000 copies in cells induced for their synthesis, are in large molar excess over the transducers. Therefore, it is not surprising that the binding proteins, even in their closed conformation, have a relatively low affinity for the transducers.

Random distribution on the cell surface gives near-optimal efficiency of detection for a chemoreceptor. Only a thousandth of the surface must be covered for molecular capture to be half as good as if the entire surface were receptive, because diffusion causes molecules to undergo multiple collisions with a localized area of the surface. Experimental data indicate that transducers are not localized near flagellar basal bodies, a result consistent with the membrane-bound receptors being randomly dispersed.

D. Intracellular Signal Transduction

Because the cytoplasmic domain of the transducers is the source of the intracellular signal, conformational changes initiated by binding of chemoeffectors to the external receptor domain must be propagated through the membrane to the cytoplasmic domain. The transducers exist as homodimers, but no cooperativity has been seen, so signal propagation across the membrane may occur via conformational changes in only one subunit. The mechanism of transmembrane signaling is one of the least understood aspects of bacterial chemotaxis.

The products of six genes are required for chemotaxis to all attractants and repellents in *E. coli* or *S. typhimurium*. Null mutations (those leading to a complete knockout of functional gene product) in the *cheA*, *cheW*, and *cheY* genes result in cells that cannot tumble because their flagella turn only CCW, even after the addition of repellents. Cells with null mutations in *cheR* are smooth swimmers until they encounter a repellent, which causes them to tumble incessantly; they are defective in adaptation because they lack the ability to methylate the transducers. Null mutations in the *cheB* and *cheZ* genes lead to a CW-biased, tumbly phenotype; *cheB* mutants cannot demethylate the transducers and are also defective in adaptation. Both *cheB* and *cheZ* mutants respond to large jumps in attractant concentrations with CCW flagellar rotation. The *che* genes are organized into two tandem operons at 42 mintues on the *E. coli* chromosome and are expressed under the same control as late genes of the flagellar regulatory system. [*See* FLAGELLA.]

The *che* genes have been sequenced and their protein products purified. Analysis of *che* gene sequences led to the discovery of homologies with two-component signaling systems, many of which regulate the expression of multigene complexes. The components consist of a modulator, which receives and integrates environmental input, and a response regulator, which produces an output activity that is controlled by signals received from the modulator. In the chemotactic system, CheA is the modulator and CheY and CheB are response regulators.

The signaling state of transducers is determined by whether they are occupied by attractants or repellents and by their level of methylation. The signal produced by the transducers controls the rate of ATP-dependent autophosphorylation of CheA at the N-3 position of a histidine residue near its N terminus. This control requires CheW, which forms a complex with CheA and CheY. Genetic and biochemical data suggest that CheA and CheW probably interact directly with the transducers as well. Phosphorylated CheA can transfer its phosphate group to an aspartate residue in CheY. The entire phosphorylation cascade, including stimulation of CheA autophosphorylation by transducers and reversal of this stimulation by attractants, has been carried out *in vitro* with purified proteins.

The phosphorylated form of CheY (CheY-P) binds to the switch complex of the flagellar motor to stimulate CW rotation. CheZ is a phosphatase that accelerates the attractant response by removing phosphate from CheY, thereby diminishing CW rotation. CheZ may also antagonzie CheY-P action by binding to the switch to promote CCW rotation. The measured response latencies (about 200 msec) of tethered cells after the onset of attractant or repellent stimuli indicate that significant increases and decreases in the intracellular CheY-P level can occur within that time. The CCW-locked behavior of strains with null mutations in *cheA*, *cheW*, and *cheY* is due to the absence of CheY-P.

Relatively little is known about signaling in systems that do not involve the methyl-accepting signal transducers. However, it has recently been proposed that HPr (histidine protein) of the PTS is the critical link between the PTS-dependent and methylation-dependent signaling pathways. Also, in species exemplified by the gram-positive *Bacillus subtilis*, methyl-group transfer seems to be involved in signaling as well as in adaptation.

E. Adaptation

Temporal sensing of spatial gradients requires a memory with a retention time of seconds. This memory is also reflected in the adaptation that occurs over a period of tens of seconds to minutes after large step changes in chemoeffector concentration. Such stimuli are seldom encountered in the natural environment but have proven extremely useful for amplifying the adaptation process to an experimentally manageable scale.

L-methionine is required for tumbling and for adaptation to many attactants in *E. coli* because it is a precursor of *S*-adenosyl methionine, which is used as a methyl donor by CheR. Four or five specific glutamate residues in the cytoplasmic domain of the transducers are methylated to form γ-carboxyl methyl esters. Variable patterns of methylation generate the different bands seen with each transducer during sodium dodecyl sulfate–polyacrylamide gel electrophoresis. The two regions in which the methylated glutamates are located flank the highly conserved central core of the signaling domain of the transducers. The extent of methylation modulates the conformation, and therefore activity, of the signaling region. The methylated positions occur at intervals of seven residues, so they will be in position on the same face of an α-helix to serve as substrates for the methylating and demethylating enzymes.

CheR is a methyltransferase with constitutive activity. CheB is a methylesterase that releases methyl groups as methanol. CheB also serves as a deamidase, and two of the glutamate residues that are methylated are derived from glutamine by deamidation. These glutamines may balance the signaling state of transducers that are newly inserted into the membrane and have not yet equilibrated with the methylation–demethylation system.

Methylation or demethylation restores the signaling state of transducers to the prestimulus level after attractant or repellent stimuli, respectively. The time required to reach a new level of methylation after the addition of an attractant or repellent is the same as the time required for the cells to adapt and return to normal run–tumble behavior. Changes in the extent of methylation are determined by changes in the demethylation rate. Attractants transiently inhibit demethylation and deamidation by CheB, and repellents stimulate demethylation and deamidation. Adaptation is complete when receptor occupancy and methylation are in balance. In an adapted cell, the ratio of CW–CCW flagellar rotation returns to a value that produces run–tumble behavior. Methylation is not involved in adaptation of responses to oxygen and PTS substrates, which must occur by another, as yet uncharacterized, process.

Two mechanisms modulate methylation of the transducers: availability of the glutamate residues for demethylation and control of the activity of CheB. Activation of CheB in response to repellents is accomplished by phosphorylation via CheA. Inhibition of CheB activity by attractants does not require CheA but may involve CheW. The aspartate residue that is phosphorylated in CheB is in an N-terminal regulatory domain, whereas the catalytic domain of CheB is at its C terminus. There is no CheZ-like phosphatase for CheB, but spontaneous dephosphorylation is rapid.

Changes in methylesterase activity affect methylation of all transducer species during the adaptation process. Once the adapted state has been reached, methylation of attractant- or repellent-bound transducer remains higher or lower, respectively, whereas the level of methylation of other transducers returns to the unstimulated level of about one methyl group per transducer. Thus, transient changes in methylation during adaptation are caused by feedback regulation of CheB activity. The level of methylation in the adapted state, however, is determined by substrate-induced changes in the conformation of the transducers in the cytoplasmic domain, in which the glutamyl methyl esters are located.

F. Integrated Model for Chemotactic Sensing

A model incorporating the known features of chemotactic signaling in *E. coli* and *S. typhimurium* is presented in Fig. 2. In the absence of attractants or repellents, transducers maintain CheA autophosphorylation at a level that generates an intermediate level of CheY-P. As a result, cells run and tumble in a random pattern. Accumulation of excess CheY-P is blocked by phosphatase activity of CheZ. CheA also maintains as intermediate level of CheB-P, whose methylesterase activity balances methyl-

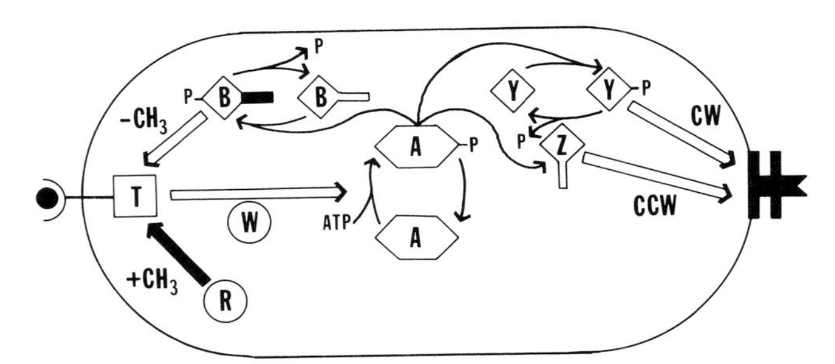

Figure 2 Schematic diagram of intracellular signaling during bacterial chemotaxis. CheA, CheB, CheR, CheW, CheY, CheZ, and a generalized transducer are indicated by the large capital letters A, B, R, W, Y, Z, and T, respectively. Phosphate is symbolized by a small capital P. The active, phosphorylated form of CheB is designated by a solid tail. Activities of CheB, CheW, CheY-P, and CheZ that are thought to be involved in regulating the activity of other components of the chemotactic system are represented by large open arrows, and the constitutive methylating activity of CheR is shown as a large filled arrow. Phosphorylation–dephosphorylation reactions are indicated by the thin, curving arrows, as is the possible regulatory interactions between CheA and CheZ. ATP, adenosine triphosphate; CCW, counter-clockwise; CW, clockwise.

group transfer by CheR so that a basal level of transducer methylation (one out of four or five possible sites) is maintained. CheB-P accumulation is limited by autodephosphorylation.

Upon addition of an attractant, ligand-bound transducer blocks autophosphorylation of CheA. CheY-P levels fall rapidly, the flagella turn CCW, and the cells run. At the same time, however, the level of CheB-P falls, although probably more slowly than CheY-P. The resulting inhibition of methylesterase leads to a global rise in the level of transducer methylation that restores an intermediate level of transducer signaling and of CheA autophosphorylation. Concentrations of CheY-P and CheB-P also return to intermediate levels, and normal run–tumble behavior resumes. Increased methylation of the attractant-bound transducer species persists to cancel the attractant-induced signal, whereas methylation of other transducer species returns to the prestimulus level.

Removal of the attractant converts the overmethylated transducer species into a CW signaler that stimulates CheA autophosphorylation. Levels of CheY-P rise to induce CW rotation and tumbling, but the overmethylated transducers are rapidly demethylated by CheB-P to return the transducers to a signaling state that maintains an intermediate level of CheA activity.

Addition of a repellent has essentially the same effect as removal of an attractant, with adaptation occurring because the transducer species mediating the response is demethylated to a level of less than one methyl group per transducer. Similarly, removal of a repellent acts much like addition of an attractant.

The combined activities of all of the transducers determine the behavior of a cell swimming in a complex chemical environment. Gradually increasing or decreasing concentrations of attractants and repellents modulate the net signal that is produced. The cell compares the integrated state of receptor occupancy, sampled over about 1 sec, with its memory of the prior state of receptor occupancy, measured over the previous 3 sec and stored as the methylation level of the transducers. If attractant occupancy outstrips methylation, or if methylation levels are low with respect to repellent occupancy, the cell extends its run until the CheR methyltransferase, aided by temporarily diminished activity of the CheB methylesterase, brings methylation into balance.

II. Amoeboid Chemotaxis

A. Functions of Amoeboid Chemotaxis

Amoeboid chemotaxis is the oriented movement of amoeboid cells in response to differences in the concentrations of specific chemicals. Although the direction of cell motion can be affected by a variety of different external agents, including temperature (thermotaxis), electric fields (galvanotaxis), light (phototaxis), and surface adhesion (haptotaxis), the response of amoeboid cells to soluble compounds (chemotaxis) has been studied the most extensively. The two amoeboid cell types whose chemotactic responses are best characterized are *D. discoideum* (a soil amoeba) and the neutrophil (a phagocytic white blood cell). This discussion is based on results obtained from studies on these two types of cells.

For cellular slime molds such as *D. discoideum,* amoeboid chemotaxis has multiple functions. The natural food source of slime molds is bacteria, and during the growth phase, sensitivity to folate allows amoebae to find bacterial colonies. This response provides a mechanism for finding food. Such a function is probably also important for other free-living amoebae. Neutrophils, on the other hand, do not use bacteria as a food source. Rather, they are involved in killing bacteria and other infectious agnets. Neutrophils and other white blood cells use chemotaxis to move rapidly to sites of infection and to carry out their various immune functions. Cellular slime molds also use chemotactic responses to organize various stages of a developmental program involving aggregation of up to 100,000 cells. Aggregation is followed by a number of morphogenetic stages ending in formation of a fruiting body. Such coordination of cell movement during development also occurs in most higher eukaryotes, including movement of neural crest cells and movement of the growth cones of axons.

However, amoeboid chemotaxis can also be important in a number of human diseases. Amoeboid parasites such as *Entamoeba histolytica* can invade the intestinal system, crawling through intestinal tissues and digesting them. Other amoeboid parasites such as *Naegleria* or *Acanthamoeba* can invade the nervous system, causing meningoencephalitis. Movement of human tumor cells may also be important in cancer. Some of the most dangerous cancers are metastatic, leaving the site of the original tumor to spread throughout the body to establish new tumors. Metastasis involves the movement of tumor cells into the circulation and subsequent movement of these cells out the circulation and into new sites.

B. Amoeboid Motility

1. Cell Motion

Amoeboid cells range in size from the very large *Amoeba proteus* (up to 400 μm long) down to white blood cells and *D. discoideum* (10–20 μm long). Unlike motility based on stable structures such as cilia or flagella, amoeboid motion utilizes structures that are transient. The cell protrudes a pseudopod, which is an extension of the cell cytoskeleton covered by the plasma membrane (Fig. 3). As the cell interior flows into the pseudopod, the rear of the cell is brought forward, resulting in a net movement of the center of mass of the cell. Multiple pseudopods can be present at the same time, competing for dominance, and eventually one wins out, with the cell interior flowing into it and new pseudopods forming from it. Pseudopods can arise from any side of the cell, and what was formerly the rear of the cell can become the front. The net result is an irregular, dynamic cell periphery.

Specific pseudopodial shapes are characteristic for particular species. Large amoebae, such as *A. proteus,* produce lobopods (large cylindrical pseudopods) in which prolonged flows of intracellular vesicles are seen. Such cytoplasmic streaming is also seen in the extended networks of the acellular slime molds of the genus *Physarum*. These species have a solid, gellike ectoplasm underlying the cell membrane. As the cell extends a pseudopod, the more fluid endoplasm from the center of the cell flows into the pseudopod and solidifies to form ectoplasm. At the rear of the cell the ectoplasm becomes fluid and flows forward. The large size of these cells facilitates clear visualization of the gel and sol states of the cytoplasm that are correlated with movement.

Smaller species, such as *D. discoideum,* display much smaller lobopods, as well as long, slender pseudopods termed filopods. These cells also show an ectoplasm–endoplasm division of the cytoplasm, but on a reduced scale. A hyaline layer of ectoplasm (which appears homogeneous in the light microscope) is directly under the cell membrane, and the internal vesicles and nucleus are surrounded by the soluble endoplasm. As a new lobopod is extended, the region directly under the membrane is hyaline.

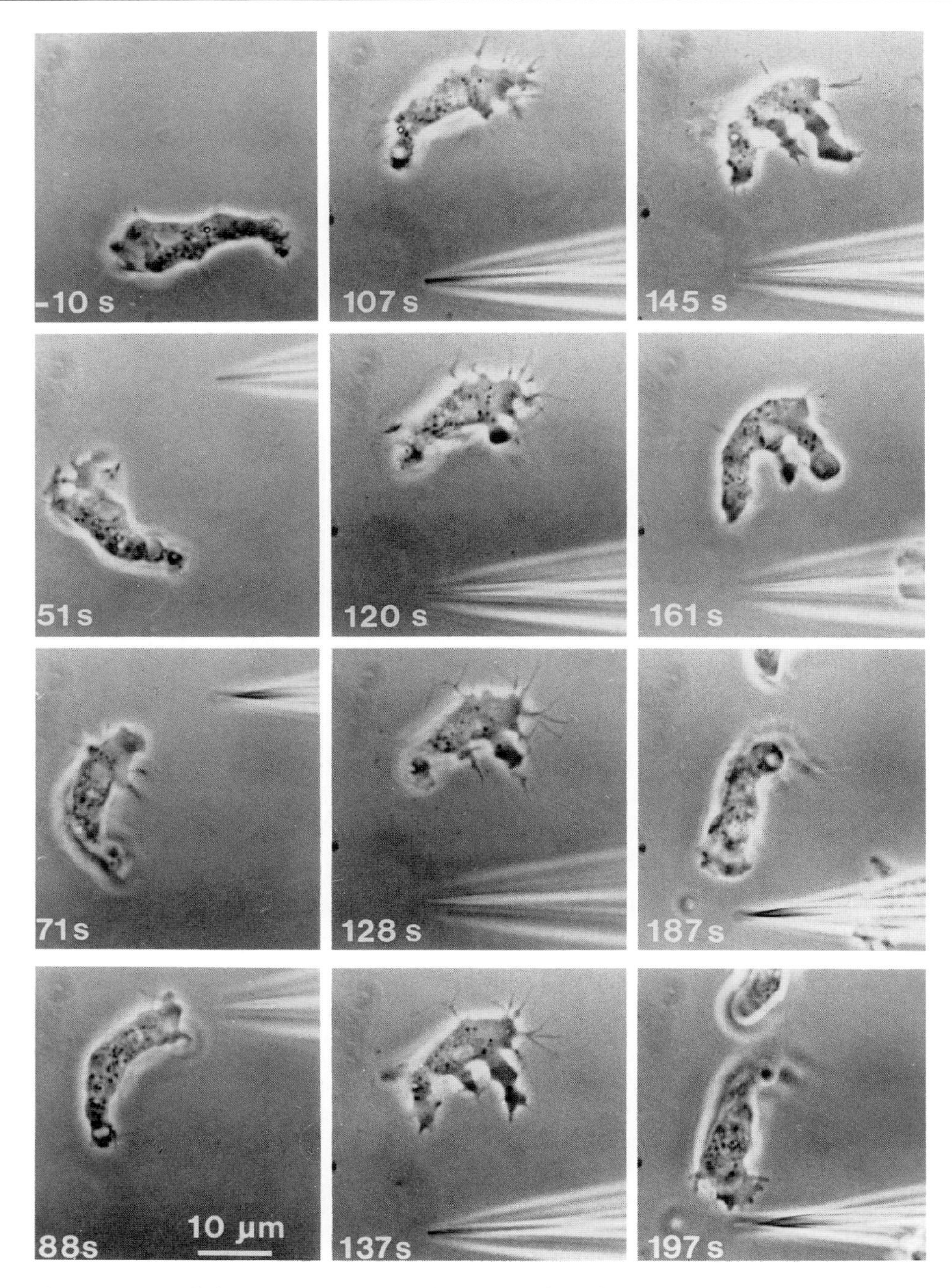

Figure 3 Typical chemotactic responses of a *Dictyostelium discoideum* cell to a micropipette filled with 100 μM cyclic adenosine monophosphate. At the beginning, the cell was moving toward the left. The micropipette was introduced at 0 sec, and the cell turned toward the tip maintaining its initial front. At 100 sec, the pipette was moved to a new position, and the cell now responded by producing multiple pseudopods. After a phase of competition, the original rear end became established as the new front. In this particular case, an aggregation-competent cell of mutant HG1132, defective in the F-actin cutting protein severin was used. The response of this cell was indistinguishable from wild-type chemotaxis. [From Segall, J. E., and Gerisch, G. (1989). *Curr. Opin. Cell Biol.* **1**, 44–50.]

Filopods can be found at both the front and rear of the cell, sometimes being extended before lobopod formation begins, and forming the final visible attachment of the cell to its previous position before the cell moves on.

When isolated and placed on artificial substrates, mammalian cells (ranging from white blood cells to neurons) tend to form relatively stable, broad, and flat lamellipods that often exhibit ruffling movements at the leading edge. Lamellipods are devoid of intracellular vesicles and organelles when viewed in the light microscope. The lamellipod is present at the front of the cell.

2. Cytoskeletal Proteins

Current evidence suggests that an actomyosin network plays a central role in generating amoeboid movement. In *D. discoideum,* actin and myosin make up 8 to 0.5%, respectively, of the total cell protein. Actin filaments are localized to the hyaline layer under the plasma membrane, whereas actin monomers are present in the endoplasm as well. In oriented, locomoting cells, actin filaments are present at high density in the leading pseudopod and are observed in thinner layers along the inner surface of the membranes lining the side and rear of the cell (Fig. 4). Myosin type II, the double-headed myosin that forms thick filaments, lines the sides and rear of the cell. Of the several isoforms of myosin type I, which is a single-headed myosin that does not form thick filaments, at least one is localized at the tips of pseudopods. The microtubule organizing center is located near the nucleus, and about 30 microtubules radiate outward from it toward the periphery.

Modulation of cell movement probably requires careful regulation of the polymerization state of actin in different regions of the cell. A number of actin-binding proteins have been isolated that could be important in controlling both the number and localization of actin filaments. Actin molecules are in equilibrium between globular, monomeric G actin and filamentous, multimeric F actin. F actin interacts with myosin to generate contraction via ATP hydrolysis. G actin-binding proteins, such as profilin, may inhibit filament formation by sequestering actin monomers.

Proteins that might accelerate actin assembly include membrane proteins such as ponticulin. Other proteins that might stabilize actin filaments under the membrane (leading to formation of the hyaline zone) include cross-linkers such as filamin, fodrin, and the 120-kDa gelation factor. Other proteins lead to bundling of actin filaments and are probably involved in the formation of filapods (which are filled with actin bundles). In addition, there are proteins that destabilize actin filaments, either by severing them (e.g., severin) or by capping the growing end (cap32/34). Finally, the actual force-generating molecules, the myosins, include both the conventional double-headed myosin type II, which can form bipolar thick filaments, and the monomeric myosin type I molecules.

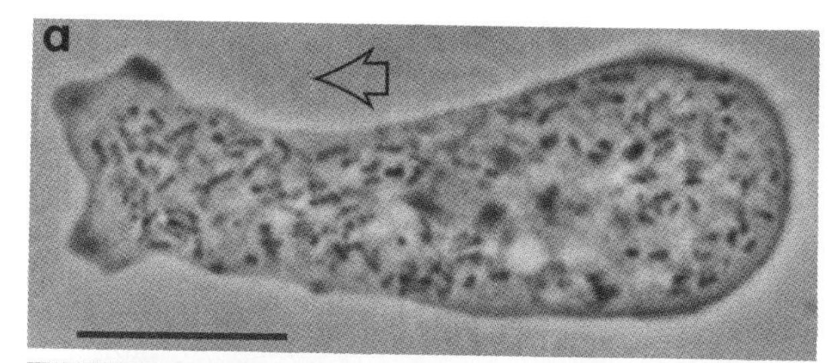

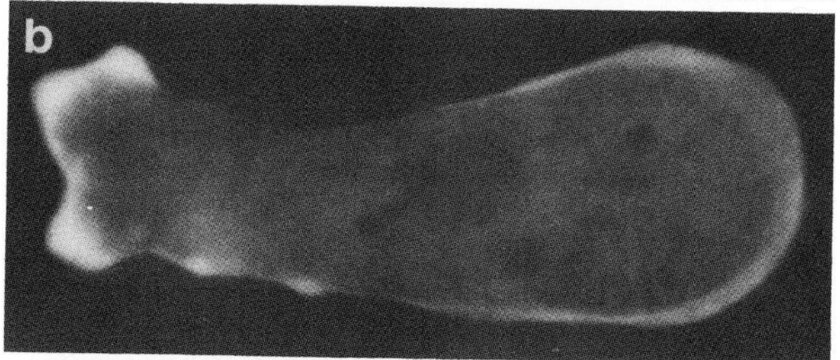

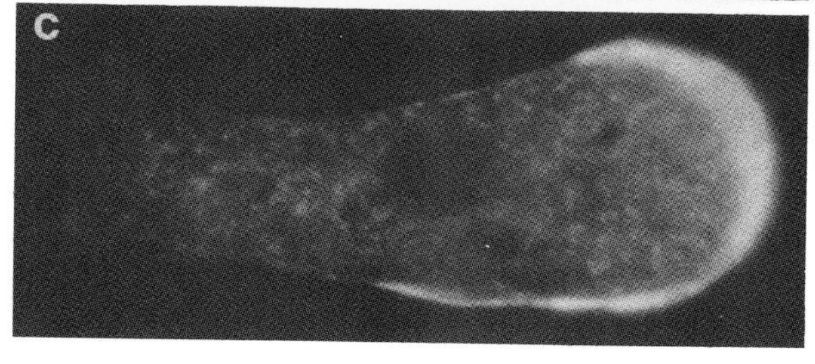

Figure 4 Double immunofluorescence of *Dictyostelium discoideum* amoebae. (a) Phase-contrast micrograph. Arrow indicates the direction of locomotion. (b) Direct immunofluorescence using TMRITC-labeled anti-*D. discoideum* actin IgG. (c) Indirect immunofluorescence using anti-*D. discoideum* myosin IgG followed by FITC-labeled goat anti-mouse IgG. Fluorescent rods in the cytoplasm may represent individual myosin-thick filaments. [From Fukui, Y., Shigehiko, Y., and Yumura, T. K. (1987). *Meth. Cell Biol.* **28,** 347–356.]

3. Cell-Substratum Adhesion

In addition to the cytoskeletal machinery necessary for the shape changes that generate motion, amoeboid cells need to attach to a substratum. The region of the cell anchored to the substratum provides the support necessary for pushing or pulling the cell mass forward. Adhesion molecules that are involved in cell motility have been best characterized for mammalian cells. Such molecules can be involved in cell–cell adhesion as well as adhesion to the substratum. For example, the integrin family consists of membrane proteins made up of dimers of α- and β-subunits. The extracellular domain contains a ligand-binding region, and the intracellular domain binds to the actin cytoskeleton via an indirect linkage involving talin, vinculin, and possibly other cytoskeletal proteins. Neutrophils from patients lacking the white blood cell integrins LFA-1, Mac-1 and p150,95 are defective in cell-substratum attachment, cell movement, and chemotaxis. The observed defects in cell movement and chemotaxis seem to be due to the inability of the cells to attach firmly to the substratum, since other responses induced by chemoattractants, such as actin polymerization, are normal.

C. Chemotactic Responses

The placement of a micropipette containing a solution of chemo-attractant in the vicinity of an amoeboid cell induces a reorientation of the cell, resulting in directed movement toward the micropipette (Fig. 3). For many species, this involves the formation of new pseudopods at the points on the cell surface that are closest to the pipette. Pseudopods already present in other regions of the cell are retracted. Cell movement is also oriented toward higher concentrations in chambers utilizing stable, linear gradients of attractants. Perhaps the most dramatic example of chemotaxis is the oriented movement of thousands of *D. discoideum* cells in streams toward an aggregation center in response to pulses of cyclic adenosine monophosphate (cAMP) relayed by the cells as they move forward.

The chemical nature of the chemoattractant varies with the species studied. For some species, only cell extracts have been used as attractants, without further characterization. Many cellular slime mold species respond to folate during the growth phase, probably as a way of finding their bacterial prey, whereas the chemoattractant used to direct aggregation in starving populations is species-dependent (cAMP for *D. discoideum*, the modified dipeptide glorin for *Polysphondelium pallidum*). Neutrophils and *A. proteus* respond to the tripeptide formylmethionine–leucine–phenylalanine, which is often found at the beginning of the signal sequences of bacterial proteins. A large number of chemoattractants, which are probably used to coordinate immune responses, have been isolated for the various white blood cell types. For example, neutrophils respond to fragment 5a of complement, and eosinophils and T cells also have specific chemoattractants. Thus, no obvious chemical characteristics identify a compound as a chemoattractant.

A feature that distinguishes bacterial and amoeboid chemotactic responses is the pattern of cell movement. In response to a strong stimulus, neutrophils or *D. discoideum* amoebae move in virtual beelines toward the source: The paths are nearly straight, with the net velocity toward the source being 80% of the total speed of the cell. Bacterial movement, on the other hand, is more diffusive, with the maximum drift rate toward the source being around 20% of the swimming speed. These distinctions in path morphology reflect inherent differences in the sensing mechanisms that operate in the two organisms. Bacteria use a purely temporal sensing mechanism, as described in the first section, whereas amoeboid cells can coordinate responses from different regions of the cell to determine the direction of motion.

Although the coordination of the entire cell is inherently a spatial mechanism, the actual measurements occurring on the cell surface can also incorporate temporal changes. Exposure of *D. discoideum* cells to rapid, uniform changes in chemoattractant concentration, which generate a spatially uniform temporal increase in chemoattractant binding to cell-surface receptors, produces changes in speed as well as changes in cell morphology. The speed drops briefly, reaching a minimum 20–30 sec after the stimulus, and recovers within 60 sec. The area of the cell first decreases, then it increases above the prestimulus value to reach a maximum 60 sec after the stimulus, and then it returns to the prestimulus value. Such changes could reflect the generation of intracellular signals responsible for cell orientation in a spatial gradient.

D. Signal Processing

1. Receptors and G proteins

Specific cell-surface binding sites for folate and cAMP (in *D. discoideum*) and f-MetLeuPhe and fragment C5a (in neutrophils) have been identified. At least some of these binding sites correspond to the receptors that detect the external chemoattractant concentration. Genes encoding a cAMP receptor from *D. discoideum* and an f-MetLeuPhe receptor from neutrophils have been cloned and sequenced. Although there is no strong homology between the two receptors, both show the motif of seven membrane-spanning α-helices that is characteristic of G protein-coupled receptors.

The C termini of the receptors have stretches of serine and threonine residues that are potential phosphorylation sites. The *D. discoideum* cAMP receptor has been shown to be phosphorylated when cells are exposed to cAMP. Phosphorylation is probably important for adaptation to background levels of stimulus.

Coupling of receptors to the cytoskeleton occurs through G proteins. *In vitro* measurements show that chemoattractant binding in isolated membranes displays dependence on guanine nucleotides, which is indicative of a receptor–G protein interaction. In *D. discoideum, fgdA* mutants are defective in chemotactic responses to cAMP and contain mutations in the gene encoding a Gα-subunit. This subunit, like the cAMP receptor, is phosphorylated when cells are exposed to cAMP. Pertussis toxin blocks chemotactic responses in neutrophils, presumably by inactivating a G protein.

2. Intracellular Signals

The cell-surface receptors measure the local concentration of chemoattractant, and this information is transduced into the local activation of G proteins. These activated G proteins are involved in the first step in a signal transduction chain that ends in the rearrangement of the cytoskeleton and cell movement. The steps that are activated by the G proteins should constitute the integration and comparison events that allow the cell to determine the orientation of the chemoattractant gradient.

In both *D. discoideum* and neutrophils, phospholipase C is activated by chemoattractants. Inositol trisphosphate and diacylglycerol are generated by the cleavage of phosphotidylinositol. Inositol trisphosphate may in turn release calcium from intracellular stores. Chemoattractants stimulate influxes of calcium ions and effluxes of protons in *D. discoideum* and neutrophils. The further consequences of these signals are not yet known.

In cellular slime molds, addition of chemoattractant stimulates synthesis of intracellular cyclic guanosine monophosphate (cGMP). Mutants of the *strF* complementation group are defective in the cGMP-specific phosphodiesterase that hydrolyzes the cGMP. In these mutants, the duration of the increase in intracellular cGMP that is stimulated by addition of chemoattractant is prolonged from 30 sec in the wild type to over 3 min. This prolonged increase in intracellular cGMP is accompanied by a prolonged drop in cell speed. Intracellular cGMP may function as a global suppressor of cell movement. In neutrophils, chemoattractants activate synthesis of cAMP, but blocking cAMP synthesis does not affect the chemotactic response.

3. Changes in the Cytoskeleton

Addition of chemoattractant to *D. discoideum* or neutrophils leads to a rapid increase in actin polymerization. In *D. discoideum,* the amount of F actin peaks at about 10 sec, drops down to near prestimulus levels at 30 sec, and then increases in a broader peak at 60 sec before returning to the prestimulus level. In neutrophils, the initial increase in F actin persists for several minutes. The mechanism of control for actin polymerization is not known, but it involves the generation or uncapping of new filaments that then increase in length. Because F actin makes up the meshwork of the hyaline layer that is present in newly formed pseudopods, F actin probably forms a support for the pseudopod and may be responsible for pseudopod extension.

Both the regulatory light chain and the heavy chain of myosin type II are phosphorylated in response to chemoattractants. Phosphorylation of the light chain should increase the forces generated by actomyosin complexes. Phosphorylation of the heavy chain should cause disassembly of thick filaments and reduction in the force generated by actomyosin complexes, which consist of myosin-thick filaments interacting with actin filaments. These counteracting effects could be useful if they are spatially separated. For example, phosphorylation of myosin light chain may occur in regions of the cell where contraction takes place and phosphorylation of myosin heavy chain may occur in regions that should expand.

Myosin type II most likely plays a role in contraction at the rear of the cell, because it is located in that region and because mutants of *D. discoideum* lacking myosin type II can still extend pseudopods. Such mutant cells move at about 50% the speed of wild-type cells, are more rounded in shape, and show 20% of the chemotactic orientation of wild-type cells. Myosin type II and actomyosin generate forces that significantly improve cell movement and orientation but are not absolutely necessary for them to occur.

The roles of actin-binding proteins and myosin type I in cell movement and chemotaxis are unclear. *Dictyostelium discoideum* mutants lacking severin (an F-actin severing protein), or α-actinin or 120-kDa gelation factor (two cross-linking proteins), appear normal. Mutants lacking certain myosin type I isoforms show only subtle, if any, defects. Parallel signal pathways may act on the cytoskeleton to generate spatial coordination that is important for amoeboid chemotaxis. Defects in any one pathway may provide only a quantitative reduction in chemotactic performance, as in the myosin type II mutants.

III. Conclusion

The two model systems described provide complementary information regarding signal transduction in chemotaxis. The bacterial system has supplied information about receptor structure and function, intracellular signaling, and second-messenger action. Chimeras between the Tar protein of *S. typhimurium* and mammalian insulin receptor exhibit signaling function, suggesting a common motif for transmembrane signaling between prokaryotes and eukaryotes. No direct eukaryotic analogs of the two-component systems have been described yet, but activation of CheY by phosphorylation and of H-*ras* by GTP binding exhibit structural and functional similarities.

The signal transduction machinery of *D. discoideum,* on the other hand, utilizes a sophisticated receptor-signaling mechanism mediated by G proteins. The distinguishing feature of this system is that the transduction chain must preserve and possibly amplify the differences in the signals coming from different parts of the cell. The part of the cell that is on the high side of an attractant gradient is more likely to extend a pseudopod and direct cell movement up the gradient. Multiple intracellular signals are probably integrated to provide this capability.

The differences in the design of the two signal transduction pathways reflect the different requirements imposed by the motility of the cells. Bacteria are small and move rapidly, whereas amoebae are large and move slowly. In *E. coli,* signals from multiple receptors converge to modulate the concentration of a single component—phosphorylated CheY protein. Bacterial motility is also simpler in that the flagellar motor has a binary CW or CCW output. A single pathway that measures the average occupancy of the receptors over the entire cell and compares it with the methylation state of the transducers is sufficient to determine how the switching probability, and thus the CW–CCW ratio, of the motor should be set.

In contrast, control of amoeboid motility appears to involve a receptor producing multiple outputs, which involve at least actin and myosin. The localization of these two proteins is different, although overlapping, and reflects the orientation of the cell in a spatial gradient. Temporal changes in attractant concentration also lead to multiple responses and adaptation in amoebae. The interplay of responses to spatial and temporal gradients must generate localized extensions of pseudopods up an attractant gradient.

Microbial behavior has provided insights into the cellular and molecular mechanisms by which all cells respond to their chemical environments. It has also produced some major surprises, like the reversible rotary motor of the bacterial flagellum and the myosin-independent machinery of amoeboid cell movement. Further revelations of common themes in signal transduction, and the discovery of other unique structures and capabilities involved in motility and chemotaxis, will continue to enliven studies of the sensory biology of microorganisms.

Bibliography

Devreotes, P. N., and Zigmond, S. H. (1988). *Annu. Rev. Cell Biol.* **4,** 649–686.

Luna, E. J. and Condeelis, J. S. (1990). *Dev. Genet.* **11,** 328–332.

Macnab, R. M. (1987). Motility and chemotaxis. *In* "*Escherichia coli* and *Salmonella typhimurium.* Cellular and Molecular Biology," Vol. 1, (F. C. Neidhardt, J. L. Ingraham, K. B. Low, B. Magasanik, M. Schaechter, and H. E. Umbarger, eds.), pp. 732–759. American Society for Microbiology, Washington, D.C.

Manson, M. D. (1992). *Adv. Microb. Physiol.* **33,** 277–346.

Segall, J. E. (1990). Mutational studies of amoeboid chemotaxis using *Dictyostelium discoideum. In* "Biology of the Chemotactic Response" (J. P. Armitage and J. M. Lackie, eds.), pp. 241–272. Cambridge University Press, Cambridge, England.

Spudich, J. A. (1989). *Cell Reg.* **1,** 1–11.

Stewart, R. C., and Dahlquist, F. W. (1987). *Chem. Rev.* **87,** 997–1025.

Stock, J. B., Ninfa, A. J., and Stock, A. M. (1989). *Microbiol. Rev.* **53,** 450–490.

Stockem, W., and Klopocka, W. (1988). *Int. Rev. Cytol.* **112,** 137–183.

Vandekerckhove, J. (1990). *Curr. Top. Cell Biol.* **2,** 41–50.

Chlamydial Infections, Human

Abdallah M. Isa
Tennessee State University

Glossary

Chlamydia Bacterial organisms that cause infection in humans

Conjugate Antibody labeled with fluorescein or enzyme

Substrate Chemical producing color upon action of an enzyme

CHLAMYDIAL INFECTIONS are a host of clinical syndromes caused by infection with any of the various chlamydial serotypes. Chlamydiae are a group of obligate intracellular parasites that cause a variety of clinical syndromes and, until recently, were considered viruses because of their small size and because of the difficulty associated with growing such organisms on synthetic media. For the first time ever, *Chlamydia* were grown in the laboratory in the mid-1950s in China in embryonated chicken eggs. Currently, however, Chlamydiae can be grown in the laboratory using an array of established cell lines. The nomenclature and classification of Chlamydiae has been a problem and the terminology has ranged from Bedsoniae, to the psittacosis–lymphogranuloma venereum–trachoma group, to the trachoma–inclusion conjunctivitis group. Recently, they have been recognized as a separate group and were assigned the genus *Chlamydia*. The genus *Chlamydia* is made up of three distinct species known to date: *Chlamydia psittaci, Chlamydia trachomatis,* and *Chlamydia pneumoniae*.

I. Disease Syndromes Associated with *Chlamydia*

A. Psittacosis

Psittacosis is a syndrome caused by *C. psittaci* that primarily infects avian and lower mammalian species. However, humans may develop psittacosis infection by acquiring the agent from such infected animals. Psittacosis in humans leads to pneumonia with high fever and ultimate death.

B. Trachoma

Trachoma is an ocular infection that afflicts hundreds of millions of humans worldwide. The organism is acquired by contact between the eye and infected matter, primarily discharge from infected eyes. Upon establishment of ocular infection in the upper conjunctiva, the infection is characterized by eye irritation and discharge. If not treated early, pannus develops and ultimate blindness occurs. Trachoma is caused by *C. trachomatis* serotypes A, B, Ba, and C.

C. Inclusion Conjunctivitis

Inclusion conjuctivitis infections are caused by *C. trachomatis* serotypes D–K. These infections are acquired by direct or sexual contact between normal and infected individuals. In inclusion conjunctivitis of the newborn, the organism is acquired by the newborn while passing through the birth canal of infected mothers. The disease is characterized by swelling of the eyes and purulent discharge. However, it is self-limiting and the eyes recover within 2 wk. Inclusion conjunctivitis in adults is more severe than in the newborn. The organism is acquired by direct contact between eyes of normal individuals and contaminated material from infected individuals. Piror to chlorination of public swimming pools, inclusion conjunctivitis was a common occurrence.

D. Genital Chlamydial Infections

Chlamydia are transmitted by sexual contact in which the etiologic agent is any of the serotypes D–K. This is the leading cause of nongonoccal urithritis and epidymitis in the male. In the female, however, cervicitis and acute salpingitis may occur, which may lead to sterility.

Lymphogranuloma venereum is a systemic sexually transmitted disease caused by any of the *C. trachomatis* serotypes L2 and L3. The disease in women is characterized by proctitis, rectal stricture, or rectovaginal fistula. [*See* SEXUALLY TRANSMITTED DISEASES.]

E. Chlamydial Pneumonia

Infants born to infected mothers may develop the classic inclusion conjunctivitis syndrome, chlamydial pneumonia, or both by acquiring the organism while passing through the birth canal of the infected mother. Chlamydial pneumonia is caused by the newly recognized chlamydial species (*C. pneumoniae*) and is characterized by a relatively mild but prolonged illness. Chlamydial pneumonia symptoms are more severe in young children and elderly patients and may follow a chronic afebrile course in infants <6 mo of age and may be accompanied by a cough. Many of the infants with chlamydial pneumonia may also have conjunctivitis.

II. Diagnosis

A. Culture

Chlamydial organisms may be cultured in the laboratory in embryonated chicken eggs or in cyclohexamide-treated cell lines. Infection can be ascertained by immunofluorescence (IFA).

B. Serologic Tests

There is no one single serologic assay that is useful in diagnosing all chlamydial infections. Basically, two systems are currently available: IFA and enzyme immunoassay (EIA).

1. Immunofluorescence

a. Antigen Detection

Clinical specimens suspected of having the organism are smeared onto wells in Teflon-coated slides, allowed to air dry, and then fixed with acetone. Fluorescein isothiocyanate (FITC)-labeled monoclonal antibodies to *Chlamydia* are added to the wells and incubated for 30 min at 37°C. The walls are washed with phosphate-buffered saline (PBS), blotted, and examined using a fluorescence microscope. The appearance of two or more fluorescing elementary bodies per field is considered positive for chlamydial infection.

b. Antibody Detection

i. Microimmunofluorescence The microimmunofluorescence (MIF) assay is performed by fixing purified elementary bodies in dots onto microscope slides. Serial dilutions of patient specimen (serum, tears, or exudate) are added to the fixed dots on the slide and incubated for 30–60 min in a moist chamber at 37°C. After washing with PBS, an FITC-labeled anti-IgM, -IgG, or -IgA is added and incubation is carried further. After washing, slides are examined using a fluorescence microscope. Dots showing bright fluorescence are considered positive. This method requires the laboratory to be equipped to grow the organism and to prepare the antigen slides.

ii. Indirect Immunofluorescence The principle of indirect IFA is the same as in the MIF earlier with the exception that these kits are commercially available. The chlamydial organism is grown in cell culture in monolayers on wells of Teflon-coated microscope slides. Chlamydial organisms form intracytoplasmic inclusions on the periphery of the nucleus of infected cells. Most commercially available kits employ the broad spectrum serotype L2 as the infecting agent. Serial dilutions of patient specimens (serum, tears, or exudate) are placed on individual wells and incubated to permit formation of primary antigen–antibody complex (if antibody is present in patient specimen). Positive and negative controls are tested with the patient specimens. After incubation and washing of the slides, an FITC-labeled anti-human globulin (anti-IgM, -IgG, or -IgA) is added to each well and slides reincubated. After washing and blotting, slides are examined using a fluorescence microscope. Positive tests appear as fluorescing perinuclear inclusions. The endpoint (titer) of the specimen tested is the reciprocal of the dilution used, showing bright apple-green fluorescence.

2. Enzyme Immunoassays

The EIAs employ an enzyme-labeled antibody and an enzyme substrate that produces color upon the action of the enzyme on the substrate.

a. Antigen-Capture Assay

The clinical specimen, normally on a swab or urine sediment, is boiled for 15 min in transport medium, cooled, and applied to microtiter plate wells that have already been coated with antibody to *Chlamydia*. The wells are incubated to permit formation of primary antigen–antibody reaction. After washing, an enzyme-labeled antichlamydial antibody is added to the wells and incubation is carried further. At the end of incubation and washing, an enzyme substrate is added to each well, and at the end of incubation a soluble color develops, the intensity of which is proportional to the amount of antigen present in the patient specimen. The color can be quantitated by reading spectrophotometrically. Depending on the enzyme-substrate system used, a stop reagent may be required to stop the enzyme reaction. The antigen-capture assay for *Chlamydia* described earlier is tedious and time-consuming.

b. Antibody-Capture Assay

As opposed to the previously mentioned antigen-capture assay for *Clamydia,* a simpler and less demanding assay requiring less time will be described here. Microtiter plate wells are coated with purified elementary bodies (antigen). Diluted patient specimen (serum, tears, or exudate) are added to the antigen-coated wells and incubated for 30 min at room temperature. Wells are washed and enzyme–antibody conjugate (anti-IgM, -IgG, or -IgA) is added to the wells and incubation is carried for 30 min at room temperature. After washing, an enzyme substrate is added and wells incubated for 30 min at room temperature. Development of color indicates the presence of antibodies to *Chlamydia* in the patient specimen. The intensity of color is proportional to the amount of antibody present in the patient specimen. The color intensity is quantitated spectrophotometrically. Depending on the enzyme substrate used, stop reagent may not be needed. [*See* ELISA TECHNOLOGY.]

Bibliography

Amico Laboratories Inc. (1991). Chlamydia product insert.

Goh, B. (1988). *The Practioner* **232,** 813–818.

Miller, S. T., Hamerschlang, M. R., Chirgwin, K., Rao, S. P., Roblin, P., Gelling, M., Stilerman, T., Schachter, J., and Cassell, G. (1991). *J. Ped.* **118,** 30–33.

Schachter, J. (1986). Chlamydiae. *In* "Manual Clinical Laboratory Immunology," 3rd ed. (N. R. Rose, H. Friedman, and J. L Fahey, eds.), pp. 587–592. American Society for Microbiology, Washington, D.C.

Chromosome, Bacterial

Karl Drlica
The Public Health Research Institute, and New York University

Glossary

DNA supercoiling Occurs in covalently closed, circular DNA molecules when the number of duplex turns differs from the number found in DNA molecules of the same length but containing an end that can rotate; creates strain in closed DNA molecules, and they coil; a deficiency of duplex turns generates negative supercoiling, and a surplus generates positive supercoiling

DNA topoisomerases Enzymes that change DNA topology by a breaking–rejoining mechanism; introduce and remove supercoils, tie and untie knots, and catenate and decatenate circular DNA molecules

Nucleoid Bacterial chromosome when in a compact configuration either inside a cell or as an isolated structure

Origin of replication Location on the chromosome (*oriC*) where initiation of replication occurs; for *Escherichia coli, oriC* is about 250 nucleotides long, and during initiation it specifically interacts with several proteins to form an initiation complex

THE BACTERIAL CHROMOSOME is the large circular DNA molecule present in all bacterial cells. A variety of proteins are associated with the DNA, and they are responsible for activities such as DNA replication, transcription, recombination, and packaging (the DNA is often 1000 times longer than the cell in which it resides). Replication begins at a precise point and continues bidirectionally until the two forks meet in a region 180° from the origin. Genes for related functions are often clustered on the DNA, and they tend to be oriented in the same direction as replication fork movement. Chromosomal DNA has fewer duplex turns than would be found in linear B-form DNA under the same conditions, and this places torsional (superhelical) strain on the DNA. The strain lowers energy barriers for strand separation, and in this sense the chromosome is considered to be energetically activated.

I. Historical Introduction

It was not always obvious that bacteria contain chromosomes, because their chromosomes are not separated from the cellular cytoplasm by a membrane. Nor do the chromosomes condense prior to mitosis, so they are not easy to visualize. Nevertheless, suggestions had been made as early as 1912 that bacteria might undergo mutation and that bacterial populations might respond to selective pressure. Both ideas would be consistent with bacteria having mutable genes, the functional elements of chromosomes. However, this view did not gain common currency until the mid-1940s, and even then some biologists continued to attribute changes in bacterial populations to the nongenetic phenomenon of "adaption." Evidence that bacteria undergo classical spontaneous and undirected mutation finally emerged in the late 1940s and early 1950s, thus establishing that bacteria must have chromosomes of some sort.

The chemical nature of genetic material was first uncovered by Avery and his associates with a transformation experiment in which a character for polysaccharide synthesis was transferred from one strain of pneumococcus to another using extracted DNA. At the time, the result was not universally accepted as evidence for genetic exchange, partly because the so-called transforming principle, the DNA, exerted its effect intracellularly after an unknown number of steps and partly because the identity of the ultimate target of action for the transforming principle was unknown. Moreover, there were concerns about potential residual biological activity due to minute quantities of protein contaminating the DNA preparation, and, perhaps most important, no conceptual framework explained in molecular terms how DNA could function as genetic material.

The discovery of sex in bacteria, the transfer of DNA from one bacterium to another, was more easily accepted as a form of genetic exchange. Lederberg and Tatum, building on their work with nutritional mutants, mated two different, doubly auxotrophic *Escherichia coli* strains by simple mixing. They then detected recombinant progeny as rare prototrophs able to grow on minimal medium. The requirement for cell contact convinced contemporary biologists that this mating system was an example of "true recombination" and was not simply due to "transforming substances" that mysteriously induced specific hereditary properties in bacteria. Because bacterial mating entailed transfer of large segments of genetic material from donor to recipient, Jacob and Wollman were able to establish linkage relationships among distant mutant loci by interrupting mating after various extents of transfer. It quickly became clear that the genetic information of bacteria is arranged in a linear array.

During the late 1940s and early 1950s, Chargaff's nucleotide analyses made it evident that DNA is not simply a monotonous repetition of nucleotides. This rekindled the focus on DNA as the genetic material. In 1952, Hershey and Chase announced that it is phage DNA, not protein, that is injected into bacterial cells during infection. At this point, DNA became widely accepted as the carrier of hereditary information. A year later, Watson and Crick provided a structural framework for understanding the activities of DNA. The central dogma of molecular biology emerged, a concept that still guides the study of how the chromosome replicates and expresses information.

By 1956, nucleoids, as we prefer to call bacterial chromosomes today, could be seen in living cells. Moreover, genetic analyses had established that there was no unique beginning or end to the linear linkage map of *E. coli* genes—there must be just one chromosome in a genetic sense, and it was likely to be circular. The concept of circularity was made graphically clear by Cairns. He visualized an entire DNA molecule following labeling with ^{3}H-thymidine, cell lysis, and examination by autoradiography. Occasionally, a DNA molecule remained intact through this procedure, and the resulting picture revealed circular molecules >1 mm long (about 4×10^6 base pairs). The length of the DNA of a single bacterial chromosome turned out to be about 1000 times the length of the cell from which the DNA was extracted, making it obvious that the long DNA molecule must be folded and compacted to fit inside the cell.

Our current knowledge of the chromosome involves many levels of understanding, but, due to space limitations, this article focuses on only three: the primary structure of the DNA (i.e., the arrangement of genetic information), the secondary and tertiary structure of the DNA, and selected chromosomal activities (replication, recombination, and transcription).

II. Primary Structure of Chromosomal DNA

A. Genetic Maps

In physical terms, the primary structure of chromosomal DNA is its nucleotide sequence. But a sequence is sterile information without knowledge of physiological function. Thus, in terms of information content, both the nucleotide sequence and the genetic map are essential aspects of primary structure.

Gene mapping in bacteria was originally based on the ability of an externally derived, genetically marked fragment of DNA to recombine with the homologous region of a recipient's resident DNA. The frequency at which two nearby markers recombine is roughly proportional to the distance between them. As mutations were collected for a variety of purposes, characterization of a mutation usually included determining its map position on the chromosome. The resulting genetic maps reveal relationships among genes such as operon clusters, show orientation preferences that may reflect chromosomal activities, and suggest that some chromosomal information may have been derived from plasmids and phages.

Genetic maps have been developed for a number of bacteria. The map of *E. coli* K-12 is the most nearly saturated: >1400 loci out of an estimated 3000 have been identified. *Salmonella typhimurium* LT2, another enteric organism, is a close relative of *E. coli,* and its most recent map lists 699 genes. Comparison of genetic maps of bacteria provides information on features of arrangement essential for function and helps to identify features that have evolved in different ways in distantly related organisms. For example, a comparison of the maps of *E. coli* and *S. typhimurium* shows an overall conservation of gene order even though the two species are believed to have diverged between 120 and 160 million yr ago. The maps have major differences at about 15 loci, where moieties of DNA have apparently been either inserted into or deleted from one genome or the other. The little gross rearrangement of genes in these two bacteria suggests that rearrangement, such as insertion or deletion, occurs only rarely in bacterial chromosomes or else change in gene arrangement is often deleterious. [*See* Map Locations of Bacterial Genes.]

As with *E. coli* and *S. typhimurium*, the genetic map of *Bacillus subtilis* reveals the grouping of many genes for biosynthetic or degradative pathways, presumably for purposes of coordinated regulation. Some adjacent genes have been shown to produce polycistronic messages, but whether or not *B. subtilis* has operons whose mechanisms of regulation are the same as those of *E. coli* remains to be determined.

Genetic maps of *Pseudomonas* and *Streptomyces* have revealed several intriguing relationships between gene function and location. *Pseudomonas putida* and *Pseudomonas aeruginosa* are gram-negative rods only distantly related to the enteric bacteria, and their genetic maps exhibit little similarity to the genetic arrangement found in *E. coli*. A peculiarity of gene location is shared by both *Pseudomonas* species: The great majority of the genes for biosynthetic functions and essential central metabolism are located in only about half of the total map. In contrast, genes for catabolic function are scattered throughout the genome and often occupy different loci in the two species. At least some of the catabolic clusters may have arisen by integration of plasmids that carried catabolic genes. Apparently, the loci of integration differed in the two species.

A different relationship between gene function and location is found in *Streptomyces* species. The map of the best-studied actinomycete, *Streptomyces coelicolor* A3, can be divided into four quadrants. Two contain the mapped genes, and they alternate with two genetically empty quadrants. Genes for sequential enzymes of biosynthetic pathways are often located in opposite quadrants. Whether the 180° location of metabolically related genes is relevant to genetic history, gene expression, or gene regulation is not known.

B. Physical Maps

The discovery of restriction endonucleases led to a quantum advance in genetic mapping, because these enzymes allow the accurate construction of maps in terms of nucleotide distances. The DNA of an organism can be divided into a number of large restriction fragments. Sets of discrete fragments have been cloned and used to generate encyclopedias in which the fragments are linearly ordered. In the case of *E. coli*, where many genes have been cloned, it has been possible to align the physical and genetic maps. Thus, a new mutation can now be mapped by cloning the gene followed by restriction analysis and hybridization to a set of ordered restriction fragments. This method is simpler and more accurate than the older methods employing recombination frequencies.

It is important to emphasize that there will not be a single restriction map for a given strain of bacterium, much less for a given species, because differences occur within different isolates of a single strain. For example, the species *E. coli* includes many laboratory strains (K-12, B, C, $15T^-$, and W) as well as numerous natural variants and sublines with different genetic histories. Genetic differences, as small as the alteration of a single nucleotide, can result in restriction fragment-length polymorphisms through abolition or creation of restriction sites.

It is now possible to generate large DNA fragments using restriction endonucleases that cut infrequently, and these fragments can be separated by pulsed-field gel electrophoresis (PFGE). This makes it possible to accurately determine genome size. Bacterial genomes vary over a wide range as evidenced by the following measurements: *S. aureus*, 2750 kilobase pairs (kbp); *E. coli*, 4700 kbp; *B. subtilis*, about 5000 kbp; *P. aeruginosa* PAO, 5300 kbp; and *Streptomyces* species, 6000–9000 kbp. Although different species show large differences, different strains generally vary within only 20%. Indeed, some *Streptomyces* mutants that have suffered large deletions often contain amplifications that tend to preserve a characteristic genome size.

Practical sequencing methods became available in the late 1970s, and bacterial nucleotide sequences began emerging at an ever-increasing rate. One-fifth of the entire *E. coli* genome has now been sequenced and entered into Genbank. Before long, information on the primary structure of the chromsome should be available as a flexible, integrated system providing data at three levels: (1) the genetic map, with the genes and their map locations correlated with the role of the gene product in cell metabolism, structure, or regulation, (2) the physical map in terms of locations of restriction sites, and (3) the nucleotide and amino acid sequences.

III. Secondary and Tertiary Structure of DNA

DNA secondary structure deals with the relationship of one strand of DNA to the other. The B-form of DNA proposed by Watson and Crick has dominated thinking about DNA; however, a variety of alternate forms have been constructed, and left-handed structures can be detected in plasmids in living cells. Because the alternate forms are often generated by specific nucleotide sequences, it is appealing to think that they may act as signals for specific chromosomal activities.

Once it was recognized that the chromosome is considerably longer than the cell, it became obvious that there must be tertiary structure just to fit the DNA into the cell. Early cytological studies of living cells using phase-contrast microscopy and proper adjustment of the refractive index of the external medium established that bacterial DNA is constrained into compact structures. When the DNA in living cells is stained, confocal scanning laser microscopy indicates that the nucleoid has a relatively smooth border (Fig. 1A). Fixation alters the structure (Fig. 1B), making more detailed electron microscopic studies difficult to interpret. However, a rapid freezing method has been developed for electron microscopy that also allows the DNA to be seen as distributed over most of the cell. The crucial issue is how chromosome structure (or movement) allows genes in the interior of the nucleoid to have access to the coupled transcription–translation machinery. If

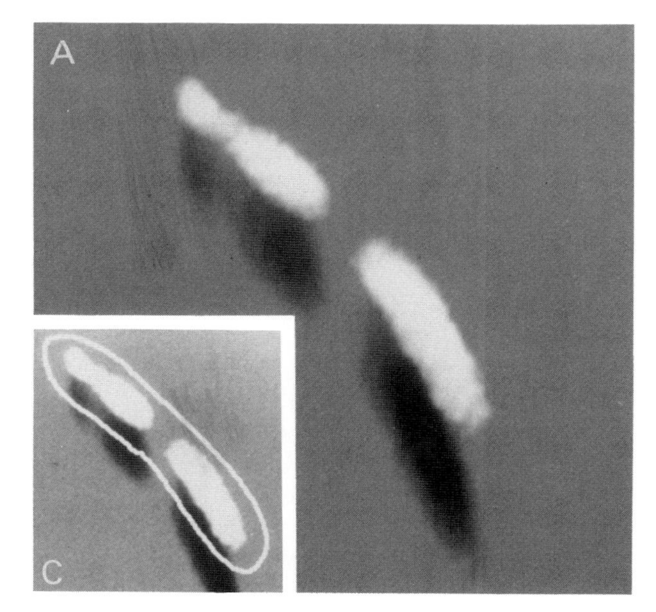

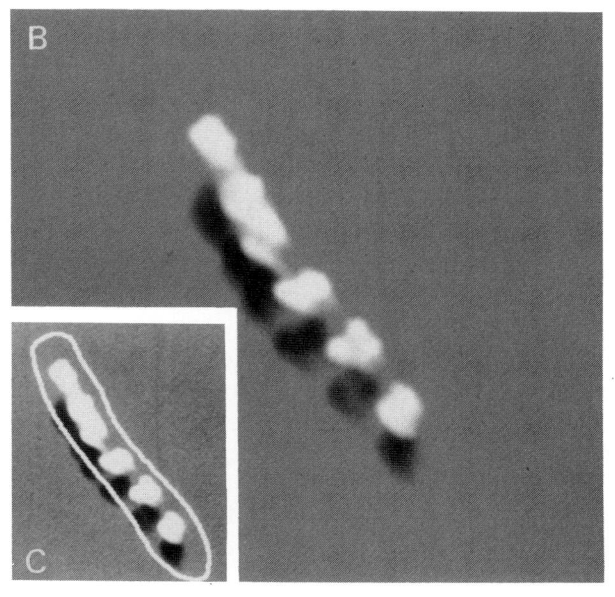

Figure 1 Bacterial nucleoids of *Escherichia coli* K-12 were visualized in a confocal scanning laser microscope. Elongated cells were obtained by growth in broth containing a low concentration (15 m*M*) of NaCl. Then, the nucleoids were stained with the DNA-specific fluorochrome DAPI (0.1 μg/ml) added to the growth medium. Under these conditions the stain had no effect on growth. The cells were observed either alive (A) or after fixation with 0.1% osmium tetroxide (B). Because the cell boundary is not easily visualized, it has been sketched in for reference (C). Multiple nucleoids were present because these fast-growing cells contain DNA in a state of multifork replication. In live cells, the nucleoid has a cloudlike appearance and a smooth boundary with the cytoplasm (protuberances, if present, would be <200 nm). (A, B) Magnification: 9000×. [Photos courtesy of Dr. Conrad Woldringh, Department of Molecular Cell Biology, University of Amsterdam, The Netherlands.]

there are deep clefts, then virtually all of the DNA would be exposed to the cytoplasm.

A. DNA Supercoiling

The bacterial chromosome can be isolated, and the resulting nucleoid contains intact DNA. An example is shown in Fig. 2. To maintain the DNA in a compact, manageable state, the isolation methods have involved high counterion concentrations; consequently, they have not proven very practical for the study of protein–DNA interactions. However, isolated nucleoids have been useful in the study of chromosome topology because extraction methods that do not break DNA molecules leave linking relationships unaffected.

When circular DNA molecules are extracted from bacterial cells, they have a deficiency of duplex turns relative to linear DNAs of the same length. This deficiency places strain on the DNA, causing it to coil (see Fig. 2). Coiling arising from a deficiency of duplex turns is loosely referred to as negative supercoiling (an excess of duplex turns would give rise to positive supercoiling). The strain, and thus supercoiling, is spontaneously relieved (relaxed) by nicks or breaks in the DNA that allow strand rotation; consequently, supercoiling is found only in DNA molecules that are circular or are otherwise constrained so the strands cannot rotate.

Because processes that separate DNA strands relieve negative superhelical strain, such processes will tend to occur more readily in supercoiled than in relaxed DNA. Among these are activities of DNA such as replication and transcription. Negative supercoiling also affects the three-dimensional configuration of DNA, facilitating loop formation and wrapping around proteins (DNA wrapped into a left-handed toroidal coil is topologically equivalent to DNA containing a negative supercoil). In a sense, negatively supercoiled DNA is energetically activated.

Soon after nucleoids were first isolated, it was discovered that they contain negatively supercoiled DNA. This feature was revealed by the ability of intercalating dyes to alter the hydrodynamic properties of nucleoids. Negative supercoils can be removed by intercalating dyes, and most comparisons of supercoiling in isolated DNA involve titration by these dyes. Because multiple nicks are required to relax the supercoils, it soon became apparent that the DNA is constrained into topologically independent domains. At first, these concepts suffered

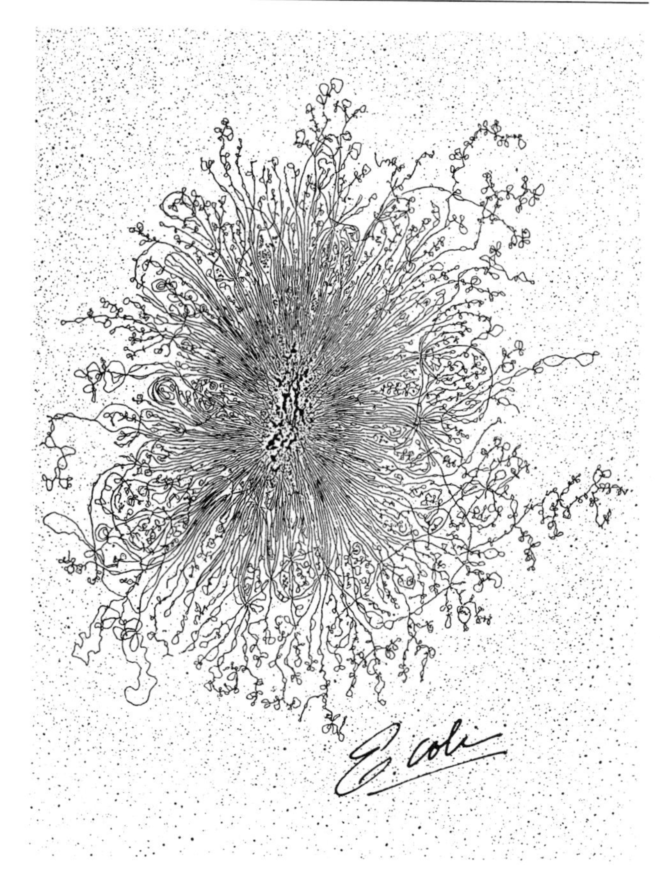

Figure 2 Electron micrograph of a purified, surface-spread *E. coli* chromosome by Ruth Kavenoff and Brian Bowen. Line under *E. coli* signature represents 2.5 μm. [© with all rights reserved by DesignerGenes Posters, Postcards, T-shirts etc, P.O. Box 100, Del Mar, California 92014.]

from uncertainties inherent in cell-fractionation strategies: Supercoiling and loops could have arisen from changes in protein-DNA and RNA–DNA interactions occurring during nucleoid isolation. Eventually, however, superhelical tension and topological domains were also detected in living cells. Supercoiling measured inside cells is probably less than half that which is measured *in vitro,* probably because proteins constrain some of the intracellular coils. The identity of these proteins is still unknown, as is the mechanism responsible for establishing the topologically independent domains.

B. DNA Topoisomerases and the Control of Supercoiling

Topoisomerases, enzymes that alter supercoiling, were discovered in the 1970s. They alter the number of turns in DNA by a strand-breaking and -rejoining process. An example of topoisomerase action is shown in Fig. 3, where both strands of the double helix are broken, a duplex is passed through the break, and then the break is sealed. Such a reaction mechanism also allows DNA knots to be tied or untied and DNA circles to be catenated or decatenated. Mutations and inhibitors became available for perturbing topoisomerase activities in living cells, and these perturbations were seen to affect many chromosomal activities.

1. Gyrase and Topoisomerase I

Two bacterial topoisomerases are involved in the control of supercoiling. One is DNA gyrase, the product of the *gyrA* and *gyrB* genes. *In vitro,* this enzyme hydrolyzes adenosine triphosphate (ATP) and introduces negative supercoils by the double-strand passage mechanism outlined in Fig. 3. In the absence of ATP, gyrase removes supercoils. Two classes of antibiotic have been discovered that inhibit gyrase. Members of the quinolone group trap a reaction intermediate in which the DNA is broken and gyrase is covalently bound to the DNA. They are useful for studying the location of gyrase–DNA interactions. The coumarins interfere with the gyrase–ATP interaction, and at moderate to high concentration they cause gyrase to relax DNA. They have been useful for examining how the loss of supercoils affects activities of DNA such as replication, transcription, and recombination. Soon after gyrase was discovered, the inhibitors were found to block the introduction of supercoils into bacteriophage λ DNA upon superinfection of a lysogen. This clearly established gyrase as a source of negative supercoiling.

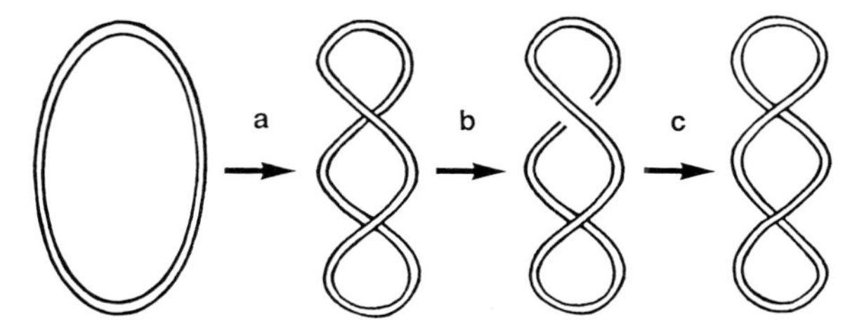

Figure 3 Negative supercoils. Gyrase generates negative supercoils by passing one duplex strand through the other. In this scheme, the enzyme binds to the DNA and creates a positive and a negative node (a). At the upper, positive node, the duplex is broken (b). The bottom strand is then passed through the break, which is then sealed (c). [From *Trends in Genetics* **6,** 433–437 (1990) as adapted from N. R. Cozzarell, *Science* **207,** 953–960 (1980).]

The second enzyme is topoisomerase I, which is encoded by *topA*. *In vitro* this enzymes relaxes negative supercoils using a strand-passage mechanism in which only one strand of the DNA is broken. A point mutation in *topA* leads to abnormally high levels of supercoiling, consistent with the idea that topoisomerase I modulates the supercoiling effect of gyrase. In *E. coli,* deletion of *topA* causes cells to grow poorly. Normal cell growth is restored by compensatory mutations, many of which map in the gyrase genes and lower supercoiling. Thus, there is little doubt that supercoiling is physiologically important.

Two aspects of the topoisomerases tend to reduce variation in supercoiling under stable growth conditions. One is the response of the enzymes to the topological state of the DNA. Gyrase is more active on a relaxed DNA substrate and topoisomerase I on a more negatively supercoiled one. Another is the homeostatic effect of supercoiling on expression of the *gyrA, gyrB,* and *topA* genes. Lowering negative supercoiling raises gyrase expression and lowers topoisomerase I expression. Raising supercoiling raises levels of topoisomerase I expression.

Bacteria also contain two other topoisomerases, topoisomerase III (the product of the *topB* gene) and topoisomerase IV (the product of the *parC* and *parE* genes). Both apparently can relax DNA *in vitro,* but at present no physiological evidence indicates that either plays a role in the control of DNA supercoiling. Topoisomerase III has a potent decatenating activity, which might be useful in recombination. Mutations in the *par* genes appear to prevent proper segregation of replicated daughter chromosomes, suggesting a role for topoisomerase IV in that process.

2. Cellular Energetics and DNA Supercoiling

Because ATP is required for gyrase to introduce negative supercoils into DNA and because in the absence of ATP gyrase removes supercoils, cellular energetics also probably affects supercoiling. *In vitro,* the ratio of ATP to adenosine diphosphate (ADP) strongly influences the level of supercoiling reached in the presence of purified gyrase, regardless of whether gyrase is introducing or removing supercoils. Higher values for the ratio generate greater negative superhelix density. Thus, ADP appears to interfere with the supercoiling interaction of gyrase with DNA while allowing a competing relaxing reaction to occur.

Several physiological correlations between supercoiling and ATP/ADP suggest that intracellular DNA supercoiling might also depend on ATP/ADP, probably through its effect on the balance between the supercoiling and relaxing activities of gyrase. One correlation was observed shortly after lowering oxygen tension: A parallel decline in ATP/ADP and supercoiling occurred. Later, both ATP/ADP and supercoiling increased in a nonparallel manner, eventually reaching a situation in which both were higher than under aerobic conditions. The quantitative relationship between DNA supercoiling and ATP/ADP before and after the drop induced by anaerobiosis was similar to the relationship observed with purified gyrase when ATP/ADP was varied. Another correlation was seen when cells were suddenly exposed to high concentrations of sodium chloride. Both ATP/ADP and supercoiling increased rapidly. After a few minutes, both gradually declined to a steady-state level that was higher than at low-salt concentration.

No evidence indicates that topoisomerase I plays a role in the influence of adenine nucleotide concentration on DNA supercoiling. However, it could be that normal intracellular levels of ATP/ADP lead to unacceptably high levels of supercoiling. Then an additional relaxing activity, effective at high levels of supercoiling, would be necessary to modulate the effect of gyrase. Because *topA* mutants can exhibit elevated levels of supercoiling, topoisomerase I is the most likely enzyme to perform such a relaxing function. At low levels of supercoiling, topoisomerase I is rather ineffective *in vitro,* and its relaxing activity is not apparent *in vivo* (the presence of topoisomerase I has little effect on the rate of relaxation induced *in vivo* by the coumarin inhibitors of gyrase).

3. Transcription and Supercoiling

The interaction between supercoiling and transcription is complex, and each influences the other. Negative supercoiling is expected to facilitate the DNA strand separation associated with initiation of transcription, and supercoiling should enhance expression from some genes. Many biochemical studies support this idea. However, moderate and high levels of supercoiling frequently inhibit initiation of transcription, presumably by interferring with proper promoter recognition by RNA polymerase. Thus, there are optimal levels of supercoiling for transcription, and they vary from one gene to another. This means that perturbations of gyrase that relax DNA inside cells will cause expression to increase, decrease, or remain unchanged depending on the gene examined.

Transcription also affects supercoiling. This was first observed with plasmid DNA isolated from *topA* mutants: The supercoiling of pBR322 DNA was much more negative than that of a closely related plasmid unable to transcribe the *tet* gene. It was subsequently proposed that tracking of the transcription complex along DNA generates positive supercoils ahead of the complex and negative supercoils behind. Topoisomerase I would normally remove the negative supercoils and gyrase the positive ones. Thus, negative supercoils will accumulate in a *topA* mutant, and positive supercoils will accumulate when gyrase is inhibited.

If the topoisomerases are balanced and suitably active, transcription and similar translocation processes would have only a very transient and local effect on supercoiling. However, cases have been found with wild-type cells where induction of very high levels of transcription from a plasmid results in abnormally high levels of negative supercoiling. Thus, the corrective action of gyrase and topoisomerase I can be overwhelmed.

It is expected that transcriptional effects on supercoiling are more pronounced for some genes than for others. One contributing factor is anchorage of the transcribed gene so that rotation of the transcription complex around the DNA is retarded. Anchorage effects may be particularly evident with genes encoding membrane proteins where the nascent polypeptide chains may be simultaneously bound to membrane and ribosomes. For example, with pBR322 the *tet* gene encodes a membrane protein, and deletion of the translation signals from *tet* lowers the abnormally high level of negative supercoiling found in a *topA* mutant. Another factor is the orientation of the gene relative to adjacent transcription units. Convergent or divergent units will tend to focus their relaxing or supercoiling effects on the intervening regions. Experimentally, changing the orientation of transcription units in plasmids does affect supercoiling. Thus, in special situations it is likely that transcription influences local supercoiling, and now good evidence indicates that this is actually the case in a plasmid.

Transcription-mediated changes in supercoiling provide two potentially important elements to the control of supercoiling. First, they provide a way for specific regions of a DNA molecule to have levels of supercoiling that differ greatly from average values.

Second, they provide a way for supercoiling to change in one region of the chromsome without affecting another.

4. Interaction of Gyrase with DNA

Knowing the rules determining when and where the topoisomerases interact with DNA is important for understanding how they control supercoiling. Gyrase–DNA interactions have been investigated in living cells by using the quinolone class of gyrase inhibitor to trap gyrase on DNA. Subsequent treatment with detergent reveals broken DNA, and analysis of the DNA fragments provides inferences about the relative distribution of gyrase on DNA. There are many sites on a DNA molecule where gyrase interacts, perhaps as many as 10,000 on the *E. coli* chromosome. This would give gyrase access to virtually all of the genome. Nucleotide sequence recognition by gyrase plays a role in determining the locations of the interactions, but DNA conformation is also likely to be important. For example, in a plasmid model system, induction of transcription enhanced the gyrase–DNA interactions located between the 3′ ends of divergent transcription units, a region where transcription was expected to cause local relaxation (or positive supercoiling).

A small number of high-affinity sites that serve special purposes may be superimposed on a background of low-specificity sites. One example of such a site has been found in the genome of bacteriophage Mu, which contains a strong, centrally located gyrase-binding site. Mutants of Mu, which could grow in an *E. coli gyrB* mutant that normally prevents Mu replication, had single nucleotide changes that increased the efficiency of gyrase interaction with the site. Deletion of the site delayed the onset of phage replication. Another example has been found in the *par* locus of plasmid pSC101. This 400-bp locus is required for plasmid stability, and it contains a site for which gyrase has a 20–30-fold higher affinity than for the rest of the plasmid. Mutation of this site lowers plasmid supercoiling in some hosts. Thus, strong sites can affect supercoiling; whether or not they are also important for DNA folding has yet to be determined.

High-affinity sites on the chromosome are not as clearly defined. Saturating concentrations of quinolone cleave the chromosome into about 50 pieces, and analysis by PFGE suggests that some of these pieces may arise from cleavage at specific sites. Unrelated experiments have led to the identification of a family of high-affinity gyrase sites called REP sequences. Members of this family are slightly >30 bp long and tend to occur in clusters called REP elements. REP elements, which contain 2–4 REP sequences, are present in 100–200 copies per chromosome, always in intergenic regions. Gyrase preferentially binds to DNA containing an REP element, and in the presence of a quinolone, gyrase generates DNA cleavage near and within the element. The function of the REP sequences and other high-affinity sites on the chromosome is still undefined.

5. Extracellular Environment and Supercoiling

Because DNA supercoiling can have a profound effect on the expression of numerous genes, it has been attractive to imagine that DNA supercoiling might be sensitive to the environment in a way that would facilitate bacterial adaptation to changing environments. The first hint that this might be the case came from a report that some mutations that prevent anaerobic growth map at or near the gyrase genes. Subsequently, it was shown that supercoiling differs in plasmids extracted from cells grown overnight under aerobic or anaerobic conditions and that supercoiling is more negative in chromosomes extracted from cells growing anaerobically. Plasmid supercoiling also becomes more negative when growth medium osmolarity is increased. Thus, certain environmental conditions apparently affect the regulation of DNA supercoiling.

C. Histonelike Proteins and Constrained Supercoils

Inside cells, the deficiency of duplex turns is partitioned into at least two components. In one component, which represents about 40% of the total, the deficiency is free to exert superhelical tension that can be detected by supercoiling assays. In the other, no supercoiling is detected, presumably because the DNA is either denatured by proteins (denaturation removes negative supercoils) and/or wrapped around proteins (wrapped DNA constrained by proteins could be unresponsive to probes of supercoiling). The success of studies with eukaryotic chromatin spurred searches for nucleosomes, the beadlike structures in which about 200 bp of DNA is wrapped by protein. Nucleosomelike particles have not been easily isolated from bacteria, so if they are present, they may be quite labile under ordinary extraction procedures.

Among the proteins that wrap DNA *in vitro* is the histonelike protein called HU. This protein has attracted attention because it is abundant and because it forms nucleosomelike structures *in vitro*. It appears to wrap DNA and facilitate formation of specific DNA structures involved in site-specific recognition of DNA sequences. If protein–DNA interactions of this type are dynamic, as *in vitro* studies have suggested for HU, they could have important effects on controlling the level of free superhelical tension. So far, cell fractionation strategies have failed to establish a general role for HU analogous to that of eukaryotic histones. Indeed, electron microscopic studies suggest that HU may not be involved in packaging the whole chromosome. Instead, its primary role may be to facilitate transient bending and looping. Short-range DNA looping is emerging as an important factor, particularly with respect to the control of transcription.

IV. Chromosome Activities

A. DNA Replication

The major features of chromosome replication have been established for many years. Semiconservative replication was demonstrated by density-shift experiments in 1958, and a few years later the autoradiograms prepared by Cairns revealed a partially replicated circle containing a large replication "eye." The possibility arose that the entire chromosome might be replicated by two forks moving in opposite directions. In the early 1970s, it became clear that replication begins at a fixed origin (*oriC*) and that the two forks do indeed proceed in opposite directions until they reach a terminus 180° around the chromosome. For *E. coli,* this was established both by a declining gradient of gene dose as markers became farther from the origin and by the sequence of replication of prophage DNA located at different chromosomal positions. [*See* DNA Replication.]

Bacterial chromosomes can contain more than one pair of replication forks, and the presence of multiple forks allows *E. coli* cells to double at 20-min intervals even though the forks require 40 min to traverse the chromosome. At rapid growth rates, there can be four chromosomes replicating to form eight at the time of cell division. Thus, *E. coli* can be thought of as a polyploid organism, even though it is genetically haploid.

Initiation of replication has long been a focus of attention, for the understanding of its control is expected to lead to new knowledge about the regulation of the cell cycle. Early in the study of initiation, heat-sensitive mutations were obtained in genes called *dnaA*, *dnaB*, and *dnaC*. They made it possible to uncouple initiation from the elongation phase of replication. Then, the origin was cloned by its ability to confer replication proficiency to an *ori*-minus plasmid. The availability of the origin on a small piece of DNA and the availability of purified initiation proteins allowed the development of an *in vitro* initiation system. Based on this, we have learned that initiation involves the specific binding of the DnaA protein to *oriC*. If the origin DNA is negatively supercoiled, the DNA becomes wrapped around a complex of proteins that includes the histonelike protein HU. Strand separation then occurs, and the replication bubble enlarges as priming and DNA synthesis begin.

Initiation of replication requires gyrase activity and negative supercoiling, and that requirement provides a *raison d'etre* for the chromosome being divided into many topologically independent domains. If the chromosome were but a single domain, the nicks and gaps that follow the replication forks would prevent the introduction of supercoils necessary to initiate the next round of replication: Multifork replication could not exist and minimal bacterial doubling times would be increased to the time required for a complete round of replication (about 40 min for *E. coli* at 37°C).

As replication forks approach the terminus of replication, they slow drastically. This feature was used to locate the terminus in *E. coli* by determining the direction of replication of a set of prophages of known orientation and map position. There are nucleotide sequences that halt fork movement in an orientation-specific manner, apparently acting as binding sites for a termination protein encoded by a nearby gene. In *E. coli,* the terminus seems to act as a large trap for the two replication forks. The traps are flanked by zones containing additional terminators arranged to block replication forks that pass through the terminus and head back toward the origin.

At the end of a round of replication, the chromosomes must segregate to the daughter cells. Catenanes appear to be unlinked by gyrase, but what happens next is largely unknown. Segregation has long been thought to involve attachment to the membrane, and indeed DNA extracted from bacteria usually has a membrane fragment bound to it. The origin of replication has a particularly high affinity for the membrane, especially when hemimethylated. Methylation of only one DNA strand is an indicator of newly replicated DNA, and this property could be important in the attachment of the chromosome to a piece of membrane that might pull the daughter chromosomes apart. The forces involved in separating the chromosomes have not been clearly defined, and ideas involving passive diffusion or charge density of the DNA are still being considered.

B. Genetic Recombination

As pointed out earlier, recombination played an important role in determining the arrangement of genetic information on the chromosome. Soon after the conjugation system was established, transduction was demonstrated in *S. typhimurium*. In this process, a small piece of donor DNA is delivered into the recipient by a bacteriophage vector. By the mid-1950s, three systems of genetic exchange, transformation, conjugation, and transduction had been discovered, and the study of bacterial genetics was well under way. [*See* RECOMBINANT DNA, BASIC PROCEDURES.]

Two types of recombination have been identified, homologous recombination and site-specific recombination. Homologous recombination entails crossover between DNA molecules of similar nucleotide sequence and occurs throughout the chromosome; thus, it has been particularly useful for genetic mapping. Once homologous recombination could be measured, it became possible to screen for mutants that failed to carry it out. The gene called *recA* was identified, and soon a number of related genes were discovered whose products participate in several pathways of homologous recombination as well as in pathways of DNA repair. In contrast, site-specific recombination occurs only at specific nucleotide sequences that are recognized by particular recombination proteins. Examples of site-specific recombination include bacteriophage integration, phase variation, and certain types of transposition. Distinguishing whether homologous or site-specific recombination drives a particular genetic rearrangement is central to understanding the rearrangement and the context within which it occurs.

A variety of infrequent events lead to chromosomal rearrangement, and they vary in frequency of occurrence and in effect on cell growth. By appreciating the underlying reasons for the difference between neutral and deleterious rearrangements, we expect to define more clearly aspects of genetic organization important to cell function. On the smallest scale are short stretches or repeated nucleotides that provide the opportunity for slippage and the generation of short internal duplications following replication and recombination. Sequence comparisons suggest that slippage may play a role in the diversification of individual genes.

Recombination between repeated sequences within and between bacterial chromosomes can generate deletions, duplications, transpositions, and inversions. Duplications and deletions arise when the repeated sequences are in direct orientation; inversions arise when the repeat is inverted. Examples are found with the seven *rrn* clusters, sets of similar assemblies of ribosomal and transfer RNA genes that undergo recombination and generate a variety of rearrangements. Other repeated sequences that function to generate rearrangements are the *rhs* loci (recombination hot spot), duplicate IS5 sequences, and experimentally introduced copies of Tn10. A cascade of sequential rearrangements has even been identified. These involve large transpositions and inversions in *B. subtilis,* which are now interpreted in terms of recombination at specific junction points in the chromosome.

Rates of inversion tend to be quite low and tend to vary in frequency with location in the chromosome. Several ideas have emerged that help explain variations in viability in rearranged strains and the stability of the wild-type genetic organization of bacterial genomes. In one, the gradient of gene dosage generated by the replication of DNA (ranging up to a factor of 4 when two rounds of replication are in progress) may constrain genes to certain locations relative to distance from the origin. In another, the symmetry of the locations of the origin and terminus of replication may be important to keep the replication arms of equal length. In a third, the direction of transcription of certain genes relative to the direction of replication may be important to preserve. A fourth idea concerns regions flanking the terminus in *E. coli*. The entire region can be inverted, but inversions having an endpoint within the region do not occur. In one set of experiments, the product of the inversion was generated by other means, and the cells were viable. Thus, there seems to be a structural barrier to inversion in this region rather than a functional prohibition against the product of inversion. The nature of the barrier is unknown. For some unknown reason, these seemingly general ideas derived from studies with *E. coli* do not seem to apply to many of the large inversion mutations in *B. subtilis*, mutations that are generated at high frequency (as high as 10^1), which are genetically stable and may involve a major disturbance in the symmetrical location of the origin and terminus of replication. Clearly, there is much to learn about factors that constrain nucleotide sequence locations in bacterial genomes.

C. Gene Expression

The factors that affect expression of individual genes have been under intense study ever since the operon model was advanced in 1961. Protein–DNA interactions have dominated this area, and recently the role of distant nucleotide sequences and local DNA configuration has been attracting attention. At the chromosomal level, the control of large sets of related genes has become particularly interesting because these control circuits may reveal principles also operative in the development of higher organisms. In some cases, a single repressor controls a large number of genes in a circuit. In others, changes in promoter recognition by RNA polymerase appear to be key events. Tertiary structure in the form of supercoiling may even have a global effect on transcription, based on the knowledge that this aspect of chromosome structure responds to changes in extracellular environment.

Under a particular situation, a gene may be active, inactive but activable, or cryptic. Cryptic genes approximate working coding sequences, but they are defective in ways that generally require one or a few mutations to become active. How and why nonfunctioning cryptic genes are retained in the genome is not well understood. Calculations show that a cryptic gene would be retained by a population if a deleterious environmental condition arises from time to time in which the cryptic gene would nullify if reactivated. Cases have been found in which changes in supercoiling allow expression of cryptic genes.

V. Concluding Remark

We now have considerable information about the primary structure of the chromosome, including the nucleotide sequence and relative orientation of many genes, and we know something about the enzymes that manipulate the DNA to store, copy, and extract information. But we have few clues to explain how the DNA is compacted to fit inside the cell. It seems reasonable that bacteria have nucleosomelike structures. Such structures could provide some compaction of the DNA (in eukaryotic organisms, nucleosomes pack the DNA by sevenfold), and they could provide a way to constrain the supercoils that are not relaxed by the introduction of nicks into intracellular DNA. But the presence of nucleosomelike structures has not been unequivocally demonstrated in bacteria. Proteins such as the histonelike protein HU constantly assemble and disassemble complexes with DNA *in vitro;* if these proteins are responsible for nucleosomelike structures, cell lysis might easily disrupt the structures. This would explain why nucleosomelike structures are so difficult to detect. But how could a rapid assembly–disassembly process prevent supercoils from being relaxed by nicks in the DNA?

Acknowledgments

I thank Monica Riley and Conrad Woldringh for critical comments on the manuscript. This work was supported by a grant from the National Science Foundation (PMB 8718115).

Bibliography

Bachmann, B. (1990). *Microbiol. Rev.* **54,** 130–197.

Drlica, K., and Rouviere-Yaniv, J. (1987). *Microbiol. Rev.* **51,** 301–319.

Drlica, K., and Riley, M. (eds.) (1990). "The Bacterial Chromosome." 469 pp. American Society for Microbiology, Washington, D.C.

Krawiec, S., and Riley, M. (1990). *Microbiol. Rev.* **54,** 502–539.

Pruss, G., and Drlica, K. (1989). *Cell* **56,** 521–523.

Chromosome Segregation

Olga Pierucci
Roswell Park Cancer Institute

Charles E. Helmstetter
Florida Institute of Technology

Glossary

Chromatid age Determined by the age of the template DNA strand: Young chromatids contain a template formed during the preceding round of chromosome replication, and old chromatids contain a template formed during some earlier round of replication

Chromatid pair Products of chromosome duplication, each containing a newly polymerized DNA strand and a template strand formed during an earlier round of DNA replication

Disjunction Separation of sister chromatids at anaphase

Equipartition Allocation of one chromatid from each pair into each progeny cell

Mother/daughter cells Two progeny cells formed by asymmetrical cellular reproduction, such as by budding in the yeast *Saccharomyces cerevisiae*

Nonrandom segregation Cosegregation of chromatids of like age; cosegregation of chromosomes and conserved cell structures; oriented distribution of chromosomes into progeny cells

Sister cells Two progeny cells formed by symmetrical cellular reproduction, such as by binary fission in the bacterium *Escherichia coli*

CHROMOSOME SEGREGATION is the process by which a cell partitions its duplicated chromosomes between the two progeny cells formed by a division. Equipartition of genetic information is essential for the maintenance of a viable population: Unequal chromosome distribution can generate nonviable cells lacking chromosomes (aploid cells) and/or cells with aberrant chromosome complements (aneuploid cells). The error in the equipartition of the genetic complement is a function of the microorganism, prokaryotic or eukaryotic, but is of the order of 10^{-5}, or less, per cell division. The high chromosomal segregation accuracy is consistent with a precise control of the events leading to the formation of two new, viable cells at division.

In principle, the equipartitioning process could yield either a random or nonrandom assortment of chromatids between the progeny cells, where the differentiation between chromatids in each pair is based on the age of the template DNA strand. In the case of random segregation, either progeny cell could inherit the chromatid with the oldest DNA template strand, independent of the asymmetrical inheritance of any other cellular element, such as the old polar cap in bacteria or old cell wall in yeast. To achieve this, chromosomes would interact individually with the segregation apparatus such that their orientation with respect to any identifiable, conserved cell structure would be independent of the age of the DNA strands in the chromatids. On the other hand, nonrandom segregation implies the oriented transmission of conserved subcellular components, including the chromosomes, into the new progeny cells. This would infer the existence of a preferential interaction among cosegregating elements, either because they are an integral part of a structure of higher complexity or because the specific interactions are a function of the age of the elements.

The purpose of this article is to describe the chromosomal segregation process in microorganisms, both prokaryotes and eukaryotes, with emphasis on the extent to which it proceeds by random or nonrandom means. The primary questions to be addressed are as follows. (1) Does the allocation system specify which chromosome enters a particular progeny cell? (2) If it does, how might this chromosomal discrimination be achieved?

I. Chromosome Segregation in Prokaryotic Microorganisms

A. Random versus Nonrandom Segregation

1. Theoretical Considerations

For prokaryotes, the chromosomal segregation process will be described by referring to the bacterium *Escherichia coli* growing with a generation time greater than C + D min (Fig. 1). C + D is the time between initiation of chromosome replication and the subsequent cell division, and is about 60 min in *E. coli* growing at 37°C. The newborn cells in such a culture (Fig. 1a) contain a single chromosome. The chromosome is a circular DNA molecule that replicates bidirectionally from a single replication origin called *oriC*. For convenience, it is represented in the figure by a short, linear double-stranded portion of the molecule, with *oriC* at the right end. This *oriC*-containing portion of the chromosome is positioned asymmetrically in the newborn cell for reasons that will be explained later. One of the DNA strands of the chromosome was formed during the most recent round of DNA replication (the "young" strand), and the other (the "old" strand) was formed during some earlier round of replication. Upon replication and segregation, the younger and older chromosomes separate from one another to eventually reside within the new daughter cells (Fig. 1b, c). The directions in which the chromosomes are partitioned can be discriminated by the fact that every cell has a "young" polar cap, formed at the previous division, and an "old" polar cap, formed during some earlier division. If segregation were totally nonrandom, the chromosome with the older template strand would always segregate toward the older pole. If segregation were random, it would be equally likely to segregate in either direction. The question therefore is: Does the cell distinguish between the two chromosomes in terms of the direction of partition? [*See* CHROMOSOME, BACTERIAL.]

2. Experimental Findings

The chromosomal segregation process in prokaryotes has generally been studied by following the distribution of radioactively or density-labeled chromosomal DNA strands between progeny cells as a function of time. If the postlabeling incubation conditions permit determination of ancestral relationships, then the mode of segregation can be identified. For many studies on bacterial cells, the DNA strands have been identified by pulse-labeling cells with radioactive thymidine and observing the distribution of radioactivity among progeny cells during growth of the cells in chains, usually in a viscous medium. In experiments of this general type, per-

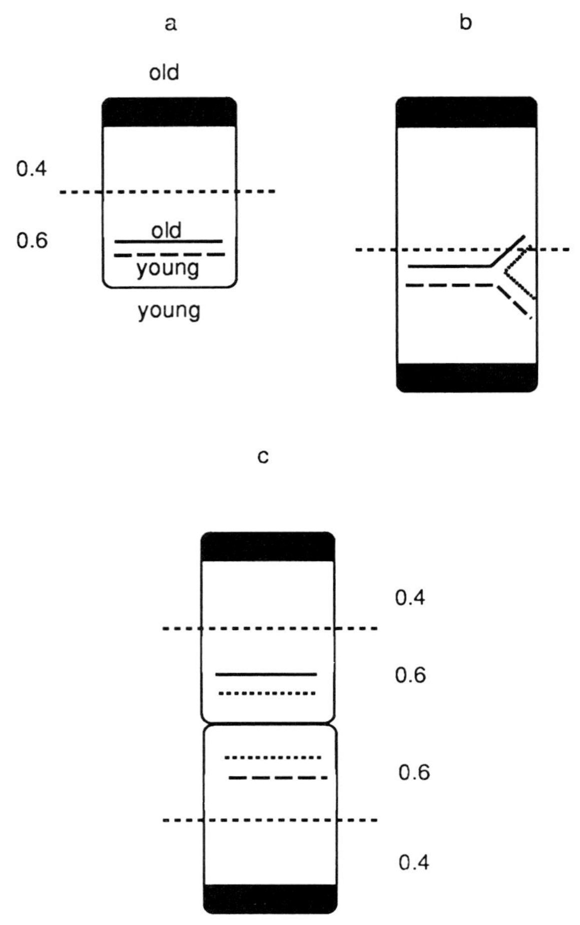

Figure 1 Chromosome segregation in slowly growing *Escherichia coli*. A cell growing such that C + D is less than the generation time is shown at ages 0 (a), 0.5 (b), and 1.0 (c) in the division cycle. The chromosome is shown schematically as a linear structure, for convenience, with *oriC* at the right end, and replication in progress in the cell of age 0.5. The "old" DNA strand has been used previously as a template, and the "young" DNA strand is being used for the first time as a template. The horizontal line within each cell identifies the center of the cell, which will be the site of formation of the next septum. [From Helmstetter, C. E., and Leonard, A. C. (1990). *Res. Microbiol.* **141**, 30–39.]

formed over the past 25 years, both random and nonrandom segregation have been reported for various bacterial species. It has now become evident that one of the explanations for these differing observations is that they were due to differences in the growth conditions of the bacterial cells. Chromosome segregation appears to be essentially random during rapid growth and becomes nonrandom during slower growth. Although there are apparent exceptions to this conclusion (e.g., random segregation in *Streptococcus faecium* at all growth rates), it has been consistently observed in *E. coli* and *Bacillus subtilis*, the most thoroughly examined bacteria. Thus, in these organisms, partitioning must be accomplished in such a way that the nonrandomness of segregation is observable only at lower growth rates.

The observation of growth rate-dependent nonrandom chromosome segregation eventually led to the idea that segregation could be described as a probabilistic process. For instance, in *E. coli* growing rapidly with a doubling time of 25 min at 37°C, the probability the older strand will segregate toward the older pole is close to 0.5 (random), but at lower growth rates the probability increases to >0.5. It has never been detected to reach 1.0 in *E. coli*, i.e., totally nonrandom. The actual probability value depends on the cell strain, but as an examle, it reaches a maximum of 0.65 for *E. coli* B/r growing with a doubling time of 60 min at 37°C.

B. Models for the Segregation Mechanism

Based on the preceding description, we can conclude that slow-growing bacteria can discriminate between chromosomes and segregate them accordingly. Before analyzing how this discrimination might take place, it is necessary to consider the mechanisms that could push or pull the chromosomes apart. The actual mechanism has not been identified, but there are two basic types to consider: those that involve an active partitioning system and those that do on. Considering the latter category first, chromosomal separation could be achieved through movement of the DNA into the space made available by growth of the cell in volume. One newly formed chromosome might remain in place while the other moves away during replication, or the two replicating chromosomes might separate coordinately, as more cytoplasmic space became available. The driving force for the separation could be affected by the transcription/translation of genes along the chromosome. As one example, if the genes for ribosomal RNAs and proteins located near *oriC* were transcribed and translated on both new chromosomes in a "ribosomal assembly microcompartment," expansion of the microcompartment could separate the origin regions and set in motion the separation of the complete nucleoids.

Most models for the chromosome segregation mechanism propose an attachment between DNA and the cell envelope (membrane and/or wall) as a means to separate the chromosomes. The well-known "replicon" model proposed in 1963 suggested that separation of these attachment sites for DNA by growth of the envelope between them would also separate the DNA attached to the sites. This model, as stated, required zonal envelope growth between the attachment sites to affect the separation. Subsequently, it was noted that this basic idea was valid even if the cell envelope followed a pattern of diffuse synthesis, as appears to be the case. If chromosomal attachment sites were located at the junctions between the polar caps and the lateral envelope of rod-shaped bacteria, they could serve as the anchor points for chromosome separation, independent of the mode of lateral wall synthesis, as long as the poles were conserved. According to one detailed proposal of this type, the replication origin would be attached to a junction site at the older pole and the terminus to the junction site at the newer pole. After initiation of a round of replication, one of the new origins is displaced and transported to the junction site at the newer pole, possibly by remaining associated with the processing replication complex, displacing the terminus. Growth of the lateral envelope, taking place only between the junction sites, would separate the chromosomes. [*See* DNA Replication.]

The putative chromosomal origin attachment sites could coincide with, or be demarcated by, the "periseptal annuli," which have been detected recently in gram-negative bacilli. The annuli consist of two concentric rings of membrane–wall adhesion, which straddle the sites at which division septa develop. Thus, upon completion of septation and division, an annulus might define the junction between the lateral envelope and each newly formed pole. Nascent annuli are formed in proximity to the central division sites (annuli) and then move to the centers of the domains destined to be the complementary sister cells. It is conceivable that this parting of the nascent annuli could be responsible for the separation of the chromosomes either because the chromosomes are attached to them or because the annuli delineate the domains of the cell in which chromosomal attachment and movement are permissible. Some evidence for these concepts will be presented in subsequent sections of this article.

C. Evidence for DNA–Membrane Attachment

If chromosomal segregation is mediated by attachment of the DNA to envelope components, then such association should be experimentally detectable at least during some portion of the cell division cycle. Indeed, there have been numerous reports of such DNA–envelope associations. DNA–membrane complexes have been isolated from bacterial cell lysates, generally by differential centrifugation. These complexes were found to be enriched for chromosomal DNA at *oriC*, suggesting that *oriC* is membrane-bound at some point in the cycle, but it has also been reported that the chromosome is bound at numerous (20–80) additional sites. It must be emphasized, however, that the presence of DNA associated with membrane after fractionation of cell lysates does not necessarily confirm that the attachment was actually present *in vivo*. Nonetheless, the apparent specificity of the *oriC* attachment is strongly suggestive of such an *in vivo* relationship. Recently, *in vitro* reconstructions have also been used to identify DNA–membrane interactions. Upon mixing membrane and chromosomal DNA, it has been found that *oriC* binds with high specificity. The most striking finding is that *oriC* DNA binds to membrane only during the restricted interval after initiation of replication when the newly polymerized DNA strand has not yet been modified by methylation. *In vivo, oriC* remains hemimethylated for about 8 min after initiation of a round of replication at 30°C. These findings indicate that *oriC* binding exists transiently after initiation of replication, and, thus, if it is involved in segregation, the chromosomal separation process must at least begin at this time.

Presumably, the binding of DNA to membrane is mediated by adapter proteins that connect the DNA to the membrane. In the case of bacterial plasmid DNA, convincing evidence indicates that these adapters are products of plasmid-encoded genes involved in partitioning (*par* genes). In the case of the chromosomal *oriC*, there is also evidence indicating that proteins are involved in membrane-binding. Origin-binding proteins, some found in membrane fractions, have been isolated, and these could participate in the association of the hemimethylated origin with membrane. Additionally, some proteins known to be required for initiation of chromosome replication interact with membrane. A protein required for initiation of replication in *B. subtilis* is also required for *in vitro* membrane attachment of DNA. Similarly, the DnaA protein, required for initiation of replication in *E. coli*, appears to be part of a DNA–membrane complex. Antibodies to DnaA inhibit DNA–membrane binding *in vitro*. In summary, good evidence indicates that chromosomal replication origins bind to the cell envelope, but the participation of the binding in segregation has yet to be demonstrated experimentally.

D. Mutants Defective in Segregation

One approach to the identification of the components of the bacterial DNA segregation system has been to isolate mutants defective in some aspect of partitioning. In mutants of this type, the chromosomes do not separate properly into daughter cells, resulting in the appearance of cells lacking chromosomes, cells containing unusually large numbers of chromosomes, or cells with division defects. A number of temperature-sensitive mutants have been isolated that display the behavior expected of a partitioning defect at elevated temperature. Upon characterization, it was found that most of these mutants were defective in topoisomerases, enzymes involved in the decatenation or folding of chromosomes. Chromosomeless cells also appear in mutants with defects in DNA replication or recombination, or with deletions in the chromosomal replication terminus. The formation of chromosomeless cells in mutants that cannot complete the replication process is usually a consequence of continued division in the absence of replication, rather than defective segregation. A similar defect in replication/division coordination may also account for the behavior of recombination-deficient mutants. The chromosomeless cells formed by mutants with terminus deletions could be related specifically to segregation if this region were involved in partitioning. Recently, a new mutant has been isolated that may have a specific defect in segregation. This mutation, called *mukA*, causes the production of chromosomeless cells of normal size at a frequency of 0.5–3% of the cells in an exponentially growing population. The chromosomeless cells are formed because division of some cells results in the production of one daughter with both chromosomes. This finding is consistent with a defect in the partitioning system, and the *mukA* gene was found to correspond to the *tolC* gene, which encodes an outer membrane protein. This may be the first gene to be identified that participates in the putative membrane-based partitioning process. Many more such mutants will have to be isolated and characterized to determine if partitioning is truly mediated by DNA–membrane associations.

E. Segregation and the Cell Cycle

The timing of chromosome segregation in the division cycle is undoubtedly determined in large measure by the timing of chromosome replication. In bacteria, cells divide C + D min after initiation of a round of replication. In slowly growing cells, these C + D intervals follow one after the other, similar to the S + G2 period in eukaryotes. In rapidly growing cells, the times between successive initiations of replication are shorter than C + D min, so that the intervals overlap. Nevertheless, cells always require the C + D interval to replicate their chromosomes and separate them into the daughter cells. Obviously, the chromosomes cannot be fully separated until replication has been completed, but segregation could start at initiation of the round. If the attachment of the replication origin to membrane were involved in the process, and it occurred transiently after initiation of replication, as suggested earlier, then it would be expected that segregation would begin at initiation. The balance of the chromosomes could separate gradually behind the origins as replication progressed during the C period (the time for a round of chromosome replication), until the chromosomes were completely separated in the D period (the time between the end of a round and cell division). There is both cytological and molecular evidence for such gradual separation of daughter chromosomes during the C period. Once replication is completed, there is then rapid movement (or coalescence) of the chromsomal DNA toward the regions of the cell destined to become the centers of the new sister cells. Recently, it has been suggested that this final, complete separation of the chromosomes requires protein synthesis at or after the end of the round of replication. It has also been suggested that completion of chromosomal separation initiates division by lifting an inhibition of septum formation. Viewed in this light, replication, segregation, and division form a coordinate, continuous process that ultimately leads to the formation of two essentially identical sister cells.

F. Physiological Basis for Nonrandom Chromosome Segregation

Based on the preceding, it can be concluded that bacterial chromosome segregation is nonrandom and that it probably involves a DNA–membrane attachment. Recently, a model has been proposed to explain the physiological basis for the observed nonrandom segregation of chromosomes in *E. coli*. This proposal states that *oriC* cannot be located within the domains of the polar caps when replication initiates and that this restriction in initiation locale is the sole determinant of the nonrandom nature of chromosome segregation. The basis for this conclusion will be explained by considering the behavior of *oriC*, because the rest of the chromosome will follow *oriC* into the same daughter cell. Referring again to Fig. 1a, the newborn cell has just been formed by a division and has not yet initiated chromosome replication. Therefore, it contains only one chromosome. The *oriC* on this chromosome must be located at some finite position within the volume of the cell. According to the proposal, it can be located anywhere in the cell, probably at random in the population as a whole, but not within the existing, older polar cap (indicated in black, Fig. 1a). Based on simple geometric consideration, there is a greater probability that it will be located in the half of the cell that contains the new pole rather than in the half cell that contains the old pole. Because a polar cap occupies about one-sixth of the cell volume, 0.6 of the newborn cells in a population will have *oriC* in the half cell with the young pole (as drawn in the example in Fig. 1a), and 0.4 of the cells will have *oriC* in the older cell half.

When replication initiates, *oriC* is attached to the membrane at sites along the lateral envelope of the cell that are external to the polar caps. The attachment takes place somewhere within the cell half in which the *oriC* resides. As described earlier, the boundaries of this attachment/replication zone, which cannot include the polar cap, could be demarcated by the polar and centrally located periseptal annuli. Once polymerization begins (Fig. 1b), the *oriC* with the younger template remains sequestered within its domain, so that it, and the entire chromosome subsequently formed on the younger template, stay within the developing daughter cell (lower cell half in this example, Fig. 1b). The *oriC* with the older template strand is *always* translocated away from the younger one, to eventually take up residence in the equivalent domain of the complementary sister cell (upper cell half, Fig. 1b). Throughout the ensuing completion of replication and cell division, each chromosome remains located (or can move freely) within the attachment/replication zone of the developing daughter cell (lateral cylinder *and* septum). Because there was a probability of 0.6 that the *oriC* was located in the younger half of the cell at the time of initiation of replication, this means that the chromosome with the old template strand has a 0.6 probability of being partitioned toward (and segregating with) the older cell pole. This is equivalent to saying that a given DNA strand has a 0.6 probability of segregating in the same direction in two successive division cycles. The end result (Fig. 1c) is that the chromosomes will be distributed within the newborn cells in the culture with the indicated probabilities. Finally, if replication/segregation proceeds by an attachment mechanism, the *oriC*s must release from the sites between initiation events so that the entire process can be repeated. This requirement is also consistent with the finding, described previously, that *oriC* only binds to membrane for a limited interval of time after initiation of replication.

The preceding proposal accounts quantitatively for the experimentally observed probabilistic, nonrandom nature of chromosome segregation in *E. coli,* with slight differences expected for cells with different shapes. It also accounts for the observation that segregation becomes more random as the growth rate increases because the proportion of the cell that is permissive for initiation increases dramatically as the growth rate increases. This is due to a decrease in the fraction of polar caps bracketing a replicating chromosome that are "old," causing segregation to appear to be more random in the population as a whole. Thus, the nonrandom aspect of chromosome segregation is apparently due to the asymmetry of the zone in the cell permissive for initiation of replication, whereas the partitioning system itself is entirely deterministic and must possess a mechanism to discriminate between template strands of different ages. The putative attachment sites for DNA probably reside external to the cell poles because they coincide with the growing portions of the envelope, and the polar caps are generally conserved. Finally, it should also be noted that the concept of "attachment," although appealing, is not a requirement for this explanation of nonrandom segregation because the asymmetric chromosomal distribution could be based equally well on the volume available for the chromosome, with the proviso that *oriC* cannot reside within the volume of a polar cap once replication begins.

II. Chromosome Segregation in Eukaryotic Microorganisms

While the end result of chromosome segregation is, for all normal cells, the equipartition of genetic information during division, the partitioning apparatus is more highly differentiated and more complex in eukaryotes than in prokaryotes. The description of the process in eukaryotic microorganisms will again be aimed at answering questions regarding the presence or absence of nonrandom segregation. To achieve this aim, chromosome partition will be presented in some detail for the two yeasts, *Saccharomyces cerevisiae* and *Schizosaccharomyces pombe*. Examination of the phenotyes of mutants possessing partitioning defects in these organisms has led to an understanding of both the structures involved in DNA partitioning and the extent to which the process is nonrandom.

A. Segregation and the Mitotic Cycle

Saccharomyces cerevisiae grows by budding; the rod-shaped *S. pombe* by extension, primarily from the old end, and fission at the cell center. In both yeasts, growth is asymmetrical. At division, one of the progenies inherits old wall and a wall scar at the division site, whereas the other harbors the newly synthesized wall and has no scar. Chromosomes duplicate in the S phase of the cycle, segregate in mitosis, and return to the interphase state upon exit from mitosis. The onset of these and other stages in the mitotic cycle are associated with periodic phosphorylation and dephosphorylation of proteins involved in the control and execution of each stage. The formation of the chromosomal segregation apparatus initiates with the duplication of the organizers of nuclear and cytoplasmic microtubules, namely the nuclear membrane-embedded spindle pole bodies (SPBs). Nuclear microtubules (composed of α- and β-tubulin proteins) radiate from the SPB into the nucleus, whereas cytoplasmic microtubules emanate from it toward the cytoplasm. During early G1 in the budding yeast, the SPB is oriented toward the site where the bud emerges, and cytoplasmic microtubules are directed toward the bud site. Once the SPB duplicates, the two bodies separate to opposite nuclear sites and a spindle of microtubules forms between them. In *S. cerevisiae*, the SPB duplicates in G1, and SPBs remain at the nuclear site and then separate late in S phase. In *S. pombe*, the SPB duplicates in G2, and then the spindle forms and elongates at the start of mitosis.

Interaction of spindle microtubules with the centromere region of each chromatid precedes partitioning. Centromere DNA sequences and the protein–DNA complexes at the centromere region may be the determinants of the bipolar orientation of the chromatids. Protein–DNA complexes at the centromere are inferred by the nuclease resistance of DNA sequences that are essential for centromere function. In *S. cerevisiae*, about 130 base pairs (bp) of a nuclease-resistant 220–250-bp centromere core are required for function. There are sequences of homology among the centromeres of the chromosomes: 8 bp in centromere DNA element I (CDE I) and 25 bp in CDE III. These two DNA elements are separated by a third, CDE II, which harbors a 78–86-bp sequence that is >90% A+T-rich. Limited changes in the length, or some substitutions of base pairs, in the CDE II region might not affect centromere function, but the conserved nucleotide sequences of the CDE I and CDE III elements must be maintained. Changes in centromere functionality are caused by single base changes in CDE III, whereas deletions in CDE I might affect proportionally the nuclease resistance of the core and the fidelity of segregation. As with the chromosome, the presence of a centromere in plasmid DNA ensures the equipartition of that plasmid at division, but more than one centromere per chromosome or plasmid usually leads to malsegregation, i.e., the DNA between the centromeric regions might be torn by the attempted migration of the chromatid toward opposite nuclear poles.

While the DNA content of the two yeasts is similar, the chromosome numbers are not: There are 17 and 3 chromosomes in the budding and fission yeast, respectively. The centromeres of the *S. pombe* chromosomes are relatively large, about 50 kb. In chromosome II, the centromere is characterized by four 14-kb tandem repeats and a 7-kb core sequence. The core has no centromeric function. Centromere function resides in the sets of repeats. Studies of specific deletions in the repeated sequences differentiate centromeric regions involved in the maintenance of sister chromatids at meiosis I from others involved in chromatid partition during mitosis and meiosis II.

At metaphase, the oriented sister chromatids align at the equatorial plane of the spindle. The chromatids disjoin at anaphase and separate along the nuclear microtubules toward each SPB. Upon dissolution of the spindle, the nucleus divides, in contrast to higher eukaryotes in which the previously disrupted nuclear membrane reforms around each chromosomal DNA complement. Cytokinesis separates the nuclei in the progenies.

B. Mutants Defective in Segregation

Information on the mechanism of chromosome segregation, and the extent to which it is nonrandom, have often arisen from studies on mutants defective in an aspect of chromosomal partitioning. The properties of a few of these mutants, which relate directly to the question of segregation nonrandomness, are summarized in this and the following sections.

1. Defects in SPBs and Microtubules

The duplication and function of the SPBs require several gene products. Mutants with a defective $p34^{cdc2}$ protein kinase, which is required for initiation of both the S phase and mitosis, are uninucleate and harbor a modified SPB with an attached satellite. Mutants defective in a calmodulinlike protein, the CDC31 and *cal1* gene products in *S. cerevisiae* and in *S. pombe*, respectively, have a single large SPB, unseparated chromatids, and more abundant microtubules. Aploid and diploid cells are produced upon division of haploid cells. The calmodulinlike protein might be a SPB protein, controlling the Ca^{2+}-dependent SPB duplication and separation. Negative controls on SPB duplication have also been detected since incubation of some mutants under nonpermissive conditions results in formation of multiple SPBs and multipolar spindles.

Once the SPB duplicates and the new pair separate to opposite nuclear poles, the spindle and the nuclear envelope expand so that they extend from the pole of the mother cell to the bud in budding yeast, and from one to the other pole in fission yeast. The migration of the chromatids within the nucleus and the motion of the nuclei within the cells depend on the nuclear and cytoplasmic microtubules, respectively. Microtubule destabilization by drugs, such as Nocodazole, induces spindle and cytoplasmic microtubule loss. Defective at the nonpermissive temperature in the β-tubulin gene, *tub* mutants of *S. cerevisiae* duplicate DNA but remain uninucleate and chromosome separation is blocked. In *tub* mutants lacking function in both nuclear and cytoplasmic microtubules, or in cytoplasmic microtubules alone, the nucleus is positioned at random within the mother cell, consistent with a lack of orientation in nuclear migration. In mutants in which the cytoplasmic but not the nuclear microtubules are functional, the majority of the nuclei migrate into the neck between the mother and the bud. In *S. pombe,* the nucleus is normally found at the cell center, except for the stage prior to nuclear division when the nucleus is elongated from one pole to the other. After nuclear and cell division, the newly formed nuclei migrate to the centers of the new sister cells. Random placement of the nucleus within the cell body is observable in mutants defective in α_1-tubulin (*nda2*) and β-tubulin (*nda3*): Spindle formation is anomalous, chromosome movement toward the SPBs is blocked, and nuclear division is inhibited.

2. Defects in Chromosome Separation

DNA topoisomerases are needed for sister chromatid decatenation and for resolution of recombined homolog chromosomes. A frequent phenotype of mutants in topoisomerase II is defective nuclear division, with the uninucleated cells having the nucleus positioned at the cell center. A similar phenotype is seen in a series of mutants designated cut: the sister chromatids are unable to be separated by the spindle, either because of a defective spindle or a defective chromosome structure. They may have a short spindle and unseparated, duplicated chromosomes. In these mutants, the septa, by traversing the undivided nucleus, cut the cells, that is, the cells are untimely torn.

Defective disjunction of chromatids at anaphase results in the production of aneuploid cells. This is the case in a class of mutants called dis (defective in sister chromatid disjoining). The mutants are defective in chromosome partitioning. The chromosomes duplicate, align at metaphase, and then migrate to the SPBs in the absence of disjunction, leading to an unequal distribution at the two cell poles.

C. Extent of Nonrandom Chromosome Segregation

1. Segregation between Spindle Poles

To evaluate the extent of nonrandomness of chromosome segregation in eukaryotic microorganisms, which contain more than one chromosome, two questions must be addressed. First, do chromatids of "like age" (e.g., those with the oldest template strand) cosegregate toward the same spindle pole? Second, if cosegregation exists, are the cosegregants distributed randomly or nonrandomly into progeny cells? With regard to the first question, conflicting evidence in support of either random segregation or chromosomal cosegregation. A few examples are as follows. In a recent study, *S. cerevisiae* chromosomes were labeled with 5-bromodeoxyuridine (BUdR), and the fate of the BUdR-labeled DNA was followed by immunofluorescence during subsequent growth of the cells in BUdR-free medium. After three generations of growth, 90–95% of the cells contained labeled DNA in their nuclei. This is consistent with random chromatid segregation, since a significant proportion of the cells would have been unlabeled if chromatids of like age cosegregated. On the other hand, cosegregation of chromatids harboring DNA strands of like age was detected in an earlier study on *S. cerevisiae* with a labeling protocol. The 17 chromosomes were fully labeled with radioactivity, and the cells were incubated further in nonradioactive medium. At the third generation, about 50% of the cells were not radioactive, indicating that the radioactive DNA strands were not dispersed at random in the progenies. The number of nonradioactive cells was about equivalent to the value expected if the complementary parental DNA strands were cosegregating together, namely if instead of 17 replicating chromosomes there were a single DNA molecule in the cells. The data were interpreted as indicative of the cosegregation of DNA strands of like age toward the same spindle pole. Cosegregation of chromatids of like age has been reported in *Aspergillus nidulans*. Conidia, radioactively labeled in their DNA, were germinated, and the radioactivity associated with each nucleus in the hyphae was determined by autoradiography. In hyphae containing four and eight nuclei, all the radioactivity was concentrated in two nuclei, consistent with the cosegregation of chromatids harboring DNA strands of like age.

Similar conclusions concerning chromosome cosegregation have been reached in studies on mutants with segregation defects of the type described in the previous section. When a cold-sensitive mutant of *S. cerevisiae,* designated *ndc1-1,* was grown at 13°C, the nonpermissive temperature, the spindle was normal, the chromosomes replicated, and the SPBs segregated at division, but all the chromatids were attached to only one of the spindle poles. At division, a chromosomeless cell and a diploid cell were generated. Production of chromosomeless cells continued for several generations, due to the cosegregation of the unseparated chromatids to the same SPB, resulting in the production of multiploid and aploid cells. Chromosome partitioning is also defective in this mutant during meiosis. Normally, diploid cells form four progeny haploid cells during meiosis, which become encapsulated as spores within a thick-walled body called an ascus. Meiosis I involves duplication of the SPB and chromosomes, spindle formation, chromosome pairing, and resolution of the homologous chromosomes within the nucleus. The meiosis I spindle is then dissolved. Following this, the two SPBs duplicate, two new spindles are formed, and the chromatids are separated and partitioned in meiosis II into the four nuclei that are each incorporated into a spore. In the mutant, sporulation of diploids results in the production of asci containing two spores, each of which is diploid due to defective chromosome segregation at meiosis II. Chromosome partition in mitosis and meiosis II is through the association of replicated centromeres to the spindle, whereas chromosome partition in meiosis I is in the absence of centromere separation. The lack of effect of the *ndc1-1* mutation on meiosis I, and its effects on both mitosis and meiosis II, suggest that the NDC1 gene product may be needed for a necessary function of the newly synthesized centromere region.

Cosegregation of chromsomes has also been detected in mutants of *S. pombe*. One instance involves mutants in the *nuc-2* gene, whose product, a p67 protein, is enriched in the nuclear scaffold and may be a centromeric DNA-binding protein, needed for centromere separation in anaphase, or for centromere–microtubule association. In temperature-sensitive mutants incubated at the nonpermissive temperature, all the duplicated chromosomes become located at only one of the spindle poles. Thus, in a situation in which chromosomes are unable to separate, but the spindle is functional, all the chromatids cosegregated to one or the other pole. In a *dis* mutant, defective in sister chromatid disjunction, the frequency of cosegregation of the three chromosomes in either progeny (3 : 0, 0 : 3 configurations) is equal to the frequency of segregation of one and two chromosomes (2 : 1 and 1 : 2 configuration) in either cell. A frequency of 50% for the cosegregation of the three chromosomes is significantly higher than expected if each chromosome segregated independently into the two cells at division.

2. Segregation between Progeny Cells

The cosegregation of joined chromatids to the same spindle pole and the deviation from a random distribution of the older DNA strands at the nuclear poles suggest that there could be a polarization in chromatid–microtubule binding in yeast, dictated either by the structure of the centromeres, the polarity of the microtubules, or both. The orientation in chromosome partition, when it has been observed, might indicate that the entire chromosomal complement of the cell has the capability to behave, as far as segregation is concerned, as a single chromosome. However, the progeny cell into which these conserved units are segregated appears to be chosen at random. In the *ndc-1* mutant of *S. cerevisiae,* the nucleus containing the entire chromsomal complement of the nondisjoined chromosomes is found with equal probability in the mother or daughter cells. Similarly, in the *dis* mutants of *S. pombe,* each sib has the same probability of inheriting the cosegregating, nondisjoined chromatids. Thus, even if a polarity in mircotubule–DNA binding might result in nonrandom segregation of two or more chromatids within the nucleus, the cosegregation would be randomly distributed between the cells at division.

In summary, evidence indicates similarities in the segregation properties of prokaryotes and lower eukaryotes. In both cell types, the segregation machinery might be capable of distinguishing between chromosomes based on the age of the template DNA strand. In prokaryotes, the chromosome with the older DNA is moved away from the locale of the younger chromosome, and in the eukaryotes, chromosomes of like age have been reported to cosegregate in some experiments, but not all. Furthermore, similar nonrandom chromatid segregation has also been detected in mammalian cells, although it is only partially nonrandom and again not always observed. Despite any deterministic aspect of segregation, however, the specific progeny cell into which a specific chromosome or chromosome complement is partitioned apparently is chosen essentially at random. Cosegregated eukaryotic chromosomes have an equal probability of residing in either progeny, while the older prokaryotic chromosome has a probability of not >65% of residing in the sister cell with the older polar cap.

Bibliography

Byers, B. (1981). Cytology of the yeast life cycle. *In* "The Molecular Biology of the Yeast *saccharomyces:* Life Cycle and Inheritance" (J. N. Strattern, E. W. Jones, and J. R. Broach, eds.), pp. 59–96. Cold Spring Harbor Laboratory, Cold Spring Harbor, New York.

Hiraga, S., Niki, H., Imamura, R., Ogura, T., Yamanaka, K.,

Feng, J., Ezaki, B., and Jaffe, A. (1991). *Res. Microbiol.* **142,** 189–195.

Leonard, A. C., and Helmstetter, C. E. (1989). Replication and segregation control of *Escherichia coli* chromosomes. *In* "Chromosomes: Eukaryotic, Prokaryotic, and Viral," Vol. III (K. Adolph, ed.), pp. 65–94. CRC Press, Boca Raton, Florida.

Murray, A. W., and Szostak, J. W. (1985). *Annu. Rev. Cell Biol.* **1,** 289–315.

Neff, M. W., and Burke, D. J. (1991). *Genetics* **127,** 463–473.

Newlon, C. S. (1988). *Microbiol. Rev.* **52,** 568–601.

Ohkura, H., Adachi, Y., Kinoshita, N., Niwa, O., Toda, T., and Yanagida, M. (1988). *EMBO J.* **7,** 1465–1473.

Robinow, C. F., and Hyams, J. S. (1990). General cytology of fission yeasts. *In* "Molecular Biology of the Fission Yeast" (A. Nasim, P. Young, and B. F. Johnson, eds.), pp. 273–330. Academic Press, New York and London.

Schaechter, M. (1990). The bacterial equivalent of mitosis. *In* "The Bacterial Chromosome" (K. Drlica and M. Riley, eds.), pp. 313–322. American Society for Microbiology Press, Washington, D.C.

Complement and Inflammation

M. Kathryn Liszewski and John P. Atkinson
Washington University School of Medicine

Glossary

Alternative pathway Branch of the complement system that serves as an innate defense mechanism capable of recognizing "foreignness" in the absence of antibody

Anaphylatoxins Peptide fragments of complement proteins (C5a, C3a, and C4a) that mediate inflammation by causing increased blood vessel permeability, contraction of smooth muscle, and/or directed migration of phagocyte cells to a site of tissue injury

Classical pathway Branch of the complement system that is initiated by antibodies of the IgM or IgG isotype binding to antigen

Complement system Recognition and effector system of plasma consisting of at least 30 proteins, which promotes inflammation, provides an innate defense mechanism, and "complements" the humoral (antibody) immune system to protect against foreign particles

Opsonization Coating of a substance with components (such as complement proteins) that are ligands for specific host cell receptors to promote adherence and, in some cases, ingestion of the particle

THE COMPLEMENT SYSTEM consists of at least 30 proteins that interact to provide an innate defense against microbes, an adjunct (i.e., a complement) to the antibody-directed (humoral) immune system and a potent means of promoting inflammation. Inflammation is a response by the host to injury and is characterized by swelling, redness, heat, and pain. This article will examine the critical role played by the complement system in the inflammatory process. First, a brief overview of inflammation will be given. This will be followed by a description of the complement system in general and its proinflammatory components in particular.

I. Overview of Inflammation

Many conditions may provoke an inflammatory reaction: microbial infections such as those from the flu or "strep" throat; mechanical body damage such as that from a scratch, sunburn, or bruise; allergens such as pollens or poison ivy; and autoimmune disorders in which the body attacks itself. Although the trigger may differ, the inflammatory response is similar. The four basic hallmarks of inflammation (redness–rubor, swelling–tumor, heat–calor, and pain–dolour) were described by Celsus in the first century A.D. We now understand that these overt signs are produced by the release of mediators that cause capillary dilation and influx of fluid and cells.

When a condition elicits such a reaction, mediators are synthesized, released, and activated. Some of these include proteins (complement protein fragments), chemicals (histamine, serotonin), and fatty acids (prostaglandins, leukotrienes). These mediators activate other cells, produce vasodilation, and induce mobile cells to migrate to the site of inflammation. Adherence-promoting factors are also expressed on cells and the vascular endothelium. As a result, white blood cells (e.g., neutrophils, lymphocytes, and monocytes–macrophages) adhere to the endothelium and then migrate out of the vascular space and into the area of injury.

Inflammatory reactions may occur rapidly, as in certain allergic reactions such as asthma or hayfever; within several hours, as in certain types of arthritis; or not until after several days, as with tuberculosis. Most inflammatory episodes resolve expeditiously, although in some cases recovery is variable and depends on the extent and location of the injury and the facility with which the foreign element can be eliminated. In the case of autoimmune syndromes and certain microorganisms, a state of chronic inflammation may be produced with exacerbations and remissions.

II. Categories of Inflammation

Inflammatory reactions fall into four basic categories: allergic or immediate hypersensitivity reactions, acute and subacute inflammation mediated by cytotoxic antibodies, subacute inflammation mediated by immune complexes, and chronic inflammation (delayed hypersensitivity reactions). The complement system can be involved in the first three.

A. Allergic Reactions

Inflammation may result from an allergic trigger. Antibodies of the IgE isotype mediate this response, which may begin within seconds and generally subsides after about 30 min. However, if the antigenic stimulus continues to be present, the reaction may persist (as in asthma). These antibodies bind to receptors on mast cells and special white blood cells called basophils. If these bound antibodies react with an allergen, the cell releases vasoactive components such as histamine and bioactive compounds termed leukotrienes.

Antibody isotypes of the IgM and IgG variety can also produce allergic-type inflammatory reactions by activation of the complement cascade. In this case, small peptide fragments of the complement proteins bind to receptors on mast cells or basophils and cause the cells to fire.

Examples of allergic reactions include urticaria (hives), seasonal rhinitis (hayfever), asthma, and anaphylaxis.

B. Inflammation Mediated by Cytotoxic Antibodies

When circulating antibodies of the IgG and IgM variety attach to a cell surface, the cell is marked for destruction by two processes. First, there are receptors for F_c (nonantigen binding portion of IgG) on phagocytes. Second, the complement system becomes activated. When this occurs, complement components also become bound to the cell and promote its phagocytosis while other fragments elicit the inflammatory response. Incoming cells (such as neutrophils) release degradative enzymes and generate toxic oxygen products. Consequently, the binding of cytotoxic antibodies can cause much tissue damage. More importantly, emigrating cells attempt to phagocytose the antibody and complement-coated target. Antibody- and complement-mediated destruction of red blood cells, or platelets occur in autoimmune syndromes in which the host synthesizes antibodies to antigens on the membranes of these blood elements.

C. Inflammation Mediated by Immune Complexes

As just described, antibodies to cell-surface components may initiate an inflammatory response. Antibodies (and complement) may also latch onto soluble antigens and form immune complexes (ICs). Normally, these aggregates are promptly eliminated by host cells. However, the formation of ICs may lead to pathologic states if the production of ICs outstrips their removal or if the complexes formed are inappropriately deposited. When ICs lodge in organs, the subsequent inflammatory response may lead to tissue damage.

As a rule, detrimental ICs may be produced by three means: persistent infections, autoimmune illnesses, and recurrent extrinsic antigen exposure.

In some *persistent infections,* certain bacterial, parasitic, or viral illnesses produce a low-grade, continuous infection that generates chronic antigenemia (persistance of antigen). Antigen combines with antibody to form ICs that may then deposit in tissue such as the kidney. An example of this situation is chronic viral hepatitis. [*See* HEPATITIS.]

In *certain autoimmune illnesses,* ICs may deposit in tissues, especially the skin, joints, and kidney. In most of these conditions, antibody is made to intracellular antigens (e.g., autoantibodies to ribonuclear proteins) that have been released after cell death. Systemic lupus erythematosus and rheumatoid arthritis are examples of immune complex-mediated syndromes.

Following *extrinsic antigen exposure,* ICs may develop in organs that are repetitively challenged from an exogenous antigen. Thus, if certain materials are repeatedly inhaled, an inflammatory reaction will develop in the lung. Molds, plants, or animals commonly evoke such a reaction. The antibody response eliciting such inflammatory reactions is secondary to IgG rather than IgE response to the antigen. Examples of these diseases include "farmer's lung" and "pigeon breeder's lung."

D. Chronic Inflammation—Delayed Hypersensitivity

As its name implies, delayed hypersensitivity takes longer to develop. It is mediated by T lymphocytes and monocytes–macrophages. A familar example is the contact allergic reaction of poison ivy. Typically, delayed hypersensitivity reactions take several hours to appear, peak at 24–48 hr, and then regress over a period of a few days. This type of an inflammatory response also mediates graft rejection.

III. The Complement System

A. Overview

The complement system consists of at least 30 plasma and membrane proteins. It can be compared to a cascade, waterfall, or domino type of reaction in which one protein assists in activating the next. There are two divisions of this cascade. One pathway is activated by antibodies; the other is a surveillance system ready for immediate defense. Although each pathway is initiated differently, they merge in the later steps.

One route of complement activation is termed the classical pathway. This system becomes activated if antibody binds to its antigen such as a virus or bacterium. In this case, an antibody defines the target and the complement system facilitates the attack.

The second route of complement activation is termed the alternative pathway. Contrary to its name, the alternative pathway probably evolved before the classical pathway. It serves as an initial line of defense against certain types of infections. This pathway is a surveillance system, forming its own primitive immune system, because it can recognize and destroy foreign substances. As an example of the potency of this system, if a gram-negative bacterium is placed in serum, it becomes rapidly coated with ~2 million complement protein fragments in a few minutes. This reaction is independent of antibody (see Fig. 1).

B. Function

The complement system has two major functions.

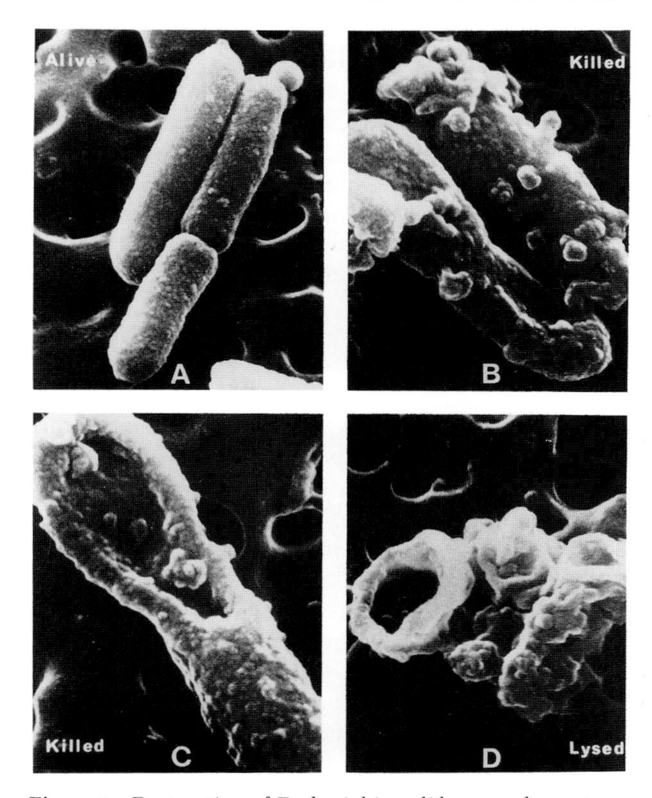

Figure 1 Destruction of *Escherichia coli* by complement. (A) Intact *E. coli*. (B, and C) Killing of *E. coli* by purified complement components. (D) Combined action of complement and a plasma enzyme (lysozyme) on destruction of particle. [Reproduced from Schreiber, R. D., Morrison, D. C., Podack, E. R., and Müller-Eberhard, H. J. (1979). *J. Exp. Med.* **149,** 870., by copyright permission of the Rockefeller University Press, New York.]

1. Membrane Modification

The primary function of complement is to alter microbial membranes by lysis or opsonization. The latter is more important in host defense and refers to the coating of microbes with complement factors, which promotes their ingestion by phagocytic cells or clearance by cells that shuttle them to sites enriched with immune cells including phagocytes.

2. Promotion of Inflammation

The complement system provides potent mediators of the inflammatory response. For example, activated components can trigger the release of compounds such as histamine from mast cells. Histamine causes contraction of smooth muscles and dilation of blood vessels, which, in turn, facilitate the exudation of plasma and cells into an area of inflammation. Additionally, certain complement components are chemotactic factors and thereby attract cells to an inflammatory site.

In these two general ways, the complement system enhances the destruction of foreign particles. However, in autoimmune diseases, complement may act as a well-meaning but destructive robot in contributing to *pathologic* inflammation and destruction of its *own* tissues. In most of these instances, antibody is inappropriately synthesized to a self-antigen.

C. Nomenclature

Although complement was first identified in the 1880s, the magnitude of its biochemical complexity was not initially appreciated until the 1960s. Components have been named as they have been elucidated, which has led to some inconsistencies in terminology.

Components of the classical pathway are designated by the letter C with a number, usually in the order of activation (e.g., C1, C2). Components of the alternative pathway are designated as *factors* followed by letters (e.g., factor B, factor D).

Components circulate in an inactive form. During activation, a peptide fragment is proteolytically generated. This fragment is usually designated by a small letter. For example, if C3 is cleaved, it releases a small fragment termed C3a and the remaining larger protein is termed C3b.

D. Classical Pathway

To facilitate understanding of the rather complicated and interactive nature of the classical pathway, four general steps need to be appreciated (see Fig. 2).

1. Attachment

If the host is infected (e.g., by a virus), the immune system synthesizes antibodies to the pathogen. Once these bind to their target, component C1 of the classical pathway *attaches* to the other end of antibody. Thus, the first "domino" in the classical pathway, component C1, is activated when antibody binds to its antigen.

2. Activation

In this stage, complement components with specific enzymatic activity are produced. These subsequently activate other complement proteins to continue the cascade. After C1 attaches to antibody, it undergoes a conformational change, which allows

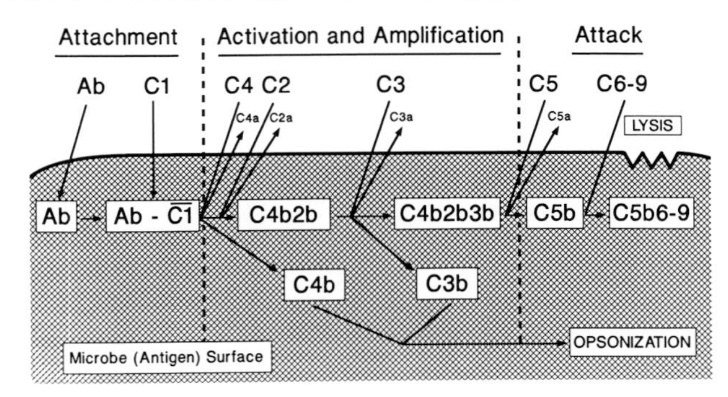

Figure 2 Engagement of classical pathway of complement on a microbial membrane. Attachment: Binding by C1 to antibody initiates the cascade. Activation and amplification: Activated C1 cleaves circulating C4 and C2 components, which form an enzymatic complex (C4b2b, the "C3 convertase") that activates C3. Both C4b and C3b may also individually bind to the membrane to serve as opsonins (see text). If C3b binds to C4b2b, it forms the "C5 convertase" and initiates the attack phase. Attack: The complement-coated microbe is "attacked" by host cells that opsonize (bind and ingest) it or by the membrane lytic components (C5–C9) [Adapted from Liszewski, M. K., and Atkinson, J. P. (1990). The complement system and immune complex diseases. *In* "Internal Medicine," (J. H. Stein, ed.). 3rd ed., pp. 1635–1642. Little Brown and Company, Massachusetts.]

part of it to become a serine protease and activate components C4 and then C2 by proteolytic cleavage. A small peptide, C4a, is liberated and the remaining larger segment, C4b, attaches nearby on the foreign cell surface. C2 is also activated by C1 through proteolytic cleavage and, if the larger C2 fragment binds to C4, a "convertase" is formed. It is the function of this bimolecular enzyme complex, C4b2b, deposited on a microbial surface, to rapidly activate (convert) many molecules of the next component, C3. Because of the enzymatic nature of C1, many activated C4 and C2 molecules are produced. As a result, clusters of these complement components form near the site where antibody is bound.

3. Amplification

As just described, following its attachment and activation, C1 becomes a protease and cleaves thousands of C4 and C2 molecules—an amplification step. These complement proteins are rapidly deposited on the foreign surface. A second amplification step involves C3. The convertase C4b2b forms on the membrane near the site where C1 is bound to antibody. This enzyme complex cleaves C3 releasing a small fragment, C3a. C3a then binds to a receptor on mast cells and triggers these cells to discharge their contents. The larger fragment, C3b, attaches covalently to the foreign membrane. Because of the enzymatic nature of the convertase, many C3b's are formed and become clustered on the microbial surface. C3b and, to lesser extent, C4b bound to the foreign membrane promote the phagocytosis of the foreign element (see later).

4. Attack

Deposited C3b serves several purposes. Some molecules interact with C4b2b to form the complex C4b2b3b. This is the "C5 convertase," which cleaves C5 into two fragments. The larger C5 fragment, C5b, associates with the membrane and triggers the so-called membrane attack complex (MAC), while the smaller fragment, C5a, is released to act as a potent inflammatory agent. C5b incorporates C6 and C7. This complex (MAC) may insert into the foreign membrane, bind C8 and C9, and lyse the organism.

The majority of deposited C3b (and many C4b) serve another purpose during the attack phase. These components are involved in the process called opsonization. The literal definition of this term is "to prepare for ingestion." With opsonization, foreign particles are coated with complement components C3b and/or C4b. These "beacons" serve as ligands for specific host cell receptors. Such receptors either assist in the ingestion of the particle or help shuttle complexes for elimination. Thus, through opsonization, a foreign element is coated with complement components that subsequently facilitate the pathogen's ingestion or elimination via specific host cell receptors.

E. Alternative Pathway

As already mentioned, the alternative pathway almost certainly developed before the classical pathway. Its purpose is to serve as a first line of defense (native immunity). Because it requires some time for an antibody to be made against a foreign element, other means of initially dealing with the particle are necessary. The alternative pathway provides an immediate recognition and effector pathway capable of neutralizing the infectious element.

The four primary constituents of the alternative pathway are C3, factor B, factor D, and properdin (sometimes called factor P). A most remarkable aspect of the alternative pathway is its ability to recognize "foreignness" and quickly amplify on such targets. The end result of the activation of the alternative pathway is the coating of the particle by C3b, which serves as an opsonin or initiates the lysis by terminal components C5–C9 (the MAC). It is at the level of C3 where both the alternative and classical pathways merge (see Fig. 3).

F. Control of the Complement System

Strict regulation of the complement system is essential in preventing host cells and tissues from being destroyed. The fact that nearly half of the members of the complement system function in its regulation attests to the absolute necessity of tight control. To understand how the complement system regulates itself, we may refer back to the basic steps involved in its activation. Control is exerted at several levels, i.e., at the level of the pathways' activation, amplification, and attack steps. In addition, there is also control of its pro-inflammatory peptides.

1. Control during Activation

The classical pathway is initiated when C1 binds to antibody. *C1 inhibitor* controls activated C1. It does not prevent appropriate activation but blocks

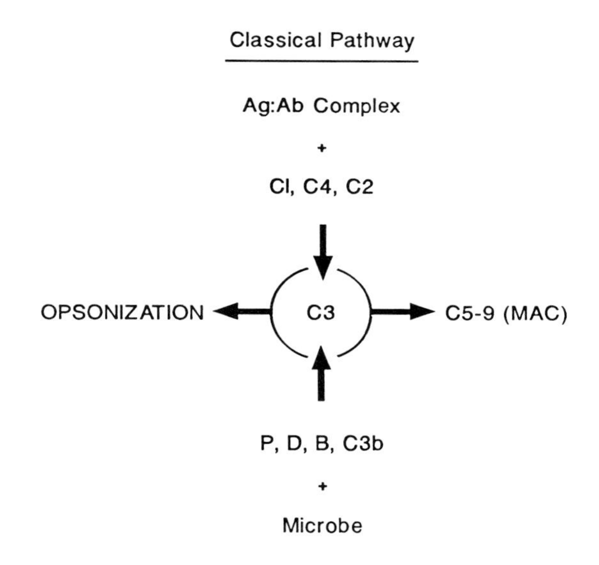

Figure 3 Simplified scheme of complement pathway. The central and pivotal role of C3 is illustrated. Although initiated differently, both the classical and alternative pathways interact at the level of C3. Note: Ag:Ab complex refers to antibody (Ab) binding to the foreign target [antigen (Ag)]. MAC, membrane attack complex.

chronic C1 activation or C1 activation in the absence of antibody.

2. Control during Amplification

Both the classical and alternative pathways are quickly and powerfully amplified as the convertases are generated. As a result, the opportunity exists for destruction of host cells as well as foreign cells. To prevent this, a series of functionally, structurally, and genetically related regulators have evolved. These proteins block inadvertent convertase activity on host cells, thereby allowing its activity to be focused on pathologic substrates. Six such regulatory proteins that exist either in plasma or on host membranes have been identified (see Table I).

These molecules exert regulation in three ways: inhibition of formation of C3 convertases, disruption of formed C3 convertases (decay accelerating activity), and/or inactivation of C3 and C4 components in convertases (cofactor activity). Some accomplish this task in plasma while another set is expressed on most cells to prevent damage by complement to self-tissue.

Table I Complement System Regulatory Proteins

Plasma components	Action
C1 inhibitor	Prevents inappropriate or excessive activation of C1
Factor I	Regulates C3 and C4 by proteolytic cleavage
Factor H	Regulates C3
C4 binding protein	Regulates C4
S protein	Regulates fluid phase MAC
Anaphylatoxin inactivator (carboxypeptidase N)	Regulates C3a, C4a, and C5a
Membrane components	
Decay accelerating factor	Regulates C4 and C3 by preventing formation of and disassociating enzyme complexes that cleave C3
Membrane cofactor protein	Regulates C4 and C3 by serving as a cofactor for their cleavage by factor I
Complement receptor type 1	Regulates C3 convertase
Membrane inhibitor of MAC (CD59)	Regulates MAC on cells

MAC, membrane attack complex.

3. Control during "Attack" Step

Plasma and membrane regulatory proteins that modulate the MAC have been identified (see Table I). One protein couples to the binding site on the C5b67 complex to prevent further assembly. In addition, on cells a membrane protein binds to C8 and prevents its attachment to C9. Thus, MAC inhibitory proteins inhibit the development of both fluid-phase and membrane-bound MAC on host but *not* on foreign tissue.

4. Control of Inflammatory Activity

The pro-inflammatory peptides C3a, C4a, and C5a, generated during complement activation, are potent mediators of inflammation. An anaphylatoxin inactivator of serum, termed carboxypeptidase N, removes a carboxyl-terminal arginine from each anaphylatoxin. This effectively inactivates C3a and C4a and markedly reduces the activity of C5a.

IV. Inflammatory Activities of Complement Components

A. Anaphylatoxins

Chief among the complement components involved in the process of inflammation are the anaphylatoxins C3a, C4a, and, especially, C5a (see Table II). The term was coined because of the old observation that guinea pigs undergo fatal shock, resembling anaphylaxis, if injected with serum containing various substances known to activate the complement system. The effect was attributed to the appearance of "anaphylatoxins," now known to be C4a, C3a, and C5a.

These peptides bind to specific receptors on the membranes of mast cells and basophils, causing them to "fire." Biologically active mediators such as histamine and serotonin are released, producing enhanced vascular permeability, contraction of smooth muscle, and local influx of fluid (edema). C4a is only mildly active, whereas C3a is moderately and C5a is strongly active.

In addition, C5a possesses chemotactic activity. Chemotaxis is the process by which cells are induced to migrate into an area because of the chemical concentration gradient of a substance. C5a is a potent chemotactic agent for cells such as neutrophils. These cells migrate into the area, release

Table II Proinflammatory Activities of Complement Components

Activity	Complement component	Description
Vasodilation	C5a, C3a, and C4a	Stimulate cells to release inflammatory mediators such as histamine
Chemotaxis	C5a	Attracts immune cells; increases cell movement
Opsonization	C3b/C4b, C3bi, and C3dg	Coating of foreign particles to facilitate their adherence and ingestion by serving as a ligand for receptors on pro-inflammatory cells
Cell lysis	C8/C9	Alteration of foreign cell membrane

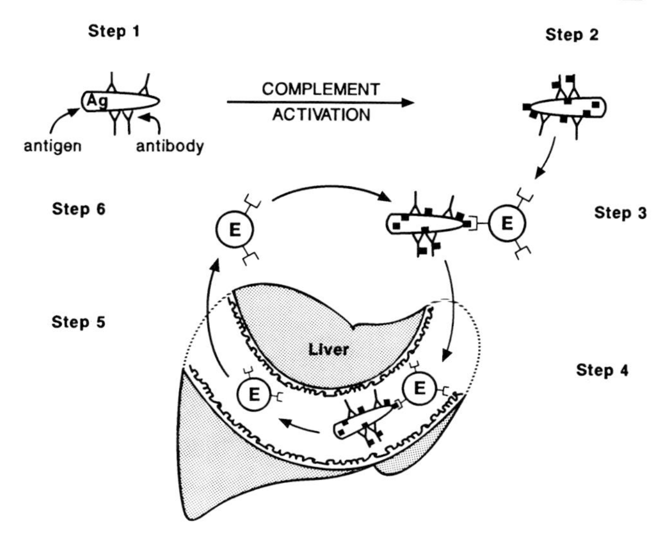

Figure 4 Erythrocyte processing of immune complexes: The EPIC mechanism is a metabolic pathway for immune complex (IC) processing. Complement-coated ICs are bound by complement receptor type 1 (CR1) on erythrocytes (E) and shuttled to the liver or spleen. Organ cells (macrophages) remove the ICs and the E returns to circulation ready for another round. Step 1, IC forms in the circulation; step 2, IC is opsonized with C3b/C4b; step 3, IC binds to E CR1; step 4, IC becomes bound to Hepatic macrophage; step 5, bond between IC and E is broken, E returns to circulation; step 6, E may recycle to bind more IC. [Adapted from Herbert, L. A., and Cosio, F. G. (1987). *Kidney Intl.* **31,** 877–885.]

molecules that damage a microorganism, and attempt to phagocytose the microbe. In summary, the activation of the complement system results in the production of small peptides known as anaphylatoxins, which bind to cellular receptors and cause the release of inflammatory mediators. The latter subsequently produce capillary permeability, contraction of smooth muscle, and local edema. Chemotaxis of cells into the area follows complement activation. C5a is the most important complement-derived anaphylactic and chemotactic factor and has a central role in the process of inflammation. [*See* CHEMOTAXIS.]

B. Opsonins

Integral to the process of inflammation is the elimination of the foreign element. One way the body accomplishes this task is by cell lysis through complement components of the MAC. However, a more important way in which this goal is met is through opsonization (see Table II; see Section III.D.4).

Among the most important opsonins are the bound fragments of C3 and C4 (i.e., C3b and C4b). Activation of both the alternative and classical pathways results in the deposition of C3b on the target. Activation of the classical pathway causes deposition of C4b as well.

Specific complement receptors on host cells bind these activation fragments. Complement receptor type 1 (CR1) is found primarily on erythrocytes and leukocytes. It binds C3b and C4b.

Complement receptor type 2 (CR2) is present on human B lymphocytes, where it binds a breakdown product of C3 termed C3dg. Interestingly, CR2 also binds to Epstein–Barr virus (EBV; the agent causing infectious mononucleosis). Binding of EBV to CR2 allows the virus to enter and infect the human cell. This is an interesting example of utilization of a human receptor by a virus to gain entrance to a cell.

Complement receptor types 3 and 4 are closely related and found on leukocytes. They are efficient at promoting the ingestion of foreign particles coated with a cleavage fragment of C3b known as C3bi.

Recently, a pathway has been delineated in which the erythrocyte is a central player in a process for elimination of complement-coated foreign complexes. The process of binding of a C3b-coated target to an erythrocyte is called immune adherence. The shuttling of such bound complexes has been termed the EPIC (erythrocyte processing of immune complexes) mechanism (see Fig. 4). The activation of the complement system results in C3b and C4b deposition on the foreign target. CR1 on erythrocytes binds to such a complex (immune adherence). As the erythrocyte flows through the liver or spleen, its immune complex cargo is transferred to resident macrophages that ingest, degrade, and thereby eliminate the particle. Subsequently, the erythrocyte returns to the circulation ready to latch onto another complex. The EPIC mechanism provides an efficient means to dispose of immune particles and to prevent their injurious deposition in tissues of the body.

Bibliography

Davis, B. D., Dulbecco, R., Eisen, H. N., and Ginsberg, H. S. (eds.) (1990). "Microbiology," 4th ed. J. B. Lippincott Company, Philadelphia.

Paul, W. E. (ed.) (1989). "Fundamental Immunology," 2nd ed. Raven Press, New York.

Ross, G. D. (ed.) (1986). "Immunobiology of the Complement System." Academic Press, Boca Raton, Florida.

Conjugation, Genetics

Jack A. Heinemann
NIH, NIAID, LMSF, Rocky Mountain Laboratories

Glossary

Conjugant Donor or recipient cell involved in conjugation

Exconjugant Donor or recipient cell that has uncoupled following conjugation

Merozygote Recipient exconjugate that has received transferred DNA

Mobilizable plasmid Conjugal plasmid that carries *oriT*

Replicon Molecule capable of autonomous replication

Self-transferable plasmid Conjugal plasmid that directs its own transfer between cells, i.e., encodes *tra*, *mob*, and *oriT* functions

tra*, *mob*, and *oriT Groups of genes specific to the mobilization and transmission of conjugal elements

Transconjugant Merozygote that has inherited transferred DNA and segregates the molecule to daughter cells

Transmission Inheritance of genetic information transferred between cells

BACTERIAL CONJUGATION is the term applied to the transmission of genetic material from a bacterial cell to another cell by a process that requires cell–cell contact and shields exchanged material from exposure to the environment. These properties differentiate conjugation from viral transduction and DNA transformation. Conjugation is mediated by plasmids. Plasmids may encode all functions specific to the transmission of genetic material between cells. Conjugal plasmids have been found to exist both extrachromosomally and in association with the host chromosome, sometimes tranferring both themselves and the chromosome between cells. Conjugation can direct DNA transfer among a broad spectrum of organisms and may be a mechanism capable of transferring DNA from bacteria to any living creature.

I. Description of Conjugation

A. Introduction to the History and Study of Conjugation

Conjugation between bacteria was first reported in 1946 as the culmination of a series of experiments performed by J. Lederberg and E. M. Tatum. Capitalizing on their unique collection of biochemical mutants in *Bacterium coli* (now *Escherichia coli*), they demonstrated genetic recombination between strains. At a low frequency, a mixture of two strains, each with as many as three unshared auxotrophies, would give rise to wholly wild-type derivatives or derivatives with two of three markers reverted. Control monocultures of each strain treated similarly gave rise to derivatives with at most one marker reverted. Reversion was the result of the conjugal exchange of chromosomal genes between strains. Lederberg and Tatum predicted that recombination was mediated by a bacterial fertility factor (F), now known to be an example of one member of one family of conjugal elements.

The work of Lederberg and Tatum illustrates the contribution of serendipity to fundamental discoveries. It was fortunate that they had introduced genetic markers into *E. coli* strain K-12 because many other available *E. coli* isolates were infertile, and mixtures of those strains would not have produced recombinants. Furthermore, strain K-12 harbors an element with rare attributes. First, regions of sequence homology with the host's chromosome allow F to integrate into the chromosome and mobilize chromosomal markers in addition to itself, necessary in this case to detect genetic recombination. Second, F transmits as much as 1000-fold higher than other similar plasmids because its conjugation-specific functions are derepressed. Had different strains or a different F factor been used to assay the complementation of chromosomal markers, it is likely that the assay would have failed.

Since the days of Lederberg and Tatum, progress has been made in determining the genetic organization of plasmids, especially plasmids of two groups, F and P, and characterizing plasmid-encoded functions that metabolize DNA. In the following discussion about genes carried by conjugal elements, the F and P plasmids will be considered typical unless compelling evidence indicates notable exceptions. In addition, more recent characterizations of both plasmid and chromosomal genes involved in conjugation, and other DNA transmission systems not mediated by F- and P-type elements, will be included. [*See* PLASMIDS.]

B. Gene Transmission

DNA transmission is the culmination of a series of events in conjugation (Fig. 1). The transmission process is divided into four steps proceeding from largely donor to recipient localized events. First, the donor conjugant, a bacterial cell harboring a conjugal element, contacts a recipient conjugant. Second, DNA transfer occurs, followed by resolution of the cells into exconjugants. Third, the newly acquired molecule establishes itself within the recipient exconjugant, or merozygote. Establishment may require a strategy for evasion of host immune response (e.g., DNA restriction by endonucleases) and does require a means of replicating, either by integration into a resident replicon or by self-replication. Fourth, the new molecule segregates during division of the merozygote to each daughter transconjugant.

C. Plasmid Families

Donor and recipient conjugants became known as "male" and "female," respectively, following W. Hayes' demonstration that transmission occurs unidirectionally from a donor to a recipient. Interactions between two males or a male and female were considered compatible when transconjugants resulted. Two female cells were incompatible, therefore, because their mixture failed to promote intercellular recombination. Today, compatibility has a plasmid-centric definition. Crosses between two F^+ donors would now be considered unproductive because the two F factors are incompatible.

Plasmids are compatible when they can coexist in the same cell and segregate to both daughter cells during cell division. Compatible plasmids have fewer DNA sequences in common and fewer interchangeable gene products than incompatible plasmids. A single cell can be infected by a number of compatible plasmids. Two plasmids that are incompatible, however, tend to compete for stable replication and segregation. Incompatible plasmids are considered to be closely related because of their overlapping replication and segregation requirements, similarity in their DNA sequences, and interchangeability of their conjugal functions. Plasmids are placed in *incompatibility* (Inc) groups, such as IncF or IncP, based on these criteria.

D. Nomenclature

Genes required to transmit a self-transferable plasmid are called *tra,* for transfer (or *trb,* when the number of transfer genes exceeds 26). Plasmid genes necessary for conjugation can be subdivided into three subsets: (1) trans-acting genes involved in cell–cell interactions and DNA transit; (2) trans-acting, and (3) cis-acting genes required to mobilize the molecule for transit. The first subset has been referred to as the mating pair formation system (Mpf) and the latter two subsets as the DNA transfer and replication system (Dtr). A plasmid such as F encodes a complete set of functions for DNA mobilization and transfer (Mpf + Dtr), whereas the mobilizable plasmid colE1, for example, can mediate its own mobilization (Dtr) but requires a self-transferable plasmid to effect its transfer in trans. Trans-acting functions encoded by and required for the transmission of mobilizable plasmids are called *mob* (DNA mobilization). The Dtr genes of a self-transferable plasmid involved in DNA metabolism are analogous to the *mob* genes of a mobilizable plasmid: Both groups act at a specific DNA sequenced called *oriT* (origin of transfer), so both groups will be referred to as mobilization functions. Mobilization genes are plasmid-specific; they are interchangeable only among very similar plasmids, sometimes not among all plasmids within the same incompatibility group. This is probably due to the coevolution of *mob* functions and their target DNA sequence, *oriT,* the only known cis-acting function required for plasmid transmission. Strikingly similar *oriT* loci, such as carried by the IncP plasmids RP4 and R751, are nonetheless distinguishable by their respective mobilization functions.

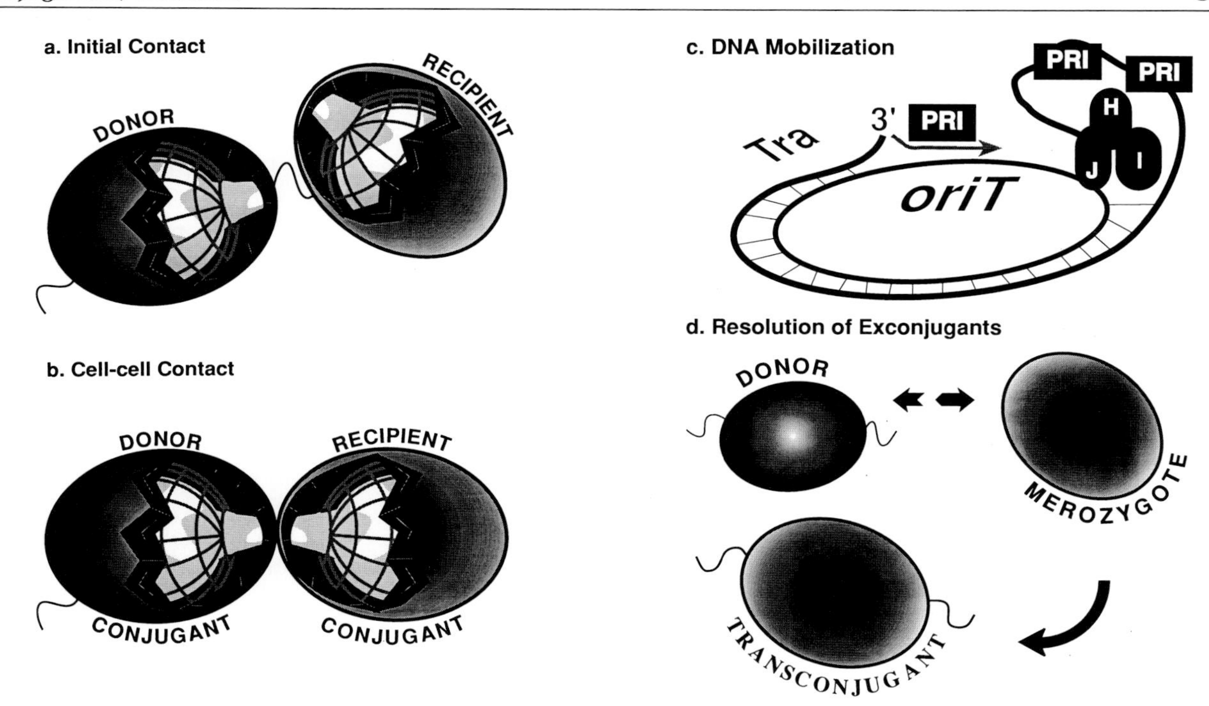

Figure 1 Cell–cell interactions and DNA transfer. (a) Cutaway views of bacterial conjugants, each pair representing a donor and recipient. Squiggles decorating donor cells are pili. The dark sphere surrounding the cells represents the outer cell membrane. The gray lattice represents the cell wall. The inner black outline represents the volume bordered by the cytoplasmic or inner cell membrane. The contact point between inner and outer membranes is a site of adhesion, from which pili are thought to emerge. One pilus bridges the cells. (b) Possible consequences of pilus contact with a recipient cell. Following contact, the pilus depolymerizes, perhaps drawing cells together into aggregates. Pili may also align cells, juxtaposing functions required for DNA transfer. Because the pilus probably makes contact with the recipient's cytoplasmic membrane, the possibility that cells are tethered at their respective zones of adhesion is illustrated. (c) Enlargement of a conjugative element at initiation of mobilization. The black ovals portray the nucleosome, part of which, at least in the F system, is thought to be embedded in both the inner and outer membranes. H, I, and J represent the RP4 *traH, traI,* and *traJ* gene products, which generate nicks at *oriT in vitro.* The double-stranded ellipse depicts strands of the conjugal plasmids. One strand has been interrupted by a nick at *oriT*. TraI remains covalently attached to the 5′ end of the cleaved strand. The IncP *pri* gene product, indicated by the black box, may prime replacement DNA strand synthesis in the donor, indicated by the gray arrow growing 5′ to 3′, and appears to accompany the transferred strand into the recipient, perhaps to prime discontinuous strand synthesis. The relative position of the *tra* region to *oriT* is indicated. (d) Exconjugants immediately after DNA transfer. Cells resolve into donors and merozygotes; the latter develop into transconjugants (indicated by pili) after the transferred plasmid is inherited.

II. Plasmid and Host

A. The Conjugal Donor

1. Cell Level Interactions

The *tra* functions of a conjugal plasmid confer upon its host a donor phenotype. In F, most *tra* genes are clustered in one macro-operon transcriptionally activated by the *traJ* product, which is repressed in most F-like plasmids by the fertility inhibition (*fin*) genes *O* and *P*. FinOP are likely active as RNA molecules. FinP may encode a complementary (or antisense) RNA sequence to the *traJ* transcript, preventing ribosome binding. FinO may promote interaction of *finP* and *traJ* RNA, but its mode of action is not understood. Plasmids of other incompatibility groups display a different organization. The *tra* functions of the $IncI_1$ plasmid ColIb and the IncP plasmid RP4, for example, are clustered into three separate regions called Tra1, Tra2, and Tra3. The position of *oriT* relative to the *tra* functions, however, is conserved between incompatibility groups.

Thirteen of the approximately 31 conjugation-specific genes in F are involved in the production, assembly, and erection of pili, cell surface structures generally required for plasmid transmission. Three types of pili have been described: thin flexible, thick flexible, and rigid. The F pilus, composed of many subunits of a single polypeptide in a helical array, is thick and flexible. Both pilus structure and serological reactivity are similar for pili encoded by plasmids in the same incompatibility group.

Pili have been proposed to participate in as many as three processes: signal transduction, cell–cell aggregation, and DNA transport. Nevertheless, functions attributed to them lack direct experimental support. Evidence for these hypothetical functions is summarized below.

An attractive assumption is that pili head a signal tranduction pathway. Because donor cells lacking pili or prevented from contacting recipients fail to initiate subsequent conjugation-specific events, pili may "sense" contact between conjugants and trigger such events, including pilus depolymerization and DNA metabolism. The type of sensory apparatus that pili might represent is unknown. Beyond the fact that pili are depolymerized shortly after contact, with their subunits returning to the donor cell membrane, no evidence indicates a signal transduction pathway.

Pilus depolymerization may be a mechanism by which conjugates aggregate (Fig. 1b). In this model, the pilus is required for establishing contact with a recipient and then drawing it to the donor. Because Zn^{2+} ions and certain pilus-specific phage both bind to the F-pilus tip and inhibit cell aggregation, the first interaction between pilus and recipient cell may involve the tip of the pilus. Subsequent depolymerization of the pilus might be responsible for drawing conjugants together into aggregates that include up to 80 individual cells.

The stability of these aggregates varies with time. Initially shear-sensitive aggregates become shear-insensitive. The more stable association has been attributed to bridging of the juxtaposed cells by the DNA transit apparatus. No molecular details of stabilization are known, making an intepretation problematical. The relevance of these observations to conjugation in general is also unclear. The transition from shear-sensitive to shear-resistant aggregation has only been demonstrated for F-mediated crosses between conjugants in liquid growth medium. Even in F-mediated matings where conjugants fail to establish shear-insensitive aggregates in liquid, mating is accomplished at normal frequencies on a solid substrate. Shear-insensitive contact may be a secondary attribute of F-mediated aggregation that permits DNA transfer under marginal conditions, such as in liquid, but not a functional aspect of DNA transit per se.

The efficiency of crosses in liquid has been ascribed to F's thick, flexible pilus. Plasmids of other incompatibility groups (such as $IncI_1$ and C) that determine thick or thin flexible pili also generally transmit equally well in liquid or on solid environments, although some exceptions exist. Plasmids that determine rigid pili, such as IncP and W, transmit relatively poorly in liquid. The $IncI_1$ plasmids R144 and R864a both elaborate two pilus types, thin flexible and thick rigid, and transmit equally well in both environments. However, mutant derivatives of R144 and R864a that constituitively express only thick pili transmit relatively poorly in liquid culture. It is possible that the thin pili stabilize conjugants during an encounter in liquid and are otherwise dispensible, but flexible pili have yet to be shown as either necessary or sufficient to account for high-frequency transmission in liquid. Demonstrating that indeed the only defect in plasmids expressing only thick pili was the absence of thin pili would settle the issue. Moreover, that the pilus itself stabilizes conjugating cells has yet to be determined. Swapping a rigid pilus for the thick, flexible F pilus by introducing a known, nonpolar mutation in the F-pilus structural gene (*traA*) and culturing a compatible but rigid pilus plasmid along with the mutant F plasmid may provide insight into this matter.

Pili may be the passageways routing DNA from the cytoplasm of one cell to the cytoplasm of the other. F pili are hollow and of adequate dimension to pass a single-strand of DNA. Indeed, DNA transmission between cells separated by a filter is indirect evidence that DNA traverses the pilus. Because these filters can be penetrated by pili but not whole cells, the pilus could be a DNA conduit.

The experimental data, however, are not consistent with an extended pilus being the channel. Although gene transfer occurs between cells separated by a filter, the apparatus previously described does not exclude all possibility of cell–cell contact, only cell passage, through the filter's pores. Furthermore, intact pili are not required for DNA transfer. When pili were disassociated by treatment with the detergent SDS after mating pair stabilization but before DNA transfer, transfer was not inhibited. Nonetheless, the pilus base, which survives detergent treatment, may suffice to provide a channel for DNA transit, or a pilus may suffice as an alternate route for DNA delivery.

Rather than DNA conduits, pili may be indirectly involved in transfer. While cells are tethered, pili might gather the membrane components of the transfer apparatus at each of their ends, allowing heterologous functions to diffuse away. Pili are anchored at membrane zones of adhesion in donor cells, points of contact between cytoplasmic and outer membranes, and may reach analogous sites in recipients where the cytoplasmic membrane is probably most accessible (Fig. 1b). Indirect evidence indicates actual contact with the cytoplasmic membrane of the recipient. When the TraS protein, which discourages transmission of additional F plasmids into occupied cells (see Section II.B.1.b), is present in the cytoplasmic membrane of recipients, both DNA transmission to the recipient and plasmid DNA synthesis in the donor are reduced, suggesting that contact with the inner membrane is required to initiate transfer. Furthermore, transient fusion of conjugants' membranes has been observed in F-mediated transmission. Certain heterologous envelope proteins, such as phage λ receptor sites, are reported to exchange bidirectionally between conjugants, but membrane exchange itself has not been seriously tested, in part because the transfer of certain heterologous cytoplasmic proteins during conjugation has not been detected. However, these proteins may be excluded from the recipient either passively, by size restrictions, or actively, by barring proteins not normally associated with membranes or DNA. Because many *tra* gene products are localized to membranes and at least some plasmids are replicated in association with the cytoplasmic membrane, membrane exchange between cells remains a viable alternative mechanism of DNA delivery. An intriguing possibility is that F transfer results from the transfer of *mob* factors, such as TraYI (see Section II.A.2), that may anchor the plasmid to the cytoplasmic membrane.

Of the many genes necessary for the elaboration of F pili, some also appear to affect cell aggregation and DNA transfer. Products of *traG* and, to a lesser extent, *traN* may be necessary for shear-resistant aggregation. The range of *traG*$^-$ phenotypes has led to the suggestion that this protein has two roles, one in pilus biosynthesis and the other in aggregate stabilization. Mutations in either *G* or *N* inhibit DNA transmission; however, the effects of *traN* mutations can be partially alleviated in matings performed on a solid substrate. Piliated donors of *traG*$^-$ plasmids form stable mating aggregates less frequently than donors of wild-type plasmids but more frequently than the frequency of plasmid transmission reflects. These observations do not distinguish between TraG having a structural role in aggregates or having an activity that leads to the stabilization of cell–cell contact.

The F *traD* and *traU* gene products remain the best candidates for components of the putative transit apparatus. Temperature-sensitive *traD* mutants have been used to demonstrate that *traD* acts after mating aggregates are formed but prior to complete DNA transmission. TraD is also required by RNA phage f2 to penetrate F$^+$ cells. Cells bearing plasmids with amber mutations in *traD* support f2 infection through the adsorption and ejection stages (characterized by sensitivity to RNase) but are resistant to the entry of phage RNA. This result suggests that TraD may serve as a membrane pore for nucleic acid transit. TraU also has a dual role, supporting phage infection and conjugation. Plasmids carrying point mutations in the *traU* gene were previously thought deficient in pilus biosynthesis because of their inability to support infection of pilus-specific phages. New studies have shown that cells with plasmids carrying *traU* deletions still present pili, albeit fivefold fewer than the wild-type number. The presence of pili on cells resistant to male-specific phages argues that TraU may be a constituent of the nucleic acid transit apparatus, transferring both the F plasmid and the phage genome.

Excepting the structure of the DNA molecule, nothing is known about the mechanics of tranfer. It may be misleading to assume that all *tra* genes are either actually involved in donor specific events or limited to DNA transit. Products of the *tra* genes may function after transfer to establish the new molecule in the recipient cell. Mutations in these genes would confer the same phenotype as mutations in genes that prevented DNA transit. Because separating genetically DNA transfer from transmission is difficult, data concerning "transfer mutants" should be interpreted cautiously, so that the possible contribution of cofactors accompanying transferred DNA molecules is not overlooked.

2. DNA Metabolism

a. Plasmid Genes

Concurrent with pilus depolymerization, DNA is prepared for transit by the concerted actions of the *mob* gene products. Mobilization begins with cleavage of the plasmid at *oriT*. Heterologous molecules carrying *oriT* receive strand- and site-specific cleavages at *oriT* when grown with a self-transferable plasmid. Therefore, it has been proposed that conjugal plasmids in unstimulated donor cells oscillate between covalently closed circular and nicked circular forms (Fig. 1c). After donor–recipient contact, the equilibrium is shifted to favor the nicked circular form, and DNA unwinding is initiated. In the F plasmid system, endonuclease activity is associated with the *traY* and *traI* (formerly *traZ*) gene products. IncP plasmid RP4-specified TraI and TraJ proteins relax supercoiled molecules carrying a compatible *oriT in vitro* and are probably complexed with TraH in a nucleoprotein complex *in vivo*. After nicking, TraI remains covalently attached to the 5′ end of the cleaved strand.

Apparently, *oriT* consists of two or more cis-acting domains. Studies on three plasmids, representing three different incompatibility groups, suggest that part of *oriT* is required for strand-specific cleavage and a second part is required for transfer or establishment following transfer. An analysis of different deletions in the F plasmid *oriT* showed that some sequences required for plasmid transmission were not necessary for strand cleavage. The *oriT* of IncP plasmid RP4 was separated into two functional domains. One domain allowed transmission at a reduced frequency while another had no apparent function alone but enhanced the transmission frequency of a plasmid also carrying the other. This suggests that the first half contains the "nicking" domain and the second half functions in later steps of transmission. Finally, an examination of various mutant *oriT*s from the IncQ plasmid R1162 provided compelling evidence for a "nicking" domain and a "resolution" domain within *oriT*, based on the way individual mutants interacted with a linked wild-type *oriT*.

Completion of the conjugal cycle in the donor requires replacing the transferred strand. DNA synthesis occurs simultaneously with helix unwinding (Fig. 1c). The displaced strand is mobilized 5′ to 3′, making replication-driven displacement consistent with the polarity of F plasmid transmission. However, a temperature-sensitive allele of *dnaE*, a host-encoded gene necessary for both chromosomal and plasmid DNA synthesis, did not prevent DNA transmission at the nonpermissive temperature.

TraY and TraI proteins of F may be involved in orienting the plasmid and nick site for transfer. Subcellular localization techniques have revealed that TraY resides in the cell envelope and TraI in the cytoplasm. TraY alone or in conjunction with TraD may anchor the plasmid to the cell membrane in alignment with the transit apparatus. TraI has been characterized as a DNA helicase, with a DNA-dependent adenosine triphosphatase (ATPase) activity. The ATPase activity of TraI may provide the energy to propel the displaced strand into the recipient, a mechanism consistent with the polarity of transmission. TraYI is plasmid-specific, being unable to recognize the *oriT*s of different plasmids even within the same incompatibility group. TraI displays no sequence preferences for helicase activity *in vitro* and, therefore, may be plasmid-specific due to a conserved interaction with its TraY partner or, alternatively, because it regains sequence dependence *in vivo*. Helicase activity may be further modulated by TraM, another plasmid-specific function. TraM is located in the inner membrane and reportedly binds in the area of *oriT*. A possible role for the *traM* (and/or *traY*) gene product is to open gaps in the duplex, allowing entry and activation of TraI. Evidence for such a role comes from the finding that *traM* is not required for pilus biosynthesis, cell aggregation, or DNA nicking but is required for DNA transmission and for synthesis of a replacement strand in the donor. These latter functions depend on DNA unwinding, a property of the helicase.

TraYI are also required for the stimulation of recombination observed between homologous regions of F plasmids and λ phage. The observed hyperrecombination depends on the host *recBCD* pathway unless the phage and plasmid "infect" a recipient cell simultaneously. These observations can be reconciled given the proposed role of *recBCD* in recombination, which is, in part, to unwind strands of a broken DNA molecule. The single-stranded DNA generated is a substrate for RecA, which seeks out and pairs complementary homologous strands. TraYI probably introduces a break in the double-stranded plasmid, allowing RecBCD entry. RecBCD activity may be unnecessary to stimulate recombination between newly transferred molecules if they persist transiently as single strands in the recipient.

b. Chromosomal Genes

Conjugation is itself a type of replication, in which daughter molecules are segregated to neighboring cells rather than to daughter cells. Although not specific to conjugation, chromosomally encoded replication functions may participate in conjugal transmission and vegetative replication of the chromosome and plasmid. Discussion here will be limited to genes and gene products with demonstrated involvement in DNA transmission. Reported so far are integration host factor (IHF), gyrase, *cpxA*, *sfrA*, *cpxB*, *sfrB*, and *rho*.

IHF, whose subunits are the product of the *himA* and *hip* genes, helps regulate a number of cellular activities—such as phage λ integration, transposition, and gene expression—and has at least two effects on F-mediated conjugation. Both effects may be understood in terms of the proposed role of IHF, which is to bend DNA into higher-order structures. First, IHF may promote conjugal DNA synthesis and helix unwinding from *oriT*. IHF-binding in the *oriT* region of F has been demonstrated *in vitro*, and *himA*$^-$ or *hip*$^-$ donors transmit plasmids at a reduced frequency. Disruption of some IHF-binding sites in the *oriT* region results in an accumulation of plasmid multimers in transconjugants. A reduced transmission frequency from IHF$^-$ donors may reflect aberrant processing of DNA intermediates or a decrease in initiation of unwinding. Second, the *in vivo* effects of IHF mutations may be a consequence of differences in gene expression. Transcription of the F genes *traM*, gene *X* (function unknown), and *traJ* is diminished in *himA*$^-$ cells and the size of *traJ* and gene *X* transcripts is reduced. Attenuated *traJ* expression in turn reduces expression of other conjugal functions because *traJ* controls expression of most conjugation-specific genes on F.

The IHF effect is not universal. IHF does not affect IncP plasmids. Curiously, the *in vivo* effect on transmission seen in *him* or *hip* mutants is suppressed when F is replicated at a higher copy number than the wild-type plasmid. A low level of *tra* gene expression from a larger pool of genes may suffice to accumulate a critical concentration of transfer functions.

A role for gyrase has been inferred from the observation that nalidixic acid, a gyrase inhibitor, also inhibits mobilization and transmission of F. However, a direct role for gyrase in DNA transfer has not been demonstrated.

Host genes *cpxA*, *sfrA*, *cpxB*, and *sfrB* all have pleiotropic effects on the cell surface and appear to affect the levels of *tra* gene products found in the outer membrane. Homologous to *envZ* and *ompR* are *cpxA* and *sfrA* (also called *seg*, *dye*, *arcA*, *fexA*, *msp*, and *cpxC*), respectively, and they thus display a "two-component regulator" system motif. The "sensor" component (CpxA) resides in the cell envelope, where it may receive specific environmental signals. Signal transduction proceeds through modification of the "regulator" component (SfrA), a cytoplasmic protein. Consistent with its hypothesized role, SfrA appears to regulate the transcription of the genes for a particular class of proteins. Donors with mutant *sfrA* alleles transcribed less *traJ* messenger RNA (mRNA) than *sfr*$^+$ donors, but this condition could be reversed by directing *traJ* transcription from a heterologous promoter. In concert with TraJ, SfrA controls expression of *traY*. It has yet to be demonstrated, however, that *cpxA* and *sfrA* interact, and therefore they are probably members of two different modulator–effector pathways. Although not a component of the regulator system, SfrB may also affect gene expression. Mutations in the *sfrB* gene reduce mRNA levels of *traYI*, perhaps by permitting premature *rho*-dependent termination of the transcript, since mutations in *rho* suppress the *sfrB*$^-$ phenotype.

B. The Conjugal Recipient

1. Cell Level Interactions

Although many functions are both required for and specific to donor activity, no functions unique to and required of recipients have been recognized. It was expected that recipient cells actively identify themselves to donors, perhaps by presenting a conjugal receptor on their surface. This expectation has no experimental support. In fact, mating may be an unprovoked response that can be inhibited but not further encouraged. Conjugal elements, on the other hand, do encode surface exclusion proteins that discourage conjugation between two cells bearing incompatible plasmids. These factors are expressed in donor cells but apparently regulate their host's participation as a recipient.

a. DNA Transfer

Putative conjugal receptors have been identified. Searches for recipients deficient in conjugal activity, or Con$^-$ cells, had implicated two classes of molecles normally found on the cell surface. The first class was lipopolysaccharide (LPS), a major component of cell outer membranes. The second class was protein, particularly the product of the *ompA* (outer membrane protein A) gene. Because mutations that alter LPS synthesis often alter the protein composition of the outer membrane, the conjugation defect of LPS mutants may be indirect—for example, these mutants may fail to present the true pilus receptor. Mutations of *ompA* are also pleiotropic, making it difficult to assess the protein's role in conjugation. Even different plasmids within the F group display distinct stringencies in their requirement for *ompA* product during mating in liquid growth media. However, in support of a specific role for OmpA and LPS in conjugation is the finding that purified *ompA* product inhibited conjugation in the presence of LPS, whereas LPS alone had little effect.

Con^- strains were uniformly compared to test the intriguing possibility that a particular LPS moiety or the *ompA* protein served as a receptor. Con mutants were isolated as one of two phenotypes: cells defective as recipients in conjugation or cells resistant to phage infection and defective in conjugation. The attractive assumption about phage-resistant mutants was that conjugal "infection" is similar to phage infection, at least as the level of cell recognition and transfer of nucleic acids. Four independent *E. coli* conjugation mutants initially selected for resistance to either phage K3 or ST-1, or the three phages T3, T4, and T7, and seven indepenent *E. coli* mutants directly selected for a conjugal defect revealed (1) as a group, the 11 isolates failed to display a uniform defect in either LPS or *ompA*, (2) all but one were competent to mate on a solid substrate (the phenotype of the exception was inconclusive), suggesting that *ompA* and LPS may affect the establishment of a conjugal relationship in liquid but are unnecessary for the relationship itself and are not involved in DNA transfer, (3) all were capable of binding purified F pilin to the same degree as their parental strains (whereas *Bacillus* or *Pseudomonas* strains did not bind pilin), and (4) the one mutant amenable to examination by electron microscopy dislayed a propensity to form shear-sensitive aggregates. Interestingly, microscopy could only be used on one of the 11 mutants because the Con^- clones clump in the absence of F^+ donors. The Con^- phenotype may be simply due to an exclusion of donors from aggregates of Con^- cells. The survey concluded that if a specific pilus receptor exists, it cannot be a single form of LPS or the *ompA* gene product.

The broad host range for DNA transfer is also inconsistent with the existence of a specific pilus receptor. IncP plasmids bridge large evolutionary distances by transferring among essentially all gram-negative bacteria, to and from gram-negative and gram-positive bacteria, and to the yeast *Saccharomyces cerevisiae* (discussed in Section III). Given the broad range of species susceptible to DNA transfer by a single conjugation system, it seems unlikely that all these recipients would carry a common specific receptor.

b. Surface Exclusion

The exchange of related plasmids between donors is discouraged but not prevented by surface exclusion. Donor cells that harbor plasmids belonging to different incompatibility groups usually act like virgin recipients for the plasmid of the other donor, making exclusion of related elements plasmid-specific. In F, the genes determining surface exclusion are called *traS* and *traT*. The *traS* protein probably resides in the inner membrane, whereas the *traT* product is found in the outer membrane. TraT is a doughnut-shaped protein that may trap the pilus tip in its central hole. TraT and TraS may stifle the mating signal required for initiation of conjugal DNA synthesis or instead may inhibit molecular transfer. The actual mechanism of surface exclusion remains unknown.

2. DNA Metabolism

Like donor chromosome-encoded functions, no known recipient chromosome-encoded functions are specific to conjugation. Those functions known to be involved in conjugation are concerned with chromosomal DNA replication and gene expression. Upon entry, a linear, single-stranded molecule must be recircularized and made double-stranded. Recircularization may be the consequence of plasmid-encoded functions that accompany the transferred DNA, perhaps in concert with host-encoded functions. In the IncP system, it appears that the product of the *pri* gene, a combination helicase and primase, is cotransferred with plasmid DNA. These functions are consistent with the proposed roles for a transferred protein in DNA synthesis. The transfer of other cytoplasmic proteins during conjugation may be inferred from genetic evidence; transconjugant cells fathered by F^+ λ lysogens display transient immunity to λ phage, possibly due to the transfer of the cI protein.

Cis-acting sites on transferred DNA may act to specifically bind proteins to the transferred strand of DNA. Proteins synthesized in the donor, or recruited from the recipient, may convert the transferred DNA to a replicon (repliconation). The similar plasmids pBR322 and colE1 possess sequences that allow conversion of single-stranded molecules to double-stranded replicons by a ΦX174 mechanism. These sequences have been identified as factor Y-binding sites and permit the assembly of a multiprotein primosome complex in ΦX174. Putative factor Y-binding sequences confer upon heterologous single-stranded phage the ability to convert to double-stranded replicative forms. Because the factor Y-binding site is close to the putative *oriT* region of pBR322 and lacks a detectable function in vegetative replication, it was suggested that factor Y sites are required for converting conjugally transferred single-stranded DNA into a replicon. Nonspecific cis-acting sites may also function to recruit necessary proteins of either the donor or recipient for repliconation. Proteins that bind to the ends of DNA molecules may associate efficiently with the nick site at *oriT*, such as TraI of RP4 and the *virD2* gene product of the *Ti* plasmid in *A. tumefaciens*. Nonspecific single-stranded DNA binding proteins could also play roles in transmission. Four such proteins are *ssb* of F, *virE2* of Ti, and *ssb* and *recA* proteins of the *E. coli* chromosome.

III. Host-Range Determinants

Recipients provide a number of challenges to incoming nucleic acids, which may potentially affect their replication and segregation to daughter cells. Elaborate modification–restriction systems threaten a recently transferred molecule with degradation. Incoming elements must be compatible with resident plasmids to establish as replicons. A molecule must replicate using the host replication machinery, express its own replication genes, or integrate into a resident replicon. In crosses involving *Hfr* (high frequency of recombination) donors, transmission of a gene is the consequence of recombination. *Hfr*s are usually formed by homologous recombination between the host chromosome and a segment of the F plasmid to link plasmid and chromosome. Integrated plasmids can still direct conjugal transfer. However, conjugation is often aborted before an entire chromosome with integrated F is transmitted. In these cases, establishment of transferred DNA requires recombination with the recipient chromosome. As most bacteria—perhaps most organisms—prevent recombination between nonhomologous molecules, replication and segregation strategies based on transfer and integration usually require the donor and recipient to be related.

Generally, conjugal plasmids can be divided into those with a broad host range and those with a narrow host range. In contrast to members of the IncP group, the F plasmid displays a classic narrow host-range phenotype, being transmitted routinely only between various Enterobacteriaceae. Evidence indicating that host range is limited by the inheritance requirements of different species rather than by *tra* functions comes from two types of experiments. First, plasmid mutations affecting host range alter but do not expand host range or uniformly affect host range; each mutation limits the plasmid's transmission to a different set of species. Such mutations map to the replication region of the plasmid.

Second, swapping replication, but not *tra*, functions alters host range. The host range of the F plasmid can be expanded to include *Pseudomonas aeruginosa* by increasing the variety of replication sequences it carries. The most dramatic example of such an analysis is that conjugal plasmids mediate DNA transfer betweeen a prokaryote (*E. coli*) and a eukaryote (the yeast *S. cerevisiae*). When the replication region of a yeast replicon was inserted into either an IncP or IncF plasmid, each of their respective conjugal functions promoted transmission of the plasmids to yeast.

In addition to DNA, conjugating cells transfer proteins. The particular molecular cofactors transferred during conjugation may influence the range of species to which a plasmid can be transmitted. Mutations in the *pri* gene of IncP plasmid RP4 affects the frequency of transmission of the mutant plasmid differentially depending on strain and species of recipient. Although a speculative proposition, the much broader host range of IncP plasmids over IncF plasmids may in part be due to the repertoire of functions available to IncP plasmids immediately after transfer.

IV. Other DNA Transmission Systems

A diversity of DNA transmission systems have been identified in the Eubacteria. The systems discussed here are similar to the conjugal systems discussed earlier but are native to *A. tumefaciens*, *Enterococ-*

cus faecalis, Bacillus subtilis, and *Spiroplasma citri* rather than to *E. coli.*

Agrobacterium tumefaciens is the causative agent of plant crown gall disease. Pathogenicity is mediated by the transmission of plasmid DNA from the bacterium to the target cell of a susceptible host plant. The transferred sequences cannot autonomously replicate in a plant cell and indiscriminately integrate into the chromosomes, reminiscent of transposable elements. Genes necessary for virulence are located on the Ti plasmid and activated by exudates released from the wounds of susceptible plants. Ti can transmit itself in its entirely to other strains of *A. tumefaciens* by conjugation, and interbacterial transmission is provoked by plant exudates. Ti also transmits a subset of its sequences to plants by a mechanism that exhibits parallels to conjugation. Putative transfer intermediates that have the same physical profile as intermediates in conjugation can be recovered from virulent strains of bacteria. The parallel between conjugation and this form of pathogenic transfer is strengthened by the finding that the *mob* and *oriT* regions of a mobilizable bacterial plasmid can replace certain virulence functions in *A. tumefaciens*. DNA intermediates of a conjugal *mob–oriT* interaction presumably substitute for DNA mobilized by virulence functions and are suited for transfer by virulence functions, perhaps in a way analogous to transfer of mobilizable plasmids by self-transferable plasmids.

A plasmid-mediated conjugative system identified in *E. faecalis* is like F in its requirement for cell contact and its insensitivity to exogenous nucleases; it resembles Ti in being regulated by diffusible compounds, in this case pheromones. Pheromone-regulated plasmids of *E. faecalis* are of two types, self-transferable and mobilizable, and are capable of transmitting antibiotic resistance determinants as well as chromosomal markers. Pili have not been reported in the *E. faecalis* system. Instead, donor cells present a proteinaceous aggregation substance that facilitates conjugant aggregation. The *E. faecalis* recipient initiates mating by secreting a peptide pheromone to which donors respond. Their response is mediated by a conjugative plasmid-encoded pheromone receptor that, when bound to pheromone, initiates events that culminate in DNA transmission. Transconjugants no longer exhibit pheromone activity and therefore do not induce the donor phenotype in cells harboring related elements. Different plasmids detect different pheromones, and each specifically stifles the pheromone activity of that peptide in its host without affecting other pheromones. As with the F system, donor and recipient cells pass through two levels of aggregate stability to achieve effective contact. An additional similarity to the F system is that aggregation-deficient recipients will mate on a solid substrate.

A conjugal system native to *E. faecalis, Streptococcus pneumoniae,* and *Streptococcus sanguis* is apparently not mediated by plasmids but by the conjugative transposons Tn916, Tn918, Tn920, and Tn1545. Conjugative intercellular transposition is resistant to exogenous deoxyribonuclease and requires cell contact. These transposons presumably exist transiently as extrachromosomal elements when they excise from the host chromosome to effect either conjugal transfer or intrachromosomal transposition. As conjugal elements, transposons have minimized the stage of their life cycle requiring autonomous replication, being dedicated to recombination with resident replicons. Work with Tn 1545 suggests further similarities between transposons, phage, and conjugal plasmids. Tn1545 encodes two proteins, designated Xis-Tn and Int-Tn, that, like phage λ's Xis-λ and Int-λ, promote site-specific integration of DNA with minimum DNA homology requirements.

In *B. subtilis,* chromosomal DNA is transmitted between cells in the absence of autonomous plasmids. DNA transfer in *B. subtilis* requires cell contact, initiates at a single site (transfer origin), and correlates with chromsomal replication, like *Hfr*-mediated transfer in *E. coli.* A possible difference between the *Hfr* phenotype of *E. coli* and contact-dependent chromosomal transfer in *B. subtilis* is that the latter is limited to a small segment of the life cycle. Contact-dependent DNA transfer between *B. subtilis* cells has been observed only during the first replication of the chromosome of a germinating spore. Transfer either does not occur at other stages or is obscured by the contact-independent natural transformation system in *B. subtilis*.

A system of chromsomal transmission has been reported in the mycoplasma *S. citri,* which is thought to be a descendant of gram-positive bacteria. Consistent with the presumed lineage of the mycoplasma, conjugation between *E. faecalis* and *Mycoplasma hominis* has been demonstrated, and chromosomal transmission between *S. citri* cells resembles the *B. subtilis* system in that there is an absence of plasmids. Chromsomal transmission in

S. citri requires cell contact and is insensitive to exogenous nucleases. As in the *Hfr* mechanism of chromosome transmission, recombination between transferred DNA and homologous sequences in the recipient apparently is required. Because centrifugation or incubation in polyethylene glycol, both of which promote protoplast fusion, increases the frequency of transmission, transient membrane fusion has been implicated in this system.

V. Conclusion

A. Parallels in DNA Transmission Processes

Conjugation, transposition, and phage infection are kindred processes insofar as the elements engaged in such activities have analogous or interchangeable functions. Conjugation and transposition are linked by the identification of elements displaying hybrid qualities. Phage λ and Tn1545, which utilize analogous integration functions, and conjugative transposons are examples of hybrid elements. Conjugal elements resemble phage that have minimized the extracellular stage of their life cycles. Both conjugation and phage infection are mediated by extrachromosomal elements that sometimes sport regions of chromosomal DNA or integrate into the chromosome. Recent demonstration of genetic complementation of S^- λ phage by the *srnB* gene of F and the *kil* gene of colE1, all encoding proteins that appear to promote bacteriolysis, provides another example of potential relatedness. Both viruses and plasmids encode protection against related elements through replication incompatibility and surface exclusion. Some plasmid–virus and virus–virus combinations are also incompatible at the level of replication or exclusion by a process called abortive infection. For example, phages T7 and ΦII adsorb and transfer DNA to F^+ and F^- hosts, but infections in F^+ cells differ from infections in F^- cells. In F^+ hosts, phage mRNA for a group of proteins is reduced, synthesis of those proteins is virtually eliminated, and DNA synthesis is inhibited. These effects are due to the *pifA* and *pifB* gene products of F. Expression of these genes is regulated by *pifC*, which is also involved in the apparent inhibition of IncP transfer genes by F. Also, at least one prophage inhibits transmission of P plasmids. *Pseudomonas aeruginosa* harboring prophage B3 inherit P plasmids less frequently than do nonlysogenic cells. Phage and plasmids may employ similar or identical mechanisms to guard against superinfection of like phage, like plasmids, and like plasmids and phage.

Phage and some conjugal elements may share sequence homology, like factor Y binding sites in pBR322 and ΦX174. Extensive sequence homology exists between the *oriT* nicking domains of IncP plasmids (RP4, R751) and the nick sites of the border sequences within the Ti plasmid. There is a 7 of 12 basepair homology between the consensus nick sequence derived from a comparison of the IncP and Ti sequences and the nick sites of ΦX174 and plasmids C194 and UB110.

B. Conjugal Network

DNA exchange mechanisms may constitute a network whereby subscribers are updated with the genetic advances of neighbors. In some cases, such as in *A. tumefaciens*, the release of genetic information may benefit one species at the expense of another. In other cases, however, such as the dissemination of antibiotic resistances among microbes, the sharing of genetic information may prove advantageous to both. The bidirectional exchange between bacteria and the unidirectional exchange from bacteria to plants and fungi are evidence supporting the concept that genetic information is shuttled among all organisms regardless of their distances from a common ancestor. In one scenario, genetic information may flow from plants, animals, fungi, and archaebacteria back to eubacteria by natural transformation, a mechanism of nucleic acid uptake exhibited by some eubacteria.

C. Adaptations of Conjugation

Conjugation may primarily benefit autonomously replicating nucleic acids that multiply faster than an asexual host by dispersing to other individuals. But conjugal transfer has clear benefit to microbes too, providing an avenue for distributing antibiotic resistance determinants and other novel traits. Conjugal transmission of chromosomal DNA may further aid related strains by providing homologous sequences with which to repair damaged chromosomal DNA. Finally, conjugation can be involved in pathogenesis, i.e., crown gall, and perhaps some organisms use conjugation to establish a symbiosis.

D. Evolutionary Impact

The simple acquisition of new traits is one way that conjugation benefits species. Conjugation provides a means for disseminating novel markers of selective advantage, like resistances to antibiotics, quickly through the microbial community and even the infrequent inheritance of useful markers could alter the development of a species. The full evolutionary impact of genetic transfers may be difficult to determine. What, for example, could be the consequence of transferring proteins or DNA molecules unable to replicate? Transferred molecules may have a heritable effect on the recipient whether or not that particular molecule persists in the population. This has recently been demonstrated for RecA protein transfer between cells and the subsequent induction of λ prophage by the transferred protein. Transfer of the RecA or, as suggested earlier, the cI protein alters a heritable epigenetic stage. Transferred DNA molecules can deliver more than just novel genes: They can provide templates for the repair of chromosome damage. To an individual that has suffered from a transient but lethal assault on its genome, a transferred template for DNA repair might extend its life long enough to reproduce. Information-sharing may make a much larger contribution to evolution than has been appreciated by using changes in nucleotide or amino acid sequence as a yardstick.

Acknowledgments

I thank Drs. K. Tilly, P. Rosa, P. Policastro, J. Swanson, B. Ankenbauer, and S. Fischer for critical review of the manuscript. For help in preparing the manuscript, I am grateful to L. Hamby, J. Steele, S. Smaus, and C. Smaus. I am indebted to the participants of the 1990 Banff meeting on Gene transfer organized by L. Frost and W. Paranchych.

Bibliography

Bernstein, H., Hopf, F., and Michod, R. (1987). *Adv. Genet.* **24,** 323–370.

Guiney, D. G., and Lanka, E. (1989). Conjugative transfer of IncP plasmids. *In* "Promiscuous Plasmids of Gram-Negative Bacteria" (C. M. Thomas, ed.), pp. 27–56. Academic Press, London and San Diego.

Heinemann, J. A. (1991). *Trends in Genetics* **7,** 181–185.

Ippen-Ihler, K. (1989). Bacterial conjugation. *In* "Gene Transfer in the Environment" (S. B. Levy and R. V. Miller, eds.), pp. 33–72. McGraw-Hill, New York.

Porter, R. D. (1988). Modes of gene transfer in bacteria. In "Genetic Recombination" (R. Kicherlapati and G. R. Smith, eds.), pp. 1–41. American Society for Microbiology, Washington, D. C.

Willetts, N., and Wilkins, B. (1984). *Microbiol. Rev.* **48,** 24–41.

Continuous Culture

Jan C. Gottschal
University of Groningen

Glossary

Auxostat Any continuous culture system in which a growth-dependent parameter is maintained constant and all other parameters, including the dilution rate and, hence, the specific growth rate, adjust accordingly

Chemostat Any continuous culture system in which the dilution rate and, hence, the specific growth rate is set externally and all other parameters adjust accordingly

Dilution rate (D) Flow rate of incoming fresh medium (F) divided by the actual volume (V) of the culture in the continuous culture vessel

Half-saturation constant for growth (K_s) Substrate concentration at which the specific growth rate equals half the maximum specific growth rate (μ_{max})

Specific growth rate (μ) Rate of increase of biomass relative to the biomass already present ($1/x \cdot dx/dt$)

Steady state Condition of a continuous culture in which changes in density and physiological state of the cells are no longer detectable

Yield coefficient (Y) Quantity of cells produced per substrate consumed

AS OPPOSED TO the "closed" batch-type of cultivation, a continuous culture is a typical "open" system in which a well-mixed culture is continuously provided with fresh nutrients and the volume is usually kept constant by continuous removal of culture liquid at the same rate at which the fresh medium is supplied. Several types of continuous cultures can be constructed. The most common type surely is the chemostat, in which the growth rate (μ) of the cells is fixed by the *external* control of the rate at which medium is fed to the culture and the cell density is set by the concentration of the limiting nutrient (S_r) present in the reservoir medium. The dilution rate of the culture (D) can be varied over a large range of values from nearly 0 to 10–20% below the maximum specific growth rate (μ_{max}) of the organism. Over this range of dilution rates, steady states can be obtained during which the growth of microorganisms is exactly balanced by the rate of removal of cells from the culture. At dilution rates approaching the maximum specific growth rate, cell densities become progressively lower, and growth at μ_{max} is not possible at all. The only way to maintain cultures at (near) maximum rate of growth is to switch over to some kind of *internal* control of the supply rate of fresh medium. Such a control mechanism could be based on growth-dependent parameters (e.g., biomass or metabolic end-products) or on feedback control of a nutrient concentration. In comparison to the vast number of studies on the behavior of microorganisms grown in chemostats, very little information is available on the use of these nonsubstrate-limited continuous culture systems. Thus, although in theory continuous cultures are not synonymous to chemostats, in practice nearly all continuous culture work has been done in chemostats.

I. Principles of Continuous Culture

A. Single Substrate-Limited Growth

1. Theoretical Background

In a substrate-limited continuous culture, a chemostat, the fresh medium is supplied at a constant rate (F). Because the culture liquid, including the cells, is removed at the same rate, a constant volume (V) is maintained. The dilution rate (D) of the chemostat is defined as F/V. The composition of the inflowing medium is such that, in the most simple case, growth in the chemostat results in (almost) complete consumption of only one of the nutrients, the growth-limiting substrate, with a concentration of S_r in the feed. The specific rate at which growth proceeds will at first be maximal, but upon depletion of the limiting nutrient this rate will rapidly decline, which may conveniently be described by the well-known empirical relationship proposed by Monod (as early as 1942):

$$\mu = \mu_{max} s/(K_s + s), \tag{1}$$

where μ (the specific growth rate) $= (1/x)(dx/dt)$, with x representing biomass, s the actual concentration of the growth-limiting nutrient in the culture, and K_s the half-saturation constant for growth, numerically equal to the substrate concentration at which $\mu = \frac{1}{2}\mu_{max}$. Due to the supply of fresh medium, growth does not stop completely but continues at a rate determined by the feed rate. The combined effect of growth and dilution eventually leads to a steady state in which $\mu = D$:

change in biomass with time = growth − output

$$dx/dt = \mu x - Dx = (\mu - D)x = 0 \tag{2}$$

From this formulation, it will be clear that if $\mu = D$, the biomass concentration in the culture remains constant with time ($dx/dt = 0$) and a steady-state situation will be obtained. Such a steady state appears to be, to some extent, self-adjusting with respect to small changes in dilution rate. If $\mu > D$, the substrate consumption rate will exceed its supply rate, leading to a drop in the residual substrate concentration in the culture. According to Equation 1, this will cause a fall in the specific growth rate, a process that continues until $\mu = D$. A similar situation exists if initially $\mu < D$. In this case, an increasing concentration of growth-limiting substrates results in an increase in the specific growth rate. Steady states can be obtained for values of D below the critical dilution rate (D_c), which in most cases is close to μ_{max} because S_r is usually much larger than K_s:

$$D_c = \mu_{max}[S_r/(K_s + S_r)]. \tag{3}$$

Operation of a chemostat at dilution rates above this critical dilution will, of course, result in complete washout of the biomass and in a substrate concentration equal to that in the fresh medium.

Under steady-state conditions, the change in substrate concentration is given by the followig balance:

change in substrate concentration with time = substrate input − substrate output − substrate consumed

$$ds/dt = DS_r - Ds - \mu x/Y = 0 \tag{4}$$

where s represents the limiting substrate concentration in the culture and Y a yield coefficient defined as the quantity of cells produced per substrate consumed. From Equation 4, solved for x, it follows that

$$x = Y(S_r - s), \tag{5}$$

and from Equation 1, solved for s, it can be seen that

$$s = K_s D/(\mu_{max} - D) \tag{6}$$

From Equation 5, it is immediately clear that the steady-state cell density is determined primarily by the concentration of the growth-limiting substrate in the reservoir medium (S_r), whereas Equation 6 indicates that s is a direct function of the dilution rate, of course with the assumption that the parameters Y, μ_{max}, and K_s are constant for a given organism. This is most clearly illustrated in Fig. 1, in which arbitrarily chosen values for Y, K_s, and μ_{max} have been used. Although the assumption that the μ_{max} and K_s are constant over a wide range of dilution rates is

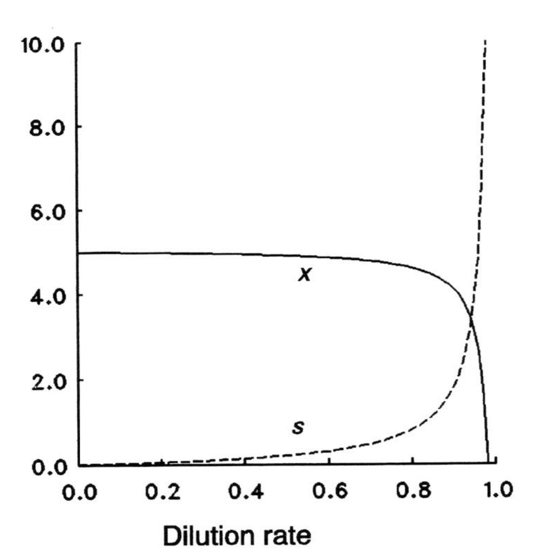

Figure 1 Theoretical relationship of the dilution rate and the steady-state values of bacterial concentration (x) and substrate concentration (s). Data were calculated assuming the following growth parameters: $\mu_{max} = 1.0\ hr^{-1}$, $Y =$ 0.5 g/g, and $K_s = 0.2$ g/liter and a substrate concentration in the inflowing medium of $S_r = 10$ g/liter.

certainly warranted, this is less so for the yield coefficient. This parameter may vary to some extent not only due to growth in different substrates, but also due to accumulation of intracellular products, changes in cell composition and viability, and variability in the maintenance energy requirements, especially at very low dilution rates. Most significantly, Fig. 1 demonstrates that at dilution rates well below D_c for a given organism relatively high cell densities can be obtained in the presence of very low, growth-limiting substrate concentrations. This aspect of continuous cultivation plus of course the fact that cells remain in an actively, relatively well-controlled physiological state has made chemostats extremely useful tools for physiological and ecological studies.

2. Determination of Some Growth Parameters

The chemostat is an excellent tool for the accurate determination of the maximum specific growth rate (μ_{max}). It will often give a more reliable result than a batch culture because (1) prior to the determination of μ_{max}, the culture can be grown at a rate close to μ_{max}, ensuring that all cells are optimally adapted to growth at their near-maximum rate, (2) lag-phases will not interfere with the measurement, and (3) possible influences of changing substrate and product concentrations are minimized. For the actual measurement, the dilution rate is increased (in one step) to a value of 20–50% above the critical dilution rate. This results at once in alleviation of the substrate limitation and in gradual washout of the culture. The rate at which this proceeds is according to the following expression:

$$-\mathrm{d}x/\mathrm{d}t = (\mu - D)x, \qquad (7)$$

which after integration gives

$$\ln x = (\mu - D)t + \ln x_0, \qquad (8)$$

where x_0 represents the cell density at the start of the washout period. Because the substrate is no longer limiting, the culture grows at μ_{max} and a plot of ln x versus t yields a slope of $\mu_{max} - D$. Because D is fixed at a known value, μ_{max} can be determined.

Another important growth parameter, the yield coefficient, can be obtained very easily from cultures in steady state at any dilution rate. According to Equation 5, the steady-state cell density is a function of Y, S_r, and s. Thus, Y can be obtained directly by dividing the cell density by the reservoir substrate concentration, assuming that $s << S_r$. This gives a yield value at a particular dilution rate, and it is important to remember that its value may differ among various rates of dilution and often will be significantly different from values obtained in batch culture.

First of all, bacteria tend to accumulate reserve materials when grown with an excess of substrates (batch culture), so it is important to decide how to express the biomass quantity (as protein, dry weight, total nitrogen, total organic carbon, etc.). Second, in most cases, the cell yield will drop with decreasing dilution rate (especially under carbon and energy limitations) as a result of increasing "maintenance-energy" requirements. (A discussion of these factors is outside the scope of this article but can be found in some of the specialized literature given in the Bibliography.) The third growth parameter, which is particularly decisive for the ability of microorganisms to grow successfully at very low substrate concentrations, is the half-saturation constant for growth (K_s), as used in Equation 1, describing the specific growth rate as a function of the substrate concentration. With most organisms and for many substrates, this parameter is very small (typically in the micromolar range). For many substrates, this causes significant problems when its values must be measured and often results in relatively inaccurate measurements. In analogy with the well-known Michaelis–Menten constant for half-saturation of enzymes, K_s is often obtained from a Lineweaver–Burk plot, a linearized form of the "Monod" equation (Equation 1). Unfortunately, this is not a very accurate plot, because those points obtained at the very low end of the measured substrate concentration range (which usually are the most *in*accurate values) put most weight on the position of the line in such a plot. For this reason, the so-called direct linear plot has been introduced and has generally been accepted as far superior (more accurate) to the Lineweaver–Burk plot. Again, the Monod equation is converted into a linearized form as follows:

$$\mu_{max}/\mu = K_s/s + 1. \qquad (9)$$

Subsequently, substrate concentrations in steady-state cultures at two or more different dilution rates (=specific growth rates) are measured and plotted as shown in Fig. 2. Note that the specific growth rates corresponding to the various substrate concentrations are plotted at the Y axis through $s = 0$. From the crossing point of the plotted lines, the projection

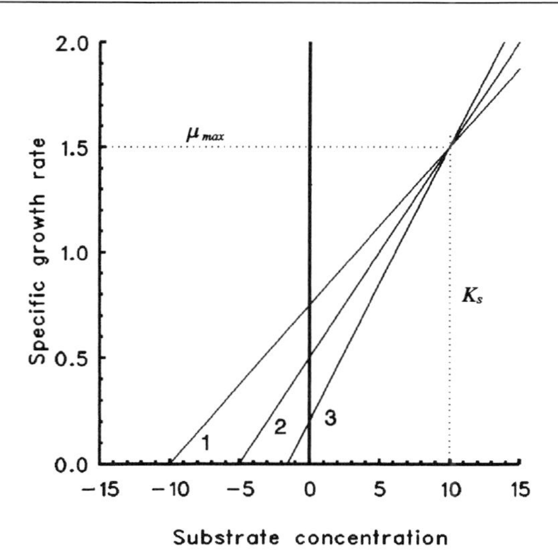

Figure 2 Example of a direct linear plot assuming the following arbitrary combinations of dilution rates (D) and steady-state substrate concentrations: line 1: $s = 10\ \mu M$, $D = 0.75\ \mathrm{hr}^{-1}$; line 2: $s = 5\ \mu M$, $D = 0.5\ \mathrm{hr}^{-1}$; line 3: $s = 1.54\ \mu M$, $D = 0.2\ \mathrm{hr}^{-1}$.

is made on the X axis, which directly gives the value of K_s in the same units as s. Moreover, from this same plot one can estimate the μ_{max} as the projection of the crossing point on the left Y axis. It is good practice to measure the steady-state substrate concentrations at relatively high specific growth rates, because this gives higher values for s and, thus, usually more accuracy.

B. Multiple-Substrate–Limited Growth

When microorganisms are grown under substrate-sufficient conditions in batch culture in a medium with, for instance, two utilizable carbon sources, the cells often adapt themselves in such a way that the substrate that supports the highest growth rate is used preferentially while synthesis of enzymes involved in the utilization of the second substrate remains repressed. As a result, sequential utilization of the substrates occurs and so-called diauxic growth of the organism takes place. However, in continuous culture in which two carbon substrates serving similar physiological functions are both present at very low (growth-limiting) concentrations, repression of the synthesis of one set of enzymes by the presence of very low concentrations of a second substrate is not likely to occur. Indeed, in chemostats that were fed with two or more substrates at the same time, truly multiple substrate limitation may occur. Usually this can be easily proven by increasing the concentration of each of the substrates in the reservoir separately and demonstrating that the cell density increases in each case.

The situation is less obvious in the case of substrates, which do not fulfill similar physiological functions (e.g., oxygen + carbon substrate, carbon source + phosphate, or phosphate + nitrogen, etc.). In such cases, one would expect only one substrate to be truly growth-limiting. Of course, the concentration of the second substrate might be made so low that it does affect normal functioning of a specific metabolic process (e.g., the synthesis of phospholipids in the cell membrane). However, although this might influence the precise composition of the membranes, it does not necessarily affect the overall rate of growth. In other words, the culture could be limited by more than one distinctly different substrate, but the specific growth rate remains limited by just one of them. In such cases, increasing the concentration of only one substrate (the primary growth-limiting one) will lead to an increase of the cell density (until the second substrate becomes the primary growth-limiting substrate!).

Obviously, description of the specific growth rate under truly multiple-substrate limitation will be different from that under conditions of single nutrient limitation. Under such circumstances, μ is best described as a function of all limiting substrates, i.e., $\mu = \mu(s_1, s_2, \ldots, s_n)$. In the simplest case, in which none of the functionally similar substrates interferes significantly with the consumption of the others, the most straightforward description is summation:

$$\mu = \mu(s_1) + \mu(s_2) + \ldots \mu(s_n). \tag{10}$$

All substrates thus contribute proportionately to the overall specific growth rate. Evidently, this approximation will be valid only for growth at very low, far from saturating substrate concentrations; otherwise, unrealistically high ($>\mu_{max}$) values will be obtained. But due to its relative simplicity, it has been the method of choice in several mathematical simulation studies of mixed substrate-limited growth. However, in many cases, the real situation is probably much more complicated because the utilization of one of the substrates is influenced strongly by the presence and consumption of the others. In fact, these substrate interactions may account for changes in the yield coefficients, the steady-state substrate concentrations, and the maximum uptake capacities for the individual substrates. In such cases, the specific growth rate should be expressed interactively, for example, as the product of the growth rates on the individual substrates:

$$\mu = \mu(s_1)\cdot\mu(s_2)\cdot \ . \ . \ . \ \mu(s_3). \tag{11}$$

This kind of expression can be tuned further to the specific properties of the organism by including, in the growth rate terms of the individual substrates, parameters describing a degree of interference by the presence of the other substrates. However, in most cases, it will not be possible to determine the values of such factors with any degree of accuracy, in which case their use will not contribute to a better understanding of the properties of the organisms.

C. Substrate-Limited Growth in Mixed Cultures

If two (or more) different bacteria are grown in a chemostat under limitation of a nutrient (or nutrients), and if they share the ability to metabolize one or more of these limiting nutrients, these organisms will compete for these nutrients, and only those bacteria that under this competitive pressure manage to maintain a specific growth rate equal to D will remain in the culture. Initially, when the chemostat has been inoculated with a mixture of bacteria, unrestricted growth is possible for some time until the primary substrate is (nearly) exhausted. During this initial period, the specific growth rate of each species will approach its maximum value because $s >> K_s$. Competition for the available amount of substrate is thus dominated by the value of μ_{max} of the individual species. Upon reduction of the substrate concentration, the specific growth rate will drop according to the respective μ–s relationships. When at this point the supply of fresh medium is started at a rate lower than the maximum rate of growth of at least one member of the mixed culture, a substrate-limited mixed continuous culture will be established. Evidently, the competitiveness of each species will depend on its μ_{max}, its K_s, and the imposed dilution rate (D).

This is seen most clearly by examining Equation 6 and by recalling that the organism that attains the highest specific growth rate at a given substrate concentration will eventually eliminate its competitor. This may be visualized by comparing the μ–s relationships of the competing species (Fig. 3). In Fig. 3A, an arbitrary μ–s relationship of two organisms, A and B, is presented, indicating that at any given substrate concentration (thus also at any D in continuous culture) the specific growth rate of organism A will be higher than that of organism B. However, in Fig. 3B, the two μ–s relationships intersect, which means that the specific growth rate of organism A exceeds that of organism B to the left of the intersection, whereas organism B exhibits the higher specific growth rate to the right of the intersection. Thus, in the example of Fig. 3A, organism A will, in due time, outcompete (i.e., outnumber) organism B at any dilution rate, whereas in the second example (Fig. 3B) this is true only for dilution rates resulting in s values below the intersection. At higher dilution rates, organism B will take over.

Two important comments must be made here. First, it should be emphasized that because the cell yield does not appear in the equations describing the μ–s relationships it does not affect the outcome of the competition in these mixed cultures. Second, this discussion of competition for one limiting substrate assumes cases of *pure* competition: No other interactions between bacterium A and B should occur. In practice, this will not very often be the case and, as a result, elimination of one species by the other is often incomplete or may not comply very accurately with the predictions based on the μ–s relationships. In fact, this latter situation may be the rule rather than the exception, because many examples of coexistence of species grown with just one primary limiting nutrient have been reported. In some cases, this is explained by inhibitory or stimulatory interactions between the competing species, and often it can be attributed to the presence of secondary growth substrates, resulting from the metabolism of the competing organisms. This phenomenon of growth of mixed cultures at the expense of more than one limiting nutrient in continuous culture can of course be studied in a more controlled fashion by using culture media that contain known mixtures of one or more substrates. Some theoretical studies appearing in the literature lead to the conclusion that to sustain a stable mixed culture of a certain number of species, at least that same number of growth-limiting substrates is required. Yet this is only a minimum requirement, because the actual growth parameters of the competing organisms pose further restrictions on both the dilution rate and the relative concentrations of the growth-limiting substrates in the medium. One of these species-dependent param-

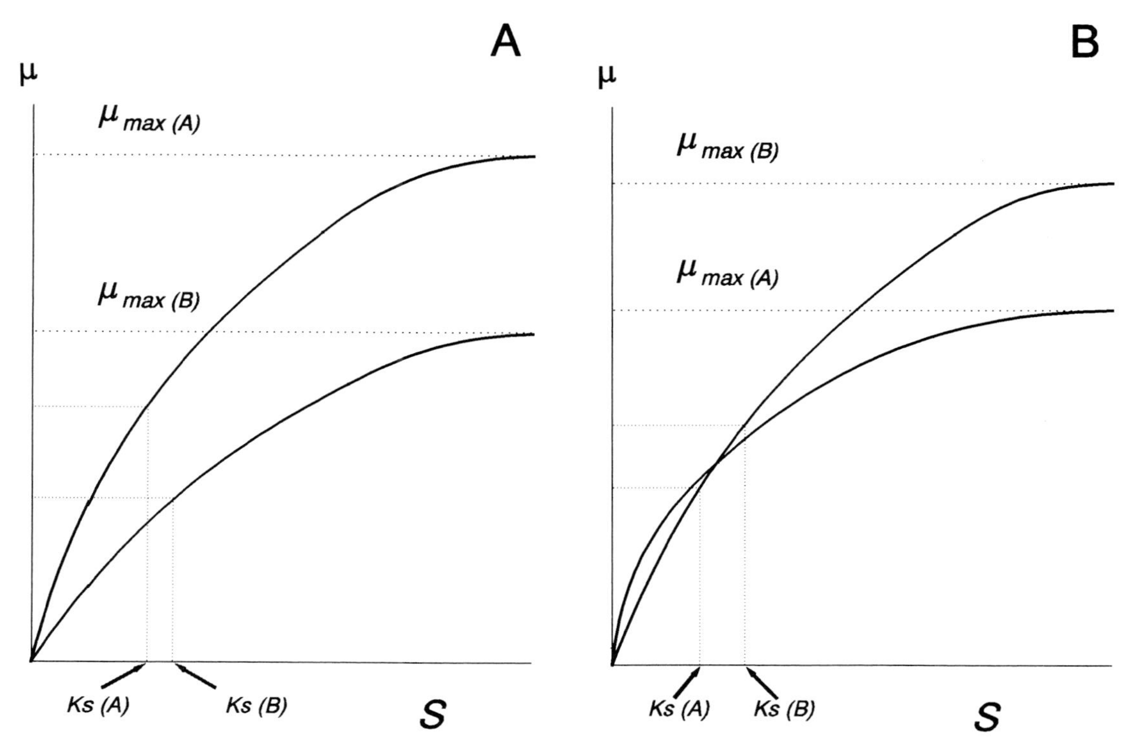

Figure 3 μ–s relationships of two organisms. *A* and *B*. (A) $K_s(A) < K_s(B)$ and $\mu_{max}(A) > \mu_{max}(B)$; (B) $K_s(A) < K_s(B)$ and $\mu_{max}(A) < \mu_{max}(B)$.

eters is the cell yield (Y). For two substrates (*1* and *2*) and two competing species (*A* and *B*), both capable of using substrates *1* and *2*, with the assumption that *A* is the faster growing species on limiting concentrations of one substrate and *B* outcompetes *A* at the other substrate, the following relationship defines the range of input concentrations that will permit stable coexistence of *A* and *B*:

$$Y_{A1}/Y_{A2} < S_{r2}/S_{r1} < Y_{B1}/Y_{B2}.$$

D. Nonsubstrate-Limited Growth in Continuous Culture

In continuous cultures operating as chemostats, the growth rate of bacteria is limited by the external control of the rate at which fresh medium is fed to the culture. Although this rate can be varied over a large range of values, growth at a rate close to μ_{max} is difficult to achieve in conventional chemostats. The cell density drops steeply when the critical dilution rate is approached, making cultivation under such circumstances inherently unstable. To overcome this problem, one must switch over to some kind of internal control of the supply rate of fresh medium. In other words, using feedback control the dilution rate is used to fix the concentration of a growth-dependent product or a substrate. The best-known example of such a feedback system is based on continuous measurement of the turbidity as a measure of cell density. Unfortunately, reliable construction of such a "turbidostat" has always been difficult due to problems of fouling of the sensors used to measure turbidity. Therefore, over the years several other growth-dependent parameters have been proposed and tested as alternatives for biomass itself such as CO_2, inhibitory substrates (i.e., sulfide), light, oxygen, carbohydrates, simple ions (i.e., potassium), and, last but not least, protons. Any continuous culture, based on feedback regulation of a growth-dependent parameter, may be called an "auxostat." The pH auxostat particularly has proven to be extremely useful because it relies on a feedback control based on the proton concentration, which is affected by almost any growing microorganism. Moreover, its measurement is very easy and usually very stable using standard pH probes. Very briefly, the basic theory of a pH auxostat is as follows. Assuming that during growth protons are excreted into the culture liquid, the change in concentration can be expressed as

$$dH^+/dt = \mu x h + D[H^+_R] - D[H^+_C] - DB_R, \quad (12)$$

where x = the culture density (g dry weight of cells per liter), h the stoichiometry of proton formation per g dry weight of cells, $[H^+_R]$ the proton concentration in the reservoir medium, $[H^+_C]$ the proton concentration in the culture, and B_R the buffer capacity of the reservoir medium. For the simplest situation, in which only a slight difference exists between the pH of the medium and that of the culture, Equation 12 reduces to

$$dH^+/dt = \mu x h - DB_R \quad (13)$$

and in steady state with $\mu = D$ results in the following expression for the cell density in the culture:

$$x = B_R/h, \quad (14)$$

demonstrating that the steady-state cell density is a linear function of the buffering capacity of the medium, assuming that h is independent of B_R. Combining Equation 14 with the general nutrient balance for continuous culture, Equation 4 gives in steady state

$$s = S_r - B_R/hY. \quad (15)$$

Finally, solving the conventional Monod Equation 1 for the obtained steady-state values of s allows a plot of the specific growth rate (μ), the steady-state substrate concentration (s), and the steady-state cell density (x) as a function of the buffering capacity (B_R) (see Fig. 4). As can be seen from this illustration, the specific growth rate of the cells remains near μ_{max} over a large range of buffering capacities but, of course, will drop at high buffering capacities due to the decreasing concentration of remaining growth substrate. Obviously, this effect will be most prominent with cells possessing relatively high K_s values for the substrate used. In principle, this will offer the opportunity to choose the buffering capacity such that the substrate concentration becomes strongly growth rate-limiting, thus creating an overlap with the conventional mode of growth in a chemostat.

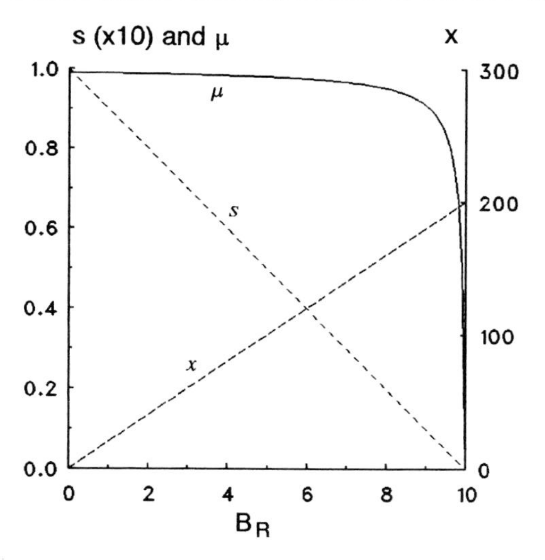

Figure 4 Major parameters for growth in a pH auxostat. s (substrate concentration in the culture), μ (specific growth rate), and x (cell density) as a function of the buffering capacity in the reservoir medium. Arbitrarily chosen values: $Y = 20$, $h = 0.05$, $K_s = 0.1$, $S_r = 10$, and $\mu_{max} = 1$.

II. Basic Continuous Culture Equipment

From the preceding theoretical treatment of continuous culture systems, it is evident that the only fundamental design requirement of a continuous culture system is that the culture is kept growing by continuous input of fresh medium and an output of culture liquid at the same rate. As already explained, this can be accomplished either through external control of a set pump rate (chemostat-type operation) or through internal control by growth-dependent parameters (turbidostat-type operation). Over the years, an enormous number of more or less useful concepts of such fermenter systems has appeared, with sizes ranging from just a few milliliters to several thousands of liters. It is obviously beyond the scope of this article to elaborate on the actual design of continuous culture systems (see Bibliography); nevertheless, several general considerations are important for the construction of most general-purpose (research) chemostats. In Fig. 5, an example is given of a bench-scale (ca. 500 ml working volume) continuous culture system that has proven its effectiveness and flexibility of operation over many years of laboratory practice. Some of its design characteristics are as follows:

- It can be used for anaerobic and for aerobic cultivation. Because the lid is made of black

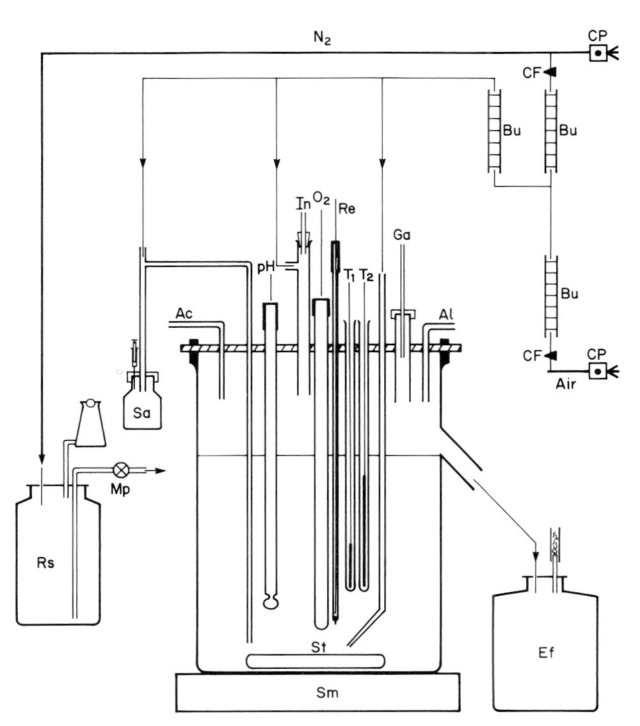

Figure 5 Schematic drawing of a small-scale (500 ml working volume), low-cost glass chemostat. All gases pass cotton wool filters (not shown) before entering the fermenter. The N_2 is freed of traces of oxygen by passage over heated copper turnings. Ac and Al, titration inlets for acid and alkali, respectively; Bu, burette with soap solution for measuring the gas flow; CF, constant flow regulator; CP, constant pressure regulator; Ef, effluent vessel; Ga, possible gas outlet for analysis; In, medium inlet; Mp, peristaltic medium pump; O_2 , oxygen electrode; pH, autoclavable pH electrode; Re, redox electrode; Rs, medium reservoir; Sa, sampling bottle; Sm, stirring motor; St, magnetic stirrer bar; T_1 and T_2, temperature sensor and heating element, respectively. [Reprinted, with permission, from J. C. Gottschal (1990). Different types of continuous culture in ecological studies. *In* "Methods in Microbiology," Volume 22 pp. 87–124. Academic Press, San Diego.]

neoprene rubber (at least 10 mm thickness, resulting in very low permeability for most gases), holes can be drilled easily to fit glass tubing and probes of various sizes.

- Sterility is easily maintained, because the entire setup can readily be autoclaved and the stirring is carried out with a magnetic stirring bar, thus avoiding difficulties with sealing of a stirring shaft through the lid of the fermenter.
- Rubber tubing is made of butyl rubber for anaerobic cultivation and silicon tubing is used for aerobic cultures for more flexibility and durability.
- The medium is supplied by peristaltic tube pumps using marprene rubber for anaerobic cultivation. The culture volume remains constant as it flows out freely through a side arm together with the outflowing gas (e.g., air, N_2).
- The medium inlet is designed such that small drops fall freely into the culture and the inside of the inlet glass tube is kept dry by a continuous flow of gas, to avoid back-growth from the culture into the medium tubing.
- Depending on the actual size of the lid, there is room to fit several probes such as oxygen probes, pH and redox electrodes, specific product/nutrient probes, and biomass probes.

Clearly this type of continuous culture will need to be adapted to any specific needs. For example, for very volatile or hydrophobic substrates, an open gas phase is not appropriate. Due to the rubber lid, the vessel cannot be pressurized, which for some applications will be absolutely necessary. Moreover, although neoprene rubber is poorly permeable for most gases, it is quite feasible that a number of substrates or products may pass this barrier. Finally, aeration may become insufficient at very high densities, especially in combination with high dilution rates, and this problem will become evident also at low cell densities if this type of fermenter is scaled up to volumes well above 1 liter. In general, special attention must be paid not only to sufficient aeration but also to proper mixing.

III. Ecological and Physiological Applications

A. Pure Culture Studies

In most natural environments, microorganisms will experience serious limitation of one or more essential nutrients, particularly carbon, nitrogen, phosphorus, and oxygen but sometimes also certain metal ions, vitamins, or any specific precursor of cell components that cannot be synthesized entirely by the organism itself. For this reason, if one wants to understand and predict the behavior of microorganisms in nature, it is absolutely necessary to study the mechanisms adapted to cope with nutrient deprivation. It is thus not surprising that the chemostat—a continuous culture system in which growth is continuous but limited by one or more nutrients—is most often the preferred tool both for ecological and for physiological studies of bacteria. In addition, compared with feedback-controlled continuous culture systems (i.e., turbidostats, auxostats, etc.), chemostats are constructed and maintained most easily. [*See* LOW-NUTRIENT ENVIRONMENTS.]

From the large number of studies using chemostats with many different microorganisms, including aerobes and anaerobes, chemolithotrophic and chemoorganotrophic species, phototrophic algae, and cyanobacteria, it has appeared that, in most cases, these microorganisms are able to adapt themselves very effectively to the imposed nutrient-deprived conditions. Adjusting the dilution rate over a large range from near μ_{max} to well below 0.01 hr^{-1} can result in changes in nutrient availability from near-saturation levels to far below the K_s value of the organism (i.e., in many cases to nM concentrations, depending on the limiting nutrient involved). In response to such dramatic decreases in substrate availability, microorganisms apparently maintain the highest possible rate of metabolism and growth by increasing their nutrient uptake potential through an increase in the V_{max} of the uptake system. Similarly, severalfold increased levels and activities have been observed in many microorganisms for many different intracellular enzymes involved in the initial steps of the catabolism of the growth-limiting nutrient(s). Interestingly, this kind of enzyme response has also been observed with extracellular enzymes—for example, in some thermophilic *Clostridia* capable of hydrolyzing starch. Both α-amylase and pullulanase increased with decreasing dilution rates as a result of derepression of their synthesis with an optimum at about one-third of the maximum growth rate of the organism. Two more examples illustrate the fact that derepression also occurs with enzymes that are not even involved in the metabolism of the growth-limiting substrate! In *Alcaligenes eutrophus,* a facultatively autotrophic hydrogen-oxidizing organism, hydrogenases appear at low dilution rates under succinate-limiting conditions, which do not require hydrogenase activity. Even ribulose–biphosphate carboxylase (RubPcase), a key enzyme in the CO_2-fixing pathway in this organism during autotrophic growth, appears at increasing activity when the dilution rate goes down during heterotrophic, succinate-limited growth. In another bacterium, *Pseudomonas oxalaticus,* grown heterotrophically (on sodium-oxalate), RubPcase is not usually detectable. But when the dilution rate becomes very low (<0.02 hr^{-1}), substantial derepression is observed, and the activity eventually (at a dilution rate of 0.005 hr^{-1}) becomes very similar to that found during autotrophic growth (on formate), in which case the cells indeed depend on CO_2 fixation.

The generally accepted interpretation of such results is that certain metabolites (which may of course be different for different enzymes) reach such low concentrations intracellularly that they no longer exert any repressive effect on the synthesis of a given enzyme, whether that enzyme is required or not!

Another response commonly observed with chemostat cultures in response to decreasing nutrient supply are changes in morphology resulting in increased surface : volume ratios. For example, *Arthrobacter* species have been shown to change from spheres into rods, *Caulobacter* species are known to produce long stalks when grown under severe nutrient limitation, and most rod-shaped organisms tend to grow longer and thinner with decreasing dilution rates. Increased surface : volume ratios are believed to facilitate nutrient flux into the cell due to an increased ratio of solute uptake systems in the cell surface to a relatively small cell volume. Yet another well-documented mechanism of enhancing metabolic activity at decreasing nutrient levels is the transition to different pathways with similar functions but with much higher overall substrate affinities (lower K_m values). This type of phenotypic adaptation is known in various organisms for the metabolism of sugars, glycerol, and nitrogen sources. Anaerobic fermentative bacteria, lactic acid bacteria, and several clostridia in particular are well known for their ability to change their fermentation pattern so that the overall energy yield increases with decreasing substrate availability. Finally, a pronounced response of microorganisms to the presence of growth-limiting concentrations of an essential nutrient other than the carbon and energy sources is that of a redirection of fluxes of metabolites containing the limiting element. Under these conditions, the carbon conversion efficiency is generally low, and this may lead to a significant accumulation of carbon in intracellular reserve materials, extracellular polymers, or a variety of low-molecular weight metabolites.

Chemostats evidently have proven extremely valuable in studying substrate-limited growth because they make it possible to cultivate microorganisms for long periods of time under well-defined and very reproducible conditions of uninterrupted growth, thus allowing for optimal adaptation to the imposed circumstances. This unique combination of advantages of chemostat cultivation has been exploited especially in studies focusing on the use of not just one limiting nutrient but of two and, in some cases, even more simultaneously limiting substrates (see also Section I.B). For example, in one of the pioneering studies (in the late 1960s) it was shown that *Escherichia coli,* which in batch culture strongly preferred glucose over lactose (glucose repressed the synthesis of β-galactosidase), consumed both sugars simultaneously at low dilution rates; however, at high dilution rates, repression of the β-galactosidase still occurred, again resulting in the use of glucose only. After prolonged cultivation under these same conditions, they found that β-galactosidase-constitutive mutants were strongly selected for. These mutants eventually dominated the culture and utilized glucose and lactose simultaneously up to the washout rate (0.9 hr^{-1}), demonstrating the strong selective power of prolonged chemostat cultivation for mutants (or contaminants!) that grow slightly faster under the prevailing conditions than the original strain (see also later). This phenomenon of simultaneous consumption of two and more sugars has since been demonstrated many times and apparently is the rule rather than the exception. Moreover, further experimentation along these lines has shown that also substrates, which, if present separately, support quite different modes of metabolism and in batch culture are used sequentially, are consumed simultaneously in chemostats. For example, substrate combinations such as formate + acetate, formate + oxalate, formate + mannitol, thiosulphate + acetate, methylamine + glucose, and methanol + glucose all can be used to generate energy and, if present alone, result in either autotrophic, methylotrophic, or heterotrophic growth. The microorganisms that were capable of utilizing these different substrates simultaneously under carbon and/or energy limitation can grow "mixotrophically." In other words, they can combine autotrophic (CO_2 serves as the carbon source), methylotrophic (C_1 compounds other than CO_2 are the carbon source), and heterotrophic (C_{2-n} compounds are used as carbon source) modes of metabolism. In various studies, this mixotrophic metabolism was studied in chemostats by supplying the cultures with variable mixtures of these different substrates. The most important conclusion to be drawn from these studies is that such metabolically flexible bacteria can accurately tune their enzymic machinery to the nutritional demands. This was most apparent for the oxidation potential of the cultures for the substrates supplied and for the capacity to convert CO_2 or formaldehyde (in the case of some of the methylotrophic organisms) into cellular material. Both activities clearly responded to the ratio of the "autotrophic" or "methylotrophic" and "heterotrophic" substrates in the inflowing medium in such a way that the organisms could optimize their biomass yield. In these particular cases, the observed increase in growth efficiency is explained by the energy-saving effect of the utilization of an organic substrate on the synthesis of biomass as compared with biosynthesis on the basis of CO_2 fixation. Obviously, similar positive effects on cell yields can be expected in general when additional growth-limiting substrates are present and if metabolic regulation permits the use of energetically "cheaper" pathways for biomass synthesis.

One of the major advantageous characteristics of a chemostat culture—its continuity of growth conditions—also yields the greatest conceptual problem for using it in ecological research, because it is very unlikely to find many natural habitats that show a degree of constancy with respect to nutrient supply and physicochemical conditions that are as strict as those established in chemostats at steady state. But fortunately, due to another very important property—precise control of the growth environment—the chemostat offers excellent possibilities to study microbial cultures under controlled *dis*continuous conditions. Of course, the standard chemostat equipment needs to be extended with some sort of timing and switching device to make changes to dilution rate, type of medium, light irradiance, aeration, pH, temperature, etc., in a controlled manner. In fact, sophisticated, gradual changes in growth conditions are possible using (micro)computer-controlled equipment in combination with proper software specifically designed for this purpose. Such computer-controlled fermenters and chemostats have become quite common in industry, and for laboratory use most of the hardware can be obtained from many specialized firms. However, although very few studies so far have been reported in which these new facilities have been exploited in fundamental studies on the ecology or physiology of microorganisms, some interesting examples exist of the use of chemostats in studies on the effects of short-term environmental changes. Such studies have revealed that many microorganisms can adapt phenotypically to a large range of variations in their environment, and this has led to the recognition that a given genotype is very much a product of its environment. But the ability to respond adequately to such changes is of course ultimately restricted by its genotype. Thus, some organisms apparently can grow on many different substrates over a wide range of temperatures, oxygen partial pressures, and pH values, whereas others really are to be considered specialists with respect to just one type of energy source under very specific physicochemical conditions.

The ability to store reserve materials intracellularly during periods of relative abundance of carbon and energy resources is often seen as a useful adaptation to survival of periods of very strong nutrient deprivation. Not only does it allow a certain degree of continued endogenous metabolism to maintain cellular integrity, it may also provide sufficient energy, at the moment nutrients become available again, to allow rapid uptake and to synthesize new enzymes. Studies with *Streptococcus cremoris* in lactose-limited chemostats subjected to starvation periods of various lengths and subsequent renewed supply of lactose have indeed shown that a sustained intracellular pool of phosphoenolpyruvate may be responsible for the rapid reestablishment of growth following the periods of starvation. In another study, using *Thiobacillus versutus* grown in thiosulfate and/or acetate-limited chemostats, the metabolic flexibility of this bacterium is clearly demonstrated. The organism is capable of growth with acetate or thiosulfate alone but also with mixtures of these two substrates when both are growth-limiting. Switching between acetate and thiosulfate limitation revealed that as long as enzymes required for catabolism of the new substrate following a switch-over were still present in the cells, growth resumed without a detectable lag. Moreover, the same was true when growth-limiting concentrations of the previous growth substrate were present along with the new substrate after complete loss of these enzymes. The survival value of such metabolic properties are revealed most clearly in experiments with chemostat-grown mixed cultures.

B. Mixed Culture Studies

The ability to capture nutrients from their immediate surroundings surely represents one of the most important properties of microorganisms to survive in nature. Because in most microorganisms energy sources and other inorganic nutrients will often be present in growth-limiting quantities, competition for limiting nutrients is considered a major interaction between microbes. Because chemostat cultivation represents an excellent way of growing microorganisms continuously under conditions of nutrient limitation, mixed culture studies have long been dominated by studies in the competitiveness of many different microorganisms under a great variety of environmental conditions. In most of these early studies, examples of "pure competition" with one growth-limiting nutrient between two bacteria have been studied. In the majority of these cases, in which no additional interactions were involved, the outcome of the competition clearly supported the competitive exclusion principle. Thus, the organism with the higher specific growth rate at a given substrate concentration (i.e., dilution rate) completely outcompeted the slower-growing species. In some cases, the outcome of the competition clearly depended on the actual dilution rate and the μ–s relationships crossed, indicating that some species are much better adapted to growth at very low substrate concentrations whereas others do better at higher rates in the presence of higher substrate concentrations. This conclusion appears true for many microbial species as such observations have been made for a large range of physiologically different types of microorganisms: aerobic and anaerobic heterotrophs, chemolithotrophic bacteria, and phototrophic organisms. However, despite these seemingly straightforward results of pure competition, a significant number of studies have appeared in which stable mixed cultures were observed in chemostats limited by only one primary substrate. In these cases, in which competition may not have been so "pure," the occurrence of additional relationships between the competing organisms does usually account for the observed coexistence. Classic examples are those in which fermentative bacteria produce secondary substrates (the fermentation products of the primary limiting substrate) or in which vitamins or other requirements excreted by two species establish commensalistic or mutualistic relationships. A good example of reciprocal stimulation of two species in mixed culture was demonstrated during attempts to produce yoghurt by continuous cultivation. In this case, *Lactobacillus bulgaricus* produced amino acids (from casein) that stimulated the growth of *S. thermophilus*, which in turn produced small quantities of formic acid, thus stimulating *L. bulgaricus*. Of course, these latter types of interactions may also be studied in batch culture systems, but the significant advantage of doing it in continuous culture is the degree of constancy of the growth conditions that can be obtained and the avoidance of interfering lag-phases, substrate exhaustion and variable product accumulation, and changes in biomass, so characteristic of batch culture experimentation. These factors also make continuous culture the method of choice in some detailed studies on interspecies hydrogen transfer. For example, the sulfate-reducing bacterium *Desulfovibrio vulgaris* growing at the expense of lactate or pyruvate requires sulfate as an electron acceptor in batch culture. However, stable mixed cultures of *D. vulgaris* and *Methanosarcina barkeri* (a hydrogen-consuming methanogen) could be obtained in chemostats with lactate (or pyruvate) as the limiting substrate in the absence of sulfate. The methanogen acted as "the electron acceptor" for *D. vulgaris* and kept the hydrogen partial pressure below a detectable level. Finally, bacteria with seemingly conflicting physicochemical requirements have been grown in stable mixed chemostat cultures. Thus, a strictly aerobic species, *Pseudomonas* sp., and an obligate anaerobe, *Veillonella* sp., were grown with lactate as the growth substrate under controlled oxygen-limiting conditions. The anaerobic organisms fermented lactate to acetate, propionate, hydrogen, and carbon dioxide, whereas the aerobic species metabolized part of these substrates to carbon dioxide and water, using oxygen as its electron acceptor. Because care was taken that oxygen remained limiting for growth of the aerobe, the oxygen partial pressure remained sufficiently low so as to prevent significant growth inhibition of the anaerobe. This kind of sensitive mutualistic interacting mixed cultures would be very difficult to maintain in batch culture systems, in which the oxygen demand is changing significantly with time.

Whereas the preceding examples all represent well-defined mixed cultures of a very limited number of species, it is of course quite feasible to maintain multispecies, undefined mixed populations in well-controlled continuous culture systems. This has indeed been done, often in an attempt to reproduce entire communities in the continuous culture as they occur in their natural habitats. However, this type of investigation will probably not do much to obtain a better understanding of the respective ecosystems. Continuous cultures are very useful tools for studying certain aspects of microbial life in the laboratory, but they are entirely inappropriate to simulate *in vivo* systems.

C. Enrichment and Selection

In selective enrichment cultures, a (natural) sample, containing many different microorganisms, is kept under conditions that favor a particular physiological type of organism or group of organisms, thus allowing them to increase in number relative to physiologically different types of microorganisms. Invariably, the mechanism involved is that the chosen growth conditions allow the favored type to grow faster than other types of microorganisms. The "art" of enriching for certain bacteria has been exploited for many years. In batch culture enrichments, the selection pressure is directed toward organisms that exhibit the highest specific growth rate during repeated transfers in media containing relatively high substrate concentrations. This is still the most widely employed technique for obtaining microbes with specific physiological properties and a high μ_{max} . Although repeated transfers offer some degree of continuity during the selection process, the actual growth conditions are continuously changing due to changes in biomass, (inhibitory) product formation, substrate concentration, and, unless accurately controlled batch fermenters are used, significant changes in physicochemical conditions. The selective pressure of this type of enrichment culture can be improved enormously by using continuous cultures based on the principles of turbidostats or various types of auxostats. Because the dilution rate of such cultures is set by the μ_{max} of the population present in the culture, it will be evident that very strong selection takes place against any organism growing at a lower rate. Moreover, this type of approach also permits a more precise definition of the actual selective conditions, because a specific set of growth conditions can be maintained constant over a long period of time. Particularly for industrial applications, this would seem the method of choice for selecting highly productive strains for very specific purposes from natural samples and for further strain improvements. So far, however, this technique has been employed only incidentally.

On the other hand, if the objective is to obtain organisms that are not so specialized in high rates of growth, the obvious choice is to use substrate-limited chemostats. Indeed, a significant number of examples exist of successful isolation of this type of organisms using chemostat enrichments. Especially in ecologically oriented research, this is considered one of the best possibilities for obtaining microorganisms more representative of those prevailing in natural (aquatic) environments. In addition to the important fact that this technique selects for organisms that are optimally adapted to growth under severe nutrient limitation, many other physical and chemical parameters and, hence, the overall selection pressure can be maintained constant over prolonged periods of time. Alternatively, these conditions may be changed in a controlled way, resulting, for example, in a defined pattern of continued environmental changes such as light–dark cycles, pH, temperature, or aerobic/anaerobic or osmotic transitions. This clearly adds enormously to the precision of the selective pressure that may be applied.

Another particularly interesting aspect of chemostat enrichments is the possibility to select organisms under conditions of multiple-substrate limitation. Chemostat enrichment under such circumstances has demonstrated that metabolically versatile bacteria could be selected that are specialized in using several different substrates at the same time if present at growth-limiting concentrations. Because such bacteria usually display lower maximum growth rates relative to more specialized species, they will, in most cases, be missed by employing batch-type enrichment techniques, even if mixtures of several substrates were used.

Finally, chemostat enrichments provide excellent possibilities for selective enrichment of entire microbial communities. Mixed cultures of bacteria, representing simple microbial food chains, or more complex food webs, in batch cultures usually appearing as successional changes in populations, will in continuous flow systems all grow simultaneously for extended periods of time. Fascinating examples of enrichment of such complex microbial communities can be found especially in the literature concerning microbial degradation of (man-made) recalcitrant compounds such as herbicides and various halogenated aromatics. Particularly with respect to this latter category of substrates, it should be recognized that in some cases the substrates involved are to some extent toxic, even for the organisms capable of degrading them. This surely adds to the advantages of using substrate-limited selections in chemostats. Of course the interesting possibility exists to consider further strain or population "improvement" by employing auxostat selection as a subsequent technique to obtain useful xenobiotic degrading microbes. [*See* BIODEGRADATION.]

IV. Concluding Remarks

Continuous culture systems represent excellent tools for studying microorganisms under accurately controlled growth conditions. In particular, the possibility to maintain cells for long periods of time under rigorously controlled environmental conditions ensures that the physiology of these cells is accurately and reproducibly tuned to the chosen conditions, permitting detailed studies of the way microorganisms respond to particular environmental constraints. Moreover, exploiting the possibilities of using continuous culture systems in a controlled *dis*continuous mode greatly adds to its usefulness in studying microbial adaptability and flexibility. In addition to this introduced heterogeneity in time by (ir)regular changes in growth conditions, further heterogeneity, so typical for most natural ecosystems, has in some cases been introduced by connecting two or more continuous cultures to each other. This has been done, for example, to study predation by protozoa. In this case, the first fermenter is used to grow a bacterial culture, and the outlet of this fermenter serves as the inlet (the bacterial feed) of the second fermenter containing the bacterivorous protozoa. In some applications, two or more fermenters are connected in the same way but with the purpose of transferring bacteria, substrates, and products to the next fermenter, in which the growth conditions are kept different (i.e., aerobic/anaerobic, denitrification/sulfate reduction/methanogenesis, etc.). In yet another application, two or three chemostats were connected by dialysis membranes, allowing solute exchange but no cells.

As evident from the preceding survey of the basic applications of continuous culture, chemostats have always dominated the field, and, thus, most of our knowledge of the physiological state of microorganisms growing in steady state is based on substrate limitation. Surely this represents for most microorganisms their natural state for most of the time. Notwithstanding, it is not their only physiological state and surely from a practical point of view not always the most desirable condition of growth. In many cases, it is of interest to learn more about the properties of cells growing at very high rates in the presence of known relatively high substrate concentrations. With the development of new, reliable, and accurate probes, exciting new ways of continuous culturing are being developed both at the laboratory-bench scale and at the industrial scale. Only relatively few applications of the various possible auxostats (or ''nutristats,'' if the growth-substrate concentration is used as the feedback signal sustaining a flow of fresh nutrients) have appeared in the literature. The most significant examples of control parameters used so far are pH, NH_3 , H_2S, redox potential, O_2, CO_2, glucose, and, of course, biomass itself, either by continuous turbidity measurement or by a recently developed probe based on dielectric spectroscopy through which a measure for the total electric capacitance of the suspended cells is obtained.

Proper choice and combination of the different types of continuous culture offer a very powerful tool for physiological and ecological studies of microbes and also allows for ''tailor-made'' selection and strain improvement of microorganisms for biotechnological applications.

Bibliography

Dykhuizen, D. E., and Hartl, D. L. (1983). *Microbiol. Rev.* **47**(2), 150–168.

Fraleigh, S. P., Bungay, H. R., and Clesceri, L. S. (1989). *Trends Biotechnol.* **7,** 159–164.

Gottschal, J. C. (1986). Mixed substrate utilization by mixed cultures. *In* ''Bacteria in Nature,'' Vol. 2 (E. R. Leadbetter and J. S. Poindexter, eds.), pp. 261–292. Plenum Press, New York.

Gottschal, J. C. (1990). Different types of continuous culture in ecological studies. *In* ''Methods in Microbiology,'' Vol. 22 (J. R. Norris and R. Grigorova, eds.), pp. 87–124. Academic Press, London.

Wimpenny, J. W. T. (ed.) (1988). ''Handbook of Laboratory Model Systems for Microbial Ecosystems, ''Vols. I and II. CRC Press, Boca Raton, Florida.

Coronaviruses

Michael M. C. Lai
Howard Hughes Medical Institute, and University of Southern California School of Medicine

Glossary

Complementation Interaction of gene products from two different viruses to enhance the growth of the parental viruses in a mixed infection; the genotypes of the parental viruses are not altered

Nonstructural protein Virus-specific protein made from the viral genome in the infected cells but not incorporated in the virion particles; typically, these proteins are involved in the regulatory role in the replication cycle of viruses

Open reading frame (ORF) Continuous stretch of genetic region in which there is no termination codon; each ORF has the potential to make a protein

Peplomers Spikes present on the surface of virion particles

RNA recombination Exchange of genetic sequence between nonsegmented RNAs as a result of cross-overs between RNAs; this is in contrast to reassortment of RNA segments, in which genetic exchanges occur by mixing RNA fragments

CORONAVIRUSES are a group of animal viruses that are characterized by a distinct morphology with prominent spikes on the surface of the virus particles, resembling the corona of the sun. This group of viruses includes many important human and animal pathogens, causing a variety of respiratory, gastrointestinal, and neurological diseases. All of them contain a lipid envelope, a positive-sensed RNA genome of huge size and a complex transcription–replication strategy.

The first coronavirus isolated (in 1937) was avian infectious bronchitis virus (IBV) from chickens with the gasping disease, which is a devastating, lethal respiratory infection with neurological involvement, resulting in chickens "curled in balls." In 1949, a murine coronavirus was isolated from a paralyzed mouse. This virus has the capacity to cause encephalomyelitis with extensive destruction of myelin and was named JHM, after a distinguished Harvard professor, J. H. Mueller. A few years later, a mouse hepatitis virus (MHV) was isolated from a mouse developing hepatitis. MHV was later found to be related to JHM. In the 1960s, several human coronaviruses including 229E and OC43 ("OC" stands for "organ culture" of human embryonic trachea used to isolate the virus) were isolated from school children and medical students with common colds. These viruses had a similar morphology and were later recognized to be also antigenically related. Subsequent biochemical studies showed that all these viruses contained similar biochemical characteristics. The name "Coronavirus" was proposed in 1968 to include all these viruses, and Coronaviridae was established as a new virus family in 1975.

I. Classification

The current known members of Coronaviridae are listed in Table I. There is only one single genus. The viruses can be grouped into four antigenic groups: Group 1 includes porcine transmissible gastroenteritis virus (TGEV), human coronavirus (HCV) strain 229E, feline infectious peritonitis (FIPV), feline enteric coronavirus (FEV), canine coronavirus (CCV), and rabbit coronavirus (RbCV); Group 2 includes MHV, bovine coronavirus (BCV), human coronavirus strain OC43, rat coronavirus (RCV), sialoda-

Table I Members of Coronaviridae and Their Diseases

Virus	Designation	Natural host	Diseases
Avian infectious bronchitis virus	IBV	Chicken	Tracheobronchitis, neurological involvement
Bovine coronavirus (neonatal calf diarrhea coronavirus)	BCV	Cow	Gastroenteritis
Canine coronavirus	CCV	Dog	Gastroenteritis
Feline enteric coronavirus	FECV	Cat	Enteritis
Feline infectious peritonitis virus	FIPV	Cat	Peritonitis, meningoencephalitis, pleuritis, pneumonia, disseminated granulomatous diseases
Hemagglutinating encephalomyelitis virus	HEV	Pig	Respiratory and enteric infections, encephalomyelitis
Human coronavirus	HCV	Human	Common colds
Human enteric coronavirus	HECV	Human	Diarrhea
Mouse hepatitis virus	MHV	Mouse	Hepatitis, encephalomyelitis
Porcine transmissible gastroenteritis virus	TGEV	Pig	Gastroenteritis
Porcine respiratory coronavirus	PRCV	Pig	Respiratory diseases
Rabbit coronavirus	RbCV	Rabbit	Myocarditis, pleuritis
Rat coronavirus	RCV	Rat	Respiratory diseases, pneumonia
Sialodacryoadenitis virus	SDAV	Rat	Sialodacryoadenitis, keratoconjunctivitis
Turkey coronavirus	TCV	Turkey	Enteritis

cryoadenitis virus (SDAV), and porcine hemagglutinating encephalomyelitis virus (HEV); Group 3 includes IBV; and Group 4 includes turkey coronavirus (TCV). More recently, TCV has been shown to be closely related to BCV. Thus, Group 4 may no longer exist as an independent group.

II. Growth and Host Range of Coronavirus

Coronaviruses in general have relatively restricted host range, infecting only cells of their own natural host species. However, cross-species infections occasionally occur. In tissue culture, most coronaviruses can be grown in established cell lines, but some infect only primary tissue cells. Many coronaviruses have fastidious requirements for tissue culture growth. Thus, many of the human coronaviruses were initially isolated using organ culture of human embryonic trachea, intestine, or nasal mucosa. The growth in tissue culture requires extensive adaptation by serial passages of virus in culture.

In animal infections, most coronaviruses cause diseases also only in the species of their natural hosts. However, transmission to other animal species has been reported. For instance, MHV can infect rats and cause encephalomyelitis. BCV can also infect humans, causing diarrhea. Furthermore, suckling mice have been used for transmission of several different coronaviruses, including human coronaviruses. In animals, viruses have high tissue specificity.

III. Structure and Composition of Virus

A. Morphology and Physical Properties of the Virus

Coronavirus virions are spherical, enveloped virus particles, ranging from 80 to 160 nm in diameter, that are characterized by the presence of widely spaced and prominent, bulbous surface projections of up to 20 nm in length, covering the entire virion surface and giving it the corona appearance. These projections are called peplomers, and there are about 200 peplomers per virion. In some coronaviruses, a second layer of small, granular projections located at the base of the larger bulbous projections can also be seen (Fig. 1). The presence of the small

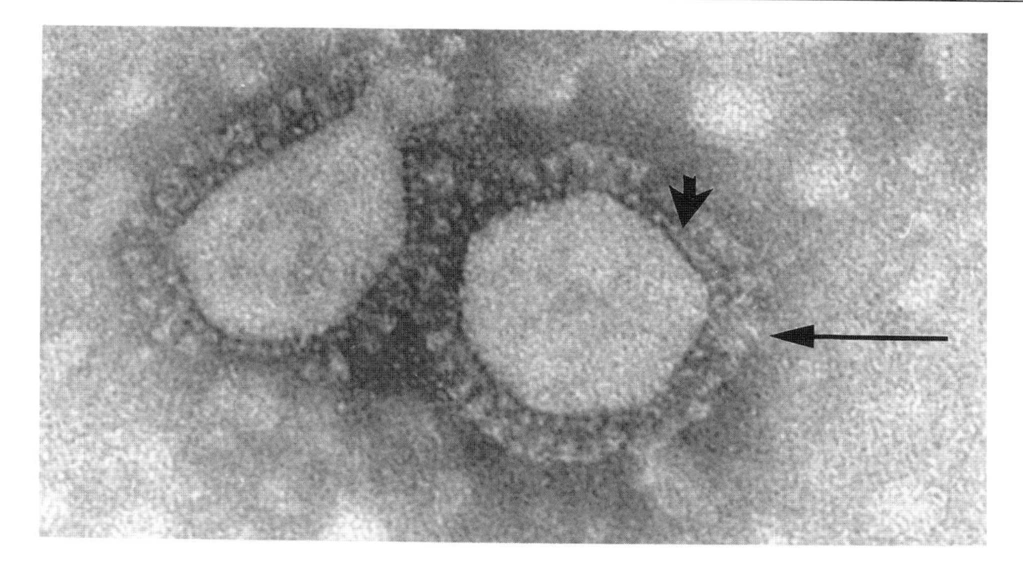

Figure 1 Electron micrographs of turkey coronavirus. Arrowhead denotes short projections, whereas the arrow indicates long spikes. (Photos courtesy of P. Tijssen, Institut Armand-Frappier, Quebec, Canada.)

projections appears to correlate with the presence of the hemagglutinin-esterase protein, which is present only in some coronaviruses (see later). Inside the envelope resides a helical nucleocapsid of 6 to 8 nm in diameter, which can be seen in thin sections of virions. The virus particle has a buoyant density of 1.17 to 1.20 g/cm^3 in sucrose solution.

B. Structural Proteins

Coronavirus contains three or four structural proteins (Fig. 2), depending on the virus species (Table II).

1. The first protein is the spike protein, designated as S (or E2) protein. S is a glycoprotein of approximately 150 to 200 kDa and forms an oligomer structure, which constitutes the large bulbous spikes on the viral surface. It is frequently cleaved into two proteins of nearly equal size (termed S1 and S2) in most coronaviruses. However, in TGEV, FIPV, and CCV, no cleavage occurs. The extent of cleavage also depends on the host cells in which the virus is grown. This cleavage is required for the induction of cell fusion by some coronaviruses. S1 (derived from the N-terminal half of the S protein) forms the outer bulbous parts of the spikes, while S2 makes up the stalk, which anchors via the C-terminal hydrophobic domain in the viral envelope. S2 is relatively conserved; in contrast, S1 is highly diverged in length and sequence among different coronaviruses. The protein is glycosylated by N-linked glycans and also acylated by palmitic acid. This protein is responsible for virus binding to receptors on target cells, induction of cell-to-cell fusion, elicitation of virus-neutralizing antibodies and cell-mediated immunity, and also hemagglutination in some

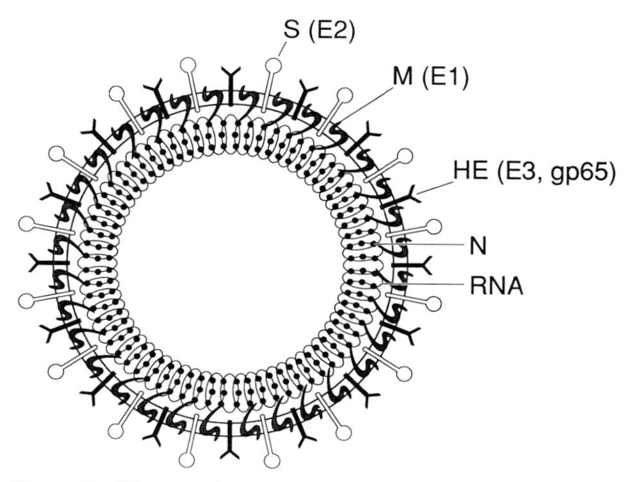

Figure 2 Diagram of coronavirus structure. S forms long spikes and HE forms short spikes. M is embedded in the envelope, but its C terminus interacts with N. N and the viral genomic RNA form nucleocapsid, which is helical.

Table II Structure and Functions of Coronavirus Structural Proteins

Protein	Molecular mass (kDa)	Modification	Functions
Spike protein (S, E2)	150–200	N-linked glycans palmitic acid cleavage	(1) Binds to receptor on target cells (2) Facilitates virus penetration and uncoating (3) Induces cell-to-cell fusion (4) Elicits neutralizing and nonneutralizing antibodies (5) Elicits cell-mediated immunity (6) Causes hemagglutination
Membrane protein (M, E1)	20–30	O- or N-linked glycans	(1) Interacts with nucleocapsid during virus assembly—determines the site of virus maturation (2) Elicits humoral and cellular immunity
Hemagglutinin-esterase protein (HE, E3, gp65)	65	N-linked glycans	(1) Causes hemagglutination and hemadsorption (2) Binds to receptors on target cells? (3) Facilitates elution of virus from cell surface? (4) Elicits humoral and cellular immunity
Nucleocapsid protein (N)	50–60	phosphorylation	(1) Forms nucleocapsid (2) Regulates viral RNA synthesis?

coronaviruses (e.g., IBV and TGEV) that lack the HE protein (see later). The monoclonal antibodies specific for the S protein can neutralize virus infectivity and inhibit fusion-inducing ability of the virus.

2. The second protein is the M or membrane protein (also E1 or matrix protein in older literature). This protein is an integral membrane protein with a molecular mass of 20 to 30 kDa. It is glycosylated by O-linked glycans at serine or threonine residues in MHV and BCV; however, in IBV and TGEV, glycans are N-linked to asparagine residues. The amino terminus of the M protein is exposed to the outside of the virus particle, whereas the carboxy terminus is exposed to the inside and may interact with the nucleocapsid. The middle contains three hydrophobic domains, which span the viral envelope three times. The M protein determines the site of virus assembly by interacting with the nucleocapsid in the infected cells.

3. The third protein is N protein, which, together with virion RNA, constitutes the nucleocapsid. This is a phosphoprotein of 50 to 60 kDa, with phosphorylation occurring on serine residues. N may interact with the M protein during viral assembly. It may also participate in the regulation of viral RNA synthesis.

4. The fourth protein is the hemagglutinin-esterase (HE) protein (also E3 or gp65). This protein is unique in that only some coronaviruses, including BCV, MHV, HEV, TCV, and HCV-OC43, have this protein. It is a glycoprotein of 65 kDa, with N-linked glycans. The protein forms dimer through disulfide bonds and constitutes the smaller of the spikes observed in some coronaviruses. The protein contains hemagglutinin and esterase activities, although the former is weak in some viruses. Hemagglutinin allows the virus to bind to red blood cells (to 9-*O*-acetylated neuraminic acids), while the esterase activity removes the acetyl group from the sialic acid. This protein shares some sequence homology with the hemagglutinin protein of influenza C virus. The role of this protein in the biology of virus is unclear. The monoclonal antibodies specific for the HE protein of BCV can inhibit the viral infectivity, indicating that HE is required for viral infectivity. However, the HE protein of MHV is not necessary for viral infection, because it is absent in many MHV isolates.

C. Viral Genomic RNA

The coronavirus genome is a single-stranded, linear RNA of 27 to 31 kb, which is by far the largest known viral RNA. It has a 5′ cap structure and a 3′-poly(A) and represents the sense strand (i.e., it can be used directly for translation). The RNA is infectious on transfection into cells. Thus, it is a positive-sensed RNA. The complete sequence of viral RNA has been determined for IBV and MHV. Both of these RNAs contain 10 open reading frames (ORFs), some of which partially overlap each other. The protein products of most of these ORFs are made from separate mRNAs in the infected cells, but a few are made together from a single mRNA. In the latter case, these ORFs are considered to belong to one gene. Based on this criterion, the genomic organization of IBV and MHV RNAs are shown in Fig. 3. Thus, IBV has six genes, whereas MHV has eight. Although the overall genomic organization of MHV and IBV RNAs is similar, there are notable differences (e.g., in IBV, gene 5, which encodes nonstructural proteins, is inserted between genes encoding M and N proteins, while, in MHV, the latter two genes are immediate neighbors). MHV contains two genes, 2 and 2-1, which are not present in IBV. At the 5′-end of the RNAs, there is a leader sequence of 60 to 70 nucleotides, which is also present at the 5′-end of mRNAs (see below). Following the leader RNA are approximately 200 to 500 nucleotides of untranslated sequence. In regions between the genes, there is a small stretch of consensus sequence of approximately 8 to 10 nucleotides, which shows homology with the 3′-end of the leader sequence. This sequence is important for the transcription of the downstream genes. In some virus preparations,

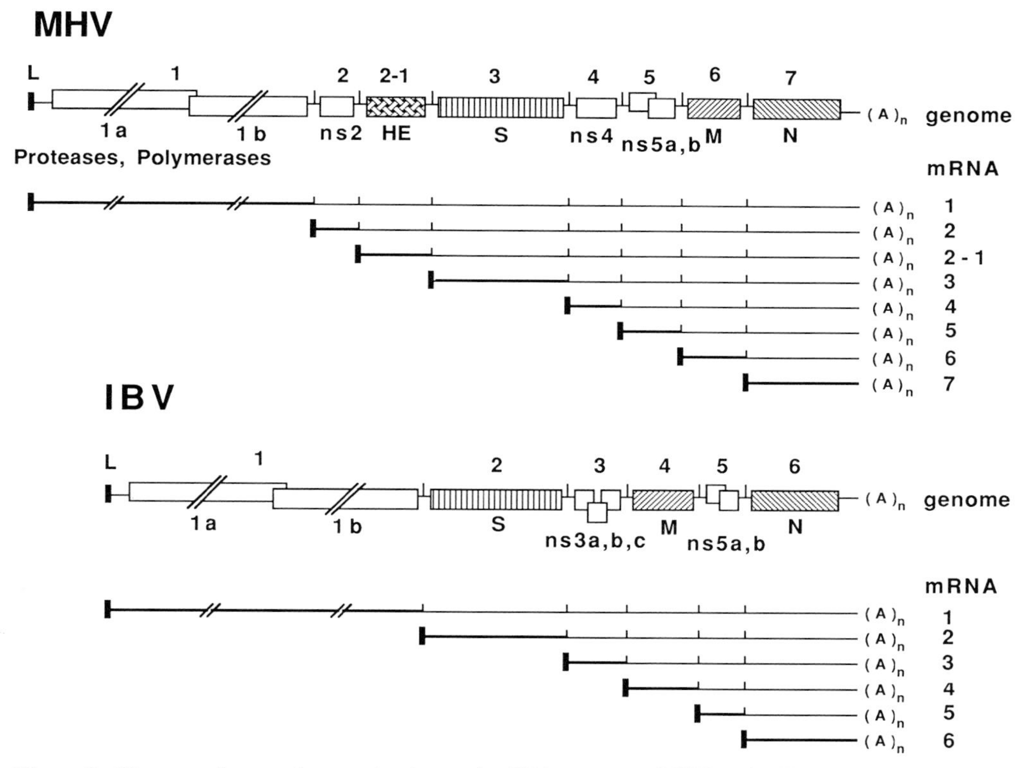

Figure 3 Diagram of genomic organization and mRNA structure. MHV and IBV are two coronaviruses whose RNA genomes have been completely sequenced. Boundaries of each gene are marked by vertical lines. Various boxes indicate open reading frames (ORFs), which have potential to encode a protein. Open boxes encode nonstructural proteins and the rest structural proteins. Boxes in the same region between two vertical lines are considered to belong in the same gene, and their gene products are made from the same mRNA. Gene 1 consists of two ORFs, which are not drawn to scale because they together constitute 20–22 kb, approximately equivalent to two thirds of the entire RNA genome (31 kb for MHV and 28 kb for IBV). Note that genes 2 and 2-1 of MHV do not have their counterparts in IBV RNA. Gene products of each gene are indicated. L represents the leader RNA and $(A)_n$ represents polyA. ns represents nonstructural proteins. mRNAs are represented by straight lines. Only the 5′-most ORF (thick lines) of each mRNA is functional.

a small amount of subgenomic mRNAs (see below) can also be detected. These mRNAs are most likely packaged nonspecifically into the virion. Some virus preparations also contain defective-interfering RNA of various sizes.

Some other coronavirus RNAs have been partially sequenced. Their genomic organization is similar to that of either IBV or MHV RNAs.

D. Lipid

The envelope of virion particles is composed of lipid bilayer, the composition of which is similar to that of the membrane of the host cells. The lipid is derived from the membrane of the Golgi complex or rough endoplasmic reticulum during virus assembly.

IV. Replication Cycle of Viruses

A. Viral Adsorption to Receptors on the Cells

The first step of viral infection is the attachment of virus particles to the target cells. This step is mediated by the S protein. However, in viruses containing HE protein, both S and HE proteins may mediate virus attachment. The attachment site is a specific receptor protein on the plasma membrane of the target cells. For MHV, the receptor is a 110-kDa protein present on the brush-border membrane of the intestine and hepatocytes. It is a member of the carcinoembryonic antigen (CEA) protein family, and the monoclonal antibody specific for this protein can prevent MHV infections *in vitro*. The transfection of the cDNA clone for this receptor protein into a resistant cell line rendered the latter susceptible to MHV infection. The receptor proteins for other coronaviruses (e.g., HCV or RCV) appear to be different from that for MHV. It is not clear whether HE and S bind to the same receptor protein. [*See* MICROBIAL ATTACHMENT, MOLECULAR MECHANISMS.]

B. Virus Penetration and Uncoating

Once the virus attaches to the target cells, the virus enters the cells by either one of the following two mechanisms. One is endocytosis or "viropexis" (i.e., the virus is internalized through the plasma membrane via coated pits) (Fig. 4). The vesicles containing virus particles are then fused with endosomes inside the cells. In the low pH environment within the endosomes, the viral envelope fuses with the endosomal membranes, releasing viral nucleocapsid into the cytoplasm. The second mechanism is the direct fusion between the viral envelope and plasma membrane on the cell surface, releasing the viral nucleocapsid into the cytoplasm as a result. The relative importance of these two mechanisms is unknown. It is not clear what triggers the virus entry. The removal of the S1 from IBV rendered the virus noninfectious although the virus still attached to the target cells, suggesting that virus attachment alone is not sufficient for virus entry. The penetration step requires active cellular metabolism.

C. Macromolecular Synthesis

Once the viral nucleocapsid is released into the cytoplasm, the viral RNA will be replicated by a virus-specific RNA polymerase. Replication occurs via RNA-dependent RNA synthesis. The entire process of viral replication takes place in the cytoplasm. Thus, viral growth is resistant to treatment with actinomycin D, which inhibits DNA-dependent RNA synthesis, and can take place in cells from which the nuclei have been experimentally removed. The first step of viral replication (i.e., primary translation in Fig. 4) is the synthesis of an RNA-dependent RNA polymerase from the incoming viral genomic RNA, using normal cellular ribosomes and translational machinery. The polymerase then transcribes the viral RNA into negative-stranded RNA. Thus, the viral genome serves as both the mRNA for the synthesis of RNA polymerase and the template for the synthesis of negative-strand RNA. The latter is in turn used as a template for the synthesis of mRNAs.

1. mRNA Structure

In virus-infected cells, mRNAs are the predominant viral RNA species, outnumbering negative-strand RNA by at least 20-fold. Depending on virus strains, the number of viral mRNAs varies from six to eight. The largest of these mRNAs is equivalent to the genomic size; the rest of them are subgenomic in size. The mRNAs are numbered in the order of decreasing size (Fig. 3). The mRNAs discovered after the viral mRNAs were initially named are given an inserted number (e.g., mRNA 2-1). These mRNAs have a nested-set structure (i.e., the sequence of the smallest mRNA is included in the next larger mRNA), and all the mRNAs have sequences starting from the 3′-end of the genome and extending for various distances in the 5′-direction. Thus, except for the smallest mRNA, all the mRNA species contain at least two genes; however, only the 5′-most gene of each mRNA can be used for translation; therefore, each mRNA, in principle, can encode only one protein.

Another distinct feature of mRNAs is that they contain a stretch of 60 to 70 nucleotides of leader sequence, which is transposed from the 5′-end of the genomic RNAs. This sequence is not present inside the genomic RNA at the initiation points of mRNAs.

2. Strategy of RNA Synthesis

The incoming genomic RNA is first transcribed by viral RNA polymerase into negative-strand RNA, which includes both genomic and subgenomic sizes corresponding to those of mRNAs. The positive-strand RNAs are then transcribed from these negative-stranded RNA templates, which contain an anti-leader sequence derived from the 5′-end of the genome, resulting in various mRNA species containing a leader sequence. The issue that is not clear at the present time is how these subgenomic negative-stranded RNAs with an anti-leader sequence are synthesized. There are several possible mechanisms. One is that the genomic RNA is transcribed first into a full-length, negative-strand RNA, which is then spliced into subgenomic negative-strand RNAs. This mechanism would use a modified form of RNA splicing, which has to occur in the cytoplasm. Alternatively, the full-length (−)-strand RNA is used as the template to synthesize a leader RNA, which then binds to the intergenic sites on the template RNA and used as primer for the transcription of subgenomic mRNAs. This mechanism is termed "leader-primed transcription." Once the mRNAs are made, they will in turn be transcribed into subgenomic negative-strand RNA, which then becomes the templates for additional mRNA synthesis. Still another possibility is that subgenomic (−)-strand RNAs are transcribed directly from the small amount of mRNAs nonspecifically incorporated into the virus particles. Currently, these mechanisms cannot be distinguished. [*See* RNA SPLICING.]

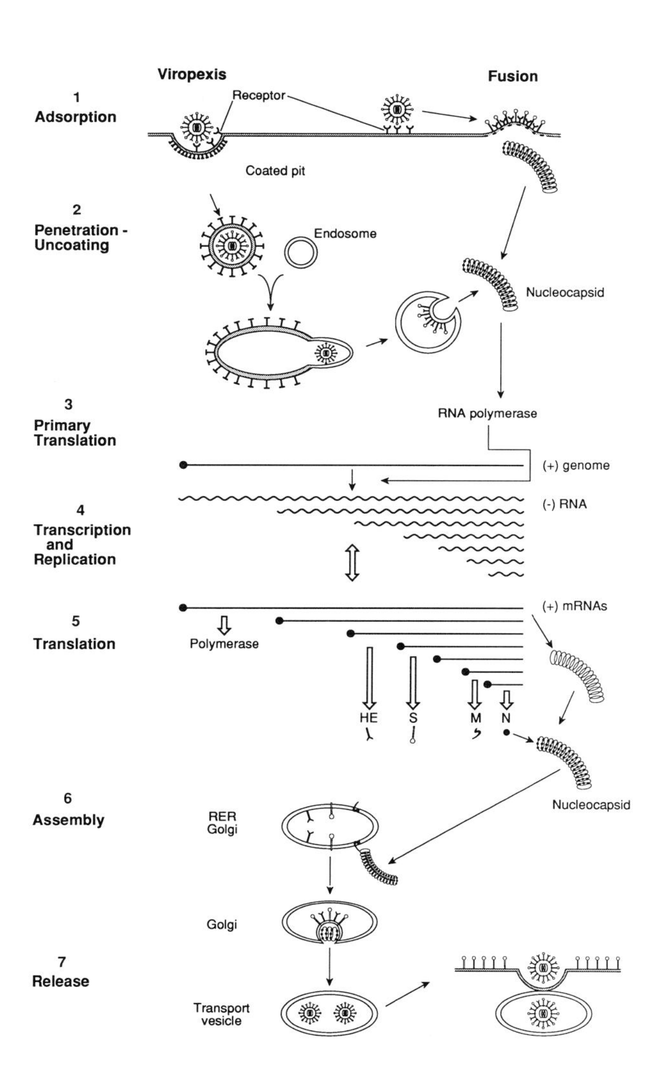

Figure 4 Strategy of coronavirus replication. Model is based on MHV. (1) Viral adsorption to receptors; (2) viral penetration is by either viropexis (endocytosis) or direct virus–cell membrane fusion. The former process requires fusion between virus-containing vesicles and endosomes. Viral nucleocapsid is released by fusion between viral envelope and endosomal membranes. Both processes require the activity of S protein. (3) Primary translation: Freed nucleocapsid is used for translation of RNA polymerases. (4) Transcription and replication: RNA polymerases are used for the synthesis of (−) strand and (+) strand RNA. The mechanism of synthesis of subgenomic (−) strand RNAs is not yet clear. They can be replicated from the (+) strand mRNAs. (5) Translation: mRNAs are used for the translation of structural as well as nonstructural proteins. The latter are probably used for RNA synthesis and gene regulation. HE, S, and M proteins are translated by membrane-bound ribosomes and inserted into rough endoplasmic reticulum (RER), where they undergo initial glycosylation. N is synthesized by free ribosomes and phosphorylated immediately in the cytosol. (6) Virus assembly: Genomic-sized RNA complexes with N to form nucleocapsid, which interacts with M on the RER or Golgi compartment. Virus particles then bud into the lumen of RER or Golgi and acquire all three envelope proteins. (7) Virus particles travel through the Golgi complex, and glycoproteins undergo further modifications. The virus is transported by transport vesicles and released from cells. The entire process takes place in the cytoplasm.

The relative amounts of mRNAs vary, with the smaller mRNAs being more abundant. The genomic-sized mRNA1 serves dual functions as the mRNA for the synthesis of polymerase as well as the genomic RNA packaged into the new virus particles. Whether there is a switch of transcription machinery from the synthesis of subgenomic mRNAs to the synthesis of genomic-sized RNAs late in the infection to prepare for virus assembly is not clear. There appears to be such a switch in BCV-infected cells. It has been speculated that the N protein may be involved in such a transcriptional switch.

3. Translation Strategy

Each coronaviral mRNA is functionally a monocistronic mRNA (i.e., each mRNA makes only one protein), despite the fact that most of the mRNAs include the sequences of more than one gene. Presumably the ribosome binds to mRNA and travels until it reaches the first or second AUG (initiation codon) and starts translation. Once the termination codon is reached, the ribosome is separated from mRNAs, and translation cannot be reinitiated from the downstream AUG. Thus, only the 5′-most gene is translated. This translation strategy holds true for most of the viral mRNAs; however, there are exceptions. For instance, in IBV, the 5′-most gene of mRNA3 contains three ORFs that slightly overlap each other, and all three of them are translated from this mRNA species. All three proteins have been detected in the IBV-infected cells. mRNA5 of MHV also contains two ORFs. The downstream ORF is preferentially translated. These proteins are the results of internal initiation. Another exception is gene 1, which contains two ORFs with slight overlap. These two ORFs are translated into one protein via a ribosomal frameshift mechanism (i.e., the ribosome and translation complex shifts reading frame near the end of the first ORF, thus allowing translation to continue on the second ORF). As a result, a single protein encompassing both ORFs is made. The ribosomal frameshifting is triggered by a pseudoknot RNA structure and a consensus sequence motif at the junction of the two ORFs.

4. Posttranslational Modifications and Processing of Viral Proteins

a. Structural Proteins

i. S Protein The S protein is synthesized on membrane-bound ribosomes as a precursor protein, which is cotranslationally glycosylated into a 150-kDa protein. The N-linked glycans are added in the rough endoplasmic reticulum. The protein forms a trimer in the Golgi and is then transported into the *trans*-Golgi compartment, where secondary glycosylation and fatty acylation (palmitic acid) take place. At least some of the S protein is cleaved into S1 and S2 by cellular proteases in the *trans*-Golgi compartment. Additional carbohydrate residues, including sialic acids, are added to the uncleaved and cleaved S proteins in the *trans*-Golgi network, and the protein matures to become the S1 and S2 proteins of 90 kDa. Some of the S protein is transported to the cell surface.

ii. M Protein The M protein is synthesized as a precursor 21-kDa protein and undergoes the first step of O-linked glycosylation in the endoplasmic reticulum. The remaining glycosylation takes place in the *trans*-Golgi compartment or the *trans*-Golgi network. After glycosylation, the mature M protein is 23 to 25 kDa in molecular mass and is retained in the Golgi in the perinuclear region, in contrast to the S protein. In viruses that have N-linked carbohydrates, the modification pathway of M protein is presumably similar to that of S protein.

iii. HE Protein The HE protein is synthesized as a precursor 45-kDa protein and is cotranslationally glycosylated by N-linked carbohydrates to become a mature 65-kDa protein. This glycosylation takes place in the rough endoplasmic reticulum. When the protein is transported into *cis*-Golgi, it undergoes some trimming of the existing carbohydrate chains. The protein may be transported to the cell surface.

iv. N Protein The N protein is synthesized on the free ribosomes and is immediately phosphorylated in the cytosol. The protein stays in the cytosol, where it complexes with viral RNAs. Some N protein may be transported to the nuclei under certain circumstances. Its functional significance is unknown.

b. Nonstructural Proteins

Not very much is known about the metabolic pathways of the nonstructural proteins. The only nonstructural protein shown to undergo posttranslational processing is gene 1 product. Sequence analysis of this gene shows that it contains two or three protease domains, which are similar to papain-like proteases and poliovirus protein 3C protease, respectively. The protein has a predicted size of 700 to 800 kDa, which is cleaved into multiple proteins. One of the cleavage events is at the N-terminal portion of the primary translation product, generating a p28 product. This cleavage in MHV is carried out by the papain-like protease within the gene 1 product itself. Other processing events have not been precisely determined.

D. Virus Assembly

The packaging of viral genomic RNA into virions requires a specific packaging signal, which is located at the 3′-end of gene 1. The newly synthesized genomic RNA first complexes with the N proteins to form a nucleocapsid. The nucleocapsid then interacts with the carboxyl terminus of M protein, which is embedded in the membrane of the endoplasmic reticulum or, in other cell types, the early Golgi compartment. The virus buds into the lumen of the endoplasmic reticulum or Golgi, using the cellular membrane as its envelope and incorporating S, HE, and M proteins into the envelope. The virus particles then travel through the Golgi complex, and the viral glycoproteins undergo further modifications during the transport. The S and HE proteins are not essential for virus assembly, but M is required for virus assembly although its glycosylation is not. The release of virus particles may occur after the cells are lysed. However, some virus particles can be released using cellular secretory pathway.

V. Effects of Viral Infections on Cells

Viral infections often cause cytopathic effects (CPE), which is characterized by syncytia formation as a result of cell-to-cell fusion. Some infected cells develop vacuolization and spindling. At the molecular level, cellular RNA and protein synthesis are gradually inhibited as viral infection proceeds. The infection eventually results in cell lysis and death. However, in some cells, coronaviruses can establish persistent infection without significant effects on the cellular metabolism. The persistent infection *in vitro* is often maintained as a state of "carrier culture," in which only a small fraction of cells are infected and release virus particles. The mechanism of persistent infection is not clear.

VI. Genetics

A. Temperature-Sensitive Mutants and Genetic Complementation

Two different types of temperature-sensitive (ts) mutants of MHV have been isolated: one [RNA(−) mutants] is defective in RNA synthesis at the nonpermissive temperature, and the other [RNA(+) mutants] synthesizes RNA but does not produce infectious virus particles. Most of the RNA (−) ts mutants have lesions in gene 1, which encodes RNA polymerase. Many of the ts mutants also have altered pathogenicity in animals (i.e., in contrast to the wild-type virus, which causes encephalitis in mice, many ts mutants of MHV have attenuated virulence and cause mainly demyelination).

These ts mutants have been grouped into five to seven groups, according to their ability to complement each other. Only the mutants with defects in different viral proteins can complement each other. Viruses that cannot complement each other are considered to affect the same gene function and, thus, belong to the same "complementation group." Four to six of the complementation groups include RNA (−) mutants, suggesting that four to six different viral proteins are required for RNA synthesis. Using recombination techniques (see below), the genetic locations of these different complementation groups have been mapped within gene 1 of the MHV genome.

B. RNA Recombination

When two coronaviruses are co-inoculated into culture cells or co-infect an animal, viruses containing part of genomic sequences from both parental viruses can be isolated among the virus progeny. This is due to homologous RNA recombination (genetic cross-over). Recombination has so far been demonstrated only in MHV.

RNA recombination in MHV can occur almost anywhere in the viral RNA genome and occurs at a very high frequency, which has been calculated to be approximately 25% for the entire MHV genome, meaning that 25% of virus progeny derived from a mixed infection would have at least a genetic crossover somewhere in the genome. This recombination frequency is astonishingly high for viruses with a nonsegmented RNA genome. RNA recombination probably occurs by a copy-choice mechanism (i.e., RNA transcription pauses at certain sites on the template RNA; the nascent RNA transcript dissociates from the template RNA and later rejoins a different template to continue RNA synthesis). Occurrence of RNA recombination in coronaviruses has also been demonstrated in experimental and natural viral infections of animals. The latter was suggested from the sequence relationship in the M and S genes between different IBV strains during natural evolution. Furthermore, the HE gene of MHV is closely related to the hemagglutinin protein of influenza C virus, suggesting a nonhomologous recombination event between coronavirus and influenza C virus. The propensity of coronavirus to undergo a high frequency of homologous recombination is unusual among RNA viruses. Only one other RNA virus family (i.e., picornavirus) has the ability to undergo homologous recombination.

C. Defective-Interfering Particles

When coronaviruses are passaged in tissue culture at high multiplicity of infection, the virus progeny often includes defective-interfering (DI) particles, which are defined as a virus containing a defective (often much smaller) viral RNA and having an ability to interfere with the replication of the wild-type virus. It needs a helper virus for replication. Different types of DI RNA have been detected in MHV-infected cells after different rounds of viral passages. Some of these DI RNAs are incorporated into virus particles, whereas others are not. The difference between these two types of DI RNA is the presence or absence of an RNA-packaging signal, which is located at the 3′-end of gene 1. DI RNA replicates at a very high efficiency and interferes with the synthesis of most viral mRNA species, except for the smallest mRNA, which encodes N protein. Thus, N protein synthesis may be required for the replication of DI RNA. All the DI RNAs that have been sequenced so far consist of sequences derived from several discontiguous regions of the wild-type viral RNAs, but they always contain both the 5′- and 3′-end sequences of the genome, which contain the signal for RNA replication. When viruses containing DI RNAs are passaged serially in tissue culture, the species and sizes of DI RNA continue to change on each passage. Thus, similar to other RNA viruses, different sets of DI RNAs have evolutionary advantages under different conditions. The DI RNA is most likely generated by polymerase jumping during RNA synthesis. The significance of DI RNA in natural viral infection is not yet clear.

VII. Clinical Manifestation and Pathogenicity

Coronaviruses cause a variety of diseases in humans and animals (Table I). The most common coronaviral infections involve the respiratory tract. Human coronaviruses are responsible for nearly 30% of common colds in the winter and is the second most common cause (after rhinovirus) of common colds. The virus infects and kills ciliated epithelial cells of the upper respiratory tract and, occasionally, the lower respiratory tract. In animals, IBV, RCV, and PRCV also cause respiratory infections, but IBV could spread from the respiratory tract to the reproductive system. RCV could also infect the salivary and lacrimal glands, similar to the disease caused by SDAV. Thus, coronaviruses have a very wide disease spectrum.

The other common coronaviral infections involve the gastrointestinal tract. Coronavirus-like particles have been isolated from neonatal necrotizing enterocolitis and other diarrheas of humans, although their etiological relationship has not been established. In animals, TGEV, BCV, CCV, and FECV also cause diarrhea of various severity. Usually, severe enteric diseases are demonstrated only in newborns or younger animals. These viruses have tropism for the epithelial cells of the intestinal tract.

The molecular basis of the coronaviral tropism for the epithelial cells of respiratory and gastrointestinal tracts could be due to the S protein. Mutations in the S protein (e.g., in the neutralization-escape mutants of MHV, which were selected with neutralizing monoclonal antibodies) altered the viral pathogenicity of the virus. PRCV, which was derived from TGEV in natural evolution and has a deletion in the S-protein gene, causes respiratory infection instead of gastrointestinal infection, as seen in TGEV. Other viral proteins, however, may also contribute to the viral pathogenicity.

Other coronaviruses cause less common illnesses; for examples, MHV causes hepatitis, encephalitis, and demyelination, FIPV causes peritonitis, and RbCV causes myocarditis. These systems provided the best studied models of coronaviral pathogenesis. The better known systems are described below.

A. Murine Coronavirus

Many MHV strains, including the JHM and A59 strains, cause either acute, lethal encephalomyelitis or subacute and chronic demyelination, resulting in hindleg paralysis and extensive damage in the white matter of brain, when inoculated intracerebrally or intransally into mice. The diseases caused by MHV in mice can be altered by several different factors.

1. Genetic makeup of virus: Alterations of the viral genome changed viral pathogenicity. For example, neutralization-escape virus variants selected by using S protein–specific monoclonal antibodies generally lost the ability to cause acute encephalitis but increased the ability to cause demyelination. Different natural MHV strains also show different cellular tropism *in vitro*. For instance, the JHM strain has a predilection for glial cells, whereas MHV-3 shows tropism for eppendymal cells, neurons, and meningeal cells.
2. Genetic makeup of the host animal: Balb/C mice are susceptible to JHM infection whereas SJL mice are resistant. The susceptibility and resistance of mouse strains can be demonstrated using *in vitro* neuron cultures and macrophage cultures. Thus, the resistance of SJL mice is evident at the level of target cells, which may have a defective virus receptor. The genetic resistance in mice is controlled by several autosomal recessive and dominant genes.
3. Age of the animals: In general, newborn mice are very susceptible to JHM infections. As the animal ages, its susceptibility to JHM infections gradually decreases. At least one of the factors is the age-dependent differentiation of the target cells.
4. Host immune responses also play a significant role in subacute and chronic demyelination. Passive immunization of mice with monoclonal antibodies against any of the viral structural proteins can prevent lethal acute encephalitis with MHV and cause demyelination instead. Similarly, adoptive transfer of T cells can also alter the pathogenesis of MHV infections.

Another well-studied model of MHV pathogenesis is MHV-3-induced hepatitis. This virus causes fulminant hepatitis in C57 Bl/6 mice but not in A/J mice. The genetic resistance is not due to restriction of virus replication in the liver but is dependent on the immune cells. The viral pathogenesis also is not solely dependent on direct viral cytopathic effects but is dependent on the host immune response.

B. Feline Infectious Peritonitis Virus

Feline Infectious Peritonitis Virus (FIPV) replicates in the small intestine, and its primary target cells are macrophages. There is a correlation between the pathogenicity of FIPV *in vivo* and the ability of the virus to infect macrophages *in vitro;* more virulent FIPV strains infect macrophages at higher efficiency. Macrophages carrying viruses also allow the virus to disseminate outside the intestine. The humoral immunity does not protect but rather enhances the FIPV pathogenicity, a phenomenon termed "antibody-dependent enhancement" of virus infection because the virus–antibody complexes attach to macrophages via Fc receptors more efficiently than direct virus infection. However, the cell-mediated immunity is an important factor in eliminating viruses and protecting cats against this disease.

VIII. Epidemiology

Human coronavirus infection is very common. More than 90% of adults have antibodies against human coronaviruses OC43 and 229E, which are responsible for nearly 30% of common colds. A similarly high prevalence rate has also been noted with coronavirus infections in animals. Currently, live attenuated vaccines are available for IBV, TGEV, BCV, and FIPV.

Bibliography

Boursnell, M. E. G., Brown, T. D. K., Foulds, I. J., Green, P. F., Tomley, F. M., and Binns, M. M. (1987). *J. Gen. Virol.* **68,** 57–77.

Cavanagh, D., and Brown, T. D. K. (eds.). (1990). "Coronaviruses and Their Diseases." Plenum Press, New York.

Kyuwa, S., and Stohlman, S. A. (1990). *Semin. Virol.* **1,** 273–280.

Lai, M. M. C. (1990). *Annu. Rev. Microbiol.* **44,** 303–333.

Lee, H.-J., Shieh, C.-K., Gorbalenya, A. E., Koonin, E. V., La Monica, N., Tuler, J., Bagdzhadzhyan, A., and Lai, M. M. C. (1991). *Virology* **180,** 567–582.

McIntosh, K. (1990). Coronaviruses. *In* "Virology" (B. N. Fields and D. M. Knipe, eds.), pp. 857–864. Raven Press, New York.

Spaan, W., Cavanagh, D., and Horzinek, M. C. (1988). *J. Gen. Virol.* **69,** 2939–2952.

Corrosion, Microbial

Marianne Walch
Center of Marine Biotechnology, University of Maryland, and Naval Surface Warfare Center

Glossary

Anode Site on a metal surface at which oxidation of metal atoms is the predominant reaction

Biofilm Microbial community that develops at the interface of two phases (e.g., at a solid–liquid interface)

Cathode Site on a metal surface at which electrons produced from metal oxidation reactions are reduced

Depolarization Acceleration of corrosion by removal of corrosion reaction products that have accumulated at the anodic and/or cathodic sites on a metal

Differential concentration cell Electrochemical potential differences established on a metal surface by variations in the concentrations of chemical species in contact with the metal; results in a flow of electrons and corrosion of the metal

Tubercle Raised hard deposit on a metal surface, largely composed of ferric or manganic oxides, which results from the activity of iron or manganese-oxidizing bacteria

MICROBIOLOGICALLY INFLUENCED CORROSION (MIC) is the deterioration of metals that is initiated, accelerated, or otherwise influenced by the presence and/or metabolic activities of bacteria and other microorganisms. Biocorrosion is another term frequently used for this. MIC occurs most frequently in aqueous systems in which microorganisms have attached to the metal surface and accumulated in a surface-associated community known as a biofilm. All metallic materials that are in contact with nonsterile aqueous or moist environments are susceptible to fouling by microorganisms. The resultant biofilms are complex, dynamic communities of microorganisms composed of living cells, their metabolic products and exopolymers, trapped detritus, and inorganic corrosion products (Fig. 1). The presence of a biofilm on a metal can markedly change the nature of a metal surface and affect both the kinetics and types of corrosion reactions that occur there.

I. Description of Microbiologically Influenced Corrosion

A. Importance

The involvement of microorganisms in metal corrosion was suggested as early as 1891, when it was hypothesized that the corrosive action of water on lead could be due to ammonia, nitrates, and nitrites produced by bacterial action. In the early twentieth century, researchers began to conclude that the corrosion of underground iron and steel structures was in part due to bacterial activity. The iron-oxidizing bacterium *Gallionella* actually was isolated from the corrosion products on a buried steel conduit, and indications of the presence of sulfur bacteria also were detected. In the 1920s, reports appeared suggesting that hydrogen sulfide production by bacteria

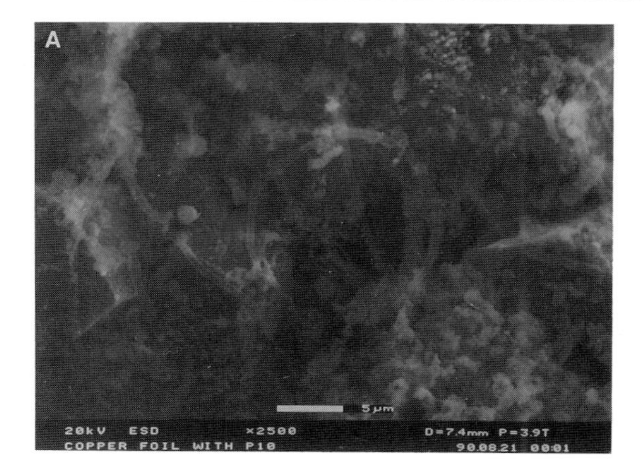

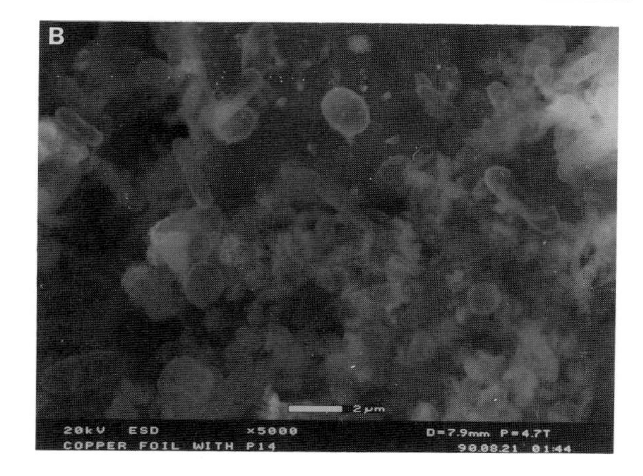

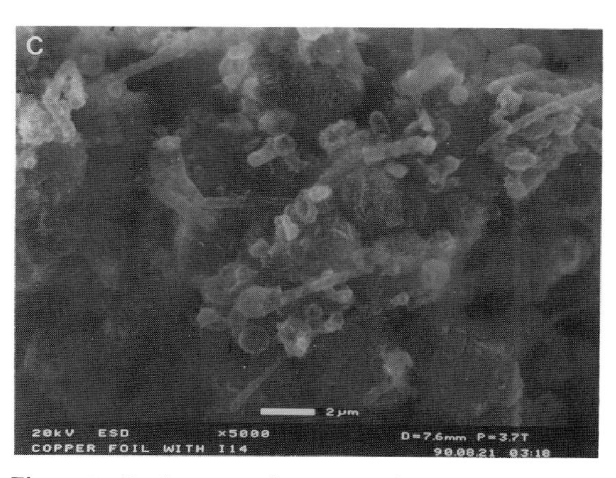

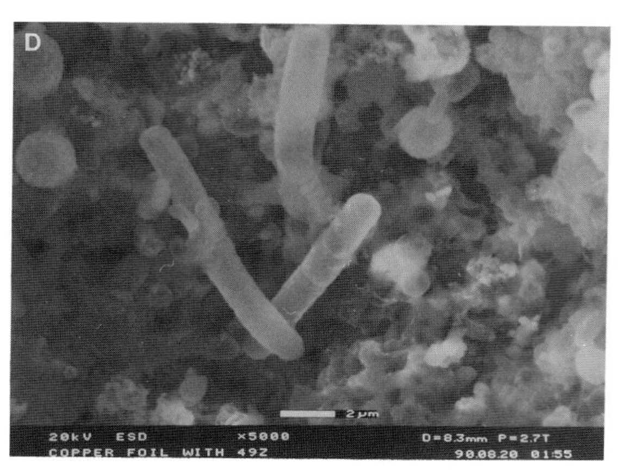

Figure 1 Environmental scanning electron micrograph of mixed species biofilms, including sulfate-reducing bacteria, growing on the surfaces of copper foils. Bacteria are growing enmeshed in a complex biofilm matrix of corrosion products and exopolymeric material. (A) Bar = 5 μm; (B–D) bar = 2 μm. [Courtesy of Richard Ray, Naval Oceanographic and Atmospheric Research Laboratory, Stennis Space Center, Mississippi.]

could profoundly affect metal corrosion and that condenser tube failures could be attributed in some cases to bacterially produced ammonia. In 1922, a series of investigations led the Dutch scientist von Wolzogen Kuhr to postulate that sulfate-reducing bacteria promote corrosion of iron under anaerobic conditions by removing hydrogen that accumulates on the surface. His classic publication is regarded by many as marking the true beginning of the "science" of microbiological corrosion.

Since that time, a growing number of reports have appeared describing the effects of microorganisms on metal corrosion, and MIC has emerged as a major economic problem. The influence of microbial activities on corrosion processes has become increasingly apparent with greater use of metals and metal alloys in situations in which they are exposed to microbial colonization. A wide range of corrosion phenomena in which microorganisms are involved has been recognized. Biocorrosion is of considerable concern to the power-generating, chemical-processing, oil, and shipping industries as well as to the military. Bacterial activity, for instance, is thought to be responsible for >75% of the corrosion in productive oil wells and for >50% of the failures of buried pipelines and cables. The economic costs of MIC are not known, but fouling by microorganisms can greatly alter the corrosion behavior of metals as well as impair the efficient operation of engineering systems.

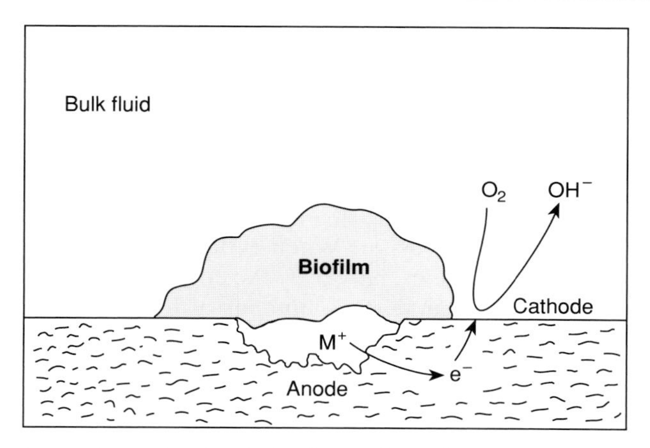

Figure 2 Electrochemical cell established beneath a patch of biofilm on a metal surface, resulting in the formation of local anodes and cathodes.

B. Mechanisms

Corrosion is an electrochemical phenomenon in which differences in potential between two sites on a metal cause a movement of electrons, resulting in oxidation and dissolution of the metal (Fig. 2). Sites on the metal surface at which oxidation of metal atoms is the predominant reaction are known as anodes. Anodic reactions produce metal cations or metal oxides plus electrons:

$$M \longrightarrow M^{2+} + 2\,e^-$$

Electrons produced from the oxidation reactions are simultaneously driven to other sites on the metal where reduction is the predominant reaction. These sites are known as cathodes. Electrons often are removed at cathodes by reduction of oxygen or, under acidic conditions, hydrogen ions:

$$\tfrac{1}{2}\,O_2 + H_2O + 2\,e^- \longrightarrow 2\,OH^-$$

$$2\,H^+ + 2\,e^- \longrightarrow 2\,H \longrightarrow H_2$$

Microorganisms may initiate or accelerate corrosion reactions by creating conditions that establish differences in electrical potential on the metal or by stimulating either the anodic or cathodic reactions. Identification of specific mechanisms for MIC has been difficult due to the complexity of both electrochemical and microbiological processes in biofilms. A number of interdependent corrosion reactions may occur simultaneously in a given situation, and some, but not others, of these may involve bacteria. A variety of MIC processes are, however, clearly recognized and their mechanisms at least partially understood.

Microorganisms can contribute to the deterioration of metals in one or more of the following ways, each of which depends on the physiological and ecological characteristics of the organisms involved:

1. Direct influence on the rate of anodic or cathodic reactions.
2. Breakdown of passive metal oxide surface films by metabolic activity or products of metabolism.
3. Generation of corrosive metabolites such as sulfide or acids, leading to the creation of a corrosive environment.
4. Creation of differential concentration cells on the metal surface by uneven distribution of microbial colonies or extracellular polysaccharide "slimes."
5. Degradation of corrosion inhibitors and protective paints and coatings.

In most cases, MIC results in localized corrosion of materials (i.e., pitting or crevice corrosion) rather than uniform loss of material over the entire metal surface.

II. Role of Sulfate-Reducing Bacteria in Microbiologically Influenced Corrosion

Many microorganisms play a role in corrosion processes, but by far the bulk of attention has focused on sulfate-reducing bacteria (SRB). SRB are obligate anaerobes, with relatively restricted nutritional requirements, which function in microbial communities, or consortia, as the terminal link in the oxidative biodegradation of primary nutrients. Although inhibited by oxygen, these bacteria thrive in anaerobic microenvironments beneath inorganic or organic deposits on metals, within colonies, or under biofilms, where release of secondary mutrients and scavenging of oxygen by the respiration of aerobic bacteria favors growth of SRB. Furthermore, hydrogen sulfide (H_2S) produced by SRB is a reducing agent that removes oxygen, thereby protecting established colonies of sulfate reducers from its harmful effects. Thus, even in aerated waters, highly active populations of SRB frequently are established on metal substrata. The last decade or so has seen a tremendous increase in our knowledge and understanding of the nutrition, physiology, and ecological importance of these organisms. [*See* SULFUR METABOLISM; ANAEROBIC RESPIRATION.]

Anaerobic corrosion by SRB is one of the most important types of MIC observed in the field. There are many documented cases of SRB-induced corrosion in the paper, petroleum, chemical-processing, water distribution, and general engineering industries. Open pitting and gouging of the metal are characteristic of stainless steel failures attributed to SRB. The pits typically are filled with soft black iron sulfide corrosion products with bright active metal beneath. Under deposits, these pits develop with small entrance holes and deep, cavernous cavities that penetrate far into the metal. SRB attack on cast iron results in graphitization of the metal; the iron dissolves away, leaving a soft graphite residue. Sulfate reducers also cause pitting of aluminum and copper alloys.

A number of hypotheses and considerable controversy exist over the mechanism(s) of anaerobic corrosion by SRB. The first and, until recently, most widely accepted hypothesis was that of von Volzogen Kuhr and Van der Vlught, who, in 1934, proposed a cathodic depolarization mechanism to explain the accelerated corrosion of iron caused by sulfate reducers. Corrosion reactions in general tend to slow down as reduction products accumulate at the cathode. The metal then is said to be cathodically polarized. Any effect that accelerates removal of these products and increases the corrosion rate is known as depolarization. SRB were postulated to depolarize corroding iron alloys by removing cathodic hydrogen that accumulates on the metal surface (Table I). This shifts the electrochemical equilibrium toward removal of electrons and causes more iron to be dissolved at the anode. The enzyme hydrogenase, synthesized by many species of SRB, was thought to be involved in this process.

Table I Reactions Proposed in the Cathodic Depolarization Theory of Iron Corrosion by Sulfate-Reducing Bacteria

Reaction	
$4\,Fe \longrightarrow 4\,Fe^{2+} + 8\,e^-$	Anodic reaction
$8\,H_2O \longrightarrow 8\,H^+ + 8\,OH^-$	Water dissociation
$8\,H^+ + 8\,e^- \longrightarrow 8\,H^0$	Cathodic reaction
$SO_4^{2-} + 8\,H^0 \longrightarrow S^{2-} + 4\,H_20$	Bacterial consumption of H^0
$Fe^{2+} + S^{2-} \longrightarrow FeS$	Corrosion products
$3\,Fe^{2+} + 6\,OH^- \longrightarrow 3\,Fe(OH)_2$	
Overall reaction	
$4\,Fe + So_4^{2-} + 4\,H_2O \longrightarrow 3\,Fe(OH)_2 + FeS + 2\,OH^-$	

Evidence from pure culture experiments conducted in the absence of sulfate have demonstrated that SRB that synthesize hydrogenase are indeed capable of depolarizing iron and steel, leading to increased corrosion. When sulfate is present in the growth medium, however, the corrosion mechanism is less clear; even a number of hydrogenase-negative strains induce high rates of corrosion. This accelerated attack seems to depend on the ferrous sulfide corrosion product and may be significantly influenced by its physical and chemical forms. Other proposed mechanisms include depolarization by the cathodic activity of H_2S, which receives electrons from the metal to form H_2, and the production of corrosive volatile phosphorous compounds. The consensus among researchers now is that sulfide production by SRB probably is of greater quantitative importance than cathodic depolarization in the overall corrosion process. It is important to note that several or all of these mechanisms may operate simultaneously in conjunction with other biocorrosion activities, and the predominant reactions likely are determined by interactions of material characteristics, environmental chemistry, and the ecology of the biofilm community. [*See* SULFIDE-CONTAINING ENVIRONMENTS.]

III. Production of Corrosive Metabolites

A wide variety of organic acids are secreted by heterotrophic microorganisms during fermentation of organic substrates. The types and amounts of these compounds found in a biofilm depend on the organisms present and the available substrates. Organic acids initiate or accelerate corrosion of many types of metals by removing or preventing the formation of protective oxide films. Some species of bacteria, particularly those of the genus *Thiobacillus,* also produce mineral acids, such as H_2SO_4, which are extremely corrosive. The corrosive activity of metabolites such as sulfides and acids are intensified when they are trapped and accumulate in microenvironments beneath colonies, deposits, and biofilms.

Thiobacillus ferrooxidans has been implicated in the corrosion of iron and steel pipes and pumping machinery used in microbiological mining of low-grade ores. *Thiobacillus* species also are responsible for considerable damage to stonework and concrete. MIC is the principal cause of deterioration of concrete infrastructures in the wastewater collection and treatment industry. Corrosion of reinforced concrete sewer pipes by the highly acidic environment created by species of *Thiobacillus* has become an extraordinarily expensive problem for municipalities all over the world. Destruction of concrete sewer pipes is caused by the sulfuric acid produced by thiobacilli living on the crown of the sewer pipes, which is not in contact with the sewage. Anaerobic bacteria in sewer biofilms generate hydrogen sulfide from sulfate in the sewage. The sulfide enters the air space in the sewer where it is oxidized by *Thiobacillus* species. This acidic environment, with pH levels as low as 1, results in severe corrosion of the concrete pipes.

IV. Biofilm Patchiness and Concentration Cells

Corrosion currents are established when dissimilar metals are in electrical contact or when differences in electrochemical potential exist between different sites on the same metal. Potential differences can occur due to imperfections in the material, inclusions in the metal matrix, or differential concentration cells. Differences in concentrations of chemicals in contact with a metal surface occur beneath surface deposits or growths that limit diffusion to or away from the metal surface. Differential aeration cells frequently are established, for example, when patches of microbial colonies or slimes limit access of oxygen to the underlying metal surface. In this case, the areas of low oxygen become anodic, and localized dissolution of the metal occurs at these sites.

Biofilms typically are nonuniform or patchy in their distribution, and areas under respiring colonies are quickly depleted of oxygen relative to surrounding uncolonized areas, even in aerated environments. As little as 15 μm of accumulated biofilm material may limit oxygen diffusion through the film to the point that the underlying metal interface is virtually anaerobic. In addition to creation of anodic sites on the metal, such environments also favor growth of SRB and loss of adherent protective metal oxide films, both of which further accelerate localized corrosion reactions. [*See* BIOFILMS AND BIOFOULING.]

All microorganisms that are attached to surfaces secrete extracellular polysaccharides (EPS). Among other functions, these materials serve to anchor cells to the substratum. Mature biofilms contain copious quantities of these hydrated exoploymers, which form a gellike matrix on the surface. The EPS in biofilms have many effects on interfacial processes, including the following:

1. Immobilization of water at the biofilm–water interface.
2. Entrapment of metals and corrosion products at the substratum.
3. Decrease in rates of diffusion of chemical species to and away from the substratum.
4. Immobilization and inactivation of corrosion inhibitors and/or biocides.

Metal deposits have been reported in association with bacterial exopolymers in biofilms. These may be accumulated from the bulk water or from the corroding metal. The ability of EPS to bind metal cations can protect bacteria on surfaces from the toxic effects of those metals. Bacteria on copper alloys, for example, have been observed to secrete greater amounts of EPS than those growing on less reactive alloys, presumably as a protection mechanism against copper ion toxicity. Metal deposits in biofilm exopolymers also may contribute to corrosion reactions by establishing galvanic (or dissimilar metal) currents between the deposits and the metal of the substratum.

V. Bacterial Oxidation and Reduction of Metals

Certain bacteria are known to catalyze the oxidation of iron or manganese to obtain energy for growth. Microorganisms also may be involved in the oxidation of chromium species. Other organisms accumulate and precipitate abiotically oxidized metals. The activity of these organisms on iron and steel surfaces results in the deposition of ferric and/or manganic oxides in discrete, raised, hard mounds known as tubercles. These deposits typically are brown or reddish brown in color, often with rust-colored streaks emanating from them. They are generally hemispherical or conical in shape and can be up to several inches in diameter. The iron-oxidizing bacteria most often cited in formation of tubercles are the stalked organism *Gallionella* and filamentous forms such as *Sphaerotilus*, *Crenothrix*, and *Leptothrix*.

The tubercles deposited by metal-oxidizing organisms result in anaerobic microenvironments beneath the deposits, creating oxygen concentration cells and providing conditions for growth of obligately anaerobic bacteria such as sulfate reducers. They also create environments for the accumulation of chloride ions (to maintain charge neutrality), which combine to form acidic ferric and manganic chlorides, compounds that are highly corrosive to stainless steel. Deep cavernous pits often are found in the metal beneath these tubercles.

Bacteria that reduce ferric iron have been isolated from environments such as oil production facilities and have also been implicated in corrosion processes. It is suggested that the reduction of iron disrupts or prevents formation of passive oxide layers that normally protect metals from attack.

VI. Microbial Influence on Metal Embrittlement and Fracture

Hydrogen embrittlement is a corrosion phenomenon in which the entry of atomic hydrogen into the matrix of some types of metals results in a loss of ductility and tensile strength, leading to cracking or fracture of the materials when they are subjected to mechanical stress. Weakening of materials by hydrogen embrittlement can cause catastrophic failures with little or no prior warning. Hydrogen damage of metals can occur when hydrogen atoms are absorbed during fabrication processes or during service in a hydrogen-rich environment. Sources of environmental hydrogen may be hydrogen gas (H_2), if it is dissociated into atomic hydrogen (H^0), or electrolytic hydrogen, if recombination of H^0 into H_2 is prevented by the presence of sulfides or other hydrogen evolution poisons. Hydrogen sulfide has a severe effect on the formation and propagation of cracks in susceptible materials.

The role of microorganisms in hydrogen embrittlement and fracture of metals is unclear. Bacteria involved in hydrogen transformations in biofilms may be important both in accelerating and in retarding hydrogen damage. Several mechanisms have been proposed for microbial influence on embrittlement. These include

1. production of hydrogen gas during fermentation of organic substrates,
2. consumption of hydrogen,
3. production of hydrogen sulfide, which may stimulate adsorption of hydrogen atoms by preventing their recombination into molecular hydrogen,
4. production of hydrogen ions (H^+) via organic and mineral acids, which may be reduced to atomic hydrogen at cathodic sites,
5. destabilization of metal oxide films, and
6. acceleration of corrosion rate, resulting in an increase in production of cathodic hydrogen.

VII. Microbial Ecology of Microbiologically Influenced Corrosion

Biofilms almost always are comprised of many different species and populations of microorganisms that live together in a complex, dynamic ecosystem characterized by microenvironments and physical and chemical gradients. Although it is traditional to discuss biocorrosion mechanisms in terms of particular environmental conditions and individual mechanisms for specific microorganisms, in reality biofilm microorganisms grow in synergistic communities, or consortia, that interact nutritionally and conduct combined processes that individual species alone cannot perform. Aerobic and anaerobic microorganisms, for example, proliferate in the same biofilm, with anaerobic heterotrophs forming colonies in lower portions of the film and feeding on the diffusable secondary nutrients secreted by aerobes near the biofilm–water interface (Fig. 3). The interactions among populations of bacteria within biofilm communities are an important consideration in analysis of MIC problems. All microorganisms colonizing a metal surface have the potential for influencing corrosion processes, and very seldom does a single mechanism operate in isolation. [*See* ECOLOGY, MICROBIAL.]

Conversely, corrosion processes and metal substratum characteristics influence the rate, extent, and distribution of colonizing microorganisms. Relatively nonreactive metals such as titanium and stainless steel foul rapidly and develop highly diverse biofilm communities. On reactive metals such as copper and aluminum large amounts of loosely adherent and sometimes toxic corrosion products are formed. Microbial colonization of such materials tends to be slower, and the communities often are less diverse, as is characteristic of extreme environ-

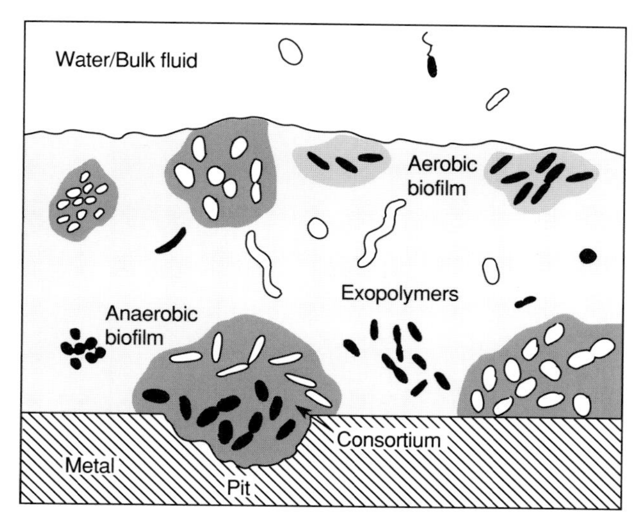

Figure 3 Model of a mature biofilm on a corroding metal surface, illustrating the complex and interactive nature of the microbial community.

ments. Nonuniform corrosion may lead to patchy colonization and growth. And certain species of bacteria may preferentially attach to welds or be attracted to particular inclusions or metal species. Several studies have demonstrated that the diversity and species composition of adherent microbial films may be markedly different on different types of alloys that have been exposed to the same aqueous environment. [*See* MICROBIAL ATTACHMENT, MOLECULAR MECHANISMS.]

VIII. Control of Microbiologically Influenced Corrosion

It is often difficult to determine the degree of involvement of microorganisms in a corrosion failure by simple observation of the damage, because appearance of the corrosion damage may be similar whether or not bacteria are involved. Treatment and prevention of abiotic corrosion and MIC, however, require distinctly different approaches. Furthermore, cell death does not necessarily stop the influence of microorganisms on corrosion. Deposits and corrosive metabolites may continue to exert their influence, even in the absence of metabolic activity. Prevention of biocorrosion therefore requires some knowledge of the mechanisms involved. Choice of protective methods, including protective coatings, MIC-resistant alloys, biocides and corrosion inhibitors, and new design criteria depends on the specific microbial processes that are likely to occur in each particular situation. Further understanding of how materials, environment, and microorganisms interact is required for development of truly effective methods for controlling MIC.

Bibliography

Cord-Ruwisch, R., Kleinitz, W., and Widdel, F. (1987). *J. Petroleum Technol.* **39**(1), 97–106.

Dexter, S. C. (ed.) (1986). "Biologically Induced Corrosion." National Association of Corrosion Engineers, Houston, Texas.

Dowling, N. J., Mittleman, M. W., and Danko, J. C. (eds.) (1990). "Microbially Influenced Corrosion and Biodeterioration." Institute for Applied Microbiology, University of Tennessee, Knoxville.

Ford, T., and Mitchell, R. (1990). *Annu. Rev. Microbial Ecol.* **11**, 231–262.

Hamilton, W. A. (1985). *Annu. Rev. Microbiol.* **39**, 195–217.

Hamilton, W. A. (1987). Sulphate-reducing bacteria and the mechanism of corrosion in the marine environment. *In* "Microbes in the Sea" (M. A. Sleigh, ed.), pp. 190–202. Ellis Horwood, Ltd., Trichester, England.

Little, B. J., Wagner, P. A., Characklis, W. G., and Lee, W. (1990). Microbial corrosion. *In* "Biofilms" (W. G. Characklis and K. C. Marshall, eds.), pp. 635–670. John Wiley and Sons, New York.

Walch, M. (ed.) (1990). *Mar. Technol. Soc. J.* **24**(3), 3–44.

Widdel, F. (1988). Microbiology and ecology of sulfate- and sulfur-oxidizing bacteria. *In* "Biology of Anaerobic Microorganisms" (A. J. B. Zehnder, ed.), pp. 469–585. John Wiley and Sons, New York.

Cosmetic Microbiology

Daniel K. Brannan
Abilene Christian University

Glossary

Adulteration Addition of any harmful substance that may make a product harmful to users under usual conditions of use; adulterated products contain filthy, putrid, or decomposed substances; products are adulterated if packed and held under insanitary conditions or if the container is unsafe

Antimicrobial Compounds that kill or inhibit the growth of microbes—used as disinfectants or preservatives

Class II Recall Recall of products classified as a priority situation where the consequences may be immediate or long range and possibly hazardous to health or life-threatening (e.g., contamination of a cosmetic with a potential pathogen); Class I is an emergency situation where the consequences are life-threatening

Cosmetics Articles that someone applies to, sprinkles on, or rubs into their body to cleanse, beautify, or promote attractiveness or alter their appearance; the product must not affect normal bodily function or structure; this definition excludes soap, products whose detergent properties are due to alkali-fatty acids and are labeled and sold only as soap

Disinfect Killing microbes on surfaces to levels that are not harmful to health or the quality of the product

Preservative Chemical agent used to prevent microbial growth in finished products; it prevents their multiplication or kills them to prevent spoilage or contamination of the product

Sanitizer Chemical agent used to disinfect equipment

COSMETIC MICROBIOLOGY is a subdiscipline of microbiology. It studies how to produce cosmetics free of pathogens and to prevent spoilage due to microorganisms. The cosmetic microbiologist's goal is to improve the safety and maintain the aesthetic quality of cosmetics. The cosmetic microbiologist must understand microbial physiology, pathogenic microbiology, and microbial ecology. In addition to microbiology, the cosmetic microbiologist understands organic and physical chemistry, toxicology, engineering, and regulatory/environmental laws.

I. Background and Importance of Cosmetic Microbiology

A. Regulations and History

Cosmetic manufacturers invest considerable effort to reduce the risks of microbial contamination in their products since the economic effects are great. Contamination requires having to scrap spoiled product, conduct Class II recalls from the Food and Drug Administration (FDA), and handle litigation from harmed consumers.

Some cosmetics firms become incredulous when they discover their cosmetic caused an infection or contains contaminating microorganisms. Thus, cosmetic microbiology is often a reactive rather than a proactive field. Microbiologists solve problems that are preventable. The joy of solving the problem, however, becomes one of kinship with the microbiological masters of old. Like Pasteur solving the spoilage problems of French wines, the techniques of microbiological sleuthing are still the same. Get all information on the subject, formulate a hypothesis for why the problem exists, test it, and then provide a practical solution for the problem.

Even the novice can identify and quantify the microbial contaminant in a cosmetic. The expert cosmetic microbiologist eliminates the contaminant and prevents it from occurring again. He or she also determines how to use or scrap the bad product in an environmentally safe manner. The novice just adds more biocide, a guaranteed way to adapt the contaminant to the biocide and compound the problem.

In 1938, U.S. Congress passed an administrative bill regulating cosmetics. This law was the Food, Drug and Cosmetic Act of 1938. This action was the culmination of a sequence of events. The existing 1907 Drug Act was considered too weak to ensure effective drug and food safety. A stronger bill was proposed in 1933, but it did not pass both House and Senate at that time. In 1937, a company marketed an oral tonic made with a poisonous ingredient. The resulting deaths prompted Congress to pass the Act in 1938. It gave the FDA a means of regulating and defined cosmetics. The act also defined adulteration and allowed the FDA to request recalls. If a manufacturer fails to conduct the recall, the Agency can seize the product. Finally, Congress passed tighter regulatory control on label claims, ingredient listing, and product safety warnings—the Fair Packaging and Labeling Act of 1973–1975.

Cosmetics industries thrived in the 1930s. The only microbial problems were preventing visible mold growth with parabens. When high-volume manufacture of cosmetics in the 1940s increased, so did bacterial and mold spoilage. Companies began including bacteria in their preservative challenge tests and using bactericidal preservatives.

In 1943, the Toilet Goods Association [later called the Cosmetics, Toiletries and Fragrance Association (CTFA)] established its Scientific Section. Member companies also founded The Society of Cosmetic Chemists to discuss formulation and preservation on a scientific level.

Physicians used antibiotics indiscriminately in the 1950s. Industry's use of antimicrobials in cosmetics ranged from deodorants and soaps to toothpastes and shampoos. Our preoccupation with germs made the market ripe for such products. Often, there was an unclear distinction between whether the biocide was used to provide a functional aspect in the cosmetic or used as a preservative. Soon, *Staphylococcus* spp., *Streptococcus* spp., *Pseudomonas* spp., *Serratia* spp., *Enterobacter* spp., and *Klebsiella* spp. caused contamination problems and became resistant to biocides. Kallings, in 1965, frequently found contaminated cosmetics on store shelves.

So the FDA conducted 25 drug and cosmetic product recalls during 1966–1968. In a 1969 sampling, the FDA found contamination in 20% of 169 cosmetics tested. Thus, the least regulated of all consumer products came under fire. Rather than face impractical regulations, the industry launched a cooperative CTFA/FDA relationship of self-regulation.

In 1967, a CTFA formed the Microbiology Committee from member companies to address these contamination problems. The committee conducted a survey of almost 4000 products to show that cosmetics were free of objectionable microorganisms. This committee developed test methods and conducted collaborative studies to improve manufacture of microbially free cosmetics. They issued technical guidelines covering good manufacturing and microbiological practices. With microbiological test methods and good manufacturing practices in place, the industry satisfied one of the FDA's chief concerns.

However, in 1974, instances of blindness occurred from use of mascaras contaminated by the user with *Pseudomonas* spp. The next concern of the FDA was whether or not a cosmetic product could withstand microbial insults added during consumer use. Recently, the FDA rephrased this concern to ask if preservative challenge tests predicted the risks of in-use contamination. Preservative challenge tests are methods used to see how well the biocide in a product kills microorganisms. The CTFA Microbiology Committee arranged for university and industry collaboration to compare laboratory methods of assessing the preservative efficacy of eye area cosmetics. They found that a variety of laboratory methods for measuring preservative adequacy in eye area cosmetics were satisfactory.

This testing occurred after the FDA gave a contract to Georgia State University in 1975 to develop eye area challenge tests. The "double membrane technique" resulted and was further tested in another FDA contract given to the University of California at San Francisco in 1977. Both contracts had little effect on the industry. In 1985, the FDA gave a contract to Schering-Plough and the University of North Carolina. They tried developing preservative challenge testing methods for creams and lotions that were predictive of consumer contamination. The FDA has not published the data from this study.

In 1987, the CTFA published the results of a survey to determine if companies tried to correlate their challenge test data with consumer use data. Nearly all companies claimed they had correlation programs in place. None presented their data, however. There are only two challenge test methods that have published validation data proving their ability to predict the in-use potential for consumer contamination. These are those of Brannan, published in 1987, and Lindstrom, published in 1986. Both tests are modifications of the CTFA test.

B. Future Expectations

Government regulation of cosmetics is increasing. Development of standard preservative efficacy tests is desired by the FDA. In 1990, it established a joint program with the CTFA and Association of Official Analytical Chemists (AOAC) to develop standard preservative challenge tests that are predictive of consumer contamination. In a 1989 House Appropriations Committee progress report, the FDA emphasized the "development and use of microbial testing procedures which ensure that cosmetic preservation systems effectively protect products from hazardous microbial contamination during . . . consumer use."

Any standardized preservative challenge test developed as a result of this effort will not approximate the rigorous methods with which most reputable companies already test their products. Additionally, it is naive to think that one standard preservative challenge or microbial content method will apply to all product types. Too many interacting factors occur to permit only a few standard tests. Finally, no laboratory challenge test can consider every conceivable formula or product type. It cannot consider every consumer use condition, every organism that mutates to grow in the product, every storage condition, or every combination of these events.

II. Sanitary Manufacture

A. Good Manufacturing Practices

Sanitary manufacture of cosmetics is critical to prevent microbial spoilage and to protect the consumer from potential pathogens. It allows plants to voluntarily meet the appropriate sections of current food and drug Good Manufacturing Practices (GMPs) (21 CFR part 110, 210, 211). Congress established the Requirement for GMPs in 1969. It defines FDA expectations of manufacturers of food, drugs, and medical devices who follow the Food, Drug and Cosmetic Act of 1938. The Act states that companies must pack, hold, or prepare cosmetics under conditions where they cannot become contaminated. This requirement typically implies suitable premises, equipment, raw materials, record keeping, sample retention, stability testing, environmental control, and trained personnel who operate using approved procedures. The microbiological deficiencies most cited by the FDA when doing a cosmetic facility inspection are citations for water contamination, inappropriate or absent testing of final product, and lack of validated methods or validation records.

B. Water

Cosmetic plants classify water as "raw" or processed. Raw water comes from either the city or the plant's own well for use in personal hygiene, cooling, toilets, or drinking. Processed water is water for making products. This water is either softened, deionized, distilled, or reverse osmosis-treated.

A variety of organisms such as *Escherichia coli* and *Pseudomonas* spp. can contaminate raw water. In storage tanks, the numbers of bacteria may rapidly increase to 1×10^6 colony-forming units per milliliter (CFU/ml). This occurs due to chlorine depletion, particularly during warmer months. Softened and deionized water also contain gram-negative bacteria. In brine-regenerated resins, *Bacillus* and *Staphylococcus* spp. also may grow. These bacteria are in raw water at low levels but then multiply rapidly in the ion-exchange resins that serve as fluidized bed bioreactors. Distilled and reverse osmosis water is free of microbes as it leaves the still or membranes. It rapidly becomes contaminated with gram-negatives in storage and distribution. This contamination occurs because of microorganisms that grow back from the various outlets and enter the storage tank.

Biofilms then form on the surfaces of the tanks, pumps, and pipes of the water distribution system. This colonization provides a microbial reservoir that contaminates the water passing through. The shear force of the passing water causes intermittent biofilm sloughing. Nonattached microorganisms also may grow in unused sections of the pipelines called dead-legs. Turbulence of the water passing by carries the organisms into the main water line. [*See* BIOFILMS AND BIOFOULING.]

C. Raw Materials

Most raw materials used in cosmetics are dry powders, natural gels, or surfactants. A few examples of these are talc and quaternized clay, aloe vera, and ammonium lauryl sulfate. These illustrate the key concerns and ways to handle raw materials. In dry powders and natural gels such as aloe vera, the primary contaminant is a spore-forming *Bacillus* or *Clostridium* spp. In surfactants, a wide variety of gram-negative bacteria may grow.

In thickening agents and talc, the spore-formers begin growing during wet portions of their manufacture. Fortunately, most of the spore-forming organisms involved are nonpathogenic *Bacillus* spp. Washing the talc provides an ideal setting for spore-forming bacteria to grow in the moist powder. Drying of the talc preserves the spores. The microbial content will be low if one reduces the process time. One should cool the moist powder to <25°C or heat it >45°C until the drying step. One highly effective means of eliminating spores is use of gamma irradiation. Manufacturers refer to irradiated quaternized clays and talc as "cosmetic grade." Although the only issues associated with use of irradiation are for foods, one should be cautious of the regulatory considerations surrounding the use of irradiated raw materials in cosmetics.

Aloe vera is also notorious for harboring spore-formers. Manufacturers harvest aloe in dry dusty areas. So it has a high bioburden of spore-laden dust. Other organisms that contaminate aloe are *Erwinia* and *Pectobacterium* spp. Manufacturers pasteurize aloe once. This only destroys vegetative cells. Since food manufacturers also use aloe, they use food-grade biocides. These are not sporicidal. Therefore, it is common to receive aloe with counts as high as 10^5–10^6 CFU/ml.

The solution is Tyndallization—double or even triple pasteurization. However, slow cooling between each pasteurization provides a warm environment (25°–45°C) permitting growth for periods of 4–8 hr. This can actually compound the problem because the spores that survive can germinate and grow forming even more spores. Germination without growth is desirable. This allows the next pasteurization to work against vegetative cells. Thus, one should flash pasteurize the aloe, cool it rapidly, and repeat within 24 hr to get a microbially free product.

Surfactants are difficult to keep uncontaminated. The exceptions are ammonium xylene sulfonate and sodium lauryl sulfate. These are hostile due to pH extremes and usually require no preservation. The only precautions are keeping the domes of the storage tanks free of condensation, because this is where microbes grow. When the condensate drips down onto the surfactant, it spreads microbes as a thin film. Use of circulating fans on the tank domes controls this condensation. The fans force air across dust filters and ultraviolet lights and then circulate it across the top of the dome.

Other surfactants, such as ammonium lauryl sulfate (ALS), are highly susceptible to microbial attack and require preservation. Some companies use isothiazolinones (Kathon®) at 5–10 ppm. Formaldehyde at 100–150 ppm is also effective. This preservative can fail, however, since many organisms easily develop formaldehyde resistance. If the manufacturing stream is not scrupulously clean, surfactants can become contaminated.

Surfactants support a succession of microbes similar to the ecological succession of microbes in milk. *Pseudomonas* spp. are the primary invaders. They have an inducible formaldehyde dehydrogenase. However, *Enterobacter* or *Klebsiella* spp. can also be primary invaders because of their capsule-producing capacity. These organisms eliminate formaldehyde either by enzymatic means or nonspecific reaction with polysaccharide capsular material. Several organisms succeed *Pseudomonas* or *Enterobacter*. *Serratia* sp. is the first. The surfactant may even turn pink. *Proteus* sp. then follows. It reduces the sulfate portion of ALS to hydrogen sulfide.

Thus, one should know what the primary invader species is to solve these problems. If it is *Pseudomonas* spp., one will have to use a completely different preservative in the surfactant. If it is *Enterobacter*, simple addition of EDTA at 0.05–0.1% will be enough to destroy the capsule. This allows formaldehyde to penetrate the capsule and kill the organism. A simple capsule stain and oxidase test can save countless hours of preservative development.

D. Personal Hygiene

Topics to cover when training for hygienic manufacture include personal hygiene, operator-borne contamination, and ways of preventing cross-infection. Personal hygiene includes washing hands, wearing clean clothing, and keeping hair and beards covered. For aseptic manufacturing, the manufacturer may use presterilized single-piece suits, foot coverings, hair and beard covers, face masks, and gloves.

Washing the hands often is the most important and inexpensive way to prevent contamination. Employees should not wear jewelry. They should wash their hands every time they leave and return to the process area. Even a brief 15-sec hand washing period will cause significant drops in numbers compared to unwashed hands. However, a thorough scrubbing with warm water (32°–43°C) and bactericidal soap for at least 1 min is best.

It is aesthetically displeasing to the consumer to find human hair in a product. To prevent such contamination, hair and beard coverings should be used in all areas of manufacture. Employees should wear disposable hair covers that completely block all hair from exposure. Personnel should put on their hair cover first, then wash their hands. The employee replaces the hair cover each time he or she leaves and returns to the work area.

Clothing must be clean and without adornment. Uniforms are a good means of controlling compliance and actually help promote good attitudes toward sanitation. Light-colored uniforms or lab coats show a need for cleaning much earlier than dark-colored ones. Pockets on the uniforms should have button down flaps. This prevents materials in them from falling into the product or process machinery.

E. Sanitary Design of Buildings

Roominess, simplicity, bright lighting, no clutter, and even a fresh clean-smelling environment give a positive perception of cleanliness. This perception creates the right attitude within the organization to promote compliance with sanitation rules. The FDA regulations written for food and drug plant design can be used as a model for cosmetic plants since there are no cosmetic GMPs. These regulations require adequate and separate space for equipment and materials storage. They require separate areas for operations that could contaminate the cosmetic. Also required are adequate lighting, ventilation, plumbing, and protection from pests.

The grounds surrounding a well-designed plant should be neat. Keep decorative landscaping features such as ponds, fountains, and sites that provide nesting for birds and rodents at least 30 ft away from the building. A 3-ft perimeter of gravel should surround the building. Trim all trees, shrubs, and lawns to avoid insects that may find harborage in the long grass and unkempt shrubbery. Keep the driveways, docks, refuse sites, and parking lots free of debris. Also, keep these areas drained. Use a perimeter fence around the plant grounds to filter paper and debris.

The building should be a simple, box design with no adornments, ledges, or architectural details to encourage bird nesting. Instead, it should have coved or sloped ledges. The exterior walls and foundations should have no cracks or holes. Avoid porous and cracked walls. Avoid or at least caulk baseboards so there are no spaces that harbor insects. Roofs should have positive slopes of 1 in/8 ft and have exterior drainspouts to take water away from the perimeter of the building. Screen all vents on the roof.

Ceilings should have limited overhead pipes. These trap dust and provide pathways for insect movement. Either cove, round, or slope overhead beams to ease cleaning. Give inside and outside ledges on windows a 20° slope. The ceilings should be easy to clean, nonporous, and painted with epoxy paint. A plant may use suspended ceilings if inspected regularly. Use load-bearing walls instead of columns on the interior of the building. If using columns, they should be cylindrical or at least sloped at the floor to help cleaning.

Floors should be impervious to water, free of cracks and crevices, and resistant to chemicals. Tiled floors are desirable but expensive. Concrete is satisfactory but must be water-sealed. Surround edges of upper mezzanines and wherever pipes pass through the floor with 4-in curbs or sleeves. This prevents water from passing to the processes below. Keep floors dry. In areas where water is frequently on the floors, use epoxy or urethane coatings on the concrete to prevent water saturation. Provide adequate drainage by using trench drains set in floors sloped at $\frac{1}{8}$–$\frac{1}{4}$ in/ft When using circular floor drains, place them every 400 ft^2. However, it is typically difficult to direct water toward circular drains. Drains and troughs should drain well since moist areas provide ideal harborage of insects. Screen all floor drains to keep rodents from entering via sewers.

Loading docks should be at least 3 ft above grade and entrances into the plant opened only when needed. Screen all windows and doors with 16-mesh screen or keep them closed. Doors should fit well, have automatic closures, and be made of metal. Protect large openings with air curtains. These should sweep air from the top to the bottom of the opening at a rate of 4400 ft^3/min and at a 25–30° angle. Equip ventilation systems with HEPA (high-efficiency particulate air) filters capable of removing 90% of particles that are 1–5 μm in diameter.

Provide hand-washing facilities in the restrooms and near entrances to the manufacturing area itself. Surgical-style washing facilities are showy but superfluous and expensive. Instead they should be easy to use, easy to find, convenient, and accessible. They should be at least 8 ft away from any process stream to prevent contamination of the product.

F. Sanitary Design of Equipment

Manufacturers of cosmetics rely on sanitary design of equipment developed by food sanitarians. Their standards for sanitary equipment design apply to cosmetic equipment design. Engineers should design equipment with sanitation in mind.

The material used for pipelines should be smooth to prevent biofilm formation. Stainless steel such as AISI 302, or type 316 with sanitary pipe junctions (dairy fittings) should be used. Pipe interiors should be smooth without rough seams. Keep bends smooth and rounded without sharp right angles. The center-line radius of pipe bends should never be less than the outside diameter of the pipe. Slope all pipes $\frac{1}{8}$ in/ft away from tanks to permit proper drainage. Avoid sags or depressions that will trap stagnant fluids. Plastic or polyvinylchloride (PVC) piping is undesirable.

Cosmetic plants should use only diaphragm, plug-cock, or butterfly valves. Each of these has advantages and disadvantages. The key concern is that disassembly is easy to permit cleaning. Washers, O-rings, and diaphragms provide microenvironments particularly susceptible to microbial colonization. Those made of silicone rather than natural rubber are more resistant to microbial growth.

Use tanks made of stainless steel. Interiors should have no sharp corners to complicate cleaning. Welds should be flush and ground to 120 grit compatible with the surface finish of the rest of the tank. Tank roofs should be domed and their bottoms rounded. Openings into tanks should have protective lips surrounding them. Hatches, covers, and lids should be overlapping to prevent debris from entering into the tank. Protrusions into the tank such as thermometers, pressure sensors, and spray balls should form smooth welded junctions. Viewing ports are flush with the tank interior. Never place a well or depression anywhere in a tank. These trap stagnant product.

Seal all bearings on stirring devices and keep them oil-free. Drains should be flush with the tank wall and ground smooth. Design inlet pipes to the tank with air gaps to prevent back siphonage. Fit them with flared protective shields directly above the gap to prevent contamination from the environment. Never seal pipes into the top of the tank; they will serve as reflux columns to permit microbial growth in the condensate that drips back into the tank.

Peristaltic and diaphragm pumps are more sanitary than rotary positive displacement or reciprocating pumps. Pumps should be self-draining and free of pockets or crevices that trap product. The most important criterion for pumps is that they are easy to disassemble and clean. Pumps bolted together require tools to take them apart so personnel will clean them less. Clamped pumps are more amenable to cleaning. Product contact parts should be stainless steel and without bearings that directly contact the product stream. Rotors attached to their shafts only by pressure contact are superior to those that use bolts, cotter pins, or hexagonal screws. Pipes entering and leaving pumps should be in smooth curves. Allow plenty of space around the pump to help cleaning and maintenance.

G. Gaining Employee Commitment

The only way to get employees involved in sanitary manufacture is to get the managers involved. If the employees see the managers wearing hair coverings and washing their hands, then the employees also will obey the rules. People's communal instincts will compel them to cooperate when their superiors set the pace. One achieves cooperation through education, training, and dedication to a sanitary program.

The ideal situation is to have everyone cooperate at the start. The ongoing commitment of managers, and their employees, to proper sanitation is facilitated by an active and knowledgeable quality assurance (QA) manager. Typically, the most successful QA functions are those where the QA manager's reporting relationships are independent of day-to-day manufacturing needs. This allows the QA manager to direct appropriate activities such as employee practices and training and product testing with the primary objective of maintaining product quality.

A corporate microbiologist is also needed to enforce quality assurance and to audit a plant's conformance to GMP. The corporate microbiologist is responsible for problem-solving support, and developing test methods and preservation systems for products. A company concerned with total quality allows the plant microbiologist and the corporate microbiologist to communicate directly with each other. Some management may think this communication leads to whistle-blowing about sanitation problems. As a result, the microbiologist may be accused of "not being a team player." This is one of the finest accolades for a cosmetic microbiologist. You know you are doing the job right when you are not popular for your stands.

The corporate microbiologist is also responsible for training and education. This can be through classes with outlines and books, or clever use of visual aids such as posters and films. Management and employees should attend these training sessions. Training should cover the regulatory requirements for clean manufacture and the illnesses that can result from products contaminated with microorganisms. This training includes personal hygiene, good housekeeping and sanitation, and sanitary equipment design. Training is effective only if the employee is given valid reasons for the hygienic principles taught. This means he or she will need a basic understanding of microbiology without a scientific, esoteric approach. The rapid nature of microbial growth, their ability to adapt, their small size, and their ubiquity should be taught in easy to understand terms.

III. Sanitization

Several elements of effective cleaning and sanitizing are important in cosmetic microbiology. These include the types of cleaning and sanitizing agents used, the kinds of equipment, the type of soil to be removed, when to clean, and how frequently to clean. Typically, cleaning and sanitizing are done at the end of a production run. Other times for cleaning and sanitizing may be after a set number of manufacturing hours or between shift changes.

Base the choice of when to clean and sanitize on facts rather than convenience. Such facts are gained by monitoring the system for microbial contamination during production. The first detection of microbial counts is the maximum length of time to go between sanitizations.

A. Cleaning

Immediately before washout, one should pump the system free of all product. Sometimes a "pig" (a foam rubber bullet) is air blown through long lines to remove traces of product. Detergent-based cosmetics such as shampoos rarely use cleaning agents. Typically, the system is only washed out with hot water followed by sanitization. Use detergents or caustic materials for cosmetics with an oily nature, emulsions, and powders. These lift and suspend the oily portions by reducing interfacial and surface tension.

Factors that govern choice of cleaning agent are whether or not it is safe and effective, nondamaging, and compatible with the formulation. Alkaline cleaning agents remove the lipid portions of cosmetics. Other cleaning agents for mineral scale or very fatty materials may be acid- or solvent-based, respectively. For heavy duty cleaning, use concentrations as high as 2000–3000 ppm. Concentrations of solutions for clean-in-place (CIP) detergents range from 1000 to 1500 ppm. Chlorinated cleaners enhance cleaning but are not sanitizers. Many companies sell what they call chlorinated sanitizing cleaners. Cleaning agents work well at a pH >9. Chlorine works well at a pH range of 4–7. Mix them and you have either an ineffective cleaner or an ineffective sanitizer.

The water used for cleaning is as critical as the cleaner used. It should be potable, low in hardness, and hot (70°C) for fatty residues or at least warm (43–54°C) for most other cleaning operations. Control of hardness can be done using softeners or ion-exchange columns. Check and service these routinely so high bacterial counts (>100 CFU/ml) are avoided. Alternatively, one can use detergents with organic chelators or phosphates. Regardless of the cleaning agent used, make routine checks for microbial growth. Quaternary detergents are especially susceptible to pseudomonads, but all detergents can support growth if not frequently changed.

Use physical methods to clean especially hard to clean systems. Probably the most effective physical method is scrubbing by hand with a brush and detergent-in-water solution. Unfortunately, this is labor-intensive, costly, and dependent on the attitude of the one doing the scrubbing. Use plastic brushes with synthetic fibers, never sponges and rags. Dry using air rather than drying cloths.

Alternatives to hand scrubbing are use of high-pressure cleaning and CIP methods. High-pressure spray systems can deliver from 500 to 1000 psi. CIP systems are by far the most popular methods. These are line loops connecting the various tanks and equipped with a CIP pump that can deliver 2–3 gal/min at 400–800 psi. Cleaning agent, sanitizer, or rinse water should be pumped through the system. High flow rates (four to five times the product flow rate) provide shear that will strip biofilms from the inner pipe surfaces. Locate spray balls in the tops of tanks. Assure that they rotate with the pressure of the solution so the spray reaches all points inside the vessel. The critical points to check are the shear flows and that the cleaning agent and the sanitizer contacts every surface.

B. Sanitizing

For some cosmetics, a thorough cleaning of the manufacturing equipment will be adequate to provide microbial control. Microorganisms are controlled by physical removal or by removing the nutrients required for growth. Some cosmetics require that the manufacturing equipment is sterile.

The perfect sanitizing agent should be safe for use by the plant personnel and act rapidly under conditions of use. It should not interact with the cosmetic or leave a residue that could interact with it. The agent should be easy to use and inexpensive. Of the various types of chemical sanitizers, none meet all aspects. The major types of chemical sanitizers available for use in cosmetics are halogens, quaternary ammonium compounds, phenolics, aldehydes, and alcohols.

The most commonly used halogen in the cosmetics industry is chlorine in the form of hypochlorous acid (HOCl). A variety of chlorine sources exist including sodium hypochlorite, calcium hypochlorite, chloramines, chlorocyanurates, and gaseous chlorine. Use these at concentrations that will provide from 100 to 200 ppm available chlorine. The pH of the chlorine solution should be ≤ 6.5. A pH <3.5 may corrode the metal if concentrations exceed 200 ppm and contact time exceeds 30 min. One should assay used solutions after a sanitization for free chlorine. Significant drops to <20 ppm show the system was improperly cleaned and contained materials that provided a chlorine demand. In this case, repeat the cleaning and sanitization before production begins.

Cosmetic plants use glutaraldehyde as a disinfectant rather than formaldehyde due to the potential carcinogenicity of formaldehyde vapors. Glutaraldehyde used at 2% and buffered to pH 7.5–8.5 is active against gram-positive and gram-negative bacteria, fungi, viruses, and spores. It acts within 10 min. Formaldehyde used as a preservative is not carcinogenic, but some consumers have a negative perception of its use.

Quaternary ammonium compounds (quats) are cationic surface-active agents that are particularly effective against gram-positive bacteria but ineffective against gram-negatives. In fact, dilute in-use solutions serve as selective media for *Pseudomonas* spp. Plants use quats to sanitize small coupling pieces, associated piping, and exterior surfaces of tanks. Make and use the solutions daily to avoid adaptation problems. One advantage to quats is that they are substantive and so provide residual antibacterial activity on treated surfaces. Apply quats at 500–1000 ppm without rinsing. They are most effective at high temperatures and a pH of 10.

The cosmetics industry uses phenolics only for floor, wall, and ceiling disinfection. The activity of phenolics diminishes markedly in the presence of organic material on surfaces. Therefore, clean the floors with an alkaline detergent and rinse before sanitizing. Phenolics are most effective at acid pH. They are effective against vegetative bacteria and molds.

Use alcohols as surface disinfectants for tank tops and other working surfaces around packing lines. The most commonly used alcohols are ethanol (70% aqueous) and isopropanol (50% aqueous). Due to volatility, their action is brief and limited to vegetative cells.

Disinfectant suppliers have tried developing combination detergent–sanitizers. Mixtures of quats and nonionic detergents, or solutions of anionic detergents plus chlorine-releasing compounds are most common. More often than not, the combination results in either an ineffective cleaner or an ineffective sanitizer. Rarely are the two completely compatible with each other. The better approach is to apply a cleaning compound to the system followed by a rinse and then the sanitizing agent.

The most effective sanitizing agent is heat. Exposure of the making system to 82°C (180°F) for 15–30 min is effective. The advantage in using heat over chemical sanitizers is that heat penetrates biofilms where chemical sanitizers do not.

C. Water System

It is critical to design water systems that permit effective sanitization. Holding tanks and associated piping should be stainless steel. Avoid using pipes made of copper or galvanized piping. Never use PVC or black iron pipe. Water lines entering making tanks should not permit back-siphonage of the tank contents into the water system. This is easily done by providing a shielded air gap. Prevent back-siphonage in the system when pressure drops in the water line by avoiding cross-connections. Construct filters in the water system in parallel. This allows a filter needing service to be isolated from the rest of the system.

Supply water treatment systems with chlorinated city water. This treated water should go directly into the stainless steel storage tanks and be periodically heated to 82°C (180°F). Circulate the heated water through the system and into the storage tank. An alternative to heat sanitation of water is ozonation. Use ultraviolet lights at the point of use to remove the ozone and further sanitize the water.

IV. Preservation

Both clean manufacture and preservation should compliment each other. Do not use preservation to mask unsanitary manufacture. Instead, preservation is primarily a way to protect the consumer during use of the cosmetic product. In addition, preservation is an aid to extend the shelf-life of the product.

Cosmetics are such complex products that their preservation is more of a subjective art than an objective science. There is such a complexity of interacting factors in the cosmetic that using one's intuition to select a preservative is often more successful than relying on isolated facts. The formulation factors to consider when choosing a preservative include pH, oil–water partitioning, and interaction with raw ingredients. In addition, preservatives may localize in an emulsion and even react with the container.

Some of the more common and effective preservatives in use today are isothiazolinones (Kathon ®), hydantoins (Glydant ®), imidazolidinyl or diazolidinyl urea (Germall®), oxazolidines (Nuosept C®), formalin, and parabens. Choose preservatives based on effectiveness as determined in a preservative challenge test (Section V). Other considerations should be safety, compatibility with the aesthetics of the product, cost, ease of formulation, and availability. Several choices of preservative systems for a cosmetic should be available in case adaptation to the preservative occurs during manufacture. Some microbiologists like to use combinations of several different preservatives; the logic is to prevent microbial adaptation. This practice is unnecessary in plants where sanitation standards are high. It also can create more safety risks and be expensive.

V. Test Methods

There are only two test methods for the cosmetic microbiologist to know: the Preservative Challenge Test (PCT) and the Microbial Content Test (MCT). All the other test methods are variations of these two tests. In addition, the microbiologist needs to know the basic microbiological techniques that support the conduct of these two tests.

A. Preservative Challenge Tests

The basic way to conduct a PCT is simple. Inoculate a product and see how long it takes to eliminate or reduce the inoculum to an acceptable level. Variations of the PCT include its adaptation for use in water-miscible and -immiscible products and eye area products. Additional variations are those designed by several groups such as the AOAC and the United States Pharmacopaeia (USP). These groups set these variations based on collective anecdotal experiences rather than data from rigorous scientific principles.

There are many variations to this simple technique. Some of these variations are which organisms to use, concentration of the organisms, single versus multiple inoculation, and how often or how long to follow the elimination process. Other variations include whether or not the product should be diluted, whether or not the challenge organisms should be a mixed or pure inoculum, and what diluents/plating media are best.

The most used PCT tests are summarized here; they are those of the CTFA and the Society of Cosmetic Chemists of Great Britian methods for cosmetics. For pharmaceuticals, the USP and the British Pharmacopoeia (BP) have developed PCT methods. Other PCT methods are those being developed by the AOAC and a variety of rapid test methods that are simple adaptations of *D*-value methods to cosmetics.

Nearly all the methods include *Pseudomonas aeruginosa*, *Staphylococcus aureus*, *Candida albicans*, and *Aspergillus niger*. In addition to these, the USP includes *E. coli*. CTFA includes *E. coli*, *Bacillus subtilis*, *Penicilleum luteum*, and spoilage isolates. Only the Society of Cosmetic Chemists of Great Britain (UK test) makes specific inoculum recommendations for each type of cosmetic product.

The levels of inoculum range from 10^5 to 10^6 CFU/ml or g and a single challenge for the USP test to 10^6–10^7 CFU/ml with several challenges for the UK test. Interpretations of the tests also vary. The CTFA test makes no formal recommendations but requires that the test continue for at least 28 days. The USP test requires a 3-log reduction of bacteria within 2 wk. Yeasts and molds should remain at or below the initial count. Over the 28-day period, all organisms are to remain at these levels.

The BP test requires that a 3-log reduction of the bacterial load occurs. It requires a 2-log reduction of yeasts and molds within 7–14 days. The BP test expects shampoos to be "self-sterilizing" in 7 days, creams and lotions to "show drastically reduced counts," and eye cosmetics to be "bactericidal to *P. aeruginosa*."

There are criticisms of these tests: improper selection of challenge organism, microbial load of the challenge, number of challenges, and the end-points stipulated. The FDA's chief criticism is that PCT methods do not predict whether or not consumers can contaminate the product during use. As a result, in 1986 and 1987, two researchers developed methods that were predictive of consumer contamination potential. These are the only methods with published data to validate their ability to predict the potential for consumer contamination.

B. Microbial Content Tests

MCTs detect the numbers and sometimes the types of organisms present in the cosmetic. Most tests for microbial content are plate-counting methods; however, one can use other techniques.

Plate-counting methods are also known as aerobic plate counts or total plate counts. In plate count methods, use an appropriate diluent to neutralize the preservative and any other antimicrobial ingredient in the cosmetic. A variety of neutralization agents are available. Not all are effective or nontoxic to the organism needing detection. If neutralization via some agent is not possible, then one should use physical dilution or membrane filtration. Similarly, the medium for plating the product is critical for accurate recovery of the microorganisms potentially present in the product.

Use either spread plates or pour plates. Use as low a dilution as possible. For example, a 1 : 10 dilution of the product will yield a better detection limit than a 1 : 100 dilution. Once inoculated into the agar plate, incubate for 1–5 days. Count the viable colonies that form and record the number as CFUs. There are several sources of error using this method. The biggest error is that no single medium will detect all types of microbes in all types of products.

Other methods for enumerating microbes in cosmetics are the direct microscopic count, the most probable number (MPN), radiometric methods, and impedance measurements. In the direct microscopic count, one spreads a known amount of cosmetic on a slide and counts it. The advantage of this technique is that it is fast. It counts both living and dead cells and is only good for heavily contaminated products. Also, the products interfere with seeing the microbes and the technician will be highly susceptible to fatigue.

In the MPN technique, the technician places various dilutions of the product in replicate tubes of nutrient medium. He or she then compares the number of replicates showing growth to a standard MPN table. From this, the number of bacteria in the sample is estimated. This method is useful when the organisms do not grow well using standard plating techniques.

Radiometric methods rely on the organisms present in the product to degrade a ^{14}C-labeled substrate into $^{14}CO_2$. The $^{14}CO_2$ evolved is related to total microbial levels in the product. The radiolabeled substrate used is typically glucose. Use ^{14}C-glutamate or ^{14}C-formate instead of ^{14}C-glucose to detect *Pseudomonas*. The major negative with this technology is the expense; however, it offers rapid results (6–8 hr) compared to traditional plating methods.

Impedance measurement is a widely used method for detecting microorganisms in cosmetics. It is especially useful for following microbial activity during preservative challenge testing. One also can use impedance methods to approximate microbial numbers in finished product. The method uses the ability of the growing microbial culture to produce changes in impedance of electrical current as it metabolizes the nutrients in the medium. This method will estimate microbial activity within 5 hr. The major negative with this method is the cost and poor detection limits: 10,000 CFU/g of product.

Plate counts will always be the benchmark of quality control in the cosmetic microbiology lab. They are easy to do, are inexpensive, and result in numbers of microbes to which most people can easily relate. However, even plate counts do not accurately quantify the numbers of microbes present. First, microbes typically clump together. A CFU is probably not a result of a single microbe but, instead, anywhere from one to several thousand microbes.

Reliance on plate count methods should be reevaluated. Regardless of the method used, the key for microbial monitoring is to show relative microbial levels and trends during the process. One should base such monitoring on a system in control. The data should be used to keep the system within total quality limits. Because total quality concepts and just-in-time production require rapid turnaround, the use of rapid methods may one day replace time-consuming plate-counting methods.

Bibliography

Brannan, D. K., and Dille, J. C. (1990). *Appl. Environ. Microbiol.* **56,** 1476–1479.

Brannan, D. K., Dille, J. C., and Kaufman, D. J. (1987). *Appl. Environ. Microbiol.* **53,** 1827–1832.

Geis, P. A. (1988). *Dev. Indus. Microbiol.* **29,** 305–315.

Hugo, W. B., and Russell, A. D. (1983). "Pharmaceutical Microbiology," 3rd ed. Blackwell Scientific Publications, Oxford.

Kallings, L. O. (1966). *Acta Pharm. Suecica* **3,** 219.

Lindstrom, S. M., and Hawthorne, J. D. (1986). *J. Soc. Cosmet. Chem.* **37,** 481–488.

Russell, A. D. (1980). "Neutralization Procedures in the Evaluation of Bactericidal Activity." Society for Applied Bacteriology, Technical Series No. 15. Academic Press, London.

Russell, A. D., Ahonkhai, I., and Rogers, D. T. (1979). *J. Appl. Bacteriol.* **46,** 253–260.

Russell, A. D., Hugo, W. B., and Ayliffe, G. A. J. (1982). "Principles and Practice of Disinfection, Preservation, and Sterilization." Blackwell Scientific Publications, Oxford.

Troller, J. (1983). "Sanitation in Food Processing." Academic Press, Orlando, Florida.

Crystalline Bacterial Cell Surface Layers (S Layers)

Uwe B. Sleytr and Paul Messner
Centre for Ultrastructure Research and Ludwig Boltzmann Institute for Molecular Nanotechnology

Glossary

Crystalline surface layers (S layers) Regular crystalline surface layers in prokaryotic organisms, composed of protein or glycoprotein subunits

Glycoproteins in bacteria Present as crystalline arrays of macromolecules on surfaces of eubacteria and archaebacteria

Two-dimensional protein crystals Regular arrays of (glyco)proteins present as outermost envelope component in many prokaryotic organisms

S LAYERS are surface envelope components on prokaryotic cells, consisting of two-dimensional crystalline arrays of (glyco)protein subunits. S layers have been observed in species of nearly every taxonomical group of walled eubacteria and represent an almost universal feature of archaebacterial envelopes. As porous crystalline arrays covering the cell surface completely, S layers have the potential to function (1) as protective coats, molecular sieves, and molecule and ion traps; (2) as a structure involved in cell adhesion and surface recognition; and (3) in archaebacteria, which possess S layers as exclusive wall components, as a framework that determines and maintains cell shape.

I. Introduction

The different cell wall structures observed in prokaryotic organisms, particularly the outermost envelope layers exposed to the environment, reflect evolutionary adaptations of the organisms to a broad spectrum of selection criteria. Crystalline cell surface layers (S layers) are now recognized as common features of both eubacteria and archaebacteria.

Most of the presently known S layers are composed of a single (glyco)protein species endowed with the ability to assemble into two-dimensional arrays on the supporting envelope layer. S layers, as porous crystalline membranes completely covering the cell surface, can apparently provide the microorganisms with a selective advantage by functioning as protective coats, molecular sieves, molecule and ion traps, and as a structure involved in cell adhesion and surface recognition. In those archaebacteria which possess S layers as exclusive envelope components outside the cytoplasmic membrane, the crystalline arrays act as a framework that determines and maintains the cell shape. They may also aid in cell division.

S layers, as the most abundant of bacterial cellular proteins, are important model systems for studies of structure, synthesis, assembly, and function of proteinaceous components and evolutionary relationships within the prokaryotic world. The isolated crystalline arrays have also shown considerable potential in biotechnological applications.

II. Location and Ultrastructure

Although considerable variation exists in the complexity and structure of bacterial cell walls, it is possible to classify cell envelope profiles into three main groups on the basis of structure, biochemistry, and function (Figs. 1 and 2).

1. Cell envelopes formed exclusively of a crystalline surface layer (S layer) composed of (glyco)protein subunits external to the cytoplasmic membrane (most halphilic, thermophilic and acidophilic, and gram-negative archaebacteria) (Figs. 1a and 2a). In some organisms (e.g., *Methanospirillum, Methanosaeta, Methanothrix*), a regularly arranged sheath is present in addition to the S layer (Fig. 1b)

2. Gram-positive cell envelopes of eubacteria with a rigid peptidoglycan containing sacculus of variable thickness outside the cytoplasmic membrane, and gram-positive cell envelopes of archaebacteria (e.g., *Methanothermus*) with a rigid sacculus composed of pseudomurein (Figs. 1c and 2c,d)

3. Gram-negative envelopes of eubacteria with a thin peptidoglycan sacculus and an outer membrane (Figs. 1d and 2b). [*See* CELL WALLS OF BACTERIA.]

Although not a universal feature as in archaebacteria (Fig. 3), crystalline arrays of (glyco)proteins have been detected as outermost envelope components in organisms of most major phylogenetic branches of gram-positive and gram-negative eubacteria (Fig. 3).

At present the most useful electron microscopical preparation procedure for detecting S layers on intact cells is freeze-etching (Fig. 4). High-resolution studies on the crystalline arrays are primarily performed on negatively stained specimens. Both two- and three-dimensional computer image reconstruction techniques are applied for the evaluation of electron micrographs, but the accuracy of information on the mass distribution in the crystal lattices is generally limited to about 2 nm. These studies revealed that most S layers have oblique (p2), square (p4), or hexagonal (p6) symmetry. The morphological units of these lattices consist of two, four, or six identical monomers (Fig. 5). Depending on the type of S layer, the center-to-center spacing between the adjacent morphological units can range from 3 to 35 nm.

Some bacteria assemble more than one S layer on their surface (see Fig. 1c,d). The superimposed lattices can show identical or different crystallographic lattice types and can be composed of different

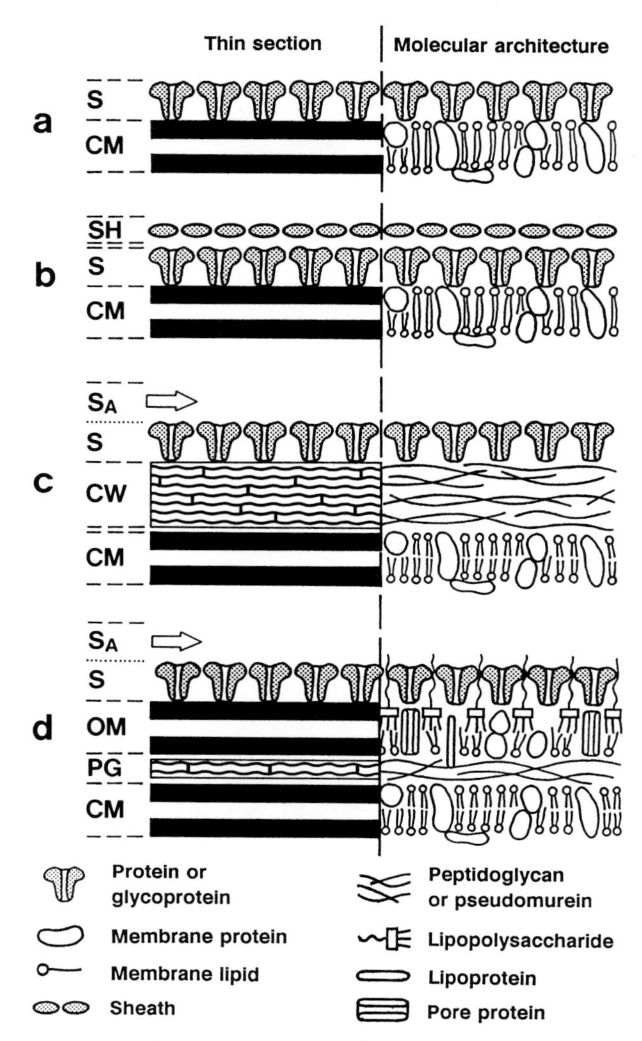

Figure 1 Schematic illustration of major classes of prokaryotic cell envelopes containing crystalline cell surface layers (S layers). *Left:* thin-section profile as revealed by electron microscopy; *right:* molecular architecture showing major components. (a) Cell envelope structure of gram-negative archaebacteria with S layers as exclusive cell wall components and (b) with an additional regularly arranged sheath; (c) gram-positive cell envelope as observed in eubacteria and archaebacteria; (d) gram-negative cell envelope as observed in eubacteria. S, crystalline surface layer; S_A, location for possible additional S layer; SH, sheath; CW, rigid cell wall layer composed primarily of peptidoglycan in eubacteria or pseudomurein in archaebacteria; PG, peptidoglycan layer; CM, cytoplasmic membrane, OM, outer membrane. [Modified from Sleytr, U. B., and Messner, P. (1989). Self-assembly of crystalline bacterial cell surface layers (S layers). *In* "Electron Microscopy of Subcellular Dynamics" (H. Plattner, ed.), pp. 13–31. CRC Press, Boca Raton, Florida.]

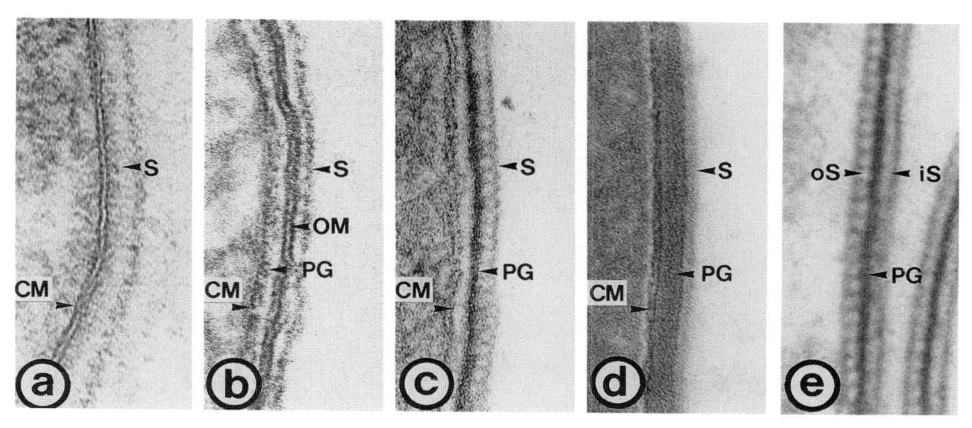

Figure 2 Ultrathin sections of different types of prokaryotic cell envelopes possessing crystalline surface layers. (a) Envelope of the gram-negative archaebacterium *Pyrodictium occultum* consists exclusively of an S layer and a cytoplasmic membrane (CM). (Micrograph courtesy of H. König, unpublished material.) (b) The gram-negative cell envelope of *Aeromonas salmonicida* A450 is composed of a thin peptidoglycan layer (PG), an outer membrane (OM), and an S layer. (Micrograph courtesy of L. L. Graham and T. J. Beveridge, unpublished material). (c) and (d) are examples of cell envelopes of gram-positive eubacteria. (c) *Clostridium thermohydrosulfuricum* L77-66 reveals only a thin peptidoglycan layer, which is characteristic for most mesophilic and thermophilic Bacillaceae possessing crystalline surface layers. See also Fig. 4b and c. (d) *Lactobacillus buchneri* 41021/251 shows a thick peptidoglycan containing sacculus with a thin S layer attached to it (see also Fig. 4a). (e) Cell wall preparation of *C. thermohydrosulfuricum* L77-66. An additional S layer (iS) has assembled on the inner surface of the peptidoglycan layer from a pool of S-layer subunits present in the meshwork of the peptidoglycan layer on removal of the cytoplasmic membrane [compare with (c)]. oS, outer S layer. Magnification of all micrographs: 182,500 ×.

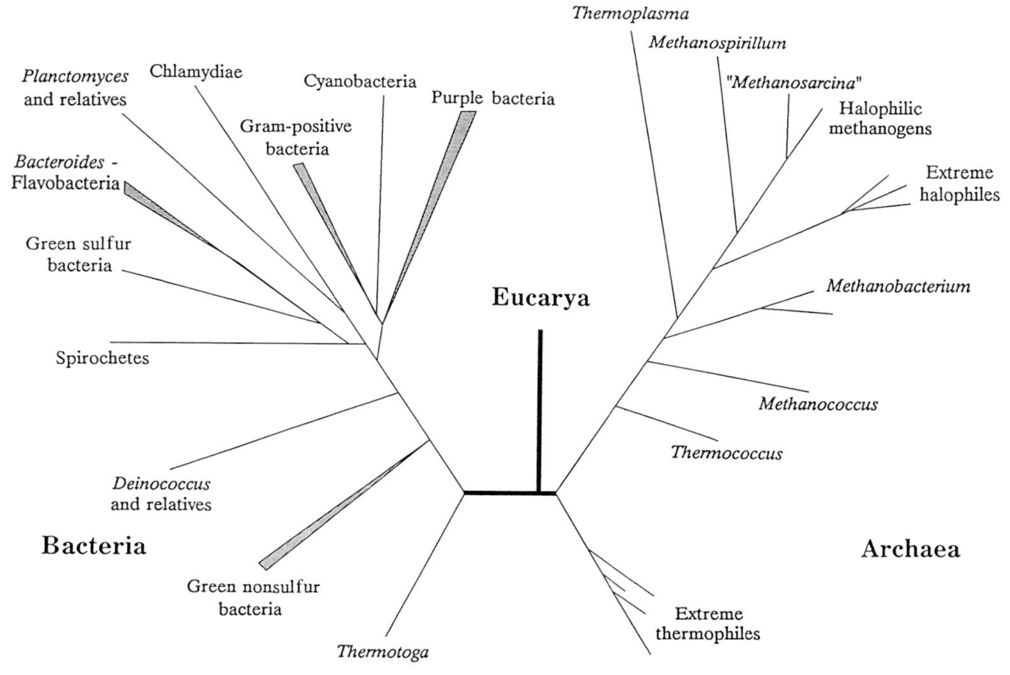

Figure 3 Universal phylogenetic tree of prokaryotic organisms showing the major groups of eubacteria (bacteria) and archaebacteria (archaea). With the exception of *Thermoplasma*, which exists as naked protoplasts, S layers are an almost universal characteristic of archaebacterial cell envelopes. Among eubacteria, organisms possessing S layers have been detected in all major phylogenetic branches. [From Woese, C. R. (1987). *Microbiol. Rev.* **51**, 221–271.]

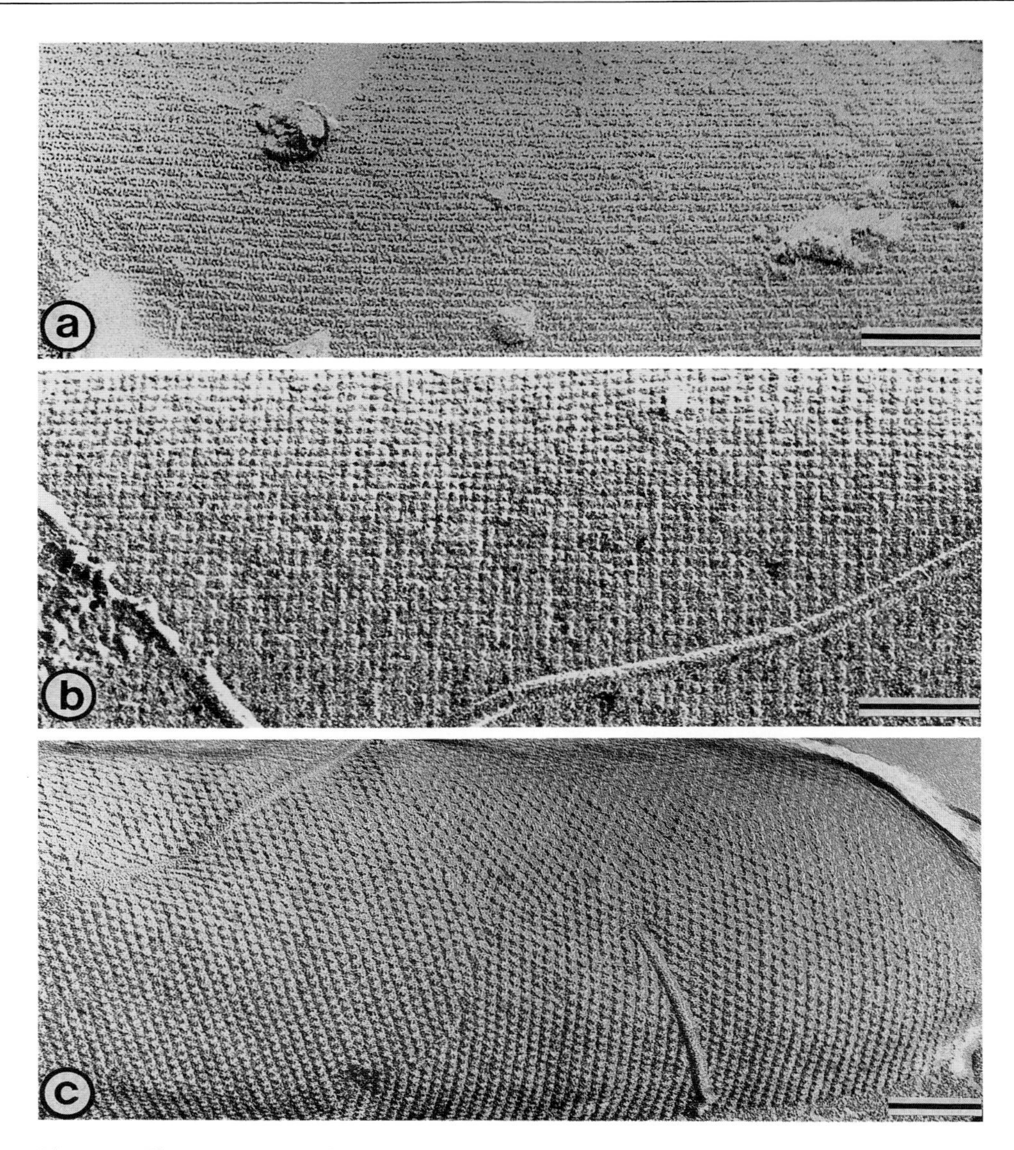

Figure 4 Electron micrographs of freeze-etched preparations of intact bacteria. (a) *Lactobacillus acidophilus* SH1; (b) *Clostridium thermosaccharolyticum* D120-70; (c) *Clostridium thermohydrosulfuricum* L111-69. The oblique (a), square (b), and hexagonal (c) S layer completely covers the cell surface. Bars = 100 nm.

proteinaceous monomers. S layers of most archaebacteria have hexagonal symmetry whereas in eubacteria, square and oblique lattices are also frequently observed.

Comparative studies on the distribution and uniformity of S layers have revealed that, in some species, individual strains can show a remarkable diversity regarding the geometry and lattice constants of the arrays and the molecular weight and glycosylation patterns of the constituent subunits.

A characteristic feature seen with many S layers is a smooth outer and a more textured inner surface. Because of their crystalline nature and composition of identical macromolecules, S layers exhibit uniform pore diameters. But individual S-layer lattices can display more than one type of pore (Fig. 5b). From high-resolution electron micrographs, pore sizes in the range from 2 to 6 nm and a porosity of the protein meshwork between 30% and 70% have been estimated.

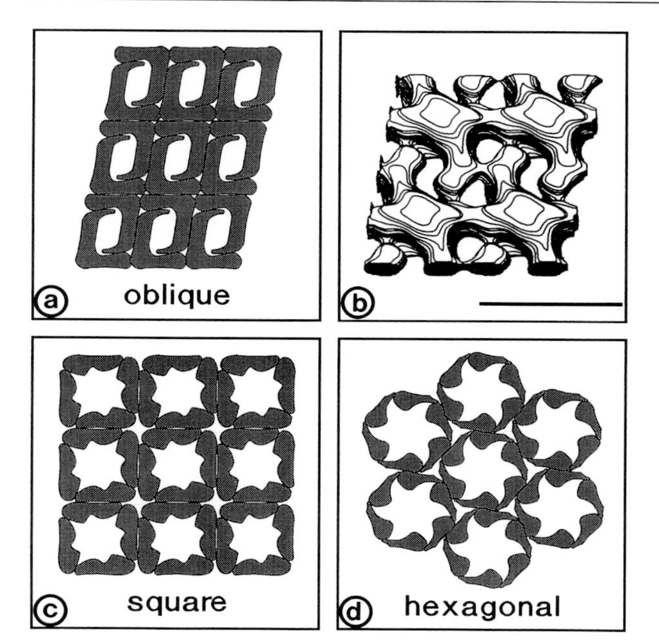

Figure 5 Schematic drawing of most common types of S-layer lattices observed on prokaryotic cells. (a) Oblique lattice; (b) top view of the three-dimensional structure of the oblique S-layer lattice from *Bacillus coagulans* E38-66, revealing two types of pores. Bar = 100 nm. [From Pum, D., Sára, M., and Sleytr, U. B. (1989). *J. Bacteriol.* **171**, 5296–5303.]; (c) square lattice; (d) hexagonal lattice.

III. Isolation and Chemical Characterization

S layers of different bacteria may vary considerably with respect to their resistance to disruption into their monomeric subunits, and a wide range of methods has been applied for their isolation and purification. The subunits of most S layers interact with each other and with the supporting envelope layer through noncovalent forces. Most commonly, in gram-positive organisms, a complete disintegration of S layers into monomers can be obtained by treatment of intact cells or cell walls with high concentrations of H-bond breaking agents (e.g., urea or guanidine hydrochloride). S layers from gram-negative bacteria frequently disrupt upon application of metal chelating agents (e.g., EDTA, EGTA), cation substitution (e.g., Na^+ to replace Ca^{2+}), or pH changes (e.g., pH <4.0). From extraction and disintegration experiments it can be concluded that the bonds holding the S-layer subunits together are stronger than those binding the crystalline array to the supporting envelope layer. There are some indications that S layers of some archaebacteria (e.g., *Thermoproteus* species and *Methanospirillum hungatei*) are stabilized by covalent bonds between adjacent subunits.

Most S layers are composed of a single, homogeneous protein or glycoprotein species. SDS-polyacrylamide gel electrophoresis reveal apparent molecular masses for subunits in the range of approximately 40 to 220 kDa.

Comparison of amino acid analyses and genetic studies on S layers from both archaebacteria and eubacteria has shown that the crystalline arrays are usually composed of weakly acidic proteins. The content of hydrophobic amino acids is generally high and the cystein or methionine content low.

A remarkable feature of both archaebacteria and eubacteria is their ability to glycosylate their S-layer proteins. Whereas among archaebacteria, most S layers appear to be glycosylated, evidence for S-layer glycoproteins in eubacteria has only recently obtained, and it seems that glycosylated eubacterial S layers are rather rare. [*See* ARCHAEBACTERIA (ARCHAEA).]

So far, the glycan chains and linkages of these eubacterial and archaebacterial glycoproteins are significantly different from those of eukaryotes. A broad spectrum of methods for carbohydrate characterization, including ^{1}H and ^{13}C nuclear magnetic resonance measurements, have been applied for elucidating the primary structures of the glycan chains of selected organisms. As summarized in Fig. 6, all glycan chains studied are polymers of linear or branched repeating sequences of two to six monosaccharide units, which include a wide range of hexoses, deoxy or amino sugars, uronic acids, and even sulfate or phosphate residues as constituents.

IV. Biosynthesis, Genetics, and Morphogenesis

An intact, "closed" S layer on an average-sized, rod-shaped cell consists of approximately 5×10^5 monomers. Thus, at a generation time of about 20 min, at least 500 copies of a single polypeptide species with a M_r of approximately 100,000 have to be synthesized, translocated to the cell surface, and incorporated into the S-layer lattice per second.

During the past decade a substantial amount of information on the biosynthesis and genetics of S layers has accumulated. Insights into the molecular

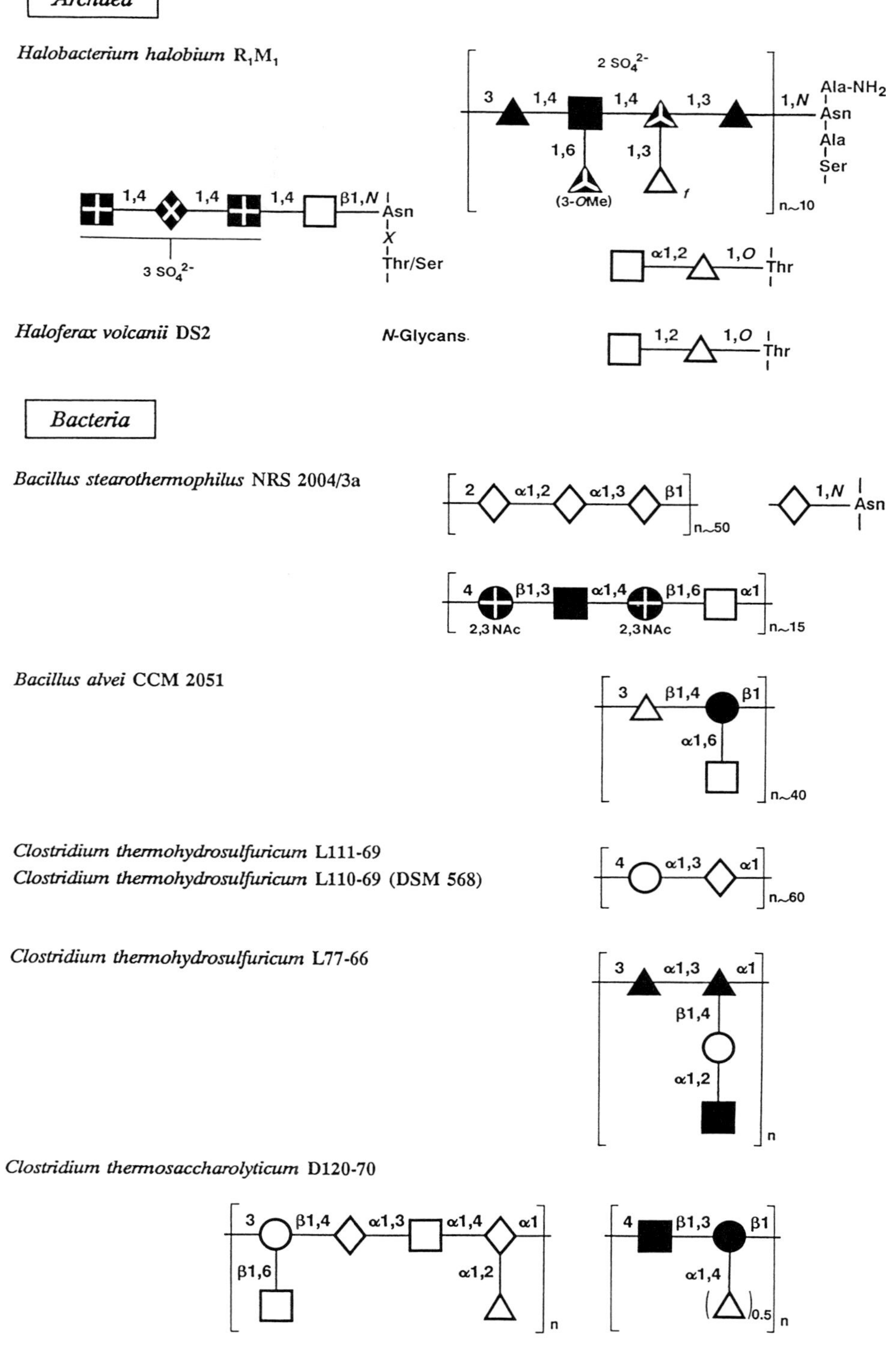

Archaea
Halobacterium halobium R1M1
2 SO4 2-
3
1,4
1,4
1,3
1,N
Ala-NH2
Asn
Ala
Ser
1,6
1,3
(3-OMe)
f
n~10
1,4
1,4
1,4
β1,N
Asn
X
Thr/Ser
3 SO4 2-
α1,2
1,O
Thr
Haloferax volcanii DS2
N-Glycans
1,2
1,O
Thr
Bacteria
Bacillus stearothermophilus NRS 2004/3a
2
α1,2
α1,3
β1
n~50
1,N
Asn
4
β1,3
α1,4
β1,6
α1
2,3NAc
2,3NAc
n~15
Bacillus alvei CCM 2051
3
β1,4
β1
α1,6
n~40
Clostridium thermohydrosulfuricum L111-69
Clostridium thermohydrosulfuricum L110-69 (DSM 568)
4
α1,3
α1
n~60
Clostridium thermohydrosulfuricum L77-66
3
α1,3
α1
β1,4
α1,2
n
Clostridium thermosaccharolyticum D120-70
3
β1,4
α1,3
α1,4
α1
β1,6
α1,2
n
4
β1,3
β1
α1,4
0.5
n
Clostridium symbiosum HB25
6
α1,4
β1,3
α1,4
α1
P
n~15

organization of S-layer biosynthesis were obtained from cloning experiments and sequencing studies. S layers characterized by genetic studies have shown to be synthesized with an N-terminal signal sequence, which is then cleaved off during translocation of the polypeptide across the cytoplasmic membrane. Although many S-layer proteins are synthesized as secretory precursors, usually the mature proteins are not liberated into the surrounding medium but are assembled into regular arrays on the cell surface. An exception is certain strains of *Bacillus brevis*, which, dependent on the growth conditions, produce considerable excess amounts of S-layer material, which are then shed into the medium.

Unlike purely proteinaceous S-layer protomers and despite the efforts of several research groups to elucidate the biosynthetic pathways of the S-layer glycoproteins, not very much detailed information could be gathered until now. In certain gram-positive archaebacteria (see Fig. 1c) two different types of lipid carriers are involved in the biosynthesis of the cell envelope components. Undecaprenol is the common lipid carrier for pseudomurein biosynthesis, whereas dolichol is involved in the biosynthesis of the S-layer glycoproteins.

Homology comparisons between S-layer proteins of different bacteria show that little overall homology exists. These findings confirm the diversity of S layers as observed by biochemical and ultrastructural investigations and are in accord with the nonconservative character of other cell surface macromolecules (e.g., lipopolysaccharides in gram-negative eubacteria).

On removal of the disintegrating agent (e.g., by dialysis), isolated S-layer subunits from numerous gram-positive and gram-negative bacteria have shown the ability to assemble into two-dimensional crystalline arrays with the same lattice dimensions as those observed on intact cells. Depending on the S layer type and assembly conditions (e.g., pH, ionic strength, ion composition), S-layer subunits may aggregate either into flat sheets (Fig. 7a), open-ended cylinders (Fig. 7b), or vesicular structures. Both single- or double-layer assembly products have been obtained. The latter frequently are generated in the presence of bivalent cations, whereby the two constituent monolayers can aggregate either with their inner or their outer faces.

Isolated S-layer subunits of gram-positive and gram-negative eubacteria have also shown the ability to recrystallize on the cell envelope fragments from which they had been removed, on those of other organisms, or on untextured charged or uncharged inanimate surfaces.

So far, the most detailed self-assembly and reattachment experiments have been performed with S layers from Bacillaceae. S layers of these organisms reveal a high anisotropic charge distribution. The inner surface is negatively charged, whereas the outer face is charge-neutral around pH 7. This characteristic of the S-layer subunits appears to be essential for the proper orientation during local insertion in the course of lattice growth.

Detailed studies have been performed to elucidate the dynamic process of assembly of S layers during cell growth. Freeze-etching preparations of rod-shaped cells generally reveal a characteristic orientation of the lattice with respect to the longitudinal axis of the cylindrical part of the cell (see Fig. 4). For maintenance of such a good long-range order during cell growth, S layer protomers must have the ability to recrystallize on the supporting envelope layer. Labeling experiments with fluorescent antibodies and colloidal gold/antibody marker methods indicated that different patterns of S-layer growth exist for gram-positive and gram-negative bacteria. In gram-positives, growth of the S-layer lattice primarily occurs on the cylindrical part of the cell by insertion, at multiple bands or helically arranged bands; in gram-negatives, incorporation of new subunits occurs at random. In both types of organisms entirely new S-layer material also appears at regions of incipient cell division and the newly formed cell poles. Little information is available on the growth of S-layers in those gram-negative archaebacteria (se Fig. 1a), which possess S layers as the exclusive cell wall component. Lattice faults such as dislocations and disclinations have been suggested to function both as incorporation sites for new subunits and as initiator sites for cell division.

Figure 6 Composite diagram to illustrate diversity of glycan structures of prokaryotic S-layer glycoproteins. Symbols are □, D-glucose; ○, D-mannose; △, D-galactose; ◇, L-rhamnose; ■, *N*-acetyl-D-glucosamine; ●, *N*-acetyl-D-mannosamine; ▲, *N*-acetyl-D-galactosamine; ◆, *N*-acetyl-D-bacillosamine (2-acetamido-4-amino-2,4,6-trideoxy-D-glucose); ⊞, glucuronic acid; ⊕, 2,3-diacetamido-2,3-dideoxymannuronic acid; ▲, galacturonic acid; ◆, iduronic acid; *f, furanose;* Ⓟ, *phosphate;* SO_4^{2-}, *sulfate;* 3-OMe, 3-*O*-methyl; Asn, asparagine; Ser, serine; Thr, threonine, Ala, alanine; *X*, interchangeable amino acid; n, degree of polymerization. For constituent sugars of archaebacteria, no steric configurations were determined. [From Messner, P., and Sleytr, U. B. (1991). *Glycobiology* **1**, 545–551.]

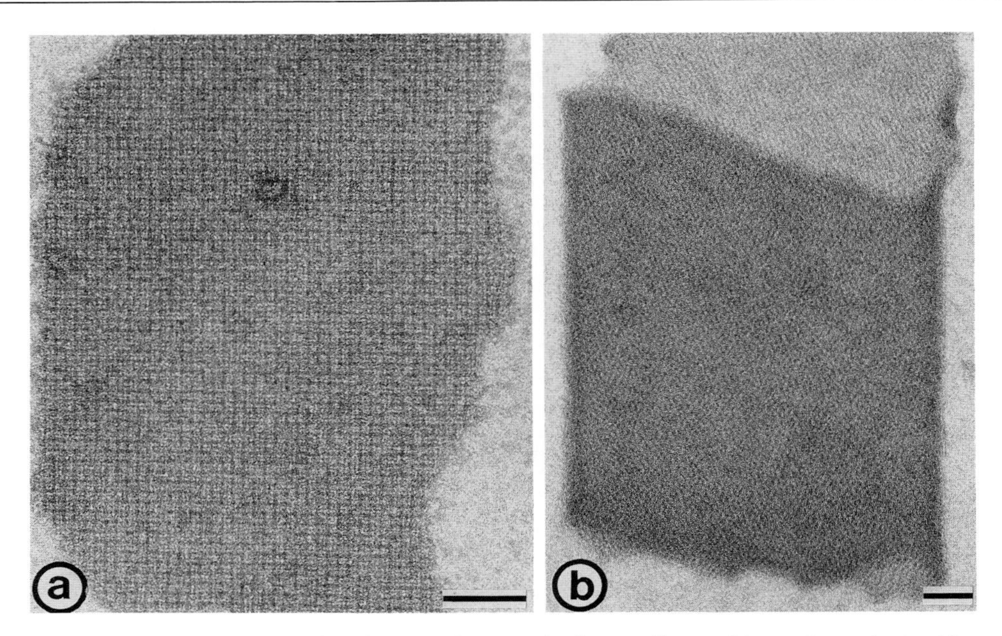

Figure 7 Electron micrographs of negatively stained S-layer self-assembly products obtained by recrystallization of isolated subunits. (a) Sheet-like self-assembly product obtained with S layers from *Clostridium thermosaccharolyticum* D120-70 revealing a square lattice. Bar = 100 nm; (b) open-ended cylindrical self-assembly product obtained with the oblique S-layer lattice from *Bacillus stearothermophilus* NRS 2004/3a. Bar = 100 nm.

V. Cellular Function and Application Potential

Because prokaryotic organisms possessing S layers are ubiquitous in the biosphere (see Fig. 3), it can be expected that the porous network of regularly arranged (glyco)proteins has evolved as the result of quite diverse interactions of the bacteria with specific environmental and ecological conditions.

Although relatively few data are available on specific functions of S layers, there is strong evidence that the crystalline arrays have the potential to function as (1) cell shape determining/maintaining framework; (2) protective coats, molecular sieves, and molecule or ion traps; and (3) promoters of cell adhesion and surface recognition.

A. Cell Shape Determination and Maintenance

Because of their structural simplicity and from a morphogenetic point of view, it was suggested that S layers, like membranes, could have fulfilled barrier and supporting functions as required by self-reproducing systems (progenotes) during the early period of biological evolution. Evidence for such a function still exists in those archaebacteria that synthesize S layers as the only rigid cell wall component (e.g., in square bacteria, some methanogens, and almost all sulfate reducers, halophilic and extremely thermophilic S°-metabolizers) (see Fig. 3).

B. S Layers Related to Pathogenicity

One of the best studied S-layer systems as it pertains to pathogenicity is the crystalline arrays present on members of the genera *Aeromonas* and *Campylobacter*. In *Aeromonas salmonicida*, which causes the fish disease known as furunculosis, generally an S layer is required for virulence. The S layer physically protects the infecting cells against proteolysis and complement, and is essential for macrophage infiltration and resistance.

In *Campylobacter fetus* subspecies *fetus*, the agent that causes abortion in sheep and cattle and various systemic infections or acute diarrheal illness in humans, S layers appear to make the cells resistant to phagocytosis and to bactericidal activity of serum.

Crystalline surface layers identified on other pathogens of humans and animals including species of *Bacteroides, Brucella, Chlamydia, Cardiobac-*

terium, Rickettsia, Wolinella, Treponema, Clostridium, and *Bacillus* may be of similar functional relevance as virulence factors.

C. S Layers as Molecular Sieves and Promoters for Cell Adhesion and Surface Recognition

Data obtained by high-resolution electron microscopy indicate that S layers are highly porous structures with pores of defined size and morophology (see Fig. 5). Permeability studies on S layers of mesophilic and thermophilic Bacillaceae provided information on their molecular sieving properties. For example, S layers of *B. stearothermophilus* revealed sharp exclusion limits for molecules larger than 45,000, indicating a limiting pore diameter in the crystalline protein meshwork of about 4.5 nm. With some mesophilic Bacillaceae, pore sizes as small as about 2.5 nm have been determined. These permeability studies also showed that the pores in the S-layer membranes have a low tendency for fouling, a feature regarded essential for an unhindered exchange of nutrients and metabolites up to a defined molecular size.

S layers acting as molecular sieves have the potential to function not only as barriers preventing molecules from entering the cell (e.g., lytic enzymes, complements, antibodies, surfactants, biocides). They can also generate a functional equivalent to the periplasmic space of gram-negative eubacterial envelopes in preventing the release of molecules (e.g., enzymes, toxins) from the cell.

Cells of *Aquaspirillum serpens* and other gram-negative species were shown to be resistant to predation by *Bdellovibrio bacteriovorus* when they are covered by an S layer. However, S layers of different organisms (e.g., Bacillaceae) were demonstrated to act as specific sites for phage adsorption. Analysis of phage-resistant mutants of *Bacillus sphaericus* showed that although the crystalline arrays were present on all mutants, the molecular weight of the S-layer subunits had been changed.

With regard to cell adhesion and cell recognition properties of S layers, an important aspect is the frequently observed anisotropic distribution of charged groups on both faces of the (glyco)protein lattice. Most Bacillaceae possessing S layers reveal a charge-neutral outer surface, which physically masks the net negatively charged peptidoglycan sacculus. In comparison with S-layer-deficient strains, S-layer-carrying Bacillaceae have also shown a much greater ability to adsorb to positively charged or hydrophobic surfaces.

D. Application Potential

Based on the data obtained by fundamental studies on S layers, a considerable potential in biotechnological and nonbiological applications became evident. Applications for S layers have been found in the production of isoporous ultrafiltration membranes, as supports for a defined covalent attachment of functional molecules (e.g., enzymes, antibodies, antigens, protein A, biotin, avidin) as required for affinity and enzyme membranes or the development of biosensors. S-layer membranes have also been used as support for Langmuir Blodgett films, mimicking the molecular architecture of gram-negative archaebacteria (see Fig. 1a). Finally, most recently it was demonstrated that S-layer fragments or self-assembly products can be used for a geometrically well-defined covalent attachment of haptens and immunogenic or immunostimulating substances. These haptenated S-layer structures act as strong immunopotentiators. Finally, the fundamental studies on the biosynthesis, regulation of synthesis, and secretion of S-layer proteins can be expected to be of great relevance for recombinant-DNA technology, exemplifying the transfer of recombinant-DNA products out of bacterial cells.

Bibliography

Baumeister, W., Wildhaber, I., and Engelhardt, H. (1988). *Biophys. Chem.* **29,** 39–49.

Hovmöller, S., Sjögren, A., and Wang, D. N. (1988). *Prog. Biophys. Mol. Biol.* **51,** 131–163.

Koval, S. F. (1988). *Can. J. Microbiol.* **34,** 407–414.

Messner, P., and Sleytr, U. B. (1991). *Glycobiology* **1,** 545–551.

Messner, P., and Sleytr, U. B. (1992). Crystalline bacterial cell surface layers. *In* "Advances in Microbial Physiology," Vol. 33, (A. H. Rose and D. W. Tempest, eds.). Academic Press, London (in press).

Pum, D., Sára, M., and Sleytr, U. B. (1990). *J. Vac. Sci. Technol. B* **7**(6), 1391–1397.

Sára, M., and Sleytr, U. B. (1988). Membrane biotechnology: Two-dimensional protein crystals for ultrafiltration purposes. *In* "Biotechnology," Vol. 6b (H.-J. Rehm and G. Reed, eds.), pp. 615–636. VCH, Weinheim.

Sleytr, U. B., and Messner, P. (1988). *J. Bacteriol.* **170,** 2891–2897.

Sleytr, U. B., and Messner, P. (1989). Self-assembly of crystalline bacterial cell surface layers (S-layers). *In* "Electron Microscopy of Subcellular Dynamics" (H. Plattner, ed.), pp. 13–31. CRC Press, Boca Raton, Florida.

Sleytr, U. B., Messner, P., and Pum, D. (1988). Analysis of crystalline bacterial surface layers by freeze-etching, metal shadowing, negative staining and ultrathin sectioning. *In* "Methods in Microbiology/Electron Microscopy in Microbiology," Vol. 20 (F. Mayer, ed.), pp. 29–60. Academic Press, London.

Sleytr, U. B., Messner, P., Pum, D., and Sára, M. (eds.) (1988). "Crystalline Bacterial Cell Surface Layers." Springer Verlag, Berlin.

Culture Collection, Functions

Robert E. Stevenson and Harold Hatt
American Type Culture Collection

Glossary

Cryopreservation Freezing and storage of cells at very low temperatures (−150 to −180°C)

Cryoprotectant Substance added to cell suspensions to reduce damage during freezing and thawing

Freeze-drying Removal of water from frozen cell suspensions through sublimation under vacuum; cells are rehydrated to produce viable cultures

Lyophilization Term for freeze-drying

Public collection Supplies cultures on demand, usually for a fee; publishes catalogs; may have government support; broad collection of strains; sometimes called service collection

Seed stock Batch, usually early passage, held in reserve by a collection and from which distribution stocks are prepared

Specialist collection Usually a research collection; often highly specialized and may be the only source of certain strains

Type culture Strain upon which the name description of a species or subspecies is based (i.e., the nomenclatural type); often called type strain; rules of nomenclature, which govern the designation of a type, are published for various disciplines

A CULTURE COLLECTION is an officially constituted organization performing the functions of acquiring, preserving, authenticating, and distributing microorganisms or *in vitro* cultured cells to qualified scientists.

I. Historical Development

The beginning of fermented foods as an item of diet is lost in the mists of antiquity, but models from Egyptian tombs of early dynasties (2000–3000 B.C.) show baking and brewing as common practices. Cultures, or "mothers," were jealously guarded and handed on to new brides establishing their households or transmitted as a trade secret to a young brewmaster.

At the end of the nineteenth century, when the findings of the great European bacteriology pioneers laid the basis for the science, the need to collect, preserve, and compare cultures from different sources became obvious.

At the Institute of Hygiene at the Faculty of Medicine in Prague, František Král began a collection of bacteria, yeasts, and filamentous fungi. In 1890, he struck out on his own and established a private laboratory called "Král's Bakteriologisches Laboratorium in Prague." According to Czechoslovakian sources, he supplied cultures and media to users in diagnostic laboratories and for teaching. The first catalog of holdings was published in 1902. This is the first public, or service, culture collection known.

Following Král's death, E. Pribram, in 1914–1915, transferred the collection of several hundred cultures to the University in Vienna. He too published a catalog of the holdings in 1919. In 1930, when Pribram joined the faculty of Loyola University School of Medicine in Chicago, he brought half of the collection with him. The part remaining in Vienna was destroyed near the end of the World War II. When Pribram died, some of his cultures were transferred to the American Type Culture Collection (ATCC).

In 1911, the Society of American Bacteriologists (SAB) started a collection at the American Museum of Natural History in New York City. In 1925, the effort was enlarged by a committee of scientists representing the National Research Council, the SAB, the American Phytopathological Society, the

American Zoological Society, and the John McCormick Institute for Infectious Diseases. This public collection—the ATCC—was first housed at the McCormick Institute in Chicago. The depression of 1932–1937 diminished the resources of the institute greatly, forcing the ATCC to move to the Georgetown University School of Medicine. In its 10 years there, the collection added strains of protozoa and algae. In 1947, the ATCC was incorporated as a nonprofit organization and a rented facility was obtained in Washington, D.C.'s Foggy Bottom; a building in the neighborhood was purchased in 1956.

By 1960, rapid growth in biomedical research in the United States and a concomitant need for research resources expanded the ATCC's role to acquire and distribute animal and plant viruses and animal cell lines. Government and industry combined to pay for a custom-designed facility built in Rockville, Maryland, which was occupied in 1964. Continuing growth required the purchase of two adjacent buildings in the 1980s.

In other countries, parallel collection activities generally were supported by government agencies related to health, agriculture, or education, and collections tended to specialize with respect to the host (human, veterinary, agriculture) or discipline (mycology/botany, yeast, bacteria, etc.). For example, in 1904 the Centraalbureau voor Schimmelcultures was founded in The Netherlands by the International Association of Botanists. The National Collection of Type Cultures was established in 1920 at the Lister Institute in London. After work in the 1940s with *Neurospora crassa* biochemical mutants and their genetic basis, it became apparent that stock centers for genetic strains were needed. In 1960, the Fungal Genetics Stock Center was established at Dartmouth College.

II. Different Kinds of Collections

The 1989 list of culture collections of microorganisms, published by the World Data Center on Microorganisms, contains addresses of 345 organizations that consider themselves in that category. A survey of 356 culture collections, reported in 1986, revealed over half a million cultures are preserved in the world's culture collections.

In a larger sense, many laboratory scientists with their freezers full of specimens are also managing "working collections," so it is difficult to draw the line.

In its recently published "Guidelines for the Establishment and Operation of Collections of Microorganisms," the World Federation for Culture Collections speaks to attributes of "public service," "long term commitment by the sponsoring organization," and "appropriate standards." For the purposes of this article, culture collections will be defined as "an officially constituted organization performing the functions of acquiring, preserving, authenticating and distributing microorganisms or *in vitro* cultured cells to qualified scientists."

Many of the guiding principles outlined here will be useful for the individual scientist who wishes to preserve small numbers of cultures under investigation, and, hopefully, the useful products of such research will be deposited eventually in a public service collection to make them readily available to others.

"Public collections" publish catalogs and make these holdings available for exchange or fees and perform a role comparable to that of a Bureau of Standards for microbiologists.

Some collections have specialized genetic stock center goals and serve a more restricted clientele of experts. Certain popular research model systems enjoy this level of resource. Thus, there are the *Escherichia coli* Genetic Stock Center at Yale University, the *Bacillus* Genetic Stock Center at Ohio State University, the *Fusarium* Genetic Stock Center at Penn State University, and the Phage Collection at Montreal, to name a few.

The Coriell Medical Research Institute (Camden, New Jersey) specializes in human genetic mutant cell lines, and the National Institute of Allergy and Infectious Disease maintains a resource of materials relating to human immunodeficiency virus, including cells and virus strains.

Because the patent laws of the United States and most countries require a deposit of biological material to support a patent application, either for enablement, best mode, or to avoid undue experimentation, patent depositories are necessary. They provide an objective, neutral resource for the deposit, safe-keeping, and eventual distribution of cultures involved in patented processes.

The ATCC accepted its first deposit for patent purposes in 1949, long before it was a formal requirement of any patent office. The ATCC was approved in 1981 as the first International Depository Authority (IDA) under the International Budapest Treaty for deposits, which meets patent office requirements in many countries. Currently 24 countries are signatory to the treaty.

The Budapest Treaty was established in 1977 but became operational in 1981 with the approval of the IDAs. The Treaty's official name is the Budapest Treaty on the International Recognition of the Deposit of Microorganisms for the Purpose of Patent Procedure. The Treaty requires signatory countries to recognize a deposit with any IDA that has been approved by the World Intellectual Property Organization. No country may impose requirements different from or additional to those that are provided in the Treaty.

III. Finding an Organism

To find specific categories of materials, the World Data Center on Microorganisms, located at the RIKEN Institute (Wako, Japan), publishes a catalog/directory that cross-indexes culture holdings and culture collections. Another service, The Microbial Strain Data Network, also has this directory capability and can provide information about the associated information or known attributes of the holdings of collections. Should these not reveal the desired material, inquiries can be made to researchers currently publishing in the field or to scientific societies having such types of members. The American Society for Microbiology conducted a survey some years ago that revealed a large number of nonpublic collections with very specific focus.

The ATCC, besides having a very diversified and large collection, also has cross-references to similar materials and may be able to provide equivalent strains or locate the specific one needed.

IV. Functions of Culture Collection

The 1990 World Federation for Culture Collections publication "Guidelines for the Establishment and Operation of Collections of Cultures of Microorganisms" gives helpful information on the organizing and operating principles of collections and suggests minimal requirements for resources and management. Because collections represent significant investments of time and money, the commitment to establish and maintain one is not trivial and should represent judicious planning and realistic estimation of resources required.

The variety of reasons for collecting and the ephemeral nature of some goals make it difficult to generalize, but there is an abiding need for some permanent collections that serve a museum, Library of Congress, or Bureau of Standards-like function in perpetuity as the ultimate resting place for material obtained and studied at great cost and labor. Just as many books pass from current novelty, strains once popular may fade from popularity. Regardless, the literature cites them, they serve for comparison and reference, and they must be preserved.

The functions of the collections are many and go beyond the mere assembly of curiosities or items to exhibit, much as a museum or art gallery would acquire artifacts for display.

A. Scope

The scope of the material collected depends on the interests to be served, but collections are based on an ordering or selection by recognized criteria or scientific significance. One could have a systematics collection with the goal of having all recognized type strains of particular groupings. Another collection might consist of clinical isolates from Asian flu victims between 1960 and 1980 on the North American continent.

The exercise of the selectivity is the responsibility of the curator, who should be a scientist knowledgeable in the field. This expert will consult with colleagues and should seek items to add and compare with present holdings.

Descriptions, attributes, properties, etc., that are associated with holdings become permanently linked in collections. It is important that collection standards are established for the determination of these data and that the historical information on isolation or source, publications, and conditions of storage and handling are recorded accurately.

B. Preservation

The short generation time of many microorganisms and the resultant increased opportunity for genetic drift makes preservation processes that suspend viability attractive. Carrying this one step farther, a concept of seed stock underlies a strategic management concept central to culture collections.

Simply stated, one prepares a well-characterized batch of multiple aliquots as "seed" and then goes back to this material to expand for distribution stock. Using this principle, a collection could provide essentially comparable material to scientists for 50–100 years after the original seed was laid down. Even afterward, the next seed batch would only be several generations later, thus minimizing the genetic drift tendency. The accessioning scheme in Fig. 1, used by the ATCC Cell Culture Department, is an example of a seed stock concept in practice.

Preservation processes used in culture collections include freeze-drying (or lyophilization), controlled rate freezing with or without cryoprotectants such as glycerol, dimethylsulfoxide, skim milk, sucrose, honey, etc., layering under mineral oil, desiccation in sterile soil or sand, and supercooling. A full description of procedures is beyond the scope of this article, but abundant literature is available.

C. Distribution

Culture collections have a societal responsibility when distributing cultures to make reasonable requirements that recipients of the cultures are trained to handle infectious agents and that the use of them will pose no risk to public health. Names and addresses of recipients as well as what they were sent should be recorded. As technical improvements and better quality-control testing become available, previously unsuspected contaminants, or passenger agents, have been discovered. Having good records of recipients enables a collection to inform them of the new findings, which may have an important bearing upon their research.

The shipping of etiologic agents is controlled by national and international bodies. These range from U.S. Commerce Department export licenses to prescribed forms of packaging and labeling to conform to postal and Department of Transportation regulations. Culture collections must comply with all of these rules, and obtaining the necessary permits often results in an unavoidable delay in fulfilling requests for material. Also, infectious agents are prohibited in the postal services of Canada and France, so more expensive forms of shipment must be used in these countries. [*See* CULTURE COLLECTIONS: METHODS AND DISTRIBUTION.]

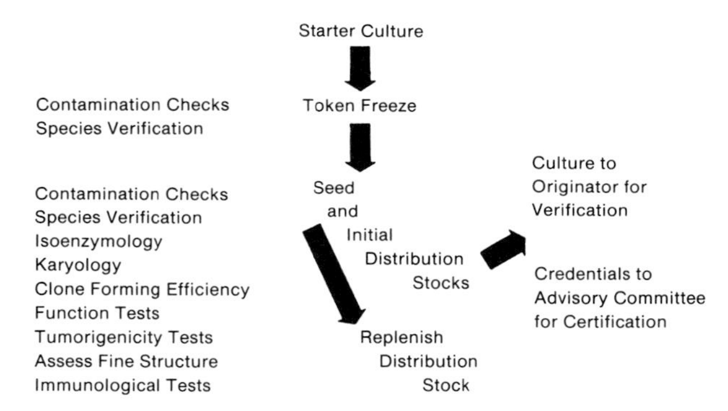

Figure 1 Accessioning scheme. From the starter culture, a token freeze is prepared and screened for contamination and verified as to species. From the token freeze, seed stock and initial distribution stock are prepared and characterized or tested as indicated. Samples of this stock are returned to the depositor for approval, and documentation is reviewed. When new distribution stock is needed, the highly characterized seed stock is used.

D. Identification

Identification is the description of an organism that differentiates it from other organisms. Culture collections must have the capability to identify the organisms represented in their holdings. This implies that trained personnel and adequate equipment resources are available.

Morphological, biochemical, staining, immunological, and (where appropriate) pathogenicity tests are used to identify organisms that are grown in pure culture.

Some culture collections will identify unknown organisms as a service, but prior arrangements should be made. Unknown or poorly labeled material arriving at a collection is more often than not autoclaved at once.

Relatedness between organisms is of both practical and theoretical importance, and besides the enumeration of properties, sets of descriptors may be treated mathematically in numerical taxonomy to explore relationships.

Codes of nomenclature specify the procedures for validly and legitimately publishing new names. They also specify, for the bacteria, the requirements that a nomenclatural type be deposited in a major culture collection and that the strain designation used therein be cited in the publication describing the new taxon.

Unfortunately, the standards for description of organisms and cell lines in the materials section of scientific papers are not uniformly enforced. The Council of Biology Editors Style Manual gives authoritative advice on proper forms of citation and should be consulted on these matters.

Use of improperly documented or generally unavailable organisms is a waste of everyone's time and money. If the material is newly isolated, care should be taken to create a cache of it and ensure its availability for at least 5 years for comparisons and validation of the published work.

Strains should be characterized as completely as

current techniques allow. This serves as a baseline for future comparison as new stocks are prepared from seed stock.

E. Catalog (Information)

A necessary function, cataloging may be as simple as a list of the names of the holdings in the collection. To most, cataloging also includes pertinent information about the date, place, and circumstances of isolation of the original material, the primary bibliographic citation, and a history of possession or provenance of the culture.

More elaborate catalogs provide an annotated bibliography or lists of strains with special applications summarizing organisms that exhibit similar properties (restriction enzymes, thermophiles, assay strains, etc.).

The catalog is the major entree to a collection, and its setup and indexing are important considerations for the usefulness of a collection.

Strain data in the catalog reflect the strain's current taxonomic position, isolation history, and subsequent custodial care and lineage. The importance of a good history is often overlooked but may be critical to understanding later why some things work and others do not. For example, cultures obtained from patients before clinical use of antibiotics are likely not to contain plasmids conferring resistance to antibiotics. An anthrax strain believed to have come from Pasteur's laboratory does not have a phage that induces hematoxin production, whereas most natural isolates do. Did Pasteur pasteurize his strain? We do not have enough information to say for sure.

Some collections now have electronic data bases that provide catalog information. Online, personal computer floppy disks, and compact disk versions exist and are presently being evaluated by users. Versions of these provide for rapid searching for characteristics or varieties of interest.

The combination catalog/directory, produced by the World Data Center for Microorganisms, that lists the microorganisms/cell lines in one section and cross-indexes them to where they are held in the collections section is a very valuable tool.

Other services inform the user what each collection knows about its holdings (Microbial Strain Data Network) and can help to focus searches for information.

V. Future

The advances of molecular biology and biotechnology have been laid on the foundations provided by microbiological tools—genetically engineered strains, hybridomas, DNA vectors, etc. These materials are a large part of the future for culture collections.

Endangered species are being helped as zoos and culture collections collaborate in collecting, handling, freezing, and storing sperm, eggs, embryos, and tissue cells.

As pointed out by L. I. Sly, "Culture collections are entrusted with the conservation of an invaluable resource which is part of our natural world heritage. The significance of the irreplaceable gene pool represented in culture collections may only be fully appreciated in the light of new scientific discoveries and technological developments."

Bibliography

Ashwood-Smith, M. J., and Farrant, J. (eds.) (1980). "Low temperature Preservation in Medicine and Biology." University Park Press, Baltimore, Maryland. 323 pp.

Cohn, J. P. (1989). *Zoogoer* **18,** 25–29.

Porter, J. R. (1976). The world view of culture collections. *In* "The Role of Culture Collections in the Era of Molecular Biology" (R. Colwell, ed.), pp. 62–72. American Society for Microbiology, Washington, D.C.

Sly, L. I. (1986). Culture collection technologies and the conservation of our microbiol heritage. *In* "Applied Microbiology" (H. Doelle and C.-G. Heden, eds.), pp. 1–37. D. Reidel, UNESCO, Paris.

Takishima, Y., Shimuira, T., Udagawa, Y., and Sugawara, H. (1989). "Guide to World Data Center on Microorganisms with a List of Culture Collections in the World." World Data Center on Microorganisms, Saitama, Japan. 249 pp.

WFCC Standards Committee. (1990). "Guidelines for the Establishment and Operation of Collections of Cultures of Microorganisms" (D. L. Hawksworth, ed.). World Federation for Culture Collections, Secretariat, Brazil. 16 pp.

Culture Collections: Methods and Distribution

C. P. Kurtzman
U.S. Department of Agriculture

Glossary

Cryoprotectant Substance added to cell suspensions to maintain viability during freezing and thawing
Culture Tube or flask of medium containing the growth of an organism
Curator Scientist responsible for a culture collection or a portion of a collection
Lyophilization Freeze-drying process for preserving organisms in a metabolically arrested state
Patent culture Culture of an organism that is the subject of a patent
Preservation Means for maintaining organisms in a living state, usually in a condition of metabolically arrested growth
Strain Single microbial isolate

ANYONE who has looked at seed and flower catalogs soon develops an appreciation of the need for a reliable source of seeds for both common plants and exotic forms. Microbial culture collections have a role rather similar to suppliers of plant germplasm because they preserve and distribute the microorganisms essential for teaching, research, and manufacturing the large number of antibiotics, enzymes, and other organic compounds that we have come to rely on in our daily lives. Most of the key microorganisms in use today trace their origins to many different parts of the world. Because culture collections maintain strains from all over, the user of these collections can profit from the efforts of literally thousands of scientists who have isolated unique microorganisms worldwide.

I. Culture Collections as Resource Centers for Science and Technology

Culture collections generally fall into two categories: service collections and in-house/research collections. The main goal of service collections is to collect and preserve cultures and to distribute them to the scientific and industrial public. The scope of the germ plasm maintained may be quite diverse, or it may focus on a specialized group of microorganisms. Fees are charged for strain distributions and other services, but the fees collected are usually insufficient to cover all operating costs, and, thus, public funds are required to make up the difference. For the collection to remain effective, curators must constantly acquire new strains, and these are obtained from various researchers, other culture collections, and isolations by the collection staff. Other services provided generally include safe storage of cultures for individual clients, accessioning of patent cultures, and identification of unknown organisms. An example of a large service collection is the American Type Culture Collection, Rockville, Maryland, which maintains many types of microorganisms as well as plant and animal cell lines.

In-house/research collections may represent the personal collection of an individual scientist, or they may provide the service/research needs of a particular laboratory or organization. These collections were not established to provide all of the options available from a service collection, but most, nonetheless, distribute cultures. An example of a large in-house collection (80,000 strains) is that of the Agricultural Research Service (ARS) Culture Collection, Peoria, Illinois. The ARS Culture Collection was established to provide strains for U.S. Department of Agriculture scientists but, over the years, the collection's prominence led to strain distributions on a worldwide basis. Distributions are limited to 12 strains per request and a catalogue is not presently issued. The ARS Culture Collection also serves as an International Patent Culture Depositary Authority. Holdings are comprised of yeasts, molds, bacteria, and actinomycetes of agricultural and industrial importance.

The locations and holdings of most culture collections may be found in the second edition of the World Federation for Culture Collection's "World Directory of Collections of Cultures of Microorganisms (V. F. McGowan and V. B. D. Skerman, 1982) available from the Secretary, UNEP/UNESCO/ICRO Panel on Microbiology, Swedish University of Agricultural Sciences, S-750 07 Uppsala, Sweden, in book form or microfiche. The third edition (1986) is available from the United Nations Environment Programme, Information Service, P.O. Box 30552, Nairobi, Kenya.

The identification of microorganisms is another service often provided by culture collections. Such work usually requires careful microscopic observations, results from various types of growth tests, and, increasingly, reliance on comparisons of nucleic acids and other macromolecules. Correct identification is needed for a variety of reasons, including the following: control of diseases in humans, animals, and plants; sustaining patent applications; predicting which microorganisms might yield novel products; food safety; and an understanding of microbial ecology. Because the identification process may take considerable time and materials, collections often charge fees for the work or limit the number of unknowns that they will accept. [*See* CULTURE COLLECTION, FUNCTIONS.]

II. Methods for Preserving Microorganisms

Periodic transfer of microbial strains to fresh culture media is one means for maintaining a collection, but the technique has some major drawbacks. Periodic transfer is time-consuming and, therefore, costly; in addition, if one waits too long, the strains may die. More importantly, continually growing microorganisms adapt to their laboratory culture media and may eventually lose the characteristics for which they were originally deposited.

Preservation methods have been developed that allow storage of cultures in an arrested, nongrowing state. The method in longest use is lyophilization, or freeze-drying, which was first used on a large scale in the early 1940s. Cells to be lyophilized are first suspended in a cryoprotective medium such as bovine serum or skim milk. The cells are then transferred to glass ampoules and frozen, and a vacuum is applied to cause the frozen cells to dry. Once the cells are dried to around 1% moisture, the ampoules are sealed with a gas-oxygen torch while still under vacuum. Ampoules are then stored in a refrigerator until needed. Although not all microorganisms can survive the lyophilization process, the majority that do are often alive for >50 yr, the longest period for which there are significant data.

Freezing without drying represents the second commonly used method for long-term storge of microorganisms. Storage at −20°C, the temperature of common household freezers is sometimes effective, but reliance is usually on the vapor phase of liquid nitrogen (ca. −165°C). Cells are usually suspended in a cryoprotectant such as 5–10% glycerol or 5–10% dimethylsulfoxide and transferred to polypropylene ampoules, which are placed in numbered spaces in the liquid nitrogen freezer. Fastidious organisms may require a programmed slow rate of freezing to survive. To revive frozen cultures, ampoules are usually thawed quickly in a 37°C waterbath to prevent elongation of ice crystals, and the cells are then transferred to an appropriate growth medium. Maximum storage times in liquid nitrogen freezers are unknown, but many organisms survive for decades and one might reasonably expect even longer periods. In recent years, ultracold freezers (−80°C) have become commonplace in many laboratories, and they may represent an alternative to liquid nitrogen for long-term frozen storage of cultures.

III. Distribution of Cultures

The strains maintained in culture collections range from those used for the manufacture of foods to those that are serious pathogens of humans, animals, and plants. Shipping of pathogens or recombinant strains with uncertain properties are regulated in the United States by the U.S. Department of Agriculture, the U.S. Public Health Service, the U.S. Environmental Protection Agency, and the U.S. Department of Transportation. These agencies have two concerns: (1) whether or not the organisms should be introduced into new areas, and (2) that they be properly contained during shipment. Additionally, the Postal Service, International Civil Aviation Organization, and the International Air Transport Association have regulations for safe containment during transport. Examples of microorganisms and their shipping requirements are given in Table I.

To obtain cultures, one must first determine where the strains of interest are located. This can usually be done from culture collection catalogues or by contacting the collection. Requests are then made by letter, telefax, or sometimes by telephone. Any required permits must be obtained before the cultures can be shipped. Catalogues of the American Type Culture Collection generally indicate which, if any, permits are required for transport into and within the United States.

IV. Patent Protection for Economically Significant Microorganisms

In 1949, the United States Patent and Trademark Office implemented its new recommendation that cultures be deposited in conjunction with patent applications concerning microbiological inventions. The reasoning was that for chemical, electrical, or mechanical patents, a diagram or formula can sufficiently describe the invention, whereas in a microbiological patent, illustrations and narrative descriptions are generally inadequate to define sufficiently the microorganism used and therefore comply with the requirement for a full and complete disclosure of the invention.

In keeping with U.S. Patent Law, the subject strain of the patent application must be deposited with a culture collection that is to maintain this strain for at least the life of the patent (17 yr). The depositor has the option of making the patent culture freely available from the date of deposit or requesting that it not be distributed until issuance of the patent, at which time the culture must be freely available. The depositor is obligated to resupply the culture should the collection find its stocks changed or nonviable.

Other industrial countries have implemented similar requirements, and many have also established a national patent culture depository. Culture collections are obligated to maintain detailed accession records on their patent cultures and to keep records of all distributions.

Although the patent laws of most nations have the same basic aims, requirements for filing can differ considerably. In the event that an applicant wishes to apply for foreign patents, the filing, translation, and attorney fees can be costly. During the late

Table I Classification of Microorganisms According to Biological Hazard and Their Shipping Requirements

	Class I: Agents of no recognized hazard under ordinary conditions
Examples	*Saccharomyces cerevisiae, Trichoderma reesei, Lactobacillus casei*
Shipping	Culture tube in fiberboard or other container. Permits as required.
	Class II: Agents of ordinary potential hazard
Examples	*Aspergillus fumigatus, Candida albicans, Cryptococcus neoformans, Staphylococcus aureus*
Shipping	Culture tube wrapped in absorbent material, placed in metal screw-cap can, placed in fiberboard container. Permits as required.
	Class III: Pathogens involving special hazard
Examples	*Coccidioides immitis, Histoplasma capsulatum, Bacillus anthracis, Yersinia pestis*
Shipment	Culture tube heat-sealed in plastic, wrapped in absorbent material, placed in hermetically sealed can, placed in sturdy cardboard box. Permits as required. Etiologic agent warning label necessary.
	Class IV: Pathogens of extreme hazard
Examples	*Arthroderma simii, Pasteurella multocida,* certain animal and plant viruses
Shipment	Culture tube heat-sealed in plastic, wrapped in absorbent material, placed in hermetically sealed can, placed in sturdy cardboard box. Required permits. Etiologic agent warning label necessary.

[U.S. Department of Health, Education and Welfare, 1972; U.S. Department of Health and Human Services, Public Health Service, 1983.]

1970s, several international agreements were reached, allowing a single application to be recognized in a number of countries.

A. Patent Cooperation Treaty

The Patent Cooperation Treaty (PCT) became effective in 1970, was amended in 1978 and 1979, and has been signed by all industrially important countries. The treaty allows an applicant to file a single application in a standard format through the applicant's national patent office and have the application recognized as a valid filing in as many PCT countries as selected. This procedure results in extra costs, but these are much less than for the collective cost of numerous individual filings. One disadvantage of this system is loss of secrecy. In the United States, the application is kept secret until approved. If rejected by the Patent Office or abandoned by the applicant, the procedure can still be practiced as a trade secret. In most other countries, patent applications are published, thus disallowing the trade secret option.

B. European Patent Convention

The European Patent Convention (EPC), which is restricted to European countries, became effective on 7 October 1977. Applications may be filed with the European Patent Office (EPO) in either The Hague (P.B. 5818, Patentlaan 2, 2280-HV-Rijswijk, ZH, The Netherlands) or Munich (Motorama-Haus, Rosenheimer Strasse 30, D-8000 Munich 80, FRG). The official languages recognized by this treaty are English, French, and German. Applicants who file under the PCT may list the combined EPC countries as a "selected country." Members of the European Patent Organization are Austria, Belgium, Germany, France, Italy, Liechtenstein, Luxembourg, The Netherlands, Sweden, Switzerland, and the United Kingdom.

The protocol for a deposit under the EPC requires that the depositor state in the letter accompanying the culture that the strain is being deposited under EPC Rule 28. Once the strain number is received, the depositor then completes the patent application with the EPO. To receive a restricted culture under the EPC, the request must include the culture name and strain number as well as note that the deposit is under the EPC. The culture collection then sends a form that the requester must complete and forward to the EPO. The collection is then notified by the EPO if the strain is to be released.

C. Budapest Treaty on the International Recognition of the Deposit of Microorganisms for the Purposes of Patent Procedure

The Budapest Treaty was signed on 28 April 1977 and provides that a single deposit in an approved culture collection satisfies the patent application disclosure requirements of all member countries of the Union established by the treaty. The major advantage of this treaty is that a single culture deposit in an approved collection will satisfy all countries selected in multicountry filings under the PCT or the EPC. Additionally, applicants filing from a country with an approved collection may deposit there and not be concerned whether or not import permits might be required for deposit in another country. Information concerning the Budapest Treaty may be obtained from the World Intellectual Property Organization, 32 chemin des Columbettes, 1211 Geneva 20, Switzerland.

The basic protocol for a deposit under the Budapest Treaty, which automatically comes under the conditions of the EPC is as follows. The depositor states in a letter with the culture that the strain is being deposited under the Budapest Treaty. The depositor must complete the culture collection's accession form, and the collection must perform viability tests on the culture and fill out a Budapest Treaty form that includes the number of the newly accessioned strain. This is returned to the depositor who is responsible for completing the filing. The depositor may make the culture freely available at the time of deposit or allow it to be handled under the conditions of Rule 11 of the Treaty, which specifies regulations concerning distribution of cultures. The depositor may receive a culture of the deposit at any time. The depositor may also authorize distribution to a third party. Any industrial property office (patent office) to which the Treaty applies may receive a culture if necessary in the patent evaluation procedure. Other persons may obtain a culture if an industrial property office to which the Treaty applies certifies that, under the applicable law, that person has the right to the subject culture. In practice, outside parties usually make their request directly to the depositary authority. However, request forms should be obtained from the property office of the contracting state wherein the patent application has been submitted. Following processing, the depositary authority is notified by the appropriate industrial property office concerning availability of the culture. Decisions on release of cultures under both the Budapest Treaty and the EPC are made by industrial property offices and not by the depositary authority.

In addition to the preceding international treaties, governments may select certain foreign culture collections as suitable for their national deposits. In that event, the collection must be knowledgeable of those particular foreign laws. Some publications are quite helpful in explaining these laws; the journal *Industrial Property* is also an excellent source of information on contemporary patent law.

V. Conclusions

Culture collections have a significant impact on our daily lives because of the need for pure cultures of microorganisms in medicine, agriculture, and industry. This impact will further increase under the impetus of biotechnology as the search quickens for organisms that produce novel metabolic products.

Bibliography

Cooper, I. P. (1982). "Biotechnology and the Law." Clark Boardman, New York.

Fritze, D., and Kocur, M. (1990). "The Král Collection—The First Recorded Service Culture Collection." *Proc. Král Symp. Inst. Ferment.*, Osaka, Japan.

Haynes, W. C., Wickerham, L. J., and Hesseltine, C. W. (1955). *Appl. Microbiol.* **3**, 361–368.

Kirsop, B. E., and Kurtzman, C. P. (1988). "Living Resources for Biotechnology: Yeasts." Cambridge University Press, Cambridge.

Saliwanchik, R. (1982). "Legal Protection for Microbiological and Genetic Engineering Inventions." Addison-Wesley, Reading, Massachusetts.

Cyanobacteria, Molecular Genetics

Stephanie E. Curtis
North Carolina State University

Glossary

Blooms Dense growths of cyanobacteria that form on the surface of bodies of water

Oxygenic photosynthesis Form of photosynthesis found in cyanobacteria and plant chloroplasts of which oxygen is a by-product

Phycobilins Light-harvesting pigments characteristic of cyanobacteria and certain eukaryotic algae

Phycobilisomes Highly ordered aggregates of phycobilins

Plankton Microbial organisms that live in open waters

Thylakoids Membranes in which photosynthesis occurs

Trichomes Chains of filaments of cyanobacterial cells

Vegetative cell Undifferentiated cyanobacterial cells

CYANOBACTERIA are photosynthetic bacteria that comprise the largest subgroup of the gram-negative bacteria. Among photosynthetic bacteria, the cyanobacteria are distinguished by their ability to perform an oxygenic, or plantlike photosynthesis. The cyanobacteria are an ancient group of bacteria with a fossil record dating to nearly 3 billion years ago. Photosynthesis by early cyanobacteria is believed to be largely responsible for the conversion of the earth's atmosphere from anaerobic to aerobic. Today, cyanobacteria exhibit diverse forms and are adapted to a broad spectrum of ecological niches. Recent advances in the application of molecular genetics to cyanobacteria has led to a better understanding of some of the interesting biological properties of these organisms.

I. Properties of Cyanobacteria

The cyanobacteria are a very large and diverse group of photosynthetic bacteria. They are unique among prokaryotes in that they perform an oxygen-evolving photosynthesis characteristic of plants. Based on this feature and others, the cyanobacteria were classified as algae for nearly 200 years and were commonly referred to as the blue-green algae. As the prokaryotic nature of these organisms became evident, the name cyanobacteria came to replace the term blue-green algae as a means to distinguish the cyanobacteria from the eukaryotic algae. The name cyanobacteria derives from the Greek word for blue (*kyanos*), because many cyanobacteria exhibit a characteristic blue-green color.

The literature on cyanobacteria can be confusing because different taxonomic names for the same strain are sometimes used by different authors. This situation arose because the original classifications of cyanobacteria were developed by botanists using morphological and ecological characteristics of samples from nature. This classification system proved very difficult to use for the identification of pure cultures of cyanobacteria and led in recent years to the increased application of criteria used in the classification of bacteria. Many of the strains discussed in this article are designated by their nomenclature in the Pasteur Culture Collection (PCC), a group of pure reference strains classified using prokaryotic criteria.

A. Habitats and Habits

Cyanobacteria are obligate photoautotrophs, organisms that must obtain energy for cellular metabolism from photosynthesis; thus they are found in nature only where light is available. Within this restriction, cyanobacteria have a wide distribution in freshwater, marine, and terrestrial environments. Cyanobacteria have been called the creatures of "earth, wind, and fire," a reference to their ability to live in very diverse and harsh habitats.

In freshwater environments cyanobacteria are very abundant and geographically widespread. They may be found in freshwater habitats of extreme salinity, temperature, and pH. Lakes with high concentrations of nutrients can promote cyanobacterial blooms that can have serious ecological consequences. Cyanobacteria are also found widely in marine environments and often are abundant in rocky shore areas, intertidal zones, salt marshes, and marine muds. In warm shallow marine bays, cyanobacteria can aggregate to form extensive mats. Although cyanobacteria are generally rather inconspicuous in open marine waters, the red filamentous cyanobacteria of the genus *Trichodesmium* are an exception. These organisms, from which the name Red Sea is derived, can aggregrate and form easily visible bundles near the surface of the water. In terrestrial habitats, cyanobacteria occur primarily on the surface of soils, rocks, and trees and can be found in the extreme temperatures of polar and desert environments.

A number of cyanobacterial species occur in symbiotic association with a wide variety of other organisms. The symbiotic partners include angiosperms, gymnosperms, ferns, bryophytes, and the fungal component of lichens. The biological advantage of these associations to the cyanobacteria is not clear, because many of the relationships are nonobligate and the cyanobacteria can be cultured as free-living. In many associations with photosynthetic organisms, the cyanobacteria are capable of nitrogen fixation and probably provide the partner with fixed nitrogen. Cyanobacteria can also form associations with algae and with other bacteria, although whether or not these relationships are truly symbiotic is not clear.

Cyanobacteria do not have flagellar structures and are incapable of "swimming"; however, they do exhibit two interesting forms of motility. Some cyanobacteria are capable of "gliding," a poorly understood type of movement that occurs when cells are in contact with a solid surface or other cells. The second form of motility is mediated by intracellular structures termed gas vesicles which are common to plankton. Cyanobacteria can regulate buoyancy by means of gas vesicles, thus enabling them to remain in, or move to, water depths where light is most abundant.

B. Morphology

The cyanobacteria exhibit great morphological diversity. They include unicellular, colonial, and filamentous forms with wide variations within these types (Fig. 1). The cyanobacteria range in size from 0.5–1 μm in diameter (typical of other bacteria) up to 60 μm in diameter.

1. Intracellular Structures

The vegetative cells of cyanobacteria have a distinctive structure. The cell wall is a bilayered structure similar to that of other gram-negative bacteria. Outside the cell wall, cyanobacteria are frequently surrounded by mucilaginous sheathes or envelopes that can bind groups of cells or filaments together. The most extensive internal structures viewed by electron microscopy are thylakoids, a series of flattened membrane sacs where photosynthesis takes place. Thylakoids can be found at the periphery of the cells as parallel concentric circles or in convoluted arrangements where they traverse most of the cell. Chlorophyll and carotenoid photosynthetic pigments are localized in the thylakoid membranes. In contrast to most eukaryotic algae and plants that use chlorophyll *a* and *b*, cyanobacteria contain only chlorophyll *a*. The major accessory pigments used for light harvesting are the phycobilins, which are organized in phycobilisomes found on the outer surface of the thylakoid membranes. The phycocyanin class of phycobilins are blue and, together with chlorophyll, give many cyanobacteria a characteristic blue-green color. However, some cyanobacteria also use phycoerythrin, a red phycobilin, and these bacteria appear red or brown in color. Each phycobilin absorbs a specific range of wavelengths, and the composition of phycobilins can vary among cyanobacteria. A subset of the filamentous cyanobacteria can alter their phycobilisome composition to adapt to different wavelenths of light, a process termed chromatic adaptation.

A number of cytoplasmic inclusions are visible by light or electron microscopy in all cyanobacterial vegetative cells. Several of these inclusion bodies represent stored nutrients in the form of insoluble polymers. These include carbon reserves (glycogen granules), nitrogen reserves (cyanophycin granules), and phosphate reserves (polyphosphate granules). Cyanobacteria use glycogen as the primary nonnitrogenous organic reserve material and store it as small granules between thylakoids. A nitrogen storage compound unique to cyanobacteria is cyanopohycin, a polymer of arginine and aspartic acid.

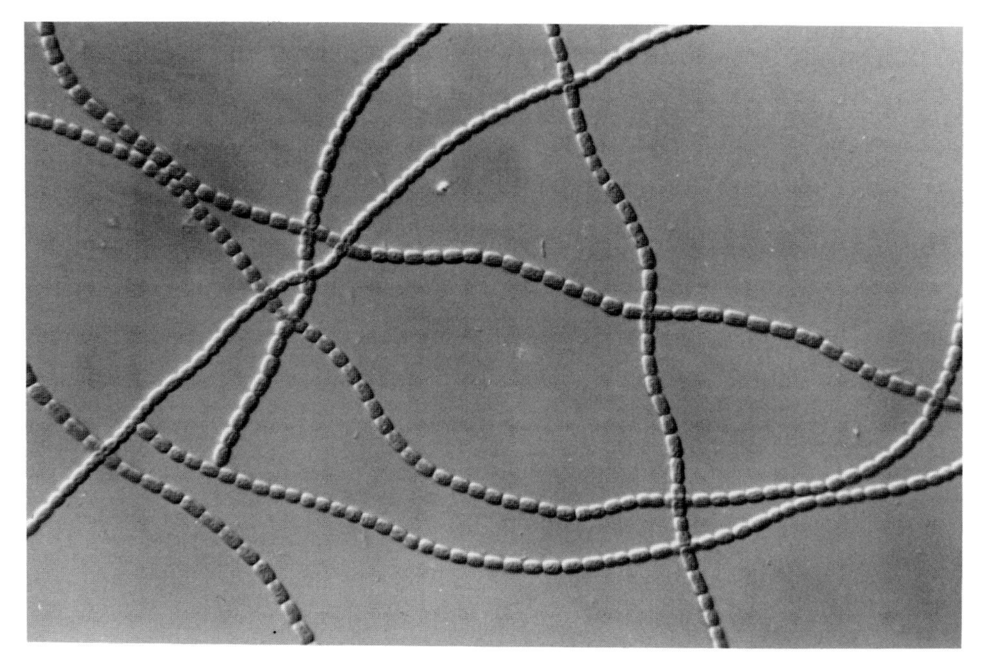

Figure 1 *Anabaena* sp. strain PCC 7120, an example of a filamentous cyanobacterium. The filaments are composed entirely of vegetative cells. Magnification: 400 ×. [Differential interference microscopy. Photograph by the author.]

Cyanophycin polymers can be observed as large granules that accumulate as cultures approach stationary growth. During nitrogen deprivation, cyanophycin as well as phycobilins are broken down and used as sources of nitrogen. A striking form of inclusion body in cyanobacteria is the carboxysome, or polyhedral body. Carboxysomes have a characteristic polyhedral structure and contain compartmentalized ribulosebisphosphate carboxylase, the enzyme responsible for CO_2 fixation in photosynthesis.

Gas vesicles are not found in all cyanobacteria but, as mentioned earlier, are common to planktonic cyanobacteria. These are hollow cylindrical structures that contain a gas-filled space. Cyanobacteria use these structures to regulate buoyancy and, thus, depth in the water column.

2. Specialized Cells

Three types of specialized cells formed by filamentous cyanobacteria in response to environmental conditions are hormogonia, akinetes, and heterocysts. Not all filamentous strains produce hormogonia or heterocysts, and a subset of heterocystous strains produce akinetes. Heterocysts are involved in the fixation of atmospheric nitrogen and are discussed in detail in Section III. Hormongonia and akinetes are cell types associated with dispersal and survival of cyanobacteria. Hormongonia are the major forms of reproduction in certain filamentous cyanobacteria. They consist of short filaments of smaller cells that differentiate and fragment from the trichome. Akinetes are enlarged and thickened single cells analogous to spores that are involved in tolerance to unfavorable conditions. Both akinetes and hormongonia can germinate to produce vegetative filaments.

C. Genomes

Cyanobacterial genomes can be observed by electron microscopy as fibrils in the central portion of the cell, and their structural characteristics have been shown to be similar to those of other prokaryotic genomes. Much of our knowledge of cyanobacterial genomes is derived from biochemical analyses of genome base compositions and genome sizes from over 100 strains representing a large number of genera. As has been observed with other bacterial groups, cyanobacterial genomes thus studied exhibit a wide variation in base compositions from 35 to 71% mol G+C. However, cyanobacterial genomes are much more diverse in size (genetic information) than those of other bacteria. The smallest cyanobacterial genomes (2.4×10^6 bp) fall in the narrow range seen with most other bacteria ($1.5–5.4 \times 10^6$ bp); however, cyanobacterial genomes range in size up to 1.3×10^7 bp, similar to those of eukaryotic fungal genomes. In general, genome sizes correlate with the morphological or physiological complexity of the organisms. In certain strains of both unicellular and filamentous cyanobacteria, the chromosome has been shown to present in multiple copies, the number of which can be affected by growth condition.

Because no natural genetic recombination systems are known to exist for cyanobacteria (see Section II), it has not been possible to use classical genetics to construct the genetic maps that have proven so useful in studies of other bacteria. Recently, the first physical map for a cyanobacterium was constructed using very large genomic DNA fragments and a separation technique known as pulsed-field gradient electrophoresis. In these studies, the chromosome of *Anabaena* sp. strain PCC 7120 was shown to be a circular molecule of 6.3×10^6 bp. The location of a number of genes was placed on the physical map by DNA hybridization techniques, thus providing the first cyanobacterial genetic map. [*See* Map Locations of Bacterial Genes.]

Many cyanobacterial strains contain extrachromosomal replicons or plasmids, and usually these strains carry several different types of plasmids. The copy number of different plasmid types has been shown to vary with growth condition. The plasmids thus characterized exhibit a wide range of sizes, and some are very large ($>1.5 \times 10^5$ bp). At present, no genes or functions have been attributed to any plasmid, and little is known about the exchange of plasmids between strains. Despite this lack of knowledge, plasmids have been important sources for origins of replication used in the construction of experimental gene transfer systems in cyanobacteria (see Section II). [*See* Plasmids.]

A characteristic feature of prokaryotic genomes is the existence of mobile pieces of DNA or transposons. The simplest transposons are insertion sequence elements that encode only their own transposition functions. Insertion elements may be important in bacterial genome evolution as agents that can cause mutations or chromosomal rearrangements that may be deleterious or advantageous. Insertion sequence families have been recently identified and characterized from several strains of filamentous cyanobacteria, and they have been detected in some unicellular strains. In general, the cyanobacterial insertion sequences are similar in size and sequence arrangement to those of other bacterial genomes. Although not yet unexploited for genetic manipulations, the cyanobacterial transposons may at some point prove useful in mutagenesis techniques (see Section II).

II. Genetics of Cyanobacteria

In both prokaryotes and eukaryotes, systems of genetic analysis and manipulation have been very powerful tools in the study of biological processes. For many years, genetic analysis in cyanobacteria was limited by the absence of classical systems for genetic exchange such as conjugation (mating) or transduction. Within the last decade, technologies have been developed that allow the routine transfer of DNA into cyanobacteria. These technologies have opened up new avenues for both creating and analyzing mutations and have greatly increased the facility and sophistication with which cyanobactrial properties of interest can be studied.

A. Gene Transfer

Different strains of cyanobacteria vary in the way in which DNA is most easily introduced into them; however, the basic approaches to transfer are similar with all methods. The DNA of interest is usually transferred into the cyanobacterial cell after being cloned in an *Escherichia coli* recombinant plasmid vector, sometimes referred to as a cargo vector. In addition to origins of replication that allow propagation in *E. coli,* these plasmids contain antibiotic resistance markers that can be used to select for cells that receive the transferred DNA.

For some studies, it is desirable to have the introduced DNA maintained on an independent replicon within the cyanobacterial cell. Examples include experiments in which foreign genes or extra copies of an endogenous gene are introduced. As *E. coli* origins of replication do not function in cyanobacteria, such studies make use of shuttle or biphasic cargo vectors that contain both *E. coli* and cyanobacterial origins of replication. The cyanobacterial origins are derived from endogenous cyanobacterial plasmids.

For other applications such as gene inactivation or replacement studies, integration of the donor DNA into the chromosome is desirable. Cargo plasmids that lack the cyanobacterial origin of replication (suicide vectors) are utilized in these studies. If the plasmid contains DNA with homology to the chromosome, recombination may occur between the donor DNA and the chromosome. This can lead to integration of the entire plasmid from a single recombination event, or exchange of homologous DNA between the donor plasmid and chromosome from a double recombination event. Nonreplicating plasmids that do not integrate are lost from the recipient cell during successive cell divisions.

Many bacteria have evolved restriction-modification systems as defense mechanisms against the introduction of foreign DNA. In these systems, restriction enzymes (sequence-specific endonucleases) are produced that cleave foreign DNA. The host DNA is protected from cleavage by methylation of the sequences recognized by the enzymes produced by the host cell. Many strains of cyanobacteria have been shown to produce one or more restriction enzymes. The existence of these enzymes poses a barrier for DNA transfer, because incoming DNA that has been propagated in *E. coli* is likely to lack appropriate modification against the host restriction enzymes and is, thus, subject to cleavage. As discussed later, this problem is circumvented in some transfer methods by methylation of the DNA prior to transfer.

1. Unicellular Cyanobacteria

Transformation, or the uptake of exogenous DNA, has been an important method of experimental gene transfer in the study of many bacterial species. Transformation of cyanobacteria was first reported in 1970 and is routinely used today. The technique is primarily limited to unicellular strains of the genera *Synechococcus* and *Synechocystis* that exhibit natural competence, the ability to take up exogenous DNA without artificial treatment. For such strains, the transformation procedure is a relatively simple one in which cyanobacterial cells are mixed with the donor DNA. After a period of time to allow uptake and expression of the exogenous DNA, transformed cells are selected by their ability to grow on the antibiotic to which the donor DNA encodes resistance. [*See* GENETIC TRANSFORMATION, MECHANISMS.]

In some unicellular strains, the restriction systems of recipient cells apparently do not interfere with plasmid transformation. In other strains, the efficiency of transformation is greatly reduced as the number of sites in the plasmid recognized by the host's restriction enzymes increases. In one cyanobacterium, this has been circumvented by using a mutated strain in which the restriction enzyme activity is reduced.

2. Filamentous Cyanobacteria

The ease and efficiency with which transformation occurs in naturally competent unicellular cyanobacteria had made these strains of choice for studies of many biological properties including photosynthesis (see Section III). Unfortunately, some biological properties of interest are exhibited only by filamentous cyanobacteria, and, as yet, transformation (either natural or artificial) has not been reliably achieved with these strains.

A major breakthrough in the study of filamentous cyanobacteria was the development of a conjugal system for gene transfer. This procedure is based on the ability of *E. coli* to conjugate with a broad range of gram-negative bacteria including cyanobacteria. In this system, the DNA of interest is cloned on a cargo vector containing sequences that allow conjugal transfer from *E. coli*. The cargo vector may be replicating or nonreplicating in cyanobacteria depending on the application. Other necessary plasmids are those that encode DNA transfer (helper plasmid) and conjugal structures (conjugal plasmid). *Escherichia coli* cells bearing the three plasmid types are mated with cyanobacterial cells, during which the cargo vector is transferred into the recipient cyanobacterial cells. As with transformation, cells that received the plasmid (exconjugants) are selected by their ability to grow on the antibiotic to which the plasmid encodes resistance. [*See* CONJUGATION, GENETICS.]

Cleavage of transferred DNA by restriction enzymes does pose a major problem in the commonly studied filamentous cyanobacteria. To circumvent this, the conjugal system includes procedures for modifying the cargo plasmid in *E. coli*. This is accomplished through genes on the helper plasmid for methylases that modify restriction sites cleaved by the host restriction enzymes.

An alternative form of gene transfer that has proven useful in filamentous cyanobacteria is electroporation. In this method, purified DNA is moved into cells via an electrical current that reversibly permeabilizes the cell membrane. As with conjugal transfer, the efficiency of transfer is increased by methylation of the DNA prior to introduction. Both conjugal transfer and electroporation are methods that are also applicable to unicellular strains.

B. Mutagenesis

Mutants impaired in particular cellular functions or developmental pathways have been important tools in the analysis of both eukaryotes and prokaryotes. Cyanobacterial mutants may appear spontaneously or can be generated experimentally by either classical or gene transfer techniques.

1. Classical Mutagenesis

Many classical mutagens have been successfully employed with cyanobacteria, although strains vary greatly in the degree to which different mutagens are effective. Both ultraviolet irradiation and a spectrum of chemical mutagens have been used to create random mutations. These techniques have successfully been applied to the isolation of antibiotic-resistant, temperature-sensitive, and auxotrophic mutants as well as mutants impaired in photosynthesis, nitrogen fixation, and cellular differentiation.

2. Gene Transfer-Based Mutagenesis

The ability to transfer DNA opens alternative avenues for creating mutations. Transposon tagging provides a means of creating random mutations and then isolating the mutated genes. The use of this technique to study heterocyst development is discussed in Section III. Mutations can also be made in a directed manner to modify specific genes of interest. These methods are most often used to inactivate or alter genes and determine the effect on a specific function or pathway. This has been applied very successfully to the study of photosynthesis in cyanobacteria (see Section III).

a. Random Mutagenesis by Transposon Tagging

Transposon-based mutagenesis has been widely used to create random mutations in both eukaryotic and prokaryotic systems. The technique is based on the ability of a transposon to move from one segment of DNA to another and cause a null mutation when a recipient gene is interrupted. Organisms that carry random transposon insertions can be screened for phenotypes that result from mutations in the function of interest. The mutated genes can subsequently be identified and isolated through the transposon "tag" that it carries.

Transposon mutagenesis was first reported in a unicellular cyanobacterium a decade ago, but the procedure was not widely used because of the low frequency of transposition. Recently the *E. coli* transposon Tn*5* and derivatives of it have been used successfully for the mutagenesis of filamentous cyanobacteria. Tn*5*, which is not native to cyanobacteria, contains a gene encoding resistance to the antibiotic kanamycin. In the mutagenesis procedure, Tn*5* on a suicide vector is transferred into cyanobacterial cells by conjugation. Because the plasmid cannot replicate, kanamycin-resistant exconjugants derive from cells in which the Tn*5* has jumped from the suicide vector into the genome. Insertion of Tn*5* into the genome should be random such that different cells will carry Tn*5* in different chromosomal locations.

b. Directed Mutations

i. Inactivation of Specific Genes Null mutations can also be created in a directed fashion, by targeting to specific genes that have been previously cloned. The general strategy is to replace a wild-type gene with one that is mutated to render it inactive. This technique has been applied most widely in the transformable cyanobacteria but has also been successfully used with filamentous cyanobacteria. The cloned gene under study is modified *in vitro* such that it is interrupted with an antibiotic resistance marker, sometimes referred to as an inactivation cassette. The inactive gene is transferred into cyanobacterial cells on a suicide cargo vector. Cyanobacterial cells that are selected for antibiotic resistance have the resistance marker in the chromosome from recombination between the plasmid and chromosomal DNA. In the unicellular cyanobacterial strains, double cross-overs between the plasmid and chromosome leading to gene replacements occur at a higher frequency than single cross-overs in which the entire plasmid becomes integrated. In filamentous cyanobacteria, double cross-overs occur at a lower frequency than single cross-overs. To facilitate experiments in which double cross-overs are desirable, a system for selection of double cross-overs has recently been described.

ii. Site-Directed Mutagenesis Site-directed mutagenesis techniques are used to change specific sequences in genes; the method is similar to gene inactivation. Specific nucleotide changes are made in a cloned gene using recombinant DNA techniques *in vitro*. The altered gene is then introduced into cyanobacterial cells on a suicide cargo plasmid. Several methods of selection for replacement with the mutated gene are available depending on the particular experiment. This technique has been used routinely in the transformable strains but has not been applied to any degree of filamentous strains.

III. The Use of Molecular Genetics in Studies of Cyanobacteria

Molecular genetics can be defined as the study of molecular processes underlying gene structure and function. The advent of recombinant DNA technology and subsequent technologies made possible the cloning of genes and detailed analyses of gene structure and function. These technologies coupled with gene transfer systems provide a means to study the molecular basis of many interesting biological properties of cyanobacteria. This section describes the application of molecular genetics to the study of photosynthesis, nitrogen fixation, and evolution in cyanobacteria.

A. Photosynthesis

Photosynthesis, the process by which light energy is converted to chemical energy, can be considered the most important biochemical reaction on earth because it has made possible the existance of life as we know it. The oxygenic photosynthesis performed by plants and cyanobacteria differs from that of other photosynthetic bacteria in using two photosystems with water as an electron donor. The reactions in which light energy is captured and converted occur in the thylakoid membranes. There are four major multisubunit complexes in thylakoids: photosystem II (PSII), the cytochrome f/b_6 complex, photosystem I (PSI), and the adenosine triphosphate (ATP) synthase. The complexes are functionally connected by small electron transport proteins. The absorption of light by PSII catalyzes the removal of electrons from water, which results in oxygen evolution. The energized electrons move through an electron transport chain to the cytochrome f/b_6 complex and then to PSI, where they are used to produce reduced nicotinamide adenine dinucleofide phosphate (NADPH). As a result of the electron transfer, a proton gradient forms across the membrane, and this gradient is used in the generation of ATP by the ATP synthase. The products of the reactions in the thylakoid membranes, ATP and NADPH, are fed into the CO_2 fixation cycle, in which complex carbon compounds are produced.

1. The Use of Cyanobacteria in Studies of Oxygenic Photosynthesis

Plant growth and productivity are tightly linked to photosynthetic capability and efficiency. The alteration of gene-encoding proteins involved in photosynthesis represents one approach for the genetic engineering of crop plants for increased productivity. This approach requires a detailed understanding of photosynthesis and the ability to manipulate photosynthesis genes.

In the plant cell, photosynthesis takes place in chloroplasts. These organelles have small genomes that encode a limited number of proteins. The majority of chloroplast proteins are encoded by the nuclear genome of the plant, translated in the cytoplasm and then transported into the chloroplast. Many of the multisubunit complexes found in the chloroplast are composed of proteins encoded by both the nuclear and chloroplast genomes. The formation of these complexes thus requires coordinate expression of these gene products and assembly within the chloroplast.

As discussed earlier, the ability to generate specific mutants in genes encoding particular functions has been very important in understanding such functions. The difficulty with probing photosynthesis in this manner in plants is that most of the photosynthetic complexes are composed of subunits encoded by both the nuclear and chloroplast genomes. Although it is possible to transform the nuclear DNA of some plants, until very recently it has not been possible to transform chloroplasts. Cyanobacteria provide attractive alternatives to plants in the study of photosynthesis because this process is encoded by one small genome and certain strains are relatively easy to manipulate genetically. An additional and essential feature is the availability of strains that can grow in the absence of photosynthetic activity because the introduction of mutations in photosynthesis can be lethal to cells dependent on this process. Given the similarity of photosynthesis between plants and cyanobacteria, information gained about photosynthesis from cyanobacteria should be applicable to photosynthesis in plants. [*See* PHOTOSYNTHESIS AND CHLOROPLASTS.]

2. Isolation of Cyanobacterial Photosynthesis Genes

Photosynthesis genes have been isolated by a variety of molecular genetics techniques commonly employed in the isolation of genes from other biological systems. These include the use of heterologous gene probes, oligonucleotide probes, and screening of expression libraries with antibody probes.

The first genes were isolated by taking advantage of the evolutionary relationship between cyanobacteria and chloroplasts. Several photosynthesis genes from chloroplasts were cloned and characterized before the homologous genes were isolated from cyanobacteria. Because some chloroplast gene sequences share a high degree of similarity with cyanobacterial genes, they were successfully used as heterologous probes to isolate analogous genes from recombinant libraries of cyanobacterial DNA. Many genes in cyanobacteria are arranged in operons, and a large number of genes have been identified by close linkage to previously characterized photosynthesis genes. In some cases, these were in turn used to isolate plant photosynthesis genes that displayed a different linkage arrangement in the chloroplast genome.

Oligonucleotide probes have been used extensively in the isolation of genes in cases where a portion of the amino acid sequence of the product was known. In this method, the protein sequence is used to generate a corresponding DNA sequence that is then synthesized as an oligonucleotide and used as a probe to screen a recombinant library of cyanobacterial DNA. If two oligonucleotides can be generated that are separated by a short sequence (0.1–2 kb), they can be used to amplify sequences of the gene using the polymerase chain reaction.

Immunological screening of expression libraries has also been applied to the isolation of cyanobacterial photosynthesis genes. In this method, libraries in which the cloned cyanobacterial DNA is expressed as protein products in *E. coli* are screened with antibodies to the gene product of interest. Clones that bind the antibody carry a DNA segment that encodes either part or all of the target gene.

Currently genes are available for many components of the major photosynthetic complexes and pathways including phycobilisomes, PSI, PSII, the ATP synthase, the cytochrome f/b_6 complex, electron transport, and CO_2 fixation. The sequences of these genes have provided a great deal of information about the structure of the gene products and provided a data base for studies on photosynthetic functions and cyanobacterial evolution (see Section III. C).

3. The Use of Mutagenesis to Probe Photosynthesis in Cyanobacteria

The availability of cloned photosynthesis genes and cyanobacterial gene transfer systems has allowed studies on the structure and function of photosynthetic components by directed mutagenesis techniques. In such studies, genes are inactivated or altered, and the effect on photosynthetic activity and structure of the photosynthetic complex are assessed. This approach has been applied most extensively to studies of PSII and associated phycobilisomes because much is known about their structure and function. Phycobilisomes are easily dissociated from the membrane and the phycobilin proteins are water-soluble; consequently, they have been characterized in detail from a number of perspectives. PSII is functionally and structually analgous to the photosynthetic reaction center of purple bacteria, for which a crystal structure has been solved. In addition, transformable strains are available that are capable of growth without PSII activity. Thus a great deal of structural information is available on which to shape mutagenesis strategies, and appropriate strains exist in which to perform the experiments. Studies on PSI have not been as extensive because less is known about this complex and some of the strains used for PSII mutagenesis are not appropriate for PSI mutagenesis. In recent years, rapid progress has been made in the cloning of PSI genes and in the understanding of PSI subunit composition and structure. This together with the use of alternative strains for PSI mutagenesis now make feasible PSI studies of the type described in the following section for PSII.

4. Studies on Photosystem II

PSII of cyanobacteria consists of a membrane intrinsic core involved in electron transfer, and a group of peripheral proteins that form a water-splitting complex extrinsic to the membrane in the intrathylakoid space. Light is funneled to PSII by phycobilisomes that are present on the membrane surface. The PSII core is composed of at least six proteins. Two proteins termed CP43 and CP47 bind chlorophyll, and two others, D1 and D2, form the binding region for reaction center components. The best-characterized protein of the water-splitting complex (MSP) is involved in stabilization of the water-splitting complex and the PSII core.

The functions of the individual components of PSII have been examined in a number of mutagenesis experiments. These include studies in which specific genes have been inactivated, mutated at specific sites, or substituted with genes from other species. The cyanobacteria in which these experiments were performed, *Synechococcus* sp. strain PCC 7002 and *Synechocystis* sp. strain PCC 6803, are capable of growth of carbon sources in the light when PSII is inactive.

Several studies have been conducted on cyanobacterial strains in which genes for PSII core proteins and the MSP have been inactivated. The general conclusions from these studies are that inactivation of any of the PSII core proteins or MSP causes a loss in ability to grow photoautotrophically. Analysis of thylakoid proteins from mutants in which the gene for one PSII core protein had been inactivated showed that there were decreased levels of other PSII proteins. Recent studies have shown that the other proteins are synthesized in a relatively normal fashion but do not accumulate to normal levels in the membrane. Thus, the presence of each of the PSII core proteins and MSP is probably required for the assembly and stability of the PSII complex. Currently a number of laboratories are using site-directed mutagenesis to alter specific amino acids within different PSII proteins. Analysis of the affect of the mutations on photosynthesis should provide considerable information on the role of specific amino acids residues in PSII structure and function.

Most PSII proteins are highly conserved between plants and cyanobacteria. Interspecific gene replacement studies have been conducted to test the ability of plant chloroplast proteins to substitute for cyanobacterial proteins. The D1, D2, and CP43 genes have been replaced with the corresponding genes from plants. Cyanobacterial strains that have plant D1 substitutions assemble a functional PSII core complex and exhibit photoautrophic growth, although at a slightly lower rate than strains without the substitution. However, strains with plant substitutions of the D2 and CP43 genes are nonphotosynthetic. These results suggest that the plant genes are not indiscriminately interchangable with their cyanobacterial counterparts. Experiments have also been conducted with hybrid genes in which one part of the gene is derived from a cyanobacterial gene and the complementary portion is derived from a plant gene. Strains bearing a CP47 hybrid gene were able to grow photoautotrophically, although at lower rates than strains with the native cyanobacterial gene. Experiments of this type should allow the identification of regions of the PSII proteins that are important for function specifically in cyanobacteria.

Although not detailed here, the types of experiments described for PSII including gene interruptions, interspecific gene substitutions, and site-directed mutagenesis have also been applied to the study of phycobilosomes. These studies have led to considerable information on the role of individual subunits of the phycobilisome in the function and structure of the complex as well as information on how chromatic adaptation is mediated.

B. Nitrogen Fixation

Nitrogen is an element to all forms of life, but for nitrogen to be utilized by most biological systems it must be "fixed" or combined with other elements such as oxygen or hydrogen. Biological nitrogen fixation is confined to a few diverse genera of bacteria. Nitrogen-fixing bacteria use the enzyme nitrogenase to convert dinitrogen gas to ammonia. Because nitrogenase activity is extremely sensitive to oxygen, an anaerobic environment for nitrogenase must be provided. In addition, nitrogen fixation is an energy-intensive process, and large amounts of energy must be available for the fixation reaction. [*See* BIOLOGICAL NITROGEN FIXATION.]

The machinery and biochemical reactions of nitrogen fixation are highly conserved among all nitrogen-fixing organisms. The greatest variations in adaptations for nitrogen fixation occur in the manner in which the requirement for anaerobiosis is met. This requirement is easily fulfilled for obligate anaerobes such as *Clostridium pasteurianum*. Other nitrogen-fixing bacteria have mechanisms not only to provide an anaerobic environment for nitrogenase activity but, in some cases, mechanisms for the protection of nitrogenase upon exposure of oxygen. Among free-living bacteria, these mechanisms include faculative anaerobic growth (ex: *Klebsiella pneumoniae, Rhodospirillum rubrum*), a high respiration rate (ex: *Azotobacter vinelandii*), and protection of nitrogenase by complexing with special proteins (ex: *Azotobacter vinelandii*). Bacteria of the genus *Rhizobium* fix nitrogen through a symbiotic relationship. *Rhizobium* and specific host plants participate in the formation of a nodule on the plant root within which nitrogen fixation occurs. Within the nodule, an anaerobic environment is produced through proteins that scavenge oxygen. [*See* ANAEROBIC RESPIRATION.]

A subset of cyanobacteria that includes both unicellur and filamentous forms are also capable of nitrogen fixation. These cyanobacteria face a special problem in providing an anaerobic environment for nitrogenase. The energy for nitrogen fixation is derived from photosynthesis, but because this process results in oxygen evolution, it is incompatible with nitrogen fixation. Cyanobacteria that fix nitrogen aerobically circumvent this problem by several interesting mechanisms. In the marine planktonic genus *Trichodesmuium*, nitrogen fixation depends on the formation of photoactive aggregates of filaments in which internal O_2-reduced microzones are present. The ability to fix nitrogen is thus related to the sea state, with becalmed waters favorable for aggregate formation. At present, the spatial organization of photosynthesis and nitrogen fixation within the *Trichodesmium* aggregrate or individual filaments is not known. In some unicellular marine genera such as *Gloethece*, nitrogen fixation and photosynthesis are temporally separated. These strains limit nitrogen fixation to periods of dark, while photosynthesis occus in the light. The best-studied mechanism of facilitating nitrogen fixation is a spatial separation of photosynthesis and nitrogen fixation in different cell types. Certain filamentous cyanobacteria can generate differentiated cells called heterocysts that fix nitrogen while the neighboring vegetative cells of the filament perform photosynthesis.

1. Heterocyst Development

Nitrogen starvation of a subset of the filamentous cyanobacteria induces the differentiation of heterocysts at regular intervals of about every tenth cell along the vegetative filament (Fig. 2). The differentiation of heterocysts from vegetative cells is a 24–36-hr process that involves many changes that facilitate nitrogen fixation. These changes include the induction of nitrogenase, the elimination of major sources of oxygen, and alterations that support the energy requirements of nitrogen fixation. Early in heterocyst development, several new envelope layers are synthesized that reduce the entry of oxygen and other gases from outside the cell. Within the cell, the oxygen-evolving portion of photosynthesis (PSII) is inactivated. The enzyme nitrogenase is induced exclusively within heterocysts late in the developmental pathway. In the nitrogen fixation reactions, ATP and electrons donated from NADPH are used by the nitrogenase enzyme to convert nitrogen gas to ammonia. ATP is synthesized in the heterocyst through PSI activity, and NADPH is provided by the metabolism of carbon compounds produced through photosynthesis and transported into the heterocyst from vegetative cells. In turn, fixed nitrogen produced by heterocysts is transported to neighboring vegetative cells. The heterocyst is a terminally differentiated cell and does not divide. As vegetative cells of the filament divide, new heterocysts are formed between preexisting ones at the appropriate spacing.

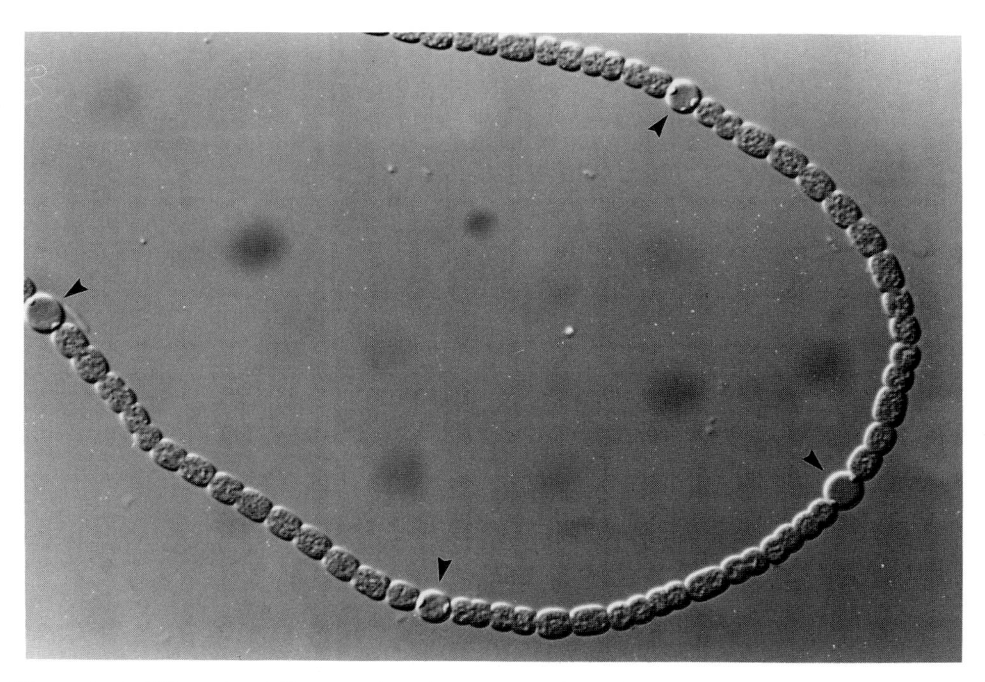

Figure 2 A differentiated filament of *Anabaena* sp. strain PCC 7120. Heterocysts are marked by arrowheads. The refractile granules at the ends of the heterocysts are cyanophycin plugs. Magnification: 1000 ×. [Differential interference constrast microscopy. Photographs by the author.]

2. The Isolation and Characterization of Genes Involved in Nitrogen Fixation in Heterocystous Cyanobacteria

In heterocystous cyanobacteria, heterocyst development is a prerequisite for nitrogen fixation. Thus there are two types of genes involved in nitrogen fixation: (1) genes required for the process of nitrogen fixation and (2) genes required for heterocyst differentiation.

a. Nitrogen-Fixation Genes

Genes required for nitrogen fixation (*nif*) genes were first characterized in *K. pneumoniae*. At least 19 *nif* genes in *K. pneumoniae* are arranged in eight contiguous operons. The *nif*H, D, and K genes encode polypeptides of the nitrogenase complex. Other *nif* genes encode regulatory functions, electron transport to nitrogenase, the maturation of certain gene products, and the synthesis of the molybdenum-iron cofactor of nitrogenase.

Because nitrogen fixation is highly conserved among nitrogen-fixing bacteria, *nif* genes from one organism can be used as heterologous probes to isolate analogous genes from other organisms. Using this strategy, the *nif*H, D, K, and S genes from *K. pneumoniae* were used to isolate the corresponding genes from *Anabaena* sp. strain PCC 7120 (*Anabaena* 7120). These were the first *nif* genes isolated from cyanobacteria, and their characterization revealed a puzzling gene arrangement. In *K. pneumoniae,* the *nif*H, D, and K genes are tightly linked and form an operon. While the *nif*H and D genes were found next to each other in *Anabaena* 7120, the *nif*K gene mapped 11 kb away. Also, *nif*S mapped next to *nif*H, a different arrangement than that in *K. pneumoniae*. In subsequent experiments with DNA from purified heterocysts, two DNA rearrangements were shown to occur in the *nif* region during heterocyst development. In each rearrangement, a specific DNA segment is excised from the vegetative cell DNA by site-specific recombination.

One of the rearrangements occurs in the *nif*D region. In heterocyst DNA, the *nif*H, D, and K genes are contiguous, resulting from the excision of an 11-kb element that interrupts the end of *nif*D. The excision is mediated by a protein encoded by the *xis*A gene at one end of the 11-kb element. Excision occurs by recombination between an 11-bp sequence at each end of the element. The *nif*D rearrangement is first detected about 18 hr into development, after envelope synthesis is complete and the heterocyst is believed to be internally anaerobic. The rearrangement results in the formation of a *nif*-HDK operon as in *K. pneumoniae*. In experiments designed to examine the role of *xis*A in heterocyst development, the wild-type *xis*A gene of *Anabaena* 7120 was replaced with a mutated gene carrying an inactivation cassette. The *nif*D rearrangement and nitrogen fixation were blocked in the *xis*A mutant, but heterocyst development was normal. This experiment demonstrated that rearrangement is required for nitrogen fixation but that heterocyst development is an independent process.

The second characterized rearrangement in *Anabaena* 7120 involves a 55-kb element that interrupts a gene called *fdx*N that maps next to *nif*S. Recombination also occurs through repeated sequences at the element ends; however, the repeats are different than those used in the *nif*D rearrangement. The *fdx*N rearrangement is not mediated by *xis*A was as shown by normal excision of the 55-kb element in the strain with the *xis*A inactivation. Recently the *fdx*N excision has been shown to occur via a second recombinanse encoded by a gene (*xis*F) within the 55-kb element. Excision of the 55-kb element forms an operon between *fdx*N and three *nif* genes.

Surveys of many other nitrogen-fixing cyanobacteria have revealed that in general the unicellular cyanobacteria and filamentous nonheterocystous cyanbacteria have contiguous *nif*H, D, and K genes. It was originally thought that the gene rearrangements observed in *Anabaena* 7120 represented part of a commitment to the differentiated state of the heterocyst, because the rearrangements are irreversible. However, heterocystous strains of cyanobacteria exist that lack the 11-kb and 55-kb elements and, thus, do not undergo rearrangment. The absence of these elements in some heterocystous cyanobacteria and the *xis*A inactivation experiments in *Anabaena* 7120 demonstrate that the element rearrangements are not required for heterocyst development. A current idea for the origin of the 11-kb and 55-kb elements in *Anabaena* is that they represent the vestiges of viral genomes that integrated into the *nif* region of an ancestral heterocystous cyanobacterial strain.

Many of the other *nif* genes of *Anabaena* 7120 have been identified by linkage to the *nif*K and S genes in heterocyst DNA. The arrangement of characterized *Anabaena nif* genes differs from that of *K. pneumoniae* in at least two respects. Some genes such as *nif*S are in different arrangements in the two bacteria. Also, the *Anabaena nif* region contains genes that are absent in *K. pneumoniae*. An example is the *fdx*N gene, which encodes a ferredoxin that possibly participates in electron transfer to nitrogenase. There are also unidentified genes in the *Anabaena nif* region that are absent in *K. pneumoniae*. At least one of these has been shown to be required for nitrogen fixation by gene inactivation experiments.

b. Heterocyst Development Genes

Heterologous *nif* genes provided a powerful means of isolating analogous genes from cyanobacteria. Because heterocyst development is unique to cyanobacteria and heterologous probes from other bacteria are not available, genetic strategies have been exclusively employed in the isolation of heterocyst-specific genes. These strategies make use of mutants that are impaired in the ability to fix nitrogen. Such mutants carry mutations either in heterocyst formation or in nitrogen fixation. The former mutants can be distinguished by their ability to fix nitrogen under anaerobic conditions. Nitrogen-fixation mutants of either type can be maintained on growth medium that contains fixed nitrogen.

Mutants in nitrogen fixation induced by chemical mutagens or ultraviolet irradiation were first used to isolate heterocyst-specific genes from *Anabaena* 7120. The genes were identified by the complementation of mutants with libraries of *Anabaena* DNA sequences cloned on plasmid vectors. The libraries were conjugated into the mutant strains, and cells that received DNA containing a wild-type copy of the mutated gene were identified by the ability to grow without fixed nitrogen. The complementing clones were then isolated and the gene of interest narrowed down by complementation studies with progressively smaller fragments of DNA. The best-characterized genes isolated by these methods are *het*A and R. The *het*A gene was identified by its ability to complement a mutant deficient in the synthesis of one heterocyst envelope layer. The mutant is impaired in nitrogen fixation because the defective envelopes apparently do not prevent oxygen from entering the heterocyst. Expression of the *het*A gene is first detected at about 7 hr of development, consistent with a role for *het*A in envelope synthesis. The *het*R gene was identified by its ability to complement a mutant that fails to differentiate heterocysts. A low level of *het*R expression in vegetative cells increases by 6 hr after nitrogen starvation. Wild-type cells containing wild-type *het*R on a plasmid show increased heterocyst frequency, even in the presence of media containing fixed nitrogen. These results suggest that *het*R is a regulator of heterocyst development. Many other heterocyst-specific genes have been isolated by mutant complementation and are currently under study.

An alternative genetic strategy that shows great potential for the isolation of heterocyst-specific genes is tagging by transposon Tn*5*. As discussed earlier, genes that carry Tn*5* insertions can be readily identified and isolated by means of the transposon tag. The isolated Tn*5*-bearing gene can in turn be used to isolate a wild-type copy of the gene from a recombinant library of wild-type DNA sequences. A sophisticated derivative of Tn*5* has recently been described and used to isolate genes that are activated within several hours of nitrogen starvation. Characterization of these genes should provide insight into the early steps of heterocyst differentiation.

c. Evolution

The cyanobacterial lineage extends far back in the history of the earth. Cyanobacteria probably appeared first in the Late Archean Age, 2.8–2.5 billion years (Ga) ago, a time when the earth's atmosphere was anaerobic. Evidence from this period indicates cyanobacteria in stromatolites, macroscopic structures of calcium carbonate that are remnants of microbial communities. Stromatolites closely resemble intertidal deposits that form today from the aggregation and precipation of sediment by communities of cyanobacteria and other bacteria in certain regions of the world. There is a very plentiful fossil record of cyanobacteria from the Proterozoic Era, which lasted about 2 billion years (2.5–0.5 Ga ago). During this era, cyanobacteria occupied vast areas of earth and inhabited a broad spectrum of soil and aquatic environments. Evidence of their abundance can be found in the remnants of extensive stromatilic reefs formed by cyanobacterial communities in the earth's oceans.

Cyanobacteria are credited with the conversion of the earth's atmosphere from anaerobic to aerobic by the release of oxygen derived from photosynthesis. This conversion led to the appearance of obligate aerobes about 1.7 Ga ago and their evolution into higher organisms of today, all of which require oxygen. Later in the Proterozoic Era (by 0.6 Ga ago), multicellular organisms, or metazoans, appeared on earth. The florishing metazoan population of the Late Proterzoic period was probably supported by the cyanobacteria-dominated microbial mats that gave rise to stromatolites. The ecological importance of cyanobacteria had diminished by the end of the Proterozoic, perhaps due to the destruction of stromatolite habitats by metazoans.

The fossil and geologic records suggest that the cyanobacteria were the first organisms on earth capable of performing an oxygenic form of photosynthesis. Observations on the prokaryotic nature of plant chloroplasts and the similarity in photosynthesis between cyanobacteria and chloroplasts led in part to symbiotic theories of eukaryotic cell evolution put forth in the early twentieth century. The endosymbiotic theory proposes that the chloroplasts and mitochondria of modern eukaryotic cells derived from bacteria that came to reside symbiotically within a primitive eukaryotic cell.

Since the advent of molecular biology, a wealth of molecular sequence information has accumulated supporting the endosymbiotic theory, and this hypothesis is now widely favored. It is generally accepted that chloroplasts evolved from endosymbiotic events between primitive eukaryotes and ancestral photosynthetic prokaryotes such as cyanobacteria. This scenario fits well with the origin of chloroplasts of eukaryotic red algae, which like cyanobacteria contain chlorophyll *a* and phycobilin pigments. However, the chloroplasts of green algae and higher plants lack phycobilins and contain both chlorophyll *a* and *b*. Since the discovery in the 1960s of prochlorophytes, photosynthetic prokaryotes that perform oxygenic photosynthesis using chlorophyll *a* and *b*, it has been suggested that prochlorophytes are the evolutionary intermediates between cyanobacteria and green chloroplasts.

A number of studies that address the role of prochlorophytes in chloroplast evolution have been conducted in the last few years. In these studies, specific DNA sequences from several genera of cyanobacteria, three species of prochlorophytes, and chloroplasts from several genera were compared. The genes studied included those for ribosomal RNA and two different proteins. The degree of similarity of these organisms was calculated from the sequence data, and evolutionary trees were constructed. The interesting conclusions from all of these studies are that none of the identified prochlorophyte species is directly related to chloroplasts and that the prochlorophytes represent a highly diverged group that has evolved from more than one ancestor. These findings suggest that the use of chlorophyll *b* as a light-harvesting pigment has developed independently several times in evolution, leading to the green chloroplast lineage and each of the prochlorophyte lineages.

Bibliography

Alam, J., Vrba, J. M., Cai, Y., Martin, J. A., Weislo, L. J., and Curtis, S. E. (1991). *J. Bacteriol.* **173,** 5778–5783.

Bancroft, I., Wolk, C. P., and Oren, E. V. (1989). *J. Bacteriol.* **171,** 5940–5948.

Bryant, D. (1987). *Can. Bull. Fish. Aquat. Sci.* **214,** 423–500.

Bryant, D. A., and Tandeau de Marsac, N. (1988). *Methods Enzymol.* **167,** 755–766.

Buikema, W. J., and Haselkorn, R. (1991). *Genes Dev.* **5,** 321–330.

Elhai, J., and Wolk, C. P. (1988). *Methods Enzymol.* **167,** 747–755.

Golden, S. S. (1988). *Methods Enzymol.* **167,** 714–728.

Haselkorn, R. (1989). Excision of elements interrupting nitrogen fixation operons in cyanobacteria. *In* "Mobile DNA" (D. Berg and M. Howe, eds.), pp. 735–742. American Society for Microbiology, Washington, D.C.

Palenik, B., and Haselkorn, R. (1992). *Nature* (*London*) **355,** 265–267.

Porter, R. (1986). *CRC Crit. Rev. Microbiol.* **13,** 111–132.

Williams, J. G. K. (1988). *Methods Enzymol.* **167,** 766–779.

Wolk, C. P., Cai, Y., and Panoff, J.-M. (1991). *Proc. Natl. Acad. Sci. USA* **88,** 5355–5359.

Cytokines in Bacterial and Parasitic Diseases

Dennis L. Stevens
Veterans Affairs Medical Center, and University of Washington

Glossary

Autocrine action of a cytokine Cytokine produced and secreted by a specific cell regulates the function of the same cell

Colony-stimulating factor Cytokines produced by lymphocytes and mononuclear phagocytes that stimulate the growth and differentiation of immature leukocytes in bone marrow

Cytokine Intercellular regulatory protein produced by cells of the immune system that are induced by specific stimuli and that enhance, or inhibit, other effector cells

Endocrine function of a cytokine Cytokine produced by a cell affects the function of distant cells, which are reached through cardiovascular circulation

Interleukin Cytokine produced by leukocytes that acts on other leukocytes

Lymphokine Effector cytokine produced by activated lymphocytes

Monokine Effector cytokine produced by mononuclear phagocytes

Paracrine action of a cytokine Cytokine produced by a cell affects the function of adjacent or nearby cells

Pleiotropism Property of a given cytokine whereby it may have effector functions on a variety of cells

CYTOKINES are intercellular regulatory proteins that mediate a multiplicity of immunologic as well as nonimmunologic biological functions. These proteins are induced by specific stimuli to modify and enhance the function of effector cells and are regulated by feedback inhibition or by the production of other cytokines. Under normal conditions, minute amounts of the individual cytokine are sufficient to mediate the desired response. The inability to generate specific cytokines in experimental or natural disease states greatly alters the course of infection, usually to the detriment of the host. In other conditions, massive release of specific cytokines induced by infectious agents may induce shock, multiorgan failure, and death. At the present time, modern technological advance have provided recombinant cytokines as well as potent neutralizing antibodies against each cytokine. Thus, physicians and researchers have at their disposal promising new agents to treat infectious diseases. However, the cascade of cytokine elaboration creates a dilemma for clinicians since selection of an appropriate therapy will depend on the ability to rapidly determine the cytokine-associated stage of the disease process.

Despite the diverse nature of cytokines, they all share the following properties:

1. Cytokines are produced during acute inflammation and serve to mediate and regulate immune and inflammatory responses.
2. Cytokine secretion requires production of new messenger RNA (mRNA) and protein synthesis.
3. Cytokines of the same type may be synthesized and secreted by a variety of cells.

4. Cytokines of a specific type may affect the function of a variety of cells types. This property is referred to as pleiotropism.
5. Cytokine actions are redundant. The function of a given cytokine may be identical to that of a different cytokine.
6. Cytokines may facilitate the production of a second cytokine.
7. Cytokines may have similar actions upon a target cell such that the effect of different cytokines acting together on a cell are either additive or synergistic.
8. Cytokines effect other cells through their ability to bind to specific receptors. The specific receptor functions can be either autocrine, paracrine, or endocrine.

This article will describe the different types of cytokines, the characteristics and known functions of individual cytokines, the role that various cytokines play in various disease states, and, finally, the interaction of specific cytokines in the amplification of the disease process.

I. Introduction

Striking advances in our understanding of inflammation have occurred in the last century. Prior to 1850, new information in the field of medicine was largely descriptive and based on visual examination of signs of disease in patients as well as anatomical drawings of postmortem examinations. The discovery of the microscope opened the eyes of scientists and physicians to the histology of tissue, detection of microbes, parasites and fungi, and primitive descriptions of the cell in general. The age of cell function was next and the structure, function, and biochemistry of cells and subcellular organelles was extensively studied and cataloged. In the last 10 years, the discovery of cytokines has launched intensive research into the unraveling of the means by which cells communicate and control one another.

I. Interleukin-10

Recently, IL10 (cytokine synthesis inhibitory factor) has been shown to inhibit mononuclear cell synthesis of TNF, IL1, and IL6 at the transcriptional level. Interleukin-10 synthesis is not constitutive and is induced in mononuclear cells within 7 to 8 hr with maximum production at 24 hr by stimuli such as LPS. Interleukin-10 synthesis, in turn, in suppressed by IL4. Interleukin-10 has both autocrine and paracrine functions and affects mononuclear cell function in a variety of ways. First, IL10 down regulates expression of Class II MHC expression on macrophages and may further alter T-cell responses through inhibition of TNFα, a known T-cell growth and differentiation factor. Similarly, IL10 inhibited macrophage production of reactive oxygen intermediates but not reactive nitrogen intermediates. The latter are lethal to intracellular parasites, and are also potent suppressors of T-lymphocyte functions.

III. Specific Cytokines

A. Interleukin-1

Interleukin-1 (IL1) is secreted by a variety of cells including monocyte–macrophage, natural killer (NK) cells, Langerhans cells, endothelial cells, neutrophils, fibroblasts, and adult T-cell leukemia cells. Two forms of IL1 are found in such cells and each is encoded by a separate gene located on the long arm of chromosome 2. There is a 26% homology between amino acid sequences and 45% homology in nucleotide sequences, of human IL1α and IL1β. The homologous region of IL1α and IL1β, termed the CD region, consists of 150–186 amino acids and contains the minimal recognition site for the IL1 receptor (IL1R).

IL1 synthesis is induced by lipopolysaccharide (LPS), by certain bacterial toxins, and by the process of phagocytosis. These stimuli result in transcription, translation, and processing of the IL1 precursor. Transcription is short-lived and can be suppressed by corticosteroids and by a putative endogenous repressor protein. Translation is increased by calcium ionophore and leukotrienes but is inhibited by prostaglandin-induced cyclic adenosine monophosphate (cAMP) production. Within the cell, IL1 is associated with lysosomes and not endoplasmic reticulum. The majority of IL1β is secreted, whereas IL1α is more membrane-associated.

Both forms of IL1 bind to a common receptor and mediate their biological effects through classical receptor-controlled events. IL1 binding is associated with fluxes of both Na/K and calcium. Receptors are constitutively present on T cells, but they are down-regulated following ligand–receptor binding and up-regulated by T-cell activation. [*See* INTERLEUKINS.]

B. Tumor necrosis factor

Tumor necrosis factor (TNF) was first described as a protein induced by LPS that was capable of producing hemorrhagic necrosis of tumors *in vivo*. TNF is produced primarily by monocytes and macrophages and is structurally and functionally related to lymphotoxin, a protein produced by lymphocytes. The macrophage product is designated TNFα; lymphotoxin is known as TNFβ. The gene for TNFα is located on the short arm of chromosome 6 and is linked to the major histocompatibility complex (MHC). TNFα is synthesized as a large precursor molecule (233 amino acids) that is cleaved to the functional molecule (157 amino acids) prior to release from cells. It is not produced constitutively but is induced by a variety of stimuli such as phorbol esters, LPS, interferon-γ (IFN-γ) and bacterial toxins such as pyrogenic exotoxin A, toxic shock toxin-1 (TSST-1), and staphylococcal enterotoxins.

The *de novo* production of TNFα is complex since accumulating evidence indicates that it originates from two sources: a small intracellular pool of TNFα mRNA, which is maintained in an untranslated state, as well as newly transcribed TNFα mRNA. Stimuli such as LPS trigger an acceleration of gene transcription within minutes. This is followed by rapid translation of the newly synthesized mRNA as well as the stored mRNA. IFN-γ controls TNFα synthesis at the translational level and is capable of overcoming the translational block of the LPS-resistant C3H/HEJ mouse, which cannot synthesize TNFα in response to endotoxin. TNFα mRNA synthesis peaks at 90 min after monocytes are treated with LPS and then declines over several hours to baseline values even in the continued presence of the inducer. Existing evidence suggests that LPS-induced cytokine production utilizes a protein kinase C pathway. In addition, substances that increase cAMP levels suppress TNFα production, whereas elevation of cyclic guanosine monophosphate has the opposite effect. Because TNF is proinflammatory and cycloxygenase inhibitors are antiinflammatory, the control of TNFα synthesis by indomethacin and/or PGE_2 are of major importance. PGE_2 activates PGE_2 receptors, which increase intracellular cAMP through the conversion of adenosine triphosphate to cAMP by the enzyme adenylate cyclase. The net effect is suppression of TNFα synthesis. Conversely, indomethacin enhances both the transcription and synthesis of TNFα as well as IL1. Whether this is due to a direct effect of indomethacin or the fact that indomethacin inhibits synthesis of PGE_2, thereby preventing an accumulation of cAMP, is not clear at the present time. It is clear that indomethacin can inhibit the effector function of synthesized or exogenous TNFα [(i.e., TNFα-induced production of arachidonic acid metabolites such as PGE_2 and prostacyden (PGI_2)] and, in so doing, prevents shock and tissue destruction.

C. Interferon-γ

IFN-γ is a potent activator of mononuclear phagocytes produced by antigen activated T-lymphocytes. It is produced both by IL2-secreting $CD4^+$ helper T cells and by nearly all $CD8^+$ T cells. Antigen activation initiates transcription of IFN-γ, and this response is augmented by IL2. IFN-γ induces the maturation of mononuclear phagocytes into macrophages and directly induces synthesis of the enzymes that mediate the respiratory burst. These macrophages are much more efficient in phagocytosis and killing of bacteria, although other responses such as tumor killing are only partially enhnaced by IFN-γ. Other effects of IFN-γ are (1) enhancement of cellular and humoral immune responses through increased MHC Class I and II expression on a variety of cell types; (2) promotion of differentiation of T- and B-lymphocytes; (3) activation of neutrophils; (4) activation of NK cells; (5) activation of vascular endothelium with promotion of $CD4^+$ T-lymphocyte adhesion, which facilitates lymphocyte extravasation at the site; and (6) enhancement of TNFα synthesis at the transcriptional level. The importance of this latter finding was substantiated in studies demonstrating that (1) pretreatment of mice with rIFN-γ increased TNF production fivefold after injection of endotoxin, (2) pretreatment of mice with rIFN-γ increased mortality induced by endotoxin, and (3) treatment with anti-IFN-γ reduced mortality due to endotoxin. [*See* INTERFERON.]

D. Lymphotoxin

Lymphotoxin (TNFβ) is synthesized by T-lymphocytes in response to stimuli such as LPS, although in smaller quantities than the TNFα made by mononuclear phagocytes. TNFβ is usually produced coordinately with IFN-γ and, like IFN-γ, activates both polymorphonuclear leukocytes (PMNLs) and vascular endothelial cells, causing increased leukocyte adhesion. This activity is further enhanced by the presence of IFN-γ. TNFα and

TNFβ have nearly identical biological effects and both bind to the same receptor. TNFβ is not readily detected in the circulation, supporting the hypothesis that TNFβ is usually a locally acting paracrine factor and not a mediator of systemic injury or shock.

E. Interleukin-6

IL6 (also known as IFN-β_2 or B-cell growth factor) consists of 184 amino acids (secreted form). It is produced in T cells, monocytes, fibroblasts, endothelial cells, and neoplastic cells. IL6 can be induced by IL1, TNFα, and platelet-derived growth factor. Although IL6 is produced by macrophages in the presence of LPS, data suggest that it is induced by TNF and IL1 rather than LPS itself.

The actions of IL6 are not completely established, yet it appears that two of the main functions are induction of the acute phase response and growth stimulation for activated B cells late in the sequence of B-cell differentiation. In response to noxious stimuli such as trauma, burns, and infection, C-reactive protein (a nonspecific opsonin), α_2-macroglobulin, fibrinogen, serum amyloid A protein, and several antiproteases all increase dramatically in the serum. In contrast, the concentrations of albumin and the iron-binding protein, transferrin, decrease in serum. Although such stimuli are all associated with increased plasma concentrations of IL1, TNF, and IL6, it is primarily IL6 that controls the dynamics of the acute phase response. Although plasma concentrations of IL6 are extremely high in patients with septic shock, it is unclear at the present time whether IL6 contributes to the pathogenesis of shock or its effects are for the most part beneficial due to its feedback inhibition of TNF synthesis and its induction of the acute phase reactants, both of which would tend to favor the host and the reestablishment of homeostasis.

F. Interleukin-2

T-lymphocyte activation occurs when the antigen-MHC Class II complex on the surface of an antigen-presenting cell (mononuclear phagocyte) interacts or binds to the T-cell receptor (TCR). Within minutes of the binding, T cells begin transcribing a variety of genes whose protein products are known to be essential for functional activation. Three main categories of genes are expressed early during T-cell activation: cellular protooncogenes, cytokine genes, and cytokine receptor genes. Transcripts of two protooncogenes, *c-fos* and *c-myc*, are first detectable within 15 min and 1 hr, respectively, after T-cell stimulation. The *fos* protein is thought to be involved in the transcriptional regulation of other genes including the IL2 gene. Transcription of the genes for IL2 and IFN-γ begins within 1 hr following TCR-mediated stimulation of human lymphocytes and peaks at around 4 hr. Simultaneously, the transcription of IL2 receptor genes begins and results in the up-regulation of surface IL2 receptors. In this way, secreted IL2 produced by an activated T-cell functions in an autocrine fashion to stimulate growth and proliferation of antigen-specific T-lymphocytes. In addition, IL2 stimulates the synthesis of other T cell-derived cytokines such as IFN-γ and lymphotoxin (TNFβ). IL2 also stimulates the growth of NK cells and enhances their cytolytic functions. Such cells are also called lymphokine-activated killer cells. Similarly, IL2 stimulates the growth of B-lymphocytes and increases antibody synthesis. It should be emphasized that the major function of IL2 is serving as an autocrine growth factor for T-lymphocytes. Although IL2 is produced by CD8$^+$ T-lymphocytes, it is produced in much greater quantities by CD4$^+$ T cells. [*See* T LYMPHOCYTES.]

G. Interleukin-8

IL8 is a 72-amino acid polypeptide originally isolated from human mononuclear phagocytes stimulated with bacterial lipopolysaccharide. IL8 is identical to neutrophil-activating factor and monocyte-derived neutrophil chemotactic factor. IL8 is also induced rapidly by the cytokines IL1 and TNF. IL8 induces chemotaxis, degranulation, and respiratory burst activity in human PMNLs at nanomolar concentrations by interacting with a specific plasma membrane receptor. Some studies have demonstrated that IL8 also induces PMNL adhesion to unstimulated human endothelial cells by a mechanism that involves enhanced expression of CD11/CD18 integrins. In contrast, other studies demonstrate that IL8-stimulated PMNLs do not bind well to endothelial cells stimulated with TNF.

H. Colony-Stimulating Factors

The cytokines that stimulate expansion and differentiation of bone marrow progenitor cells are collectively called colony-stimulating factors (CSFs). Some of the CSF molecules are capable of causing bone marrow stem cells to proliferate into erythroid, myeloid, monocyte, and megakaryocyte cell lines. Others affect only specific cell lines. Some of the actions of CSFs are inhibited by cytokines such as TNFα, TNFβ, and IFN-γ, whereas others are stimulated or enhanced by IL1 and IL6. The function of the various CSFs are discussed below.

1. Interleukin-3

The IL3 CSF is also known as multilineage CSF because it acts on the most immature bone marrow precursor cells and stimulates production of all the aforementioned cell lines. It is a 20–26-kDa product of CD4$^+$ cells.

2. Granulocyte–Macrophage Colony-Stimulating Factor

Granulocyte–macrophage CSF (GM-CSF) is a 22-kDa protein produced by activated T cells and by activated mononuclear phagocytes, endothelial cells, and fibroblasts. It acts more distally in the bone marrow maturation process on cells that have already differentiated. The net effect is to increase the production of both mononuclear phagocytes and granulocytes. GM-CSF is not found circulating in blood and is, therefore, thought to act locally at sites of production to activate mature leukocytes. Infusion of TNFα or TNFβ in mice has been associated with increased GM-CSF transcripts in tissue and increased circulating levels of GM-CSF in serum of animals leading to the concept of a cascade of mediators that affect bone marrow production starting with TNFα and TNFβ followed by GM-CSF and IL1. This cascade is responsible, in part, for the hematopoietic response to inflammation. GM-CSF does not influence *in vitro* adherence, directed migration, or *in vivo* infiltration of neutrophils. *In vivo* pretreatment of neutrophils with either GM-CSF or granulocyte colony-stimulating factor (G-CSF) enhanced production of superoxide anion, induced degranulation and enhanced phagocytosis.

Finally, eosinophilia is a prominent feature in many patients with parasitic diseases. Although GM-CSF and IL3 can be associated with increased eosinophilia in marrow, IL5 appears to be the major cytokine mediating the selective eosinophilia in filarial and other helminthic diseases.

3. Granulocyte Colony-Stimulating Factor

G-CSF is a 175-amino acid protein produced by monocytes, fibroblasts, and endothelial cells and regulates the production of neutrophils within the bone marrow. The effects of G-CSF on PMNL function are similar to those already described for GM-CSF. G-CSF is currently being used in a variety of neutropenic states, as described later.

IV. The Role of Cytokines in the Development of Fever: Leukocytosis, Polymorphonuclear Leukocyte Priming, and the Acute Phase Response

Fever has remained one of the earliest and most interesting responses to infection. Workers in the previous centuries diligently studied the patterns of fever and their descriptions have stood the test of time. In the 1950s, one researcher recognized that a substance found in the serum of infected animals could induce fever in other animals. This substance was heat-labile, required protein synthesis, and was not directly related to "pyrogenic" substances produced by bacteria. This substance was named endogenous pyrogen, or leukocyte endogenous mediator (LEM), and was thought to originate from peripheral blood PMNLs. Subsequently, the monocyte has been shown to be the major producer of LEM in peripheral blood, and LEM activity has since been attributed primarily to IL1 and TNF. In patients with fever, the function of cells and even organs is also altered. Specifically, zinc, serum iron, and transferrin levels are dramatically decreased. Serum albumin synthesis is slowed and production of IgM, IgG, complement, fibrinogen, and other acute phase reactants, such as C-reactive protein, are accelerated. Although such events are associated with increased plasma concentrations of IL1, TNF, and IL6, it is primarily IL6 that controls the dynamics of the acute phase response. The "resting" activity of circulating cells such as PMNLs are increased in febrile patients. It is clear that this is related, in part, to the direct effects of elevated temperature since *ex vivo* PMNL activity in blood of normal patients increases linearly from ~31° to 40°C. Thereafter, cell activity becomes reduced, likely due to the denaturing effect of high temperature. It now appears that the enhanced basal activity of circulating PMNLs from patients with fever is also related to priming induced *in vivo* by the cytokines TNF, IL1, and probably IL8. It is interesting that TNF and IL1 induce mononuclear cells to produce IL8. Increased production of TNF shuts off bone marrow synthesis of both erythroid and myeloid cell lines. Yet, despite this effect increased white cell counts, left shifts, and increased numbers of immature PMNLs are observed *in vivo* during febrile episodes in patients with infection. Because both TNF and IL1 suppress bone marrow stem cell proliferation, but also stimulate monocytes to synthesize G-CSF, a potent inducer of bone marrow synthesis of granulocytes, it would appear that G-CSF may be capable of overcoming the myeloid suppressing effects of TNF. Similarly, G-CSF decreases the time necessary for bone marrow release of myeloid cells. The net effect is to increase the peripheral white blood cell count and to shift the cell population to more immature forms such as band form neutrophils, metamyelocytes, and myelocytes. Thus, in significant infections, there is evidence of leukocytosis with a left shift, and these circulating white cells are activated. The elevated white count and left shift are primarily due to G-CSF, whereas *in vivo* activation or priming is regulated by or the consequence of TNF, IL1, and IL8.

V. The Role of Cytokines in Specific Infections

A. Gram-Negative and Gram-Positive Bacterial Infections: Mechanisms of Shock

TNF has been implicated as a mediator of endotoxin-induced shock based on the following findings:

1. Increased serum levels of TNF were found in animals infused with endotoxin and the time course correlated with development of hypotension.
2. Endotoxin induced synthesis of TNFα by human mononuclear cells.
3. Infusion of low-dose endotoxin in humans resulted in peak serum TNF concentrations, which occurred 90 min postinfusion and then fell to baseline within 3–4 hr.
4. The quantity of TNFα in serum was directly related to the severity of shock.
5. Passive immunization of experimental animals with a neutralizing monoclonal antibody against TNF protected animals from endotoxin infusion.
6. Passive immunization with neutralizing antibody against TNF protected baboons from lethal challenge with viable *Escherichia coli*.

Based on both *in vitro* and *in vivo* studies, it has become clear that LPS stimulates a cascade of cytokine synthesis with TNF peaking in serum by 60–90 min and persisting for ~3 hr; next IL1 levels peak at 2 hr and fall to baseline at 3.5–4 hr; lastly, IL6 levels peak at 3.5 hr and become negative at about 5 hr. *In vitro* data suggest that toxins from gram-positive bacteria such as streptolysin O and pyrogenic exotoxin A can also stimulate a similar cascade of cytokines. The significance of these cascades is that IL1 and TNF in concentrations found in blood are capable of interacting in a synergistic manner to induce both shock and tissue injury. Clearly, infusion of large amounts of TNF alone can induce the manifestations of shock, whereas IL1 does not by itself induce shock. The mechanisms whereby TNF induces shock are not well understood, but the following specific actions of TNF can, at high concentrations, contribute to its lethal effects:

1. reduction in myocardial contractility,
2. reduction in vascular smooth muscle tone resulting in vasodilation,
3. induction of profound hypoglycemia,
4. intravascular thrombosis due to enhanced PMNL–endothelial adherence mechanisms leading to vascular occlusion and decreased tissue perfusion, and
5. alteration of endothelial cell integrity leading to extravasation of fluid into the interstitium.

In addition to its role in shock, TNFα has also been implicated as a major mediator of complications associated with meningitis caused by *Streptococcus pneumoniae* and *Hemophilus influenza*.

B. Viral Infections

1. Natural Immunity to Viruses

Interferon was initially discovered as a protein released from virus-infected cells that prevented viral replication in uninfected cells. Recently, interferons have been classified as either type 1 (IFN-α and IFN-β) or type 2 (IFN-γ). Both type 1 interferons are produced in response to active viral replication *in vivo* as well as in cell cultures. Although the structure and sequence of IFN-α and IFN-β are vastly different, all type 1 IFN molecules bind to the same cell surface receptor and induce the following cellular responses:

1. inhibition of viral replication,
2. inhibition of cell proliferation,

3. enhanced lytic function of NK cells,
4. enhanced expression of Class I MHC molecules, and
5. inhibition of expression of Class II MHC molecules.

2. Acquired Immunity to Viruses

Humoral or mucosal immunity generated as a result of prior active infection or vaccination provides the first line of defense against viral infections. However, the major mechanism of specific immunity is mediated by virus-specific cytotoxic T-lymphocytes (CTLs). CTLs can be either $CD8^+$ cells that recognize viral antigens in association with expressed Class I MHC molecules on a variety of cell types (see type 1 IFN earlier) or $CD4^+$ cells that recognize viral antigens presented in association with Class II MHC molecules. In either case, synthesis of IL2, TNFβ, and IFN-γ by $CD4^+$ cells is of major importance in the mediation of the destruction of virus-infected cells. The importance of the host response to virally infected cells is substantiated by the clinical observation that immunocompromised patients with hepatitis B infection are most likely to become asymptomatic carriers of the virus. As one might expect, hepatitis B infection in patients with HIV-1-induced AIDS are also more likely to have asymptomatic infection due to a lack of $CD4^+$ lymphocytes. In contrast, immunocompetent individuals are more likely to develop either acute hepatocellular destruction of virus-infected hepatocytes with rapid clearing of the infection, satisfactory resolution, and long-term immunity, or the immunological response to viral-infected cells may be so violent that acute hepatic necrosis and death ensue.

C. Intracellular Parasites

A number of infectious agents that are capable of replicating within unstimulated monocytes or other cells of the human host also require a T-cell response to eradicate infection. Cytokines are crucial to limit and clear these intracellular parasites; however, excessive cytokine production is also important in the development of the specific pathological effects of each agent. Table I lists the intracellular parasites that require cell-mediated immunity for resolution. IFN-γ therapy given to animals infected with these pathogens has resulted in resolution of infection if given before or at the time of infection. In addition, in *Mycobacterium avium-intracellulare* and *Leishmania donovani* infections, IFN-γ enhanced antimicrobial activity even if given after establishment of infection. The importance of specific cytokines in *Leishmania* infections is further substantiated by the demonstration that pretreatment of animals with neutralizing monoclonal antibody against IFN-γ and TNF rendered animals more susceptible to infection. Specifically, animals demonstrated impaired acquired resistance, increased severity of infection, or increased mortality after sublethal challenge. In addition, in a model of *Mycobacterium bovis* infection, granuloma formation was also disrupted by anti-TNF monoclonal antibody, suggesting that the cellular immune response was dramatically impaired. Thus, IFN-γ and TNF appear to play important roles in resistance against infection by these intracellular pathogens by increasing macrophage killing of parasites and by enhancing granuloma formation. In contrast, in cerebral malaria, TNF may play the opposite role. Specifically, in *Plasmodium falciparum* infection, cerebral malaria is associated with high levels of TNF in serum and the highest levels were associated with a worse prognosis. In experimental infections, administration of both anti-IL3 and anti-GM-CSF monoclonal antibody prevented both the appearance of cerebral malaria in animals infected by *Plasmodium berghei* and a rise in serum TNFα. Administration of anti-IFN-γ monoclonal antibody had a similar effect, leading to the hypothesis that activated T cells secrete IL3, Gm-CSF, and IFN-γ, which in turn activated monocytes to secrete TNFα. TNF could then contribute to the vascular injury through its ability to upregulate the expression of endothelial adherence molecules.

Table I Intracellular Pathogens against which Cellular Immunity Appears to be Important

Protozoa	*Toxoplasma gondii, Toxoplasma cruzi, Leishmainia* sp.
Bacteria	*Listeria monocytogenes, Brucella* sp., *Legionella pneumonphila, Salmonella* sp.
Mycobacteria	*Mycobacterium tuberculosis, Mycobacterium avium-intracellulare, mycobacterium kansasi, mycobacterium leprae*
Fungi	*Histoplasma capsulatum, Cryptococcus neoformans, Coccidioides immitis, Paracoccidiodes brasiliensis, Blastomyces dermatitidis*
Viruses	*Herpes simplex, Cytomegalovirus, Varicella zoster, Human Immunodeficiency Virus, Epstein–Barr, Rubeola*
Others	*Chlamydia* sp.

The interplay of cytokines in the pathogenesis of an infectious disease is no better defined that in a mouse model of *Trypanosoma cruzii* infection. In *Trypanosoma cruzii* infections, γ-interferon endogenously produced in resistant strains of mice but not susceptible ones prevents infection. Because the γ-interferon produced in both strains is biologically active, an inhibitor of γ-interferon in the susceptible strains was suspected and found to be IL10. Thus, IL10 production by spleen cells in susceptible mice challenged with *T. cruzii* inhibits the production and function of γ-interferon as well as other cytokines, allowing proliferation of the parasite within macrophages. Recently, Heinzel and Locksley have demonstrated that in response to acute challenge with T. cruzii, CD4+ cells from resistant mice expressed γ-interferon and IL2, whereas CD4+ cells from susceptible mice expressed IL4 and IL10 M-RNA. Anti-CD4+ antibody given to susceptible animals at the time of *T. cruzii* challenge, resulted in depletion of CD4+ cells expressing IL4 and IL10, and repopulation with CD4+ cells expressing IL-2 and γ-interferon. Such animals were then resistant to *T. cruzii* infection. Thus, two populations of CD4+ cells exist (TH1 and TH2 and these cells determine resistance or susceptibility based upon their dominant phenotypic expression of cytokines (TH_1= IL2 and γ-interferon) or (TH_2= IL4 or IL10), respectively.

VI. A Look to the Future: The Potential Role of Cytokines and Anticytokine Antibodies in the Treatment of Specific Infections

Two recombinant cytokines have been used to treat active infection in humans in two situations, both with exciting results. First, rIFN-γ has been used to treat patients actively infected with lepromatous leprosy, AIDS-related *M. avium-intracellulare* bacteremia, cutaneous leishmaniasis, and diffuse cutaneous and visceral leishmaniasis. Although small numbers of patients have been treated, results demonstrate partial to dramatic improvements in the clinical courses of each type of infection. IL2 has also been used intradermally in patients with lepromatous leprosy and facilitated marked reductions in the number of acid-fast bacilli at the site. The study also suggested that local production of IFN-γ due to the injected IL2 accounted at least partially for the observed effect.

Two cytokines have also been used prophylactically in patients with qualitative or quantitative PMNL deficits. Specifically, patients with chronic granulomatous disease have received IFN-γ subcutaneously with the effect that (1) monocytes and neutrophils demonstrated increased respiratory burst activity and antibacterial capacity and (2) patients had fewer febrile and infectious complications. In addition, patients with congenital and cyclic neutropenia have received subcutaneous injection of G-CSF with resultant increase in PMNL counts for 14–21 days and fewer infectious complications. Currently, G-CSF is also being investigated in the treatment of patients with febrile neutropenia as a complication of chemotherapy. Lastly, several clinical trials are currently underway to investigate the efficacy of anti-TNFα antibody and IL1 receptor antagonists in the treatment of septic shock. Phase I trials indicate that both agents are well tolerated and look very promising, particularly if patients receive initial treatment early in the course.

Bibliography

Abbas, A. K., Lichtman, A. H., and Pober, J. S. (1991). Cytokines. *In* "Cellular and Molecular Immunology" (A. K. Abbas, A. H. Lichtman, and J. S. Pober, eds.), pp. 226–243. W. B. Saunders Company, Philadelphia.

Cannon, J. G. (1991). Cytokines and shock. *In* "Cytokines and Inflammation" (E. S. Kimball, ed.), pp. 307–329. CRC Press, Boca Raton, Florida.

Gupta, S. (1988). *Scand. J. Rheumatol.* **76**(suppl.), 189–201.

Heinzel, F. P. (1990). *J. Immunol.* **145,** 2921–2924.

Heinzel, F. P., Sadick, M. D., Mutha, S. S., and Locksley, R. M. (1991). *Proc. Natl. Acad. Sci, U.S.A.* **88,** 7011–7015.

Kaushansky, K., Broudy, V. C., Harlan, J. M., and Adamson, J. W. (1988). *J. Immunol.* **141,** 3410–3415.

Lewis, D. B., and Wilson C. B. (1990). *Pediatr. Infect. Dis. J.* **9,** 642–651.

Limaye, A. P., Abrams, J. S., Silver, J. E., Ottesen, E. A., and Nutman, T. B. (1990). *J. Exp. Med.* **172,** 399–402.

Malefyt, R. W., Abrams, J., Bennett, B., Figdor, C. G., and de Vries, J. E. (1991). *J. Exp. Med.* **174,** 1209–1220.

Murray, H. W. (1990). *Diagn. Microbiol. Infect. Dis.* **13,** 411–421.

Ohlsson, K, Bjork, P., Bergenfeldt, M., Gageman, R., and Thompson, R. C. (1990). *Nature* (*London*) **348,** 550–552.

Saukkonin, K., Sande, S., Cioffe, C., Wolfe, S., Sherry, B., Cerami, A., and Tuomanen, E. (1990). *J. Exp. Med.* **171,** 439–448.

Steinbeck, M. J., and Roth, J. A. (1989). *Rev. Infect. Dis.* **11,** 549–568.

Sullivan, G. W., and Mandell, G. L. (1991). *Curr. Opinion Infect. Dis.* **4,** 344–349.

ISBN 0-12-226891-1